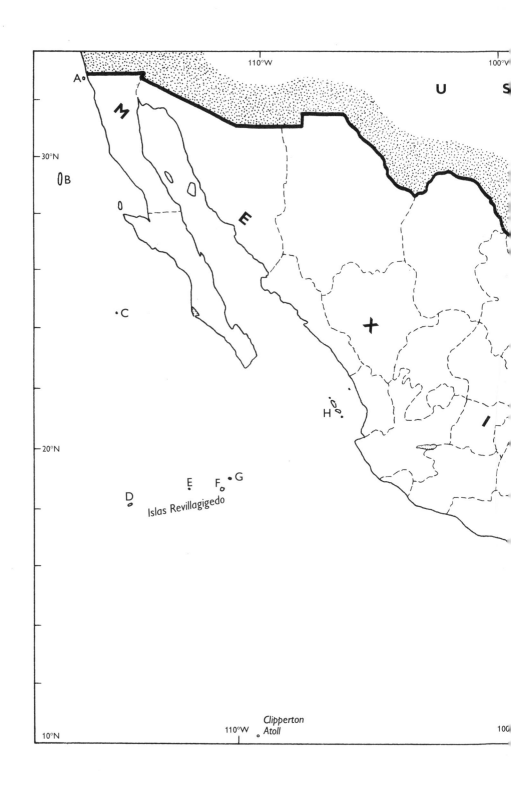

110°W   100°W

U S

A•

M

30°N

0B

E

•C

20°N

H

+

I

E   F   •G

D

Islas Revillagigedo

Clipperton
Atoll

10°N   110°W   100

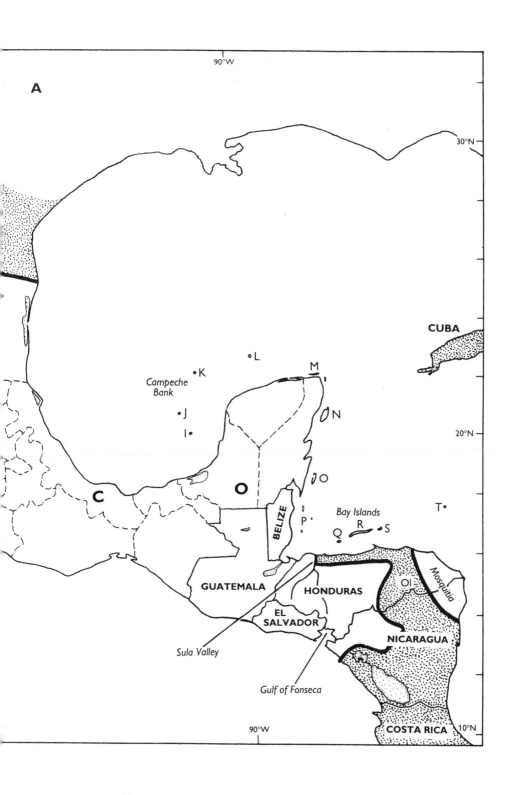

A GUIDE TO
The Birds of Mexico
and
Northern Central America

Steve Howell first visited Mexico from November 1981 to April 1982, armed with a strong background of birding in Britain and the Western Palearctic. He found the great diversity of often unfamiliar birds fascinating, but soon learned that relatively little was known about these birds. For the next eleven years, he spent up to eight months each year studying birds in Mexico and Central America and, in March 1986, while visiting John Guarnaccia and Mercedes Villamil in Panama, decided to write the sort of guide he would have liked to have had on his first visit to Mexico. Having taken the plunge, there remained the major problem of illustrating the guide. Some years earlier he had met **Sophie Webb**, a biologist and talented artist who also had field experience in the Neotropics. Webb's skill at sketching birds in the field was an attribute that Howell considered essential. Webb agreed to illustrate the guide and there followed over seven years of painstaking labor, in the field, in museums, and in libraries, as Howell and Webb worked on what became a much larger project than either had anticipated.

After more than 74 months of virtually daily field work in all corners of Mexico and Central America, the authors have gained field experience with over 99 per cent of the species treated by this guide, including all but six of the extant breeding species.

Sponsored by
POINT REYES BIRD OBSERVATORY
Stinson Beach
California 94970, USA

A GUIDE TO

The Birds of Mexico
and
Northern Central America

Steve N. G. Howell
Sophie Webb

OXFORD
UNIVERSITY PRESS

OXFORD

UNIVERSITY PRESS

Great Clarendon Street, Oxford OX2 6DP

Oxford University Press is a department of the University of Oxford.
It furthers the University's objective of excellence in research, scholarship,
and education by publishing worldwide in

Oxford   New York

Auckland Cape Town Dar es Salaam Hong Kong Karachi
Kuala Lumpur Madrid Melbourne Mexico City Nairobi
New Delhi Shanghai Taipei Toronto

With offices in

Argentina Austria Brazil Chile Czech Republic France Greece
Guatemala Hungary Italy Japan South Korea Poland Portugal
Singapore Switzerland Thailand Turkey Ukraine Vietnam

Oxford is a registered trade mark of Oxford University Press
in the UK and in certain other countries

Published in the United States
by Oxford University Press Inc., New York

First published 1995
Reprinted 1995, 1999, 2000, 2001, 2004, 2005, 2005, 2007

A catalogue record for this book is available from the British Library

Library of Congress Cataloging in Publication Data
Howell, Steve N. G.
A guide to the birds of Mexico and northern Central America /
Steve N. G. Howell, Sophie Webb.
Includes bibliographical references and index.
1. Birds—Mexico.   2. Birds—Central America.   3. birds—Mexico—
Indentification.   4. Birds—Central America—Indentification.
I. Webb, Sophie.   II. Title.
QL686.H68   1994   598.2972—dc20   93-31659
ISBN 978-0-19-854012-0 (Pbk)

10

Printed in China

Dedicated to our parents

Ewart and Charmian Howell

and

Dwight and Nancy Webb

and to

William J. Schaldach Jr.

a remarkable mentor without whom
this book never would have happened

CONTENTS

The plates are to be found between pp. 400 and 401

COLOR PLATES

ACKNOWLEDGEMENTS

This work would not have been possible without the help of many people. First and foremost, we thank the small core of believing individuals whose encouragement and support in the early stages were critical: John Guarnaccia, Mercedes Villamil, Dodge and Lorna Engleman, Frank A. Pitelka, Allan R. Phillips, Lina J. Prairie, Peter Pyle, Richard G. Wilson, Jack and Paula De Groot, Ken and Marci Collis, Leonora Klein, and Violet L. McIvor.

A special thanks also goes to Point Reyes Bird Observatory (PRBO), in particular to David G. Ainley, Grant Ballard, Patrice Daley, Daniel C. Evans, Geoffrey R. Geupel, Susan Goldhaber, Karen Hamilton, B. Denise Hardesty, Janet Kjelmyr, Gary W. Page, Larry B. Spear, Dan L. Reinking, W. David Shuford, Lynne E. Stenzel, Liz Tuomi, Laurie Wayburn, Oriane Williams, and numerous field biologists at the PRBO Palomarin Field Station, for logistical support and help from the earliest stages through to the final product.

Important support for field work was provided by the following. The Kelton Foundation sponsored an expedition to Isla Guadalupe and Las Islas Revillagigedo in January–March 1988, and another expedition to pelagic waters off western Mexico in May 1992. Larry B. Spear, David G. Ainley, and the officers and crew of the National Oceanic and Atmospheric Administration vessel 'Discoverer R102' enabled SNGH, during October–December 1989, and SNGH and SW during October–December 1990, to visit Clipperton Atoll and also to gain field experience with virtually all of the Pacific seabird species treated in this guide. The personnel and fishermen of the Cooperativa de Pesca in Progreso, Yucatán, and the Mexican naval authorities in Progreso, enabled SNGH to visit the islands of the Campeche Bank in October 1984. Other logistical support for field work was provided generously by William J. Schaldach Jr. and his family (Sierra de los Tuxtlas, Veracruz); Richard G. and Elda Wilson (Mexico City and surrounding areas); Barbara MacKinnon de Montes (Quintana Roo); George and Janet Cobb (Yucatán); Peter Hubbell (Antigua, Guatemala); Rose Ann Barnhill (Puerto Barrios, Guatemala); and Ing. Cesar A. Alvarado B. (Lancetilla Botanical Gardens, Honduras). Miguel Alvarez del Toro and Rafael Solis G. assisted with an expedition to El Triunfo, Chiapas, Jack Bucklin and FUNDAECO helped SNGH with a trip to El Golfete, Guatemala, and Miguel Dominguez helped with a visit to the Sierra de La Laguna, BCS. In addition, Will Russell and WINGS Inc., Lina J. Prairie and Golden Gate Audubon Society, Kathleen Dickey and Natural Excursions (PRBO), and Richard G. Wilson and Ecotours were of major help to SNGH with travel and field work in Mexico and Belize.

An important precursor to SNGH's study of status and distribution was a three-month trip with Peter Pyle, from December 1983 to February 1984, driving through virtually all of Mexico's states. This memorable trip highlighted how little was known about Mexican bird distribution.

The generous assistance afforded us by personnel at the following museums and other institutions was essential and greatly appreciated: American Museum of Natural History (AMNH; Allison Andors, George F. Barrowclough, Christine Blake, Robert W. Dickerman, Wesley E. Lanyon, Mary LeCroy, Richard A Sloss, François Vuillemier, and Jennifer Williams); California Academy of Sciences (CAS; Stephen F. Bailey, Louis F. Baptista, James B. Cunningham, and Betsey Cutler); Museum of Vertebrate Zoology, University of California, Berkeley (MVZ; Ned K. Johnson and Barbara Stein); United States National Museum (USNM; Richard C. Banks, Bonnie B. Farmer, Gary R. Graves, Joe T. Marshall Jr. and Richard L. Zusi); British Museum (BM; Peter Colston and Graham S. Cowles); Western Foundation of Vertebrate Zoology (WFVZ; Ed N. Harrison, James T. Jennings, Lloyd F. Kiff, Manuel Marin A. and Clark Sumida); Moore Laboratory of Zoology, Occidental College (MLZ; John C. Hafner); University of California, Los Angeles (UCLA; James R. Northern); Los Angeles County Museum (LACM; Kimball L. Garrett); Colección del Instituto de Historia Natural, Tuxtla Gutierrez (IHN; Gerardo de J. Cartas H.); VIREO (Geoff LeBaron and Doug Wechsler); Academy of Natural Sciences, Philadelphia (ANSP; Mark B. Robbins); Colección de Ornitología, Instituto de Biología, Universidad Nacional Autonoma de México (UNAM), Mexico City (IBUNAM; Gonzalo Gaviño de la T. and Alejandro Melendez H.); Museo de Zoología 'Alfonso L. Herrera', Facultad de Ciencias, UNAM, Mexico City (FCUNAM; Adolfo G. Navarro S.); Museum of Comparative Zoology, Harvard University (MCZ; Raymond A. Paynter Jr.); El Salón de las Aves, Saltillo (Aldegundo Garza de Leon); Carnegie Museum of Natural History, Pittsburgh (CM; Kenneth C. Parkes and D. Scott Wood); University of British Columbia, Vancouver (UBC; Richard J. Cannings); Delaware Museum of Natural History (DMNH; David M. Niles); Field Museum of Natural History (FMNH; David E. Willard); Southwestern College, Kansas (SWC; Max. C. Thompson); University of Kansas (Tristan J. Davis and Richard O. Prum); Louisiana State University (Steven W. Cardiff); Royal Ontario Museum (ROM; Jon C. Barlow and James A. Dick); and University of Michigan (Robert B. Payne).

SW would like to thank the numerous people who aided her during the labor of painting the plates. Of these, special acknowledgement must go to Guy Tudor whose encouragement, comments, and generous loan of photographic material were invaluable. Thanks also to Jonathan L. Atwood, Stephen F. Bailey, Robert A. Behrstock, David F. DeSante, Robert W. Dickerman, Jon L. Dunn, Albert Earl Gilbert, Keith Hansen, Joe T. Marshall Jr., Allan R. Phillips, Peter Pyle, Richard G. Wilson, and Dale A. Zimmerman for criticisms of certain plates.

The loan of AMNH specimens to SW at PRBO allowed this work to be completed in a reasonable time frame, and special acknowledgement is due to Christine Blake and Mary LeCroy for their smooth and efficient handling of the loans.

The following people generously made available photographs which aided in painting the plates: Kenneth P. Able, Peter Alden, Philip W. Atkinson, Robert A. Behrstock, Rick Bowers, Aldo Brando, Ron Branson, A. D. Brewer, Mark Brigham, Allen Chartier, William S. Clark, Debby Cotter, Barbara A. Dowell, Jack C. Eitniear, John Gerwin, James D. Gilardi, Ed Greaves, Dick Hoffman,

Inga Kellogg, Greg W. Lasley, Sebastian Lousada, Curtis Marantz, Simon Perkins, Peter Pyle, Robert S. Ridgely, Patricio Robles Gil, Andres M. Sada, Luis Santaella, David A. Sibley, Mark J. Whittingham, Allan Wofchuck, and Tom and Sheri Wood.

We thank Steve E. Cornelius, Juan C. Martinez-Sánchez, Barbara M. de Montes, Andres M. Sada, Byron Swift, Sherry Lynn Thorn, Walter A. Thurber, and Paul Wood for constructive criticism of the conservation chapter. Opinions expressed therein, however, remain our responsibility.

Cynthia Kevorkian and Amacker Bullwinkle took SNGH's rough drafts of bird distribution and, with considerable skill, patience, and commitment, turned them into accurate and presentable range maps. Any errors remaining are SNGH's responsibility.

Others, too numerous to mention, helped in many ways. Of this multitude, a few should be thanked here: Juliet Bloss, Christopher Kales, Michelle le Marchant, Eleanor Munro, Andrew Taylor, Patrick Webb, and Janet Wessel.

Finally, we thank the staff of Oxford University Press for their care and patience with this project.

This is contribution number 564 of the Point Reyes Bird Observatory.

Sources of information

To the best of our knowledge, the text includes accounts for all species reliably recorded in Mexico and northern Central America through the end of 1992.

Information for the sections on status and distribution was derived from numerous sources. Published works, mostly based on specimens, constitute the backbone of the distributional information. However, many of the finer points of distribution were unpublished previously and result from our own field work and that of numerous others (see below). Most of the data relating to species' abundance comes from our own field work.

As well as crediting unpublished sources of information, we cite specific sources for (i) information published after the sixth edition of the *American Orthithologists' Union (AOU) checklist of North American birds* (1983), (ii) information published earlier but overlooked by, or contrary to, that work, and (iii) all species recorded five or fewer times in the region.

Specimen data have to be used carefully. Whenever possible we examined and/or verified critical specimens, and at the same time used caution when dealing with the few notoriously unreliable collectors, mainly M. del Toro Avilés (see Binford 1989) and G. F. Gaumer (see Parkes 1970).

Published information was found to be erroneous, or at least questionable, so frequently that we tried to verify all critical records. Errors perpetuated from mislabeled specimens of the 1800s and early to mid-1900s are still fairly common in recent published works (for example, AOU 1983, see Wilson and Ceballos-L. 1986). We have tried to cull as may of these errors (some from recent sources) as possible, while at the same time drawing attention to other records that we consider doubtful. Our uncertainty is often indicated by phrases such as 'requires verification', or 'possibly also occurs in . . . (reference)'. We hope that, rather than condemn such records, this will prompt correction or confirmation of our decisions. We also realize that before this guide appeared, there were few if any sources to indicate what really occurred where, and when, or what might be considered unusual. We hope that the information in the status and distribution accounts, some of it certainly incomplete or tentative (for example, altitudinal limits, seasonal occurrences), will form a baseline for future observations.

From the numerous sight reports that we evaluated in the course of this work, it became clear that many birders consider that once they have crossed into Mexico, anything is possible, and it doesn't necessarily need to be documented. There is, however, no reason why criteria for records from Mexico or Central America, and this includes unusual regional (state, department, etc.) records, should be any different from standards expected in North America or Europe. We have tried to evaluate reports accordingly. We have accepted numerous single-observer non-specimen records (including first country records), when confirmed by written descriptions, sketches, photographs, tape recordings, etc. Many other records, mostly of regional importance and some of them no doubt valid, were rejected if they lacked conclusive documentation; these included difficult-to-identify species for which inadequate descriptions were provided and, far more commonly, reports with no attempt at documentation.

For sight records we tried to apply strict judgement criteria, but we are as guilty as anyone of being biased towards competent observers personally known to us; this is inevitable. When previously unpublished records are supported by photographs or tape recordings known to us, this is indicated by noting *photo* or *tape* following the relevant observer's initials.

As noted above, we owe much of the data in the sections on status and distribution to many people who kindly took time to review lists of critical questions and who, together, provided a wealth of information that could not otherwise have been obtained. The following people contributed unpublished information, including tape recordings of species unfamiliar to us and/or clarified published material relating to status, distribution, and taxonomy: Robert P. Abrams, David G. Ainley, Peter Alden, Miguel Alvarez del Toro, John C. Arvin, Stephen F. Bailey, Richard C. Banks, Luis F. Baptista, Robert A. Behrstock, Chris Benesh, Mauro Berlanga C., Laurence C. Binford, Rick Bowers, Dawn Breese, Kelly Brough, Jack Bucklin, John L. Butler, Richard J. Cannings, Gerardo Cartas H., Hector Ceballos-Lascurain, Allen Chartier, George and Janet Cobb, Kathleen Collins, John Coons, Jorge Correa S., Alan Demartini, David F. DeSante, Robert W. Dickerman, Barbara A. Dowell, Hugh Drummond, Charles A. Ely, Richard A. Erickson, Ted L. Eubanks Jr., Jules G. Evens, C. Craig Farquhar, Tim Fenske, David J. Fisher, Aldegundo Garza de León, Hector Gomez de Silva, Brad Goodhart, Mary E. Gustafson, J. W. Hardy, Jack Holloway, Mark A. Holmgren, Jim and Karen Horton, Thomas R. Howell, Peter Hubbell, Eugene Hunn, Douglas A. James, Stuart Johnston, Kenn Kaufman, Joe Keenan, William H. Keener, Adam Kent, Jeff Kingery, Oliver Komar, Greg W. Lasley, Brian Lavercombe, Ron LeValley, Adrian J. Long, Sebastian Lousada,

Charlie Lownes, Cindy Ludden, Oddvin Lund, James F. Lynch, Guy McCaskie, Alejandro Melendez H., Greg Meyer, Bruce W. and Carolyn M. Miller, Burt L. Monroe Jr., Gale Monson, Barbara M. de Montes, Joseph Morlan, Eduardo Palacios, Dennis R. Paulson, Gary Perless, Bruce G. Peterjohn, A. Townsend Peterson, Allan R. Phillips, Robert L. Pitman, Lina J. Prairie, William L. Principe, Peter Pyle, Robert L. Pyle, Kurt A. Radamaker, Robert S. Ridgely, Don Roberson, Gerardo, Javier, and Patricio Robles Gil, Chandler S. Robbins, Gary H. Rosenberg, Rose Ann Rowlett, Stephen M. Russell, Richard Ryan, Andres M. Sada, Javier Salgado O., William J. Schaldach Jr., Peter E. Scott, Robert C. Self, Charles G. Sibley, David A. Sibley, J. Robert Singleton, Thomas B. Smith, Larry B. Spear, Rich Stallcup, David Stejskal, John Sterling, F. Gary Stiles, C. Rick Taylor, Bernie Tershy, Stuart Tingley, Peter D. Vickery, Joan M. Walsh, Roland H. Wauer, Richard E. Webster, David C. Wege, Dora Weyer, Bret Whitney, Sartor O. Williams III, Richard G. Wilson, David E. Wolf, D. Scott Wood, Paul Wood, Thomas E. Wurster, and Richard L. Zusi.

We are indebted to all of these people and to those who have published critical information about the birds of Mexico and Central America. While we made every effort to check all relevant published sources, it is likely (inevitable?) that somewhere along the way we missed something, and we apologize to any authors whose work we have overlooked inadvertently. At the same time, it would be impossible to check and include all of the unpublished information that exists for the region. We are aware of significant records omitted because the observers were unable to respond to our requests for information. We hope that they, and others who see gaps in our knowledge, will be prompted to publish their findings.

In 1992 the journal *The Euphonia* was founded and deals specifically with Mexican birds. We recommend this as an outlet to publish findings that supplement and correct information given in this guide for Mexico.

ABBREVIATIONS

Periods are used only when confusion might otherwise occur; for example, MA is often used for Massachusetts, hence we use M.A. for Middle America. Compass directions are written in upper case: E, N, SW, etc. The standard symbols ♂ (male, ♂♂ males) and ♀ (female, ♀♀ females) are used.

alt	alternate plumage	M.A.	Middle America
Baja	Baja California	N.A.	North America
basic	basic plumage	Plateau	Mexican Plateau
cen	central	SL	sea level
C.A.	Central America	S.A.	South America
imm(s)	immature(s)	spec.	specimen
Isthmus	Isthmus of Tehuantepec	ssp.	subspecies
		Yuc Pen	Yucatan Peninsula (including N Guatemala and N Belize)
juv(s)	juvenile(s)		
Los Tuxtlas	Sierra de los Tuxtlas, S Veracruz		

Mexican states

Standard abbreviations are used for Mexican states (see Figs 2(a) and 2(b)) though BCN is used instead of the ambiguous BC. Look at car license plates and the abbreviations soon will become ingrained.

Ags	Aguas Calientes	Mor	Morelos
BCN	Baja California Norte	Nay	Nayarit
BCS	Baja California Sur	NL	Nuevo Leon
Camp	Campeche	Oax	Oaxaca
Chis	Chiapas	Pue	Puebla
Chih	Chihuahua	Quer	Queretaro
Coah	Coahuila	QR	Quintana Roo
Col	Colima	SLP	San Luis Potosí
DF	Distrito Federal	Sin	Sinaloa
Dgo	Durango	Son	Sonora
Gto	Guanajuato	Tab	Tabasco
Gro	Guerrero	Tamps	Tamaulipas
Hgo	Hidalgo	Tlax	Tlaxcala
Jal	Jalisco	Ver	Veracruz
Mex	Mexico	Yuc	Yucatan
Mich	Michoacan	Zac	Zacatecas

Museums and institutions

AMNH	American Museum of Natural History
ANSP	Academy of Natural Sciences, Philadelphia
BM	British Museum of Natural History
CAS	California Academy of Sciences
CM	Carnegie Museum of Natural History
DMNH	Delaware Museum of Natural History
FCUNAM	Facultad de Ciencias, University Nacional Autonoma de México
FMNH	Field Museum of Natural History
IHN	Instituto de Historia Natural
LACM	Los Angeles County Museum of Natural History
MCZ	Museum of Comparative Zoology, Harvard University
MLZ	Moore Laboratory of Zoology, Occidental College
MVZ	Museum of Vertebrate Zoology, University of California, Berkeley
ROM	Royal Ontario Museum
SWC	Southwestern College, Kansas
UBC	University of British Columbia, Vancouver
UCLA	University of California, Los Angeles
USNM	United States National Museum of Natural History
WFVZ	Western Foundation of Vertebrate Zoology

INTRODUCTION

AREA COVERED

This guide treats the approximately 1070 species of birds that occur in Mexico and northern Central America.

Northern Central America, as defined in this book (see Figs 1, 2(a) and (b)), includes the Atlantic Slope lowlands south to the Ulua–Comayagua drainage (or Sula Valley) in western Honduras; the interior and highlands south to north-central Nicaragua; the Pacific Slope lowlands south to western Nicaragua in the vicinity of the Gulf of Fonseca; all Mexican islands, the cays of Belize, the Honduras Bay Islands; and offshore waters to a distance of 200 nautical miles (or agreed international or median line borders) from habitable land. We also include the lowland pine savannas (the Mosquitia) of eastern Honduras and northern Nicaragua, which are biogeographically a part of northern Central America. Clipperton Atoll (10°18′N, 109°13′W), a French possession, is also included as it has been claimed by Mexico, and is unlikely to be treated in any other guide.

Appendix E treats an additional 50 species known from the humid Atlantic Slope lowlands (or Olancho rain forests) of eastern Honduras. The Swan Islands, claimed by Honduras, lie some 200 km offshore in the Caribbean Sea and have a West Indian avifauna; they are treated by Bond (1985).

We consider it important to recognize biogeographic divisions rather than political boundaries, especially in so complex a region as Middle America (Mexico and Central America), the meeting place of the Nearctic and Neo-tropical zones.

In general, tropical species of humid evergreen forest (rain forest) range north on the Atlantic Slope to eastern Mexico. Few species that range west of the Sula Valley do not reach Mexico. Those species reaching their northern limit in the Olancho rain forests have affinities with southern Central America and South America.

Tropical species of arid deciduous forest (thorn forest) occur on the Pacific Slope. Here a division is harder to draw, and we chose the Gulf of Fonseca, although many species typical of the arid Pacific Slope of Middle America range from Mexico to northern Costa Rica.

Temperate zone species range south in the highlands and interior. Most reach their southern limit in or north of the highlands of Honduras and western Nicaragua, which are also the southern limit of native pines in the New World. The next highland block to the south (Costa Rica and western Panama) has its affinities with South America and also hosts a number of endemic species.

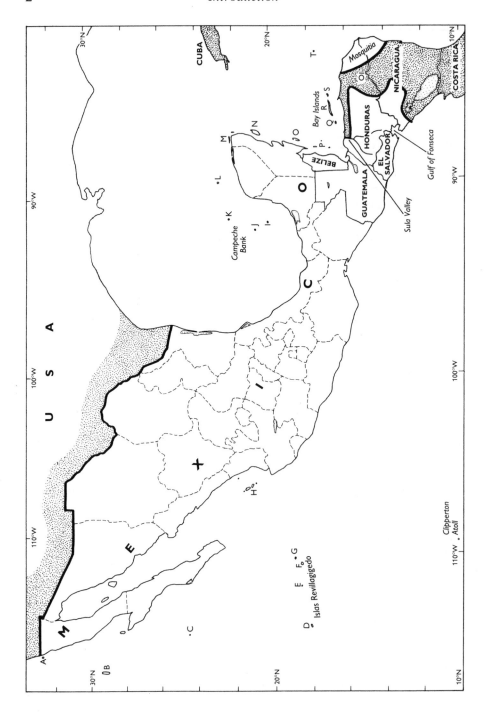

Fig. 1 Mexico and northern Central America: the area covered by this guide. **A:** Islas Los Coronados (BCN); **B:** Isla Guadalupe (BCN); **C:** Alijos Rocks (BCS); **D:** Isla Clarión (Col); **E:** Roca Partida (Col); **F:** Isla Socorro (Col); **G:** Isla San Benedicto (Col); **H:** Islas Tres Marias (Nay); **I:** Cayo Arcas (Camp); **J:** Arrecifes Triángulos (Camp); **K:** Cayo Arenas (Yuc); **L:** Arrecife Alacrán (Yuc); **M:** Isla Holbox (QR); **N:** Isla Cozumel (QR); **O:** Chinchorro Bank (QR); **P:** Belize Cays; **Q:** Utila Island (Honduras); **R:** Roatan Island (Honduras); **S:** Guanaja Island (Honduras); **T:** Swan Islands (Honduras); **Ol:** Olancho Rain Forests. The stippled areas show the limits of the guide region.

In addition to the mixing of temperate and tropical avifaunas, a remarkably high degree of endemism characterizes the region. Over one-third of all species breeding in the region covered by this guide are endemic to Middle America. Most of the endemics occur on the Pacific Slope and in the adjacent interior valleys and highlands.

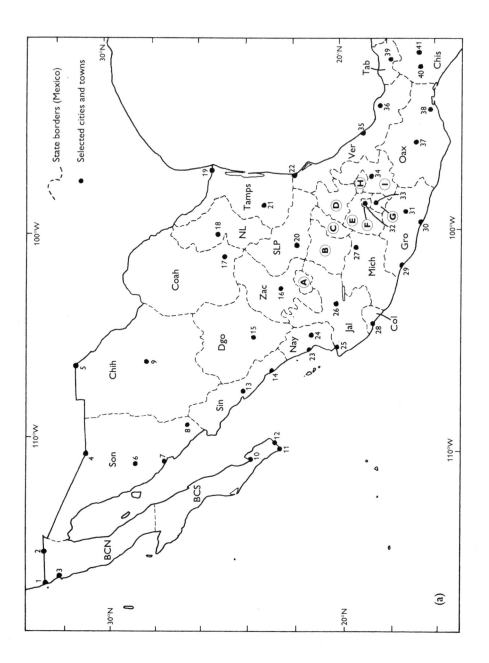

(a)

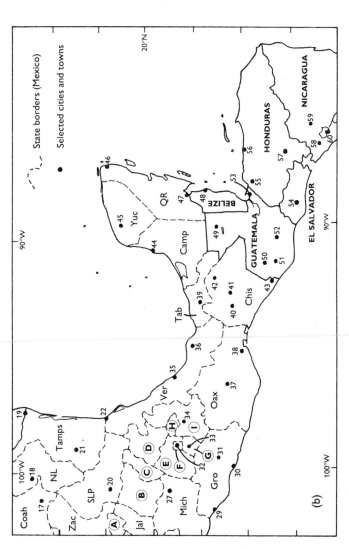

Fig. 2 (a) Political divisions of north and central Mexico; (b) political divisions of southern Mexico and northern Central America. **A**: Aguas Calientes; **B**: Guanajuato; **C**: Queretaro; **D**: Hidalgo; **E**: Mexico; **F**: Distrito Federal: **G**: Morelos; **H**: Tlaxcala; **I**: Puebla. **1**: Tijuana/San Diego (USA); **2**: Mexicali/Calexico (USA); **3**: Ensenada; **4**: Nogales/Nogales (USA); **5**: Ciudad Juarez/El Paso (USA); **6**: Hermosillo; **7**: Guaymas; **8**: Alamos; **9**: Chihuahua; **10**: La Paz; **11**: Cabo San Lucas; **12**: San José del Cabo; **13**: Culiacan; **14**: Mazatlan; **15**: Durango; **16**: Zacatecas; **17**: Saltillo; **18**: Monterrey; **19**: Matamoros/Brownsville (USA); **20**: San Luis Potosi; **21**: Ciudad Victoria; **22**: Tampico; **23**: San Blas; **24**: Tepic; **25**: Puerto Vallarta; **26**: Guadalajara; **27**: Morelia; **28**: Manzanillo; **29**: Zihuatanejo; **30**: Acapulco; **31**: Chilpancingo; **32**: Mexico City; **33**: Cuernavaca; **34**: Puebla; **35**: Veracruz; **36**: Catemaco; **37**: Oaxaca; **38**: Tehuantepec; **39**: Villahermosa; **40**: Tuxtla Gutierrez; **41**: San Cristobal de las Casas; **42**: Palenque; **43**: Tapachula; **44**: Campeche; **45**: Mérida; **46**: Cancún; **47**: Chetumal; **48**: Belize City; **49**: Tikal; **50**: Huehuetenango; **51**: Quezaltenango; **52**: Guatemala City; **53**: Puerto Barrios; **54**: San Salvador; **55**: San Pedro Sula; **56**: La Ceiba; **57**: Tegucigalpa; **58**: León; **59**: Matagalpa; **60**: Managua.

GEOGRAPHY AND BIRD DISTRIBUTION

In this chapter we outline the physical geography of Mexico and northern Central America, the main regions of bird distribution (see Figs 3(a) and (b)), and species characteristic of each region. Our divisions are based primarily on biogeographic considerations and are somewhat broader than biotic divisions such as those of Goldman and Moore (1945). Biogeography relates the distribution of living organisms to factors such as topography, climate, and vegetation. In addition, one must consider temporal aspects such as glaciations and changing sea levels. A basic knowledge of avian distribution helps one to appreciate the reasons underlying species distribution, and may simplify potential identification problems. Related to this, one should realize that much of the region's avifauna is sedentary, unlike in North America or Europe where the majority of species are migratory.

As a starting point, one may consider the lowlands of the USA/Mexico border a reasonable division between tropical and temperate avifaunas. While a number of temperate species range south into Middle America in the cooler highlands, tropical species, such as Tropical Kingbird and Tropical Parula, range only a short distance north of the USA border. In addition, several North American species barely reach northern Mexico as breeders, for example Red-breasted Nuthatch and Ruby-crowned Kinglet (which, in the region covered by this guide, breed only on Isla Guadalupe, off north Baja). In addition, the Mexico/USA border straddles the arid interior and western desert regions whose affinities cannot be placed with either country.

The topography of Mexico and northern Central America is highly complex, which often results in extremely localized bird distributions. There are two coastal plains (Atlantic and Pacific) of varying width which, sooner or later, run into often imposing mountain ranges, or sierras, inland. These sierras serve to separate the coastal slopes from the interior regions, for example the Mexican Plateau, which is bounded to the west by the Sierra Madre Occidental, to the east by the Sierra Madre Oriental, and to the south by the Central Volcanic Belt. The Isthmus of Tehuantepec is the only low-lying connection between the two coastal slopes in the region, and is therefore an important biogeographic feature. It has acted and still acts both as a barrier to highland species to the north and south, and as a bridge for species to cross between Pacific and Atlantic slopes. The ranges of many species end in the mountains and interior of Oaxaca 'north' (more properly west) of the Isthmus.

Glaciations have played a major role in the avian endemism seen today in Mexico and Central America. With an overall cooling of climate, much water became locked up as ice, resulting in a drier climate. Humid forest remained only in the wettest areas, such as the Selva Lacandona, or the lower Río Dulce/Río Motagua drainage. Birds of arid regions were forced into the few remaining pockets (or refugia) of warm conditions, such as the Balsas drainage or the Pacific side of the Isthmus of Tehuantepec. In isolation, new forms often developed, and some dispersed later following glacial retreat. On the other hand, species typical of temperate conditions ranged more widely during glaciations. With a warming climate and glacial retreat, however, they were forced to find pockets of suitable habitat, usually on higher slopes, such as the Sierra Madre del Sur of Guerrero and Oaxaca or Los Tuxtlas in southern Veracruz. There, isolated from similar pockets, species evolved and diverged. In addition, for long periods in the geologic past, higher sea levels separated Central America from the vast avian reservoirs of both North and South America, again allowing species to develop in isolation.

Three broad terrestrial divisions can be identified in the region: the Pacific Slope, the Interior, and the Atlantic Slope, each of which can be divided further. Sierras may fall within any of these divisions. For example, the western slopes of the Sierra Madre Occidental are part of the Pacific Slope, while the eastern slopes have their affinities with the interior. The offshore Pacific Ocean constitutes yet another distinct region treated here under the Pacific Slope.

Pacific Slope

A. Offshore Pacific islands

These include Isla Guadalupe, the Islas Revillagigedo (San Benedicto, Socorro, Roca Partida, and Clarión), and Clipperton Atoll. Like many offshore islands, Mexican islands have a high percentage of endemic species and subspecies which, inevitably (?), have been decimated by introduced mammals.

Isla Guadalupe rises to 1200 m at its northern end where remnant groves of pines, oaks, and cypress survive. Its breeding birds are of temperate origin. Of the eight endemics, the Guadalupe Storm-Petrel is presumed extinct and the Guadalupe Junco persists only in small numbers; the endemic subspecies of Crested Caracara, Bewick's Wren, and Rufous-sided Towhee are extinct. Guadalupe's breeding seabirds include Black-vented Shearwater, Leach's Storm-Petrel, Western Gull, Xantus' Murrelet, Cassin's Auklet, and the recently arrived Laysan Albatross.

The Islas Revillagigedo (specifically the larger islands of Socorro, rising to over 1000 m, and Clarión) have an interesting mix of breeding species. Some, such as Socorro and Clarion wrens, are of temperate origin; others, like Socorro Parakeet, are of tropical origin. Of the 15 endemic breeders (13 of them landbirds), the Socorro Dove is presumed extinct, and Townsend's Shearwater and Socorro Mockingbird are endangered; the San Benedicto subspecies of Rock Wren was wiped out by a volcanic eruption in 1952. The unique location of the Islas Revillagigedo is reflected in their varied breeding seabird fauna which includes Laysan Albatross, Wedge-tailed Shearwater, Red-billed and (probably) Red-tailed tropicbirds, and Great and Magnificent frigatebirds.

Clipperton Atoll, about 1000 km southwest of the coast of Guerrero, is sufficiently remote that no landbirds have colonized it. Although only 1.8 square km in area, it constitutes an important breeding ground for thousands of seabirds, including vast colonies of Brown and Masked boobies. Smaller numbers of other seabirds also breed there, including Black Noddy and White Tern

Fig. 3(a) Biogeographic divisions of north and central Mexico.

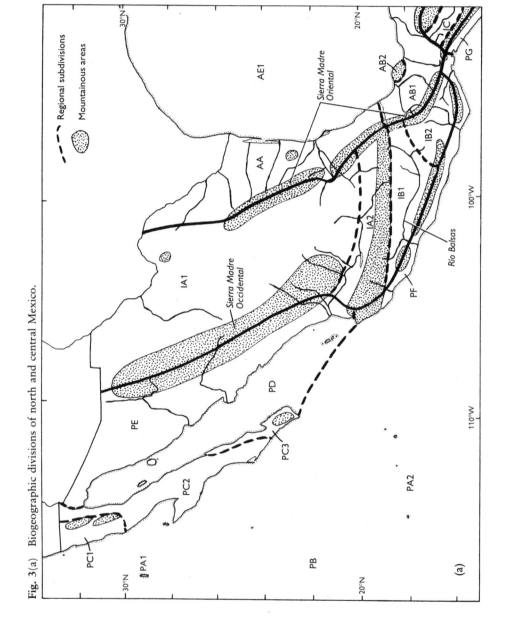

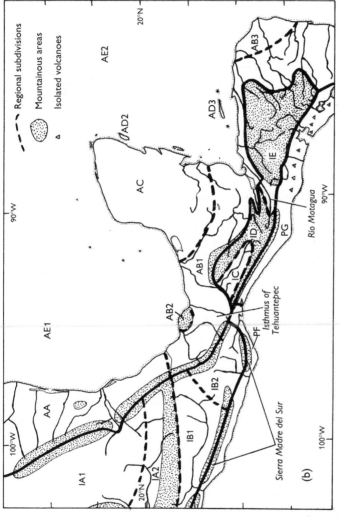

Fig. 3(b) Biogeographic divisions of southern Mexico and northern Central America.

Bold lines separate the three major regions (P: Pacific Slope; I: Interior; A: Atlantic Slope). The second letter of the abbreviation refers to the geographic area under that alphabetical heading in the text. PA1: Isla Guadalupe; PA2: Islas Revillagigedo; PB: offshore pacific waters; PC1: northwest Baja; PC2: Baja deserts; PC3: Cape District; PD: Gulf of California; PE: northwest Mexico; PF: southwest Mexico; PG: Central American Pacific Slope. IA1: Mexican Plateau; IA2: Central Volcanic Belt; IB1: Río Balsas drainage; IB2: interior Oaxaca; IC: central Valley of Chiapas; ID: Chiapas–Guatemala Highlands; IE: Interior highlands and valleys of northern Central America. AA: northeast Mexico; AB1: humid southeast; AB2: Sierra de Los Tuxtlas; AB3: Mosquitia; AC: Yucatan Peninsula; AD1: Campeche Bank; AD2: Isla Cozumel; AD3: Honduras Bay Islands; AE1: offshore Gulf of Mexico; AE2: offshore Caribbean.

which do not breed elsewhere in the guide region. Clipperton, not surprisingly, acts as a magnet to lost migrants and has hosted breeding populations of American Coot and Common Moorhen, presumably derived from vagrants.

B. Offshore Pacific waters

These extend to 200 nautical miles from habitable land, and contribute a distinctive, yet poorly known element to the Mexican and Central American avifauna. The offshore region can be divided into the temperate waters off northwest Baja, subtropical waters south to the vicinity of the Islas Revillagigedo, and tropical waters south to Clipperton Atoll. An interesting oceanographic feature is the Middle American Trench which attains depths of 4000–5000 m as close as 50 km from shore between Jalisco and Oaxaca.

Pelagic species off northwest Baja are much like those off adjacent California, with regular non-breeding visitors including Black-footed Albatross, Cook's Petrel, and Sooty Shearwater, but things soon change south from there. Some species (for example, Galapagos Storm-Petrel) occur commonly offshore north to waters off BCS throughout the year. The majority of species, however, appear to occur mostly from May to October, ranging to these subtropical and tropical waters from their nesting islands in the central and south Pacific. The abundance and seasonal occurrence of many species almost certainly varies from year to year in response to oceanographic conditions. The commoner non-breeding pelagic visitors known to this area include Dark-rumped, Juan Fernandez, Tahiti, and Kermadec petrels, and Pink-footed, Christmas, and Audubon's shearwaters.

C. Baja California

The peninsula of Baja California can be divided into three regions: the northern mountains (rising to 2000 m and 3000 m in the Sierra de Juarez and Sierra San Pedro San Martir respectively) and their western slopes, the deserts which constitute most of the peninsula, and the tropical Cape District at the southern tip.

The northern sierras are closely allied to the mountains of southern California and support several species that occur nowhere else in Mexico. Species found at higher altitude, mostly in the conifer forests, include California Condor (extinct), Mountain Quail, Calliope Hummingbird, Williamson's Sapsucker, Pinyon Jay, Clark's Nutcracker, Mountain Chickadee, and Cassin's Finch. The lower slopes of chaparral, oaks, and riparian groves host breeding species such as Red-shouldered Hawk, Anna's Hummingbird, Nuttall's Woodpecker, Wrentit, California Thrasher, Lazuli Bunting, and Lawrence's Goldfinch. Ashy Storm-Petrel, Pelagic Cormorant, and Black Oystercatcher, which breed off and along the mostly rocky coasts north of 30° N, and winter visitors such as Red-breasted Sapsucker, Varied Thrush, and Fox and Golden-crowned sparrows, emphasize the similarity between northwest Baja and California.

Most of Baja comprises rugged inhospitable deserts which host California Quail, California Gnatcatcher, Grey Thrasher, and California Towhee, as well as a variety of more widespread desert species, some of which also range into the northwestern chaparral and into the Cape District. Belding's Yellowthroat is endemic to isolated freshwater marshes in the deserts of south Baja. The northeastern deserts and adjacent Río Colorado delta are more closely allied to the Sonoran Desert than to peninsular Baja. Replacement species there include Gambel's Quail, Black-tailed Gnatcatcher, and Abert's Towhee. Except for a few bays and estuaries, the hot sandy beaches and rocky headlands of the Pacific coast are distinctly different from the coast of northwest Baja, and mangroves (and Mangrove Warblers) start appearing at about 27° N. The inshore islands of Baja's central Pacific coast hold important colonies of several seabirds, including Black-vented Shearwater, which breeds only in Mexico.

The tip of Baja extends just across the Tropic of Cancer and, together with the Sierra Victoria which rises to 2000 m, helps give the Cape District its distinctive avifauna. The lower slopes of the Sierra are covered with arid deciduous scrub, while the upper slopes support an outpost of pine–oak woodland. The endemic Xantus' Hummingbird has its center of abundance in the Sierra, although it also occurs widely in BCS, while Cape Pygmy-Owl, San Lucas Robin, and Baird's Junco are endemic to the Cape District sierras. As one might expect, the tip of Baja is an excellent place to find migrants and vagrants, and has produced, among other species, the first North American record of White Wagtail.

D. Gulf of California

The Gulf of California and its numerous desert islands comprise an important area for seabirds. A high percentage of the world populations of Black (70%) and Least (90%) storm-petrels, Blue-footed Booby (40%), Heermann's (90% +) and Yellow-footed (100%) gulls, Elegant Tern (95% +), and Craveri's Murrelet (90%) breed in the Gulf, as well as large numbers of Brown Boobies, Brown Pelicans, and Royal Terns.

E. Northwest Mexico

This region extends in the lowlands from the Sonoran Desert to southern Nayarit, where the coastal plain is squeezed out by mountains extending almost down to the sea. It also includes the Sierra Madre Occidental south to the Río Grande de Santiago/Río Lerma drainage in southern Nayarit and northern Jalisco.

Birds of the Sonoran Desert (extending south to northern Sinaloa) are much like those in northeast Baja California with a few additions such as Bendire's Thrasher and Rufous-winged Sparrow, and the loss of Río Colorado specialities like Abert's Towhee. As one moves south and inland, the desert gives way to deciduous thorn scrub and woodland. Species characteristic of this habitat in northwest Mexico include Rufous-bellied Chachalaca, Elegant Quail, Black-throated Magpie-Jay, Purplish-backed Jay, Black-capped Gnatcatcher, and Five-striped Sparrow. In southern Sinaloa and Nayarit, the thorn forest may grade into taller, tropical semideciduous forest, and into oak scrub, pine–oak, and pine forest in the Sierra Madre. Fingers of tropical forest extend north and inland along river valleys to southern Sonora and southwestern Chihuahua. Birds of the semideciduous forest provide the first real taste of the Tropics on the Pacific Slope, and include widespread tropical species, for example Laughing Falcon, Squirrel Cuckoo, and Masked Tityra, as well as endemics such as Mexican Parrotlet. The forests of the Sierra Madre and its canyons, or barrancas, host many typical pine—oak forest birds, and specialities such as Thick-billed Parrot, Eared Quetzal, Imperial Woodpecker (extinct?), Sinaloa Martin, and Tufted Jay.

Extensive irrigation schemes in the Río Colorado delta have greatly reduced the attractiveness of what was once a major wintering area for waterfowl. South from the delta, the coast to southern Sonora is mostly arid and rocky, but thence to central Nayarit are numerous shallow bays, coastal lagoons, and marshes. These areas, including most notably the Marismas Nacionales of southern Sinaloa and northern Nayarit, support large wintering populations of waterfowl and shorebirds.

The Islas Tres Marias, 80–100 km off the coast of Nayarit, have an avifauna similar to the adjacent mainland but several species have diverged at least to the subspecific level. Birds characteristic of the islands include Mexican Parrotlet, Yellow-headed Parrot, Lawrence's [Broad-billed] Hummingbird, Happy Wren, Grayson's Thrush, Tropical Parula, and Streak-backed Oriole.

F. Southwest Mexico

This region extends from southern Nayarit to the Isthmus of Tehuantepec, with marked endemism occurring in the Sierras which back the narrow coastal plain. The Río Balsas, an important biogeographic feature, cuts down to the coast where it forms the border between the states of Michoacan and Guerrero. The two main mountainous areas are the coastal-facing slopes from southern Nayarit to Colima (including the Volcanes de Colima), and the Sierra Madre del Sur of Guerrero and Oaxaca.

The coastal plain supports mostly arid deciduous woodland, or thorn forest, with an avifauna similar to that of northwest Mexico, but typical species include West Mexican Chachalaca, Doubleday's Hummingbird (east of the Balsas), White-throated Magpie-Jay, San Blas Jay, White-lored Gnatcatcher, and Orange-breasted Bunting. Rosita's Bunting and Sumichrast's Sparrow are endemic to thorn forest on the Pacific side of the Isthmus.

In the mountains from southern Nayarit to Colima, humid pine—oak forest with dense understory covers the upper slopes, while lower down is found fairly tall, tropical, semideciduous forest which, in shady barrancas, often retains an evergreen aspect. The avifauna includes an interesting mix of highland species at the western end of the Central Volcanic Belt, and Pacific Slope endemic species and subspecies. The avifauna is strikingly different from that of the Sierra Madre Occidental and includes Ornate Hawk-Eagle, Long-tailed Wood-Partridge, Singing Quail, Mexican Hermit, Mexican Woodnymph, Olivaceous Woodcreeper, Scaled Antpitta, Grey-breasted Wood-Wren, Chestnut-sided Shrike-Vireo, and Cinnamon-bellied Flowerpiercer.

All but the Mexican Woodnymph also occur in the Sierra Madre del Sur where bird diversity is much greater owing to the Río Balsas which has apparently prevented many species from spreading farther north on the Pacific Slope. Like the mountains from Nayarit to Colima, the Sierra Madre del Sur consists of several smaller sierras. These may be lumped biogeographically into the Sierra de Atoyac of Guerrero and the Sierra de Miahuatlán of Oaxaca, separated by the Río Verde drainage. The middle elevations of these sierras are covered with often extensive tracts of humid evergreen forest (cloud forest), which there reaches its northern limit on the Pacific Slope. Birds of the Sierra Madre del Sur are a mixture of endemics (both Mexican and of the Sierra itself), Central American cloud forest species, and a few more widespread tropical species. These include Barred Forest-Falcon, Pheasant Cuckoo, Garnet-throated Hummingbird, Collared Trogon, Spectacled and Ruddy foliage-gleaners, Strong-billed and Barred woodcreepers, White-throated and Unicolored jays, Black Thrush, and Slate-blue Seedeater; in addition, Short-crested Coquette and White-tailed Hummingbird are endemic to the Sierra de Atoyac, and Blue-capped Hummingbird to the Sierra de Miahuatlán.

The coast of southwest Mexico is mostly arid and rocky. Mangroves occur wherever conditions are favorable, and there are large lagoons in Colima and Guerrero which support good populations of wading birds and wintering waterfowl. Numerous rocky islets dot the inshore waters and are used for roosting and nesting by tropicbirds, boobies, pelicans, and terns.

G. Central American Pacific Slope

The Pacific Slope of northern Central America is, in many ways, much like that of southwest Mexico, but it has a greater diversity of tropical species. Sadly, the Pacific coastal plain from the Isthmus to the Gulf of Fonseca has been cut, burned, and converted mostly to an open, agricultural wasteland, with little trace remaining of the natural forest.

The range of a number of thorn forest species of the Mexican Pacific Slope ends at, or north of, the Isthmus. This is due in part to the more humid nature of the Pacific lowlands of Chiapas which are unsuitable for birds adapted to arid thorn forest. The Isthmus also marks the northern Pacific Slope limit of numerous tropical lowland species, many of them declining, for example Red-throated Caracara (extirpated?), White-bellied Chachalaca, Scarlet Macaw (extirpated), Orange-chinned Parakeet, and Yellow-naped Parrot. Species of open habitats, however, such as Giant Wren, endemic to the Pacific lowlands of Chiapas, and Striped Cuckoo, Rufous-breasted Spinetail, and Common Tody-Flycatcher, may have increased with forest clearing.

The arid Pacific Slope resumes in eastern Guatemala and consequently one finds species such as Spot-bellied Bobwhite, White-lored Gnatcatcher, and Streak-backed Oriole in that area. Hoffmann's Woodpecker barely enters into the guide region at the Gulf of Fonseca.

The vegetation of the foothills and highlands of the Sierra Madre de Chiapas (extending from eastern Oaxaca to Guatemala) and its extensions in the isolated volcanic peaks of El Salvador, resembles that of the Sierra Madre del Sur. The avifauna of the foothills and tongues of gallery forest suggests a poorer version of the Atlantic Slope humid forests mixed with Central American endemics. It includes White Hawk, Spectacled and Black-and-white owls, Lesser Swallow-tailed Swift, Blue-throated Sapphire, Blue-tailed Hummingbird, Collared Araçari, White-necked Puffbird, Royal Flycatcher, Long-tailed Manakin, Cabanis' Tanager, and White-eared Ground-Sparrow.

The higher cloud forest of the Sierra Madre de Chiapas is closely allied to the interior highlands of Chiapas and northern Central America, and its avifauna is included with the Interior regions. In the relatively narrow bottleneck of northern Central America, the complexities of topography make it difficult to distinguish between Pacific Slope and Interior. Many species of the Pacific Slope are also typical of the Interior, though the humid Atlantic Slope lowlands remain relatively distinct. In addition, some species typical of the Pacific Slope occur in arid interior valleys on the Atlantic Slope of Guatemala and Honduras.

The coast of northern Middle America is mostly low-lying, with numerous estuaries and extensive lagoons which are important for breeding and migrant wading birds and for migrant shorebirds. Relatively few wintering ducks occur south of Chiapas. The arid rocky coastline and islets of the Gulf of Fonseca resemble the Gulf of California and support populations of Red-billed Tropic-bird, Blue-footed and Brown boobies, and Bridled Tern.

Interior

A. Mexican Plateau and Central Volcanic Belt

The rolling plains of the Plateau slope gradually from about 500–1000 m in the north to 1500–2000 m in the south where they merge with the Central Volcanic Belt whose highest peaks rise to over 5000 m. The southern edge of the Central Volcanic Belt drops into the arid Río Balsas drainage while to the west and east the volcanoes merge with the coastal-facing sierras.

The Plateau is similar overall to deserts of the southwestern USA. Much of the land is used for ranching and, in the warmer southern sections, for intensive cultivation. The higher slopes are covered with arid oak and pine–oak woodland. Worthen's Sparrow is the only Plateau endemic; otherwise, the avifauna consists mostly of widespread desert and grassland species such as Scaled Quail, Chihuahuan Raven, Verdin, Black-tailed Gnatcatcher, and Cassin's and Black-throated sparrows. With the onset of typically cold winters, many species breeding on the Plateau withdraw south or to warmer lowlands. During this season, the extensive lakes of the Plateau hold thousands of migrant waterfowl and Sandhill Cranes.

The peaks of the Central Volcanic Belt, from Pico de Orizaba in the east to the Volcanes de Colima in the west, rise as forested islands from the dusty agricultural plains at the south end of the Plateau. From oaks on the lower slopes, one climbs into pine–oak and then pine forest, with firs on the cooler and wetter slopes. The peaks of the highest volcanoes are barren and rocky, covered with snow throughout the year. Both the highland forests and the scattered lakes and marshes of the Volcanic Belt are important centers of endemism, although many of the 'central highland endemics' also range north into the Sierra Madre Occidental and south into the sierras of Guerrero and Oaxaca.

Endemics and isolated populations of wide-ranging species found in the Volcanic Belt marshes include American Bittern, Yellow Rail (extirpated?), King Rail, Marsh Wren, Black-polled Yellowthroat, and Slender-billed Grackle (extinct); the Mexican Duck also ranges north to lakes in the western Plateau. Birds typical of the highland forests include Strickland's Woodpecker, Grey-barred Wren, Russet and Ruddy-capped nightingale-thrushes, Red Warbler, and Olive-backed Towhee, while Striped and Sierra Madre sparrows inhabit bunch-grass clearings in open pine woods.

B. Southwest Interior

This region may be divided into the Río Balsas drainage and the interior valleys of central Oaxaca. Between them, Mexico's southwest Pacific Slope and arid interior constitute the richest area of avian endemism in northern Middle America. Owing to its very hot climate, the Río Balsas basin was an important refuge during glaciations. The avifauna of the interior valleys of central Oaxaca is similar in many ways to that of the Balsas, but the higher elevations in central Oaxaca support mostly subtropical and temperate scrub and less Pacific Slope thorn forest. As a consequence, many Pacific Slope species that extend inland along the Balsas are absent from central Oaxaca, for example Orange-fronted Parakeet, Lesser Ground-Cuckoo, Russet-crowned Motmot, Golden-cheeked Woodpecker, Happy Wren, and Orange-breasted Bunting.

Birds characteristic of the arid Southwest Interior include Dusky Hummingbird, Grey-breasted Woodpecker, Pileated Flycatcher, Boucard's Wren, Ocellated Thrasher, and Slaty and Dwarf vireos. Some of these range north locally into the Central Volcanic Belt. More locally, the Banded Quail, Balsas Screech-Owl, and Black-chested Sparrow are endemic to the Balsas drainage, while Beautiful Hummingbird, White-throated Towhee, and Bridled and Oaxaca sparrows are typical of central Oaxaca.

C. Central (interior) Valley of Chiapas

Although the Río Grijalva, which drains the Central Valley of Chiapas, flows into the Gulf of Mexico, the valley is basically isolated from the Atlantic Slope by mountains through which the Grijalva cuts its exit via the spectacular Sumidero Canyon. Tropical deciduous forest and scrub used to cover the Central Valley but most has long since been cleared for agriculture. The foothills still support semideciduous woodland and, higher up, cloud forest, much of which has been converted to coffee plantations. The cloud forest avifauna is treated with the Chiapas–Guatemala highlands. The Central Valley's avifauna includes a mix of arid Pacific Slope species, some at the southern limit of their range, and Central American endemics typical of the subtropical interior from the Isthmus to northern El Salvador and western Honduras. Species in the former category include Flammulated Flycatcher and Red-breasted Chat, while Rufous Sabrewing, Slender Sheartail, Belted Flycatcher, and Bar-winged Oriole are characteristic of the latter.

D. Chiapas–Guatemala highlands (including southeastern Oaxaca)

While closely allied to the highlands of Honduras and north-central Nicaragua, the Chiapas–Guatemala highlands, rising to over 3000 m, have their own distinctive avifauna. Humid pine–oak forest and cloud forest once covered much of these highlands but today are restricted to the steeper, higher, and more remote slopes since much of the land has been cleared for agriculture. The highland avifauna includes northern, endemic, and Central American species. Pine Flycatcher, Grey Silky, Rufous-sided Towhee, Yellow-eyed Junco, and Hooded Grosbeak all are at the southern end of their range here. The Chiapas–Guatemala highland endemics are Atitlan Grebe (extinct?), Horned Guan, Bearded Screech-Owl, Pink-headed Warbler, and Black-capped Siskin, while Green-throated Mountain-Gem, Wine-throated Hummingbird, Blue-throated Motmot, Black-capped Swallow, Black-throated Jay, Blue-and-white Mocking-bird, and Rufous-collared Thrush range no farther south than central Honduras.

E. Interior highlands and valleys of northern Central America (from eastern Guatemala to western Nicaragua)

The highlands of Honduras and north-central Nicaragua, rarely rising above 2400 m, are separated from the Chiapas–Guatemala highlands by low-lying hills and depressions which run roughly along the eastern border of Guatemala. The Río Motagua Valley, which cuts inland from the Caribbean coast of Guatemala, helps emphasize the break between Honduras and most of Guatemala.

The Honduras–Nicaragua highlands support mostly semiarid pine–oak forest, with more humid forest, including cloud forest, on the higher slopes. As there is no well-defined sierra along the Pacific Slope in this region, species typical of the arid Pacific Slope spill into the interior of Honduras, including valleys draining north to the Caribbean. Atlantic Slope species range inland along these same valleys so one may find, for example, Berylline, Rufous-tailed, and Cinnamon hummingbirds feeding side-by-side.

Of the interior valleys, the upper Río Motagua has the greatest Pacific Slope component, with birds such as Spot-bellied Bobwhite, Lesser Ground-Cuckoo, Buff-collared Nightjar, White-throated Magpie-Jay, Rufous-naped Wren, White-lored Gnatcatcher, and Streak-backed Oriole, as well as Russet-crowned Motmot and Varied Bunting at the southern end of their range. The arid interior valleys of Honduras have fewer Pacific Slope species but their varied avifauna includes Atlantic Slope species such as Aztec Parakeet, and localized populations of Green Jay and White-bellied Wren. In addition, the recently rediscovered endemic Honduran Emerald has evolved in isolation in these valleys.

Species typical of the semiarid interior highlands include Red-throated Parakeet, Elegant Trogon, and Bushy-crested Jay. The higher pine–oak and cloud forests are similar to those of Chiapas and Guatemala, with species such as White-breasted Hawk, Highland Guan, Buffy-crowned Wood-Partridge, Fulvous Owl, Resplendent Quetzal, Mountain Elaenia, and Mountain Thrush. The Green-breasted Mountain-Gem is endemic to the Honduras–Nicaragua highlands east of the Sula Valley, and the highlands of north-central Nicaragua support an outpost of Three-wattled Bellbirds, a species otherwise restricted to Costa Rica and Panama. In addition, the Honduras–Nicaragua highlands mark the southern extent of native pines in the New World, and with them the southern limit of birds such as Mexican Whip-poor-will, Steller's Jay, Brown Creeper, Eastern Bluebird, Painted Redstart, and Olive Warbler.

Atlantic Slope

A. Northeast Mexico

The Atlantic Slope of northeast Mexico extends from the USA border along the Río Grande to the vicinity of Jalapa, Veracruz, where the Central Volcanic Belt nearly squeezes out the coastal plain. In the north, the plains bordering the Río Grande merge into the Mexican Plateau, and birds typical of the Plateau (Greater Roadrunner, Cactus Wren, etc.) occur on the Atlantic Slope in Tamaulipas. As on the Pacific Slope, desert scrub gives way to deciduous woodland which may grade into tropical semideciduous forest as one moves south and inland. The rugged Sierra Madre Oriental backs much of the coastal plain, and may be divided into the drier, more interior-like ranges from Coahuila to southern Nuevo León, and the wetter, coastal-facing ranges from central Tamaulipas to central Veracruz.

The native vegetation of the Atlantic Slope has been drastically changed by the activities of man, far more so than the Pacific Slope, and little untouched forest remains. Most of the lowlands have been converted to ranching and arable land; today it is hard to believe that the lowlands of southern Tamaulipas and northern Veracruz were once forested. A few corridors of gallery forest survive along some rivers, and one still can find tracts of thorn forest in places. Most of the forest birds are now restricted to the foothills and highlands where forest persists on steep limestone ridges, or where coffee plantations have allowed trees to remain.

Species characteristic of the lowlands and foothills of northeast Mexico include Green Parakeet, Red-crowned Parrot, Tawny-collared Nightjar, Tamaulipas Crow, Carolina Wren, Altamira Yellowthroat, and Crimson-collared Grosbeak, as well as many widespread tropical species such as Red-lored Parrot, Blue-crowned Motmot, Lineated Woodpecker, Boat-billed Flycatcher, and Golden-crowned Warbler. Thus, on the Atlantic Slope, one only has to venture 300 km south of the USA border to get truly into the Neotropics. The southern limit of the northeast coastal plain is an interesting pocket of arid thorn forest and scrub, now mostly reduced to small remnant patches, between Veracruz City and Jalapa. Birds characteristic of this area include Buff-collared Nightjar, Mexican Sheartail, Rufous-naped and White-bellied wrens, and Varied Bunting.

Spectacular ridges and volcanic peaks form the northern sierras. Desert vegetation around their bases gives way to chaparral, then oaks, and pine–oak forest, with firs in cooler, wetter areas. Many species are the same as in northwest Mexico, and the northern Sierra Madre Oriental is an outpost for Mexican highland endemics such as Russet Nightingale-Thrush, Hooded Yellowthroat, and Rufous-capped Brushfinch, and for two western North American species, Clark's Nutcracker and MacGillivray's Warbler. Other birds typical of these sierras include Maroon-fronted Parrot, Pine Flycatcher, Grey Silky, and Colima Warbler.

The avifauna of the coastal-facing ranges from central Tamaulipas to central Veracruz is closely linked to that of the tropical foothills mentioned above, but higher elevations support stands of cloud forest. While a few birds such as Bumblebee Hummingbird and Black Thrush range north to Tamaulipas, the valley of the Río Santa Maria in southeast San Luis Potosí has served as a barrier for most cloud forest species. South of this river, the remaining patches of cloud forest in the Sierra Madre Oriental host species such as Barred Forest-Falcon, White-faced Quail-Dove, Garnet-throated Hummingbird, Emerald Toucanet, Tawny-throated Leaftosser, Scaled Antpitta, Azure-hooded Jay, and Slate-colored Solitaire, which all range south into Central America, in addition to Bearded Wood-Partridge and White-naped Brushfinch, both endemic to eastern Mexico.

Vast lagoon complexes protected by barrier beaches characterize the coast of northeast Mexico. The lagoons are fringed by mangroves and constitute important nesting areas for waterbirds in general, as well as important wintering grounds for migrant waterfowl and shorebirds.

B. The humid southeast

The 'humid southeast' refers to the region formerly covered by humid evergreen forest (rain forest) in the Atlantic Slope lowlands from northern Oaxaca and southern Veracruz to northern Honduras; the Yucatan Peninsula is treated separately. Sadly, the forests of the humid southeast have been all but cleared by man, mostly to provide poor grazing land for cattle. Today, species such as Cattle Egret, Great-tailed Grackle, and White-collared Seedeater are usually associated with an area that, fifty or so years ago, supported Harpy Eagles and Scarlet Macaws. The mountains backing the southeast coastal plain support cloud forest and pine–oak forest, although much has been cleared.

The northernmost extensions of neotropical rain forest are Los Tuxtlas in southern Veracruz, and the foothills of northern Oaxaca in the vicinity of Tuxtepec. Species at their northern limit here include Great Tinamou, Spotted Wood-Quail, Scaled Pigeon, Brown-hooded Parrot, White-necked Jacobin, Slaty-tailed Trogon, Rufous-tailed Jacamar, Chestnut-colored Woodpecker, Plain Xenops, Lovely Cotinga, Green Shrike-Vireo, Green Honeycreeper, and Crimson-collared Tanager, all of which range south at least to southern Central America.

The diversity of neotropical birds then increases in steps southward, significant steps in Mexico being the Isthmus of Tehuantepec (adding species such as

White-necked Puffbird, Scaly-throated Leaftosser, Cinnamon Becard) and the Selva Lacandona (adding species such as Short-tailed Nighthawk, Great Potoo, Purple-crowned Fairy, Plain Antvireo). The lower Río Dulce/Río Motagua valleys of eastern Guatemala mark another important biogeographic break (the northern limit for species such as Olivaceous Piculet, Buff-throated Woodcreeper, Grey-headed Piprites), with several species (for example, Band-tailed Barb-throat, Crowned Woodnymph, Bare-crowned Antbird) also entering southern Belize where the northern limit of their range is defined by the Maya Mountains.

These are all fairly small steps in comparison to the Sula Valley in western Honduras, which marks the Atlantic Slope boundary for this guide. Crossing the Sula Valley, one finds 50 species in eastern Honduras that do not occur north, or more correctly west, of the valley (see Appendix E).

In the humid southeast there are some natural clearings in the form of grassy savannas where characteristic birds include Double-striped Thick-knee, Plain-breasted Ground-Dove, and Fork-tailed Flycatcher, while the Spot-tailed Night-jar occurs in savannas in the Isthmus. Pine savannas occur locally in the humid southeast from northern Guatemala to the Mosquitia of northeastern Honduras and adjacent Nicaragua. These interesting features, known as pine ridges in Belize, support a distinctive avifauna with characteristic breeding birds such as Azure-crowned Hummingbird, Acorn Woodpecker, Vermilion Flycatcher, Grace's Warbler, Hepatic Tanager, and Botteri's, Rusty, and Chipping sparrows.

The cloud forests of the sierras backing the coastal lowlands from central Veracruz to the Isthmus support an avifauna similar to that of northeast Mexico, while the higher pine—oak forests are much like the Central Volcanic Belt highlands. Some additions to the cloud forest list include Emerald-chinned and Stripe-tailed hummingbirds, Black-crested Coquette, and Blue-crowned Chlorophonia. The Dwarf Jay is endemic to the humid pine—oak forests of central Veracruz and northern Oaxaca, and Sumichrast's and Nava's wrens are endemic to forested limestone outcrops from central Veracruz to northwest Chiapas.

As well as being the northern limit of humid evergreen forest, Los Tuxtlas, rising from the coastal plain to 1600 m, are a minor center of endemism. Birds characteristic of the Sierra include Purplish-backed Quail-Dove, Long-tailed Sabrewing, and Plain-breasted Brushfinch, with northern outposts of Yellowish Flycatcher and Yellow-backed Oriole.

South, or more correctly east, of the Isthmus, the Atlantic-facing sierras of Chiapas and Guatemala support a mixture of Atlantic Slope cloud forest species and Central American species (dealt with under the Interior). Some of the humid evergreen forest species gained by the step into the Selva Lacandona appear there to be more typical of foothills (for example Slaty Antwren, Nightingale Wren, and Shining Honeycreeper), although farther south they are as typical of lowland forest.

From southern Veracruz to Belize there are extensive freshwater marshes and lagoons which support large colonies of wading birds, including a small popula-tion of Jabirus.

C. The Yucatan Peninsula

The Yucatan Peninsula, or simply the Yucatan, is a low-lying limestone shelf protruding north into the Caribbean. It includes the states of Yucatan, Campeche, and Quintana Roo plus the northern parts of the Petén Department of Guatemala and the northern parts of Belize. The vegetation grades from rain forest in the south to arid coastal scrub along the north coast. There is a gradual decrease in species diversity as one moves north into southern Campeche and southern Quintana Roo. This decrease continues almost imperceptibly north through Quintana Roo where it is offset slightly by the occurrence of Yucatan endemics and Caribbean species. Species diversity drops off and the avifauna changes quickly as one moves into the drier northwest corner of the Yucatan, a region that corresponds roughly to the state of Yucatán.

The Yucatan is a center of avian endemism, and characteristic species of the peninsula include Ocellated Turkey, Yucatan Bobwhite, Yucatan Parrot, Yucatan Poorwill, Yucatan Nightjar, Yucatan Woodpecker, Yucatan Flycatcher, Ridgway's Rough-winged Swallow, Yucatan Jay, Black Catbird, Grey-throated Chat, Rose-throated Tanager, and Orange Oriole; a few of these species range southwest and southeast into adjacent regions. Mexican Sheartail and Yucatan Wren are specialities of the arid northwest where one also finds Lesser Roadrunner, Cinnamon Hummingbird, and White-lored Gnatcatcher, species usually associated with the Pacific Slope.

The low-lying coasts of the Yucatan, particularly the Caribbean coast, are characterized by long sandy beaches which hold only small populations of shorebirds, gulls, and terns, mostly in winter. Along the north coast of the Yucatan, however, extensive lagoons support large numbers of nesting and migrant waterbirds, including a sizeable breeding population of American Flamingos.

D. Caribbean islands and cays (pronounced 'keys') from the Campeche Bank to the Honduras Bay Islands

Most of the islands and cays are low and coral-based, but the Honduras Bay Islands, rising to 365 m on Guanaja, are of volcanic origin and represent a continuation of the Sierra de Omoa of northwestern Honduras. A notable feature of this region is the barrier reef of Belize, the longest in the Western Hemisphere, consisting of over 400 mangrove, sand, and coral islands and islets.

The Campeche Bank comprises four main island groups lying 120–160 km off the northwest Yucatan. Nine seabird species nest on the bank including large colonies of Masked Boobies, Magnificent Frigatebirds, and Sooty Terns; no landbirds breed there.

The avifauna of the offshore islands and cays from Isla Cozumel to the Honduras Bay Islands has a distinct Caribbean flavor. The more widespread, yet often still local, species include White-crowned Pigeon, Caribbean Dove, Smooth-billed Ani, Caribbean Elaenia, and Yucatan Vireo. Approximately 50 species breed on Isla Cozumel and endemics include Cozumel Emerald, Cozumel Wren, Cozumel Thrasher, Cozumel Vireo, and subspecies of Rufous-browed Peppershrike, Bananaquit, and Stripe-headed Tanager. About 30 species breed on the Bay Islands, including Grey Hawk, Yellow-naped Parrot, Northern Potoo, Canivet's Emerald, and Yucatan Woodpecker. In winter, the Caribbean islands and cays host northern migrants such as Cape May, Black-throated Blue, and Prairie warblers, all rare on the adjacent mainland.

Many of the islands and cays in this region support no breeding birds, but locally one finds colonies of seabirds such as Brown Pelican, Magnificent Frigatebird, Roseate, Bridled, and Sooty terns, and Brown Noddy. Off Belize, a large colony of Red-footed Boobies breeds on Lighthouse Reef, and Black Noddies formerly bred on Glover's Reef.

E. Offshore Gulf of Mexico and Caribbean waters

As well as being an important feeding ground for locally nesting seabirds, these offshore waters host a number of migrants. Regular visitors include Audubon's Shearwater and Pomarine and Parasitic jaegers. Species with only single records (although further work may show some to be of regular occurrence) are Cory's, Great, and Manx shearwaters, White-tailed Tropicbird, and Great Skua.

CLIMATE AND HABITAT

Climate

Climate, both global and local, plays a major role in determining the distribution of vegetation types. The two main climatic factors in this respect are temperature and rainfall. Temperature in Mexico and northern Central America is determined largely by latitude and altitude, while precipitation depends on air currents, topography, ocean currents, and ocean temperature. However, while the climatic regimes of Middle America are easily described, their mechanics are far from understood.

Temperature increases as one moves south in the Northern Hemisphere, but at the same time decreases with increasing altitude. South from southern Sonora and central Tamaulipas, the lowlands are tropical. Northwest and northern Mexico, and the highlands from central Mexico south, are temperate. The foothill zone between tropical lowlands and temperate highlands is often referred to as subtropical and includes habitats such as cloud forest. The subtropical zone in Mexico and northern Central America varies in extent, but lies mostly between 900 m and 2100 m, although the lowlands of northern Tamaulipas and the foothills of central Sonora may also be considered subtropical. Frosts and snow occur regularly from November to March in the interior and highlands south to the Central Volcanic Belt, with frosts occurring irregularly (mostly December to February) south to the highlands of Honduras. The mean monthly coldest temperatures throughout the region covered in this guide occur during December–January, and the highest temperatures during May–September (mostly May–June, that is, at the end of the dry season).

Most of the region in this guide lies in the latitude of the Northeast Trade Winds, though northwest Mexico extends into the belt of mid-latitude westerlies. Off the Atlantic coasts, prevailing winds from east and northeast blow in across the Caribbean and Gulf of Mexico. Off the north Pacific coast the prevailing winds are northwesterly, swinging to northeasterly from around Jalisco south.

The Doldrums, or equatorial calms, are a belt of warm, moist, calm air between the Northeast and Southeast Trades. The Doldrums migrate north in the northern summer, reaching their maximum latitude (about 12° N, that is, just south of the guide region) in August, then move back south to near the Equator by February. Their northward movement appears to be the main factor causing the wet season in Middle America.

Most of Mexico and northern Central America experiences a basic regime of 'summer' wet (or rainy) and 'winter' dry seasons, though the timing of these seasons varies geographically and also may vary by up to a month from year to year.

Atlantic Slope

The moisture-laden Northeast Trades move in almost continuously across the warm waters of the Caribbean and Gulf of Mexico. As the Doldrums shift north, they bring rising warm air which forces the Northeast Trades to rise, cool, and drop their moisture, causing a distinct wet season from June to October; rain is far from continuous, however, with most falling from mid-afternoon into evening. As the Doldrums shift south again, precipitation becomes less regular and a dry season occurs from November to May. In the months of July and August (often mid-July to mid-August) rains usually lessen slightly on both slopes, mostly from the Isthmus south.

The 'dry' season is not pronounced on the Atlantic Slope from the Isthmus south due to *Nortes*. These are irregular incursions of cold air masses, which are typically forced south in association with strong cold fronts over the Great Plains of interior North America. As these fronts push south, their cold air masses encounter warm air over the Gulf of Mexico and Caribbean, and rain storms result, often accompanied by a marked drop in temperature. Most Nortes occur between November and early March, with fewer and generally weaker fronts from October to early May.

In addition, the coastal-facing sierras of the Atlantic Slope intercept the Northeast Trades throughout the year, causing the Trades to rise and drop their moisture. The mountains themselves also produce convective rains.

Another factor influencing rainfall is the irregular passage of cyclones (tropical storms) which develop off both Atlantic and Pacific coasts, mostly from August to October. The low-pressure systems associated with such storms may be so large that effects can be felt hundreds of kilometers away from the storm centers. Cyclones with winds in excess of 120 km per hour are termed hurricanes, and their devastating effects on making landfall are well known.

Belize and northern Guatemala also experience frequent early-morning fogs in the dry season which provide considerable moisture and relief from potentially long, dry periods.

Hence, the dry season in the humid southeast occurs mostly from February to May, and is more prolonged in the lowlands. Average annual rainfall in much of the humid southeast is around 2000–2500 mm (80–100 in), with 80–90% falling during June–December; locally, up to 3500–5000 mm (140–200 in) of rain falls, with as much as 90% from June to January. Mean monthly low temperatures are 21–23 °C (low 70s °F), and highs 27–30 °C (low to mid-80s °F). The unusually low winter temperatures (for the Tropics) are due to Nortes and for most of the year mean monthly temperatures vary by only a few degrees. Average daily temperature in humid tropical climates, however, often varies by 8–10 °C (15–18 °F).

By virtue of their geographic position, the northeast Atlantic Slope lowlands of Mexico escape the rain from Nortes and so experience a marked dry season from November to May. Average annual rainfall is 625–1000 mm (25–40 in), 70–80% of which falls from May–October. Mean monthly low temperatures are 16–19 °C (low to mid-60s °F), with mean highs 27–28 °C (low 80s °F).

The northwest corner of the Yucatan also experiences an arid climate similar to that of northeast Mexico. Average annual rainfall there is 450–900 mm

(18–35 in), 70–80% falling from May–October. Mean monthly low temperatures are 21–23 °C (low 70s °F), and mean highs 27–29 °C (low to mid-80s °F). As one moves east and south, rainfall increases, so that in Quintana Roo and at the base of the Yucatan average annual rainfall is 1200–1500 mm (48–60 in), again with 70–80% during May–October.

Pacific Slope

Winds moving in from the Atlantic rarely reach the Pacific coast with any great moisture, and this rainshadow effect helps explain the generally arid nature of the Pacific Slope. The warm air brought north by the Doldrums, however, heats the land, causing onshore breezes and convective rains. The result is the Pacific Slope wet season. This pattern may be supplemented in September and October by tropical cyclones off the Caribbean coast. These storm systems are so large that their influence extends across the Central American isthmus and causes onshore westerly winds along the Pacific coast.

Rains mostly occur as heavy showers and thunderstorms from afternoon into early evening. They may begin by mid-May in Central America but usually do not reach northwest Mexico until June or July. October generally marks the end of the Pacific Slope wet season and the prolonged dry season is from November to May or June. Pacific Slope rains can be heavy, and for a few months the vegetation's dead grey aspect is exchanged for a lush green appearance. Thus, it is not low precipitation but a prolonged dry season that creates the Pacific Slope's arid climate.

North of the Isthmus, average annual rainfall in the lowlands is 750–1500 mm (30–60 in), with 80–90% falling in May–October. The deserts of Baja California and Sonora, however, average less than 250 mm (10 in) of rain a year and many months may pass without precipitation. In the lowlands south of the Isthmus, yearly rainfall averages 1750–2000 mm (70–80 in), again with 80–90% falling in May–October. Mean monthly low temperatures vary from 16–19 °C (low to mid-60s °F) in northwest Mexico, to 21–25 °C (low to mid-70s °F) elsewhere. Mean monthly highs are generally 27–29 °C (low to mid-80s °F), but may reach 32–35 °C (low 90s °F) in the northwest deserts.

The humid foothill and highland forests from southern Nayarit to El Salvador are maintained in much the same way as those on the Atlantic Slope, with locally generated convective rains being a critical factor. Annual rainfall in the mountains from the Sierra Madre del Sur south averages 3000–4000 mm (120–150 in) with some parts of the Sierra Madre de Chiapas receiving up to 5000 mm (200 in); most falls during April–November.

Interior

Moisture from winds moving in off the Atlantic or Pacific oceans rarely reaches the extensive interior north of the Isthmus which therefore experiences an arid, desert-like climate. Rain in the arid interior regions is mostly from June to October, the majority convectional in association with warmer air penetrating from the coasts. The narrower land mass south of the Isthmus means that moisture from the Northeast Trades carries farther into the interior but there it falls mostly on the higher parts of the sierras and still allows the existence of desert-like interior valleys.

On most of the Mexican Plateau and into the Central Volcanic Belt, average annual rainfall is 250–750 mm (10–30 in), with 70–80% falling in June–September; the northern Plateau deserts receive less than 250 mm (10 in) of rain annually. Mean monthly low temperatures are mostly 9–15 °C (high 40s to 50s °F), but often drop to around or below freezing (0 °C; 32 °F) in the higher mountains. Mean monthly highs are 18–26 °C (mid-60s to 70s °F), reaching 32–35 °C (low 90s °F) in the northern deserts.

In the southwest Mexican interior, average rainfall is 450–750 mm (18–30 in) with about 80% falling during May–September. Mean monthly low temperatures are 16–19 °C (low to mid-60s °F), and highs 21–25 °C (low to mid-70s °F), reaching at least 27–30 °C (low to mid-80s °F) in the depths of the Balsas Basin. The Central Valley of Chiapas is similar but slightly wetter, with 750–1000 mm (30–40 in) of rain annually, and slightly hotter (mean monthly highs 27–29 °C, or low 80s °F).

Most of the Chiapas–Guatemala highlands have an average annual rainfall of 1500–2500 mm (60–100 in), with 70–80% falling in May–October. The higher peaks, however, intercept more moisture-laden winds and may receive up to 3000–4000 mm (120–150 in), while the rainshadowed valleys receive only 1000–1500 mm (40–60 in). Mean monthly lows for the Chiapas–Guatemala highlands are 7–11 °C (mid-40s to low 50s °F), though on higher peaks the temperature routinely drops to freezing (0 °C; 32 °F); mean monthly highs are 16–18 °C (low 60s °F) although daytime temperatures may reach 25–26 °C (high 70s °F).

The interior highlands of Honduras and Nicaragua receive 1000–2000 mm (40–80 in) of rain, most of it falling May–October. Greater amounts fall locally at higher elevations and as little as 500–1000 mm (20–40 in) falls in the arid interior valleys where rain may be concentrated into only 3–4 months (June/July–September). Mean monthly low temperatures in the highlands are 16–19 °C (low to mid-60s °F), colder on the higher peaks, and 21–25 °C (low to mid 70s °F) in the arid interior valleys. Mean monthly highs are 18–30 °C (mid-60s to 80s °F).

Baja California
Although the cold, south-flowing California Current swings offshore as it approaches Mexico, its influence still reaches northwest Baja California. Consequently, this corner of Mexico has a Mediterranean climate distinct from the rest of the region in this guide, and much like that of coastal California. The wet season is from November to March and the summers are long and dry. In summer, however, fog often forms along the coast as warm air from the land condenses over the cold offshore waters.

In general, coastal northwest Baja California (north of about 30° N) receives about 250–400 mm (10–15 in) of rain a year, virtually all of it falling during

November–April; the mountains may receive 500–1000 mm (20–40 in), again most of it November–April, with some falling as snow. Mean monthly low temperatures along the coast are around 10 °C (50 °F), while high up in the sierras, lows average around or just below freezing (0 °C; 32 °F). Maximum temperatures occur in August–September, with mean monthly highs of 21–26 °C (70s °F) along the coast, and 27–30 °C (80s °F) inland.

Habitat

While several systems are used to classify animal distributions with respect to ecological situations, 'life zones', etc., bird distribution is basically related to habitat. Some birds are habitat-specific, others have a broader habitat tolerance, and some species are adaptable generalists. The main requirements a habitat must satisfy are food, shelter, and, in the case of resident birds, nest sites and nest materials.

In the species accounts we have tried to use readily understood terms to describe habitats. The following explanations cover the main habitats in Mexico and northern Central America. It should be noted that habitats are rarely sharply defined; instead, they tend to merge with one another, often over an elevational gradient.

A simplified overview of habitats in the region covered by this guide may be useful. In general, on both slopes, desert scrub in the north gives way to tropical deciduous forest, or thorn forest. Deciduous may become semideciduous and then evergreen forest depending on the amount and frequency of precipitation. Climbing into the mountains from the coastal slopes, deciduous forest gives way to oak scrub which becomes oak, and then pine–oak woodland, or forest. Cloud forest occurs locally on the coastal-facing slopes, and fir forest or pine–fir (conifer) forests occur in cooler, wetter areas. The dry interior regions mostly support arid scrub, deciduous forest, and arid oak and pine–oak woodland.

Tropical forests do not contain pure or extensive stands of a single species and, while most forests have characteristic plant species, a list of these (few of which have standardized common names) will mean little to most people. Instead, we have opted for broader descriptive terms for the main habitats. In general, forest is denser than woodland; that is, in woodland the canopy is not closed and light readily penetrates to the understory; often the trees are smaller and shorter than in forest. Pines rarely form forest *per se* but mature old stands of pines may be referred to as forest, particularly when mixed with oaks.

Humid evergreen forest

This is mature tropical and subtropical broad-leaf forest which, although all the trees lose their leaves at some time or other, retains a lush evergreen aspect throughout the year. It includes rain forest and cloud forest. In Mexico and northern Central America, rain forest occurs mostly from sea level to 1000 m, montane rain forest at 750–1500 m, and cloud forest at 900–2100 m.

Rain forest is the classic 'jungle' that many people associate with the tropics. In the guide region it is restricted (locally due to extensive clearing by man) to the Atlantic Slope from the Isthmus to northern Honduras, with extensions north to Los Tuxtlas in southern Veracruz and to the foothills of northern Oaxaca. Rain forests host the greatest diversity of plant and animal species on Earth, although the forests in the guide region are peripheral and thus relatively poor in species diversity. Mature rain forest is supported by plentiful rain throughout the year and is characterized by a canopy of tall trees (to 30 m, many with buttressed roots), a wealth of lianas and epiphytes, and a relatively open understory rich in palms. Montane rain forest is a general term for the transition between rain forest and cloud forest. It is also restricted to the Atlantic Slope.

Typical rain forest birds include Great Tinamou, Short-billed Pigeon, Mealy Parrot, Long-tailed and Little hermits, Purple-crowned Fairy, Slaty-tailed Trogon, White-whiskered Puffbird, Scaly-throated Leaftosser, Russet Antshrike, Dot-winged Antwren, Sepia-capped, Ruddy-tailed, and Sulphur-rumped flycatchers, Rufous Piha, White-breasted Wood-Wren, Green Honeycreeper, Olive-backed Euphonia, and Black-throated Shrike-Tanager, with birds such as Slaty-breasted Tinamou, Barred Forest-Falcon, Violet Sabrewing, Mexican Antthrush, and Lovely Cotinga being equally typical of montane rain forest (and some also of cloud forest).

Cloud forest has a humid subtropical climate and owes its existence to plentiful moisture throughout the year, much of it derived from low clouds or fogs which envelop and shroud the forest. Cloud forests occur, often very locally, where rising warm air regularly meets cool air and thus tend to be restricted to coastal-facing slopes. They are found on both Atlantic and Pacific slopes, often within or just below larger areas of pine–oak forest. Cloud forest canopy may reach 30 m and the trees are far more laden with epiphytes and mosses than in rain forest, at times to the extent that branches collapse under the weight. Tree ferns often are considered classic cloud forest indicators, though they range into montane rain forest and occur locally near sea level, as in Los Tuxtlas. On windswept ridge tops a lower stunted forest (Elfin Forest) occurs, and at edges and in clearings a dense shrubby growth prevails. Elfin Forest is rare and not a significant habitat in the guide region, though it is important farther south. Bamboo is also more important farther south but occurs locally in the guide region; it may be associated with lowland or highland forests.

Typical cloud forest birds include Highland Guan, Barred Parakeet, White-faced Quail-Dove, Emerald-chinned and *Eupherusa* hummingbirds, Green-throated and Green-breasted mountain-gems, Resplendent Quetzal, Spectacled and Ruddy foliage-gleaners, Tawny-throated Leaftosser, Spotted Woodcreeper, Azure-hooded Jay, Slate-colored Solitaire, and Blue-crowned Chlorophonia, with Bumblebee and Wine-throated hummingbirds typical of edges and low shrubby growth, and Blue and Slate-blue seedeaters and Slaty Finch associated with seeding bamboo.

Pine–oak forest

Pine–oak and oak forests, or woodlands, are a dominant feature of the high-lands throughout Mexico and northern Central America (Mexico alone has over

170 species of oaks). As the name suggests, such forests are composed primarily of various species of pines and oaks, and the canopy may be as high as 20–30 m. Humid pine–oak forest occurs mostly on coastal-facing slopes and ridge tops, while arid pine–oak forest occurs mostly on the drier interior slopes. The main distinction between humid and arid pine–oak is the dense shrubby understory of the former. Intermediate areas may be termed semiarid or semihumid pine–oak. Humid pine–oak forest is sometimes enshrouded by clouds and often heavily laden with epiphytes, suggesting cloud forest. Pine and pine–oak forests sometimes merge with cloud forest for which we use the term pine–evergreen forest.

Species typical of pine–oak forest include Band-tailed Pigeon, Whiskered Screech-Owl, Green Violet-ear, White-eared Hummingbird, Hairy Woodpecker, Spot-crowned Woodcreeper, Pine Flycatcher, Steller's Jay, Mexican Chickadee, Grey Silky, Chestnut-sided Shrike-Vireo, and Hepatic Tanager, with Long-tailed Wood-Partridge, Mountain Trogon, Green-striped Brushfinch, and Collared Towhee more typical of humid pine–oak, and Montezuma Quail, Bridled Titmouse, White-breasted Nuthatch, and Painted Redstart typical of arid oak and pine–oak. Garnet-throated Hummingbird, Black-throated, Dwarf, and White-throated jays, and Black Thrush are typical of pine–evergreen forest.

Within humid pine–oak forests one finds stands of firs in the coldest, most humid areas. Few bird species are restricted to this habitat which, south of the Central Volcanic Belt, occurs in very local and usually small stands, but the breeding distribution of Golden-crowned Kinglet is closely tied to fir forests.

In open pine–oak and pine woodland, park-like clearings and grassy areas occur. Birds typical of these habitats include Eastern and Western bluebirds, American Robin, Rufous-collared Thrush, Chipping Sparrow, and Yellow-eyed Junco, and, in bunch grass areas, Striped Sparrow. The lowland pine savannas of northern Central America are discussed under savannas.

Semidecidous forest

Semidecidous forest occurs in drier areas than humid evergreen forest. It may have the appearance of rain forest for much of the year, although during the distinct dry season many trees lose most or all of their leaves. Semideciduous forest occurs locally on the Pacific Slope from southern Nayarit south, and on the Atlantic Slope in northeast Mexico and in the eastern and southern parts of the Yucatan Peninsula. The avifauna suggests that of rain forest but is less diverse. On the Pacific Slope, this habitat often occurs along streams in deciduous forest.

Species typical of this habitat include Thicket Tinamou, Mexican Hermit, Mexican Woodnymph, Green Jay, Carolina and White-browed wrens, Yellow-green Vireo, Rufous-browed Peppershrike, Grey-throated Chat, Rose-throated Tanager, and Crimson-collared Grosbeak.

Deciduous forest

Deciduous forest experiences a marked dry season when most or all trees lose their leaves and the forest has a dead grey appearance. A striking feature of deciduous forests is that several trees flower in the dry season. At that time their bright yellow, pink, and white blossoms can be seen from long distances against an otherwise bleak canopy, allowing pollinators to locate them easily. Tropical deciduous forest canopy may reach 20–30 m and the understory is usually fairly dense and thorny. This habitat is sometimes referred to as thorn forest and occurs (although now extensively cleared) along much of the arid Pacific Slope of Middle America, and in the northern Yucatan Peninsula.

Birds typical of tropical deciduous forest (mostly from the Pacific Slope) include Orange-fronted Parakeet, Lesser Ground-Cuckoo, Cinnamon Hummingbird, Citreoline Trogon, Nutting's and Flammulated flycatchers, Magpie-Jays, San Blas and Yucatan jays, Happy, Banded, and White-bellied wrens, Black-capped and White-lored gnatcatchers, Red-breasted Chat, and Streak-backed Oriole.

Gallery forest
Gallery forest or woodland refers to corridors of forest or woodland which grow along watercourses, usually in otherwise open or scrubby habitats. In temperate areas, the term riparian woodland is usually used. On the Pacific Slope of eastern Oaxaca and adjacent Chiapas, Long-tailed Manakins are typical of gallery forest; in northern Baja California, Bell's Vireos are associated with riparian groves.

Second growth
Second growth refers to successional-stage, shrubby and wooded habitats which differ from mature forest by having smaller trees and lower, more open canopy which lacks emergent trees. Second growth along edges and roadsides is characterized by thickets, tangles, and plants such as *Heliconia* and *Cecropia*. Forest edge, especially where altered by man (such as around clearings or along roads) provides more habitat for species that formerly were restricted to naturally occurring second growth such as light gaps, stream sides, etc. Typical second growth and forest-edge species include Little Tinamou, Blue Ground-Dove, Rufous-tailed Hummingbird, Violaceous Trogon, Rufous-tailed Jacamar, Great and Barred antshrikes, Dusky Antbird, Slate-headed Tody-Flycatcher, Tropical Pewee, White-collared Manakin, Crimson-collared and Scarlet-rumped tanagers, Blue-black Grosbeak, and Yellow-tailed Oriole.

Plantations
Plantations represent a special type of man-made woodland. The resultant monoculture is usually much poorer in bird diversity than the original vegetation, as exemplified by the coconut and banana plantations which have replaced tropical deciduous forest and scrub on much of the Pacific coastal plain from Nayarit south. Other common, but uninteresting plantations, or groves, are those of mangos, papayas, and citrus fruits, mostly oranges.

One culture which is noted for birds is coffee, where the shade trees help retain much of the aspect and avifauna of the original forest. Among other species, trees of the genus *Inga* are used commonly for shade and their filamentous whitish flowers (blooming mostly during March to June, depending on location) are favored by hummingbirds. In addition, the coffee bushes themselves provide at least a surrogate understory where birds such as quail-doves, Pheasant Cuckoo, and ant-tanagers can live.

Scrub

In general, scrub is the equivalent of deciduous or semideciduous woodland where the climate cannot support such tall growth. Scrub, therefore, is found mostly in arid areas and is characterized by scraggly, often thorny bushes (to 1–2 m), small trees such as *Acacia* (to 3–5 m), and cacti. Scrub may be open (allowing easy access in most directions), semiopen (allowing less ready access), or relatively dense and impenetrable, approaching thorn forest in appearance. Often the ground is rocky or stony with sparse grass cover or, as on the Mexican Plateau, scrub may alternate with rolling grasslands.

Temperate to subtropical areas of open and semiopen arid scub are often termed deserts, and they usually support a variety of cacti; true deserts receive less than about 250 mm (10 in) of rain annually. One also finds arid scrub in tropical climates from the interior of central Mexico south. Birds typical of arid scrub include Banded, Scaled, and Gambel's quail, Dusky, Broad-billed, Violet-crowned, Lucifer, and Beautiful hummingbirds, Grey-breasted and Gila wood-peckers, Scrub Jay, Verdin, several *Campylorhynchus* wrens, several thrashers, Dwarf Vireo, and most *Aimophila* sparrows.

In northwest Baja and on mountains in north and central Mexico, one finds *chaparral*, a dense, semiarid yet evergreen scrub, supported by moisture from fog or clouds. Characteristic chaparral birds include Wrentit, California Thrasher, and Colima Warbler (summer).

Savanna

Savanna refers to natural grasslands with scattered trees and bushes, especially palms and oaks. Most savannas exist on flat or gently rolling, poor soils and are often maintained by periodic burning. They may be dry or, at least seasonally, marshy with scattered ponds, and are found locally in the lowlands on both slopes. The native grasses are usually less than 0.5 m tall, but in most areas savannas are heavily grazed by livestock.

Pine savannas are a specific type of savanna found locally on sandy soils from northern Guatemala to northern Nicaragua, and are grasslands with often fairly extensive woods of the tropical pine *Pinus caribaea*. They are traversed by streams and rivers lined with humid gallery forest and, at least in the wet season, tend to be dotted with ponds and marshes.

Extensive clearing of forest for grazing land has produced pseudo-savannas in many areas. These tend to be covered with tangled non-native grasses and are referred to as open areas with scattered trees and hedges. Elsewhere, savannas have been ploughed and drained for cultivation or simply fenced and converted to grazing land.

Birds characteristic of savannas include Aplomado Falcon, Double-striped

Thick-knee, Plain-breasted Ground-Dove, Common Nighthawk, Fork-tailed Flycatcher, and Grassland Yellow-Finch. Pine savannas support a distinctive avifauna which includes Azure-crowned Hummingbird, Acorn Woodpecker, Grace's Warbler, and Hepatic Tanager, species usually associated with pine—oak highlands.

Grasslands

Grasslands are similar to savannas but are more open with fewer scattered bushes and usually no trees; on the Mexican Plateau, they often alternate with desert scrub. Most (all?) of the native grasslands of the region have been altered by man, usually by being enclosed for grazing land. Like savannas, grasslands would traditionally be maintained by seasonal burning but many are now maintained by grazing, or overgrazing.

Typical grassland breeding species include Savannah and Grasshopper sparrows, and Eastern and Western meadowlarks. This habitat is particularly important for wintering North American birds such as Mountain Plover, Sprague's Pipit, several sparrows, and longspurs.

Marshes

Marshes are non-forested areas that are flooded for much or all of the year and are thus characterized by aquatic vegetation. Salt marshes tend to be flooded daily by tides but often are not permanently flooded; they are typical of temperate regions and tend to be replaced by mangroves in tropical areas. Salt marshes in the region covered by this guide are characterized by Cordgrass (*Spartina*) and Pickleweed (*Salicornia*), and host birds such as Clapper Rail and Savannah Sparrow.

Freshwater marshes and associated lakes have been drained and 'reclaimed' by man in most of the region so that few extensive marshy areas survive today. The famous and formerly extensive Río Lerma Marshes in central Mexico now consist of small and shrinking patches, while the Río Lerma itself has been reduced to a narrow, polluted trickle. In some areas, seasonally flooded fields, such as for rice, may provide a substitute habitat for numbers of marsh and waterbirds. Freshwater marshes are characterized by reeds, rushes, and sedges which usually form dense masses of vegetation such as reedbeds. Open water tends to be partly or fully covered with floating plants, and scrub and small trees such as willows and acacias often border marshes. Typical freshwater marsh birds include bitterns, several rails and crakes, Marsh and Sedge wrens, most yellowthroats, and several sparrows and blackbirds.

Mangroves

Mangroves form a specialized plant community which grows in tropical areas inundated permanently or periodically with salt or brackish water. They occur commonly along both coasts and on numerous islands from central Baja California, Sonora, and Tamaulipas south through the region. The trees usually are 3–10 m tall, locally attaining 15–20 m, and mangrove swamps or forests are characterized by an often dense 'understory' of stilt roots. Mangrove swamps are dominated by Red Mangroves (*Rhizophora*) growing in deeper and usually permanently flooded areas, Black Mangroves (*Avicennia*) in shallower areas, and White Mangroves (*Laguncularia*) growing in drier areas which may be flooded only irregularly. Button Mangroves (*Conocarpus*) tend to occur in low flooded areas inland of the coast.

Mangroves often provide relatively safe sites for breeding colonies of wading birds such as egrets and Agami and Boat-billed herons. They also are home to species such as Clapper Rail, Rufous-necked Wood-Rail, Mangrove Cuckoo, kingfishers, Mangrove Vireo (typical of mangroves only on the Pacific coast), and Mangrove [Yellow] Warbler.

Lakes, ponds, and reservoirs

These are freshwater habitats widely distributed throughout the guide region, with a notable concentration in the interior of north and central Mexico, and relatively few in the southwest interior of the country. They generally provide habitat for resident birds only when surrounded by marshes or woodland. Many lakes have been or are being drained, and those remaining are becoming increasingly polluted; witness, for example, the sad state of the once great Lake Chapala. Despite degradation, these habitats remain important stopovers and wintering areas for thousands of migratory ducks and shorebirds.

Coastlines

Coastlines in northern Middle America vary from rocky, as along much of the Pacific coast, to sandy, as on most of the Atlantic coast. Estuaries with extensive mud-flats are restricted to the temperate areas of northern Mexico. Large coastal lagoons, usually bordered by mangroves, occur along both coasts.

Coastal breeding species include Collared, Snowy, and Wilson's plovers, American Oystercatcher, Laughing Gull, several terns, and Black Skimmer, many of which prefer to nest on protected spits, sand bars, or islets in lagoons. Rocks and islets, mostly along the Pacific coast, provide nesting and roosting sites for seabirds such as Brown and Blue-footed boobies, and Bridled and Sooty terns.

The estuaries of northern Mexico support large numbers of wintering geese, ducks, and shorebirds, with a large percentage of the world population of Black Brant wintering in northwest Mexico. Elsewhere, except around river mouths and lagoons, the coasts support mostly small and scattered populations of Black-bellied Plover, Willet, Whimbrel, Ruddy Turnstone, and Sanderling.

Pelagic

Pelagic refers to offshore ocean waters. The categorization of open ocean habitats, based on characters such as temperature and salinity, is poorly understood and we have not attempted any distinctions in this guide. Inshore waters in this book correspond roughly to waters within sight of land. True pelagic species are rarely seen from the mainland. Offshore islands and their birds are discussed in the chapter on geography and bird distribution.

The high diversity of seabirds in the Pacific might be considered compensation for the paucity found in the Gulf of Mexico and Caribbean. Typical pelagic species occurring off the coasts of northern Middle America include tropicbirds, boobies, and several species of petrels, shearwaters, and storm-petrels.

Ranchland and agricultural land

These terms should be self-explanatory. Ranchland refers to areas grazed by cattle and may be brushy or grassy. Agricultural land refers to all arable crops such as grains, beans, and agaves. Large areas of Mexico and northern Central America, especially in temperate regions, have been converted to open, seasonally barren, agricultural land. Most of the flat or gently rolling land in many areas has long since been cleared for arable practices so that, particularly from central Mexico south, one sees corn planted on cleared hillsides with slopes of 60° or more. Agricultural land includes hedges, weedy areas along fences, etc. No birds are endemic to such areas but many birds use them for feeding, such as large flocks of geese and blackbirds in winter.

MIGRATION

Migration may be long-distance or local. The former refers to birds regularly traveling considerable distances between their breeding and non-breeding grounds. These migrations are usually oriented north–south. Local migrations involve birds moving shorter distances, often less predictably or regularly, and include altitudinal migration, seasonal withdrawal from peripheral breeding areas, and post-breeding dispersal.

One thing that may be unfamiliar to observers from North America or Western Europe is the large proportion of species in Mexico and northern Central America that are sedentary. Of the 800 or so species that breed in the guide region, about 440 have little or no migratory tendencies. Thus, in their lives the majority of individuals of these species move no more than a few kilometers from where they hatched.

Long-distance migration

About 225 species of long-distance migrants occur in Mexico and northern Central America but are not known to breed there regularly. As a result of its proximity to continental North America, where the majority of breeding species are migratory, Mexico hosts more migrants than any other Latin American country. Of 160 winter visitors in the guide region, 75 winter regularly no farther south than Mexico, and a further eight species typically range no farther south than western Nicaragua.

Some species that reach northern Mexico in winter, such as Rough-legged Hawk and Red-breasted Nuthatch, exhibit marked fluctuations in annual abundance related to climate or food supply to the north.

During migration or winter, the populations of many species breeding in the guide region (for example, Common Moorhen, Mourning Dove, Yellow Warbler) are supplemented greatly by northern migrants.

Other long-distance migrants are species with populations breeding in northern or northeast Mexico (often at the southern limit of a wider range in North America), and wintering mostly on the Pacific Slope from western Mexico south. This group includes Black-chinned Hummingbird, Western Kingbird, Scissor-tailed Flycatcher, Black-capped Vireo, and Colima Warbler.

The remaining long-distance migrants are: 29 species that also breed locally in northern Mexico, 22 summer residents, 36 transients, 22 pelagic visitors, and 30 vagrants. There is a single austral (southern hemisphere) landbird migrant species: Blue-and-white Swallow.

The status of some species, based on present knowledge, is subjective, for example, it is not clear whether Wilson's and Red-necked phalaropes winter irregularly or regularly in the guide region. In such cases we have opted for the more conservative status, at least until sufficient data indicate otherwise.

Winter visitors (Tables 1 and 2)

This category includes all northern migrants that occur regularly in northern Middle America during the northern winter (mostly August to May, which

Table 1. Winter visitors. a: winters north and west of Isthmus of Tehuantepec (except irregular or vagrant occurrences); b: winters mostly or entirely from Isthmus south and east; c: has bred/may breed irregularly in Mexico (present status often unclear).

Red-throated Loon[a]	Whimbrel
Pacific Loon[a]	Long-billed Curlew
Common Loon[a]	Marbled Godwit
Horned Grebe[a]	Ruddy Turnstone
Northern Fulmar[a]	Black Turnstone[a]
Northern Gannet	Surfbird
American Bittern[a,c]	Red Knot
Tundra Swan[a]	Sanderling
White-fronted Goose[a]	Semipalmated Sandpiper
Snow Goose[a]	Western Sandpiper
Ross' Goose[a]	Least Sandpiper
Brant[a]	Dunlin[a]
Canada Goose[a]	Stilt Sandpiper
Wood Duck[a]	Short-billed Dowitcher
Green-winged Teal	Long-billed Dowitcher
Northern Pintail	Common Snipe[c?]
Blue-winged Teal[c]	Red Phalarope
Northern Shoveler	Pomarine Jaeger
Gadwall	Parasitic Jaeger
Eurasian Wigeon[a]	Bonaparte's Gull[a]
American Wigeon	Mew Gull[a]
Canvasback[a]	Ring-billed Gull
Redhead[c]	California Gull[a]
Ring-necked Duck	Herring Gull
Greater Scaup[a]	Thayer's Gull[a]
Lesser Scaup	Lesser Black-backed Gull[a]
Black Scoter[a]	Glaucous-winged Gull[a]
Surf Scoter[a]	Black-legged Kittiwake[a]
White-winged Scoter[a]	Common Tern
Common Goldeneye[a]	Black Tern
Bufflehead[a]	Common Murre[a]
Hooded Merganser[a]	Ancient Murrelet[a]
Common Merganser[a,c]	Rhinoceros Auklet[a]
Red-breasted Merganser[a]	Short-eared Owl[a]
Broad-winged Hawk	Chuck-will's Widow
Ferruginous Hawk[a]	Ruby-throated Hummingbird
Rough-legged Hawk[a]	Rufous Hummingbird[a]
Merlin	Allen's Hummingbird[a]
Sandhill Crane[a]	Belted Kingfisher
Black-bellied Plover	Lewis' Woodpecker[a]
Pacific Golden Plover[a]	Yellow-bellied Sapsucker
Semipalmated Plover	Red-naped Sapsucker[a]
Piping Plover	Red-breasted Sapsucker[a]
Mountain Plover[a]	Yellow-bellied Flycatcher
Greater Yellowlegs	Least Flycatcher
Lesser Yellowlegs	Hammond's Flycatcher
Solitary Sandpiper	Grey Flycatcher[a]
Wandering Tattler	Eastern Phoebe[a]
Spotted Sandpiper	Great Crested Flycatcher[b,c]

Table 1. (*continued*)

Tree Swallow	Black-and-white Warbler
Winter Wren[a]	American Redstart
Mountain Bluebird[a]	Worm-eating Warbler
Swainson's Thrush	Swainson's Warbler[b]
Wood Thrush[b]	Ovenbird
Varied Thrush[a]	Northern Waterthrush
Grey Catbird	Louisiana Waterthrush
Sage Thrasher[a]	Kentucky Warbler[b]
American Pipit[a,c]	Hooded Warbler
Sprague's Pipit[a]	Wilson's Warbler
Cedar Waxwing	Rose-breasted Grosbeak
Yellow-throated Vireo[b,c]	Indigo Bunting
Philadelphia Vireo[b]	Dickcissel
Blue-winged Warbler	Clay-colored Sparrow[a]
Golden-winged Warbler[b]	Brewer's Sparrow[a]
Tennessee Warbler[b]	Field Sparrow[a]
Nashville Warbler	Vesper Sparrow[a]
Virginia's Warbler[a]	Lark Bunting[a]
Northern Parula	Baird's Sparrow[a]
Chestnut-sided Warbler[b]	Sharp-tailed Sparrow[a]
Magnolia Warbler	Fox Sparrow[a]
Cape May Warbler[b]	Lincoln's Sparrow
Black-throated Blue Warbler[b]	Swamp Sparrow[a]
Myrtle Warbler	White-throated Sparrow[a]
Townsend's Warbler	Golden-crowned Sparrow[a]
Hermit Warbler	White-crowned Sparrow[a]
Black-throated Green Warbler	McCown's Longspur[a]
Golden-cheeked Warbler[b]	Chestnut-collared Longspur[a]
Yellow-throated Warbler	Baltimore Oriole
Pine Warbler[a]	Purple Finch[a]
Prairie Warbler[b]	American Goldfinch[a,c?]
Palm Warbler	

includes migration periods) but are not known to breed regularly in the region. Several species (for example, Broad-winged Hawk, Semipalmated Sandpiper) occur much more commonly as transient migrants. Non-breeding individuals (usually immatures) of numerous species of waterbirds oversummer frequently, including loons (northwest Mexico), ducks (mostly northern Mexico), shorebirds, gulls, and terns, as well as species that breed in the region such as pelicans, boobies, cormorants, and herons. No non-breeding landbird is known to oversummer in northern Middle America.

Four groups account for over 60% of the regular northern winter visitors. These are wildfowl, shorebirds, wood-warblers, and sparrows, with most wildfowl and sparrows being at the southern edge of their winter ranges.

Two additional species are extirpated as regular winter visitors: Trumpeter Swan and Whooping Crane. In addition, many of the vagrants (see Table 6)

Table 2. Winter visitors with small and/or very local breeding populations in northern Mexico. a: breeds in Baja California only.

American White Pelican	Northern House Wren[a]
Mallard[a]	Ruby-crowned Kinglet[a]
Bald Eagle	Hermit Thrush[a]
Northern Harrier[a]	Orange-crowned Warbler[a]
Red-shouldered Hawk[a]	Lucy's Warbler
Sora[a]	Black-throated Grey Warbler
Willet	MacGillivray's Warbler
Forster's Tern	Western Tanager
Long-eared Owl	Lazuli Bunting[a]
Calliope Hummingbird[a]	Green-tailed Towhee[a]
Williamson's Sapsucker[a]	Dark-eyed Junco[a]
Olive-sided Flycatcher[a]	Yellow-headed Blackbird
Willow Flycatcher	Brewer's Blackbird[a]
Dusky Flycatcher[a]	Cassin's Finch[a]
Red-breasted Nuthatch[a]	

occur during the winter months and two species (Eurasian Wigeon, Lesser Black-backed Gull) apparently occur regularly enough to be considered winter visitors.

Table 2 lists an additional 29 species and one subspecies that occur mostly as winter visitors, although small and/or very local populations breed in extreme northern Mexico. Seventeen of them breed only in Baja California.

Summer residents (Table 3)

Twenty-two species breed in the guide region but migrate south for the winter. Some, like Swainson's Hawk, Yellow-billed Cuckoo, and Bank Swallow, are much commoner as transients when North American breeding populations pass through the region. Six of the summer residents do not breed north of Mexico.

Table 3. Summer residents. a: winters irregularly and/or rarely in guide region; b: except Clipperton Atoll, where probably resident.

Swallow-tailed Kite	Spot-tailed Nightjar
Plumbeous Kite	Black Swift[a?]
Swainson's Hawk[a]	Western Pewee
Elegant Tern[a]	Streaked Flycatcher
Roseate Tern	Sulphur-bellied Flycatcher
Least Tern[a]	Piratic Flycatcher
Bridled Tern[a]	Purple Martin
Sooty Tern[b]	Sinaloa Martin
Brown Noddy[b]	Bank Swallow[a]
Yellow-billed Cuckoo	Cliff Swallow
Common Nighthawk	Yellow-green Vireo

Transient migrants (Table 4)

The 36 transient migrants breed in North America and winter in or offshore of southern Central America, the Caribbean, and South America. Their main periods of passage are April–May and August–October. Six of these transients winter irregularly or rarely in the region, five at the northern edge of their winter range, and one (Prothonotary Warbler) at the periphery of a circum-Caribbean winter range.

Table 4. Transient migrants. a: winters irregularly and/or rarely in region; b: may breed in region.

Mississippi Kite	Acadian Flycatcher
American Golden Plover	Alder Flycatcher
Upland Sandpiper	Eastern Kingbird
Eskimo Curlew	Grey Kingbird
Hudsonian Godwit	Veery
White-rumped Sandpiper	Grey-cheeked Thrush
Baird's Sandpiper	Red-eyed Vireo[b]
Pectoral Sandpiper[a]	Black-whiskered Vireo
Buff-breasted Sandpiper	Blackburnian Warbler
Wilson's Phalarope[a]	Bay-breasted Warbler
Red-necked Phalarope[a]	Blackpoll Warbler
Long-tailed Jaeger	Cerulean Warbler
Franklin's Gull[a]	Prothonotary Warbler[a]
Sabine's Gull[a]	Connecticut Warbler
Arctic Tern	Mourning Warbler
Black-billed Cuckoo	Canada Warbler
Chimney Swift	Scarlet Tanager
Eastern Pewee	Bobolink

Pelagic visitors (Table 5)

Of the 22 pelagic visitors, 17 occur off the Pacific coast, one off the Atlantic coast, and four may occur off both coasts. Of the Pacific species, two (Stejneger's Petrel, Markham's Storm-Petrel) have been recorded only around Clipperton Atoll and, were it not for breeding populations on Clipperton, Black Noddy and White Tern would be only pelagic visitors (or vagrants?) in the guide region.

Vagrants (Table 6)

Thirty species have occurred as vagrants in Mexico and northern Central America, although this number probably could be doubled easily if there were half the number of active observers in Mexico as in most states of the USA. Of the total, 20 are North American breeding species (though a few barely qualify as such), all of which have turned up in migration or winter; four are wide-ranging seabirds; four are of Eurasian origin; and two are of South American origin.

Table 5. Pelagic visitors. a: Clipperton Atoll only.

Short-tailed Albatross	Pink-footed Shearwater
Black-footed Albatross	Flesh-footed Shearwater
Parkinson's Petrel	Buller's Shearwater
Tahiti Petrel	Short-tailed Shearwater
Dark-rumped Petrel	Sooty Shearwater
Juan Fernandez Petrel	Christmas Shearwater
Kermadec Petrel	Audubon's Shearwater
Herald Petrel	Harcourt's Storm-Petrel
Cook's Petrel	Galapagos Storm-Petrel
Stejneger's Petrel[a]	Markham's Storm-Petrel[a]
Cory's Shearwater	South Polar Skua

Table 6. Vagrants. a: Clipperton Atoll only.

Yellow-billed Loon	Glaucous Gull
Great Shearwater	Great Black-backed Gull
Manx Shearwater	Grey-backed Tern[a]
White-tailed Tropicbird	Pigeon Guillemot
Garganey	Crested Auklet
Oldsquaw	Dark-billed Cuckoo[a]
Harlequin Duck	Dusky Warbler
Bar-tailed Godwit	Arctic Warbler
Ruff	Northern Wheatear
American Woodcock	Brown Thrasher
Great Skua	White Wagtail
Little Gull	Red-throated Pipit
Black-headed Gull	Le Conte's Sparrow
Black-tailed Gull	Lapland Longspur
Kelp Gull	Rusty Blackbird

Local migration

Most local migrants tend to be insectivores, frugivores, or otherwise dependent on a food supply linked to warmer climates. Local migration may be divided into four categories: altitudinal migration, peripheral withdrawal, post-breeding dispersal, and other.

Altitudinal migration

This refers mostly to species moving downslope in winter to avoid cold and/or rainy winter seasons, and also may be related to greater abundance of food at lower elevations in winter. Thus, depending on the harshness of a winter, or perhaps seasonal abundance of food, altitudinal migrations vary from year to year. Obviously, the number and variety of altitudinal migrants varies regionally.

Table 7. Altitudinal migrants in the Gomez Farias region, Tamaulipas

Mountain Trogon	Crescent-chested Warbler
Spot-crowned Woodcreeper	Hooded Yellowthroat
Tufted Flycatcher	Painted Redstart
Greater Pewee	Rufous-capped Warbler
Grey-collared Becard	Blue-hooded Euphonia
Brown-backed Solitaire	Hepatic Tanager
White-throated Thrush	Flame-colored Tanager
Orange-billed Nightingale-Thrush	Black-headed Siskin
Blue Mockingbird	Hooded Grosbeak

Enough work has been done in only three areas to indicate which species undertake altitudinal migrations; such movements may be only 100–200 m down to a different habitat, or from 1500 m to near sea level.

For the Gomez Farias region of central Tamaulipas, Arvin (1990) listed 18 species of altitudinal migrants (Table 7); small numbers (usually) of each species move downslope in the non-breeding (dry) season.

In the Sierra de los Tuxtlas of southern Veracruz, long-term studies by W. J. Schaldach Jr., supplemented by observations by SNGH, show that at least 12 species (Table 8) regularly move downslope, mostly between November and February. This corresponds with strong Nortes that bring cold and rain to the region.

In the Valley of Mexico, long-term observations by R. G. Wilson indicate that in winter, at least some individuals of 12 species (Table 9) move down into the oak woodlands from the higher pine and fir forests, sometimes supplementing the smaller breeding populations of the lower woodlands. High aerial feeders such as swifts also mostly or entirely leave the Valley in winter.

Another form of altitudinal migration is that exhibited by some birds nesting on the Mexican Plateau. The Plateau typically experiences severe winters so that insectivores in particular may withdraw south and/or to the adjacent lowlands. A few hardy individuals, however, may remain, mostly around lakes and ponds. Species in which most of at least the northern Plateau breeding population moves out in winter include Vermilion and Ash-throated flycatchers, Cassin's Kingbird, and Varied Bunting.

Table 8. Altitudinal migrants in Los Tuxtlas, Veracruz

Emerald Toucanet	Green Shrike-Vireo
Yellowish Flycatcher	Tropical Parula
Grey-collared Becard	Slate-throated Redstart
Slate-colored Solitaire	Blue-crowned Chlorophonia
Black-headed Nightingale-Thrush	White-winged Tanager
White-throated Thrush	Common Bush-Tanager

Table 9. Altitudinal migrants in the Valley of Mexico, DF. a: only rarely recorded at lower elevations in winter.

Hairy Woodpecker	White-throated Thrush
Strickland's Woodpecker[a]	Grey Silky
White-breasted Nuthatch	Crescent-chested Warbler
Pygmy Nuthatch[a]	Red Warbler
Grey-barred Wren[a]	Slate-throated Redstart
Golden-crowned Kinglet[a]	Olive Warbler[a]

Peripheral withdrawal

The most common type of peripheral withdrawal in Mexico occurs when most or all individuals in a population leave the extreme northwestern and/or northeastern parts of their range in winter; in northwestern Mexico, for example, withdrawing species are resident by the time one reaches southern Sonora. Species that withdraw from extreme northwestern Mexico and northeastern Mexico (and often the adjacent southwestern USA) are listed in Table 10, although this list probably is incomplete. For example, Sutton and Burleigh (1941) provide suggestive data on other candidates in northeastern Mexico. Table 10 does not include species that withdraw more extensively or are more strongly migratory, such as Broad-billed Hummingbird and Buff-breasted Flycatcher. Whether withdrawing individuals retreat only just to or beyond the northern edge of the species' winter range remains unknown.

It appears that some species (for example, Northern Jacana, Groove-billed Ani, Blue-black Grassquit) may withdraw in summer from the upper edges of their range along valleys on the western Plateau.

An interesting phenomenon, apparently due to Nortes, is the winter withdrawal of several species from most or all of the Atlantic Slope north of the Isthmus, and to a lesser extent from the Yucatan Peninsula (Table 11). Often the Gulf Slope populations winter on the warmer and drier Pacific Slope from the Isthmus to northern Central America.

Table 10. Species that withdraw from northern Mexico in winter. a: northwestern Mexico only; b: northeastern Mexico only; c: see Table 11.

Common Black Hawk[a]	Tropical Kingbird[a]
Blue-throated Hummingbird	Couch's Kingbird[b]
Magnificent Hummingbird	Thick-billed Kingbird[a]
Elegant Trogon[a]	Rose-throated Becard
Northern Beardless Tyrannulet	Tropical Parula
Dusky-capped Flycatcher[a]	Slate-throated Redstart[b]
Brown-crested Flycatcher[a,c]	Hepatic Tanager

Table 11. Species that withdraw mostly or entirely from the Atlantic Slope north of the Isthmus in winter. a: withdraw also from Yucatan Peninsula.

Green-breasted Mango[a]	Brown-crested Flycatcher
Yellow-bellied Elaenia[a]	Grey-breasted Martin[a]
Tropical Pewee	Red-legged Honeycreeper[a]

Post-breeding dispersal

Post-breeding dispersal occurs in all birds to some extent, although detecting only slight movements or the dispersal of inconspicuous species requires long-term local study. Well-known post-breeding dispersals include northward movements along the Pacific coast by Black Storm-Petrel, Brown Pelican, Heermann's Gull, Elegant Tern, and Craveri's Murrelet, and northward dispersal by Yellow-footed Gulls to the inland Salton Sea in southern California. Other waterbirds, including Roseate Spoonbill, Caspian Tern, and Black Skimmer, wander to lakes on the Plateau.

Other

Perhaps the least studied local movements are those of hummingbirds, which somehow seem to 'know' when flowers will be blooming, and accordingly find them. Thus, for example, when *Inga* trees at sea level start blooming in northern Honduras, foothill and highland hummingbird species such as Black-crested Coquette and Sparkling-tailed Woodstar appear as if by magic. When the *Inga* bloom peaks in the Sierra de Atoyac foothills in Guerrero, 10 or more hummingbird species appear, from both highlands and lowlands, and then vanish when the bloom dies off. Hummingbirds of highlands and temperate habitats are more prone to such movements than species in less seasonally affected habitats such as rain forest or cloud forest. Other species whose local movements would repay study include several parrots and orioles.

Another form of local movement is that of waterbirds which rely on ephemeral water-level conditions. Thus, species such as Pinnated Bittern and King and Spotted rails have relatively restricted ranges in the dry season but disperse widely to breed when the rains come.

Nomadic wandering is another poorly understood aspect of migration. In the guide region it is most apparent with swallows, and perhaps swifts, which wander widely in search of food and do not seem tied to a particular area or set migration path or schedule.

Undoubtedly, with more interest and long-term studies we will learn more about local migration. We hope that this brief discussion will spur increased interest in the subject.

A HISTORY OF ORNITHOLOGY IN MEXICO AND NORTHERN CENTRAL AMERICA

The history of ornithology in Middle America justifies a book in itself and here we provide only a brief outline of the subject. The early development of ornithological exploration is best understood within the frame of world history. Thus little, if anything, was achieved while Middle America remained under Spanish rule, science being anathema to the prevalent religious fervor of the conquistadors.

Mexican independence, however, dating from 1821, made it possible for all Europeans and Americans to settle in or visit Mexico and send home whatever they wanted. During the same year, the dissolution of the Viceroyalty of Guatemala into its five provinces (which then became the five Central American countries) opened Central America to foreigners.

The novelty of the New World brought numerous pioneers and explorers to Mexico and northern Central America. Many were sponsored by European noblemen who wished to obtain and exhibit curiosities of the little-known Neotropics. Most of the early visitors were French or English, and while some spent several years traveling in the countries, others visited for shorter periods, taking excursions from ports of call where sailing vessels were anchored. Rarely were birds the main focus of these collectors who were interested in anything new, from plants and butterflies to birds and mammals.

The exploration of the mid-1800s is studded with names such as Botteri, Boucard, Bullock, Cabot, Craveri, De Lattre, Deppe, Grayson, Morellet, Prévost, Sallé, and Xantus. Specimens were sent back to Europe or the eastern USA where ornithologists puzzled over and named new species, often in honor of the collectors. Most of this early work, however, was characterized by at best imprecise labeling of specimens, many of which arrived simply from 'Mexique' or 'Oaxaca', etc., and misinterpretations of localities from these early specimens still plague analyses of distribution patterns today.

Notable among the early collectors were Ferdinand Deppe and Andrew Jackson Grayson. Deppe spent much of five years (1824–29) traveling and collecting in Mexico, and sending most of the specimens back to Germany. Temascaltepec, in the state of Mexico, and the adjacent village of Real de Arriba are famous Deppe localities whence originate the type specimens of many Mexican and western North American birds. Grayson was a naturalist and artist who moved from California to Mazatlán in 1859 to include Mexican birds in his portfolio, *Birds of the Pacific Slope*, which he hoped to publish as a western counterpart to Audubon's *Birds of America*. He lived in Mexico until his death in 1869 and was the first naturalist to visit and describe birds from the Islas Tres Marias and from Isla Socorro in the Islas Revillagigedo.

Here we have skipped a little ahead, though it is hard to provide a strictly chronological account of the smorgasbord of exploration in different parts of the region during this period. In December 1857 the Englishman Osbert Salvin landed at Belize and worked his way overland to Guatemala. During the next seven years, Salvin, accompanied on one journey by his sponsor Frederick Ducane Godman, traveled widely in Guatemala as well as elsewhere in Central America. Wherever he went, Salvin aroused interest and trained numerous collectors, both natives and resident Europeans. In addition, other collectors visited Mexico and Central America under the auspices of Salvin and Godman. The culmination of Salvin's efforts was the legendary *Biologia Centrali-Americana*, a 63-volume treatise on the flora and fauna of Mexico (excluding Baja California) and Central America. Although, necessarily, various specialists wrote many of the chapters, Salvin himself wrote the ornithological sections (published 1879–1904) with insight and ability well ahead of his day. As recently as 1938, Griscom wrote that 'ninety percent of what we know today about the ornithology of Guatemala is due to . . . Salvin [and] . . . Godman.'

Two other naturalists who spent long periods in the region during the late 1800s were A. L. François Sumichrast and George F. Gaumer. Sumichrast resided in Orizaba in the 1860s before moving to the Pacific Slope of the Isthmus where he continued to work into the 1880s. Gaumer based himself in the Yucatan, though he also visited Belize and the Honduras Bay Islands. Most of his work was in the 1880s. Both men published natural history observations, including Sumichrast's notable *The geographical distribution of the native birds of the department of Vera Cruz, with a list of migratory species* (1869), and collected numerous specimens which were sent to North American and European museums. Gaumer specimens, however, are notorious for their inaccurate labeling.

Toward the end of the 1800s, American museums and private collectors began to take a strong interest in Mexico and Central America and, by the early part of the twentieth century, European dominance of ornithological exploration in Middle America had waned. Collectors active in this period included A. W. Anthony, F. B. Armstrong, J. H. Batty, L. Belding, Ned Dearborn, Colonel N. S. Goss, M. H. Peck, W. B. Richardson, P. W. Shufeldt, and C. H. Townsend.

The most important feature of the turn of the century, from which one might date modern work on the distribution and taxonomy of Mexican birds, was the work of E. W. Nelson and E. A. Goldman. Under the auspices of the old US Division of Economic Ornithology and Mammalogy, these two men traveled throughout every state of Mexico, during nearly continuous field work from 1892 to 1906. Unfortunately, other duties prevented Nelson and Goldman from publishing fully their wealth of data, but the more significant discoveries found their way into the literature. An interesting overview of their work is provided in Goldman's *Biological investigations in Mexico* (1951).

Another important mark was the pioneering work for distributional summaries in Middle America. Dickey and Van Rossem's *The birds of El Salvador* (1938) was the result of A. J. Van Rossem's field work in that country in 1912 and 1925–27 and represents a remarkable study in which little-known subjects such as local migrations, breeding seasons, and molt were examined. Few subsequent works have treated the birds of an area in such depth.

Concurrent with increasing field work and the accumulation of specimens, American ornithologists began to compile taxonomic and distributional data. The most prominent is Robert Ridgway's monumental work *The birds of North and Middle America* (1901–50) which summarized the taxonomy and distribution of birds in the region through the early part of this century, including most of Nelson and Goldman's specimens. Other notable compilations were J. Grinnell's *A distributional summation of the ornithology of Lower* [=Baja] *California* (1928), L. Griscom's *The distribution of bird-life in Guatemala* (1932), and A. J. Van Rossem's *A distributional survey of the birds of Sonora, Mexico* (1945).

Although several American museum personnel made field trips to Middle America dating from the end of the nineteenth century, the bulk of specimens collected through the middle of the twentieth century was still obtained by professional collectors, although often in the employ of American museums. The 1920s to 1940s mark the period when professional ornithologists began to supercede professional collectors, though the latter also tended to be experienced field naturalists in their own right.

The most famous collectors of this period were W. W. Brown and C. C. Lamb in Mexico, and A. P. Smith and C. F. Underwood in Central America. Lamb continued working into the 1950s and is particularly noteworthy. He collected the first examples of species still much sought-after today, namely Maroon-fronted Parrot, Short-crested Coquette, and Tufted Jay. In addition, the Mexican collector Mario del Toro Aviles worked extensively in Mexico, mostly in the 1940s and 1950s. He is notorious, however, for his careless and often blatantly inaccurate labeling of specimens which, to this day, confuse the interpretation of many species' distributions.

Ornithologists who worked, or began their long careers, in Mexico and Central America in the 1930s and 1940s include E. R. Blake, P. Brodkorb, A. S. Leopold, J. T. Marshall Jr., A. R. Phillips, A. F. Skutch, G. M. Sutton, A. C. Twomey, J. Van Tyne, H. O. Wagner, D. W. Warner, and A. Wetmore. One remarkable man who emerged in this period is Miguel Alvarez del Toro. A native of Colima, Alvarez del Toro moved to Chiapas where he has studied birds with an avid interest and keen intellect. He has become a legendary symbol in the fight to save Chiapas' natural heritage and his important book *Las aves de Chiapas* appeared in 1971.

A major event in the development of Mexican ornithology was the compilation of the *Distributional checklist of the birds of Mexico* by H. Friedmann, A. H. Miller, L. Griscom, and R. T. Moore. This two-volume checklist appeared in 1950 and 1957 and enabled ornithologists to evaluate what was known, and unknown, and to appreciate the wealth of Mexico's avifauna.

From the 1950s, concomitant with an improvement in the infrastructure for travel, and perhaps helped by information in the distributional checklist, interest in Mexico by North American ornithologists increased greatly, much of it in connection with various universities.

One publication in particular perhaps spurred this growing interest. What readers may have failed to note is that the work described thus far was undertaken *without* the aid of field guides. The first bird guide for anywhere in the region was Emmet R. Blake's *Birds of Mexico* (1953). This pioneering work is primarily a manual for in-hand identification which, as recently as the 1960s, was how most identifications were made. Another work that stimulated interest in Mexico at this time was George M. Sutton's *Mexican birds: first impressions* (1951).

The main driving force for field work was taxonomy, that is, the 'need' to name and classify all species and subspecies of birds in Mexico and northern Central America. Distributional work went hand in hand with taxonomy, and ornithology from the 1950s to the 1970s focused on these two subjects. Within these fields one may distinguish between numerous dedicated field workers who often published their findings, and a smaller number of museum ornithologists who compiled taxonomic and distributional data, often for places they had barely, if ever, visited. These two groups working hand in hand filled out our knowledge of birds in Middle America.

Ornithologists active in the guide region in the 1950s and 1960s included R. F. Andrle, D. H. Baepler, R. C. Banks, J. Davis, L. I. Davis, R. W. Dickerman, E. P. Edwards, J. W. Hardy, H. C. Land, W. E. Lanyon, R. T. Orr, K. C. Parkes, A. R. Phillips, J. S. Rowley, W. J. Schaldach Jr., R. K. Selander, C. G. Sibley, F. B. Smithe, R. E. Tashian, W. A. Thurber, J. D. Webster, and D. A. Zimmerman. Some are famous for their work on particular subjects. For example, Dickerman is synonymous with marshes, Hardy with jay behaviour, Lanyon with *Myiarchus* flycatchers, Rowley with nests and eggs, and Webster with Zacatecas. The name of Allan R. Phillips stands out as an overall authority on the taxonomy and distribution of Middle American (and North American) birds. Phillips' *The known birds of North and Middle America* (two volumes, 1986 and 1991) set the record straight on many matters of taxonomy and distribution.

A trend that also developed in this period, especially in association with Louisiana State University, was that of regional studies of distribution and taxonomy. Thus developed works on the Mexican states of Oaxaca (L. C. Binford), Veracruz (F. W. Loetscher), Tabasco (D. G. Berrett), and Colima (W. J. Schaldach Jr.), and on the Yucatan Peninsula (R. A. Paynter Jr.), Belize (S. M. Russell), Honduras (B. L. Monroe Jr.), and Nicaragua (T. R. Howell, unpublished).

So far, people had managed with Blake's guide, even though it lacked color plates; but then one doesn't miss what one doesn't have. The next guide to Middle American birds was Frank B. Smithe's admirable work, *The birds of Tikal* (1966), which treated the birds of this famous Mayan site in northern Guatemala and provided color plates for many 'unfamiliar' tropical species. A spate of guides followed: Hugh C. Land's *Birds of Guatemala* (1970), Ernest P. Edwards' *A field guide to the birds of Mexico* (1972), L. Irby Davis' *A field guide to the birds of Mexico and Central America* (1972), and Roger T. Peterson and Edward L. Chalif's *A field guide to Mexican birds* (1973).

The publication of these guides no doubt contributed to the feeling that the birds of Middle America were more-or-less known. The 1970s and 1980s certainly saw a lapse in interest in the region by ornithologists. Distributional work is not complete, however, and much still remains to be learned. Notable publications of recent years are Robert L. Pitman's *Atlas of seabird distribution and relative abundance in the eastern tropical Pacific* (1986), Richard G. Wilson and Hector Ceballos-Lascurain's *The birds of Mexico City* (1986), and Walter A. Thurber *et al.*'s *Status of uncommon and previously unreported birds of El Salvador* (1987).

After 'all the birds were classified', attention has sometimes turned to other aspects of ornithology, most notably behaviour and ecology. These subjects receive considerable attention in North America but their surface has been barely scratched in Middle America. The most important contributor to our knowledge of Middle American bird behaviour, particularly nesting habits, is undoubtedly Alexander F. Skutch. Although most of Skutch's work has been centered in Costa Rica, his earlier neotropical work (dating from 1930!) began in Honduras and Guatemala. The nests and eggs of a number of species in the guide region remain undescribed.

Birdwatching, or birding, really hit Mexico and Central America in the 1970s, although some pioneering birders were active from the 1950s. The appearance of several guides in the early 1970s coincided with the desire of North American birders to venture farther afield, and numerous individuals and organized tours have ranged over much of Middle America ever since. Notable among the early birders were Peter Alden, who published a guide to finding birds in western Mexico in 1969, and John C. Arvin and Kenn Kaufman who, between them, have visited most corners of Mexico.

Our field work in Middle America began in 1981. Hence, this guide is effectively a product of the work of all of the above-mentioned individuals, and we hope that it will help all those who watch and study birds in Mexico and northern Central America in the future.

CONSERVATION

'Conservation of the environment in Mexico and Central America is a major problem' (Ramos 1985). More specifically, the problem is uncontrolled human population growth combined with widespread low standards of living and education, and governments that have other priorities and may even be corrupt and uninterested in conservation.

There is no short-term solution, and that is a problem in itself. Most politicians, not simply those in Latin America, are elected for short periods and rarely, if ever, have the foresight, or interest, in any long-term planning, let alone for the environment. Since much of the forests in North America and Western Europe has already been destroyed, however, the finger is not there to be pointed. Instead, one can only hope that something may be learned from the mistakes that have been made in 'developed' countries. At the same time, much of the existing degradation, and hence poverty, in Latin America is due to foreign 'aid' transposing unsuitable, temperate technologies on tropical ecosystems.

Both general and specific environmental conservation problems must be viewed in terms of cultural differences. There is often a lack of conservation awareness among people living in all Latin American countries. Latin cultures are based on large families which need to be fed. Hence, more land is cleared, more beans and corn planted, more children produced. The cycle continues and wholesale habitat destruction results. Habitat destruction is the single largest threat to the maintenance of biological diversity in the Neotropics, with much of the forest being cleared for poor cattle land to supply an unnecessary beef demand in the USA. In addition, American companies continue to produce and export harmful pesticides to Latin America where, consequently, wildlife as a whole, including North American migrants, suffers.

More specific problems that are often found in 'developed' countries are no less common in Latin America. Birds (of almost any size) are hunted for food. Most small boys living in the country and small towns possess slingshots and kill birds for 'sport', typifying a lack of aesthetic appreciation of birds and lack of respect for animal life. There is also the common tendency in rural areas to shoot any bird of prey on sight. Paradoxically, keeping parrots and songbirds in cages about the house is considered an appreciation of nature. Sadly, the desire to have birds, particularly parrots, as pets has spread far beyond Middle America. The illegal international bird trade has effectively extirpated several species from much of their former range in the guide region. Within 20 years, or less, we predict that wild Scarlet Macaws and Yellow-headed Parrots may be things of the past in Mexico.

Other birds which have all but vanished from most of the guide region, due to uncontrolled hunting (in some cases combined with habitat loss), are mostly large game birds such as Great Curassow, Crested Guan, and Wild and Ocellated turkeys. Populations of Brown-backed and Slate-colored solitaires have been locally extirpated by capture for cage birds—it is a sad morning indeed that one experiences in forests deprived of these fine songsters. The spectacular Imperial Woodpecker is probably extinct; Military Macaws have been wiped out from most of the Pacific Slope; Yellow-naped Parrots are disappearing. More depressing examples would be simply wasting paper.

On a brighter note, Mexico signed on as a member of CITES (the Convention on International Trade in Endangered Species) in 1991 and, with El Salvador's accession in 1987, all Central American countries are now CITES signatories. While signing a document has never solved an issue by itself, recognizing the problem is an important first step.

With a few notable exceptions the numerous parks and reserves in Mexico and northern Central America exist only on paper. There is rarely (ever?) sufficient funding or human resources to protect such areas and enforce legislation on hunting and trapping birds, cutting trees, etc. Until recently, it was unclear in some countries which arm of government was responsible for parks and reserves, and some agencies were faced with potentially contradictory missions; for example, federal reserves in Mexico were, up until 1992, administered by SEDUE (the Secretary for Urban Development and Ecology).

Conservation in the guide region depends ultimately on political decisions, and politicians need data. How many species are in an area, what are their ecological requirements, how big an area should be set aside; these are all questions that we rarely have the information to answer.

Fortunately, the need to compile databases has been recognized. The International Council for Bird Preservation (ICBP) completed recently its analysis of global biodiversity through patterns of bird endemism: Mexico ranked in the top 10 countries with the highest numbers of restricted-range species, but also ranked in the top 10 countries with the highest number of *threatened* restricted-range species (ICBP 1992). The ICBP also published recently the benchmark Red Data Book for the Americas (Collar *et al.* 1992). In 1987, Conservation International (CI) created two ecosystem data centers: a national biotic resources information center established at the Center of Ecology at the National University of Mexico, and a state-level database for Chiapas run by the Instituto de Historia Natural (IHN). These are important resources which can provide critical information for federal agencies involved in policy-making and planning.

Designating and then protecting reserves or *habitats* efficiently is the key to maintaining species diversity. Using 'sexy' species, such as the Resplendent Quetzal, as symbols is a useful way to highlight the predicament of threatened areas. However, it is the whole ecosystem, not just one or two species, that needs to be protected. Captive breeding may be an acceptable last resort in some cases, such as the Socorro Dove which is presumed extinct in the wild. Often, however, it is abused; if the amount of money raised for raptor breeding and reintroduction programs could be spent on purchasing and protecting habitat, or fighting for the worldwide ban of harmful pesticides, then perhaps something useful might be achieved.

Local biologists face many technical problems that may not be apparent to outsiders. Most scientific literature is in English, and Latin American universities rarely have the funds to subscribe to all the necessary journals. Consequently, local biologists are often unequipped to keep up with developments in their field. Outside ornithologists frequently know little more, as exemplified by much that is published in North American journals. Foreign involvement *can* help, but too many studies are short-term ones, undertaken by eager North American or European students who lack adequate knowledge of their subject. *Long-term* research projects are essential, but require commitment. Individuals who have made long-term avian studies in Middle America are few—M. Alvarez del Toro and A. R. Phillips (Mexico), A. F. Skutch and F. G. Stiles (Costa Rica), A. Wetmore (Panama). The list does not go on.

It has to be communicated somehow that development and conservation are not incompatible and that our future depends upon wise management of the environment. Education at all levels is needed, from government planning to efficient farming practices, and birth control. The environment's long-term economic value (for watersheds, medicine, selective logging, tourism, etc.), and the need for its management, have to be made evident to politicians and local farmers alike.

As an old proverb says: 'one cannot love what one does not know.' Educational programs must help people understand the natural environment, the role humans play in it, and that they are a part, not the center, of it.

Mexico

In 1992 SEDUE was divided into the Secretaria de Desarrollo Social and the Instituto Nacional de Ecología; it is not yet clear whether this will be better for the environment, but early signs are encouraging.

Mexico's 64 federally administered *areas naturales protegidas* (protected natural areas) comprise 44 National Parks, 13 Special Biosphere Reserves, and 7 Biosphere Reserves (SEDUE 1989). The criteria for establishing each have varied, as have the uses and management procedures.

National Parks are defined as 'areas that are biogeographically representative on a national level, with one or more ecosystems important for their scenic beauty and their scientific, educational, recreational, or historic value' (SEDUE 1989). It is evident that most national parks have been created for recreation: 27 of the 44 parks are in the highly populated Central Volcanic Belt between Guadalajara (Jal) and Xalapa (Ver). Many parks encompass spectacular natural formations such as the Cascada de Basaseachic (Chih), or the Cañon de Sumidero (Chis), and three are centered on famous Mayan ruins.

Special Biosphere Reserves (formerly Ecological Reserves) are 'areas with one or more natural ecosystems, not significantly altered by humans, which contain species that are endemic, threatened, or in danger of extinction' (SEDUE 1989). These mostly comprise vulnerable ecosystems such as Isla Rasa (BCN), or Río Lagartos (Yuc), many of which are small enough to be protected reasonably well.

Biosphere Reserves are extensive areas (of more than 10 000 hectares) that incorporate representative and unaltered ecosystems as well as areas altered by man. They are set up to incorporate the goals of habitat conservation, scientific investigation, education, and integrated human exploitation. Thus, they recognize that people cannot simply be excluded from using an area. The five Biosphere Reserves include Montes Azules in the Selva Lacandona (Chis), and Sian Ka'an (QR).

Like several governments elsewhere in Latin America, the Mexican government is increasingly aware of the role of the private sector in conservation. It is also becoming more willing to consider creative financing mechanisms for environmental issues. In February 1991, the government signed with CI, Mexico's first debt-for-nature agreement. This represents an important step forward by Mexico in its growing commitment to support major conservation work.

Many Mexican universities have an increasing number of qualified young biologists who are greatly needed and involved in their country's age of environmental awareness. Working with external support organizations such as CI, ICBP, the Smithsonian Institution, and Point Reyes Bird Observatory (PRBO), Mexico's new generation of biologists is working on a wide array of biomonitoring and conservation-oriented projects from Baja California and the Gulf of California to the Selva Lacandona and the Yucatan.

An outstanding regional example of environmental awareness may be seen in the state of Chiapas. Much of this can be attributed to the work of Miguel Alvarez del Toro, a remarkable man who has written many books on the natural history of this rich state. The Tuxtla Gutierrez Zoo, named in honor of Alvarez del Toro, is renowned as the finest zoo in Latin America and exhibits only regional fauna. A visit reveals the wonders of Chiapas' natural heritage, with intelligent and useful interpretation, to many thousands of people a year, including the all-important school children. The zoo operates under the auspices of the IHN, which also helps manage the state's official reserves, such as the famous El Triunfo, and plays an important part in environmental education in Chiapas. People who wish to support conservation can become members of FUNDAMAT, a private non-profit organization established to assist the IHN. Another conservation organization in Chiapas is ECOSFERA, a non-profit private organization founded in 1988 to promote the conservation of natural resources through their sustainable use.

The Yucatan Peninsula is another area of notable conservation awareness. Much forest still remains in the region, due in large part to the large population of native Mayan people who traditionally have learned to live with nature rather than against it. Sian Ka'an Biosphere Reserve, established in 1986, covers 528 147 hectares (1.3 million acres) of diverse habitats in central Quintana Roo. Much of the initiative for education and research in relation to the reserve comes from the private organization, *Amigos de Sian Ka'an A.C.*, founded in 1986 by concerned and interested persons. Recently *Amigos* produced a book, *100 common birds of the Yucatan Peninsula* (MacKinnon Vda. de Montes 1990), the sale of which has provided funding for a Spanish translation of the book. The informative text and colorful photos are a model example of what is needed to generate interest at local levels. In 1989, PRONATURA-Yucatan, A.C. (independent of PRONATURA A.C.) co-hosted the First International Workshop on Ecotourism on the Yucatan Peninsula. Other work of PRONATURA-Yucatan includes baseline studies in the Calakmul Biosphere Reserve, and environmental education.

PRONATURA A.C. is a private non-profit organization, created in 1981, which aims to develop activities that further conservation and an appreciation of Mexico's natural resources as a whole. PRONATURA-Chiapas is involved in education at a grass-roots level and also manages small reserves in the state.

DUMAC (Ducks Unlimited of Mexico, A.C.), founded in 1974, is a non-profit organization dedicated to the conservation, development, and improvement of wildlife resources, especially for waterfowl. DUMAC recently formulated a 10-year master plan to evaluate threats to Mexico's most important wetlands and to design and implement conservation and management strategies. DUMAC has also established three field research centers which provide logistical bases for both Mexican and foreign students, and has completed more than 70 projects, including the rehabilitation of over 250 000 hectares of wetlands.

The *Agrupación Sierra Madre* was founded in 1989 to promote awareness of Mexico's richness of natural resources with a view to conservation. *Sierra Madre* has an annual campaign (United for Conservation) which involves prominent Mexican companies and institutions in raising funds for specific conservation projects. In recent years, these have included Isla Rasa, the Lacandon rain forests, and Mexico's northern deserts. In addition, *Sierra Madre* has edited three art books on important themes of Mexican natural history.

Notable among other private conservation associations is *Pro Esteros*, an international grass-roots organization formed in 1988. *Pro Esteros* aims to promote education and awareness in respect of preserving Baja California's internationally important, yet little-studied, coastal wetlands. In conjunction with SEDUE, *Pro Esteros* recently completed the complicated application to request reserve status for Estero Punta Banda on the Pacific coast of northern Baja California.

The Island Endemics Foundation, established in 1990, is working in conjunction with CIB (Centro de Investigaciones Biologicas) to save the fragile ecosystems of Las Islas Revillagigedo. Major concerns are the endangered Socorro Mockingbird and a reintroduction program for the Socorro Dove.

Another recent development is the *Club para la Conservación y Observación de Aves, Distrito Federal* (CCOA-DF), a birding club formed in 1990 in Mexico City. The club now counts more than 100 members, including many Mexicans and foreigners living in the city. Working with SEDUE, the club has already made an impact in one of its main areas of concern: halting the illegal cage bird trade. CCOA-DF also offers an active field trip program. Hot on the heels of CCOA-DF has come the *Club de Observadores de Aves del Noreste* which conducts regular field trips and is promoting an appreciation of birds in northeast Mexico.

Mexico is such a large country that regional approaches, whether as part of a national system or as separate, often private organizations, may be the best way to effect conservation projects. Environmental awareness has grown exponentially in Mexico in the past 10 years and it should continue to do so, giving hope to protection of the diverse natural wealth with which the country is blessed.

Guatemala

In February 1989, in a push to protect the nation's rapidly dwindling forests, the government of Guatemala passed the *Ley de Areas Protegidas* (Protected Areas Law) which proposed 44 new nature conservation areas and established boundaries for the six existing *biotopos* (biological reserves managed by the University of San Carlos). The new law brought all new and existing parks and reserves under the auspices of the *Consejo Nacional de Areas Protegidas* (National Council of Protected Areas). In addition, President Vinicio Cerezo signed an executive order designating all Guatemalan territory north of latitude 17°10'N (about 1 000 000 hectares) as the Mayan Biosphere Reserve.

This good news reflects the growing awareness for conservation in Central America and potentially puts over 15% of Guatemala's territory under protection. The proposed areas include many 'islands' of cloud forest on peaks throughout the country, all of which constitute important watersheds that are among the most threatened habitats in Middle America.

The new law named the proposed 'protected areas' and created the legal mechanism to transform them into national parks, biotopos, wildlife reserves, etc. The reserve boundaries, however, have yet to be determined specifically. Technical studies and boundary specification are already underway in some areas. In this respect, the work of private organizations such as the Nature Conservancy and FUNDAECO (a non-profit foundation created by out-going president Cerezo) is all important.

Like other Central American countries, Guatemala has realized that involvement of the private sector can be important in conservation. Thus, in 1990, *Defensores de la Naturaleza* (Defenders of Wildlife; founded in 1983) was assigned the management of the Sierra de las Minas Biosphere Reserve. Other non-profit organizations in Guatemala, such as the Guatemala Audubon Society (chartered by National Audubon in 1988), are working toward environmental education and the purchase and protection of reserves.

The Maya Project, begun in 1988 by the Peregrine Fund, aims to study neotropical raptors as environmental indices to develop management strategies for protected areas. The project is developing census techniques for tropical forest raptors, studying the nesting biology of little-known species, and most importantly, actively involving Guatemalans in the work. Most of it is centered at Tikal National Park.

Thus environmental awareness in Guatemala is snowballing but it is important to keep the momentum going and carry through the proposed parks to well-defined, protected, and managed reserves. This will require time, money, dedicated commitment, and an understanding of local community development needs around reserves.

Belize

Blessed by a still small (yet rapidly growing) population, Belize may have as much as 85% of its native vegetation (60% of its forest) relatively intact. This is at striking odds with Mexico or other Central American countries and has helped make Belize a prime attraction for international conservation organizations.

The conservation movement in Belize began in the 1960s when a few interested persons formed the Belize Audubon Society (BAS) in 1969. From modest beginnings, the now independent BAS has grown to become an influential body in Belizean conservation. The BAS' first success story was to promote the Jabiru from being viewed as an opportunistic meal to a legally protected symbol of national pride. Environmental education continues to be a major goal of the Society.

In 1981, the Belizean Government passed the National Parks System Act and the BAS received a mandate from the government to help it develop and run the parks. The first area designated under the new act, in 1982, was Half Moon Caye, and BAS has kept working steadily with the government to add others: by 1986, Crooked Tree, Cockscomb, Guanacaste, and Blue Hole had all been designated as protected areas.

In 1987, the Programme for Belize was initiated as an organization sponsored by societies such as Massachusetts Audubon, National Audubon, and the Nature Conservancy. The Programme's objectives are to acquire land for conservation and, in cooperation with the Belize government, to encourage environmentally sustainable economic development. The first major project was the creation of the Río Bravo Conservation Area, a mostly forested area of about 60000 hectares that will be open to multiple use.

The Belize Zoo and Tropical Education Center, established in 1983, exhibits only local fauna and has become famous as a conservation movement. Every school child in Belize has visited the zoo as part of a long-term plan of environmental education.

Since 1987, Manomet Bird Observatory (MBO) has carried out intensive research into Belize's tropical forests. In particular, MBO was instrumental in conducting research leading to a proposal that helped the Bladen Branch, a rich area of rain forest in southern Belize, to be designated a nature reserve in 1990.

Ecotourism in Belize has flourished in recent years and tourism now vies with agriculture for the country's number one foreign income source. Private reserves in the country include Chan Chich and Monkey Bay Wildlife Sanctuary, with others appearing regularly. However, like any country with such rich natural resources, Belize faces potentially unsustainable development pressures and care must be taken if all is not to be damaged by misguided planning.

El Salvador

An unstable political climate combined with economic problems and over-population (the country is one of the most densely populated in the world) means that conservation is low on the list of immediate concerns in El Salvador.

As of 1986, none of the parks and reserves was protected by specific laws. All, however, came under the *Ley Forestal* (Forestry Law) of 1973 which provided for some protection of wildlife. In 1986 six areas were still administered and maintained as parks (two existing and one proposed national park, two wildlife refuges, and one biological reserve) and 40 other areas had been studied, six of them proposed as national parks (Thurber *et al.* 1987). Lack of funding and sporadic civil unrest have made protection of natural areas at best difficult. In January 1992, however, a peace agreement was signed that has lasted up to the publication of this book. This may be an encouraging sign for the environment.

Unlike other Central American countries, El Salvador has retained its parks and wildlife agency under the jurisdiction of the agriculture ministry. Encouragingly, however, two articles mandating natural resource conservation were included in El Salvador's new constitution, and the country has recently embarked upon a plan to entrust municipalities and agricultural cooperatives with direct management of new parks and reserves being established with its agrarian reform program (CENREN 1987). Whether this latter move will bode well for the environment remains to be seen.

Honduras

In 1987 the Government of Honduras declared 37 areas of cloud forest as national parks, wildlife refuges, and biological reserves. The main drive behind this far-seeing move was to protect the remaining islands of forest as watersheds. These forests on mountains above 1800 m are the ecosystems which have the greatest capacity to generate drinking water at the lowest cost. The need to protect the biological diversity of the reserves was also recognized.

The new law means that there are now 54 'areas protegidas' in Honduras: 13 National Parks, 9 Wildlife Refuges, 21 Biological Reserves, a Biosphere Reserve, and 10 other reserves (Asociación Hondureña de Ecología 1990). Additional proposed reserves bring the total to 79 protected areas that comprise the *Sistema de Areas Protegidas de Honduras*, the responsibility for which was removed from the ineffective RENARE (Renewable Natural Resource Agency) to COHDEFOR (the State Forestry Corporation) in 1991. This is hoped to have the dual benefit of strengthening the latter and eliminating overlapping authority between the two bodies.

Honduras has also recognized the potential importance of the private sector in conservation. For example, the Cuero y Salado Foundation has been given legal authority to manage a national wildlife refuge. However, as in other Central American countries, it is not yet clear how non-governmental organizations will handle law enforcement and management of protected areas.

Since the late 1980s (and mostly since 1990), the birth of over 20 organizations dedicated to conservation in Honduras reflects the rapidly growing environmental awareness in the country. Many of the new organizations are involved in environmental education in schools and local communities. As in other countries, funding remains a problem, and conservation organizations, at both regional and international levels, must work together if either is to be successful in the long term.

Nicaragua

Almost 20 years before any other Central American country, Nicaragua established a small department to specifically address wildlife conservation. As in neighboring countries, however, the department has led an unstable existence, and recent social and economic problems have not helped. For example, the vigorous conservation movement of the early 1980s was crippled by the economic embargo imposed by the US government, and later by the contra war.

With a change of government in 1989, Nicaragua embarked upon an ambitious plan to restructure its natural resource management authority (IRENA), along the lines of a superagency for the environment (Gutierrez 1990). Of immediate concern to IRENA is trying to secure financial and technical support for the conservation of three major areas: Cayos Miskitos in the Caribbean, and the last two extensive areas of rain forest, Boswás (in the north-northeast) and Río Indio (in the southeast).

As of 1989, Nicaragua had 19 protected areas and 18 other potential new reserves (Cornelius, personal communication). These include all volcanic peaks above 800 m in the Pacific region, which were given legal protection in 1987 by the Nicaraguan government. Although the 19 protected areas comprise only 3% of Nicaragua's national territory, one must remember that the country is the largest in Central America.

Nicaragua shares with the rest of Latin America the problem of an advancing agricultural frontier. This has been compounded recently by settlement of the contras in forested areas within the limits of Boswás and Río Indio, where they are seeking permits to cut the forest to sell to private companies.

Like most other Central American countries, Nicaragua has university undergraduate programs in ecology and natural resource management, and the Central American University of Nicaragua is restructuring its programs to make them more responsive to the socio-economic realities of natural resource management. Environmental awareness is growing rapidly in all sectors of life, and potential political stability in the forseeable future should help Nicaragua to work at evaluating and protecting its vast natural resources.

Conclusions

While there is much to be commended about conservation awareness and efforts in Mexico and northern Central America, much remains to be done for the region to enter the next century with a semblance of its wildlife resources ecologically and genetically intact. Efficient and educated management, and protection of existing reserves are priorities. It must also be recognized that efforts at conserving wildlife cannot be directed exclusively at existing parks and reserves since many critical forested areas and wetlands remain outside protected areas. The value of tropical agricultural systems to wildlife conservation has also been largely ignored. For example, many coffee plantations in Mexico and Guatemala are managed under native shade trees and occupy an elevational range that is poorly represented in protected areas. The establishment of private reserves to complement federally protected areas, and innovative management of economically important habitat for successful coexistence between man and wildlife, will be essential to conserve the region's biodiversity.

In order for wildlife to survive in Mexico and northern Central America, a restructuring of each country's economic policy, and of outside countries' foreign policy, is needed. This must be accompanied by commitments of funding and technical assistance that far surpass what has been available thus far from international conservation organizations and bilateral aid agencies. Most importantly, it must be recognized that improving the quality of life of people in the region, and ensuring the survival of wildlife, are intertwined.

The best way you can help with conservation in countries of the guide region is to visit them, on your own or with organized tour groups, and contribute to the economy in a way that supports conservation. For anyone wishing to go beyond this stage, the following addresses may be of use, and donations to any of these organizations will be invaluable in helping to save a rapidly vanishing heritage. Addresses of Mexican and Central American organizations may change frequently; those given here are correct as of 1992, to the best of our knowledge.

Conservation International
1015 18th Street NW, Suite 100
Washington, DC 20036, USA

BirdLife International
  (formerly ICBP)
Wellbrook Court, Girton Road
Cambridge CB3 0NA, UK

The Nature Conservancy
1815 N. Lynn Street
Arlington, VA 22209, USA

World Wildlife Fund
1250 24th Street NW
Washington, DC 20037, USA

Amigos de Sian Ka'an, A.C.
Apartado Postal 770
Cancun 77500
Quintana Roo, MEXICO

Instituto de Historia Natural/
FUNDAMAT
Apartado Postal 6, C.P. 29000
Tuxtla Gutierrez,
Chiapas, MEXICO

DUMAC, A.C.
Apartado Postal 776
Monterrey, N.L.
MEXICO

ECOSFERA
Apartado Postal 219, C.P. 29200
San Cristóbal de las Casas
Chiapas, MEXICO

Pro Esteros
1825 Knoxville Avenue
Long Beach, CA 90815
USA

PRONATURA-Chiapas
Apartado Postal 219
San Cristóbal de las Casas
Chiapas, MEXICO

PRONATURA-Yucatán
Calle 1D #254-A (x 36 y 38)
Col. Campestre, 97120 Merida
Yucatán, MEXICO

Programme for Belize
PO Box 385 M
Vineyard Haven
MA 02568, USA

Defensores de la Naturaleza
7a. Avenida y 13 Calle, Zona 9
Edificio La Cúpula
Ciudad de Guatemala
GUATEMALA
Central America

Belize Audubon Society
PO Box 1001
Belize City, BELIZE
Central America

Belize Zoo and Tropical Education
   Centre
Mile 30, Western Highway
Belize District, BELIZE
Central America

FUNDAECO
25 Calle 2–39, Zona 1
Ciudad de Guatemala
GUATEMALA
Central America

BIRDING IN MEXICO AND NORTHERN CENTRAL AMERICA

Rather than providing a list of specific localities where one can expect to find various species, we consider it more useful to provide some information on *how* to look for birds. A major problem with giving specific sites in the region is that, even within parks, habitat alteration and 'development' are so frequent and rapid that reliance on directions, even recent ones, can be frustrating when a new road has been built, a forested area cleared for cattle pasture, etc.

Most species can be found without too much difficulty by following this procedure.

1. Check the range maps and read the status and distribution accounts to make sure you are or will be in the right place at the right time.
2. Read the habitat and habits section to learn more specifically where to look and what behavior to expect from the bird (for example, skulking low in brush, foraging high in canopy).
3. Read-up on the vocalizations to be aware of what to listen for.
4. Study the plates, descriptions, and notes on similar species to know what to look for and at when you see the bird.

Then, you just have to find the habitat! Recognizing habitat is perhaps the single most important trick to finding birds, although birders intent on finding regional specialities almost invariably have a network of friends and contacts that can provide current specific site information. For birders who simply enjoy birding, getting out into the field early and taking binoculars and a notebook is about all there is to it, although once a quiet trail or side road has been found, the following advice may be useful to those unfamiliar to traveling or birding in the region covered by this guide.

The people of Mexico and northern Central America are generally friendly and helpful if one is polite and friendly to them. The commonest reaction to birders is curiosity. Private property generally is not viewed as it is in North America, and surprise (at being asked) is often the reaction obtained when one asks permission to enter or cross someone's land. It is always best, however, to err on the side of courtesy. If one speaks even a little Spanish it will be much appreciated, but universal sign language and smiling usually work well.

Many rural people carry impressive-looking machetes, and in some areas rusty-looking rifles slung across the back are not infrequent. These are carried in the same way one might carry a pocket knife or binoculars and should not be viewed as threatening—although one's first impression on meeting a group of machete-wielding people on a remote trail might be otherwise. The group of people will pass, usually smiling, and bid you 'buenos días' (good morning) or 'buenas tardes' (good afternoon), and you should reply in kind. In fact, rural people will often be scared of you and breaking the ice by getting in the first 'buenos días' is often a good idea, although many women and children may be too frightened to answer.

Contrary to popular belief, local drivers are not out to kill all and any pedestrians. At the same time, driver awareness of pedestrians may not be what one is used to at home. Drivers often honk their horns at pedestrians (and not just foreigners) which may be considered rude in some cultures and can take a while to get used to. It is simply a way of saying 'don't step into the road now', or even just 'hello'. Wave at the drivers and almost invariably they will wave back.

In our experience, crime is rarer in Mexico and northern Central America than in much of the USA, and one of us (SNGH) feels much safer in Guatemala City, for example, than in New York. Having said this, it is always sensible to be aware of one's surroundings and prevention is invariably the best 'cure'. Carrying around flashy photographic or recording gear and being loud and conspicuous are behaviors to avoid. Having a pair of binoculars and a small tape recorder has never caused us any problems.

Birding in Mexico and Central America far surpasses anything we have experienced in North America or Europe. Having binoculars and being polite and aware of one's surroundings are things that should be innate and are all that is needed.

USING THIS BOOK

TAXONOMY

Systematics and theory

The purpose of taxonomy (the science of classification) is to describe and name divisions of living organisms and to establish relationships between them. Most people are aware of the major division of living things into two kingdoms—plants and animals. Birds are a group at the level called a **class** within the Animal Kingdom and members of the Class Aves are characterized by having feathers.

Divisions within the Class Aves bring us to the level of **species**. A group of similar species forms a **genus**. Every known species possesses a unique name based on a binomial system based upon Latin, an internationally accepted common language. The specific name comprises the genus name (first letter capitalized), followed by the specific name (all lower case). Hence, the American Robin (*Turdus migratorius*) can be classified as follows:

Class: Aves
Order: Passeriformes
Family: Turdidae
Genus: *Turdus*
Species: *migratorius*

The major taxonomic divisions that concern most birders and field ornithologists are species, subspecies, and genus (plural: genera).

Species

For many professionals, as well as amateur ornithologists and birders, the question of what constitutes a species, and why, is often puzzling. We follow the biological species concept (BSC) which is defined as follows: *a species can be defined as a population, or group of populations, of actually or potentially interbreeding individuals, reproductively isolated from all other such populations. The members of a species should be able to interbreed freely and produce fertile offspring.* Problems often arise, however, in interpreting this definition (see below).

Some biologists have proposed alternatives to the BSC which, like any definition, will have some problems. The main challenge to the BSC is the phylogenetic species concept (PSC) which defines a 'species' as the 'smallest diagnosable cluster of individual organisms in which there is a parental pattern of ancestry and descent' (Cracraft 1981), and in which subspecies are *taxa non grata*. The

PSC often relies on trivial differences in morphology or plumage and overlooks important biological information; such an approach seems to derive from the inability of persons to understand a complex natural world rather than the inability of birds to conform to our attempts at classification. We believe that a species is a biological entity, that the BSC is as good as any definition, and that subspecies are an important part of avian taxonomy. At the same time, hopefully the BSC will benefit from the PSC challenge, and avian taxonomy as a whole will advance.

Subspecies

The term subspecies (or race) is applied to groups within a species that can be distinguished by virtue of certain characters, often slight differences in measurements or plumage tones. Subspecific differences may be clinal (that is, those which occur with intergradation in a continuous population) or allopatric (those which occur between separate populations).

A subspecies' name is appended to its specific name. For example, the American Robin is split into eight subspecies; usually, the first subspecies described is the nominate subspecies: *Turdus migratorius migratorius*, and the distinct population in the mountains of southern Baja California, for example, is *T. m. confinis*, the San Lucas Robin.

Genera

As mentioned above, a genus is a group of closely related species. Traditionally, morphological factors (such as plumage, bill shape, skull structure, etc.) have been important in identifying genera. Clearly the degree to which species are considered similar, or dissimilar, is highly subjective. Newer lines of thought prefer to classify genera as monophyletic groups between the levels of family and species. Monotypic genera, defined as those which have a single extant species, are naturally commoner in older families and relatively rare in passerines.

Evolution and practical problems

How do species evolve?

When a population becomes divided, the resultant separate groups will tend to develop differently for various reasons, such as slight differences in genetic composition or in environmental conditions. The taxonomist is faced with the problem of deciding whether, given sufficient time in isolation, the separated groups have diverged enough to be considered separate species. However, although species evolve in isolation, proof of their status as species is often indicated only through loss of isolation. That is, should the two forms meet again (a phenomenon known as secondary contact), will they retain their differences and demonstrate reproductive isolation (that is, truly constitute two species) or will they merge back together into one freely interbreeding population (that is, constitute but a single species)?

The role of the taxonomist

The taxonomist is a biologist trained in the principles of evolutionary biology. He/she should evaluate all factors relating to speciation for a given case and decide as best possible. Even a simple list of all factors involved in species determination would be too long for this chapter. The most obvious factors for birds are plumage (coloration, pattern, and sequence), voice (songs and call notes), behavior (including courtship displays), egg patterns, nest site and structure, and habitat preference. More recently, biochemical techniques (such as DNA–DNA hybridization) have been used in an attempt to determine more precisely taxonomic relationships. While we recognize that biochemistry can, and should, contribute another facet to understanding taxonomic relationships, such techniques are still in their infancy and most results have yet to be widely accepted.

The degree to which each factor contributes to defining a particular species varies greatly. For example, ritualized courtship displays seem more important than plumage in albatrosses, plumage pattern and display more important than voice in parrots, and voice more important than plumage in many flycatchers.

It must be recognized that speciation is a dynamic evolutionary process. A population does not attain specific status overnight. Drawing lines to indicate that certain birds form a species rather than subspecies is a purely human designation. It is the birds themselves that decide the reproductive isolation.

Taxonomic problems

There are two main types of problems in trying to determine specific relationships. First, if two populations have separate breeding ranges (that is, are **allopatric**), it is a real problem to determine whether they are different species. All factors should be analyzed carefully before any decision is reached. In many instances, however, especially in the Neotropics, relatively little is known about numerous allopatric populations and any decision is at best tentative. Examples of allopatric problems include the Sharp-shinned and White-breasted hawks, Montezuma and Ocellated quails, and Green-throated and Green-breasted mountain-gems. Second, in areas where populations have come into secondary contact and share, at least in part, a common breeding range (that is, are at least partly **sympatric**), the taxonomist is presented with a natural laboratory in which to study evolutionary processes. The case of two or more sympatric species may sound simple at first—either they interbreed or they don't—but in reality it is riddled with complexities. Examples of secondary contact may be found in many familiar North American birds such as Red-shafted and Yellow-shafted flickers, Rose-breasted and Black-headed grosbeaks, Eastern and Western meadowlarks, and Baltimore and Bullock's orioles. Each case of secondary contact is different. There will almost always be periods when isolating mechanisms are tested in newly reunited populations. Interbreeding can vary from being common and extensive (flickers) to rare and local (meadowlarks). In these two cases, most taxonomists agree that the Red-shafted and Yellow-shafted flickers are representatives of a single species whereas the meadowlarks constitute two species. Problems arise in more complex cases, but to analyze every individual in an area of sympatry is essentially impossible. Consequently, decisions are based on statistical analyses of samples and, as such, are open to individual interpretation. In addition, since evolution is a dynamic process, subsequent analyses of a given situation may produce different results and perhaps different conclusions.

Approximately 1070 species (or about 11% of the world's 9600 or so bird species) occur today in the region covered by this guide. There has been, and often still is, considerable debate among taxonomists concerning the specific relationships of over 15% of these 1070 'species'. In many cases, critical field work is still needed to delimit ranges and study breeding biology, vocalizations, etc.

In cases where incipient species appear to be involved (that is, forms which appear to be attaining reproductive isolation but which may not yet have fully achieved specific status), we have indicated that uncertainty exists. If the forms are known, or thought to be, sympatric, we have used a question mark, for example Clark's Grebe *Aechmophorus* (*occidentalis?* = Western?) *clarki*, or Long-tailed Sabrewing *Campylopterus* (*curvipennis?* = Wedge-tailed?) *excellens*. In cases of clear or apparent allopatry, an option has been given, for example, White-breasted Hawk *Accipiter chionogaster* or *A. striatus* (in part, = Sharp-shinned Hawk?), Ocellated Quail *Cyrtonyx montezumae* (in part, = Montezuma Quail?) or *C. ocellatus*, etc.

In this evolving natural world there will always be cases which might be best placed in a 'don't know' or borderline category. To keep pace with the rapidly changing opinions of taxonomists is difficult. The methods of naming incipient species used here enable one to be aware of taxonomic problems and, hopefully, will prompt further study. For the birder who simply enjoys observing birds, it should not matter if the bird is a species in its own right or not. The pleasure is in being able to recognize a bird, regardless of its taxonomic status, learn about it, and enjoy it for what it is.

MOLT AND PLUMAGE

Although birders often consider molt and plumage rather confusing, a basic understanding of these two intertwined subjects can be important for bird identification. The molt terminology we use follows that proposed by Humphrey and Parkes (1959, 1963). They pointed out correctly that terms such as 'summer plumage', 'breeding plumage', 'postnuptial molt', etc., can be inaccurate and confusing. This is never truer than in tropical regions. For example, distinct seasonal plumage differences are most pronounced in migratory, temperate-breeding species, but for most tropical birds 'breeding' and 'non-breeding' plumages are the same. Other problems with traditional nomenclature include loons and most terns which attain their 'first winter' plumage in their first summer, or immature birds which may have a 'postnuptial' molt even though they have not yet bred.

Humphrey and Parkes proposed a plumage terminology independent of a bird's life cycle, one that can be adapted for all birds, which follow a fundamentally similar plumage succession regardless of family.

Molt is the process by which birds replace their feathers. This requires considerable energy expenditure, so most molting occurs before or after breeding, and before or after migration, the two other major energy demands in a bird's life. Factors which affect the extent and timing of molt are many and varied, and include the following.

Breeding chronology, and consequently molt timing, varies considerably within Mexico and northern Central America such that some individuals of a species may be completing their molt after breeding while others are just beginning to breed.

The feathers of birds in harsh desert environments may wear more quickly than those of birds in protected woodland. Molts within a species may thus vary from partial to complete in response to environmental conditions.

Birds that are highly dependent on flight feathers for feeding, such as hummingbirds, have necessarily a gradual molt. An alternative approach is that of many ducks which, having a high wing-loading, drop all their flight feathers simultaneously and risk a short flightless period in preference to several months of labored flight.

In some sexually dimorphic species, males may resemble females for much of the year but attain a brightly colored plumage for breeding, for example, Ruddy-breasted Seedeater. In several passerines, first-year males attain the adult male breeding' plumage only partially, for example, Black-headed Grosbeak.

Regardless of these variations, molt can be described in terms of three general strategies. A molt may be partial, incomplete, or complete. Partial molt involves replacement of all or some contour (head and body) feathers only. Incomplete molt involves replacement of contour and some flight feathers (remiges and rectrices). A complete molt is the replacement of all feathers.

Newly hatched birds are described as altricial or precocial. The former refers to birds that hatch in a helpless condition, usually with their eyes closed, and a naked body, although several wading birds, hawks, and owls are covered with down on hatching. Precocial young hatch with their eyes open and can move about shortly after drying; they are downy and only partly, or not at all, dependent on their parents for food. The first true (non-downy) plumage of a bird is its **juvenal** plumage, which is attained upon or shortly after leaving the nest. Birds in juvenal plumage are known as juveniles. After loss of the juvenal plumage, plumages are attained in cyclic fashion, mostly on an annual basis.

The **basic** plumage of a bird is typically the plumage worn for most (or all) of the year. This corresponds to winter or non-breeding plumage. The plumage following juvenal is called first basic plumage. In species that take more than one year to assume adult plumage, such as large gulls, the terms first basic, second basic, etc., are used until the plumage no longer changes in appearance (= adult plumage). We use the more widely understood term **adult** rather than 'definitive basic' suggested by Humphrey and Parkes. Unless otherwise specified, 'basic' in the species accounts means adult (or adult-like) basic, which also includes first basic when similar to adult basic. In many tropical species that do not change their appearance for breeding, basic plumage is replaced once a year by a complete molt.

Some species attain a different, **alternate** plumage for breeding, corresponding to summer or breeding plumage. Again, this may be termed first alternate, second alternate, etc., until adult alternate plumage (definitive alternate of Humphrey and Parkes) is attained.

Molts are described in terms of the incoming plumage. That is, the molt preceding breeding is the prealternate molt, but that following breeding is the prebasic molt.

Typically the prebasic molt of adults is complete. The *first* prebasic molt in most species is partial: juvenal flight feathers are retained for almost one year until the second prebasic molt, which is complete. In the species accounts, this is often indicated by a phrase such as 'quickly attains adult-like plumage' for birds that soon lose their distinctive juvenal contour feathers. The first prebasic molt may take from less than a week (some warblers) to several months (Yellow-bellied Sapsucker). In general, larger birds take longer to molt than smaller birds, and in some species such as certain seabirds and raptors, molt may be extremely protracted such that it is more or less continuous throughout the year. Prealternate molt, if it occurs, typically is partial.

The extent and timing of the prebasic molt are highly variable, and timing in particular can be of use in field identification. For example, most Western Sandpipers molt one month or so earlier than Semipalmated Sandpipers. A juvenile found in early October would thus most likely be Semipalmated, as a Western would then be mostly or entirely in basic plumage. A very worn plumage Dusky/Hammond's Flycatcher seen on autumn migration would be a Dusky since adult Duskies molt on their winter grounds after migration, while Hammond's molt before they migrate and thus would be in fresh plumage. However, juveniles of both Hammond's and Dusky appear fresh in autumn migration, so an overall picture of a species' molt strategy is important.

The molt strategies of many neotropical birds are poorly known. Some of the information presented here is tentative and based on examination of series of museum specimens. Hummingbird molts are notably little known and unstudied. Molt strategies that are particularly poorly known are mentioned in the species accounts in the hope of prompting study.

OUTLINE OF THE SPECIES ACCOUNTS

With the publication of the sixth edition of the *American Ornithologists' Union (AOU) checklist of North American birds* (1983), a taxonomic and nomenclatural base was established for the species of North and Middle America. In some cases our opinions differ from those of the AOU, including the placement of species within genera or families, usually based on our own field and museum studies. Where our views concerning a species' taxonomic status differ from those of the AOU, this may be stated at the end of the relevant species' account.

Family accounts

The number of species in each family occurring in the guide region is given in parentheses following the family heading; an asterisk (*) indicates that the total includes recently extinct species (Appendix A). Family accounts precede the species accounts. They summarize general biological information and field characters common to the family as represented in the guide region. Included are notes about habitat, structure, plumage, molt, voice, diet, and breeding. If all members of a family share common characters such as age/sex-related plumages, number and color of eggs, etc., this is stated in the family account and not repeated in every species account.

At the beginning of a genus we often include a brief note that includes more specific information than the family accounts. These notes should be read carefully as an appreciation of generic characters can assist in field identification.

Species accounts

If a widely used alternative English name exists, it is given in parentheses, for example, Mexican (Blue-rumped) Parrotlet. Brackets indicate that the form is or has been considered part of a larger taxonomic grouping, for example, Mexican [Black-faced] Antthrush. An asterisk (*) following the scientific name indicates that more than one subspecies occurs in the region; absence of an asterisk indicates that we consider the species monotypic. An appended subspecies name gives the single subspecies that occurs in the region. See the discussion of taxonomy (pp. 61–4) for methods of indicating uncertain specific status.

Spanish names are given in the species account headings and are derived from a survey of the available literature (Friedmann *et al.* 1950; Miller *et al.* 1957; Wetmore 1965–1984; Smithe 1966; Land 1970; Alvarez del Toro 1971*c*; Edwards 1972; Cramp and Simmons 1977–1990; Meyer de Schauensee and Phelps 1978; Birkenstein and Tomlinson 1981; Sada *et al.* 1987; Peterson and Chalif 1989; Robles Gil *et al.* 1989; Stiles and Skutch 1989). They represent, in our opinion, the most suitable names for each species. In some cases we had to create names, for example, for recently split species that are not given Spanish names in any other text. We have attempted to standardize the format and principles for Spanish names with a view to reasonably short, unambiguous names (Howell, unpublished).

The species accounts contain a maximum of five headings, one of which (identification) is split into three subheadings. A heading or subheading may be omitted if not relevant, or headings may be combined when appropriate.

Identification (ID)

The primary purpose of this guide is to facilitate field identification. Characters important for identification are noted in italics in the species' descriptions.

The first character given under ID is a species' length, in inches and centimeters. The length of every species was measured (by SNGH) from museum skins, from tip of bill to tip of tail with the bird laid on its back. The different preparation methods of numerous collectors were, as far as possible, taken into account and the measurements provide comparable, if not exact, lengths for all species. We avoid giving only average length as this obscures size variation within species (sexual, geographic, or simply individual) which can be important for identification. Geographic variation in length, when noticeable, is indicated. For example, N > S means that northern birds of a species average larger than southern birds. It should be noted that *length and size are not the same thing*, as volume (and effectively size, if the proportions are similar) increases by the cube of increase in length. Thus, a bird that is 24 in (61 cm) long can appear *markedly larger* than a bird that is 20 in (51 cm) long.

Total lengths may be followed by supplementary measurements in parentheses such as bill length (B, exposed culmen), tail length (T, measured from the point of insertion of central rectrices), and wingspan (W). Larry B. Spear and David G. Ainley of PRBO kindly allowed us to use wingspans taken from freshly dead specimens collected as part of their long-term study of Pacific seabirds. In addition, measurements from Spear's specimens allowed us to calculate a ratio of wing chord to wingspan for many seabirds, raptors, and owls. These ratios were used, when other data were lacking, to determine wingspans of structurally similar congeneric species. Wingspans for Diomedeidae, Procellariidae, Hydrobatidae, Phaethontidae, Sulidae, Pelecanidae, Fregatidae, Stercorariidae, and Lariidae are all derived from unpublished data of Spear and Ainley. Manuel Marin A. kindly provided data on swift wingspans.

After length, there is usually a short note about one or more points important to field identification, such as status, distribution, structure, etc. If subspecific variation is likely to be noticed in the field, the main subspecies or subspecies groups (and usually their geographic ranges) are noted briefly. A subspecies 'group' is a subspecies or group of similar subspecies that may be distinguished in the field from other groups; this concept can be useful in appreciating biogeographic affinities. The group name is that of the first-named subspecies that occurs in the region. It was not always feasible to evaluate subspecific variation outside the guide region and some of our subspecies or groups may be equivalent to larger groups. Slight subspecific differences may be simply noted in the descriptions. Alternatively, the most widespread subspecies is described, and local subspecies are compared with it after the main description. The complexities of subspecific variation preclude a consistent treatment for all species.

Following the introductory note are the three main identification criteria: Description, Voice, and Habitat and habits.

Description (Descr.) The presence or absence of age, sex, or seasonal variation is stated first. (If there is no variation among all members of a family or genus, this is noted in the family account or genus note.) Age differences are described from adult (male and female) through the most commonly encountered plumages. Thus, with raptors, adult is usually followed by juvenile and subsequent immature stages; but for orioles, adult is followed by immature, then juvenile, since juvenal plumage is worn for only a brief period. Seasonal plumages are generally described in an order related to their frequency of encounter in the region. (For a discussion of molt and plumage, see pp. 64–7.) Descriptions often build upon preceding descriptions; thus, for Little Tinamou, ♀ resembles ♂ except for those characters noted. Eyes (considered synonymous with irides) are dark unless otherwise stated. Legs and feet are the same color as each other unless otherwise stated; whichever are more conspicuous on a particular species is noted.

Most of the terms we use to describe birds are fairly standard and widely understood. Figure 4 illustrates terms that may require explanation. Note that many anatomical terms are explained in the glossary.

One factor that relates directly to the usefulness of the species descriptions is understanding what is meant by a given color. Red, yellow, blue, and such are familiar colors. Other colors, however, may be less familiar and unavoidably, are interpreted differently by different people. Several carefully reproduced color charts exist but often are not readily available, and color descriptions between any two charts usually differ. The following colors are used frequently and are now explained with examples given in parentheses.

chestnut: dark reddish brown (belly of Agami Heron, body of Northern Jacana, underbody of ♂ Orchard Oriole).

rufous: brownish red (Rufous-necked Wood-Rail, crown of Russet-crowned Motmot, Rufous-backed Robin).

cinnamon: brownish orange (underparts of Cinnamon Hummingbird, underparts of Tufted Flycatcher, tail of Sumichrast's Sparrow).

tawny: orangish brown to yellowish brown (Tawny-crowned Greenlet, Tawny-winged Woodcreeper, rump of Bright-rumped Attila).

ochre: brownish yellow (underbody of Ochre Oriole, Ochre-bellied Flycatcher, crest of Chestnut-colored Woodpecker (pale ochre)).

buff: pale yellowish (Buff-breasted Flycatcher, underparts of Mangrove Cuckoo, underparts of Happy Wren (ochraceous buff)).

olive: dull yellowish green; (upperparts of Greenish Elaenia, Common Yellowthroat, and Chestnut-capped Brushfinch).

dusky: dirty greyish or brownish, often applied to markings that are not sharply distinct.

dark: a generic term applied to dark plumage markings which, although having color, do not show any obvious hue and where specifying a color would not aid in field identification.

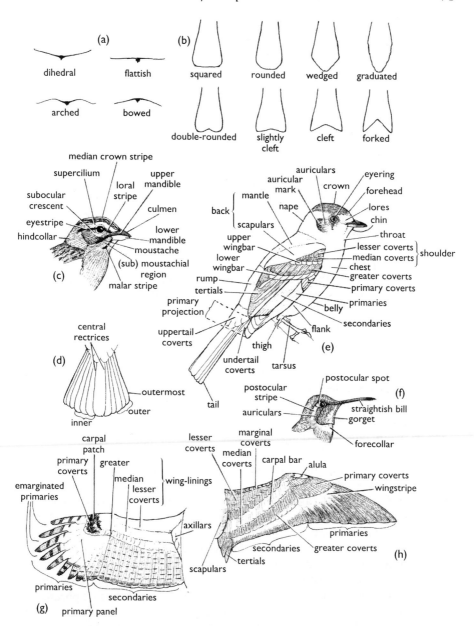

Fig. 4 Bird topography (also see Glossary). (a) Head-on flight shapes; (b) tail tip shapes; (c) head of a sparrow; (d) tail; (e) parts of a flycatcher; (f) head of a hummingbird; (g) underwing of a raptor; (h) upperwing of a shorebird.

Three colors are used only for bare parts.

horn: pale brownish yellow (bills of Western and Summer tanagers).
flesh: pale pinkish (lower mandible of Western and Yellowish flycatchers, legs of
    Striped Sparrow).
amber: reddish orange (eyes of many adult parrots, eyes of adult Double-
    toothed Kite).

Compound colors should be interpreted as the second color tinged or washed
with the first; for example brownish red is red tinged with brown.

Voice   In general the vocalizations described in the voice sections represent the
songs and calls given most frequently by birds in the region, with emphasis on
those relevant to identification. Thus, chip notes rather than songs of migrant
North American wood-warblers are given, and the voices of many seabirds are
omitted.

While some birds, such as pigeons, tend to have relatively stereotyped voices
that lend themselves to transcription, this is an exception rather than the rule.
The songs of most species vary considerably; some species have more than
one song, and different calls serve different functions such as contact, alarm,
aggression, etc. Song may vary individually or regionally, sometimes to such an
extent that hummingbird songs can differ noticeably between singing assemblages
of the same species within the same patch of forest!

The voice transcriptions are those of SNGH whenever possible. Although bird
voices are heard differently by virtually everyone, this method may allow comparisons to be made once the user is familiar with the interpretation of a few
vocalizations. In some cases, the vocalizations given are those transcribed by
SNGH from tape recordings and these are indicated by the source and a
notation, *tape*.

Rather than attempting to transcribe numerous variations of songs, we have
often opted to use descriptive adjectives Transcriptions, however, may be given
for shorter songs and calls when their components and phrases can be distinguished. In most cases, two (or more) transcriptions are given, which represent how SNGH wrote down the same song (or call) on two different occasions
and/or how the song varies. Two or more transcriptions of a song or call
followed by 'etc.', indicate that (numerous) other variations may occur.

The frequency with which a song or call is given can be important. Often we
give the time required to utter 10 phrases, from the start of the first phrase to the
end of the pause after the tenth phrase, for example 10/32–38 s. This translates
into about one phrase every 3.5 s. In cases where phrases may not be repeated as
many as 10 times, the range is given for the common number of phrases, for
example 6/9–10 s, or simply the interval between single phrases. These times
indicate only the typical range of frequency, which can change with individuals
and their mood; for example, excited birds may sing faster.

Comparisons between species are often made, usually comparing uncommon
with common or more familiar species. We recommend learning thoroughly the

songs and calls of a number of commoner species and then building upon this base by comparison; this will also help in appreciating regional variation in vocalizations.

Voice transcriptions are indicated in italics. The terms high, low, rising, and falling all apply to pitch. An apostrophe (') indicates a noticeably rising inflection and/or a fairly abrupt break, or ending. *Wh-ooo'* means that the call rises toward the end of the second phrase and/or ends abruptly. Capital letters indicate strikingly loud or louder phrases, for example *chee-o'REER*. Rather than taking up space, two full stops, or periods, (..) are used to indicate when phrases may continue. This is usually accompanied by an indication of how often they are repeated, for example 'a deep booming hoot *oot oot..*, 3–12×'.

In some cases, voice may be the only safe way to distinguish between similar species in the field, for example Willow and Alder flycatchers. At the other extreme, we may be unable to distinguish what are presumably species-specific vocalizations, and many calls and songs are specific to individuals, as are human voices.

The primary function of song is to proclaim territory and advertise for mates. Hence, most song is restricted to the period immediately preceding and during breeding. A secondary period of song after nesting may teach young birds the complexities of songs that are not fully innate. Some birds, however, particularly in the Tropics, sing throughout the year. True songbirds, or oscines, have a specialized syrinx to help with sound production. All passerines in the taxonomic sequence from larks to finches are oscines; other landbirds from ovenbirds through manakins are known as suboscines. Many other birds advertise by what is usually known as song. Often one finds a relationship between plumage and singing ability; for example dull brown birds such as wrens are accomplished singers, while the dazzling honeycreepers have poorly developed and rarely heard songs.

Many species sing from prominent perches or even in display flights, while others sing from cover. A few species, notably some wrens, sing in duet, at times contributing different parts of a song. In temperate-breeding species, it is mostly ♂♂ that sing, but in many tropical species both sexes sing. The dawn chorus is a well-known phenomenon and is a prominent feature of tropical forests; many tyrant-flycatchers have 'dawn songs' rarely heard other than at or even before first light. Dawn in tropical forests is a memorable experience, as sounds slowly build up into an almost overwhelming chorus and then gradually fade away. By 9 or 10.00 it is hard to believe that an hour or two earlier 40–50 species were singing!

Habitat and habits (Habs.) The major habitats of a species are given first (see discussion of habitats, pp. 26–33). Arid, semiarid, semihumid, and humid are used as relative terms. If no climatic range is given, then a species occurs in habitats not strongly affected by climate (for example, many waterbirds) or occurs throughout the range from arid to humid (for example, many widespread landbirds).

Habits include notes about a species' behavior, especially if relevant to identification. In particular, knowing where a bird occurs in the forest can be useful.
We have divided forests into three levels: low levels are roughly those at or
below 'eye level' (about 2 m or 6 ft); upper levels are those in or just below the
canopy; and mid-levels cover the intermediate areas, irrespective of tree height.
Behavior may also be described in comparative terms under the similar species'
heading.

Nesting data. At the end of the Habitat and habits section is a brief description (when data are available) of the nest and eggs of species that breed, have
bred, or probably breed in the region. Alternatively, this is summarized in the
family account or genus note; for example, all hummingbirds (as far as known)
lay two white eggs. We mention if the nest and/or eggs of a species appear to be
undescribed. We attempted to examine eggs of all species that breed in the
region so that descriptions of colors, patterns, etc., would be comparable (the
museum collection, usually WFVZ, where we examined the eggs is indicated in
parentheses). We also tried to take into account post-mortem egg color change
when possible. For species whose eggs may be highly variable, both geographically
and individually, the country of origin of eggs described is noted (if outside the
guide region). The eggs (and nests) of several species remain undescribed, and
for others we found either literature descriptions of the eggs or a reference (Kiff
and Hough 1985) to their existence in collections that we were unable to
examine. A lack of reference following the description of a species' eggs indicates that the information comes from one of the main references of nesting
habits: Skutch, or Stiles and Skutch (mostly Central American species); Rowley,
Alvarez del Toro (Mexican species), and Bent (North American species). References other than these are cited. We have assumed the eggs listed by Kiff and
Hough (1985) to be correctly identified. However, if the species involved are not
known (or assumed) to have white eggs (such as owls, woodpeckers), we have
indicated this as 'Eggs undescribed (Kiff and Hough 1985)', since we have found
no published description of that species' eggs.

Nest descriptions come mostly from the literature but we have examined or
seen nests of representatives of almost all families of birds that breed in the
region. In a few cases, previously undescribed nests are noted and credited to the
relevant observers.

Most species of temperate regions begin breeding between March and June
and end between July and September. Tropical species often have more protracted breeding seasons although most rain forest and cloud forest species
breed from February to September, the majority breeding from March to August.
A few rain forest species start as early as January. The onset of breeding for
some northern desert species, notably *Aimophila* sparrows, is tied to the start of
the rainy season in July. Several waterbirds exploit ephemeral water-level conditions and may breed in any month. Hummingbirds are an exception to most
'rules'. While most migratory northern species breed from March or April to
August or September, hummingbird breeding cycles are tied closely to food
abundance, such that in the highlands from central Mexico south, most humming-

birds breed from August to February during the peak flowering period. Nesting periods are rarely noted in the species accounts except for species that do not fit the general 'spring–summer' schedule mentioned above.

Similar species (SS)

For each species, this section notes the distinguishing features of similar species which follow in the text, or refers to the SS accounts of preceding species. For vagrants and other very rare species, all similar species may be discussed under the rare species. If two similar species are allopatric this may be noted since geography can be an important factor in the identification of very similar species that do not occur together; for example, Bumblebee and Wine-throated humming-birds. Many species in Mexico and Central America are sedentary. This should be noted especially by visitors from North America and Europe where the majority of birds are migrants and hence likely to turn up out of their usual range. Thus, if Pileated Flycatcher occurs in arid scrub in interior southwestern Mexico, it will not be found in the humid Atlantic Slope lowlands of eastern Mexico. The SS section may be omitted if a bird is unmistakable.

Status and distribution (SD)

This section gives the status of a species (which may vary within the region), its abundance *in its main habitat(s)*, and its distribution within the region. The SD accounts should be used in conjunction with the species range maps. (For an explanation of data sources, see Acknowledgements.) Months are divided into three periods of approximately 10 days: early (1–10th), mid (11–20th), and late (21st–end of month). Months may be given without more specific periods; for example, May–Aug indicates a species occurs from early May to late August.

Status Status is defined by the following seven categories, with months of occurrence given in parentheses.

resident: Breeds and resides within its range throughout the year; for example, Great Tinamou.

summer resident: Breeds in the region but is present only for a period during the northern summer; for example, Streaked Flycatcher (late Mar–Sep).

breeder: A local summer resident, that is a species that does not leave the region in winter but has different breeding and non-breeding ranges (months in parentheses indicate presence in the nesting area, not necessarily the period of active nesting); for example, Barn Swallow (Feb–Oct).

winter visitor: Non-breeding visitor present during the northern winter; for example, Snow Goose (Oct–Mar).

transient: Non-breeding visitor only present during spring and/or autumn migra-tion; for example, Alder Flycatcher (Apr–early Jun, Aug–Sep).

visitor: Non-breeding visitor present for varying periods, up to all year; for example, Sooty Shearwater (Jan–Dec).

vagrant: Bird outside its normal range; for example, Arctic Warbler.

A few points should be considered regarding the status of birds in Mexico and northern Central America, particularly in terms of seasons. Within such a large and diverse region (lying between approximately 13° and 33° north of the equator), seasons differ from those in the USA, Canada, and Europe. (See the discussion of breeding seasons in the Habitats and habits section above.) Migration is mainly from March to May and from July to November, although some migrants are still heading south in December when others are already returning north! Spring, summer, autumn, and winter are terms that may be applied to nearctic species but should be applied with caution to neotropical species.

Abundance   Most birds tend to be 'common' in their optimum habitat and at the right season. We have tried to make abundance ratings comparable between closely related species, recognizing that a common hawk is numerically less abundant than a common sparrow. We attempted to determine each species' abundance throughout its range in the region, but the abundance codings still may be biased by our own field experience.

Abundance is indicated by the following four categories. Proper season refers not only to when migrants are present but also to when otherwise inconspicuous resident birds are readily detectable by song.

C : Common. Usually recorded daily *in proper habitat and season*, often in large numbers; for example, Grey Hawk, Couch's Kingbird, Red-throated Ant-Tanager.
F : Fairly common. Recorded at least on most days *in proper habitat and season*, usually in small numbers; for example, Double-toothed Kite, Royal Flycatcher, Tawny-crowned Greenlet.
U : Uncommon. Generally not recorded daily, *even in proper habitat and season*, but usually recorded at least once a week in small numbers; for example, Grey-headed Kite, Ruddy-tailed Flycatcher, Grey-headed Tanager.
R : Rare. Unlikely to be seen, *even in proper habitat and season*, and then only in small numbers; for example, Solitary and Harpy eagles, Slaty Finch.

Even though an observer is making a specific effort to locate rare species, seeing such birds usually depends on spending considerable time in a suitable area and/or luck. Many species, especially vagrants, are considered rare but this simply may be a function of our present (lack of) knowledge.

Several factors that affect a species' apparent or real abundance, particularly in the Tropics, should be considered.

1. Abundance usually varies within a species' range. Codings (C, U, etc.) are given in order of a species' abundance over most of its range in the region. Thus, C to F indicates that a species is widely common, more locally fairly common. A species is likely to be less abundant in suboptimal habitat and also may be less common at the edge of its range. The abundance of transient migrants refers to their greatest abundance in their period of occurrence which, for several reasons, may vary in timing from year to year. Migrants are unlikely to occur in their greatest abundance at the extremes of their migration periods.

2. Some birds are rarely seen but often heard (such as tinamous and rails). They are still common although their presence may be missed completely by observers unfamiliar with the species' vocalizations. In addition, certain birds are somewhat seasonal in their vocalizations, such as several owls and nightjars, and may appear absent from suitable areas for much of the year.

3. Tropical forests contain a much greater species diversity than temperate forests although individuals of tropical forest species tend to be less numerous than temperate forest species. Thus few tropical forest species might be considered common by North American or European observers. Abundance in tropical forests is somewhat relative since a common bird may be missed completely one day but 20–30 individuals may be seen the next.

4. Species which apparently are absent from suitable areas, occur in a very specific niche within a broader habitat, or are local due to widespread extirpation by man are termed *local*. Examples include Orange-breasted Falcon, Sumichrast's Wren, Horned Guan. Species that occur in naturally localized habitats, such as marshes and cloud forest, are not termed local solely for this reason.

5. Some species, not commonly considered migrants, move around within their range during the year. The movements may vary from year to year depending on food supply. Thus, a species may be resident throughout its range for much or all of the year but the bulk of the population may wander in response to local conditions. In these cases, the term *locally/seasonally* may be used; for example, Bar-winged Oriole is a locally/seasonally F to C resident in its range.

6. Some species are prone to cyclic variations in their abundance. The term *irregularly* is used for migrant species whose occurrence in the field guide region varies from year to year as a function of weather patterns or food abundance outside the region; for example, Rough-legged Hawk is an irregularly U to R winter visitor. In the Neotropics one also notices cyclic trends (apparent or real?) in the abundance of some species. As we have little or no data on such trends we have 'averaged out' abundance based on our experience in over 10 years in the region.

Distribution   Distribution is generally described from north and west to south and east, primarily in terms of Baja, Pacific Slope, Interior, and Atlantic Slope (see Figs 3(a) and (b)). Approximate elevations are given in meters above sea level but should be taken only as guidelines; birds will occur in their favored habitat(s) regardless of elevation. Any species occurring up to 3000–3500 m should be considered likely to occur as high as available habitat exists.

The discussion of geography and bird distribution (pp. 6–21) should help in understanding the principles of avian distribution in the guide region. 'South to central Mexico' means south to the vicinity of the Central Volcanic Belt, that is a line drawn roughly from Manzanillo to Veracruz City; 'interior from cen Mexico S' means interior regions extending south from just south of the Central Volcanic Belt, that is from the Balsas drainage south.

Range (RA)

The world range of each species is outlined to give a broader picture of the birds in the guide region. When the global breeding range of a species falls within the region, RA is combined with SD.

Note (NB)

This section may include comments about the taxonomic status of a species, and brief descriptions of related or similar species that may be expected in the region.

Plates

The color plates are a key part of this guide and are usually one's first reference in identifying an unfamiliar bird. Species are portrayed to scale on a given plate, or a line divides the plate into two scales. The illustration and facing page notes will often be sufficient to identify a bird but we recommend a thorough reading of the full species' account. Several guides cover North American birds and we have opted to devote plate space to Middle American species. However, several 'North American' species are shown on the plates. These are mainly similar species illustrated for comparison with non-North American species (for example, Blue-grey and other gnatcatchers) and species that are typically Mexican or Middle American although their ranges may just extend into the southwestern USA (for example, Mexican Chickadee, Olive Warbler).

Generally a sense of taxonomic sequence (AOU 1983 Checklist) has been followed in the order of the plates. When a species is significantly out of taxonomic sequence, its location is given on the page facing the plate where it would be expected.

Facing each plate are short captions which include a species' English and scientific names; its *average* length (or wing span for flying birds) in inches and centimeters; the page number for the relevant species accounts; and a note summarizing identification criteria, often with an indication of range, habitat, or habits.

Maps

Distribution maps are included for all species (except those known only from Clipperton Atoll). This is the first attempt at doing so for all species known from Mexico and northern Central America. Basic ranges were compiled by SNGH from a review of the existing literature (major references are listed in the bibliography) and a wealth of unpublished information (see Acknowledgements). Much information is specifically credited in the species accounts. A large amount of our field work in the past six years has been aimed at filling gaps in our knowledge of species' distributions, but much remains to be learned. Range limits often had to be interpolated based on SNGH's working knowledge of Mexico's complex biogeography and habitat distribution. In cases where records of a species are few (such as Solitary Eagle, Slaty Finch), localities of specific records may be plotted within the interpolated range. Thus the maps indicate where a species can be found or be expected to occur *in its habitat* and *during the right season*. Reference to the SD sections of the species accounts will help fill in details of abundance, habitat, and seasonal occurrence.

A map immediately highlights gaps in present knowledge and provides a visual catalyst for correcting and updating bird distribution. Often a question mark (?) is employed on maps to indicate uncertainty. Every effort has been made to make the ranges as accurate as possible given the constraints of the scale used. In particular, the northern range limits for many species in Mexico (mainly Sonora, Sinaloa, Tamaulipas, and the Yucatan Peninsula) are not precisely known. Stephen M. Russell (Sonora), John C. Arvin (northeast Mexico), and Thomas R. Howell (Nicaragua) were particularly helpful in determining range limits for numerous species in the region. Any mistakes, however, remain those of SNGH who would much appreciate hearing of corrections to the range maps of any species.

Figure 5 shows the symbols used to convey information on the maps.

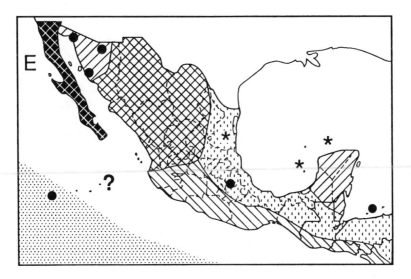

Fig. 5 Range map symbols.

⬡ Resident breeder

⬡ Summer resident (breeder)

⬡ Winter (non breeding) visitor

⬡ Former (resident breeder) range

⬡ Transient migrant

⬡ Non-breeding visitor

● Breeding colony (mainly used for water birds), or a presumed resident breeding record in an extensively interpolated range (mainly land birds), or a disjunct, presumed or proven breeding population.

★ Migrant (non-breeding) occurrence, includes transient, vagrant, and winter records (see text). If superimposed on transient range indicates rare winter occurrence.

? Status uncertain (see text)   E Extirpated population (islands)

GLOSSARY

alar bar: pale diagonal bar on outer coverts of upper wing, that is, a short carpal bar.

allopatric: having separate (breeding) ranges; usually refers to closely related or similar forms. See pp. 61–4.

alternate (alt): the plumage attained by many species before or during the breeding season, though not necessarily by breeding adults; equivalent to 'summer', breeding, or nuptial plumage. See basic; see also pp. 65–7.

altricial: hatching in a helpless state, usually with eyes closed. See precocial.

alula: a small feathered projection at the bend, or joint, most distal on the leading edge of a bird's wing.

arched: as viewed head-on, wings held raised near the body and bowed distally; also a decurved bill that rises initially.

auricular patch: a contrasting patch on the auriculars.

axillars: the innermost feathers lining the underside of a bird's wing; the 'wing pits'.

band: a horizontal marking, subjectively broader than a bar.

bar: a horizontal marking, subjectively narrower than a band.

bare parts: the naked areas of a bird, that is, bill, legs, feet, and any naked areas on the face, throat, etc. Also known as soft parts.

basic: the plumage worn by a bird usually for most (or all) of the year; equivalent to 'winter' or non-breeding plumage. See alternate; also see pp. 65–7.

bib: a contrasting, usually dark area, on the throat and/or chest, such as on titmice and orioles.

bowed: as viewed head-on, wings held curved downward, slightly below the body plane.

braces: a contrasting, usually pale, pair of stripes on either side of the back, formed by pale edges to scapulars or outer mantle. Also sometimes referred to as mantle V or scapular V, depending on location.

brood parasite: a species that lays its eggs in the nests of other birds, leaving them to care for and raise its young.

cap: a contrasting patch on the crown.

carpal: relating to the wrist (carpus), the bend along the leading edge of a bird's wing.

carpal bar: a contrasting diagonal bar extending from the carpus across the secondary wing coverts.

carpal patch: a patch, usually dark, on the underwing at the carpus, that is, on the primary coverts.

Central America: a geographic term encompassing Belize, Guatemala, El Salvador, Honduras, Nicaragua, Costa Rica, and Panama, but *not* Mexico. Politically, however, Central America does not include Belize or Panama.

cere: a bare, leathery patch of skin into which the nostrils open. Only certain birds, such as raptors and pigeons, have this feature, and it may be brightly colored.

chick: precocial young.

cline: a gradual geographical change, usually in terms of size or plumage.

complete molt: a molt involving the replacement of all feathers (see pp. 65–7).

conspecific: of the same species.

contour feathers: the feathers that make up the outline, or contours, of a bird's body; includes the wing coverts but not the flight feathers.

cooperative breeder: species in which several individuals, usually closely related, help with nest building, incubation, feeding young, etc.

cosmopolitan: of worldwide distribution.

coverts: feathers that cover, and thus protect; for example, secondary coverts cover the bases of the secondaries, ear coverts (or auriculars) cover the ear region.

crèche: a gathering of young of certain colonially nesting birds.

crepuscular: active at dawn and dusk.

culmen: the ridge of the upper mandible on a bird's bill; in terms of bare part colors, often used to refer to the upper part of the upper mandible.

dihedral: as viewed head-on, wings held raised in a V shape; a slight dihedral refers to wings only slightly raised above the horizontal.

dimorphic: having two distinct morphs or variants, often color-related as in light and dark morphs.

distal: farthest from the center, or midline, of a bird; opposite of proximal.

dumping: laying eggs in other birds' nest, thus a dump nest is one which contains eggs of more than one female.

ear-spot: a small contrasting mark on the auriculars, particularly of certain gulls.

endemic: restricted, at least as a breeding bird, to a specific region.

extirpated: no longer occurring in a region although existing elsewhere; 'locally extinct'.

eye-crescents: contrasting, usually pale, crescents above and below the eye; that is, an eyering broken in front and behind.

eyering: a contrasting feathered ring around the eye. See orbital ring.

eyestripe: a contrasting, usually dark, line through the eye.

face: an anthropomorphic term, encompassing the forehead, lores, superciliary region, auriculars, and moustachial region or any combination of these which might be considered the face.

facial disks: saucer-shaped disks of feathers surrounding the eyes, mainly used in reference to owls.

feral: a free-flying bird derived from domesticated or captive stock.

filoplumes: specialized hair-like or bristle-like feathers.

fledge: to leave the nest.

fledgling: typically an altricial bird out of the nest but still dependent on its parents.

flight feathers: the main feathers of the wings and tail used for flight; that is, the remiges and rectrices.

forecollar: a contrasting band, or collar, across the foreneck.

form: a general term for distinguishable entities, including species, subspecies, etc. Often used where uncertainty regarding specific status exists.

frugivorous: feeding on fruit.

frontal bar: a contrasting dark bar, or band, across the forecrown, as on plovers.

gonys: the ventral ridge of the lower mandible from its tip to the separation of the mandibular rami, or gonydeal angle.

gular pouch (also gular sac): a throat pouch of naked skin.

heterodactyl: form of a foot with the inner two toes pointing back, outer two toes forward, as in trogons.

Holarctic: a faunal region combining the northern parts of the Old World (Palearctic) and New World (Nearctic). Often used in reference to birds with high northern latitude breeding distribution.

hindcollar: a contrasting band or collar on the hindneck.

hood: a contrasting area covering all or most of the head and neck, as in certain gulls.

immature: a general term for a non-adult bird.

incomplete molt: a molt in which all or most contour feathers and some flight feathers are replaced (see pp. 65–7).

intergrade: a bird that is intermediate between two closely related parental forms which are not necessarily species. (A cross between two accepted full species is a hybrid.)

iridescent: the effect of changing colors caused by reflected light from specially structured (iridescent) feathers; for example, on hummingbirds.

juvenal: the first feathered, or non-downy, plumage of a young bird.

juvenile: a bird in its first feathered (or juvenal) plumage.

lek: an assemblage of displaying and/or singing males.

lobed feet: feet with stiff flaps of skin between the toes to aid in swimming; only seen in grebes, coots, phalaropes, and finfoots.

loral stripe: a contrasting, usually dark, stripe in the lores.

lores: the area between the base of a bird's upper mandible and eyes.

M pattern: a contrasting, M-shaped dark pattern visible on the upperparts of certain birds in flight, formed by the outer primaries, primary coverts, and carpal bar, and often joined across the rump.

malar stripe: a contrasting stripe or line extending back from the lower corner of the lower mandible. See moustache.

mantle: the central back (or interscapular area) of a bird which, together with the scapulars forms the back; in other works mantle refers to the back, scapulars, and upperwing coverts.

melanistic: having excessive dark pigmentation.

Middle America: a geographic term encompassing Mexico and Central America.

mirror: a white spot or mark on or near the tips of the primaries of gulls.

monotypic: not having subspecies.

morph: a distinct variation, usually color-related; color phase has sometimes been used for color morph but implies a changing or transitory state.

mottled: plumage marked with coarse spots or irregular blotches.

moustache: a contrasting stripe or line extending back from the gape, sometimes extending down to the malar region. The area between the moustache and malar is the submoustachial or moustachial region. See malar stripe.

nail: the hardened, horny tip of the upper mandible; mostly used for ducks.

Nearctic: a faunal region comprising North America and the temperate interior of Mexico south to the Isthmus of Tehauntepec.

Neotropics: a faunal region comprising South and Central America and the tropical areas of Mexico.

nestling: altricial young confined to the nest. See fledgling.

nuchal: of, or pertaining to, the nape, as in nuchal collar.

occipital plumes: modified feathers, usually slender, projecting from the hind-crown or nape.

orbital ring: the naked ring of skin surrounding the eye, often brightly colored. See eyering.

Palearctic: a faunal region comprising Europe, Africa north of the Sahara, and Asia north of the Himalayas.

pantropical: of tropical regions across the world.

partial molt: a molt involving the replacement of most or all contour feathers but no flight feathers (see pp. 65–7).

passerine: a perching bird of the order Passeriformes characterized by a perching foot, with three toes pointing forward, and one toe backward. Includes all species in the taxonomic sequence from Ovenbirds (Furnariidae) onward.

pectinate: tooth-edged, refers to the central claw of certain birds, such as Nightjars.

pelagic: of the open ocean.

polygamous: having more than one mate.

polymorphic: having more than two distinct plumage morphs. See dimorphic.

postocular spot/stripe: a spot or stripe beginning at the posterior edge of the eye.

prealternate: the molt by which alternate plumage is attained (see pp. 65–7).

prebasic: the molt by which basic plumage is attained (see pp. 65–7).

precocial: able to move about and feed soon after hatching and drying. See altricial.

primaries: the outer flight feathers of a bird's wing, attached to the hand, or manus.

primary projection: on a closed wing, the projection of the tips of the primaries beyond the secondaries and/or tertials.

raptor: a bird of prey, used mainly for hawks, eagles, and falcons

rectrices (singular rectrix): the flight feathers of the tail.

remiges (singular remex): the flight feathers of the wing; that is, the secondaries and primaries.

rictal bristles: stiff, bristle-like feathers around the gape of certain birds, such as nightjars, flycatchers, and swallows.

ruff: an area of elongated, often erectile feathers on the neck.

scapulars: the shoulder feathers of a bird, often covering the lesser upperwing coverts at rest.

secondaries: the inner flight feathers of a bird's wing, attached to the forearm, or ulna. See tertials.

semipalmated: having partial webbing between the toes.

semiprecocial: capable of leaving the nest soon after hatching (hatched with eyes open, covered with down) but remaining there to be fed, as in gulls, terns, alcids.

shaft-streak: a contrasting streak along the shaft of a feather.

shield: a naked, shield-like plate on the forehead.

shorebird: any bird in the suborder Charadrii (thick-knees, plovers, oyster-catchers, avocets, sandpipers, and allies). Synonymous with wader in the Old World.

shoulder: lesser, or lesser and median, upperwing coverts, usually when forming a contrasting patch.

spectacles: facial markings formed by a combination of an eyering, supraloral stripe, and/or postocular line.

speculum: a patch of metallic-looking feathers; mostly used for the patch on the secondaries of ducks.

straightish: not quite straight; used for bills that have a slight, often almost imperceptible, decurve, such as hummingbirds.

streak: a short, longitudinal mark.

stripe: a longitudinal mark, longer and more continuous than a streak.

subocular: below the eye. The lower eye crescent is also a subocular crescent. See supraocular.

supercilium: a stripe over the eye, typically from the base of the bill back to the hind auriculars. Also known as eyebrow.

supraloral spot/stripe: a spot or stripe over the lores.

supraocular: above the eye; see subocular.

sympatric: sharing, at least in part, the same (breeding) range; usually refers to closely related or similar forms. See pp. 61–4.

tail streamers: elongated, usually attenuated, rectrices, typically either central rectrices (for example, tropicbirds) or outer rectrices (for example, terns).

talons: sharp, usually strong, and slightly decurved claws of raptors.

tarsus: the lower part of a bird's 'leg', above the toes; technically, the upper part of a bird's foot.

temperate: refers to the parts of the Earth between the Tropics and the Arctic and Antarctic circles; generally used for regions with temperate, non-tropical, climate, thus includes many highland areas south of the Tropic of Cancer. See tropical.

tertials: the inner three secondaries which, typically, are modified as coverts for the other secondaries.

tibia: the 'upper' part of a bird's 'leg', above the tarsus; technically, between a bird's heel and knee.

tropical: refers to parts of the Earth between the Tropics of Cancer and Capricorn; generally used for regions with a tropical climate, that is, where the mean low temperature does not fall below about 22 °C, or 70 °F. See temperate.

tomia (**singular tomium**): the cutting edges of a bird's bill.

totipalmate: fully webbed, with all four toes interconnected by webbing.

vent: the region around the opening of the cloaca.

vermiculations: fine, often wavy, bars that generally create an overall effect rather than stand out as marks in themselves (as do bars).

wattle: naked, fleshy, pendant skin on the head or neck of a bird (for example, throat wattle); usually brightly colored.

web: the vane or broad surface of a feather, distinct from the shaft; inner web is closest to the midline of a bird.

wingbar: a bar formed by the contrasting, usually pale, tips of the secondary wing coverts when wing is closed; the upper wingbar is formed by the tips of the median coverts, the lower wingbar by the tips of the greater coverts.

wing-linings: underwing coverts, usually when contrastingly colored.

wing-loading: a value obtained by dividing a bird's wing area by its body mass. Birds with relatively small wing area and a heavy body (such as Sooty Shearwater) have a high wing-loading; birds with relatively large wing area and light body (such as Buller's Shearwater) have low wing-loading. Thus birds with low wing-loading have a more leisurely, or buoyant, flight.

wing-margins: the contrasting dark edges to the underside of the wings of certain seabirds, such as *Pterodroma* petrels, formed by the dark trailing edge of the remiges, leading edge of the outer primaries, carpal mark or bar, and, on some species, dark leading edge of the inner wing.

wing panel: a contrasting, usually pale, panel on the wing; at rest, usually refers to a panel on the secondary coverts, secondaries and/or tertials; in flight, usually refers to a panel across the primary bases.

wingstripe: a contrasting, often white, stripe visible on the spread wing, particularly of certain shorebirds. Formed by the bases of the flight feathers and/or tips of the greater coverts.

zygodactyl: form of a foot with the outer and inner toes pointing backward, the two middle toes forward, as found in cuckoos, parrots, owls, and woodpeckers.

FAMILY AND SPECIES ACCOUNTS

See pp. 68–79 for explanation of the contents of the species accounts. See p. 79 for a key to patterns and symbols used on the range maps.

For explanation of names in parentheses and brackets, e.g. Mexican (Blue-rumped) Parrotlet, or Mexican [Black-faced] Antthrush, see p. 68, Species accounts.

* an asterisk following the number of species in the family heading: see p. 68, Family accounts.
* an asterisk after the scientific name means that two or more subspecies occur in the guide region. Only subspecific variation likely to be noticed in the field is discussed. See p. 69, Identification.

Abbreviations used in the species accounts
See p. xv for Mexican state names and other geographic abbreviations.
See p. xiii–xiv for names of those credited by their initials in the species accounts.
See p. xvi for names of museums and other institutions credited by acronyms in the species accounts.

ID	Identification
Descr.	Description
Habs.	Habitat and habits
SS	Similar species
SD	Status and distribution
RA	Range
NB	Note
SL	Sea level
N, S, E, W, SE, etc.	Compass directions
cen	Central
alt	Alternate plumage (see pp. 65–7, Molt and plumage)
basic	Basic plumage (see pp. 65–7)
♂	Male ♀ Female
juv(s)	Juvenile(s) imm(s) Immature(s)
spec.	Specimen ssp. Subspecies

The following abbreviations are further explained on p. 76.
C   Common
F   Fairly common
U   Uncommon
R   Rare

TINAMOUS: TINAMIDAE (4)

A neotropical family of fowl-like, terrestrial birds of forest and second growth, tinamous are large-bodied and almost tail-less, with slender necks and small heads. Bills slender and slightly decurved, dark above, pale below; legs and feet fairly stout. ♀♀ often brighter than ♂♂ but both sexes cryptically colored in browns and greys, often marked with darker and lighter barring. Young precocial, molting into adult-like plumage when about half-grown and perhaps undergoing a second molt into adult plumage when full-grown. Tinamous are mostly detected by voice, which is generally quite loud, hollow, and quavering.

Food vegetable matter and small invertebrates. Nests are scrapes on the ground; ♂♂ incubate and raise the young, one ♀ may lay for several ♂♂, and more than one ♀ may lay in the same nest. Eggs notably lustrous, usually 2–7 per nest.

GREAT TINAMOU

Tinamus major*   Plate 15
Tinamú Mayor

ID: 15–18 in (38–46 cm). SE humid forest. **Descr.** Sexes similar. *Legs greyish.* Greyish face with white eyering contrasts with darker crown, throat white; often shows a dark auricular patch. Neck and upperparts brownish, marked with black bars and spots. Grey-brown underparts, paler on belly, dark vermiculations become distinct dark bars on flanks, thighs, and undertail coverts.

    Voice. A haunting call, evocative of tropical lowland forests, and given day or night, mainly at dawn and dusk. Powerful, eerie, quavering paired whistles with a variable introduction, typically of 1–4 notes: *whi hoooor-ooo hooor-ooo hooor-ooo...,* or *whoo hoo-hoo hoo ooohr-ooo ooohr-ooo...,* etc. Usually 1–6 pairs (rarely more) are given in intensifying series. Flushed birds make a rapid whistling twitter as they fly off.

Habs. Humid evergreen forest. Prefers relatively open forest floor. Often flushes explosively from 6–10 m away, or runs a short distance and freezes, at times in full view. Eggs: 3–5, glossy blue (WFVZ).

SS: Other tinamous markedly smaller, the two species closest in size have bright red legs. Tinamou shape and relatively uniform plumage should preclude confusion with other forest gamebirds.

SD: F to C (where not heavily hunted) resident (SL–1200 m) on Atlantic Slope from S Ver and N Oax to Honduras; reports from SE Pue (AOU 1983) may be in error. U in much of range due to hunting.

RA: SE Mexico to N Bolivia and cen Brazil.

LITTLE TINAMOU

Crypturellus soui meserythrus   Plate 15
Tinamú Menor

ID: 8.5–9.5 in (20–24 cm). *SE humid second growth.* **Descr.** Sexes differ. ♂: *legs olive to olive-yellow.* Head and neck dark grey, throat whitish. Upperparts dark rich

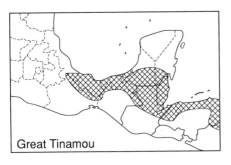

Great Tinamou

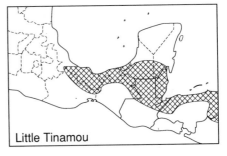

Little Tinamou

brown. Chest brown becoming *greyish cinnamon on belly*, undertail coverts barred buff. ♀: wing coverts and tertials edged rufous, foreneck grey, *underparts cinnamon-rufous*.

Voice. Clear, tremulous whistles given singly or in series up to 8× (rarely more), may call at night. Single whistles, and the 1st of a series, tend to swell then fade, successive whistles rise slightly, and a series often intensifies overall, ending abruptly: *wheeee-eeer, wheeeeer, wheeeeer..*, etc. Strongly recalls introductory phrases of Singing Quail song. May suggest Great Tinamou but higher, less haunting, and series not paired.

Habs. Humid second growth, forest edge, rarely forest interior. Secretive and rarely seen except crossing trails. Usually runs when alarmed. Eggs: 2, vinaceous grey (WFVZ).

SS: Thicket and Slaty-breasted tinamous larger with bright red legs, prefer different habitats; Uniform Crake has bright red legs, stout yellowish bill.

SD: F to C resident (SL–1400 m) on Atlantic Slope from N Oax to Honduras; U to R, N to S Ver and S Yuc Pen.

RA: SE Mexico to Bolivia and SE Brazil.

THICKET TINAMOU

Crypturellus cinnamomeus*   Plate 15
Tinamú Canelo

ID: 10–11.5 in (25.5–29 cm). Both slopes. Two groups: *occidentalis* (W Mexico), *cinnamomeus* (E Mexico to C.A.). **Descr.** Sexes differ. *Occidentalis* ♂: *legs bright red to orange-red. Head and neck greyish*, crown darker, throat whitish. *Upperparts grey-brown*, wing coverts and tertials sparsely spotted and barred buff and black, rump and uppertail coverts barred black. *Underparts pale greyish, washed*

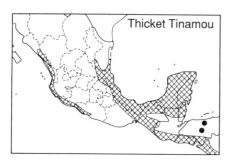

Thicket Tinamou

buff on belly and flanks, undertail coverts barred black. ♀: *upperparts boldly barred black and whitish to buffy cinnamon*, flanks barred dark brown and buff to cinnamon. *Cinnamomeus*: *more richly colored and browner overall. Upperparts of* both sexes *boldly barred dark brown and buff to whitish*, or (in NE Mexico) ♂ has barring restricted to posterior upperparts. *Face, sides of neck, and underparts cinnamon to greyish cinnamon*, becoming paler to whitish on belly. Dark barring on ♀ often extends across chest.

Voice. A loud, clear, hollow, fairly abrupt, 2–3-syllable whistle: *hoo-oo* or *hoo-oo-oo* or *hoo oo-oop'*, etc., which slurs up consistently (?) at the end in *occidentalis*: *whooo'r ew'* or *hooo o oop'*.

Habs. Second growth woodland and thickets, forest edge. Rarely flies, prefers to run or freeze if alarmed. Eggs: 2–7, pale vinaceous grey (WFVZ).

SS: Slaty-breasted Tinamou (humid evergreen forest) darker than all forms of Thicket Tinamou, especially on upperparts, ♀ less contrastingly barred cinnamon-rufous on upperparts. Grey-necked Wood-Rail larger and unbarred, with long yellow bill, grey neck, and black abdomen.

SD: C to F resident (SL–1800 m) on Pacific Slope from Sin to Gro (*occidentalis*), on Atlantic Slope from cen Tamps to Yuc Pen, and on Pacific Slope and locally in interior from Isthmus to El Salvador and Honduras.

RA: Mexico to NW Costa Rica.

NB: Reported hybrids of Thicket × Slaty-breasted Tinamou from Honduras (Monroe 1968) appear to pertain to brightly marked *C. b. costaricensis*.

SLATY-BREASTED TINAMOU

Crypturellus boucardi*   Plate 15
Tinamú Jamuey

ID: 10–11 in (25.5–28 cm). SE humid forest. **Descr.** Sexes differ. ♂: *legs bright red. Head and neck slaty grey*, crown blackish, throat whitish. *Upperparts dark brown*, slightly paler wing coverts usually sparsely spotted buff. *Underparts grey*, washed buffy brown on flanks, belly, and thighs, flanks and undertail coverts barred dark brown and buff. ♀: *upperparts extensively barred blackish and cinnamon*, flanks barred dark brown and buffy cinnamon. *C. b.*

costaricensis (E of Sula Valley) has greyish-cinnamon underbody, brighter in ♀.

Voice. A far-carrying, low, mournful, slightly tremulous, drawn-out whistle, *ooo-o-oooo* or *whoo-oo-ooo*, less often a shorter *hoo-ooo*; recalls the sound made by blowing across the top of a bottle.

Habs. Humid evergreen forest. Favors densely vegetated forest floor where hard to see; rarely flies, prefers to run or freeze when alarmed. Eggs: 2–7, pale mauve-grey (WFVZ).

SS: See other tinamous; Purplish-backed Quail-Dove (S Ver) has more horizontal build and longer tail, more likely to flush if disturbed; Grey-necked Wood-Rail has long yellow bill, rufous sides.

SD: C to F resident (near SL–1500 m) on Atlantic Slope from S Ver and N Oax to Honduras.

RA: SE Mexico to N Costa Rica.

NB: See Thicket Tinamou.

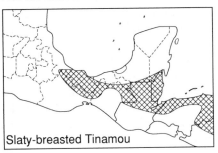

Slaty-breasted Tinamou

LOONS: GAVIIDAE (4)

Loons, often known as divers in the Old World, are large aquatic birds that breed in high northern latitudes. They occur almost exclusively at sea in the region and only in Mexico. Long-bodied and low-swimming, their legs with webbed feet are set well back on the body for strong underwater propulsion. Their long dagger-like bills are used to catch fish and other aquatic animals. Ages differ slightly; molt into adult-like 1st basic plumage begins in early to late winter. Eyes red in adults, brownish in juvs. Sexes similar, plumage varies seasonally. Loons fly with their neck extended and drooped distinctively below their body plane, feet projecting beyond their short tail. Mostly silent in winter but may give occasional honks or yelps.

RED-THROATED LOON
Gavia stellata　　　　Not illustrated
Colimbo Gorjirrojo

ID: 21–25 in (53–63 cm). *Bill relatively slender with tapered lower mandible.* **Descr. Basic:** bill grey to blackish. *Grey crown and hindneck do not contrast strongly with white face and foreneck (dark eye tends to stand out on pale face).* Upperparts grey, speckled white, underparts white. **Juv:** head and sides of neck more extensively dusky, attains 1st basic from early winter on. **Alt** (rarely seen in Mexico): head and neck grey, hindneck has vertical white lines, *foreneck dull red (often looks blackish). Grey-brown upperparts finely spangled white.*
　Habs. Coastal bays, estuaries, open ocean. Often holds head slightly raised, accentuating up-tilted bill. Rarely in small groups.
SS: Pacific Loon slightly larger and more heavily built with heavier straight bill; darker in all plumages, typically shows marked contrast between upperparts and underparts in basic. Western and Clark's grebes slimmer and longer-necked, black-capped with yellow bills, often in flocks.
SD: U to F winter visitor (Nov–Mar) along Pacific coast of Baja, U to R in N Gulf of California; winter vagrant to Hgo (spec., per ARP).
RA: Holarctic breeder; winters S in New World to NW Mexico, SE USA.

PACIFIC [ARCTIC] LOON
Gavia arctica pacifica　　Not illustrated
Colimbo Artico

ID: 22–26 in (56–66 cm). *Bill straight.* **Descr. Basic/juv:** bill grey, often with darker tip. *Blackish-grey crown and hindneck contrast with white face and foreneck,* dark often extends narrowly under chin (eye does not stand out on face). *Blackish upperparts darker than hindneck,* with subdued pattern of paler checkers (adult) or scaling (juv) visible at close range; underparts white. Juv attains 1st basic from mid-winter on. **Alt:** head and neck pale grey, sides of neck striped white, *foreneck black with purple sheen. Black upperparts boldly checkered white.*

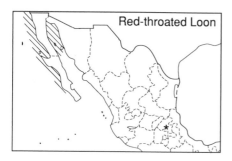

Red-throated Loon

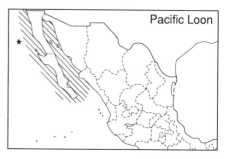

Pacific Loon

Habs. Open ocean, coastal bays and estuaries. Often in flocks up to several hundred birds, occurs farther offshore than other loons, often migrates in flocks and tends to fly low over the ocean.

SS: See Red-throated Loon; oversummering Pacifics often in faded and worn basic, best identified by bill shape and any retained patterning. Larger and bulkier Common Loon has heavier bill, more angular head shape, basic has smudgier and less neatly demarcated face and neck pattern; wing-beats slower than Pacific Loon, larger feet project more beyond tail. Imm cormorants rarely show such black and white contrast, often hold their hooked bills tilted up (bills held level in Pacific and Common loons), usually jump from the water as they dive, and have gular markings.

SD: C to F transient and winter visitor (Nov–May) along coasts of Baja and Son (probably also to Sin); R to U through the summer.

RA: Holarctic breeder; winters S in New World to NW Mexico.

NB1: *G. a. pacifica* (Pacific Loon), breeding N.A. and E Siberia, sometimes considered specifically distinct but see Storer (1978).
NB2: Known as Black-throated Diver in the Old World.

COMMON LOON
Gavia immer   Not illustrated
Colimbo Común

ID: 26–33 in (66–84 cm). Large and bulky with *stout and straightish bill, culmen slightly decurved.* **Descr. Basic/juv:** bill grey to greyish white with *dark culmen.* Blackish crown and hindneck contrast with white face and foreneck. *Upperparts blackish, slightly paler than hindneck,* with subdued pattern of paler checkers (adult) or scaling (juv); underparts white. Juv attains 1st basic from late winter on. **Alt:**

head and neck glossy blackish green (appearing black) with 2 cross bars of black and white striping on neck. Black upperparts boldly checkered white.

Habs. Harbors, bays, estuaries, open ocean, rarely inland lakes. Often flies high (100+ m) over the ocean.

SS: See Pacific Loon; also imm cormorants under Pacific Loon SS.

SD: C to F winter visitor (Nov–Apr) along coasts of Baja, Son, and Sin; F to U to Col (JCA, SNGH), possibly R and irregular to Gro (RR) and Chis (ARP). U to R in winter on Atlantic coast from Tamps to N Ver. Irregular R to U winter visitor (Nov–Mar) on inland lakes S to Mex (PDV), DF (RGW), Hgo (SNGH, SW), and S Ver (WJS). U to R in summer in NW Mexico.

RA: Holarctic breeder; winters S in New World to N Mexico, SE USA.

NB: Known as Great Northern Diver in the Old World.

YELLOW-BILLED LOON
Gavia adamsii   Not illustrated
Colimbo Piquiamarillo

ID: 28–34 in (71–86 cm). *Vagrant.* **Descr/ SS:** *Resembles Common Loon but greyish-white to creamy bill very long with straight pale culmen, at least distally, tapered lower mandible gives up-tilted aspect.* Plumage resembles Common Loon but paler overall in basic, *whitish face typically with contrasting dark smudge on auriculars.*

Habs. Pelagic, but vagrants may occur in harbors, bays, etc.

SD: Vagrant, two records from BCN: Islas Los Coronados (Nov 1968: Jehl 1970), and N Gulf of California (Jun 1973: Simon and Simon 1974).

RA: Holarctic breeder; winters S in New World to NW USA, E Canada.

NB: Known as White-billed Diver in the Old World.

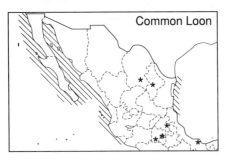
Common Loon

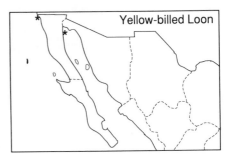
Yellow-billed Loon

GREBES: PODICIPEDIDAE (6)

A cosmopolitan family of medium-sized to small aquatic birds, grebes superficially resemble small, more slightly built loons but have lobed toes and usually more slender bills. Ages differ; precocial young of most species have dark-striped whitish heads and apparently go through more than one downy plumage before their molt into adult-like basic in 1–3 months. Sexes similar but plumage varies seasonally. Most species have white wing patches visible in flight. Vocalizations are mainly coos, clucks, and reedy trills.

Food small fish and invertebrates. Grebes breed on fresh water; their platform nests of vegetable matter often are floating and anchored to vegetation. Eggs whitish when fresh, quickly become soiled; 3–10 (usually 4–6) per nest, larger clutches may result from dumping by more than one ♀.

LEAST GREBE

Tachybaptus dominicus *   Figure 6
Zambullidor Menor

ID: 8.5–9.5 in (21.5–24 cm). *Slender bill appears slightly up-tilted* **Descr. Adult:** *eyes golden yellow, black bill* tipped whitish. Crown blackish, *face and neck slaty greenish grey*, throat black (alt) to dusky (basic). Upperparts brownish black, sides greyish brown, undertail coverts and underparts white. *Broad white stripe on secondaries and inner primaries obvious in flight.* **Juv:** eyes duller, bill pale below at base. Throat whitish, striped head may be retained to mid-winter.

Voice. Varied purring and chattering calls, including a rolled, purring trill, often starting hesitantly, *pc, pc, purrrr...*; suggests Ruddy Crake but softer and lower, less whinnying, usually shorter; also a quacking *kwrek*, and a rough, slightly shrill, clucking *eh kehr*, 2nd note slightly emphatic. Juvs give a high, insistent *sii sii ... or ssip ssip ...*

Habs. Freshwater lakes, ponds, and marshes, mangroves; prefers water with vegetated edges for cover. In pairs or

Fig. 6 (a) Atitlan [Pied-billed] Grebe; (b) Pied-billed Grebe; (c) least Grebe

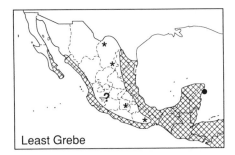

Least Grebe

small groups, rarely up to 30 birds; associates loosely with Pied-billed Grebes.

SS: Pied-billed Grebe larger and bulkier with much deeper pale bill, overall brownish plumage.

SD: C to F resident (SL–1500 m) on both slopes from Sin and Tamps, and locally in interior from Isthmus, to El Salvador and W Nicaragua; U to F and local in BCS (Howell and Webb 1992f) and S Son, R and irregular in interior (to 2400 m) N of Isthmus.

RA: S Texas, Mexico, and Greater Antilles to cen Argentina.

PIED-BILLED GREBE

Podilymbus p. podiceps   Figure 6
Zambullidor Piquipinto

ID: 11–13 in (28–33 cm). *A stocky grebe with notably thick bill.* **Descr. Alt:** *bill pale blue-grey with broad black vertical bar,* orbital ring whitish. Grey-brown overall, face and neck greyer with black throat. Upperparts dark brown, undertail coverts and underparts white. **Basic/juv:** bill creamy to pale grey without black bar. Head and neck cinnamon-brown with whitish throat; whitish-striped head of juv may be retained for 1–2 months.

   Voice. A series of hollow, clucking *cohw* or *cow* notes, often accelerating, then tail-

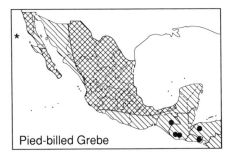

Pied-billed Grebe

ing off, may begin with whinnying calls. Also single clucks.

   Habs. Much as Least Grebe, but more often on open water, rarely in estuaries and bays. At times in flocks of 50+ birds.

SS: Stocky proportions and bill shape distinguish Pied-billed from other small grebes and diving ducks. See Atitlan Grebe (below).

SD: F to C resident (SL–2500 m) in Baja and locally from N Mexico to El Salvador and Honduras. C to F winter visitor (Oct–Mar) throughout the region.

RA: N.A. locally through M.A. to S.A., mainly in temperate areas; N N.A. populations largely migratory.

ATITLAN [PIED-BILLED] GREBE

Podilymbus (podiceps?) gigas   Figure 6
Zambullidor de Atitlán   No map

ID: 13–16 in (33–40.5 cm). Lake Atitlan, Guatemala. Extinct? **Descr./SS:** *Resembles Pied-billed Grebe but larger, bill proportionately deeper and stouter, white* to pale grey *with black vertical band* in juv/basic as well as alt, orbital ring bright whitish. Plumage darker overall on head and neck.

   Voice. Much like Pied-billed Grebe but lower pitched (La Bastille 1974).

   Habs. Much as Pied-billed Grebe, favors reed-fringed shore.

SD/RA: Restricted to Lake Atitlan, Guatemala. From maximum population of 200+ birds in 1930s through early 1970s decreased to 50–60 birds in early 1980s (after water level of lake began to drop following 1976 earthquake). Present status unclear (see NB).

NB: Although Hunter (1988) declared the Atitlan Grebe extinct, it is not clear whether she noted maintenance of black band on whitish bill throughout the year or in juv which would have confirmed Atitlan identification. In addition, she did not capture and measure all birds on Lake Atitlan and it is unclear whether she distinguished size differences *within* the *Podilymbus* present. In May 1989 we observed and compared directly two very large *Podilymbus* grebes (*gigas?*) with an obviously smaller, normal Pied-billed Grebe on Lake Atitlan. We suggest that isolating mechanisms were not complete (i.e. Atitlan Grebe was simply a large, dark ssp of Pied-billed and that re-invading Pied-billed Grebes are swamping (have now swamped?) Atitlan Grebes.

HORNED GREBE

Podiceps auritus   Not illustrated
Zambullidor Cornudo

ID: 11.5–13 in (29–33 cm). Rare in winter, NW Mexico. *Bill straight, crown flattish.* **Descr. Basic:** eyes reddish, bill greyish with small white tip. *Black cap (through level of eye) and hindneck contrast strongly with white face and sides of neck.* Upperparts blackish, underparts white. White secondaries striking in flight. **Alt** (rare in Mexico): bill black with pale tip, head and upperparts blackish, bushy golden tufts ('horns') extend back from eyes; *neck and sides above waterline chestnut.*

    Habs. Bays, estuaries. Singly or in small groups, often associates with more numerous Eared Grebe.

SS: Eared Grebe has steep forehead with profile peaking in mid- rather than hindcrown, bill appears slightly up-tilted; basic Eared has face and neck sides dusky, not contrasting strongly with black cap which extends below level of eye; alt Eared has black neck and extensive golden 'ears' over sides of head. Western and Clark's grebes much larger and longer-necked with yellowish bills.

SD: U to R winter visitor (Nov–Mar) along Pacific coast of N Baja; R (to U?) in N Gulf of California.

RA: Holarctic breeder; winters S in New World to NW Mexico, SE USA.

NB1: Known as Slavonian Grebe in the Old World. **NB2: Red-necked Grebe *P. grisegena*** (17–20 in; 43–51 cm) may occur as vagrant to N Mexico, especially NW coasts (see Appendix B). Fairly stocky with flat head, yellow-based blackish bill. In basic, *whitish head sides contrast with black cap through level of eye and with dusky neck sides.* Hindneck and upperparts blackish, underparts white. White secondaries and leading edges of inner wings visible in flight. Alt (unlikely in Mexico) resembles basic but sides of neck chestnut.

EARED GREBE

Podiceps nigricollis californicus
Zambullidor Orejudo   Not illustrated

ID: 11–12.5 in (28–31.5 cm). *Bill appears slightly up-tilted, forehead steep.* **Descr. Basic/juv:** eyes reddish (dull in juv), bill greyish. *Blackish cap (to below eye) and hindneck do not contrast strongly with dusky face and sides of neck*, though throat and face often are whitish. Upperparts blackish. White secondaries obvious in flight. **Alt:** bill black. Head, neck, and upperparts black, *broad fan of golden-yellow plumes ('ears') spreads out behind eye*, sides of body above waterline chestnut.

    Voice. Mostly silent except when nesting: a plaintive, slightly reedy, rising *hreep* or *hoo-reep.*

    Habs. Lakes, marshes, coastal bays, estuaries; fresh water with emergent vegetation for nesting. Often in small groups, in winter in flocks up to a few hundred birds.

SS: See Horned Grebe.

SD: Irregular and local breeder in cen volcanic belt (Wilson *et al.* 1988) and, at least formerly, in BCN and N Plateau. C to F winter visitor (Nov–Apr) S to cen Mexico; U to R to Guatemala and El Salvador (Thurber *et al.* 1987). Nonbreeding birds oversummer locally, S to El Salvador.

RA: Breeds W N.A. to cen Mexico; N populations migratory, wintering to N C.A. Also breeds in much of Old World and (formerly) in S.A.

NB1: Known as Black-necked Grebe in the Old World. **NB2:** The forms in S.A. (extinct?) are sometimes considered a separate species, *P. andinus.*

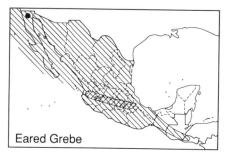

Horned Grebe   Eared Grebe

WESTERN GREBE

Aechmophorus occidentalis*   Not illustrated
Achichilique Piquiamarillo

ID: 19–24 in (48–61 cm). Neck long and
slender, bill long and narrow. **Descr.** Eyes
red (dull in juv), *bill greenish yellow to
yellow. Black cap (through level of eye)
and hindneck contrast strongly with white
head and neck sides. Lores and feathers
around eye blackish* in alt, dusky in imm/
basic with whitish lore spot. Upperparts
grey, underparts white. White to pale grey
secondaries and base of primaries often
striking in flight. In studied N.A. populations, adult-like pattern appeared 20–50
days after hatching but bill was black, becoming yellow by 80 days (Ratti 1979).
   Voice. Reedy and scratchy notes, including a 2–3-syllable mate-advertising call,
kree-kree or *kree-kree-kreet*, and an up-
slurred reedy *h-weep h-weep* ...
   Habs. Freshwater lakes with surrounding
reedy vegetation (breeding and winter),
also coastal bays and estuaries (winter).
Locally in flocks up to a few hundred
birds.
SS: See Clark's Grebe (below).
SD: The relative status and distribution of
Western and Clark's grebes in Mexico
have yet to be clarified. Western Grebes
breed in interior Mexico (750–2100 m) S
to Gro and Pue; they are less numerous
than Clark's in the N and cen areas of this
range (where Westerns may be only winter
immigrants?) but are at least as common as
Clark's at the S outpost of both forms in N
Gro. Western Grebes breed irregularly in
BCN (REW). Western is C to F (Oct–May,
U to Jun, non-breeding birds resident?)
along the Pacific coast of BCN, less common along the coasts of BCS and Gulf of
California coasts of Baja, Son, and possibly Sin (unspecified Western/Clark's).
Western is R winter visitor (Nov–Dec) E
to Coah and NL (SNGH, AMS photo).
RA: W N.A. to cen Mexico.
NB: Studies of Mexican populations would
help resolve the taxonomic status of Western and Clark's Grebes; up to 30–33% of
populations in Mich and Gro have been
considered intermediate (Feerer 1977).

CLARK'S [WESTERN] GREBE

Aechmophorus (occidentalis?) clarki*
Achichilique Piquinaranja   Not illustrated

ID/Descr.: 17–22 in (42–55 cm). *Resembles
Western Grebe but bill bright yellow to
yellow-orange. In alt, white facial feathering extends to above eye; in basic/imm
these areas may be dusky.* In studied N.A.
populations, downy Clark's (20–50 days
after hatching) were generally snowy white;
the bill at this age was black but attained
adult color by 80 days (Ratti 1979).
   Voice. Much like Western but mate ad-
vertisement call (at least in N.A.) is a
single-noted *kreeet*.
   Habs. Similar to Western Grebe but seems
to occur less often (or is at least less common) on salt water.
SS: See Western Grebe.
SD: F to C resident (750–2100 m) in interior
Mexico S to Gro and Pue, may wander
locally but seems mostly sedentary; also
breeds BCN (REW). U to R (Oct–Jun,
year-round?) along Pacific coast of Baja,
locally inland in BCS, and probably also in
Gulf of California (needs further study).
RA: W N.A. to cen Mexico.
NB: See Western Grebe.

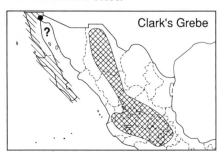

Western Grebe

Clark's Grebe

ALBATROSSES: DIOMEDEIDAE (3)

Albatrosses are very large, long-winged, pelagic birds characteristic of southern oceans, with three species occurring in the North Pacific. Bills long and stout with external tubular nostrils fused at the base of the culmen. Ages similar or different, young altricial; several years required to maturity but molt sequences poorly known; adults appear to have one complete molt per year after breeding, and 1st prebasic molt may also be protracted and complete. Sexes similar, patterned in dark browns, black, and white. Flight slow and leisurely when calm, but graceful with much effortless-looking gliding and soaring when windy. Vocal mainly during courtship displays.

Food squid, fish, and other marine animals taken near the sea surface. Albatrosses breed colonially on offshore islands and typically pairs mate for life, bonded by complex courtship rituals. A single white egg is laid in a shallow scrape on the ground and rearing young requires 3–5 months.

SHORT-TAILED ALBATROSS
Diomedea albatrus   Not illustrated
Albatros Rabón

ID: 33–37 in; W 84.5–93 in (74–84 cm; W 214–235 cm). Very rare. **Descr.** Ages differ. **Adult:** *large bill pink,* tipped bluish, *feet flesh. Head and body white, crown and nape washed golden yellow. Upperwings: outer half black, inner half white* with black patch on inner secondary coverts. Underwings white with black tip and narrow trailing edge. Tail black. **Juv/imm:** 1st year lacks bluish tip to bill. *Dark sooty brown overall* with whitish outer primary shafts on upperwings. Adult plumage attained over several years; underbody, then upperbody and upperwing coverts become progressively whiter. **Habs.** Sometimes accompanies ships and fishing boats.
SS: Juv resembles Black-footed Albatross but larger and more heavily built; note larger,

bright pink bill of Short-tailed. Occasional aberrant Black-footed Albatrosses have dull flesh bills but lack golden-yellow head and white patches on upperwing coverts.
SD: Formerly numerous off Pacific coast, at least S to 25° N; recorded throughout the year till late nineteenth century when heavy exploitation of birds at nesting islands brought the species close to extinction. Now protected, there are signs of a gradual recovery, with one recent Mexican record: an adult near Isla San Benedicto in Apr 1990 (Santaella and Sada 1991*a*); also recent Nov–Dec sightings of imms off California (PP, RS).
RA: Breeds on islands off Japan; ranges at sea in N Pacific.

BLACK-FOOTED ALBATROSS
Diomedea nigripes   Figure 7
Albatros Patinegro

ID: 28–32 in; W 81–90.5 in (71–81.5 cm; W 206–230 cm). All dark. **Descr.** Ages differ. **Adult:** *bill blackish* to dull dusky flesh with dark tip, *feet black. Dark sooty brown overall with white feathering around base of bill, white tail coverts,* and white outer primary shafts on upperwings; from below, darker head and chest contrast with paler underbody. **Juv:** tail coverts dark, becoming white with age. Some birds (aberrant or with Laysan characters) have pale head and underbody.
Habs. Pelagic, unlikely to be seen from land; often accompanies ships.

Short-tailed Albatross

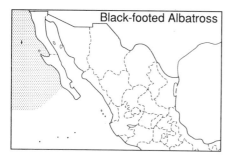

Black-footed Albatross

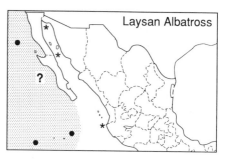

Laysan Albatross

SS: See Short-tailed Albatross.

SD: F to C visitor (Mar–Oct) off Baja; R to U (Nov–Feb).

RA: Breeds W and cen Pacific; ranges widely in N Pacific.

NB: Hybridizes rarely with Laysan Albatross.

LAYSAN ALBATROSS

Diomedea immutabilis   Figure 7
Albatros de Laysan

ID: 28–31 in; W 79–86.5 in (71–79 cm; W 200–220 cm). Black and white. **Descr.** Ages similar. *Bill flesh pink* with dusky tip, legs pale flesh. *White overall with black back and upperwings*; note dark smudge through eye, whitish outer primary shafts on upperwings. Underwings white with black margins and blackish markings on coverts.

Voice. On or near land in courtship displays, gives high, whinnying calls which

Fig. 7 (a) Laysan Albatross; (b) Black-footed Albatross

may suggest Whimbrel, also a braying *aaahhr*, and bill-clapping, etc.

Habs. Does not usually follow ships. Breeds (Nov–May) on rocky or grassy-topped islands.

SD: U to F (commonest Nov–Mar) offshore from Baja to vicinity of Islas Revillagigedo. Colonized islands off W Mexico in 1980s (Howell and Webb 1992c) with breeding confirmed on Islas Guadalupe (in 1986), Clarión (in 1988), and San Benedicto (in 1992). R visitor (Apr–May) to N Gulf of California (Newcomer and Silber 1989) where first noted 1982.

RA: Breeds W and cen Pacific, recently in E Pacific off W Mexico. Ranges widely in N Pacific.

NB: See Black-footed Albatross.

PETRELS AND SHEARWATERS: PROCELLARIIDAE (22)

Procellarids are cosmopolitan, medium-sized to large pelagic birds with long wings, webbed feet, and external tubular nostrils fused at the base of the culmen into a characteristic 'tubenose'. Ages/sexes similar, young altricial. First pre-basic molt appears to be protracted and complete, starting toward end of first year, gradually synchronizing with adult cycle of one complete molt per year after breeding. Several species polymorphic. Plumages colored in black, browns, greys, and white, often in striking patterns. Braying and chattering calls are usually heard only at or near breeding grounds.

Food fish, squid, and other marine organisms; some species scavenge from fishing boats. Nests are in crevices, burrows, or in dense vegetation, with one (rarely two) white eggs laid.

NORTHERN FULMAR

Fulmarus glacialis rodgersii   Not illustrated
Fulmar Norteño

ID: 16–18 in, W 39–44 in (40.5–45.5 cm, W 99–112 cm). *A large, stocky, dimorphic petrel with a notably thick neck, 'melon-headed' appearance, and thick bill.* **Descr.** *Bill and feet flesh pink to yellowish,* tubenose often dark. **Dark morph:** *smoky grey to grey-brown overall,* rarely appearing dark sooty brown; flight feathers usually darker with *whitish flash across base of primaries.* **Light morph:** head and underparts white, underwings white with dark margins. *Upperparts grey, rarely silvery, remiges darker with white flash across base of primaries.* Intermediate morphs also occur, dark morphs commonest off Mexico. **Habs.** Pelagic but may be seen from land during storms. Follows and scavenges from fishing boats. Flight steady and strong with powerful flapping; in moderate winds glides frequently on slightly bowed wings held stiffly and straight out from body.

SS: Dark shearwaters more slender with narrower wings, darker plumage lacks white flashes on primaries; only rare Flesh-footed Shearwater has pale bill. White-bodied gulls have white rumps and tails, different flight, lack tubenose bills.
SD: Irregularly U to F winter visitor (mainly Nov–Apr) off Baja; some birds over-summer, particularly after 'invasion' winters; vagrant (Jun 1985) to Gulf of California (DB, BT).
RA: Breeds N Atlantic and N Pacific oceans; winters in Pacific S to NW Mexico and Japan.
NB: *Fulmarus* might best be merged with *Procellaria.*

PARKINSON'S (BLACK) PETREL

Procellaria parkinsoni   Not illustrated
Petrel de Parkinson

ID: 16–18 in, W 43–47 in (40.5–45.5 cm, W 110–120 cm). An all-dark, fulmar-like petrel. **Descr.** *Bill pale horn, tipped black, feet dark. Blackish brown overall,* under-

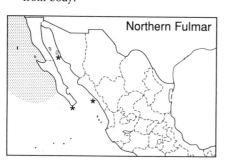

Northern Fulmar

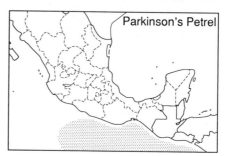

Parkinson's Petrel

side of primaries may appear paler, silvery grey.

Habs. Pelagic. Flight and habits much like Northern Fulmar, scavenges from boats etc.

SS: Flesh-footed Shearwater less thickset, slender bill and feet pink.

SD: Irregularly (?) U to R visitor (Apr–Oct) N to waters off Gro; recorded off Gro, Oax, Guatemala, and El Salvador (Jehl 1974; RLP).

RA: Breeds off New Zealand; ranges to waters off W S.A. and M.A.

NB: Bulwer's Petrel *Bulweria bulwerii* (10.3–11.5 in, W 26.7–28.3 in, 26–29 cm, W 68–72 cm) may reach southern Pacific waters of region (see Appendix B). It is a small dark petrel with long, pointed wings and long, wedge-shaped tail, typically held closed in a point. *Blackish brown overall, slightly paler below, with paler upperwing bar* (across greater secondary coverts) *visible at close range*; bill and feet black. Flight leisurely, weaving low over water with fairly quick, shallow wingbeats and prolonged glides, often banks in low, wheeling arcs, bowed wings pressed slightly forward and angled back at carpals. In moderate to strong winds may bank and wheel higher, like a *Cookilaria*.

GENUS PTERODROMA

Pterodroma petrels (also called gadfly-petrels) are highly pelagic birds with steeper foreheads and thicker bills than shearwaters. In the region, all species are found in the Pacific and are very unlikely to be seen from mainland. Cook's Petrel and other similar small *Pterodroma* are collectively referred to as *Cookilaria* petrels. Bills black, feet often bicolored (flesh or grey-blue basally, black distally). From below, long undertail coverts cover most of tail. Typical flight is fast and wheeling, with little or no flapping, though more leisurely with flapping in calm conditions. Many species show a contrasting **M** pattern across their upperparts formed by darker outer primaries, secondary coverts, and rump. Species with mostly white underwings have black wing-margins formed by the dark trailing edge, primary tips, and a carpal bar from the primary coverts across the median coverts. *Pterodroma* are notoriously difficult to identify at sea. The most

important points to check are head/body contrasts and wing patterns; shape and flight are often the best characters but require considerable comparative experience. *Pterodroma* taxonomy is poorly understood and the present grouping might be better considered as more than one genus. Feed by picking from the sea surface, in flight or sitting on water; some species catch flying-fish in flight.

JUAN FERNANDEZ [WHITE-NECKED] PETREL

Pterodroma externa   Figure 8
Petrel de Juan Fernandez

ID: 16.5–17.5 in; W 40.5–43.3 in (42–45 cm; W 103–110 cm). A large *Pterodroma* with long, narrow, pointed wings often slightly crooked back; long tail. Plumage suggests a large *Cookilaria*. **Descr.** *Crown, nape, and upperparts grey with black eye-patch*, dark cap, *blackish M pattern*, and dark tail; uppertail coverts often mottled white. *Underparts white, underwings with dark carpal spot, narrow black trailing margins*. Some imms and molting birds have extensively white napes, much like White-necked Petrel and, in worn plumage, the crown and upperparts become darker, sooty-brown (cap often looks blackish). In bright light, the upperwings may appear dark brownish with head and back paler and greyer. Feet black and flesh.

Habs. Pelagic. In light winds, low wheeling glides on bowed wings interspersed with easy, often languid wingbeats; in moderate to strong winds, flight fast and wheeling with high arcs and little or no flapping. Often over tuna schools in mixed-species feeding flocks.

SS: In strong or poor, low-angle light when upperparts appear brown, may resemble light morph Wedge-tailed Shearwater or Dark-rumped Petrel, both of which have thick black underwing-margins; note also

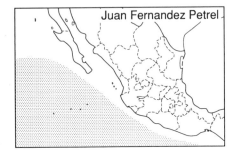
Juan Fernandez Petrel

more slender brownish head and dark tail below of Wedge-tailed Shearwater, extensive blackish hood and bold white forehead of Dark-rumped Petrel. *Cookilaria* petrels markedly smaller with faster wingbeats, more erratic flight in moderate to strong winds. Buller's Shearwater has more leisurely, buoyant flight without high wheeling, note dark cap, pale panel from secondaries across outer secondary coverts. See White-necked Petrel (NB below).

SD: F to C visitor (Jan–Dec, commoner Apr–Oct) N to waters around Clipperton (Pitman 1986; RLP, LBS); U to F (late May–Aug at least, probably Apr–Oct) N to waters off Col and vicinity of Islas Revillagigedo (Pitman 1986; RLP); U to R in these waters Nov–Dec (SNGH, LBS), probably Nov–Mar.

RA: Breeds on islands off Chile; ranges at sea into N Pacific, mostly to 5–15° N.

NB: White-necked Petrel *Pterodroma cervi-*

Fig. 8 (a) Juan Fernandez Petrel ((i) white-necked variation). (b) Dark-rumped Petrel; (c) Tahiti Petrel; (d) Kermadec Petrel

calis (16–17 in, W 40–42 in; 40.5–43 cm, W 102–107 cm) may occur off S Mexico and Clipperton (see Appendix B). Much like Juan Fernandez Petrel with broad, clean-cut white nape contrasting strongly with blackish cap and darker back, slightly less distinct blackish **M**; *told safely from Juan by bolder dark underwing-margins with distinct blackish carpal bar*; tail slightly paler, more grey-brown than Juan.

DARK-RUMPED PETREL

Pterodroma p. phaeopygia　　　Figure 8
Petrel de Galapagos

ID: 15.5–16.5 in; W 39.5–41 in (39.5–42 cm; W 100–104 cm). A fairly large *Pterodroma*, structurally similar to Juan Fernandez Petrel. **Descr. Adult/worn plumage:** *upperparts dark brown with no **M** pattern, white forehead contrasts strongly with blackish hood*; may show white mottling on uppertail coverts. *Underparts white, wings with thick black margins* rarely hard to see because of bright light. Feet black and flesh. **Juv** (and fresh-plumage adult?) *upperparts dark brownish grey with indistinct darker **M** pattern*.

Habs. Much as Juan Fernandez Petrel.

SS: See Juan Fernandez Petrel. Light morph Wedge-tailed Shearwater has brownish head without bold white forehead, tail black from below.

SD: R to U offshore visitor (late Feb–Oct) N to waters off Gro and vicinity of Islas Revillagigedo (Pitman 1986; RLP), probably from the Galapagos.

RA: Breeds in Hawaii and Galapagos Islands; ranges at sea in the tropical Pacific.

KERMADEC PETREL

Pterodroma neglecta juana　　　Figure 8
Petrel de Kermadec

ID: 14.5–15.5 in; W 39–40.5 in (37–39.5

cm; W 99–103 cm). A medium-large, polymorphic *Pterodroma* with *long, fairly broad wings and relatively short tail*. **Descr. Dark morph:** *dark brown overall*, underbody often paler brownish. *White primary shafts form white flash on upperwings; underwings dark with bold, skua-like white flash across base of primaries*, and often a narrow white crescent on under primary coverts. Some birds have white mottling on face and throat but this, and white basal rectrix shafts, rarely visible at sea. Underside of primaries rarely silvery grey without skua-like white flashes. Feet black. **Light morph:** head and underparts white, contrasting with dark brown to grey-brown upperparts and underwings; head and chest often mottled dusky, especially noticeable as darker chest band, throat white. *Wings as dark morph or underwings may show whitish mottling along median and greater coverts.* Feet black and flesh. Intermediate morphs include birds with dark head and chest and white throat and underbody, to essentially all-dark birds with some white mottling on underbody.

Habs. Pelagic. In light winds flight relatively leisurely, usually low over water with deep strong wingbeats, easy glides on bowed wings; in moderate winds prolonged gliding with low wheeling arcs, wings often slightly angled back at carpals.

SS: South Polar Skua chunkier with shorter, broader wings, bolder white wing flashes above, and more powerful, flapping flight. Pomarine Jaeger lacks gliding and wheeling *Pterodroma* flight, adults usually have long central rectrices, juvs often barred below; plumage, however, can be remarkably similar to Kermadec Petrel. Herald Petrel smaller and more slender with narrower wings, relatively longer tail, more

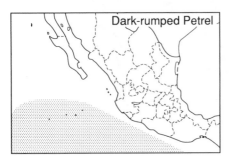

Dark-rumped Petrel

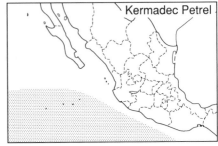

Kermadec Petrel

buoyant and wheeling flight, lacks white primary shafts above; in bright light, upperparts of dark morph Herald may show indistinct dark **M**, underside of primaries flash dull silvery, rarely bright whitish. Tahiti Petrel larger with longer narrower wings lacking white primary shafts above, underside of primaries rarely appears dull silvery. Shearwaters and petrels in molt may show white wing flashes, superficially suggesting Kermadec Petrel.

SD: U visitor (late Feb–Oct at least) N to waters off Col and vicinity of Islas Revillagigedo (Murphy and Pennoyer 1952; Pitman 1986; RLP).

RA: Breeds in S Pacific; ranges at sea N to 40° N in cen Pacific.

HERALD PETREL

Pterodroma arminjoniana ssp.
Petrel Heráldico   Not illustrated

ID: 13.5–14.5 in; W 35.8–37.5 in (34.5–37 cm; W 91–95 cm). *A medium-sized, relatively lightly built* Pterodroma *with long, fairly narrow, pointed wings.* **Descr. Dark morph:** *blackish brown overall with in-distinct* **M** *pattern visible above in bright light. Underwings show dull silvery flash across base of primaries,* rarely a bright whitish flash and whitish crescent on under primary coverts. White mottling on throat and whitish bar on lesser underwing coverts rarely visible at sea. Paler dark morphs have paler brownish underbody, whitish mottling on greater underwing coverts. Feet black. **Light morph:** head and chest pale brownish with white throat patch, underbody white. Upperparts grey-brown to brown with indistinct dark **M**; *underwings with brighter white flash at base of primaries and more white mottling on greater coverts than dark morph.* Feet black and flesh.

Habs. Pelagic. In light winds, flight buoyant with loose, flicking wingbeats, wings pressed slightly forward and slightly angled back at carpals, prolonged leisurely glides on bowed wings; in moderate to strong winds, flies with buoyant, fairly high wheeling arcs.

SS: See dark morph Kermadec Petrel. Christmas Shearwater best told by flight: long, buoyant, shearing glides on bowed wings interspersed with quick bursts of stiff flapping; note slender bill and shorter tail.

SD: Probably a R but irregular visitor to waters around Clipperton and S of Islas Revillagigedo, few records to date: three light morphs 300–400 km S of Isla Socorro (Oct 1991, LBS); one dark morph 100 km SE of Clipperton (Oct 1989, SNGH, LBS).

RA: Breeds tropical S Pacific, W Atlantic, and Indian oceans; ranges into N Pacific, at least to 25° N in cen Pacific.

NB: Murphy's Petrel *Pterodroma ultima* (13.5–14.5 in, W 35.5–37.5 in; 34.5–37 cm, W 90–95 cm) may occur as U visitor (Mar–Apr) offshore extreme NW Baja. Dark brownish grey to grey overall, head often appears darker than greyer upperparts which typically show a fairly distinct dark **M**. Underwings dark with usually distinct silvery flash across base of primaries. Higher wing-loading than Herald Petrel results in less buoyant flight in light winds and high, steep wheeling arcs in stronger winds.

TAHITI PETREL

Pterodroma rostrata ssp.   Figure 8
Petrel de Tahiti

ID: 15.7–17 in; W 39.7–42.5 in (40–44 cm; W 101–108 cm). *A large* Pterodroma *with long, narrow, pointed wings; bill large and heavy.* **Descr.** *Head, upper chest, and upperparts dark brown; white underbody contrasts with dark chest and dark under-*

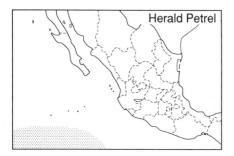
Herald Petrel

Tahiti Petrel

wings. Uppertail coverts and base of tail often contrastingly paler brownish. Underwings may show paler stripe toward trailing edge. Feet black and flesh.

    Habs. Pelagic. In light winds, long, shearing glides on bowed wings interspersed with loose, powerful flapping; wing tips may appear to curl up when gliding. In moderate to strong winds, flight fast and powerful, with relatively low wheeling arcs and little or no flapping. Typically does not associate with feeding flocks.

SD: F to C visitor (May–Dec at least) N to waters around Clipperton (Pitman 1986; SNGH, LBS); U to F visitor (late May–Nov at least) N to waters off Col and S of Islas Revillagigedo (Pitman 1986; SNGH, LBS, RLP).

RA: Breeds in cen and W tropical Pacific; ranges widely in S Pacific and N to 20° N.

NB: Sometimes placed in the genus *Pseudobulweria*.

COOK'S PETREL
Pterodroma cookii   Not illustrated
Petrel de Cook

ID: 12–13 in; W 29.5–31.5 in (30.5–33 cm; W 75–80 cm). Common off Baja. **Descr.** *Upperparts grey* (browner when worn) *with blackish M pattern; pale grey head* with dark eye-patch *does not contrast with back*; tail blackish distally, whitish outer rectrices rarely visible at sea. *Underparts white*, narrow black wing-margins inconspicuous. Feet grey-blue.

    Habs. Pelagic. In light winds flies with fast, usually low wheeling arcs interspersed with rapid wingbeats; in moderate to strong winds flight more erratic with steep wheeling arcs and little or no flapping.

SS: See Black-winged Petrel (NB below). Stejneger's Petrel has dark cap (especially noticeable in profile or from below),

brighter **M** pattern above, slightly more distinct black wing-margins.

SD: F to C visitor (Mar–Oct) off Baja S to about 22° N, U to R (Nov–Feb); probably U to F transient (mainly Mar–Apr, Sep–Nov) S to Clipperton, recorded Oct–early Dec off Islas Revillagigedo and Clipperton (SNGH, LBS), and possibly Apr off Islas Revillagigedo (Santaella and Sada 1991*b*).

RA: Breeds on islands off New Zealand; disperse to waters off W N.A.

NB1: Black-winged Petrel *Pterodroma nigripennis* (12–13 in, W 28.7–30.3 in; 30.5–33 cm, W 71–75 cm) may occur in waters off southern Mexico and Clipperton (see Appendix B). Resembles a chunky Cook's Petrel but blackish **M** often less distinct on darker upperwings, grey from nape extends down onto sides of chest, underwings show striking thick black margins; feet black and flesh. **NB2: Mottled Petrel** *Pterodroma inexpectata* (13–14 in, W 33–36 in; 33–35.5 cm, W 84–92 cm) occurs as regular transient (Nov–Dec) of California, with recent records adjacent to BCN (PP). Slightly larger but distinctly heavier-bodied than *Cookilaria*. Head and upperparts smoky grey with blackish **M** pattern and contrasting paler hindwing panel (secondaries and inner primaries); underparts white with blackish belly and thick black carpal bar on silvery-white underwings. Flight strong with long low glides and bursts of quick flapping in calm conditions; high bounding arcs in moderate to strong winds.

STEJNEGER'S PETREL
Pterodroma longirostris   Not illustrated
Petrel de Stejneger   No map

ID: 11.5–12.5 in; W 28–29.5 in (29–31.5 cm; W 71–75 cm). Rare transient. **Descr.** Upperparts grey with distinct blackish **M** pattern; *slaty-blackish cap contrasts with grey upperparts*; tail blackish distally. Underparts white, *note dark grey extension from neck onto sides of chest forward of wings*, narrow black wing-margins. Feet grey-blue.

    Habs. Much as Cook's Petrel.

SS: See Cook's Petrel, White-winged Petrel (NB below).

SD: Probably a R to U transient (Nov–Dec) off Pacific coast, one record to date: one bird 320 km S of Clipperton (Dec 1990, SNGH, SW).

Cook's Petrel

RA: Breeds on islands off Chile; disperses to waters off Japan.

NB: White-winged Petrel *P. leucoptera* (12–13 in, W 29–31 in; 30.5–33 cm, W 74–79 cm) may occur (Apr–Oct?) S of Clipperton (see Appendix B). Closely resembles Stejneger's but slightly larger; convex blackish hood with white forehead, suggesting Dark-rumped Petrel (versus concave cap of Stejneger's), contrasts with grey back, black underwing-margins thicker.

GENUS PUFFINUS: Shearwaters

Shearwaters typically are more slender than petrels, with less steep foreheads and more slender bills. Wingbeats stiff, varying from relatively leisurely (larger species) to relatively fluttery (smaller species). Size, flight, and pattern/contrast are important field characters. Several species are often seen from land. Feed by shallow diving, from flight or sitting on water.

CORY'S SHEARWATER

Puffinus diomedea   Not illustrated
Pardela de Cory

ID: 18–20 in; W 44–48 in (45.5–50.5 cm; W 112–122 cm). Rare off Atlantic coast. **Descr.** Bill flesh horn with dark tip, feet flesh. *Upperparts grey-brown, dusky sides of head do not contrast sharply with white throat*, may show whitish band at base of tail. Underparts white with dark wing-margins.

 Habs. Pelagic. Flight often distinctly languid with easy wingbeats, prolonged glides.

SS: Great Shearwater has contrasting dark cap, pale hindcollar, dark markings on basal underwing coverts.

SD: Probably R but regular (?) off Atlantic coast of region (recorded off S Texas, Aug–

Oct), one Mexican record to date: one bird 25 km SE of Isla Cozumel, 31 Oct 1991 (Santaella and Sada 1992).

RA: Breeds E Atlantic Ocean and Mediterranean Sea; ranges in Atlantic Ocean to waters of N.A. and NE S.A.

NB: Often placed in the genus *Calonectris*.

PINK-FOOTED SHEARWATER

Puffinus creatopus   Figure 9
Pardela Patirrosada

ID: 18–19 in W 43–46 in; (45.5–48 cm; W 110–117 cm). A large, bicolored Pacific shearwater. **Descr.** *Bill pink with dark tip, feet pink. Upperparts dark grey-brown,* dusky sides of head do not contrast sharply with white throat; upperparts may show slightly contrasting M pattern in fresh plumage and bright light. Underparts white with dark wing-margins and variable dusky mottling on wing coverts and flanks. *On water looks stocky, large-headed, with dusky head, neck sides, and sides along waterline.*

 Habs. Pelagic, but may be seen from shore with large movements of Sooty Shearwaters. Flies with fairly leisurely wingbeats and frequent glides, may bank fairly high in moderate to strong winds. Often in feeding flocks with other shearwaters, boobies, etc.

SS: Light morph Wedge-tailed Shearwater has slightly more buoyant flight with bowed wings pressed forward and angled back at carpals; note longer, narrower tail (shorter and broader in molting birds), grey bill, dark cap often cleaner-cut. On water, Wedge-tailed appears smaller with longer, more slender neck, smaller head, and contrasting white foreneck, chest, and sides along waterline (neck and sides dusky in large-headed Pink-footed). Black-vented Shearwater smaller, with quicker wingbeats and short glides.

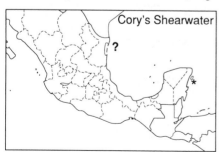

Cory's Shearwater

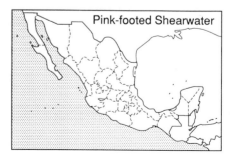

Pink-footed Shearwater

Fig. 9 (a) Wedge-tailed Shearwater ((i) dark morph); (b) Pink-footed Shearwater

SD: F to C visitor (Feb–Nov) offshore N to Jal, U to R (Dec–Jan), becoming F N to BCN (Mar–Oct) where U to R (Nov–Feb), (Pitman 1986; RLP); irregularly U to F (Feb–Oct) in Gulf of California (DB, RL, BT).

RA: Breeds on islands off Chile; ranges N in E Pacific to waters off Alaska.

FLESH-FOOTED SHEARWATER
Puffinus carneipes   Not illustrated
Pardela Patipalida

ID: 18–19 in; W 43–47 in (45.5–48 cm; 110–120 cm). A large, dark Pacific shearwater. **Descr.** *Bill pink with dark tip, feet pink. Plumage blackish brown overall,* underside of remiges paler, at times flashing dull silvery.

Flesh-footed Shearwater

Habs. Pelagic, unlikely to be seen from land. Flight much like Pink-footed Shearwater; associates with other large shearwaters, may follow ships.

SS: Heavier build, broader wings, pale bill, dark underwings, and leisurely flight distinguish Flesh-footed from Sooty and Short-tailed shearwaters. Dark morph Wedge-tailed Shearwater has slightly more buoyant flight with bowed wings pressed forward and angled back at carpals; note longer, narrower tail (shorter and broader in molting birds), grey bill (can appear pale), pale ovals at base of bill. See dark morph Northern Fulmar and Parkinson's Petrel.

SD: U to R visitor (mainly Jan–Jul?) offshore from Baja to Col and vicinity of Islas Revillagedo (Pitman 1986; RLP, AMS); R in Gulf of California (May 1982, RL).

RA: Breeds in S Indian and SW Pacific oceans; birds from latter area range to waters off W N.A. from Alaska to W Mexico.

GREAT SHEARWATER
Puffinus gravis   Not illustrated
Pardela Mayor

ID: 18–20 in; W 42.5–45.3 in (45.5–50.5 cm; W 108–115 cm). *Atlantic vagrant.*

1 Great Shearwater
2 Manx Shearwater

Descr. Bill dark, feet pink. *Blackish cap (to just below eye) separated from dark grey-brown upperparts by narrow white hind-collar (may be hard to see); white uppertail coverts form contrasting band at base of black tail.* Underparts white with dark mottling on wing coverts and belly, wings with blackish margins.

Habs. Pelagic. Flight strong, fairly stiff wingbeats interspersed with glides on bowed wings.

SS: See Cory's Shearwater.

SD: Vagrant, one record: a recently dead bird found at Tulum, QR, in Jul 1978 (Ash and Watson 1980); should be looked for (occurs rarely off Texas, Jul–Sep).

RA: Breeds S Atlantic; ranges into N Atlantic.

WEDGE-TAILED SHEARWATER

Puffinus pacificus cuneatus   Figure 9
Pardela Colicuña

ID: 17–18 in; W 39–43 in (43–45.5 cm; W 99–109 cm). A large, polymorphic Pacific species with long, wedge-shaped tail usually held closed; tail strikingly shorter and broader-looking in molt. **Descr.** *Bill grey* (often looks pale in bright light), feet pink. **Light morph:** *upperparts, including sides of head to below eye, dark brown to sooty brown;* upperwing may show paler bar on greater coverts. Underparts white with

thick dark underwing-margins, often some dark mottling on inner wing coverts. On water, note white foreneck, chest, and sides along waterline. Intermediate morphs (rare) have white areas washed or mottled dusky. **Dark morph:** *dark sooty brown overall,* paler below; underparts may appear warm brown in bright light. Upperwing may show paler bar on greater coverts, and underside of primaries a pale basal flash. *Note pale ovals at base of bill.*

Voice. Nocturnal at colonies. From burrow gives eerie, drawn-out, rising and falling, wailing moans in series of 2 and 3: *ouwhh 'oahh ouwhh* etc.

Habs. Pelagic. Flight typically unhurried with wings pressed forward and angled back at carpals, glides with wings held bowed slightly above body plane, recalling Buller's Shearwater. May glide and wheel for long periods in moderate to strong winds, but typically lower than *Pterodroma.* Often in feeding flocks over tuna schools with other shearwaters, boobies, and terns.

SS: Long tail and buoyant flight distinctive, but molting birds in particular may be mistaken for Flesh-footed and Pink-footed shearwaters. Black-vented Shearwater smaller with rapid stiff wingbeats and short glides. Sooty, Short-tailed, and Christmas shearwaters smaller, and shorter tailed, with faster wingbeats; Sooty has silvery-white underwing flashes.

SD: About 1000 pairs (70–90% dark morph) breed (Mar–Oct) on San Benedicto, Islas Revillagigedo. F to C visitor (Nov–Jun) off M.A. coast N to BCS, U during Jul–Oct (Pitman 1986; RLP); light morphs commoner than dark morphs.

RA: Breeds widely in tropical Pacific Ocean, off Australia and New Zealand, and in Indian Ocean. Ranges at sea off M.A., S.A., Japan, and S Red Sea coasts.

BULLER'S SHEARWATER

Puffinus bulleri   Not illustrated
Pardela de Buller

ID: 17–18 in; W 38–40 in (43–45.5 cm; W 96–102 cm). *A smartly patterned Pacific shearwater.* **Descr.** Bill dark, feet pink. *Upperparts grey with blackish cap through level of eye, contrasting blackish M pattern, and striking, broad, whitish band from secondaries across outer secondary coverts;* tail black. *Bright white underparts*

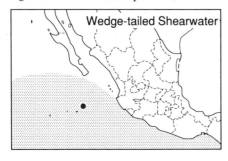

Wedge-tailed Shearwater

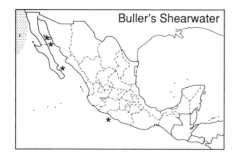

Buller's Shearwater

contrast sharply with dark cap, narrow black underwing-margins visible only at close range.

Habs. Pelagic, unlikely to be seen from land. Flight unhurried with leisurely wing-beats and prolonged, buoyant gliding, wings pressed forward and held bowed above body plane. In moderate to strong winds, flight lower and less erratic than *Pterodroma*.

SS: Unlikely to be confused with other shearwaters but see Juan Fernandez Petrel.

SD: Irregular R to U transient (Sep–Nov) off Baja; also R (Jun–Sep) offshore from BCS to Gro and near Clipperton, and in Gulf of California (Gisiner *et al.* 1979; Pitman 1986; Tershy *et al.* 1993).

RA: Breeds on islands off New Zealand, ranges at sea to Pacific coasts of N.A. (Alaska to NW Mexico) and S.A.

SOOTY SHEARWATER
Puffinus griseus   Not illustrated
Pardela Gris

ID: 17–18 in; W 37.7–41.2 in (43–45.5 cm; W 96–105 cm). A fairly large, dark shearwater with narrow, pointed wings. **Descr.** *Bill and feet grey.* Plumage *dark sooty grey-brown overall,* head and upperparts darker; *underwings usually show distinct silvery-white flash across median and greater coverts.*

Sooty Shearwater

Habs. Pelagic, but often seen from land. Flight strong, wingbeats powerful and often fast, with wings usually slightly angled back at carpals; notably fast with high banking in moderate to strong winds. At times in flocks of hundreds off Baja. Ones and twos join feeding flocks off SW Mexico and C.A.

SS: Short-tailed Shearwater smaller, flight less powerful, with quicker flicking wing-beats, feet may project more noticeably beyond shorter tail, underwings with whitish flashes but rarely, if ever, bright silvery white; some Short-taileds have noticeably smaller bills and can show whitish chin, rarely seen in Sooty; some birds, however, may not be safely told from Sooty at sea. Smaller Christmas Shearwater has dark underwings, different flight. See Flesh-footed and Wedge-tailed shearwaters.

SD: C to F visitor (Apr–Oct) off BCN where U to R (Nov–Mar); U to F transient and visitor (mainly Feb–Jun, Sep–Dec) S off coast to C.A. Irregularly F to C (Feb–Nov) in Gulf of California (DB, RL, BT). Vagrant off Tamps (Jan 1992, per GWL).

RA: Breeds on islands off Australia, New Zealand, and S S.A., ranges at sea into N Atlantic and N Pacific oceans.

SHORT-TAILED SHEARWATER
Puffinus tenuirostris   Not illustrated
Pardela Colicorta

ID: 16–17 in; W 35.8–37.8 in (40.5–43 cm; W 91–96 cm). A medium-sized, dark Pacific shearwater, *very similar to Sooty Shearwater.* **Descr.** Bill and feet grey. Plumage *dark sooty grey-brown overall,* darker on head and upperparts, *underwings with pale grey flash across median and greater coverts;* chin often whitish.

Habs. Pelagic, but may be seen from land in storms. Habits much as Sooty Shear-

Short-tailed Shearwater

water but flies with slightly faster, less strong wingbeats; not in flocks in the region, at times follows ships.

SS: See Sooty, Flesh-footed, and Wedge-tailed shearwaters. Christmas Shearwater smaller with more buoyant flight and dark underwings.

SD: U to R visitor (Nov–Mar) off Pacific coast of Baja (SNGH, RLP, PP), R and irregular in Gulf of California (LBS, spec.) and S to Nay (JM, PP); beached carcass found in Gro (Oct 1972, ARP).

RA: Breeds S Australia and offshore islands, ranges at sea to waters off W N.A. (Alaska to Mexico).

CHRISTMAS SHEARWATER

Puffinus nativitatus Not illustrated
Pardela Pardo

ID: 13–14 in; W 32.5–35.3 in (33–35.5 cm; W 83–90 cm). A medium-sized, dark Pacific shearwater. **Descr.** Bill and feet dark. *Plumage uniform, dark sooty brown overall*, although underside of primaries can show dull silvery flash.

    Habs. Pelagic, unlikely to be seen from shore. Flight typically low with bursts of rapid wingbeats and long buoyant glides on bowed wings; may bank fairly high in strong winds. Often in feeding flocks of other shearwaters, boobies, etc.

SS: See Sooty, Short-tailed, Flesh-footed, and dark morph Wedge-tailed shearwaters, dark morph Herald Petrel.

SD: Irregularly (?) F to U offshore visitor (Apr–Sep) from Mich to Guatemala (Pitman 1986; Howell and Engel 1993; LACM, spec.); also recorded off Clipperton (Pitman 1986).

RA: Breeds in W cen Pacific, ranges in tropical Pacific to waters off SW Mexico.

BLACK-VENTED SHEARWATER

Puffinus opisthomelas Figure 10
Pardela Mexicana

ID: 14–15 in; W 31–33 in (35.5–38 cm; W 79–84 cm). *A small Pacific shearwater*, endemic Mexican breeder. **Descr.** Bill grey, feet dusky to dusky flesh. *Upperparts, including sides of head, dark brown* (rarely appearing blackish), *dusky sides of head do not contrast sharply* with white throat. *Underparts white with broad dark wing-margins* and variable dusky mottling on wing coverts and flanks; undertail coverts dark. Underbody rarely entirely dusky.

Voice. Undescribed (?).

Habs. Inshore waters, often seen from land. Flight fairly rapid with stiff wingbeats and short, rarely prolonged, glides. Often in flocks up to several hundred birds.

SS: Townsend's and Audubon's (southern) shearwaters are smaller with quicker wingbeats, clean-cut black-and-white plumage; Townsend's has white flank patches, Audubon's smaller.

SD/RA: C to F resident (breeds Feb–Jul) off Pacific coast of Baja, breeding on Islas Guadalupe, San Benito, Cedros, and Natividad, probably also on Isla San Martín. Post-breeding dispersal N (Aug–Feb) to California (rarely British Columbia) and S to Col, R to Gro, possibly to Oax (Feb 1983, SNGH, PP); also irregularly U to C (Jun–Feb, mainly Jul–Nov) in Gulf of California (Tershy *et al.* 1993, DB, BT).

NB: Traditionally considered conspecific with Manx Shearwater.

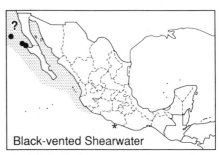

MANX SHEARWATER

Puffinus p. puffinus Not illustrated
Pardela Manx See Great Shearwater for map

ID: 14–15 in; W 31.5–33.5 in (35.5–38 cm; W 80–85 cm). Atlantic vagrant. **Descr.** Bill black, feet flesh. *Upperparts, including sides of head to below eye, blackish; throat*

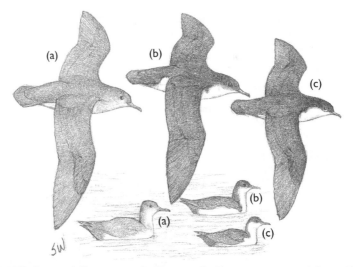

Fig. 10 (a) Black-vented Shearwater; (b) Townsend's Shearwater; (c) Audubon's Shearwater (*P. l. subalaris*)

and underparts *(including undertail coverts) white* with blackish wing-margins.
   Habs. Pelagic. Flight fairly rapid with stiff wingbeats and short, rarely prolonged, glides.
SS: Audubon's Shearwater smaller, with proportionately shorter wings, longer tail, more rapid wingbeats; note dark undertail coverts.
SD: Vagrant, one record: a beached carcass found near Dangriga, Belize, in Feb 1990 (Howell *et al.* 1992*b*)
RA: Breeds N Atlantic Ocean; wintering in S Atlantic.

TOWNSEND'S SHEARWATER
Puffinus auricularis   Figure 10
Pardela de Townsend

ID: 12.5–13.5 in; W 30.3–32.3 in (32–34.5 cm; W 77–82 cm). *A small, black-and-white Pacific shearwater*, endemic to Mexico. **Descr.** Bill dark, feet flesh. *Upperparts, including sides of head to below eye, blackish, dark extends down as partial collar on sides of neck, white flanks extend up to form patches (usually conspicuous) on sides of rump. Underparts clean white* with blackish wing margins and tail coverts. Worn plumage notably browner above, white sides of rump may be poorly defined. On water, note white sides along waterline.
   Voice. In flight over breeding grounds (at night), a slightly intensifying series of 2–6

gruff, throaty, slightly braying 2–3-syllable phrases: *ahr eh ahr eh* . . . or *ahr ah-ah ahr ah-ah* . . ., etc. (LFB tape); calls at colony may suggest chachalacas chorusing.
   Habs. Pelagic, near land only at breeding grounds. Flight typically fast and low, with bursts of rapid wingbeats interspersed with glides, at times banks high in strong winds; wingbeats more rapid than Black-vented Shearwater. Flocks stage offshore in late afternoon near breeding islands. Off W Mexico often in feeding flocks with other shearwaters, boobies, terns, etc.
SS: See Black-vented Shearwater. Audubon's Shearwater smaller and shorter-winged; note clean-cut cap through eyes, dark sides along waterline of sitting birds; lacks dark patches on sides of neck and white flank patches.
SD/RA: F but threatened breeder in Islas Re-

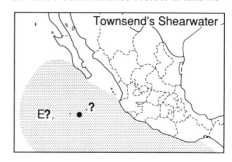

villagigedo (Nov–May) on Clarión (extirpated?), Socorro, and (at least formerly) San Benedicto; San Benedicto may be recolonized but suitable habitat was destroyed by volcanic activity in 1952. On Socorro, Townsend's Shearwater is preyed upon by feral cats (Jehl and Parkes 1982), and on Clarión by feral pigs (Howell and Webb 1989*b*; Santaella and Sada 1991*b*). Ranges at sea N (Dec–Jun) to BCS and Nay and (Jan–Dec) off W Mexico S to El Salvador (Jehl 1982; Pitman 1986; Howell and Engel 1993).

NB: Often considered conspecific with Newell's Shearwater (*Puffinus (a.?) newelli*), breeding in Hawaii. Newell's may occur in waters S of Clipperton (Pitman 1986; Dec 1990, SNGH). Rarely distinguishable from Townsend's Shearwater at sea but proximal undertail coverts white, black and white in face cleaner-cut, slightly longer-tailed.

AUDUBON'S SHEARWATER
*Puffinus lherminieri**   Figure 10
Pardela de Audubon

ID: 11.5–13 in; W 26–29 in (29–33 cm; W 66–74 cm). Two groups: *lherminieri* group (Caribbean) and *subalaris* (E Pacific). A small, short-winged, black-and-white shearwater; Caribbean birds relatively long-tailed. **Descr. *Lherminieri***: bill dark, feet flesh to dusky. *Upperparts, including cap through or just below eye, dark brown to blackish*, dark from hindneck extends down to sides of chest as partial collar. Underparts white with blackish wing-margins and undertail coverts; flanks mottled dusky. ***Subalaris***:

smaller, shorter-tailed. Lacks dark half-collar. On water, appears mostly dark along waterline.

Habs. Flight rapid and fluttery, with rapid, stiff wingbeats, low over the water with only short glides. Gregarious, often at feeding flocks with other shearwaters, terns, etc.

SS: See Townsend's and Black-vented shearwaters.

SD: F to C visitor (Jan–Dec) off Pacific coast of M.A. N to Gro, U (irregular?) N to waters off Jal (Howell and Engel 1993); originating from Galapagos population. Reports from Islas Revillagigedo (Santaella and Sada 1991*b*) are not credible. R to U visitor to Atlantic waters: recorded Mar, Jul, and Oct off QR (SNGH, OL, AMS).

RA: Breeds widely in tropical oceans; ranges at sea in E Pacific from Galapagos N to W Mexico; in W Atlantic from Caribbean to Gulf of Mexico and SE USA.

NB: More than one species may be included in the presently accepted Audubon's Shearwater, and relationships of some forms of Audubon's with *P. assimilis* (Little Shearwater) also are unclear.

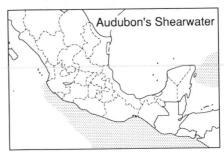

STORM-PETRELS: HYDROBATIDAE (8*)

Small to very small pelagic birds found over the world's oceans. Like procellarids, the external tubular nostrils are fused at the base of the culmen. Feet webbed. Ages/sexes similar, young altricial. First prebasic molt appears to be protracted and complete, starting toward end of first year, synchronizing gradually with adult cycle of one complete molt per year, after breeding. One species polymorphic. Plumage predominantly dark brown to blackish with paler upperwing bars (across greater secondary coverts); upperwing bars duller when fresh, brighter when worn. Several species have white uppertail covert patches (often erroneously referred to as rump patches). Bills and legs black. Rarely vocal away from breeding grounds where give purring calls from nest crevices, and more varied calls in flight.

Food zooplankton and other small marine animals; many species feed with wings raised, feet pattering on the sea surface. Storm-Petrels nest in crevices on rocky islands, visiting their colonies at night. All species lay single white eggs.

Identification of storm-petrels at sea is frequently problematic, and most similar species *do* overlap in size (*contra* many field guides) due to geographic variation. Judging shape correctly is important, particularly wing and tail shape, and the shape and extent of the white uppertail covert patch (if present). Judging flight is also important but requires comparative experience. Flight may vary with wind speed, wind direction relative to flight direction, bird's behaviour, etc.

LEACH'S STORM-PETREL
*Oceanodroma leucorhoa**   Figure 11
Paíño de Leach

ID: 6.5–8.5 in; W 15.7–19.3 in (16.5–21.5 cm; W 40–49 cm). *Common offshore Pacific species, polymorphic. Longish wings, usually angled back distinctly at carpals, tail forked* (appears notched when closed). In the Pacific, a cline occurs from larger, northern breeders with white uppertail coverts, to smaller, all-dark southern breeders; Atlantic birds have white uppertail coverts. **Descr.** Dark sooty brown overall with *band- to U-shaped white uppertail covert patch (divided by dusky median stripe which may be hard to see), through all intermediate stages to birds with all-dark uppertail coverts.* White often extends slightly onto lateral tail coverts. *Pale upperwing bar conspicuous;* underwings may flash silvery in bright light.
Voice. Burrow call a hard, rolled purring, typically rapid bursts of 2–5 s duration punctuated by abrupt, plaintive, gulping inhalations, *urrrrr...*, *ieh'urrrrr...*, etc.; flight calls are gruff, chuckling chatters, typically a clipped introductory note followed by 2 accelerating, chuckling phrases, *krruh ti-ti-krruh-kuh kuh-huh-huh-huh*, or *krrih pih-pih-pih-pih pyuh piupiupiu*, etc., at times ending slightly emphatically (DGA tape). Isla Guadalupe winter breeders have a rougher purring call, and overall shorter, low, slightly rasping or scratchy flight calls with little or no chuckling quality, *krreh keh-eh-eh-eh-ehr*,

Leach's Storm-Petrel

Fig. 11 (a) Leach's Storm-Petrel (dark morph below); (b) Black Storm-Petrel; (c) Galapagos Storm-Petrel; (d) Least Storm-Petrel

or *krreh keh-eh-eh-eh rreh eh-rr*, etc. (DGA tape).

Habs. Pelagic, rarely seen from land except near breeding grounds. In light winds, erratic bounding flight varies from fairly fast to leisurely, with fairly deep, often rapid, wingbeats and short glides suggesting a medium-sized bat; in stronger winds may glide long distances, banking like a small shearwater. When alarmed, such as close to ships, flight may be much more erratic, almost frenetic, with fast deep wingbeats and no gliding, much like Galapagos Storm-Petrel. Smaller S birds (including all-dark birds) have quicker wingbeats, more erratic and fluttery flight. Often foot-patters when feeding.

SS: Galapagos Storm-Petrel smaller and usually darker, with longer, bright white uppertail covert patch; flight quicker and more erratic, in light winds rarely glides. Harcourt's Storm-Petrel typically darker overall with less distinct upperwing bar, white uppertail covert band brighter without dusky median stripe, less pointed wings held straighter out from body, tail only slightly cleft, flight often different. See Wilson's Storm-Petrel (NB2 below). Ashy Storm-Petrel (BCN) smaller than N Leach's and relatively longer-tailed, flight more direct and fluttery with little or no gliding. Least Storm-Petrel smaller with more erratic flight and little or no gliding; note shorter, wedge-shaped tail. Black Storm-Petrel larger and longer-winged with leisurely flight, more languid wingbeats.

SD: F to C breeder (May–Oct, also Oct–Apr on Isla Guadalupe) off Pacific coast of Baja on Islas Los Coronados, Guadalupe, and San Benito, and Alijos Rocks (RLP). C to F visitor (Jan–Dec) off Pacific coast throughout region, more common southward in non-breeding season. All-dark birds mainly inshore and U to R S of Islas Revillagigedo and Col. May occur off Atlantic coasts (R off Texas, May–Jul).

RA: Breeds N Pacific and N Atlantic oceans. In Pacific, non-breeding birds occur S to 15°S; all-dark birds mainly off Mexico, R to Panama (Pitman 1986).

NB1: On Isla Guadalupe it appears that two populations breed, one in summer (late May–early Sep), the other in winter (mid-Oct–Apr). Most of the latter birds have white uppertail coverts, an exception to the N–S cline. **NB2: Wilson's Storm-Petrel** *Oceanites oceanicus* (6.5–7.5 in, W 14.5–17.2 in; 16.5–19 cm, W 37–44 cm) could occur off both coasts (see Appendix B). Relative to Leach's, wings not markedly angled back, tail squarish; yellow-webbed feet project beyond tail in flight. Blackish brown overall with pale upperwing bar; white uppertail covert patch squarish to band-shaped, extending noticeably onto lateral tail coverts. Flight usually low and fluttery, fairly direct, often with fairly stiff wingbeats (may recall Barn Swallow). **NB3: Fork-tailed Storm-Petrel** *Oceanodroma furcata* (8–8.5 in, W 17.3–19 in; 20–21.5 cm, W 44–48 cm) of NE Pacific probably occurs irregularly off Baja, at least in winter (see Appendix B). Tail long and forked (fork can be hard to see). Flight typically fast and fluttery with little gliding. Plumage smoky blue-grey overall, paler below, with blackish eye-patch. Pale grey upperwing bar contrasts with blackish lesser and median coverts and outer primaries; underwings grey with black axillars and lesser and median coverts.

GUADALUPE STORM-PETREL
Oceanodroma macrodactyla Extinct. See Appendix A.

HARCOURT'S (BAND-RUMPED) STORM-PETREL
Oceanodroma castro ssp.   Not illustrated
Paíño de Harcourt

ID: 8–8.5 in; W 17–19 in (20–21.5 cm; W 43–48.5 cm). Probably occurs off both coasts. *Wings fairly long and slightly angled back at carpals, tail slightly cleft.* **Descr.** *Sooty blackish brown overall, clean-cut white uppertail covert band* extends onto lateral tail coverts, bases of outer rectrices white. Paler upperwing bar usually indistinct.

Habs. Pelagic, unlikely to be seen from land. In light winds wingbeats deep and fairly floppy with little gliding; in stronger winds flight fairly buoyant with frequent, long, shearwater-like glides interspersed with fairly shallow wingbeats; foot-patters when feeding (PP).

SS: See Leach's and Wilson's (NB2 under Leach's) storm-petrels. Galapagos Storm-Petrel smaller with deeper wingbeats, more erratic flight, and little gliding; note long, wedge-shaped, white uppertail covert patch.

SD: Offshore Pacific visitor, perhaps irregularly/seasonally F (Apr–Aug?) to waters around Islas Revillagigedo. The only documented record from the region is one bird photographed near 19°N 115°W (about 75 km NNW of Isla Claríon) on 2 Aug 1989 (DR). Other reports from Islas Revillagigedo and BCS (Santaella and Sada 1991*b*) are not credible. May occur in Atlantic waters (R off Texas May–Jul).

RA: Breeds in Pacific on Galapagos Islands, Hawaii, and islands off Japan; also in E Atlantic. Pelagic distribution poorly known, ranges in warm waters of Atlantic and Pacific oceans.

NB: Known as Madeiran Petrel in Europe.

GALAPAGOS (WEDGE-RUMPED) STORM-PETREL
*Oceanodroma tethys** 			Figure 11
Paíño de Galapagos

ID: 6–6.5 in; W 13–15.7 in (15–16.5 cm; W 33–40 cm). *Small Pacific species with relatively long, narrow, pointed wings, notched tail.* **Descr.** Sooty blackish brown overall; *very long, bright white uppertail covert patch typically imparts gleaming white-ended appearance,* but often has hint of dusky median stripe distally; white patch shorter, recalling Leach's Storm-

Harcourt's Storm-Petrel

Galapagos Storm-Petrel

Petrel, in molting birds. Pale upperwing bar indistinct in fresh plumage.

Habs. Pelagic, unlikely to be seen from land. In light winds, flight erratic with strong, deep wingbeats and little or no gliding, recalling a small bat. In moderate to strong winds glides more and flight may resemble Leach's Storm-Petrel in lighter winds, but wingbeats quicker, flight more erratic. Often foot-patters when feeding; associates readily with Leach's Storm-Petrel.

SS: See Leach's and Harcourt's storm-petrels.

SD: C to F visitor (Jan–Dec) off Pacific coast N to Baja (regularly to 23° N; Pitman 1986; SNGH, LBS); U to F (May–Jul at least) in S Gulf of California (Howell and Webb 1992*f*).

RA: Breeds on Galapagos Islands and off Peru, ranges at sea N to W Mexico, S to Chile (20° S).

ASHY STORM-PETREL
Oceanodroma homochroa   Not illustrated
Paino Cenizo

ID: 7.3–7.7 in; W 15.3–17.3 in (18.5–19.5 cm; W 39–44 cm). *An all-dark Pacific storm-petrel, tail long and cleft.* **Descr.** Dark sooty grey overall, more brownish when worn, with pale greyish upperwing bar; note pale greater underwing coverts.

Voice. Burrow call much like Leach's Storm-Petrel: prolonged series of hard, rolled, clicking, and purring churrs, usually 1.5–3.5 s long, punctuated by abrupt, plaintive, gulping inhalations; flight calls shrieky to slightly gruff with emphatic barking or yelping phrases, often ending with a gruff phrase, *kri-ih whee-pu', ki-krr,* or *krieh whii pi-pi-peu' k'k'kirr,* etc. (DGA tape).

Habs. Pelagic, unlikely to be seen from shore. Flight somewhat fluttery, fairly

direct to slightly weaving, low over water, with fairly rapid and often shallow wingbeats, little or no gliding.

SS: Black Storm-Petrel larger with leisurely, rangy wingbeats and frequent glides; Least Storm-Petrel smaller and darker with shorter, wedge-shaped tail, more erratic flight with deeper wingbeats. See dark morph Leach's Storm-Petrel.

SD: R breeder (Apr–Oct?) on Islas Los Coronados, BCN; pelagic range off Mexico poorly known, recorded S (Apr) to vicinity of Islas San Benitos, BCN (Townsend 1923). Reports from farther S require verification.

RA: Breeds on inshore Pacific islands from cen California to N BCN, ranges at sea in vicinity of nesting islands.

BLACK STORM-PETREL
Oceanodroma melania   Figure 11
Paiño Negro

ID: 8.5–9 in; W 19–21 in (21.5–23 cm; W 48–53 cm). *A large, dark Pacific storm-petrel with long, angled wings and long forked tail.* **Descr.** Dark sooty grey-brown overall, *pale upperwing bar does not extend to leading edge of wing.*

Voice. Burrow call is a prolonged, strong, rapid, rattling churr up to 10–12 s in duration, swelling rapidly from a soft start and ending abruptly, with 1 s or more between churrs; flight calls rough and screechy with an accelerating ending, suggest Least Storm-Petrel but fuller, more emphatic, and often with a shrieky quality, *kreeih kreehr kree-kree-kree-kreehr,* or *kriih krri, krri ki-ki-kihr,* etc. (DGA tape).

Habs. Mainly inshore. May be seen from land, especially during or after storms. Flight often more languid and rangy than other storm-petrels, may suggest Black Tern. Direct flight with easy, fairly fast,

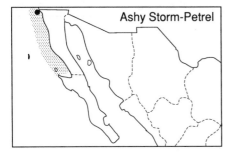
Ashy Storm-Petrel

Black Storm-Petrel

deep wingbeats and often prolonged glides; wingbeats may be quicker, shallower, and more flicking in light winds. In light to moderate winds glides suggest Leach's Storm-Petrel but are longer. Often accompanies ships; foot-patters briefly when feeding.

SS: See Ashy, dark Leach's, and Markham's storm-petrels.

SD: C breeder (Apr–Oct) on islands in N Gulf of California and off Pacific coast of Baja. F to C visitor (Jan–Dec) along Pacific coast throughout region, commonest southward in non-breeding season.

RA: Breeds off S California and both coasts of Baja. Ranges at sea S to 15° S (off Peru) and N to cen California.

MARKHAM'S STORM-PETREL

Oceanodroma markhami   Not illustrated
Paíño de Markham   No map

ID: 8.5–9 in; W 19.3–21.3 in (21.5–23 cm; W 49–54 cm). Clipperton. A large, dark storm-petrel, very similar to Black Storm-Petrel. *Wings long, tail long and forked.* **Descr.** Dark sooty grey-brown to brown overall, *pale upperwing bar extends to leading edge of wing.*

SS/Habs: Pelagic, well offshore. Flies with shallower wing-beats and more gliding than Black Storm-Petrel but field separation of the two species is problematic; Markham's often appears paler and browner than Black (LBS). Note that pale upperwing bar stops short of leading edge of wing in Black Storm-Petrel.

SD: Irregularly (?) U to R visitor (Jul–Aug at least) N to waters E of Clipperton (Loomis 1918; LBS).

RA: Breeds along Peruvian coast. Recorded at sea off E S.A. S to 26° S and regularly N to 5° N (Pitman 1986; LBS), rarely to 13° N.

NB: Sometimes considered conspecific with *O. tristrami* of Hawaii.

LEAST STORM-PETREL

Oceanodroma microsoma   Figure 11
Paíño Mínimo

ID: 5.5–6 in, W 12.5–14 in (13.5–15 cm, W 32–36 cm). Mexican endemic breeder. *A small, dark Pacific species with relatively long, narrow, pointed wings and medium-short, wedge-shaped tail.* **Descr.** Sooty blackish brown overall with poorly contrasting upperwing bar.

Voice. Burrow call a relatively soft (?), rapid, rolling churr, bursts of 3–4 s punctuated by abrupt, plaintive inhalations; flight calls surprisingly rough, harsh to burry phrases, accelerating toward a slightly emphatic ending but lacking the strong and often screechy quality of Black Storm-Petrel, *krrih krrih krrih-krri-krri-krri,* or *krruh kuh-uh krr-krr-krr-krr,* etc. (DGA tape).

Habs. Mainly inshore. May be seen from land, especially in Gulf of California, and during or after storms; often associates loosely with Black Storm-Petrel. In light winds, flight fast and fairly direct with deep wingbeats and no gliding, recalls small bat; in stronger winds may glide briefly. When feeding, flight erratic with dashing changes in direction, foot-patters.

SS: See Ashy, Black, and dark morph Leach's storm-petrels.

SD/RA: C breeder (Apr–Sep) on islands in N Gulf of California and on Islas San Benitos off Pacific coast of BCN; C to F at sea (Jan–Dec) over inshore waters S to 5° S, U to R to 10° S (Pitman 1986); ranges N irregularly (mainly Aug–Oct) to California; commonest southward in non-breeding season.

Least Storm-Petrel

TROPICBIRDS: PHAETHONTIDAE (3)

Spectacular, highly aerial seabirds of tropical oceans. Bills stout and pointed, feet totipalmate. Ages differ, young altricial; sexes similar. Molts poorly known; adults appear to have a complete annual molt after breeding, although tail streamers may be replaced at any time; 1st prebasic undescribed (?). Bills red to yellow in adults, duller in juvs; feet black with blue-grey to yellowish bases. Plumage white overall, marked with black masks and dark flank streaks. Adults have extremely long and attenuated central tail streamers; central rectrices barely project in juvs. Tropicbirds typically fly high above the sea and often sit on the water, tails held cocked. Vocal mostly at or near breeding islands.

Food fish and squid obtained by plunge-diving. Nest in crevices, burrows, shady places on rocky islands; single egg creamy to pinkish white, typically densely flecked and speckled with purplish browns and greys (WFVZ).

WHITE-TAILED TROPICBIRD
Phaethon lepturus *   Not illustrated
Rabijunco Coliblanco

ID: 14–17 in + 10–15 in streamers; W 34–38 in (35.5–43 cm + 25.5–38 cm streamers; W 86–97 cm). Vagrant. A small and slender tropicbird. **Descr. Adult:** *bill yellow to reddish orange.* White overall with black eyestripe, black outer primaries, and *diagonal black bar from tertials across median upperwing coverts. Tail streamers white.* **Juv:** bill dark blue-grey, becoming yellow in 1st year. Lacks tail streamers, tail tipped black; *crown extensively spotted blackish, back and inner upperwing coverts coarsely barred black.* May attain adult-like plumage (streamers shorter?) by 1st prebasic molt when a few months old, but bill yellow to yellow-green through 1st year.
  Habs. Pelagic and coastal. Flight buoyant and at times erratic, with relatively grace-ful, often fairly deep wingbeats.
SS: Red-billed Tropicbird larger with rel-atively labored flight, juv has mostly white crown, upperparts more narrowly barred black, black eyestripes usually meet around nape, adult has finely barred upperparts, red bill. Red-tailed Tropicbird larger and bulkier, appears all white, adult has wire-like red tail streamers. Large terns have deeper wingbeats and more relaxed flight, forked tails.
SD: May be regular off Caribbean coast (Mar–May?), only record is an adult on Isla Cozumel (Mar 1983, per GWL, photo); unverified reports from Caribbean coast of Belize (Wood *et al.* 1986) and Guatemala (Wetmore 1941*b*). Tideline carcass found on Pacific coast of Chis (Jan 1975, ARP, spec.). Vagrant to Clipperton (Nov 1987, PP).
RA: Breeds widely in tropical oceans; ranges mostly in vicinity of breeding islands.

RED-BILLED TROPICBIRD
Phaethon aethereus mesonauta   Figure 12
Rabijunco Piquirrojo

ID: 17–19 in + 16–20 in streamers; W 38–43 in (43–48 cm + 40.5–50.5 cm; W 97–110 cm). *The common Pacific coast tropicbird of the region.* **Descr. Adult:** *bill red* to orange-red. White overall with black eyestripe, *narrow black barring on upperbody and inner upperwing coverts,* and *black outer primaries.* Long tail streamers white. **Juv:** bill dull yellow. Lacks tail streamers, tail tipped black; *crown white* with variable, but rarely heavy, black spotting, *black eyestripes*

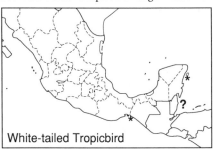

White-tailed Tropicbird

Fig. 12 Red-billed Tropicbird

meet around nape. Imm (1st basic?) resembles adult but bill orange, tail streamers may be shorter.

Voice. Rough to reedy screeches and shrill chatters, *krreea* and *krri-krri-krri-krri-krrriik*, etc.

Habs. Pelagic and coastal, typically around rocky coasts and islands with steep cliffs. Flies with surprisingly ungraceful, hurried wingbeats. Often in pairs or small groups, locally to a few hundred birds around nesting islands. Usually solitary at sea, associating rarely with feeding flocks.

SS: Red-tailed Tropicbird more heavily built with blunter wings, appears all white, adult's wire-like red tail streamers hard to see; see White-tailed Tropicbird. Large terns have deeper wingbeats and more relaxed flight, forked tails.

SD: F to C breeder (chronologies vary) along Pacific coast from Alijos Rocks and Gulf of

California to at least Col and Gro (Howell and Engel 1993), possibly Oax (SNGH, PP) and in Gulf of Fonseca; seasonal movements need study. U to F visitor (Jan–Dec) off Pacific coast from BCS to El Salvador and Honduras, U to R to BCN and Clipperton.

RA: Breeds and ranges along Pacific coast from Mexico to Peru, in Caribbean, and in tropical Atlantic and Indian oceans.

RED-TAILED TROPICBIRD
Phaethon rubricauda rothschildi
Rabijunco Colirrojo   Not illustrated

ID: 17–19 in + 11–15 in streamers; W 42–46.5 in (43–48 cm + 28–38 cm streamers; W 107–119 cm). Offshore Pacific. **Descr. Adult:** bill bright red. *Predominantly white,* with black eyestripe, tertial centers, and outer primary shafts. *Wire-like red tail streamers often difficult to see.* **Juv:** bill dark grey, becoming orange in 1st year. Lacks tail streamers, tail tipped dusky. *Crown, nape, upperbody, and inner upperwing coverts spotted and barred black*; belly mottled dusky. Subsequent molts and plumages undescribed (?). **Imm (1st basic?):** bill orange. Crown, nape, and upperparts more sparsely marked with black spots and bars, belly speckled dusky, tail streamers short, less wire-like, and whitish or washed pink. 2nd basic (?): resembles adult but bill orange-red, tail streamers shorter and paler, rump and

Red-billed Tropicbird

median upperwing coverts finely spotted black.

Voice. Calls (mostly in display flight) are slightly gruff, barking clucks, *kweh* and *gwehk* or *wahk*, etc.

Habs. More pelagic than other tropicbirds, often far offshore. Flight less hurried than Red-billed Tropicbird due to larger surface area of more rounded wings. In 'bicycling' flight display, pair hovers into wind, one bird above the other, top bird drops as lower bird climbs and both hover again; position change repeated, often several times.

SS: See other tropicbirds.

SD: U to R visitor (Jan–Dec) offshore from Baja to Col. Probably breeds on Isla San Benedicto (Howell and Webb 1990a) and Clipperton (RLP).

RA: Breeds widely in tropical Pacific and Indian oceans; ranges to waters off N M.A.

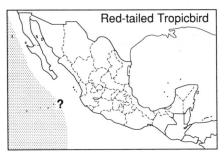

BOOBIES AND GANNETS: SULIDAE (5)

Large, long-bodied seabirds with long, pointed wings and tapered tails. Boobies are smaller than gannets and breed in tropical oceans; gannets breed in temperate oceans. Bills long, stout, and pointed; feet totipalmate. Ages differ, young altricial; molts (including 1st prebasic) protracted and poorly known. Sexes similar but ♀♀ larger. Adult plumages patterned with white, black, and dark brown; juvs duller. Bills and feet often brightly colored in boobies. Flight powerful, deep wingbeats interspersed with glides. Rarely vocal except on breeding grounds, occasionally when feeding.

Food mostly fish and squid, often caught by spectacular plunge-dives. Breed colonially, most nest on bare ground. Eggs: 1–3, chalky whitish to pale green (WFVZ).

MASKED BOOBY

*Sula dactylatra** Figure 13
Bobo Enmascarado

ID: 29–32 in, W 60–68 in (73.5–81 cm, W 152–173 cm). Largest booby, occurs both coasts. **Descr. Adult:** eyes and bill yellow, facial skin (mask) blackish; feet yellowish to olive-grey. In *granti* of SE tropical Pacific bill pinkish orange (orange-yellow by 2nd year), feet grey. *White overall with black flight feathers*, greater upperwing coverts, and tips to longest scapulars. **Juv/imm:** eyes duller, bill dusky yellowish, feet greyish, bill attains adult color by 20 months. *Dark brown head and neck separated from dark brown upperparts by whitish hindcollar.* Underparts white, with dark flight feathers and dark bar on lesser wing coverts. Attains adult plumage over 3 years; mostly like adult by 2 years of age.

Voice. ♂ gives shrill, piping to slightly wheezy whistles, including a *whi whiii'ooo* suggesting Black Hawk-Eagle, ♀ makes gruff brays.

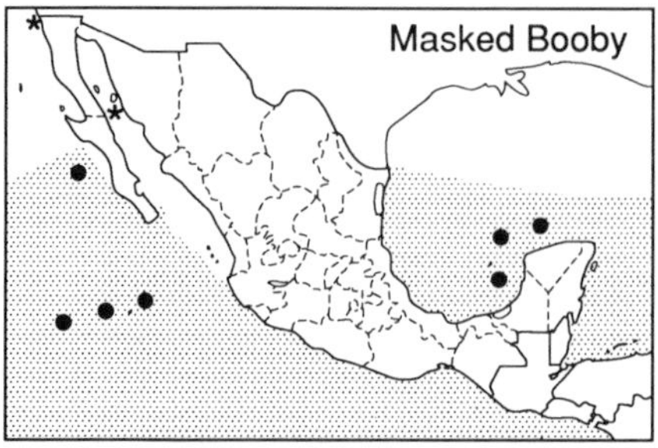

Masked Booby

Habs. Pelagic, rarely seen from mainland. Flight steady and powerful, fairly high to low over water, chases flying-fish. Nests on barren ground, prefers flat or gently sloping terrain.

SS: Adult might be taken for older imm Northern Gannet but head white, bill yellow, more black on secondaries. Smaller and slimmer white morph Red-footed Booby has white inner secondaries, white-tipped to all-white tail, dark patch on under primary coverts; note red feet and pale blue bill. Brown Booby lacks whitish hindcollar, dark brown extends onto chest, under primary coverts dusky. Blue-footed Booby (inshore) has paler, streaked head and neck, white-tipped tail (mostly white from above), dark underwings with large white axillar patches.

SD: C to F resident (breeding chronologies vary) in Pacific on Alijos Rocks, Islas Revillagigedo and Clipperton; small numbers of *granti* breed (usually mating assortatively) at last two sites (Howell and Webb 1990*a*; RLP). U to F visitor (Jan–Dec) off Pacific coast from BCS to C.A., vagrant to Islas Los Coronados (Apr 1988, Everett and Teresa 1988), and Gulf of California (Jun 1992, DB, BT). In Atlantic region, C resident (breeds Feb–Nov at least) on Campeche Bank; U visitor (Jan–Dec) to Tamps (JCA), Ver (WJS), Belize, and Honduras.

RA: Breeds and ranges widely in all tropical oceans.

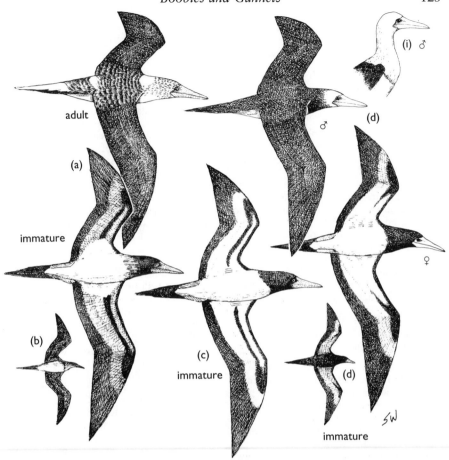

Fig. 13 (a) Blue-footed Booby; (b) Red-footed Booby (immature/dark morph); (c) Masked booby; (d) Brown Booby *S. l. brewsteri*; (i) extreme variant, Clipperton

BLUE-FOOTED BOOBY

Sula n. nebouxii   Figure 13
Bobo Patiazul

ID: 28–31 in, W 58.5–65.5 in (71–78.5 cm, W 148–166 cm). Large Pacific booby. **Descr. Adult:** eyes pale yellow, bill and facial skin greyish, *feet bright blue. Head and neck whitish with spiky, dark brown streaking,* upperparts dark brown with white patch across lower hindneck, short white bars on back and innerwing coverts, white patch across upper rump; *tail appears white from sides and above* with dark corners, *dark from below, tipped white.* Underparts white, *underwings mostly dark with large white axillar patch and two pale bars along coverts.* **Juv/imm:** eyes dusky, bill darker, feet dull bluish.

Head and neck brownish, contrasting more with white underbody; white bars above indistinct. Attains adult pattern over 2 years.

Voice. ♂ has whistles, ♀ brays, similar to but slightly deeper than Brown Booby (DB, BT).

Habs. Inshore waters, often seen from mainland. Groups often feed in shallow water off beaches, also circles and dives from high up. Nests on fairly flat ground.

SS: Adult distinctive; imm told from smaller Brown Booby by mostly whitish tail, less contrasting dark/white border on chest, underwing pattern, feet color. See Masked Booby.

SD: C to F resident (breeds Jan–Aug) from cen Gulf of California S to Tres Marietas, ranging commonly into N Gulf; probably

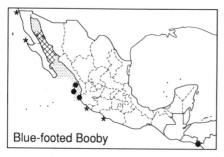

Blue-footed Booby

also breeds in Gulf of Fonseca. Mostly sedentary, irregularly R to U visitor (Mar–Jun at least) S to Gro (SNGH photo; RAE), possibly to Oax (Jehl 1974) and N (Jul–Nov) to Islas Los Coronados. Reports from Islas Revillagigedo (Jehl and Parkes 1982) require verification.

RA: Breeds in Pacific off NW Mexico, Panama, Ecuador, and Peru. Ranges near breeding islands throughout year.

BROWN BOOBY
*Sula leucogaster**   Figure 13
Bobo Vientre-blanco

ID: 27–30 in, W 53–60 in (68.5–76 cm, W 135–153 cm). Both coasts. **Descr. Adult:** eyes brownish, feet greenish to yellow; in ♂ deep blue face blends abruptly with pale glaucous bill, in ♀ pale greenish-yellow face blends into pale pinkish bill; colors brightest when nesting. *Dark brown overall with sharply contrasting white lower chest, belly, undertail coverts, and underwing coverts.* ♂ *brewsteri* (Pacific) has *crown and often whole head and neck whitish* (most extensive at Clipperton) with dark chest band. **Juv/imm:** eyes pale grey, bill greyish, feet dusky to fairly bright orangish flesh. *Brown overall,* lower chest, belly, and undertail coverts flecked whitish, i.e. with hint of adult pattern; *note contrast between dark underbody and*

whitish underwing coverts. Resembles adult in 2 years, often less clean-cut until 3rd year.

Voice. ♂ gives hoarse, hissing whistles, ♀ makes harsh brays.

Habs. Inshore and pelagic, often seen from mainland but feeds less often off beaches than Blue-footed Booby. Flight easy but strong, often low over water. A common member of feeding flocks off W Mexico. Often nests on steeper and more broken ground than other boobies.

SS: Brown-plumaged Red-footed Boobies told by paler body contrasting with dark underwings, whitish-tipped tail; note bill and feet colors, more crooked wings, slighter build. See Masked and Blue-footed boobies.

SD: C to F resident (breeding chronologies vary) along Pacific coast, breeding from Gulf of California locally to Gro (Goldman 1951; SNGH, SW), possibly also in Gulf of Fonseca; U resident on Islas Revillagigedo, C on Clipperton. R (Jan–Mar, Jul–Aug, irregular?) N along Pacific coast of Baja (RL). F to C breeder (Mar–Sep) in Atlantic region on Arrecife Alacrán and Isla Contoy (BM); R to U in Gulf of Mexico (Sep–Jan) whence post-breeding dispersal presumably into Caribbean; U to F visitor (Jan–Dec) to Belize and Caribbean Honduras. Vagrant to Tamps (Dec 1987, per GWL).

RA: Breeds widely in all tropical oceans; ranges at sea in vicinity of breeding grounds.

RED-FOOTED BOOBY
*Sula sula**   Figure 13
Bobo Patirrojo

ID: 26–29 in, W 52.5–59 in, (66–83.5 cm, W 134–150 cm). A polymorphic booby with rangy build, angled wings, low wing-loading. **Descr. Adult. White morph:** eyes

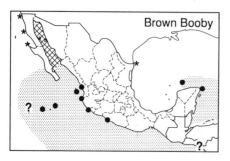

Brown Booby

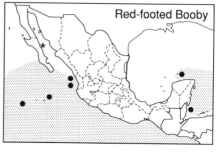

Red-footed Booby

brown, *bill pale blue*, facial skin pale blue and pink, *feet bright red*. *White overall* with yellow wash on crown and nape, *remiges black except for white inner secondaries*; tail blackish, tipped white, in most Pacific birds. **Brown morph:** brown overall; *head, neck, and underbody usually paler in contrast to dark underwings, tail tipped whitish.* Many birds show dark necklace across chest. Rump, tail coverts, and tail white in so-called white-tailed brown morph. **Intermediate morph:** highly variable appearance between white and brown morphs. **Juv/imm** (all morphs): bill and face dark grey, feet dusky to orange-red; bill becomes dusky flesh to pinkish with dark tip, and face mauve-blue, then colors 'reverse' to blue bill and pink face of adult. *Juv plumage as brown morph adult.* After 8 months white morphs become increasingly like adult, first on head and neck. Attains adult plumage over 3 years.
Voice. Both sexes give low, guttural brays and chatters.
Habs. Highly pelagic, rarely seen near mainland. Flight often graceful and agile, skims low over water and chases flying-fish. Nests in trees and bushes, exceptionally on ground.
SS: See other boobies; Northern Gannet (Gulf of Mexico) larger and heavier, adult has white secondaries, juv speckled white.
SD: In Pacific, F to C resident (breeding chronologies vary) on Islas Revillagigedo and Clipperton, U on Tres Marías and Isla Isabel, Nay. F to U offshore visitor (Jan–Dec) from BCS to C.A. (RLP), commonest from Mich S; R (Apr–May) to Gulf of California (RL). In Atlantic region, C resident on Half-Moon Cay, Belize, and 1–2 pairs recently found nesting at Arrecife Alacrán (Tunnell and Chapman 1988). R to U visitor (Jan–Dec) off QR and Belize. Revillagigedo breeders 95% white morph,

brown morphs predominate on Clipperton; in Caribbean only white and white-tailed brown morphs reported.
RA: Breeds and wanders widely in all tropical oceans.

NORTHERN GANNET
Morus bassanus   Not illustrated
Bobo Norteño

ID: 32–36 in, W 71–84.5 in (81.5–91.5 cm, W 180–215 cm). *Winter visitor to Gulf of Mexico.* **Descr. Adult:** eyes bluish white, bill blue-grey, facial skin and feet blackish. *White overall with yellow wash on head and nape, black primaries and upper primary coverts.* **Juv/imm:** eyes dusky. *Dark brown overall, paler below, with narrow white uppertail covert V and white speckling on upperparts.* Head, rump, and underparts become white first, a *patchwork pattern of black and white secondaries and rectrices follows*; 3rd year birds much like adult. Attains adult plumage in 4–5 years.
Habs. Pelagic, may be seen from land during storms. Flight powerful, often soars fairly high.
SS: See Masked and Red-footed boobies.
SD: U to R winter visitor (Oct–May, most records Jan) off Tamps and Ver (JCA, AMS, WJS).
RA: Breeds N Atlantic; ranges S in New World to SE USA and Gulf of Mexico.

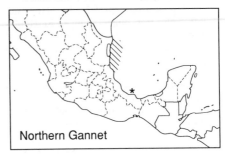

Northern Gannet

PELICANS: PELECANIDAE (2)

A cosmopolitan family of very large aquatic birds which inhabit freshwater lakes, lagoons, sea coasts, etc. Bills very long with large distensible gular pouch along lower mandible. Legs short and thick, feet totipalmate. Wings long and broad, tails short. Ages differ, young altricial; sexes similar, slight seasonal variation. Molts protracted, partial 1st prebasic occurs within a few months; first complete molt begins toward end of first year. Plumage patterned in white, greys, browns, and black. Pelicans fly with deep, powerful wingbeats and often glide and soar. Mostly silent, rarely uttering grunts and hisses.

Food mostly fish (although amphibians and crustaceans at times taken into dip-net bills), caught by diving or scooped-up while swimming. Nest colonially in low bushes or on ground. Eggs: 1–3, whitish.

AMERICAN WHITE PELICAN
Pelecanus erythrorhynchos   Not illustrated
Pelícano Blanco Americano

ID: 57–65 in, W 95–115 in (145–165 cm, W 242–292 cm). Very large with very long bill (13–14 in, 33–35.5 cm). **Descr.** Ages differ slightly. **Basic:** *bill*, pouch, facial skin, *and feet orange. White overall with black outer secondaries, primaries, and upper primary coverts.* **Alt:** bill and face brighter, bizarre fibrous horn grows near end of upper mandible. Chest and short occipital crest pale yellow, nape becomes sooty grey after eggs laid. **Juv/1st basic:** bill flesh, feet duller orange. Resembles basic but upperwing coverts washed brown. Attains adult-like plumage by 2nd prebasic molt.
  Habs. Freshwater lakes, marshes, coastal lagoons, estuaries. In small groups up to flocks of a few thousand. Swimming groups feed by rounding up fish in shallow water. Often seen soaring or, during migration, flying in long lines.
SS: When soaring in the distance, can be con-

fused with Wood Stork which flies with neck outstretched, long legs trailing, and has black inner secondaries and tail.
SD: Breeds locally in coastal Tamps and, in winter, in cen Dgo (Knoder *et al.* 1980). F to C but nomadic transient and winter visitor (Sep–Apr) in coastal lowlands from Son to Isthmus, and from Tamps to N Yuc Pen; irregularly U to R winter visitor on Pacific coast of Baja and from Chis to Honduras (Thurber *et al.* 1987), in interior S to cen volcanic belt, and U to C in interior BCN. Irregularly U to F through the summer in winter range.
RA: Breeds W N.A. to N Mex; winters to S Mexico and N C.A.

BROWN PELICAN
Pelecanus occidentalis *   Not illustrated
Pelícano Café

ID: 44–54 in, W 75–100 in (112–137 cm, W 190–254 cm). W > E. Coastal. *Bill very long* (11–15 in, 28–38 cm). **Descr. Basic:** eyes pale blue, bill greyish, pouch, facial skin, and feet blackish to grey. *Head and*

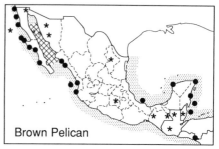

American White Pelican

Brown Pelican

neck white with yellow wash on crown. *Body and wings overall silvery grey, underparts darker, with blackish flight feathers*; underwings dark with broad whitish median stripe. **Alt:** bill and pouch become partly reddish. Nape and hindneck chestnut (darker in Pacific birds), base of foreneck pale yellow. **Juv/1st basic:** eyes brownish. *Dirty grey brown overall with whitish belly.* **2nd basic:** resembles adult but duller, less silvery. **2nd alt:** hindneck mottled chestnut. Adult plumage attained by 3rd prebasic molt.

Habs. Coastal, but may follow large rivers inland during winter storms, rare far offshore. Often in small groups, flying low over the waves; feeds by plunge-diving. Nests and roosts on cliffs and bushes.

SD: C to F resident along both coasts, but breeds only locally, with colonies mainly along Pacific coast of Baja, from Gulf of California to Jal, and along Caribbean coast of Yuc Pen. Irregular U to R visitor inland (Jun–Oct) in NE BCN, Son, and Sin (Anderson *et al.* 1977), and (mainly Dec–Mar) from Isthmus (SNGH, LJP) to Guatemala (Beavers *et al.* 1991; SNGH, SW). Vagrant (Jul 1990, after storms) to DF (per RGW), Coah, and NL (TLE).

RA: S USA to Panama and Venezuela; ranges generally in vicinity of breeding areas, rarely N to W Canada, S to N Brazil.

128

CORMORANTS: PHALACROCORACIDAE (4)

A cosmopolitan family of large aquatic birds with long necks and fairly long tails. Bills long and hooked, feet totipalmate. Feathers not fully waterproof, birds often perch with wings outstretched to dry. Ages differ, young altricial; sexes similar, slight seasonal variation. Molts protracted, partial 1st prebasic begins within a few months, first complete molt occurs toward end of first year. Eyes glassy green to blue-green (brownish in juvs), bills dark or (mainly juvs) dark above and pale below, bare facial and gular skin often brightly colored, feet black. Adults blackish overall, often with a green sheen, juvs dark brown. Flight typically fairly low over the water with necks outstretched. Rarely vocal, occasional grunts uttered when nesting.

Food mostly fish caught by diving, also crustaceans and amphibians. Cormorants breed colonially, often with other species, nests are bulky platforms of sticks and vegetation in trees, on cliffs or rocky islands. Eggs: 2–7 (usually 3–5), chalky bluish white to greenish white (WFVZ).

DOUBLE-CRESTED CORMORANT

Phalacrocorax auritus*   Not illustrated
Cormorán Bicrestado

ID: 28–32 in (71–81 cm). *Both coasts. Hind edge of gular skin rounded.* **Descr. Adult basic:** *facial and gular skin bright orange-yellow.* Blackish overall; head, neck, and underparts with green sheen, upperparts browner, scaled black. **Alt:** brighter green gloss, white supercilium crests present up to nesting. **Juv/1st basic:** dark brown overall, paler below, *face, foreneck, and chest whitish.* **1st alt:** may have white filoplumes on head briefly before molt into **2nd basic:** resembles basic adult but duller, with head, neck, and underparts mottled brown. **2nd alt:** attains partial white crests. Attains adult plumage by 3rd prebasic molt.

Habs. Estuaries, coastal bays and lagoons, also large lakes and rivers in N Mexico.

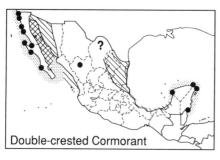
Double-crested Cormorant

On Pacific coast rarely in large flocks, but in Yuc Pen forms flocks up to several hundred birds. In flight, head held slightly above neck plane.

SS: Neotropic Cormorant smaller and relatively longer-tailed, posterior edge of gular skin pointed and often bordered whitish. Brandt's and smaller Pelagic cormorants (both Baja) strictly coastal and pelagic; both have dark faces and gular areas, fly with head and neck held straight, Brandt's relatively shorter-tailed; imms of both species more uniform.

SD: F to C resident along both coasts of Baja S to 24°N, and locally on coast of N Son, R in winter (Oct–Apr) S to Nay. U and local resident (2000 m) in Dgo (Williams 1977). C to F resident from NW Yuc (AC photo) to N Belize. F to C winter visitor (Oct–Apr) along Gulf coast of Tamps and Camp, U to R to Ver and Tab, locally inland in N Mexico to Son, NL, and SLP. Non-breeders U to F in summer (May–Sep) in winter range in E Mexico. Reports from Islas Revillagigedo and Gro (AOU 1983) require verification.

RA: Breeds N.A. to N Mex and Belize; winters to N Mexico and Caribbean.

NB: Great Cormorant *Phalacrocorax carbo* (34–38 in; 86.5–96.5 cm) is increasing in N.A., could occur in N Yuc Pen in midwinter. It is larger and more heavily built than Double-crested, adult has broad, whitish-feathered border to yellowish gular skin, 1st basic dark brown overall with whitish throat, brownish neck and chest, and white belly.

NEOTROPIC (OLIVACEOUS) CORMORANT

Phalacrocorax brasilianus   Not illustrated
Cormorán Neotropical

ID: 25–27 in (63.5–68.5 cm). Widespread. *Tail relatively long, posterior edge of gular skin pointed.* **Descr. Adult basic:** *facial and gular skin brownish yellow to flesh-orange.* Blackish overall with little or no green sheen, upperparts browner, scaled black, indistinct narrow *whitish border to gular skin.* **Alt:** green sheen to head, neck, and underparts; bright white gular border and variable white filoplumes on head and neck sides. **Juv/1st basic:** dark sooty brown overall but face, foreneck, and underparts soon become whitish to pale brown. Subsequent molts and plumages undescribed (?); may attain adult plumage by 2nd prebasic molt.
   Habs. Lakes, marshes, coastal lagoons, mangroves, small roadside ponds, also regularly on sea, close to shore. Often in large concentrations, perches readily on fence posts, in trees, etc. Flies with head slightly higher than neck plane.
SS: See Double-crested Cormorant. Brandt's and Pelagic cormorants rarely overlap in range; note face and throat patterns. Anhinga longer and more slender with narrow, dagger-like bill, scapulars and upperwing coverts patterned white, flies with long neck outstretched, often soars.
SD: C to F resident (SL–600 m) on both slopes from S Son, NL, and Tamps to El Salvador and Honduras, locally U to F in interior (to 2000 m) of N and cen Mexico (breeds Dgo and Jal, Williams 1977), and in Bahia Magdalena, BCS (RL photo). U

to C visitor (commoner and more widespread in winter) in interior from NE Mexico to Honduras, and in S BCS (SNGH, PP).
RA: SW USA, Mexico, and Caribbean to S S.A.
NB: Formerly *P. olivaceus.*

BRANDT'S CORMORANT

Phalacrocorax penicillatus   Not illustrated
Cormorán de Brandt

ID: 29–33 in (73.5–83.5 cm). A large, pelagic, *relatively short-tailed* cormorant of NW Mexico. **Descr. Adult basic:** *facial and gular skin bluish.* Dull blackish overall, upperparts slightly browner, scaled black; *pale buff band borders gular skin.* **Alt:** brighter with dark green sheen overall, white filoplumes on sides of neck and on back. **Juv/1st basic:** dark brown overall, slightly paler below, especially a *broad paler V across chest.* **1st alt:** may have white filoplumes on neck briefly before molt into **2nd basic:** resembles adult but less glossy, with head, neck, and underparts mottled brown. **2nd alt:** white filoplumes on neck; attains adult plumage by 3rd prebasic molt
   Habs. Sea coasts, large bays. Where common, often in large flocks. Flies with head held in same plane as outstretched neck.
SS: See Double-crested Cormorant. Pelagic Cormorant smaller with dark face, smaller and more slender bill, longer tail, alt has large white flank patches, juv more uniform.
SD: C to F resident along Pacific coast of Baja S to about 24° N, U to F in Gulf of California S to 28° N; in winter (Oct–Mar) U to R to tip of Baja and to N Nay (LACM spec.).
RA: W N.A. to NW Mexico.

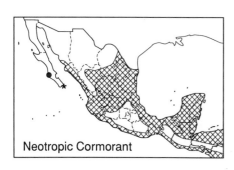

Neotropic Cormorant

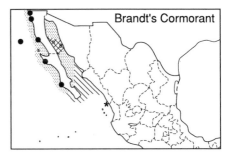

Brandt's Cormorant

PELAGIC CORMORANT

Phalacrocorax pelagicus resplendens
Cormorán Pelágico   Not illustrated

ID: 25–28 in (63.5–71 cm). *NW Baja.*
Descr. Adult basic: *facial and gular skin
dark reddish.* Black overall with dull green
sheen, no crest. **Alt:** strong green and purple
sheens, *conspicuous white flank patch,* and
short bushy crests on forehead and nape.
Juv/1st basic: dark brown overall, lacks
any obvious paler contrast. **1st alt:** may
have white filoplumes on head briefly before 2nd prebasic molt when attains adult
plumage.
  Habs. Rocky sea coasts and islands. Singly
  or in groups up to 10 birds. Flies with
  head in same plane as thin neck.
SS: See other cormorants.

SD: U resident along Pacific coast of NW
  BCN, breeds on Islas Los Coronados (5–
  10 pairs), possibly also on mainland; more
  widespread in winter when U to 27°N, R
  to tip of Baja (Feb 1988, SNGH).
RA: Bering Sea and N Pacific S to N Japan
  and NW Mexico.

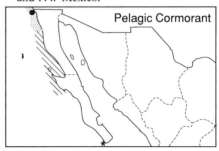
Pelagic Cormorant

ANHINGAS: ANHINGIDAE (1)

Anhingas are aquatic, long-necked, and long-tailed cormorant-like birds of pantropical distribution. Bills long, narrow, and pointed, with serrated tomia; feet totipalmate. Feathers not fully waterproof, birds often perch with wings outstretched to dry. Ages/sexes differ, young altricial. First prebasic molt begins within a few weeks of fledging, subsequent molts may be similar to those of cormorants (?). Plumage blackish overall, marked with white and browns. Mostly silent but may utter clicking chatters.

Food fish, amphibians, and aquatic reptiles caught by diving. Nest colonially, often with other waterbirds; nest is a stick platform. Eggs: 3–5, chalky greenish white (WFVZ).

ANHINGA

Anhinga anhinga leucogaster Not illustrated
Anhinga Americana

ID: 33–37 in, W 43–47 in (84–94 cm, W 109–119 cm). Fairly common in lowlands. **Descr. Basic ♂:** eyes red, bill yellow to yellow-orange with dusky culmen and tip, bare facial skin bluish, feet yellow-olive. Black overall with green gloss. Elongated scapulars and lesser and median upper-wing coverts white, edged black, *white greater upperwing coverts form bold wing-bar.* **Alt:** has white filoplumes on neck. **♀:** *head, neck, and upper chest pale greyish buff.* **Juv/1st basic:** eyes duller, bill dark. Resembles ♀ but browner, with reduced white markings, head and neck cinnamon-buff to greyish buff. Resembles adult of respective sex by 2nd basic but ♂ still has less white on upperparts, some brown mottling on neck. Attains adult plumage by 3rd prebasic molt.
Habs. Lakes, marshes, slow-moving rivers, mangroves. Swims low in water, often with only head and neck above the surface. Often perches with wings outstretched. Flies with neck extended, deep wingbeats interspersed with glides; often soars, wings held flattish. Migrates in flocks, soaring on thermals, at times mixing loosely with hawks.
SS: See Neotropic Cormorant. Flocks may suggest hawks at a distance.
SD: F to C resident (SL–900 m), breeding locally, on both slopes from S Sin and Tamps to El Salvador and Honduras; U in Yuc Pen and Belize. In E Mexico, commoner in winter (Sep–Apr) when N migrants present.
RA: Breeds SE USA and Mexico to Ecuador and N Argentina; northernmost breeders withdraw S in winter.

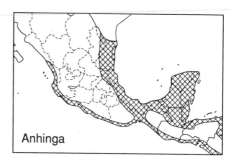

Anhinga

FRIGATEBIRDS: FREGATIDAE (2)

Large and spectacular, highly aerial seabirds of pantropical distribution, frigate-birds have very long and narrow wings, and long, deeply forked tails. Bills long and hooked, feet totipalmate. ♂♂ have highly distensible bright red gular sacs used in display; usually the sacs are deflated and appear as a narrow reddish throat patch. Frigatebirds have the lightest body weight (2–3 lb; 1–1.3 kg) in relation to wing area of any living bird. Ages/sexes differ, young altricial; molt sequences undescribed but juvenal or juvenal-like plumage retained for about 2 years; probably first breed at 6–7 years of age. Imm plumages complex and specific identification often problematic. Plumage overall black in ♂♂, marked with white and greys in ♀♀ and imms. Frigates often soar very high, almost effortlessly, for long periods and do not settle on water. Mostly silent at sea, but bill-clacking noises and quacking, chippering, and wheezy screams given on nesting grounds.

Food fish, jellyfish, and crustaceans, picked from at or near sea surface; also chase and pirate other seabirds such as terns, gulls, and boobies, forcing them to disgorge their food, and earning frigates the name 'man o' war birds'. Nest colonially, at times with other species; nests are stick platforms in bushes or low trees, rarely on grass tussocks. Successful breeders nest only every two years due to slow growth of young (which first fly at about 6 months of age) and prolonged post-fledging dependency (6–10 months). Eggs: 1–2, white.

MAGNIFICENT FRIGATEBIRD
Fregata magnificens　　　　　　Figure 14
Fragata Magnífica

ID: 35–42 in; W 82–94 in (89–107 cm; W 210–240 cm). Common and widespread.
Descr. ♂: orbital ring and *feet blackish*, bill greyish. *Plumage wholly lustrous black*, with purple and green sheens to upper-parts; paler inner secondaries form brown-ish wedge on trailing edge of inner wing.
♀: *orbital ring blue*, feet pink. *Black over-all with broad white chest band in the form of an inverted U*, pale brownish alar bar.
Juv: orbital ring and *feet blue-grey*, bill blue-grey with flesh tip. Blackish overall with white head, chest, and median belly; *note pointed black chest side wedges* (rarely meeting across center) helping enclose *white diamond-shaped belly patch*; alar bars pale brownish. 1st stage imm: head and under-body mostly or all white, *may show dusky collar*. Head and belly become marked with black as molts to 2nd stage imm: head and body variably marked with black. Axillars show wavy, whitish scal-lops. Feet may be pink in ♀, gular area pinkish in ♂. Subadult: *much like adult* (lacks pale alar bar) but ♂ *has underparts flecked whitish*, ♀ has white-streaked throat.

Habs. Pelagic and coastal. Roosts on trees, cliffs, rigging of ships, etc. Flight easy and buoyant with deep wingbeats, effortlessly soars high on bowed and crooked wings, tail usually held closed. See family account.

SS: Great Frigatebird (offshore Pacific) subtly stockier with somewhat barrel-chested appearance. Plumages similar to Magni-ficent but adult ♂ has pale brownish alar bar, axillars scalloped pale greyish, note pink feet; adult ♀ has grey throat, red orbi-tal ring, broader black belly patch. Juv Greats have cinnamon head and chest, and large oval white belly patch; older imms much like Magnificent but chest usually appears clouded or stained cinnamon, belly becomes mostly black before head becomes marked with black; some birds lack pale axillar scallops; note colors of bare parts on older imms.

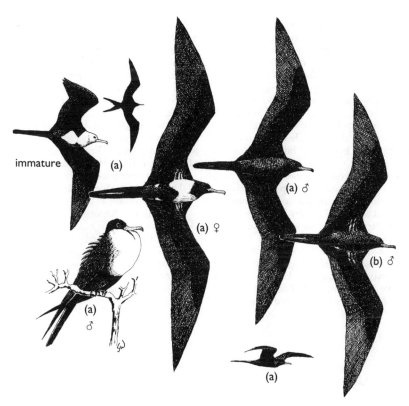

immature (a)

(a) ♂

(a) ♀

(b) ♂

(a) ♂

(a)

Fig. 14 (a) Magnificent Frigatebird; (b) Great Frigatebird

SD: C resident, but breeds only locally, along both coasts from BCS, Gulf of California, and Ver, to El Salvador and Honduras; also Islas Revillagigedo where breeds sympatrically with Great Frigatebird on San Benedicto (Howell and Webb 1990*a*). R to U visitor to Pacific coast of BCN, U to F to N Tamps. Rare inland in S Mexico and N C.A., usually after storms; birds probably cross between oceans via Isthmus.

RA: Breeds in Pacific from NW Mexico to Ecuador, in Atlantic from SE USA to S Brazil and in Cape Verde Islands; ranges at sea in vicinity of breeding grounds.

GREAT FRIGATEBIRD
Fregata minor ridgwayi   Figure 14
Fragata Pelágica

ID: 32–39 in, W 78–87 in (81.5–99 cm, W 198–220 cm). Offshore Pacific. **Descr. ♂:** orbital ring black, bill greyish, *feet pinkish*

red. Lustrous black overall with purple and green sheens to upperparts; paler inner secondaries form brownish wedge on trailing edge of inner wing. *Note pale brownish alar bar, axillars scalloped pale greyish.* ♀: *orbital ring red,* bill pale blue-grey to pale flesh, feet pink. Blackish overall with *grey throat,* broad, white U-shaped chest band, and pale brownish to whitish alar bar; chest usually clouded pale cinnamon. **Juv**: orbital ring blue-grey, bill blue-grey to flesh-grey with flesh tip, feet flesh. *Head and median chest cinnamon, blunt-tipped, black chest side patches* help enclose large, oval white belly patch; axillars may be blotched white; alar bars bright whitish. Head often bleaches to whitish but throat and chest retain cinnamon wash. **1st stage imm**: head and underbody mostly or all white, *cinnamon clouding on throat and chest* may be hard to see. Alar bars whitish to pale brownish. *Belly becomes marked with black* as molts to **2nd stage imm**: *belly becomes mostly black before black appears on head,* ♂ attains ♀-like plumage but chest flecked black. Axillars may show wavy whitish scallops. Gular area may be pinkish in ♂. **Subadult**: *much like adult but* ♂ *has U-shaped chest band,* flecked whitish; ♀ may have stronger cinnamon mottling on chest than adult.

Habs. Pelagic and highly aerial. Habits much as Magnificent Frigatebird.

SS: See Magnificent Frigatebird.

SD: C to F resident in Islas Revillagigedo, breeding confirmed only on San Benedicto (Howell and Webb 1990*a*); C non-breeding resident (500–1000 birds) on Clipperton. Reports from mainland Mexican coast (for example, Knoder *et al.* 1980) are not credible.

RA: Breeds widely in tropical Pacific and Indian oceans, also on islands in W Atlantic off Brazil; ranges at sea in vicinity of breeding grounds.

HERONS: ARDEIDAE (16)

A cosmopolitan family of typically long-necked and long-legged birds found near water, where they hunt for fish with long, dagger-shaped bills. Feet unwebbed. Wings broad and rounded, short tails concealed at rest by closed wings. Elongated neck feathers often lay over the chest when the neck is retracted. Ages similar or different, sexes usually similar. Several species dimorphic. Plumage varies from cryptically patterned browns, blacks, and buffs (bitterns) to entirely white (several egrets). Many species attain greatly elongated plumes on their head, back, and lower foreneck for courtship displays; these plumes usually are dropped during nesting. Colors of bare parts often intensify or even change completely for a short period at the onset of breeding [alternate colors indicated in square brackets], but usually fade soon after eggs are laid. Young altricial and downy; juv plumage usually worn for a few months and replaced by similar 1st basic. Molts, especially of imms, are protracted, with distinct alt and basic plumages difficult to recognize; alt often best indicated by the presence of head plumes. Herons fly with their necks retracted in a deep neck bulge, legs projecting beyond the tail. Voices mostly harsh grunting and deep croaking calls.

Food varied; many species mostly eat fish, but diet of the group as a whole includes amphibians, reptiles, crustaceans, insects, small mammals, and nestling birds. Many herons are social, and breed and roost colonially (except bitterns and tiger-herons), often with other waterbirds; this is assumed unless stated otherwise in the species accounts. Nests are platforms of vegetation, often sticks; eggs unmarked, pale green to pale blue, unless stated otherwise (WFVZ).

GENUS BOTAURUS: Bitterns

Bitterns are large, brown, cryptically patterned herons that usually skulk in dense marsh vegetation, especially reeds. Eyes yellow, bills stout and long, yellowish with dusky culmen and tip, lores yellow, legs yellowish to yellow-olive. Nests built up out of water or lodged in vegetation. Eggs: 2–5, brownish to olive-buff.

PINNATED BITTERN

Botaurus pinnatus caribaeus   Plate 1
Avetoro Neotropical

ID: 25–30 in (63.5–76 cm). *Neotropical.*
  Descr. Ages/sexes similar. *Buff crown and neck narrowly barred blackish brown;* auriculars buffy brown with narrow pale supercilium. *Upperparts rich buff, heavily striped and barred blackish producing a bold pattern; tail blackish brown.* Throat white, foreneck and underparts creamy to pale buff, foreneck and sides of chest streaked brown. White tufts on sides of chest usually concealed but conspicuously displayed in threat (and courtship?). In flight, blackish-brown remiges and buff-tipped primary coverts contrast with brown innerwings.
  Voice. When flushed, a gruff *owhk* or *owk owk owk*, etc. ♂ booms in breeding season, often at dusk or night, a deep, booming *poonk* or *poonkoo* (Stiles and Skutch 1989).

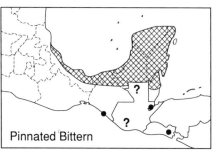

Pinnated Bittern

Habs. Freshwater habitats, from small roadside ditches with rushes to extensive marshes with tall reeds. Usually singly but small loose groups gather at favored feeding areas; hunts mainly by slow stalking. Often in fairly open situations; when alarmed may retract neck and crouch down in shorter vegetation, or freeze with neck up-stretched in taller reeds. Flight heavy with wingbeats mainly above body plane, more reminiscent of a tiger-heron than of American Bittern.

SS: Smaller American Bittern has shorter bill, upperparts more uniform (tail not contrastingly dark), underparts more heavily striped brown, neck lacks barring, black lateral neck stripes often conspicuous. Immature tiger-herons have barred upperparts, especially bold across closed wings, and are often arboreal whereas bitterns are almost strictly terrestrial. Immature nightherons are smaller, grey-brown overall, spotted and streaked whitish and buff above.

SD: F to C but local resident (around SL) on Atlantic Slope from Ver locally to Belize; F (?) but local on Pacific Slope in Chis (MAT spec.) and El Salvador. One record for Guatemala (Izabal, May 1992, DJF) where probably also a local resident, although records in Beavers (1992) require verification; unrecorded in Honduras.

RA: SE Mexico locally to N Argentina.

AMERICAN BITTERN
Botaurus lentiginosus   Plate 1
Avetoro Americano

ID: 20–24 in (51–61 cm). *Mainly a winter visitor in the region.* **Descr.** Ages differ slightly, sexes similar. **Adult:** *dark brown crown and nape contrast slightly with tawny-brown auriculars and narrow whitish supercilium; neck tawny brown with black malar extending down sides of*

upper neck (can be hard to see). Upperparts brown with narrow buff edgings and vermiculations, appearing fairly uniform, wing coverts often slightly paler. Throat and foreneck whitish, throat with narrow median stripe of brown streaking, *foreneck heavily striped tawny brown;* buff underparts coarsely streaked brown. In flight, blackish brown remiges and ochre-tipped primary coverts contrast with brown innerwings. White tufts on sides of chest usually concealed. **Juv:** lacks black malar stripes (appear by 1st prebasic molt, Aug–Nov).

Voice. When flushed may give a hoarse, gruff *wok* and *woh-wok*, etc. Boom (N.A.) is a deep, plunging *glun'k-lung*, repeated steadily.

Habs. Freshwater marshes with large reedbeds; in winter also salt marshes. Habits similar to Pinnated Bittern but often more skulking and harder to see; deeper wingbeats more reminiscent of a night-heron.

SS: See Pinnated Bittern. Immature nightherons grey-brown overall, spotted and streaked whitish and buff above.

SD: U to F winter visitor (Oct–Apr; SL–2500 m) S to cen Mexico, U to R in S Mexico, R and irregular to El Salvador and Honduras. Resident (at least formerly) in cen volcanic belt.

RA: Breeds N.A., locally in cen Mexico; winters S to Mexico, rarely to West Indies and S to Panama.

LEAST BITTERN
*Ixobrychus exilis**   Not illustrated
Avetorito Americano

ID: 11–12 in (28–30.5 cm). *Small and skulking.* **Descr.** Ages/sexes differ. ♂: eyes pale lemon, bill orange with dusky culmen,

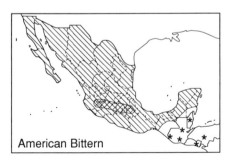
American Bittern

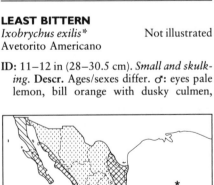
Least Bittern

lores yellow-green [scarlet], legs yellow [orange]. *Black crown contrasts with ochraceous-buff auriculars* and rufous hindneck, sides of neck ochraceous buff. *Upperparts black with narrow white braces; ochraceous buff lesser and median upperwing coverts form conspicuous pale panel in flight*; greater upperwing coverts and tertials rufous, primary coverts blackish, tipped rufous. Flight feathers blackish. Throat whitish, foreneck and underparts pale buff, black tufts on sides of chest usually concealed, belly and undertail coverts whitish. ♀: *black replaced by dark brown*, sides of neck and upperwing panel greyish buff, braces buff, *paler underparts streaked brown*. Juv: resembles ♀ but bill dusky pink to yellowish with dark tip, upperparts broadly edged pale buff.

Voice. Call may be given by flushed birds: a sharp, slightly explosive, at times accelerating *kak-kak-kak-kak-kak-kak* or *ka-ka-ka-ka…*, reminiscent of Clapper Rail, lacks rhythmic or pumping quality of King Rail. 'Song' is an accelerating then slowing series of 6–8 low, moaning clucks, *cowh-cowh…* or *puhk-puhk…*

Habs. Mainly freshwater but also brackish marshes. Usually skulks in tall vegetation, feeds by careful stalking, often through reeds in deep water. When flushed, flies up weakly with legs dangling. At times loosely colonial: nest usually lodged in vegetation over water. Eggs (N.A.): 2–5, bluish white, often with sparse faint brown flecks (WFVZ).

SS: Larger and stockier Green Heron darker overall, lacks pale buff wing patches.

SD: F to U but local resident (SL–2500 m) in Baja, coastal Son, and from cen and E Mexico locally to Honduras and El Salvador. F transient and winter visitor (Sep–Apr; SL–2500 m) throughout most of region; status unclear in many areas due to skulking habits.

RA: Breeds from N.A. locally to N Argentina. N migrants winter S to Panama and Colombia.

BARE-THROATED TIGER-HERON

*Tigrisoma mexicanum**   Plate 1
Garza-tigre Gorjinuda

ID: 28–32 in (71–81.5 cm). *Throat naked, bill long and stout, culmen slightly decurved*, legs fairly short. **Descr.** Ages differ, sexes similar. **Adult:** eyes yellow, bill

blackish with paler tomia, lores greenish yellow, *throat and upper foreneck naked greenish yellow to yellow-orange*, legs dark greyish olive. *Black crown contrasts with grey face*, neck vermiculated dark brown and buff; foreneck with narrow dark median stripe bordered white and black. Dark olive-brown upperparts vermiculated buff. Underparts cinnamon with grey thighs, axillars, and undertail coverts. Flight feathers blackish, tail often vermiculated whitish. **Juv/1st basic:** *head, neck, and chest barred cinnamon-buff and dark brown, throat naked yellow*, some birds with trace of white-edged dark foreneck stripe. *Dark brown upperparts broadly barred cinnamon, boldest on greater upperwing coverts and tertials*, uppertail coverts blackish, tipped white. Underparts pale buff with sparse to heavy, broad dark brown barring, axillars barred black and white. Remiges dark grey with wide-spaced narrow whitish barring, tail blackish with 4–5 narrow white bars. **2nd basic:** *darker overall. Dark brown crown narrowly barred tawny-buff, neck narrowly barred dark brown and buff with white-edged dark median stripe. Dark brown upperparts vermiculated buff.* Underparts pale buff with variable dusky barring. **3rd basic:** resembles adult but crown and auriculars with some buff and dark brown barring, underparts duller cinnamon, sparsely barred buff. 4th basic may retain dusky bars on auriculars and underparts.

Voice. When flushed may give a deep, throaty *woh woh woh woh* or *wok woh-woh…*, etc. In breeding season, especially at dusk or night, gives far-carrying, deep, hoarse grunts or roars, at times paired, *ohrrr ohrrr…* or *rrohr rrohr…*, etc., 10/10–15 s.

Habs. Varied freshwater habitats and mangroves, typically with wooded edges or trees nearby but rarely inside heavy forest. Usually solitary but several may gather in small areas, often in fairly open situations at marsh edges. If disturbed, takes flight heavily and usually perches in trees. Hunts by waiting or slow wading. Partly nocturnal. Nests in trees. Eggs: 2–3, greenish white, rarely flecked brown.

SS: See Rufescent Tiger-Heron (NB below) and *Botaurus* bitterns.

SD: F to C resident (SL–1000 m) on both slopes from S Son and S Tamps to El Salvador and Honduras.

RA: Mexico to Panama and NW Colombia.
NB: **Rufescent Tiger-Heron** *Tigrisoma lineatum* (Plate 1) (25–29 in; 63.5–73.5 cm) may occur from lower Río Dulce, NE Guatemala, S (see Appendix B). Illustrated to faciliate confirmation (?) of its presence in region. A retiring heron of swampy lowland forest; shape and habits much like Bare-throated Tiger-Heron but bill relatively stouter and shorter. **Adult:** eyes yellow to amber, bill greyish above, dusky horn below with dark tip, *bare yellow skin at gape borders feathered white throat*, legs yellow-olive. *Head and neck rufous*; blackish-brown upperparts vermiculated rufous to tawny-buff. Throat and foreneck white with dark median foreneck stripe; underparts dusky pale cinnamon. **Juv/1st basic:** *note greyish-flesh bill with dark culmen, yellowish base below, yellowish to yellow-olive legs. Rufous crown and nape marked with blackish chevrons; auriculars and neck buffy-cinnamon with dark chevrons on neck. Dark brown upperparts boldly and heavily spotted cinnamon and pale buff, becoming distinct broad bars only on greater coverts and tertials*; uppertail coverts blackish, tipped white. *Throat white*, foreneck often has trace of dark median stripe. Pale buff underparts marked with short dark brown bars. Blackish remiges have narrow, whitish to buff bars, black tail has 3–4 narrow white bars. **2nd basic:** *darker overall*. Head and neck barred brownish buff to cinnamon and dark brown, *blackish upperparts have narrower brownish buff to cinnamon barring*. Foreneck resembles adult, buff underparts usually have dark bars; remiges may be all dark. **3rd basic:** resembles adult but sides of head and neck narrowly barred blackish, buffier underparts may have dusky bars. May show some dark bars at base of neck in 4th year.

Voice. A harsh, groaning, far-carrying, long-drawn-out *kwooohh* or *kwawwh*, mostly at night (Stiles and Skutch 1989).

GREAT BLUE HERON
Ardea herodias *   Not illustrated
Garzón Cenizo

ID: 40–50 in (101.5–127 cm). Widespread, dimorphic. **Descr.** Ages differ, sexes similar. **Dark morph. Adult:** eyes yellow, *bill yellow [orange] with dusky culmen*, lores greyish [bright lime green], *legs dusky flesh to brownish yellow. Head white with elongated black supercilium*, neck dusky vinaceous with white throat and black-streaked foreneck. *Upperparts grey*, flight feathers darker. *Black sides often appear as black 'shoulders'*; median underparts whitish with heavy black streaks, flanks grey, thighs cinnamon, undertail coverts whitish. **Alt:** attains elongated slaty black occipital plumes, and silvery white lower foreneck and back plumes. **Juv/1st basic:** eyes duller, bill blackish above. Crown slaty grey, neck duskier, underparts often buffier, lacks black sides of chest; underparts broadly streaked whitish and dusky, thighs paler cinnamon. **1st alt:** may have short occipital plumes. **2nd basic:** crown mixed white and grey, occipital plumes short, sides of chest mixed black and white. Adult plumage attained by 3rd prebasic molt. **White morph:** legs paler and yellower than dark morph. *Plumage entirely white*, alt adult has elongated white occipital, lower neck, and back plumes. Intergrades resemble dark morph but have white head and neck, black supercilium reduced or lacking.

Voice. A deep, throaty *rrok* or *rroh*.

Habs. Wide variety of freshwater and saltwater habitats, most commonly at lakes and coastal lagoons (white morph exclusively coastal). Singly or in small loose

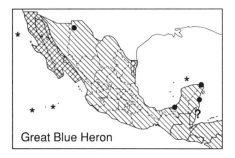

Bare-throated Tiger-Heron   Great Blue Heron

groups. Hunts mainly by standing and waiting, or stealthy stalking. Nest usually in trees, also reedbeds and cliffs; 3–5 eggs.

SS: Sandhill Crane (winter visitor to N Mexico) grey overall; note red crown patch, dark bill and legs, flies with neck outstretched. White morph larger than other white herons; told from Great Egret by pale legs, less uniformly yellow bill.

SD: Dark morph: F but local breeder (SL–2000 m) in NW Mexico S to Sin, and on Atlantic Slope from Tamps to Yuc Pen; C to F transient and winter visitor (Sep–Apr; to 2500 m) throughout region; nonbreeders oversummer widely. White morph: F to U resident (breeds locally), in coastal N and E Yuc Pen.

RA: Breeds N.A. S to Mexico, disjunctly in Galapagos; white morph breeds throughout Caribbean to N Venezuela. N populations migratory.

GREAT EGRET

Egretta alba egretta   Not illustrated
Garza Grande

ID: 33–39 in (84–99 cm). Common and widespread. **Descr.** Ages differ slightly, sexes similar. Eyes yellow, *bill bright yellow* [yellow-orange], lores yellow-green [lime-green], *legs black*. **Adult:** *entirely white; has elongated back and lower foreneck plumes in alt.* **Juv:** yellow bill tipped dusky, eyes paler.
  Voice. A low, guttural, rasping or creaky, drawn-out *ahrr-rr* or *owhh-uh*, and *owhh owhh*, etc.
  Habs. Wide variety of freshwater and brackish habitats. Singly or in loose flocks. Hunts by standing and waiting, or slow stalking. Nests in trees and bushes; 2–5 eggs.

SS: See white morph Great Blue Heron; other white herons smaller; Reddish Egret (coas-

tal) closest in size but note bicolored or dark bill, often feeds actively.

SD: F to C but local breeder (SL–2000 m), mainly in coastal lowlands, from Son and Tamps to El Salvador and Honduras. C to F transient and winter visitor (Aug–Apr; to 2500 m) throughout region; nonbreeders oversummer widely.

RA: Breeds N.A. to S S.A., also much of Old World; N New World populations withdraw S in winter.

NB1: Traditionally placed in the genus *Casmerodius*; may form a link between the genera *Egretta* and *Ardea*, thus its generic placement is equivocal. **NB2:** Known as Great White Egret in the Old World.

SNOWY EGRET

Egretta thula*   Not illustrated
Garza Nivea

ID: 19–23 in (48.5–58.5 cm). Common and widespread. **Descr.** Ages differ slightly, sexes similar. **Adult:** eyes yellow, *bill black, lores bright yellow* [orange], *legs black, feet (and often hind tarsi) yellow* [orange]. *Plumage entirely white.* Attains elongated occipital, back, and lower foreneck plumes in alt. **Juv:** bill often yellow, tipped black, to 5 weeks of age; hind tarsi extensively yellow, eyes paler.
  Voice. A low, gruff *eh-ahrr* or *rah-ahrr*, etc.
  Habs. Often at beaches and river mouths as well as freshwater and brackish marshes, lakes, etc. Singly or in loose groups. Often hunts with much dashing and chasing. Nests usually low in trees and bushes; 2–4 eggs.

SS: Imm Little Blue Heron has bicolored bill, legs and feet greenish yellow, outer primaries tipped dusky; feeds less actively. Larger white morph Reddish Egret has heavier bill often with pink base, dark legs and feet.

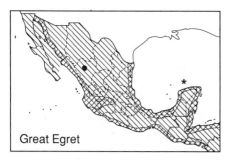

Great Egret

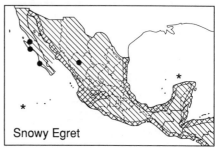

Snowy Egret

Cattle Egret stockier with stout yellow bill, legs and feet dark most of year.

SD: F to C but local breeder (SL–2000 m), mainly in coastal lowlands, from Baja, Son, and Tamps to El Salvador and Honduras. C to F transient and winter visitor (Aug–Apr; to 2500 m) throughout region; non-breeders oversummer widely.

RA: Breeds N.A. to S S.A.; N.A. populations mostly withdraw S in winter.

NB: Western Reef-Heron *Egretta gularis* (20–24 in, 51–61 cm) is expanding in Old World and has been recorded in E N.A. and West Indies in recent years; should be looked for on Atlantic coast (see Appendix B). Bill heavier than Snowy Egret and often appears slightly decurved. Bill brownish horn or with some yellow at base below, lores pale greenish [pinkish], legs greenish black, feet (and often hind tarsi) yellowish [orange]; imm often has feet and hind tarsi duller, olive-yellow. Dark morph overall dark slaty blue-grey with contrasting white throat and upper foreneck; may show some white areas in wings, usually only a few primary coverts. White morph (very unlikely to occur) entirely white. Alt adults have elongated occipital, back, and lower neck plumes.

LITTLE BLUE HERON
Egretta caerulea   Not illustrated
Garza Azul

ID: 20–24 in (51–61 cm). Common and widespread. Descr. Ages differ, sexes similar. Adult: eyes yellow, *bill blue-grey with distal 30–50% black, lores dull greenish* [blue], *legs and feet dark greenish grey to greenish yellow* [black]. Dark slaty blue-grey overall, head and neck darker, more purplish blue, becoming brighter reddish purple in alt when also has elongated crown, back, and lower foreneck plumes.

Juv/1st basic: *legs and feet greenish yellow*. Plumage *entirely white except for dusky tips to outer 6–8 primaries*; may have greyish or rusty wash on crown and back. Attains adult plumage in 1st summer when birds with mixed slaty and white plumage are often seen.

Voice. A harsh, gruff, complaining, drawn-out *rraa-aahh* or *grah-ah-h*.

Habs. Wide variety of freshwater and brackish habitats; often singly and in areas where other herons are rare or absent, such as isolated pools and small drainage ditches. Often loosely gregarious. Typically hunts by standing and waiting, less often by active stalking. Often flies with neck outstretched, at least for short periods after take-off. Nests in bushes and trees; 2–6 eggs.

SS: See Snowy Egret. Dark morph Reddish Egret larger and paler overall, head and neck cinnamon, bill dark or with pink base; often feeds much more actively. Cattle Egret stockier with stout yellow bill.

SD: F but local breeder (SL–1700 m), mainly in coastal lowlands, from BCS and S Son to Isthmus, and from Tamps to N Ver, and NW Yuc Pen; possibly in C.A. F to C transient and winter visitor (Aug–Apr; to 2500 m) from cen Mexico S; non-breeders oversummer widely.

RA: Breeds N.A. to S Peru and Uruguay; N.A. populations mostly withdraw S in winter.

TRICOLORED HERON
Egretta tricolor   Not illustrated
Garza Tricolor

ID: 22–26 in (56–66 cm). Mainly coastal. Descr. Ages differ, sexes similar. Adult: eyes red, *bill yellowish with dark culmen and tip* [bluish with black tip], lores yellowish [violet], *legs and feet greenish yellow*

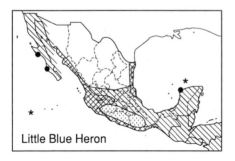
Little Blue Heron

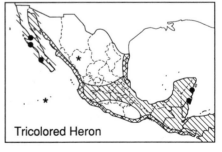
Tricolored Heron

[pinkish red]. Head, neck, and upperparts slaty blue-grey, throat and foreneck white with narrow median stripe of rufous flecking; white rump and uppertail coverts usually visible only in flight. Flight feathers slaty grey. Underparts, including underwing coverts, white. Has elongated white occipital plumes and creamy-buff back plumes in alt. **Juv/1st basic:** eyes yellowish, bill greyish, legs and feet greyish green. Crown grey, face and neck cinnamon-rufous; upperwing coverts edged rufous. Often has short occipital plumes in 1st summer. Attains adult plumage by 2nd prebasic molt.

Voice. A guttural, drawn-out aahhrr or aaahhr, etc.

Habs. Brackish and freshwater marshes. Singly or in loose groups. Hunts by active running and chasing or stealthy wading, often in water up to its belly. Nests low in bushes, at times on ground; 2–4 eggs.

SS: Contrasting white rump and belly distinctive.

SD: F to U but local breeder (SL–300 m) mostly near coast, from BCS and S Son to El Salvador, and from Tamps to Yuc Pen. F to C transient and winter visitor (Aug–Apr; SL–1500 m), on both slopes throughout; U to R and local (to 2500 m) in interior from cen Mexico S, and in BCN; non-breeders oversummer widely.

RA: Breeds W Mexico and SE USA to cen Peru and N Brazil; N.A. populations mostly withdraw S in winter.

REDDISH EGRET
Egretta rufescens*   Not illustrated
Garza Rojiza

ID: 26–30 in (66–76.5 cm). Coastal. Dimorphic. **Descr.** Ages differ, sexes similar. **Adult. Dark morph:** eyes whitish, bill pinkish [bright pink] with distal 30–50% black, lores pale pinkish [violet], legs and feet dark blue-grey [cobalt]. Slaty blue-grey overall with cinnamon head and neck; may show white patches in plumage. Alt has brighter head and neck, shaggy head and neck plumes, and elongated back plumes. **White morph:** entirely white; attains shaggy head and neck plumes, and elongated back plumes in alt. **Juv/1st basic:** bill and lores dark greyish. Dark morph grey overall, may show a pinkish-cinnamon hue on neck; white morph white. Often has shaggy neck plumes in 1st summer. Attains adult plumage by 2nd prebasic molt.

Voice. A low, moaning or complaining owah oahh-oahh or owh oahh, etc.

Habs. Coastal lagoons, beaches, estuaries; rarely away from salt water. Singly or in small groups. Often feeds actively with much wing-spreading and dashing around. Nests in bushes or on ground; 2–7 eggs.

SS: See Little Blue Heron, Great and Snowy Egrets.

SD: F to U but local breeder on Pacific coast of BCS, in Gulf of California S to Sin, and in Isthmus; on Atlantic coast S to N Ver (Loetscher 1941) and in Yuc Pen. U to R visitor (Jan–Dec, mostly Aug–Mar) along Pacific coast to BCN, El Salvador, and presumably Honduras, on Atlantic coast S to Tab. Vagrant to cen Oax (1500 m, Dec 1991, SNGH); other reports from interior (Friedmann et al. 1950) require confirmation.

RA: Breeds locally in W Mexico, SE USA, Caribbean, and E Mexico; non-breeding birds occur S to Costa Rica on Pacific coast and to N Venezuela in Atlantic region.

CATTLE EGRET
Bubulcus i. ibis   Not illustrated
Garza Ganadera

ID: 18–21 in (45.5–53 cm). Stocky with relatively short and stout bill; often away

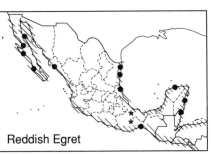

Reddish Egret

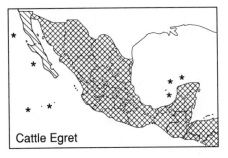

Cattle Egret

from water. **Descr.** Ages/sexes similar.
Eyes pale yellow [orange], *bill yellow*
[orange], lores yellowish, *legs and feet
dark olive-grey [yellow-orange].* Basic
plumage *entirely white. Attains elongated
buff plumes on crown, back, and lower
foreneck in alt;* faint buff wash often retained into winter.

Voice. Gruff clucks and grunts, a low *kuh*
or *kyuh,* and *kwoh-kwoh...,* etc.

Habs. Typically feeds around cattle or
other livestock, usually 1–2 birds per
animal; also follows ploughs, rarely in
marshes unless cattle present. Feeds by
walking, interspersed with a dash and
stab. Nests mainly in trees and bushes,
less often in reedbeds; 2–5 eggs.

SS: See Snowy Egret and imm Little Blue
Heron.

SD: F to C but local breeder (SL–2100 m)
throughout, from NE BCN and Tamps S.
C to F transient and winter visitor (Sep–
Apr) throughout; non-breeders oversummer
widely. The spread of this species throughout the region, and the Americas as a whole,
is remarkable: first noted in the New
World in Surinam in 1877, then spread
over the N half of S.A. This apparently was
followed by a jump N over the Caribbean
to the SE USA in the early 1940s and a
secondary wave moved N through M.A. in
the 1950s (Crosby 1972). The main expansion along both slopes and through the
interior of Mexico was in the 1960s, and
by the 1970s Cattle Egrets were well established throughout the region. The species'
range is probably still increasing and its
new-found niche alongside introduced
cattle is (unfortunately) not threatened.

RA: Cosmopolitan and increasing.

GREEN [GREEN-BACKED] HERON
Butorides virescens *   Not illustrated
Garza Verde

ID: 15–17 in (38–43 cm). A widespread,
small dark heron. **Descr.** Ages differ, sexes
similar. **Adult:** eyes yellow [orange], *bill
blackish with yellowish base below* [glossy
black], lores yellow [black], *legs yellowish
[reddish orange].* Crown and erectile crest
glossy blackish green, *face, neck, and chest
rufous to dark chestnut,* white stripe from
throat down foreneck is streaked dark
brown. *Upperparts blackish green,*
elongated scapular and mantle feathers
with pale grey cast, wing coverts narrowly

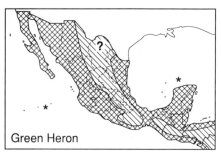

Green Heron

edged buff; flight feathers blackish. Underparts greyish. **Juv/1st basic:** bill yellowish
with dusky culmen and tip. Blackish crown
streaked rufous; *face, neck, and chest grey-
brown to rufous-brown, streaked buff;*
underparts whitish, streaked dark brown.
Wing coverts broadly edged tawny and
buff. Face and sides of neck more solidly
rufous in 1st summer. Attains adult plumage by 2nd prebasic molt.

Voice. A complaining, sharp bark, *kyow!*
or *kyah,* and *kah-kah-kah-kah-kah,* when
flushed. Also agitated clucking, *kweh-
kweh...* or *kuh-kuh...*

Habs. Marshy areas with wooded edges,
mangroves, small ponds; rarely in open
areas or large reedbeds. Usually solitary.
Hunts by standing and waiting, or slow
stalking. At times loosely colonial; nests
low in bushes and trees; 2–4 eggs.

SS: Unlikely to be confused with other species;
see Least Bittern.

SD: F to C but local breeder (SL–1500 m),
mainly in coastal lowlands, from Baja,
Son, and Tamps to El Salvador and Honduras. C to F transient and winter visitor
(Sep–May: SL–2500 m) throughout region.

RA: Breeds N.A. to cen Panama; N populations migratory, moving S in winter as far
as N S.A.

NB1: Often considered conspecific with
Butorides striatus of S.A. and much of Old
World. Critical field studies are needed in
areas of possible sympatry (cen Panama?).
NB2: *B. sundevalli* of the Galapagos
Islands may also be conspecific with the
virescens/striatus complex.

AGAMI (CHESTNUT-BELLIED) HERON
Agamia agami   Plate 1
Garza Agami

ID: 26–30 in (66–76 cm). *Uncommon and
retiring. Bill very long and slender, legs fairly*

short. **Descr.** Ages differ, sexes similar. **Adult:** eyes amber to red, bill black, yellowish below at base, lores yellowish, legs olive to yellow. Head and hindneck black with short silvery grey occipital crest. *Sides of neck chestnut, lower neck covered with a spray of silvery-grey plumes, throat and foreneck white with chestnut median stripe. Upperparts dark glossy green* with indistinct chestnut braces; remiges blackish. *Underparts deep chestnut,* thighs often slightly paler. Has elongated silvery-white occipital and back plumes in alt. **Juv/1st basic:** eyes pale greyish to amber-yellow, bill black above, greenish yellow below. Crown and nape black, slightly crested, *face and neck grey-brown*; median stripe slight or lacking on throat and upper foreneck. Upperparts dark grey-brown. *Underparts creamy to pale buff, coarsely streaked dark brown on chest.* **2nd basic:** resembles adult but silvery neck plumes less extensive, underparts mottled cinnamon and chestnut. Attains adult plumage by 3rd prebasic molt.

Voice. Usually silent but may give a low *guk* in alarm (Stiles and Skutch 1989).

Habs. Shady pools and streams inside humid forest, wooded lagoons, mangroves. Solitary and retiring, at least for most of the year; forms small breeding colonies (only in mangroves?) which may associate with other species. Hunts by waiting and spearing fish with its long bill. Nests low in trees; 2–4 eggs.

SS: Unlikely to be confused with other species, especially in its habitat.

SD: U to R resident on Atlantic Slope (SL–150 m) from S Ver to Honduras; also in mangroves on Pacific coast of Chis (and Guatemala?). A report from SE NL (Contreras 1988) is remarkable.

RA: SE Mexico to Ecuador and Amazonian Brazil.

BLACK-CROWNED NIGHT-HERON

Nycticorax nycticorax hoactli Plate 1
Garza-nocturna Coroninegra

ID: 22–25 in (56–63.5 cm). *A stocky and compact, mainly nocturnal heron. Bill stout and dagger-like, neck usually hunched; feet project slightly beyond tail in flight.* **Descr.** Ages differ, sexes similar. **Adult:** eyes red, bill black, lores yellow-olive [black], legs yellow [bright pink]. *Crown and nape black* with two white occipital plumes (longer in alt), white forehead and supraloral stripe; *face, neck, and underparts whitish. Back black* with green sheen, rump and tail grey. Wings pale grey above and below. **Juv/1st basic:** eyes yellow to amber, *bill greenish yellow with dark culmen and tip*, legs greenish yellow. Crown and nape blackish brown, streaked brown to whitish, face and neck streaked pale buff and dark brown; throat whitish. *Upperparts dark brown with cinnamon to buff streaks on mantle and fairly long, broad, pale buff to white wedge-shaped streaks on scapulars and upperwing coverts; greater secondary coverts broadly tipped white.* Flight feathers dark grey-brown. Whitish underparts heavily streaked grey-brown, vent and undertail coverts whitish, usually unmarked. In late winter begins protracted molt. **2nd basic:** may show short white occipital plumes. Forehead whitish to pale brown, crown blackish brown, face and neck grey, often with brownish cast; throat and foreneck white. Underparts pale grey. Upperparts slaty grey with green sheen, wing coverts grey. Flight feathers grey to grey-brown. **2nd alt:** often has longer occipital plumes. Adult plumage attained by 3rd prebasic molt when about 3 years old.

Voice. A hoarse or gruff, barking *wok!* or *owhk*, often given in flight.

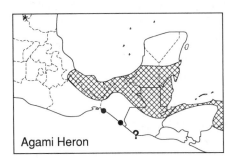

Agami Heron

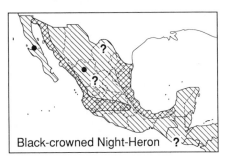

Black-crowned Night-Heron

Habs. Freshwater and brackish marshes. Mainly nocturnal but may be active during day; in loose flocks or solitary. Hunts by standing and waiting, and slow stalking. Nests in trees and bushes, rarely on ground; 3–5 eggs.

SS: Yellow-crowned Night-Heron less compact, typically stands taller, neck not hunched, note thicker bill; in flight, feet project noticeably beyond tail; juv/imm has black bill, finer pale spotting on upperparts. Boat-billed Heron superficially similar but note bill shape, flight shape even more compact, feet barely project beyond tail and, if seen before too dark, adults show black wing-linings. See bitterns.

SD: F to C but local breeder (SL–2200 m) in Baja, on Pacific Slope and in interior from Son and Dgo to cen volcanic belt, and on Atlantic Slope from Tamps to SW Camp; possibly S to N C.A. C to F transient and winter visitor (Sep–Apr; to 2500 m) almost throughout region; U to F in Yuc Pen; non-breeders oversummer widely.

RA: Virtually cosmopolitan (absent from Australasia); in New World breeds from S Canada to S S.A., N.A. populations mostly withdraw S in winter.

YELLOW-CROWNED NIGHT-HERON
*Nycticorax violaceus**   Plate 1
Garza-nocturna Coroniclara

ID: 20–23 in (51–58.5 cm). *Mainly nocturnal. Bill thick and heavy, neck often held outstretched; in flight, feet project noticeably beyond tail.* **Descr.** Ages differ, sexes similar. **Adult:** eyes amber [scarlet], *bill black (may be greenish yellow below at base),* lores greyish yellow [dark green], legs yellow [scarlet-orange]. *Head black with creamy to buff crown, broad white auricular stripe,* and 2–6 elongated, black and white occipital plumes; neck and underparts grey. Upperparts have spiky

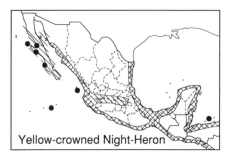

Yellow-crowned Night-Heron

pattern of dark brown feathers edged pale grey. Flight feathers blackish; *underwings grey.* Attains elongated back plumes in alt. **Juv/1st basic:** eyes orange-yellow to amber, legs greenish yellow. Crown and nape blackish, streaked buff, face and neck streaked pale buff and dark brown; throat whitish. *Upperparts dark brown, marked with short, cinnamon to buff, wedge-shaped streaks and spots, wing coverts narrowly edged whitish.* Flight feathers dark grey. Pale buff underparts heavily streaked dark brown, vent and undertail coverts unmarked whitish. **1st alt:** forehead and crown buffy brown with 2–4 short, white occipital plumes, broad postocular stripe and nape blackish, auriculars dull whitish. Upperparts more uniform through wear. **2nd basic:** resembles adult but neck and underparts washed brown, upperparts edged pale brownish grey. **2nd alt:** attains occipital and back plumes. Attains adult plumage by 3rd prebasic molt when about 3 years old.

Voice. A gruff, barking *kwehk* or *kyowk,* higher and more nasal than Black-crowned Night-Heron; often silent in flight.

Habs. Mangroves and coastal areas, including open beaches; also freshwater marshes. Habits similar to Black-crowned Night-Heron but seen more often during day. Usually nests fairly high in trees; 2–5 eggs.

SS: See Black-crowned Night-Heron and bitterns.

SD: F to C but local breeder (SL–1000 m) on Pacific Slope from BCS and Son to El Salvador (at least formerly), on Isla Socorro, and on Atlantic Slope from Tamps to Honduras. C to F transient and winter visitor (Sep–Apr) on both slopes; U to R (to 2500 m) in cen volcanic belt; non-breeders oversummer widely.

RA: Breeds NW Mexico and SE USA to N Peru and E Brazil; N.A. populations mostly withdraw S in winter.

NB: Often placed in the monotypic genus *Nyctanassa.*

BOAT-BILLED HERON
*Cochlearius cochlearius**   Plate 1
Garza Cucharón

ID: 18–21 in (45.5–53 cm). *A stocky and compact nocturnal neotropical heron. Bill remarkably broad and heavy. In flight, feet* *barely project beyond tail.* **Descr.** Ages differ, sexes similar. **Adult:** eyes dark brown, bill black above, yellow below [all black], lores greyish, legs greenish yellow [yellow]. *Black crown and strongly crested nape contrast with whitish forehead and creamy to pale vinaceous face;* throat white, neck pale vinaceous. *Upperparts pale grey* with broad blackish band across upper mantle. Flight feathers grey. *Underparts vinaceous; black flanks, axillars, and lesser and median underwing coverts* contrast with pale grey greater coverts and remiges. Has extremely long and exuberant crest of broad black plumes in alt. **Juv/ 1st basic:** forehead grey-brown, *black crown slightly crested; face, neck, and upperparts grey-brown,* often with some cinnamon on scapulars and mantle. Flight feathers brownish grey. *Underparts pale grey with buff or pinkish wash,* flanks duskier. **1st alt:** crest slightly longer, upper mantle often mottled dark brown. **2nd basic:** forehead greyish, crown black with moderate crest, face and neck dusky vinaceous, foreneck pale grey. Upperparts grey with blackish-brown upper mantle; flight feathers grey. Underparts cinnamon, flanks dull blackish, underwing coverts mottled buff and blackish. **2nd alt:** some birds show paler back and broad elongate crest. Adult plumage attained by 3rd prebasic molt.

Voice. Usually silent in flight at dusk. At roosts and colonies makes a low, accelerating clucking *cu-cu-cu-cu-kah'*, a more laughing *ah-ah-ah-ah-ah-ah cu-ah*, etc.

Habs. Mangroves, freshwater marshes, riverside trees, etc. More strictly nocturnal than night-herons, unlikely to be seen in day away from roosts or colonies. Roosts in denser areas than night-herons and hence more easily overlooked. Solitary feeder at night; function of bill still not fully understood. Nests in trees. Eggs: 2–4, bluish white, faintly speckled cinnamon (WFVZ).

SS: See Black-crowned Night-Heron.

SD: F to C but local resident (SL–900 m), mainly near coasts, on both slopes from S Sin and cen Tamps to El Salvador and Honduras.

RA: Mexico to W Ecuador and N Argentina.

NB: Sometimes placed in the monotypic family *Cochleariidae*.

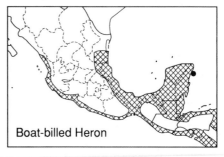

Boat-billed Heron

IBISES AND SPOONBILLS: THRESKIORNITHIDAE (4)

A cosmopolitan family of large wading birds with fairly long necks and long legs. Ibises have long decurved bills; spoonbills have long flattened bills that expand into spatulate tips. Feet slightly webbed. Ages differ, sexes similar. Young altricial; molt strategies vary. Color of bare parts can change at onset of breeding [alternate colors are noted in square brackets]. Flight strong, flocks often in V formation. Usually silent, mostly making low grunts and clucks.

Food mostly crustaceans, amphibians, molluscs, small fish, insects, etc. Typically social and gregarious, breeding in large colonies, often with other waterbirds. Nests are platforms of vegetation, especially sticks.

WHITE IBIS
Eudocimus albus　　　　Not illustrated
Ibis Blanco

ID: 21–25 in (53.5–63.5 cm). Mainly coastal. **Descr. Adult:** eyes bluish white, *bill, bare face, and legs pink [scarlet]. Entirely white except for black tips to outer 4 primaries, usually only noticeable in flight.* **Juv:** eyes grey, bill initially dusky at base, face feathered. Head and neck whitish, streaked dusky, *upperparts dark brown with white rump and uppertail coverts usually only noticeable in flight;* tail dark brown. *Underparts white,* underside of remiges dark brown. In 1st winter begins protracted molt into **1st basic:** in 1st summer upperparts mottled white; by 2nd winter resembles adult but may show some brown feathers in upperparts.

Habs. Mangroves and brackish lagoons, less often freshwater marshes. Social and gregarious, feeds by probing and picking. In flight, neck and legs held extended; wingbeats interspersed with short glides. Nests in reedbeds or low trees and bushes. Eggs: 3–4, greenish white to pale buff, spotted brown (WFVZ).

SS: Told from herons by decurved bill; other ibises all dark. Wood Stork larger with stout darker bill, naked grey head and neck, black remiges obvious in flight.
SD: F to C resident (SL–150 m), breeding locally, from BCS, Son, and Tamps along both slopes to Chis (possibly to El Salvador) and Honduras; more common and widespread (Aug–Apr; to 1000 m) with post-breeding dispersal and arrival of N migrants. Vagrant (Sep–Dec; to 1500 m) to Coah and NL (AGL, AMS).
RA: W Mexico and SE USA to N Peru and French Guiana.

GLOSSY IBIS
Plegadis falcinellus　　　　Not illustrated
Ibis Lustroso

ID: 21–25 in (53.5–63.5 cm). *Local but increasing.* **Descr. Basic/imm:** *eyes brown, lores slaty with narrow whitish to pale blue edging* above and below, bill and legs greyish (often looking slightly pinkish when wet). Head, neck, and underparts greybrown, head and neck streaked whitish, undertail coverts dull metallic green. *Upperparts dark, metallic bronzy green;*

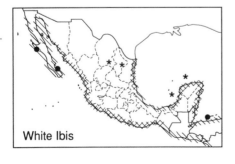

White Ibis

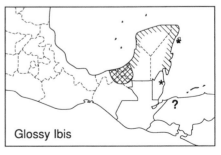

Glossy Ibis

remiges metallic green, tail bronzy purple. **Alt:** lores flush bright blue, legs may become reddish. *Head, neck, and underparts chestnut*, undertail coverts metallic greenish purple. Upperbody burnished purplish chestnut becoming purple-green on rump and uppertail coverts, lesser upperwing coverts chestnut, median and greater upperwing coverts bronzy green. **Juv:** resembles basic but head and neck unstreaked grey-brown; attains basic within a few months, adult plumage by 2nd prebasic molt.

Habs. Brackish and freshwater marshes. Associates loosely with herons, egrets, etc., when feeding but single birds likely to fly off separately. Associates readily with White-faced Ibis. Flies with neck outstretched and slightly drooped, legs extended; wingbeats interspersed with short glides. Nest lodged above water in marsh vegetation. Eggs: 1–4, pale greenish blue to blue (WFVZ).

SS: White-faced Ibis has red eyes (brownish in juv), plain slaty-grey lores of basic/imm become red in breeding birds. See imm White Ibis.

SD: Presumed local breeder in Usumacinta delta (N Chis to SW Camp) where locally/seasonally C to F resident; U to R visitor (Jan–Jul at least, probably Jan–Dec) to Yuc Pen, especially Caribbean coast of QR, Belize (Jan–Feb 1993, SNGH), and probably to N Honduras (Howell and de Montes 1989). Other records from Belize (Wood *et al.* 1986) require verification.

RA: In New World, where may have spread from Old World in 1800s, breeds E and SE USA, Greater Antilles, and N Venezuela; recently S Mexico. Also locally in Old World; N populations migratory.

WHITE-FACED IBIS
Plegadis chihi Not illustrated
Ibis Cariblanco

ID: 20–24 in (51–61 cm). *Common and widespread.* **Descr. Basic/imm:** *eyes bright red, lores slaty*, bill and legs greyish (often looking slightly pinkish when wet). *Plumage identical to Glossy Ibis*; note colors of bare parts. **Alt** (Apr–Aug): lores flush bright red, often appearing reddish from late winter, legs may become reddish, *band of white feathers borders lores and extends behind eyes.* **Juv:** eyes grey-brown, plumage as Glossy Ibis.

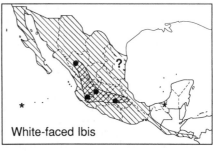

White-faced Ibis

Habs. Freshwater marshes, less often brackish areas. Habits, voice, nest, and eggs much like Glossy Ibis with which it associates readily.

SS: See Glossy Ibis, imm White Ibis.

SD: F to U but local breeder (May–Aug; 1500–2200 m) in interior from Dgo to cen volcanic belt (SOW) possibly on Atlantic Slope in Tamps (Sprunt and Knoder 1980). C to F winter visitor (Sep–Apr; SL–2500 m) on Pacific Slope from Son to Guatemala, U to F in Baja, R to El Salvador; C to F in interior S to cen volcanic belt, and on Gulf Slope to S Ver, R to SW Camp (Howell and de Montes 1989). Non-breeding birds oversummer in winter range, mainly in NW and cen Mexico.

RA: Breeds W and S USA to cen Mexico, locally in S.A. from N Colombia to cen Argentina. Northern birds winter from S USA to El Salvador.

ROSEATE SPOONBILL
Platalea ajaja Not illustrated
Espátula Rosada

ID: 28–31 in (71–79 cm). *A distinctive pink wading bird with a long spatulate-tipped bill.* Timing of molts poorly known. **Descr. Adult:** eyes red, bill glaucous with blackish mottling, naked head glaucous [golden buff] with black nape band, legs red, feet (and often hind tarsi) blackish. *Head naked, neck and upper mantle white, upperparts pink* with reddish-pink lesser upperwing coverts and uppertail coverts; tail orange. Upper chest white with tawny-buff sides and, in alt, a central patch of stiff curly reddish-pink feathers; underparts pink, underwing coverts and tail coverts reddish pink. **Juv:** eyes yellow, bill pale flesh to yellowish, head feathered, legs dusky reddish. *White overall, underwing coverts and tail pale pink*, outer 3–6 primaries, primary coverts, and alula edged dusky.

1st basic: attained quickly. *Pale pink overall*, rarely a few reddish-pink flecks on lesser wing coverts, tail pale buff. **2nd basic:** much like adult but duller, little or no reddish pink on upperwing coverts and tail coverts. Attains adult basic by 3rd prebasic molt when about 3 years old.

Habs. Brackish coastal marshes and lagoons, less often in fresh water. Often in small flocks, feeds by sweeping bill from side to side through shallow water. Flies with neck and legs extended; wingbeats shallow and stiff, interspersed with glides. Nests in trees and bushes. Eggs: 2–5, whitish, marked with browns (WFVZ).

SS: Distinctive bill shape and pink plumage should preclude confusion; local people often refer to spoonbills as flamingos.

SD: F to U resident (SL–500 m), breeding locally on both coasts, from S Son and Tamps to El Salvador and Honduras. R to F post-breeding and winter visitor (Jan–Dec, mostly May–Aug; to 2200 m) to cen volcanic belt and Plateau (SNGH, AGL, SOW, RGW), and R to NE Baja (at least formerly).

RA: NW Mexico and SE USA to cen Argentina.

NB: Often placed in the monotypic genus *Ajaia*.

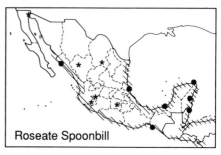

Roseate Spoonbill

STORKS: CICONIIDAE (2)

A mainly Old World family with three species occurring in the New World. Large to very large wading birds with long legs and long, stout, pointed bills. Feet semipalmated. Wings long and broad, used for soaring, tails short. Ages differ, sexes similar. Young altricial, juv plumage lost within a few weeks of leaving nest, imm much like adult. Plumage predominantly white, head and neck naked. Flight powerful, neck held extended and legs trailing. Typically silent but utter low grunts and hisses; also communicate by bill-rattling.

Varied diet includes fish, amphibians, reptiles, snakes, small mammals, and birds. Storks breed colonially or in scattered pairs; breeding may depend on water levels and vary in timing from year to year, with birds in some areas not breeding in certain years. Nests are stick platforms, usually high in trees. Eggs: 2–4, whitish (WFVZ).

JABIRU

Jabiru mycteria   Figure 15
Jabirú

ID: 51–60 in (129.5–152.5 cm). *The largest flying bird in the Americas. Bill massive* (12–13 in; 30.5–33 cm). **Descr. Adult:** bill and legs black. *Naked head and neck black with broad red band around base of neck* and down-like white feathering on nape; *plumage entirely white.* **Juv:** downy dusky head and neck feathering mostly lost within a few weeks, upperparts pale grey, feathers edged silvery grey-brown. Remiges whitish, primaries pale brownish on inner webs; tail white. Underparts whitish. **1st basic:** head and neck naked as adult but red duller, plumage white overall with some brownish at base of neck below bare skin, scattered brownish feathers on upperparts. **2nd basic:** resembles adult but may retain a few brownish upperwing coverts. Successive plumages undescribed (?).

Habs. Extensive marshy savannas, rice fields, swampy woodland. Singly or in small groups, feeds by slow stalking, stabs and picks at prey. Breeds in scattered pairs, at times loosely colonial.

SS: Should not be confused with other species; Wood Stork much smaller with black flight feathers.

SD: U resident (around SL), breeding locally (Dec–May) on Atlantic Slope from E Tab through S Yuc Pen to Belize, also in the Mosquitia; most Belize breeders move to Usumacinta drainage (Jul–Nov); R and irregular visitor (Dec–Jan) to N coast of Honduras. R and irregular visitor (mostly Dec–Mar) on Pacific Slope from E Oax (GC, WJS) to El Salvador.

RA: SE Mexico to N Argentina and Uruguay.

WOOD STORK

Mycteria americana   Figure 15
Cigüeña Americana

ID: 35–40 in (89–101.5 cm). Widespread. *Bill stout and slightly decurved.* **Descr. Adult:** bill blackish to mixed black and horn, *head and neck naked* (forehead and

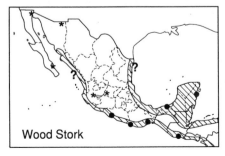

Jabiru   Wood Stork

Fig. 15 (a) Jabiru; (b) Wood Stork

crown creamy horn, nape band black, rest of head and bare neck scaly dark grey), legs blackish to blue-grey, feet pink. *Plumage white overall except black primaries, secondaries* (tertials white), primary coverts, alula, *and tail.* In alt shows elongated undertail coverts which project beyond tail in flight. **Juv:** bill yellowish, naked face blackish. Head and upper neck feathered dusky grey with whitish-feathered patch on crown. **Imm:** loses white feathering on crown, neck sparsely feathered, bill becomes mottled black. Successive stages underscribed (?) but does not attain completely naked head and neck of adult until about 4 years old (JMW).

Habs. Freshwater, brackish, and salt marshes, often wooded. In small groups to flocks of a few hundred birds. Feeds by wading and probing. Breeds and roosts colonially, sometimes with other species.

SS: See White Ibis. When soaring, especially at a distance, may recall White Pelican (white tail, lacks long projecting legs) or King Vulture (short neck and short tail without projecting legs).

SD: F to C resident (SL–500 m), breeding locally, on both slopes from Sin and Tamps to El Salvador and Honduras. Postbreeding dispersal nomadic; R visitor (Jan–Dec, mostly May–Oct) to Baja and (to 1800 m) to SW Plateau (SOW).

RA: W Mexico and SE USA to W Ecuador and N Argentina.

FLAMINGOS: PHOENICOPTERIDAE (I)

Flamingos are locally distributed in warm climates in the Americas, Africa, and Eurasia. They are large wading birds with very long necks and legs, and bills that are laterally flattened and bent in a 'broken nose' profile, the lower mandible being shaped into a shovel-like scoop. Feet webbed. Ages differ, young precocial; molts poorly known, protracted (?). Sexes similar but ♂♂ larger. Calls are honking and cackling noises.

Flamingos feed in water of any depth accessible to their long legs and necks, and suck water and sediments from which plankton, tiny fish, and brine fly larvae are filtered with the tongue and bill. Social and gregarious, colonies of several hundred to several thousand pairs breed in shallow lagoons; raised mud cone nest has cup-like depression at top; 1 white egg laid (rarely 2). Young at a few days of age gather into crèches which are attended communally by parents.

AMERICAN [GREATER] FLAMINGO
Phoenicopterus ruber   Figure 16
Flamenco Americano

ID: Adult 45–48 in, juv 35–40 in (adult 89–101 cm, juv 115–122 cm). *Unmistakable.* **Descr. Adult:** eyes pale lemon, bill tri-colored (base pale creamy, mid-section pink, tip black), legs pink with darker joints. *Bright pink overall,* palest on scapulars, brightest on axillars and underwing coverts; *black primaries and secondaries* (tertials pink) *usually only noticeable in flight.* **Juv:** eyes dusky, bill tricolored (greyish, creamy yellow, and black-tipped), legs grey with dark joints. *Whitish to pale grey overall;* upperparts, chest, and sides streaked dusky; greater secondary coverts and *remiges black,* primary coverts tipped black; axillars pale pink. Tail edged blackish. **1st basic:** attained at a few months of age; *whitish overall, faintly tinged pink,* lesser wing coverts finely streaked dusky, median wing coverts pink, greater wing

American Flamingo

Fig. 16 American Flamingo

coverts and remiges as juv, axillars pink. May become pinker in 1st alt. **2nd basic:** resembles adult but slightly paler overall. Attains adult plumage by 3rd prebasic molt (?).

Voice. Honking and cackling, a low *onk* and *ohrn*, and a higher *aah aah ahnk*; the overall sound of a flock very goose-like.

Habs. Saline lagoons and estuaries. Typically social and gregarious, in small groups up to flocks of thousands. Flight strong, flocks usually fly in V-shaped formation, with long necks and legs extended; does not soar and rarely glides.

SD: Locally C resident (breeds Apr–Aug) in N Yuc (lagoons E of Río Lagartos), rarely small groups build nests as far W as extreme N Camp (JRS). Post-breeding and non-breeding dispersal (Jan–Dec) is mostly W to Celestún, Yuc, where 50–75% of the population winters (Nov–Mar); U visitor E to Caribbean coast of QR (Bahia Ascensión), R and irregular to N Belize. Population has apparently increased from 12 000 birds in 1971 to 26 000 in mid-1980s, perhaps due to protection of main nesting and wintering grounds (Espino-Barros and Baldassare 1989).

RA: New World; disjunct populations in Caribbean, Yuc Pen, and Galapagos.

NB: Often considered conspecific with Greater Flamingo (*P. roseus*) of Old World.

SWANS, GEESE, AND DUCKS: ANATIDAE (40)

This diverse cosmopolitan family of aquatic birds, collectively known as water-fowl, is well known throughout the world as a traditional food source. Size varies from the relatively small teal to large swans; all are swimming birds with webbed feet. Bills broad and somewhat flattened with serrated tomia, rounded tip, and a slightly hooked nail. Necks fairly short to very long, wings fairly broad and pointed and tails short in most species. Plumage dense and water-proof, underlain with down. Most waterfowl are only or primarily winter visitors to the region (33 of 40 species), arriving between Sep and Dec and leaving Mar to May; a few oversummer in N Mexico.

Two distinct subfamilies occur in the region, both divided into a number of tribes. Subfamily Anserinae (swans, geese, and allies): ages usually differ, sexes similar; young precocial and cared for by both parents. Subfamily Anatinae (ducks): ages/sexes differ; young precocial and cared for by ♀. Juvs usually resemble ♀. Basic plumage of ♂♂ is often known as eclipse plumage; it usually resembles the dull ♀ plumage and is worn for a few months during wing molt. Molt into full breeding colors occurs in late autumn, and courtship and pair formation take place through the winter and early spring.

TRIBE DENDROCYGNINI: Whistling-Ducks

Whistling-Ducks are named for their high-pitched whistling calls. They are gregarious, mostly tropical ducks with long necks and legs. Wings fairly long and broad; in flight feet project beyond tail and neck held drooped. Molts poorly known, young soon resemble adults. Patterned mainly in browns and black. Primarily nocturnal feeders, main food is grains. Nest not lined with down as in most waterfowl; 6–18 eggs, white (WFVZ).

FULVOUS WHISTLING-DUCK
Dendrocygna bicolor   Not illustrated
Pijiji Canelo

ID: 19–21 in (48.5–53.5 cm). Local and nomadic. **Descr.** Ages differ slightly, sexes similar. **Adult:** bill and legs dark grey. *Head, neck, and underparts ochraceous tawny*, narrow hindneck stripe blackish, sides of neck with diagonal whitish furrows. *White tips to elongated flank feathers form stripe along sides*, undertail coverts white. Blackish upperbody broadly edged tawny-cinnamon, *white uppertail coverts obvious in flight. Wings look all dark in flight:*

lesser and median wing coverts chestnut, greater wing coverts and flight feathers black. **Juv:** duller overall, poorly contrasting uppertail coverts dusky cinnamon-buff for a few weeks.
Voice. A high, slightly hoarse, 2-syllable whistle *ch-hehw* or *pi-hew*, and a more drawn-out *chi-heahw*, usually given in flight.
Habs. Open marshes, flooded fields, lagoons; roosts by day in dense vegetation, rarely in trees. Often in flocks, rarely in thousands; small groups may associate with Black-bellied Whistling-Ducks. Nest platform of vegetation usually over water.
SS: Juv Black-bellied Whistling-Duck has subdued adult pattern, most noticeably grey face and neck, white wingstripes. At a

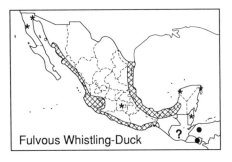

Fulvous Whistling-Duck

distance on water, Fulvous may recall dabbling ducks, especially ♀ Pintail; note white side stripe.
SD: Nomadic and unpredictable, populations appear to fluctuate widely but this may be due in part to difficulty of detecting roosting birds. Locally and irregularly F to C resident (SL–1500 m) on Pacific Slope from S Son to Chis (possibly Guatemala, per PH), disjunctly in El Salvador where colonized in 1970s (Thurber *et al.* 1987), on Atlantic Slope from Tamps to Tab and SW Camp (SNGH); disjunctly in NW Honduras. Irregular R to U visitor (Jan–Dec, mainly Sep–Apr, to 2300 m) to Yuc Pen, Belize (RGW), interior cen Mexico, and BCN.
RA: S USA, Mexico, and Caribbean locally to cen Argentina; also Africa and SW Asia.

BLACK-BELLIED WHISTLING-DUCK
Dendrocygna a. autumnalis   Not illustrated
Pijiji Aliblanco

ID: 18–20 in (45.5–50.5 cm). Common and widespread. **Descr.** Ages differ, sexes similar. **Adult:** *bill and legs bright pinkish red.* Head and upper neck grey with dark brown crown and hindneck stripe, white eyering. Lower neck and chest tawny-brown, *belly and flanks black*, undertail coverts mottled black and white. Upperparts darker tawny-brown to cinnamon-brown, becoming black on rump and uppertail coverts, *broad white upperwing stripe* (across median and greater coverts, and base of primaries) striking in flight; flight feathers otherwise black. Underwings blackish. **Juv:** bill and legs grey. Resembles washed-out adult, belly and flanks sooty brown.
Voice. High, piping whistles, usually given in flight, *pi-yih pyi-pyi-pyi* or *chee twee WEE wee-wee*, etc. Also high, rapid, piping, and twittering whistles near nest.

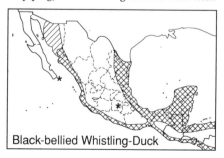

Black-bellied Whistling-Duck

Habs. Wooded marshes, swampy forest, lagoons, mangroves, flooded fields, often roosts in trees. Nests typically in tree cavities, less often on ground. Communal 'dump' nests may contain up to 100 eggs.
SS: See Fulvous Whistling-Duck. Muscovy Duck more heavily built, feet do not project beyond tail, adult has entire forewing (above and below) white, plumage black overall. Downy young (dive when alarmed) may suggest Ruddy or Masked ducks in head pattern and often are left unattended.
SD: C to F resident (SL–750 m) from S Son (to N Son in summer) and Tamps along both slopes to El Salvador and Honduras; N populations somewhat migratory; elsewhere wanders in winter. R visitor (Aug–Mar at least, to 2300 m) to interior cen Mexico. Vagrant to BCS (Jun 1991, Howell and Webb 1992*f*). Often domesticated, wild birds wary and not around habitation.
RA: S USA and Mexico, to Ecuador and N Argentina.

TRIBE ANSERINI: Swans and geese

Swans and geese are winter visitors to the region. Ages similar or different, sexes similar. Juvs attain adult plumage by a protracted molt through first year and family units are maintained through the winter. Swans are very large waterfowl with very long necks that enable them to reach food (aquatic plants) in deeper water than dabbling ducks and geese; often feed with head and neck submerged for long periods, also feed on grains. Adults white, juvs brownish grey. Geese are large, long-necked waterfowl, plumage varies from overall white to dark grey-brown. Social and noisy, often in large flocks and faithful to traditional wintering areas. Geese spend much time walking, feed on vegetable matter, including grains, and often make long daily flights to feeding areas, returning at night to roost on the safety of lakes.

WHISTLING [TUNDRA] SWAN
Cygnus c. columbianus   Not illustrated
Cisne de Tundra

ID: 48–58 in (122–147 cm). Rare winter visitor to N Mexico. **Descr.** Ages differ.

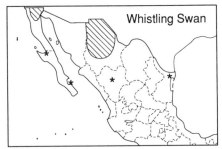

Whistling Swan

Adult: *bill black, usually with a small* (rarely fairly large) *yellow basal spot* near eye, legs black. *Plumage entirely white,* head and neck may be stained rusty by ferrous pigments in marshes. **Imm:** bill pinkish with dark tip, often mostly black by mid-winter. Plumage smoky grey-brown overall, becoming increasingly white through 1st winter; by spring usually only head and neck greyish.
   Voice. Mellow, honking calls, *whooh* or *hooh* and *huh, huh*, etc.; slower paced, less coarse than Snow Goose.
   Habs. Large lakes, rarely estuaries. See tribe account. Unlikely to occur in groups of more than 8–10 birds.
   SS: Snow Goose smaller and chunkier with pink bill (adult), dark bill (juv), black primaries obvious in flight. See larger Trumpeter Swan.
   SD: Irregular R winter visitor (Dec–Feb; SL–2000 m) to N Mexico, recorded S to BCS, Dgo, and Tamps.
   RA: Holarctic breeder; New World populations winter to S USA, rarely N Mexico.

TRUMPETER SWAN
Cygnus buccinator   Not illustrated
Cisne Trompetero   No map

ID: 60–70 in (152–178 cm). Former winter visitor. **Descr.** Ages differ. **Adult:** bill and legs black. Plumage entirely white, *eye stands out at base of narrow black loral area.* **Imm:** bill pinkish with dark tip. Overall smoky grey-brown plumage retained through spring.
   Voice. Nasal to trumpeting honks, unlike Whistling Swan.
   Habs. Much like Whistling Swan.
   SS: Told from Whistling Swan by larger size and larger bill, more prominent black eye, and voice.
   SD: Formerly wintered in coastal N Tamps, last recorded 1909. Numbers increasing

with protection and reintroduction programs in N.A., may return to winter in Mexico: recent report from the Rio Grande, Texas (Dec 1989/Jan 1990, per GWL).
   RA: NW N.A.; formerly to E and cen N.A., wintering S to NE Mexico.

WHITE-FRONTED GOOSE
*Anser albifrons**   Not illustrated
Ganso Careto Mayor

ID: 27–31 in (68.5–78.5 cm). N Mexico in winter. **Descr.** Ages differ. **Adult:** *bill pink, legs orange.* Grey-brown overall, *band around base of bill white,* belly mottled black. *White tail coverts contrast with black, broadly white-tipped tail.* **Imm:** lacks white bill base and black belly mottling, traces of which usually appear by spring.
   Voice. Varied honking calls, *ahng* and *yow-a-yowng*, etc.; higher and more musical, or laughing, than Snow Goose calls.
   Habs. Extensive marshes, arable areas, lakes, estuaries, etc. See tribe account.
   SS: White front and black belly mottling distinctive; juv dark morph Snow Goose has dark grey bill and legs, lacks white uppertail coverts.
   SD: F to C winter visitor (Oct–Mar; SL–2500 m) S in coastal lowlands to Sin and N Ver, on Plateau to cen Jal; U to F in Baja; R and irregular (formerly regular?) to S Mexico; vagrant to Belize (Jan 1973, AMNH spec.).
   RA: Holarctic breeder; winters S in New World to cen Mexico.
   NB: Also known as Greater White-fronted Goose.

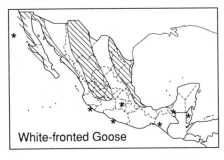

White-fronted Goose

SNOW GOOSE
Anser c. caerulescens   Not illustrated
Ganso Blanco

ID: 25–29 in (63.5–73.5 cm). N Mexico in winter. Dimorphic. **Descr.** Ages differ.

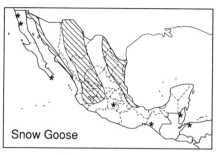

Snow Goose

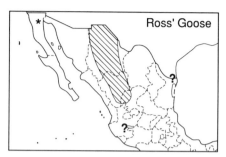

Ross' Goose

White morph. Adult: *bill pink with noticeable black tomia*, legs pink. *Plumage white overall, black primaries striking in flight*; upper primary coverts pale grey. **Imm:** bill and legs dark grey, often dull pinkish by mid-winter. Head, neck, and underparts dirty whitish, upperparts duskier, pale grey-brown. Much like adult by spring. **Dark morph** (uncommon in region). **Adult:** *white head and upper neck contrast with blackish-grey lower neck and body*, rump and tail coverts white (belly white on some birds). *Upperwing coverts pale grey, remiges blackish.* Tail blackish, tipped white. **Imm:** smoky grey-brown overall, upperwing coverts paler; much like adult by spring.

Voice. Fairly coarse, nasal to laughing honks, *ahnk* or *ahn*, and *yah'nk* or *yahnk*, etc.

Habs. Extensive marshes, arable areas, lakes, estuaries, etc. See tribe account. Traditional food marsh plants, but moving increasingly to arable fields.

SS: Ross' Goose distinctly smaller and shorter-necked, stubby bill lacks black tomia. See Whistling Swan, White-fronted Goose.

SD: F to C winter visitor (Oct–Mar; SL–2500 m) S in coastal lowlands to Son (U to Sin) and N Ver, on Plateau to cen Jal, R to DF (RGW). Vagrant (late Nov–Feb) S to Belize (Howell *et al.* 1992b) and NW Honduras.

RA: Breeds N.A.; winters S to cen Mexico.

NB1: Sometimes placed in the genus *Chen*. **NB2:** Hybrids (rare) between Snow and Ross' geese appear intermediate in most characters but apparently have the black tomia of Snows.

ROSS' GOOSE

Anser rossii   Not illustrated
Ganso de Ross

ID: 21–24 in (53.5–61 cm). Winter visitor to

N Mexico. Dimorphic, rare dark morph unrecorded in region. **Descr.** Ages differ. **White morph. Adult:** *bill and legs pink. Plumage white overall, black primaries striking in flight*; upper primary coverts pale grey. **Imm:** bill and legs duller. Dirty whitish overall; much like adult by spring. **Dark morph. Adult:** similar to dark Snow Goose but neck blackish, often extending in narrow line up nape onto crown, underbody mostly white, only chest and sides blackish. **Imm:** blackish areas duller and more brownish, head mottled brown.

Voice. A honking *ownk* and other notes; mellower than Snow Goose.

Habs. Much as Snow Goose with which it usually occurs. Family groups of Ross' keep together within large flocks of Snows.

SS: See Snow Goose.

SD: U to F winter visitor (Nov–Mar; 1200–2500 m) to N Mexico, mainly on Plateau S to Dgo and Zac, R to BCN (SNGH, PP); possibly R to Jal (RB), and N Tamps (Saunders and Saunders 1981).

RA: Breeds cen Canadian Arctic; winters mainly in California, in low density E to N Mexico and S USA.

NB1: Sometimes placed in the genus *Chen*. **NB2:** See NB2 under Snow Goose.

BRANT

Branta bernicla nigricans   Not illustrated
Branta

ID: 22–26 in (56–66 cm). Winter visitor to NW Mexico. **Descr.** Ages differ. **Adult:** bill and legs black. *Head and neck black with narrow white collar on upper neck.* Whitish flanks contrast with blackish-brown chest and belly, and with slaty grey-brown upperparts; tail coverts white. Flight feathers blackish. **Imm:** white collar indistinct or lacking, pale upperpart edgings distinct; much like adult by spring.

Brant

Voice. Low, grunting, and rolling calls, unlike other geese; in flight a gruff, clipped *rruh* and *rruhk*, the overall sound of a flock being a slightly disjointed, low chuckling.

Habs. Coastal bays and estuaries. Feeds mostly on eelgrass (*Zostera*); very unlikely to be seen away from immediate vicinity of coast.

SS: Most ssp of Canada Goose much larger, all told by broad white chinstrap, paler upperparts.

SD: C to F winter visitor (Nov–Apr) along Pacific coast of Baja S to Bahia Magdalena; since 1950s has expanded its winter range to coasts of Son and N Sin where locally F; vagrant to Nay (Jan–Feb 1985, PP) and Yuc (Feb 1983, Gatz *et al.* 1985), the latter bird also thought to be of the Pacific ssp *nigricans*. A report from Gro (Saunders and Saunders 1981) requires verification.

RA: Holarctic breeder; winters S in New World to NW Mexico and E USA.

NB1: Known as Brent Goose in the Old World.

CANADA GOOSE
Branta canadensis*   Not illustrated
Ganso Canadiense

ID: 22–40 in (56–101.5 cm). Winter visitor to N Mexico. *Marked size variation related to ssp* (see below). **Descr.** Ages similar. Bill

Canada Goose

and legs black. *Head and neck black with broad white chinstrap.* Underparts brownish white to dusky brown, undertail coverts white. Upperparts dark brown becoming black on rump, white uppertail coverts contrast with black tail. Four ssp occur in Mexico. Cackling Goose *B. c. minima*: small (22–24 in; 56–61 cm), body dusky brown overall, little contrast between upperparts and underparts; Baja. Richardson's Canada Goose *B. c. hutchinsii*: small (24–26 in; 61–66 cm), pale brown underparts, narrow white ring at base of black neck; NE Mexico. Lesser Canada Goose *B. c. parvipes*: medium (28–30 in; 71–76 cm), underparts paler than upperparts; N and NE Mexico. Western Canada Goose *B. c. moffitti*: large (36–40 in; 91.5–101.5 cm), underparts only slightly paler than upperparts; N Mexico.

Voice. Varies from lower, more drawn-out honking of large birds to higher, shorter yelping calls of small birds.

Habs. Extensive marshes, arable areas, lakes, etc. Typically in flocks, separate from other geese; *minima* also likely to occur singly and can be fairly tame.

SS: See Brant.

SD: F to U winter visitor (Oct–Feb; around SL) to NE Mexico (Tamps and N Ver); U to R in Baja, Son, and on Plateau (to 2000 m) S to Dgo, and (at least formerly) to cen Jal. Vagrant (at least formerly) S to Chis (Saunders and Saunders 1981).

RA: Breeds N.A.; winters to S USA and N Mexico.

TRIBE CAIRININI: Perching ducks

Perching ducks are heavily-built ducks with relatively broad wings and fairly long, broad tails. Legs are set farther forward than in dabbling ducks, hind toe well developed and claws strong and sharp. Prefer wooded areas and nest in tree cavities. Varied diet includes grains, aquatic vegetation, fruits, nuts, small fish, insects, crabs.

MUSCOVY DUCK
Cairina moschata   Figure 17
Pato Real

ID: 26–34 in (66–86.5 cm) ♂ > ♀. *A large neotropical duck, often domesticated.* **Descr.** Sexes similar. **Adult:** bill pinkish,

Fig. 17 Muscovy Duck

mottled black (♂ has warty protuberances at base of bill and bare blackish facial skin), legs blackish. *Plumage black overall, glossed green and purple, white wing coverts and axillars usually only visible in flight.* **Juv/imm:** all black, often with small white spot on upperwing, visible in flight. Less iridescent than adult, attains white wing coverts in 1st year. Domesticated birds usually show at least some white patches in plumage.

    Voice. Usually silent; ♂ makes a high hissing, ♀ a low quack.

    Habs. Wooded lakes and rivers, marshes, mangroves. Heavily hunted in most areas and typically very wary, rarely seen other than in flight. Singly or in small groups, flight heavy and goose-like. ♂♂ polygamous. Nest in tree cavities. Eggs: 8–9, greenish white (WFVZ).

SS: See Black-bellied Whistling-Duck.

SD: U to R resident, locally/seasonally F (SL– 1200 m), from S Sin and cen NL (Leopold 1959) along both slopes to El Salvador and Honduras; probably extirpated locally by hunting. Nestbox programs may help populations to recover.

RA: Mexico to W Colombia and N Argentina.

WOOD DUCK

Aix sponsa   Not illustrated
Pato Arcoiris

ID: 17–19 in (43–48.5 cm). Rare visitor. *Head has bushy pointed crest; tail fairly long and wedge-shaped.* **Descr.** Sexes differ. **Alt ♂:** eyes red, bill mostly bright red with black culmen and tip, legs brownish yellow. *Head and crest dark glossy green with 2 narrow white lateral stripes, white throat forks into chinstrap* and partial collar. *Purplish neck and chest separated from cinnamon-buff sides and flanks by bold black-and-white stripe;* belly and vent white, undertail coverts purplish. Upperparts glossy blackish, white secondary tips and outer primary webs form *bold white stripe on closed wing. In flight wings appear dark* (speculum and greater upperwing coverts bluish) with *white trail-*

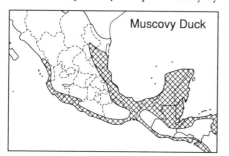

Muscovy Duck

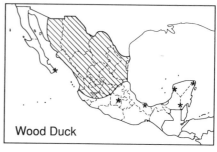

Wood Duck

ing edge to secondaries. ♀/**imm:** eyes brown, bill pinkish grey. Mottled grey-brown and brown overall, *thick teardrop-shaped white eyering distinctive*, belly and undertail coverts whitish; *wings similar to* ♂. Attains adult-like plumage by early winter. **Basic** ♂ (Jun–Sep): resembles ♀ but note red eyes, red bill.

Voice. An up-slurred squeal, *wheeah* or *whee-uk* and *whee-u-uk*, and a lower rolled *whur-rruh-ruk*, etc.

Habs. Wooded ponds and lakes, slow-moving rivers, marshes, mangroves. Singly or in pairs; may associate with other ducks on water but likely to fly off separately.

SS: ♂ unmistakable, ♀ told from brownish ♀ dabbling ducks by crest and white eyering; in flight told by long wedge-shaped tail, white trailing edge to secondaries.

SD: U to R but increasing winter visitor (Nov–Mar, SL–2300 m) S to cen Mexico (Williams 1987; SNGH, RGW), R to S Ver (Pashley 1987) and (Dec–Feb) to Yuc Pen (Howell and de Montes 1990). Irregular and R S to Jal in Jun–Jul (Williams 1987). Vagrant to BCS (Nov 1987, TEW).

RA: Breeds N.A.; winters to N Mexico.

TRIBE ANATINI: Dabbling ducks

Dabbling ducks are small to medium-sized waterfowl found on fresh water and in estuaries. Hind toes lack the lobe of webbing characteristic of diving ducks (following 3 tribes); take-off from water nearly vertical. ♂♂ brightly colored and strikingly patterned, ♀♀ cryptically marked in browns. The secondaries typically form a distinct speculum on the upperwing; underwings usually whitish with dusky remiges, and narrow dark leading edge to coverts in teal. Unless stated otherwise, basic ♂♂ resemble ♀♀ except for colors of bare parts and wing patterns. Social and noisy, ♂♂ and ♀♀ generally have distinct calls. Dabblers, as the name suggests, feed at or near the water surface, also on land; they dive rarely except to elude predators. Food mostly aquatic vegetation and grains. Nest platforms of vegetation lined with down.

GREEN-WINGED TEAL
Anas crecca carolinensis   Not illustrated
Cerceta Aliverde

ID: 14–15 in (35.5–38 cm). Widespread winter visitor, rare S of Isthmus. **Descr.**

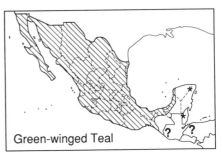

Green-winged Teal

Ages/sexes differ. **Alt** ♂ (Oct–Jun): bill blackish, legs greyish. *Head and upper neck chestnut with broad dark green post-ocular stripe.* Chest buff with dark spotting, body grey overall, browner above, with vertical white stripe on sides of chest, black-margined buff undertail coverts, and whitish belly. *Upperwings greyish, speculum green at base, black distally, with buff leading edge*, underwings grey with whitish coverts. ♀/**imm:** bill grey with orangish sides below. Mottled brown and pale brown overall, with darker crown and narrow eyestripe; *wings similar to* ♂. Resembles adult by early winter.

Voice. ♂ has a piping, reedy *krreek* or *krriip*, ♀ a low quack.

Habs. Estuaries, marshes, lakes. Often in large flocks, flight fast and flocks tightly packed.

SS: Blue-winged and Cinnamon teal have pale bluish forewings and yellowish legs; when swimming, however, the 3 teal species can be very similar. Cinnamon has a relatively long and broad bill, relatively warm brown plumage; Blue-winged relatively grey-brown, face has distinct whitish spot at base of bill.

SD: C to F winter visitor (Oct–Apr; U to R from mid-Jul and into May; SL–2500 m) S to cen Mexico; U to R in S Mexico; R and irregular to Belize (Wood *et al.* 1986), possibly to Guatemala (Land 1970), Honduras (Monroe 1968), and El Salvador (Thurber *et al.* 1987).

RA: Holarctic breeder; New World populations winter S to N C.A. and Caribbean.

MALLARD
Anas p. platyrhynchos   Not illustrated
Pato de Collar

ID: 20–23 in (51–58.5 cm). *N Mexico.* **Descr.** Sexes differ. **Alt** ♂ (Oct–Jun): bill yellow with black nail, legs orange. *Dark*

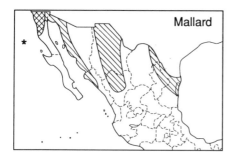
Mallard

glossy green head and upper neck separated from dark chestnut-brown chest by
narrow white collar. Body pale grey overall with black rump and tail coverts, tail
white. Wings grey-brown above, speculum
blue to violet-blue with white leading and
trailing edges; silvery-white wing-linings
contrast strongly with dusky remiges. ♀/
juv: bill orange, mottled black. Chest mottled brown and pale brown; head and upper
neck paler, buff to greyish buff, with dark
flecking except on throat, and dark crown
and eyestripe; belly usually paler. *Tail
and undertail coverts edged whitish; wings
similar to ♂.* Resembles adult by early winter. **Basic ♂** (Jul–Oct): resembles ♀ but bill
olive-yellow, chest rufous-brown.
   Voice. A full, often loud quacking *wahk-
wahk* or *wahnk-wahnk . . .*, and a lower
rrah rrahk, etc.
   Habs. Lakes, marshes, estuaries, arable
fields. Usually in pairs or small flocks,
often among Mexican Ducks, rarely in
flocks up to a thousand birds in N Chih.
Flight heavy with fairly shallow wingbeats. Nest platform of reeds, grasses,
etc., on ground near water. Eggs: 6–12,
pale greenish to greenish white (WFVZ).
SS: Mottled and Mexican ducks both distinctly
darker and more uniform than ♀ Mallard
(obvious in direct comparison) without
paler belly, tail and undertail coverts edged
brown, faces have distinct unstreaked buff
area at bill base; Mottled Duck lacks white
leading edge to speculum. ♀ Gadwall smaller, forehead steeper and head more angular; note white square on speculum.
SD: U breeder (SL–1000 m) locally in N
BCN (where non-breeders also oversummer). U winter visitor (Nov–Mar; SL–
2000 m) in N Mexico (locally C in N
Chih, S to BCN, N Sin, Dgo, and N
Tamps; formerly more numerous and
widespread, occurred regularly to cen

Mexico (Saunders and Saunders 1981),
rarely to N C.A.
RA: Holarctic breeder; winters S in New
World to N Mexico.

MEXICAN DUCK [MALLARD]
Anas (platyrhynchos?) diazi   Not illustrated
Pato Mexicano

ID: 20–23 in (51–58.5 cm). *A ♀-plumaged
Mallard of the Plateau.* **Descr.** Sexes similar. Bill olive-yellow (♂) to yellow-orange
(♀) with black nail, ♀ sometimes has dark
mottling on culmen; legs orange. Mottled
dark brown and warm brown overall,
head and upper neck contrastingly paler,
buff with dark flecking except in fairly
large area at base of bill and on throat and
foreneck, dark brown crown and eyestripe.
*Speculum green-blue to blue with white
leading and trailing edges. Silvery-white
wing-linings contrast strongly with dusky
remiges.*
   Voice. Loud quacking much like Mallard.
   Habs. Lakes, marshes, arable fields, drainage ditches. Usually in pairs or small
flocks, at times up to a few hundred birds.
Nest similar to Mallard. Eggs: 4–9, pale
greenish (WFVZ).
SS: See ♀ Mallard. ♀ Gadwall more slightly
built, paler, forehead steeper and head
more angular; note white square on speculum.
SD/RA: C to F resident (900–2500 m) on
Plateau from Chih and Dgo (and locally
in SW USA) to cen volcanic belt; U to
R in winter (Nov–Mar; locally resident?)
E to W Tamps and SE SLP (JCA), and S to
Mor.

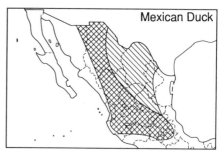
Mexican Duck

MOTTLED DUCK
Anas fulvigula maculosa or *A. platyrhynchos*
(in part)   Not illustrated
Pato Tejano

ID: 20–23 in (51–58.5 cm). *A ♀-plumaged*

Mallard of NE Mexico. **Descr.** Sexes similar. Bill olive-yellow with black nail (♂) to orange-yellow, mottled black (♀), *note black spot at base below*; legs orange. Mottled dark brown and warm brown overall, *head and upper neck contrastingly paler, buff with faint dark streaking on hind auriculars* and sides of neck, dark brown crown and eyestripe. *Greenish-blue speculum bordered white* (often indistinctly) *only on trailing edge. Silvery-white wing-linings contrast strongly with dusky remiges.*

Voice. Loud quacking much like Mallard.

Habs. Freshwater and brackish marshes and ponds, arable fields. Gregarious but rarely in sizeable flocks, often with other ducks in winter. Nest similar to Mallard. Eggs: 8–12, greenish white (WFVZ).

SS: See ♀ Mallard. ♀ Gadwall more slightly built, paler, forehead steeper and head more angular; note white square on speculum.

SD: F to U resident (around SL) in coastal Tamps (possibly extreme N Ver); R to U winter visitor (Oct–Mar; SL–500 m) inland to S NL (Howell and Webb 1990d), S to Ver, possibly to Tab (Saunders and Saunders 1981).

RA: S USA and NE Mexico.

NB: Black Duck A. (platyrhynchos?) rubripes (21–24 in; 53.5–61 cm) may occur in N Tamps in winter (see Appendix B). Much like Mottled Duck but darker overall, blackish brown with poorly contrasting brown edgings; head and neck contrastingly paler, greyish with extensive dark flecking (only median throat unstreaked), and dark crown and eyestripe. Speculum blue to violet-blue, edged white only on trailing edge. Bill yellowish (♂) to greenish yellow (♀) with black nail, ♀ often has blackish mottling on culmen, greyish bill of imm may be kept until at least Jan; legs orange-red to orange.

NORTHERN PINTAIL

Anas a. acuta Not illustrated
Pato Golondrino Norteño

ID: 20–23 in (51–58.5 cm). A fairly large but slender, longish-necked duck; common and widespread in winter. *Tail fairly long and pointed, finely elongated central rectrices of ♂ add 3 in (7.5 cm) to length.* **Descr.** Sexes differ. Alt ♂ (Oct–Jun): bill blue-grey with black culmen and nail, legs grey. *Head and hindneck dark brown, white of chest and foreneck extends as stripe into dark sides of neck.* Body pale grey overall, elongated black scapulars edged white, median underparts and broad band on hind flanks white. Tail white with black central points. Upperwings grey, speculum dark bronzy green with buff leading edge, white trailing edge, underwings pale grey. ♀/imm: bill grey with black nail. *Plumage mottled brown and pale grey-brown; speculum brown with white trailing edge.* Resembles adult by mid-winter.

Voice. ♂ has a high, rolled *wirrr-p* or *quirrr-k'*, ♀ a slightly reedy quacking *kwerr'k.*

Habs. Estuaries, marshes, lakes, arable fields. Often in large flocks, locally up to several thousands.

SS: ♀ longer-necked, more elegant, and often paler than other ♀ dabbling ducks; note white trailing edge to speculum. See Fulvous Whistling-Duck.

SD: C to F transient and winter visitor (Sep–May, R from Jul; SL–2500 m) almost throughout region; U to R on E coast of Yuc Pen and in Belize. Small numbers oversummer in BCN.

RA: Holarctic breeder; winters S in New World to N S.A.

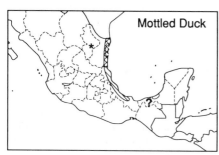

Mottled Duck

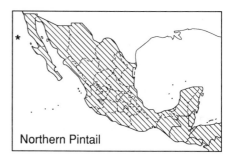

Northern Pintail

BLUE-WINGED TEAL
Anas discors   Not illustrated
Cerceta Aliazul

ID: 15–16 in (38–40.5 cm). Common and widespread winter visitor. **Descr.** Sexes differ. **Alt** ♂ (Oct–Jun): bill black, legs yellowish. *Head and neck blue-grey with broad white crescent in front of eye.* Body brownish (chest and sides spotted blackish), with broad white band on hind flanks, black tail coverts. *Pale grey-blue forewing sometimes visible at rest, striking in flight;* speculum green with white leading edge. **Basic** ♂ (Jun–Sep): much like ♀ but forewing brighter. **♀/imm:** bill greyish. Mottled brown and grey-brown, head and upper neck paler with *dark crown and eyestripe, whitish spot at base of bill; bluish forewing* duller than ♂. Resembles adult by midwinter.

Voice. ♂ has a slightly reedy, piping *pseep;* ♀ gives low quacks.
Habs. Marshes, lakes, coastal lagoons. Often in flocks, locally in thousands. Nest platform of reeds, grass, etc., in dense vegetation near water. Eggs: 6–11, creamy (WFVZ).
SS: See Green-winged Teal. ♀ Cinnamon Teal has slightly spatulate bill suggesting Northern Shoveler, color richer overall, face more uniform with whitish eyering, but juv often greyer with more distinct face pattern, perhaps not always safely told from Blue-wing.
SD: C to F transient and winter visitor (Sep–May; SL–2500 m) throughout most of region, commonest in S and E; U in Baja. Small numbers oversummer locally in N and cen Mexico; has bred DF (RGW), possibly Chih (Stager 1954).
RA: Breeds N.A.; winters S to cen Peru and cen Argentina.

GARGANEY
Anas querquedula   Not illustrated
Cerceta Cejiblanca

ID: 15–16 in (38–40.5 cm). Vagrant. **Descr.** Sexes differ. **Alt** ♂ (Feb–May): bill blackish, legs grey. *Head dark brown with broad white postocular stripe;* body brown overall with pale grey sides and elongated, white-striped hind scapulars. *Wings resemble* ♂ Blue-winged Teal but paler and greyer overall, bluish grey, speculum also bordered white on trailing edge. ♀: bill grey. Mottled brown and grey-brown, head and upper neck paler with *paler supercilium set off by dark crown and eyestripe; note whitish spot at base of bill continuing posteriorly as narrow pale stripe below dark eyestripe, short dusky auricular stripe* often indistinct. Upperwings greybrown, speculum dull greenish brown with white leading and trailing edges. **Imm** ♂/**basic** ♂ (Aug–Feb): resembles ♀ but note bluish-grey forewing, duller in imm.

Voice. ♂ has a dry, crackling rattle, ♀ a low quack much like Green-winged Teal.
Habs. Lakes, marshes, usually fresh water. Likely to be found with other ducks, especially Blue-winged Teal.
SS: Alt ♂ unmistakable; note face and wing pattern of ♀/imm/basic ♂, pale blue-grey forewings of ♂.
SD: Vagrant; one record: a ♂ near Escuinapa, Sin, on 22 Mar 1973 (PA).
RA: Breeds Eurasia; winters Africa and S Asia to Australia. Regular vagrant to W N.A.

Garganey

CINNAMON TEAL
Anas cyanoptera septentrionalium
Cereta castaña   Not illustrated

ID: 15–16 in (38–40.5 cm). Local in summer, widespread in winter. *Bill slightly spatulate.* **Descr.** Sexes differ. **Alt** ♂ (Oct–Jun): eyes red, bill blackish, legs yellowish.

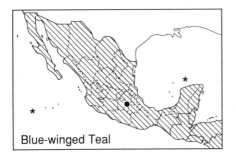
Blue-winged Teal

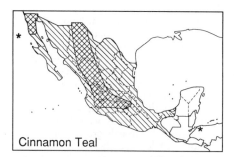

Cinnamon Teal

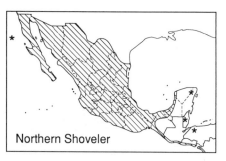

Northern Shoveler

Bright chestnut overall, darker and browner on upperparts. *Pale grey-blue forewing sometimes visible at rest, striking in flight*; speculum green with white leading edge. Basic ♂ (Jun–Sep): much like ♀ but forewings brighter. ♀/juv: eyes brown, bill grey. Mottled brown and warm brown, head and upper neck paler with *dark crown and whitish eyering*, indistinct whitish spot at base of bill; *bluish forewing* duller than ♂. Juv may be duller and greyer with more distinct whitish spot at bill base and darker eyestripe, much like Blue-winged Teal. Resembles adult by mid-winter.

Voice. Much like Blue-winged Teal; ♂ has a slightly reedy, piping *pseep*, ♀ low quacks.
Habs. Marshes, lakes, coastal lagoons. Often in flocks but rarely of more than a few hundred birds. Nest platform of reeds, grasses, etc., in dense cover near water. Eggs: 6–12, creamy to buffy cream (WFVZ).
SS: See Blue-winged and Green-winged teal. Northern Shoveler larger with heavier and longer bill, orange legs.
SD: U to F summer resident, breeding locally (May–Aug; SL–2500 m) in BCN and on Plateau S to cen volcanic belt. C to F transient and winter visitor (Sep–May; SL–2500 m) S to cen Mexico, U to R to Guatemala, R and irregular to NW Honduras. Reports from Yuc Pen (Paynter 1955a) and Belize (Wood *et al.* 1986) require verification.
RA: Breeds W N.A. to cen Mexico; winters to N S.A.

NORTHERN SHOVELER
Anas clypeata   Not illustrated
Pato Cucharón Norteño

ID: 17–20 in (43–50.5 cm). Common and widespread winter visitor. *Bill slightly longer than head and markedly spatulate.* Descr. Sexes differ. Alt ♂ (Oct–Jun): eyes

yellow, bill blackish, legs reddish orange. *Dark glossy green head and neck contrast with white chest, underbody reddish chestnut* with broad white band on hind flanks. Upperparts blackish, elongated scapulars edged white, tail white. *Pale grey-blue forewing sometimes visible at rest, striking in flight*; speculum green with white leading and trailing edges. ♀/imm: eyes brown, bill orange with dark culmen. Mottled brown and dark brown overall, head and upper neck paler, outer rectrices whitish; forewings duller, more greyish than ♂. Imm attains ♂ plumage via supplemental plumage (like dull alt ♂, often with whitish crescent in front of eye) by spring.

Voice. Usually silent; ♂ has a low, gruff *ahrr rrah*, ♀ a quiet quacking *rehk*.
Habs. Marshes, lakes, estuaries. Often in flocks, at times in thousands. Typically feeds by filtering surface water with bill, less often up-ends.
SD: C to F transient and winter visitor (Sep–May; U to R from Jul; SL–2500 m) almost throughout region; U to R on E coast of Yuc Pen and in Belize; oversummers locally in N Mexico.
RA: Holarctic breeder; winters S in New World to NE S.A.

GADWALL
Anas strepera   Not illustrated
Pato Pinto

ID: 18–21 in (45.5–53.5 cm). A fairly large, heavily built duck, common in winter N of Isthmus. Descr. Sexes differ. Alt ♂ (Oct–Jun): bill dark grey, legs orangish. Grey-brown head and neck, and *black rump and tail coverts contrast with overall grey body*; elongated scapulars edged cinnamon, belly white. Upperwings grey with chestnut median coverts, black greater coverts, *white inner square on dark speculum*. ♀/imm: bill orange with black

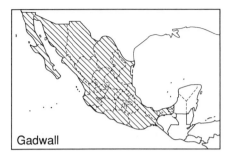

Gadwall

culmen. Mottled brown and pale brown, head and neck paler, belly whitish; *wings similar to ♂ but chestnut and black areas reduced, often lacking in imm ♀.* Resembles adult by mid-winter.

Voice. Quacking *raahk rrahk* or *kwah . . .* of ♂ slightly rougher than Mallard; ♀ has a quieter quack.

Habs. Lakes, marshes, coastal lagoons. Rarely in flocks of more than a few hundred birds.

SS: See Mallard, Mottled and Mexican ducks. ♀ Wigeon smaller, more reddish brown overall with blue-grey bill.

SD: C to F winter visitor (Oct–May; SL–2500 m) S to cen Mexico, U to S Mexico; reports from Yuc Pen (Saunders and Saunders 1981) require verification. Unrecorded C.A. but may occur at least in Pacific lowlands of Guatemala. Small numbers oversummer locally in N Mexico.

RA: Holarctic breeder; winters in New World to S Mexico, N Caribbean.

EURASIAN WIGEON
Anas penelope   Not illustrated
Pato Silbón

ID: 17–20 in (43–51 cm). *Rare winter visitor to N Mexico.* **Descr.** Sexes differ. **Alt ♂** (Oct–Jun): bill blue-grey with black nail, legs grey. *Head and upper neck rufous-chestnut with broad creamy-yellow fore-*

head stripe. Chest pinkish, *body pale grey overall,* belly and broad band on hind flanks white, tail coverts black. *White forewing striking in flight,* speculum dark green with black trailing edge. *Underwings, including axillars, smoky grey.* ♀/ **imm**: *mottled brown and cinnamon-brown overall,* head and upper neck rarely paler and greyer; belly whitish. Forewing grey-brown (♀) to pale grey (imm ♂). Resembles adult by early winter but imm forewing pattern retained through 1st winter. Hybrid Eurasian × American Wigeon are fairly common in California and may occur in Mexico; some closely resemble ♂ Eurasian Wigeon but show green postocular stripe in rufous-chestnut head.

Voice. Whistles of ♂ and low quacks of ♀ much like American Wigeon.

Habs. Much like American Wigeon with which it usually occurs.

SS: ♀ American Wigeon has head and neck contrastingly paler and greyer, typically with more distinct dusky eye-patch (♀ Eurasian Wigeon head more uniform, warmer brownish), axillars white. See note on hybrids (above).

SD: R to U winter visitor (Oct–Apr; SL–250 m) to BCN (RS, REW), R to BCS (PA), Son (DS) and Tamps (GWL); vagrant S to Jal (Feb 1993, SNGH) and to Clipperton.

RA: Breeds N Eurasia, recently has spread to N.A.; winters in small numbers along coasts of N.A.

AMERICAN WIGEON
Anas americana   Not illustrated
Pato Chalcuán

ID: 17–20 in (43–51 cm). Common and widespread in winter. **Descr.** Sexes differ. **Alt ♂** (Oct–Jun): bill blue-grey with black nail, legs grey. *Head and upper neck grey with broad whitish forehead stripe, and broad dark green postocular stripe. Body*

Eurasian Wigeon

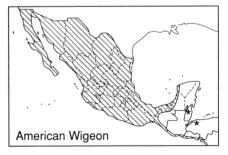

American Wigeon

pinkish overall, belly and broad band on hind flanks white, tail coverts black. *White forewing striking in flight*, speculum dark green with black trailing edge. Underwings pale grey, *axillars white*. ♀/**imm**: mottled brown and cinnamon-brown overall, *head and upper neck contrastingly paler and greyer*, typically with dusky eye-patch, belly whitish. Forewing grey-brown, adult ♀ and imm ♂ usually with greater coverts broadly edged white, imm ♂ may have pale grey to mostly white forewing in 1st winter. Resembles adult by early winter, but imm forewing pattern usually retained through 1st winter.

Voice. ♂ has a high, thin, slightly hoarse, whistled *whew-heu* and *whi-hew-hew*, etc.; ♀ gives low quacks.

Habs. Estuaries, lakes, marshes. Often in large flocks, locally in thousands. Mostly feeds by grazing.

SS: See Eurasian Wigeon, Gadwall.

SD: C to F transient and winter visitor (Oct–May, U to R from Jul; SL–2500 m) N of Isthmus, irregularly F to U to N and W Yuc Pen, U to Honduras; may oversummer locally.

RA: Breeds NA; winters to N S.A.

TRIBE AYTHYINI: Scaups

Scaups are medium-sized ducks of fresh and salt water. Legs set far back and hind toes lobed to aid in underwater propulsion; unlike dabblers, scaups have to run along the water surface to gain momentum for take-off. Sexes differ. Basic (eclipse) ♂♂ much like ♀♀. Boldly patterned ♂♂ appear black and white, ♀♀ dark brown overall; paler remiges often form broad contrasting stripe along the trailing edge of the upperwings; underwings whitish with dusky remiges. Imms attain adult plumage through first winter, ♂♂ much like dull adults by mid-winter. Mostly silent, ♂♂ have cooing or mewing whistles given in courtship; ♀♀ rarely give low growls. Feed by diving (to depths of 6 m), often at night, with much of day spent sleeping, bills tucked into scapulars. Food aquatic vegetation, insects, molluscs, and crustaceans.

CANVASBACK

Aythya valisineria   Not illustrated
Pato Coacoxtle

ID: 20–23 in (51–58.5 cm). *Long sloping forehead and long bill.* **Descr.** Alt ♂ (Oct–

Canvasback

Jun): eyes red, bill blackish, feet grey. *Chestnut head and neck and black chest contrast with silvery-grey body*, rump and tail coverts black. *Upperwings pale grey.* ♀: eyes brown. Tawny brown head (often with whitish spectacles) and neck, chest slightly duskier, *body pale grey overall*, white below. *Upperwings grey with paler remiges.*

Habs. Lakes, coastal bays, and lagoons. Mostly concentrated in flocks at favored sites, rarely up to a few thousand birds.

SS: Redhead smaller with steeper forehead, shorter blue-grey to grey bill has broad black tip, ♂ has yellow eyes; in flight note pale remiges above.

SD: F to C winter visitor (Nov–Apr; SL–2500 m) S to cen Mexico, U in Baja, U to R in S Mexico (possibly including Yuc Pen, Saunders and Saunders 1981), R and irregular to NW Honduras.

RA: Breeds NW NA; winters to Mexico, rarely N C.A.

REDHEAD

Aythya americana   Not illustrated
Pato Cabecirrojo

ID: 18–20 in (45.5–51 cm). *Steep forehead.* **Descr.** Alt ♂ (Oct–Jun): eyes yellow, bill blue-grey with whitish subterminal band and black tip, feet grey. Chestnut head and neck and black chest contrast with pale

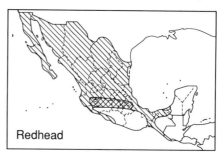
Redhead

grey body, rump and tail coverts black. *Upperwings dark grey with paler grey remiges.* ♀: eyes brown, *bill grey with black tip and pale subterminal band. Tawny brown overall* with whitish belly; lores usually paler, often shows narrow whitish spectacles. *Wings similar to ♂.*

Voice. ♂ gives a low, mewing *mowh* or *me-owh* in courtship.

Habs. Lakes, coastal bays and lagoons; marshes and lakes with emergent vegetation for nesting. Nest platform of reeds etc., in marshy vegetation near or over water. Eggs: 6–12, pale olive-buff to pale greenish (WFVZ); notorious for dumping eggs in several nests, often of other species.

SS: See Canvasback. Scaups have bold white wingstripes, ♀♀ darker brown overall, typically with bold white band around base of bill, ♂♂ have blackish heads. ♀ Ring-necked Duck has peaked nape, contrasting grey face, bolder bill pattern.

SD: Irregular U and local breeder (May–Aug; 1000–2300 m) in cen volcanic belt, at least in Jal (Williams 1975) and DF (Wilson and Ceballos-L. 1986). C to F winter visitor (Nov–Apr; SL–2500 m) S to cen Mexico; U to R to S Guatemala, possibly to Yuc Pen (Saunders and Saunders 1981). Small numbers oversummer locally in N Mexico.

RA: Breeds N.A., locally to cen Mexico; winters to N C.A.

RING-NECKED DUCK
Aythya collaris   Not illustrated
Pato Piquianillado

ID: 16–18 in (40.5–45.5 cm). *Nape strongly peaked.* **Descr.** Alt ♂ (Oct–Jun): eyes yellow, *bill blue-grey with white subterminal band, broad black tip, and narrow white band at base,* feet grey. Black overall with *pale grey sides bordered by contrasting white vertical stripe at sides of chest,* belly whitish. *Upperwings dark grey with paler remiges.* ♀: eyes pale brown, bill grey with black tip and whitish subterminal band. Dark brown overall with *pale grey face and sides of neck,* becoming whitish on lores, *narrow, white, teardrop-shaped eyering usually distinct,* belly whitish. *Wings similar to ♂.*

Habs. Wooded ponds and lakes, lagoons, marshes, rarely on salt water. Often at smaller and more wooded water bodies than other diving ducks, rarely in flocks of 100+ birds.

SS: See Redhead. ♀ scaups lack pale grey face, white around base of bill bolder; in flight note their bold white wingstripes.

SD: F to C winter visitor (Oct–Apr; SL–2500 m) S to cen Mexico, U in S Mexico, R and irregular to NW Honduras.

RA: Breeds N.A.; winters S to Panama.

GREATER SCAUP
Aythya marila mariloides   Not illustrated
Pato-boludo Mayor

ID: 17–20 in (43–51 cm). Rare winter visitor to NW Mexico. *Forehead steep and crown fairly flat.* **Descr.** Alt ♂ (Oct–Jun): eyes yellow, bill blue-grey with black nail, feet grey. *Head and upper neck black with green sheen,* chest black, upperparts pale grey becoming black on rump and tail coverts, sides and belly white. *Upperwings grey with dark-tipped white remiges.* ♀: bill grey with black nail. Dark brown overall with broad white band around base of bill, belly whitish. *Wings similar to ♂.*

Habs. Coastal bays and estuaries, open ocean, rarely adjacent freshwater lagoons. Singly or in small groups up to 10 birds; may associate with Lesser Scaup.

SS: Lesser Scaup has more sloping forehead and peaked nape, narrower bill, in flight white wingstripes extend only to outer

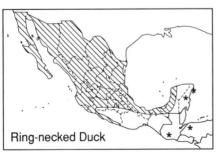

Ring-necked Duck

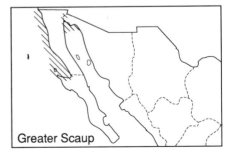

Greater Scaup

secondaries, becoming pale grey on primaries; ♂ has purple sheen to head (but can appear green at times), ♀ best told by head and bill shape. See Ring-necked Duck, Redhead.

SD: R to U winter visitor (Nov–Mar) along Pacific coast of Baja S to Bahia San Ignacio, R (irregular?) in N Gulf of California. Reports from Sin (Friedmann *et al.* 1950) are unsatisfactory.

RA: Holarctic breeder; winters S in New World to NW Mexico, SE USA.

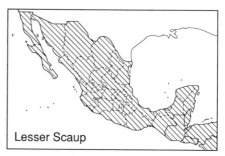
Lesser Scaup

LESSER SCAUP
Aythya affinis　　　　Not illustrated
Pato-boludo Menor

ID: 15–17 in (38–43 cm). Common and widespread winter visitor. *Forehead fairly sloping, crown peaks at rear.* **Descr.** Alt ♂ (Oct–Jun): eyes yellow, bill blue-grey with black nail, feet grey. *Head and upper neck black with purplish (rarely green) sheen,* chest black, upperparts pale grey becoming black on rump and tail coverts, sides and belly white. *Upperwings grey with dark-tipped white secondaries, pale grey primaries.* ♀: bill grey with black nail. Dark brown overall with broad white band around base of bill, belly whitish. *Wings similar to ♂.*
Habs. Lakes, ponds, coastal lagoons, estuaries, rarely open ocean. Often in large flocks, locally in thousands.
SS: See Greater Scaup, Ring-necked Duck, Redhead.
SD: C to F winter visitor (Oct–Apr; SL–2500 m) throughout most of region, U in Yuc Pen and Belize; small numbers may oversummer locally.
RA: Breeds N.A.; winters to N S.A.

TRIBE MERGINI: Sea ducks and saw-billed ducks

Sea ducks and saw-billed ducks, all winter visitors to the region, are small to medium-large ducks that feed by diving (to depths of 15 m or more) and may stay submerged for 1–2 minutes. Like scaups they have lobed hind toes and run along the water before take-off. Generally silent except during courtship. Sea ducks (7 species), as the name suggests, frequent salt water (at least in winter), though some species also occur on fresh water. Typically require 2 years to mature and attain adult plumage in 2nd winter. ♂♂

vary from almost all black (scoters) to strikingly patterned black and white (goldeneyes); ♀♀ overall dark brown or brown and white. Several species have large white wing patches obvious in flight. Food mainly molluscs, crustaceans, aquatic insects. Saw-billed ducks (3 species), also known as mergansers, have fairly long, slender bills with strongly serrated tomia used for grasping fish. Occur on fresh and salt water. ♂♂ strikingly patterned, appear black and white overall; ♀♀ grey-brown overall; all have white wing patches obvious in flight. Food mostly small fish, crustaceans, amphibians, and aquatic insects.

HARLEQUIN DUCK
Histrionicus histrionicus　　Not illustrated
Pato Arlequín

ID: 15–17 in (38–43 cm). Vagrant. *Bill small, forehead steep.* **Descr.** Sexes differ. Alt ♂ (Oct–May): bill and legs grey. *Slaty blue overall, strikingly patterned with white. Broad white crescent in front of eye* continues over eye to nape as narrow chestnut line; note white spot and vertical bar on auriculars, white collar, and black-bordered white stripe on sides of chest. *Chestnut sides and flanks,* white stripe along scapulars, white tertial edgings, and small white spot at sides of black undertail coverts complete harlequin pattern. *Wings*

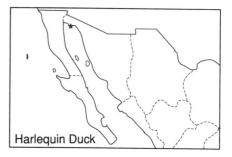
Harlequin Duck

appear all dark above and below. Basic ♂ duller and browner with trace of alt pattern. ♀/imm: dark brown overall with *small round white spot on auriculars, larger, less sharply defined white spot below eye,* and often a third small whitish patch in front of eye, belly whitish. Imm ♂ shows traces of adult plumage by midwinter.

Habs. Rocky sea coasts, rarely coastal lagoons, estuaries, harbors. Lone birds may associate with other sea ducks.

SS: ♀ similar to scoters, especially Surf Scoter; note small bill, steep forehead. ♀ Bufflehead has single elongated white patch on auriculars, chest and sides paler.

SD: Vagrant; two records, both at Puerto Peñasco, Son: one ♂ in Mar–Apr 1977 (Kaufman and Witzeman 1979), 2–3 ♀♀ in Jan–Mar 1990 (TEW and others).

RA: Holarctic breeder; winters S in New World to California and NE USA.

OLDSQUAW
Clangula hyemalis   Not illustrated
Pato Colilargo

ID: 16–18 in (40.5–45.5 cm). Vagrant. Bill small, forehead steep. Elongated central rectrices of adult ♂ add 4–5 in (10–12.5 cm) to length. **Descr.** Sexes differ. **Basic. ♂:** bill blackish with pink subterminal band, feet grey. *Looks white overall,* note pale brown-grey face with *large dark brown auricular patch*; chest, broad Y on mantle, rump, uppertail coverts, and central rectrices blackish brown. *Wings dark brown.* ♀/imm: bill grey, ♂ often with partial pink band. *Head and neck whitish with dark brown crown, nape, and large hind auricular patch,* chest greyish brown, rest of underbody white. Upperparts dark brown overall. Imm ♂ develops elongated silvery-white scapulars through 1st winter. **Alt** (unlikely in region). ♂: head, neck, and chest dark brown with large whitish face patch. ♀: face variegated dark and pale brown with little white.

Habs. Open ocean, bays, harbors, estuaries, small lakes near coast. Dives well and may stay under water for long periods.

SD: Vagrant (late Nov–early Apr) to NW Mexico; 4 records: all from Gulf of California (BCN, Son, Sin) S to Jal (Huey 1927; Russell and Lamm 1978; Kramer 1982; Howell and Pyle 1987).

RA: Holarctic breeder; winters S in New World to California, NE USA.

NB: Known as Long-tailed Duck in the Old World.

BLACK SCOTER
Melanitta nigra americana   Not illustrated
Negreta Negra

ID: 18–20 in (45.5–51 cm). Rare winter visitor to NW Baja. *Head shape slightly domed.* **Descr.** Sexes differ. ♂: *bill black with swollen orange-yellow knob at base,* feet blackish. Plumage entirely black; *in flight underside of remiges appears contrastingly paler.* ♀/imm: bill dark grey. Dark brown overall with *contrasting pale greyish face and upper neck, and darker crown and hindneck. Underside of remiges contrasts as in ♂.* Imm has whitish belly, ♂ becomes mottled with black from midwinter onward.

Habs. Open ocean, coastal bays, rarely estuaries and harbors. Often with other scoters.

SS: All-black head and bill pattern of ♂ distinctive; some ♀ Surf Scoters, particularly oversummering birds, appear to have pale face and dark cap, at close range whiter areas corresponding to face spots usually visible; note squarer head, heavier bill, orange feet. Basic ♂ Ruddy Duck has whiter

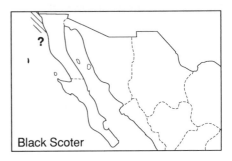

Oldsquaw

Black Scoter

face, body sides paler, bill blue-grey, rarely on salt water.

SD: R winter visitor (Nov–Apr?) along Pacific coast of BCN S to 32°N (SNGH, PP). May oversummer in NW BCN but reports of Hubbs (1955, 1956) require verification.

RA: Holarctic breeder; winters S in New World to NW Mexico, E USA.

NB: Known as Common Scoter in the Old World, although New World Black Scoter sometimes considered specifically distinct.

SURF SCOTER

Melanitta perspicillata   Not illustrated
Negreta de Marejada

ID: 18–20 in (45.5–51 cm). Common winter visitor to NW Mexico. *Forehead short and steep, head shape fairly squared, bill relatively heavy.* **Descr.** Sexes differ. ♂: eyes white, *bill mostly orange with white base almost filled by black spot,* feet pinkish red. Black overall with *bold white patch on forehead and hindneck.* ♀/imm: eyes dark, bill dark grey, feet duller pinkish, often dusky in imm. Dark brown overall with *white spot on auriculars and lores.* Imm has whitish belly, ♂ becomes mottled with black from mid-winter onward.

   Habs. Open ocean, coastal bays, estuaries, harbors. May be in large flocks, up to hundreds, often closer inshore than other scoters.

SS: White-winged Scoter has longer, more sloping forehead, white secondaries (often hard to see when swimming); ♂ lacks ♂ Surf's bold white head patches; ♀'s white lore spot round versus vertical and oblong in Surf. See Black Scoter.

SD: C to F winter visitor (Nov–Apr) along Pacific coast of Baja S to Bahia Magdalena and in N Gulf of California; small numbers oversummer locally; R and irregular in winter S to Nay, and (Nov–Dec) inland

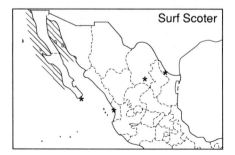

in NE Mexico S to Coah (SNGH, SW); probably also occurs on coast of N Tamps.

RA: Breeds N.A.; winters S to NW Mexico, SE USA.

WHITE-WINGED SCOTER

Melanitta fusca deglandi   Not illustrated
Negreta Aliblanca

ID: 20–23 in (51–58.5 cm). Uncommon winter visitor to NW Mexico. *Forehead sloping.* **Descr.** Sexes differ. ♂: eyes white, *bill yellow-orange with black basal knob,* feet pinkish red. Black overall with *white teardrop-shaped patch through eye, white secondaries striking in flight.* ♀/imm: eyes dark, bill blackish, feet duller pinkish, often dusky in imm. Dark brown overall with *whitish spot on auriculars and lores, white secondaries striking in flight.* Imm has whitish belly, ♂ becomes mottled with black from mid-winter onward.

   Habs. Open ocean, coastal bays, estuaries. Often with Surf Scoters.

SS: See Black and Surf Scoters.

SD: U winter visitor (Nov–Apr) along Pacific coast of BCN, R to BCS and in N Gulf of California; R in summer (May–Oct).

RA: Holarctic breeder; winters S in New World to NW Mexico, E USA.

NB: Known as Velvet Scoter in the Old World.

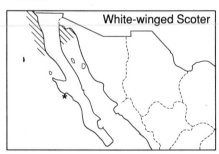

COMMON GOLDENEYE

Bucephala clangula americana
Ojodorado Común   Not illustrated

ID: 17–20 in (43–51 cm). Winter visitor to N Mexico. *Forehead slopes into high peaked crown.* **Descr.** Sexes differ. ♂: eyes yellow, bill blackish, feet orangish. *Head glossy blackish green with large white spot on lores;* neck, chest, and underbody white. Upperparts black, elongated white scapulars merge with white sides. Wings blackish with *large white patch on trailing edge*

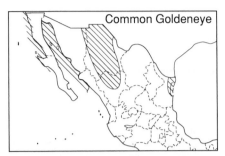

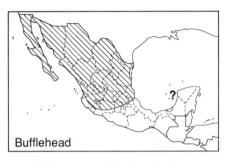

of innerwing. ♀: bill blackish, with narrow
pink subterminal band in adult. *Head dark
brown to tawny-brown, white neck appears
as narrow collar when neck retracted.* Body
grey overall, often with white stripe visible
on closed wing, belly whitish. *Wings similar to ♂ but 2 dark bars split white patch.*
Imm: resembles ♀ but chest paler, white
wing patch smaller in ♀s; ♂ has darker
head, black bill, white lore spot appears
from mid-winter onward.

Habs. Lakes, estuaries, coastal bays.
Often in small groups. In flight, wings
make a hollow, hoarse whistling sound.

SD: U to F winter visitor (Nov–Apr) on
Pacific coast of Baja and in Gulf of Califor-
nia S to Sin; U to R in interior (to 2000 m)
S to Dgo, and on Atlantic coast in Tamps.

RA: Holarctic breeder; winters S in New
World to N Mexico, SE USA.

NB: Barrow's Goldeneye *Bucephala islandica*
(17–20 in; 43–51 cm) may occur (Dec–
Mar) as vagrant to NW Mexico (see
Appendix A). Differs from Common
Goldeneye in steep forehead, flatter crown.
♂ has glossy blackish-purple head with
large white crescent on lores; white tips to
inner scapulars form band of short white
bars separate from white sides. Wings
similar to ♂ Common Goldeneye but black
bar splits white square. ♀ has orange bill
(or dark with orangish subterminal band
in imm), dark brown head; wings similar
to ♀ Common but lesser upperwing
coverts greyer so white patch appears
smaller. Often with Common Goldeneyes.

BUFFLEHEAD

Bucephala albeola   Not illustrated
Pato Monja

ID: 14–16 in (35.5–40.5 cm). Winter visitor
to N Mexico. **Descr.** Sexes differ. ♂: bill
grey, feet orangish. Head black with green
and purple sheen, *large triangular white*

*patch flares out behind and above eye.
Neck, chest, and underbody white,* upperparts blackish. Wings blackish with *large
white square on trailing edge of innerwing.*
♀: feet dusky flesh. *Head dark brown with
broad white stripe on auriculars,* otherwise
grey-brown overall, darker above, belly
whitish. Wings dark brown with *small
white square on inner secondaries.* **Imm**:
resembles ♀ with smaller white auricular
stripe, ♂ shows trace of white head patch
by mid-winter, ♀ has smaller white wing
patch.

Habs. Lakes, coastal lagoons, estuaries,
harbors. Often in flocks, at times up to a
few hundred birds.

SS: ♂ Hooded Merganser has erectile crest,
black stripes at sides of chest, flanks
greyish cinnamon. See Harlequin Duck
and scoters.

SD: F to C winter visitor (Nov–Apr) in Baja
and on Pacific Slope from Son to Sin, F to
U in interior (to 2500 m) S to cen volcanic
belt; U to R on Atlantic Slope from Tamps
to N Ver, possibly R to NW Yuc Pen
(Saunders and Saunders 1981).

RA: Breeds N.A.; winters to cen Mexico, SE
USA.

HOODED MERGANSER

Mergus cucullatus   Not illustrated
Mergo de Caperuza

ID: 16–18 in (40.5–45.5 cm). Rare winter
visitor to N Mexico. *Bill fairly long and
slender, erectile fan-like crest often striking.*
Descr. Sexes differ. ♂: eyes yellow, bill
black, feet yellowish. Head and neck
black, *large white patch flares out behind
eye.* Chest and belly white with *2 vertical
black stripes on sides of chest,* flanks
greyish cinnamon. Upperparts blackish,
elongated tertials edged white. Wings
blackish, *inner half of upperwings whitish,
split by 2 dark bars.* ♀: eyes dark, *bill dark*

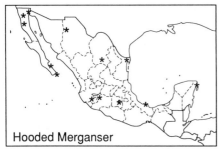

Hooded Merganser

with orangish base below. Grey-brown overall with cinnamon crest, belly whitish. Wings blackish, speculum black with whitish-tipped inner secondaries and white inner leading edge. **Imm** ♂: resembles ♀, shows traces of ♂ plumage by mid-winter.
Habs. Lakes and ponds, especially with wooded and reedy edges. Singly or in pairs, often with other waterfowl but usually flies off separately.
SS: ♀ nondescript but distinctive; note small size, cinnamon crest, slender bill. See Bufflehead.
SD: R but probably regular winter visitor (Nov–Mar; SL–1500 m) to N BCN, R and irregular S (formerly regular?) to cen volcanic belt; vagrant (Dec) to S Ver (Pashley 1987) and QR (Lopez O. *et al.* 1989).
RA: Breeds N.A.; winters to S USA, rarely N Mexico.
NB: Sometimes placed in the genus *Lophodytes*.

COMMON MERGANSER
Mergus merganser americanus
Mergo Mayor   Not illustrated

ID: 23–26 in (58.5–66 cm). N Mexico in winter. *Bill long and slender, head angular with bushy crest.* **Descr.** Ages/sexes differ. ♂: bill red with black tip, feet orangish red. *Dark green head and upper neck contrast with white chest* and underbody. Mantle

and inner scapulars black, becoming pale grey on rump and uppertail coverts, white outer scapulars usually merge with white sides. *Inner half of upperwings white* with indistinct dark bar on median coverts, outer half blackish, underwings whitish. ♀/**imm**: feet duller. *Dark brown head and upper neck contrast sharply with whitish chest* and belly; throat white. Flanks and upperbody grey overall. Upperwings dark with white patch on inner secondaries, smaller on imm ♀. ♂ shows traces of adult plumage by mid-winter, especially blackish collar at base of neck.
Habs. Lakes, mountain rivers, rarely salt water. Often in small groups, usually apart from other species. Nests in tree cavities, usually near water. Eggs: 9–12, buffy white (WFVZ).
SS: Red-breasted Merganser (mainly coastal) has ragged double crest, red eyes, darker overall; ♂ has brownish chest, grey flanks; brownish-cinnamon head and neck of ♀ do not contrast sharply with pale brownish sides of chest.
SD: U to F winter visitor (Nov–Apr; SL–2000 m) to N tier of Mexican states (but mainly Chih and Coah), from BCN to Tamps; formerly F to C, occurring S to cen volcanic belt; apparently bred Chih (Van Rossem 1929).
RA: Holarctic breeder; winters S in New World to N Mexico, SE USA.
NB: Known as Goosander in the Old World.

RED-BREASTED MERGANSER
Mergus serrator   Not illustrated
Mergo Copetón

ID: 20–23 in (51–58.5 cm). N Mexico in winter. *Bill long and slender, ragged double crest.* **Descr.** Sexes differ. ♂: eyes red, bill reddish with dark tip, feet reddish orange. *Dark green head contrasts with white neck. Chest cinnamon-brown* with

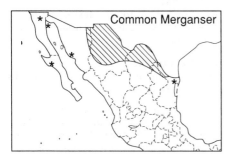

Common Merganser

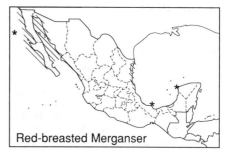

Red-breasted Merganser

black-and-white patch on sides, belly whitish, flanks pale grey. Mantle and scapulars blackish, becoming pale grey on rump and uppertail coverts, white upperwing coverts often form white stripe along sides. *Inner half of upperwings white*, outer half blackish, underwings whitish. ♀/imm: bill orangish with dark tip, eyes dull in imm. *Cinnamon-brown head and neck blend with grey-brown sides of chest*; throat, foreneck, and belly whitish. Flanks and upperparts brownish grey overall. Upperwings dark with white patch on inner secondaries, smaller in imm ♀. ♂ shows traces of adult plumage by mid-winter.

Habs. Coastal bays, estuaries, lakes. Often in small groups, may associate loosely with other species.

SS: See Common Merganser.

SD: F to U winter visitor (Nov–Apr) along coasts of Baja (rarely oversummers), Son (R inland), and N Sin, R to Nay; F to U on Atlantic coast in Tamps, R to S Ver (PP) and NW Yuc Pen (SNGH, RGW). A report from Chih (Hubbard and Crossin 1974) requires verification.

RA: Holarctic breeder; winters S in New World to N Mexico, SE USA.

TRIBE OXYURINI: Stiff-tailed ducks

Stiff-tailed ducks are small, compact, freshwater ducks with fairly long broad tails. They are very capable divers, well equipped with a rudder of stiff rectrices and legs set far back for strong propulsion. Unlike other ducks, stiff-tails can sink slowly below the surface, scarcely leaving a ripple. Alt ♂♂ are handsome, chestnut overall with bright blue bills, ♀♀ and basic ♂♂ are brownish overall. Rarely vocal. Food mostly vegetable matter and aquatic insects.

RUDDY DUCK
Oxyura j. jamaicensis   Figure 18
Pato Tepalcate

ID: 14–16 in (35.5–40.5 cm). Local in summer, widespread in winter. **Descr.** Ages/sexes differ. **Alt ♂** (Jan–Dec): bill blue, feet grey. *White face contrasts with black crown and nape and chestnut body*, belly whitish. Tail black. Wings dark. **Basic ♂** (Jan–Dec, mainly Sep–Mar): bill blue-grey, body grey-brown. ♀: bill blue-grey.

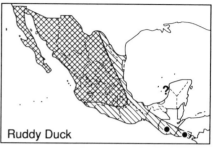

Ruddy Duck

Dark brown cap (to lower edge of eye) and nape, *face pale grey with dusky horizontal stripe*. Body dark brown overall (greyer in basic), belly whitish. Tail blackish. Wings dark. **Juv/imm ♂:** resembles ♂, white face becomes apparent after 3–4 months, attains ♂ plumage when about 1 year old.

Voice. ♂ gives low clucks during courtship.

Habs. Lakes, marshes, coastal lagoons, rarely salt water. Social, often in flocks up to a few hundred birds in winter. Runs across water prior to take-off. ♂ strikes bill against inflated chest in display to dispel air bubbles into surrounding water. Nest platform of reeds etc., in emergent marshy vegetation. Eggs: 5–10, whitish to buffy white (WFVZ). Unlike other ducks, ♂ may help with care of young.

SS: Masked Duck typically skulks in emergent vegetation and rarely seen, ♀/imm/basic ♂ warmer and buffier overall with double dark stripes on face, alt ♂ has black face. See Black Scoter (coastal NW Baja).

SD: F to C but local breeder (SL–2500 m) in Baja, and in interior and on Pacific Slope S to cen volcanic belt and S Mor (RWD); irregularly (near SL–1500 m) in Guatemala (per PH) and probably El Salvador (Thurber *et al.* 1987); may breed in any month. C to F S to Chis in winter (Oct–Apr), U to R and irregular to Guatemala and Honduras, possibly to Yuc Pen (Saunders and Saunders 1981).

RA: Breeds N.A. to cen Mexico and Caribbean; winters to N C.A. Disjunct populations in S.A.

MASKED DUCK
Oxyura dominica   Figure 18
Pato Enmascarado

ID: 13–15 in (33–38 cm). *A rarely seen, secretive neotropical duck.* **Descr.** Ages/sexes differ. **Alt ♂:** bill blue with black tip, feet grey, narrow orbital ring pale blue.

Fig. 18 (a) Masked Ducks; (b) Ruddy Duck

Black face and crown contrast with chestnut nape, neck, and chest. Body mottled chestnut and blackish overall, belly whitish. Tail black. Wings dark, *white square on trailing edge of innerwing* rarely visible except in flight. ♀/imm/basic ♂: bill blue-grey to grey, lacks blue orbital ring. *Buff supercilium contrasts with dark brown crown and eyestripe, face buff with dark horizontal stripe.* Body mottled dark brown and tawny-brown overall, belly whitish. Tail blackish. White wing patch smaller in ♀.

Voice. ♂ rarely gives a throaty *oo-oo-oo* or *kir-roo-kirroo-kiroo*, ♀ a short, repeated hiss (Stiles and Skutch 1989).

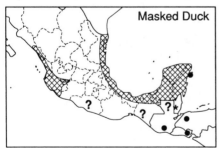

Masked Duck

Habs. Freshwater ponds and marshes with emergent vegetation, especially rushes; often on small bodies of water such as roadside ditches. Skulking and hard to see but emerges onto open water at night to feed. In pairs or small groups, not associating with other ducks. Take-off from water nearly vertical like dabbling ducks. Displays may be similar to Ruddy Duck but little known. Nest platform of reeds etc., well hidden in emergent marshy vegetation. Eggs: 3–6, buffy white.

SS: See Ruddy Duck.

SD: Poorly known. Locally/seasonally U to R resident (SL–1000 m) on Pacific Slope from S Sin to Col, possibly in Chis (Alvarez del Toro 1971*c*), on Atlantic Slope from Tamps to Yuc Pen (RAB, G and JC, PW); possibly to 1500 m in Lerma drainage (Lake Chapala, at least formerly, Nordhoff 1922); disjunctly in NW Honduras and, since 1970, on Pacific Slope in Guatemala (PH) and El Salvador (Thurber *et al.* 1987). Somewhat nomadic and irregular in occurrence due, at least in part, to fluctuating water levels.

RA: Mexico and Caribbean locally to S Peru and N Argentina.

NEW WORLD VULTURES: CATHARTIDAE (5*)

This New World family consists of large, superficially hawk-like birds with long, broad wings used for soaring. Heads and upper necks are naked and often brightly colored, with warty protuberances. Bills hooked for tearing into flesh but feet relatively weak. Ages similar or different, sexes similar; altricial young initially naked but quickly acquire down, fledge in 2–3 months. Plumage predominantly black or white. Molts protracted, and complete replacement of flight feathers may take one year or more. Mostly silent but may utter grunts and hisses.

Food almost exclusively carrion. 'Nests' are rudimentary, eggs being laid on bare surfaces such as the ground, cliffs, in caves, hollow logs, ruined buildings, etc.

Recent research indicates that New World Vultures are closely related to storks and not to raptors nor Old World Vultures.

BLACK VULTURE

Coragyps atratus *   Plate 7
Zopilote Negro

ID: 22–26 in; W 55–62 in (56–66 cm; W 140–158 cm). Common and widespread, social. *Wings long and broad, tail short, broad, and squared.* **Descr.** Ages similar. *Naked head and upper neck grey and wrinkled, bill dark grey with pale horn tip (all dark in juv), legs whitish. Plumage black overall, white shafts above and webs below of outer 6 primaries form contrasting whitish wing panels.*
Habs. Open areas, towns, rubbish dumps, river mouths, lake shores, rarely dense forest. Typically soars higher than Turkey Vulture, wings slightly raised, tail slightly spread. Wingbeats shallow, stiff, and rapid, interspersed with glides. Commonly in groups up to a few hundred birds, often on the ground where hops and ambles readily. Roosts communally, often with Turkey Vultures, in tall trees,

on pylons, etc. Eggs: 1–3, greenish white to greenish, usually marked with browns (WFVZ).
SS: Juv King Vulture soon shows white mottling on underparts, lacks white flashes in primaries. Adult Common Black and Great Black hawks and Solitary Eagle have similar flight shape but head and neck larger, wingbeats deep and powerful, not hurried, tails show broad white band; all lack large white flashes in primaries.
SD: C to F resident (SL to 2000 m, R to 3000 m) on both slopes and locally in interior virtually throughout region; absent from Baja. Appears to have decreased locally, at least in W and cen Mexico (ARP), and locally extirpated (for example, from Valley of Mexico since 1950s). Migratory status uncertain, appears resident in most areas.
RA: S USA and Mexico to cen Chile and cen Argentina.

TURKEY VULTURE

Cathartes aura *   Plate 7
Aura Cabecirroja

ID: 26–30 in; W 65–72 in (66–76.5 cm; W 165–183 cm). Common and widespread. *Wings long and broad, tail fairly long and slightly rounded. At rest, wingtips reach to or slightly beyond tail tip.* **Descr.** Ages differ slightly. **Adult:** eyes greyish, *naked head red to purplish red* with semicircle of whitish to green warts in front of and below eye, bill whitish, *legs pale flesh to reddish.* Blackish overall, *upperbody and*

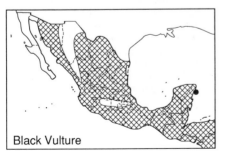

Black Vulture

upperwing coverts edged brown; pale brownish to whitish shafts of outer 6 primaries form paler panel on upperwing. Silvery grey underside of remiges, and to a lesser extent rectrices, *contrast with blackish coverts.* **Juv**: naked head greyish, bill dusky; upperwing coverts more neatly edged brown.

Habs. Open, semiopen, and wooded country, rarely dense forest. Highly aerial, soars low to high, wings held in marked dihedral. Wingbeats deep and easy, soars and glides for long periods with little flapping, tilting characteristically from side to side. Singly or in small groups, but may occur in flocks of thousands when migrating. Roosts communally (see Black Vulture). Eggs: 1–3, whitish, heavily marked with browns and greys (WFVZ).

SS: Lesser Yellow-headed Vulture (favors marshes) very similar but slightly smaller, wings project farther beyond tail at rest; blacker overall, outer primary shafts whiter (often quite striking), legs whitish, adult has overall yellow-orange head without contrasting pale bill (compared to whitish bill contrasting with dark head of Turkey Vulture). Zone-tailed Hawk smaller; note feathered black head, yellow cere, bold white tail band of adult. Golden Eagle has larger, feathered head and neck, lacks strongly two-tone underwings, flies with less marked dihedral.

SD: C to F resident virtually throughout region (SL–2100 m, U to R to 3500 m where may be present only in winter). Much of N.A. population winters and migrates through region; main migration periods Feb–Apr and Sep–Nov.

RA: N.A. to S S.A.; northernmost populations wintering S to S.A.

LESSER YELLOW-HEADED (SAVANNA) VULTURE

Cathartes b. burrovianus   Plate 7
Aura Sabanera

ID: 22–24 in; W 58–65 in (56–61 cm; W 148–165 cm). *S savannas, very similar to Turkey Vulture.* Wings long and broad, tail fairly long and slightly rounded. *At rest, wingtips project noticeably beyond tail tip.* **Descr.** Ages differ slightly. **Adult:** eyes red, *naked head multicolored (orange-yellow overall* with broad bluish-purple band at base of bill, pale blue crown, blood red nape band), bill pale flesh, *legs whitish.* Blackish overall, *white shafts of outer 6 primaries form contrasting panel on upperwing, often surprisingly striking,* at times suggesting Black Vulture; underside of flight feathers contrastingly silvery grey, similar to Turkey Vulture. **Juv:** undescribed (?). Probably has dusky eyes and bill, greyish head, wing coverts may be edged paler.

Habs. Marshes, savannas, open grasslands, mangroves. Behavior and flight much as Turkey Vulture but also hunts actively, quartering marshes like a harrier in search of small aquatic animals; rarely soars high. Usually perches low or on ground. May be somewhat migratory, at least nomadic, in response to fluctuating water levels. Nests in tree hollows, under grass clumps, etc. Eggs: 2, whitish, heavily marked with browns and greys (WFVZ).

SS: See Turkey Vulture.

SD: F to C resident (SL–150 m) on Atlantic Slope from S Tamps to N Honduras. U to F on Pacific Slope from Oax to Guatemala, possibly to El Salvador (Thurber *et al.* 1987).

RA: SE Mexico to N Argentina.

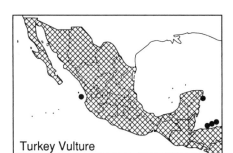

Turkey Vulture

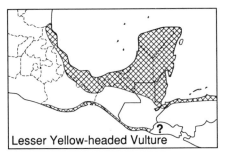

Lesser Yellow-headed Vulture

CALIFORNIA CONDOR *Gymnogyps californianus* Extinct. See Appendix A.

KING VULTURE

Sarcoramphus papa   Plate 7
Zopilote Rey

ID: 28–32 in; W 69–76 in (71–81.5 cm; 176–193 cm). *Rare lowland forest species, decreasing in numbers. Wings long and very broad, tail short, broad, and squared.* **Descr.** Ages differ. **Adult:** eyes white; *head and upper neck naked and multicolored,* head mainly dark grey with bright orange orbital ring and yellow wattles at base of bill, neck orange-red, bill red, legs pale grey. *Plumage white overall* (often with buffy or pinkish wash) *with black flight feathers,* alula, and upper primary coverts; dusky grey neck ruff at base of naked neck. **Juv:** eyes and bill dark, initially downy-feathered head and upper neck grey overall but orbital ring, bill, and nape soon show orange to orange-red. Plumage *slaty grey overall, white mottling on axillars, underwing coverts and underbody* increases in prolonged molt into **1st basic:** head and neck colors resemble adult but duller, wattles smaller. Neck ruff and upperparts slaty grey, underbody white with some dusky mottling. Subsequently, underparts become clean white and white mottling appears on upperwing coverts and back. By 3rd basic (?) much like adult but upperwing coverts mottled blackish. Full adult plumage may not be attained until 5th or 6th year.

Habs. Humid and semihumid forest, less often in drier and more cleared areas whence vanishing. Usually seen singly, associates with other vultures. Soars low to high, wings held flat with tips raised or held in slight dihedral, wingbeats deep and strong, rarely flaps when soaring and gliding. Eggs: 1, chalky white, unmarked (WFVZ).

SS: See Black Vulture. Adult possibly confusable at a distance; see Wood Stork; White Hawk smaller, mostly white in Mexico.

SD: U to F resident (SL–1500 m) on Atlantic Slope from S Ver and N Oax to Honduras, U to R on Pacific Slope from Gro (SNGH, SW) to El Salvador. Formerly ranged N on Pacific Slope to Sin, on Atlantic Slope to N Ver, possibly S Tamps (Arvin 1990). Increasingly threatened by habitat destruction but locally F to C in heavy forest on Atlantic Slope.

RA: SE Mexico to N Argentina.

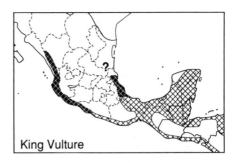

King Vulture

KITES, HAWKS, EAGLES, AND ALLIES: ACCIPITRIDAE (40)

A large cosmopolitan family of diurnal raptors. Size ranges from the small ♂ Sharp-shinned Hawk (105 g) to the massive ♀ Harpy Eagle (7.5 kg). Eagles are simply large hawks. Bills sharply hooked and feet strong with sharp talons. Nostrils open in a soft, leather-like cere and often the lores and orbital skin are naked. Wings typically are broad and rounded, tails short to long and usually broad. Long wings and a shortish broad tail are well adapted for soaring, short wings and a long tail give maneuvrability inside forests. Ages differ, young altricial and downy, fledge in 3—4 weeks (small accipiters) to 3—4 months (large eagles). Sexes similar or different, and ♀♀ often markedly larger. Molt strategies of raptors, particularly in the Neotropics, are poorly known. Juvenal plumage is usually worn through most of the first year, and juv flight feathers are often noticeably longer than those of adults. Many species attain adult plumage by completion of the 1st prebasic molt when about 1 year old, but larger species require up to 5 years. Successive imm stages (1st basic, 2nd basic, etc.) indicate passage of about one year unless stated otherwise. Molts are generally completed once a year but they may be protracted with contour feather molt well underway, at times almost complete, before flight feather molt begins. Some species, notably large eagles, require more than a year to completely replace their remiges. Plumages are usually somberly colored though patterns may be striking. Several species are dimorphic. Upperparts mostly dark grey or brown, underparts usually paler, often with dark barring or streaking. Tails typically have contrasting dark and pale bands which usually vary in number with age, adults having fewer bands than juvs and imms. Often the cere, lores, orbital skin, and legs are brightly colored; bills blackish unless stated otherwise.

Hawk nests typically are stick platforms, some notably bulky, others surprisingly flimsy, usually lined with green leaves; most species nest in trees. Eggs: 1—5, generally whitish, often attractively marked with browns and reds.

OSPREY

*Pandion haliaetus**   Figure 19
Gavilán Pescador

ID: 22—26 in; W 60—68 in (56—66 cm; W 153—173 cm). A unique fish-eating hawk often placed in its own family. Differs from other hawks in its heavily scaled tarsi and feet; rough soles and a reversible hind toe help grasp wriggling prey. Long wings held crooked in flight. Head slightly crested. **Descr.** Sexes similar. **Adult:** eyes yellow, cere and legs blue grey. *Head and underparts white*, crown streaked dark brown, *broad dark brown eyestripe* merges with dark brown nape (dark eyestripe almost lacking in 'white-headed' Caribbean ssp *ridgwayi*); ♀♀ especially may have slight brownish chest band. Upperparts dark brown. Tail tipped white, upperside barred

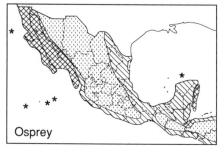

Osprey

Fig. 19 Osprey (*P. h. ridgwayi* perched; *P. h. carolinensis* in flight)

grey-brown and dark brown with 5–6 dark bars; underside pale grey with 4–6 dark bars, distal bar widest. Underwings: *white coverts contrast with blackish carpal patch,* remiges dusky with *broad pale panel across primaries.* **Juv:** white-tipped upperparts and remiges produce neatly scaled effect, tail more broadly tipped white. **Voice.** Clear to slightly hoarse, piping to ringing whistles, typically in loud, slightly ascending series, *kyew-kyew ...* or *hiuw hiuw ...,* etc.

Habs. Lakes, rivers, estuaries, coasts, and islands, widely in migration. Often circles over water, soaring and gliding with wings crooked. Perches on conspicuous snags, posts, etc. Fishes by spectacular feet-first plunges, and carries off fish, aligned head-forward, in talons. Nests are bulky stick structures in trees, on cliffs, even on ground, locally in loose colonies.

Eggs: 2–3, whitish to pinkish cinnamon, heavily mottled with reds and browns (WFVZ).

SS: Can be confused in flight at a distance with large gulls which have more pointed wings, lack black carpal patches and dark eyestripe.

SD: F to C but local breeder (around SL) on Pacific coast and numerous islands of Baja, coasts and islands of Gulf of California S to Tres Marías, and on Caribbean coast of QR and Belize; may breed elsewhere but no confirmed records (?). F to C on both slopes in migration and winter (Sep–Apr; SL–1500 m); non-breeders oversummer widely. Regular in interior on migration (Sep–Oct and Mar–May; to 2500 m).

RA: Almost worldwide except S.A. In New World breeds S to Mexico and Caribbean; N populations migratory, wintering to S.A.

KITES

A diverse group, separable into several genera by virtue of structural characters but best divided for field purposes into two types: rounded-winged and pointed-winged. Kites inhabit warm climates and feed mostly on reptiles, insects, and snails, for which their relatively weak legs and feet suffice. Ages often differ strikingly, sexes similar or different. Most attain adult plumage by a complete molt when about 1 year old. Three of the pointed-winged species are long-distance migrants, wintering in S.A.

GREY-HEADED KITE

Leptodon cayanensis   Plates 5, 8, 12
Milano Cabecigris

ID: 18–21 in; W 36–42 in (45.5–53 cm; W 92–107 cm). Uncommon SE forest species. A fairly large kite with *fairly long, broad, rounded wings, and long, slightly rounded tail.* Juv dimorphic. **Descr.** Sexes similar.

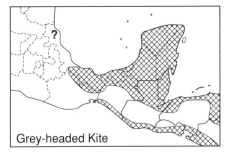

Grey-headed Kite

Adult: eyes dark brown, cere, lores, orbital ring, and feet blue-grey. *Pale grey head contrasts with slaty black upperparts. Throat and underparts white.* Tail black with narrow whitish tip, 2–3 pale grey bars on upperside, 2 broad white bands on underside. Underwings: *coverts black, primaries boldly checkered black and white* (especially when backlit), *secondaries pale grey with narrow dark barring; dark underwings contrast with pale body.* **Juv. Light morph:** cere, lores, orbital ring, and feet yellowish. *Head and underparts white with black crown patch* (auriculars rarely also dark), sides of neck streaked blackish. Upperparts dark brown, edged cinnamon. Tail upperside blackish with 2 broad, pale grey-brown bands, underside pale grey with 2–3 dark bars. Underwings: coverts creamy white, *pale grey remiges barred black, boldly on outer 5–6 primaries* which may form checkered panel when backlit. **Dark morph:** resembles light morph (including tail pattern) but *head blackish, throat and underparts streaked blackish* (can be almost solidly dark on chest), white throat often has dark median stripe. Underwings: similar to light morph but black barring on primaries usually broader, coverts often with some, usually sparse, dark streaks and blotches.
Voice. A fairly rapid, often excited-sounding clucking or laughing *kyuh-kyuh-kyuh . . .* or *keh-keh-keh . . .*, 10/1.5–2.5 s, up to 40+ × in series; may suggest Lineated Woodpecker but steadier, less laughing. Calls from canopy, including exposed perches; ♂ and ♀ (higher versus harder and lower voices) may call to one another.
Habs. Humid evergreen to semideciduous forest and edge, mangroves, gallery woodland. Flies with wings flattish to slightly bowed, and slightly pressed forward, tail closed to slightly spread. Often soars, at times high. Wingbeats when soaring slow and easy, but more rapid, even hurried, when climbing or in direct, low flight. May be identified at long range by distinctive display flight: interrupts glides with short, steep climbs using high, floppy, rapid wingbeats followed by a brief glide down with wings held in strong dihedral. Nest a shallow platform of twigs high in trees. Egg (reportedly of this species; Trinidad): 1, whitish with faint fine darker markings (WFVZ).

SS: Adult distinctive. Juv light morph suggests rare Black-and-White Hawk-Eagle which has black lores and orange cere, longer wings white below without bold black outer primary bars, note white leading edge to innerwing. Juv Hook-billed Kite has more oval wings with trailing edge pinched in at body, larger bill, often some dark barring on underparts. Collared Forest-Falcon has long legs, long graduated tail, juv barred below.
SD: U to F resident (SL–1500 m, mainly below 500 m) on Atlantic Slope from S Ver (possibly from S Tamps, at least formerly) to Honduras, on Pacific Slope locally from E Oax to Honduras.
RA: S Mexico to Ecuador and N Argentina.

HOOK-BILLED KITE
*Chondrohierax uncinatus**
Milano Piquiganchudo   Plates 2, 8, 10, 12

ID: 15–18 in; W 32–37 in (38–45.5 cm; W 81–94 cm). A medium-sized, *highly variable kite. Bill heavy and strongly hooked. Wings broad and rounded, pinched in at body, tail long and squared to slightly rounded.* **Descr.** Sexes differ. **Adult. Light morph.** ♂: *eyes whitish*, bill black with horn to yellow-green base below, orbital ring, *lores, and cere greenish,* often with brighter lime-green patch at gape; bare crescent in front of eye orange-yellow to orange, feet yellow to yellow-orange. Head and upperparts slaty grey to slaty black, uppertail coverts tipped white. *Throat and underparts grey to slaty grey,* barred white or with white barring reduced to absent; tail coverts whitish, rarely with a few dark bars. *Tail black* with narrow whitish tip, *upperside with a broad pale greyish band distally, a narrower whitish band basally, underside with 2 broad whitish bands* (basal band often covered by tail coverts).

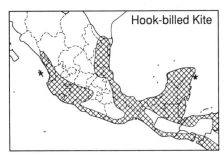

Hook-billed Kite

Underwings grey overall, bold white barring on outer primaries often appears as translucent panel when backlit, coverts narrowly barred whitish. Some ♂♂ have cinnamon-brown and white barring on underparts, partial cinnamon hindcollar. ♀: grey face contrasts with dark crown and rufous hindcollar. Upperparts dark brown. Throat and underparts coarsely barred rufous and cream, rufous bars often narrowly edged blackish; throat and chest may be solidly rufous. Tail as ♂ or pale bands washed cinnamon. Underwings: coverts whitish, usually barred and washed rufous, remiges pale grey, barred darker, inner primaries washed rufous, outer 5–6 primaries boldly barred black and white, often appearing as translucent panel when backlit. Some ♀♀ have more slaty-blackish upperparts, poorly marked hindcollar, underparts barred dark brown and cream, thus much like some ♂♂. Dark morph (sexes similar): almost entirely dark slaty grey to brownish black, tail coverts usually tipped whitish. Tail black with 1–2 broad white bands on upper and underside and narrow whitish tip, upper bands often washed grey distally. Underwings uniformly dark slaty or with white to pale grey spots across primaries. Juv. Light morph: eyes initially dusky, bare facial colors paler. Crown blackish, face, hindcollar, and underparts creamy white; underparts unmarked, or with sparse dusky barring on sides, or with extensive dark barring. Some juvs have grey face and rufous and cream barring on underparts like adult ♀. Dark brown upperparts edged rufous except white-tipped tail coverts. Tail blackish brown with narrow whitish tip, 3–4 pale grey-brown to whitish bands, pale bands may be washed cinnamon. Underwings whitish, coverts may be flecked and barred rufous to brown, pale grey to grey remiges barred black, inner primaries usually washed rufous, more boldly barred outer 5–6 primaries often appear as translucent panel. Dark morph: blackish to blackish brown overall, upperparts edged paler as in light morph. Tail black with 2 pale bands, above pale grey-brown to whitish, below whitish (basal band usually hidden), and narrow whitish tip. Underwings slaty, remiges sparsely barred white to pale grey, usually boldest on outer primaries.

Voice. A rapid, slightly clipped, clucking chatter, weh keh-eh-eh-eh-eh-eh-eh-eh-eh-eh, or w-kehehehehehehehehehe (Moore 1992, tape).

Habs. Forest and edge, mangroves, thorn forest. Wingbeats often distinctly loose and floppy; soars with wings flattish or slightly raised, and slightly pressed forward, tail slightly to fully spread; glides with wings slightly bowed and curled up at tips, tail closed. Soars often but rarely for long periods. Spends much time perched in forest where searches for land snails, its main prey. Nest is a cupped platform of sticks at mid- to upper levels in trees. Eggs: 2, buffy white, marked with dark reddish browns (WFVZ).

SS: The floppy flight and paddle-shaped wings, with pinched-in bases and spread-fingered 'hands', are distinctive; flight and shape readily separate Hook-billed Kite from superficially similar buteos which have stiffer wingbeats, different tail patterns, etc.; note also the translucent, boldly barred primaries of Hook-billed Kite. In flight can be mistaken easily for hawk-eagles, especially Ornate, when its smaller size may be hard to judge; best told by squarer tail, bolder dark flight feather barring, fewer and usually bolder tail bands. Juv Snail Kite told in flight by arched wings, tail pattern. Juv Bicolored Hawk has longer legs and tail, narrower hindcollar, underparts unbarred. Collared Forest-Falcon has much longer legs and tail, dark crescent on auriculars. Black morph Hook-billed Kite can be puzzling, especially in flight: Black Hawks have broader wings and shorter tails; Crane Hawk has longer tail, long red legs, white band across primaries, dark lores. See Grey-headed Kite.

SD: U to F resident (SL–2000 m) on Atlantic Slope from Tamps to Honduras, on Pacific Slope from Isthmus to Honduras. R to U on Pacific Slope (to 2500 m) N of Isthmus to Sin, and inland locally in W cen Mexico to Mex (SNGH, PP); formerly perhaps to DF and Gto (Wilson and Ceballos-L. 1986).

RA: Mexico (and S Texas) to N Argentina.

SWALLOW-TAILED KITE
Elanoides forficatus*　　　　Not illustrated
Milano Tijereta

ID: 22–24 in; W 46–53 in (56–61 cm; W 117–135 cm). An unmistakable and grace-

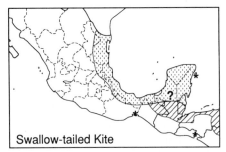

Swallow-tailed Kite

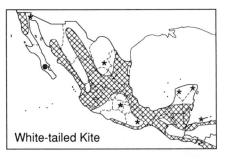

White-tailed Kite

ful raptor with long, pointed wings and very long, deeply forked tail. **Descr.** Sexes similar. **Adult:** eyes red, cere and feet blue-grey. White head contrasts with black upperparts, much of which show a strong blue to purplish sheen. *White underparts and wing-linings contrast with blackish underside of flight feathers.* **Juv:** eyes brownish. Flight feathers and upper primary coverts tipped white, tail shorter and less deeply forked. May attain adult-like tail by 1st prebasic molt on winter grounds, and adult plumage by 2nd prebasic molt when just over 1 year old (?).

Voice. Shrill, piping, and ringing whistles, given mainly in flight, at times in song-like series, accelerating and slowing, *k-leep k-leep k-leep k-leep* ... or *whee-whee-whee* ..., etc.

Habs. Humid evergreen and pine–evergreen forest, often near water; widely during migration. Flight graceful and buoyant with deep, easy wingbeats and leisurely soaring, wings flat with tips curled up; glides with wings slightly bowed. Often in pairs or small groups; flocks to 50+ bird in migration. Nest platform of twigs high in trees. Eggs: 2–3, whitish, marked with browns (WFVZ).

SD: F to C summer resident (Feb–Aug, near SL–1500 m) on Atlantic Slope from E Chis to Honduras. F transient (Feb–Mar, Aug–Sep) along Atlantic Slope throughout, including Yuc Pen, R transient on Pacific Slope S of Isthmus.

RA: Breeds SE USA and S Mexico to N Argentina; winters in S.A.

WHITE-TAILED [BLACK-SHOULDERED] KITE

Elanus leucurus majusculus   Not illustrated
Milano Coliblanco

ID: 15–16 in; W 35–39 in (38–40.5 cm; W 89–99 cm). An increasingly common kite

of open country. *Wings long and pointed, tail long and squared to slightly cleft.* At rest, wingtips reach tail tip. **Descr.** Sexes similar. **Adult:** eyes red, cere and feet yellow. Head and underparts white with black patch in front of eye. Nape and upperparts pale grey, *lesser and median upperwing coverts black*, primaries darker than back. *Tail white with pale grey central rectrices.* Underwings white with black spot on primary coverts, primaries dark grey. **Juv:** eyes brownish. Head and underparts white, crown and nape streaked dark brown, chest mottled cinnamon. Upperparts grey-brown with rufous and whitish edgings, lesser and median upperwing coverts blackish; wing coverts and remiges tipped white. Tail pale grey with dusky subterminal bar. **1st basic:** attained quickly; resembles adult by early winter except tail which is replaced by spring.

Voice. Mostly silent; at times utters low, rasping notes, *rrehh-rr* or *hi shhrrrr*, etc.

Habs. Open country with scattered trees, savannas, marshes, irrigated agricultural areas. Flies with easy wingbeats, wings held in a dihedral during gliding and infrequent soaring. Often hovers and perches on roadside wires like a kestrel. Nest platform of twigs, low to high in bush or tree. Eggs: 3–5, whitish to pale brown, heavily mottled with browns (WFVZ).

SS: ♂ Northern Harrier has broader wings, black wingtips, white uppertail covert band, hunts by quartering low over ground. Mississippi and Plumbeous kites dark, soar frequently on flattish wings but hover rarely, found mostly in forested areas.

SD: C to F resident (SL–1500 m) in BCN (first noted NE BCN in 1987, TEW) and on both slopes, from Son (where first recorded 1979, DS) and Tamps to Honduras and W Nicaragua. U and local (non-breeding wanderers in some areas) in BCS (Howell and Webb 1992*f*), on Plateau (to 2500 m) to which it spread in late 1970s and 1980s; recorded Chih, Coah, Dgo, Zac, Ags, Gto, Mex, and DF (SNGH, AGL, PP, SW, per RGW), and N Yuc Pen. N.A. and Mexican populations have expanded rapidly since the 1950s; first recorded El Salvador, Honduras, and Nicaragua as recently as the early 1960s (Eisenmann 1971) but now declining, at least in El Salvador, perhaps due to pesticides (Thurber *et al.* 1987).

RA: SW USA to cen Chile and cen Argentina.

SNAIL KITE

Rostrhamus sociabilis*   Plates 4, 10, 11
Milano Caracolero

ID: 18–20 in; W 41–47 in (45.5–51 cm; W 104–119 cm). A social and gregarious kite of SE freshwater marshes. *Wings broad and rounded, tail fairly long and squared to slightly cleft, rounded when spread. At rest, wingtips extend beyond tail. Bill slender and strongly hooked.* **Descr.** Sexes differ. ♂: eyes red, *cere, lores, orbital ring, and feet bright orange-red. Slaty blue-black overall* with blacker flight feathers; *distal uppertail coverts, base of tail, and undertail coverts white,* tail tipped whitish. Underwings slaty black, remiges often barred pale grey, most noticeable as pale primary panel. ♀: *cere, lores, orbital ring, and feet yellow to orange. Head and upperparts dark brown to blackish brown, forehead and supercilium usually streaked creamy;* upperparts edged cinnamon when fresh. *Throat whitish to pale buff, underparts dark brown, mottled buff to cinnamon;* thighs dark brown, barred cinnamon, undertail coverts buff. *Tail as* ♂. Underwings: coverts mottled blackish and cinnamon, smoky grey secondaries barred darker, primaries pale grey with dark bars,

often appearing as broad pale panel when backlit. **Juv:** eyes brownish, *cere and facial skin blue-grey, feet yellowish.* Resembles ♀ but *face, throat, and underparts pale buff to ochraceous buff with dark postocular stripe,* dark crown streaked buff, *underparts coarsely streaked dark brown;* thighs and undertail coverts spotted and streaked dark; whitish primary panel below when backlit. **1st basic** ♂: duller than adult, often with an overall brownish cast, throat streaked whitish, underparts with some cinnamon mottling; possibly with more extensive pale underwing areas. **1st basic** ♀: resembles ♀ but underparts streaked and mottled dark brown, cinnamon, and buff, head has distinct buff supercilium, auriculars may be streaked creamy buff. Attains adult plumage by 2nd prebasic molt when about 2 years old.

Voice. A rasping *ahrrrr,* a creaky *ehrr-rrr,* a dry, crackling chatter, and a rasping cackle *kah-ah-ah-ah-ah;* mostly silent.

Habs. Freshwater lakes and marshes. Flight agile, wings held arched and tail often spread, wingbeats leisurely and floppy. Quarters low over water and picks snails from near surface with talons; at times soars high. Perches low on posts, bushes, and ground. Usually breeds and roosts communally. Nest is a stick platform in reeds or low in bush, over or near water. Eggs: 2–4, whitish to greenish white, heavily marked with browns (WFVZ).

SS: Shape, slender hooked bill, and white tail base distinctive; note bright facial skin and legs of adults. Black Hawks much more heavily built, wings fall short of tail tip at rest, juvs and imms have boldly barred tails; juv Black-collared Hawk has contrasting pale head, upperparts mottled rufous-orange, feet pale flesh; in flight all have very broad wings and short tails. Harris' Hawk has shorter wings, longer rounded tail, rufous shoulders and thighs. See Hook-billed Kite.

SD: C to F but local resident (SL–1500 m) on Atlantic Slope from cen Ver to NW Honduras but only locally/seasonally U in N Yuc Pen (Aug–Dec at least); U to F but local visitor (resident?) on Pacific Slope in Gro (SNGH, SW), Chis (AM), and Guatemala (RWD spec.). Locally nomadic due to fluctuating water levels; vulnerable to pollution and drainage.

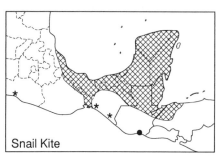
Snail Kite

RA: SE USA, Cuba, and S Mexico to W Ecuador and N Argentina.

DOUBLE-TOOTHED KITE

Harpagus bidentatus fasciatus Plates 2, 8, 13
Milano Bidentado

ID: 12.5−14 in: W 25−29 in (32−35.5 cm: W 63.5−73.5 cm). *Fairly small, rather accipiter-like kite. Wings fairly long, broad and tapered, tail long and squared to slightly rounded. In flight, fluffy white undertail coverts spread to sides of rump.* At rest, wingtips reach about half-way down tail. **Descr.** Sexes differ slightly. ♂: eyes amber, cere and lores green to yellowish, feet yellow to orange-yellow. Head and upperparts slaty grey, often with white mottling on scapulars. *Throat white with blackish median stripe, chest rufous* (often broken with grey and white bars, mainly in center), *lower chest and belly barred dark grey and white*, grey bars with narrow rufous edges basally; undertail coverts white. Tail blackish with narrow whitish tip, 3 whitish bars above, and 2−3 whitish to pale grey bands below. *Underwings white, primaries barred black*, secondaries with indistinct dusky bars most distinct distally, base of secondaries may be unbarred. ♀: chest more solidly rufous, rufous bars on underparts wider than grey bars. **Juv:** paler eyes can be yellowish. Head and upperparts dark brown, often with whitish mottling on scapulars. *Throat as adult*, underparts whitish to buff, unmarked to lightly streaked dark brown on chest, faintly barred dusky on flanks and thighs, or heavily streaked dark brown on chest and belly, flanks and thighs barred dark brown: undertail coverts white, rarely with a few dark spots. *Tail* blackish brown with narrow whitish tip and with 3−4 whitish bars, basal band may be hidden by tail coverts. *Underwings as adult.*

Voice. High, thin whistles, *tsee up*, a more excited *tswee-u tswee-u*, and a faster *t-sup-sup-sup-sup-sup* and *tsip tsii-yiip*, etc., given in flight by pairs. Also high, slightly shrill whistles, *shiep* or *seeh*, repeated, given by birds in canopy.

Habs. Humid evergreen forest and edge, locally in semideciduous forest. Often soars, sometimes quite high, but rarely for long periods. Usually soars and glides on distinctly bowed wings, tail closed; rarely, however, soars with flattish wings, tail slightly spread, when looks much like an *Accipiter*. Display flight an accipiter-like series of stoops interspersed with short climbs. Stick platform nest usually high in trees. Eggs (Trinidad): 2−3, whitish, marked with browns and greys (WFVZ).

SS: Flight, combined with extensive white undertail coverts, unmistakable, enabling identification at long range. Sharp-shinned and Cooper's hawks soar on shorter, flattish wings, tails slightly spread, lack such conspicuous white undertail coverts. Juv Barred Forest-Falcon does not soar, is barred below, has bright yellow bare facial skin. Small buteos are larger and more heavily built; adult Broad-winged Hawk has brownish eyes, thick dark malar, 1−2 broad pale tail bands.

SD: F to C resident (SL−1500 m) on Atlantic Slope from S Ver to Honduras, U to R in Yuc Pen and locally on Pacific Slope in Jal and Col (SNGH), Gro (Dixon and Davis 1958; SNGH, SW), E Oax, and El Salvador (Dickey and Van Rossem 1938).

RA: S Mexico to Bolivia and Brazil.

MISSISSIPPI KITE

Ictinia mississippiensis Plates 2, 13
Milano de Misisipi

ID: 13.5−15 in: W 33−37 in (34−38 cm; W 84−94 cm). Transient in S and E. *Wings long and pointed, tail long, squared*

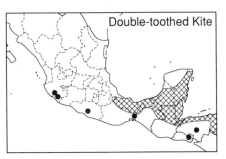

Double-toothed Kite

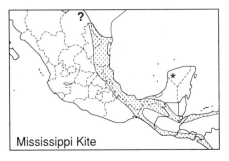

Mississippi Kite

to slightly cleft. At rest, wingtips project slightly beyond tail. **Descr.** Sexes similar. **Adult**: eyes red, cere grey, feet yellow. *Whitish head contrasts with slaty grey upperparts, small black patch in front of eye. Pale grey secondaries form conspicuous broad band on closed wing and contrasting patch on upperwing in flight. Tail blackish.* Underparts and underwings grey, slightly paler than upperparts. Primaries blackish, usually with slight rufous edging and spots on inner primaries. **Juv**: eyes brownish, cere yellow. Head whitish with dark streaking and *bold white supercilium.* Upperparts slaty grey, edged rufous; white-tipped *primaries contrastingly darker than back. Underparts whitish* with coarse reddish streaking. *Tail black with 2–3 narrow white bars visible from below.* Underwings: coverts mottled rufous and dark brown, remiges dark grey, *primaries lack rufous* but often show pale basal flash. **1st basic**: attained on winter grounds; resembles adult but retains juv flight feathers and most underwing coverts, underparts may show slight rufous and whitish mottling. Attains adult plumage by 2nd prebasic molt on winter grounds.

Habs. Aerial, usually in small flocks, rarely up to a few hundred birds, often flying high; small numbers associate loosely with other migrating raptors. Soars and glides on flattish wings, wingtips often slightly curled up, wingbeats deep and easy, tail slightly spread. Usually silent.

SS: Plumbeous Kite has shorter tail and shorter outermost primary noticeable in flight, wingtips project well beyond tail at rest; adult darker overall, lacks pale secondaries, note rufous flash in primaries and 2–3 white tail bars; juv has underparts streaked dark greyish, primaries not contrastingly darker than upperparts, rufous flash in primaries; imm overall darker than Mississippi Kite with rufous flash in primaries, bolder tail bands (as seen from below); note structure.

SD: F to C transient (Apr–early May, late Aug–mid-Oct; SL–1500 m) along Atlantic Slope, U to R on Pacific Slope S of Isthmus, and R (Sep–Oct) in Yuc Pen (G and JC). Increasing in SW USA and breeds adjacent to region in W Texas.

RA: Breeds S USA; winters S.A.

PLUMBEOUS KITE

Ictinia plumbea   Plates 2, 13
Milano Plomizo

ID: 13–14.5 in, W 33–37 in (33–37 cm, W 84–94 cm). Summer visitor to S and E. *Shape similar to Mississippi Kite but tail shorter, outermost primary shorter; at rest, wingtips project well beyond tail.* **Descr.** Sexes similar. **Adult**: eyes red, cere grey, *feet salmon-red to yellow-orange. Slaty grey overall,* upperparts blacker, head paler but not strongly contrasting, small black patch in front of eye. Flight feathers blackish overall, *inner webs of outer primaries rufous, tail shows 2–3 narrow white bands below* (rarely visible from above unless tail fully spread). *From below, rufous flashes in primaries usually obvious.* **Juv**: eyes brownish, cere and feet yellowish. Head whitish, heavily streaked blackish, at times with white supercilium. Upperparts slaty black, white-tipped *primaries not contrastingly darker than back. Tail similar to adult but white bars narrower. Underparts whitish, heavily streaked dark grey.* Underwing coverts mottled grey and white, *rufous on primaries* less extensive than adult, primaries often show pale basal flash. **1st basic**: attained on winter grounds; resembles adult but retains juv flight feathers and most underwing coverts, underparts may be mottled whitish. Attains adult plumage by 2nd prebasic molt on winter grounds.

Voice. Mostly silent; at times gives shrill, thin whistles, *si-yoo* and *siseeeoo*, etc.

Habs. Humid evergreen and semideciduous forest and edge, often near water, mangroves. Highly aerial, spends much time soaring and gliding, catches insects on the wing. Soars on flattish wings, wingtips often curled up slightly, wingbeats deep and easy, tail slightly spread. Nest is

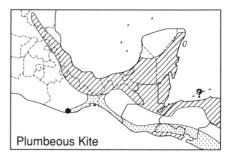

Plumbeous Kite

a stick platform at mid- to upper levels in trees. Eggs: 1–2, whitish with faint sparse cinnamon markings (WFVZ).

SS: See Mississippi Kite.

SD: F to C transient and summer resident (Feb–Aug; SL–1200 m) on Atlantic Slope from Chis to Honduras, U N to S Tamps and on Pacific Slope locally from cen Oax to El Salvador. No verified winter records.

RA: Breeds SE Mexico to N Argentina; winters in S.A.

BALD EAGLE

Haliaeetus l. leucocephalus   Not illustrated
Aguila Cabeciblanca

ID: 30–34 in, W 72–83 in (76–86.5 cm, W 183–211 cm). N Mexico. *Wings long and broad, tail fairly short and broad, squared to slightly wedge-shaped.* **Descr.** Sexes similar. **Adult:** eyes pale yellow, bill, cere, and feet yellow. *Dark brown overall with white head, tail,* and undertail coverts. **Juv:** eyes brownish, bill and cere blackish. Dark brown overall, belly slightly paler, flight feathers and greater upperwing coverts blacker. Tail usually with some whitish mottling at base, *underwing coverts mottled whitish.* **1st basic:** resembles juv but *belly whitish,* back and upperwing coverts mottled whitish; whitish supercilium and throat often apparent. **2nd basic:** similar to adult but head dirty whitish, underwing coverts retain some whitish mottling, tail mottled dark brown at base. **3rd basic:** like adult but tail usually shows dusky tip. Attains adult plumage by 4th prebasic molt.

Voice. Mostly silent except near nest; a rising whistled *reep reep reep,* and a more screechy *kreer kreer kreer,* etc.

Habs. Lakes, estuaries, sea coasts, extensive marshes. Spends much time sitting, in trees, on ground, etc. Scavenges at carcasses and catches fish with talons by plunging into water. Wingbeats deep and powerful, soars and glides with wings flat to slightly raised, tail slightly spread. Nest is a bulky stick structure in tree, on cliff. Eggs: 1–2, whitish, unmarked (WFVZ).

SS: Golden Eagle has longer tail and narrower wings, flies with wings in slight dihedral, underwing coverts darker than remiges.

SD: R resident (SL–200 m) in Baja (vicinity of Bahia Magdalena, Henny *et al.* 1978) and Son (Brown *et al.* 1987); formerly more widespread in Baja. R to U in winter (Nov–Mar) S in lowlands to Nay and N Ver, on Plateau U to F to Dgo, Coah, and probably Zac (Knoder *et al.* 1980; SNGH, AGL, SW).

RA: Breeds N.A. to NW Mexico; winters to N Mexico.

NORTHERN HARRIER

Circus cyaneus hudsonius   Not illustrated
Gavilán Rastrero

ID: 18–22 in; W 39–46 in (45.5–56 cm; W 99–117 cm). Widespread winter visitor to open country. *Wings long and slightly tapered,* tail long and squared. *Owl-like facial disks distinctive.* **Descr.** Ages/sexes differ. ♂: eyes pale yellow, cere yellow, feet orange-yellow. *Head, upper chest, and upperparts grey* to brownish grey, black-tipped outer 5 primaries form *contrasting black wingtips,* secondaries tipped dusky. *White uppertail coverts form conspicuous band.* Tail grey with 4–5 indistinct dusky bars. Underbody white, chest and flanks flecked cinnamon. *Underwings white with black wingtips and dusky trailing edge to inner primaries and secondaries,* coverts flecked cinnamon. ♀: head and upperparts dark brown with pale upperwing bar; *white uppertail covert band striking in flight.*

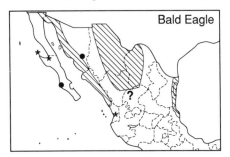

Bald Eagle

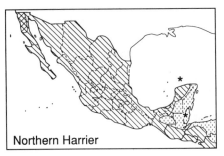

Northern Harrier

Tail dark brown above with narrow pale tip, central rectrices with 3–4 pale grey-brown bands, outer rectrices with 3 cinnamon to pale buffy-cinnamon bands; tail underside banded pale cinnamon to pale grey and dark, with 2–3 dark bands. Underparts pale buff, heavily streaked dark brown. Underwings: coverts mottled whitish and brown, remiges barred dark brown and greyish, primaries often show contrasting pale panel when backlit. **Juv:** eyes brownish. *Resembles* ♀ *but underparts bright cinnamon* (fading to pale buff by spring), dark brown streaking only on upper chest and flanks, underside of secondaries darker so pale primary panel more contrasting. Attains adult plumage by 1st prebasic molt.

Voice. Rarely vocal except around nest where may give a rough, rapid, yelping chatter, *kyeh-kyeh* ...

Habs. Open country in general, especially marshes, lake edges. Usually quarters low over ground but soars high at times. Wings held in distinct dihedral, rarely almost flat, wingbeats fairly deep and easy, flight generally leisurely but notably quick and agile when hunting; feeds mostly on small mammals. Often roosts communally in winter, up to 20 birds at a site. Nest is a flimsy platform of grass and sticks on or near ground. Eggs: 4–6, whitish, unmarked or with sparse pale brown flecking (WFVZ).

SS: Buteos are more heavily built and broader-winged, wingbeats stiffer and less leisurely, usually soar fairly high. See White-tailed Kite.

SD: F to U but local breeder (SL–300 m) in BCN, S to 30°N. C to F transient and winter visitor (Sep–Apr; SL–2500 m) S to cen Mexico; U to F to Honduras and W Nicaragua.

RA: Holarctic breeder. In New World breeds S to NW Baja; winters to N S.A.

NB: Known as Hen Harrier in the Old World.

GENUS ACCIPITER: Accipiters

Accipiters live mostly in forests and are seen relatively infrequently. Their fairly short and rounded wings and long tails provide good maneuvrability while chasing small birds. Legs and toes long. Sexes similar, ♀♀ markedly larger than ♂♂. Adult plumage attained at end of 1st year.

SHARP-SHINNED HAWK
*Accipiter striatus**   Plate 13
Gavilán Pajarero

ID: 11–14 in; W 20–25 in (28–35.5 cm; W 51–63.5 cm). *Small with relatively small head and squared to slightly notched tail.* In flight, head projects only slightly beyond leading edge of wings. **Descr. Adult:** eyes orange to red, cere and legs yellow to orange-yellow. *Crown, nape, and upperparts slaty blue-grey* (♀ more brownish grey), whitish face and throat streaked rufous. *Underparts rufous* (darkest in N.A. migrant *velox*, palest in *madrensis* of Gro and Oax), *coarsely barred whitish* except for all-rufous thighs (barred white in *velox*); undertail coverts white. Tail tipped white, upperside broadly banded blackish brown and pale brownish grey, with 3 pale bands; underside pale grey with 3–4 dark bands. Underwings whitish, coverts flecked brownish, pale grey remiges barred dark grey. **Juv:** eyes yellow. Head and upperparts dark brown, face streaked paler, typically with distinct narrow whitish supercilium; upperparts edged rufous, scapulars often mottled whitish. Throat and *underparts whitish, heavily streaked and spotted rufous-brown* except clean white undertail coverts. Tail similar to adult but 3–4 pale bands above browner. Underwings as adult.

Voice. A sharp, hard, yelping *kyew-kyew-kyew* ...; rarely vocal in winter.

Habs. Breeds mainly in pine–oak woodland. In winter, wooded and forested areas in general, widely during migration. Spends much time perched, flies with bursts of quick stiff wingbeats interspersed with short glides, soars with wings flattish and pressed forward slightly, tail slightly spread. Nest is a bulky stick platform at mid- to upper levels in trees.

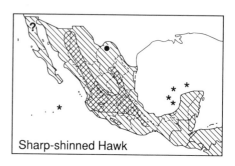

Sharp-shinned Hawk

Eggs: 3–5, whitish, marked with browns (WFVZ).

SS: Cooper's Hawk larger (but large ♀ Sharp-shinned close in size to small ♂ Cooper's) with relatively larger head and longer, rounded tail, wings not pressed forward while soaring, wingbeats less hurried, favors more open country; adult shows marked contrast between dark crown and paler nape, juv often lacks whitish supercilium. Juv White-breasted Hawk has dark crown and nape contrasting with greyish upperparts, white underparts with inconspicuous fine dark streaking. See Double-toothed Kite.

SD: F resident (1000–3000 m) in interior and on adjacent slopes from Son (and probably BCN) and Coah to Oax. F to C and widespread transient and winter visitor (Oct–Apr; SL–3000 m) almost throughout region, though U in Yuc Pen where may be commoner as a transient.

RA: Breeds N.A. to Mexico; N.A. populations largely migratory, wintering S to Panama.

WHITE-BREASTED [SHARP-SHINNED] HAWK
Accipiter chionogaster or *A. striatus* (in part)
Gavilán Pechiblanco   Plates 2, 13

ID: 11–13 in; W 20–25 in (28–33 cm; W 51–63.5 cm). Endemic; replaces Sharp-shinned Hawk in highlands S of Isthmus. **Descr. Adult:** eyes amber to red, cere and feet yellow to orange-yellow. *Crown, nape, and upperparts blackish grey* (♀ slightly browner), auriculars whitish to slaty grey. *Throat and underparts white, thighs pale buff.* Tail tipped white, upperside broadly banded blackish and pale greyish, with 3 pale bands, underside pale grey with 3–4 dark bands. Underwings white, pale grey remiges barred dark grey. **Juv:** eyes yellow (?). Crown and nape blackish with narrow whitish supercilium and

broad dark eyestripe; upperparts blackish brown, narrowly edged rufous when fresh. Auriculars, throat, and *underparts white; auriculars, throat and chest with fine dark streaks,* rarely extending onto belly; *thighs often washed pale cinnamon.* Tail similar to adult but 3–4 pale grey to grey-brown bands on upperside. Underwings white, coverts with sparse dark flecking, pale grey remiges barred blackish brown.

Habs. Pine and pine–oak woodland, semideciduous and evergreen forest. Habits much as Sharp-shinned Hawk. Voice, nest, and eggs undescribed (?).

SS: See Sharp-shinned Hawk, Double-toothed Kite.

SD/RA: F to U resident (600–3000 m, rarely ranging lower), mainly in interior and on adjacent Pacific Slope, from E Oax (Binford 1989) and Chis to Honduras and N cen Nicaragua.

BICOLORED HAWK
Accipiter bicolor*   Plate 2
Gavilán Bicolor

ID: ♂ 13–15 in, ♀ 17–19 in; W 24–31 in (♂ 33–38 cm, ♀ 43–48 cm; W 61–79 cm). *Uncommon hawk of SE humid lowland forest.* Marked sexual size difference. *Plumage lacks dark streaking or barring.* **Descr. Adult:** eyes amber, cere yellowish, legs yellow. *Crown and nape blackish, upperparts dark slaty grey* (♀ slightly browner). Auriculars, throat, and *underparts pale grey* (mainly ♂♂) *to dusky grey, with contrasting rufous thighs* and whitish tail coverts. Tail upperside blackish with narrow white tip, 2–3 poorly contrasting grey bands, underside with 3 white bands (basal band often covered by tail coverts). Underwings: coverts whitish or mottled with grey, remiges dusky grey, secondaries faintly barred darker, primaries broadly barred whitish. **Juv:** eyes yellow. *Upper-*

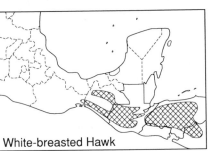

White-breasted Hawk

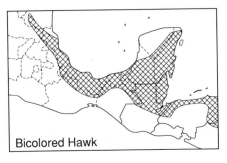

Bicolored Hawk

parts dark brown, crown and nape blacker, often with partial (rarely complete) hindcollar same color as underparts. Underparts cream to ochraceous buff (rarely cinnamon in some ♀♀), thighs may be spotted dark brown. Tail similar to adult but bands on upperside paler brown, often mixed with white. Underwings: coverts same color as underparts, remiges as adult but barring on primaries may be washed buff.

Voice. A fairly hard, slightly gruff, rapid, barking *keh-keh-keh . . .*, higher in ♂, suggesting Cooper's Hawk; rarely vocal except when nesting.

Habs. Humid evergreen and semideciduous forest and edge. A rarely encountered species of deep forest, most likely to be seen perched quietly in forest whence it makes dashing hunting flights after its main food, small birds. Flies over canopy or across open areas with hurried wingbeats and little or no gliding; does not soar. Nest a bulky cup of sticks, lined with leaves, at mid- to upper levels in trees. Eggs: 2, whitish with faint brownish flecks (WFVZ).

SS: Juv Barred Forest-Falcon usually has some dark barring on underparts, strongly graduated tail has 3–6 narrow whitish bars. Collared Forest-Falcon larger with longer legs and tail, dark crescent on auriculars.

SD: U resident (SL–1400 m) on Atlantic Slope from S Tamps (Martin 1951*a*) and Ver to Honduras; possibly on Pacific Slope in Guatemala (Griscom 1932).

RA: E Mexico to Chile and Argentina.

COOPER'S HAWK

Accipiter cooperi   Not illustrated
Gavilán de Cooper

ID: 15–20 in; W 27–34 in (38–51 cm; W 68.5–86.5 cm). Medium-large, *relatively large-headed accipiter. Tail long and*

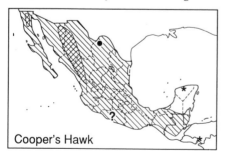

Cooper's Hawk

slightly rounded. In flight, head projects noticeably beyond leading edge of wings. **Descr. Adult:** eyes orange to red, cere and legs yellow to orange-yellow. *Crown, nape, and upperparts slaty blue-grey* (♀ more brownish grey), *crown contrastingly darker*, whitish face and throat streaked rufous. Underparts rufous with coarse whitish barring, undertail coverts white. *Tail boldly tipped white*, upperside broadly banded blackish brown and pale brownish grey, with 3 pale bands, underside pale grey with 3–4 dark bands. Underwings whitish, coverts flecked rufous, pale grey remiges barred dark grey. **Juv:** eyes yellow to glaucous. Head and upperparts dark brown, face streaked paler, sometimes with narrow whitish supercilium; crown and nape coarsely streaked rich brown, upperparts edged rufous, scapulars usually mottled whitish. *Throat and underparts whitish, streaked dark brown except clean white undertail coverts.* Tail similar to adult but 3–4 pale bands on upperside browner. Underwings as adult but coverts flecked brown.

Voice. A hard, rapid, barking *keh-keh-keh . . .*; rarely vocal in winter.

Habs. Breeds mainly in pine–oak woodland. In winter, open woodland and edge, semiopen country with scattered trees; more often in open situations than Sharp-shinned Hawk. Flies with stiff, strong wingbeats alternating with glides; soars with wings held flat or slightly raised, leading edge fairly straight out from body. Nest is a bulky stick platform at mid- to upper levels in trees. Eggs: 3–5, whitish, sometimes finely marked with browns (WFVZ).

SS: See Sharp-shinned Hawk. Northern Goshawk (NW Mexico) has relatively broader wings and shorter tail, proportions in flight more buteo-like, juv more heavily streaked below (including tail coverts), tail bands wavier, dark tail bands narrowly edged whitish. Juv Roadside, Grey, and Red-shouldered hawks may appear surprisingly accipiter-like: Roadside has streaked chest contrasting with barred belly, Grey has boldly marked face, Red-shouldered has more heavily marked underparts (including tail coverts), narrow pale tail bands, and in flight shows pale crescent across outer primaries.

SD: U to F resident (1000–3000 m) in BCN

and, mostly in interior, from Chih to Dgo and Coah, possibly to Gro. F winter visitor (Oct–Apr) S to Isthmus; U to R to N C.A.
RA: Breeds N.A. to N Mexico; N populations migratory, wintering to N Costa Rica.

NORTHERN GOSHAWK
Accipiter gentilis apache   Not illustrated
Gavilán Azor

ID: 20–25 in; W 37–44 in (51–63.5 cm; W 94–112 cm). *Large accipiter of NW Mexico. Flight shape more buteo-like than other accipiters: relatively long and broad wings may appear tapered, tail slightly rounded and relatively short.* **Descr. Adult:** eyes red, cere and legs yellow. *White supercilium contrasts with blackish crown and auriculars,* upperparts dark slaty grey (may be mixed with brown in 1st basic). *Throat and underparts dusky grey* with narrow blackish shaft streaks and fine whitish vermiculations, *undertail coverts white.* Tail upperside grey with narrow whitish tip, 3–5 poorly contrasting dark bars (often appears uniform grey above); underside pale grey with 3–5 indistinct dark bands. Underwings grey, coverts marked as underparts, primaries often show some whitish barring. **Juv:** eyes yellow. Head and upperparts dark brown, usually with distinct whitish supercilium; upperparts with variable paler brown mottling. *Throat and underparts pale buff, heavily streaked and spotted dark brown from chest to undertail coverts.* Tail tipped white, upperside banded dark brown and pale brownish grey, the 3–4 pale *bands often wavy and with narrow whitish margins,* underside pale grey with 3–4 dark bands. Underwings: pale buff coverts flecked brown, whitish remiges barred dusky grey, more contrasting on primaries.
Voice. Generally silent except around nest

where calls are loud and screaming; a hard *wheeu wheeu* ... or *whee'a whee'a,* and *kaah kaah kaah* ... (California).
Habs. Conifer and pine–oak forests. Usually stays in forest and seen infrequently. Hunts from perches or while flying through forest. Soars infrequently, mainly in late winter and spring during displays. Wingbeats powerful and stiff, interspersed with short glides; soars with wings held flat to slightly raised, and pressed forward slightly. Nest is a very bulky platform of sticks at mid- to upper levels in trees. Eggs: 2–4, whitish, unmarked (WFVZ).
SS: See Cooper's Hawk. Buteos have shorter tails, less tapered wings, at rest their wings extend more than half-way to tail tip.
SD: R (to U?) resident (1000–3000 m) from Son and Chih to Jal, and in Gro (Blake 1950a). A report from Coah (Urban 1959) requires verification.
RA: Holarctic breeder; N populations migratory. In New World breeds S to W Mexico.

CRANE HAWK
Geranospiza caerulescens*   Plates 4, 8, 10
Gavilán Zancudo

ID: 18–21 in; W 36–41 in (46–53.5 cm; W 92–105 cm). *A lanky, fairly small-headed hawk usually found near water. Legs very long, wings broad and rounded, tail long and squared to slightly rounded.* **Descr.** Sexes similar. **Adult:** eyes orange-red to red, bill black, cere grey, *legs bright orange-red.* Plumage blackish slate overall; lower belly, thighs, and undertail coverts vermiculated whitish. *Tail blackish with narrow white tip and 2 broad white bands, from below the broader basal band often covered by tail coverts.* Underwings blackish with *narrow white band across outer 4–7 primaries* conspicuous (rarely in-

Northern Goshawk

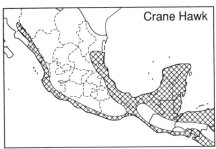

Crane Hawk

distinct) in flight from above and below, whitish vermiculations and bars on coverts and base of remiges. **Juv:** eyes amber, bill grey, legs orange. Overall slightly paler than adult, with whitish forehead, short supercilium, auriculars, and throat; pale buff mottling on chest, belly, and thighs; undertail coverts mottled buff, rarely entirely creamy buff. *Tail as adult. Underwings as adult* but coverts and base of primaries more distinctly barred and vermiculated whitish.

Voice. A clear, plaintive whistle, quite loud, *wheeo-oo* or *wheeeoo.*

Habs. Mangroves, swampy woodland, thorn forest, semiopen areas with scattered trees and forest patches, mostly near water. Usually seen perched in trees or clambering around on trunks, poking into crevices with its double-jointed legs, and flapping wings for balance. Wingbeats distinctively loose and floppy, usually soars only for short periods, wings flattish and pressed slightly forward, tail slightly spread (usually closed in steady flight or glide). In display flight, flap-flap-flap-glide progression low over canopy interrupted by brief climb with quicker loose flaps, then glide back down to original level. Nest platform of sticks at mid- to upper levels in trees. Eggs: 2, white, unmarked (WFVZ).

SS: Lanky appearance, small head, long red legs, and floppy wingbeats distinctive. Stockier black hawks have broader wings and much shorter tails, yellow ceres and legs. See black morph Hook-billed Kite.

SD: U to F resident (SL–1500 m) on both slopes from S Son and S Tamps to El Salvador and Honduras.

RA: Mexico to Peru and N Argentina.

BLACK-COLLARED HAWK
Busarellus n. nigricollis   Plates 4, 11
Aguililla Canela

ID: 18–22 in; W 45–53 in (45.5–56 cm; W 114–134 cm). A distinctive, fish-eating raptor. *Wings very broad and rounded, tail short and squared.* At rest, wingtips about equal with tail tip. In flight appears almost tail-less when tail spread. **Descr.** Sexes similar. **Adult:** eyes dark brown, cere and lores grey, legs pale flesh. *Head creamy white,* crown streaked dark brown, *underparts orange-rufous with black collar across upper chest. Upperparts orange-rufous* with black shaft streaks; tertials

orange-rufous, usually with some black barring and broad black tips; *secondaries rufous with broad black trailing edge; primaries black.* Tail blackish with narrow whitish tip and 2–3 narrow orange-rufous bands at base. Underwings: coverts orange-rufous with sparse black bars most distinct on greater coverts, primaries black, secondaries orange-rufous with sparse dark bars, broad blackish trailing edge. **Juv:** *whitish head* usually shows dark eyestripe. *Underparts creamy to rich buff, heavily streaked and mottled dark brown with collar, and often belly, solidly blackish brown*; thighs and undertail coverts barred dark brown. *Upperparts mottled and barred dark brown and orange-rufous,* nape and upper mantle mottled and coarsely streaked buff. Tail has 3–4 orange-rufous bars basally. Underwings: coverts dull cinnamon-rufous, barred blackish: primaries blackish, subtly barred grey; secondaries paler than adult, more densely barred dark and without bold dark trailing edge. **1st basic:** resembles adult but underparts and underwing coverts flecked whitish and dark, thighs barred dusky, secondaries and tail with less broad dark tips, secondaries with broader dark bars. Attains adult plumage by 2nd prebasic molt.

Voice. Short, hoarse rasps, *eh-rrr* or *eh'rrrr,* etc., and a slightly piercing, clear, screaming whistle, *hileeee* or *ki-leeeee*; usually silent.

Habs. Freshwater marshes, lakes, slow-moving rivers, mangroves. Usually perches fairly low, at times on ground. Wingbeats deep and powerful, often soars, wings in shallow arch with tips slightly curled up, tail spread; glides with wings bowed and tips curled up, tail often closed. Nest is a bulky stick platform, low to high in bush or tree near water. Eggs (Paraguay): 1–2,

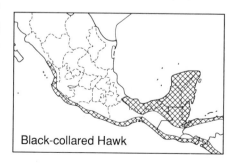

Black-collared Hawk

whitish, marked with browns and greys (WFVZ).

SS: See Snail Kite. In silhouette told from black hawks by smaller and shorter tail, less broadly based wings than Common Black Hawk.

SD: U to F resident (SL−100 m) on Atlantic Slope from S Ver to Honduras; R and local in Yuc Pen and on Pacific Slope from S Sin to El Salvador.

RA: Mexico to N Argentina.

NB: Traditionally considered a buteonine hawk, Olson (1982) provided evidence suggesting that *Busarellus* may be more closely related to kites.

WHITE HAWK
Leucopternis albicollis*   Plates 5, 11
Aguililla Blanca

ID: 19−22 in; W 45−52 in (48.5−56 cm; W 114−132 cm). Striking and unmistakable SE rain forest hawk. Wings very broad and rounded, tail broad, fairly short, and squared. Noticeable geographic variation. Descr. Sexes similar. Adult: eyes dark, cere and lores blue-grey, legs pale yellow. *White overall with broad black tips to outer 6 primaries*, black alula, and large subterminal black spots on rectrices which form *black tail band*. Juv: outer 7 primaries more broadly tipped blackish; outer webs of outer primaries, secondaries (but not tertials), and greater secondary coverts mottled and barred black, primary coverts black with white tips; tail with broader, solid black subterminal band. Underwings may show dusky subterminal spots on secondaries. 1st basic: resembles adult but with slight black mottling on bases of secondaries, more extensive black on outer primaries, black streaking on upperwing coverts. Attains adult plumage by 2nd pre-basic molt when about 2 years old. Adults on Pacific Slope and in S Belize and NE

Guatemala tend to show more extensive black markings on upperwings, resembling *L. a. costaricensis* (E of Sula Valley): adult has black remiges, upper primary coverts, and greater upper secondary coverts, tipped white, some black spotting on outer lesser and median upperwing coverts, and broader black tail band. Juv similar but with more extensive black spotting on lesser and median upperwing coverts.

Voice. A hoarse, drawn-out, whistled scream, *whii-ii-eeihhr* or *hweeiiihrr*, etc., frequently given while soaring; suggests a pig squealing.

Habs. Humid evergreen forest. Often soars, singly or in pairs, low to high over forest. Wingbeats deep and often floppy, soars with wings flattish and pressed slightly forward, tail spread. Usually perches in forest, less often above canopy or at edges. Nest platform of twigs high in trees. Eggs (Trinidad): 1, bluish white, marked with browns (WFVZ).

SS: See King Vulture.

SD: F to C resident (SL−1000 m) on Atlantic Slope from S Ver to Honduras; U to R and local on Pacific Slope from Chis (MAT) to El Salvador (Thurber *et al.* 1987).

RA: SE Mexico to Amazonian Brazil.

COMMON BLACK HAWK
Buteogallus anthracinus*   Plates 4, 8, 10, 11
Aguililla Negra Menor

ID: 18−21 in; W 43−50 in (45.5−53 cm; W 109−127 cm). Large, heavily built hawk with fairly long legs. Wings very broad and rounded, tail short, broad and squared (*juv has noticeably longer tail than adult*). At rest, *wingtips reach nearly to tail tip, primaries project noticeably* (2−3.5 in; 5−9 cm) *beyond tertials*. In flight, appears almost tail-less when tail spread. Descr. Sexes similar. Adult: eyes dark, *cere, lores, bill base, and legs bright yellow* to orange-

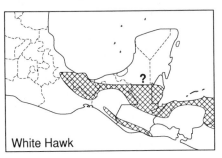

White Hawk

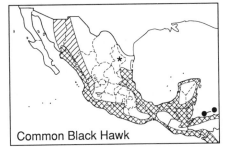

Common Black Hawk

yellow. Dark slaty grey overall, *not* black (can look dark slaty blue-grey in bright light), belly and thighs finely barred white, tail coverts tipped white. *Tail black with white tip and single broad white band* (narrow white basal band usually covered by tail coverts). *Underwings: coverts blackish, remiges often contrastingly paler with whitish flash on base of outer 2–5 primaries,* greyish to buff mottling on base of inner primaries, greyish to rufous mottling across base of secondaries; *remiges broadly tipped blackish. Whitish flash on base of outer primaries* usually distinct in N Mexico (and SW USA), but *often indistinct or lacking in S Mexico and C.A. where many birds have slaty grey remiges poorly contrasting with underwing coverts.* **Juv:** *face pale buff with dark eyestripe and dark malar.* Crown, nape, and upperparts dark brown with crown, nape, and mantle streaked cream to pale buff; scapulars and upperwing coverts often with some whitish to pale buff mottling; lateral uppertail coverts barred buff. Throat and underparts whitish to rich buff, streaked dark brown, often with solid dark patches on sides of chest; thighs and undertail coverts barred and spotted dark brown. *Tail* narrowly tipped whitish, *upperside barred blackish brown and white to pale buffy grey-brown (4–7 pale bars), underside whitish with 4–7 dusky bars often poorly defined,* the distal dark bar being widest. *Broad, whitish to buff panel across base of primaries visible from above in flight.* Underwings: coverts whitish to rich buff, flecked dark brown, *whitish outer primaries (forming contrasting panel, translucent when backlit)* broadly tipped blackish and narrowly barred dark brown (rarely unmarked creamy), inner primaries and secondaries dusky cinnamon to greyish buff, barred dark brown with broader dark trailing edge. Attains adult plumage by 1st prebasic molt although 2nd-year birds may have buff-flecked underparts, more boldly bicolored underwings than adults.

Voice. Often calls when soaring and in display. An ascending, at times prolonged, series of loud, ringing whistles, the last note often lower, or short series of fuller, slower whistles alternated with more rapid series: *yeep yeep ...* or *pyiih pyiih ...,* and *yiih-yiih-yiih-yiih-yiih yeep yeep yeep ...,* etc.

Habs. Marshes, mangroves, forested areas, usually near water. Commonly seen soaring or perched conspicuously on trees, fence posts, at times on ground; runs after crabs on beaches. Wingbeats deep, powerful, and fairly slow, at times floppy. Soars on flattish wings, tail usually spread; in display flight climbs with exaggerated slow wingbeats, calling excitedly, circling and gliding with legs dangling. Stick platform nest at mid- to upper levels in trees. Eggs: 1–3, whitish, usually marked with browns (WFVZ).

SS: Can be confused easily, especially in flight, with larger Great Black Hawk and Solitary Eagle. Great Black Hawk has narrower wingbase, relatively longer tail (especially juv) often less spread when soaring and gliding. At rest note longer legs and short primary projection; adult has two white tail bands (usually only one visible from below), more uniformly dark underwings, white uppertail coverts, and grey lores; juv and imm usually have whitish head that lacks strong dark malar stripe, note more numerous dark tail bars of juv with very broad distal dark band or narrow dark bars to tail tip. Solitary Eagle has relatively longer, less broadly based wings, with longer outer primary. At rest, wingtips may project slightly beyond tail; also note very thick legs and heavy bill; adult grey above; juv has solidly dark brown thighs and overall pale greyish tail with diffuse dusky subterminal band but no distinct barring. See Snail Kite, Crane Hawk, and black morph Hook-billed Kite.

SD: C to F resident (SL–1800 m) on both slopes from S Son and E NL (AMS) and S Tamps to El Salvador and Honduras. U to F summer resident (Mar–Sep; to 2000 m) inland locally (mostly along rivers) to N Son; R (winter status unclear) inland along Balsas drainage to Mor; U to F but local in interior from Chis S.

RA: Mexico (SW USA in summer) to N S.A.

NB: Mangrove-inhabiting birds of the Pacific coast from Chis (MAT spec.) are sometimes treated as a separate species, Mangrove Black Hawk, *B. subtilis* (Aldrich and Bole 1937; Dickey and Van Rossem 1938; Amadon 1961; Monroe 1963a, 1968). Monroe (1963a) reported that *subtilis* in El Salvador and Honduras differed from *anthracinus* by the 'lack of rufous on the secondaries' but this feature is variable: a ♂ *subtilis* (AMNH 813147) collected in mangroves on the Pacific coast of E Guatemala, has more extensive rufous mottling on the secondaries than many *anthracinus*. Also, many *anthracinus* have so little rufous on their secondaries that it is almost invisible. The main difference between the two forms in the region appears to be size. Wing chord, measured to the nearest 5 mm, averages 325–360 mm in ♂ *subtilis* compared with 360–390 mm in ♂ *anthracinus*. Similarly, wing chord of ♀ *subtilis* is smaller (335–370 mm) than that of ♀ *anthracinus* (370–410 mm) (*n* = 31 *subtilis*, from Monroe 1963*b*, 1968; *n* = 16 ♂, 17 ♀ *anthracinus*).

The form in the Honduras Bay Islands appears intermediate in size between *anthracinus* and *subtilis* (♂ wing chord 345–385 mm, ♀ 365–380 mm; from Monroe 1968) and was first considered as *B. subtilis utilensis* by Monroe (1963*a*), then as *B. anthracinus utilensis* (Monroe 1968). We see no reason to consider *subtilis* as other than a mangrove-inhabiting ssp of Common Black Hawk. Voice and behaviour are very similar and many *anthracinus* on the Atlantic Slope live in mangroves where, in the manner of *subtilis*, they run after crabs on beaches and mud-flats.

GREAT BLACK HAWK

Buteogallus urubitinga ridgwayi
Aguililla Negra Mayor   Plates 4, 8, 10, 11

ID: 20–24 in; W 47–54 in (51–61 cm; W 120–137 cm). Large, heavily built hawk with *long legs*. Wings broad and rounded, tail fairly short, broad, and squared (*juv has noticeably longer tail than adult*). At rest, *wingtips do not reach tail tip, primaries project only slightly* (1–2 in; 2.5–5 cm) *beyond tertials*. In flight appears almost

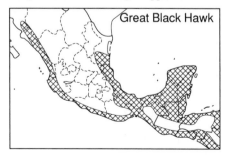

Great Black Hawk

tail-less when tail spread. **Descr.** Ages differ, sexes similar. **Adult:** eyes dark brown, cere yellow, *lores grey to dull pale yellowish*, legs yellow. Blackish slate overall with *white uppertail coverts and coarse white barring on thighs. Tail black with white tip, a single broad white median band, and a narrow white basal band (from below, basal band usually covered by tail coverts; from above, upper band merges with white tail coverts and may produce the effect of a white tail with a broad black median band). Underwings blackish*, coverts may have whitish vermiculations, base of *outer 1–3 primaries whitish, remiges indistinctly barred grey;* may show small pale panel on trailing edge of middle primaries when backlit. **Juv:** *face pale creamy overall with no dark malar stripe.* Crown, nape, and upperparts dark brown with crown, nape, and mantle streaked cream to buff; median and greater secondary coverts barred and mottled tawny to rufous, scapulars often mottled pale buff; uppertail coverts barred buffy white to mostly pale buff. Throat and underparts creamy buff to rich buff, underparts streaked dark brown, often with solid dark patches on sides of chest; thighs and undertail coverts barred and spotted dark brown. *Tail narrowly tipped whitish, upperside narrowly barred blackish brown and creamy to pale buffy grey-brown (10–13 pale bars), underside buff with 9–12 dark bars, narrow dark bars continue to tail tip* or *distal dark band may be much broader. Broad, whitish to buff panel on base of primaries visible from above in flight.* Underwings: coverts creamy buff to rich buff, flecked dark brown; buff outer primaries (forming contrasting panel, translucent when backlit) broadly tipped blackish and narrowly barred dark brown; inner primaries and secondaries cinnamon to dusky rufous, narrowly barred dark brown, with broader dark trailing edge. **1st basic:** resembles juv but may show indistinct dark malar stripe; upperparts more uniform with uppertail coverts barred whitish to mostly white underparts generally less heavily streaked. Tail tipped whitish, upperside pale grey-brown to whitish with 5–7 coarse dark bars, underside whitish to buff with 4–6 dark bars, distal dark band broadest. Buff to whitish primary panel less distinct, underside of inner primaries and secondaries greyish with sparser dark barring. **2nd basic:** resembles adult but head, chest, and underwing coverts usually show some buff flecking, inner primaries often include 2–3 buff, darkly barred feathers forming a distinct panel; underside of remaining remiges barred slightly paler grey than adult. *Tail variable* (2nd basic tail sometimes shown by birds in mostly 1st basic): *boldly banded black and white with 2–4 often wavy pale bands* (white on upperside, whitish to dirty creamy buff on underside); uppertail coverts mostly white to pale buff. Attains adult plumage by 3rd prebasic molt.

Voice. A shrill, long-drawn-out, screamed whistle, given while perched or in flight. In display flight, an excited, high, piping, or bubbling whistle run into an accelerating, rolled series or trill, *whi' whi hi-i-i-i-i-i ... or whi' pi-pi-pi-pi-pi-pi-pi ...*

Habs. Forested areas, often near water, marshes, mangroves. Habits much as Common Black Hawk but rarely seen on ground; soars with wings flattish and pressed slightly forward, tail typically less spread than Common. In display soars in circles, calling, with legs dangling. Nest platform of sticks usually high in trees. Eggs (S.A.): 1, whitish, usually marked with browns (WFVZ).

SS: See Common Black Hawk. Solitary Eagle larger with more massive legs and bill; at rest, wingtips extend to or beyond relatively shorter tail; adult grey, tail has broad white median band; juv and imm have solidly dark brown thighs, juv has pale greyish tail with diffuse dusky subterminal band but no distinct dark barring, imm tail similar but with broad paler median band.

SD: U to F resident (SL–1800 m) on both slopes from S Son and S Tamps, and in interior locally from Chis, to El Salvador and Honduras.

RA: Mexico to N Peru and N Argentina.

SOLITARY EAGLE

Harpyhaliaetus solitarius * Plates 4, 8, 10, 11
Aguila Solitaria

ID: 25–31 in; W 60–74 in (63.5–79 cm; W 152–188 cm). Rare and little-known eagle of mountains and foothills. *Bill massive, legs thick and powerful.* Wings long, very broad, and rounded; tail short, broad and squared (juv has noticeably longer tail than adult). *At rest, wingtips reach to or slightly*

beyond tail tip, primaries project slightly (2–2.5 in; 5–6 cm) beyond tertials. In flight appears almost tail-less when tail spread. Nape feathers slightly elongated but rarely, if ever, appears crested in the field. **Descr.** Sexes similar. **Adult:** eyes dark, cere, lores, and legs yellow. *Slaty blue-grey* to dark slaty *overall*, tail coverts tipped white. *Tail blackish with narrow white tip and single broad white median band* (narrow white basal band usually covered by tail coverts). Underwings dark grey, outer primaries broadly tipped black with paler flash on base of outer 2–4 primaries, inner primaries and secondaries with dark bars and broad dark trailing edge. **Juv:** *face buff with broad, dark brown postocular stripe continuing down sides of neck to merge with solidly dark sides of chest; no dark malar stripe.* Crown, nape, and upperparts dark brown, nape and mantle coarsely streaked rich buff. Throat and underparts buff, streaked and mottled dark brown, *thighs solidly dark brown. Tail grey-brown with narrow whitish tip, upperside with broad but poorly defined dark subterminal band, indistinct dark spotting and narrow broken barring; underside appears pale grey, finely mottled dusky, subterminal band as upperside.* Underwings: coverts rich buff with coarse, usually sparse dark spotting, *creamy-white outer 6 primaries (forming contrasting pale panel, translucent when backlit) broadly tipped blackish and unbarred*; inner primaries and secondaries pale grey, marked like undertail. **1st basic:** resembles juv but tail grey with broad dark subterminal band, variable dark mottling, and wavy bars at base. **2nd basic:** resembles adult but browner, head, body, and underwing coverts sparsely streaked and mottled buff, tail resembles adult but broad median band pale grey. Attains

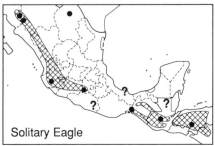

Solitary Eagle

adult plumage by 3rd prebasic molt when about 3 years old (?).

Voice. In flight, a rapid series of short, powerful, whistled notes on same pitch *ple ple ple ple ple ple* ...; while perched, a long-drawn-out, powerful whistle, the second half on lower pitch *keeeeeer-loooooo* (Stiles and Skutch 1989).

Habs. Pine and pine–oak forest, humid evergreen forest. Habits little known; in Costa Rica soars frequently. Nest is a bulky platform of sticks high in trees. Eggs: 1, whitish, unmarked (WFVZ).

SS: See Common Black and Great Black hawks, both of which are frequently mis-identified as Solitary Eagle. Golden Eagle has narrower wings and longer tail, lacks broad white tail band. Imm Bald Eagle lacks pale primary panel, younger imms have all-dark heads, older imms have pale bills.

SD: R resident (600–2100 m) on Pacific Slope and in adjacent interior, recorded Son, Chih, Jal, Mex (SNGH, SW), and E Oax; also N Coah (AGL spec.) and single old specs from Guatemala and Honduras. Probably R in foothills and highlands throughout region. However, many recent reports from Ver (Los Tuxtlas; Winker *et al.* 1992c), Chis (Palenque etc.), Guatemala (Vannini 1989a; Beavers *et al.* 1991), and Belize (Wood and Leberman 1987; Eitniear 1991) require verification or are not credible; nesting reports from Oax (Smith 1982) and Chis (GC, photo) are equivocal (photos examined).

RA: Mexico to N Argentina.

HARRIS' HAWK

Parabuteo unicinctus * Plates 3, 11
Aguililla de Harris

ID: 19–21 in; W 40–47 in (48.5–53.5 cm; W 102–120 cm). A dark, buteo-like hawk. *Wings fairly long, broad and rounded, tail long and slightly rounded.* At rest, wing-

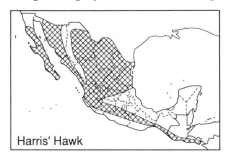

Harris' Hawk

tips extend about half-way to tail tip.
Descr. Ages differ, sexes similar. **Adult:** eyes dark, cere, lores, orbital ring, and legs yellow to orange-yellow. *Blackish brown, overall,* lesser upperwing coverts and edges of median upperwing coverts rufous, *thighs solidly rufous,* tail coverts white. *Tail blackish with broad white tip and broad white band at base* (mostly covered by tail coverts). Underwings dark overall, axillars and lesser and median coverts rufous. **Juv:** cere, lores, and legs paler. Head browner, rufous on wing coverts paler. Throat and underparts mottled and streaked pale buff (often with a broad pale band on lower chest), thighs rufous with white barring. *Whitish tail tip less striking,* uppertail much like adult but browner, may show 8–10 indistinct paler bars (mostly on inner webs), *undertail greyish with 2–12 narrow dark bars most distinct distally.* Underwings: coverts mottled rufous, *whitish outer 6 primaries (forming contrasting pale panel) broadly tipped black* and often barred dark brown, inner primaries and secondaries dusky grey with narrow darker barring. Attains adult plumage by 1st prebasic molt.

Voice. Generally silent; at times gives a gruff, rasping *ehhk* and *eh'hk eh'hk eh'hk*, etc., usually near nest; also a shrill, whistled *whieh whieh* ..., at least given by juvs, suggesting Red-tailed Hawk.

Habs. In N and cen Mexico, arid to semi-arid areas with brush, large cacti, and Joshua trees, thorn scrub. In S Mexico and C.A., arid to semiarid savannas and marshes with scattered trees. A social and cooperative nester, usually seen in twos and threes; often perches on large cacti and telegraph poles. Usually hunts from perches, soars with wings flat or slightly bowed and pressed forward, tail spread. Wingbeats often floppy, unlike buteos. Nest is a compact stick platform, low to high in tree or large cactus. Eggs: 2–4, whitish, may be lightly marked with browns (WFVZ).

SS: Shape distinctive. Dark morph buteos lack rufous shoulders and distinctive black and white tail pattern; juv Red-shouldered Hawk paler overall, lacks bold tail pattern. See Snail Kite.

SD: F to C resident (SL–1800 m) in cen and S Baja, on both slopes S to Col and N Ver, and on Plateau S to Zac and SLP; U to R and local S in interior to Oax. R and local on Pacific Slope from Isthmus to El Salvador. Often kept in captivity and extralimital distributional records (such as DF, Yuc Pen) may refer to escapes.

RA: SW USA and Mexico to cen Chile and cen Argentina.

GENUS BUTEO: Buteos

Buteos, known as Buzzards in the Old World, are often considered to represent typical hawks. Wings broad and rounded, tails short to medium-long. Spend much time soaring and prey mainly on mammals. Ages differ, sexes similar, several species dimorphic. Attain adult plumage in 1–2 years.

GREY HAWK
Buteo nitidus *　　　　　　　Plates 3, 9
Aguililla Gris

ID: 16–18 in, W 32–37 in (40.5–45.5 cm, W 81–94 cm). Widespread. *Relatively long-tailed,* wingtips rest about half-way down tail. **Descr. Adult:** eyes dark, cere and legs orange-yellow to yellow. *Head and upperparts grey,* paler on head, distal uppertail coverts white. *Underparts pale grey to grey, coarsely barred whitish,* undertail coverts white. *Tail white-tipped, blackish on upperside with 2 white bands, basal band narrower;* underside dark grey with 3 white bands, basal band usually covered by tail coverts. Underwings: whitish overall, coverts lightly spotted and barred grey, remiges barred dusky, outer 5 primaries tipped black, inner primaries and secondaries with dusky distal band. **Juv:** *face whitish to buff with dark eyestripe and malar stripe.* Crown, nape, and upperparts dark brown; crown and nape often streaked buff; back and *upperwing coverts usually mottled rufous-brown. White-tipped distal uppertail coverts form white*

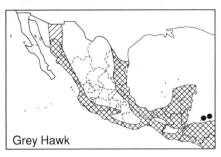

Grey Hawk

band at tail base. Throat and underparts whitish to creamy buff (rich buff when fresh, fades by winter) with dark brown spotting and streaking; *thighs barred dark brown,* undertail coverts whitish. *Tail tipped whitish, upperside grey-brown with 6–9 dark brown bars,* underside pale grey with 6–9 narrow dark bars, often indistinct; *tail bars often wavy, progressively wider distally.* Underwings: coverts whitish, flecked and barred dark brown, remiges barred dusky, outer 5 primaries narrowly tipped dark brown, inner primaries and secondaries with dusky distal band. In flight often appears uniformly pale below, especially when backlit. Attains adult plumage by 1st prebasic molt.

Voice. Loud, mournful, rich whistles in series, typically 3–7 ×, *h'lee h-weeoo h-weeoo ...* or *weeeooo wi-weeeooo wi-weeeooo ...,* etc. Also a clear, whistled scream, *wheeeeu* or *wheeeuu,* not as rough as Roadside Hawk scream; juv gives a hoarser *wheeeih.*

Habs. Wooded and forested habitats, less often semiopen areas with scattered trees. Perches conspicuously on telegraph poles and roadside trees. Soars with wings flat to slightly raised, tail slightly to fully spread. Nest is a stick platform at mid- to upper levels in trees. Eggs: 2–4, whitish, may be lightly marked with browns (WFVZ).

SS: Adult distinctive. Juv Broad-winged Hawk told by longer primary projection, shorter tail; bolder dark trailing edge to underwing; less distinct white uppertail covert band; fewer, more even-width, and less wavy tail bands with broad distal band; upperside of tail often less contrastingly marked; grey-brown rather than rufous-brown mottling on upperwing coverts; spotted not barred thighs; and usually less boldly marked face. Juv Red-shouldered Hawk has less boldly marked face, underparts usually more heavily marked (including tail coverts); note tail pattern. See Cooper's Hawk.

SD: C to F resident (SL–1800 m) on both slopes from cen Son, NL, and Tamps, and locally in interior from Balsas drainage, to El Salvador and Honduras; in summer (Mar–Sep) to N Son.

RA: SW USA and Mexico to Ecuador and N Argentina.

NB: Sometimes placed in the genus *Asturina.*

ROADSIDE HAWK
Buteo magnirostris *   Plates 3, 9, 69
Aguililla Caminera

ID: 13–16 in; W 27–31 in (33–40.5 cm; W 68–79 cm). Common small buteo of both slopes. Three groups: *griseocauda* group (most of Mexico), *direptor* group (SE Chis and N C.A.), and *gracilis* (Isla Cozumel). **Descr. Griseocauda. Adult:** *eyes whitish, cere, lores, and legs bright orange-yellow to orange.* Head and upperparts grey-brown, upper tail coverts whitish to buff, coarsely barred rufous. Throat whitish, streaked grey-brown, *chest dusky grey-brown, usually mottled and streaked whitish to buff, belly rufous, coarsely bar-red whitish to buff;* thighs whitish to pale cinnamon with narrow rufous barring, undertail coverts whitish to pale buff, sparsely spotted dark brown. *Tail tipped whitish, upperside boldly and evenly banded blackish brown and pale greyish, with 3–4 pale bands,* underside pale grey with 3–5 narrow dark bands. Underwings: coverts creamy to buff, flecked dark brown, remiges pale grey with dark bar-ring, *inner webs of inner primaries and outer secondaries washed rufous,* appear-ing as translucent panel when backlit. **Juv:** eyes brown to amber, legs yellow. Head and upperparts grey-brown, often with pale buff streaking in face and especially supercilium; crown and nape streaked whitish to buff. Throat and underparts whitish to pale buff, *chest coarsely streaked* dark grey-brown to rufous-brown, *belly barred* rufous and whitish (lower belly may be unbarred). *Tail similar to adult* but usually with 4–5 pale grey bands above, 4–6 narrow dark bands be-low. Wings often with little (sometimes no) rufous. Attains adult plumage by 1st pre-basic molt. *Direptor.* **Adult:** overall paler

and greyer above with pale tail bands washed rufous, chest often mottled rufous, rufous in primaries brighter. *Gracilis* smaller. **Adult:** pale buff thighs sparsely spotted dark brown. **Juv:** *whitish face with dark eyestripe, underparts white to pale buff, chest streaked dark brown, belly and thighs sparsely spotted dark brown.*

Voice. Common call, usually from perch, a complaining, drawn-out scream, *rreeeaew* or *meeeahhh.* In display flight (less often from perch), an often persistent, fairly rapid, nasal, laughing or barking series, at times with one or more introductory notes, *heh-heh-heh...* or *keh heh-heh-heh* or *reh, reh, heh-heh-heh*, etc.; may suggest Lineated Woodpecker.

Habs. Open and semiopen country, es-pecially humid areas; less often in forest. Perches on roadside fence posts, telegraph wires, trees. Flies with rapid stiff wingbeats interspersed with short glides; soars in-frequently (except in noisy display), wings flattish or slightly arched (in marked dihed-ral during display), tail closed or only slight-ly spread. Nest platform of sticks at mid- to upper levels in trees. Eggs: 2, whitish, speckled to mottled with brown (WFVZ).

SS: Broad-winged Hawk shorter tailed, underparts either barred (adult) or streaked (juv), tail usually spread while soaring, prefers more wooded areas. Juv Red-shouldered Hawk larger, usually shows some rufous on upperwing coverts, has narrower pale bands on uppertail, and in flight shows narrow pale panel across outer primaries. See Cooper's Hawk, Double-toothed Kite.

SD: C to F resident (SL–1500 m, rarely to 2000 m) on Atlantic Slope from E NL and Tamps to Honduras, on Pacific Slope U to F from Jal to Isthmus, C to F from Isthmus to El Salvador and Honduras.

RA: Mexico to Ecuador and N Argentina.

RED-SHOULDERED HAWK
Buteo lineatus *   Plate 9
Aguililla Pechirroja

ID: 18–22 in; W 36–43 in (45.5–56 cm; W 91–109 cm). E > Baja. N Mexico, mainly in winter. A fairly large, relatively slender buteo. Wings fairly long and broad, tail fairly long (especially juv) and squared. Two groups: *elegans* (Baja) and *lineatus* group (N and cen Mexico, mainly winter). **Descr. Adult.** *Elegans*: eyes dark, cere and

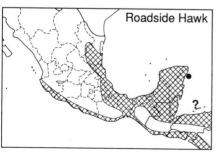

Roadside Hawk

legs yellow. Head and upperparts dark brown, with dusky malar stripe, overall streaked and narrowly edged rufous, *lesser upperwing coverts rufous; rest of upperwings boldly barred and checkered blackish and white*; narrow white panel visible across base of primaries from above in flight. *Tail* tipped white, *upperside blackish with 3–4 narrow white bands*, underside dusky with 3 narrow whitish bands (basal band may be covered by tail coverts). Throat whitish to pale cinnamon, streaked dusky, *underparts bright rufous*, chest with sparse dark streaks, *belly and thighs barred whitish*, undertail coverts creamy buff, barred cinnamon. Underwings: coverts rufous, barred whitish, remiges whitish, barred dark brown. *From below, narrow pale crescent (translucent when backlit) across outer primaries distinctive.* **Lineatus**: underparts typically paler, cinnamon to cinnamon-rufous, upperwings less boldly and brightly patterned; uppertail has 3–5 narrow white to whitish bands undertail, 3–4 narrow whitish bands; underwings less strikingly patterned. **Juv. Elegans**: head and upperparts dark brown, streaked buff to ochraceous buff, dark malar stripe often bolder than adult; *lesser upperwing coverts dark brown, edged rufous, rest of wings resembles adult but less striking*, dark brown and whitish. *Tail* tipped whitish, upperside *blackish brown with 3–4 (rarely 5) evenly spaced whitish to pale grey bars*, underside dusky with 3–4 whitish bars. Throat and underparts pale buff to creamy, throat and chest streaked dark brown (often with dark median throat stripe), markings become coarse teardrops and bars on belly, flanks, and undertail coverts; buff thighs barred dull rufous. Underwings: coverts creamy to buff, spotted and barred dark brown and rufous-brown, remiges whitish

with narrower, sparser dark barring than adult; *note narrow pale crescent across outer primaries*. **Lineatus**: head and upperparts have paler, whitish to pale buff streaking, often forming a whitish supercilium. Upperwings dark brown with indistinctly rufous-edged lesser coverts; median and greater coverts poorly patterned with pale grey-brown to dusky cinnamon barring. *Uppertail has 4–5 evenly spaced, narrow pale greyish bands*, undertail has 4–6 poorly contrasting whitish bands. Creamy to whitish throat and underparts more sparsely marked than *elegans*, some birds with only slight dark median throat stripe and coarse teardrop spots on chest and belly, unmarked creamy thighs, and sparse dark spots on undertail coverts. Underwings: coverts creamy to whitish, spotted dark brown, remiges sparsely barred dark brown. Juv *texanus* (of NE Mexico) has more heavily marked underparts, resembling elegans.

Voice. An often persistent, screamed *kyeeahh, kyeeahh ... or kyaah kyaah ...*, etc., suggesting Brown Jay.

Habs. Semihumid to arid riparian woodland, wooded and semiopen country with scattered trees and hedges, often near water. Perches on fence posts, telegraph poles, or trees, from which it hunts. Wingbeats fairly rapid and stiff, interspersed with short glides; soars with wings flat to slightly bowed; glides with wings bowed, tail slightly spread to spread. Nest is a stick platform at mid- to upper levels in trees. Eggs: 2–4, whitish, marked with browns (WFVZ).

SS: See Cooper's, Grey, and Roadside hawks. Juv Broad-winged Hawk has shorter tail, flies with wings flattish, less heavily marked below than *lineatus*, pale tail bands broader than dark bands (reverse of Red-shouldered), shows large pale primary panel in flight. Juv Red-tailed Hawk more heavily built, flies with wings in slight dihedral, usually has contrasting dark head and belly band; note large pale primary panel in flight.

SD: F to U resident (near SL–1000 m) in Baja S to 26°N, R resident in N Tamps (at least formerly, WFVZ). U to R winter visitor (Oct–Apr; SL–2500 m), mainly in interior, S to Jal, Mex, and cen Ver. Erroneously reported as breeding S to cen Mexico (AOU 1983).

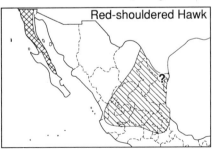
Red-shouldered Hawk

RA: Breeds N.A. to N Mexico; winters to cen Mexico.

BROAD-WINGED HAWK

Buteo p. platypterus Plates 3, 9
Aguililla Aluda

ID: 15–17 in; W 32–36 in (38–43 cm; W 82–92 cm). *Small migrant buteo. Wings relatively pointed, tail fairly short; at rest, wingtips project more than half-way down tail.* Dimorphic (dark morph rare). **Descr. Adult. Light morph**: eyes amber to brownish yellow, cere and legs yellow. Head and upperparts dark brown with thick blackish malar stripe; upperwing coverts and scapulars mottled paler grey-brown, uppertail coverts broadly tipped whitish to pale brown. *Throat and underparts whitish, underparts coarsely barred and mottled greyish rufous to rufous,* concentrated on chest; undertail coverts white. *Tail white-tipped, upperside blackish with 2 whitish to pale grey bands* (distal band broad, basal band often partly covered by tail coverts), *underside dusky with broad white band, narrower basal band usually covered by tail coverts.* Underwings: coverts white to creamy, sparsely spotted and barred rufous, *remiges whitish with broad dusky trailing edge,* secondaries and inner primaries with narrow dark barring. **Dark morph:** blackish brown overall, including underwing coverts, flight feathers as adult light morph so that *silvery underside of remiges contrasts with coverts.* **Juv. Light morph:** *head and upperparts dark brown with whitish supercilium and dark malar stripe,* crown streaked whitish to buff, nape mottled white; upperwing coverts and scapulars edged rufous and mottled paler brown, uppertail coverts broadly barred and tipped whitish. Throat and underparts whitish to pale buff, often with dark median throat stripe, underparts

spotted and streaked dark brown, *thighs spotted dark brown,* undertail coverts unmarked. *Tail whitish-tipped, upperside grey brown with 4–5 poorly contrasting, dark brown bands, distal band notably broader, underside pale greyish with 4–5 dark bars.* Underwings whitish, coverts flecked and barred dark brown, *remiges similar to light adult but dusky trailing edge less distinct,* outer primaries may be faintly barred and spotted dark brown. **Dark morph:** usually shows some whitish and rufous mottling on underparts and underwing coverts. *Flight feathers as light juv.* Attains adult plumage by 1st prebasic molt. **Voice.** A high, thin, drawn-out whistle, *s-eeeeeeu* or *ssiiiiiiu,* given from perch and in flight.

Habs. Forested areas, clearings, open woods. Often soars, wings held flattish and slightly pressed forward, tail spread to slightly spread; glides on flattish wings. Usually perches in forest or at edges, rarely in open situations. Occurs in flocks of several thousand birds during migration when often mixes with other migrating raptors.

SS: See Grey, Red-shouldered, and Roadside hawks, Double-toothed Kite. Dark morph Short-tailed Hawk (common) often has whitish forehead and lores, tail less distinctly barred, underside of secondaries dusky relative to pale primaries. Zone-tailed Hawk has relatively longer and narrower wings and tail, flies with wings in strong dihedral.

SD: C transient (Mar–mid-May, Sep–Oct, R into Nov), main migration route apparently along Atlantic Slope and adjacent foothills, with smaller numbers along Pacific Slope S of Isthmus, U to R W to NL. R (to U?) migrant on Pacific Slope N of Isthmus. U to F winter visitor (Oct–Mar; near SL–2000 m, mainly above 250 m) on Pacific Slope from Col (SNGH), and on Atlantic Slope from N Chis to Honduras; R and irregular in winter N to BCS (Howell and Pyle 1988), Nay (Clow 1976), DF (JKe, RGW), Tamps, and S QR (JKi).
RA: Breeds E N.A.; winters S Mexico to S.A.

SHORT-TAILED HAWK

Buteo brachyurus fuliginosus Plates 3, 8, 9
Aguililla Colicorta

ID: 16–18 in, W 34–40 in (40.5–45.5 cm, W 86.5–102 cm). *Widespread small buteo.*

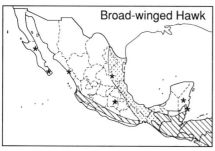

Broad-winged Hawk

Wings relatively long and broad, slightly tapered toward tips, tail medium length and squared. At rest, wingtips reach about to tail tip. *Dimorphic, both morphs common.* **Descr. Adult. Light morph:** eyes dark, cere and legs yellow. *Head and upperparts dark brown,* forehead and lores often whitish, uppertail coverts tipped pale grey. *Dark hood contrasts strongly with creamy-white throat and underparts,* sides of chest often with a few dark streaks. Tail tipped white, upperside grey brown with 4–5 dark bars, underside pale grey with 3–4 darker bars, distal band broadest; tail may appear washed cinnamon when backlit. *Underwings white overall,* outer 5 primaries tipped blackish, inner primaries and secondaries barred dark brown and tipped dusky (*secondaries often appear contrastingly dusky*). **Dark morph:** *blackish brown overall, including underwing coverts,* forehead and lores often whitish. *Flight feathers as adult light morph.* **Juv. Light morph:** as light adult but crown, nape, and auriculars streaked buff to whitish, *sides of chest with more extensive dark streaking;* in flight may appear dirtier white below. Uppertail grey-brown with 4–6 dark bars, undertail pale with 3–5 dark bars, distal band slightly wider; tail often looks fairly uniform unless spread. Underwings similar to adult but secondaries duskier. **Dark morph:** as dark adult but *underparts and underwing coverts flecked cinnamon to whitish.* Flight feathers as juv light morph, remiges duskier and less contrasting than dark adult. Attains adult plumage by 1st prebasic molt.

   Voice. A clear, whistled *clee-u, clee-u,* or *wheeu wheeu,* or *klee-ee klee-ee,* etc., and a slightly accelerating *whee whee whee wheeu.*

   Habs. Wooded and forested country in general, plantations, mangroves. Usually seen soaring, rarely perched. Soars with wings flattish, tips of primaries slightly curled up, tail slightly spread. Often hangs in wind and hovers. Stick platform nest usually high in trees. Eggs: 1–3, whitish, often marked with browns (WFVZ).

SS: See juv Broad-winged and Grey hawks which have much more heavily marked underparts than light morph juv Short-tailed. Red-tailed Hawk flies with wings in slight dihedral, dark morph adult has rufous tail; dark juv longer-tailed with more numerous and distinct dark tail bars; note large pale panel on base of primaries above and below. Zone-tailed Hawk has relatively longer and narrower wings and tail, flies with wings in strong dihedral, adult tail pattern distinctive, juv has more strongly two-toned underwings and paler underside to tail. Adult White-tailed Hawk overhead shows single black band on shorter white tail, wings more broadly based and more tapered.

SD: C to F resident (SL–2000 m, locally to 3000 m) on both slopes from cen Sin and Tamps to El Salvador and Honduras; U N to cen Son (SNGH, SW). Both morphs common though dark morphs may be slightly commoner in W Mexico; light morphs commoner in E Mexico, including Yuc Pen.

RA: Mexico (locally S USA) to Ecuador and N Argentina.

SWAINSON'S HAWK
Buteo swainsoni   Plate 9
Aguililla de Swainson

ID: 19–22 in; W 47–52 in (48.5–56 cm; W 119–132 cm). Large migratory hawk with *long, noticeably tapered wings;* tail medium length and squared. *At rest, wingtips reach to or beyond tail tip.* Dimorphic (dark morph uncommon), plumage variable. **Descr. Adult. Light morph:** eyes

Short-tailed Hawk

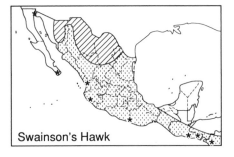

Swainson's Hawk

brown, cere and legs yellow. Head and upperparts dark brown to grey-brown, distal uppertail coverts barred white. *White throat contrasts with rufous-brown to grey-brown chest* (median chest may be mottled whitish), underparts otherwise white, often with brown spots and bars on sides and belly. Tail white-tipped, upperside dark grey with 6–8 blackish bars, underside whitish with 4–6 narrow dark bars, the subterminal bar widest (tail often looks uniform). Underwings: *white coverts*, sometimes sparsely spotted dark brown, *contrast with dark grey remiges*, remiges faintly barred blackish, base of outer primaries may be paler grey. **Dark morph:** Head, upperparts, and flight feathers as light morph. *Throat and underparts dark brown with dark-barred buff undertail coverts*; underwing coverts mottled dark brown and rufous. A variation ('rufous morph') has median throat whitish, belly dark rufous, undertail coverts whitish with some rufous barring, underwing coverts mottled whitish and rufous. **Juv:** *face pale buff with dark eyestripe and malar stripe.* Crown, nape, and upperparts dark brown, nape and upper mantle coarsely streaked pale buff, scapulars mottled pale buff. Underparts rich buff (soon fading to pale buff or whitish); some birds sparsely spotted dark brown on sides and flanks, others heavily spotted and streaked dark brown on chest; belly mottled darker, thighs coarsely barred dark brown, undertail coverts may be barred dark brown. Tail similar to adult but with 8–10 narrower dark bars above. Underwings: coverts as underparts, remiges typically dusky grey with paler inner primaries. **1st basic** (dimorphism becomes apparent). **Light morph:** as light adult but whitish-mottled chest rarely forms complete bib, belly and underwing coverts mottled dark brown; base of primaries paler. **Dark morph:** as dark adult but crown and nape streaked pale buff, chest and underwing coverts mottled pale buff, belly and thighs mottled rufous; base of primaries paler. Attains adult plumage by 2nd prebasic molt on breeding grounds.

Voice. Usually silent; near nest may give a clear to slightly rough, drawn-out, mewing scream, *wheeeooo.*

Habs. Savannas, grassy plains, and farmland with scattered trees and bushes, widely during migration. Hunts mainly in flight, wingbeats deep and flexible; soars with wings in marked dihedral, less pronounced when gliding, tail slightly spread. Hangs in wind and hovers, especially when windy. Typically perches low, often on ground. Occurs in flocks of several thousands during migration when often associates with Turkey Vultures and Broad-winged Hawks. Stick platform nest low to high in trees. Eggs: 2–3, whitish, often marked with browns (WFVZ).

SS: White-tailed Hawk has broader wings, adult shorter-tailed with bold black tail band, lacks dark bib; juv has blackish-brown head and upperparts, underside of remiges paler than coverts; imm has less contrasting dark remiges below with inner primaries darker than secondaries, adult-like tail. On perched birds, note Swainson's small bill, slighter build; light juv has more boldly marked face than White-tailed, dark morph has all-dark thighs, lacks white chest patch of juv White-tailed. Juv Red-tailed Hawk soars with less-marked dihedral, wings less pointed, underwing often has broad pale primary panel.

SD: F to U summer resident (Mar–Aug; 500–2100 m) from Son to E Dgo and Coah, possibly N NL (TLE); formerly BCN. C to F transient (Mar–Apr, Sep–Oct; U to R to Nov), main migration route along Pacific Slope S of Isthmus, Atlantic Slope N of Isthmus. U to R transient on Pacific Slope and in interior N of Isthmus. Irregular R to U winter visitor (Dec–Feb) on Pacific Slope N to Nay, possibly also on Atlantic Slope. Reports from Belize (Wood and Leberman 1987) require verification.

RA: Breeds W N.A. to N Mexico; winters mainly in S.A., irregularly N to M.A.

WHITE-TAILED HAWK
Buteo albicaudatus hypospodius
Aguililla Coliblanca   Plates 3, 8, 9

ID: 19–23 in; W 49–54 in (48.5–58.5 cm; W 124–137 cm). A large buteo of open country. *Wings long, broad, and tapered, tail short and squared in adult*, markedly longer in juv. At rest, wingtips extend beyond (adult) or almost to (juv) tail tip. **Descr. Adult:** eyes dark, cere and legs yellow. Head, sides of neck, *back, and upperwings slaty blue-grey* (remiges darker), *with rufous lesser upperwing* coverts and rufous mottling on scapulars. *Rump, uppertail coverts, and tail white; uppertail with broad black subterminal band* and several fine dusky bars. Throat and underparts white, sides and belly usually sparsely vermiculated greyish to rusty. Underwings: whitish, coverts vermiculated greyish to rusty, remiges faintly barred blackish with dark trailing edge, inner primaries often appear as dusky panel. **Juv:** *head and upperparts blackish brown,* often with buff supercilium and streaking on auriculars, upperparts edged rufous when fresh; creamy white uppertail coverts often barred dark brown. *Underparts buff, quickly fading to whitish, with variable dark brown mottling, or mostly dark with whitish patch on central chest*; thighs and whitish undertail coverts typically barred dark brown. *Tail pale grey to pale grey-brown* (rarely tinged rufous) with numerous (10–15) indistinct dusky bars and often a broader dark distal band. Underwings: coverts as underparts, remiges pale grey with darker barring, outer primaries tipped blackish. **1st basic:** resembles adult but throat dark, head and upperparts darker and browner with less rufous, underparts and underwing coverts with dark mottling and stronger dark barring, often forming a dark belly band contrasting with white chest. *Tail similar to adult but with 6–8 narrow black bars on upperside, or pale grey with fairly broad black subterminal band.* Attains adult plumage by 2nd prebasic molt.

Voice. Usually silent; near nest may give a screaming, slightly rough to slightly clipped *keh keh-eh keh-eh keh-eh keh-eh,* or *kre keh-heh keh-heh ...,* etc. (CCF tape, Texas).

Habs. Savannas, open farmland with scattered trees and bushes, brushy ranchland. Soars with wings in marked dihedral and slightly pressed forward, tail slightly spread. Often hangs in wind and hovers, particularly over fields being burned where 20–30 birds may gather. Nest is a bulky stick platform usually in crown of bush or low tree. Eggs: 2–3, whitish, often marked with browns (WFVZ).

SS: See Swainson's Hawk, Short-tailed Hawk. Juv Red-tailed Hawk has less tapered wings, primaries often appear as broad paler panel on underwing, flight feathers more coarsely barred, light morph has dark lesser underwing coverts, dark morph has dark undertail coverts. Other large dark buteos have dark undertail coverts, bolder bands on tails, and often more contrasting whitish underside to remiges.

SD: F to U but local resident (SL–1500 m) on Pacific Slope in Son, in interior and on adjacent Pacific Slope from Dgo and S Nay to Honduras, and on Atlantic Slope from Tamps to Belize and the Mosquitia; C locally in Tab and SW Camp.

RA: SW USA and Mexico to cen Argentina.

ZONE-TAILED HAWK

Buteo a. albonotatus   Plates 3, 8
Aguililla Aura

ID: 18–21 in; W 48–54 in (45.5–53 cm); W 122–137 cm). A widespread black hawk with long wings and tail; *in flight strongly recalls Turkey Vulture.* At rest, wingtips reach to tail tip. **Descr. Adult:** eyes brown, cere and legs yellow. *Dark slaty grey overall* with whitish lores. *Tail* white-tipped, *upperside with 2 pale grey bands, underside with 2–3 white bands* (the 1–2 narrower basal bands often covered by tail coverts). *Two-tone underwings: coverts black, pale grey remiges barred blackish.* **Juv:** blackish grey-brown overall (looks black), with variable white flecking on nape, mantle, and underparts. *Tail*

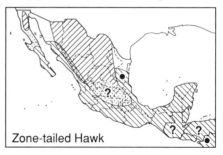

White-tailed Hawk   Zone-tailed Hawk

whitish-tipped, *upperside grey-brown with 5–7 dark bars, underside pale grey with 4–7 narrow dark bars, distal dark bar widest.* Underwings more contrastingly two toned than adult, with paler, silvery grey remiges. Attains adult plumage by 1st prebasic molt.

Voice. Usually silent; near nest may give a slightly rough, piercing, drawn-out scream, *reeeeah* or *meeeeeahhr*.

Habs. Wooded and semiopen country, mainly in arid to semiarid areas when nesting, more widely in winter, including humid evergreen forest. Flies with wings in marked dihedral, tail closed to slightly spread. At times soars with Turkey Vultures which it appears to mimic in plumage and flight. Nest is a stick platform at mid- to upper levels in trees. Eggs: 2–3, whitish, rarely with sparse dusky markings (WFVZ).

SS: See Turkey Vulture. Dark morph Red-tailed Hawk has relatively shorter and broader wings and tail, adult has rufous tail, juv has duskier secondaries, pale primary panel. See dark morph Short-tailed and Broad-winged hawks. Dark morph Rough-legged and Ferruginous hawks (both N Mexico in winter) are more heavily built, fly with slight dihedral, silvery-white underside of remiges contrast more, tail patterns distinct.

SD: F to U breeder (500–3000 m; Mar–Sep) in BCN and N Mexico, mainly in interior and on adjacent slopes; probably also in Guatemala (Vannini 1989a) and N cen Nicaragua (TRH). U in winter in N Mexico; U to F but widespread transient and winter visitor (Sep–Apr) S to Oax; U to R to El Salvador and Honduras, including Yuc Pen (Lopez O. *et al.* 1989). Despite frequently breeding reports from cen Mexico to N C.A., most reports pertain to winter (Dickey and Van Rossem 1938; Monroe 1968; Land 1970; Thurber *et al.* 1987; Binford 1989); status in much of the region needs clarification.

RA: SW USA and Mexico to S Brazil.

RED-TAILED HAWK
Buteo jamaicensis *   Plates 8, 9
Aguililla Colirroja

ID: 19–23 in; W 46–54 in (48.5–58.5 cm; W 117–137 cm). Widespread large buteo. *Wings long and broad, tail medium length, broad and squared. Juv noticeably longer-*tailed than adult. At rest, wingtips approach or just reach tail tip. *Plumage variable*, also dimorphic (dark morph uncommon). **Descr. Adult. Light morph:** eyes yellow to brown, cere and legs yellow. Head and upperparts dark brown, *whitish to pale brown mottling on scapulars and median upperwing coverts forms contrasting V at rest and in flight*; uppertail coverts whitish to pale rufous, barred dark brown. Throat and underparts whitish or washed pale rufous, with dark streaking heaviest on belly where typically forms contrasting dark band (ill-defined or lacking in paler *kemsiesi* of Chis and N C.A. highlands, and in paler forms of *calurus* in N Mexico). *Tail rufous*, paler on underside, with dark subterminal bar and often 1–8 narrower dark bars. *Underwings: whitish overall with dark bar along leading edge of inner wing*, dark-flecked coverts may be washed pale rufous; remiges barred dark brown. **Dark morph:** dark brown overall (including underwing coverts), little or no paler mottling on upperparts, undertail coverts often slightly paler rufous. *Flight feathers as light morph adult.* Rufous morph resembles dark morph but underparts dark rufous with dark brown belly band. *B. j. harlani* may reach N Mexico in winter, resembles dark morph but blackish overall, usually with variable whitish mottling on underparts, tail pale grey with dusky mottling and dark subterminal band. **Juv. Light morph:** *resembles light adult but pale mottling above bolder, primaries paler (visible in flight as pale panel above and below), tail grey-brown with whitish tip, upperside with numerous narrow dark bars, underside with 5–8 narrow dark bars.* **Dark morph:** resembles dark adult but usually underparts and underwing coverts mottled whitish, *flight feathers as light juv.* Rufous morph resembles juv light

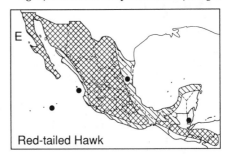

E

Red-tailed Hawk

morph but underparts more heavily streaked dark brown. Attains adult plumage by 1st prebasic molt.

Voice. A hoarse, drawn-out, whistled scream *wheee-eahr* or *wheeiihr*, often given when soaring; also a shorter, hoarse mewing *wheeir* which may be repeated steadily.

Habs. Breeds in temperate regions, mostly pine and pine–oak forest, deserts; widespread in winter but avoids heavy forest, in open areas of humid SE. Perches commonly on telegraph poles and fence posts; soars frequently. Soars and glides with wings in slight dihedral, tail spread when soaring; may glide with wings nearly flat. Often hangs in wind and at times hovers. Stick platform nest at mid- to upper levels in trees, rarely low in desert vegetation. Eggs: 2–3, whitish, usually marked with browns (WFVZ).

SS: Adult told by rufous tail. Juv can be confused with several species, see Red-shouldered, Broad-winged, Short-tailed, Swainson's, and White-tailed hawks. Ferruginous and Rough-legged hawks (both N Mexico in winter) have relatively longer wings and tails, the former has larger bill, the latter has smaller bill; dark morphs have more contrasting silvery-white underside of remiges; note tail patterns.

SD: C to F resident (SL–3000 m) in Baja and, mostly in interior, S to Oax; U to F (750–3000 m) to Honduras and N cen Nicaragua, also in Mountain Pine Ridge, Belize, and (around SL) in the Mosquitia. C winter visitor (Oct–Mar; SL–3000 m) S to cen Mexico; U in winter on Atlantic Slope from S Ver to N Yuc Pen; altitudinal migrant elsewhere, descending locally to foothills and lowlands.

RA: N.A. to W Panama.

FERRUGINOUS HAWK

Buteo regalis   Not illustrated
Aguililla Real

ID: 21–24 in; W 51–58 in (53.5–61 cm; W 129–147 cm). *N and cen Mexico in winter. Wings long and slightly tapered, tail fairly long and squared.* At rest, wingtips reach almost to tail tip. *Bill large and aquiline, tarsi feathered.* **Descr.** Dimorphic (dark morph uncommon). **Adult. Light morph:** eyes yellow to brown, cere and feet yellow. *Head and underparts whitish,* head

streaked brown, thighs (and rarely belly) barred rufous. Upperparts grey-brown, nape and upper mantle coarsely streaked white, *back and upperwing coverts mottled rufous;* uppertail coverts whitish to rufous, barred dark brown. *In flight from above, pale base of primaries forms distinct panel. Tail greyish white to pale rufous without dark bars. Underwings overall white,* coverts sparsely marked rufous, narrow dark carpal crescent usually obvious; outer primaries tipped blackish; trailing edge of inner primaries and secondaries dusky. **Dark morph:** *dark brown to dark rufous-brown overall* (including underwing coverts), brighter rufous on lesser upperwing coverts and tail coverts. *Tail white basally, pale greyish distally, with narrow dark subterminal band.* Remiges as light adult; *note narrow whitish carpal crescent on underwing.* **Juv. Light morph:** *as light adult* but rufous on upperparts less distinct, whitish band across base of primaries more striking, underparts and underwings white with only sparse dark spotting. *Tail pale grey with broad diffuse dusky subterminal band.* **Dark morph:** as dark adult but browner, less rufous. *Uppertail greyish with 3–4 indistinct dark bars, undertail whitish with diffuse dusky subterminal band.* Attains adult plumage by 1st prebasic molt.

Habs. Open grassland and farmland with scattered trees and bushes. Often perches on telegraph poles, fence posts, on ground. Flies with wings in dihedral, sometimes only slightly raised, tail slightly spread while soaring. Commonly hunts by flying low, harrier style; also soars high; hangs in wind and hovers occasionally.

SS: See Red-tailed and White-tailed hawks. Dark morph Rough-legged Hawk has small bill, lacks white carpal crescent on underwing.

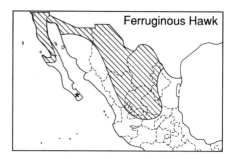

Ferruginous Hawk

SD: F to C winter visitor (Oct–early Apr; SL–2000 m) to N Mexico S to BCN, Son, Dgo, and Coah; irregularly U to R to BCS, N Tamps, and S over Plateau to Gto and Hgo. A report from DF (Wilson and Ceballos-L. 1986) is questionable.

RA: Breeds W N.A.; winters to N Mexico.

ROUGH-LEGGED HAWK
Buteo lagopus sancti-johannis Not illustrated
Aguililla Artica

ID: 20–23 in; W 50–58 in (51–58.5 cm; W 127–147 cm). *Rare winter visitor to N Mexico. A large, long-winged, long-tailed buteo.* At rest, wingtips reach to tail tip. *Bill small, tarsi feathered.* Dimorphic (dark morph rare). **Descr. Adult. Light morph:** eyes yellowish to brown, cere and feet yellow. *Head and underparts whitish, coarsely streaked dark brown, heaviest on upper chest and belly (where often solidly dark),* lower chest often contrastingly whitish, undertail coverts whitish. Upperparts dark grey-brown, *mottled whitish. Tail white with broad blackish subterminal band* and (mainly ♂) often 1–3 narrow dark bars. Underwings: coverts mottled dark brown and whitish, *large blackish carpal patch contrasts with dark-tipped whitish remiges.* **Dark morph:** blackish brown overall (including underwing coverts), head and chest rarely mottled whitish. *Tail dark grey with narrow whitish tip, upperside with 2–4 narrow whitish basal bars (mainly ♂), or dusky, blacker subterminally, underside silvery white with broad blackish subterminal band.* Remiges as light adult. **Juv. Light morph:** as light adult but head and chest paler, less heavily streaked dark brown, *belly solidly dark brown. Upperwings with broad whitish band across base of primaries obvious in flight. Tail whitish with broad blackish subterminal band.* Underwings whitish with less dark mottling, dark carpal patch more striking. **Dark morph:** as dark adult but more brownish, some whitish mottling on chest. *Uppertail dark grey-brown above with indistinct, broad, darker terminal band, undertail as light juv.* Remiges as light juv. Attains adult plumage by 1st prebasic molt.

Habs. Open grassland, farmland, and marshes with scattered trees and bushes. Often perches on telegraph poles, fence posts, ground. Wingbeats at times almost floppy; flies with wings in dihedral; frequently hangs in wind and hovers.

SS: See Red-tailed and Ferruginous hawks.

SD: Irregular U to R winter visitor (Nov–Mar; near SL–2000 m) to N Mexico from BCN (GMcC) and Son to W Coah; most records from Chih.

RA: Holarctic breeder; in New World winters S to USA, rarely to N Mexico.

GOLDEN EAGLE
Aquila chrysaetos canadensis Not illustrated
Aguila Real

ID: 31–36 in; W 72–84 in (79–91.5 cm; W 183–213 cm). N Mexico. *Wings long and broad, tail fairly long and squared to slightly rounded, tarsi feathered.* **Descr.** Ages differ, sexes similar. **Adult:** eyes brown, bill black-tipped with grey base, cere and feet yellow. *Blackish brown overall, crown and nape golden brown;* paler median upperwing coverts form contrasting panel on upperwing. Tail blackish brown with 2–3 faint paler wavy bars. Underwings: remiges dark grey with faintly paler mottling, contrasting slightly with darker brown coverts. **Juv:** *tail white with broad black subterminal band, white patch on underside of outer secondary and inner primary bases* usually obvious from below, sometimes visible from above; pale upper-

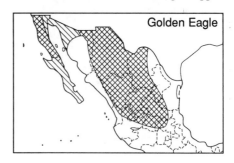

Rough-legged Hawk   Golden Eagle

wing bar less distinct. **1st basic:** resembles juv but white tail base mottled dark brown, white on underwings reduced. **2nd basic:** resembles adult but usually has some white mottling at tail base. Attains adult plumage by 3rd prebasic molt when about 4 years old.

Voice. Usually silent, may give quiet, chippering calls.

Habs. Arid to semiarid mountainous country, deserts, open grassland, and farmland with scattered trees and bushes. Soars and glides with wings in slight dihedral, rarely flat; at times hangs in wind but does not hover. Nest is a bulky mass of twigs on cliff or in tree. Eggs: 1–3, whitish, marked with browns and often greys (WFVZ).

SS: See Turkey Vulture, Bald Eagle.

SD: U to R resident (600–3000 m; locally to SL in winter) in BCN (to BCS in winter) and in N Mexico S to Dgo and SLP (per PRG), probably to N Mich and Hgo (Mengel and Warner 1948; Davis 1953), possibly to Oax (Winker *et al.* 1992a); more widespread in winter.

RA: Holarctic, in New World S to N Mexico.

CRESTED EAGLE
Morphnus guianensis　　　　Plate 6
Aguila Crestada　　　　　　　Figure 20

ID: 28–34 in; W 60–73 in (71–86.5 cm; W 152–185 cm). *Lowland rain forest in C.A., should be looked for in Mexico. Large and powerful, bill heavy, legs thick. Long, erectile, single-pointed crest. Wings*

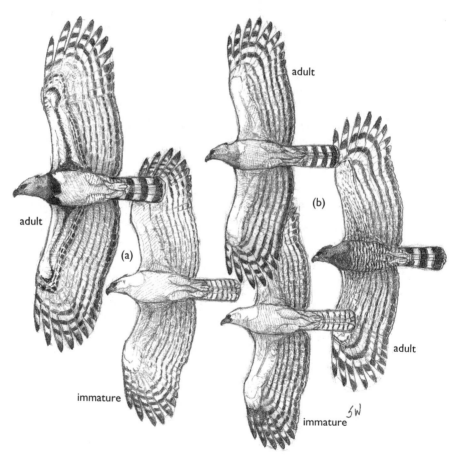

Fig. 20 (a) Harpy Eagle, adult and first-year; (b) Crested Eagle, light and dark morph adults and first-year

broad and rounded, tail broad and long, *slightly rounded.* Dimorphic, dark morph rare. **Descr.** Ages differ, sexes similar. **Adult. Light morph:** eyes brown, bill, cere, and lores blackish, legs yellow. *Head and upper chest pale grey* to dusky grey, crest black. *Underparts white with sparse cinnamon barring, boldest on thighs.* Upperparts blackish, typically with pale mottling on median and greater upperwing coverts, uppertail coverts tipped white. Tail boldly banded black and pale grey with narrow whitish tip; uppertail with 3 broad pale grey bands, undertail with 3 broad whitish bands (basal band often covered by tail coverts). Underwings white, *primaries boldly barred black,* secondaries barred dusky. **Dark morph**; head darker than light morph, upperparts blacker with little or no pale mottling on wing coverts; *upper chest blackish, white underparts and underwing coverts barred black.* **Juv** (both morphs): whitish overall, longest crest feathers tipped dusky, upperparts vermiculated pale grey. Remiges blackish above, mottled and barred paler. Uppertail greybrown with numerous (7–10) narrow blackish bars and whitish tip; undertail whitish with 6–8 narrow dark bars. *Underwings white with narrower blackish barring on primaries than adult.* **1st basic:** *resembles juv but upperparts mottled darker with contrasting pale wing covert panel, crest tipped black, head and chest washed pale grey.* Tail variable: upperside typically blackish with 3–4 pale grey to grey-brown bands, rarely grey-brown with 5–7 dark bars; underside boldly banded black and whitish with 3–4 whitish to pale grey bands, or whitish with 4–5 dark bars. Dark morph may have duskier head and chest, less contrasting pale wing covert panel. **2nd basic. Light morph:** resembles adult but head and chest paler, underparts

with little or no cinnamon barring, upperwing coverts with more pale mottling. Uppertail has 3–4 pale grey to grey-brown bands, undertail 3–4 whitish to pale grey bands. **Dark morph:** resembles adult but head and chest paler, upperwing coverts with more pale mottling, underparts and underwing coverts with only sparse black bars and spots. Tail as light 2nd basic. Attains adult plumage by 3rd prebasic molt when about 3 years old.

Voice. Shrill, high whistles, sometimes 2-parted with 2nd part higher (Stiles and Skutch 1989).

Habs. Humid evergreen forest. A bird of forest canopy and interior, rarely seen but may perch conspicuously in tall trees; rarely soars. Nest is a bulky stick structure high in trees. Eggs: 2, creamy white, unmarked.

SS: Harpy Eagle larger and more heavily built with broader wings, shorter tail, and thicker legs; note forked crest, less distinct dark barring on underwings. Adult Crested has unmarked white underwing coverts. Juv and imm Harpies may have coarser tail barring relative to similar age Cresteds. Juv Ornate Hawk-Eagle smaller, upperparts dark brown; note dark thigh bars, fewer bands on tail from below.

SD: R to U resident (SL–500 m) on Atlantic Slope from N Guatemala and Belize (Miller and Miller 1992*b*) to Honduras. Should be sought in E Chis and S Camp.

RA: Guatemala to N Argentina.

HARPY EAGLE

Harpia harpyja　　　　　　　　　Plate 6
Aguila Arpia　　　　　　　　　　Figure 20

ID: 34–42 in; W 72–88 in (86.5–107 cm; W 183–224 cm). *Huge powerful eagle, rare and endangered. Bill very heavy, legs and feet massive. Fairly long, erectile crest forked. Wings very broad and rounded, tail broad and fairly long, tip squared to*

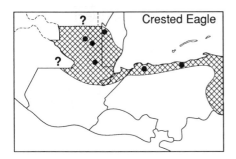

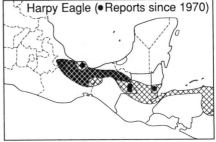

slightly rounded. **Descr.** Ages differ, sexes similar. **Adult**: eyes brown, bill, cere, and lores blackish, legs yellow. *Head pale grey with blackish crest. Black upper chest contrasts with white underparts, thighs barred black,* axillars black. Upperparts blackish with white-tipped tail coverts. Tail boldly banded black and pale grey; upperside with 3 broad pale grey bands and whitish tip, underside with 3 broad whitish bands (basal band usually covered by tail coverts). Underwings white with black bar along median coverts, blackish barring on primaries, dusky barring on secondaries. **Juv**: whitish overall, upperparts vermiculated pale grey, with slight black mottling on greater upperwing coverts. Remiges blackish above, mottled and barred paler. Uppertail greyish with 7–8 blackish bars, undertail whitish with 6–8 narrow black bars. Underwings white, remiges less boldly barred than adult. **1st basic**: resembles juv but upperparts mottled blackish, crest tipped dusky, chest and axillars pale grey. Tail variable: upperside grey with 1–5 narrow, often broken, black bars distally (mainly ♂♂?), or with 6–7 narrow black bars (mainly ♀♀?); underside whitish with 2–6 narrow black bars. Underwings have bolder blackish bars on primaries than juv. **2nd basic**: upperparts with more black mottling, crest tipped black, some black mottling on chest, thighs often with slight black barring. Tail boldly banded black and pale grey, upperside with 4–6 black bars, underside whitish with 4–5 black bars. **3rd basic**: resembles adult but with some pale mottling on upperparts, chest band mottled pale grey, uppertail with 3–4 slightly narrower grey bands. Attains adult plumage by 4th prebasic molt when about 4 years old.

Voice. Plaintive, piping whistles, *whee whee wheeu*, with emphasis on last note; and quiet clucking *chuk chuk*, etc. (captive birds, SNG).

Habs. Humid evergreen forest, requires large undisturbed areas where it and its prey (monkeys, curassows, etc.) can avoid persecution. Stays mostly in forest, but hunts over and may perch in canopy; does not soar. Nest is a large stick structure, high in tree. Eggs: 1–2, whitish, unmarked.

SS: See Crested Eagle.
SD: R resident (SL–500 m) on Atlantic Slope

from S Ver and E Oax to Honduras. Extirpated from much of original range; formerly U to R resident N to S cen Ver; possibly also on Pacific Slope of Chis (MAT). A report from Campeche (Friedmann *et al.* 1950) is unsatisfactory.
RA: SE Mexico to N Argentina.

BLACK-AND-WHITE HAWK-EAGLE
Spizastur melanoleucus   Plates 5, 8, 12
Aguila Blanquinegra

ID: 20–25 in; W 46–56 in (51–63.5 cm; W 117–142 cm). The rarest hawk-eagle in the region. Crest bushy, without prominent long feathers, tarsi feathered. *Wings broad and relatively long,* tail medium length and squared to slightly rounded. *At rest, wingtips reach about half-way down tail.* **Descr.** Ages differ, sexes similar. **Adult**: eyes yellow, *cere and gape bright yellow-orange, black lores and eyering form bold mask through eye,* feet yellow. *Head and underparts white with black crown and crest. Upperparts blackish,* uppertail coverts dark brown; *note bold white leading edge to inner wing, often striking in flight.* Tail pale-tipped, upperside grey to grey-brown with 3–4 dark bars, underside whitish with 3–4 dark bars, distal band broadest; *tail may appear dusky when closed, in contrast to white underparts. Underwings white,* outer primaries tipped dark, inner primaries and secondaries dusky; may show cinnamon-washed panel on outer primaries when backlit. **Juv**: upperparts blackish brown, underside of outer primaries barred blackish. Uppertail has 4–5 dark bars. Attains adult plumage by completion of 1st prebasic molt when about 1 year old.

Voice. Piping whistles given in flight, suggesting Ornate Hawk-Eagle in quality, but clearer, more piping. An introductory note followed by 4 accelerating whistles,

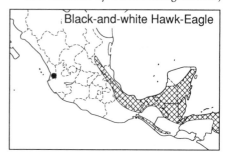
Black-and-white Hawk-Eagle

ending with a more emphatic, at times slightly disyllabic whistle: *whee whi-whi-whi-whi whee-eer*, or *whii whi-whi-whi-whi whiih*. Calls less often than other hawk-eagles.

Habs. Humid evergreen forest and edge. Glides and soars with wings held flattish though tips may be curled up; tail closed when gliding, slightly spread when soaring; hunts by stooping from considerable heights. In apparent courtship, ♂ and ♀ soar together over wide area, staying close then separating and chasing each other, stooping together and at one another, etc. Nest is a stick platform high in trees. Eggs (Guyana): 1, whitish with sparse, coarse dark brown splotches (WFVZ).

SS: See light morph juv Grey-headed Kite. Juv Ornate Hawk-Eagle has relatively shorter wings and longer tail, long crest, usually some black barring and spotting on white underparts with black-barred thighs, underwings have more distinct dark barring on remiges. White leading edge to Black-and-white's wings often striking, head-on suggests King Vulture.

SD: U to R resident (SL–1000 m) on Atlantic Slope from S Ver to Honduras; on Pacific Slope (300–1500 m) from E Oax to Guatemala, R N to S Tamps (at least formerly); apparently also locally (2000 m) in S Nay (MVZ spec.).

RA: Mexico to N Argentina.

BLACK HAWK-EAGLE

Spizaetus tyrannus serus   Plates 5, 8, 12
Aguila Tirana

ID: 24–29 in; W 50–61 in (61–73.5 cm; W 127–155 cm). *Erectile crest bushy, without prominent long feathers*, tarsi feathered. *Wings long, broad, and rounded, tail long and squared to rounded.* At rest, wingtips reach base of tail. **Descr.** Ages differ, sexes similar. **Adult:** eyes golden yellow, cere

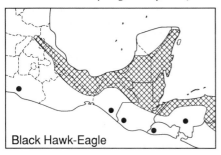

Black Hawk-Eagle

and lores greyish, feet yellow. *Black overall; crest white basally* (often obvious), uppertail coverts tipped white, thighs and undertail coverts coarsely barred white. Tail boldly banded black and pale with narrow whitish tip; upperside with 3 pale grey-brown to grey bands, underside with 3 whitish bands (basal band often covered by tail coverts). *Underwings: coverts black* (at close range some white flecking visible), *remiges checkered black and white.* **Juv:** eyes paler. *Creamy-white forehead and broad supercilium contrast with black-and-white mottled crown and crest, and black auriculars.* Throat and median upper chest creamy white, streaked blackish, underparts otherwise dark brown; sides of chest coarsely streaked black, thighs and undertail coverts barred white. Nape and upperparts blackish brown, rump sides and uppertail coverts tipped white. Tail more broadly tipped white, with 4 fairly broad pale grey to grey-brown bands on upperside (basal 2 bands often mottled white), 4 whitish bands on underside (basal band often covered by tail coverts). *Underwings checkered and barred blackish and white overall.* Within a few months (?) begins head and body molt into **1st basic:** similar to juv but upper mantle black, *chest creamy white, heavily streaked black, belly boldly spotted and barred white.* Toward end of body molt (age about 1 year?) the very worn flight feathers are replaced by adult-like feathers. Shortly after this, or while flight feathers are being replaced, begins molt into **2nd basic:** *resembles adult but with indistinct whitish-streaked supercilium, belly often sparsely spotted white.* Probably attains adult plumage by completion of 3rd prebasic molt when 2–3 years old.

Voice. Loud, clear whistles, often given while soaring, a drawn-out *whi, whi-wheeooo* or *wh' whee-hoo*, and *wheeoo*, often preceded by several introductory notes, *whi whi ... wheeoo* or *wh wh ... weee-oo*, etc. Main whistle rolled into 3 syllables in Gro birds: *wh-whi wh-whee-hee-hoo*, or *wh-whee-ee-oo*; thus may suggest Ornate Hawk-Eagle. In flight, imm may simply give steady, ringing whistles, *whee whee ...* or *wheep wheep ...* Adults also give a single *weeoo* when perched.

Habs. Humid evergreen and semi-deciduous forest. Frequently soars, often quite high, mainly in mid- to late morning. Wingbeats deep and powerful, wings held flattish and slightly pressed forward, tail slightly spread. Usually perches in forest or at edges. Stick platform nest high in trees. Eggs (captivity): 1, whitish, finely marked with reddish browns (WFVZ).

SS: Ornate Hawk-Eagle similar in shape but often appears to have slightly shorter and broader wings, with more pronounced bulge in trailing edge of secondaries; note voice, plumage if visible. See Hook-billed Kite, Zone-tailed Hawk.

SD: U to F resident (SL–1000 m) on Atlantic Slope from S Ver to Honduras; R N to SE SLP and locally (500–1500 m) on Pacific Slope in Gro (Webb and Howell 1993), Chis (AJL), Guatemala (Vannini 1989*a*), and El Salvador (Thurber *et al.* 1987).

RA: E Mexico to Ecuador and N Argentina.

ORNATE HAWK-EAGLE

Spizaetus ornatus vicarius   Plates 5, 8, 12
Aguila Elegante

ID: 22–27 in; W 46–56 in (56–68.5 cm; W 117–142 cm). *Long erectile crest often striking,* tarsi feathered. *Wings broad and rounded, tail long and squared to slightly rounded.* At rest, wingtips reach base of tail. **Descr.** Ages differ, sexes similar. **Adult:** eyes golden yellow, cere, lores, and orbital ring dull greyish-olive, lores may be darker grey, feet yellow. *Crown and crest black; face, sides of neck, and chest rufous, bordered by bold black malar stripe. Throat and median chest white, rest of underparts white, boldly barred black,* undertail coverts with sparser and narrower black bars. Nape reddish brown, mantle and white-tipped lesser upperwing coverts blackish, rest of upperparts dark brown, tail coverts tipped white. Tail boldly

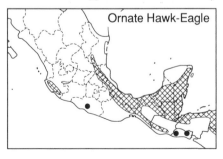

Ornate Hawk-Eagle

banded black and pale with pale greyish tip; upperside with 3 broad pale grey to brownish-grey bands, underside with 3 broad whitish to pale grey bands (basal band may be covered by tail coverts). *Underwings whitish, often contrasting with dark body:* coverts flecked dark, remiges pale grey distally with blackish barring boldest on outer primaries, indistinct on secondaries; *outer primaries may appear as translucent panel when backlit.* **Juv:** eyes paler, lores and orbital skin blue-grey. *Head and underparts white, longest crest feathers tipped black,* crown streaked blackish, dark malar stripe indistinct or lacking, *sides and flanks sparsely marked with coarse black spots and bars, thighs barred black.* Upperparts dark brown, white-tipped lesser upperwing coverts and remiges blacker. Tail tipped whitish, upperside grey to grey-brown with 4–5 black bands, underside appears whitish with 3–5 narrow black bands, basal band often covered by tail coverts. Underwings whitish, barring of remiges less distinct than adult, outer primaries barred blackish. Within a few months (?) begins head and body molt into **1st basic:** resembles adult but *face and chest paler rufous, black malar less distinct,* upper mantle mottled black. *Chest white, sparsely spotted black at sides, belly coarsely spotted and barred black;* underparts overall whiter than adult. Towards end of body molt, flight feathers are replaced. New tail has 3–4 broad pale bands on upperside, 3–4 black bands on underside. Attains adult plumage by completion of 2nd prebasic molt when about 2 years old.

Voice. Loud, piping whistles, often repeated tirelessly while soaring, *whi whee-whee-wheep* or *wh' whee-whee-wheep,* and *whi whee-whee . . .,* the *whee* repeated 2–9 ×, and *whee-a' whee whee whee whee-a'* or *wee-hu wi-wi-wi-wee-hu,* etc.; unlike Black Hawk-Eagle, introductory series more hurried and last note not drawn out. Perched juv has a loud, clear whistle, repeated irregularly, *wheeeu* or *wheeee.*

Habs. Humid evergreen and semi-deciduous forest. Often soars, mainly in mid- to late morning, wings held flattish and pressed slightly forward, tail closed to slightly spread. Wingbeats deep and powerful; in display flight climbs with deep, floppy wingbeats, and stoops with wings closed, almost somersaulting at times. Usually perches in forest or at edges. Stick platform nest high in trees. Eggs: 1, white, faintly marked with brownish red (Lyon and Kuhnigk 1985).

SS: See Hook-billed Kite, juv Grey-headed Kite, Black-and-White Hawk-Eagle, Black Hawk-Eagle, juv and imm Crested Eagle.

SD: U to F resident (SL−1500 m) on Atlantic Slope from S Tamps to Honduras, locally on Pacific Slope in Jal and Col (Schaldach 1963, 1969), Gro (Webb and Howell 1993), and from E Oax to El Salvador.

RA: Mexico to N Peru and N Argentina.

FALCONS AND ALLIES: FALCONIDAE (12)

A fairly diverse, cosmopolitan family of raptors differing from hawks and allies mainly in internal structural characters. Usually ages differ and sexes are similar. Young altricial and down-covered, fledge in 1–2 months. Juvenal plumage of most species retained through first year; adult plumage attained by 1st prebasic molt when about one year old, unless stated otherwise.

Falcons feed on birds, mammals, reptiles, and insects; Crested Caracaras often scavenge carrion; Red-throated Caracaras are noted for feeding at wasp nests and sometimes eat fruit. Some species are cavity nesters, others build stick nests in trees, others nest on cliffs.

RED-THROATED CARACARA

Daptrius americanus   Figure 21
Caracara Comecacao

ID: 21–25 in; W 38–45 in (53.5–63.5 cm; W 97–115 cm). *Extirpated* (?). Wings and tail long, broad, and slightly rounded. Descr. Ages/sexes similar. Eyes red (brown in juv?), bill horn-yellow distally, cere and base of bill blue-grey; *lores, orbital skin, and upper throat red, legs red. Plumage black overall with white belly,* flanks,

thighs, and undertail coverts.

Voice. Loud, carrying calls, quality reminiscent of Collared Forest-Falcon, *cowh cah-cowh,* or *ka-ow ka-kow,* and a longer series, *cowh cowh cowh cowh cowh ka-cow*; also in flight, a deep, throaty *rrrah,* recalling Scarlet Macaw (Costa Rica).

Habs. Humid evergreen forest and edge, plantations. Typically social, in pairs or small, often noisy groups that travel in the canopy, at times low or even on the ground. Flies with slow, deep wingbeats; does not soar. Said to lay 2–3 eggs, whitish to buff, spotted with rusty and browns (Brown and Amadon 1968).

SS: Unlikely to be confused with other species. ♂ Great Curassow crested, has bright yellow knob on bill.

SD: No reports in over 20 years from W of Sula Valley (?). Formerly, apparently U to F resident (SL–750 m) on Atlantic Slope in S Ver (Lowery and Dalquest 1951) and in N Honduras (mainly E of Sula Valley); also on Pacific Slope (near SL–750 m) of Chis and Guatemala. While deforestation may be a major reason for this species' disappearance from the Pacific Slope, it is

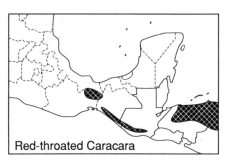

Red-throated Caracara

Fig. 21 Red-throated Caracara

unclear why it has vanished elsewhere (including most of C.A.).

RA: SE Mexico to Ecuador and cen Brazil.

CRESTED CARACARA
Caracara plancus*   Plate 7
Carcara Común

ID: 19–23 in; W 45–52 in (48–58.5 cm; W 115–132 cm). Common and widespread in open country. Wings and tail long, broad, and slightly rounded. *Head and long neck project well forward of wings in flight*; nape slightly crested. **Descr.** Ages differ, sexes similar. **Adult:** eyes pale brown to yellow, bill pale blue, cere and lores orange-red, legs yellow to reddish. *Crown and crest black, face, neck, and upper chest white*, chest barred and spotted blackish. Lower chest and belly black, undertail coverts white. Upperparts black, mantle narrowly barred whitish, uppertail coverts white. *Tail white with broad black terminal band* and numerous dusky bars. *Underwings: black with broad, dark-barred whitish panel across base of outer primaries*, also visible from above in flight. **Juv:** dark brown overall, white areas of adult dirty whitish to buff, chest and nape streaked dark brown, belly usually mottled whitish, wing coverts more boldly spotted whitish.
Voice. Mostly silent, at times utters hard, dry, clucking *chak* notes and creaky rattles.
Habs. Open and semiopen country with scattered trees and bushes. Flight slow with steady, fairly deep wingbeats; soars and glides on flat to slightly bowed wings, rarely high or for long periods. Often perches on posts or ground where walks with agility. Singly or in pairs, locally in groups up to 50 birds. Nest is a mass of sticks in bush or small tree. Eggs: 2–4, buff to brown, heavily marked with darker browns (WFVZ).

SD: F to C resident (SL–1500 m) in BCS, on Pacific Slope from Son to El Salvador, on Atlantic Slope from Tamps to SW Camp; U to F in NW Yuc Pen and locally in Honduras, in interior (to 2500 m) almost throughout except N Plateau. Reports from Belize (Wood *et al.* 1986) require verification.

RA: S USA and Mexico to S S.A.

NB: Extinct Isla Guadalupe form *C. (plancus?) lutosus* (Guadalupe Caracara) has been considered specifically distinct.

LAUGHING FALCON
Herpetotheres cachinnans*   Plate 7
Halcón Guaco

ID: 18–22 in; W 31–37 in (46–56 cm; W 79–94 cm). An unmistakable raptor with fairly short, rounded wings and long tail. **Descr.** Ages/sexes similar. **Adult:** eyes dark, cere and feet yellow. *Head and underparts pale buff with broad, dark brown mask from lores to nape*, crown finely streaked dark brown. *Upperparts dark brown*. Tail blackish with buff tip, upperside with 4–5 buff bands, underside with 3–4 buff bands. In flight, upperwings show broad cinnamon flash across base of primaries. Underwings: buff to pale rufous, coverts sparsely spotted dark brown, remiges barred dark brown, primaries often appear as translucent cinnamon panel when backlit. **Juv:** usually shows some dark spotting on thighs.
Voice. Far-carrying, often loud, laughing or crowing calls, a fairly steady *wah wah* ... or *w-hah w-hah* ..., 10/6–7 s; a more drawn-out *ha-cow ha-cow* ..., 10/12–19 s or, when dueting, *ha-cah ha ha-cah* ... or *ah-ow owh, ah-ow owh* ..., at times breaking into maniacal laughter; also a soft *ha* or *hah*, repeated irregularly.
Habs. Forest and edge, open areas with scattered trees and forest patches, savan-

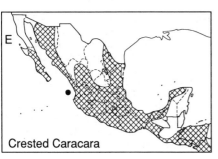

Crested Caracara

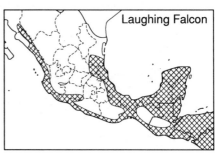

Laughing Falcon

nas, palm groves. Perches conspicuously with upright stance on dead trees, telegraph poles, etc. Wingbeats usually hurried and fairly shallow; does not soar. Nest usually in tree cavity, sometimes in old hawk nest. Eggs: 1–2, brown with heavy, darker brown mottling (WFVZ).

SD: F to C resident on Atlantic Slope (SL–1500 m) from S Tamps to Honduras; U to F on Pacific Slope (SL–1500 m) from S Son to El Salvador and Honduras.

RA: Mexico to Peru and N Argentina.

GENUS MICRASTUR: Forest-Falcons

Forest-Falcons have short, rounded wings and long, graduated tails. Legs long for clambering around in trees and even running on ground. Ages differ, sexes similar. Inhabit forest interior and heard far more often than seen. Nest in tree cavities.

BARRED FOREST-FALCON
*Micrastur ruficollis**   Plate 2
Halcón-selvático Barrado

ID: 13–15 in; W 19–23 in (33–38 cm; W 49–59 cm). Small, accipiter-like forest species. See genus note. **Descr. Adult:** eyes yellow, cere, *lores, broad orbital ring, and legs bright yellow-orange. Head and upperparts slaty grey*, wings with brownish cast (more so in ♀). Throat pale grey, *underparts white, narrowly barred blackish*. Tail blackish with white tip and 3 narrow whitish to pale grey bars (from below, basal bar often hidden by tail coverts). Underside of remiges slaty grey, barred white at base. **Juv:** *head and upperparts dark brown to blackish*, head and upper mantle darkest, *often with narrow, broken*

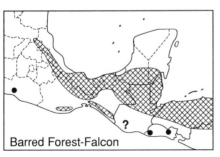

Barred Forest-Falcon

whitish hindcollar. Wings and uppertail coverts often marked with indistinct pale brown bars and spots. *Throat and underparts creamy white to pale buff, most birds with obvious barring*; some have completely barred underparts except for throat; rarely almost unmarked with a few bars on sides of chest. Tail blackish with white tip and 3–4 (6 in *oaxacae* of Sierra Madre del Sur of Oax and Gro) narrow white bars, bolder than adult. Underside of remiges as adult.

Voice. A distinctive, sharp, yapping bark, *kyak* or *kaah*, repeated steadily, 10/12–16 s, more hesitant series to 10/20–23 s, mostly around dawn; also varied, bouncing-ball or slightly laughing series, *cah cah cah-cah-cah* or *kowh kowh kowh-kowh-kowh*, etc.

Habs. Humid evergreen and semideciduous forest, rarely edges and clearings. Perches quietly in forest, usually fairly low, at times attends ant swarms to prey on small birds. Elusive when calling, at other times often confiding. Does not soar and unlikely to be seen flying other than quickly through forest or across roads. Eggs: 2–3, reddish brown to cream, with dark marbling (Thorstrom *et al.* 1990).

SS: See Double-toothed Kite, juv Bicolored Hawk. Collared Forest-Falcon much larger and lankier, light morph adult unbarred, juv has duller facial skin.

SD: F to C resident (50–1600 m) on Atlantic Slope from SE SLP to Honduras; U to F and local on Pacific Slope (100–2500 m) from Gro (Navarro 1986; SNGH, SW) to El Salvador.

RA: S Mexico to Ecuador and N Argentina.

COLLARED FOREST-FALCON
Micrastur semitorquatus naso   Plate 2
Halcón-selvático Collarejo

ID: ♂ 21–22 in, ♀ 24–25 in; W 30–37 in (♂ 53.5–56 cm; ♀ 61–63.5 cm; W 76–94 cm). *Large and lanky forest raptor.* See genus note. At rest, wingtips reach to base of tail. Dimorphic (dark morph rare). **Descr. Adult. Light morph:** eyes brown, cere, lores, and broad orbital ring dull yellowish green, legs yellow. *Face, hindcollar, and underparts white to pale buff*, rarely rich buff, *with black crown and crescent on hind auriculars*. Upperparts blackish brown, uppertail coverts tipped white. Tail blackish with white tip and 3 fairly narrow white bars. Underside of remiges dark grey, barred white. **Dark morph:** blackish brown overall or with lower chest and belly barred and spotted white; uppertail coverts tipped white, flight feathers as light morph. **Juv. Light morph:** head and upperparts dark brown, *auricular crescent and hindcollar less contrasting than adult* (collar rarely absent), upperparts often with pale buff bars and spots, mainly on wings and rump; uppertail coverts tipped white. *Throat and underparts whitish to buff, coarsely barred and scalloped dark brown.* Tail blackish with white tip and 3–4 white bars; remiges as adult. **Dark morph:** resembles dark adult but browner; belly, thighs, and undertail coverts (rarely also scapulars and upperwings) with pale brown bars and spots, tail as light juv.

Voice. A hollow, far-carrying *cohw* or *owhh*, often repeated steadily but not hurriedly, rarely faster than 1/2 s; also more varied, at times slightly laughing series, *ka how ow ow ow*, and an ascending laugh ending with a lower, drawn-out note, *hoh-hoh-hoh-hoh-hoh-hoh-hoh-hoh howh*, etc.; juvs give a more plaintive *mehow* or *kyeoh*, and a quiet, reedy chippering in alarm.

Habs. Evergreen to deciduous forest. Perches quietly in forest, high to low; attends ant swarms (Mays 1985). Elusive when calling, usually calls from high in canopy. Does not soar; flies mostly through forest or above canopy. Nests rarely (?) in ruined buildings (Cobb 1990). Eggs: 2, brown to buff with heavy darker brown mottling (Wetmore 1974; G and JC).

SS: See Hook-billed Kite, Bicolored Hawk.
SD: F to C resident (SL–1800 m) on both slopes from S Sin and S Tamps to El Salvador and Honduras.
RA: Mexico to Peru and N Argentina.

GENUS *FALCO*: True falcons

True falcons are small to fairly large raptors with long, pointed wings and longish tails. Legs fairly short. Ages differ, sexes usually similar. All species have dark brown eyes and black-tipped blue-grey bills. Eggs buff to whitish, speckled and mottled with browns (WFVZ).

AMERICAN KESTREL
*Falco sparverius** Plate 13
Cernicalo Americano

ID: 10–11.5 in; W 24–25.5 in (25.5–29 cm, W 61–65 cm). Widespread. *Wings and tail fairly long and narrow.* **Descr.** Ages differ slightly, sexes differ. ♂: cere, orbital ring, and legs yellow to orange-yellow. *Crown blue-grey with rufous central patch; face whitish with black moustache and black stripe on auriculars,* nape rufous-brown with dark spot each side and dark median stripe. Throat whitish, underparts pale cinnamon, becoming whitish on undertail coverts, sides and flanks spotted black. Upperbody rufous, back barred black, *upperwing coverts and tertials blue-grey,* spotted black. Remiges blackish, secondaries tipped blue-grey. *Tail rufous with white tip and broad black subterminal bar,* outermost rectrices white with 2–3 black bars. *Underwings whitish,* coverts spotted black, *remiges barred blackish with subterminal white checkering on primaries when backlit.* ♀: *head pattern similar to ♂ but duller; rufous upperparts barred dark brown.* Throat and underparts whitish, chest and flanks streaked rufous brown, belly more sparsely marked, thighs some-

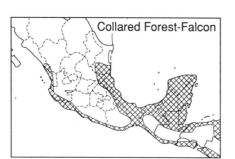

Collared Forest-Falcon

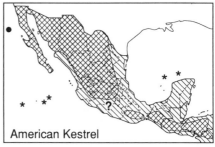

American Kestrel

times with coarse brown spotting. *Upper-tail rufous with 8–10 black bars*, subterminal bar usually widest in adult; undertail paler with 6–8 dark bars. Underwings similar to ♂ but coverts creamy buff, spotted dark brown, lacks distal white primary checkering. **Juv** ♂: resembles adult ♂ but back more extensively barred black, underparts paler with chest streaked black, flanks more heavily spotted black. Attains adult-like plumage by 1st winter. *F. c. tropicalis* group (resident S from Oax) has little or no rufous in crown, underparts of ♂ whitish.

Voice. A shrill, screaming *krieh-krieh ...* or *kree-kree ...*, and a rapid, chippering *k-ree k-ree k-ree ...*, etc.

Habs. Open and semiopen country with scattered trees and bushes. Perches conspicuously on wires and telegraph poles. Flies with fairly loose floppy wingbeats, flight rarely direct and purposeful like Merlin; often hovers and hangs in wind. Soars and glides on flattish wings. Nests in cavities in trees, old buildings; 3–5 eggs.

SS: Note distinctive head pattern, floppy flight, and hovering. Merlin and Bat Falcon are more compact and shorter-tailed, have swift and strong flight, appear darker overall.

SD: F to U resident (SL–3000 m) in Baja and in interior and on adjacent slopes to N Mich and N Mex, possibly to N Gro (Friedmann *et al.* 1950), *tropicalis* group is U resident (600–3000 m) in Oax, F to C from Chis to cen Honduras, and (around SL) in the Mosquitia. C to F transient and winter visitor (Oct–Apr) almost throughout; U to F in Yuc Pen.

RA: Breeds N.A., locally to S S.A.; much of N.A. population migratory, wintering S to Panama.

MERLIN

Falco columbarius*   Plate 13
Esmerejón

ID: 10.5–12.5 in; W 25–27.5 in (26.5–32 cm; W 63.5–70 cm). Widespread winter visitor. *A small, compact, and fast-flying falcon. Wings fairly broadly based.* **Descr.** Ages/sexes differ. ♂: cere and orbital ring greenish yellow to yellow, feet yellow. Face whitish, streaked dark, usually with *narrow pale supercilium and indistinct dark moustache.* Crown, nape, and *upperparts slaty*

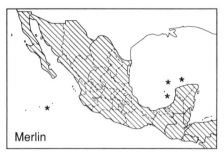

Merlin

blue-grey. Tail black with white tip, upperside with 2–3 broad greyish bars, 1–2 whitish bars visible on underside. Throat and underparts whitish, usually washed buff to pale rufous, with heavy dark streaking tending to bars on flanks. *Underwings appear dark*, checkered black and white overall. ♀/juv: *resemble ♂ but crown, nape, and upperparts dark grey-brown to dark brown*, uppertail has 2–3 narrow whitish bars, 1–2 bars on undertail. *F. c. richardsoni* (winters to cen Mexico) markedly paler, some ♀♀/juvs pale grey-brown above.

Voice. In interactions may give a chippering screaming *krih-krih ...*

Habs. Open and semiopen country, estuaries, lakes, coasts. Perches low, often on ground. Usually seen in low direct flight, wingbeats fast and powerful; does not hover. Soars and glides on flattish wings.

SS: See American Kestrel. Bat Falcon darker overall with contrasting white throat and upper chest. Peregrine Falcon larger with proportionately shorter tail, thick black moustache.

SD: U to F transient and winter visitor (Oct–Apr; SL–2500 m) throughout region, commonest in coastal areas, especially Yuc Pen.

RA: Holarctic breeder, in New World breeds S to N USA; winters to N S.A.

APLOMADO FALCON

Falco femoralis*   Plates 6, 13
Halcón Aplomado

ID: 15–18 in; 32–37 in (38–45.5 cm; W 81–93.5 cm). *A local and decreasing falcon with long wings and tail.* At rest, wingtips fall short of tail tip. **Descr.** Ages differ, sexes similar. **Adult:** cere, orbital ring, and feet yellow to orange-yellow. *Broad creamy-white supercilium contrasts with* blue-grey crown and black eyestripe, supercilia meet on nape. Face, throat, and chest whitish with black moustache, median chest often streaked blackish, *upper belly and flanks black*, narrowly barred white, *lower belly, thighs, and undertail coverts cinnamon.* Upperparts slaty blue-grey, primaries darker, secondaries and uppertail coverts tipped white. *Tail slaty grey with white tip, upperside with 5–6 narrow white bars, underside with 3–4 bars.* Underwings dark, checkered black and white. **Juv:** feet paler. Face and chest washed buff, chest more heavily streaked blackish, belly band dark brown, lower belly and thighs paler. Upperparts blackish brown, tail has 6–7 narrow white bars above, 4–5 bars below.

Voice. A full screaming *keeh-keeh-keeh* ..., and a single sharp *keeh* or *kiih*.

Habs. Savannas and open country, often marshy, with scattered trees. Often perches on bushes and bare trees, at times on ground. Hunts a wide variety of prey, mostly birds and insects, and may be seen soaring at dusk, catching and eating insects on the wing. Nest is a stick platform low to high in tree or bush; 2–3 eggs.

SS: Peregrine Falcon more heavily built with broader-based wings and shorter tail, face with bolder moustache, lacks blackish belly band. Bat Falcon (smaller) and rare Orange-breasted Falcon are built more like Peregrine, both have blackish heads and lack distinct white tail bars.

SD: U to F but local resident (SL–100 m) on Atlantic Slope from SE SLP (RAB) and Ver to W Camp, in Belize, and the Mosquitia. R and local on Plateau in Chih (JRG, PRG), on Pacific Slope in Oax. Resident (formerly) in N and W Mexico, in Yuc, and possibly Pacific Slope of Guatemala.

RA: Mexico locally to S S.A.

BAT FALCON
Falco rufigularis*   Plates 6, 13
Halcón Murcielaguero

ID: 9–11 in; W 24–29 in (23–28 cm; W 61–73.5 cm). Small, compact, tropical falcon, looks all dark with white throat. *Wings broadly based, tail fairly short.* **Descr.** Ages differ, sexes similar. **Adult:** cere, orbital ring, and feet orange-yellow to yellow. *Head and upperparts dark slaty blue-grey,* head and primaries blacker, upperparts with dark shaft streaks. *Throat, sides of neck, and upper chest white,* often washed cinnamon on chest and sides of neck; *lower chest and belly black* with narrow white to pale cinnamon barring, *thighs and undertail coverts rufous.* Tail black with narrow white tip and 3–4 narrow whitish bars. Underwings dark, checkered black and white. **Juv:** feet paler. Throat, sides of neck, and chest washed cinnamon to (rarely) bright orange, upperparts washed brown, undertail coverts barred black.

Voice. A penetrating, screaming *kree-kree-kree* ... or *hew-hew-hew* ..., etc.; also a single sharp *kik*.

Habs. Forest edge and clearings, open areas with scattered patches of trees, towns; mainly in humid areas. Typically perches conspicuously on dead snags, often in pairs. Flight rapid and dashing with fast wingbeats, reminiscent of Merlin or Peregrine; main prey birds, bats, and insects. Usually soars only for brief periods, wings held flattish. Nests in cavities in trees, buildings, etc.; 2–3 eggs.

SS: Larger and rare Orange-breasted Falcon has proportionately larger bill and feet, adult has orange-rufous chest contrasting with white throat (beware Bat Falcons with strong orange on neck sides and chest), narrower black chest band with coarser pale barring, upperparts blackish

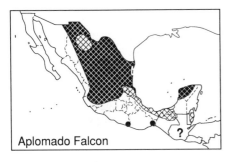

Aplomado Falcon

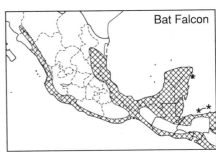

Bat Falcon

with broad blue-grey edgings, juv thighs barred black. White-collared Swift high in flight can easily be taken for Bat Falcon; note black throat, proportionately longer wings. Peregrine Falcon larger with bold black moustache on white face, rump and uppertail of adult paler. See Merlin, Aplomado Falcon.

SD: F to C resident (SL–1600 m) on Atlantic Slope from E NL (Sutton 1948*b*) and Tamps to Honduras; R to U on Pacific Slope from S Son to El Salvador.

RA: Mexico to Ecuador and N Argentina.

ORANGE-BREASTED FALCON
Falco deiroleucus　　　　　　Plates 6, 13
Halcón Pechirrufo

ID: 13.5–16 in; W 30.5–35.5 in (34.5–40.5 cm; W 77–90 cm). *Rare falcon of SE humid lowland forest. Feet proportionately large*; at rest, wing tips about equal with tail tip. *Flight shape Peregrine-like*, wings broadly based, tail fairly short. **Descr.** Ages differ, sexes similar. **Adult:** cere, orbital ring, and feet yellow to orange-yellow. *Head and upperparts slaty blackish, edged blue-grey. Throat white*, sides of neck washed cinnamon, *chest orange-rufous*, belly black with coarse whitish to cinnamon barring and spotting, thighs and undertail coverts rufous. Tail black with white tip and 3–4 narrow whitish bars. Underwings dark, checkered black and white. **Juv:** feet paler. Upperparts blacker, finely edged pale buff when fresh, *chest paler rufous with black streaks, thighs and undertail coverts cinnamon, coarsely barred and spotted black.*
Voice. A hard, often insistent screaming *kyowh-kyowh-kyowh* ... or *kyah-kyah-kyah* ..., etc.; may suggest a rapid Brown Jay screaming but more clipped. Also single, more barking screams, *kyow* or *kyowh*, etc.

Habs. Humid evergreen forest, edges and clearings, especially near cliffs. Perches conspicuously on tall dead snags. Flight fast and powerful, recalling Peregrine; main prey birds. Nests in cavities in cliff faces; 3–4 eggs.

SS: See Bat Falcon. Peregrine Falcon less dark below overall with white throat and chest, barred or streaked underparts.

SD: R to U and local resident (SL–800 m) on Atlantic Slope from S Ver (WJS) to Honduras. Also R resident (at least formerly) on Pacific Slope of Guatemala (probably also adjacent Chiapas). The only Mexican spec. of this species reportedly came from Tecolutla (cen Ver) but this locality may be erroneous.

RA: SE Mexico locally to N Argentina.

PEREGRINE FALCON
*Falco peregrinus**　　　　　　Plate 13
Halcón Peregrino

ID: 15–20 in; W 38–47 in (38–51 cm; W 96–119 cm). Large and heavily built falcon, *wings long and broad-based, tail fairly short. At rest, wingtips reach about to tail tip.* **Descr.** Ages differ, sexes similar. **Adult:** cere, orbital ring, and feet yellow to orange-yellow. Head and upperparts dark slaty grey, blackest on head; *thick black moustache contrasts with white auriculars and sides of neck*; rump and uppertail coverts paler blue-grey. *Throat and underparts white to pale pinkish buff*, upper chest often streaked blackish, rest of *underparts narrowly barred blackish*. Tail blackish with white tip, upperside with 4–6 paler grey bars, underside with 3–5 indistinct whitish bars. *Underwings pale buff to creamy, with dark barring.* **Juv:** cere, orbital ring, and feet pale yellowish to blue-grey. *Black moustache contrasts more on paler head, forehead and supercilium often whitish. Grey-brown to blackish-*

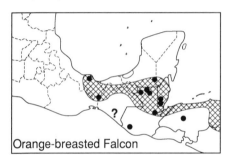

Orange-breasted Falcon

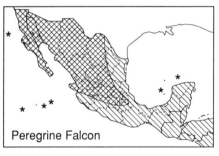

Peregrine Falcon

brown upperparts edged pale buff. Throat and underparts whitish to pale cinnamon-buff, heavily streaked dark brown, throat lightly streaked or unstreaked. *Underwings dark*, checkered and barred dark brown and cinnamon-buff. Tail bars pale buff, narrower than adult.

Voice. A hard, barking to screaming *kaah kaah ...* or *kyuh-kyuh ...*, etc.

Habs. Open and semiopen country, lakes, marshes, sea coasts, usually around cliffs when nesting. Wingbeats shallow and powerful, flight fast, often hunts by stooping on prey from considerable heights. Soars and glides with wings slightly raised to flat; rarely hovers. Nest is a stick platform on cliff ledge or in tree; 2–4 eggs.

SS: See Merlin and Bat, Aplomado, and Orange-breasted falcons. Prairie Falcon slightly longer-tailed, less thickset, wingtips fall short of tail tip at rest; upperparts brown, note black axillars and bar on underwing.

SD: U breeder (Mar–Aug; SL–2000 m) in Baja and around Gulf of California (about 70 pairs, 1965–1984; Porter *et al.* 1988); U to R in interior and on adjacent slopes from Chih and Coah S to Dgo (Hunt *et al.* 1988), possibly to cen Mexico (Jun records in Mex and Pue, SNGH, RGW). U to C transient and winter visitor (Sep–Apr; SL–2500 m) throughout, most common in coastal lowlands, especially Yuc Pen.

RA: Almost worldwide; in New World breeds from Canada to N Mexico and in S S.A.; N migrants winter to S S.A.

PRAIRIE FALCON

Falco mexicanus Not illustrated
Halcón Pradeño

ID: 16–19 in; W 38–45 in (40.5–48 cm; W 96–114 cm). *N Mexico, mainly in winter. A large falcon with long wings and tail.* At rest, wingtips fall short of tail tip. **Descr.** Ages difer, sexes similar. **Adult:** cere, orbital ring, and feet yellow. *Face whitish with*

broad dark brown auricular mark and dark moustache; narrow supercilium whitish, crown brown. Brown upperparts edged and barred paler. *Throat and underparts whitish to pale creamy buff* with often sparse dark spotting and streaking heaviest on flanks. Tail grey-brown with 3–5 poorly contrasting paler bars on underside; appears uniform above. *Underwings pale with blackish bar from axillars along greater coverts.* **Juv:** cere, orbital ring, and feet paler, sometimes blue-grey. Buff-edged upperparts produce neatly scaled effect when fresh; upperparts darker when pale tips wear off.

Voice. A screaming *kyiih kyiih kyiih ...*; usually vocal only around nesting areas.

Habs. Arid to semiarid, open and semi-open areas with scattered trees and bushes, hilly and rocky desert, often around cliffs when nesting. Flight fast with powerful, shallow wingbeats; soars and glides on flattish wings; rarely hovers. Nests on cliff ledges; 3–4 eggs.

SS: See Peregrine Falcon.

SD: U to F and local breeder (near SL–2000 m) in BCN and from Chih and Coah to Dgo and N SLP (Lanning and Hitchcock 1991). F to C transient and winter visitor (Oct–Mar; SL–2500 m) in Baja, on both slopes in Son and Tamps, and in interior S over Plateau to E Jal and N Hgo.

RA: Breeds W N.A. to N Mexico, winters to cen Mexico.

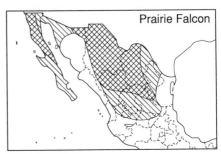
Prairie Falcon

CHACHALACAS, GUANS, AND CURASSOWS: CRACIDAE (8)

Cracids are large, primitive, neotropical gamebirds with longish necks and long broad tails. Bills fowl-like, legs fairly long and stout. Many species are crested and several have brightly colored wattles, bare facial skin, or other head and bill adornments. Sexes usually similar, but strikingly different in a few species. Precocial young able to fly when a few days old, their wings and tails grow rapidly, and they often resemble miniature adults when about half-grown; juvs typically molt into adult plumage before being fully grown. Plumage colored in black, greys, browns, and white, the black areas often with a green or blue sheen. Chachalacas are loud and raucous; other species make whistles, yelps, and low booming noises.

Food mostly vegetable matter, especially fruit. Nests are fairly flimsy platforms of twigs and leaves, often lined with green leaves, low to high in tree or bush. Eggs: 2–3, white, often becoming soiled by plant juices in the nest (WFVZ). Larger species thought to breed when 2 years old but chachalacas may breed in their 1st year.

GENUS ORTALIS: Chachalacas

Chachalacas are smallish cracids, conspicuous in much of the region. All Mexican species were at one time considered conspecific. Ages differ slightly, sexes similar. Eyes amber (grey-brown in juvs), bills and legs greyish, naked loose throat skin reddish. Juvs differ from adults in their narrower rectrices which have poorly contrasting paler tips. Calls are loud and raucous chatters, higher in ♀♀.

PLAIN CHACHALACA
*Ortalis vetula** Plate 14
Chachalaca Común

ID: 16–22 in (40.5–56 cm). *Atlantic Slope.*
Descr. Adult: orbital skin grey. Head greyish; neck, upper chest, and upperparts olive-brown (paler and greyer in *pallidiventris* of Yuc Pen). *Underparts pale buffy brown, slightly paler than upperparts* (contrastingly whitish in *pallidiventris*), undertail coverts dusky cinnamon. Tail blackish with green sheen and *fairly broad pale tips to all but central rectrices, tips whitish in* mccallei *of NE Mexico and in* pallidiventris, *pale buff to brownish in* vetula *of SE Mexico and C.A.*

Voice. Raucous, screechy to gruff, burry chattering, including a rhythmic 3-syllable *kuh-kuh-ruh* or *cha-cha-lac*, and a low, gruff, purring or growling *krrr krrr krrr krrr*, etc. Other shrieking and honking chatters may suggest large *Amazona* parrots.
Habs. Brushy woodland, second growth, and forest edge. Typically in groups of 4–20 birds, feeding low to high in trees, at times on the ground. Often seen flying across roads, rivers, small clearings, etc., with swooping glides, tail spread and neck outstretched.
SS: West Mexican Chachalaca (may be sympatric in Isthmus) larger, whiter on underbody, pale rectrix tips broader. May not overlap with White-bellied Chachalaca

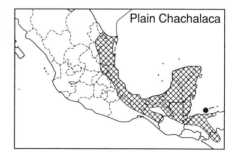
Plain Chachalaca

(despite frequently reported sympatry) which has whiter underparts and white-tipped rectrices. ♀ Highland Guan (cloud forest) darker and browner, barred blackish overall.

SD: C to F resident (SL–1800 m) on Atlantic Slope, and locally in interior, from Tamps and E NL to Honduras (including Isla Utila, where U) and N cen Nicaragua.

RA: S Texas and NE Mexico to N cen Nicaragua, disjunctly in NW Costa Rica.

NB: White-bellied Chachalaca sometimes considered conspecific with Plain Chachalaca.

RUFOUS-BELLIED [WEST MEXICAN] CHACHALACA
Ortalis (poliocephala?) wagleri   Plate 14
Chachalaca Vientre-castaña

ID: 23–27 in (58.5–68.5 cm). Endemic to NW Mexico. **Descr. Adult:** orbital skin greyish (?). Head grey, darkest on crown, with bushy frontal crest; neck, chest, and upperparts grey-brown with strong olive sheen. *Belly and undertail coverts chestnut.* Tail blackish with green sheen and *broad rufous tips to all but central rectrices.* Intergrades with West Mexican Chachalaca have intermediate belly and tail tip coloration.
Voice. Loud, gruff, and scratchy chattering, often in rhythmic chanting choruses of 4–5-syllable phrases, *kirr-i-i-kr* or *chrr-i-k-rr* or *chrr-uh-uh-rr*, etc.; also longer cackling and growling series, etc.
Habs. Arid to semihumid woodland, thorn forest, second growth. In pairs or groups up to 10 or so birds, feeding low to high in trees, at times on the ground; calls from high in canopy but can be hard to see.
SS: West Mexican Chachalaca has whitish belly, buff-tipped tail; intergrades may occur in area of overlap.

SD/RA: C to F resident (SL–2000 m) on Pacific Slope from S Son to N Jal where appears to intergrade abruptly with *O. poliocephala.*

NB1: Often considered conspecific with West Mexican Chachalaca; intergrades are known from the vicinity of Puerto Vallarta, Jal. Critical field studies would help resolve the taxonomic status of the two forms, Banks' (1990*b*) opinion that all intergrades were derived from captive birds being unverified conjecture. The apparent range break between the two forms can also be attributed to incomplete collecting and the absence, until recently, of roads in many areas; semidomesticated chachalacas are rare in Mexico. **NB2:** Sometimes known as Wagler's Chachalaca.

WEST MEXICAN CHACHALACA
Ortalis poliocephala   Plate 14
Chachalaca Mexicana

ID: 23–27 in (58.5–68.5 cm). *Endemic to W Mexico.* **Descr. Adult:** resembles Rufous-bellied Chachalaca but orbital skin pinkish red, lower chest and *belly whitish*, undertail coverts cinnamon, *rectrices broadly tipped creamy buff, often washed rufous.* Intergrades with Rufous-bellied Chachalaca.
Voice. Much like Rufous-bellied Chachalaca: gruff, throaty, rhythmic chattering, *chur-uh-uh-uhr*, etc.; also longer series, *kahrr-ahrr-ahrr-ahrr-ahrr*, etc.
Habs. Much as Rufous-bellied Chachalaca; also occurs locally in pine–oak forest.
SS: See Plain and Rufous-bellied chachalacas. White-bellied Chachalaca smaller, rectrices broadly tipped white, undertail coverts whitish, occurs more in lowlands; note voice.
SD/RA: C to F resident (SL–2400 m) on Pacific Slope and in interior from Jal to SW Chis.

Rufous-bellied Chachalaca

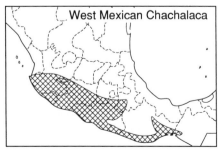
West Mexican Chachalaca

NB: See NB1 under Rufous-bellied Chachalaca.

WHITE-BELLIED CHACHALACA
Ortalis leucogastra   Plate 14
Chachalaca Vientre-blanco

ID: 19–22 in (48.5–56 cm). *Endemic to Pacific Slope S of Isthmus.* Descr. Adult: orbital skin greyish (?). Head greyish; neck, chest, and upperparts grey-brown to rich brown with slight olive sheen. *Belly and undertail coverts contrastingly whitish,* tail coverts may be washed pale buff. Tail blackish with green sheen and *fairly broad white tips to all but central rectrices.*
  Voice. Gruff and burry chattering, including a rhythmic 4-syllable *k-ku'uh-ruh'* or *ch-k-uh-urr,* repeated, and a deep, gruff *chk-chk-chk ...,* etc.
  Habs. Humid to semiarid brushy woodland, second growth and forest edge. In pairs or small groups, mostly arboreal.
SS: See West Mexican and Plain chachalacas.
SD/RA: C to F resident (SL–800 m) on Pacific Slope from Chis to W Nicaragua.
NB: See Plain Chachalaca.

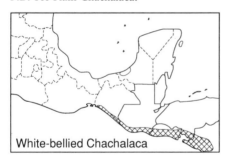

White-bellied Chachalaca

HIGHLAND GUAN
(BLACK PENELOPINA)
Penelopina nigra   Plate 14
Pajuil

ID: 23–26 in (58.5–65.5 cm). *A dark, chachalaca-like guan, endemic to SE cloud forest.* ♂ *has conspicuous red throat wattle.* Descr. Ages/sexes differ. ♂: eyes amber, bill and orbital skin red, *throat wattle and legs bright red.* Black overall, upperparts with green-blue sheen, belly duller and browner. ♀: bill greyish, sparsely feathered throat skin pink, legs dusky reddish. Dark brown overall; head, neck,

Highland Guan

chest, and *upperparts barred and vermiculated rufous.* Underparts paler grey-brown, barred buffy brown. *Tail black with numerous narrow to fairly broad rufous bars.* Imm ♂: resembles ♂ but throat wattle slight or lacking, upperparts and central rectrices sparsely vermiculated rufous, underparts barred buffy grey. Juv: resembles ♀ but eyes grey-brown; ♂ darker overall than ♀.
  Voice An ascending, clear whistle, *wheeeeeeoo.* In display a bizarre 'crashing tree' sound, often immediately following a whistle: initially 1–2 raps, then a creaking, falling, rattling crash, *ke'k'a, arrrrrrr,* apparently made by the wings as birds glide from one perch to another. Alarm call a high, slightly reedy *see-uh,* may be run into twittering.
  Habs. Humid evergreen and pine–oak forest. Mostly arboreal, usually singly at mid- to upper levels of large trees, rarely on the ground; ♂♂ seen more often than ♀♀.
SS: Crested Guan larger (though juv similar size to Highland Guan), with neck and underparts streaked white, bill and face dark; poor illustration of Crested Guan in Peterson and Chalif (1973) has led to frequent misidentification as Highland Guan. See Plain Chachalaca.
SD/RA: Threatened by hunting and habitat destruction; in undisturbed areas still F to C resident (1000–3000 m) on both slopes and in interior from SE Oax and Chis to Honduras and N cen Nicaragua.

CRESTED GUAN
Penelope p. purpurascens   Plate 14
Pavo Cojolito

ID: 32–36 in (81–91 cm). *Large species of both slopes, heavily hunted and generally scarce. Both sexes have conspicuous red throat wattle* and erectile *bushy crest.*

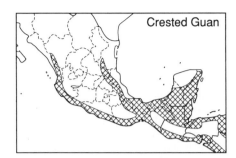

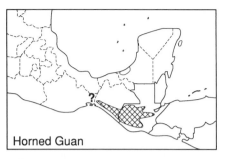

Descr. Ages/sexes similar. **Adult**: eyes red, bill black, bare facial skin deep blue-grey, throat wattle orange-red, legs dull reddish. Plumage *blackish brown overall (often looks black)*; head, neck, chest, and upperparts strongly glossed green, more bronzy on rump and uppertail coverts. *Neck, back, and underparts streaked and mottled white.* **Juv**: eyes brownish, throat wattle reduced or lacking; white streaking reduced.

Voice. Loud honking and yelping cries, often repeated tirelessly early and late in the day: *yoink yoink yoink* ... or *kyeh-kyeh-kyeh* ..., etc.; also a gruff low *urmmff*.

Habs. Humid to semihumid evergreen and semideciduous forest, locally in pine–evergreen and humid pine–oak forest. Often in pairs or small groups, usually high in trees, rarely on the ground. Flight labored with a distinctive loud rush of wings; can glide considerable distances across canyons.

SS: See Highland Guan.

SD: U to R resident (SL–2500 m) on both slopes from S Sin and S Tamps to El Salvador and Honduras, widely decimated by hunting. Still F to C where protected, such as Tikal, Guatemala, and in remote areas.

RA: Mexico to Ecuador and N Venezuela.

HORNED GUAN
Oreophasis derbianus   Plate 14
Pavón Cornudo

ID: 31–35 in (79–89 cm). *A bizarre cracid, endemic to remote cloud forest in Chis and Guatemala. Threatened by habitat destruction. Unique, striking red horn of bare skin on top of head.* **Descr.** Ages differ, sexes similar. **Adult**: bill yellow, horn, small naked gular area, and legs red. *Black overall*, upperparts with bluish sheen. *Foreneck and chest white,*

finely streaked black, tail has white band near base. **Juv**: eyes dusky, horn develops through 1st year. Wings and tail dark brown, tail vermiculated and finely spotted cream.

Voice. ♂ in display (Feb–early May) gives a low, ventriloquial, almost subliminal boom, *uhmm uh'mmm uh'mmm uh'mmm*, and variations.

Habs. Humid evergreen forest, rarely pine–evergreen forest. Mostly arboreal, runs quickly among mid- to upper levels of large trees and overlooked easily. Usually found singly, often naturally curious.

SD/RA: U to R and local resident (2000–3000 m) in Sierra Madre de Chiapas of Chis (possibly also extreme SE Oax) and W and cen Guatemala (Andrle 1967a). Recently discovered in Sierra de Las Minas, E Guatemala (Howell and Webb 1992b).

GREAT CURASSOW
*Crax rubra**   Plate 14
Hocofaisán

ID: 30–36 in (76–91.5 cm). *Very large, lowland cracid, highly prized as a table bird and consequently extirpated from much of former range. Prominent curly crest striking.* ♀ *polymorphic (barred morph uncommon).* **Descr.** Ages/sexes differ. ♂: eyes amber, bill black with yellow base and swollen knob, legs greyish. *Glossy black*

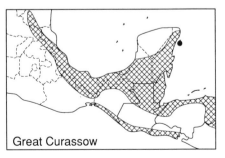

overall with white lower belly and undertail coverts. ♀: bill horn with grey base.
Dark morph (commonest form in Mexico): *head and upper neck barred and spotted black and white, crest black with broad
white median band.* Lower neck, upper
chest, and upperparts blackish, darkest on
back, wings and tail usually mottled and
vermiculated rufous and whitish; tail may
show 5–7 narrow whitish bars. Underbody cinnamon, belly and thighs may be
barred blackish. **Barred morph** (Isthmus
and Yucatan): crest more extensively white;
lower neck, upper chest, and mantle boldly
barred blackish and white, rest of upperparts boldly barred dark brown, rufous-
brown, and pale buff. Tail black with 10–
12 pale buff bars. Underbody cinnamon.
Rufous morph (mostly C.A.): resembles
dark morph but lower neck, chest, and
upperparts rufous, upperparts sometimes
sparsely barred black, uppertail coverts
barred buff. Tail rufous with buff tip and

5–6 buff bars, often edged black. **Imm ♂:**
resembles ♂ but bill knob slight or lacking,
develops through 1st year. **Juv:** eyes dusky
brown. Resembles ♀ (♂ resembles dark
morph ♀).
Voice. ♂ has a low, almost subliminal
booming *mm uhmm-ump*, and variations.
Calls include a sharp, piping *wheep!*
Habs. Undisturbed humid evergreen
forest, also mangroves on Pacific coast.
Rarely seen except in remote or protected
areas. Mostly terrestrial, but also feeds in
trees and ♂♂ may call from fairly high in
trees. Usually in pairs or small groups,
runs rather than flies when disturbed.
SD: R to U resident (SL–1500 m; F to C in
the few areas where it is protected) on
Atlantic Slope from S Tamps to Honduras,
including Isla Cozumel (at least formerly);
on Pacific Slope from E Oax to El Salvador, extirpated from much of original
range.
RA: E Mexico to Ecuador.

TURKEYS AND QUAIL: PHASIANIDAE (17)

A fairly diverse cosmopolitan family of primarily terrestrial chicken-like birds. Bills stout and short, legs fairly stout, used for running and scratching ground for food. Wings broad and rounded. Ages/sexes usually differ, ♂♂ larger than ♀♀. Young precocial and able to fly within a few days, assume juv plumage before fully grown. First prebasic molt, almost complete, occurs within 1–2 months. Imms retain outer 2 juv primaries and usually some upperwing coverts but most resemble adults; a few species have distinct imm plumage which is replaced by adult plumage within a few months. Calls are typically loud and carrying, notably clear and whistled in Wood-Partridges and some tropical quail, members of a pair often duet.

Food mostly vegetable matter, especially seeds; also berries, green leaves, insects and other small invertebrates, small amphibians, and reptiles. Nests are shallow scrapes in the ground, often somewhat hidden at base of bush or tuft of grass. Large clutches (6–18 eggs) typical of temperate quail, smaller clutches (3–6 eggs) usual for tropical species.

GENUS MELEAGRIS: Turkeys

Turkeys are large game birds with fairly long, broad tails that are spread in display. Heads and upper necks naked and brightly colored, covered with warty protuberances and wattles. Plumage strongly iridescent. Ages/sexes differ. ♂♂ have well-developed tarsal spurs; ♀♀ smaller and duller overall. ♂♂ polygamous, ♀♀ care for the young. Nest is a scrape in the ground; 8–15 eggs, creamy buff, speckled with browns (WFVZ). Turkeys are often considered as a separate family, Meleagrididae.

WILD TURKEY
Meleagris gallopavo*   Not illustrated
Guajalote Silvestre

ID: ♂ 40–44 in, ♀/imm 28–36 in (♂ 101–112 cm, ♀/imm 76–91.5 cm). Very local, extirpated from most of original range but widely and commonly domesticated. **Descr.** ♂: bill yellowish horn, *head and upper neck naked, bluish with red warts*, inflatable forehead wattle and throat wattle red, legs reddish pink. *Metallic bronzy overall with extensive black scaling*, (often looks black overall), rump metallic blue-black, scaled black, lower rump and uppertail coverts broadly tipped white. Long black tuft of feathers hangs from

chest. Remiges barred dark brown and white. *Tail vermiculated black and rufous with broader black subterminal band and fairly broad whitish tip.* ♀: lacks inflatable forehead wattle. Duller overall, black chest tuft only on birds older than 3 years. **Juv:** dark brown overall, feathers tipped rufous and buff, subterminally blackish, lacking strong metallic sheen. **1st basic:** attained before fully grown; resembles ♀, ♂ has trace of chest tuft. Attains adult plumage by 3rd prebasic molt when about 3 years old.

Voice. ♂ gives the familiar gobbling call heard in most farmyards and villages; ♀ makes low clucks.

Habs. Semihumid to semiarid pine–oak and oak woodland with grassy and brushy understory. Often in small groups but wary, runs at first sight of danger, rarely flies. Widely domesticated; wild birds unlikely to be seen anywhere near

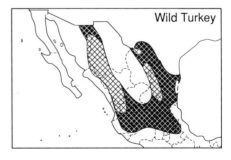

Wild Turkey

habitation but for a few private ranches in N Mexico.

SD: R (widely extirpated) to F (on private ranches) and local resident (300–2500 m) from Son and Tamps S in interior and on adjacent slopes to N Jal and SE SLP, formerly to cen Mexico.

RA: N.A. to cen Mexico.

OCELLATED TURKEY
Meleagris ocellata					Plate 14
Guajalote Ocelado

ID: ♂ 36–40 in, ♀/imm 26–33 in (♂ 91.5–102 cm, ♀ 66–84 cm). Endemic to Yuc Pen. **Descr.** ♂: bill black with horn nail, *head and upper neck naked, bright blue with orange warts* concentrated about head, red orbital ring; inflatable forehead wattle may hang down over bill, inflatable horn on crown erected in display; legs reddish pink. *Metallic blue-black overall, feathers tipped blue-green to golden with subterminal black bars*, median upperwing coverts metallic green, tipped black, *greater upperwing coverts burnished copper.* Remiges barred dark brown and white, *secondaries mostly white* on outer webs. *Tail and uppertail coverts vermiculated grey and blackish, tips patterned with copper-tipped blue-green to violet-blue eyespots.* ♀: bill flesh with grey base, orbital ring orange-red. Plumage duller overall. **Juv:** Grey-brown overall without metallic sheen, feathers tipped whitish. Tail vermiculated grey and dark brown with broader black subterminal band. Molts to adult plumage unknown (?), may parallel those of Wild Turkey.

Voice. ♂ in display makes a curious-sounding, slightly nasal, hesitant, pumping *puhk, puhk-puhk, puhk* ... or *puht-puht-puht* ..., accelerating into a roll or gobble; may suggest a motor scooter starting up; ♀ gives low, clucking notes.

Habs. Humid to semihumid semi-deciduous and deciduous forest and clearings, overgrown milpas, brushy woodland. Usually in small groups but elusive and wary in most of range. Not domesticated, though in its range birds derived from Wild Turkeys are common.

SD/RA: U to R resident (SL–300 m) in Yuc Pen, but C and conspicuous where protected such as Tikal, Guatemala. Reports from E Chis have been questioned (Wagner 1953).

NB: Traditionally placed in the genus *Agriocharis.*

GENUS DENDRORTYX:
Wood-Partridges

Wood-Partridges are medium-sized forest-dwelling quail with long tails and erectile bushy crests. Bare parts often brightly colored. Ages differ, juvs have white shaft streaks on back and underparts and attain adult-like plumage when about half-grown. Sexes similar, plumage cryptically patterned in browns and greys.

LONG-TAILED WOOD-PARTRIDGE
*Dendrortyx macroura**					Plate 15
Gallina-de-monte Coluda

ID: 12–15 in (30.5–38 cm). Endemic to cen Mexico. **Descr. Adult:** bill, orbital ring, and legs bright red to orange-red. *White supercilium and moustache contrast on black head.* Nape and upper mantle coarsely streaked red-brown and blue-grey, rest of upperparts cryptically patterned with grey-brown, buff, and dark brown. *Underparts grey to blue-grey,* chest and sides spotted and streaked rufous, thighs and flanks mottled brown, dark undertail coverts tipped whitish. Remiges

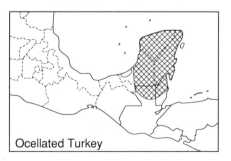

Ocellated Turkey

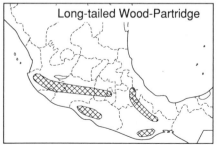

Long-tailed Wood-Partridge

and outer rectrices grey-brown to rufous-brown. *D. m. oaxacae* of N Oax often has indistinct white face stripes.

Voice. Varied, loud, rollicking whistles, often given in duet, *ohrr ohrr ohrr* ..., breaking into *kee-ohrr kee-ohrr* ..., and *whee-a-huck-u* ..., or *quee-a-ruc* ..., or *kee-a'ohr* ..., etc. Alarm call a shrill, squeaky *quid-it*, often repeated rapidly.

Habs. Humid pine–oak, pine–evergreen, and evergreen forest with dense underbrush. Terrestrial, rarely takes to trees except to roost. Skulking and elusive, mainly detected by voice. Runs low and swiftly through dense cover, tail held cocked; flushes with a loud whirr of wings. Usually seen singly or in pairs but forms small groups in non-breeding season (Aug–Mar). Eggs: 4–6, whitish, sparsely speckled brown (WFVZ).

SS: Bearded Wood-Partridge may occur sympatrically in cen Ver (Cofre de Perote, Orizaba) and perhaps Pue, but favors lower elevation cloud forest; in flight, remiges and outer rectrices rufous-brown (versus grey-brown in sympatric Long-tailed), head pattern distinctive. Singing Quail smaller, with stubby tail, rarely flushes.

SD/RA: F to C resident (1200–3300 m; R to U where heavily hunted and/or underbrush overgrazed) on Pacific Slope from Jal to Oax, in cen volcanic belt from Col to Ver, and on Atlantic Slope from cen Ver to Oax.

BEARDED WOOD-PARTRIDGE

Dendrortyx barbatus　　　　　Plate 15
Gallina-de-monte Veracruzana

ID: 13–14 in (33–35.5 cm). Endemic to NE Mexico; *endangered by habitat loss.* **Descr. Adult:** bill, orbital ring, and legs bright red to orange-red. *Crown and crest grey-brown; face, throat, neck, and upper chest blue-grey, with hindneck, mantle,*

and sides of chest streaked rufous. Rest of upperparts cryptically patterned with buff, grey-brown, blackish, and pale grey. *Underparts cinnamon,* thighs and flanks mottled grey-brown, dark undertail coverts tipped whitish. Remiges and outer rectrices rufous-brown.

Voice. Loud, rollicking whistles, repeated rapidly and often given in duet: ♂ has a 3–4-syllable *ko-orrr-EE-EE* and *ko-or-EE*; ♀ a quieter *ko-or-ee-ee-ee-eee*, usually preceded by softer notes audible only at close range (Johnsgard 1973).

Habs. Humid evergreen forest with dense underbrush. Habits much as Long-tailed Wood-Partridge. Nest undescribed (?). Eggs: 3–4, whitish (Johnsgard 1973).

SS: See Long-tailed Wood-Partridge. Singing Quail smaller, with stubby tail, rarely flushes.

SD/RA: U to F but local resident (1000–1500 m; widely extirpated), from SE SLP (extirpated?) through Hgo and Pue to cen Ver.

BUFFY-CROWNED WOOD-PARTRIDGE

Dendrortyx leucophrys*　　　　　Plate 15
Gallina-de-monte Centroamericana

ID: 11–14 in (28–35.5 cm). Highlands S of Isthmus. **Descr. Adult:** bill black, orbital ring and legs orange-red to red. *Pale buff to whitish forehead, crest, supercilium, and throat contrast with dark brown crown and auriculars.* Hindneck and sides of neck rufous, streaked whitish, upper mantle coarsely streaked rufous and blue-grey; rest of upperparts olive-brown, mottled rufous on back. *Underparts blue-grey to grey, streaked rufous on foreneck and chest, cinnamon on belly and flanks,* undertail coverts blackish brown. Remiges and outer rectrices dark rufous.

Voice. Loud, rollicking whistles, often re-

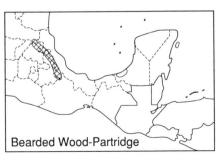

Bearded Wood-Partridge

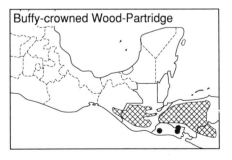

Buffy-crowned Wood-Partridge

peated rapidly, *kee-oh'rr kee-oh'rr ...*, and *kee-oh' roh-roh ...* etc., much like other wood-partridges.

Habs. Humid to semihumid pine–oak and evergreen forest with brushy undergrowth. Habits much as Long-tailed Wood-Partridge. Nest undescribed (?). Eggs: 4–5 (?), reddish buff with spots and blotches of reddish brown (Oates 1901).

SS: Singing Quail smaller, with stubby tail, rarely flushes, Spotted Wood-Quail stubby-tailed, head appears dark.

SD: F to C resident (600–3000 m) in interior and on Pacific Slope from SE Chis to El Salvador and N cen Nicaragua. Reports from N Chis (e.g. Leopold 1959) require verification.

RA: SE Mexico to Costa Rica.

SPOTTED WOOD-QUAIL

Odontophorus guttatus   Plate 15
Codorniz Bolanchaco

ID: 9–10 in (23–25.5 cm). A plump, stub-tailed forest quail; erectile crest often conspicuous. Dimorphic. **Descr.** Ages/sexes differ. ♂: bill, orbital ring, and legs grey. *Hind crown and nape rufous-orange basally, face and throat black with lines of white streaks.* Upperparts brown; scapulars, upperwing coverts, and tertials cryptically patterned with grey-brown, blackish, and buff. *Underparts dusky cinnamon* (rarely grey-brown), *spotted white*, thighs and undertail coverts buffy brown with dark barring. ♀: *lacks rufous-orange base to crest*, underparts grey-brown overall (rarely dusky cinnamon). **Juv:** resembles adult but throat whitish, blackish chest streaked white. Quickly attains adult-like plumage.

Voice. Loud, ringing, rollicking whistles, often given in duet, series lasting a minute or longer, *coo-coo-woo-loo-oo woor coo-*

coo-woo-loo-oo woor ..., or *kolliwoo cha-ku ...*, or *golli kiwoo ki cha-koo ...*, etc., hence the onomatopoeic Spanish name. Single birds may give a loud, ringing, slightly hollow *hoo-wook* or *h-wooh*, repeated.

Habs. Humid evergreen forest. Terrestrial, in pairs or small groups, often quite confiding but in areas where hunted will flush at close approach. While walking, crest often raised and lowered. Nest undescribed (?). Eggs creamy white, sometimes spotted with brown (Schonwetter 1960).

SS: See Great Tinamou, Buffy-crowned Wood-Partridge, Black-eared Wood-Quail (E Honduras, see Appendix E).

SD: F to C resident (near SL–2000 m; mostly above 600 m in Honduras) on Atlantic Slope from S Ver to Honduras; on Pacific Slope of Chis and Guatemala.

RA: SE Mexico to W Panama.

SINGING QUAIL

*Dactylortyx thoracicus**   Plate 15
Codorniz Silbadora

ID: 8–9 in (20.5–23 cm). Endemic; a small stub-tailed forest quail; erectile crest often noticeable. Feet fairly large and claws long. *Song striking.* **Descr.** Ages/sexes differ. ♂: bill and legs grey. *Cinnamon face and throat contrast with dark brown crown*; supercilium often fades to pale buff at sides of nape. Upperparts grey-brown, mantle coarsely streaked darker brown, scapulars and upperwing coverts cryptically patterned with blackish, rufous, and pale grey-brown, and marked with *sharp buff streaks, broadest on tertials. Underparts greyish to grey-brown* (palest in *sharpei* of Yuc Pen), *finely streaked white on chest and flanks*, flanks and undertail coverts with dark spotting. ♀: face and throat pale grey, underparts cinnamon. **Juv:** resembles

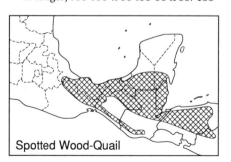
Spotted Wood-Quail

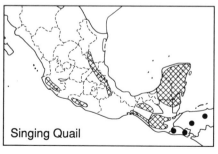
Singing Quail

adult but throat paler (whitish in ♀), underparts dusky cinnamon-buff (♂) to cinnamon (♀), spotted dark brown on chest and flanks. Quickly attains adult-like plumage.

Voice. An arresting series of loud, quavering whistles, initially hesitant single whistles 3–6 × (rarely more), breaking into a rollicking chorus repeated 2–8 ×, typically 3 ×: *wheerr, wheerr, wheerr ki'ki-i-weer ki'ki-i-weer* or *choo-oo, choo-oo, choo-oo, choo, choo-choo-churry-chewt choo-choo-churry-chewt* ... Introductory notes suggest Little Tinamou and may be given singly throughout day. At times, choruses may be repeated with no introductory notes. Also a liquid twittering when nervous.

Habs. Humid to semihumid pine–evergreen and semideciduous forest, locally (*sharpei*) in semiarid deciduous forest. Terrestrial, often detected by scratching sounds made by birds feeding in dry leaf litter; presence known mostly by loud and carrying song. In pairs or small groups which may be very confiding, flies rarely and prefers to run or freeze if alarmed. Nest presumably on ground. Eggs: 5, creamy with sparse fine dusky flecks (MCZ).

SS: See Wood-Partridges. ♂ Bobwhites (usually in more open areas) have either white throat and supercilium or black head and chest (Northern), or black-and-white-striped head and black throat (Yucatan), ♀♀ whitish below, scalloped black. Ocellated Quail usually in more open areas, upperparts less contrastingly patterned; note face pattern (striking in ♂).

SD/RA: F to C resident (250–3000 m) on Pacific Slope from Jal to Gro, on Atlantic Slope (250–2000 m, locally to 100 m) from S Tamps to cen Ver, S of Isthmus on both slopes and in interior (800–3000 m) from E Oax and Chis to El Salvador and Honduras, and (SL–300 m) in Yuc Pen.

MONTEZUMA QUAIL

*Cyrtonyx montezumae**   Plate 16
Codorniz de Moctezuma

ID: 8–9 in (20.5–23 cm). Plump quail of grassy pine–oak woodland. *Both sexes have bushy nuchal crest.* Two ssp groups told in ♂ plumage: *montezumae* group (N and cen Mexico; including *mearnsi*, shown) and *sallei* group (SW Mexico).

Descr. Ages/sexes differ. ♂ *Montezumae*: bill blackish above, blue-grey below, legs grey. *Harlequin-like head pattern:* face and throat white with 2 wedge-shaped black stripes on auriculars (lower stripe often blue-grey), black median throat stripe; neck sides and throat edged black; black forehead stripes merge with greyish-buff to tawny crest. Upperparts grey-brown to rufous-brown, barred and spotted blackish, streaked cream to buff. *Sides and flanks purplish grey, boldly spotted white, median underparts chestnut* becoming black on vent and undertail coverts. Intergrades with *sallei* group in cen Ver. *Sallei*: streaks on posterior upperparts cinnamon-rufous, *sides and flanks blue-grey, anterior spots whitish or washed buff, posterior spots larger, dark rufous to chestnut, median underparts paler.* ♀: *head shows trace of ♂ pattern (especially dark stripe below eye and dark border to sides of neck), but washed vinaceous and streaked blackish*, crest spotted black. Upperparts with less distinct black marks on tertials and wing coverts. Underparts dusky vinaceous to vinaceous cinnamon, sparsely marked with black spots and bars, especially on sides of chest. **Juv:** resembles ♀ but underparts greyish with black bars and white shaft streaks. **Imm ♂:** resembles ♂ but ♀-like head pattern kept into early winter.

Voice. Territorial call a far-carrying, descending, quavering whinny; a twittering *whi-whi whi-hu* when alarmed.

Habs. Semiarid to semihumid pine–oak woodland with grassy understory, hilly grassland with scattered trees, especially oaks and junipers. In small groups, rarely flies till nearly stepped on, prefers to crouch rather than run. Eggs: 6–12, whitish, unmarked (WFVZ).

SS: Northern Bobwhite rarely in same habitat, lacks distinct face pattern and bushy

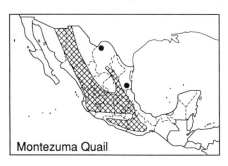

Montezuma Quail

crest, usually runs for cover when alarmed and flushes from greater distances.

SD: U to F resident (1000–3000 m) in interior and on adjacent slopes from Son and Coah to Oax (N of Isthmus). Decreasing in much of range due to overgrazing of understory and hunting.

RA: SW USA to S Mexico.

NB: *Sallei* group sometimes considered a distinct species, Salle's Quail.

OCELLATED [MONTEZUMA] QUAIL

Cyrtonyx montezumae (in part) or
C. ocellatus   Plate 16
Codorniz Ocelada

ID: 8–9 in (20.5–23 cm). *Endemic; counterpart of Montezuma Quail S of Isthmus.* **Descr.** Ages/sexes differ. *Similar to Montezuma Quail (sallei group) but ♂ has median chest tawny-cinnamon, becoming dark cinnamon-rufous on belly; anterior flank spots washed buff, posterior spots expand into broad cinnamon-rufous scallops.*

  Voice. Undescribed (?); presumably much like Montezuma Quail.

  Habs. See Montezuma Quail. Nest and eggs undescribed (?).

SS: Spot-bellied Bobwhite rarely in same habitat, lacks distinct face pattern and bushy crest, usually runs for cover when alarmed and flushes from greater distances.

SD/RA: U to F resident (1000–3000 m) in interior and on adjacent slopes from SE Oax and Chis to El Salvador and Honduras. Decreasing in much of range due to overgrazing of understory and hunting.

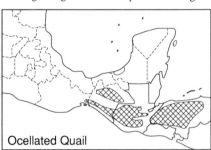

Ocellated Quail

SPOT-BELLIED [CRESTED] BOBWHITE

*Colinus leucopogon** or *C. cristatus* (in part)
Codorniz-cotui Centroamericana   Plate 70

ID: 7–8.5 in (19–23 cm). C.A.; *♂ has single-pointed crest.* Two groups: *hypoleucus* group (E Guatemala to cen El Salvador) and *leucopogon* group (cen El Salvador and Honduras to Nicaragua). **Descr.** Ages/sexes differ. *Hypoleucus.* ♂: bill blackish, legs dusky flesh to greyish. *White supercilium contrasts with dark crown and broad eyestripe, crest dark brown. Throat and underparts white,* sides greyish, hind flanks often dark brown, boldly spotted white; undertail coverts usually barred black. Upperparts patterned with grey, pale brown, whitish, and black. ♀: *buff supercilium and throat contrast with dark crown and broad eyestripe. Underparts whitish with coarse dark scalloping,* brownish marks on flanks. Intergrades with *leucopogon* group in cen El Salvador. *Leucopogon.* ♂: *greyish-cinnamon underparts heavily and coarsely spotted whitish to buff,* chest unmarked or with a few fine spots. *Throat white (leucopogon of E El Salvador and Pacific Slope of Honduras)* to *blackish, bordered white.* ♀: *supercilium and throat buff, throat streaked blackish.* **Juv** (both forms): resembles ♀ but back and chest have whitish shaft streaks.

  Voice. Territorial call a slightly hoarse to sharp *hu-wheet!* or *h hoo wuiit!,* much like Northern Bobwhite, etc.

  Habs. Arid to semiarid brushy woodland, scrub, overgrown fields. In pairs or small groups which run for cover when alarmed but also flush if surprised or if away from cover. Eggs: 11, white, unmarked (Costa Rica, Leber 1975).

SS: See Ocellated Quail.

SD: C to F resident (SL–1800 m) in interior and on Pacific Slope from E Guatemala to Honduras and W Nicaragua.

RA: E Guatemala to NW Costa Rica.

NB1: Sometimes considered conspecific with *C. cristatus* of Pacific Slope from S Costa

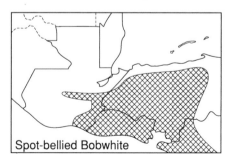

Spot-bellied Bobwhite

Rica to N S.A. **NB2:** Some authors consider all Bobwhites as variable populations of a single species.

NORTHERN BOBWHITE

*Colinus virginianus** Plate 16
Codorniz-cotui Norteña

ID: 7.5–9 in (19–23 cm). N > S. Widespread in Mexico except Yuc Pen. *Marked geographic variation complicated by range expansion due to clearing of forest and localized introduction. Four ssp groups: texanus* (NE Mexico), *graysoni* (cen Mexico), *pectoralis* group (Atlantic Slope from Ver to Tab, disjunctly in S-cen Mexico), and *coyolcos* group (Pacific Slope in Son, and from Gro to Chis and NW Guatemala). **Descr.** Ages/sexes differ. ♂. *Texanus*: bill blackish, legs dusky flesh to greyish. *White supercilium and throat contrast with blackish crown and broad eyestripe.* Upperparts patterned with blue-grey, pale brown, buff, whitish, and black. *Black forecollar borders white throat; upper chest cinnamon, rest of underparts white with coarse black scalloping,* flanks broadly striped cinnamon. Intergrades with *graysoni* (from S Tamps to N Ver) have underparts mostly cinnamon-rufous with coarse black scalloping and striping on chest and flanks. *Graysoni*: resembles *texanus* but mantle mottled cinnamon-rufous, *underparts solidly cinnamon-rufous. Pectoralis*: resembles *graysoni* but *black forecollar extends into broad black chest band,* black may extend onto belly as coarse striping. *Coyolcos*: resembles *graysoni* but *head and throat black, often with narrow whitish supercilium; chest often blackish in southernmost birds.* ♀: resembles ♂ *texanus* but *supercilium and throat buff,* cinnamon upper chest mixed with blackish and white, underparts less extensively scalloped blackish. S and SW forms often pale

buff below, lack cinnamon wash on upper chest, southernmost. Pacific coast ssp (*salvini*) markedly darker and greyer than other ssp. **Juv:** resembles ♀ but underparts dusky grey, white shaft streaks on back, chest, and flanks.

Voice. Much like Spot-bellied Bobwhite, including a sharp *hwuik! hwuik! hwuik!* or *hwuip! ...*, etc.

Habs. Brushy woodland, fields, overgrown pastures, savannas. Habits much like Spot-bellied Bobwhite. Eggs: 7–16, whitish, unmarked (WFVZ).

SS: Scaled Quail has pale grey-brown head with prominent tufted crest, prefers more arid areas. See Montezuma Quail. In range of rare *ridgwayi* ('Masked Bobwhite') of Son, Elegant Quail has prominent crest, underparts boldly spotted, Gambel's Quail has prominent crest, fairly uniform pale grey overall.

SD: C to F resident (SL–2500 m) on Atlantic Slope from Coah to Tab; in interior from SLP to SE Nay and Jal, from S Mor and S Mex to SW Pue and N Gro, and from Chis to NW Guatemala; on Pacific Slope from cen Gro to Chis. R resident in Son. May spread to base of Yuc Pen and Pacific lowlands of Guatemala with clearing of forest.

RA: E N.A. to S Mexico and NW Guatemala.
NB: See NB2 under Spot-bellied Bobwhite.

YUCATAN (BLACK-THROATED) BOBWHITE

*Colinus nigrogularis** or *C. virginianus* (in part) Plate 16
Codorniz-cotui Yucateca

ID: 7.5–8 in (19–20.5 cm). Yuc Pen and N C.A. **Descr.** Ages/sexes differ. ♂: bill blackish, legs dusky flesh to greyish. *White supercilium, black eyestripe, and white-bordered black throat create striking head pattern.* Crown, nape, and upperparts

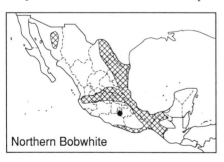

Northern Bobwhite

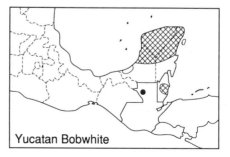

Yucatan Bobwhite

patterned with pale grey, cinnamon-rufous, blackish, and white. *Underparts white, boldly and coarsely scalloped black,* vent and undertail coverts cinnamon-rufous; underparts may look broadly striped black and white in the field. ♀: *buff supercilium and throat contrast with dark crown and broad eyestripe.* Upperparts patterned with grey, pale brown, whitish, and black. *Whitish underparts coarsely scalloped black; underparts appear boldly spotted whitish overall.* **Juv:** resembles ♀ but underparts dusky grey, white shaft streaks on back, chest, and flanks.
Voice. Much as other bobwhites, including a shrill *kreh-kreh ...* when flushed.
Habs. Arid to semiarid brushy woodland, overgrown fields, beach scrub. Habits much as Spot-bellied Bobwhite. Clutch in wild undescribed (?), eggs white (MCZ).
SS: See Singing Quail.
SD/RA: C to F resident (SL–200 m) in Yuc Pen S to cen Camp and cen QR (SNGH, SW); disjunct populations in N Guatemala, Belize, and the Mosquitia.
NB: See Spot-bellied Bobwhite.

BANDED (BARRED) QUAIL
Philortyx fasciatus   Plate 16
Codorniz Barrada

ID: 7.5–8.5 in (19–21.5 cm). *Mexican endemic; replaces Bobwhite in Balsas drainage. Adult has single-pointed crest.* **Descr.** Ages differ, sexes similar. Bill blackish, legs flesh. **Adult:** *head and upper chest cinnamon-brown with black crest,* whitish throat; chest indistinctly barred whitish. Upperparts brownish grey with short, thick, blackish bars and narrow buff bars. *Underbody whitish, sides and flanks boldly barred black.* **Juv:** throat whitish, back and chest with whitish shaft streaks. Within 2–3 months molts into **imm:** resembles adult but with black face and throat.

Attains adult plumage at 4–5 months of age.
Voice. Territorial call a ringing *keeah!* or *k-yah',* repeated, sometimes in rapid, sharp series. Also a squealing *wheer whee'ar* and *wheer-pee-pee pee-peer,* etc., when flushed, and a quiet growl in alarm.
Habs. Arid to semiarid thorn forest, scrub, overgrown fields. In groups of 6–20 birds, runs quickly when alarmed and also flushes short distances into cover. Nest and clutch in wild undescribed (?); breeds Jul–Sep.
SS: See Northern Bobwhite.
SD/RA: C to F resident (SL–1500 m), mainly in interior, from S Jal and Col through Balsas basin to SW Pue and cen Gro.

SCALED QUAIL
*Callipepla squamata**   Plate 16
Codorniz Escamosa

ID: 9–10 in (23–25.5 cm). *Deserts of N and cen Mexico. Adults have tufted crest.* **Descr.** Ages differ, sexes similar. Bill blackish, legs grey. **Adult:** head and upperparts grey-brown, *crest tipped whitish,* tail grey. *Neck and chest blue-grey, boldly scaled black,* grey-brown flanks streaked whitish; belly and undertail coverts buff, scaled black, central belly dark chestnut in ♂ *castanogastris* of NE Mexico. **Juv:** grey-brown overall without bold black scaling, crest shorter without whitish tip. Upperparts streaked white, wing coverts and tertials vermiculated blackish. Tail barred whitish. Underparts have indistinct brownish bars, chest streaked whitish.
Voice. Territorial call a hoarse, wheezy *rrehh!,* repeated; may be alternated with bouts of dry, nasal, rhythmic clucking *chow-chowk' chow-chowk' ...* or *cow-cow-h cow-cow-h ...,* also a dry, clucking *chek-ah,* and a sharp, metallic, ringing *ching* in alarm.

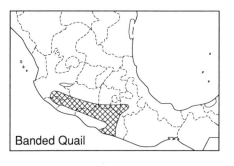

Banded Quail

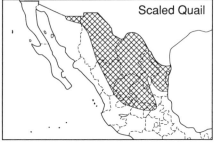

Scaled Quail

Habs. Arid brush and grassland with scattered bushes. In pairs or small groups which run for cover when alarmed but also flush if surprised or if away from cover. Eggs: 9–16, whitish, usually speckled brown (WFVZ).

SS: Tufted crest and heavily scaled appearance distinctive.

SD: C to F resident (1000–2000 m) in interior from NE Son and NL S over Plateau to NE Jal and Hgo, also on Atlantic Slope (to near SL) in Tamps.

RA: SW USA to cen Mexico.

ELEGANT QUAIL
*Callipepla douglasii**   Plate 16
Codorniz Elegante

ID: 9–10 in (23–25.5 cm). *Endemic to NW Mexico. Adult has straight crest,* longer in ♂. **Descr.** Ages/sexes differ. Bill dark grey, legs grey. ♂: face and throat whitish, streaked and spotted black, with *tawny crest*; blue-grey hindneck mottled chestnut. *Upperparts grey to blue-grey, scapulars and tertials mottled chestnut and edged whitish,* wing coverts and remiges greyish brown. Underparts grey to blue-grey, *lower chest, belly, and flanks boldly spotted white,* flanks and sometimes sides of chest mottled rufous. ♀: *browner overall, crest dark brown,* white throat has finer dark spotting. Upperparts grey-brown, scapulars and tertials cryptically patterned blackish, buff, and rufous-brown, wing coverts edged and vermiculated buff; tail vermiculated pale grey-brown. Underparts washed brown, whitish spotting usually extends onto upper chest. **Juv:** resembles ♀ but whitish throat unspotted, back streaked whitish. Underparts whitish, chest mottled grey-brown and streaked white, flanks barred dark brown.

Voice. Territorial call a hollow, ringing *whoi!* or *hoik!,* repeated. Also sharp, nasal clucking, *keh teh-wek keh teh-wek …* or *whirr ki-dik …,* which may be given in short, rollicking series; an emphatic *hui'ka hui'ka,* and an excited *huit huit …* when nervous.

Habs. Arid to semihumid brushy woodland and scrub, overgrown fields. Habits much as Scaled Quail. Eggs (captivity): 4–11, whitish, speckled with pale browns (WFVZ).

SS: Gambel's Quail prefers more arid desert habitat, has recurved black crest, lacks bold spotting below. ♂ has black-bordered face and throat.

SD/RA: C to F resident (SL–1000 m) on Pacific Slope from cen Son and SW Chih to N Jal.

GAMBEL'S QUAIL
*Callipepla gambelii**   Not illustrated
Codorniz de Gambel

ID: 9.5–10.5 in (24–26.5 cm). *NW Mexican deserts. Adults have blob-tipped recurved crest,* longer in ♂. **Descr.** Ages/sexes differ. Bill blackish, legs grey. ♂: *white-bordered rufous crown and white-bordered black face and throat striking; forehead* and crest *black.* Hindneck, chest, and upperparts blue-grey, hindneck with slight dark scaling. Whitish edges to inner tertial webs form contrasting stripes. *Belly buff with blackish central patch, chestnut flanks coarsely streaked white;* undertail coverts buff with coarse dark streaks. ♀: *browner overall with plain greyish face and throat,* crown brownish. Belly lacks dark central patch but is finely streaked dark brown. **Juv:** resembles ♀ but upperparts cryptically patterned with pale grey-brown, white, and blackish, lacks boldly patterned flanks.

Voice. A loud, crowing *ka-kya, ka kah-ka,* and *ka ka-ka,* etc., accent usually on

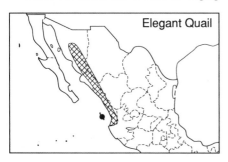

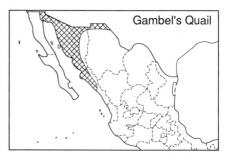

2nd syllable; a single crowing *ka-owh* or *ka-aah*; also sharp clucking and more liquid chattering calls, etc.

Habs. Arid desert scrub, brushy riparian woodland. Habits much as Scaled Quail. Eggs: 9–12, creamy to pale buff, spotted with browns (WFVZ).

SS: California Quail (allopatric?) told by heavy black scaling on belly, brown flanks, fine white speckling and dense blackish scaling on hindneck, ♂ has tawny-brown crown. See Scaled Quail, Northern Bobwhite.

SD: C to F resident (SL–1500 m) from NE Baja to cen Sin and N Chih.

RA: SW USA and NW Mexico.

CALIFORNIA QUAIL

*Callipepla californica**   Not illustrated
Codorniz Californiana

ID: 9–10 in (23–25.5 cm). *Baja only. Adult has blob-tipped, recurved crest, longer in* ♂. **Descr.** Ages/sexes differ. Bill blackish, legs grey. ♂: *white-bordered tawny-brown crown and white-bordered black face and throat striking; forehead pale,* crest black. Hindneck, chest, flanks, and upperparts grey-brown, *hindneck densely speckled white and scaled black,* flanks streaked white; tail grey. Whitish edges to inner tertial webs form contrasting stripes. *Belly buff to whitish with heavy black scaling* and dark rufous central patch, undertail coverts buff with coarse dark streaks. ♀: *browner overall with greyish face and throat,* crown brownish, belly lacks dark central patch. **Juv:** resembles ♀ but upperparts cryptically patterned with pale grey-brown, white, and blackish, lacks boldly patterned flanks.
Voice. Much like Gambel's Quail. A loud, crowing *ka-kwa* and *ka ka-kwah*, etc., accented on 2nd syllable, a single crowing

k-yowh, and sharp, clucking and liquid spluttering calls.

Habs. Arid to semiarid brush and scrub, chaparral, riparian woodland, brushy understory of open pine forest. Habits much as Scaled Quail. Eggs: 6–17, creamy to pale buff, spotted with browns (WFVZ).

SS: See Gambel's Quail. Mountain Quail locally sympatric in N Baja, adult has long straight crest, white-bordered chestnut face and throat, chestnut flanks boldly barred white.

SD: C resident (SL–2800 m) throughout Baja except extreme NE where replaced by Gambel's Quail.

RA: W N.A. to NW Mexico.

MOUNTAIN QUAIL

Callipepla picta confinis   Not illustrated
Codorniz de Montaña

ID: 9.5–10.5 in (24–26.5 cm). *Mountains of NW BCN. Adult has striking, very long straight crest.* **Descr.** Ages differ, sexes similar. **Adult:** bill black, legs dusky flesh. *Face and throat chestnut, bordered white at sides,* crest black; rest of head, chest, and mantle blue-grey. Upperparts olive-brown. Whitish edges to inner tertial webs form contrasting stripes. Belly whitish, *flanks chestnut with broad, black-bordered white bars,* undertail coverts blackish. **Juv:** grey-brown overall, lacks bold face and throat markings, upperparts cryptically patterned with grey-brown, buff, and blackish, lacks boldly patterned flanks.
Voice. Territorial call a far-carrying, loud, ringing *queeoh!* or *kee-oh,* often repeated steadily; also a squealing twitter when flushed.
Habs. Brushy understory of semiarid oak and pine–oak woodland, chaparral. Habits much as Scaled Quail. Eggs: 6–10,

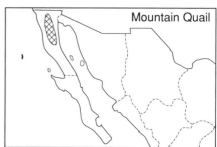

California Quail

Mountain Quail

white to pale pinkish, unmarked (WFVZ).
SS: See California Quail.
SD: F to C resident (1000–2800 m) in Sierra Juarez and Sierra San Pedro Martír, BCN.

RA: W N.A. to N Baja.
NB: Traditionally placed in the genus *Oreortyx*.

RAILS, GALLINULES, AND ALLIES: RALLIDAE (15)

This cosmopolitan family of very small to medium-sized marsh-haunting birds may be divided into rails, and gallinules and allies.

Rails are usually skulking and secretive, rarely seen except by luck, patience, or getting more than one's feet wet! Adaptations for living in marshes are long legs and toes for wading and walking on floating vegetation, and laterally compressed bodies for running quickly through reeds and grasses. Bills either fairly short or long; short-billed rails are known as crakes. Wings short and rounded, short tails often held cocked. Ages differ, sexes usually similar though ♂♂ larger. Young precocial, downy chicks of all species are black; juv plumage attained before birds are fully grown and kept up to a few months in some species. Plumage typically colored in greys and browns, often with black and white barring or spotting.

Gallinules and allies are more conspicuous than rails and spend much time swimming. Bills extend into horny shields on the forehead, and bills and legs are often brightly colored. Ages differ, sexes similar. Adults overall blackish, juvs paler.

Food mostly vegetable matter and aquatic insects but larger species also take small fish, frogs, tadpoles, and snails. Nests are platforms or cups of reeds and grasses placed in marshy vegetation; some crakes build domed nests. Eggs typically light-colored and spotted with browns.

YELLOW RAIL
Coturnicops noveboracensis goldmani
Polluela Amarilla   Not illustrated

ID: 6–6.5 in (15–16.5 cm). *Río Lerma marshes, Mexico*; extirpated? A small plump rail with *short and fairly stubby bill* (0.4× head). **Descr.** Ages differ, sexes similar. **Adult:** eyes brown, bill yellowish, legs olive-grey. *Buff supercilium, auriculars, and sides of neck contrast with dark brown crown and broad dark smudge through eye. Chest warm buff*, indistinctly scalloped dusky, contrasting slightly with white belly. Flanks dark brown, barred white, undertail coverts dull cinnamon. *Upperparts dark brown with broad buff striping and narrow white cross-barring*, wing coverts spotted white. *In flight, note white patch on inner secondaries*; remiges grey-brown. **Juv:** bill dusky. Darker overall, upperparts less contrastingly marked, crown and nape finely spotted white; juv plumage retained for a few months.

Voice. In N.A., a hard ticking *tick-tick, tick-tick-tick* or *kik-kik* . . . , often in series of 2 and 3, mainly given at night, sounds like two pebbles being knocked together.

Habs. Sedge marshes in ungrazed areas with vegetation < 0.5 m tall (Hardy and Dickerman 1965). Skulking and hard to see unless flushed by chance. Nest (N.A.) built on a tussock of grass near ground. Eggs (N.A.): 8–10, creamy buff, speckled with rusty and greys (WFVZ).

SS: Juv Sora (plumage kept till early winter) twice as large with larger and stouter bright yellow bill, upperparts lack buff striping, lacks white wing patch in flight.

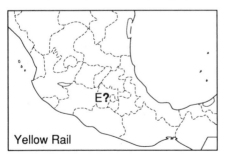

Yellow Rail

SD: Formerly F but local resident (2500 m) in upper Río Lerma Valley, Mex. Present status unknown, last recorded 1964 (RWD; Hardy and Dickerman 1965) but most of the area has since been drained. Could occur elsewhere in Río Lerma drainage W to Lake Chapala, and should be looked for.
RA: N.A. and cen Mexico; N populations winter S to S USA.

RUDDY CRAKE
Laterallus ruber   Plate 17
Polluela Rojiza

ID: 5.5–6 in (14–15.5 cm). The common crake of S lowlands. Bill straight, fairly stout (0.7× head). **Descr.** Ages differ, sexes differ slightly. ♂: eyes red, bill black, *legs slaty olive-grey. Slaty blue-grey head* contrasts with rufous throat and neck, median throat often whitish. Back rufous, rest of upperparts dark grey-brown. *Underparts cinnamon-rufous, slightly paler than upperparts, becoming rufous on flanks and undertail coverts.* ♀: rump and uppertail coverts rufous-brown. **Juv:** eyes dusky, bill dusky horn. Dark slaty grey overall, blacker above, throat and belly often whitish; quickly attains adult-like plumage.
Voice. Common call, given all year, a rising and falling, churring or purring trill, typically preceded by quiet, piping notes audible at close range, *whiit whiit ... urrrrr-rrrr ...*, persistent series may last 30 s or more. Also a hard, clipped ticking *chk* or *tek*, often doubled *chk-chk*, at times followed by quiet low rasps, *chk-ehrr chk ...*, etc.
Habs. Varied freshwater habitats from reedbeds to damp grassy fields and roadside ditches. Best detected by voice but feeds in open situations at marsh edges or on floating mats of vegetation, mainly early and late in the day. Difficult to flush;

flies strongly, often with legs dangling. Nest is a domed structure of grasses near the ground. Eggs: 4–6, whitish to cream, marked with rusty and greys (WFVZ).
SS: Larger Uniform Crake mostly found in damp forest and swamps, lacks grey hood, legs bright red. Grey-breasted Crake from above (e.g. flushed) less ruddy on back, grey chest and black-and-white-barred flanks distinctive. In E Honduras, see White-throated Crake (Appendix E).
SD/RA: C to F resident (SL–1500 m) on both slopes from S Tamps and Col (SNGH tape) to N Nicaragua (Mosquitia) and El Salvador; locally in interior from Isthmus S. Possibly in NW Costa Rica.

GREY-BREASTED CRAKE
Laterallus exilis   Plate 17
Polluela Pechigris

ID: 5.5–6 in (14–15.5 cm). Local in N C.A. Bill 0.5–0.7× head. **Descr.** Ages differ, sexes similar. **Adult:** eyes red, *bill blackish with bright lime-green base below, legs yellowish. Head and chest grey*, throat whitish, *hindcrown and nape rufous*, rest of upperparts dark olive-brown, some birds have white bars on wing coverts. Remiges grey-brown. Tail blackish. *Belly whitish, flanks and undertail coverts boldly barred black and white.* **Juv:** eyes dusky, bill duller. Sooty black overall with whitish throat and median chest; quickly attains adult-like plumage.
Voice. An overall descending, fairly rapid series of 2–8 (usually 5–8) piping notes, suggests a Great-tailed Grackle or, when close, a frog; carries well; at close range, a soft introductory note audible, *tik dee-deedeedeedeedee*, or *tk duiduiduiduidui, duiduiduiduiduiduiduiduidui*, etc. When repeated steadily, 3–4 s between series. Less often a churring rattle suggesting Ruddy Crake but lower and drier (and usually

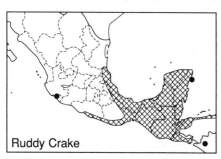
Ruddy Crake

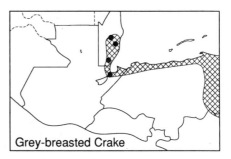
Grey-breasted Crake

shorter?); a low, gruff *chk*, and other soft chuks and churs.

Habs. Wet grassy swales and marshy pastures; habits, nest, and eggs similar to Ruddy Crake.

SS: See White-throated (Appendix E) and Ruddy crakes. Juv Black Rail has upperparts spotted white, underparts less extensively barred.

SD: F to C but local resident (around SL) in Belize and NE Guatemala (Howell and Webb 1992*b*); also E Honduras; should be looked for elsewhere.

RA: C.A. to Ecuador and Brazil.

BLACK RAIL
*Laterallus jamaicensis** Not illustrated
Polluela Negra

ID: 5–5.5 in (12.5–14 cm). Local and secretive, status poorly known. Bill straight, 0.5× head. **Descr.** Ages differ, sexes similar. **Adult:** eyes red, *bill black*, legs dusky flesh (becoming orange-red in breeding season?). *Head, chest, and upper belly dark slaty grey*, rest of underparts darker with narrow white barring. Nape and upper mantle chestnut-brown, rest of *upperparts blackish, extensively flecked white*. Remiges dusky grey, sparsely spotted white. **Juv:** eyes dusky. Head and upperparts have trace of brownish wash on nape, sparser white spotting. Throat and underparts pale grey becoming whitish on throat and belly, flanks spotted whitish.

Voice. One or 2 slightly hard, piping notes followed by a gruff, rolled note, often given at night: *hii'kii-durr* or *kee-kee-durr*, or simply *kii-kurr*; carries well; when repeated steadily, given every 4–8 s (Belize). Also a gravelly churring, often prolonged, *gurrrr-rrrr . . .*, and a quiet, low *kuk* or *cwuc*.

Habs. Grassy freshwater marshes and seasonally dry savanna, also coastal salt marshes in BCN. Secretive and easily overlooked. Flies strongly when flushed, legs quickly tucked in, shape compact and Sora-like. Nest similar to Ruddy Crake. Eggs (N.A.): 6–8, whitish, speckled with browns and greys (WFVZ).

SS: See Grey-breasted Crake.

SD: Formerly F but local breeder (SL–1000 m), presumed resident, in NW BCN; mostly extirpated due to habitat loss, and only one recent record (Erickson *et al.* 1992). Local (F to C?) resident in Belize. May breed along Colorado River in BCN and Son. Also recorded (resident?) in SE SLP, N Ver, and Guatemala (extirpated?); a report from NW Honduras (Monroe 1968) requires verification.

RA: Breeds N.A., locally C.A. and W S.A; N.A. populations winter in S USA, Caribbean, and possibly E Mexico and N C.A.

CLAPPER RAIL
*Rallus longirostris** Plate 17
Rascón Picudo

ID: 12–15 in (30.5–38 cm). *Large rail of salt marshes. Bill long (1.2–1.5× head) and straight or slightly decurved toward tip.* Three groups: *obsoletus* group (NW Mexico; including *beldingi*, shown), *pallidus* group (Yuc Pen), and *saturatus* (NE Mexico). **Descr.** Ages differ, sexes similar. **Adult:** eyes reddish, bill blackish above, orange to red below, tipped dusky, *legs brownish flesh to orangish. Obsoletus:* head, hindneck, and upperparts olive-brown to grey-brown, face paler with dusky lores and pale supraloral stripe, throat whitish; upperparts streaked and mottled dark brown, upperwing coverts rufous-brown. *Foreneck and chest cinnamon-rufous, belly white, flanks and undertail coverts barred dark brown and white.* Remiges dark grey-brown. *Pallidus:* paler overall, upperparts greyer, *foreneck and*

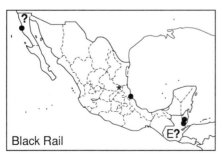

Black Rail

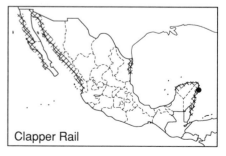

Clapper Rail

chest pale pinkish buff to cinnamon, white flank bars wider. *Saturatus:* similar to *pallidus* group but darker overall, especially upperparts, *sides of neck washed grey.* **Juv:** eyes brownish. Upperparts darker than adult, more blackish; underparts mottled grey and whitish, flank bars absent or indistinct. Juv plumage mostly attained when about half-grown, adult-like plumage attained within a few months.

Voice. Much like King Rail but often drier and less pumping, a dry, laughing *kah-kah-kah kah-kah* ... or *cah-cah-cah-cah* ...; suggests Least Bittern; also a single sharp *kek!* repeated.

Habs. Salt-water and brackish marshes with *Spartina* grass, mangroves. Often seen feeding at marsh edges or in open channels; runs quickly but, like most rails, flushes if surprised and flies a short distance, legs dangling. Nest (N.A.) is a platform of grasses, usually near ground. Eggs: 6–14, buff, flecked and spotted with browns (WFVZ).

SS: King Rail (allopatric?) favors freshwater marshes, black and white flank bars bolder than *obsoletus* group, neck and chest brighter than *saturatus.* Virginia Rail is a third the size of Clapper Rail, darker overall with blue-grey auriculars, bright rufous upperwing coverts.

SD: C to F resident coastally in Baja, from Son to Nay, and in N Tamps and N and E Yuc Pen.

RA: N.A. through M.A. and Caribbean to Ecuador and S Brazil.

KING RAIL

Rallus longirostris* (in part) or *R. elegans**
Rascón Real   Plate 17

ID: 13–16.5 in (33–42 cm). Large rail of freshwater marshes. *Bill long (1.2–1.5×
head) and straight or slightly decurved toward tip.* **Descr.** Ages differ, sexes similar.

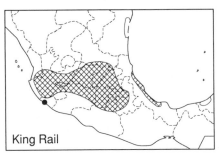

King Rail

Adult: eyes reddish, bill blackish above, orange to red below, tipped dusky, *legs brownish flesh.* Head, hindneck, and upperparts olive-brown, face paler with dusky lores and pale supraloral stripe, throat whitish; upperparts streaked and mottled dark brown, upperwing coverts rufous-brown. *Foreneck and chest cinnamon-rufous, belly white, flanks and undertail coverts barred dark brown and white.* Remiges dark grey-brown. *R. e. elegans* (E Mexico): smaller, *overall richer in color, flank bars bolder.* **Juv:** eyes brownish. Upperparts darker than adult, more blackish; underparts mottled grey and whitish, flank bars absent or indistinct. Juv plumage mostly attained when about half-grown, adult-like plumage attained within a few months.

Voice. A gruff *ki ka-ka-ka ka-ka-ka* ... with a rhythmic or pumping quality; also a sharp, gruff *keh-keh-kehrr* and *keh-kerrr,* a single sharp *kek!* which may be repeated steadily, and *kek-kek-kek-kek* or *kehk* ... in flight when flushed, suggesting Least Bittern.

Habs. Freshwater marshes with tall reeds and cattails. Habits, nest, and eggs like Clapper Rail.

SS: Virginia Rail a third the size (or smaller) than King Rail, darker overall with blue-grey auriculars, bright rufous upperwing coverts. See Clapper Rail.

SD: C to F resident (1200–2500 m) in cen Mexico from S Nay and Jal to S SLP and Pue *(tenuirostris);* and (near SL) on Pacific Slope in Col (SNGH), on Atlantic Slope in Ver (Dickerman 1971; SNGH, SW).

RA: E N.A., disjunctly in cen and E Mexico.

NB: *Tenuirostris* sometimes considered specifically distinct, Mexican Rail.

VIRGINIA RAIL

Rallus limicola*   Plate 17
Rascón de Virginia

ID: 8.5–9.5 in (21.5–24 cm). Medium-sized rail, *bill long* (1.2–1.5× head) and straight or slightly decurved. **Descr.** Ages differ, sexes similar. **Adult:** eyes red, *bill orange-red with blackish culmen and tip, legs brownish to orange-red.* Blue-grey *face contrasts with blackish crown and cinnamon-rufous sides of throat,* supraloral stripe and median throat whitish to pale cinnamon. Hindneck and upperparts

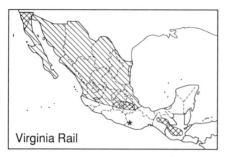

Virginia Rail

tawny-brown, streaked blackish, *rufous upperwing coverts obvious in flight*. Foreneck, chest, and upper belly cinnamon-rufous, lower belly, flanks, and undertail coverts barred black and white. Remiges dark grey-brown. **Juv**: eyes grey-brown, bill duller. Head and upperparts blacker overall, underparts blackish except whitish throat and belly, flanks slightly barred whitish.

Voice. A pumping or chortling *rreh-rreh-rreh...* or *nyuh-nyuh ...*, etc.; short, sharp calls, *kriik* and *riik* or *reek*, etc., rougher and screechier than Sora calls.

Habs. Freshwater marshes, especially with reeds and rushes, less often in brackish marshes, mainly those with *Spartina* grass. Typically furtive and hard to observe, flushes at close range with legs dangling and quickly drops back to cover. Nest platform of grasses, sometimes domed, built in low marshy vegetation. Eggs: 5–12, whitish to pale buff, spotted rusty (WFVZ).

SS: See much larger King and Clapper rails. Juv Spotted Rail larger, upperparts spotted whitish.

SD: F to U but local resident (near SL–2500 m) in NW BCN (at least formerly), and from cen volcanic belt (Mex to Pue, at least) to Guatemala. F to U winter visitor (late Aug–Apr) S to cen Mexico.

RA: Breeds N.A. to S Mexico, also N S.A.

SPOTTED RAIL

Rallus maculatus insolitus   Plate 17
Rascón Pinto

ID: 10–11 in (25.5–28 cm). A striking, fairly large rail; *bill long and straight, fairly stout* (1.3× head). *Juv dimorphic.* **Descr.** Ages differ, sexes similar. **Adult:** eyes red, *bill bright yellow-green with orange-red spot at base below, legs bright pinkish red.* Slaty

blackish overall; upperparts broadly edged olive-brown, head and *upperparts boldly spotted white*, scapulars also streaked white. *Underparts boldly barred white*, undertail coverts white. Remiges dark grey-brown. **Juv**: eyes brownish, bill and legs duller. **Dark morph**: *dark brown overall*, sootier below with whitish throat, sparse white flecks on back and underparts. **Barred morph** (unrecorded in region): resembles adult but browner overall, white spotting and barring less contrasting, tinged brown.

Voice. A loud, rough, rasping, slightly emphatic screech, which may be preceded by a short grunt, *kr'krreih kr'krreih ...* or simply *grrr* or *kehr*, repeated; and a deep, gruff pumping series, accelerating toward the end, *wuh-wuh ...* or *um-um ...*, may suggest a motor starting up (Reynard 1981, tape). Also a sharp *gek*, often repeated, when disturbed (Stiles and Skutch 1989).

Habs. Freshwater marshes with reeds, rushes, tangled second growth, flooded fields, overgrown wet ditches. Often seems unconcerned by human presence, feeds in fairly open situations at any time of day. Nest is a cupped platform of grasses, low in marshy vegetation. Eggs: 3–7, brownish buff, spotted brown (Ripley 1977).

SS: See juv Virginia Rail.

SD: Locally/seasonally F to U resident (SL–1200 m) on both slopes from Nay and Ver to Oax, Yuc Pen, and Belize; locally (U to F?) in interior from Mich to Chis, and on Pacific Slope in El Salvador (Thurber *et al.* 1987). Almost certainly more widespread; may be increasing or has been overlooked in past.

RA: Mexico and Caribbean to Peru and Argentina.

NB: Often placed in the genus *Pardirallus*.

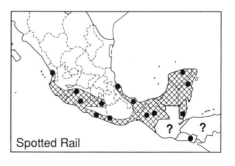

Spotted Rail

GREY-NECKED WOOD-RAIL

Aramides cajanea*   Plate 17
Rascón Cuelligris

ID: 15–17 in (38–43 cm). *Chicken-sized rail of SE forests and swamps. Bill long, stout, and straight,* 1.2× head, legs long and sturdy. **Descr.** Ages differ slightly, sexes similar. **Adult:** eyes red, bill yellow becoming green distally and often orange at base, legs bright pinkish red. *Head and neck blue-grey* with rufous nape patch and whitish throat. Olive upperparts become black on rump, uppertail coverts, and tail. *Wings bright rufous, striking in flight,* underwing coverts barred black. *Chest and sides cinnamon-rufous,* often with a whitish band between rufous chest and *black belly* (especially *albiventris* of Yuc Pen); *flanks and undertail coverts black,* thighs smoky grey. **Juv:** eyes brownish, bill and legs duller. Belly grey, mottled pinkish buff.

Voice. Song mostly at dusk and dawn, a varied, rapid, crazed-sounding, rollicking, popping, and clucking series, *coo-coo-coo-coo' ki-ki-kik' cococo,* or *koo koo koo koo whiwho wee-wee-oo,* or a shorter *koo koo koo kri-krik,* etc. Calls include a sharp, clucking shrieking when disturbed, and low grunting clucks, *puk* or *kuk,* etc. **Habs.** Swampy woodland, forest streams, pools and marshes near wooded thickets, locally at mangrove edge. Often quite conspicuous, walks with short tail cocked and frequently twitched. Elusive when calling. Nest platform of grasses and twigs at low to mid-levels in bush or low tree, usually over water. Eggs: 3–5, whitish to buffy white, spotted with browns and greys (WFVZ).

SS: Rufous-necked Wood-Rail (mangroves) smaller with shorter bill, adult has bright rufous head, neck, and underparts; juv grey overall, becoming rufous first on neck. See Slaty-breasted Tinamou.

SD: C to F resident (SL–500 m) on Atlantic Slope from S Tamps to Honduras; on Pacific Slope from cen Oax to W El Salvador.
RA: E Mexico to N Argentina.

RUFOUS-NECKED WOOD-RAIL

Aramides axillaris   Plate 17
Rascón Cuellirrufo

ID: 11.5–12.5 in (29–31.5 cm). *Fairly large, mangrove-haunting rail of both coasts. Bill fairly long* (1× head), straight and stout. **Descr.** Ages differ, sexes similar. **Adult:** eyes red, bill greenish yellow, legs bright pinkish red. *Head, neck, chest, and upper belly bright rufous,* except whitish throat and blue-grey patch on lower hindneck and upper mantle. Rest of upperparts olive becoming black on rump, uppertail coverts, and tail. *Remiges rufous (obvious in flight),* underwing coverts barred black and white. Hind flanks and vent smoky grey, *undertail coverts black.* **Juv:** eyes brownish, legs dull reddish. *Rufous of head, neck, and underparts replaced by blue-grey,* crown and hind neck darker, washed olive. Attains adult plumage over a few months.

Voice. An incisive, loud series given at dawn and dusk, *pik-pik-pik* or *pyok-pyok-pyok,* repeated about 8× and often antiphonal, the whole series reminiscent of the style of Grey-necked Wood-Rail; also a gruff *kik* in alarm (ffrench 1973). **Habs.** Mangroves. Usually skulking and elusive but may be seen walking among mangrove roots, feeding in channels, on tidal flats, etc. Nest is a twig platform, lined with weeds, in bush or small tree over water. Eggs (Trinidad): 5, whitish to buffy white, spotted with browns and greys (WFVZ).

SS: See Grey-necked Wood-Rail. Uniform Crake smaller with shorter bill and legs,

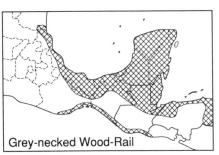

Grey-necked Wood-Rail

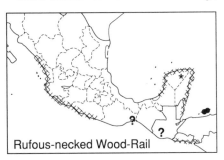

Rufous-necked Wood-Rail

rufous brown overall, rarely in mangroves.

SD: F to C (but local?) resident coastally from Sin to Gro, El Salvador to Honduras, from Camp to Belize, and on Guanaja and Roatan (Howell and Webb 1992*b*) Islands Honduras. Probably occurs elsewhere but easily overlooked.

RA: Mexico to Ecuador and Surinam.

UNIFORM CRAKE

Amaurolimnas concolor   Plate 17
Polluela Café

ID: 8–9 in (20.5–23 cm). *A plump, medium-sized rail of lowland forest and swamps. Bill stout and straight,* 0.7× head. **Descr.** Ages differ, sexes similar. **Adult:** eyes red, bill greenish yellow (rarely dark above), *legs bright pinkish red. Crown, nape, and upperparts dull rufous-brown,* greyer on crown, brighter rufous on scapulars and upperwing coverts. *Face, foreneck, and underparts rufous,* throat whitish. Remiges grey-brown. **Juv:** eyes brownish, legs duller. Slightly darker and richer above than adult with face, foreneck, and underparts dusky grey to grey-brown.

Voice. A series of up to 20 or more mellow, rich to slightly plaintive, disyllabic (the 2nd part up-slurred) whistles that may suggest Barred Woodcreeper: initially very soft, becoming loud, and then fading fairly abruptly, *tu-ih too-ih too-ih too-ih too-IH TOO-IH TOO-IH TOO-IH TOO-IH too-ih too-ih,* or *hu-wiih huu-wiih huu-wiih HU-WIIH . . .,* etc.; may be given at night, when, at a distance, only the loudest 1–5 whistles are audible. Also shorter soft series, at times soft single whistles, *tuu-ih,* etc., and, when disturbed, a sharp *plik!* or *pleek.*

Habs. Marshy and damp floor of forest and second growth thickets, locally in mangrove edge. Typically skulking and elusive, walks and runs with fairly upright

stance. Nest (Costa Rica) is a leafy cup in a hollowed stump. Eggs: 4, pale buff, marked with browns and greys (WFVZ).

SS: See Ruddy Crake, Rufous-necked Wood-Rail, Little Tinamou.

SD: F to U but local (or more widespread but overlooked) resident (SL–150 m) on Atlantic Slope from S Ver to Honduras. Vagrant (?) to Pacific coast in Chis (Alvarez del Toro 1954).

RA: SE Mexico to Peru and Brazil.

SORA

Porzana carolina   Plate 17
Polluela Sora

ID: 8–8.5 in (20.5–21.5 cm). Widespread winter visitor. *Bill stout and straight,* 0.5–0.7× head. **Descr.** Ages differ, sexes similar. **Adult:** eyes reddish, bill orange-yellow, *legs yellow-green to olive. Black lores and bib* contrast *with blue-grey face.* Crown and nape brown with black median stripe. Hindneck and upperparts olive-brown, streaked and spotted black, scapulars and upperwing coverts also streaked and spotted white. Chest and sides of *neck blue-grey,* chest flecked white. *Flanks barred black and white,* belly white, *undertail coverts cream to rich buff.* Remiges grey-brown. **Juv:** eyes brown, bill greenish yellow. Lacks black bib, *throat whitish. Face, chest, and sides buffy brown,* upperparts with less black, flank bars less distinct. Attains adult-like plumage by December.

Voice. Plaintive to slightly yelping calls, *kee-ur* or *keuh-eh,* and a bright, sharp *keek!* or *kiip* which may run into a piping, descending whinny, *keek, dee-dee-dee . . .,* etc.

Habs. Freshwater and brackish marshes, especially with reeds, flooded pastures, mangroves. Less skulking than most crakes, often feeds in open situations at marsh edge, in channels, ditches, etc.

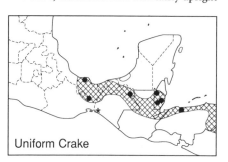

Uniform Crake

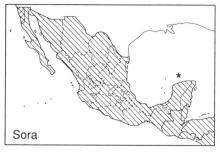

Sora

Flushes readily and usually the commonest rail underfoot (in winter) when wading in marshes. Flies well with legs tucked in, overall flight shape compact. Nest platform of grasses built up out of water or lodged in vegetation. Eggs: 6–12, buff to olive-buff, spotted with browns and greys (WFVZ).

SS: Juv superficially suggests Yellow-breasted Crake but much larger with stout paler bill, olive legs. See Yellow Rail.

SD: Local breeder (at least formerly) in BCN. C to F transient and winter visitor (mid-Aug–May, SL–2500 m) almost throughout region, but more common as a transient on N Plateau.

RA: Breeds N.A., winters from S USA to N S.A.

YELLOW-BREASTED CRAKE
Porzana flaviventer woodi   Plate 17
Polluela Pechiamarilla

ID: 5–5.5 in (12.5–14 cm). A *tiny*, little-known rail of freshwater marshes. *Bill straight and fairly heavy*, 0.7×, head. **Descr.** Ages differ, sexes similar. **Adult:** eyes reddish, *bill blackish, legs straw yellow. Narrow whitish supercilium* (often broken over eye) *contrasts with blackish crown, nape, and eyestripe.* Auriculars, *sides of neck, and chest rich buff,* auriculars washed grey, throat whitish; belly white, *flanks and undertail coverts barred black and white.* Upperparts rich dark brown, streaked and spotted white, tertials and scapulars broadly edged rich buff forming conspicuous stripes, obvious on flushed birds. Remiges dusky grey. **Juv:** neck and chest barred dusky.

Voice. A low, gruff, rolled or churring *k'kuk kuh-kurr* or *k'turr kurr-kurr*, a plaintive, squealing *kreeihr*, a Sora-like *reehr* or *krreh*, and longer series, *kree, kreee-kreeh*, etc. (Reynard 1981, tape).

Habs. Grassy freshwater marshes, flooded fields, floating aquatic vegetation such as water hyacinth. May be seen feeding at marsh edge early and late in the day but generally skulking and hard to see. Flushes up from under foot and flies weakly a short distance, legs dangling conspicuously. Nest (West Indies) loosely built in a water plant. Eggs: 3–5, creamy, lightly spotted with brown or lavender (Ripley 1977).

SS: See Sora.

SD: F to U but local resident (SL–1000 m) on Atlantic Slope from Ver to Yuc Pen and on Pacific Slope from Gro to El Salvador, locally in interior Chiapas, and undoubtedly elsewhere. The source of a report from Mich (AOU 1983) is unclear (BLM); reports from Belize (Wood *et al.* 1986) require verification.

RA: Mexico and Caribbean to N Argentina.

PURPLE GALLINULE
Porphyrula martinica   Not illustrated
Gallineta Morada

ID: 12–13 in (30.5–33 cm). A colorful marsh inhabitant. **Descr.** Ages differ, sexes similar. **Adult:** eyes red, bill red with yellow tip, *forehead shield pale blue, legs orange-yellow.* Head, neck, and underparts glossy bluish violet, becoming black on vent and thighs; *white undertail coverts form solid triangle. Upperparts glossy green, upperwings glossy turquoise-blue.* **Juv:** eyes brownish, bill duller with smaller, greyish shield, legs olive-yellow. *Face, sides of neck, chest, and flanks buff,* throat and belly whitish, *undertail coverts as adult.* Crown, hindneck, and upperparts brown, *body with olive-green sheen, wings with greenish-turquoise sheen.* Attains adult-like plumage through 1st winter.

Voice. Varied, sharp, shrill calls and gruff chatters, a sharp *kr-lik'* or *kee-k'*, a gruff

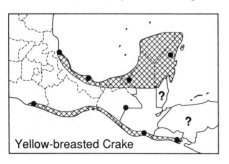
Yellow-breasted Crake

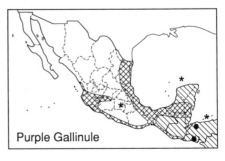
Purple Gallinule

kruk kruk kruk kruk, a wailing scream *whiehrrr* or *w'heehrr*, and a rapid, clucking series, accelerating then fading, when disturbed, *kahw cohw-cohw* ... or *keh-keh-keh* ..., etc.

Habs. Freshwater marshes with reedbeds and large areas of floating vegetation (especially water hyacinth), flooded fields. Usually near cover where it retreats quickly if disturbed. Climbs readily in marshy vegetation, at times fairly high in bushes. If flushed, flies a short distance, legs dangling. Nest is a cupped platform of grasses lodged in marshy vegetation. Eggs: 4–8, creamy to pale buff, flecked with rusty and greys (WFVZ).

SS: Common Moorhen has white stripe on flanks, central undertail coverts black, adult blackish overall with red shield, juv dusky grey without buff tones or blue-green sheen. American Coot blackish overall with central undertail coverts black, white bill and shield.

SD: F to U but local resident (SL–600 m, locally to 1600 m) on both slopes from Nay and Tamps to El Salvador and Honduras. More widespread in winter (Oct–Mar) when N migrants present; also wanders in summer, for example recorded DF (Wilson and Ceballos-L. 1986). Reports N to Son (Freidmann *et al.* 1950) appear to be erroneous.

RA: SE USA and Mexico to N Chile and N Argentina; US populations mainly migratory, wintering S to S.A.

COMMON MOORHEN

Gallinula chloropus cachinnans
Gallineta Común   Not illustrated

ID: 13–14 in (33–35.5 cm). Common and widespread. **Descr.** Ages differ, sexes similar. **Adult:** eyes red, *bill and shield red*, bill tipped yellow, *legs greenish yellow*, tibia red. Head, neck, and underparts dark slaty

grey, head and neck blacker, belly mottled whitish. *White-tipped flanks form conspicuous stripe on sides, undertail coverts white with broad black median stripe.* Upperparts olive-brown, tail blackish. **Juv:** eyes brownish, bill duller, legs yellowolive. Face, throat, and foreneck whitish, underparts pale grey, mottled whitish, belly white.

Voice. Varied sharp clucks and grunts, a sharp *keh! keh!* ... or *kep* ..., a clipped *kreik!*, and a grunting clucking *kruh-kruh-kruh* ..., etc.

Habs. Freshwater marshes and ponds with reeds, rushes, and floating vegetation, flooded fields, drainage ditches. Habits similar to Purple Gallinule but less skulking and often in loose flocks, locally to a few hundred birds. Nest is a cupped platform of grasses over water. Eggs: 4–10, buff to olive-buff, marked with browns (WFVZ).

SS: See Purple Gallinule. American Coot chunkier, more often seen swimming, lacks white flank stripe, bill and shield white.

SD: C to F resident (SL–2500 m) S to cen Mexico; F to U but local S to El Salvador and Honduras. Commoner and more widespread in winter (Oct–Mar). F resident on Clipperton where colonized since 1968.

RA: Cosmopolitan except Australia; in New World breeds from E and SW USA to N Chile and N Argentina.

NB: Formerly known in N.A. as Common Gallinule.

AMERICAN COOT

Fulica a. americana
Gallareta Americana   Not illustrated

ID: 14–16 in (35.5–40. 5 cm). Common and widespread black waterbird, toes have lobed webbing. **Descr.** Ages differ, sexes similar. **Adult:** eyes red, *bill and shield*

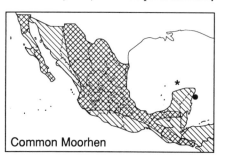

Common Moorhen

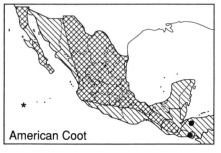

American Coot

white with broken reddish bar distally on bill and dark red oval swelling at top of shield. *Dark slaty grey overall, head and neck blacker, undertail coverts white with broad black median stripe. White trailing edge to secondaries visible in flight.* **Juv**: eyes brownish, bill greyish horn, shield develops through 1st winter. Face and sides of neck whitish, underparts pale grey mottled whitish. Attains adult-like plumage through 1st winter.

Voice. Varied nasal clucks, slightly rough quacks, and gruff chatters: *ke-yik* and *k-rrk!* or *krek!*, a longer *k-yew-r*, and *wah wahk ...* or *kuk-kuk ...*, etc.

Habs. Lakes, reservoirs, coastal lagoons, freshwater marshes with open water; rarely salt water. Conspicuous and gregarious most of the year, may skulk when nesting. In winter, flocks of several thousand birds not unusual, often associates loosely with ducks. Feeds by up-ending and diving, also grazes on land. Runs, splashing across water, to become airborne. Nest platform of grasses built up out of water or anchored to vegetation, sometimes floating. Eggs: 6–12, buffy white to olive-buff with dark and grey speckling (WFVZ).

SS: See Common Moorhen and Purple Gallinule.

SD: F to C resident (SL–2500 m), mostly in interior, to Guatemala, locally to El Salvador and Honduras. C to F and widespread in winter (Oct–Mar), small numbers oversummer widely in winter range. Formerly bred on Clipperton.

RA: Breeds N.A. to NW Costa Rica; N birds migratory, wintering S to N Colombia.

FINFOOTS: HELIORNITHIDAE (1)

Finfoots (or sungrebes) form a pantropical family of little-known aquatic birds. The New World representative is a small, superficially grebe-like bird of forested areas. Bill fairly long, straight and pointed, neck fairly long, tail fairly long, broad, and rounded. Legs short, toes have lobed webbing. Ages/sexes differ slightly, young altricial; molts undescribed (?).

Feeds while swimming, eats small aquatic animals obtained at or near water surface and from vegetation along banks. Nest is a cupped platform of twigs in vegetation low over water. Eggs: 2–4, pale cinnamon to buffy white, flecked with browns and greys. Incubation short (10–11 days), naked young are carried for an unknown time in pockets of skin under male's wings (Alvarez del Toro 1971a).

SUNGREBE
Heliornis fulica Plate 18
Pájaro-cantil

ID: 10.5–11.5 in (26.5–29 cm). See family account. **Descr. ♂:** bill flesh to horn with dusky culmen and tip, feet pale yellow with black banding on toes. White forehead and *postocular stripe contrast with black crown and broad eyestripe* which merge into black hindneck stripe; *head and sides of neck white with black lateral neck stripe.* Upperparts dark olive-brown becoming rich brown on rump and uppertail coverts. Tail black with white tip obvious in flight. Underparts white overall, washed buff on chest, flanks and undertail coverts dark greyish. **♀:** bill pinkish with dark culmen and tip, becoming bright red prior to breeding. Auriculars tawny, black of head and neck glossier. **Juv:** black stripes on head and neck duller, mottled dusky.

Voice. A series of (usually 4) sharp barks, descending overall, *wehk! wehk! wehk! w'eh* or *k'wek! k'wek! wek wek*; also a single barking *kek!* or *wek!*, and a quiet, sharp *plik*

Habs. Slow-moving rivers, streams, and lakes, mainly in humid forest, typically with overhanging vegetation along the banks, mangroves. Singly or in pairs, swims close to cover, jerking neck like a Moorhen, tail held flat on the water. Rarely, if ever, dives, prefers to swim into cover. Flies short distances if startled on open water, flight low with long, broad tail conspicuous. Nesting Dec–Jan (N Belize), Apr–May (S Chis).

SD: F to U but local resident (SL–200 m) on Atlantic Slope from cen Ver to Honduras, on Pacific Slope of Chis (Alvarez del Toro 1971a) and Guatemala (Land 1970), U N to S Tamps (GWL).

RA: E Mexico to Ecuador and SE Brazil.

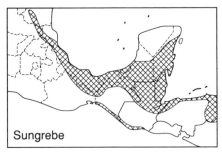

SUNBITTERNS: EURYPYGIDAE (1)

The Sunbittern, a unique neotropical species, recalls a large rail or sandpiper. It runs and walks along forested river banks, searching for prey among rocks and in small pools; does not swim. Bill long, slender, and dagger-like, legs long and toes unwebbed. Neck long and slender, body large, wings broad and rounded, tail long and broad. Ages/sexes similar; young altricial but downy. Plumage cryptically patterned in browns, greys, black, and white, but wings and tail brightly patterned, striking in flight and display. Molts poorly known (?).

Feeds on insects and other invertebrates, frogs, tadpoles, small fish. Nest is a fairly bulky cupped mass of vegetation on a branch near or over water. Eggs: 2, pale vinaceous, spotted with browns and greys (WFVZ).

SUNBITTERN

Eurypyga helias major Plate 18
Garza del Sol

ID: 17–19 in (43.5–48.5 cm). C.A. See family account. **Descr.** Eyes red (brown in juv), bill yellow-orange with black culmen, legs orange. *Appears dark grey overall with bold white scapular spots, long brownish neck, and black-and-white-striped head.* Head blackish with white supercilium and moustache, throat white. Neck and sides of chest vermiculated blackish and dull rufous, chest mottled brown and buff, underbody whitish. Upperparts, including tertials, barred blackish and dark brownish grey, scapulars boldly spotted white. Tail vermiculated dark grey and white, with 2 black-and-rufous bars. *Wings strikingly patterned:* barred grey and blackish *with bold tawny flash across outerwing marked with 2 black-tipped chestnut patches.*
Voice. A high, reedy trill (Honduras).
Habs. Fast-flowing rocky streams and rivers in humid evergreen forest, less often in swampy forest. Shy and wary, often flies at first sight of danger; also runs with agility. Wingbeats floppy, glides short distances on bowed wings. See family account.
SD: R to U resident (near SL–500 m) on Atlantic Slope from N Guatemala to Honduras. May occur (300–900 m) on Pacific Slope of Chis (MAT), and perhaps adjacent Guatemala (at least formerly). Traditionally reported from Atlantic Slope of Chis but no specimens known (?), and no recent reports.
RA: Guatemala (S Mexico?) to Brazil.

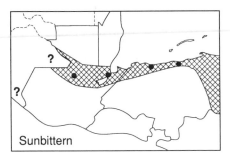

LIMPKINS: ARAMIDAE (1)

The Limpkin is a large, superficially ibis-like New World bird with a long, slightly decurved bill, long neck, and long legs. Wings fairly long and broad, tail short and covered by wings at rest. Limpkins inhabit marshy areas and walk with a high-stepping gait, picking and probing for food. Ages/sexes similar, young precocial; molts poorly known (?). Voice loud and carrying, varied wailing screams.

Food mostly freshwater snails, also mussels, crustaceans, frogs, insects. Nest platform of marsh grasses at low to mid-levels in vegetation over water. Eggs: 4–7, buff, marked with browns and greys (WFVZ).

LIMPKIN

Aramus guarauna dolosus   Plate 1
Carao

ID: 23–25 in (58.5–63.5 cm). *See family account.* **Descr.** Eyes reddish, bill pinkish orange with dark culmen and tip, legs greyish. Plumage *dark brown overall*, upperparts with olive sheen; *head, neck, and much of back and underparts coarsely streaked white*; white mottling on underwing coverts visible in flight.
Voice. Loud wailing and screaming cries, especially at dawn and dusk: a drawn-out *kyaoh kyaoh* ... or *haowh*, a rolling *krrrrh* and *krrrreeah* or *kerr-rr-rr owh'*, a shorter *kyow* or *kyowk* and *kaah* or *krrah*, and, when disturbed, a sharp, piercing *bihk, bihk* ..., etc.
Habs. See family account. Freshwater marshes and lake edges with tall reeds, swampy woodland, mangroves. Most active at dawn and dusk, partly nocturnal. Typically close to cover, heard more often than seen. Flight distinctive, like a hurried crane, with shallow, stiff wingbeats jerked on the upstroke, wings mostly above body plane.
SS: See juv White Ibis, Glossy and White-faced ibis, juv night-herons.
SD: F to C resident (SL–1500 m) on Atlantic Slope from Ver to Honduras; U to F resident (around SL) on Pacific Slope from Isthmus to El Salvador where now R, if not extirpated (Thurber *et al.* 1987).
RA: SE USA, Caribbean, and Mexico to Ecuador and N Argentina.

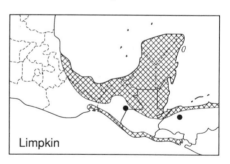

Limpkin

CRANES: GRUIDAE (2)

Cranes, mainly an Old World family, are large wading and terrestrial birds with very long necks and long legs. Superficially they resemble large herons but their elongated tertials give the rear end a distinctive bulge or 'bustle'. Bills long, straight, and pointed, forehead and crown of adults naked and brightly colored. Wings long and broad, tails short and covered by wings at rest. Cranes inhabit open marshy areas and arable plains, and often are very social. They are noted for their spectacular dancing displays which may be seen in winter. Ages differ, sexes similar. Adult plumage attained over 1–2 years. Molts protracted, with not all flight feathers replaced annually. Plumages colored in greys, browns, white, and black. Flight graceful with neck and legs outstretched, flocks often fly in lines and V formation, recalling geese. Wingbeats stiff with a relatively quick, jerked upstroke, wings mostly above body plane. Calls are loud and carrying honking cries.

Food mostly grains in winter, also invertebrates, small mammals, frogs, etc. No species breed in the region.

SANDHILL CRANE

*Grus canadensis**   Not illustrated
Grulla Canadiense

ID: 34–40 in (86.5–96.5 cm). Winter visitor to N Mexico. **Descr. Adult:** eyes amber, bill and legs blackish, naked forecrown red. *Pale grey overall*, in bright light can appear silvery grey overall, almost white. Plumage may be stained rusty brown by ferrous pigments in marshes, often noticeable on wing coverts. **Imm:** eyes dusky. Forecrown feathered, *head and neck cinnamon-brown, rest of plumage mottled cinnamon-brown and pale grey*; mostly grey by mid-winter with paler wing panel; forecrown may be dull reddish by mid-winter.
Voice. Deep, far-carrying, rolled, honking calls, *k'worrr-h, krrrow, krrrroh*, etc.; imms have strikingly different, high, slightly reedy, rolled trills.

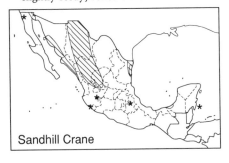

Sandhill Crane

Habs. Open arable land, lakes, marshes. Usually in noisy flocks up to few thousand birds which feed in fields and return to roost at lakes or marshes, a schedule similar to geese with which they often associate. Usually wary and take flight noisily when alarmed.
SS: Rare Whooping Crane larger and white, adult has black moustache; in flight, note black primaries. See Great Blue Heron.
SD: F to C winter visitor (Nov–Mar; SL–2500 m), mainly in interior, S to Son, Dgo, and Tamps; irregularly U to R on Pacific Slope to Col, in interior to cen volcanic belt; R in Baja. Vagrant to QR (Feb 1949).
RA: Breeds N.A.; winters to N, rarely cen Mexico.

WHOOPING CRANE

Grus americana   Not illustrated
Grulla Americana

ID: 50–55 in (127–140 cm). Former winter visitor. **Descr. Adult:** eyes pale yellow, bill olive with yellowish base, legs black; *red crown, black lores, and red-tipped moustache naked. White overall, black primaries striking in flight*; upper primary coverts and alula also black. **Imm:** crown and face feathered, *head and neck washed brown, upperparts mottled brown and white*; primaries black. Becomes whiter through 1st winter.

Voice. Loud, trumpeting, honking cries, *kr-hehh,* etc.

Habs. Coastal lagoons and marshes, extensive arable areas and lakes. Habits

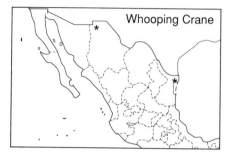

Whooping Crane

similar to Sandhill Crane with which, in Chih, it is likely to be seen.

SS: See Sandhill Crane, American White Pelican, Snow Goose.

SD: R and irregular winter visitor (Nov–Mar) to N Tamps where last reported Feb 1951 (Evenden 1952), and N Chih; birds in latter area derived from transplanted stock in N.A. species' management program. Extirpated as regular winter visitor in early 1900s; formerly occurred S to Jal and Gto.

RA: Breeds W Canada; winters coastal Texas; transplanted population summers in Idaho, winters in New Mexico.

THICK-KNEES: BURHINIDAE (1)

Thick-knees, large terrestrial shorebirds that resemble huge plovers, are nocturnal inhabitants of grasslands. They are cosmopolitan in warm climates. Bills heavy and plover-like, legs long and stout with thick intertarsal joints ('knees'). Heads large with large staring eyes, wings fairly long and bluntly pointed, tails fairly long. Ages differ slightly, sexes similar. Plumage cryptically colored in browns, greys, black, and white. Young precocial, 1st prebasic molt partial, remex molt protracted. Voice loud and carrying.

Food mostly insects, small reptiles, amphibians. Nest is a scrape on the ground. Eggs: 2, buff to pale olive-buff, heavily marked with browns and greys (WFVZ).

DOUBLE-STRIPED THICK-KNEE
Burhinus b. bistriatus Plate 18
Alcaraván Americana

ID: 18–20 in (45.5–50.5 cm). See family account. **Descr. Adult:** *eyes yellow,* bill yellowish with blackish tip, legs yellow. Face, neck, and chest greyish buff, streaked darker, with *broad whitish supercilium and broad blackish lateral crown stripes,* throat whitish. Belly whitish, undertail coverts pale cinnamon. Upperparts dark brown, edged cinnamon and buff. Tail brown, outer rectrices broadly barred black and white (rarely visible except when spread). *Upperside of remiges blackish brown with short white bar on outer 3 primaries, 2nd white bar across base of inner primaries, underwings white overall.* **Juv:** duller and greyer, forehead and lores greyish buff, dark lateral crown stripe and white supercilium extend back from eye, with a short triangular blackish stripe at sides of nape. Neck has narrower dark streaking than adult, becoming broader mottled stripes on chest. Legs duller, greenish yellow.
Voice. A loud, far-carrying, clipped barking or clucking *kah-kah-kah ...* or *kyehkyehkyeh ...,* often in prolonged, rapid series; at a distance may suggest frogs. Also a quiet, rolled to louder shrill *kwehrr kwehrr ...,* and *kreh ehr* or *kreh eh ehr,* etc.

Habs. Arid to semiarid savannas, grassland, ranchland. Easily overlooked despite its large size. In pairs or loose groups up to 20 birds which spend the day sitting or standing inconspicuously, often in the shade of small bushes or trees. Crouches or runs quickly when alarmed in preference to flying. May be domesticated (rarely in Mexico) for its watchdog qualities; calls carry over a mile (1.6 km).
SD: F to C but local resident (SL–900 m) on Atlantic Slope from cen Ver to SW Camp (SNGH, SW), and on Pacific Slope and locally in interior from Oax to El Salvador (possibly extirpated, Thurber *et al.* 1987) and Honduras. Vagrant (or escape?) in N Ver (Jan 1985, JCA photo). May spread slowly with increased clearing of forest.
RA: SE Mexico to NW Brazil.

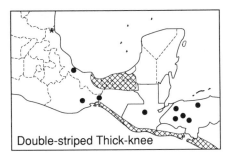

Double-striped Thick-knee

PLOVERS: CHARADRIIDAE (10)

Plovers are cosmopolitan shorebirds distinguished by their fairly short stout bills which are slightly swollen toward the tip. Legs fairly long, feet unwebbed, eyes large. Plovers typically frequent open, often fairly dry areas where they feed with a distinctive stop-start action: standing still, then running and pecking, standing still again, etc. Like most shorebirds they are often active by night. Ages differ slightly; young precocial, 1st prebasic molt partial. Juvs overall resemble basic adults and usually migrate before molting. Contour feather molt may begin during autumn migration but flight feathers are usually molted after migration. Sexes similar or slightly different, ♂♂ being blacker on the face and/or underparts than ♀♀; seasonal variation apparent in most species. First-year birds often do not breed, remain south of the nesting grounds, and retain complete or partial first basic. Plumages overall colored in browns, greys, black, and white, but patterns often striking. Calls mostly plaintive whistles and sharp chips.

Plovers feed mostly on a wide variety of invertebrates. Four species, all *Charadrius*, breed in the region. Nests are scrapes on open ground. Eggs: 2–4, pale buff to buff, spotted and scrawled with browns, greys, and black (WFVZ).

GENUS PLUVIALIS

Large to fairly large plovers of coasts and grasslands. Age/seasonal variation, sexes similar. Juvs attain basic plumage after migration, alt plumage seen briefly in Apr–May. Bills and legs black.

BLACK-BELLIED (GREY) PLOVER
Pluvialis squatarola   Not illustrated
Chorlo Gris

ID: 10.5–11.5 in (26.5–28 cm). A large plover, common on both coasts. **Descr. Basic:** face pale with whitish lores and supercilium, dusky auricular smudge. Throat and underparts whitish, streaked and mottled dusky on chest and flanks. Crown, nape,

and *upperparts grey to grey-brown, edged and spotted whitish, uppertail coverts white; tail barred black and white. Underwings whitish with black axillars, upperwings grey with* darker remiges and *bold white wingstripe.* **Alt:** *face, throat, and underparts black, broadly bordered by white* forehead, supercilium and sides of neck and chest; vent and undertail coverts white. Crown, nape, and *upperparts blackish, boldly spangled white.* **Juv:** resembles basic but browner, crown, nape, and *upperparts brightly and more finely spotted whitish to buffy yellow,* chest neatly streaked dusky. 1st basic similar but greyer, some birds mottled black on face and underparts in 1st summer.

Voice. Melancholy, slurred whistles, *kweeh-ee-ooh* or *hee-o-eeh,* and *weee-oou,* and a slightly more piping *teeeeu* or *tee-uu,* etc.

Habs. Beaches, estuaries, coastal lagoons, fields near water, rarely inland lakes. Singly or in loose flocks, at times to 100+ birds.

SS: Juv American Golden Plover smaller with long primary projection beyond tertials and beyond tail, underwings and axillars dusky; in flight, upperparts uniform with indistinct narrow whitish wingstripe.

SD: C to F transient and winter visitor (Aug–May) along both coasts; small numbers

Black-bellied Plover

oversummer widely. Irregular R to U transient and winter visitor (Sep–Jan at least; 1000–2200 m) in interior from Zac and Coah to Chis (AGL spec.; SNGH, LJP, RGW).

RA: Holarctic breeder; winters S in New World to cen Chile and N Argentina.

AMERICAN [LESSER] GOLDEN PLOVER
Pluvialis dominica   Not illustrated
Chorlo-dorado Americano

ID: 9.5–10 in (24–25.5 cm). *Transient on grasslands and coasts. Tertial tips fall short of tail tip, note long primary projection.* **Descr. Basic** (Sep–Apr): face pale with dusky auricular smudge, *whitish lores and supercilium often contrast strongly with dark crown.* Throat and underparts whitish, mottled dusky on chest and flanks. Molting birds have face and underparts scalloped and mottled black. Crown, nape, and *upperparts brown to grey-brown, edged and spotted whitish to buffy yellow. Underwings dusky, upperwings and tail uniform with upperparts,* flight feathers slightly darker with indistinct whitish wingstripe. **Alt:** *face, throat, and underparts black, with broad white border from forehead and supercilium to sides of chest,* ♀ often has whitish mottling on face and underparts. Crown, nape, and *upperparts blackish, boldly spangled buffy yellow to ochre-yellow.* **Juv** (Sep–Nov): resembles basic but crown, nape, and *upperparts spotted whitish to pale lemon, bright ochre-yellow spotting only on rump and uppertail coverts*; supercilium streaked dusky and often washed lemon; throat whitish, *underparts dusky* with fine dark streaking and mottling on chest and flanks. Some birds attain black mottling on underparts by 1st spring.
   Voice. A mellow to fairly sharp *chee-wit* or *ch-weet,* a plaintive *kl-eee'* or *k-leep*

and *klee-ee-weet,* also a longer *clee-oo-weetl* or *clee-wee-dl* in flight.
   Habs. Grasslands, fields, lake shores, rarely mud-flats. Singly or in flocks, in spring up to few thousand birds at migration stopovers; associates loosely with other shorebirds.
   SS: Pacific Golden Plover has relatively longer tertials, the tips about equal with tail tip, and shorter primary projection; juv and basic appear golden yellow overall, upperparts with bold golden-yellow spangling and spotting, crown often less contrastingly dark. Alt Pacific Golden Plover typically has white from sides of chest extending down along flanks, undertail coverts mottled white; molting birds can be hard to identify, note structure of wingtips. Mountain Plover (N Mexico in winter) plainer overall with pale legs. See Black-bellied Plover.
   SD: F to C spring transient (mid-Feb–May; SL–2500 m) on Atlantic Slope (with major concentrations in cen Ver), on Pacific Slope from Col S, and in interior; R in NW Mexico. Probably R to U autumn transient (mid-Aug–Oct) throughout, but few records.
   RA: Breeds Arctic N.A.; winters S S.A.
   NB: Traditionally considered conspecific with Pacific Golden Plover, and known as Lesser Golden Plover, *P. dominica.*

PACIFIC [LESSER] GOLDEN PLOVER
Pluvialis fulva   Not illustrated
Chorlo-dorado Asiático

ID: 9–9.5 in (23–24 cm). *Baja and offshore Pacific islands. Tertial tips about equal with tail tip, primary projection shorter than American Golden Plover.* **Descr. Basic:** face and throat whitish with dusky auricular smudge, chest and flanks mottled dusky, contrasting with whitish belly and undertail coverts. Crown, nape, and *upper-*

American Golden Plover

Pacific Golden Plover

parts *blackish brown to brown, edged and spotted golden yellow* and whitish. *Underwings dusky, upperwings and tail uniform with upperparts*, flight feathers slightly darker with indistinct whitish wingstripe. **Alt:** face, throat, and underparts black with *white border from forehead and supercilium down sides of neck to flanks*, often with some black scalloping on flanks, undertail coverts mottled white. Crown, nape, and *upperparts blackish, boldly spangled and spotted ochre-yellow to golden yellow.* **Juv** (Sep–Nov): *resembles basic but face, throat, and chest washed buffy yellow* to ochre-yellow, with neat dusky streaking on chest.

Voice. A mellow to fairly sharp, rising *ch-weet* or *ch-wit*, a longer, slightly plaintive *kleei-wee* or *klee-i-wee*, and an excited *whi-chi chi–weet'* given by territorial birds.

Habs. Beaches, grasslands, fields, open country. Singly or in small flocks, often associating with other shorebirds.

SS: See American Golden Plover.

SD: U winter visitor (Oct–Apr) on Isla Guadalupe (Jehl and Everett 1985; Howell and Webb 1992d), Islas Revillagigedo (San Benedicto and Clarión, Howell and Webb 1992g) probably also locally along Pacific coast of BCN (Wilbur 1987). U transient (Aug–Nov) on Clipperton.

RA: Breeds Arctic NE Asia to W Alaska; winters from E Africa to Pacific Ocean coasts and islands, E to W N.A.

NB: See American Golden Plover.

GENUS CHARADRIUS

Small to medium-sized plovers widespread throughout the world; most have black chest bands and are known as ringed plovers. Slight sexual variation: alt ♂♂ have more black on face and chest than ♀♀. All show a black frontal bar (at least in alt).

COLLARED PLOVER

Charadrius collaris   Plate 18
Chorlito Collarejo   Figure 22

ID: 5.5–6 in (14–15 cm). A smart petit plover of sandy coasts and river bars. Fairly slender bill often appears relatively long. **Descr.** Ages differ, sexes similar. **Adult:** bill black, legs flesh. *Large white forehead patch contrasts with black loral stripe and broad*

black frontal bar. *Crown, nape, and upperparts sandy brown* with whitish postocular stripe, *cinnamon wash on forecrown, hind auriculars, and sides of neck.* Throat and underparts white with *narrow black chestband.* Upperwings sandy brown, remiges darker, with white wingstripe. Tail dark with white sides. **Juv:** lacks black on head, chest band restricted to dark patches at sides, upperparts neatly edged pale cinnamon.

Voice. A sharp *peek*, suggesting Wilson's Plover; a shorter *kip* or *krip*, and a mellow, rolled *krip* or *kyip* which may be run into series in display (?) flight.

Habs. Sandy beaches, coastal lagoons, sand and gravel bars in rivers, rarely inland lakes. Singly or in pairs, rarely (ever?) in flocks; associates loosely with other small plovers.

SS: Snowy Plover larger, bulkier, and paler overall, with grey to dusky flesh legs, whitish hindcollar, incomplete chest band. Wilson's Plover larger with longer, heavier bill and whitish hindcollar; broad chest band black only in alt ♂. Semipalmated Plover has stubbier, thicker bill, head appears brown overall with white forehead; note white hindcollar, orange legs.

SD: F to U resident (SL–750 m) on both slopes from S Sin and S Tamps (JCA photo) to El Salvador and Honduras, but absent from Yuc Pen (reports in Edwards (1989) erroneous); occurs inland mostly from Isthmus S. Reports from Belize (Feb–Apr) may be of migrants from Caribbean populations. U to R visitor (Oct–Mar; 750–1000 m) in N Gro and Mor (SNGH, RGW); vagrant to Tlax (Jan 1984, SNGH, PP).

RA: Mexico and E Caribbean to cen Chile and cen Argentina; some populations at least partly migratory.

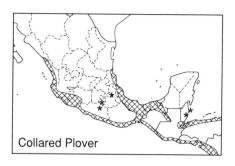

Collared Plover

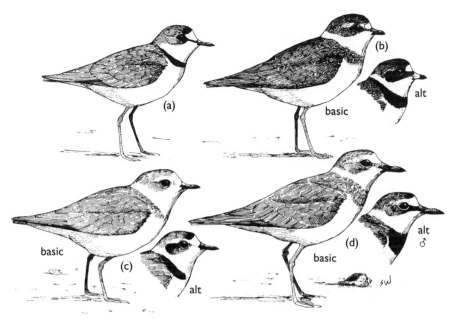

Fig. 22 (a) Collared Plover; (b) Semipalmated Plover; (c) Snowy Plover; (d) Wilson's Plover

SNOWY PLOVER

*Charadrius alexandrinus**   Figure 22
Chorlito Nivéo

ID: 6–6.5 in (15–16 cm). A widespread, overall pallid plover; bill fairly slender. **Descr.** Age/sex/seasonal variation. **Alt ♂:** bill black, *legs grey. White forehead, lores, and short supercilium contrast with black frontal bar and auricular patch,* lores may have dark stripe. Crown, nape, and *upperparts pale sandy* (paler, silvery sandy in *tenuirostris* of Atlantic coasts) with *whitish hindcollar,* cinnamon-washed crown. Throat and underparts white with *black chest side patches.* Upperwings sandy, remiges darker, with white wingstripe. Tail brown, darker distally, with white sides. **Alt ♀:** typically duller, dark head and chest side markings blackish brown. **Basic/juv:** legs may be dusky flesh. Head sandy brown with whitish forehead, lores, and supercilium, chest patches sandy brown. Upperparts of juv have fine pale edgings.

Voice. A quiet, rolled to slightly reedy *prrit* or *prip,* and a brighter *kriip* or *klip* in flight. Nesting birds give a plaintive up-slurred *hoo-reep* or *chuweep,* a low rolled *prrt* or *perrt,* and low, gravelly chatters.

Habs. Sandy beaches, salt ponds, salt-flats, lakes, etc. Often in small flocks in winter, locally up to 100+ birds; associates with other plovers, especially Wilson's Plover.

SS: Piping Plover bulkier with short, stubby bill, orange legs. See Collared Plover. Some Wilson's Plovers on Caribbean coasts have washed-out, pale sandy upperparts; note heavy black bill, flesh legs, complete chest band.

SD: F to C breeder (Mar–Jul) along Pacific coast of Baja S at least to 26° 45′ N, locally in Isthmus, and possibly elsewhere on Pacific coast; F to U breeder (Apr–Aug) on Atlantic coast from Tamps to N Ver (Coffey 1961) and along N coast of Yuc (Klaas 1968; G and JC, photo). F to U and local

Snowy Plover

breeder (May–Aug; 2000–2500 m) in interior from Zac and SLP to Mex, probably also Jal (SNGH, SW). F to C transient and winter visitor (Aug–Apr) on Pacific coast from Baja and Son to Nay; U to R to Guatemala; U to C (Aug–May) on Atlantic coast from Tamps to Yuc Pen; R to Honduras; U (Sep–Apr) in interior S to cen volcanic belt (AGL spec, SNGH, SW). Vagrant to N Guatemala (Apr 1991, Howell and Webb 1992*b*). Non-breeders oversummer locally, at least on Atlantic coast.

RA: Virtually cosmopolitan in warm climates except Australia; N populations migratory.

NB1: N New World populations sometimes considered specifically distinct, *C. nivosus*, as is the form in W S.A. (*C. occidentalis*).

NB2: Known in the Old World as Kentish Plover.

WILSON'S PLOVER
*Charadrius wilsonia**   Figure 22
Chorlito Piquigrueso

ID: 7–7.5 in (18–19 cm). A fairly large ringed plover of sandy coasts. *Bill long and heavy.* **Descr.** Age/sex/seasonal variation. **Alt ♂:** bill black, legs flesh. *White forehead and short supercilium contrast with black frontal bar and lores; postocular area and sides of hindcrown often mottled cinnamon.* Crown, auriculars, nape, and upperparts sandy brown (to pallid silvery sandy, at least in Yuc Pen) with white hindcollar. Throat and underparts white with *broad black chest band.* Upperwings sandy, remiges darker, with white wingstripe. Tail dark with white sides. **Alt ♀:** forecrown, lores, and *chest band brown*, often mottled cinnamon. **Basic/juv:** resembles ♀ but chest band paler, rarely broken in middle; juv has fine pale edgings to crown and upperparts.

Voice. A sharp *week!* and *pri-dik* or *pri-*

di-ik, a short *pik*, and dry, buzzy, bickering chatters.

Habs. Sandy beaches, coastal lagoons, salt ponds, rarely far from the coast. In winter and migration occurs locally in flocks up to few hundred birds; associates with other plovers.

SS: Semipalmated Plover smaller with shorter stubby bill, orange legs. See Collared and Snowy plovers.

SD: F to C resident along Pacific coast of BCS and from Gulf of California to Nay, in El Salvador (Thurber *et al.* 1987) and probably elsewhere; F to C breeder (Apr–Sep) on Atlantic coast from Tamps to N Ver, and resident from Yuc (WFVZ spec.) to Belize Cays (Pelzl 1969). U to F transient and winter visitor (Aug–May) from Gulf of California to Oax; F to C from Isthmus S; vagrant to Pacific coast of BCN (Aug 1992, KAR, TEW). F to C transient (mid-Aug–Oct, Mar–May), and U to F winter visitor (Oct–Mar), on Gulf of Mexico coast from Tamps to Camp; U in winter to N Honduras.

RA: Breeds Mexico and E USA to Panama and NE Brazil; N populations migratory, wintering along Pacific coast to NW Peru.

SEMIPALMATED PLOVER
Charadrius semipalmatus   Figure 22
Chorlito Semipalmado

ID: 6.5–7 in (16.5–17.5 cm). *Widespread* in winter and migration. *Bill fairly stubby.* **Descr.** Age/seasonal variation, sexes similar. **Basic/juv:** bill black, orange below at base, *legs orange* to yellow-orange. *Head brown with white forehead and typically indistinct white postocular stripe*; hindcollar white, upperparts brown (finely edged buff in juv). Throat and underparts white with brown chest band often almost broken in middle. Upperwings brown, remiges darker, with white wingstripe. Tail

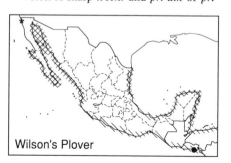

Wilson's Plover

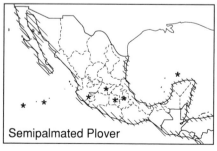

Semipalmated Plover

brown, blackish distally, edged white. **Alt:** bill orange with black tip. *Frontal bar, lores, auriculars, and chest band black,* often with black extending as narrow collar around upper mantle.

Voice. Common call, often given in flight, a slightly plaintive *ch-weeh* to a sharper *ch-wiet,* also an abrupt, slightly hollow *ky-eh'* or *kyuh',* and plaintive notes run into an accelerating, rolled chatter, *whee twee twee twee ch-uh-uh-uh-uh-uhr,* etc.

Habs. Sandy and muddy beaches, lagoons, salts ponds, estuaries, lakes, and marshes. Often in small flocks in winter, larger flocks during migration; associates with other plovers.

SS: See Collared, Wilson's, and Snowy plovers. Piping Plover overall pale silvery sandy above, bill stubbier, lores whitish, white wingstripe broader, and tail with more contrasting black distal band; basic has pale brown chest patches.

SD: F to C transient and winter visitor (Aug–May) on both coasts; R and irregular in summer. U transient and winter visitor (Sep–May; to 2500 m) in interior S at least to cen volcanic belt (SNGH, PP, RGW).

RA: Breeds N N.A.; winters coasts of S USA to cen Chile and Argentina.

PIPING PLOVER

Charadrius melodus Not illustrated
Chorlito Chiflador

ID: 6.5–7 in (16.5–17.5 cm). Winters on Gulf and Caribbean coasts. *Bill stubby.* **Descr.** Age/seasonal variation, sexes similar. **Basic:** bill black, orange below at base, *legs orange. Head and upperparts pallid silvery sandy with white lores and short supercilium,* white hindcollar. Throat and underparts white with *poorly contrasting grey-brown chest side patches.* Upperwings pale sandy, remiges darker, with broad white wingstripe. *Tail pale sandy*

with black subterminal band not extending onto white tail sides. **Alt:** bill orange with black tip. Frontal bar and chest side patches black, often with black extending narrowly around upper mantle; some birds have a full black chest band.

Voice. A clear, mellow whistle, *peep* or *peep-o* (DAS).

Habs. Sandy beaches, sand bars in estuaries, coastal lagoons. Singly or in small, loose groups; associates with other small plovers.

SS: See Snowy and Semipalmated plovers.

SD: U transient and winter visitor (Aug–Apr) along Atlantic coast from Tamps to Ver, U to F in N Yuc Pen (SNGH, PP, RGW), probably R to Belize (Russell 1964). R and irregular winter visitor (Nov–Apr at least) on Pacific coast from Son (Russell and Lamm 1978) to Nay (SNGH, DEW).

RA: Breeds interior and E N.A.; winters SE USA to Caribbean and adjacent coasts.

KILLDEER

Charadrius v. vociferus Not illustrated
Chorlito Tildío

ID: 9.5–10 in (24–25.5 cm). Widespread, noisy, and conspicuous. *Tail long and graduated, extending beyond wingtips at rest.* **Descr.** Ages/sexes similar. Orbital ring red, bill black, legs flesh. Head brown with white forehead and postocular stripe; frontal bar, lores, and auriculars blackish in alt. Throat and underparts white with *2 black chest bands,* the upper band broader. Hindneck and upperparts brown, with white hindcollar, upperparts edged cinnamon (obvious in juv and fresh basic), with *bright cinnamon rump and uppertail coverts;* tail brown with black subterminal band, edged white.

Voice. Plaintive, wailing, and screaming cries, *keeeu* or *tieeee,* and *kee-deer* or *kee-hr;* longer series when alarmed or flushed,

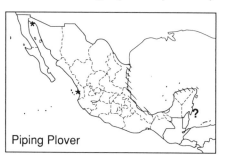
Piping Plover

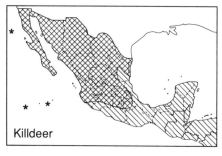
Killdeer

kee dee-dee-dee-dee ... or *tee-deet, tee dee-dee-dee* ..., etc. In display flight a repeated *kill-dee-u* or *kill-deee'*.

Habs. Fields, marshes, lakes, river bars, open habitats in general. Singly or in small groups, rarely in flocks of more than 25 birds.

SD: F to C breeder (Apr–Aug; SL–2500 m) in Baja and, mainly in interior, S to cen volcanic belt and N Gro, possibly elsewhere S to N C.A. (for example, Jun–Jul records in N Yuc, G and JC). C to F transient and winter visitor (Sep–Apr; SL–2500 m) S to Isthmus; F to U to El Salvador and Honduras.

RA: Breeds N.A. to cen Mexico, also Costa Rica; winters S to N S.A.

MOUNTAIN PLOVER

Charadrius montanus   Not illustrated
Chorlito Llanero

ID: 8.5–9 in (21.5–23 cm). A fairly elegant long-legged plover wintering in N Mexico. **Descr.** Seasonal variation, ages/sexes similar. **Basic:** bill black, legs yellowish flesh. *Whitish forehead, supercilium, and subocular crescent accentuate large-eyed expression on buff face.* Rest of head and upperparts sandy brown to grey-brown, edged cinnamon when fresh. Throat and *underparts whitish, washed buff to greyish buff on neck sides, chest, and flanks.* Upperwings brown, remiges darker, with narrow whitish wingstripe most noticeable across base of primaries; *underwings bright whitish. Brown tail with black terminal band contrasts with paler uppertail coverts.* **Alt:** frontal bar and lores black, sides of neck and chest washed pinkish cinnamon.

Voice. A slightly rasping, dry *krehrr*, a dry clipped *kep*, and a slightly reedy *krrip* or *krreek*.

Habs. Open grassland and farmland, often with sparse vegetation. Usually in small flocks.

SS: See American Golden Plover.

SD: Irregular (?) U and local winter visitor (Nov–Mar; near SL–2000 m) in BCN and from N Son to N Tamps, S in interior to Zac; R to BCS.

RA: Breeds interior W N.A.; winters SW USA to N Mexico.

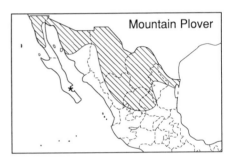

Mountain Plover

OYSTERCATCHERS: HAEMATOPODIDAE (2)

A small but cosmopolitan family of large shorebirds. Bills long, stout, straight, and laterally flattened at the tip, medium-length legs thick and fleshy. Bare parts brightly colored. Ages differ, sexes similar; young precocial, 1st prebasic molt partial. Young often stay with their parents through the first winter.

Oystercatchers feed mostly on shellfish, prised open with their long wedge-like bills. Nest is a scrape on the ground. Eggs: 2–3, pale buff, marked with dark browns and greys (WFVZ).

AMERICAN OYSTERCATCHER

*Haematopus palliatus**   Not illustrated
Ostrero Americano

ID: 16–18 in; B 3–3.5 in (40.5–45.5 cm; B 7.5–9 cm). Both coasts. **Descr. Adult:** eyes pale yellow, orbital ring red, *bill bright orange-red*, legs flesh. *Head and chest black* (with irregular mottled lower chest border in *frazari* of NW Mexico), *upperparts blackish brown* with white uppertail coverts and black tail. *Underparts white.* Upperwings black with broad white stripe from inner secondaries along greater coverts. **Juv:** eyes dusky, bill duller with dusky tip, legs dull flesh. Upperparts with fine, scaly buff edgings. Attains adult bare-part colors over 1st year.
  Voice. Loud, piping to screaming calls, *wheeh* or *whee-uh,* and *h'wheep,* often run into excited piping chatters; a sharp *keek* in alarm near nest.
  Habs. Sandy and rocky coasts, rarely estuaries. In pairs or singly, less often in small loose flocks. Associates readily with other shorebirds.
SD: U to F but often local resident from Baja and Gulf of California to Isthmus, possibly to El Salvador (Thurber *et al.* 1987). F resident on Atlantic coast from Tamps to N Ver and in N Yuc Pen; commoner in winter when also R to N Honduras.
RA: NW Mexico and E USA to cen Chile and cen Argentina.
NB1: Sometimes considered conspecific with Black Oystercatcher and also with Eurasian Oystercatcher *H. ostralegus*. **NB2:** May interbreed with Black Oystercatcher in Baja.

BLACK OYSTERCATCHER

Haematopus bachmani   Not illustrated
Ostrero Negro

ID: 16–17 in; B 2.5–3 in (40.5–43 cm; B 6.5–7.5 cm). *W Baja.* **Descr. Adult:** eyes yellow, orbital ring red, *bill bright orange-red*, legs flesh. *Blackish overall,* upperparts and belly washed brown. **Juv:** eyes dusky, bill duller with dusky tip, legs dull flesh. Upperparts with fine, scaly buff edgings. Attains adult bare-part colors over 1st year.
  Voice. Loud screams, much like American Oystercatcher, *h-eeu* and *ki-eep* or *k-eep,* a sharp *wheek!* in alarm, and excited, shrill, piping chatters.
  Habs. Rocky coasts. In pairs or family groups, associates readily with other shore-birds, mainly at high tide roosts.

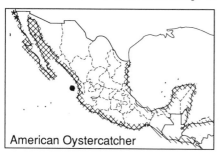

American Oystercatcher

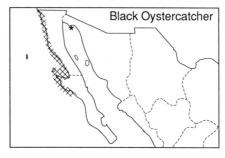

Black Oystercatcher

SD: F to U resident on W coast of Baja S to about 27°N. R visitor (Mar–May) to N Son (Russell and Lamm 1978).

RA: W N.A. to Baja.

NB: See American Oystercatcher.

STILTS AND AVOCETS: RECURVIROSTRIDAE (2)

A small cosmopolitan family of shorebirds found in warm climates. Fairly large but slender and elegant with long legs and bills; bills straight and fine in stilts, upcurved in avocets. Ages/sex/seasonal variation generally slight but noticeable; young precocial, 1st prebasic molt partial. Plumages strikingly patterned in black and white. Typically conspicuous and often noisy when nesting.

Food mostly insects, small crustaceans, and other aquatic invertebrates. Nest singly or, more often, colonially, often in association with one another. Nest is a scrape on open ground, often built up with vegetation in wetter areas. Eggs: 3–5, buff to olive-buff, marked with browns, black, and greys (WFVZ).

BLACK-NECKED STILT
Himantopus m. mexicanus   Not illustrated
Candelero Americano

ID: 14–16 in; B 2.5 in (35.5–40.5 cm; B 6.5 cm). Widespread. *Bill long, straight, and needle-like, legs extremely long* (6.5 in; 16.5 cm). **Descr.** Ages/sexes differ. ♂: bill black, *legs bright pink*. Head black with white postocular spot; hindneck and upperparts glossy black, white uppertail coverts usually visible only in flight, tail pale grey. Lores, foreneck, and underparts white, chest flushed rosy in alt. Underwings black with white axillars. ♀: back and tertials brownish, lacks rosy flush on chest in alt. Juv: resembles ♀ but upperparts have neat, scaly buff edgings.
Voice. Varied barking and clipped nasal calls, especially when nesting, *kreh* or *krek*, and *krreh*, and a steady scolding *kek-keh-keh* ... or *kih-kih* ..., and clucking *wek-wek-wek* ..., etc.
Habs. Marshes, lakes, salt ponds, coastal lagoons, river bars, rarely on open coasts. Breeds colonially and conspicuously, agitated parents scold loudly and give distraction displays. At other times, often in flocks, associating readily with other waterbirds.
SD: C to F resident (SL–1500 m) in BCN and on both slopes from Sin and Tamps to El Salvador and Yuc Pen; F to U transient and winter visitor (Aug–Apr) in Son, BCS, and S to Honduras; commoner southward in breeding range in winter. F to C but local breeder (Mar–Oct; 1000–2500 m) in interior S to cen volcanic belt; U to C but local transient and winter visitor (Aug–Mar) in interior throughout.
RA: Breeds USA to S S.A.; N populations withdraw S in winter.
NB: Sometimes considered conspecific with *H. himantopus* of Old World.

AMERICAN AVOCET
Recurvirostra americana   Not illustrated
Avoceta Americana

ID: 15.5–17 in; B 3–3.5 in (39.5–43 cm; B 7.5–9 cm). Widespread, mainly in winter. *Bill long, slender, and recurved,* more strongly so in ♀. **Descr.** Age/seasonal variation, sexes similar. **Basic:** bill black, legs grey-blue to grey. *Head, neck, and chest pale grey,* underparts white. *Upper-

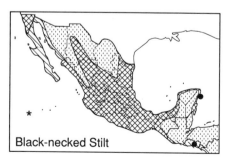

Black-necked Stilt

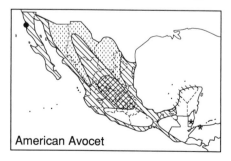

American Avocet

parts black with broad white V *on outer scapulars,* rump and uppertail coverts white, tail pale grey. In flight appears strikingly black and white with black primaries, greater upperwing coverts, and scapular stripes. **Alt:** *head, neck, and chest cinnamon* with white lores and eyering. **Juv:** resembles basic but hindneck washed cinnamon, pale buff fringes to dark areas on upperparts.

Voice. A loud, oystercatcher-like piping *kleep* or *klee-ip*, and a loud, piping, often slightly accelerating *wheep wheep whee-hee-hee* or *wheet wheet whi-whi-whi-wheeh*, etc.

Habs. Estuaries, lagoons, lakes, marshes, etc. Feeds by sweeping bill from side to side in shallow water. Usually breeds in small colonies. At other times often in flocks, locally to a few thousand birds. Readily associates with other shorebirds, especially stilts.

SD: F breeder (Apr–Aug) on coast of NW Baja; irregularly U to F but local breeder (Mar–Aug; 2000–2500 m) in interior from Coah (Van Hoose 1955) to cen volcanic belt (Wilson and Ceballos-L. 1986). C to F transient and winter visitor (Jul–May; SL–1000 m) in Baja and on Pacific Slope from Son to Isthmus; U to Guatemala; U to C but local in interior S to cen volcanic belt; F to U on Atlantic coast from Tamps to Yuc Pen (Nov–Apr, G and JC photo); R and irregular to N Honduras. Small numbers oversummer in winter range.

RA: Breeds interior N.A. to cen Mexico; winters S USA to N C.A.

JACANAS: JACANIDAE (I)

Jacanas, a small pantropical family, are superficially rail-like marsh birds with very long legs and extremely long toes and claws used for walking on floating vegetation. Brightly colored frontal shields resemble those of gallinules; wings have a horny spur at the carpal joint. Ages differ, sexes similar. Young precocial and cryptically colored, molt into juvenal plumage before fully grown and then into similar imm plumage, kept for first year. Adult black and chestnut overall, juv and imm brown above, white below; all ages have striking yellow remiges. Calls loud and raucous.

Food mostly aquatic insects, less often small fish, vegetable matter. Nest is a flimsy platform of leaves and grasses built on floating vegetation. Eggs: 3–5, glossy buffy brown with dense inky black scrawls (WFVZ).

NORTHERN JACANA
Jacana s. spinosa　　　　Plate 17
Jacana Mesoamericana

ID: 8.5–9.5 in (21.5–24 cm). Common in lowlands. **Descr. Adult:** bill bright yellow with pale blue cere, shield orange-yellow, legs greyish olive. *Chestnut overall with green-glossed black head, neck, and upper chest. Remiges bright pale yellow*, dusky-tipped, *striking in flight*. **Juv:** bill duller than adult. *Pale buff supercilium and whitish lores*, auriculars, and throat *contrast with brown crown and postocular stripe*. Hindneck dark brown, upperparts olive-brown, broadly edged cinnamon, uppertail coverts broadly edged pale chestnut. *Underparts white*, flanks and underwing coverts dark brown. Yellow remiges have broader dusky edgings than adult. Quickly molts into imm. **Imm:** resembles juv but hindneck black, with green gloss, rump and uppertail coverts chestnut, pale upperpart edgings reduced. Shield develops through 1st year; birds mottled chestnut, black, and brown often seen in summer.

Voice. Loud, screechy clicks and chatters; a shrill, raucous chatter when flushed.

Habs. Freshwater marshes and lake edges with grassy and floating vegetation, flooded fields, roadside ditches. Walks with high-stepping gait; flies with jerky, stiff wingbeats and short glides, neck outstretched, and long legs and toes trailing. Often in loose flocks, at times to 100+ birds.

SD: C to F resident (SL–1200 m) on both slopes from Sin and Tamps to El Salvador and Honduras, extending locally inland (to 1800 m; resident?) along Lerma and Balsas drainages; vagrant to DF (Jun 1984, Wilson and Ceballos-L. 1986).

RA: Mexico (rarely S Texas) and Caribbean to W Panama.

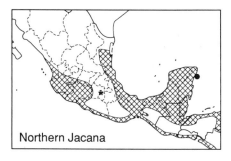
Northern Jacana

SANDPIPERS AND ALLIES: SCOLOPACIDAE (35)

Sandpipers and allies comprise a nearly cosmopolitan (none breeds in Australia), fairly diverse group of shorebirds that generally have long to fairly long bills and legs. Bills straight to slightly decurved or upturned; bill shape and length, and to a lesser extent leg length, indicate feeding methods. Wings usually long and pointed, tails fairly short. Most sandpipers breed at high latitudes, and only one species breeds in the region. Eleven others occur mostly or entirely as transients, and the rest are winter visitors; non-breeding imms of larger species often over-summer. Age/seasonal variation usually noticeable; young precocial. Most shorebirds (adults and juvs) migrate before their prebasic molt, though contour feather replacement may be well underway by the time winter grounds are reached. Sexes typically similar though ♀♀ larger with longer bills in most species. Plumage colored mostly in browns, greys, black, and white, patterns often cryptic but beautiful; flight patterns may be striking. Voices vary from loud shrieking cries to mellow whistles, quiet trills, and chips.

Sandpipers eat mostly a variety of aquatic and semiaquatic invertebrates obtained by picking, probing, or snatching. Long-legged and long-billed species feed in deeper water, shorter-billed species along edges or in drier habitats.

Many birders shirk shorebird identification, particularly when it comes to sandpipers, but they aren't really that difficult. It helps that they are often social, conspicuous, and confiding, with several species occurring together. The most important characters to note are relative size, structure, bill shape, and call; plumage is so variable that it often comes last on the list of things to check.

GENUS TRINGA

Medium-sized to fairly large sandpipers with long, straight to slightly upturned bills, long legs. Age/seasonal variation, sexes similar.

GREATER YELLOWLEGS

Tringa melanoleuca   Not illustrated
Patamarilla Mayor

ID: 11.5–12.5 in; B 2–2.3 in (29–31.5 cm; B 5–6 cm). Widespread, especially coastal. *Bill long and straightish.* **Descr. Basic:** *bill blackish, often paler olive-grey on basal half, legs orange-yellow to yellow.* Head, neck, and chest streaked whitish and grey to grey-brown, with whitish supraloral stripe, dusky lores, white eyering and throat. Rest of underparts white, flanks often with slight dusky barring. *Upperparts grey to grey-brown, spotted and flecked whitish,* with *white lower rump and uppertail coverts* and pale greyish tail with dark bars. *Upperwings dark,* underwings dusky. **Alt:** head, neck, and chest darker overall with bold dark streaking and scalloping, flanks and lateral tail coverts barred blackish. Upperparts dark grey-brown with scattered blackish feathers, spotted and spangled white. **Juv:** resembles basic but browner overall, upperparts with denser whitish spots.

Voice. A loud, shrill to ringing *tchoo-tchoo-tchoo* or *kyew kyew kyew*, less often single notes or longer series.

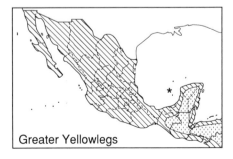

Greater Yellowlegs

Habs. Estuaries, marshes, lakes, lagoons, etc. Singly or in loose flocks, at times in hundreds. Associates readily with other shorebirds, including Lesser Yellowlegs. Feeds by sweeping bill from side to side, less often by picking.

SS: Lesser Yellowlegs smaller with relatively shorter, finer black bill, appears more delicate overall, note quieter voice. Willet larger and stockier, with thicker, more strongly bicolored bill, greyish legs, striking flight pattern.

SD: C to F transient and winter visitor (Jul–May) coastally from Baja, Son, and Tamps to Isthmus and N Yuc Pen; F to U to El Salvador and Honduras; F to U in interior (to 2500 m) S to cen volcanic belt; U to R S to Isthmus. Small numbers oversummer on both coasts.

RA: Breeds N N.A.; winters USA to S S.A.

LESSER YELLOWLEGS
Tringa flavipes   Not illustrated
Patamarilla Menor

ID: 9.5–10 in; B 1.3–1.5 in (24–25.5 cm; B 3.5–4 cm). Widespread, especially inland. Bill fairly long, slender, and straight. **Descr. Basic:** bill blackish, legs orange-yellow to yellow. Head, neck, and chest streaked whitish and grey to grey-brown, with whitish supraloral stripe, dusky lores, white eyering and throat. Rest of underparts white. *Upperparts grey to grey-brown, spotted and flecked whitish,* with *white lower rump and uppertail coverts* and pale greyish tail with dark bars. *Upperwings dark,* underwings dusky. **Alt:** head, neck, and chest darker overall with bold dark streaking and scalloping. Upperparts dark grey-brown with scattered blackish feathers, spotted and spangled white. **Juv:** resembles basic but browner overall, upperparts with denser whitish spots.

Voice. A bright to mellow *kyew kyew kyew* or *kyih kyi-kyih*, etc., not loud and ringing like Greater Yellowlegs; also shorter, mellow calls, *chew tewt* and *kyew*, etc., may suggest Short-billed Dowitcher; rarely longer series.

Habs. Marshes, lakes, lagoons, salt ponds, less often estuaries. Singly or in loose flocks, at times in hundreds. Associates readily with other shorebirds, including Greater Yellowlegs. Feeds mostly by picking.

SS: See Greater Yellowlegs. Stilt Sandpiper feeds by probing, bill slightly decurved at tip, upperparts lack fine whitish spotting; note whitish supercilium (basic) and voice. Wilson's Phalarope has finer bill, basic and juv whiter overall, tail pale grey; note different feeding habits. Solitary Sandpiper smaller and more compact with shorter, olive legs; note flight pattern and voice.

SD: F to C transient and winter visitor (Jul–May: SL–1500 m) on both slopes from Nay and Tamps to El Salvador and Honduras; U N to Baja; F to U in interior (to 2500 m), mainly from cen volcanic belt S. Small numbers may oversummer on both coasts.

RA: Breeds N N.A.; winters Mexico and SE USA to S S.A.

SOLITARY SANDPIPER
*Tringa solitaria**   Not illustrated
Playero Solitario

ID: 8–8.5 in; B 1–1.3 in (20.5–21.5 cm; B 2.5–3 cm). Widespread, especially inland. Bill straight and medium length. **Descr. Basic:** bill blackish, paler olive-grey at base, *legs olive* to olive-yellow. Head, neck, and sides of chest dusky grey-brown with *dark lores, white supraloral stripe and eyering;* rest of underparts white. *Upperparts grey-brown with fine, fairly sparse, whitish spotting. Uppertail coverts*

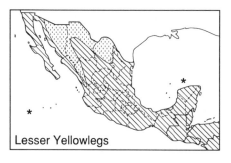

Lesser Yellowlegs

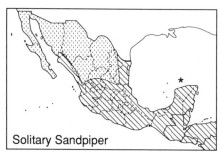

Solitary Sandpiper

and central tail dark with white, dark-barred lateral tail coverts and outer rectrices. Upperwings and underwings dark. **Alt:** head, neck, and chest streaked whitish, upperparts darker brown, brightly spotted and flecked whitish. **Juv:** resembles basic but with denser buffy-white spotting on upperparts.

Voice. A fluty, slightly piping to slightly plaintive *t-swee sweet* or *tsee-seet* and *twee-twee sweet*, and *teet, t-weet weet*, etc.; brighter and flutier than Spotted Sandpiper; may suggest alarm call of Barn Swallow; also a clipped *weit* in alarm.

Habs. Freshwater marshes, lakes, small ponds, usually not in open situations. Usually singly, separate from other shorebirds, but may form flocks of 20–30 birds in migration. Often towers when flushed, wingbeats deep, quick, and swallow-like. Bobs while feeding but less than Spotted Sandpiper.

SS: Spotted Sandpiper smaller with less upright stance, tail projects beyond wings, legs flesh to yellowish; note whitish supercilium, dusky chest side patches, white wingstripe, flight, and voice. See Lesser Yellowlegs.

SD: F to C transient (Apr–May, mid-Jul–Oct; SL–1500 m) on Atlantic Slope from Tamps to Honduras; F to U in interior (to 2500 m) and on Pacific Slope from Sin S; U to R in Baja and Son. U to R winter visitor (Sep–Apr) throughout region, except NW Mexico.

RA: Breeds N N.A.; winters Mexico and SE USA to Peru and cen Argentina.

WILLET

Catoptrophorus semipalmatus*
Playero Pihuihui   Not illustrated

ID: 12.5–14 in; B 2–2.5 in (32–35.5 cm; B 5–6.5 cm). Widespread, coastal. *Bill long,*

Willet

stout, and straight. **Descr.** Age/seasonal variation, sexes similar. **Basic:** *bill blue-grey basally, blackish distally, legs blue-grey* to olive-grey. Head, neck, and chest pale brownish grey with *white supraloral stripe and eyering*, dusky lores, and whitish throat; rest of underparts white. *Upperparts pale brownish grey* to grey-brown with white uppertail coverts and pale grey tail. *Wing pattern striking in flight:* upperwings grey-brown with *broad white stripe across secondaries and base of primaries contrasting with black primary coverts and primary tips; underwings black with broad white stripe* as above, and white lesser wing coverts.

Voice. Noisy. A nasal, laughing *kyeh-heh* and *kyeh yeh-yeh*, a shrieky *kriiehrr* and *rrieh*, a rough, yelping to screechy *wi-wi-wiih* or *krri-rri-rrit*; in alarm an often persistent, hard, rough, clipped *keh-keh-keh ... 'Song' a shrieky krr-wee-wee-weet* or *kr wi will-it*, repeated. Feeding and roosting birds give gruff, bickering, and social chatters interspersed with muted shrieks.

Habs. Sandy beaches, estuaries, coastal lagoon, salt marshes; rare inland. Singly or in loose flocks, locally in hundreds. Associates readily with other shorebirds. Noisy and wary, often the first shorebird to flush. Feeds by probing and picking. Nest is a scrape on the ground. Eggs: 4, buff to olive-buff, boldly marked with browns and greys (WFVZ).

SS: See Greater Yellowlegs.

SD: C to F breeder (Apr–Aug) in coastal Tamps (SNGH, SW). C to F transient and winter visitor (late Jul–May) along both coasts; small numbers oversummer widely. R to U transient (Apr, Aug–Oct) in interior (900–2500 m) S to cen volcanic belt (AGL spec, SNGH, RGW). Reports of breeding in Yuc (Edwards 1989) require confirmation.

RA: Breeds interior and E N.A. to NE Mexico and Caribbean; winters USA to N Chile and N Brazil.

WANDERING TATTLER

Heteroscelus incanus   Not illustrated
Playero Vagabundo

ID: 9.5–10 in; B 1.2–1.5 in (24–25.5 cm; B 3–4 cm). Rocky Pacific coasts. Bill straight, medium length, and fairly stout. **Descr.** Age/seasonal variation, sexes similar. **Basic:** bill blackish, legs yellow. *Smoky*

Wandering Tattler

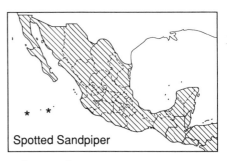

Spotted Sandpiper

grey overall, darker above, with *dark lores, white supraloral stripe, and broken white eyering*; throat, belly, and undertail coverts white. *Upperwings and tail grey, remiges darker; underwings smoky grey.* **Alt:** *face, foreneck, chest, and underparts whitish, streaked and barred blackish.* **Juv:** resembles basic but upperparts, especially wing coverts, with fine scaly pale edgings and pale flecking, chest indistinctly vermiculated paler.

Voice. A slightly plaintive piping *tee-lee-lee* and *tee-lee lee-lee-lee*, and a rougher *kree-lee* and *kree-lee-lee-lee*, etc.

Habs. Rocky coasts and islands, rarely beaches. Singly or in small loose groups; associates with other rocky coast shorebirds, especially at roosts.

SD: F to U transient (Apr–May, Aug–Oct) along Pacific coast S to Oax; apparent rarity S of Isthmus largely due to little rocky coastline. F winter visitor (Oct–Mar) along Pacific coast of Baja and on offshore islands; U to R along rest of Pacific coast and in Gulf of California. Small numbers oversummer along Pacific coast of Baja and on offshore islands, probably locally elsewhere.

RA: Breeds NW N.A.; winters W N.A. to Peru, and in cen and E Pacific.

SPOTTED SANDPIPER
Actitis macularia   Not illustrated
Playero Alzacolita

ID: 6.5–7 in; B 0.7–1 in (16.5–18 cm; B 2–2.5 cm). Widespread. Bill medium length and straight. **Descr.** Age/seasonal variation, sexes similar. **Basic:** bill dusky to flesh with dark tip, *legs dull flesh to yellow. Head, sides of chest, and upperparts grey-brown with short whitish supercilium*, dark lores, and narrow broken white eyering; upperwing coverts with dark and pale barring. *Throat, median*

chest, and underparts white. Tail dark with dark-barred white sides. Upperwings dark brown with white wingstripe, underwings white with broad black tip and trailing edge, black bars on median and primary coverts.* **Alt:** bill and legs bright orange-pink, bill tipped black. *Throat, chest, and underparts white, spotted black.* Upperparts barred and spotted blackish. **Juv:** resembles basic but wing covert barring brighter, buff, and dark brown, upperparts with fine, scaly buff edgings.

Voice. A high, slightly plaintive to piping *siit* or *seet*, and a longer *sii-wii* and *see wee-wee*, etc.

Habs. Lakes, rivers, marshes, less often estuaries and beaches. Singly, or a few birds at favored spots, usually not associating with other shorebirds. Flight typically low over water with shallow, stiff, flicking wingbeats.

SS: See Solitary Sandpiper.

SD: F to C transient and winter visitor (Jul–May; SL–2500 m) throughout; less common on N Plateau in winter.

RA: Breeds N.A.; winters S USA to N Chile and N Argentina.

UPLAND SANDPIPER
Bartramia longicauda   Not illustrated
Zarapito Ganga

ID: 11–12 in; B 1–1.3 in (28–30.5 cm; B 2.5–3 cm). Transient on grasslands. *Tall-standing with slender neck and small head, long graduated tail projects beyond wingtips at rest. Bill slender, medium length, and very slightly decurved.* **Descr.** Ages differ slightly, sexes similar. **Adult:** bill orange-yellow with blackish culmen and tip, *legs orange-yellow. Whitish eyering and pale lores accentuate big dark eye on plain buff face*; throat whitish, crown dark brown, streaked paler with pale median stripe. Neck and chest buff, streaked

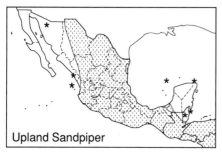

Upland Sandpiper

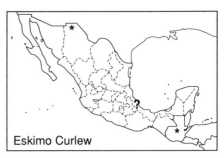

Eskimo Curlew

blackish; underparts white, scalloped blackish on sides and flanks. *Upperparts dark brown to blackish, boldly edged buff to cinnamon-buff, rump blackish.* Tail barred brown and blackish, graduated outer rectrices tipped white. *Upperwings brown, primaries and primary coverts darker.* Juv: upperparts more boldly scaly, especially wing coverts and scapulars which have dark bars.

Voice. A mellow to sharp, liquid *whi whi whuit* or *hui hui huit*, and *whi whi whi-whit*, etc.; may suggest Curve-billed Thrasher.

Habs. Grasslands, fields, open areas in general, rarely beaches. Singly or in small flocks, locally in hundreds on spring migration. Associates loosely with other grassland shorebirds.

SS: Buff-breasted Sandpiper smaller with black bill, proportionately longer yellow legs, face and underparts plain buff, wings project beyond tail.

SD: F to C transient (late Mar–mid-May, late Jul–mid-Oct; SL–2500 m) on Atlantic Slope (but R in Yuc Pen) and in interior; U on Pacific Slope S of Isthmus; R (late Jul–Aug) in NW Mexico (Grant 1964a; Russell and Lamm 1978; PH).

RA: Breeds N.A.; winters E and S S.A.

GENUS NUMENIUS: Curlews

Curlews are medium-sized to large shore-birds with slender decurved bills and long legs. Ages differ slightly. Bills blackish, flesh below at base, legs grey.

ESKIMO CURLEW
Numenius borealis   Not illustrated
Zarapito Boreal

ID: 12–13.5 in; B 2–2.5 in (30.5–34.5 cm; B 5–6 cm). *Very rare transient. Bill slender, medium length, and slightly decurved.*

Descr. Adult: *crown dark brown, streaked paler, with pale median stripe; dark eye-stripe sets off pale buff supercilium.* Face, neck, chest, and underparts pale buff to cinnamon-buff, streaked dark brown on face, neck, and chest, becoming dark scallops on flanks; throat and belly paler. Upperparts dark brown, edged and notched buff to cinnamon-buff, becoming brighter *buffy cinnamon barring on rump and uppertail coverts. Tail barred dark brown and greyish buff to buffy cinnamon.* Upperwings brown, regimes and primary coverts darker, *underwing coverts cinnamon, barred dark.* Juv: upperparts more scaly and more brightly edged.

Voice. A short, low to melodious whistle, *bee bee*, and a fluttering *tr-tr-tr* in flight (Bent 1929).

Habs. Grasslands. Likely to be found singly or in small groups, perhaps associating loosely with other grassland shorebirds. Feeds by probing and picking.

SS: Whimbrel larger, darker grey-brown overall with more boldly striped head, longer and heavier bill, brown rump and tail, underside of primaries barred paler (unbarred in Eskimo Curlew). Upland Sandpiper smaller with shorter straighter bill, bare-faced expression, yellow legs.

SD: Formerly probably a regular spring transient through interior and along Atlantic Slope N of Isthmus, but only two records, both in Apr, from Chih (Friedmann *et al.* 1950) and Guatemala (Salvin 1861). Should be looked for on coastal plains of Ver where large numbers of American Golden Plovers and Upland and Buff-breasted sandpipers stage in spring.

RA: Breeds NW N.A.; winters S S.A.

NB: Close to extinction.

WHIMBREL

Numenius phaeopus hudsonicus
Zarapito Trinador   Not illustrated

ID: 14–17 in; B 3–4 in (35.5–43 cm; B 7.5–10 cm). Widespread, mainly coastal. *Bill long and decurved distally.* **Descr. Adult:** *crown dark brown with whitish median stripe; dark eyestripe sets off pale buff supercilium.* Face, neck and chest buff, streaked dark brown, throat, belly, and undertail coverts whitish, flanks barred dark brown. Upperparts dark brown, edged and notched buff, *tail barred brown and dark brown. Upperwings brown,* remiges darker. Underwing coverts cinnamon-buff, barred dark brown, underside of remiges dusky brown, barred paler. **Juv:** upperparts with brighter and scalier cinnamon-buff edgings.
  Voice. A fairly rapid series of piping or tittering whistles, *kee hee-hee-hee-hee,* or *whee hee-hee-hee-hee-hee,* etc.; a rolled, slightly bubbling *whirr-i whirr-ip* especially in spring, etc.
  Habs. Marshes, estuaries, fields, sandy beaches, lagoons, etc. Singly or in small flocks, often associating with other shorebirds. Feeds by probing and picking.
SS: Long-billed Curlew larger with longer, more gradually decurved bill, overall cinnamon cast obvious in flight, lacks bold dark head stripes.
SD: F to C winter visitor (Sep–Apr) along Pacific coast, including Islas Revillagigedo, and in Gulf of California; F to U on Atlantic coast; commoner during migration (late Jul–Sep, Mar–early May). Small numbers oversummer along both coasts. R spring transient (Apr) in interior (SNGH, SW).
RA: Holarctic breeder; winters in New World from USA to S Chile and S Brazil.

LONG-BILLED CURLEW

*Numenius americanus**   Not illustrated
Zarapito Piquilargo

ID: 18–23 in; B 4–8 in (45.5–58.5 cm; B 10–20.5 cm). Widespread, mainly N of Isthmus. *Bill long to extremely long and decurved.* **Descr. Adult:** *crown dark brown, streaked paler, face pale buff with whitish eyering* and often a dusky loral stripe. Neck, chest, and *underparts pale buff to cinnamon, streaked dark brown on neck and chest,* barred brown on flanks. Upperparts dark brown, edged and notched cinnamon to pale buff, uppertail coverts boldly spotted and barred cinnamon; *tail barred cinnamon and dark brown. Upperwings brown with cinnamon secondaries, greater coverts, and inner primaries contrasting with dark brown primary coverts and outer primaries. Underwings cinnamon,* coverts unbarred, remiges barred dusky. **Juv:** upperparts with brighter and bolder buffy cinnamon edgings; note relatively short bill into mid-winter.
  Voice. A slightly shrieky to hoarse, whistled *ree-ip* or *hoo-reep* and a single *hoor* or *heuwr,* etc., at times run into bubbling series.
  Habs. Estuaries, salt marshes, fields, lakes, lagoons. Often in flocks, locally to a few hundred birds. Associates with other large shorebirds. Feeds by probing with long bill, often wading up to its belly.
SS: See Whimbrel. Marbled Godwit smaller with straight to slightly upturned bill.
SD: C to F transient and winter visitor (late Jul–May) in Baja and on Pacific Slope from Son to Col; U to R coastally to El Salvador (Thurber *et al.* 1987); F to U in interior (to 2500 m) S to cen volcanic belt, and on Atlantic coast from Tamps to Yuc Pen; R to Belize. A report from N Honduras

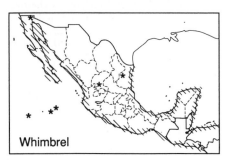

Whimbrel

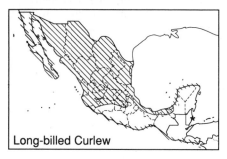

Long-billed Curlew

(Monroe 1968) is unsatisfactory. Smaller numbers oversummer on both coasts.

RA: Breeds interior W N.A.; winters S USA to N C.A., rarely to N S.A.

GENUS LIMOSA: Godwits

Godwits are medium-large shorebirds resembling curlews but with long, straight to slightly upturned bills. Bills pink basally, black distally, legs grey. Age/seasonal and often sex variation; ♀♀ noticeably larger than ♂♂.

HUDSONIAN GODWIT

Limosa haemastica   Not illustrated
Picopando Ornemantado

ID: 13.5–16 in; B 3–3.5 in (35–40.5 cm; B 7.5–9 cm). Uncommon transient. Wingtips fall beyond tail at rest. **Descr. Alt ♂:** *head whitish*, streaked dusky, with greybrown crown and white supercilium, dark loral stripe. *Foreneck, chest, and underparts reddish chestnut*, streaked and barred dark brown, flanks and belly narrowly barred whitish. Hindneck and upperparts dark brown, spotted and spangled buff and whitish on back; *white uppertail coverts contrast with black rump and tail*, outer rectrices white basally. *Upperwings blackish brown with white stripe across base of outer secondaries and inner primaries; underwings dark with white stripe.* **Alt ♀:** foreneck, chest, and underparts whitish to pale buff, scalloped chestnut on underparts. **Basic:** short white supercilium contrasts with dark crown and lores; rest of head, neck, chest, and underparts pale greyish, paler on throat and belly. Upperparts brownish grey. **Juv:** resembles basic but neck, chest, and underparts pale greyish buff, upperparts brown with scaly

buff edgings and dark subterminal scallops.

Voice. A hoarse chippering *kyik-kyik ...* or *kyeh-kyeh ...*, and a more nasal *kyehk-kyehk-kyeh-kyehk*, etc.

Habs. Coastal lagoons, estuaries, marshes. In small groups or singly. Associates loosely with other shorebirds. Feeds by probing, often wading up to its belly.

SS: Marbled Godwit larger, overall pale greyish cinnamon to buffy cinnamon with boldly spangled upperparts, lacks black-and-white wing pattern.

SD: U to F spring transient (mid-Apr–May) on Pacific coast from Guatemala to Isthmus (Binford 1989; SNGH, SW), and on Atlantic coast in Tamps and Ver (Coffey 1960); R to Yuc (PES) and in interior to DF (JKe, RGW). No satisfactory autumn records.

RA: Breeds N N.A.; winters coasts of S S.A.

BAR-TAILED GODWIT

Limosa lapponica baueri   Not illustrated
Picopando Colibarrado

ID: 14–17 in; B 3.3–4.3 in (35.5–43 cm; B 8–10.5 cm). Vagrant. **Descr. Juv:** face pale buff with buffy white supraloral stripe and dark loral stripe, crown dark brown, streaked paler. *Neck and chest buff* with dusky streaking, becoming *buffy white on thighs and undertail coverts; lateral undertail coverts barred dark brown.* Upperparts dark brown, edged and notched whitish to pale buff, *uppertail coverts and tail barred whitish and dark brown* (uppertail coverts may appear contrastingly whitish in flight). *Upperwings brown with darker remiges; underwings whitish, barred dusky;* may show whitish flash across base of primaries in flight. **Basic:** plainer and greyer overall, underparts mostly whitish. **Alt ♂:** face, foreneck, chest, and underparts chestnut. **Alt ♀:** resembles basic but chest and under-

Hudsonian Godwit

Bar-tailed Godwit

parts mottled pale cinnamon. *L. l. lapponica* (vagrant to E USA) lacks dark barring on underwings, rump and uppertail coverts white in inverted V up back.

Voice. A slightly nasal, dry *kek*, and a barking *kak-kak* (California).

Habs. Much like Marbled Godwit with which it occurs readily.

SS: Lacks cinnamon in wings of Marbled Godwit, and tail barred whitish and dark brown. Juv browner overall and basic greyer than Marbled Godwit.

SD: Vagrant, one record: a juv photographed at Ensenada, BCN, on 12 Sep 1992 (CLu, KAR). A specimen (USNM 86418, head only) from BCS has been labeled as this species. While the head matches those of numerous Bar-tailed Godwits, it also matches some Marbled Godwits. Based on equivocal plumage and bill length, we concur with Grinnell (1928) that the specimen is unidentifiable.

RA: Breeds N Eurasia to W Alaska; winters Eurasia to Africa and Australia.

MARBLED GODWIT

Limosa fedoa   Not illustrated
Picopando Canelo

ID: 15–18 in; B 3.5–5 in (38–45.5 cm; B 9–12.5 cm). Common on both coasts. **Descr.** Ages differ slightly, sexes similar. *Adult: face pale buff with paler supraloral stripe and dark loral stripe,* crown dark brown, streaked paler. Neck, chest, and *underparts pale buffy cinnamon* to pale cinnamon (pale greyish to pale buff when worn), *streaked brown on neck, barred brown on chest and flanks* (barring more extensive in alt). Upperparts dark brown, boldly spangled and notched cinnamon to pale buff, *uppertail coverts and tail barred cinnamon and dark brown. Upperwings brown with cinnamon secondaries, greater coverts, and inner primaries contrasting*

with dark brown primary coverts and outer primaries. Underwings cinnamon, unbarred. **Juv:** foreneck, chest, and underparts faintly streaked and barred darker, upperparts with neater and scalier edgings.

Voice. Nasal, slightly crowing or laughing calls, *ah-ha* or *ah-ahk*, and a single *ahk*; flocks in flight often make laughing chatters. Roosting and feeding groups make nasal, bickering chatters.

Habs. Estuaries, salt marshes, lagoons, beaches. Often in flocks, locally to a few hundred birds. Associates with other large shorebirds. Habits much as Hudsonian Godwit.

SS: See Long-billed Curlew, Hudsonian, and Bar-tailed (vagrant) godwits.

SD: C to F transient and winter visitor (late Jul–May) along coasts from Baja and Son to Col; U to R to El Salvador; F to U from Tamps to Yuc Pen, R to N Honduras. Smaller numbers oversummer. Vagrant to Coah (Aug 1980, AGL spec.).

RA: Breeds W interior N.A.; winters S USA to N Chile and Colombia.

GENUS ARENARIA: Turnstones

Turnstones are chunky shorebirds typically associated with rocky coasts and sandy beaches. Black bills medium length, stout, and slightly up-tilted at tip for turning over seaweed and stones in search of prey; legs thick and medium length. Age/seasonal variation, sexes similar. Both species show striking patterns in flight.

RUDDY TURNSTONE

Arenaria interpres*   Not illustrated
Vuelvepiedras Rojizo

ID: 8.5–9 in; B 0.7–1 in (21.5–23 cm; B 2–2.5 cm). Common on both coasts. **Descr. Basic:** *legs bright orange-red.*

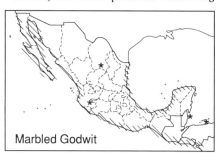

Marbled Godwit

Ruddy Turnstone

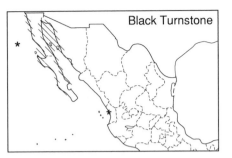

Broad, blackish-brown, double U-shaped chest band contrasts with whitish throat and underparts. Head and upperparts brownish, *head with diffuse, smudgy, whitish and dark patterning*, upperparts edged pale brown and dull rufous. *White lower back, rump, and distal uppertail coverts revealed in flight, in striking contrast to black basal uppertail coverts and white-edged black tail. Upperwings dark with white wingstripe and white bar across wing base*; underwings whitish. **Alt:** head and chest strikingly patterned black and white; upperparts rufous-chestnut with black V on mantle and scapulars. **Juv:** legs may be dull brownish yellow. Resembles basic but upperparts with neat buff to pale cinnamon edgings.

Voice. A mellow, rolled, clipped *ki-ti-tuk* or *ki-tih-tuh*, and a harder *kri-ti-tuk*, all suggesting Short-billed Dowitcher but sharper and harder; also a single *kek* and dry, rolled chatters. Feeding and roosting birds give mellower chatters and bickering.

Habs. Sandy beaches, rocky coasts, lagoons, estuaries. Often in small groups, rarely to 100+ birds. Associates readily with other shorebirds, commonly with Sanderlings on beaches.

SS: Black Turnstone darker overall with more uniform head and chest, duller legs.

SD: C to F transient and winter visitor (Aug–May) on both coasts; small numbers oversummer widely. R transient (Apr–May, Aug–Sep) in interior (to 2500 m), at least in cen Mexico (SNGH, RGW).

RA: Holarctic breeder; wintering in New World from USA to S S.A.

BLACK TURNSTONE
Arenaria melanocephala   Not illustrated
Vuelvepiedras Negro

ID: 8.5–9 in; B 0.7–1 in (21.5–23 cm; B 2–2.5 cm). NW Mexico. **Descr. Basic:** *legs dull pinkish grey to greyish flesh*. Head, chest, and upperparts sooty blackish brown, throat and postocular stripe typically paler, scapulars and wing coverts edged whitish, chest mottled darker. Belly and undertail coverts white. *White lower back, rump, and distal uppertail coverts revealed in flight, in striking contrast to black basal uppertail coverts and white-edged black tail. Upperwings dark with white wingstripe and white bar across*

wing base; underwings white. **Alt:** blacker overall, with white supraloral spot, white flecking from face extends to chest sides. **Juv:** resembles basic but upperparts with fine, pale scaly edgings.

Voice. Guttural, dry chatters and bickering calls, *weh-ka weh-ka ...* and *k-wek k-wek ...*, etc. a rolled *krrrrrrr*, and rolled, dry chatters; flight rattle higher and drier than Ruddy Turnstone.

Habs. Rocky coasts, estuaries, beaches. Often in small flocks, may associate with Ruddy Turnstones.

SS: See Ruddy Turnstone. Surfbird larger, paler grey overall, straight bill orange below at base, pattern in flight restricted to white tail with black terminal band, and white wingstripe.

SD: F to C transient and winter visitor (late Aug–Apr) on Pacific coast of BCN; U to R to BCS and in Gulf of California, and S to Nay (SNGH, PP). Small numbers oversummer in NW Baja.

RA: Breeds Alaska; winters S Alaska to NW Mexico.

SURFBIRD
Aphriza virgata   Not illustrated
Playero de Marejada

ID: 9.5–10 in; B 1 in (24–25.5 cm; B 2.5 cm). Transient and winter visitor on Pacific

coast. *A medium-sized stocky shorebird of rocky coasts; bill medium short, straight, and thick.* **Descr.** Age/seasonal variation, sexes similar. **Basic:** *bill black with yellow-orange base below,* legs olive to yellow. *Head, chest, and upperparts slaty grey, with white eyering,* whitish streaking on throat and supercilium. *Belly and undertail coverts white, spotted dark brown. Broad white band on uppertail coverts and base of tail contrasts with broad black tail tip.* Upperwings dark with *broad white wing-stripe.* **Alt:** head, chest, and underparts white, coarsely streaked and spotted black; upperparts blackish with *tortoiseshell-patterned chestnut and black scapulars.* **Juv:** resembles basic but face, throat, and chest neatly streaked and mottled whitish, dark spotting faint or absent on belly and undertail coverts, wing coverts neatly edged whitish.

> **Voice.** A quiet, mellow *whek-whek ...,* and a higher *whik,* easily lost amid the sound of crashing waves.

> **Habs.** Rocky coasts. In flocks up to several hundred birds in spring. Wintering individuals associate with other rocky coast shorebirds.

SS: See Black Turnstone.

SD: F to C spring transient (mid-Mar–Apr) on coasts and islands in Gulf of California, a major stopover for the species; U to F S to Col; U to R elsewhere on Pacific coast. U to R autumn transient and winter visitor (Sep–Apr) locally along Pacific coast.

RA: Breeds NW N.A.; winters S Alaska to S S.A.

GENUS CALIDRIS

These are often considered typical sandpipers. *Calidris* sandpipers are very small to medium-sized shorebirds; smaller species are known as peeps in North America, and as stints in Britain. Bills typically medium length, straight to slightly decurved. Age/seasonal variation, sexes similar. Bills usually black, legs black to flesh or yellowish.

RED KNOT
Calidris canutus rufa   Not illustrated
Playero Gordo

ID: 10–10.5 in; B 1–1.5 in (25.5–26.5 cm; B 2.5–4 cm). Local migrant. A *stocky,* large calidrid of muddy and sandy shores; *bill*

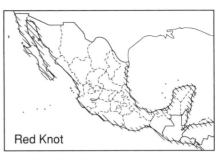

Red Knot

medium length, straightish. **Descr. Basic:** bill black, *legs dark greyish olive* to yellow-olive. Face, throat, chest, and underparts whitish, streaked and spotted dusky on face, chest, and flanks; whitish supercilium and dusky lores. Crown, nape, and *upperparts grey, white rump and uppertail coverts sparsely barred dusky, tail pale grey.* Upperwings grey with dark remiges and primary coverts, *narrow whitish wing-stripe broadest on base of primaries.* **Alt:** *face, throat, chest, and underparts brick-chestnut,* whitish on dark-barred hind flanks and undertail coverts. Crown, nape, and upperparts blackish, edged and checkered with cinnamon, whitish, and chestnut. **Juv:** resembles basic but chest washed pinkish buff when fresh, chest and flanks finely speckled dusky; upperparts with pale scaly edgings and dark subterminal scallops.

> **Voice.** Quiet, low, slightly grunting calls, *chuh* or *kuh,* and a slightly gruffer *kruh* or *gruh.*

> **Habs.** Muddy estuaries, coastal lagoons, beaches. In small, often compact flocks, associating loosely with other shorebirds. Feeds by picking and probing.

SD: U to F transient and local winter visitor (Aug–May) on both coasts; R through the summer, at least on Atlantic coast (SNGH, SW).

RA: Holarctic breeder; wintering in New World from USA to S S.A.

SANDERLING
Calidris alba   Not illustrated
Playero Blanco

ID: 7–7.5 in; B 0.7–1 in (18–19 cm; B 2–2.5 cm). Common on sandy beaches; *bill medium length and straight,* foot lacks hind toe. **Descr. Basic:** bill and legs black. *Face, throat, chest, and underparts white* with faint dusky smudge through eye, dusky

Sanderling

smudge on sides of chest. Crown, nape, and *upperparts pale grey*, wings slightly darker; *blackish lesser coverts often show as dark shoulder.* Central tail dark, lateral tail coverts and sides of tail white. Upperwings pale grey with black primary coverts and remiges, *broad white wingstripe.* **Alt:** *face, throat, and chest washed rufous and streaked dark brown.* Crown, nape, and upperparts spangled blackish, rufous, and whitish. **Juv:** face, throat, chest, and underparts white with *dark smudge through eye,* dusky streaking on sides of chest; chest washed buff in fresh plumage. Crown, nape, and *upperparts blackish, overall edged and spangled buff and whitish, nape contrastingly whitish,* wing coverts pale grey.

Voice. A bright, sharp *kip* or *kiip,* and *pliik* or *whiik.*

Habs. Sandy beaches, salt lagoons, less often estuaries, rocky coasts. In small groups, associating often with other shorebirds. Beach feeding birds noted for running out after the receding waves to pick for food, then hurrying back up the beach ahead of incoming waves. Feeds by picking and probing.

SS: Basic Red Phalarope in flight is deep-chested, with longer neck, and smaller head; note black postocular stripe, habits, call.

SD: C to F transient and winter visitor (Aug–May) along both coasts; small numbers oversummer locally. R transient and winter visitor (late Aug–early Mar; to 2500 m) in interior N and cen Mexico (SNGH, AGL spec., RGW).

RA: Holarctic breeder; wintering in New World from N.A. to S S.A.

SEMIPALMATED SANDPIPER
Calidris pusilla   Not illustrated
Playerito Semipalmado

ID: 5.5–6 in; B 0.6–0.7 in (14–15 cm; B 1.5–2 cm). *Migrant, mainly in east.*

Straightish bill, short to medium length, fairly thick throughout. **Descr. Basic:** bill and legs black. Face, throat, chest, and underparts white with dusky smudge through eye, *dusky chest side patches streaked darker; supercilium thickest at base of bill.* Crown, auriculars, nape, and *upperparts grey-brown* to brownish grey. Upperwings greyish with dark remiges, white wingstripe; median uppertail coverts and tail dark with white sides. **Alt:** face and chest streaked dark brown. Crown, nape, and *upperparts blackish, edged buff and buffy rufous,* brightest on crown, auriculars, and some scapulars. **Juv:** face and chest whitish with *dark smudge through eye,* sides of chest streaked dusky and chest washed buff when fresh. *Crown, nape, and upperparts blackish with neat pale buff and buffy-rufous edgings, at times fairly bright rufous on crown, scapulars, and tertials.*

Voice. A fairly sharp *kip* or *kyip,* a low *chrrt* or *chrit,* and a low, gruff *krrip* or *krrp.*

Habs. Coastal lagoons, estuaries, beaches, salt ponds, lakes. In flocks, in migration up to hundreds, mixing readily with other shorebirds. Feeds by picking and probing.

SS: Western Sandpiper averages larger and lankier, more Dunlin-like in shape, with longer, slightly decurved bill, longer legs, less thickset posture. Basic Semipalmated appears distinctly browner than Western in direct comparison, with streaking restricted to sides of chest; alt and juv Westerns bright rufous and grey overall, versus more uniform browner appearance of Semipalmated; alt Western has dark-spotted flanks. Western molts about one month earlier than Semipalmated; thus, by Sep, few Westerns are in full juv plumage, with most showing grey 1st basic feathers. Note calls. White-rumped Sandpiper (tran-

Semipalmated Sandpiper

sient) has long wings projecting well beyond tail at rest; note white uppertail coverts in flight. Least Sandpiper smaller, more mouse-like in its overall brown plumage, creeping gait with flexed legs; yellowish legs often covered with mud or color hard to see; note slightly decurved, tapering bill, voice.

SD: F to C transient (late Mar–early Jun, mid-Jul–Oct) on Atlantic coast from Tamps to Honduras, and on Pacific coast S of Isthmus; U to R transient (Aug–mid-Sep) on Pacific coast from BCN (Wurster and Radamaker 1992) to Isthmus. U to F winter visitor (Sep–Apr) on N coast of Yuc Pen (SNGH, PP, SW, photo), and from Isthmus (SNGH, SW) S locally along Pacific coast.

RA: Breeds N N.A.; winters S Mexico and Caribbean to N Chile and S Argentina.

WESTERN SANDPIPER

Calidris mauri   Not illustrated
Playerito Occidental

ID: 6–6.5 in; B 0.7–1 in (15–16.5 cm; B 2–2.5 cm). *Widespread and common. Bill straightish to slightly decurved, tapering slightly toward tip.* **Descr. Basic:** bill and legs black. Face, throat, chest, and underparts white with dusky smudge through eye, *dusky streaking across chest often indistinct in center.* Crown, nape, and *upperparts grey to brownish grey.* Upperwings grey with dark remiges, white wingstripe; median uppertail coverts and tail dark with white sides. **Alt:** face, chest, and *flanks streaked and spotted dark brown, auriculars washed rufous.* Crown, nape, and *upperparts blackish, broadly edged rufous and buff, rufous brightest on sides of crown and scapulars.* **Juv:** face and chest whitish with dusky smudge through eye, sides of chest streaked dusky, chest washed buff when fresh. Crown, nape, and *upper-*

parts blackish with neat rufous, buff, and whitish edgings, rufous mainly on mantle, inner scapulars, and tertials.

Voice. A high, slightly reedy *chiit* or *jiit*, a lower, more rolled *chit* or *chrit*, a hoarse *chihr*, less often *chrip*. Feeding flocks make high, excited, reedy piping and twittering.

Habs. Coastal lagoons, estuaries, beaches, salt ponds, lakes. In flocks, up to thousands, mixing readily with other shorebirds. Feeds by picking and probing.

SS: See Semipalmated Sandpiper. White-rumped Sandpiper (transient) has long wings projecting well beyond tail at rest; note white uppertail coverts in flight. Least Sandpiper smaller, more mouse-like in its overall brown plumage, creeping gait with flexed legs; yellowish legs often covered with mud or color hard to see; note voice.

SD: C to F transient and winter visitor (late Jul–May) along both coasts; U to C locally in interior N and cen Mexico (to 2500 m). R in summer locally along both coasts.

RA: Breeds N N.A.; winters S USA to N Peru and Surinam.

LEAST SANDPIPER

Calidris minutilla   Not illustrated
Playerito Mínimo

ID: 5.2–5.7 in; B 0.6 in (13.5–14.5 cm; B 1.5 cm). Widespread and common. A very small, mouse-like calidrid; *bill medium length and slightly decurved, tapering toward tip.* **Descr. Basic:** bill black, *legs olive to yellow.* Head, chest, and upperparts grey-brown with indistinct whitish supercilium, whitish throat; chest streaked darker, upperparts mottled darker. Belly and undertail coverts whitish. Upperwings grey-brown with dark remiges, narrow white wingstripe; median uppertail coverts and tail dark, edged white. **Alt:** legs brighter, yellow to orange-yellow. Face

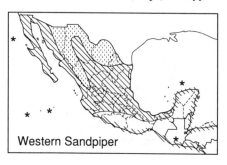

Western Sandpiper

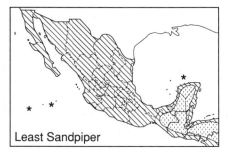

Least Sandpiper

and chest paler, strongly streaked and spotted dark brown. Crown, nape, and upperparts blackish, edged rufous and buff. **Juv:** resembles alt but chest washed buff with dark streaking mainly on sides; crown, nape, and upperparts with neat scaly rufous, buff, and whitish edgings, often forming 1–2 whitish Vs on back.

Voice. A high, reedy to slightly rolled *krreet* or *krreep*, less often a lower, reedy, rolled *krriit* or *krrip*, etc., and mellow trilling.

Habs. Marshes, coastal lagoons, lakes, estuaries, beaches; favors fresh water. In flocks, often in hundreds, mixing readily with other shorebirds but often keeping to drier and more vegetated areas than Western and Semipalmated sandpipers. Rarely wades up to its belly to feed. Feeds by picking and probing.

SS: See Semipalmated and Western sandpipers.

SD: C to F transient and winter visitor (mid-Jul–May; SL–2500 m) on both slopes throughout and in interior S to Isthmus; R in interior from Isthmus S.

RA: Breeds N N.A.; winters USA to Peru and cen Brazil.

WHITE-RUMPED SANDPIPER

Calidris fuscicollis   Not illustrated
Playerito Rabadilla-blanca

ID: 6.7–7.3 in B 0.7–1 in; (17–18.5 cm; B 2–2.5 cm). Transient on Atlantic coast. *Long wings project well beyond tail at rest*; bill medium length and straightish. **Descr. Alt:** bill and legs black, bill often with orangish spot below at base. Face, throat, and underparts whitish, streaked and spotted dark brown on face, chest, and flanks, with *whitish supercilium* and rufous-washed auriculars. Crown, nape, and *upperparts blackish, edged rufous and buff, white uppertail coverts ('rump')* con-

trast with dark rump and blackish tail. Upperwings dark with narrow white wingstripe. **Basic:** head and upperparts grey to grey-brown with whitish supercilium thickest behind eye. Throat and underparts whitish, chest streaked dusky, often some dusky spots on flanks. **Juv:** resembles alt but chest often washed buff; crown, nape, and upperparts with neat rufous, buff, and whitish edgings, often forming 1–2 white Vs on back.

Voice. A very high, thin, slightly squeaky *jiiht* or *jiht*; flocks make high thin twitters.

Habs. Coastal lagoons, beaches, estuaries, lakes. Singly or in small groups, associating readily with other shorebirds. Feeds by picking and probing.

SS: See Western and Semipalmated sandpipers. Baird's Sandpiper browner overall, upperparts with buff edgings, supercilium less contrasting, lacks white uppertail coverts; note voice.

SD: F to U transient (mid-Apr–early Jun) on Atlantic Slope from Tamps to Yuc Pen and N Honduras. U to R transient (Aug–Nov) on N coast of Yuc Pen, probably elsewhere on Atlantic coast.

RA: Breeds N N.A.; winters S.A.

BAIRD'S SANDPIPER

Calidris bairdii   Not illustrated
Playerito de Baird

ID: 6.5–7 in; B 0.8 in (16.5–18 cm; B 2 cm). Widespread transient. *Long wings project well beyond tail at rest*; bill medium length and straightish. **Descr. Alt:** bill and legs black. *Face and chest buff*, streaked dark brown, with *paler supercilium*; throat and rest of underparts white. Crown, nape, and *upperparts dark brown, edged buff to cinnamon-buff*; median uppertail coverts and tail blackish, edged white. Upperwings brown with dark remiges, narrow whitish wingstripe. **Basic** (rarely seen): greyer

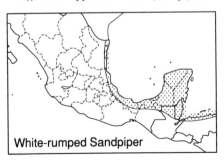

White-rumped Sandpiper

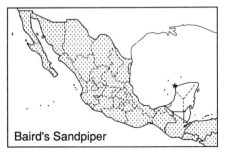

Baird's Sandpiper

overall, upperparts more uniform grey-brown. **Juv:** resembles alt but face and chest often brighter buff with less distinct streaking on chest; upperparts with neat, scaly buff and whitish edgings.

Voice. A rolled, slightly reedy to gravelly *krrrip* or *krriih*, and dry, rolled *krrit*.

Habs. Marshes, grasslands, lakes, lagoons, rarely beaches and estuaries. In flocks of hundreds, locally thousands, in N and cen Mexico; elsewhere, singly or in small groups, associating readily with other shorebirds. Feeds by picking, less often by probing.

SS: See White-rumped Sandpiper.

SD: C to F transient (mid-Mar–early Jun, Aug–early Oct; 900–2500 m) in interior S to cen volcanic belt (see Wilson and Ceballos-L. 1986); U to F on Atlantic coast, in interior from cen Mexico S, and on Pacific coast from Isthmus S; U to R in NW Mexico; R in Yuc Pen (May 1987, SJ).

RA: Breeds NW N.A.; winters from Ecuador and cen Peru to S S.A.

PECTORAL SANDPIPER

Calidris melanotus Not illustrated
Playero Pectoral

ID: 7.7–9 in; B 1–1.3 in (19.5–23 cm; B 2.5–3 cm). ♂ > ♀. Widespread transient. *Bill medium length, slightly decurved.* **Descr. Alt:** *bill black with dull ochre to olive base, legs ochre-yellow.* Face and chest whitish to pale buff, streaked and scalloped dark brown, with paler supercilium; *lower border of streaked chest clean-cut from white belly* and undertail coverts. Crown, nape, and *upperparts dark brown, edged greyish buff to buffy rufous,* brightest on crown, scapulars, and tertials; *lower rump to tail blackish with white lateral tail coverts forming ovals above at sides.* Upperwings brown with *indistinct*

Pectoral Sandpiper

narrow whitish wingstripe. **Basic** (rarely seen): greyer overall, upperparts more uniform grey-brown. **Juv:** resembles alt but face and chest buffier, finely streaked dusky; crown, nape, and upperparts with neat rufous, buff, and whitish edgings often forming 1–2 whitish Vs on back.

Voice. A reedy, low, rolled *krrih* or *krrip,* and a rolled, slurred *kreep,* etc.

Habs. Freshwater marshes, flooded fields, lakes, small pools, less often estuaries and beaches. Associates with other shorebirds, feeds by picking and probing.

SD: F to C transient (mid-Mar–early Jun, mid-Jul–Nov; SL–1500 m) on Atlantic Slope from Tamps to Honduras; F to U in interior (to 2500 m) and on Pacific Slope N of Isthmus, F to C S of Isthmus; U to R in spring in NW Mexico. Irregular R to U winter visitor (Dec–Feb) from cen Mexico S (SNGH, PP, RGW).

RA: Breeds N N.A.; winters from Peru and Brazil to S S.A.

NB: Sharp-tailed Sandpiper *C. acuminata* (7.5–8.8 in, B 1 in; 19–22 cm, B 2.5 cm; ♂ > ♀) is rare but regular autumn vagrant in W N.A., probably occurs in Baja. Resembles Pectoral Sandpiper, with which it often occurs, but posture often more hunched. Juv has face and chest bright ochre-buff to ochre, indistinctly streaked dusky at sides and with no sharp cut-off at belly; whitish supercilium (broadest behind eye) contrasts with dark rufous crown. Upperparts generally brighter and more rufous than Pectoral Sandpiper, pattern in flight similar. Alt has chest and flanks coarsely spotted and scalloped dark brown, with no clean cut-off, crown often bright rufous. Basic resembles basic Pectoral but chest washed greyish buff, with sides of chest and flanks sparsely streaked dark brown.

DUNLIN

*Calidris alpina** Not illustrated
Playero Dorsirrojo

ID: 7.7–8.5 in; B 1.2–1.5 in (19.5–21.5 cm; B 3–4 cm). Winters N Mexico. *Bill medium long to longish, slightly decurved.* **Desc. Basic:** bill and *legs black.* Head, *chest, and upperparts grey to brownish grey* with indistinct whitish supercilium; throat and rest of underparts whitish. Median uppertail coverts and tail dark with white sides. *Upperwings grey with* *darker remiges, white wingstripe.* **Juv:** plumage lost before arrival in region, but early winter birds may show some cinnamon to rufous-edged wing coverts and tertials; belly may be spotted blackish. **Alt:** face and chest whitish, streaked blackish; *large black belly patch on white underparts.* Crown, nape, and *upperparts dark brown,* edged *rufous-chestnut* and pale buff, wing coverts grey-brown.

Voice. A rough, rolled *ihrrr* or *iehrrr,* and a shorter *eihr,* all fuller than small *Calidris* calls. Feeding flocks make quiet, burry to mellow chippering.

Habs. Estuaries, coastal lagoons, beaches, etc. In flocks or singly, associating with other shorebirds. Feeds mostly by probing.

SS: Stilt Sandpiper has whiter face and chest, straighter bill, slightly decurved at tip, legs yellow, rump and uppertail coverts white, wingstripe indistinct.

SD: F to U transient and winter visitor (Oct–Apr) on coasts from Baja and Son to Nay (SNGH, PP); R to Guatemala (Dickerman 1977); U on Atlantic coast from Tamps to N Ver and in N Yuc Pen; R (irregular?) in interior S to cen volcanic belt (SNGH, PP, SW), and in Belize.

RA: Holarctic breeder; winters in New World from N.A. to N Mexico, rarely to C.A.

NB: Curlew Sandpiper *C. ferruginea* (7.5–8.3 in, B 1.3–1.5 in; 19–21 cm, B 3.5–4 cm) is rare but regular vagrant in N.A., probably occurs E Mexico. Resembles Dunlin but has more slender, decurved bill, longer legs. Note solidly white uppertail covert patch and pale grey tail; upperwings much like Dunlin. Basic resembles Dunlin but whitish chest has only slight dusky wash and indistinct streaks. Alt much like alt Red Knot in plumage but darker overall. Juv has face, chest, and underparts whitish

with white supercilium and dark eyestripe, chest washed cinnamon-buff when fresh, with indistinct dusky streaking at sides. Crown, nape, and upperparts dark brown with neat, scaly, whitish to pale buff edgings and dark subterminal scallops. Compare with Stilt Sandpiper.

STILT SANDPIPER
Calidris himantopus   Not illustrated
Playero Zancudo

ID: 8–8.7 in; B 1.5 in (20–21.5 cm; B 4 cm). Widespread. A *tall-standing, long-legged* sandpiper; *bill longish, straightish with slight droop at tip.* **Descr. Basic:** bill black, *legs yellow* (not green as often described). Face and chest whitish, streaked dusky, with *distinct whitish supercilium;* throat and rest of underparts white, flanks often streaked dusky. Crown, nape, and upperparts grey, *white uppertail coverts* often with dark flecks, *tail grey.* Upperwings grey with darker remiges, indistinct narrow whitish wingstripe. **Alt:** *auriculars and sides of crown rufous; throat streaked dusky, becoming dense dark barring on underparts.* Crown, nape, and upperparts dark brown, edged whitish to pale cinnamon. **Juv:** face, throat, chest, and underparts whitish, streaked dusky on face and sides of chest, with whitish supercilium; chest washed buff when fresh. Crown, nape, and upperparts dark brown with neat, scaly, pale buff and rufous edgings.

Voice. A quiet, low, slightly grunting *greh* or *kreh;* often silent.

Habs. Freshwater marshes, lakes, coastal lagoons, less often estuaries, beaches. In flocks or singly, associating with other shorebirds. Feeds by probing, often up to its belly in water.

SS: See Dunlin, Lesser Yellowlegs. Dowitchers are shorter-legged with longer bills, darker overall in basic, especially on chest,

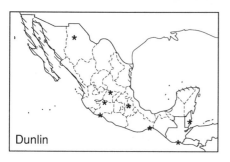

Dunlin

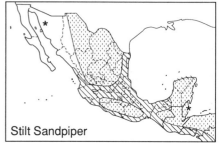

Stilt Sandpiper

legs yellow-olive; note patterns in flight, voice.

SD: F to C winter visitor (Oct–Mar; SL–1500 m) on Pacific Slope from Nay (RAB, SNGH, PP) to El Salvador, on Atlantic Slope from Tamps to N Yuc (SNGH, SW, photo), and (to 2500 m; irregularly?) in cen volcanic belt (SNGH, RGW); U N to Sin. F to C transient (late Mar–May, mid-Jul–Oct) on Atlantic Slope from Tamps to Honduras; F to U transient throughout interior and on Pacific Slope (except NW Mexico where R). Small numbers oversummer, at least on Atlantic coast (SNGH). Most references (for example, AOU 1983) have inexplicably overlooked this species as a common and widespread winter visitor in the region.

RA: Breeds N N.A.; winters Mexico to N Peru and N Argentina.

NB: Formerly placed in the monotypic genus *Micropalama*.

BUFF-BREASTED SANDPIPER
Tryngites subruficollis   Not illustrated
Playero Pradero

ID: 7–8 in; B 0.6–0.7 in (18–20.5 cm; B 1.5–2 cm). Uncommon transient. *A slender-necked and small-headed sandpiper of grasslands.* Bill medium-short, straight, and slender, legs fairly long. **Descr.** Ages differ slightly, sexes similar. **Adult:** Bill black, legs yellow. *Face, throat, chest, and underparts buff* to cinnamon-buff, *whitish eyering accentuates dark eye in blank face*; sides of neck and chest spotted blackish. Crown, nape, and *upperparts blackish, edged buff*; central rectrices dark, outer rectrices greyish buff with narrow black subterminal band. Upperwings brown with darker remiges, indistinct narrow whitish wingstripe; underwings white with dark crescent on primary coverts. **Juv:** face, throat, chest

and underparts paler, buff to pale buff, often with hint of cut-off between chest and paler belly; upperparts have neat, scaly, buff to whitish edges, wing coverts with dark subterminal crescents.

Habs. Open grasslands, usually with short grass, such as golf courses, playing fields, also sandy beaches. Singly or in small groups, may associate loosely with other shorebirds. Usually silent.

SS: See Upland Sandpiper.

SD: U to F transient (Apr–mid-May, Aug–Oct; SL–1000 m) on Atlantic Slope from Tamps to Honduras; R on Pacific Slope S of Isthmus.

RA: Breeds N N.A.; winters Paraguay, Uruguay, and N Argentina.

RUFF
Philomachus pugnax   Not illustrated
Combatiente

ID: ♂ 10.5–11.5 in, ♀ 8.5–9.5 in; B1–1.5 in (♂ 26.5–29 cm, ♀ 21.5–24 cm; B 2.5–4 cm). Vagrant. A medium-sized, fairly long-legged shorebird of highly variable appearance; *bill medium length, slightly decurved.* **Descr. Basic:** bill blackish (often paler below at base) to orangish with dark tip; legs *greyish olive* to orange to black. Face, throat, chest, and underparts whitish, streaked and mottled pale grey-brown on face and chest, typically with *white area around base of bill.* Upperparts grey-brown, edged paler; *lower rump to tail dark with white lateral tail coverts forming large ovals above at sides.* Upperwings grey-brown with darker remiges, indistinct, narrow whitish wingstripe. First basic typically retained through 1st year so spring migrants often worn and faded. **Juv:** legs typically olive. *Head and chest buff to greyish buff with paler face, darker cap*; throat, belly, and undertail coverts whitish to pale buff. *Upperparts dark brown with*

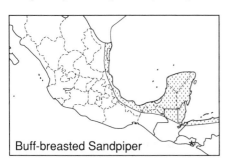

Buff-breasted Sandpiper

Ruff

neat, scaly, buff to pale buff edgings, wing coverts with dark subterminal marks. **Alt** (unlikely in region) ♀: head and chest buffy brown, streaked and mottled dark brown, belly and undertail coverts white, often with dark chevrons on flanks. Upperparts dark brown, edged buff to pale brown. ♂: resembles ♀ but in full plumage (May–Jul) attains fantastic erectile ruff on head and chest, with colors varying, rarely two birds being the same; crest often white, throat and neck ruff white, chestnut, or black, often barred and mottled with all three basic colors. Upperparts blackish, spangled and edged buff and chestnut.

Habs. Coastal lagoons, marshes, flooded fields, etc. Likely to be found feeding in association with other shorebirds but usually flies off separately. Feeds by wading or walking on dry shore, picks and probes. Flight often fairly lethargic and floppy. Mostly silent.

SS: Essentially unlike any other shorebird in the region, larger and notably longer-legged than Pectoral Sandpiper.

SD: Vagrant, one record: a basic ♀ collected on Pacific coast of Guatemala (Apr 1974, Dickerman 1975b). Reports from BCN (Warnock *et al.* 1989, description examined) and Son (Taylor 1986) are unacceptable.

RA: Breeds N Eurasia; winters S Eurasia to Africa and Australia; small numbers regular (increasing?) in N.A.

GENUS LIMNODROMUS:
Dowitchers

Dowitchers are medium-sized shorebirds somewhat resembling *Calidris*. Black bills long and straight, olive to yellowish legs medium length. The two species in the region are very similar and usually best separated by voice. Age/seasonal variation, sexes similar.

SHORT-BILLED DOWITCHER
Limnodromus griseus *　　　　　Not illustrated
Costurero Piquicorto

ID: 9.5–11 in; B 2–2.5 in (24–28 cm; B 5–6.5 cm). Widespread, mainly coastal. **Descr. Basic:** *head and chest grey to pale grey with short whitish supercilium and dark loral stripe*; throat whitish, chest often speckled dusky. Belly and undertail

coverts white, flanks mottled dusky. Upperparts grey to brownish grey, edged paler; *lower back, rump, and uppertail coverts white, spotted dusky, appearing in flight as white oval extending up back. White tail barred blackish.* Upperwings dark with narrow whitish trailing edge to secondaries. **Alt:** *face, throat, chest, and flanks cinnamon* to pinkish cinnamon with paler supercilium, dark loral stripe; neck sides, chest, and flanks spotted and barred dark brown. *Belly and undertail coverts white* or (*hendersoni,* wintering in É) cinnamon like chest. Crown, nape, and upperparts blackish, edged cinnamon to buff. **Juv:** face, throat, chest, and flanks washed cinnamon-buff, with whitish supercilium and dark loral stripe; sides of chest spotted dark brown, belly and undertail coverts whitish, undertail coverts spotted dark brown. Crown, nape, and *upperparts blackish, brightly edged ochre-buff to cinnamon-buff, tertials with broad edgings and zig-zag markings.*

Voice. A mellow *chu tu-tu* or *kyew tu-tu,* also a longer *kyew tu-tu-tu-tu-tu,* and a sharper *chi-tu;* suggests Lesser Yellowlegs or Ruddy Turnstone.

Habs. Estuaries, coastal lagoons, beaches, less often freshwater marshes. In flocks, associates readily with other species, locally with Long-billed Dowitcher. Feeds by 'sewing-machine' probing, often wading up to its belly.

SS: Long-billed Dowitcher darker overall in all plumages, basic has duskier chest, barring on flanks contrasts more with whitish belly; alt has chestnut underparts; juv has dark upperparts, including tertials, narrowly edged chestnut; note voice. See Stilt Sandpiper, Lesser Yellowlegs.

SD: C to F transient and winter visitor (mid-Jul–May) on both coasts S to El Salvador and N Yuc Pen; U to N Honduras. Small

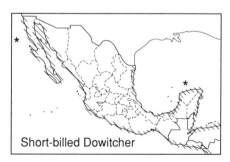

Short-billed Dowitcher

numbers oversummer locally on both coasts.

RA: Breeds N N.A.; winters S USA to cen Peru and cen Brazil

LONG-BILLED DOWITCHER

Limnodromus scolopaceus   Not illustrated
Costurero Piquilargo

ID: 10–11.5 in; B 2.3–3 in (25.5–29 cm; B 6–7.5 cm). Widespread, prefers fresh water. **Descr. Basic:** *head and chest grey with short whitish supercilium and dark loral stripe*, throat whitish; *chest usually contrasts with white belly* and undertail coverts, flanks mottled and barred grey. Upperparts grey, edged paler; *lower back, rump, and uppertail coverts white, spotted dusky, appearing in flight as white oval extending up back. Tail barred black and white.* Upperwings dark with narrow whitish trailing edge to secondaries. **Alt:** *face, throat, chest, and underparts cinnamon* to cinnamon-chestnut with paler supercilium, dark loral stripe; sides of neck and chest spotted dark brown, flanks barred dark brown. Crown, nape, and upperparts blackish, edged cinnamon to dark cinnamon. **Juv:** face, throat, chest, and flanks washed dusky cinnamon, with whitish supercilium and dark loral stripe; sides of chest spotted dark brown, belly and undertail coverts whitish, undertail coverts spotted dark brown. Crown, nape, and *upperparts blackish, edged chestnut to cinnamon-chestnut, tertials with narrow chestnut edgings.*
  Voice. A high, sharp *kiip* or *peek!*, suggests Wilson's Plover; calls may run into rapid series when flushed.
  Habs. Freshwater marshes, lakes, coastal lagoons, rarely beaches. Habits much as Short-billed Dowitcher.
SS: See Short-billed Dowitcher, Stilt Sandpiper, Lesser Yellowlegs.

SD: C to F transient and winter visitor (Aug–May, mainly from Sep; SL–2500 m) in Baja and on Pacific Slope, and in interior from Son and NL to Isthmus; U to El Salvador. F to C (Oct–May) on Atlantic Slope from Tamps to SW Camp, possibly to Belize (Wood *et al.* 1986) but no confirmed records from N or E Yuc Pen.

RA: Breeds NW N.A.; winters S USA to C.A.

COMMON SNIPE

Gallinago gallinago delicata   Not illustrated
Agachona Común

ID: 9.7–10.5 in; B 2.2–2.5 in (25–26.5 cm; B 5.5–6.5 cm). Widespread. *A skulking, cryptically colored shorebird of marshes. Bill very long and straight.* **Descr.** Ages/sexes similar. Bill blackish distally, olive basally, legs olive. *Head buff with blackish lateral crown stripe, dark eyestripe, and dark auricular stripe,* throat whitish. *Chest mottled and barred brown, rest of underparts white with dark barring on flanks* and undertail coverts. *Upperparts dark brown, cryptically patterned buff and whitish, with 2 broad buff Vs on back.* Tail blackish with rufous subterminal band and whitish tips to outer rectrices. Upperwings dark with white trailing edge to secondaries; underwings dusky, coverts barred whitish.
  Voice. A rasping *rrah-k* or *errhk* in flight; usually calls when flushed.
  Habs. Freshwater marshes, typically well vegetated, ponds, lakes, salt marshes, rarely open situations. Most often seen when flushed from close range, flies low and then climbs up before dropping back to cover. In N and cen Mexico, locally in flocks up to a few hundred birds but mostly singly or in small loose groups. Nest is a scrape under grass tussock, etc.

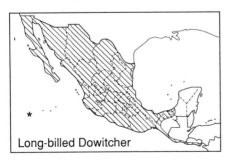

Long-billed Dowitcher

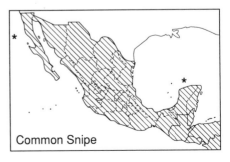

Common Snipe

Eggs: 3–4, olive-buff, blotched with browns and greys (WFVZ).

SD: C to F winter visitor (Oct–Apr; SL–3000 m) S to cen Mexico; U to F to El Salvador and Honduras. Probably bred BCN (Huey 1928); reports of breeding in cen Mexico (Jal, Gto; Friedmann *et al.* 1950) are equivocal.

RA: Holarctic breeder, also locally in S.A.; migrants winter in New World from N.A. to N S.A.

AMERICAN WOODCOCK

Scolopax minor　　　　　Not illustrated
Chocha Americana

ID: 10.5–11.5 in; B 2.3–3 in (26.5–29 cm; B 6–7.5 cm). Vagrant. *A large, plump snipe; bill long and straight, fairly thick.* **Descr.** Ages/sexes similar. Bill greyish flesh, darker at tip, legs dusky flesh. Face dusky with *long, dark loral stripe and auricular stripe, nape black with 4 cinnamon crossbars.* Chin whitish, rest of throat and *underparts pale cinnamon. Upperparts cryptically patterned* cinnamon, blackish, and grey-brown with 2 soft pale grey Vs on back; *cinnamon rump sides and lateral tail coverts striking in flight.* Tail blackish, outer rectrices tipped pale grey above, whitish below. Upperwings rufous-brown, underwings dusky with cinnamon lesser and median coverts.

Habs. Damp understory of woodland and scrub, often along shady valleys. Flushed birds erupt from underfoot making a high whistling sound, probably with their wings; flight erratic, usually soon drops back to cover. Walks hesitantly with almost constant bobbing action.

SS: Common Snipe prefers open marshy habitat, more slender overall and longer-billed, usually calls when flushed.

SD: Vagrant (Dec–Feb; SL–500 m) to Tamps (Wolf 1978; JCA) and to QR

(Lopez O. *et al.* 1989). Possibly an irregular R winter visitor in Tamps.

RA: E N.A.; N breeders withdraw S in winter.

GENUS PHALAROPUS:
Phalaropes

Phalaropes are swimming shorebirds. Bills straight and medium length, legs medium length, lobed toes aid in swimming. Age/seasonal/sex variation; alt ♂♂ duller than ♀♀. Juv plumage quickly lost and rarely seen in the region without being mixed with pale grey 1st basic feathers on upperparts.

WILSON'S PHALAROPE

Steganopus tricolor　　　　Not illustrated
Falárapo de Wilson

ID: 8–8.8 in; B 1.2 in (20.5–22 cm; B 3 cm). Widespread transient. *Bill fine and fairly long.* **Descr. Basic:** bill black, legs yellow by winter. *Face, foreneck, chest, and underparts white with dusky auriculars.* Crown, hindneck, and *upperparts pale grey,* wings darker, *with white lower rump and uppertail coverts, pale grey tail.* Upperwings grey. **Alt ♀:** legs black. *Black auricular mask extends as broad stripe down sides of neck becoming chestnut on sides of chest; lores and crown pale blue-grey* with narrow white supercilium. Throat and hindneck white, *foreneck washed chestnut,* underparts white. Upperparts blue-grey with 2 chestnut Vs on back, wings browner. **Alt ♂:** duller, crown and auriculars dark grey contrasting with bolder white supercilium; dark stripe down sides less contrasting. **Juv:** resembles basic but crown, nape, and upperparts dark brown, neatly edged pale cinnamon to whitish, foreneck washed buffy cinnamon when fresh, legs yellowish.

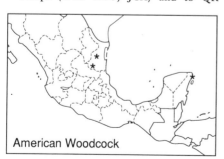

American Woodcock

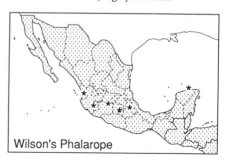

Wilson's Phalarope

Voice. Mostly silent, rarely utters quiet grunts.

Habs. Salt ponds and lagoons, freshwater marshes, lakes, rarely beaches. Singly or in flocks, at times in hundreds; associates readily with other shorebirds. Feeds by walking and picking, tail characteristically held high in the air; also by swimming and picking at water surface.

SS: Red-necked and Red Phalaropes smaller, in basic both have black postocular stripes; note white wingstripes, dark-centered rump and tail; Red Phalarope stockier with thick bill. See Lesser Yellowlegs.

SD: F to C transient (late Mar–early Jun, Jul–Oct; SL–2500 m) throughout most of region, commonest Apr–May and Aug–Sep; R to U on Atlantic Slope S of Isthmus; small numbers may oversummer locally. Irregular U to R winter visitor (Nov–Mar; SL-2500 m) on Pacific Slope and in interior from Nay and cen volcanic belt S (SNGH, PP; Wilson and Ceballos-L. 1986), also in N Yuc Pen (SNGH, LJP).

RA: Breeds N.A.; winters W S.A., rarely N to Mexico.

RED-NECKED PHALAROPE
Phalaropus lobatus   Not illustrated
Falárapo Cuellirrojo

ID: 6.7–7.5 in; B 0.7 in (17–19 cm; B 2 cm). *Bill fine and medium length.* Transient along Pacific coast. **Descr. Basic:** bill black, legs grey. *Head*, neck, and underparts *white with black postocular stripe. Hindneck and upperparts grey, whitish upperpart edgings form 2 Vs on back, visible at close range*, wings darker. Uppertail coverts and tail dark grey, lateral tail coverts white. Upperwings dark with *white wingstripe.* **Alt ♀:** *head, hindneck, and sides of chest dark grey with white supraorbital spot and white throat patch, chestnut patch on sides of neck.* Upperparts blackish with ochre-

buff to ochre edgings forming 2 broad Vs on back. Underparts white, flanks streaked dusky. **Alt ♂:** duller, head and neck pattern duller. **Juv:** resembles basic but crown, nape, and upperparts blackish, upperparts with buff edgings; foreneck and chest washed dusky cinnamon when fresh.

Voice. A slightly sharp, clipped *tlik* or *plik*, and a harder *tlk* or *tik*, usually given in flight.

Habs. Mainly inshore, coastal lagoons, lakes, less often far offshore. In flocks or singly; associates readily with other shorebirds, including Red Phalarope. Feeds by swimming and picking, rarely seen on land except to rest and preen.

SS: Red Phalarope larger and bulkier with thicker bill, basic has white head with black postocular stripe and dark nape band, uniform pale grey upperparts, broader white wingstripe; patterned upperparts appear before 'red' underparts in alt, suggesting Red-necked. See Wilson's Phalarope.

SD: C to F transient (Apr–May, Jul–Oct) off and along Pacific coast and in Gulf of California (DB, BT); irregularly U to C winter visitor (Nov–Mar) off and along Pacific coast, at least from Col S (SNGH, PP, SW, photo). Irregular R to U transient (late Nov–Dec; to 2500 m) in interior S to cen volcanic belt and in N Yuc Pen (SNGH, LJP, RGW).

RA: Holarctic breeder; winters in E Pacific from Ecuador to S S.A., rarely (?) to Mexico.

RED (GREY) PHALAROPE
Phalaropus fulicaria   Not illustrated
Falárapo Piquigrueso

ID: 8–8.5 in; B 0.7–1 in (20–21.5 cm; B 2–2.5 cm). *Bill medium length and fairly stout.* Migrant off Pacific coast. **Descr. Basic:** bill and legs black. *Head, neck, and underparts white with black postocular*

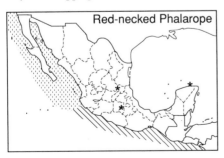

Red-necked Phalarope

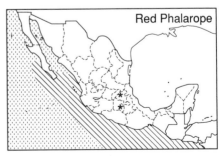

Red Phalarope

stripe, blackish nape band. Hindneck and *upperparts pale grey*, wings darker. Rump and tail dark grey, lateral tail coverts white. Upperwings dark with *broad white wing-stripe.* First basic often retains dark, buff-edged juv tertials into early winter. **Alt ♀:** bill yellow with black tip, legs brownish. *White face contrasts with black lores and crown, and deep chestnut throat, neck, and underparts.* Upperparts blackish, boldly edged ochre. **Alt ♂:** duller, whitish face contrasts less with dark lores and crown, neck and underparts mottled pale chestnut and whitish.

Voice. A high, slightly tinny *tink* or *tsik.*

Habs. Pelagic, often far offshore; also coastal, especially after storms, rarely inland lakes. Habits much as Red-necked Phalarope.

SS: See Red-necked Phalarope, Sanderling.

SD: C to F transient and winter visitor (Sep–mid-May) off Pacific coast from Baja S; at times U to F from mid-Jul in N; also in Gulf of California (DB, BT) where mostly a transient. Irregular R transient (Dec; 1500–2300 m) in interior S to cen volcanic belt (SNGH, SW, RGW).

RA: Holarctic breeder; winters in E Pacific from S California to Chile.

JAEGERS AND SKUAS: STERCORARIIDAE (5)

Jaegers and skuas breed at high latitudes and winter at sea; all are known as skuas in the Old World. Sometimes treated as a subfamily of the Laridae, they differ from gulls in their fleshy ceres, more strongly hooked bills, sharp, hooked claws, and often prominent white flashes across the base of the primaries; in contrast to gulls, ♀♀ are larger than ♂♂. Feet webbed. Wings long and fairly broad to broad, tails wedge-shaped. Skuas are large and heavily built; jaegers are smaller and more slightly built with projecting central rectrices, longest in alt adults. Ages differ strongly in jaegers, slightly in skuas; sexes similar. First basic attained by a protracted complete molt at sea, subsequent imm molts and plumages poorly known since imms remain at sea for 2–4 years; 1st basic followed by complete molt to 2nd basic, then partial prealternate and complete prebasic molts converge with adult cycles. Adults molt after autumn migration and, again, in late winter prior to and while heading north; prealternate molt includes long central rectrices in jaegers. Dimorphism confined mostly to light and dark morphs in adults, with relatively few intermediates, but juv and imm plumages of jaegers highly variable, and juv morphs not necessarily related to subsequent adult morph. Plumage colored brown and whitish overall. Mostly silent away from breeding grounds. Plumage variation in juv and imm jaegers means that identification is often based on size, structure, and flight. For more information on plumages and molts, see Cramp and Simmons (1983), Harris *et al.* (1989), Olsen (1989).

Jaegers and skuas are well known for their parasitic feeding habits at sea whereby they chase other birds, forcing them to regurgitate their food; they also, however, feed on fish picked from near the sea surface.

POMARINE JAEGER

Stercorarius pomarinus Not illustrated
Salteador Pomarino

ID: 17–20 in, W 43–49 in (43.5–51 cm, W 110–125 cm). *Large and heavily built with broad wings.* **Descr. Adult. Light morph alt** (Apr–Oct): bill dusky flesh with dark tip, legs dark. *Tail projections blunt and* twisted, spoon-like, adding 2.5–4.5 in (6.5–11 cm) to length. Head and underparts whitish, washed lemon on face, with blackish cap on forecrown, *dusky mottled chest band* (rarely lacking) *and flanks,* and dark lower belly and undertail coverts. Underwings blackish brown with *bold white flash across base of primaries.* Upperparts and tail blackish brown, white flash restricted to outer 5 or so primary shafts. **Basic** (Oct–Apr): tail projections shorter. Face, neck, and throat washed and barred dusky, cap less contrasting, tail coverts barred blackish and whitish. **Dark morph** (rare): blackish brown overall with white flashes in primaries; basic similar but underparts may appear grizzled paler. **Juv** (Aug–Nov). **Light morph:** bill greyish with dark tip, legs blue-grey to dull flesh with mostly black feet. Often a short blunt tail projection. *Brown to grey-brown overall,*

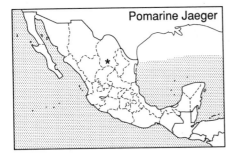

Pomarine Jaeger

upperparts darker with dull pale edgings, head and underparts paler, underparts and underwing coverts barred darker. *Note whitish base to under primary coverts, forming second white flash on underwing.* **Dark morph:** blackish brown overall, rarely distinguishable from adult except for shorter tail projections. **Imm:** in 1st year, belly often becomes whitish, upperparts more uniform. Some imms (age?) have striking whitish head and underparts. 2nd year resembles adult basic but underparts and underwing coverts strongly barred. Underwing coverts become mostly or uniformly dark by 3rd basic. Attains adult plumage in 4–5 years.

Habs. Pelagic but may be seen from land. Direct flight heavy and powerful yet buoyant, low to high over water, may glide and bank in windy conditions. Pursuit flight fast and powerful but mostly chases larger or less agile prey species (gulls, shearwaters, even boobies) than does Parasitic Jaeger. Often in small flocks in migration.

SS: Parasitic Jaeger smaller and slighter-built with less broadly based wings; alt has pointed tail projections, juv warmer brown overall, especially cinnamon wash to neck, with less extensive white flash on underwing, imms best identified by size and shape. Skuas are larger and more heavily built, lack tail projections, wings broader and flight more powerful, white flash across base of upper primaries more extensive. Imm gulls less barred or uniformly dark (except Heermann's), lack white flashes in primaries (Heermann's may show white patch on primary coverts).

SD: C to F transient and winter visitor (Aug–May) off and along Pacific Coast, including Gulf of California; U to F through the summer. U to F transient and winter visitor (Sep–Apr) off Atlantic coast from Ver (WJS) to Belize (Howell *et al.* 1992*b*), some probably oversummer. Vagrant to Coah (Nov 1986, Garza de León 1987).

RA: Holarctic breeder; winters widely in Pacific and Atlantic oceans.

PARASITIC JAEGER

Stercorarius parasiticus   Not illustrated
Salteador Parásito

ID: 16–17.5 in, W 39–44 in (40.5–44.5 cm, W 99–112 cm). Medium-sized jaeger, often looks falcon-like. **Descr. Adult. Light**

morph alt (Apr–Oct): bill and legs dark. *Sharply pointed tail projections* add 2.5–4 in (6.5–10 cm) to length. Head and underparts whitish, washed lemon on face, with blackish cap on forecrown, and dark undertail coverts; some have partial to complete dusky chest band. Underwings blackish brown with *white flash across base of primaries.* Upperparts and tail blackish brown, white flash restricted to outer 3–6 primary shafts. **Basic** (Oct–Apr): tail projections shorter (some may molt tail projections only once a year, hence same in basic as alt). Face, neck, and throat washed and barred dusky, cap less contrasting, tail coverts barred blackish and whitish. **Dark morph alt:** blackish to blackish brown overall, face often slightly paler, with white flashes in primaries similar to light morph; basic similar. **Juv** (Aug–Nov). **Light morph:** bill greyish with dark tip, legs blue-grey to dull flesh with feet mostly black. Often a *short, pointed tail projection* (0.5 in; 12 mm). *Warm brown* to dark brown *overall, nape* typically *brighter and paler, cinnamon;* upperparts darker with pale edgings (including *distinct whitish edging to tips of primaries*), head and underparts paler, head rarely whitish, underparts and underwing coverts barred darker. **Dark morph:** extreme birds hard to tell from adult except by shorter tail projection, usually some paler barring on underwing coverts and undertail coverts, imm dark morphs similar; as adult by 2nd basic. **Imm:** in 1st year, belly often becomes whitish, upperparts more uniform. Some imms (age?) have striking whitish head and underparts. 2nd year resembles adult basic but underparts and underwing coverts strongly barred. Underwing coverts uniformly dark by 3rd basic, attains adult plumage in 3–4 years.

Habs. Pelagic but may be seen from land.

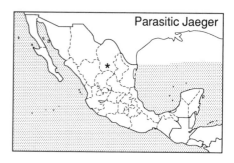

Parasitic Jaeger

Direct flight fairly buoyant and powerful, low to high over water. Often glides and banks in windy conditions. Pursuit flight fast and agile, fairly aerobatic with often prolonged chases; mostly chases terns. May occur in flocks in migration.

SS: See Pomarine Jaeger. Long-tailed Jaeger smaller with narrower wings, more graceful and buoyant flight, bill shorter and stubbier, adult and older imms show contrast between paler grey-brown upperparts and dark flight feathers; note dark underwings; juv Long-tailed has longer but blunt tail projection, light morph greyer without strong warm tones, upperparts often show adult-like two-tone contrast; note virtual lack of white in primaries above. Imm gulls less barred or uniformly dark (except Heermann's), lack white flashes in primaries (Heermann's may show white patch on primary coverts).

SD: C to F transient (Aug–Nov, Mar–May) off and along Pacific coast, including Gulf of California; U to F in winter (Dec–Feb), small numbers probably oversummer. U to R visitor (Nov–Jun at least) off Atlantic coasts from Tamps (GWL) to Honduras (Brown and Monroe 1974). Vagrant to Coah (Sep 1986, Garza de León 1987).

RA: Holarctic breeder; winters widely in Pacific and Atlantic oceans.

NB: Known as Arctic Skua in the Old World.

LONG-TAILED JAEGER

Stercorarius longicaudus   Not illustrated
Salteador Colilargo

ID: 14.5–16 in, W 37–41 in (38–40.5 cm, W 94–104 cm). Smallest jaeger, bill relatively short and stubby. **Descr. Adult. Light morph alt** (Apr–Oct): bill and feet dark, legs pale. *Long, pointed tail projections* add up to 5.5–8 in (14–20.5 cm) to length *but full streamers rarely seen in region.* Head and underparts whitish,

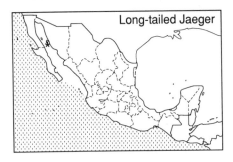

Long-tailed Jaeger

washed lemon on face, with blackish cap on forecrown, lower belly and undertail coverts dusky. *Underwings dark brown without white flash across base of primaries. Upperparts grey-brown with contrasting blackish brown flight feathers, white flash restricted to outer 2 primary shafts.* **Basic** (Oct–Apr): tail projections shorter (1.5–4 in; 4–10 cm). Face, neck, and throat washed and barred dusky, cap less contrasting, uppertail coverts barred whitish, underparts whitish with dark chest band, undertail coverts barred blackish. Dark morph adult unknown (?). **Juv** (Aug–Nov). **Light morph:** bill greyish with dark tip, legs blue-grey to dull flesh with feet mostly black. Often a *short, blunt tail projection* (1 in; 2.5 cm). Grey-brown to greyish overall, *upperparts darker with relatively bright pale edgings,* head and underparts paler (head and belly may be whitish), underparts and underwing coverts barred darker. *Upperparts may show two-tone contrast* similar to adult. **Dark morph:** blackish overall, usually with *distinct pale tips to upperparts* (except primary tips), whitish barring on underwing coverts and tail coverts. **Imm:** 1st basic (unknown in region) similar to light juv (dark morph rare) with dark chest band on paler underparts; flanks, underwing coverts, and tail coverts boldly barred. *2nd basic* (fairly common in region) *much like adult basic but axillars and underwing coverts barred.* Attains adult plumage in 3 years.

Habs. Pelagic, unlikely to be seen from land. Singly or in loose groups, often glides and banks in windy conditions. Direct flight fairly buoyant. Pursuit flight fast and aerobatic; mostly chases terns.

SS: See Parasitic Jaeger.

SD: F to U transient (Aug–Oct, Apr–May) off Pacific coast, some may oversummer (?); R transient in Gulf of California (May 1992, DB, BT).

RA: Holarctic breeder; winters in S oceans, in Pacific mostly off W S.A.

GREAT SKUA

Catharacta s. skua   Not illustrated
Págalo Grande

ID: 21–24 in, W 51–55 in (53.5–61 cm, W 129.5–139.5 cm). Vagrant. **Descr.** Ages similar. Bill and legs dark. Head and upperparts dark brown, hindneck with grizzled,

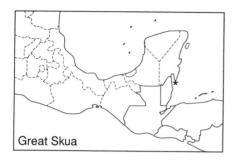

Great Skua

buff to pale lemon streaking, upperparts often mottled cinnamon to rufous and whitish. Underparts slightly paler, sooty grey-brown, often mottled and washed pinkish cinnamon.

Habs. Much as South Polar Skua.

SS: Larger than South Polar Skua, warmer brown with pale mottling and grizzling. See Pomarine Jaeger. Imm gulls have relatively longer, narrower wings, less powerful flight, more patterned overall but lack white flashes in primaries.

SD: Vagrant, one record: a tideline carcass on Ambergris Cay, Belize, Mar 1971 (Barlow *et al.* 1972, ROM spec. 109298). May be an irregular R visitor (Dec–Apr) off Caribbean coasts.

RA: Breeds on islands in NE Atlantic Ocean and in southern oceans. Ranges at sea in N Atlantic and S Pacific oceans.

SOUTH POLAR SKUA

Catharacta maccormicki   Not illustrated
Págalo Sureño

ID: 20–23 in, W 48–52 in (51–58 cm, W 122–132 cm). **Descr.** Adult dimorphic, juvs most often seen in region. **Juv:** bill blue-grey with black tip, legs dark. *Head, neck, and underparts grey to buffy grey, contrasting with slaty blackish brown underwings and upperparts, bold white flash across base of primaries* most striking below. Subsequent imm plumages poorly

understood, may resemble adult by 3rd year. **Adult.** Bill black. **Light morph:** *head, neck, and underparts buff to whitish, hindneck often paler, forming broad collar.* Upperparts and underwings blackish brown, white flashes in primaries may be bolder than juv (?). **Dark morph** (rare in region?): blackish brown overall, head and underparts slightly paler and browner, wings flashes similar to light morph.

Habs. Pelagic, unlikely to be seen from land. Flight often direct with easy powerful wingbeats of broad wings, more agile when pirating, mostly chases less agile species such as gulls and shearwaters.

SS: See Pomarine Jaeger. Imm gulls have relatively longer, narrower wings, less powerful flight, more patterned overall but lack white flashes in primaries.

SD: U to R transient and visitor (Jan–Dec; most records May–Oct, RLP) off Pacific coast; R to U (May–Nov) in Gulf of California (Tershy *et al.* 1993; SNGH).

RA: Breeds on South Shetland Islands and along coast of Antarctica; ranges at sea to N Pacific and N Atlantic oceans.

NB: Chilean Skua *C. chilensis* (20–23 in; 51–58 cm) of S S.A. may occur as rare visitor off Pacific coasts of C.A. and S Mexico (see Appendix B). Face, neck, and underparts (including underwing coverts) washed cinnamon, contrasting with blackish cap and upperparts and dusky chest band; white wing flashes as in other skuas.

South Polar Skua

GULLS, TERNS, AND SKIMMERS: Laridae (39)

A fairly uniform group of web-footed birds characteristically found near water. Ages differ, young precocial; sexes similar. First prebasic molt usually occurs within a few months and juv plumage of most migrants rarely seen in the region.

The Laridae may be divided into three subfamilies: Larinae (gulls), Sterninae (terns), and Rynchopinae (skimmers).

SUBFAMILY LARINAE: Gulls

Gulls typically breed in temperate regions and are migratory; only four species breed in the guide region, all in N Mexico. Most are coastal, at least in the non-breeding season. Bills relatively delicate in small species, stout and heavy in large species; legs medium length. Wings fairly long and broad, well-suited for flight over open water. Tails squared unless stated otherwise. ♂♂ larger, often striking in large species. From 1 to 4 years required to attain adult plumage; a 3-year gull, for example, is one that attains adult plumage in its 3rd year. Adults patterned in greys, white, and black; most imms of larger species brownish overall, at least in 1st year. Bare parts often brightly colored. Several smaller species have dark hoods in alt. Differences between alt and basic are detectable in most species, with basic often indicated by dusky streaking on the head and neck. Underwings usually show a hint of upperwing pattern and are not described unless of use in identification. Adult dark-hooded gulls usually show a variable pink flush to their underparts in spring. Colors of bare parts are given for most ages, especially 1st and 2nd years, 3rd year and older imms intermediate between 2nd year and adult; orbital ring colors for larger gulls refer to alt adults. Calls are mostly wailing, yelping, and mewing cries, deeper in large species. All large gulls (Western, Herring, etc.) give loud crowing, deep yelping, and moaning calls, rarely used for identification; imms have reedier, thin whistling, and creaky calls, deepening with age.

Larger gulls are omnivorous, eating fish, bird eggs, young and even adult birds, garbage, offal, etc.; smaller species mostly eat fish, other small aquatic animals, and may be seen hawking insects. Most gulls breed colonially, nest scrapes on the ground usually lined with grasses, seaweed, etc. Eggs: 2–4, patterned.

LAUGHING GULL

Larus atricilla   Not illustrated
Gaviota Reidora

ID: 15–17 in, W 38–42 in (38–43 cm, W 96.5–107 cm). Common and widespread 3-year gull. *Bill may appear drooped at tip.* **Descr. Adult. Basic:** bill and legs blackish. Head, neck, and underparts white with *dusky smudge behind eye. Upperparts dark grey, black wingtips lack white mirrors* but often show 1–2 white tips at rest; secondaries and inner primaries tipped white. Rump and tail white. **Alt:** bill deep red. Black hood with white eye-crescents. **Juv:** bill blackish. *Dusky grey-brown overall, belly and undertail coverts whitish,* whitish-edged upperparts appear scaly. Upperwings grey-brown with blackish leading edge to outerwing and black bar on secondaries. Rump and tail whitish, tail with broad black subterminal band. **1st basic:** head whiter, especially forehead and throat, with *dusky smudge behind eye,* back grey. *Dusky chest band usually distinct.* **1st alt:** head and underparts whitish with dusky mask or partial dark hood. **2nd year:** resembles adult but black wingtips more extensive, usually shows trace of

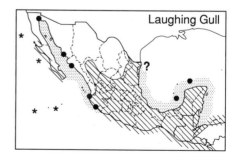

Laughing Gull

dark tail band and bar on secondaries, black hood may be mottled white.

Voice. Varied, nasal, laughing calls, *kyah-kyah* or *ka-ha*, and longer series, *ka hah-hah-hah-hah* ... or *kyah-kyah-kyah* ..., etc. First-years have a shrill *seeirr* and other reedy, screechy calls.

Habs. Beaches, estuaries, harbors, lagoons, rivers, lakes, etc. Commonest gull in the region, often in flocks of several hundred birds, at times thousands. Nests colonially on the ground. Eggs: 2–4, olive-brown to olive-buff, marked with dark browns (WFVZ).

SS: Franklin's Gull smaller with more bluntly tipped wings, smaller bill not drooped at tip, thicker eye-crescents; basic and imm have partial dark hood; 1st basic lacks dusky chest band, black tail band does not extend to sides of tail.

SD: C to F resident, breeding locally, from N Gulf of California along mainland coasts to Col, on Atlantic coast throughout. C to F winter visitor (Sep–Apr) on E coast of Baja and Pacific coast from Col S; U to F in summer (May–Aug). U to R winter visitor (Sep–Apr) N on Pacific coast of Baja (at least to 27°N), to Islas Revillagigedo and Clipperton, and in interior from S Plateau to Guatemala; U to R in interior (May–Aug). Vagrant to Isla Guadalupe (Howell and Webb 1992d).

RA: Breeds N.A. to Caribbean; winters to N S.A.

FRANKLIN'S GULL
Larus pipixcan   Not illustrated
Gaviota de Franklin

ID: 14–15 in, W 35–38 in (35.5–38 cm, W 89–96.5 cm). Transient 3-year gull, unique among gulls in the region in having 2 complete molts each year. **Descr. Adult. Basic:** bill and legs blackish. Head, neck, and underparts white with *partial dark hood*

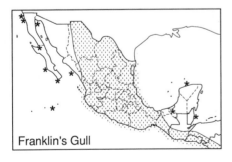

Franklin's Gull

and white eye-crescents. Upperparts dark grey, *black wingtips with large white mirrors.* In flight, note white band inside black wingtips; secondaries and inner primaries tipped white. Rump and uppertail coverts white, tail pale grey. **Alt:** bill deep red, legs may be dark reddish. Black hood with *thick white eye-crescents.* **1st basic:** head and underparts white, head with *dusky partial hood* and white eye-crescents. Back grey, upperwings grey-brown with blackish leading edge to outerwing and black bar on secondaries. Rump and tail whitish, *black subterminal tail band does not extend to outermost rectrices.* **1st alt:** similar to adult basic but black wingtips more extensive and more narrowly tipped white, often lacks white band inside black wingtips, may retain traces of tail band and bar on secondaries. **2nd basic:** wingtips generally with more black and less white than adult. Attains adult plumage by 2nd prealternate molt on winter grounds.

Voice. Particularly in spring, flocks may give pleasant, yelping, laughing calls, *kyeah kyeah* ..., and a more mewing *meeah*, etc.

Habs. Coasts, fields, lagoons, lakes, marshes, etc. In spring, flocks up to a few hundred birds move N in steady, buoyant flight, sometimes soaring and circling high. Associates readily with Laughing Gull.

SS: See Laughing Gull.

SD: C to F spring transient (Mar–early Jun) N along Pacific coast to Isthmus, thence most birds cross and continue N along Atlantic Slope; F along Pacific coast, N at least to Nay; U to R in Gulf of California and off W coast of Baja; U (to F?) in interior N of Isthmus. C to F autumn transient (Sep–Dec), spring route reversed. R on Atlantic Slope from Yuc Pen southward. R and irregular in winter (Jan–Feb) along both coasts, and in migration to Islas Revillagigedo and Clipperton.

RA: Breeds N.A.; most winter along Pacific coast of S.A.

LITTLE GULL
Larus minutus   Not illustrated
Gaviota Mínima

ID: 11–11.5 in, W 27–29 in (28–29 cm, W 68.5–73.5 cm). Vagrant 3-year gull. *Bill slender and delicate, wings fairly rounded.* **Descr. Adult. Basic:** *bill blackish* to dull

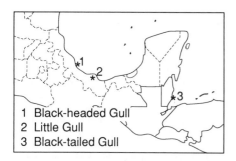

1 Black-headed Gull
2 Little Gull
3 Black-tailed Gull

red, legs dull red. Head, neck, and under-parts white with *dusky crown and ear spot*. Upperparts pale grey, trailing edge of wings white, *no black in wingtips*. Rump and tail white. *Underwings smoky grey with white trailing edge.* Alt: *black hood lacks eye-crescents.* 1st year: legs flesh, at least until spring. Head, neck, and under-parts white, with *blackish crown and ear spot. Back and upperwings pale grey with bold blackish M pattern* and inconspicuous dusky bar on secondaries. Underwings whitish. Rump and tail white, tail with black subterminal band. Usually attains partial black hood in 1st alt. **2nd year:** resembles adult but some birds retain a few black markings in wingtips.

Habs. Lakes, estuaries, lagoons, marshes. Associates often with terns and other gulls. Flight fairly buoyant and graceful, dips down to pick food from or near water surface. Usually silent.

SS: Bonaparte's Gull has more pointed wings, all ages have bold white leading edge to outerwing above and below. 1st basic Black-legged Kittiwake larger, lacks dark crown but usually has dark hindcollar, secondaries and inner primaries contrast-ingly whitish, legs black. Black-headed Gull larger with longer heavier bill, all ages have white leading edge to outerwing.

SD: Vagrant, one record: a 1st-winter bird in S Ver, Jan–Feb 1982 (Howell 1987). In-creasing in N.A. and likely to occur again.

RA: Breeds Eurasia and locally in NE N.A.; winters S in New World mainly to E USA, vagrant to W N.A.

BLACK-HEADED GULL

Larus ridibundus   Not illustrated
Gaviota Encapuchada   See Little Gull for map

ID: 15–16 in, W 35.5–39 in (38–40.5 cm, 91–99 cm). Vagrant 2-year gull, resembles large Bonaparte's Gull. *Bill fairly long and*

slender, wings poir.:ed. **Descr. Adult. Basic:** *bill and legs red.* Head, neck, and underparts white with dusky ear spot. Upperparts pale grey, *leading edge of outerwing white*, outer primaries tipped black. *From below, outer primaries form black-tipped white wedge, inner primaries dusky to blackish.* Rump and tail white. **Alt:** *dark brown hood* with narrow white eye-crescents. **1st year:** resembles basic but *bill and legs flesh to yellowish orange, bill dark-tipped*; upperwings with *dark carpal bar and bar on secondaries*, all primaries tipped black, and white primary wedge less distinct below. Usually attains nearly full hood in 1st alt.

Voice. Calls grating and shrill, typically a hard *kaahrr* and a higher, more drawn-out *kee-arrr*, etc. (Britain).

Habs. Estuaries, harbors, lagoons, lakes, marshes. Associates readily with other gulls, terns, etc. Flight notably heavier than Bonaparte's Gull.

SS: Bonaparte's Gull smaller and tern-like, all ages have delicate blackish bill, white underside to all primaries. See Little Gull.

SD: Vagrant; one record: one collected off Ver (Feb 1912), had been banded the previous summer in N Germany (Oberholser 1923; Friedman *et al.* 1950).

RA: Breeds Eurasia and locally in NE N.A.; winters S in New World to Caribbean.

BONAPARTE'S GULL

Larus philadelphia   Not illustrated
Gaviota de Bonaparte

ID: 13–13.5 in, W 31.5–34 in (33–34.5 cm, W 80–86.5 cm). Small 2-year gull; N Mexico in winter. *Bill slender and fairly delicate, wings pointed.* **Descr. Adult. Basic:** *bill black* (rarely dull red at base), legs orange-red. Head, neck, and under-parts white with dusky ear spot. Upper-parts pale grey, *leading edge of outerwing*

Bonaparte's Gull

white, outer primaries tipped black. *From below, black-tipped white primaries appear translucent when backlit*. Rump and tail white. **Alt**: *slaty-black hood with narrow white eye-crescents*. **1st year**: resembles basic but legs flesh, upperwings with *dark carpal bar and bar on secondaries*, all primaries tipped black. Usually attains nearly full hood in 1st alt.

Voice. A low, rasping *kwah*, a rasping, clipped *rehow*, and a louder, plaintive *whee-hooah*, etc.

Habs. Estuaries, harbors, coastal bays, lakes, etc. Often in flocks up to few hundred birds; flight buoyant and tern-like, dipping down to pick food from at or near surface, sometimes surface-plunging.

SS: See vagrant Black-headed and Little gulls.

SD: C to F winter visitor (Oct–May) along coasts of Baja and in Gulf of California, U to Nay and R to Col; on Atlantic coast, F to U from Tamps to N Ver, R to Yuc Pen. U to R in interior to Chih and Coah, R and irregular to Jal and Gto. Small numbers oversummer in Baja and Gulf of California. Vagrant to Isla San Benedicto (Feb 1988, SNGH, SW).

RA: Breeds N.A.; winters to N Mexico and Caribbean.

HEERMANN'S GULL

Larus heermanni   Figure 23
Gaviota de Heermann

ID: 18–20 in, W 43–48 in (45.5–50.5 cm, W 109–122 cm). *Overall dark, 4-year Pacific gull*. Bill fairly long but not heavy. Underwings dark in all ages. **Descr. Adult. Basic:** *bill red with black tip, legs blackish*. Head, neck, and underparts smoky grey, head flecked whitish and dusky. Upperparts dark grey, remiges blackish with white trailing edge to secondaries and inner primaries. *Rump and uppertail coverts pale grey, tail black with broad white tip*. Some show contrasting white upper primary covert patches. **Alt**: *clean white head striking*, orbital ring red. **Juv:** *bill flesh* to reddish pink *with black tip*. Dark brown overall, flight feathers blacker, pale-edged upperparts appear scaly. **1st year**: back uniform, faded upperwing coverts may form contrasting panel. **2nd year**: *bill orange* to orange-red *with black tip*. Grey-brown overall, rump and uppertail coverts paler greyish, scapulars and tertials tipped white; some have a few rectrices tipped white. **3rd year**: resembles adult but head often speckled dusky in alt, black tip on bill more extensive.

Fig. 23 (a) Heermann's Gull; (b) Yellow-footed Gull

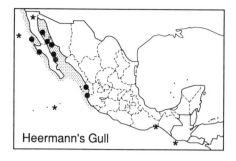

Heermann's Gull

Voice. A fairly deep, crowing *ha-ow* or *owh*, a more plaintive *kyow* or *kow*, and longer, laughing series, etc.

Habs. Rocky sea coasts, beaches, harbors, estuaries, very rarely inland. Often in flocks, regularly well offshore. Noted for pirating food from Brown Pelicans with which it often associates. Nests colonially on ground. Eggs: 2–3, greyish buff to buff, marked with browns and greys (WFVZ).

SS: See Jaegers.

SD/RA: C to F but local breeder in Gulf of California (90 per cent of population on Isla Rasa, BCN), U S to Nay, also 2 small colonies on Pacific coast of Baja, and irregular breeder along California coast. C to F post-breeding visitor (mainly Jul–Feb, some remain year-round) N along Pacific coast to W USA and S to Col; R as far N as British Columbia and S to Guatemala; vagrant to Isla Socorro.

BLACK-TAILED GULL

Larus crassirostris   Not illustrated
Gaviota Colinegra   See Little Gull for map

ID: 19–22 in, W 45–50 in (48.5–56 cm, W 114–127 cm). Vagrant 3-year gull. **Descr. Adult:** eyes pale yellow, orbital ring red, *bill yellow with black subterminal band and red tip, legs yellow.* Head, neck, and underparts white; in basic, dusky mottling on head and sides of neck concentrated as hindneck collar. *Upperparts dark grey* (similar to Laughing Gull), *wingtips black* with small white tips to outer primaries (except 2 outermost); white trailing edge to secondaries and inner primaries. *Rump and tail white, tail with broad black subterminal band,* narrower at sides, *not spreading onto outer web of outermost rectrices.* Underwing coverts pale grey. **1st year:** bill flesh with broad blackish tip, legs

flesh. Head and underparts dusky greyish with paler face and white throat, belly, and undertail coverts; distal undertail coverts barred dusky. *Rump white,* usually with sparse dark spots and bars, *tail blackish.* Upperparts mottled pale grey-brown, wing coverts darker and browner; primary coverts and remiges blackish brown. *Underwings dusky.* Head and underparts mostly white in 1st alt. **2nd year:** bill yellowish with black subterminal band, legs may be yellowish. Head and underparts white, washed and mottled dusky on crown, nape, neck, and sides of chest. *Upperparts mottled grey and grey-brown,* primary coverts and remiges blackish. *Rump white, tail blackish. Underwings dusky.* Apparently attains adult plumage by 3rd prebasic molt.

Habs. Coasts, beaches, bays, etc. The single bird sighted in Belize was on a sand bar with other gulls and terns.

SS: Lesser Black-backed Gull larger, adult has yellow bill with red gonys spot, darker grey upperparts, lacks black tail band, underwing coverts white. See Belcher's Gull (NB below).

SD: Vagrant; one record: Dangriga, Belize, on 11 Mar 1988 (RPA, photo).

RA: Breeds coastal NE USSR, Japan, and China; winters Japan to Formosa.

NB: Belcher's [Band-tailed] Gull *L. belcheri* (18–21 in, W 42–46.5 in; 45.5–53 cm, W 106–118 cm): a 3-year gull breeding on Pacific coast of S.A. in Peru and Chile; has occurred in Panama and could occur in guide region. **Adult:** eyes dark, orbital ring yellow, bill yellow with black subterminal mark and red tip, legs yellow. Head, neck, and underparts white in alt (Sep–Apr); basic (Apr–Sep) has extensive blackish-brown hood and white eyering. Upperparts brownish black, black wingtips lack white mirrors; broad white trailing edge to secondaries becomes narrower on inner primaries. Rump and tail white, broad black subterminal tail band does not extend onto outer web of outermost rectrices. Underwing coverts pale grey. **1st year:** bill and legs yellowish flesh, bill tipped black. Head, neck, and upper chest blackish brown; whitish underparts mottled dusky. Upperparts mottled grey-brown and whitish, primary coverts and remiges blackish. Rump whitish, tail black with narrow whitish tip, white basal corners to outer rectrices, and white outermost webs. Head becomes mostly white in 1st alt. **2nd year:** resembles basic adult but upperparts, especially wing coverts, browner, foreneck and chest mottled dusky, white trailing edge to wings narrower, bare parts duller; in 2nd alt head and neck become white, flecked dusky on face and throat. A less likely occurrence is **Olrog's [Band-tailed] Gull** *L. atlanticus* (20–24 in, W 50–55 in; 51–61 cm, W 127–140 cm), breeding on coast of N Argentina. Larger than Belcher's with thicker bill. Adult differs in red orbital ring, extensive black subterminal band on darker, red-tipped bill. Upperparts blacker than Belcher's with broader white trailing edge to secondaries and inner primaries, underwing coverts white, tail band narrower. Basic and imms lack the dark, sharply demarcated hood of Belcher's, and have dusky mottling mostly on hindneck and across foreneck and chest.

MEW GULL
Larus canus brachyrhynchus   Not illustrated
Gaviota Piquiamarilla

ID: 16–17 in, W 41–44 in (40.5–43 cm, W 104–112 cm). NW Baja in winter; a 3-year gull. *Bill fairly small and delicate.* **Descr. Adult:** *eyes brown* (rarely dull yellowish), orbital ring red, *bill yellowish, legs greyish green* to yellowish. Head, neck, and underparts white, streaked and mottled dusky on head and neck in basic. Upperparts grey, wingtips black with white mirrors; white trailing edge to upperwings. Rump and tail white. **1st year:** bill flesh with black tip, legs flesh. *Mottled pale greyish brown overall*, head and chest often whiter from mid-winter on; back grey. Leading edge of outerwing, bar on secondaries, and broad subterminal tail band darker, rarely blackish. **2nd year:** resembles adult but *bill*

greyish green to yellowish with dark tip or subterminal band, black wingtips more extensive with little or no trace of white mirrors, tail and secondaries usually show trace of dark bar.

Voice. Hoarse, wheezy mews, *hew* or *whew*, and a slightly reedier *hsew*, imms give a high, slightly shrill *siiir*.

Habs. Estuaries, harbors, beaches. Associates readily with other gulls at beaches, feeding off headlands, etc.

SS: Ring-billed Gull larger with larger, deeper bill, upperparts paler grey, legs and bill usually brighter; adult and 2nd year have pale yellow eyes, bill shows clean-cut black subterminal band; 1st year has contrasting whitish rump and tail with broad black tail band, more contrasting upperwing pattern. Adult and 2nd year Black-legged Kittiwake have short black legs, neat black wingtips without white mirrors, secondaries and inner primaries contrastingly whiter.

SD: U to R winter visitor (Nov–Mar) along Pacific coast of BCN.

RA: Holarctic breeder (absent from E N.A.); winters S in New World to NW Mexico.

NB: Known as Common Gull in the Old World.

RING-BILLED GULL
Larus delawarensis   Not illustrated
Gaviota Piquianillada

ID: 17–20 in, W 44.5–49 in (43–50.5 cm, W 113–125 cm). Widespread 3-year gull. **Descr. Adult:** *eyes pale yellow*, orbital ring red, *bill yellow with black subterminal band, legs yellow.* Head, neck, and underparts white, streaked and mottled dusky on head and neck in basic. Upperparts pale grey, wingtips black with white mirrors on outer 2 primaries; white trailing edge to upperwings. Rump and tail white. **1st year:** *eyes brown, bill flesh with black tip, legs flesh pink.* Head, neck, and underparts

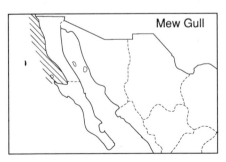

Mew Gull

Ring-billed Gull

white, streaked and mottled brown. Back pale grey, upperwings grey-brown with blackish leading edge to outerwing, black-ish bar on secondaries, and *pale grey panel on greater coverts and inner primaries. Rump and tail whitish, tail with black subterminal band.* **2nd year:** resembles adult but *bill dull yellowish with black band,* black wingtips more extensive with little or no trace of white mirrors, tail and secondaries show trace of dark bar.

> **Voice.** A yelping *ky-ow* or *y-ohw*, a mewing *kee-a kee-a kee-a kee-a*, and a thin, whining *sseeah* and *wheeah*, etc.

> **Habs.** Beaches, estuaries, lagoons, harbors, lakes, fields, etc., rarely offshore. Often in flocks, at times up to a few hundred birds.

SS: California Gull larger with relatively longer bill, darker grey upperparts, dark eyes at all ages, 2nd year and basic have greyish-green legs, bill of adult and older imms has red gonys spot; 2nd year resembles 1st year Ring-billed but less contrasting upperwings lack pale grey central panel, tail mostly black; 3rd year resembles 2nd year Ring-billed but note size and structure, colors of bare parts, darker grey upperparts. Herring Gull markedly larger with heavier bill but size of lone bird may be hard to judge; adult has pink legs, yellow bill with red gonys spot; 2nd year resembles 1st year Ring-billed (and may have dark band on bill), best told by pale eyes and brown-mottled tertials (dark with neat pale fringes in Ring-billed), broader black tail band; 3rd year Herring told from 2nd year Ring-billed by color of bare parts, etc. See Mew Gull (Baja).

SD: F to C winter visitor (Oct–May) from Baja and Gulf of California to Nay, U to R to El Salvador; F to C on Atlantic coast from Tamps to cen Ver, U to Yuc Pen; F to U in interior S to cen volcanic belt, R to Chis and Guatemala (Beavers *et al.* 1991). Small numbers oversummer, especially along N coasts.

RA: Breeds N.A.; winters S to Mexico and Caribbean, rarely to C.A.

CALIFORNIA GULL

*Larus californicus**   Not illustrated
Gaviota Californiana

ID: 18–21 in, W 46–52 in (45.5–53 cm, W 117–132 cm). Fairly large 4-year gull, winter visitor to N Mexico. *Bill relatively long but not heavy.* **Descr. Adult:** *eyes*

dark brown, orbital ring red, bill yellow to greenish yellow with red gonys spot and dark subterminal bar, legs greyish green to yellow. Head, neck, and underparts white, streaked and mottled dusky on head and neck in basic. Upperparts grey, wingtips black with white mirrors on outer 2 primaries; white trailing edge to upperwings. Rump and tail white. **1st year:** *bill flesh with black tip, legs flesh. Mottled brown overall* (head often whitish in 1st alt), pale upperpart edgings often look scaly, rump and uppertail coverts whitish, flight feathers and primary coverts blackish. **2nd year:** *bill greyish green with black tip or subterminal band,* legs greyish flesh to grey-green. Head, neck, and underparts white, streaked and mottled brown. Back grey, upperwings grey-brown with blackish leading edge to outerwing, blackish bar on secondaries, and pale panel on inner primaries. Rump and uppertail coverts whitish, *tail black with narrow white tip and outer edges.* **3rd year:** resembles adult but bill may lack red gonys spot, black wingtips more extensive with little or no trace of white mirrors, tail and secondaries show trace of dark bar.

> **Voice.** A slightly grating *whee-ahrr* or *whie ahrrr*, and a more rasping *raahrr*, etc.

> **Habs.** Beaches, estuaries, harbors, lagoons, fields, lakes, often well offshore. Locally in flocks up to a few thousand birds in Baja and N Gulf of California.

SS: See Ring-billed Gull. Herring Gull larger with heavier bill, back and upperwings pale grey; 1st year has mostly black bill, contrasting pale panel on inner primaries; 2nd year has pink legs, pale eyes (usually), upperwings have contrasting pale grey central panel; 3rd year told by pale grey upperparts, color of bare parts. Thayer's Gull differs in similar respects to Herring

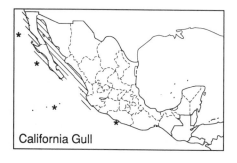

California Gull

Gull but 1st year paler overall, dark on wings and tail less contrasting, eyes dark.

SD: C to F winter visitor (Oct–May) to Pacific coast of Baja and in Gulf of California, U to F S to Nay, irregularly U to F to Gro (Howell and Wilson 1990); U to F through the summer in Baja and Gulf of California, R and irregular to Gro (SNGH); vagrant to Islas Revillagigedo. Reports from interior Mexico (Friedmann *et al.* 1950) require verification.

RA: Breeds W N.A.; winters to NW Mexico.

HERRING GULL
Larus argentatus smithsonianus
Gaviota Plateada   Not illustrated

ID: 22–27 in, W 53.5–60 in (56–68 cm, W 136–153 cm). Fairly widespread 4-year gull. *Bill stout and fairly heavy but not bulbous at tip, forehead typically fairly low and sloping.* **Descr. Adult:** *eyes pale yellow*, orbital ring yellow to pinkish, bill yellow with red gonys spot, *legs flesh-pink.* Head, neck, and underparts white, streaked and mottled dusky on head and neck in basic. Upperparts pale grey, wingtips black with white mirrors on outer 2 primaries; white trailing edge to upperwings. Rump and tail white. **1st year:** *eyes brown, bill blackish with diffuse paler base*; legs duller. Mottled brown overall (head often whitish by mid-winter), *upperparts patterned with paler scaling,* flight feathers and primary coverts blackish with *paler panel on inner secondaries.* **2nd year:** eyes usually pale, bill flesh to yellowish with black tip or subterminal band. Head, neck, and underparts whitish, streaked and mottled brown. Back pale grey, upperwings grey-brown with blackish leading edge to outerwing, blackish bar on secondaries, and pale grey panel on greater coverts and inner primaries. Rump and tail whitish, *tail with broad black subterminal band.*

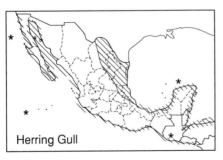

Herring Gull

3rd year: resembles adult but bill with black subterminal mark, black wingtips more extensive with little or no trace of white mirrors, tail and secondaries show trace of dark bar.

Habs. Estuaries, harbors, beaches, coastal bays, rubbish dumps, lakes. Locally in flocks up to a few hundred birds.

SS: See California Gull, Ring-billed Gull. Thayer's Gull generally smaller, head typically more domed, bill less heavy, eyes dark at all ages; adult has less black, more white on upperside of wingtips: wingtips appear similar to Herring when perched; from below, wingtips appear dusky without contrasting black tips. First year Thayer's paler overall and more uniform than Herring with less contrasting dark leading edge to outerwing, bar on secondaries, and tail band; underwings appear translucent when backlit; 2nd and 3rd years told by structure, less contrasting plumage than Herring; note dark eyes. First basic Western Gull has heavier, bulbous-tipped black bill, more domed head, brown plumage more uniform (upperparts less finely marked), lacks obvious pale panel on inner primaries. First basic Yellow-footed Gull has heavier, bulbous-tipped bill, contrasting whitish belly and white rump and uppertail coverts.

SD: F to C winter visitor (Oct–Apr) along Pacific coast of Baja, and in Gulf of California, U to Nay, R to El Salvador and Isla Clarión (Howell and Webb 1989*b*); F to C on Atlantic coast in Tamps and N Ver, U to Yuc Pen and Honduras; U to R in interior NE Mexico. U to R through the summer along both coasts, mainly in N Mexico.

RA: Holarctic breeder; winters S in New World to Mexico, Caribbean, and N C.A.

THAYER'S [ICELAND] GULL
Larus glaucoides (in part) or *L. thayeri*
Gaviota de Thayer   Not illustrated

ID: 21–24 in, W 52–57 in (53.5–61 cm, W 132–145 cm). Baja in winter; 4-year gull. *Head shape typically gently domed, bill relatively delicate for a large gull.* **Descr. Adult:** *eyes brown*, orbital ring red, bill yellow with red gonys spot, *legs flesh-pink.* Head, neck, and underparts white, streaked and mottled dusky on head and neck in basic. Upperparts pale grey, wingtips blackish above with white mirrors on outer 2 primaries, *underwings dusky at tip with blackish restricted to narrow line near tip of outer primaries*; white trailing edge to upperwings. Rump and tail white. **1st year:** *bill black,* legs flesh. *Mottled grey-brown overall,* upperparts fairly neatly patterned with paler scaling, flight feathers and primary coverts darker brown with paler panel on inner secondaries; *from below, primaries may appear translucent when backlit.* **2nd year:** bill flesh to yellowish with black tip or subterminal band. Head, neck, and underparts whitish, streaked and mottled dusky; upperparts more uniform, back grey, upperwing coverts grey-brown. Upperwings similar to 1st year but with contrasting dark bar on secondaries, darker outer primaries. Rump and tail whitish, tail with dark subterminal band. **3rd year:** resembles adult but blackish wingtips more extensive with little or no trace of white mirrors, tail and secondaries show trace of dark bar, head and underparts may be mottled with dusky through 3rd alt.

Habs. Estuaries, harbors, beaches, coastal bays, rubbish dumps. Usually found in small numbers with other large gulls.

SS: See Herring Gull. Glaucous-winged Gull larger and more heavily built with heavier, bulbous-tipped bill, from above wingtips paler in all ages, showing little or no contrast with upperparts at rest; 1st year has more uniform upperparts.

SD: U winter visitor (Nov–Apr) on Pacific coast of N Baja, U to R in N Gulf of California. Vagrant to Tamps (Dec 1986, Lasley and DeSante 1988).

RA: Breeds cen Canadian Arctic; winters mainly on Pacific coast from S Canada to NW Mexico.

NB: Thayer's Gull has been considered conspecific with Herring Gull, but appears more closely related to Iceland Gull *L. glaucoides* (Snell 1989).

LESSER BLACK-BACKED GULL
Larus fuscus graellsii   Not illustrated
Gaviota Dorsinegra Menor

ID: 21–25 in, W 52–58 in (53.5–63.5 cm, W 132–147 cm). Rare in winter, E Mexico. 4-year gull. *Overall more slightly built with narrower wings and smaller bill than Herring Gull.* **Descr. Adult:** eyes pale yellow, orbital ring red, bill yellow with red gonys spot, *legs yellow* to orange-yellow (rarely straw-flesh). Head, neck, and underparts white, streaked and mottled dusky on head and neck in basic. Upperparts *dark slaty grey,* wingtips black with white mirrors on outer 2 primaries; white trailing edge to upperwings; *underside of remiges dark.* Rump and tail white. **1st year:** eyes brown, bill black, *legs flesh.* Mottled brown overall (*head and chest usually constrastingly whitish*), remiges and *greater coverts blackish, rump and tail whitish,* tail with broad blackish subterminal band. **2nd year:** eyes pale, bill flesh to yellowish with black tip or subterminal band, legs flesh to yellowish. Head, neck, and underparts white, streaked dusky on head and neck (streaking almost absent in some 2nd alt). Back slaty grey, upperwings blackish with paler grey-brown inner coverts; underwings dark. Rump and tail whitish, tail with black subterminal band. **3rd year:** resembles adult but black wingtips more extensive with little or no trace of white mirrors, tail usually with trace dark bar.

Habs. Estuaries, harbors, beaches, rubbish dumps. Associates readily with other gulls.

SS: 1st year Herring Gull paler overall, head and underparts browner, greater secondary coverts paler than blackish second-

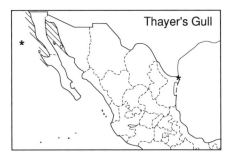

Thayer's Gull

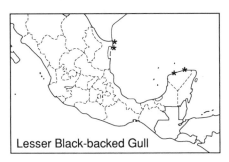

Lesser Black-backed Gull

aries; note pale panel on inner primaries. Western, Yellow-footed, and vagrant Great Black-backed gulls more heavily built with stout, bulbous-tipped bills, 1st years have brown greater secondary coverts contrasting with blackish secondaries, Great Black-backed has paler panel on inner primaries; adults have blackish backs and upperwings, clean white heads in basic, Western and Great Black-backed have pink legs. Kelp Gull has heavier bill (though small-billed ♀ similar to Lesser), adult has blackish upperparts, only one small white mirror in primaries, unstreaked white head and neck; note color of bare parts of imms, other field separation criteria poorly known.

SD: R (to U?) winter visitor (Nov–Apr at least) to Atlantic coast in N Tamps and N Yuc (Lasley 1987; Howell and Prairie 1989; Howell and Webb 1990*d*; RCS photo) Increasing in N.A. and probably will be found more widely in region; first recorded in Mexico in 1979.

RA: Breeds NW Europe; winters S to Africa, increasingly reported in New World.

KELP GULL
Larus dominicanus ssp.   Not illustrated
Gaviota Dorsinegra Sureña

ID: 22–26 in, W 52–58 in (56–66 cm, W 132–147 cm). Vagrant 4-year gull. *Bill fairly heavy to heavy.* **Descr. Adult:** eyes pale yellow, *orbital ring red*, bill yellow with red gonys spot, *legs yellowish.* Head, neck, and underparts white, *head and neck unstreaked. Upperparts slaty blackish* showing little or no contrast with black wingtips. *White mirror on outermost primary* and white trailing edge to upperwings; underside of remiges dark. Rump and tail white. **1st year:** eyes brown, bill black, *legs greyish to flesh.* Mottled grey-brown overall, paler below and on rump;

head and neck often contrastingly whitish. Remiges blackish, greater coverts dark brown; underwings dark. Tail blackish with whitish basal corners. **2nd year:** eyes paler, bill yellowish with broad black median band, legs greyish flesh to dull yellowish. Head, neck, and underparts white, streaked brownish. Upperparts grey-brown and slaty grey, may be mostly slaty black on back; underwings dark to mottled whitish on coverts. Rump whitish; more white at tail base. **3rd year:** resembles adult but bare parts duller, head and underparts may show faint dusky markings, little or no white mirror on outermost primaries, tail usually with trace black bar.

Habs. Beaches, estuaries, etc. Associates readily with other gulls.

SS: See Lesser Black-backed Gull. Yellow-footed Gull typically has more massive and deeper bill, adult has yellow orbital ring, less blackish upperparts (typically with primaries contrasting blacker); imm plumages distinct. Great Black-backed Gull larger with heavier bill, most plumages distinctive, adult has pink legs.

SD: Vagrant: up to 2 adults seen and photographed in N Yuc in four winters (Nov–May at least) since Nov 1987 (Howell *et al.* unpublished).

RA: Circumpolar in S Pacific; ranging N in W S.A. to Ecuador.

YELLOW-FOOTED GULL
Larus livens   Figure 23
Gaviota Patamarilla

ID: 22–27 in, W 56–61 in (56–67.5 cm, W 142–155 cm). Large 3-year gull endemic to NW Mexico. *Bill very heavy and bulbous at tip.* Prebasic molt Jun–Sep, markedly earlier than other gulls in region. **Descr. Adult:** eyes pale yellow, orbital ring yellow to yellow-orange, bill yellow with red gonys

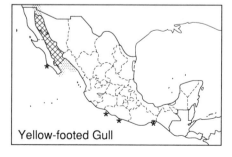

1 Kelp Gull
2 Great Black-backed Gull

Yellow-footed Gull

spot, *legs bright orange-yellow* to yellow (can appear flesh in bright light). Head, neck, and underparts white. Upperparts *blackish grey*, wingtips blacker with large white mirror on outermost primary; white trailing edge to secondaries and inner primaries. Rump and tail white. **Juv:** eyes brown, bill black, legs flesh. Mottled grey-brown overall with *whitish belly and undertail coverts*, upperparts with neat, pale, scaly edgings. *Whitish rump and uppertail coverts contrast with dark back and blackish tail*, remiges and primary coverts blackish. **1st basic:** *similar to juv but* bill may be pale at base, *head and underparts paler*, often whitish, back more uniform, rump and uppertail coverts whiter. **1st alt:** resembles 1st basic but eyes pale greyish, bill flesh to yellowish with black tip, legs may be yellowish, *back slaty grey*. **2nd year:** bill often with dark subterminal mark, red gonys spot. Head, neck, and underparts whitish, with slight dusky streaking on head and hindneck in basic, upperparts blackish grey, remiges blackish. Rump and uppertail coverts white, *tail with broad black subterminal band*.
    Habs. Rocky seacoasts, beaches, harbors, estuaries. Often in small flocks, rarely to a few hundred birds. Nests in small colonies or singly, on beaches above high-tide line. Eggs: 2–3, pale greenish buff to pale brownish, marked with browns and greys (WFVZ).
SS: See 1st year Herring Gull. Western Gull (rarely sympatric) has less massive bill, flesh-pink legs at all ages; juv and 1st year brown overall; 2nd year has less black on tail than 1st year Yellow-footed; 3rd year has less black on tail than 2nd year Yellow-footed. See vagrant Kelp Gull.
SD/RA: C to F resident in Gulf of California, mostly near or along Baja coast and on islands, F along Son coast, U to Sin. Post-breeding dispersal (mainly Jul–Sep) takes birds rarely to Pacific coast of Baja and regularly to Salton Sea, California. Imms wander in all months, at least N to Bahia Magdalena on Pacific coast of Baja; R and irregular (Nov–May) S to Isthmus (Howell and Webb 1992*a*). Many early records, including a vagrant to Isla Clarión, were not critically determined with respect to Western Gull (see NB).
NB: Formerly considered conspecific with Western Gull.

WESTERN GULL

Larus occidentalis wymani   Not illustrated
Gaviota Occidental

ID: 21–26 in, W 53.5–59 in (53.5–66 cm, W 136–150 cm). Large 4-year gull, Pacific coast of Baja. *Bill heavy and with bulbous tip.* **Descr. Adult:** eyes pale yellow to brownish yellow, orbital ring yellow, bill yellow with red gonys spot, *legs flesh-pink*. Head, neck, and underparts white, head and neck streaked dusky in basic, upperparts *blackish grey*, wingtips blacker with large white mirror on outermost primary, white trailing edge to secondaries and inner primaries. Rump and tail white. **Juv:** eyes brown, bill black, legs dusky flesh. *Mottled grey-brown overall*, upperparts with neat, pale, scaly edgings. Flight feathers and primary coverts blackish, *rump and uppertail coverts paler, barred whitish and dark brown*. **1st year:** *similar to juv but* bill paler at base, legs flesh, *upperparts more uniform*, head and neck paler in 1st alt. **2nd year:** eyes pale, bill flesh with black tip. Head, neck, and underparts whitish, streaked and mottled brown. Back slaty grey, wings and tail similar to 1st year but inner primaries paler, some slaty grey on upperwing coverts, *lateral rectrices often white at base*. **3rd year:** resembles adult but bill flesh to yellowish with black tip or subterminal bar, black wingtips more extensive with little or no trace of white mirrors, tail and secondaries show trace of dark bar. Northern breeding ssp *occidentalis* not certainly recorded in region but probably R visitor, differs from *wymani* in paler grey (grey to dark grey) upperparts of adult and older imms.
    Habs. Harbors, beaches, estuaries, rocky coasts, rubbish dumps. Often in flocks up to a few hundred birds. Nests in small to large colonies on rocky islands. Eggs: 2–

Western Gull

3, pale greenish buff to pale brownish, marked with darker browns and greys (WFVZ).

SS: See Herring and Yellow-footed gulls. Glaucous-winged Gull paler in all plumages; note that *L. o. occidentalis* and Glaucous-winged interbreed, and hybrids showing a range of intermediate characters may occur, at least in NW Baja.

SD: C to F resident along Pacific coast of Baja S to 27°N; F to U winter visitor (Oct–Apr), and U in summer, S to Bahia Magdalena, U to R to tip of Baja and (Dec–Mar) in N Gulf of California. Vagrant to Gro (Jun 1992, RAE) and Oax (Apr 1988, Howell and Webb 1992*a*). Reports from Sin and Nay (AOU 1983) probably pertain to Yellow-footed Gull; a report from Isla Clarión is similarly equivocal.

RA: W N.A. from S Canada to Baja.

NB1: Formerly considered conspecific with Yellow-footed Gull. **NB2:** Interbreeds with Glaucous-winged Gull; the two sometimes considered conspecific.

GLAUCOUS-WINGED GULL

Larus glaucescens   Not illustrated
Gaviota Aliglauca

ID: 22–28 in, W 55–62 in (56–71 cm, W 140–157 cm). Baja in winter; 4-year gull. *Bill heavy and with bulbous tip.* **Descr. Adult:** eyes brown, orbital ring reddish, bill yellow with red gonys spot, legs flesh-pink. Head, neck, and underparts white, streaked and mottled dusky on head and neck in basic. Upperparts *pale grey, wingtips duskier grey with large white mirror on outermost primary*; white trailing edge to secondaries and inner primaries. Rump and tail white. **1st year:** bill black; legs flesh. *Mottled pale grey-brown overall (may fade to creamy whitish by 1st summer), flight feathers show little or no contrast with rest of plumage*; at rest, wingtips

same shade or paler than back. **2nd year:** *blackish bill often shows paler base. Resembles 1st year but generally more uniform, back pale grey,* head and underparts paler in 2nd alt. **3rd year:** resembles adult but bill flesh to yellowish with black tip or subterminal band, wing coverts washed brown, wingtip pattern indistinct, tail with trace of dusky subterminal band.

Habs. Harbors, beaches, estuaries, rocky coasts, rubbish dumps. Usually with other large gulls.

SS: See Thayer's, Herring, and Western gulls.

SD: U to F winter visitor (Nov–Mar) and R to U in summer along Pacific coast of Baja; U to R in N Gulf of California, mainly in winter; vagrant to Isla Socorro.

RA: Breeds N Pacific; winters S in New World to NW Mexico.

NB: See Western Gull.

GLAUCOUS GULL

Larus hyperboreus barrovianus
Gaviota Blanca   Not illustrated

ID: 22–28 in, W 56–62 in (56–71 cm, W 142–157 cm). Vagrant 4-year gull. *Bill relatively long and heavy but not bulbous at tip.* Adults unrecorded in region and rare S of Canada. **Descr. 1st year:** eyes brown, *bill flesh-pink with clean-cut black tip,* legs flesh. *Mottled pale brownish grey to creamy overall* (often fades to whitish by 1st spring), *upperparts (including tail) with neat fine dusky barring. Whitish remiges paler than rest of plumage.* **2nd year:** similar to 1st year but eyes paler, bill may be yellowish with black subterminal band, *paler and more uniform overall,* back mainly pale grey in 2nd alt. **3rd year:** resembles adult but bill often with blackish subterminal mark, some brownish on upperwing coverts and tail. **Adult:** eyes pale yellow, orbital ring yellowish, bill yellow with red gonys spot, legs flesh-pink.

Glaucous-winged Gull

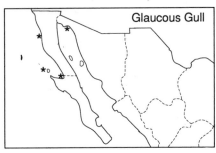
Glaucous Gull

Head, neck, and underparts white, streaked and mottled dusky on head and neck in basic. Upperparts pale grey, *trailing edge of wings and wingtips white.* Rump and tail white.

Habs. Rocky coasts, beaches, harbors, rubbish dumps. Associates with other large gulls.

SS: 1st and 2nd year Glaucous-winged Gulls rarely have such a clean-cut pink and black bill, plumage relatively uniform without fine dusky barring (especially tail), remiges rarely paler than upperparts. Thayer's Gull smaller with more delicate bill (black in 1st year), wingtips and tail darker than rest of plumage. Hybrid Glaucous × Herring gulls resemble Glaucous but show dark in wings and tail.

SD: Irregular R winter visitor (Jan–Mar) to Pacific coast and islands of BCN (Devillers *et al.* 1971), and to N Gulf of California, Son (BW photo).

RA: Holarctic breeder; winters S in New World to N USA, rarely to S USA and NW Mexico.

GREAT BLACK-BACKED GULL
Larus marinus　　　　　Not illustrated
Gaviota Dorsinegra Mayor　　See Kelp Gull
　　　　　　　　　　　　　for map

ID: 25–31 in, W 60–65 in (63.5–78.5 cm, W 152–165 cm). Vagrant 4-year gull. *Bill heavy and with bulbous tip.* **Descr. Adult:** eyes pale yellow, orbital ring red, bill yellow with red gonys spot, *legs flesh-pink.* Head, neck, and underparts white (head faintly streaked dusky in basic). Upperparts *slaty blackish, wingtips blacker with large white mirror on tips of outer 2 primaries,* white trailing edge to secondaries and inner primaries. Rump and tail white. **1st year:** eyes brown, bill black, legs duller. *Head, neck, and underparts whitish, streaked and mottled brown (head often mostly whitish),* upperparts neatly checkered blackish and white. Remiges and primary coverts blackish, paler inner primaries form slightly contrasting panel. Rump and tail whitish, *tail with wavy black bars not forming solid subterminal band.* **2nd year:** eyes pale, bill pale flesh with dark subterminal area by 2nd summer. Head, neck, and underparts white with little or no dusky streaking on head; *back blackish grey,* upperwings and tail similar to 1st year but wing coverts more

uniform. **3rd year:** resembles adult but bill often with dark subterminal mark, black wingtips more extensive, white mirror at tips of primaries reduced, tail with trace of dusky band.

Habs. Beaches, estuaries, coastal bays, etc. Associates with other large gulls.

SS: Herring Gull smaller with less heavy bill; 1st year more uniform brownish, black tail band bolder. See vagrant Lesser Black-backed and Kelp gulls.

SD: Vagrant, one record: Belize City, Belize, 11–12 Jan 1989 (Howell *et al.* 1992*b*). An earlier report from Belize (Wood *et al.* 1986) requires verification. Increasing in New World, rarely wintering S to Caribbean; should be looked for on Atlantic coast of Mexico.

RA: Breeds N Atlantic; winters S in New World to SE USA.

BLACK-LEGGED KITTIWAKE
Larus tridactylus pollicaris　Not illustrated
Gaviota Patinegra

ID: 17–18 in, W 37–41 in (43–46 cm, W 94–104 cm). Offshore winter visitor to NW Mexico; 3-year gull. *Bill fairly small and delicate, legs relatively short.* Tail squared to slightly cleft. **Descr. Adult:** *bill yellowish, legs black.* Head, neck, and underparts white, with *dusky smudging on nape and auriculars in basic;* upperparts grey with *clean-cut black wingtips;* remiges often noticeably paler than upperwing coverts. Rump and tail white. **1st year:** *bill black in 1st winter.* Head, neck, and underparts white, with *blackish ear spot and hindcollar.* Upperparts grey (*secondaries and inner primaries often appear contrastingly whitish*) with *bold black M pattern not joined across back.* Rump and tail white, tail with black terminal band. **2nd year:** resembles adult but black wingtips slightly more extensive.

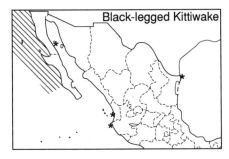
Black-legged Kittiwake

Habs. Pelagic, may be seen from land or on beaches, in harbors, etc., during or after storms. Flight fairly buoyant, low to high over water. Usually silent in the region.

SS: Sabine's Gull lacks black carpal bar, pattern of black/white/grey or grey-brown bold and striking (though Kittiwake may show bold white triangle on trailing edge of wing in some lights), forked tail distinctive but hard to see; juv Sabine's has dusky grey-brown crown, nape, and upperparts; adult has yellow-tipped black bill, dark hood in spring. See Mew Gull.

SD: Irregularly U to C winter visitor (Nov–Mar) off Pacific coast of Baja, R S to Nay and Jal (Dec 1989, DCW), vagrant to Gulf of California (Apr 1984, BT); small numbers may oversummer following invasions. Vagrant to Tamps (Jan 1992, per GWL).

RA: Holarctic breeder; winters S in New World to NW Mexico, E USA.

NB: Often placed in the genus *Rissa*.

SABINE'S GULL

Larus sabini   Not illustrated
Gaviota de Sabine

ID: 13–14 in, W 33.5–35.5 in (33–35.5 cm, W 85–90 cm). Offshore Pacific; 2-year gull. *Bill relatively delicate, tail slightly forked.* Adults and juvs complete autumn migration before molting and have complete molt in spring before moving N. **Descr. Adult. Alt** (Mar–Oct): *bill black with yellow tip,* legs blackish. *Slaty-grey hood demarcated by black collar.* Neck and underparts white. Back grey, *striking tricolored upperwing pattern: black leading edge to outerwing and grey inner coverts contrast sharply with white triangle on trailing edge of wing* formed by secondaries, greater coverts, and inner primaries. **Basic:** head white with dusky hindcollar and auricular mark, legs flesh. **Juv:** bill black, legs flesh. Lores, lower face,

foreneck and underparts white. *Crown, hindneck, and back grey-brown, upperwings patterned as adult but inner coverts grey-brown like back.* Rump and tail white, tail with black terminal band. **1st alt:** resembles adult but dark hood flecked whitish. Attains adult plumage by 2nd prealternate molt on winter grounds.

Habs. Pelagic. May be seen from shore, singly or in small flocks, but usually offshore. Flight buoyant and tern-like, low to fairly high over water. Usually silent in the region.

SS: See Black-legged Kittiwake.

SD: F to C transient (Apr–early Jun; late Aug–Nov) off Pacific coast; U to R in Gulf of California (DB, BT). Irregularly (?) U to F winter visitor (Dec–Mar) N along coast to BCS (JCA photo, SNGH, PP). Vagrant to Yuc (Feb 1989, Howell and Prairie 1989).

RA: Holarctic breeder; in New World winters in Pacific from S C.A. to cen Chile.

NB: Often placed in the genus *Xema*.

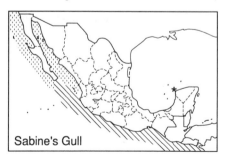

Sabine's Gull

SUBFAMILY STERNINAE: Terns

Terns resemble gulls superficially but have pointed bills, shorter legs, and typically are smaller, more slender, and graceful with noticeably forked tails. Unlike gulls, terns rarely alight on water. They are more diverse than gulls in tropical regions, and most species are migratory. The Arctic Tern is noted for traveling from the Arctic in summer to waters around Antarctica in winter, the longest known migration of any bird. Wings long and pointed, tails often long and deeply forked, alt adults of several species have long outertail streamers. Most species feed by plunge-diving for fish near the surface. Juvs of several species accompany their parents for up to several months after fledging, including during migration.

Adults of most species are white and pale grey overall with black caps in alt; juvs are usually mottled or scalloped dark above, tails lack streamers. Imms of most species resemble adults by their 2nd winter but take 2–6 years to mature. Seasonal variation obvious in most species, basic often indicated by partial black caps. Molts complex and variable, flight feather molt suspended during migration. First prebasic protracted and complete, running directly into, and often overlapping with, complete 2nd prebasic molt. Much of juv body plumage quickly molted in 1st autumn but flight feathers, including tertials, molted on 'winter' grounds, with 1st basic usually attained by end of 1st summer; by this time, however, 2nd prebasic molt has already begun, and by 2nd winter imms of most species look much like adults. It is often more convenient to talk of appearance in 1st summer than to describe transitory 1st basic plumage. Successive incomplete to partial prealternate and complete prebasic molts converge with adult cycle. Adult prebasic molt complete, mostly or all on winter grounds; prealternate molt usually includes tail and 2–6 inner primaries. The surface of new primaries is covered with a silvery grey 'bloom' which reveals dark feathers below as it wears off; hence, worn primaries are *darker* than fresh primaries. Molts of tropical terns (Sooty, Bridled, and White terns, noddies, etc.) poorly known due to their pelagic habits; not necessarily as described above. Noddies and White Tern also have distinct plumages from other terns.

Food mostly fish and marine invertebrates. Most terns nest colonially, often in mixed-species colonies. Nest scrapes may be marked with grass or shingle rims; sometimes full shallow cups of vegetation are built, often in bushes or trees by noddies. Eggs: 1–4, variable within and among species; ground color buff to olive-brown to whitish, marked with browns and greys (WFVZ).

Measurements quoted for terns in most field guides often include the tail streamers of some species which greatly increase the apparent size of a bird. Lengths given here (including tail length, T) do not include tail streamers. Bill lengths are also given in parentheses for all species.

GULL-BILLED TERN
Sterna nilotica*   Not illustrated
Golondrina-marina Piquigruesa

ID: 13–14 in, W 35–38 in, T 4 in (33–35.5 cm, W 89–96.5 cm, T 10 cm). *A fairly stocky tern with notably thick bill* (1.5 in; 4 cm) *unlike other terns; legs relatively long.* **Descr.** Age/seasonal variation. **Adult. Alt:** bill and legs black. Head, neck, and underparts white with black cap extending down nape. Upperparts, including rump and tail, silvery grey; outer primaries often blackish by late summer. Underwings white with dark trailing edge to outer primaries.

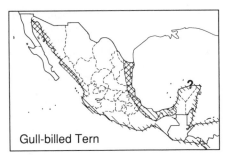

Gull-billed Tern

Basic: head white with indistinct dusky smudge through eye. **Juv:** resembles basic but legs dark reddish, upperparts slightly duskier grey with brown tips and spots, subterminal dark marks on tertials and secondaries retained to mid-winter when primaries become notably darker than adult. **1st summer:** much like adult basic; note dark outer primaries; as adult basic by 2nd winter.

Voice. Nasal, laughing and mellow, barking calls, *kek-wek* or *ku-wek* and *ket-e-wek* or *kit-i-wuk*, etc.

Habs. Beaches, salt marshes, coastal lagoons, rarely offshore. Flies with easy graceful wingbeats. Feeds by swooping down and picking insects, molluscs, small reptiles, fish, etc., near or from the ground or surface of water; often hunts over land.

SS: Stout black bill combined with relatively stocky build and ghostly pale appearance distinctive; note also long legs, feeding habits. Basic Forster's Tern has bold black mask.

SD: F to U but local breeder (Apr–Aug?) in Gulf of California from Son to Sin, probably elsewhere S at least to Isthmus; on Atlantic Slope S to cen Ver (SNGH, SW), probably also in coastal Yuc. U to F transient and local winter visitor (Aug–May) from Gulf of California to El Salvador, and on Atlantic coast from Tamps to Honduras; small numbers oversummer locally in winter range.

RA: Cosmopolitan but local in warmer areas of mid-latitudes.

NB: Often treated in the monotypic genus *Gelochelidon*.

CASPIAN TERN
Sterna caspia   Figure 24
Golondrina-marina Caspica

ID: 20–22 in, W 46.5–51 in, T 5–6 in (51–57 cm, W 118–130 cm, T 12.5–15 cm).

An unmistakable, heavily built, broad-winged tern with a stout carrot-like bill (2.5–3 in; 6.5–7.5 cm). **Descr.** Age/seasonal variation. **Adult. Alt:** *bill bright red* to orange-red with black subterminal mark and fine pale tip, legs black. Head, neck, and underparts white with black cap. Upperparts pale grey, becoming white on rump and tail; outer primaries often blackish by late summer. *Underwings white with bold blackish wedge on outer primaries.* **Basic:** forecrown streaked white from mid-summer. **Juv:** bill reddish orange to orangish with dark tip, dull orangish legs soon becoming black. Black cap streaked whitish to buff. Upperparts have brown chevrons and spots, secondaries dusky subterminally, tertials dark, edged white, tail pale grey with darker subterminal band. **1st summer:** much like adult basic but often shows dark carpal bar and bar on secondaries, dark outer primaries; as adult basic by 2nd winter. May show white flecks in black cap in 2nd summer.

Voice. A loud, low, rasping *ahhrr* or *rahrr*, and a longer *rrah-ah ah-rr* or *rrah a-aah*; begging young give a high, slightly reedy to plaintive, drawn-out *sii-iieu*.

Habs. Coastal lagoons, estuaries, beaches, inland lakes and large rivers, rarely

Fig. 24 (a) Forster's Tern; (b) Black Skimmer; (c) Elegant Tern; (d) Royal Tern; (e) Caspian Tern

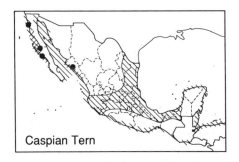

Caspian Tern

offshore. Singly or in groups, often roosts with other terns, less often with gulls. Flies with easy shallow wingbeats.

SS: Royal Tern smaller and more rangy with longer tail, less stout orange bill, lacks blackish wedge on underwing tips, forecrown extensively white much of year.

SD: Local breeder (Apr–Aug?) on W coast of Baja and Sin; probably elsewhere in NW Mexico. F to C transient and winter visitor (Aug–May) along and near coasts from Baja and Gulf of California to Guatemala, unreported El Salvador; F to U on Atlantic Slope from Tamps to Tab, coastally to Honduras; and U to R and local (1000–2500 m) S to cen volcanic belt (SNGH, PP). Small numbers oversummer locally on both coasts, especially Pacific coast, and wander (May–Jul) to W Mexican interior (Williams 1982*a*).

RA: Cosmopolitan except S.A.

NB: Formerly placed in the monotypic genus *Hydroprogne*.

ROYAL TERN

Sterna m. maxima   Figure 24
Golondrina-marina Real

ID: 17–19 in, W 42–45.5 in, T 6 in (43–48 cm, W 106–116 cm, T 15 cm). *A large, fairly rangy tern.* Bill fairly stout (2.5 in; 6.5 cm). **Descr.** Age/seasonal variation. **Adult. Alt:** *bill reddish orange to orange,*

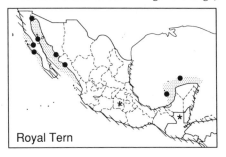

Royal Tern

legs black. Tail streamers add 2 in (5 cm) to length. Head, neck, and underparts white with *crested black cap.* Upperparts pale grey, becoming white on rump and tail; outer primaries often blackish by late summer. Underwings white with blackish trailing edge to outer primaries. **Basic:** *forecrown and lores white from mid-summer,* black postocular mask often separated from eye by white crescent. **Juv:** bill and legs orange-yellow to orange (rarely greenish yellow), legs soon become black. Resembles basic but upperparts with dark marks and spots, lesser upperwing coverts dark, secondaries dusky subterminally, tertials dark, edged white, and tail pale grey, tipped dark. **1st summer:** much like adult basic but often shows dark bar on secondaries, outer primaries and tail tips dark; as adult basic by 2nd winter. May show white flecks in black cap in 2nd summer.

Voice. Adults give a yelping *krehk*, a laughing *kweh-eh* and *kweh-eh-eh*, and soft, mellow, piping whistles; imms have a screechy, grating *rreh-eh* and *rree-ahk*, and younger birds give a high *see-ip*; also a slightly creaky *ahrr* or *aehrr*.

Habs. Beaches, coastal bays and estuaries, sandy islands, locally on lakes near coast, rarely offshore. Habits much as Caspian Tern but flies with deeper, more rangy wingbeats.

SS: See Caspian Tern. Elegant Tern smaller with markedly more slender bill which may appear drooped at tip, longer shaggy crest extends well down nape.

SD: C to F resident (mostly non-breeding) along both coasts, breeding (Apr–Aug) locally on Pacific coast of Baja, in Gulf of California, and on Campeche Bank, possibly also off Belize. R (Jan–Feb, regular?) to interior N Guatemala (Beavers *et al.* 1991); vagrant to DF (Oct 1990, JKe, RGW).

RA: Breeds NW Mexico and from E USA through Caribbean to N S.A., locally in Uruguay and W Africa; ranges along coasts S to Peru and Argentina.

NB: Royal and other similar medium-large terns sometimes placed in the genus *Thalasseus*.

ELEGANT TERN

Sterna elegans   Figure 24
Golondrina-marina Elegante

ID: 14.5–16 in, W 36.5–39.5 in, T 4.5–5 in

(37–41 cm, W 92–100 cm, T 11.5–12.5 cm). Pacific coast. A medium-large rangy tern; *slender bill* (2.5 in; 6.5 cm) *may appear drooped at tip*. **Descr.** Age/seasonal variation. **Adult. Alt:** bill orange to reddish orange, legs black (rarely orange). Tail streamers add 1.5–2 in (4–5 cm) to length. Head, neck, and underparts white with *shaggy black cap extending well down nape*; often has rosy flush on underparts. Upperparts pale grey, becoming white on rump and tail; outer primaries often blackish by late summer. Underwings white with blackish trailing edge to outer primaries. **Basic:** forecrown and lores white from late summer. **Juv:** bill and legs orange-yellow to orange (bill rarely yellow, often reddish orange by autumn). Resembles basic but upperparts with dark marks and spots, dark lesser upperwing coverts soon lost, secondaries with dusky subterminal band, tertials dark, edged white, outer rectrices tipped dark. **1st summer:** much like adult basic but often shows dark bar on secondaries, outer primaries and tail tips dark; as adult basic by 2nd winter. May show white flecks in black cap in 2nd summer.

Voice. Adults and imms give a grating, slightly screechy *rreea-h* or *rree-ahk*, and *krree-eh*, and a rough, grating *errk*; begging young give a high, piping *sii* or *seep*, and *siip-siip*, etc.

Habs. Beaches, rocky coasts and islands, coastal bays and lagoons, rarely far offshore, no inland records. Habits much as Caspian Tern, flight suggests lightly built Royal Tern.

SS: See Royal Tern.

SD/RA: C to F but local breeder (Mar–Jul) in Gulf of California, with 95% of population on Isla Rasa, BCN; local breeder on Pacific coast of Baja N to S California. Post-breeding dispersal (Jul–Oct) N to cen California, R to British Columbia, and S along Mexican coast. C to F transient (Sep–mid-Dec, mid-Feb–Apr) off and along Pacific coast, winters (Oct–Feb) off W S.A. from Ecuador to Chile; U to R and irregular in winter (Jan–Feb) N to Nay (SNGH, PP) and in summer (May–Jun) from Nay to C.A.

NB: See Royal Tern.

SANDWICH TERN
*Sterna sandvicensis**   Not illustrated
Golondrina-marina de Sandwich

ID: 13.5–14 in, W 34–36.5 in, T 4 in (34.5–35.5 cm, W 86.5–92 cm, T 10 cm). Atlantic coast. A medium-sized rangy tern with slender bill (2 in; 5 cm). **Descr.** Age/seasonal variation. **Adult. Alt:** *bill black with fine yellow tip*, legs black. Tail streamers add 1 in (2.5 cm) to length. Head, neck, and underparts white with *shaggy black cap*; may show rosy flush on underparts in spring. Upperparts silvery grey, becoming white on rump and tail; outer primaries often blackish by late summer. Underwings white with blackish trailing edge to outer primaries. **Basic:** forecrown and lores white, apparent from late summer. **Juv:** yellow tip to bill may be lacking. Resembles basic but forehead dusky, *dark chevrons* and spots *on upperparts* lost quickly, secondaries dusky subterminally, tertials dark, edged white. Rump and tail pale grey, outer rectrices tipped blackish. **1st summer:** much like adult basic but often shows dark bar on secondaries, outer primaries and tail tips dark, tail pale grey; as adult basic by 2nd winter. May show white flecks in black cap in 2nd summer. **Cayenne Tern** *S. s. eurygnatha* (of E Caribbean and S.A.) has bill mostly or entirely yellow.

Voice. A scratchy *krrii-ik* or *krri-ik*, a slightly reedy *ki-i wii-wii* given by feed-

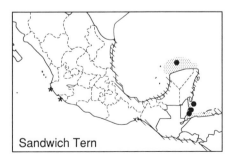

Elegant Tern   Sandwich Tern

ing groups, and high, shrill notes from imms.

Habs. Beaches, sandy islets, coastal bays and lagoons, estuaries, rarely lakes and rivers near coast. Habits much as Caspian Tern but flies with rangy wingbeats.

SS: See Gull-billed Tern.

SD: F but local breeder (Apr–Aug) on Campeche Bank and Belize cays (Pelzl 1969). F to C transient and winter visitor (Aug–May) on Atlantic coast from Tamps to Honduras (Brown and Monroe 1974); F to U on Pacific coast from Isthmus S; U to F through the summer in winter range. R (irregular?) N on Pacific coast to Col and Jal (Dec 1983, SNGH, PP). Cayenne Tern vagrant to Isla Cozumel (Dec 1992, BL).

RA: Breeds Europe and E USA to Caribbean; ranges along coasts to South Africa and S.A., on Pacific coast S to Peru.

NB: See Royal Tern.

ROSEATE TERN

Sterna d. dougallii   Not illustrated
Golondrina-marina Rosada

ID: 12–13 in, W 26–28 in, T 5–6 in (30.5–33 cm, W 66–71 cm, T 12.5–15 cm). Caribbean coast in summer. Bill 1.5 in (4 cm). Wings relatively short, tail very long and deeply forked. **Descr.** Age/ seasonal variation. **Adult. Alt:** *bill black with red base by summer*, legs red. *Tail streamers add 2 in (5 cm) to length and project well beyond wingtips at rest.* Head, neck, and underparts white with black cap; usually a rosy flush on underparts in spring. *Upperparts silvery grey becoming white on rump and tail; outer 2–3 primaries often blackish by late summer. White underwings lack dark trailing edge.* **Basic:** bill black (Sep–May), legs dull red; forehead white from late summer, tail shorter. **Juv:** *bill and legs black.* Resembles basic but forehead dusky, *upperparts with dark*

chevrons and spots, dark lesser upperwing coverts; tail pale grey with dark subterminal chevrons; *underwings white with dark subterminal area on tips of outer primaries.* **1st summer:** much like adult basic but may show dark carpal bar, outer primaries, and tail tips; from below, outer primaries may show dark trailing edge. Subsequent molts and imm plumages poorly known, may parallel Common Tern.

Voice. A slightly scratchy *krrizzik* or *kir-rik*, often doubled, suggests Sandwich Tern; a rasping *rrahk* or *ahrrrr*, a mellow *ch-dik* or *ch-weet*, a chippering *cheut cheut*, etc.

Habs. Sandy beaches, sandy islets with grassy vegetation, bays, and estuaries. Singly or in pairs, breeds in small colonies which may associate loosely with other species. Flies with fairly quick wingbeats, distinct from other medium-sized *Sterna.* Nest usually well hidden in vegetation, coral rubble, etc.

SS: Common, Forster's, and Arctic (unlikely to overlap with Roseate) terns have larger bodies, longer wings, and deeper, more graceful wingbeats, underwings with black trailing edge to outer primaries; dark chevrons on upperparts of juv Roseate distinctive, note also black bill and legs. See other species for more detailed descriptions.

SD: U to F but local summer resident (Apr–Aug) on cays off Belize (Pelzl 1969) and Honduras; U transient (late Apr–mid-May) on Isla Cozumel (SNGH, KK); also a band recovered in Sep in El Salvador (Thurber *et al.* 1987). Reports of being resident in Belize (Wood *et al.* 1986) are unfounded.

RA: Breeds E N.A. to N S.A., in W Europe, and locally from E Africa to Australasia. Winters in Atlantic from E Caribbean to Brazil, and in W Africa.

COMMON TERN

Sterna h. hirundo   Not illustrated
Golondrina-marina Común

ID: 11.5–12.5 in, W 29.5–32.5 in, T 3.5–4 in (29–32 cm, W 75–82.5 cm, T 9–10 cm). Migrant on both coasts. Bill 1.5 in (4 cm). **Descr.** Age/seasonal variation. **Adult. Alt:** *bill red with black tip* (rarely all red), legs red. Tail streamers add 1 in (2.5 cm) to length and are about equal with wingtips at rest. Head, neck, and under-

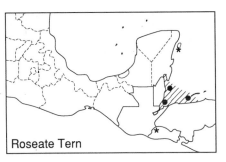

Roseate Tern

parts white with black cap. Upperparts pale grey, becoming white on rump and tail; outermost rectrices with dark outer webs; outer primaries often darker by late summer with dark wedge on trailing edge of middle primaries. *Underwings white with black trailing edge to outer primaries,* inner primaries may appear as translucent panel when backlit. **Basic:** bill black, usually with slight red at base, legs duller; forehead white from late summer; note dusky lesser upperwing coverts and subterminal bar on secondaries, tail pale grey, shorter. **Juv:** bill black with dull pinkish base, legs dull reddish. Resembles basic but forehead may be washed buff, *upperparts mottled brownish,* tertials dark, edged white, rump and uppertail coverts washed pale grey, *tail pale grey, outer rectrices edged dark. In flight note dusky secondaries and dark lesser upperwing coverts* (usually obvious at rest). **1st summer:** much like adult basic but lesser upperwing coverts and outer primaries often darker, tail tips dark; as adult basic by 2nd winter. May show white flecks in black cap in 2nd summer.

Voice. Mostly silent in the region. Calls include a sharp *cheuk*, a grating *krrrih*, a sharp *kik-kik*, and a rapidly repeated *kehrr kehrr kehrr ...*, etc.

Habs. Coastal and offshore; beaches, bays, estuaries, etc. Singly or in flocks, associates readily with other terns.

SS: Forster's Tern slightly larger with slightly heavier bill, longer legs, primaries often appear silvery above. Alt adult has orange-red bill with black tip; in basic, head white with bold black mask not joining around nape. Juv and especially 1st basic Forster's can be very similar to Common; note black mask blending with blackish hindcrown and nape, lesser upperwing coverts not contrastingly darker, legs

usually brighter orange-red. Arctic Tern (offshore Pacific migrant) has smaller body and shorter wings with more buoyant flight, neck appears shorter so head does not project as much beyond wings as in Common Tern; on standing birds (unlikely) note Arctic's markedly shorter legs. Adult Arctic has longer tail streamers, translucent primaries from below, underbody often contrastingly dusky; juv Arctic much cleaner-patterned than Common, grey and whitish, trailing edge of upperwing whitish, dark lesser upperwing coverts less distinct. See Roseate Tern.

SD: C to F transient (Aug–Nov, Apr–May) off and along Pacific coast, including Gulf of California; irregularly F to C in winter (Dec–Feb) inshore from Jal S, U to R N to Baja and Gulf of California; F to C non-breeding visitor (Mar–Sep) N to Nay (SNGH). U to F transient (Aug–Nov, Apr–May) on Atlantic coast from Tamps to Honduras; U to R in winter, mostly on Caribbean coasts; U non-breeding visitor (Mar–Sep) N to S Tamps.

RA: Breeds N.A. and across Eurasia, locally in Atlantic to N S.A. and W Africa; winters in Americas from W Mexico and Caribbean to Chile and Argentina.

ARCTIC TERN

Sterna paradisaea   Not illustrated
Golondrina-marina Artica

ID: 11.8–13 in, W 29.5–33 in, T 4.5–6.5 in (30–35.5 cm, W 75–84 cm, T 11.5–16.5 cm). *Transient offshore in Pacific.* Bill 1.2 in (3 cm), legs notably short. Tail long and forked. **Descr.** Age/seasonal variation. *Prealternate molt does not include inner primaries, hence birds do not show contrasting dark outer primaries.* **Adult. Alt:** *bill blood red,* legs red. Tail streamers add 1.2–2 in (3–5 cm) to length and project beyond wingtips at rest. Head, neck, and

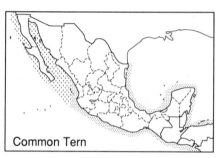

Common Tern

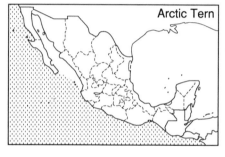

Arctic Tern

underparts white with black cap; *pale dusky underparts often contrast with whiter face.* Upperparts pale grey, becoming white on rump and tail; outer rectrices with dark outer webs. *Underwings white with narrow black trailing edge to primaries,* primaries may appear translucent when backlit. **Basic:** forehead may become white and bill blackish before heading S; basic (unlikely in region) similar to alt but tail shorter, upperwings with dusky lesser coverts. **Juv:** bill black, legs reddish. Black cap with white forehead, upperparts pale grey with whitish edgings and darker lesser upperwing coverts; *in flight, whitish flight feathers form contrasting triangle on trailing edge of wing.* Tail pale grey, outer rectrices with dark outer webs. **1st summer:** resembles adult basic (primary molt completed in 1st winter), followed by adult-like 2nd basic; forehead may be flecked white in 2nd summer.

Voice. Usually silent when migrating but may give a high, clipped *kiip* and high, shrill, grating chatters, especially when in flocks.

Habs. Pelagic, singly or in flocks; often joins mixed-species feeding flocks. Unlikely to be seen from land.

SS: See Common and Roseate terns. Forster's Tern larger and more heavily built with longer bill and legs, rarely offshore; note silvery upperwings, white underparts, and black-tipped bill of alt adult, white head with black mask of basic and juv.

SD: F to C transient (Aug–Oct, Apr–May) off Pacific coast (SNGH, RLP); no records from land.

RA: Holarctic breeder; winters around Antarctica.

FORSTER'S TERN
Sterna forsteri　　　　　　　Figure 24
Golondrina-marina de Forster

ID: 12.5–14 in, W 30–33 in, T 4.5–5 in (32–36 cm, W 76–84 cm, T 11.5–12.5 cm). Widespread in winter. Bill 1.5 in (4 cm). Tail very long and deeply forked. **Descr.** Age/seasonal variation. **Adult. Alt:** *bill orange-red with black tip,* legs orange-red to red. Tail streamers add 2.5–3 in (6.5–7.5 cm) to length and project beyond wingtips at rest. Head, neck, and underparts white with black cap. Upperparts pale grey becoming white on rump, tail pale grey with white outer webs to outer rectrices. In flight, *remiges contrastingly pale silvery* but by later summer outer primaries may be blackish. Underwings white with black trailing edge to outer primaries. **Basic:** black bill dull orangish at base, legs duller. *Head white with bold black auricular mask.* **Juv:** resembles basic but crown and upperparts washed buff (lost quickly), *black mask less distinct since hindcrown and nape blackish,* remiges less silvery, lesser upperwing coverts slightly darker, outer rectrices tipped dark; tertial centers, primaries, and tail often wear and become dark by early winter. **1st summer:** much like adult basic but mask less striking, similar to juv; often shows dark bar on secondaries, outer primaries and tail tips dark; as adult basic by 2nd winter. May show white flecks in black cap in 2nd summer.

Voice. A hard, clipped, dry *kik* or *krik,* a gruff, slightly shrill, rasping *zzhi-zzhi ...* or *krrih-krrih ...,* and a slightly grating, shrill *krrih* or *kyiih* and *kyerr kyerr ...,* etc.

Habs. Coastal lagoons, estuaries, lakes, reservoirs, beaches, etc.; rarely offshore. Singly or in flocks, associates readily with other terns.

SS: See Common, Arctic, and Roseate terns.

SD: F breeder (Apr–Aug) locally in BCN (Palacios and Alfaro 1991) and coastal N Tamps, probably also Presa Marte R. Gomez, interior Tamps (SNGH, SW). C to F transient and winter visitor (Aug–May) in Baja and on Pacific Slope from Son to Col; U to F to El Salvador (Thurber et al. 1987); U to F (1000–2500 m) in interior cen Mexico (SNGH, PP, RGW); C to F on Atlantic Slope from Tamps to Tab, coastally to NE QR, R to Honduras (Howell and Webb 1992b). Not 'common' in Belize as reported (Wood et al. 1986), with apparently no verified records from that country.

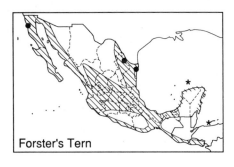

Forster's Tern

RA: Breeds N.A.; winters S USA to N C.A. and N Caribbean.

LEAST TERN

*Sterna antillarum**   Not illustrated
Golondrina-marina Mínima

ID: 8–9 in, W 19–21 in, T 3–3.5 in (20.5–23 cm, W 48–53 cm, T 7.5–9 cm). Summer visitor to both coasts. Bill 1 in (2.5 cm). **Descr.** Age/seasonal variation. **Adult. Alt:** *bill orange-yellow with black tip*, legs yellow-orange. Head, foreneck, and underparts white, *black cap with white chevron on forehead.* Upperparts pale grey fading to white on secondaries, *dark outer 2–3 primaries form wedge on leading edge of outerwing.* Underwings white with dark outer primaries. **Basic:** bill blackish, legs dusky brownish yellow. *Head white with black mask* extending around nape, hind-crown often streaked dusky; upperwings with dark lesser coverts. **Juv:** resembles basic but crown streaked brown, upperparts with brown chevrons and spots; *primary coverts and outer primaries blackish, contrasting with whitish inner primaries and secondaries.* **1st summer:** much like adult basic but outer primaries blacker, advanced birds may have yellowish base to bill, adult-like alt cap mixed with white. Like adult by 2nd basic but crown may have white flecks in 2nd summer.

Voice. High, reedy, chippering, and chattering *chi-rit* and *k-rrik*, a reedier *kree-it* or *kreet*, a clipped *ki-rit*, and a longer *kik kirvee*, etc.

Habs. Sandy beaches and islets, coastal bays and lagoons, estuaries. Singly or in groups, associates loosely with other terns. Flies with distinctive rapid wing-beats. Nests colonially on beaches, sand bars, etc.

SD: F to C but local summer resident (Apr–Aug) on coasts from Baja and Gulf of California to Isthmus, probably locally elsewhere S to Gulf of Fonseca (Brodkorb 1940; Thurber *et al.* 1987); on Atlantic coast in N Tamps (JCA), and from Yuc Pen (Lopez O. *et al.* 1989; BM) to Honduras. U to R and local non-breeder (Mar–Aug) along Pacific coast. F to C transient (Mar–May, Aug–Oct, R to mid-Dec) along both coasts. Irregular U to R winter visitor (Dec–Feb) on Pacific coast N to Nay (SNGH, PP), and R on Atlantic coast in N Yuc (SNGH, LJP). Winter reports for Baja (AOU 1983) and reports of being resident in Belize (Wood *et al.* 1986) are unfounded.

RA: Breeds N.A. to Mexico and Caribbean; winters W Mexico and S Caribbean to N S.A.

NB: Traditionally considered conspecific with *S. albifrons* of Old World.

GREY-BACKED TERN

Sterna lunata   Not illustrated
Golondrina-marina Dorsigris   No map

ID: 12.5–14 in, W 31–33.5 in, T 4.5–5.5 in (32–35.5 cm, W 79–85 cm, T 11.5–14 cm). *Clipperton.* A graceful pelagic tern with long, deeply forked tail; bill 1.5 in (4 cm). **Descr.** Ages differ. **Adult:** bill and legs black. Tail streamers add 2–2.5 in (5–6.5 cm) to length. Head, foreneck, and underparts white, *black cap with white chevron on forehead* extending narrowly back to eye. *Upperparts blue-grey* with darker remiges, outer web of outermost rectrices white. **Imm:** tail shorter, greyish. Head white with dark auricular mask, brownish streaking and wash on crown and nape, upperparts with scaly whitish to buff edgings. Subsequent plumages undescribed (?), may parallel Bridled Tern.

Habs. Pelagic; flight and habits much like Sooty Tern with which it may associate (LBS).

SS: Sooty Tern black above with white forehead patch; Bridled Tern (unlikely to overlap) darker and browner above.

SD: Vagrant; one record: 2 adults roosting with Sooty Terns on Clipperton, 19 Aug 1986 (RLP photo).

RA: Central tropical Pacific.

NB: Sometimes known as Spectacled Tern.

BRIDLED TERN

*Sterna anaethetus**   Not illustrated
Golondrina-marina Embridada

ID: 12.5–14 in, W 31–33.5 in, T 4.5–5.5 in

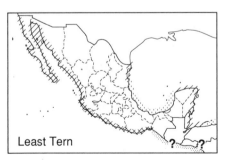

Least Tern

(32–35.5 cm, W 79–85 cm, T 11.5–14 cm). Local on both coasts. A graceful tern with long, deeply forked tail often held closed; bill 1.5 in (4 cm). **Descr.** Age/ seasonal variation. **Adult. Alt:** bill and legs black. Tail streamers add 2–2.5 in (5–6.5 cm) to length. Head, foreneck, and underparts white, *black cap with white chevron on forehead extending narrowly back to eye. Upperparts dark brownish* (may look blackish in bright light), *with paler hindneck,* darker remiges, *outer rectrices mostly white.* Underwings white with dark trailing edge. **Basic:** cap flecked white, upperparts narrowly edged whitish, outer rectrices with less white, tipped greyish. **Juv:** tail shorter, greyish. Head white with dark auricular mask, brownish streaking and wash on crown and nape, brownish wash on sides of neck; upperparts with scaly, whitish to buff edgings. **1st basic:** attained through 1st year, resembles adult basic. Much like alt adult by 2nd summer.

Voice. A mellow *kowk-kowk* or *kwawk* ..., a harder *kahrrr* around nest, etc.

Habs. Breeds on rocky islets, often close to shore, visiting adjacent beaches and headlands. Pelagic in winter, possibly not far offshore. Flight smooth and graceful; feeds by swooping down and picking food from at or near sea surface. Nest usually in crevice or rather hidden; breeds in small groups, at times in Sooty Tern colonies.

SS: Sooty Tern slightly larger and bulkier, adult black above, including hindneck, with broad white patch on forehead; no contrast between cap and back but in bright light remiges may appear paler (i.e. reverse contrast of Bridled Tern), tail shows less white at sides, underwing has broader dark trailing edge.

SD: F to U and local summer resident (late Mar–Aug) on Pacific coast at least from Nay to Gro (Howell *et al.* 1990; Howell and Engel 1993; LACM spec.), and probably in Gulf of Fonseca (Thurber *et al.* 1987); on Caribbean coast from QR (Howell *et al.* 1990) to Belize (Pelzl 1969). Apparently moves offshore or migrates S for winter, small numbers seen off Gro into Nov (RR). Reports of being resident in Belize (Wood *et al.* 1986) are unfounded.

RA: Pantropical but mostly coastal.

SOOTY TERN
*Sterna fuscata**   Not illustrated
Golondrina-marina Oscura

ID: 14–15.5 in, W 34–36.5 in, T 5.5–6 in (35.5–39.5 cm, W 86–92.5 cm, T 14–15 cm). Widespread tropical tern. Graceful with long, deeply forked tail, often held closed; bill 1.5 in (4 cm). **Descr.** Age/ seasonal variation. **Adult. Alt:** bill and legs black. Tail streamers add 2.5–3 in (6.5–7.5 cm) to length. Crown, hindneck, and upperparts black (browner when worn) with *broad white forehead patch;* remiges may appear paler in bright light, *outer web of outermost rectrices white.* Throat, foreneck, and underparts white, underparts tinged smoky grey when fresh. **Basic:** cap flecked white, upperparts narrowly edged whitish, outermost rectrices black, tipped white. **Juv:** *sooty blackish brown overall* with pale grey vent and undertail coverts; upperparts narrowly edged and tipped buff to whitish; underwing coverts pale grey, appearing silvery in bright light. Tail sooty black. Within a few months begins molt into **1st basic:** much like juv but upperparts with reduced whitish fringes, throat pale grey, belly extensively white, tail more deeply forked, short streamers tipped white. **2nd basic:** may be similar to 1st basic but upperparts more uniform, underparts whitish with dark mottling concentrated across chest. In 3rd year attains adult-like

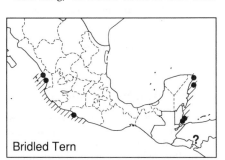

Bridled Tern

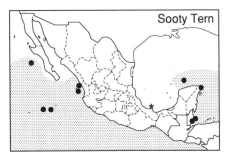

Sooty Tern

plumage but underparts usually with some dark mottling; a few birds in 5th year still show some dark speckling on underparts.

Voice. Harsh to shrill grating cries at colonies include a distinctive *ke weh-de-wek*; at sea may give a nasal to barking *ka-wake'* or *ka-a-ah*; juvs have higher reedy cries.

Habs. Breeds on sandy and rocky islets, usually with grass or other vegetation. Pelagic most of year, usually offshore over deep water. Flight and feeding much like Bridled Tern. Nests colonially, often with Brown Noddies, at times in tens of thousands; nest is scrape on open ground or on rock ledge.

SS: See Bridled Tern. Brown Noddy flies low over water with less graceful flight, tail broader and wedge-shaped.

SD: C to F but summer resident (Feb–Aug) off Pacific coast on Alijos Rocks, Islas Revillagigedo, and Clipperton (resident?), also (Mar–Aug) on Islas Tres Marías and Isla Isabel (where drastic population decline in recent years is due to predation by cats, HD), Nay; off Atlantic coast (Apr–Aug) from Arrecife Alacrán to Belize cays. U to F offshore migrant and visitor (Jan–Dec, RLP) from S BCS and Nay to El Salvador, R (Dec–Mar) off S Ver (Winker *et al.* 1992*b*). Reports of being resident in Belize (Wood *et al.* 1986) are unfounded.

RA: Pantropical.

BLACK TERN
Chlidonias niger surinamensis Not illustrated
Golondrina-marina Negra

ID: 9–9.7 in, W 23.5–25.5 in, T 3 in (23–24.5 cm, W 60–65 cm, T 7.5 cm). Widespread migrant. A small dark tern. Bill 1 in (2.5 cm). **Descr.** Age/seasonal variation.

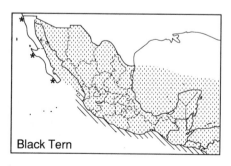
Black Tern

Adult. Basic: bill black, legs dark reddish brown. Head, neck, and underparts white with black crown patch extending down onto auriculars, *blackish patch at sides of chest. Upperparts smoky grey*, darker on mantle and lesser upperwing coverts; underwings pale grey. **Alt** (Mar–Jul): head, neck, and *underparts blackish* (blacker in ♂) *with white undertail coverts*, whitish underwings. *Upperparts smoky grey*, paler on wing coverts which may appear silvery in bright light. *In late summer and autumn, underparts blotched black and white during prebasic molt.* **Juv:** resembles basic but upperparts washed brown, upperwing coverts edged pale, underwing coverts whitish. **1st summer:** resembles adult basic but less neat, some birds with blackish patches on underparts; as adult by 2nd basic.

Voice. A piping, slightly reedy *peep* or *pseeh* and *kriih* from birds in feeding flocks; a quacking *kek* in alarm.

Habs. Widely in migration, from inland lakes to coasts and offshore waters. In winter mostly offshore, also coastal lagoons. Associates with other terns but flight and habits set it apart. Flight distinctly floppy, dipping down to water to pick food from surface.

SD: C to F transient (Apr–early Jun, mid-Jul–Oct) off and along coasts from Gulf of California (mostly in autumn?) S, U to R off Pacific coast of Baja; irregular C to F winter visitor (Nov–Mar) N to Jal (SNGH, PP); small numbers may oversummer (Jun–Jul) along Pacific coast. U to F transient (May–early Jun, late Jul–early Oct) through interior, at least in N and cen Mexico (Williams 1983; Wilson and Ceballos-L. 1986). F to C transient (Jul–Oct, Apr–early Jun) on Atlantic Slope, where U to F non-breeding visitor (Jun–Jul) at least from Yuc Pen E (G and JC, SNGH, SW); no winter records. Vagrant to Clipperton.

RA: Breeds interior N.A. and W Eurasia; winters off coasts of W Africa, N S.A., and W M.A.

GENUS ANOUS: Noddies

Noddies are blackish brown tropical terns with white caps. Tails long and graduated, held closed in flight. Bills black, legs dark. Ages differ slightly, no seasonal variation.

BROWN NODDY

Anous stolidus *   Not illustrated
Golondrina-boba Café

ID: 14–16 in, W 31–35 in, T 6 in (35.5–40.5 cm, W 79–89 cm, T 15 cm). Tropical oceans. See genus note. Bill 1.5 in (4 cm). **Descr. Adult:** *blackish brown overall with white cap* brightest on forecrown, fading out on hindcrown, and white subocular crescent. Blackish flight feathers contrast slightly with upperparts at rest. **Juv:** *whitish restricted to narrow band from forehead to eyes*, forecrown greyish brown; some may have bright whitish forecrown (as on Clipperton). Subsequent molts and plumages poorly known.

Voice. Rarely vocal except on breeding grounds when gives guttural barks and a braying *keh-eh-eh-ehr*, etc.

Habs. Pelagic, but may be seen near shore, especially on rocky islets or points, associating with other terns. Flight typically low over water, direct flight steady but flocks mill around when feeding, swooping down to pick food from at or near surface. Breeds on offshore islands and rocks, nests on rock shelves, on ground, or builds shallow cup nests in bushes or trees.

SS: Black Noddy smaller with longer and markedly more slender bill, adult has contrastingly paler tail above, juv blacker with bright white cap, primaries not obviously darker than upperparts at rest. See juv Sooty Tern.

SD: F to C but local summer resident (Apr–Aug) on islets and islands along Pacific coast from Nay to Gro (SNGH), and (Feb–Nov) on Islas Revillagigedo and Clipperton (where U, Dec–Jan). C to F but local summer resident (Apr–Aug) off Atlantic coast from Arrecife Alacrán to Belize cays. F to U transient (Mar–May, Aug–Sep) and U to R in summer (May–Jul) along Pacific coast and off Caribbean coast of region. Reports of being resident in Belize (Wood *et al.* 1986) are unfounded.

RA: Pantropical.

BLACK NODDY

Anous minutus *   Not illustrated
Golondrina-boba Negra

ID: 12–13.5 in, W 26–29 in, T 4–4.5 in (30.5–34 cm, W 66–73.5 cm, T 10–11.5 cm). Clipperton. See genus note. Bill long and notably slender (1.7 in; 4.5 cm). **Descr. Adult:** *blackish grey overall* (browner when worn) *with white cap* brightest on forecrown, fading out on hindcrown, and white subocular crescent. Remiges darker but not contrasting with upperparts at rest; *tail paler and greyer, contrasting with upperparts.* **Juv:** *blacker* overall with *bright, clearly defined white cap*, dark tail. Subsequent molts and plumages poorly known.

Voice. Rarely vocal except on breeding grounds when utters guttural growls, *ahrrr* or *garrr*, etc.

Habs. Much as Brown Noddy with which it associates readily, but flight often faster and more fluttery; nests in palm trees and on rocks at Clipperton.

SS: See Brown Noddy.

SD: Vagrant (Jul–Aug) to QR (Howell *et al.* 1990) and Honduras Bay Islands (Marcus 1983). Formerly bred (1862–1907 at least) on Half Moon Cay off Belize (Russell 1964). C summer resident (Mar–Oct) on Clipperton (most apparently leave the island Nov–Feb).

RA: Pantropical except Indian Ocean (but see NB).

NB: Often considered conspecific with *A. tenuirostris* of Indian Ocean.

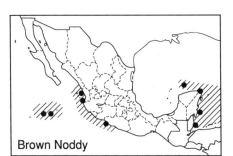

Brown Noddy

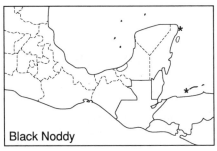

Black Noddy

WHITE (FAIRY) TERN
Gygis alba candida   Not illustrated
Golondrina-boba Blanca

ID: 12–13 in, W 28–29.5 in, T 4 in (30.5–33 cm, W 71–75 cm, T 10 cm). Clipperton. *Unmistakable and ethereal.* Bill fairly deep at base (1 in, 2.5 cm), tail forked. **Descr.** Ages differ slightly. **Adult:** bill black, paler blue-grey below at base, legs black. *Snowy white overall with big dark eyes* and narrow black eyering, dark shafts to outer primaries and rectrices visible at close range. **Juv:** upperparts mottled and tipped cinnamon-brown. Subsequent molts and plumages poorly known.

> **Voice.** Mostly silent at sea. On breeding grounds, high, thin, piping whistles, and a low, gruff scold.
> **Habs.** Pelagic. Flight buoyant with deep floppy wingbeats, usually fairly high over water, dipping down to pick food from at or near surface. Joins mixed-species feeding flocks and associates loosely with other terns. Eggs laid on branch or other flat surface.

SS: At a distance in bright light may suggest a tropicbird.
SD: Vagrant to Islas Revillagigedo (Jul 1905); R (to U?) visitor (Feb at least) off N C.A. (RLP). F breeder (30–50 birds, Feb–Oct?) on Clipperton, where R to absent Nov–Dec.
RA: Pantropical, but local in E Pacific.

White Tern

SUBFAMILY RYNCHOPINAE:
Skimmers

Skimmers form a distinct subfamily, sometimes considered a separate family. They resemble large, angular, rangy terns and are immediately distinguished by their stout, laterally compressed bills with a projecting lower mandible. Wings long and pointed, tail slightly forked. Molts are poorly known. Feed, mostly at night, by skimming low over water, lower mandible cutting through water and snapping shut on contact with food, mostly small fish and crustaceans. Usually breed colonially, nest is a scrape on open sandy ground. Eggs: 2–5, bluish white to pale buff, marked with browns and greys (WFVZ).

BLACK SKIMMER
Rynchops n. niger   Figure 24
Rayador Americano

ID: 17–18 in, W 45.5–48.5 in, B 3.5 in (43–45.5 cm, W 115–123 cm, B 9 cm). Widespread, coastal. See subfamily note. **Descr.** Age/seasonal variation. **Adult. Alt:** *bill black with red base,* legs red. *Crown, hindneck, and upperparts black with broad white trailing edge to secondaries and inner primaries,* central rump and tail blackish, broadly edged white, dusky outer rectrices broadly edged white. Lores, lower face, foreneck, and underparts white; underwings white with dark edge to outerwing. **Basic:** upperparts browner, hindneck mostly whitish. **Juv:** mandibles about equal length at fledging, adult-like by early winter. Base of bill and legs initially orange to orange-red. Head, neck, and underparts white with dark smudge through eye, brown mottling on crown and hindneck. Upperparts dark brown with bold, scaly cinnamon-buff edgings, soon fading to white. Outer rectrices edged pale grey. **1st year:** crown and upperparts brown, mottled black to mostly black in 1st summer when back contrasts with faded brown wing coverts; hindneck whitish. Molts and subsequent plumages poorly known. *R. n. cinerascens* of S.A. could occur along Pacific coast in summer (reported May–Oct in Costa Rica, Stiles and Skutch

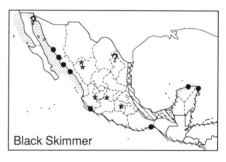

Black Skimmer

1989); adult has secondaries and sooty grey tail narrowly edged white, underwing coverts smoky grey.

Voice. Nasal, laughing to barking calls, *kyuh* or *kwuh*, and *k-nuk* or *kwuk*, etc.

Habs. Coastal lagoons, estuaries, beaches, locally inland on lakes and along rivers. See subfamily note. Singly or, more often, in flocks which fly in compact wheeling formation. Much of day spent roosting on beaches, sand bars, etc., often with terns and gulls.

SD: F to C but local breeder (Apr–Aug?) along coasts from Sin to Col (SOW); U in Isthmus (Binford 1989), and probably elsewhere; on Atlantic coast in Tamps and N Yuc Pen (Friedmann *et al.* 1950; BM). F to C but often local transient and winter visitor (Aug–May) on coasts from Son to El Salvador, on Atlantic Slope from Tamps to Tab, coastally to N Yuc Pen; U to R to N Honduras. U to R and local (Jun–Jul) in winter range when migrants from S.A. could occur on Pacific coast. F to C but local visitor (Jan–Dec) in W BCN (Palacios and Alfaro 1992c). Irregular R to U visitor (late May–Sep; 1500–2200 m) to W Plateau and cen volcanic belt (Williams 1982a; RGW).

RA: Breeds on Pacific coast in Mexico and Ecuador, on Atlantic coast from E USA to Argentina. N populations migratory.

AUKS AND PUFFINS: ALCIDAE (8)

A distinctive Holarctic family of pelagic birds, found in the region only off NW Mexico. Their center of diversity lies in the N Pacific with a secondary center from cen California to Baja. Alcids are heavy-bodied and small-winged; as a consequence, they fly fast and usually low to the water. Bills vary from slender to stout, some species growing ornamental plates in the breeding season; legs stout and strong, feet webbed. Age/seasonal variation apparent in N migrants, sexes similar; young semiprecocial. Molts poorly known, undescribed (?) in *Endomychura* murrelets. Plumage overall blackish above, white below. Rarely vocal away from nesting grounds.

Alcids feed on fish and other marine organisms caught by diving. Mexican breeding species nocturnal at nesting islands to avoid gull predation. Nest in rock crevices, burrows, under bushes, etc. Eggs: 1–2, varied.

COMMON MURRE

Uria aalge californica   Not illustrated
Arao Común

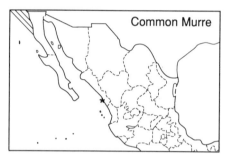

ID: 16–17 in (41.5–44 cm). Winters off NW BCN. Bill straight and dagger-like. **Descr.** Age/seasonal variation. **Imm/basic** (Jul–Mar): bill and feet black. *Crown, lores, postocular stripe, and hindneck blackish; face, foreneck, and underparts white,* streaked dark on flanks. *Upperparts blackish brown,* secondaries tipped white. Underwings white with dusky remiges. **Alt** (Dec–Aug): head and foreneck blackish brown which, along with upperparts, may become browner and faded by summer.
  Habs. Pelagic, but may be seen from land. Swims low in water and dives well. Flight strong and direct, usually low over water.
SD: R to U winter visitor (Nov–Feb) off NW BCN, S to about 30° N. Vagrant to Sin (Nov 1989, RR).
RA: Breeds N Atlantic and N Pacific oceans; winters in America S to NW Mexico and NE USA.
NB: Known as Guillemot in the Old World.

PIGEON GUILLEMOT

Cepphus c. columba   Not illustrated
Arao Paloma

ID: 13–14 in (33–35.5 cm). Vagrant. Bill straight and dagger-like. **Descr.** Age/seasonal variation. **Alt** (Feb–Aug): bill black, feet red. Sooty blackish overall with bold white wing patch on median and

greater coverts, underwings dusky. **Imm/basic** (Jul–Feb): crown and lores blackish brown; face, sides of neck, and underparts white, speckled dusky, often with dark postocular stripe. Upperparts blackish brown, white wing patch less distinct.
  Voice. High, thin whistles, mainly when nesting (California).
  Habs. Much as Common Murre but flight faster, slightly rocking.
SD: Vagrant. Three records (May–Jul), all

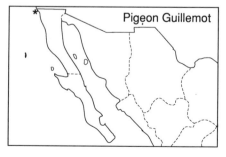

from Islas Los Coronados, BCN (Jehl 1977; Wilbur 1987).

RA: N Pacific, S to S California.

XANTUS' MURRELET

Endomychura hypoleuca *   Figure 25
Mérgulo de Xantus

ID: 9.5–10 in (24–25.5 cm). NW Mexico. Bill slender and pointed. Two ssp: *hypoleuca* and *scrippsi* (both breed off BCN). **Descr.** Ages/sexes similar. *Hypoleuca*: bill black, feet greyish. *White from lower face extends up over front of eye and onto hind auriculars so dark eye stands out on white face.* Cap (through gape), hindneck, and *upperparts blackish grey* (paler when worn). Foreneck and underparts white with dark grey flanks. *Underwings white with dusky remiges. Scrippsi: blackish cap extends below eye and across hind auriculars,* narrow white eye-crescents inconspicuous. **Voice.** At colonies, high twittering calls becoming purring choruses with many bird vocalizing. Pairs on the water may give high thin whistles. **Habs.** Pelagic, not usually seen from land. In pairs or small loose groups, dives or flushes off water (without pattering along surface) to avoid boats. Low flight fast and direct, slightly rocking. Eggs: 1–2, brownish to pale bluish, usually well marked with browns (WFVZ).

SS: Craveri's Murrelet overall brownish black above with dark spur extending onto sides of chest, underwings mostly dusky; note finer bill, blackish extends narrowly under chin. Ancient Murrelet has stubbier pale bill, white sides of neck contrast with black hood. Cassin's Auklet chunkier and shorter-necked, appears all dark on water, patters on surface before take-off, flight heavier and more direct.

SD: *Hypoleuca* is C breeder (Feb–Aug) on Isla Guadalupe, and probably Islas San Benitos, off Pacific coast of BCN (formerly elsewhere). Ranges at sea off Pacific coast of Baja and N to cen California (Aug–Sep), possibly S to Gro (Oct 1988, RR). *Scrippsi* is C breeder (Feb–Aug) on islands off Pacifice coast of BCN, from Los Coronados to San Benitos. Ranges at sea off

Fig. 25 (a) Xantus' Murrelet, (i) *scrippsi*, (ii) *hypoleuca*; (b) Craveri's Murrelet

Baja, S probably to Bahia Magdalena (Anthony 1925, ssp unspecified).

RA: S California to Baja; ranging at sea N to S British Columbia; *hypoleuca* endemic to Mexico as a breeding bird.

NB1: Formerly considered conspecific with Craveri's Murrelet. **NB2:** The two forms of Xantus' Murrelets may constitute separate species; they appear to breed sympatrically on Islas San Benitos. **NB3:** *Endomychura* sometimes merged with *Synthliboramphus.*

CRAVERI'S MURRELET

Endomychura craveri   Figure 25
Mérgulo de Craveri

ID: 9.5–10 in (24–25.5 cm). Endemic to NW Mexico. Bill slender and pointed, finer than Xantus' Murrelet. **Descr.** Ages/sexes similar. Bill black, feet greyish. *Cap (through gape), hindneck, and upperparts brownish black, dark from gape usually extends narrowly under chin*; narrow white eye-crescents inconspicuous. Foreneck and underparts white, with *dark from back extending onto sides of chest*, and dark grey flanks. *Underwings dusky*, but often with whitish median patch.

    Voice. Pairs on the water may give a rather weak, high trilling (RAE).

    Habs. Much like Xantus' Murrelet, including nest and eggs.

SS: See Xantus' Murrelet. Ancient Murrelet has stubbier pale bill, white sides of neck contrast with black hood. Cassin's Auklet chunkier and shorter-necked, appears all dark on water, patters on surface before take-off, flight heavier and more direct.

SD/RA: C breeder (Jan–Jul) on numerous islands in Gulf of California off coasts of Baja and Son; probably also along Pacific coast of Baja, at least on Islas San Benitos. Ranges at sea N to cen California (Jul–Oct) and S at least to near Islas Tres

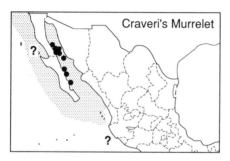

Craveri's Murrelet

Marías (Dec 1983, SNGH, PP). DeWeese and Anderson (1976) cited a record off Guatemala from Jehl (1974), although the latter did not provide conclusive details and stated that the identification should remain tentative.

NB: See NB1 and NB3 under Xantus' Murrelet.

ANCIENT MURRELET

Synthliboramphus antiquus   Not illustrated
Mérgulo Antiguo

ID: 10–10.5 in (25.5–26.5 cm). Rare in winter, NW BCN. *Bill fairly stubby.* **Descr. Adult:** *bill pale yellowish horn*, feet greyish. *White from sides of neck extends up into black hood; throat and hindneck black*, indistinct white postocular streaking more distinct in alt. *Upperparts grey*, flight feathers darker. *Underparts white* with grey flanks. Underwings white with dusky remiges. **Imm:** throat dusky to whitish.

    Habs. Pelagic, but may occur in harbors, etc., singly or in small groups. Flight fast and direct, low over water.

SS: See Xantus' and Craveri's murrelets. Cassin's Auklet has all-dark head and chest.

SD: Probably an irregular R to U winter visitor (Dec–Feb) off Pacific coast of N BCN, two records: Dec 1927 (Grinnell 1928), Feb 1980 (Wilbur 1987, per RAE).

RA: Breeds N Pacific; winters S in Americas to California, rarely Baja.

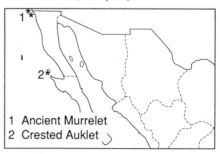

1 Ancient Murrelet
2 Crested Auklet

CASSIN'S AUKLET

Ptychoramphus aleuticus   Not illustrated
Alcita de Cassin

ID: 8.5–9 in (21.5–23 cm). W Baja. *Bill small and stout.* **Descr.** Ages/sexes similar. *Eyes whitish* (dusky in imm), bill black with pale flesh base below. Legs fleshy whitish with dark webs. *Dark smoky grey overall*, with *white supraorbital spot*, throat and underparts paler with *white*

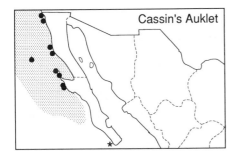

Cassin's Auklet

belly and undertail coverts usually noticeable only in flight.

Voice. On breeding grounds, a hoarse, slightly rough, purring *ruh reehr* or *uhr rreihr*, repeated rhythmically a few times, becoming full choruses when many birds involved.

Habs. Pelagic, unlikely to be seen from land. In pairs or small flocks, at times loosely to a few hundred birds. Low flight heavy and direct, patters along surface before take-off. Nests in burrows. Eggs: 1, whitish (WFVZ).

SS: See Xantus', Craveri's, and Ancient murrelets.

SD: C resident off coast of BCN, breeding (Feb–Sep) on islands S to 27° N. Ranges at sea S to 23° N.

RA: Alaska to NW Mexico.

CRESTED AUKLET
Aethia cristatella   Not illustrated
Alcita Crestada   See Ancient Murrelet for map

ID: 10–10.5 in (25.5–26.5 cm). Vagrant. *Bill thick and stubby.* **Descr.** Age/seasonal variation. Eyes whitish, bill bright red with swollen gape plate (bill smaller and duller in imm/basic), feet dark grey. Head and upperparts blackish with *erect frontal crest* and fine white postocular stripe (reduced in basic). Throat, underparts, and underwings smoky grey.

Habs. Pelagic. Flight heavy and low over water.

SD: Vagrant, one record: one near Isla Cedros, BCN (Jul 1980, Pitman *et al.* 1983).

RA: N Pacific (mainly Bering Sea).

NB: Parakeet Auklet *Cyclorrhynchus psittacula* (10 in; 25.5 cm) has been noted in offshore waters adjacent to N BCN (Mar 1991, PP) and could occur in region (Dec–Mar?). Shape, stout reddish bill, and pale eyes recall Crested Auklet. Head, neck, and upperparts blackish with white postocular stripe (reduced or lacking in imm/basic), chest mottled black and white, belly and undertail coverts white; underwings dusky. Basic/imm has throat and foreneck pale grey.

RHINOCEROS AUKLET
Cerorhinca monocerata   Not illustrated
Alcita Rinoceronte

ID: 13–13.5 in (33–34.5 cm). W Baja. *Bill deep and moderately long* (narrower in imm), with rhinoceros-like horn at base of culmen in breeding condition (unlikely in Mexico). **Descr.** Age/seasonal variation slight. Eyes pale, *bill dusky to orange*, feet yellowish to dark grey (juv). *Dark overall.* Crown, nape, and upperparts blackish, face and chest smoky grey with white postocular streaks and moustache (bolder in alt). Belly and undertail coverts whitish. Underwings dark grey.

Habs. Pelagic but may be seen from land; singly or in small groups. Flight fast and powerful, low over water.

SD: F to C winter visitor (Nov–Mar) off Pacific coast of BCN, U off BCS to about 24 °N; U to R though the summer.

RA: Breeds N Pacific; winters in Americas to NW Mexico.

NB: Horned Puffin *Fratercula corniculata* (14–15 in; 35.5–38 cm) probably occurs irregularly off BCN (May–Jul at least); see Appendix B. *Bill deep and laterally compressed* (smaller in imm). **Imm/basic:** eyes yellow, *bill dusky yellow at base, dull red distally,* legs orangish. *Head and upperparts black with whitish face patch washed dusky;* throat and foreneck black, chest and underparts white. Underwings dark. **Alt:** bill bright yellow and red, face patch white, fleshy horn above eye.

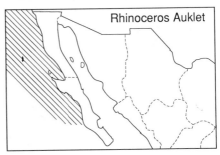

Rhinoceros Auklet

PIGEONS AND DOVES: COLUMBIDAE (23*)

A familiar group of successful, widely distributed birds. The terms pigeon and dove are interchangeable but the former usually refers to larger birds, typically of the genus *Columba*. Plump and large-bodied birds with small heads, tails vary from short and squared to long and pointed. Bills slender with distinct cere, legs relatively short and sturdy. Crops well developed for food storage and production of 'pigeon's milk', a rich milky substance fed to nestlings. Ages/sexes usually differ only slightly, ♀♀ being duller than ♂♂; juvs have extensive pale tips to contour feathers. Young altricial; adult plumage attained by complete prebasic molt within two weeks to several months after fledging. Plumage overall colored in greys, browns, and pinks, bare parts often brightly colored, legs reddish or pink. The song (advertising coo) of most species is either a series of coos or a single coo, repeated, and it is important in species recognition.

Pigeons may be divided into arboreal and terrestrial feeders, although some species bridge the division. Food mostly fruit, seeds, flowers, young leaves, at times invertebrates. Pigeons are unique among birds in being able to drink by sucking up water. Nests typically are flimsy platforms of twigs and other vegetation placed in bushes or trees; some species may nest on ground. Eggs: 1–2, white or, as in *Leptotila* and *Geotrygon*, pale buff to buffy white (WFVZ).

GENUS COLUMBA

These are large arboreal pigeons. Usually in pairs or groups, often quite wary since they are hunted for food in most of region. May be seen drinking or taking grit at quiet roadsides.

FERAL PIGEON (ROCK DOVE)
Columba livia ssp   Not illustrated
Paloma Doméstica

ID: 12–14 in (30.5–35.5 cm). Widespread non-native. **Descr.** Eyes amber (dull in juv), cere whitish, legs reddish pink. Plum-

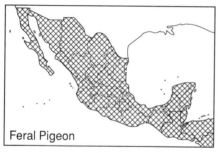

Feral Pigeon

age notably variable. Ancestral type: head, neck, and chest slaty grey with green and purple sheen on neck, belly pale grey. Upperparts grey with 2 black bars on secondary coverts and base of secondaries, rump white to slaty grey, uppertail coverts and tail slaty grey with blackish tail tip; primaries dark grey. Varies from ancestral type to washed-out pale grey and white with dark chestnut patches and purplish chest, to all white, to mostly blackish with white primaries and tail, etc.

Voice. A low, muffled cooing audible at close range.

Habs. Towns, villages, rarely far from human habitation but, in some locations, has reverted to nesting on cliffs. Singly or in flocks, often glides down with wings held in deep **V**.

SS: An annoying source of confusion with native pigeons; also may recall falcons or shorebirds in flight.

SD: C to F resident (SL–2500 m) throughout most of region but in many areas it is unclear if truly feral or simply free-flying domesticated birds.

RA: Native to Eurasia; introduced elsewhere throughout world.

PALE-VENTED PIGEON
Columba cayennensis pallidicrissa   Plate 18
Paloma Vientre-claro

ID: 11.5–13.5 in (29–34 cm). SE savannas and riverine forest. **Descr.** Ages/sexes differ slightly. ♂: eyes red, orbital ring reddish, bill blackish, legs reddish. *Head blue-grey* with purplish forehead, whitish chin, and violet-green iridescence on nape. Neck and upper chest vinaceous purple, paling to greyish vinaceous underbody and *whitish belly and undertail coverts*. Back vinaceous purple, wing coverts brownish and remiges blackish, rump blue-grey. *Tail grey-brown with broad pale distal band.* ♀: duller overall, back more brownish. **Juv:** duller, back and wing coverts narrowly edged paler.

Voice. A low *whoooo, who'koo-koo ...*, the introductory *whoooo* deep and swelling, the 2nd phrase repeated 3–4×; also a deep, hoarse, purring *whoor* or *whohr.* Song suggests Red-billed Pigeon but 2nd phrase has 3, not 4 notes, usually repeated 4×.

Habs. Marshy savannas with scattered forest patches, humid forest edge, especially along rivers. See genus note. In display flight climbs with exaggerated, slow, deep wingbeats, then glides down with wings raised in strong dihedral.

SS: Red-billed Pigeon has blue-grey belly and undertail coverts, black tail, pinkish head with whitish bill. Scaled Pigeon has similar whitish belly and undertail coverts but tail black, neck and chest boldly scaled, bill bright red. Short-billed Pigeon smaller, all dark.

SD: F to C resident (SL–700 m) on Atlantic Slope from SE Ver to Honduras.

RA: SE Mexico to Ecuador and N Argentina.

SCALED PIGEON
Columba speciosa   Plate 18
Paloma Escamosa

ID: 12–14 in (30.5–35.5 cm). SE humid forest. **Descr.** Ages/sexes differ. ♂: eyes purplish, orbital ring dark red, bill bright red with yellow tip, legs dark reddish. Head purplish brown, *neck, upper mantle, and chest whitish, boldly scaled iridescent purple* to green. Underparts whitish with heavy purplish scalloping, sparser on *belly and undertail coverts* which *appear contrastingly whitish. Upperparts rich purplish chestnut*, remiges dark brownish. *Tail blackish.* ♀: duller, scaling brownish purple, underparts with sparse brownish scalloping, *upperparts dark olive-brown.* **Juv:** resembles adult but duller, neck and upper mantle dull greyish with indistinct dark scaling, narrow buff to cinnamon edges on wing coverts and underparts, soon attains some boldly scaled adult neck feathers.

Voice. A deep, rhythmic, slightly gruff or moaning *whooo-oo, hoo'ooo ...* or *hoo, oo'hoo ...*, or *hooo hu'whooo ...* etc.; 2nd phrase repeated 2–3×. Gruffer than Red-billed or Pale-vented pigeons.

Habs. Humid evergreen and semideciduous forest and edge, clearings with tall trees. See genus note. Easily overlooked in dense canopy; may be seen in flight or perched in bare treetops at edges or in clearings. Flight slightly looser and more floppy than other *Columba.*

SS: Red-billed Pigeon has, ironically, a mainly whitish bill, appears dark overall whereas Scaled shows whitish undertail coverts contrasting with black tail, red bill and scaling on neck and chest often quite obvious, deep chestnut upperparts of ♂ distinctive. Short-billed Pigeon smaller, all

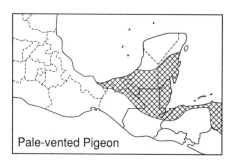

Pale-vented Pigeon

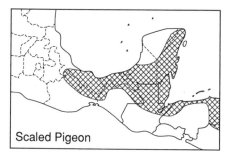

Scaled Pigeon

dark. White-crowned Pigeon all dark with white crown. See Pale-vented Pigeon.

SD: F to C resident (SL–900 m) on Atlantic Slope from S Ver to Honduras, but decreasing with clearing of forests; most apparently withdraw (Sep–Jan) from N of Isthmus and from Yuc Pen; migratory status in other areas unclear.

RA: SE Mexico to Ecuador and N Argentina.

WHITE-CROWNED PIGEON

Columba leucocephala   Plate 18
Paloma Coroniblanca

ID: 13–15 in (33–38 cm). Caribbean islands and coasts. **Descr.** Ages differ, sexes similar. **Adult:** eyes pale yellow, orbital ring whitish, bill purplish with horn tip, legs reddish. *Dark slaty blue-grey overall with bright white crown*, and black-tipped iridescent green scaling on neck. **Juv:** eyes dusky; dark brownish grey overall with dull whitish forehead, narrow pale edgings to chest and upperparts.

Voice. A deep, at times slightly burry *wh, whu-u-uh whoo, whu-u-uh whoo, ...* or *whu, cu-cu-cuu, cu-cu-cuu ...*; 2nd phrase often slightly accelarating and repeated 2–4×; also a low, strong purring, *puh'uhrr-rrr* and *purrr purr*, etc.

Habs. Humid evergreen to semideciduous forest and scrubby woodland, mangroves. See genus note. Flight steady and direct; display flight with exaggerated deep wingbeats. Breeds (Apr–Aug) on offshore islands, at times colonially, for example 400+ pairs on Crawl Cay, Belize in 1968 (Pelzl 1969).

SS: Bright white cap and overall dark grey plumage distinctive.

SD: Locally/seasonally F to C resident (around SL) on islands off Caribbean coast from N QR to Honduras, also locally/seasonally C on adjacent mainland, at least in QR. Breeding only known from

offshore islands but commutes to/from mainland to feed, and leaves some islands for the winter. Seasonal movements need study.

RA: Caribbean and adjacent areas of Mexico and Florida; wanders to mainland coasts from Honduras to Panama.

RED-BILLED PIGEON

Columba flavirostris*   Plate 18
Paloma Morada

ID: 12–14 in (30.5–35.5 cm). Widespread in lowlands. **Descr.** Ages/sexes differ slightly.
♂: eyes orangish, orbital ring dark red, *bill pale horn* with red base, legs reddish. *Head, neck, and chest dull purplish* with *slaty blue-grey flanks, belly, and undertail coverts*. Mantle and tertials olive-brown, scapulars purplish, and wing coverts and rump blue-grey; remiges dark grey. *Tail blackish.* ♀: slightly duller and paler. **Juv:** resembles adult but more rusty tone to chest and scapulars.

Voice. A deep *whoooo, oo'koo-koo-koo ...* or *whooo, hu'ku-ku-ku ...*, 2nd phrase usually repeated 2–5×; also a single swelling *whoo*, often repeated several times.

Habs. Forest and edge, including oaks, at times semiopen areas with forest patches and scattered trees; rarely in humid evergreen forest. See genus note; at times in flocks to 50+ birds. Flight fairly direct but slightly floppy. In display, climbs with exaggerated deep wingbeats, then glides in descending circles with wings held in strong dihedral.

SS: Short-billed Pigeon (rain forest) smaller and more compact with shorter tail, dark bill. Band-tailed Pigeon (pine–oak) paler overall and grey, tail with broad pale distal band. See Pale-vented, Scaled, and White-crowned pigeons.

SD: C to F resident (SL–1800 m) on Pacific

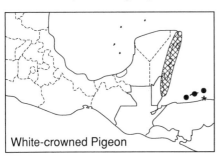

White-crowned Pigeon

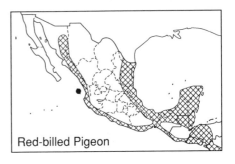

Red-billed Pigeon

Slope from S Son to Honduras, on Atlantic Slope from E NL and Tamps to Yuc Pen, locally U to F to N Honduras (SNGH); locally F to C in interior from Isthmus to Honduras.

RA: Mexico to cen Costa Rica.

BAND-TAILED PIGEON
*Columba fasciata**   Plate 18
Paloma Encinera

ID: 13–15 in (33–38 cm). Pine–oak highlands. Descr. Ages/sexes differ slightly. ♂: eyes brownish, orbital ring reddish, *bill and legs orange-yellow*, bill with black tip. Head and underparts pinkish grey becoming whitish on belly and undertail coverts; note *narrow white hindcollar* and iridescent green scalloping on hindneck. Upperparts grey, remiges blackish. *Tail grey with broad pale terminal band*; band barely present in *vioscae* of BCS. ♀: duller, underparts greyer. Juv: resembles ♀ but lacks white collar, fine pale edgings to chest and upperparts.
Voice. A deep, slightly hoarse *huh whur* or *wh'hoo*, repeated 3–4×, rarely up to 13× or more; often preceded by a deep *grrrr* and at times followed by a moaning *whorr*, longer series may end with an abrubt *wu't*.
Habs. Pine–oak and oak woodland. See genus note; at times in flocks to 50+ birds. Locally nomadic in response to acorn crops, one of its main foods.
SS: See Red-billed Pigeon.
SD: F to C resident (1000–3000 m, lower locally in winter) in BCS, and in interior and on adjacent slopes from Son and Coah to N cen Nicaragua. More widespread in winter (Sep–Apr) when recorded BCN, and when N migrants presumably occur in N interior Mexico.
RA: W N.A. to N Argentina.
NB: Birds from Costa Rica S sometimes con-

sidered specifically distinct, White-naped Pigeon (*C. albilinea*).

SHORT-BILLED PIGEON
Columba nigrirostris   Plate 18
Paloma Piquinegra

ID: 11–12 in (28–30.5 cm). *SE humid forest.* Descr. Ages differ, sexes similar. Adult: eyes and orbital ring reddish, *bill black*, legs reddish. Head and underparts dark greyish vinaceous, iridescent purple sheen on hindneck. Upperparts blackish brown. ♀ generally paler and duller. Juv: cinnamon edgings on head and body, narrower on upperparts.
Voice. A clear cooing *hoo'koo-koo-koo* or *hoo'cu-cu-hoo*, lacking an introductory note and repeated irregularly, suggests White-winged Dove but higher and clearer; also a deep purring *urrrrrrrr*, at times repeated steadily.
Habs. Humid evergreen forest. Overlooked easily in dense canopy, may be seen at edge or in relatively open fruiting trees such as *Cecropia*. Flight fairly fast and direct, low over canopy; rarely in flocks.
SS: See Red-billed, Pale-vented, and Scaled pigeons.
SD: C to F resident (SL–1000 m) on Atlantic Slope from S Ver to Honduras. Reports from N QR (Lopez O. *et al.* 1989) are erroneous and pertain to Ruddy Quail-Dove (photo examined).
RA: S Mexico to NW Colombia.

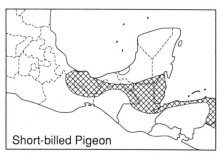
Short-billed Pigeon

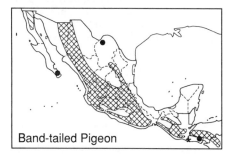
Band-tailed Pigeon

GENUS ZENAIDA

These are medium-sized doves of fairly open and lightly wooded country. Bills dark grey, orbital rings pale blue, legs pinkish.

WHITE-WINGED DOVE
Zenaida asiatica*   Plate 20
Paloma Aliblanca

ID: 10.5–12 in (26.5–30.5 cm). *Widespread and common.* **Descr.** Ages differ slightly, sexes similar. **Adult:** eyes amber. *Pale grey-brown overall,* with black stripe across lower auriculars, *bold white bar across upperwing* from outer lesser through greater coverts; remiges blackish. Central rectrices brown, *outer rectrices* grey, *broadly tipped white.* **Juv:** eyes dull, paler and greyer overall.
     Voice. A mournful, often slightly hoarse cooing, *who koo-koo-koo* or *hoo-koo'koo-koo,* and a rhythmic *hoo hoo coo, hoo oo'ooo oo'ooo oo,* etc., slowing throughout.
     Habs. Arid to semihumid woodland, brush, semiopen areas with scattered trees, ranchland. Often in flocks, may nest colonially. Migratory in N and may be seen in flocks of thousands in Apr and Sep. Flight fairly fast but slower and usually higher than Zenaida or Mourning doves, more *Columba*-like. In display, climbs with clapping wingbeats then glides down with wings slightly bowed.
SS: Bold white wingbars distinctive and usually obvious, even at rest.
SD: C to F resident (SL–2500 m) throughout most of region but absent to R in humid SE; N populations migratory, wintering mostly on Pacific Slope from Isthmus S; U to R on Plateau in winter (Nov–Mar).
RA: S USA to Caribbean and C.A.

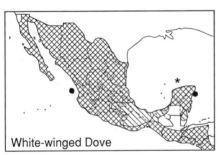

White-winged Dove

ZENAIDA DOVE
Zenaida aurita salvadorii   Plate 20
Paloma de Zenaida

ID: 10–11.5 in (25.5–29 cm). *Coastal Yuc Pen.* **Descr.** Ages/sexes differ slightly. ♂: eyes reddish. *Pinkish brown overall* with

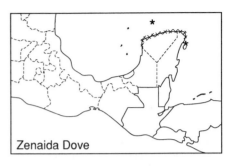

Zenaida Dove

black streak behind eye and on lower auriculars, grey-blue hindcrown and nape, iridescent violet neck patch. Underparts more violaceous pink with grey-blue flanks, upperparts more grey-brown with *bold black spots on scapulars, tertials,* and inner wing coverts. Remiges dark grey with *white trailing edge to secondaries;* central rectrices grey-brown, *outer rectrices grey with black subterminal band and broad whitish tips.* ♀: duller, buffy grey overall above, orangish pink on underbody with blue-grey flanks; lacks grey-blue on head, iridescent neck patch smaller. **Juv:** resembles ♀ but lacks iridescent neck patch, back and wing coverts edged cinnamon.
     Voice. A low mournful *hoo'ooo-oo oo-ooo,* suggesting Mourning Dove.
     Habs. Arid coastal scrub, mangroves. Singly or in pairs, rarely small groups. Flight typically low and fast, often commutes to/from barrier islands. At times nests on ground in cover.
SS: Mourning Dove has long pointed tail, lacks white trailing edge to secondaries; see White-winged Dove.
SD: F to C resident (around SL) along N coast of Yuc Pen from N Camp to N QR. Irregular R to U visitor (mostly Oct–Dec) to Isla Cozumel (RAB, AMS), vagrant to Arrecife Alacrán (Oct 1985, BM). A report from Belize (Wood *et al.* 1986) probably is in error (see Barlow *et al.* 1969).
RA: Caribbean and adjacent coasts of Yuc Pen.

MOURNING DOVE
Zenaida macroura*   Plate 20
Paloma Huilota

ID: 10–12 in (25.5–30.5 cm). Widespread and common. *Tail long and tapered.* **Descr.** Ages/sexes differ slightly. ♂: eyes brown. *Pinkish brown overall* with black

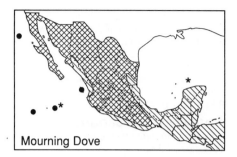

Mourning Dove

streak on lower auriculars, blue-grey hind-crown and nape, iridescent pink neck patch. More vinaceous pink below, paling to whitish on undertail coverts; upperparts more grey-brown with *bold black spots on scapulars, tertials, and innerwing coverts.* Remiges dark grey; central rectrices grey-brown, *outer rectrices grey with black sub-terminal band, broadly tipped white* to pale grey. ♀: duller, greyer overall; lacks blue-grey on head, iridescent neck patch smaller. **Juv:** resembles ♀ with extensive buff feather tips, *looks scaly and barred overall;* lacks iridescent neck patch, rectrices tipped pale grey.

Voice. A low, mournful *whoo'hoo hu-hu* or *whoo'hoo hu-hu, hu;* also a single low *whooo* which may be repeated.

Habs. Open and semiopen areas with hedges, scattered trees, brush. Often in pairs or flocks, at times in hundreds during winter. Flight fast and direct. In display, climbs with slow wingbeats and descends in glide with wings bowed.

SS: See Zenaida Dove (coastal Yuc Pen).

SD: C to F resident (SL–3000 m) throughout temperate regions of Mexico (including Islas Guadalupe, Socorro, and Clarión) S to Isthmus; winters (Oct–Apr) throughout region, F to C in interior and on Pacific Slope S of breeding range to El Salvador and Honduras, U on Atlantic Slope S from Isthmus.

RA: Breeds N.A. to Mexico, locally on Pacific Slope in Costa Rica and Panama; N breeders winter to Panama.

SOCORRO DOVE

Zenaida graysoni Presumed extinct. See Appendix A.

GENUS COLUMBINA

These are small ground-doves of fairly open country. Often in pairs or small flocks. Eyes reddish (dull in juvs), legs pinkish. Rufous flash in wings visible in flight.

INCA DOVE

Columbina inca　　　　　　Plate 19
Tórtola Colilarga

ID: 8–9 in (20.5–23 cm). Widespread. *Long graduated* tail appears square-tipped. **Descr.** Ages differ, sexes similar. **Adult:** bill dark grey. *Pinkish grey-brown overall,* paler and more pinkish below, *with extensive dark scaling,* less distinct on underparts. Remiges blackish with *bright rufous flash across primaries* striking in flight, under-wings mostly rufous with black axillars. Central rectrices grey-brown, outer rectrices black with broad white tips forming *white tail sides, striking on spread tail.* **Juv:** duller and browner, upperparts with buff sub-terminal scaling, underparts more pinkish buff, pale buff belly lacks dusky scaling.

Voice. A low, often burry cooing, *who'pu* or *whur-pu,* 10/18–27 s.

Habs. Open and semiopen habitats, including urban areas, brushy woodland and edge, mainly in drier areas. Flocks readily with other ground-doves. Flushes with distinctive loud whirr of wings.

SS: Long white-edged tail distinctive; birds missing tail told by extensive dark scaling.

SD: C to F resident (SL–3000 m) throughout most of Mexico (but U and local in N-cen Plateau) S to Isthmus, then mostly on Pacific Slope and in interior to El Salvador and Honduras. R visitor (Aug–Dec) to NE Baja where first recorded Aug 1984 (REW).

RA: SW USA to cen Costa Rica.

NB1: Formerly placed in genus *Scardafella.* **NB2:** sometimes considered conspecific with *C. squammata* of S.A.

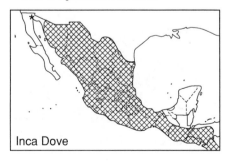

Inca Dove

COMMON GROUND-DOVE
Columbina passerina＊   Plate 19
Tórtola Común

ID: 6.2–6.7 in (16–17 cm). Widespread in relatively dry areas. **Descr.** Ages/sexes differ. ♂: *bill orange to dull flesh with dark tip*, rarely appears all dark. Face, neck, and underparts pinkish, crown and hindneck blue-grey; *dusky scaling on crown, neck, and chest* often looks mottled on chest. Upperparts grey-brown, *wing coverts and tertials paler, pinkish with dark violet spots and bars*. Remiges blackish, rufous flash across primaries striking in flight, underwings mostly rufous. Central rectrices grey-brown, outer rectrices blackish, narrowly tipped white. ♀: bill duller. Face, neck, and underparts buffy grey to pale grey-brown, crown and nape slightly darker; *dusky scaling on crown, neck, and chest* often looks mottled on chest. Upperparts darker grey-brown, *wing coverts and tertials paler and greyer, with burnished purplish-chestnut spots and bars*. **Juv:** resembles ♀ but upperparts narrowly edged buff, dusky mottling on chest less distinct. *C. p. socorroensis* (of Isla Socorro) smaller and darker overall.

Voice. A low *whuu'* or *h'woo*, 10/10–12 s. On Socorro, often a more strongly disyllabic *per'hoo*.

Habs. Open and semiopen habitats, including urban areas, brushy woodland and edge, mainly in drier areas. Flocks readily with Inca Dove and Ruddy Ground-Dove. May nest on ground.

SS: Plain-breasted (savannas) and Ruddy ground-doves lack pinkish or orange bill base (not always easy to see) and scaling on head and chest, Plain-breasted smaller with duller rufous flash on primaries, ♀ Ruddy larger and longer-tailed, black wing covert marks sparser but black bars also on lower scapulars.

SD: C to F resident (SL–2500 m) in Baja, on Isla Socorro, on Pacific Slope from Son, and in interior from S Plateau, to W Nicaragua, on Atlantic Slope from N Coah to S Ver, and in Yuc Pen (including many islands and cays) and Honduras Bay Islands.

RA: S USA to N S.A.

PLAIN-BREASTED GROUND-DOVE
Columbina minuta interrupta   Plate 19
Tórtola Pechilisa

ID: 5.5–6 in (14.5–16 cm). S savannas. **Descr.** Ages/sexes differ. ♂: bill greyish flesh with dark tip. *Face, neck, and underparts plain greyish vinaceous*; crown and nape blue-grey. Upperparts grey-brown, wing coverts paler with dark violet spots, lower scapulars with sparse dark violet marks. Remiges blackish, *rufous flash across primaries less extensive and duller than other ground-doves*, underwings mostly rufous. Central rectrices grey-brown, outer rectrices blackish, narrowly tipped white. ♀: *duller*, bill greyish. Face, neck, and *chest buffy grey*, throat and *belly paler*, rarely whitish; crown and nape greyish. **Juv:** resembles ♀ but upperparts narrowly edged buff, wing covert spots blackish.

Voice. A low, often prolonged *wuup wuup …* or *w-up w-up …*, 10/6.5–7.5 s. **Habs.** Savannas and open grassy fields. Singly or in pairs, rarely small groups. Seen most often when flushed, usually flies a short distance and drops back down; at times in open on quiet roadsides where may feed in loose association with other ground-doves but usually keeps apart. At times nests on ground.

SS: ♀ Ruddy Ground-Dove confused easily with ♀ Plain-breasted but former is larger and relatively longer-tailed, although size can be hard to judge; ♀ Plain-breasted

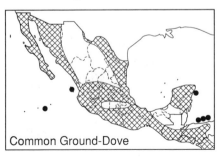

Common Ground-Dove

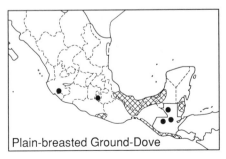

Plain-breasted Ground-Dove

slightly darker and greyer overall with less extensive rufous in wings, paler wing coverts with purple spots; black axillars of Ruddy distinctive; see Common Ground-Dove.

SD: F to C but often local resident (SL–750 m) on Pacific Slope from E Gro to El Salvador, on Atlantic Slope from cen Ver to SW Camp; also locally in N Guatemala, Belize, the Mosquitia, and (resident?) in cen Jal (WFVZ spec.) and Mor (AMNH spec.). Unrecorded in Honduras.

RA: S Mexico locally to Peru and Paraguay.

RUDDY GROUND-DOVE
Columbina talpacoti*   Plate 19
Tórtola Rojiza

ID: 6.5–7 in (16.5–18 cm). Widespread in more humid areas. *Relatively long-tailed.* **Descr.** Ages/sexes differ. ♂: bill greyish with dark tip. Face, neck, and *underparts vinaceous to ruddy vinaceous; crown and nape blue-grey.* Upperparts pinkish ruddy, *wing coverts and lower scapulars with sparse black spots and bars. Remiges extensively rufous* (rufous extends onto outer primary webs), underwings mostly rufous with *black axillars.* Central rectrices pinkish brown, outer rectrices black, outermost finely tipped whitish to buff. ♀: face, neck, and underparts greyish with paler throat. Crown, nape, and upperparts darker, grey-brown, wings marked as ♂ but less extensive rufous may not extend onto outer primary webs, outer rectrices narrowly tipped white to whitish. **Juv:** resembles ♀ but contour feathers with fine pale tips. **Imm** ♂: resembles ♀ but crown and nape greyer, underparts pale greyish vinaceous, upperparts with scattered ruddy feathers. Probably attains adult plumage in less than one year. *C. t. eluta* of W Mexico paler overall than birds in E Mexico and C.A.

Voice. A low *per-woop* or *h-woop*, 10/6–

8 s, sometimes varied to a longer *p-ter-woo'*; at a distance may sound like *wooh* or *woop*.

Habs. Open and semiopen habitats, including towns, forest edge, clearings, mainly in humid areas. At times in flocks up to 200+ birds. Flocks readily with Inca Dove and Common Ground-Dove.

SS: See ♀ Plain-breasted and Common ground-doves. ♀ Blue Ground-Dove larger and longer-tailed with bold dark rufous spots on wing coverts, uppertail rufous, underwings pale grey.

SD: C to F resident (SL–1000 m, locally to 1250 m in C.A.) on both slopes from S Son (spread N from S Sin since 1960s, SMR) and S Tamps, and locally in interior from Isthmus, to El Salvador and Honduras. Vagrant to BCS (Jun 1991, Howell and Webb 1992*f*).

RA: Mexico to Peru and N Argentina.

GENUS CLARAVIS

These are ground-doves of forest habitats, in pairs or small flocks. Eyes reddish, legs pink. Relatively arboreal. Often seen flying down forest trails or across clearings, flight fast and distinctively rocking.

BLUE GROUND-DOVE
Claravis pretiosa   Plate 19
Tórtola Azul

ID: 8–8.5 in (20.5–21.5 cm). Humid SE. **Descr.** Ages/sexes differ. ♂: bill horn. *Grey-blue overall,* paler on head and underparts, wing coverts with black spots and bars. Remiges blackish, underwings pale greyish; central rectrices grey-blue, outer rectrices black. ♀: head, chest, and upperparts brown, *uppertail coverts and central rectrices contrastingly rufous, wing coverts with bold dark rufous spots and bars;* belly pale blue-grey, undertail coverts

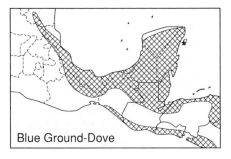

Ruddy Gound-Dove   Blue Ground-Dove

dull cinnamon. Remiges blackish, underwings grey; blackish outer rectrices noticeable if tail spread. **Juv:** resembles ♀ but upperparts with pale buff tips, dark wing covert marks dull and inconspicuous. ♂ quickly attains some adult wing coverts, then underparts and rump become bluish, lastly head, back, and flight feathers.

Voice. A clear *boop* ..., or *oop* ..., 1–6×, usually 3–6×, 6/7–9 s.

Habs. Humid to semihumid forest and edge, clearings, forest patches. See genus note. At times in flocks up to 40+ birds. Often perches at mid- to upper levels in trees whence calling birds may be hard to see until flushed.

SS: See ♀ Ruddy Ground-Dove; Maroonchested Ground-Dove (higher elevations) has bold and striking violet wing covert bars, underwings dark grey, ♂ darker and greyer with maroon chest, white tail sides conspicuous when flashed, ♀ darker and greyer, lacks contrasting rufous uppertail, outer rectrices pale-tipped.

SD: Locally/seasonally F to C resident (SL–1000 m) on Atlantic Slope from S Tamps to Honduras, on Pacific Slope from Chis to El Salvador. Engages in seasonal movements, at least in the drier parts of its range, for example absent from several sites in Yuc in Mar–May, but common in Jun–Aug (PES); Dickey and Van Rossem (1938) and Thurber *et al.* (1987) found it in El Salvador only during Jan–Jun; R to locally absent in S Ver in Sep–Jan, appearing Feb–Apr (SNGH).

RA: Mexico to Peru and N Argentina.

MAROON-CHESTED GROUND-DOVE

Claravis mondetoura salvini   Plate 19
Tórtola Pechimorada

ID: 7.5–8.5 in (19–21.5 cm). Enigmatic dove of humid montane forest. **Descr.** Ages/sexes differ. ♂: bill black. *Dark blue-grey*

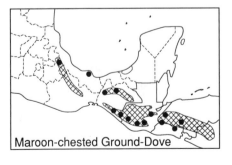

Maroon-chested Ground-Dove

overall with whitish forehead and throat, dark maroon chest, and whitish vent and undertail coverts; *bold blackish violet bars on wing coverts striking, even in flight.* Remiges blackish, underwings dark grey. Central rectrices dark grey, white outer rectrices striking on spread tail. ♀: head grey-brown with *pale tawny forehead and sides of throat*, throat whitish. Upperparts dark grey-brown, rump to central rectrices richer tawny-brown; *bold violet bars on wing coverts as* ♂. Chest and flanks greybrown, paler and greyer on belly, brighter greyish ochre on hindflanks and undertail coverts. Remiges dark grey, wing linings warm brownish; *outer rectrices blackish with bold pale tips.* **Juv:** resembles ♀ but browner, upperparts with pale buff tips, wing covert bars blackish.

Voice. Faster and more prolonged than Blue Ground-Dove. A far-carrying *woop* ..., 10/12–13 s, repeated up to 40× or more; also a slower *huh'woop* or *hwoop*, 10/15–20 s.

Habs. Humid evergreen forest and edge, second growth thickets, especially with bamboo. See genus note. At times in flocks of 10–15 birds. Calls from midlevels where may be hard to see until flushed. Nest and eggs undescribed (?).

SS: See Blue Ground-Dove (lower elevations).

SD: Locally/seasonally U to F resident (1200–2500 m), recorded at scattered localities in Ver, Chis, Guatemala, El Salvador, and Honduras. The species' 'rarity' seems linked with its cyclic abundance which may be tied to seeding bamboo; C in Sierra de las Minas, Guatemala, Aug–Dec 1958 (Land 1962*b*) and at El Sumidero, Chis, Jul 1988 (Howell 1992*b*).

RA: S Mexico locally to Peru and Bolivia.

GENUS LEPTOTILA

These birds are common but retiring, terrestrial doves of forest and woodland. Most often seen singly or in pairs as they flush off through the understory; may be seen in the open at edges, on quiet roads, and trails. When alarmed, typically bow head and slowly jerk tail. Often call from low branches and when flushed, alight at mid- to upper levels in trees. Nest at times on ground. All have yellowish eyes, blackish bills, dark reddish orbital rings, and red to pinkish-red legs.

WHITE-TIPPED DOVE

Leptotila verreauxi*   Plate 20
Paloma Arroyera

ID: 11–12 in (28–30.5 cm). *Widespread.*
Descr. Adult: *head greyish vinaceous with paler forehead, darker and greyer crown,* whitish throat; hindneck with iridescent purple sheen. *Chest and flanks pale greyish vinaceous, becoming whitish on belly* and undertail coverts. Upperparts greyish olive-brown (palest in *angelica* N of Isthmus, darkest in *bangsi* of Pacific Slope and interior from Chis S), remiges and outer rectrices darker, *outer 3–4 rectrices tipped white. Underwing coverts and axillars dark rufous.* Juv: darker overall, head and chest dusky, chest and upperparts with pale edgings.
Voice. A low, mournful, 3-syllable cooing, 1st syllable may be inaudible at a distance, *wh' whoo-ooo,* or *woo woo-oooo,* 1/9–10 s, at times preceded by a low crooning *krru* or *krrru.*
Habs. Wide variety of wooded and forested habitats from brushy thickets to humid evergreen forest (mainly edges) and pine–oak woodland with brushy understory. In display, climbs briefly with 2–3 wing claps then glides down. See genus note.
SS: Grey-headed Dove (SE) has contrasting blue-grey crown and nape, pinker chest, and whiter belly, more extensive rufous on underwings, white tips of rectrices wider and mainly on outer 2 pairs. Caribbean Dove (Yuc Pen) has whiter face, blue-grey crown and nape, brighter white belly, brighter red legs, more extensive rufous on underwings, white rectrix tips often wider and on outer 4–5 pairs. Grey-chested Dove (SE) mainly in heavy forest, darker overall with warm brown crown and nape, belly vinaceous, white rectrix tips mainly on outer 2 pairs. Quail-Doves stockier and shorter-tailed, typically flush with louder whirr of wings, all are darker overall and lack white tail tips.
SD: C to F resident (SL–2000 m) on Pacific Slope from S Son, and in interior from Balsas drainage, to Honduras and W Nicaragua; on Atlantic slope from E NL to Yuc Pen.
RA: Mexico (also S Texas) to Peru and Argentina.
NB1: Formerly known as White-fronted Dove. NB2: Mexican birds often erroneously reported to have pale blue orbital rings (Peterson and Chalif 1973) instead of dark reddish like other Mexican *Leptotila.* In S C.A. however, White-tipped doves have blue orbital rings, and the color in S.A. varies geographically from red to blue.

GREY-HEADED DOVE

Leptotila p. plumbeiceps   Plate 20
Paloma Cabecigris

ID: 10–11 in (25.5–28 cm). SE humid forest.
Descr. Adult: *crown and hindneck blue-grey, contrasting with greyish vinaceous face,* throat whitish. *Chest pale vinaceous, becoming white on belly* and undertail coverts. Upperparts dark olive-brown, remiges and outer rectrices darker, *outer 3 rectrices broadly tipped white. Underwing coverts, axillars, and underside of primaries rufous.* Juv: upperparts edged cinnamon, chest narrowly scaled buff.
Voice. A fairly short, mournful *whooo* or *huuu,* repeated steadily, 10/21–28 s; can be confused with Ruddy Quail-Dove but shorter, more abrupt.
Habs. Humid evergreen forest, second growth, and edge. See genus note.
SS: See White-tipped Dove. Grey-chested Dove darker overall with warm brown crown and nape, belly vinaceous; Caribbean Dove has whiter face, less extensive blue-grey crown, iridescent pink sides of

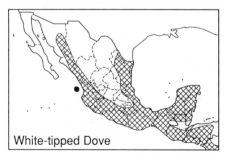
White-tipped Dove

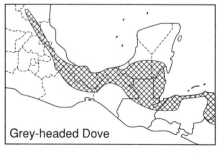
Grey-headed Dove

neck, brighter red legs. See quail-doves under White-tipped Dove SS.

SD: C to F resident (SL–900 m, locally to 1500 m in Honduras) on Atlantic Slope from S Ver to Honduras; U N to S Tamps.

RA: E Mexico to NW Colombia.

NB: Sometimes considered conspecific with *L. rufaxilla* of S.A.

CARIBBEAN DOVE

Leptotila jamaicensis gaumeri   Plate 20
Paloma Caribeña

ID: 11–12 in (29–31.5 cm). *Yuc Pen and islands*. **Descr. Adult:** *legs bright red. Face whitish, hindcrown and nape blue-grey, sides of neck iridescent rose pink*. Chest pale greyish vinaceous, becoming *white* on *belly* and undertail coverts. Upperparts greyish olive-brown, remiges and outer rectrices darker, *outer 5 rectrices tipped white*. *Underwing coverts, axillars, and underside of primaries rufous*. **Juv:** undescribed (?).

Voice. A low, 4-syllable cooing, *whu hu'oo ooo* or *hoo coo'hoo-ooo*, 1/7–9 s; may suggest Mourning Dove; in the Honduras Bay Islands, soft 1st syllable may be inaudible, *(h) woo oo-ooor*, or *h hoo-oo-oooh*, etc.

Habs. Humid to semihumid forest, dense scrubby woodland. See genus note.

SS: See White-tipped and Grey-headed doves.

SD: C to F resident (around SL) in QR (including Isla Cozumel), and S to NE Belize (Ambergris Cay, SNGH); also Isla Barbareta, Isla Roatán (Howell and Webb 1992*b*), and Little Hog Island, Honduras; F to U farther W in Yuc Pen to W and S Camp. Reports from Belize in Wood *et al.* (1986) lack documentation.

RA: Locally in Caribbean and adjacent coasts of M.A.

NB: Sometimes known as White-bellied Dove.

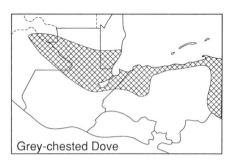
Grey-chested Dove

GREY-CHESTED DOVE

Leptotila cassinii cerviniventris   Plate 20
Paloma Pechigris

ID: 10–11 in (25.5–28 cm). SE humid forest. Relatively short-tailed. **Descr. Adult:** forehead pale greyish, *crown and nape warm brown*, throat whitish; face, neck, *chest, and flanks vinaceous greyish, becoming vinaceous on belly, whitish only on undertail coverts*. Upperparts warm dark brown, remiges and outer rectrices slightly darker, *outer 2 rectrices tipped white*. Underwing coverts, axillars, and underside of primaries dark rufous. **Juv:** upperparts extensively edged cinnamon.

Voice. A low, mournful, drawn-out *whooooo* or *coooooo*, fading away, 10/48–65 s; can be confused with Ruddy Quail-Dove.

Habs. Humid evergreen forest and adjacent second growth, *Heliconia* thickets, etc. See genus note.

SS: See White-tipped and Grey-headed doves. ♀ Ruddy Quail-Dove shorter-tailed, darker overall but belly contrastingly paler, lacks white tail tips; note face pattern, reddish bill.

SD: F to C resident (SL–500 m, locally to 1400 m in Honduras) on Atlantic Slope from N Chis and E Tab to Honduras. A report from the Pacific Slope of Guatemala (Vannini 1989*b*) is probably erroneus.

RA: SE Mexico to N Colombia.

GENUS GEOTRYGON

The quail-doves typically keep to dense shady understory where they walk quickly and quietly. Singly or in pairs, seen infrequently unless flying across roads or wandering out on quiet trails; flush with louder wing whirr than *Leptotila* and may perch on low branches.

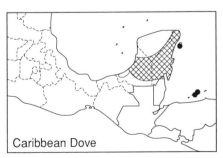
Caribbean Dove

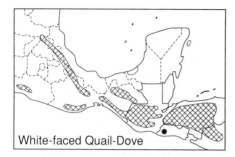

White-faced Quail-Dove

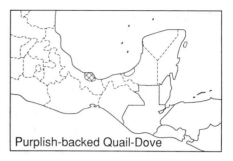

Purplish-backed Quail-Dove

WHITE-FACED QUAIL-DOVE

*Geotrygon albifacies**   Plate 20
Paloma-perdiz Cariblanca

ID: 11.5–12.5 in (29–31.5 cm). Endemic to E and S cloud forest. **Descr.** Ages differ, sexes similar. **Adult:** eyes red, orbital ring grey, bill blackish, legs red. *Looks ruddy overall with whitish head. Face whitish,* crown and nape blue-grey, sides of neck and chest greyish cinnamon; *sides of neck with distinctive dark furrows.* Rest of underparts pale cinnamon. Hindneck and *upperparts dark rufous* with vinaceous gloss on upper mantle and broad band of purplish gloss across lower mantle. Remiges and underwings dark grey. **Juv:** darker overall, underparts barred and scalloped dusky, upperparts with subterminal dusky bars and cinnamon tips.

  Voice. A low, hollow, mournful *whoooo* or *whoo'oo*, 10/25–45 s, usually 10/36–40 s; suggests Ruddy Quail-Dove but rises slightly and ends abruptly (versus fading away in Ruddy).

  Habs. Humid evergreen and pine-evergreen forest, coffee fincas. Bows and jerks tail like *Leptotila* when wary. May nest on ground (?).

SS: See White-tipped Dove. Ruddy Quail-Dove smaller, lacks contrasting whitish head; note dark face stripe, underwings rufous.

SD/RA: C to F resident (1000–2500 m) on both slopes from SE SLP and Gro, and locally in interior from Chis, to Honduras and N cen Nicaragua.

PURPLISH-BACKED QUAIL-DOVE

Geotrygon lawrencii carrikeri   Plate 20
Paloma-perdiz Morena

ID: 11.5–12.5 in (29–31.5 cm). *Endemic to Sierra de los Tuxtlas, Ver.* **Descr.** Ages dif-

fer, sexes similar. **Adult:** eyes reddish, orbital ring dull pinkish, bill dark grey, legs reddish. *Face whitish with black cheek stripe,* crown and hindneck grey-green. *Sides of neck and chest grey becoming buff-brown on flanks,* whitish on vent and undertail coverts. Upperparts greenish brown, wings browner, *broad band of glossy purple across lower mantle.* Underwings dark grey. **Juv:** darker overall, upperparts edged pale cinnamon, chest barred buff.

  Voice. A 3-syllable, slightly burry or twangy *hu' w-wohw* or *h' w-wohw*, 10/27–33 s, at a distance only the last *wohw* audible, when quality may suggest Inca Dove.

  Habs. Humid evergreen forest. See genus note. May nest on ground.

SS: Ruddy Quail-Dove smaller, ruddy or brown overall, lacks white face.

SD: F to C resident (350–1500 m) in Los Tuxtlas, S Ver.

RA: S Ver, disjunctly in Costa Rica and Panama.

NB: The song of Purplish-backed Quail-Doves in cen Panama is a single note, 'a fairly loud, nasal and somewhat hollow *cowh*, repeated steadily for lengthy periods at 3-second intervals' (Ridgely and Gwynne 1989), or 10/24 s (RAB tape). The Mexican form *carrikeri* is strikingly larger and paler than Purplish-backed Quail-Doves in S C.A. and sounds quite different from birds in Panama. This suggests that it could be specifically distinct. In NW and cen Costa Rica, however, Purplish-backed Quail-Doves give a 3-syllable *coo-ka-krrrw* or *pum-whaa-kooow*, 1/3–5 s; at a distance, only the third note is audible (Stiles and Skutch 1989; FGS, personal communication). This appears to be very like the song of *carrikeri*. Thus the taxonomic status of *carrikeri*, and indeed of the whole *G. lawrencii* complex, remains moot.

RUDDY QUAIL-DOVE

Geotrygon m. montana   Plate 20
Paloma-perdiz Rojiza

ID: 9–10 in (23–25.5 cm). Widespread.
Descr. Ages/sexes differ. ♂: eyes yellowish, orbital ring, bill, and legs reddish, bill with dark tip. *Face buff with rufous-brown cheek stripe. Crown, nape, and upperparts brownish rufous*, mantle glossed purple. Chest greyish vinaceous with vertical white bar at sides; belly, flanks, and undertail coverts buff. Underwings cinnamon-rufous. ♀: *face brownish cinnamon with dusky cheek stripe*, throat whitish. Crown, nape, and upperparts dark olive-brown. *Chest greyish cinnamon, becoming buff on belly* and undertail coverts, chest side bar pale buff. Underwings dull cinnamon-rufous. **Juv:** resembles ♀ but darker, chest and upperparts extensively edged cinnamon.

Voice. A low, mournful, slightly moaning *whoooo* or *ooooooh*, fading away slightly, 10/25–32 s (SE Mexico), 10/34–36 s (W Mexico), 10/78–98 s (NE Mexico); apparently faster where Grey-chested Dove sympatric. Compare with White-faced Quail-Dove, Grey-headed Dove.

Habs. Humid evergreen to semi-deciduous forest, coffee fincas. See genus note. May nest on ground.

SS: See *Leptotila* doves, other quail-doves.

SD: F to C resident (SL–1500 m) on both slopes from S Sin and S Tamps (Webster 1974; JCA) and SE SLP (SNGH tape) to El Salvador and Honduras.

RA: Mexico to N Argentina.

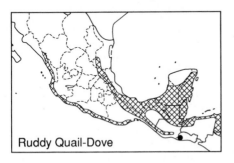

Ruddy Quail-Dove

NEW WORLD PARROTS: PSITTACIDAE (23)

A familiar group of brightly colored arboreal birds which, unfortunately, are popular as pets; as a consequence, many species are endangered in the wild and several have been extirpated from most of their natural range in the region.

Bills deep and hooked, zygodactyl feet well adapted to walk about in tree canopy. Shape varies from pointed-winged, long-tailed macaws and *Aratinga* parakeets to broad-winged and short-tailed *Amazona* parrots (or Amazons). Ages/sexes mostly similar though juvs typically duller; Amazon head patterns may show age/sex variation. Juvs often best told by dark eyes (versus amber or pale in adults). Young altricial; juvs usually associate with adults up to several months after fledging. Small species mature in 1–2 years, larger species in 3–4 years. Molts surprisingly poorly known, adults have one complete molt per year after breeding. Plumage of most species bright green overall. A trend observed in M.A. is the retention of 'neotenic' head patterns southward: southern ssp. of Red-lored, Mealy, and Yellow-headed parrots have head patterns similar to juv and imm head patterns of Mexican ssp. Voice varies from the deep throaty calls of macaws to the excited high twittering of parrotlets, but all species share a distinctive quality that says 'parrot' as soon as one hears them. In general, the smaller the species, the higher pitched and more frequently given are its calls.

Parrots feed mostly on seeds and fruits, often lifting food to their bill with their dextrous feet. Breeding data poorly known in the wild. Nest in cavities in trees, rock walls, termitaries. Eggs: 2–5, white, possibly 7 in Barred Parakeet (WFVZ). Nesting in tree cavities is assumed unless stated otherwise.

For an observer new to the Neotropics, first encounters with wild parrots may be somewhat disappointing. The typical view one gets of most species is of a pair or flock in flight, often at considerable distance. With practice, voices can be useful for identification but knowing which species occur where helps greatly; rarely do two similar parrots occur in the same range or habitat. Most identification problems occur among the Amazons; head patterns are the key but can be hard to see on flying birds and many parrots may have to go down as 'Amazon sp.' Rarely do two species associate readily other than sometimes at fruiting trees.

GENUS ARATINGA

These are 'typical' parakeets with pointed wings and fairly long, pointed tails. Plumage bright green overall, head and chest patterns important for species recognition. Ages differ slightly: adults have amber eyes, juvs have duller brown eyes; sexes similar. Calls are shrill and raucous screeching, higher and faster than parrots.

GREEN PARAKEET
Aratinga holochlora *   Plates 22, 23
Perico Verde Mexicano

ID: 11–12 in (28–30.5 cm). Mexican endemic with two recognized ssp: *holochlora* (NE and SE Mexico) and *brewsteri* (NW Mexico). **Descr.** *Holochlora:* orbital skin vinaceous grey, bill pale horn-flesh, feet pale greyish. *Bright green overall*, often with variable (usually slight) orange flecking on throat and neck. *Underside of flight* feathers metallic yellowish. ***Brewsteri:*** color of bare parts undescribed (?), bill and feet probably like *holochlora*. Differs in slight glaucous blue cast on crown and sometimes on chest.

Voice. Raucous, often shrill and shrieky screaming, *krreh-krreh* … or *krrieh krrieh*, and *kree-ik krii-krii-kriir*, also harder and deeper phrases, *kreh kreh* … or *ruh-ruh-ruh-ruh*, etc. Flight calls of Chis birds lower than NE Mexico, call structure apparently more like flight calls of Pacific Parakeet (SNGH tape, sonograms) than of Green Parakeets in NE Mexico. Voice of *brewsteri* undescribed (?).

Habs. Evergreen to semideciduous forest, plantations, locally in arid deciduous and pine–oak forest in Chis. Often in flocks up to 100+ birds which typically fly higher and with noticeably slower wingbeats than sympatric smaller parakeets. Nest in tree cavities, termitaries, and colonially in rock crevices.

SS: Aztec Parakeet and perhaps Orange-fronted Parakeet locally sympatric; both are smaller, shorter-tailed, appear to fly faster, and have different calls, useful with experience; head and chest patterns distinctive if seen well, note also remiges mostly bluish above, greyish below. See Pacific Parakeet under NB1.

SD/RA: C to F resident (SL–1500 m) on Atlantic Slope from E NL and Tamps to cen Ver; disjunctly in interior from SE Ver (HG, SNGH) and E Oax to E Chis, ranging seasonally to adjacent Pacific Slope in Isthmus. Also (1300–1800 m) in SW Chih, N Sin, and adjacent Son, where locally F to C resident (GP). Reported traditionally S to Oax on Atlantic Slope but no records from N Oax (Binford 1989); see NB1.

NB1: Pacific Parakeet sometimes regarded as specifically distinct on the basis of its pro-portionately heavier bill (Bangs and Peters 1928; Griscom 1932) and apparent sympatry with Green Parakeet in Isthmus. However, *holochlora* and *strenua* are not known to breed sympatrically and bill measurements of the two forms overlap; average mensural differences should be supported by other data before considering them separate species. See Table 1. **NB2:** Socorro Parakeet (*A. brevipes*) often considered conspecific with Green Parakeet but more distinct than *strenua* (see below); considered specifically distinct by Bangs and Peters (1928). **NB3:** Red-throated Parakeet (*A. rubritorques*) also may be conspecific with Green Parakeet.

Table 1. Bill measurements of Green Parakeets to the nearest 0.1 mm; length measured from anterior end of nostril to tip of bill (means in parentheses). *n* = sample size. Sexual differences not significant, hence sexes were combined.

	Length	Width	Depth
Holochlora (NE Mexico) (*n*=40)	22.4–26.6 (24.5)	15.1–18.9 (16.8)	24.5–30.1 (28.5)
'*Holochlora*' (E Oax–Chis) (*n*=17)	23.6–26.6 (24.9)	16.6–19.6 (18.1)	26.1–30.3 (28.4)
Strenua (Oax–Nicaragua) (*n*=40)	24.7–29.3 (26.6)	17.8–21.3 (19.4)	28.1–32.9 (30.8)
Rubritorques (*n*=30)	21.7–25.2 (23.5)	15.4–19.0 (17.0)	25.1–28.8 (27.0)
Brevipes (*n*=40)	25.0–28.4 (26.7)	17.9–21.3 (19.5)	29.7–33.5 (31.2)

PACIFIC [GREEN] PARAKEET
Aratinga (holochlora?) strenua
Perico Verde Centroamericano   Plate 22

ID: 12–13 in (30.5–33 cm). *Endemic to Pacific Slope from Isthmus S.* **Descr.** Orbital skin greyish, bill pale horn-flesh, feet pale greyish. *Bright green overall*; may have variable (usually slight) orange flecking on throat and neck. *Underside of flight feathers metallic yellowish.*

Voice. Much like Green Parakeet but flight calls lower that Green in NE Mexico

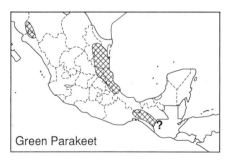

Green Parakeet

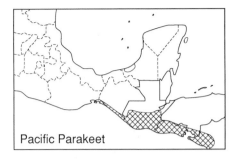

Pacific Parakeet

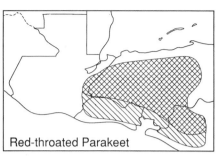

Red-throated Parakeet

(see Green Parakeet), a screaming *kreh-kreh-kreh* ..., and higher *kriih-kriih* ..., etc.

Habs. Semideciduous forest, plantations, also pine–oak highlands of Guatemala. Often in flocks up to 200+ birds which typically fly higher and with noticeably slower wingbeats than sympatric smaller parakeets. Nest in rock crevices, often colonially; also (?) in tree cavities and termitaries.

SS: See NB1 under Green Parakeet. Aztec, Orange-fronted, and Red-throated parakeets all locally sympatric but all distinctly smaller, shorter-tailed, appear to fly faster, and have different calls, useful with experience. Head and chest patterns distinctive if seen well. Olive-throated and Orange-fronted have striking pale orbital rings, remiges mostly bluish above, dark grey below; Red-throated has extensively reddish-orange mottled throat (adult), appears to have dark eyes (versus paler orbital ring of Pacific).

SD/RA: C to F resident (near SL–1500 m) on Pacific Slope from E Oax to cen Nicaragua; also in interior (to 2500 m) of Guatemala.

NB: See NB1 under Green Parakeet, and Table 1.

RED-THROATED [GREEN] PARAKEET
Aratinga rubritorques or *A. holochlora* (in part)   Plate 70
Perico Gorjirrojo

ID: 10.5–11.5 in (26.5–29 cm). *Endemic to N C.A.* **Descr.** Ages differ, sexes similar. **Adult:** *orbital skin brownish grey*, bill pale horn-flesh with dusky cere and tip, feet pale greyish. Bright green overall with variable but usually extensive *reddish-orange mottling on throat*, often a solidly orange throat; may also show some orange flecks

on sides of head. Underside of flight feathers metallic yellowish. **Juv:** red on throat reduced or absent (but may fully appear in captive juvs).

Voice. Raucous, shrill screaming in flight, *kreeah-kreeah* and *krri-krreea*, etc., often more varied and persistent when perched, *krieh-krieh*... and *krreh kreh krieh*, etc. Higher than Pacific Parakeet.

Habs. Semiarid to semihumid pine and pine–oak woodland and adjacent semideciduous forest. Pairs and small groups often seen perched in pines, rarely (?) in large flocks. Reported to roost with Green (=Pacific) Parakeets in El Salvador when behavioral differences noted (Thurber *et al.* 1987). Breeds colonially in rock walls, probably also in trees.

SS: Aztec and Orange-fronted parakeets have remiges mostly bluish above, dark grey below, both show bold pale orbital rings; head and chest patterns distinctive if seen well. See Pacific Parakeet.

SD/RA: F to C resident (600–1800 m) in interior and on adjacent Pacific Slope from E Guatemala to W Nicaragua; to near SL in non-breeding season (Nov–Apr) on Pacific Slope from El Salvador (Thurber *et al.* 1987) to Nicaragua (TRH).

NB: See NB3 under Green Parakeet.

SOCORRO [GREEN] PARAKEET
Aratinga brevipes   Plate 69
Perico de Socorro

ID: 12–13 in (30.5–33 cm). *Endemic to Isla Socorro, Mexico.* **Descr.** Ages similar (?). Orbital skin purplish brown (can look reddish), bill pale flesh-horn, feet pale greyish. Plumage bright green overall, rarely with 1–2 orange flecks on neck. Underside of flight feathers metallic yellowish. Bill stouter than Green Parakeet (see Table 1). Wing formula differs also: 10th primary usually < 7th primary (versus 10th > 7th in Green

Socorro Parakeet

Parakeet, innermost primary being 1st primary).

Voice. Notably different from Green Parakeet (*A. h. holochlora*). High, screeching or piping calls in flight, *krree-kree...*; flock calls often have a shrill, chirping quality reminiscent of small parakeets. Other shrill screams when perched, including persistent short screams, *kee-kee-kee...*, etc.

Habs. Semideciduous forest, especially on higher slopes of island, but moves around to find fruiting trees. In pairs or small flocks, fairly confiding. Nests in tree cavities (Rodriguez-E. *et al.* 1992), eggs undescribed (?).

SS: No other parrots in Islas Revillagigedo.

SD/RA: F resident on Isla Socorro, Islas Revillagigedo, population estimated at 400–500 birds (Rodriguez-E. *et al.* 1992).

NB: See NB2 under Green Parakeet.

AZTEC [OLIVE-THROATED] PARAKEET

*Aratinga astec** or *A. nana* (in part)
Perico Pechisucio   Plates 22, 23

ID: 8.5–9.5 in (21.5–24 cm). *Atlantic Slope.*
  Descr. Bill horn, may be dusky below, cere and *broad orbital ring whitish*, feet greyish. Plumage overall bright green above, *throat and chest brownish olive* (more greenish olive in *vicinialis* of NE Mexico)

paling to yellowish olive on belly. Remiges mostly dark bluish above, dark grey below, underside of tail metallic yellowish.
  Voice. Raucous screaming, *krrieh-krrieh* or *krrreh krreh*, and *krrieh krrie krreah* or *krrih rrih rrih*, etc., burrier and gruffer, less shrieky than Green Parakeet.
  Habs. Evergreen to deciduous forest and edge, semiopen areas with scattered forest patches and large trees, plantations. In pairs or flocks up to 50+ birds which fly low and fast over the canopy. Nest in termitaries.
  SS: See Green and Red-throated parakeets. Orange-fronted Parakeet may be locally sympatric in interior Honduras; note its orange forehead and bluish crown, paler and more greyish olive throat and chest.
  SD: C to F resident (SL–1000 m) on Atlantic Slope from S Tamps to Honduras.
  RA: E Mexico to W Panama.

ORANGE-FRONTED PARAKEET

*Aratinga canicularis**   Plates 22, 23
Perico Frentinaranja

ID: 9–10 in (23–25.5 cm). Pacific Slope.
  Descr. Eyes yellowish (adult), *broad orbital ring yellow*, horn bill may be dusky below, feet greyish. *Orange forehead and bluish forecrown distinctive.* Plumage bright green overall, throat and chest duller, washed greyish olive. Remiges mostly dark bluish above, dark grey below, underside of tail metallic yellowish.
  Voice. Raucous, screechy screaming, *kreer kreei-kreei*, or *rreek ree-reeh* and *krreh krrie-eh-krreh*, etc., burrier and more chattering than Pacific Parakeet.
  Habs. Arid to semihumid forest, semiopen areas with scattered forest patches and trees, plantations. Habits much as Aztec Parakeet; nests in termitaries, less often tree cavities.
  SS: See Green, Red-throated, and Aztec para-

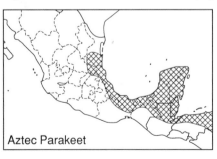

Aztec Parakeet

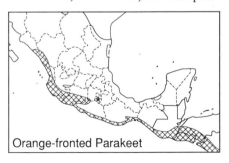

Orange-fronted Parakeet

keets. Orange-chinned Parakeet smaller with distinctive bounding flight; note voice.
SD: C to F resident (SL–1500 m) on Pacific Slope from Sin to El Salvador and Honduras, locally in interior from Balsas drainage S.
RA: NW Mexico to NW Costa Rica.

MILITARY MACAW
Ara militaris* Plates 22, 24
Guacamaya Verde

ID: 27–30 in (68.5–76 cm). E > W. *Widely extirpated by pet trade.* **Descr.** Ages/sexes similar. Eyes pale yellow (brown in juv), bare face whitish with bristle-like red and black feathers, bill all black (W) or mostly pale above (E), feet dark grey. Plumage bright green overall with turquoise rump, tail coverts, and flight feathers, central rectrices red basally. At close range note red forehead. *In flight, looks vivid turquoise and green from above, green with metallic yellow underside to flight feathers from below.*
Voice. Deep, far-carrying, raucous and throaty cries, *rrrah* and *rrr'ak*, etc.
Habs. Semideciduous forest, mainly in foothills. Flies with shallow, graceful wingbeats, long tail flowing behind; at times glides. Usually seen flying in pairs or flocks over distant ridges or valleys. Nests in tree cavities, rock crevices, formerly on rocky islets off Acapulco.
SD: U to F but local resident (near SL–2000 m) on Pacific Slope from S Son to Jal, and on Atlantic Slope from E NL to SLP; R (extirpated?) on Pacific Slope from Col to Gro. Formerly more widespread, including Pacific foothills S to W Chis.
RA: Mexico; disjunctly in S.A. from Venezuela to Argentina.
NB: Sometimes considered conspecific with

A. *ambigua* of S C.A. and S.A. (see Appendix E); the two may interbreed in W S.A. (Fjeldsa *et al.* 1987).

SCARLET MACAW
Ara macao Plates 22, 24
Guacamaya Roja

ID: 32–38 in (81–96 cm). *Unmistakable, endangered* due to relentless capture for pet trade. **Descr.** Ages/sexes similar. Eyes pale yellow (brown in juv), bare face whitish with inconspicuous, bristle-like red feathers, bill pale grey above, black below, feet dark grey. Plumage *scarlet overall* with blue rump, tail coverts, and flight feathers (central rectrices mostly red), and yellow greater upperwing coverts. Underside of flight feathers metallic golden-red.
Voice. Deep, raucous, far-carrying cries, *rrrah* and *rrrrahk*, etc., quieter at rest, *aahr* or *rah*, etc.
Habs. Humid evergreen forest. Likely to be found only in remote areas, in pairs or singly, flying over forest canopy.
SD: R to U but very local resident (SL–500 m; locally to 1000 m, at least formerly) on Atlantic Slope from E Chis to Honduras. Formerly occurred N to S Tamps and on Pacific Slope from Isthmus S. May vanish completely from region within a few years.
RA: SE (formerly E) Mexico locally to Amazonian Brazil.

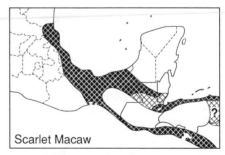
Scarlet Macaw

THICK-BILLED PARROT
Rhynchopsitta pachyrhyncha Plates 22, 24
Cotorra-serrana Occidental

ID: 15–17 in (38–43 cm). *Endemic to NW Mexico. Wings long, tail long and pointed.* **Descr.** Ages differ slightly, sexes similar. **Adult:** eyes amber, orbital ring pale greyish buff, bill black, feet dark grey. Bright green overall with dark red forehead and short supercilium, bright *red marginal upperwing coverts*, dark red thighs. *Greater*

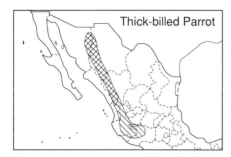

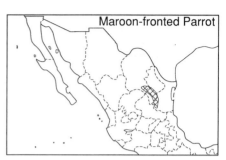

underwing coverts bright yellow, striking in flight overhead, underside of flight feathers blackish. **Juv:** eyes brownish, bill horn.

Voice. Far-carrying calls include a mellow, rolling *aahr*, and a more laughing *kah ha* and *kah ha-ha-ha-ha*, etc. Much like Maroon-fronted Parrot, and often similarly hard to locate.

Habs. Pine forest. Easy slow flight with fairly shallow wingbeats recalls macaws. Usually seen flying high, in pairs or flocks up to a hundred birds which may engage in bouts of soaring and gliding. Food mostly pine seeds. Nests in tree cavities.

SS: Lilac-crowned Parrot has shorter tail (but long for an Ámazon), red patch on upper secondaries, lacks yellow stripe across underwing, voice more raucous and screaming.

SD/RA: U to F but nomadic (1500–3000 m) resident in Sierra Madre Occidental from Chih to W Mich; bulk of population breeds (Jun–Oct) from Chih to Dgo, disperses (late Dec-Apr) S to volcanoes in Jal and Mich. Formerly ranged N to SE Arizona and S and E to cen Ver when breeding range perhaps more extensive.

NB: Sometimes considered conspecific with Maroon-fronted Parrot.

MAROON-FRONTED PARROT
Rhynchopsitta terrisi   Plates 22, 24
Cotorra-serrana Oriental

ID: 16–18 in (40.5–45.5 cm). *Endemic to NE Mexico.* **Descr.** Ages differ slightly, sexes similar. **Adult:** eyes amber, orbital ring pale greyish buff, bill black, feet dark grey. Bright green overall with *maroon forehead and short supercilium,* bright *red marginal upperwing coverts,* and dark red thighs. Greater underwing coverts yellow-olive, not contrasting, underside of flight feathers blackish. *In flight overhead*

appears bright green with contrasting dark tail. **Juv:** eyes brownish, bill horn.

Voice. Low, mellow, far-carrying calls, slightly harsher when flying, more subdued when perched, may recall laughing of distant Acorn Woodpecker; *rrahk* and *kyah-kyah*, and a laughing chatter *krrah-rrah rah-rah rah-rah*; can be frustratingly hard to locate as calls echo through canyons and off rock faces.

Habs. Pine forest. Habits, flight, etc., much like Thick-billed Parrot, but often seen about cliff faces where it nests colonially in crevices. Feeds seasonally (late Mar–Apr at least) on flowering agaves.

SD/RA: U to F but nomadic (1500–2500 m); breeds (Aug–Nov) in SE Coah and adjacent NL; ranges S to S NL (mostly May–Jul) and SW Tamps (Mar–Apr at least). Population estimated at 2000–3000 birds in 1977 (Forshaw 1981), with single roost counts up to 1600 birds (Lawson and Lanning 1980).

NB: See Thick-billed Parrot.

BARRED PARAKEET
Bolborhynchus l. lineola   Plates 22, 23
Periquito Barrado

ID: 6.5–7 in (16.5–18 cm). *SE cloud forest. Wings and tail pointed.* **Descr.** Ages/sexes

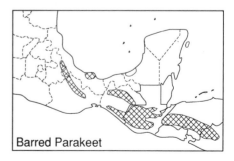

differ slightly. ♂: eyes brown, orbital ring grey, bill and feet pale flesh. Head and underparts bright green, indistinctly barred dusky on flanks and tail coverts. *Blackish upperwing covert edgings form dark shoulder and bars on median and greater coverts.* Upperparts green with sparse dark barring on nape and back, becoming bolder on rump and uppertail coverts, central rectrices black distally. ♀/juv: barring duller.

Voice. Shrill, short, screeching calls, *chirr'it* and *chirr'it chi*, etc., which carry well and are often hard to locate.

Habs. Humid evergreen and pine—evergreen forest. Usually seen flying high, calling, in pairs or small groups but often hidden in low clouds. Flight rapid and slightly bounding. Usually silent when perched and easily overlooked in canopy. Nests in tree cavities; eggs probably of this species are at WFVZ (attributed to Orange-chinned Parakeet).

SD: U to F resident (900–2400 m) on Atlantic Slope from cen Ver (no recent records?) and N Oax (Howell 1990*b*), and in interior from Chis, to N cen Nicaragua (Martinez-S. 1989), on Pacific Slope from Chis to Guatemala. Reports from Gro (such as AOU 1983) require verification. **RA:** S Mexico to Peru.

MEXICAN (BLUE-RUMPED) PARROTLET
Forpus cyanopygius*   Plates 22, 23
Periquito Mexicano

ID: 5–5.5 in (12.5–14 cm). *Endemic to NW Mexico. Tail short and squared.* **Descr.** Ages/sexes differ. ♂: eyes brown, bill flesh-horn, feet pale flesh. *Bright yellowish green overall* (body washed glaucous in *insularis* of Islas Tres Marías) *with turquoise rump,* uppertail coverts, and greater upperwing

coverts, and bluish secondaries. Turquoise underwing coverts hard to see in fast flight. ♀: *all green,* lacks turquoise and blue of ♂. **Juv** ♂: resembles ♀ but bluish on secondaries.

Voice. High, excited-sounding, screechy twitters, a reedy, rolled *kreeit* or *kree-eet*, etc.; often sounds like many more birds are present than actually are.

Habs. Semideciduous to deciduous forest, riparian woods, plantations. Usually in pairs or flocks which fly fast in compact formation. Calls often while perched but, being the size and color of leaves, hard to spot until one or more birds fly and a tree may appear to partially defoliate! Nests presumably in tree cavities; eggs undescribed (?)

SD/RA: F to C resident (SL–1400 m) on Pacific Slope from S Son to Col.

ORANGE-CHINNED PARAKEET
Brotogeris j. jugularis   Plates 22, 23
Periquito Barbinaranja

ID: 7–7.5 in (18–19 cm). *Pacific Slope from Isthmus S. Tail medium length and tapered.* **Descr.** Ages/sexes similar. Eyes brown, orbital ring whitish, bill horn, feet greyish flesh. Head and underparts bright green, washed bluish on head, *orange chin often inconspicuous.* Upperparts bluish green with *bronzy-brown lesser and median upperwing coverts* and variable bluish to violet cast on primary coverts and remiges. *Lesser and median underwing coverts bright yellow.*

Voice. Shrill, screaming phrases, typically short and clipped, *chi-chi chit* and *chree-ee chi-chit* and *krree-krreep*, etc.

Habs. Semideciduous to deciduous forest, plantations, semiopen areas with scattered forest patches and large trees. In pairs or small flocks which fly with

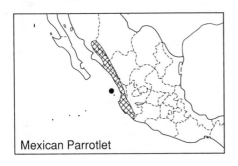

Mexican Parrotlet

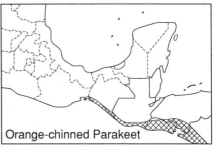

Orange-chinned Parakeet

distinctive bounding flight, several quick wingbeats interspersed with a glide. At times loosely colonial.

SS: See Orange-fronted Parakeet.

SD: C to F resident (SL–500 m) on Pacific Slope from E Oax to Honduras, locally (to 900 m) in the interior of Honduras. Reports from Gro (such as AOU 1983) probably are erroneous.

RA: S Mexico to N Venezuela.

NB: Sometimes known as Tovi Parakeet.

BROWN-HOODED PARROT

Pionopsitta h. haematotis   Plates 21, 23
Loro Orejirrojo

ID: 8–9 in (20.5–23 cm). *SE humid forest.* **Descr.** Ages differ slightly, sexes similar. **Adult:** eyes brownish, orbital ring whitish, bill and feet pale flesh. *Head dark brownish,* paler and more olive on crown, with small red auricular spot. Chest greenish olive becoming brighter green on belly. Upperparts green with violet-blue shoulders, mostly blue greater upperwing coverts and secondaries, and blackish primaries. Underside of remiges greenish blue with *bright red axillars.* Red at base of rectrices usually concealed. **Juv:** paler hood brownish olive, auricular spot brownish red, chest greenish.
Voice. Shrill, squeaky, slightly metallic calls and chatters, *kreeik* or *kreik'*, and *kreiik kreeik krri-ik*, etc.
Habs. Humid evergreen forest and edge. In pairs or small groups, rarely in flocks to 20+ birds. Flight fast and direct, low to high over canopy. Eggs undescribed (?).

SS: White-crowned Parrot larger with slower, deep wingbeats, lacks red axillars.

SD: F resident (SL–1000 m) on Atlantic Slope from S Ver to Honduras.

RA: SE Mexico to Ecuador.

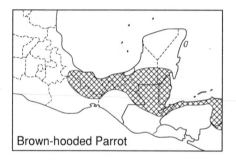
Brown-hooded Parrot

WHITE-CROWNED PARROT

Pionus senilis   Plates 21, 23
Loro Coroniblanco

ID: 9–10 in (23–25.5 cm). *Atlantic Slope.* **Descr.** Ages differ, sexes similar. **Adult:** eyes amber, orbital ring pinkish, bill pale horn, feet flesh. *Head and chest bluish with bold white forecrown,* throat whitish. Rest of *underparts blue-green with red tail coverts.* Upperparts green with *bronzy-brown shoulders* and violet-blue remiges. Central rectrices green, outer rectrices violet-blue above, blue-green below with red bases. Underwings appear blue-green overall. **Juv:** duller, white restricted to forehead, head and chest more greenish. Soon attains adult-like plumage (?).
Voice. Shrill screeches, *rreeah'k,* and a rolled *rrrie'ah,* etc.; higher and screechier than Amazons, but deeper, less shrill and metallic than Brown-hooded Parrot.
Habs. Humid to semihumid forest and edge, locally in pine–oak forest. Usually in small flocks, at times to 50+ birds. Flies with distinctive deep wingbeats, suggesting a large kingfisher.

SS: See Brown-hooded Parrot. Amazon parrots generally larger, all fly with shallow wingbeats, lack conspicuous red patch under base of tail but have red upperwing patches.

SD: F to C resident (SL–1500 m, locally to 2000 m in C.A.) on Atlantic Slope from S Tamps to Honduras, locally in interior from Chis S.

RA: E Mexico to W Panama.

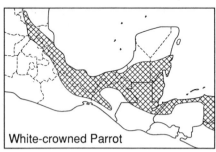
White-crowned Parrot

GENUS AMAZONA

Amazons are 'typical' parrots with rounded wings and relatively short, squared tails. Bright green overall with red upperwing patches and species-specific head markings of red, yellow, white, and blue. Calls raucous and screaming, of use in identification with practice.

WHITE-FRONTED PARROT

*Amazona albifrons**   Plates 21, 23
Loro Frentiblanco

ID: 10–11.5 in (25.5–29 cm). W > E. Widespread in relatively dry country. **Descr.** Ages/sexes differ. ♂: eyes yellowish, orbital ring greyish, bill horn-yellow, feet greyish. *Red lores and eyering, white forehead, and bluish crown.* Bright green overall (washed glaucous blue in *saltuensis* of NW Mexico) with indistinct dusky scalloping most noticeable on neck. *Alula and upper primary coverts red*, striking in flight, remiges mostly bluish violet. Red at base of tail usually concealed. ♀: white 'front' less extensive, *lacks red in wing*. **Juv** ♂: resembles ♀ but with some red on primary coverts.
    Voice. Raucous screaming, including a shrill *kyi kyeh-kyeh-kyeh ...*, a sharp, yapping *kyak-yak-yak-yak ...* or *rek rek-rek-rek ...*; may not give paired barks as Yucatan Parrot. Varied, mellow, rolled and screeching calls when perched.
    Habs. Semideciduous and deciduous forest, semiopen areas with scattered trees and forest patches, mangroves. In pairs or flocks. Wingbeats fairly fast and flight appears hurried, usually flies low over canopy; rarely seen flying high to/from roosts as are larger Amazons.
SS: Yucatan Parrot has dark auricular spot, extensive dark scalloping, at close range yellow lores and larger white forecrown patch of ♂ distinctive, ♀ has fairly plain head; from above in flight, note bright yellow-green uppertail coverts and tail of Yucatan, and ♂ has red only on primary coverts; vocal differences need study.

Lilac-crowned Parrot (W Mexico) larger and longer-tailed, flight less hurried, red upperwing patch on outer secondaries; note head pattern, voice. Other Amazons larger, slower-flying; note red secondary patches, head patterns, and voice.
SD: C to F resident (SL–1800 m) on Pacific Slope from S Son to El Salvador (where extirpated?) and Honduras (apparently absent from Col and Mich?); on Atlantic Slope from SE Ver to Yuc Pen, locally in Honduras; and in interior valleys of C.A.
RA: Mexico to NW Costa Rica.

YUCATAN (YELLOW-LORED) PARROT

Amazona xantholora   Plates 21, 23
Loro Yucateco

ID: 10–11 in (25.5–28 cm). *Endemic to Yuc Pen.* **Descr.** Ages/sexes differ. ♂: eyes brownish yellow, orbital ring pale grey, bill horn-yellow, feet greyish. *Yellow lores, white forecrown* with blue at rear, red eyering, *blackish auricular patch.* Bright green overall with *distinct dark scalloping*, small red shoulder patch; *uppertail coverts and tail contrastingly bright yellow-green, often striking from above in flight.* Upper primary coverts red, striking in flight; remiges mostly bluish violet. Red at base of tail usually concealed. ♀: *head relatively plain*, crown bluish with trace of white on forehead, yellow lores and red eyering mixed with green, generally indistinct; *lacks red in wing.* **Juv** ♂: resembles ♀ but with some red on primary coverts.
    Voice. Raucous and much like White-fronted Parrot, including a rolled *reeeah-h* and *kyeh-kyeh keeei-i-iirr*, a screechy *ree-o-rak zeek ree-o-rah*, etc.; gruff, barking calls often in pairs, *rek-rek-rek-rek rek-rek rek-rek*, or *rek-rek rek-rek rrehr*, etc.; paired calls not given by sympatric White-fronted (?).

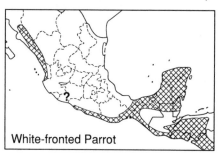

White-fronted Parrot

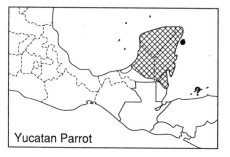

Yucatan Parrot

Habs. Semideciduous and deciduous forest and edge. Habits much as White-fronted Parrot. Large flocks commute to and from Isla Cozumel, feeding on the island, returning to roost on the mainland.

SS: See White-fronted Parrot.

SD/RA: F to C resident (SL–100 m) in Yuc Pen (including Isla Cozumel) S to N Belize; also known from Isla Roatan, Honduras (Apr 1947, spec.) where status unclear (Howell and Webb 1992*b*).

RED-CROWNED PARROT
Amazona viridigenalis　　Plates 21, 23
Loro Tamaulipeco

ID: 12–13 in (30.5–33 cm). *Endemic to NE Mexico.* **Descr.** Ages/sexes differ slightly. ♂: eyes yellowish, orbital ring pale grey, bill horn, feet pale greyish. *Crown and lores red*, supercilium and sides of neck pale bluish. Bright green overall, auriculars brighter yellow-green, with dark scalloping on back and, less distinctly, on chest. Remiges blackish to bluish violet distally and with red patch on outer secondaries; outer rectrices broadly tipped pale yellowish green. ♀/**juv:** duller, red restricted to forecrown and lores, hindcrown pale bluish.

Voice. Loud and raucous. Usually a fairly mellow, rolling *rreeoo rreeoo* or *keer-yoo keer-you*, and slightly barking *rreh-rreh-rreh-rreh* or *rrak* ..., etc.; often combined in flight, *cleeoo cleeoo, ahk-ahk-ahk-ahk,* etc. Also a quieter *rreah rreah*, and *clee-ik,* etc.

Habs. Semideciduous forest and edge, semiopen country with scattered trees and forest patches, pine–oak woodland. In pairs or loose flocks of pairs which tend to fly high to and from roost sites.

SS: Red-lored Parrot larger with darker bill,

conspicuous yellow auriculars (lacking in juv), note voice.

SD/RA: Locally/seasonally F to C resident (near SL–1000 m) on Atlantic Slope from E NL and Tamps to N Ver. Decreasing with capture for pet trade and through habitat loss.

LILAC-CROWNED PARROT
*Amazona finschi**　　Plates 21, 23
Loro Corona-violeta

ID: 12–13.5 in (30.5–34.5 cm). *Endemic to W Mexico.* Relatively long-tailed. **Descr.** Ages/sexes similar. Eyes amber (brownish in juv), orbital ring grey, bill pale horn, feet pale greyish. *Forehead and lores deep red, crown and sides of neck lilac.* Bright green overall, extensively scalloped dusky on underparts. Remiges blackish to bluish violet distally with red patch on outer secondaries; outer rectrices broadly tipped pale yellowish green.

Voice. Varied and raucous, including a shrill *krih-krih* or *kreeih-kreeh*, a rolling *krreeeih*, a deeper *kyah'ha* ..., and an almost raven-like *krra krra*. Usually calls less frequently in flight than White-fronted Parrot.

Habs. Semideciduous forest and edge, pine–oak forest, mangroves, etc. Habits much as Red-crowned Parrot.

SS: See White-fronted Parrot; Yellow-headed Parrot (R in W Mexico) told by heavier build and shorter tail, yellow on head.

SD/RA: F to C resident (SL–2000 m) on Pacific Slope from S Son and SW Chih to Oax.

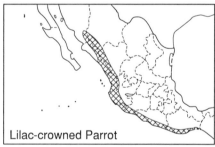
Lilac-crowned Parrot

RED-LORED (YELLOW-CHEEKED) PARROT
Amazona a. autumnalis　　Plates 21, 23
Loro Cachete-amarillo

ID: 12.5–14 in (32–35.5 cm). *Commonest large Amazon on Atlantic Slope.* **Descr.**

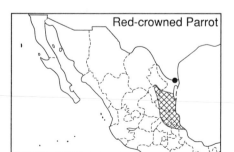
Red-crowned Parrot

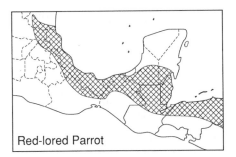

Red-lored Parrot

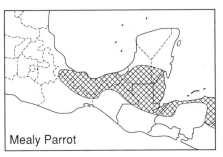

Mealy Parrot

Ages differ slightly, sexes similar. **Adult:** eyes amber, orbital ring pale grey, bill dark to horn with horn patch above, feet grey. *Lores and forehead red,* crown bluish, *anterior auriculars (cheeks) bright yellow.* Bright green overall with dark scalloping on nape and sides of neck. Remiges blackish to bluish violet distally with red patch on outer secondaries. Outer rectrices broadly tipped yellowish green. **Juv:** eyes brownish, *yellow on auriculars reduced or lacking.*
Voice. Raucous screaming, often with a fairly shrill, yapping quality *rriek-rriek, rriek-i-rrak,* and *zi-rreek* or *zee k-riek* 2–3×; also a quieter *ree-uk ree-uk,* and a mellow, slightly rolling *rieh terr-rr-riek,* etc.
Habs. Evergreen to semideciduous forest and edge, semiopen areas with scattered trees and forest patches. Habits much as Red-crowned Parrot.
SS: See Red-crowned Parrot. Mealy Parrot (rain forest) has broad whitish orbital ring, pale blue crown. Yellow-headed Parrot (semiopen country) usually has obvious yellow on head but Mexican imms and Belize adults have yellow mostly on auriculars when seen from below; note lack of red lores and forehead, voice.
SD: F to C resident (SL–750 m, to 1000 m in Honduras) on Atlantic Slope from S Tamps to Honduras.
RA: E Mexico to Brazil.

MEALY PARROT
*Amazona farinosa**   Plates 21, 23
Loro Verde

ID: 15–17 in (38–43 cm). *SE humid forest, decreasing with loss of habitat.* **Descr.** Ages differ slightly, sexes similar. **Adult:** eyes reddish, *broad orbital ring whitish,* bill blackish grey with pale horn patch above, feet grey. Bright green overall with

pale blue crown, fading to glaucous on nape. Remiges blackish to bluish violet distally with red patch on outer secondaries. Tail broadly tipped yellowish green. **Juv:** crown dull bluish. Adult *A. f. virenticeps* (E of Sula Valley) has greenish crown.
Voice. Loud and raucous but relatively mellow screaming, *churruk-churruk, chuk chuk cheeurr,* a deep, slightly gruff *rrehk-rrehk-rrehk* or *krreh-krreh,* and *rrah krrah-krrah ...,* etc. Often silent in flight.
Habs. Humid evergreen forest. Habits much as Red-crowned Parrot.
SS: See Red-lored Parrot. Juv Yellow-naped Parrot suggests Mealy but latter does not occur (except as escapes) on Pacific Slope; bold white orbital ring of Mealy distinctive.
SD: F to C resident (SL–500 m, to 1000 m in Honduras) on Atlantic Slope from S Ver to Honduras; mostly restricted to extensive tracts of forest and hence not seen readily in much of region.
RA: SE Mexico to Brazil.
NB: Sometimes known as Blue-crowned Parrot.

YELLOW-HEADED PARROT
*Amazona oratrix**   Plates 21, 23, 69, 70
Loro Cabeciamarillo

ID: 14–15 in (35.5–38 cm). *Large, formerly widespread Amazon, in danger of extinction due to capture for pet trade.* Complex plumage variation, three groups: *oratrix* group (Mexico), *belizensis* (Belize to NW Honduras), and '*hondurensis*' (N Honduras). See also SD and NB below. **Descr.** Ages differ, sexes similar. *Oratrix.* **Adult:** eyes amber, orbital ring whitish, bill and cere pale horn-flesh, feet greyish. *Head yellow.* Otherwise bright green overall with red (sometimes mixed with yellow) bend of wing, dark scalloping on sides of neck and

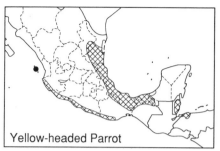

Yellow-headed Parrot

mantle, and yellow thighs. Remiges black-ish to bluish violet distally with red patch on outer secondaries. Outer rectrices broadly tipped yellowish green; red at base of tail usually concealed. *A. o. tresmariae* of Islas Tres Marías has yellow extending onto chest, underparts with glaucous cast, dark scalloping reduced. **Juv:** bill dusky, yellow restricted to crown and lores, red at bend of wing reduced or absent, thighs green. **Imm:** yellow restricted to crown, lores, anterior auriculars, and throat; attains adult plumage in 2–4 years. *Belizensis:* resembles imm *oratrix* group but orbital ring greyish white, yellow does not extend onto throat, nape may show a few yellow feathers; in Guatemalan and Honduran birds yellow usually more restricted, and may appear simply as a broad forecrown patch and broad eyering, nape often with yellow flecks and patches, cere pale horn-flesh to dusky. **Juv:** bill dusky, yellow restricted to forecrown; thighs green. **Imm:** in Belize, yellow restricted to a patch extending from crown to eyering, much like adults in Guatemala. In NW Honduras a complex transition occurs from *belizensis*-like birds to yellow-crowned and yellow-naped birds (the '*hondurensis*' group): adults have yellow crown (and sometimes a few yellow nape feathers) *or* yellow crown and full yellow nape. These birds have pale horn-flesh bills (rarely slightly dusky, imm?), pale to dusky ceres, red shoulders, etc., typical of other Yellow-headed Parrots; juvs of both types have dusky bills and yellow forecrowns (SLo).

Voice. Raucous but mellow, deep, rolled screams, often with human quality, including a rolled *kyaa-aa-aah* and *krra-aah-aa-ow*, a deep, rolled *ahrrrr* or *ahrhrrrr*, and *whoh-oh-ohr*, a throaty, rolled *rrohrr*, etc.; often silent in flight.

Habs. *Oratrix*: savannas and semiopen areas with scattered trees, gallery forest. *Belizensis*: pine savanna and adjacent evergreen forest in Belize, scrubby woodland and mangroves in NE Guatemala (JB). Sula Valley: unknown, possibly open scrubby areas (Monroe and Howell 1966). Singly, in pairs, or small groups, rarely in flocks.

SS: See Red-lored and Mealy parrots.

SD: *Oratrix*: R to U and local on Pacific Slope (around SL) from Jal (PH) to Oax; only confirmed reports in last 10 years from S Jal (KAR, TEW) and Mich (RB photo); population on Islas Tres Marías estimated at fewer than 800 birds (Konrad, unpublished). U to R and local (formerly C) resident (SL–900 m) on Atlantic Slope from E NL and Tamps to Tab and N Chis. *Belizensis*: U to F resident (around SL) in cen Belize, and locally in NE Guatemala (JB) and NW Honduras (SLo). '*Hondurensis*' R (formerly C) resident from coast of NW Honduras to Sula Valley.

RA: Mexico to Brazil (various populations).

NB1: The Yellow-headed Parrot complex is a taxonomic headache and from one to several species have been recognized. AOU (1983) considered Sula Valley yellow-crowned birds conspecific with *A. ochrocephala* (Yellow-crowned Parrot) of S C.A. based on apparent sympatry of yellow-crowned and yellow-naped birds in the Sula Valley. This scenario, however, is based on several erroneous and equivocal assumptions, for example that the yellow-naped birds were vagrant or escaped *parvipes* Yellow-naped Parrots (which have grey bills, etc.). There is a general cline in amounts of yellow on the head of 'Yellow-headed' parrots decreasing from Mexico to Belize to the Sula Valley, with a sharp step in N Honduras. The similarity of the Sula Valley yellow-napes and yellow-crowns (differing only in yellow nape patch) suggests that these could be color morphs. Indeed, the two forms interbreed readily in captivity (SLo). We consider it premature to assign specific status based on present knowledge; lumping Yellow-headed and Yellow-crowned parrots is an equally parsimonious conclusion. **NB2:** Yellow-naped Parrot *A. auropalliata* has been considered conspecific with Yellow-headed Parrot, and the Sula Valley yellow-napes have been placed with Yellow-naped Parrots although they are closer to Yellow-headed in most characters (pale bill, juv plumage, etc.). Note also that Bay Islands' Yellow-naped Parrots may have a relatively pale lower mandible (Lousada 1989), suggesting introgression of Yellow-headed characters, and that Bay Island and Mosquitia Yellow-naped Parrots have red shoulders, unlike the Pacific Slope nominate *auropalliata* but typical of the Yellow-headed group. It should be noted that Parkes' (1990*b*) critique of Lousada (1989) reflects some misunderstanding of this complex situation: *caribaea* may be a valid ssp and further study of *live* birds is needed. Monroe and Howell (1966) considered Yellow-headed and Yellow-naped parrots to be nearly sympatric in the Isthmus with no apparent intergradation. Specimens of the two forms about 100 km apart (i.e. from either side of the Isthmus) are a Yellow-headed from El Barrio (AMNH 44626) with far less yellow than typical adults (may be imm?), and a Yellow-naped from Santa Efigenia (AMNH 44624) with atypically extensive yellow on the forehead, and patches of yellow on the throat. This highly complex situation may never be resolved satisfactorily given the rarity to absence of free-flying birds in much of the region.

YELLOW-NAPED PARROT
*Amazona auropalliata** or *A. oratrix*
(in part)   Plates 21, 23, 70
Loro Nuquiamarillo

ID: 14–15 in (35.5–38 cm). Two groups: *auropalliata* (Pacific Slope S from

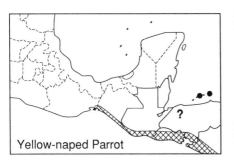

Yellow-naped Parrot

Isthmus), and *parvipes* group (?) (Bay Islands and Mosquitia). **Descr.** Ages differ, sexes similar. *Auropalliata*. **Adult:** eyes amber, orbital skin pale greyish, *bill grey*, cere dark, feet greyish. Bright green overall with *bright yellow nape*, and often some (usually slight) yellow on forehead. Remiges blackish to bluish violet distally with red patch on outer secondaries. Outer rectrices broadly tipped yellowish green; red at base of tail usually concealed. **Juv:** *lacks yellow on head* or may show a few yellow feathers on nape, *crown washed greenish blue*. *Parvipes*. **Adult:** resembles *auropalliata* but bill paler greyish (greyish horn below, at least in *caribaea* of Bay Islands), head and underparts have bluish cast, more extensive yellow forehead and nape, throat and sides of neck often mottled yellow, bend of wing red. **Juv:** *lacks yellow on head* or may show a few yellow feathers on forehead, lores, and nape, lacks red bend of wing.

Voice. *Auropalliata*: deep and raucous, though relatively mellow and rolling, often with human quality, including a gruff *rrowh* or *grrrowh*, repeated. Flight calls include a mellow, rolling *chrr-rrr uhrr-rr*, etc. *Parvipes (caribaea)*: low, far-carrying, raucous, rolled screams, *h-rrah h-rrah ...*, and *rr-aah rrowh, rr-aah rrowh, ...*, and *h' rrah rrah-rrah-rrah-rrah-rrah*, etc.

Habs. Arid to semiarid savannas with scattered trees, semiopen areas with scattered trees, gallery forest; *parvipes* group in pine woods and adjacent semideciduous forest. Habits much as Red-crowned Parrot.

SS: See Mealy Parrot.

SD: *Auropalliata*: U to F resident (SL–600 m) on Pacific Slope from E Oax to Nicaragua (but R in El Salvador, Thurber *et al.* 1987). *Parvipes*: formerly F to C resident in the Honduras Bay Islands; in 1987 estimated 50–75 birds on Roatán, 200–300 on Guanaja (SLo); in 1991 U on Roatan, F on Guanaja (SNGH); F in the Mosquitia.

RA: S Mexico to NW Costa Rica.

NB: See NB1 and NB2 under Yellow-headed Parrot.

CUCKOOS: CUCULIDAE (12)

A cosmopolitan family united by common anatomical features but strikingly diverse in appearance and habits. Most occur in forested and semiopen habitats and, although often common, some are among the hardest birds to see. Bills vary from slender and slightly decurved to deep and laterally flattened; feet zygodactyl. Most cuckoos are fairly slender and have long tails. Ages often differ slightly; young altricial, 1st prebasic molt usually partial. Sexes similar. Plumage varies from cryptically patterned browns and buffs to all black. Some species have erectile crests, and several have bold white tips to their outer rectrices which form striking patterns on the underside of their strongly graduated tails. Voices varied, songs usually series of whistles, clucks, or coos; several species mostly silent except in the breeding season.

Cuckoos feed on insects, reptiles, fruits, small mammals, eggs, and nestling birds. Nesting habits diverse; some are obligate or opportunistic brood parasites, others are loosely colonial.

GENUS COCCYZUS

These are sleek and slender arboreal cuckoos, common but often retiring and detected mostly by voice which is rarely heard other than on breeding grounds. Ages differ slightly, mainly in tail pattern. First prebasic molt partial and may be completed during migration; adult tail often attained in 1st winter. Legs grey to blue-grey. Nest is an insubstantial platform of twigs placed low in bushes to fairly high in tree forks. Eggs: 2–4, pale blue-green to bluish (WFVZ).

BLACK-BILLED CUCKOO

Coccyzus erythropthalmus   Not illustrated
Cuco Piquinegro

ID: 10.8–11.8 in (27.5–30 cm). Transient.
Descr. Adult: *orbital ring red, bill blackish,* often paler below at base. Head and upperparts grey-brown with darker auricular mask, dull rufous wash across primaries. *Throat and underparts dirty whitish,* dullest on throat, chest, and flanks. *Underside of tail grey, outer rectrices narrowly tipped whitish.* **Imm:** orbital ring greyish. Duller underparts may be washed greyish buff on throat and chest, rufous wash on primaries brighter, undertail greyish, pale tips dull and inconspicuous.

Habs. Forest and edge. Sluggish and easily overlooked. Flight swift and direct, can slip into a tree and vanish.
SS: Yellow-billed and Mangrove cuckoos have mostly yellow bills, bold white tail-spots; Yellow-billed brighter white below with bright rufous flash in wings; Mangrove buff below.
SD: U to F transient (Apr–early Jun, Aug–Nov; SL–3000 m), throughout most of region except NW Mexico; relatively few records due to inconspicuous habits.
RA: Breeds N.A.; winters in S.A.

YELLOW-BILLED CUCKOO

*Coccyzus americanus**   Not illustrated
Cuco Piquiamarillo

ID: 11–12 in (28–30.5 cm). Breeds locally, widespread transient. **Descr. Adult:** orbital ring yellow, *bill black above, orange-yellow below* with black tip. Head and upperparts grey-brown with darker auricular mask, *bright rufous flash across primaries.* *Throat and underparts white,* throat

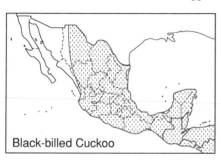

Black-billed Cuckoo

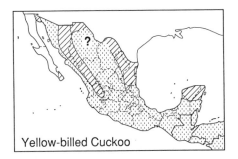

Yellow-billed Cuckoo

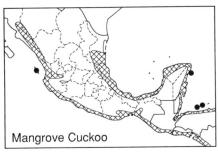

Mangrove Cuckoo

and chest may be washed greyish. *Underside of tail black, outer rectrices broadly tipped white.* **Juv:** may have blackish bill for up to 3 weeks after fledging; rufous flash on primaries brighter, *undertail pattern less contrasting,* grey and whitish.

Voice. Song (May–Aug) a dry, accelerating rattle which slows into a hollow clucking, *ka ka-ka-ka-kaka ... k'lop k'lop k'lop,* and a dry *ka'ow ka'ow ...* slowing into *cow cow cow cow.* Also a plaintive *kyow kyow ...,* and hard clucking *koh-koh-koh ...*

Habs. Much as Black-billed Cuckoo; breeds mostly in riparian woodland. Nest platform of sticks low to high in bush or tree. Breeding often tied to ephemeral abundance of tent caterpillars; a remarkably fast 18–21 days from incubation to fledging.

SS: See Black-billed Cuckoo; Mangrove Cuckoo has buff underparts, darker mask.

SD: F to C but local summer resident (May–Aug; SL–1500 m) in BCS (and Colorado delta, at least formerly), on Pacific Slope and in adjacent interior from Son and Chih locally (irregularly?) to Zac (*not* in NE Mexico as claimed by Banks 1988*b*), on Atlantic Slope from Coah to Tamps, and in N Yuc Pen; may breed elsewhere, at least irregularly; for example, Jul records from S Ver (Andrle 1967*b*), Guatemala (Wendelken and Martin 1986), and El Salvador (Thurber *et al.* 1987). F to C transient (Apr–mid-Jun, Aug–early Dec; SL–2500 m) on Atlantic Slope; U to F in interior and on Pacific Slope.

RA: Breeds N.A. to N Mexico and Caribbean; winters S.A.

MANGROVE CUCKOO
Coccyzus minor* Plate 24
Cuco Manglero

ID: 12–13 in (30.5–33 cm). Tropical lowlands. **Descr. Adult:** orbital ring yellow, bill black above, orange-yellow below with dark tip. Head and upperparts greyish with *contrasting dark auricular mask. Throat and underparts buff. Underside of tail black, outer rectrices broadly tipped white.* **Juv:** upperparts browner with indistinct mask, *undertail pattern less contrasting,* grey and whitish

Voice. Song (Apr–Aug) a hard, slightly rasping to nasal, steady series of 8–25 or more notes, *ahrk-ahrk ...* or *ahrr-ahrr ...,* often changing in quality part way through or ending with a distinct *ah-ahr.*

Habs. Forest and scrubby woodland, edge, mangroves. Habits much as Black-billed Cuckoo.

SS: See Yellow-billed and Black-billed cuckoos. Lesser Ground-Cuckoo terrestrial; note bold face pattern, cinnamon underparts, olive sheen to upperparts.

SD: F to C resident (SL–1200 m) on both slopes from S Son (SMR) and S Tamps to El Salvador and Honduras, locally in interior W Chis.

RA: Mexico and Caribbean (also S Florida) to N S.A.

DARK-BILLED CUCKOO
Coccyzus melacoryphus Not illustrated
Cuco Piquipardo No map

ID: 10–11 in (25.5–28 cm). Vagrant to Clipperton. **Descr.** Orbital ring greyish (?), *bill blackish.* Crown and nape grey with dark auricular mask, neck and sides of chest paler blue-grey. *Upperparts brown with no rufous in wings,* central rectrices olive-brown becoming black distally. Throat and underparts ochraceous buff. *Underside of tail black, outer rectrices boldly tipped white much like Mangrove Cuckoo.*

Habs. Silent and furtive much like other *Coccyzus* away from breeding grounds.

SS: See other *Coccyzus*.

SD: Vagrant, one record: a bird collected at Clipperton, 13 Aug 1958 (Stager 1964).

RA: S.A. from Colombia to N Peru, cen Argentina; partial austral migrant.

SQUIRREL CUCKOO
*Piaya cayana**   Plate 24
Cuco Ardilla

ID: 16–19.5 in (40.5–50 cm). W > E. Widespread and common tropical cuckoo. Two ssp: *mexicana* (W Mexico) and *thermophila* (E Mexico and C.A.). **Descr.** Ages differ slightly. *Mexicana:* eyes reddish, orbital ring and bill yellow-green, legs blue-grey. *Head and upperparts bright rufous,* tail tipped black. Throat and upper chest pale pinkish cinnamon, rest of underparts pale grey, darker on belly and flanks. *Underside of tail rufous, rectrices with broad black subterminal band and broad white tips.* *Thermophila:* darker overall, *underside of tail black, rectrices broadly tipped white.* **Juv** (both ssp): paler, central rectrices lack black tips above.

Voice. A sharp woodpecker-like *chik,* typically followed by a gruff snarl, *whrrr;* also *chik-chik,* a gruff, nasal *chek-e-rehr* or *sheh-k-ker,* a loud *ch'kaow,* and a rapid, chattering rattle punctuated with gruff *ch'kerr* calls. Song (mostly Apr–Aug) a steady, often prolonged series of sharp notes *pee'uk pee'uk ...* or *wheek wheek ...,* 10/4–6 s.

Habs. Humid to semiarid forest and edge, scrubby woodland. Fairly conspicuous, at mid- to upper levels of trees. Often seen crossing roads, undulating flight ends with a glide into cover. Nest platform of twigs at mid- to upper levels in bush or tree. Eggs: 2–3, white (WFVZ).

SD: C to F resident (SL–2000 m) on both slopes from S Son and S Tamps, and in interior from Balsas drainage, to El Salvador and Honduras.

RA: Mexico to Peru and N Argentina.

STRIPED CUCKOO
Tapera naevia excellens   Plate 24
Cuco Rayado

ID: 11–12 in (28–30.5 cm). S tropical lowlands. Erectile crest raised and lowered while singing. **Descr.** Ages differ. **Adult:** eyes pale brownish, orbital ring yellow-green, bill greyish horn, paler below, legs greyish. *Face buff with broad dark eye-stripe and dark malar stripe, crown and nape streaked rufous and black.* Throat and underparts pale buff, becoming ochraceous buff on hind flanks and undertail coverts. *Upperparts cinnamon to buff, back boldly striped blackish,* narrow black shaft streaks on rump and uppertail coverts. Remiges greyish; in flight, broad whitish stripe across base of primaries recalls Brown Creeper. Tail greyish, rectrices broadly edged cinnamon. **Juv:** crown blackish, spotted cinnamon, throat and chest faintly barred dusky, upperparts spotted cinnamon.

Voice. Song (Mar–Aug) distinctive and far-carrying, a clear, deliberate whistle, easily imitated, usually 2 notes, *whee whee,* the 2nd note higher, also longer variations with same quality *whee' whee' whee' whee'buh,* or *whee whee whee-boo whee-boo,* etc.

Habs. Dense second growth, scrubby woodland and edge, often near water. Rarely seen unless singing when often perches conspicuously on wires, high in trees, etc. Otherwise keeps low in cover. Brood parasite, typically of birds with domed or covered nests; eggs whitish (WFVZ).

SS: Pheasant Cuckoo larger and darker, tail longer and broader, crown and upperparts

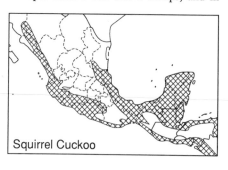

Squirrel Cuckoo

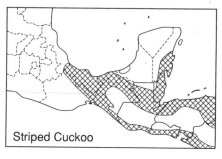

Striped Cuckoo

lack bold striping, chest with dark spotting.

SD: F to C resident (SL–1250 m) on Atlantic Slope from S Ver to Honduras, on Pacific Slope and in interior from E Oax and Chis to El Salvador and Honduras; R (resident?) N to cen QR (Lopez O. *et al.* 1989).

RA: S Mexico to Ecuador and N Argentina.

PHEASANT CUCKOO
Dromococcyx phasianellus rufigularis
Cuco Faisán   Plate 24

ID: 14–15 in (35.5–40.5 cm). Skulking and terrestrial. Tail long, broad, and strongly graduated, with long uppertail coverts. **Descr.** Ages differ. **Adult:** eyes amber, orbital ring greenish yellow, bill dark grey, paler below at base, legs greyish flesh. *Whitish supercilium contrasts with chestnut crown and nape, and broad dark brown eyestripe.* Throat and underparts whitish, *sides of neck and chest spotted dark brown.* Upperparts dark brown, scapulars and wing coverts edged buff to white, whitish bar across base of primaries visible in flight. Undertail greyish, rectrices tipped white. **Juv:** crown, nape, and upperparts darker, sparsely spotted buff, throat and chest rich buff, unmarked.
Voice. Song (Mar–Aug) a distinctive far-carrying whistle, typically *whee whee wheerr-rr,* 1st 2 notes much like Striped Cuckoo, 3rd with tinamou-like quaver; also longer, varied series accelerating toward the end, *whee-whee whee whee-whee-whee-whee,* or *whu hu hoo'hoo-hoo-hoo-hee* etc., with similar quality. Calls include a quiet growling *grrr.*
Habs. Forest and woodland, plantations with dense undergrowth. Unlikely to be detected unless heard, sings from mid- to upper levels, often conspicuously. Otherwise terrestrial and skulking, walks quietly, tail often wagged slowly; runs with much

flapping when disturbed. Brood parasite, mostly of domed or pendulous nests; only host report in Mexico is Yellow-olive Flycatcher (Wilson 1992); eggs whitish to pale buff, flecked rufous (WFVZ).

SS: See Striped Cuckoo.

SD: F resident (SL–1500 m) on Pacific Slope from Gro (SNGH, SW, tape) to El Salvador and Honduras, on Atlantic Slope from E Pue (Warner and Beer 1957) and N Oax, and in interior from Chis, to N cen Nicaragua.

RA: S Mexico to N Argentina.

LESSER GROUND-CUCKOO
*Morococcyx erythropygus**   Plate 24
Cuco-terrestre Menor

ID: 10–11 in (25.5–28 cm). Arid Pacific Slope. **Descr.** Ages differ slightly. **Adult:** *yellow orbital ring merges with pale blue teardrop behind eye, bill orange-yellow with dark culmen,* legs greyish. *Black auricular mask offsets bright orbital ring,* white postocular stripe indistinct. Crown, nape, and upperparts grey-brown, *flight feathers with strong greenish sheen,* rump and uppertail coverts (usually covered by wings) blackish, edged cinnamon. *Throat and underparts cinnamon*; underside of tail greyish, outer rectrices blackish subterminally, tipped white. **Juv:** duller, underside of tail greyish, outer rectrices indistinctly tipped buff.
Voice. Song loud, rich, rolled whistles, suggesting a referee's whistle: slow-paced single notes build into a rapid series then fade off, becoming slower and slower, *prree, prree, prree, prree-prreeprree … prree, prree, prree, prree, prree*; at times stops after 2–3 introductory notes; given through most of year. Call a burry whistle, *whirrrr,* often soft.
Habs. Arid to semiarid woodland, thorn forest, semiopen areas with brushy scrub and forest patches. Terrestrial. Walks

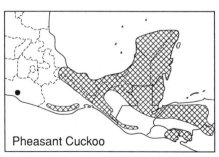

Pheasant Cuckoo

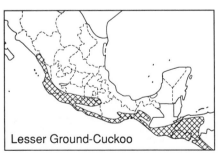

Lesser Ground-Cuckoo

slowly and deliberately, may freeze if alarmed and becomes hard to see; flushes with short low flight to cover. Nest is a shallow cup of twigs and dead leaves on ground under cover. Eggs: 1–3, whitish (WFVZ).

SS: See Mangrove Cuckoo (arboreal).

SD: C to F resident (SL–1500 m) on Pacific Slope from S Sin, and in interior locally from Balsas drainage, to El Salvador and Honduras.

RA: Mexico to NW Costa Rica.

LESSER ROADRUNNER

Geococcyx velox∗   Plate 24
Correcaminos Menor

ID: 16–20 in (40.5–50.5 cm). *Endemic; W Mexico, Yuc, and C.A.* Erectile crest often raised. **Descr.** Ages differ slightly. **Adult:** eyes brown, pale blue orbital ring extends back as blue or lilac band, becoming red on nape (red usually hidden), bill and legs greyish. *Face, throat, and underparts buff, streaked blackish on face, sides of neck, and chest.* Blackish-brown *crown spotted whitish. Upperparts dark rufous-brown,* coarsely striped pale buff, wings and upper-tail with strong green sheen, bold white spots across outer primaries visible in flight. Tail blackish below, outer rectrices boldly tipped white. **Juv:** bare facial skin greyish.

Voice. A slowing, descending series of 3–7 low, moaning, somewhat dove-like coos, *oooah* . . . or *owoah* . . .

Habs. Arid to semiarid brushy wood-land, semiopen areas, ranges into pine–oak zone. Runs strongly and quickly, cocks tail and droops wings; mainly ter-restrial but often perches in bushes or low trees. Nest is a cup of twigs and leaves in bushes. Eggs: 2–3, whitish (WFVZ).

SS: Greater Roadrunner larger, bill longer

and heavier, underparts whitish, throat and chest heavily striped blackish.

SD/RA: C to F resident (SL–3000 m) on Pacific Slope from S Son to Isthmus, and in interior from Balsas drainage to N cen Nicaragua; disjunctly in NW Yuc Pen.

GREATER ROADRUNNER

Geococcyx californianus   Plate 24
Correcaminos Mayor

ID: 20–24 in (50.5–61 cm). N and cen Mex-ico. Erectile crest often raised; bill long and heavy. **Descr.** Ages differ slightly. **Adult:** eyes pale brownish, pale blue orbital ring extends back as blue or lilac band, becom-ing red on nape (red usually hidden), bill and legs greyish. *Face, throat, and under-parts whitish, sides of throat and chest striped blackish and cinnamon.* Blackish-brown *crown spotted cinnamon. Upper-parts dark grey-brown,* coarsely striped pale buff, wings and uppertail with strong green sheen, bold white spots across outer primaries visible in flight. Tail blackish below, outer rectrices boldly tipped white. **Juv:** bare facial skin greyish.

Voice. Much like Lesser Roadrunner, a slowing series of 5–8 low, hollow, moan-ing coos, *oowh oowh* . . . or *ooah ooah* . . . Also a bill-snapping rattle, *br-rr-rrp.*

Habs. Open and semiopen arid areas, brushy woodland. Habits much as Lesser Roadrunner. Nest is a cup of twigs and leaves in bushes. Eggs: 2–6, chalky white (WFVZ).

SS: See Lesser Roadrunner.

SD: C to F resident in Baja, on Pacific Slope S to N Sin, in interior S over Plateau to N Mich, N Mex, and Hgo, and on Atlantic Slope in Tamps. Reports from DF (AOU 1983) probably are in error (Wilson and Ceballos-L. 1986).

RA: W USA to cen Mexico.

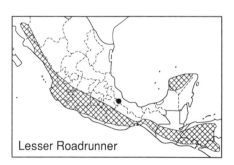
Lesser Roadrunner

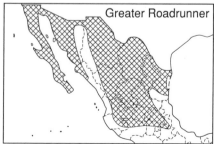
Greater Roadrunner

SMOOTH-BILLED ANI
Crotophaga ani   Plate 24
Garrapatero Piquiliso

ID: 13–14.5 in (33–37 cm). *Caribbean islands. Bill smooth, deep, and laterally flattened, culmen with raised bump near base.* **Descr.** Ages/sexes similar. Bill, bare facial skin, and legs blackish. Black overall with slight iridescence, paler scalloping on chest and back. Juv duller.

    Voice. A slurred *ree-o-rink* or *roohr-ihnk*, usually 1–4×, a mewing, almost hawk-like *reeeahh* or *reeeow*, and a slurred, mewing *reeeh yuh* or *roohr-ehw*, etc.

    Habs. Humid second growth, overgrown fields and roadsides, often near livestock. Associates with Groove-billed Ani. In pairs or small groups, clambers and flutters about in bushes, hops readily on ground, tail cocked loosely and swung about. Flight appears inept but stronger than Groove-billed Ani; rapid flaps interspersed with flat-winged glides, often crash-lands into bushes. Nests singly or communally, when 2–4 or more pairs help build a nest, in which each ♀ may lay eggs, and raise young. Nest is a bulky cup of twigs lined with green leaves, placed in dense tangles. Eggs: 4–15 or more (single ♀♀ may lay 3–5), chalky white (WFVZ).

SS: Groove-billed Ani smaller, culmen lacks bump at base, grooves on bill absent in juv; note voice.

SD: Irregular U to F resident (around SL) on Isla Cozumel; R on adjacent mainland of QR (SNGH). Probably also an irregular resident on Belize Cays, including Ambergris (SNGH), but few records. Occurrence and abundance in these areas may be linked to periodic invasions from Caribbean. F resident on Honduras Bay Islands. Reports from Isla Holbox (AOU 1983) are based on Gaumer spec. and require verification.

RA: Caribbean region and Costa Rica to Ecuador and N Argentina.

GROOVE-BILLED ANI
*Crotophaga sulcirostris**   Plate 24
Garrapatero Pijuy

ID: 12–13.5 in (30.5–34.5 cm). Common and widespread in tropical lowlands. **Descr.** Ages differ slightly. *Bill deep and laterally flattened, with 2–3 distinct grooves above and below.* **Adult:** bill, bare facial skin, and legs blackish. Plumage black overall with slight iridescence, paler scalloping on chest and back. **Juv:** duller, bill initially smooth.

    Voice. A squeaky *pi'chip* or *chi-weerp*, and varied piping and squealing noises including *chi-weeu* ..., a growling *k'rrr k'rrr* ..., and a gruff *chuh-chuh* ..., repeated up to 4–6×. Rarely heard song a rapid, mellow clucking *whiu-whiu* ... or *whiuk-whiuk* ..., several s in duration.

    Habs. Second growth, fields, roadsides, forest edge, prefers humid or wet areas; often near livestock. Habits, nesting, and eggs (WFVZ) much like Smooth-billed Ani.

SS: See Smooth-billed Ani (Caribbean islands); other black birds have 'normal' bills.

SD: C to F resident (SL–1800 m) on both slopes from S Son and N Coah to El Salvador and Honduras, also hotter interior regions from cen Mexico S; formerly S BCS (recent vagrant Nov 1985, Howell and Webb 1992f). U to F resident on Isla Cozumel, Isla Holbox, and Ambergris Cay (SNGH), occurrence may be tied to periodic invasions from mainland. A report from Half Moon Cay (Russell 1964) requires verification.

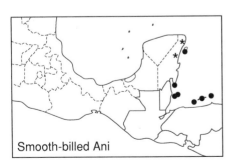

Smooth-billed Ani

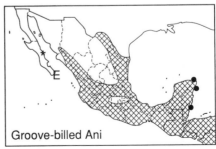

Groove-billed Ani

RA: Mexico (and S Texas) to N Chile and N Argentina.

NB: Greater Ani *Crotophaga major*, a S.A. species, has been reported from S Tamps (see Appendix B).

BARN OWL AND GRASS-OWLS: TYTONIDAE (1)

The region's single species is a cosmopolitan member of a family whose center of diversity lies in Australasia. These owls differ from other owls in their heart-shaped facial disks, relatively small eyes, short squared tails, and pectinate central claws. Bills and talons sharply hooked. Legs long and feathered, toes naked. Wings long and rounded; outer web of outermost primary has 'saw-tooth' edge to aid in silent flight. Ages/sexes similar but ♀♀ darker than ♂♂ in a given ssp; downy young molts directly into adult-like plumage. White and golden buff overall. Voice includes shrieks and hisses, does not hoot like many owls.

Barn Owls mostly eat rodents, also some birds. Nests (and roosts during the day) in old buildings, caves, hollow trees, etc. Eggs: 3–7, whitish, unmarked (WFVZ).

BARN OWL

Tyto alba*　　　　　　　Not illustrated
Lechuza de Campanario

ID: 14–16 in, W 36.5–43.5 in (35.5–40.5 cm, W 93–110 cm). *Widespread.* See family account. **Descr.** *Eyes dark brown*, bill horn, feet flesh. *Facial disks whitish to white with darker rim. Underparts whitish (♂) to buff (♀) with sparse dark spotting.* Head and upperparts mottled silvery grey and golden buff, sparsely flecked dark and white, flight feathers with dark barring. *Underwings whitish overall*, remiges with dusky bars.

　Voice. A piercing, hissing screech, often in flight; also a drawn-out, rasping screech.

　Habs. Open and semiopen country with scattered trees and forest patches, old buildings, towns. Hunts by quartering grassy areas, wingbeats shallow and easy, appearing to drift ghost-like at times; often hovers.

SS: Striped and Short-eared owls have longer wings, the latter has more floppy and buoyant flight; both lack distinctive heart-shaped facial disks and have blackish-striped underparts, ear tufts often flattened.

SD: F to C resident (SL–3000 m) throughout most of region, commonest in lowlands; may be declining with increased use of pesticides in some areas. Vagrant (?) to Isla Socorro (Apr–May 1991, HG).

RA: Virtually cosmopolitan, including many islands.

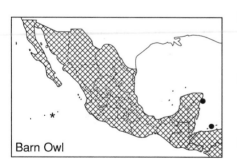

Barn Owl

TYPICAL OWLS: STRIGIDAE (30)

A familiar yet primarily nocturnal family of predatory birds, one or more of which occur in most habitats throughout the region. Owls' large forward-facing eyes are set in distinct facial disks and give them binocular vision. Their hearing is acute and some can detect and locate prey solely by sound. Bills and talons typically strong and sharply hooked but relatively weak in smaller insectivorous species. Wings rounded and fairly long; outer web of outermost primary usually has a 'saw-tooth' edge which aids in silent flight. Tails fairly short except in pygmy-owls. Several species have erectile tufts of feathers (ear tufts) at sides of the crown. Ages differ, though soft downy juv plumage is soon replaced by adult-like head and body plumage in variable 1st prebasic molt. Sexes similar though ♀ larger; several species dimorphic, ♀♀ more commonly exhibiting a reddish morph. Most species are patterned, often cryptically, in shades of brown, though facial patterns may be quite striking.

Voices vary and are often the best criteria for distinguishing otherwise similar species. Song is best heard in late winter (Dec–Mar) as many owls nest early in the year and they tend to be inconspicuous when nesting. The territorial 'song' is species-specific and the most commonly heard vocalization; it is described in the species accounts. Most owls use many different calls but, unless these are likely to be heard often, they are not noted below; consequently, 'owl noises' will likely be heard that cannot be identified and may, in some cases, be undescribed! Screech-Owls in particular have two primary songs, one used to define territories, the other in dueting.

Owls feed on mammals, birds, insects, reptiles, etc. Nest in tree cavities, old nests of other birds, on ground, etc. Eggs: 2–7, white and rounded.

Note: Although it is always nice to see owls, observers (if they care for the birds) should be content to identify and count owls by voice alone which, in many cases, is *the* critical field mark. Tape-recorded playbacks are all too often used (or abused) by people who simply wish to see another species, regardless of the undue stress caused to the birds.

GENUS OTUS: Screech-Owls

Screech-Owls are small to medium-sized, cryptically patterned owls with ear tufts; usually common in forests and woodland but strictly nocturnal. Eyes yellow to brown but night-time 'eyeshine' amber to reddish. Plumage of the group as a whole is remarkably similar: all are beautifully patterned above in soft browns (including rufous) and greys, with variable dark streaking, the dark mark-ings reduced or even absent in rufous morphs (except Flammulated); whitish-tipped scapulars and outer secondary coverts typically form contrasting lines of spots. The facial disks are generally rimmed with black and the sides of the forehead (brows) and inner edges of the ear tufts may be contrastingly whitish; at times the whitish encircles the crown in a coronal band. Underparts are whitish with dark shaft streaks and fine cross-bars. As far as is known, all juvs are whitish to buffy white overall, barred dusky to rufous (see Fig. 26). Given the remarkable similarity in overall appearance of screech-owls, the foregoing applies to all Mexican *Otus* and is not repeated in the species accounts, unless stated otherwise. Screech-Owls are best identified by voice, combined with habitat and range. They roost and nest in tree cavities, and also roost against trunks. Eggs: 2–4.

FLAMMULATED OWL
Otus flammeolus*   Not illustrated
Tecolote Flameado

ID: 6–6.7 in (15–17 cm). *Small migratory owl of mountain forests.* Ear tufts relatively short. *Long wings project slightly beyond tail at rest; feet small and weak, toes naked.* **Descr.** See genus note. *Eyes dark brown,* bill blackish. Plumage varies from overall grey mixed with rufous to mainly rufous, blackish markings distinct in all plumages; *facial disks always show rufous,* at least near eyes; *scapular spots washed cinnamon.*
   Voice. Song (Apr–Jun) a low, soft *hu* or *hoo,* repeated, 10/20–36 s; also a distinctive series *hu, h-hu hu', h-hu hu', h-hu hu'.* Ventriloquial, song does not carry far. Silent and little known most of year.
   Habs. Open pine and pine–oak forest, especially with Ponderosa Pine (*Pinus ponderosa*). Calls from close to the trunk at mid- to upper levels in pines. Feeds on insects and hunts at low to mid-levels, often near ground.
   SS: Small size, dark eyes, and voice distinctive; Whiskered and Bearded are normally the only two screech-owls in Flammulated's range and habitat.
   SD: Poorly known. F to C but local breeder (Apr–Aug; 1500–3000 m) from E Son, Chih, and Coah to Oax (Binford 1989; JCA); winters (1500–3000 m), probably F to C, from Sin and cen volcanic belt to

Guatemala, possibly to El Salvador (AOU 1983).
   RA: Breeds W N.A. to Mexico; winters W Mexico to Guatemala.

EASTERN SCREECH-OWL
Otus asio mccallii   Not illustrated
Tecolote Oriental

ID: 8–9 in (20.5–23 cm). *NE Mexico.* Dimorphic. Feet relatively large, toes bristled. **Descr.** Eyes yellow, *bill greenish.* A typical *Otus* as described in genus note; see voice and SS below.
   Voice. A quavering, toad-like trill of 3–4.5 s in duration. Mexican birds apparently lack the horse-like whinny typical in NE USA (Marshall 1967).
   Habs. Open deciduous and riparian woodland. Hunts from branch to branch, catching prey in the foliage and on the ground.
   SS: Western Screech-Owl (red morph rare) told safely by voice, also note blackish bill; Whiskered Screech-Owl (usually higher, in pine–oak) best told by voice, smaller feet.
   SD: F to C resident (near SL–1500 m) on Atlantic Slope from Coah to SE SLP and S Tamps.
   RA: E N.A. to NE Mexico.
   NB: Traditionally considered conspecific with Western Screech-Owl.

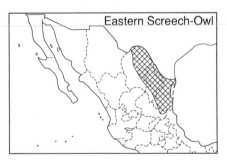
Eastern Screech-Owl

WESTERN SCREECH-OWL
Otus kennicottii*   Figure 26
Tecolote Occidental

ID: 8.5–9.5 in (21.5–24 cm). *N and cen Mexico.* Dimorphic, red morph rare. *Feet relatively large,* toes bristled. **Descr.** Eyes yellow, *bill blackish.* A typical *Otus* as described in genus note; see voice and SS below.
   Voice. Song a bouncing-ball series, typically of 5–9 hoots, *hoo-hoo hoo-oo-*

Flammulated Owl

Fig. 26 Juvenile owls. (a) Western Screech-Owl; (b) Mottled Owl; (c) Black-and-white Owl; (d) Crested Owl; (e) Spectacled Owl; (f) Fulvous Owl

oo...; also (BCS at least) a longer 13–15-note song of short hoots which rises in pitch and falls back to starting pitch (Marshall 1967), and (California) excited-sounding, fairly rapid, rolled whinnies with a slight bouncing-ball quality.

Habs. Arid to semihumid woodland (including pine–oak), semiopen areas with scattered trees and riparian woods, cardon cacti. Habits much as Eastern Screech-Owl.

SS: See Eastern Screech-Owl; Whiskered Screech-Owl best told by voice but note smaller feet, coarser black markings on underparts at comparable latitude. Not known to overlap with Balsas Screech-Owl but ranges come close in Jal; note habitat preferences, voice, brown eyes, and larger size of Balsas.

SD: F to C resident (SL–2500 m) in Baja, on Pacific Slope S to N Sin, and in interior S to cen volcanic belt.

RA: W N.A. to cen Mexico.

NB1: See Eastern Screech-Owl. **NB2:** Balsas and Pacific screech-owls are incipient species with very close affinities to Western; it may be premature to consider all three as separate species (see Marshall 1967).

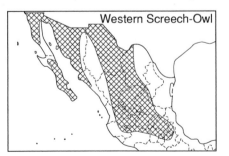

Western Screech-Owl

BALSAS [WESTERN] SCREECH-OWL

Otus seductus or *O. kennicottii* (in part)
Tecolote del Balsas   Plate 25

ID: 9.5–10.5 in (24–26.5 cm). Large screech-owl, *endemic to SW Mexico*. No red morph known. Feet large and bristled. **Descr.** *Eyes brown* (rarely golden brown), bill greenish. A typical *Otus* as described in genus note; see voice and SS below.

Voice. Bouncing-ball song recalls Western

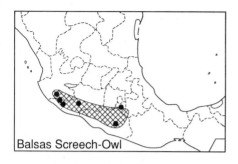

Balsas Screech-Owl

Screech-Owl but gruffer, *hooh-hooh hooh-hooh-hooh-hooh-hooh-hooh-hooh*, etc. Also a gruff, screaming whinny.

Habs. Thorn forest, arid open to semi-open areas with scattered trees. Habits much as Eastern Screech-Owl. Nest and eggs undescribed (?).

SS: Large size, brown eyes, and thorn forest habitat distinctive; smaller Vermiculated Screech-Owl occurs in dense tropical forest, eyes yellow, toes long and naked; note voice.

SD/RA: F to C resident (600–1500 m) in interior from S Jal (SNGH, SW) and Col to Mor (SNGH) and cen Gro.

NB: See NB2 under Western Screech-Owl.

PACIFIC [WESTERN] SCREECH-OWL
Otus cooperi* or *O. kennicottii* (in part)
Tecolote de Cooper   Plate 25

ID: 9–10 in (23–25.5 cm). Large screech-owl, *Pacific Slope from Isthmus S*. No red morph known. Feet large and bristled. **Descr.** Eyes yellow, bill greenish. A typical *Otus* as described in genus note, though underpart markings relatively fine; see voice and SS below.
  Voice. Gruff, bouncing-ball song of 5–15 notes recalls Balsas Screech-Owl or even Mottled Owl, *wup-wup wup-wup-wup-*

wup … or *wuh-wuh* … Also a single, low, gruff *woof* or *whuh'*.
  Habs. Arid to semiarid woodland, semi-open areas with scattered trees and cardon cacti, mangrove edge. Habits much as Eastern Screech-Owl.

SS: Not known to overlap with other screech-owls; Vermiculated prefers denser habitat; note long naked toes, voice.

SD: F to C resident (SL–1000 m) on Pacific Slope from Oax to El Salvador and Honduras.

RA: S Mexico to NW Costa Rica.

NB: See NB2 under Western Screech-Owl.

WHISKERED SCREECH-OWL
Otus trichopsis*   Plate 25
Tecolote Bigotudo

ID: 6.5–7.5 in (16.5–19 cm). *Fairly small screech-owl of pine–oak forests throughout*. Dimorphic, red morph mainly from cen Mexico S. *Feet relatively small and bristled*. **Descr.** Eyes yellow, bill greyish. A typical *Otus* as described in genus note but underpart markings relatively coarse, especially in S ssp; see voice and SS below.
  Voice. Song a rising then falling series of 4–10 hoots, *hoo-hoo-hoo-hoo hoo-hoo*; also a dueting, Morse-code-like series, *hoo-hoo hoo-t-hoo hoo-hoo hoo-t-hoo hoo-hoo* or *hoo hoo-hoo-hoo-hu' hoo-hoo-hoo-hu'* …, etc.; at times simply a 2-note repetition *hoo-hoo' hoo-hoo'* …
  Habs. Oak and pine–oak forest. Habits much as Eastern Screech-Owl.

SS: See Flammulated Owl and Western Screech-Owl; Vermiculated Screech-Owl overlaps locally in S Mexico and C.A.; told by naked toes, finer and less distinct dark streaking on underparts, and voice. Bearded Screech-Owl (S of Isthmus) has more contrasting whitish brows, more scalloped underparts (versus streaked in

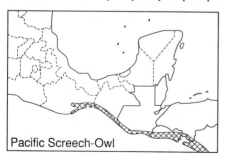

Pacific Screech-Owl

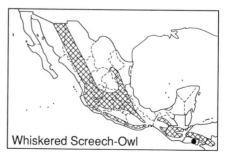

Whiskered Screech-Owl

Whiskered), naked pink toes; note voice.

SD/RA: C to F resident (750–2500 m) on both slopes from Son (also SE Arizona) and NL, and in interior from cen Mexico, to N cen Nicaragua.

NB: Sometimes known as Spotted Screech-Owl.

VERMICULATED SCREECH-OWL

*Otus guatemalae**　　　　　　　Plate 25
Tecolote Vermiculado

ID: 8–9 in (20.5–23 cm). *Tropical forest.* Dimorphic. *Naked toes* dusky flesh. **Descr.** Eyes yellow, bill greenish. A typical *Otus* as described in genus note, but *plumage relatively dark, underparts with fine dark streaks, less contrasting* than other *Otus;* note whitish brows.

　　Voice. Song a purring, toad-like trill of 2.5–19 s (usually 6–15 s) in duration; much like Lesser Nighthawk but begins softly, strengthens, and ends abruptly: *urrrrrrrr* ... Carries well but can be deceptively soft at close range. Also (S.A. at least) a rapid, rolled series of about 10 hoots, run into a purring, rolled trill, ending abruptly (RAB tape).

　　Habs. Humid to semiarid evergreen to semideciduous forest, dense scrubby woodland. Habits much as Eastern Screech-Owl; often hard to observe in dense habitat it prefers but may be seen making insect-catching flights from forest edge.

SS: See Pacific and Whiskered screech-owls.

SD: F to C resident (SL–1500 m) on Pacific Slope from S Son to Oax, on Atlantic Slope from S Tamps to Honduras and N cen Nicaragua. Should be sought on Pacific Slope S of Isthmus.

RA: Mexico to S.A. (see NB).

NB: Birds from Costa Rica to NW S.A. sometimes considered specifically distinct, *O. vermiculatus* (Vermiculated Screech-

Owl); other forms in this complex being united as *O. atricapillus* (Variable Screech-Owl), ranging from Mexico to Peru and S Argentina (Marshall *et al.* 1991).

BEARDED (BRIDLED) SCREECH-OWL

Otus barbarus　　　　　　　Plate 25
Tecolote Barbudo

ID: 6.5–7.5 in (16.5–19 cm). *Endemic to highlands S of Isthmus.* Dimorphic. Wings project beyond tail, ear tufts fairly short, *toes naked.* **Descr.** See genus note. Eyes yellow, bill greenish, *feet bright pinkish.* Whitish brows and coronal band well marked, underparts appear coarsely scalloped.

　　Voice. Song a quiet, low, cricket-like trill of 3–5 s in duration, rising throughout or at times dropping abruptly at end, repeated after a short pause. Quieter than purring, toad-like trill of Vermiculated Screech-Owl. Also a soft ventriloquial *hu.* **Habs.** Humid pine–oak forest. Habits little known. Calls from mid- to upper levels and can be elusive. Nest and eggs undescribed (?).

SS: See Whiskered Screech-Owl.

SD/RA: F to C but local resident (1800–2500 m) in N Chis and Guatemala.

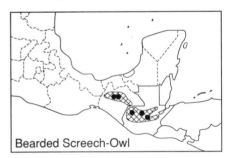

Bearded Screech-Owl

CRESTED OWL

Lophostrix cristata stricklandi　　Plate 26
Búho Corniblanco　　　　　　Figure 26

ID: 15–17 in (38–43 cm). SE humid forest. Dimorphic. **Descr.** Ages differ, sexes similar. **Adult:** eyes yellow, pale horn bill dark at sides, feet dull yellowish. *White brows lead into very long white ear tufts* (often held flattened), *facial disks tawny to chestnut,* rimmed black. Dark morph dark rich brown overall, paler tawny-brown below, with faint paler vermiculations. Light

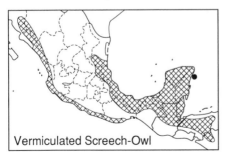

Vermiculated Screech-Owl

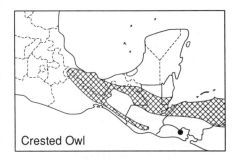

Crested Owl

morph overall cinnamon-brown above, greyish buff below with more distinct dark vermiculations, outer scapulars and upperwing coverts boldly tipped whitish. **Juv:** whitish overall with dark facial disks, short whitish ear tufts.

Voice. A deep, throaty, slightly frog-like, emphatic growl, *ohrrrr* or *gurrrr*, at close range a rapid, stuttering introduction may be audible *g'g'g'g'grrrr*, repeated every 5–10 s.

Habs. Humid evergreen forest. Roosts at mid- to upper levels, probably hunts mostly in forest but also at edges. Nests in tree cavities, eggs undescribed (?).

SD: U to F resident (near SL–1800 m) on Atlantic Slope from S Ver to Honduras; on Pacific Slope locally from E Oax to El Salvador.

RA: SE Mexico to Brazil.

SPECTACLED OWL
Pulsatrix perspicillata saturata   Plate 26
Búho de Anteojos   Figure 26

ID: 17–19 in (43–48 cm). SE humid forest, toes feathered. **Descr.** Ages differ, sexes similar. **Adult:** eyes yellow, bill horn. *Head and upper chest dark brown with white brows and broad white forecollar.* Upperparts dark brown, flight feathers indistinctly barred paler. Underparts buff, occasionally

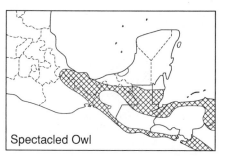

Spectacled Owl

with sparse dusky bars at sides. **Juv:** *whitish overall with blackish facial disks.*

Voice. A series of deep, resonant hoots, usually 10–20, accelerating and then fading quickly, *wuup-wuup wuup-wuup-wuup-wuup* . . ., recalls a sheet of tin being flexed quickly. Also a low, deep *whooo*, suggesting a large pigeon.

Habs. Humid evergreen forest and edge, shady plantations. Habits much as Crested Owl. Nests in tree cavities; 2 eggs.

SS: Variation of Mottled Owl hooting similar but less resonant, not as fast and less 'woofing'.

SD: F to U resident (SL–700 m) on Atlantic Slope from S Ver to Honduras; on Pacific Slope from E Oax to El Salvador where now very local due to forest destruction.

RA: SE Mexico to Ecuador and N Argentina.

GREAT HORNED OWL
*Bubo virginianus**   Plate 26
Búho Cornudo

ID: 19–22 in, W 41–49 in (48.5–53.5 cm, W 105–124 cm). *Widespread. Ear tufts thick and bushy,* toes feathered. **Descr.** Ages differ, sexes similar. **Adult:** eyes yellow, bill greyish. Facial disks grey-brown to cinnamon, rimmed black, *white foreneck often conspicuous and ruff-like.* Head and upperparts grey-brown, cryptically patterned (flight feathers barred) with buff, whitish, dark brown, and black. Underparts pale buff, *upper chest coarsely mottled dark brown, rest of underparts barred dark brown.* **Juv:** barred pale buff and dusky overall. Imm has coarse dark blotching on underparts before adult-like barring appears.

Voice. A varied series of deep, low hoots, usually 5–8, *hoo, hoo-hoo-hoo hoo-hoo,* and *hoo-hoo-hooo hoo-hoo* or *hoo h-hoo, hoo, hoo,* less often a single deep

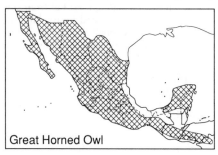

Great Horned Owl

hoo, repeated at irregular intervals; ♀ also has a wailing shriek, *rreieh*, and begging juv has a rasping shriek, often repeated insistently.

Habs. Widespread except in humid evergreen forest, often in semiopen areas, ranging from arid deserts to humid semideciduous forest and edge; mainly in temperate habitats from cen Mexico S. May be seen at dusk perching conspicuously on telephone wires and poles, etc. Nests in old crow or hawk nests, tree cavities, caves, etc.; 2–3 eggs.

SS: Other 'eared' owls are smaller, ear tufts longer and set closer together, barred underparts of Great Horned distinctive; Stygian Owl usually darker overall with dark facial disks; Long-eared Owl has black eye-patches; Striped Owl paler, whitish facial disks boldly rimmed black; all have striped underparts; Short-eared Owl has relatively longer wings and more maneuvrable flight.

SD: C to F resident (SL–3000 m) S to cen Mexico, F to U (mostly in interior) to Honduras and N cen Nicaragua; absent from much of humid SE.

RA: N.A. to Tierra del Fuego.

GENUS GLAUCIDIUM: Pygmy-Owls

Pygmy-owls are tiny but aggressive, relatively long-tailed owls which are often diurnal. Lack ear tufts, sides of hindneck have false eyespots—white-rimmed triangular black marks which resemble eyes when bird is seen from behind. Eyes yellow, bills yellowish to horn, yellow feet bristled. Ages differ. Hunt mostly at low to mid-levels for birds up to the size of American Robins or larger, also reptiles, insects; often call from mid- to upper levels, at times in tall trees. Pygmy-owls look around alertly, twitching their tail from side to side, and may be located by tracking the scolding of agitated small birds. Nest in tree cavities, often in old woodpecker holes. Eggs: 2–4.

MOUNTAIN [NORTHERN] PYGMY-OWL

Glaucidium gnoma*　　　　　Plate 25
Tecolotito Serrano

ID: 6–6.5 in (15–16.5 cm). *Highland forests. Polymorphic.* **Descr. Adult:** facial disks

brownish, flecked white, white brows narrow. Crown, nape, and upperparts greybrown to rufous-brown, *crown and back spotted whitish to buff*, boldest on crown and scapulars. *Uppertail dark brown with 5–6 narrow, fairly complete whitish bars*, upper 1–2 bars may be hidden by wings and uppertail coverts; 3–4 white bars visible on tail underside. Sides of chest brown, finely spotted buff, underparts whitish with dark brown to rufous-brown streaking, coarser and blurrier in *cobanense* (S of Isthmus). Rufous morph (mostly S Mexico to C.A.) has *head and upperparts bright rufous-chestnut with subdued paler spotting on head; tail bars pale cinnamon.* Sides of chest and underpart streaking bright rufous. **Juv:** *unspotted grey crown and nape contrast with brown upperparts*, forehead may be flecked whitish, false eyespots sootier, upperparts mostly unspotted with a few pale buffy cinnamon spots on distal scapulars, tail bars whitish to pale buffy cinnamon. No red morph juvs seen.

Voice. Song (S to Oax at least) often starts hesitantly, then runs into prolonged, fairly rapid hooting, mostly of paired notes with odd single notes thrown in: *hoo hoo hoo hoo, hoo hoo … hoo-hoo hoo-hoo hoo hoo-hoo …*, or *oot, oot oot … oot-oot oot-oot …*, 5 double notes/4.5–6.5 s; series also may start with a rapid, slightly ringing *huhuhuhuhuhuhu . . . .* Juvs give a high, passerine-like, chipping twitter (DEW tape). Voice of birds from Chis S undescribed (?).

Habs. Pine–oak, pine, and humid pine–evergreen forest. See genus note.

SS: Colima and Tamaulipas pygmy-owls shorter-tailed, safely told by 3–4 broader but slightly broken whitish bars on uppertail (2–3 on undertail) and voice; upperparts lack extensive and prominent whitish spotting often shown by Mountain but

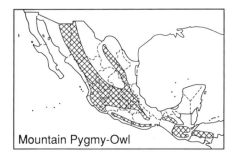

Mountain Pygmy-Owl

juvs may be extremely similar; note tail patterns. Central American Pygmy-Owl (rain forest) allopatric. Ferruginous Pygmy-Owl occurs at lower elevations; note tail pattern (which can be same as Mountain), streaked crown, voice.

SD/RA: F to C resident (1500–3500 m) in interior and on adjacent slopes from Chih (and S Arizona) and Coah to Oax; U to F from Chis to cen Honduras.

NB1: Taxonomy complex. Traditionally considered a single species, there is a distinct vocal difference among the 'Northern' Pygmy-Owls of N.A. Birds of the Pacific Slope (including Baja) give slow calls, while birds of the interior give fast calls; in addition, northern birds give single calls and southern birds double calls! Thus, it appears there are at least four distinct vocal types: slow single-hooters from British Columbia to California (*G. californicum*), slow double-hooters in BCS, fast single-hooters in interior N.A. S to Arizona, and fast double-hooters from S Arizona to Oax. N birds are generally larger but seem indistinguishable from S birds in plumage. **NB2:** The voice of birds from Chis S (in which a bright rufous morph is common) appears to be undescribed and may be different from birds N of Isthmus; if this is the case and these birds warrant specific recognition, we suggest the name Guatemalan Pygmy-Owl, *G. cobanense*.

CAPE [NORTHERN] PYGMY-OWL

Glaucidium hoskinsii Plate 25
Tecolotito del Cabo

ID: 6–6.5 in (15–16.5 cm). Endemic to mountains of BCS. **Descr.** Much like Mountain Pygmy-Owl but allopatric; note voice. Sandy grey-brown overall, averaging redder in ♀; apparently lacks grey or red morphs. Juv differs as in Mountain Pygmy-Owl (?).

Voice. Slow 'double-hoots' with occasional single hoots thrown in, thus recalls Mountain Pygmy-Owl but strikingly slower, *hoo hoo, . . . hoo hoo, . . .*, with one double-hoot in about 2 s, and 5–15 s between double-hoots. Thus 5 double-hoots/30–40 s; rarely 5 hoots in fairly steady series before running into pairs. Also gives a rapid, slightly quavering *huhuhu . . .* which often precedes bouts of hooting.

Habs. Pine and pine–oak forest, probably also deciduous woodland in winter. See genus note.

SS: Other pygmy-owls allopatric.

SD/RA: F resident (1500–2100 m, to 500 m in winter) in the Sierra Victoria, BCS; also probably resident in the Sierra de la Giganta, BCS, N to 26° 30′ N (Mar and Oct records).

NB1: See Mountain Pygmy-Owl. **NB2:** Northern Pygmy-Owl *G. c. californicum* (7–7.5 in; 18–19 cm) may be rare winter visitor (Oct–Mar?) to BCN. Indistinguishable from Mountain Pygmy-Owl by plumage but larger; note voice (unlikely to call in winter?), a single low *hoot*, repeated, 10/19–25 s. Reports of *G. (g.) californicum* from Coah pertain to *G. gnoma*, based on voice and equivocal wing length (Miller 1955) or on wing length (Ely 1962).

COLIMA [LEAST] PYGMY-OWL

*Glaucidium palmarum** Plate 25
Tecolotito Colimense

ID: 5.5–6 in (14–15 cm). Endemic to W Mexico. **Descr. Adult:** facial disks brownish, flecked white, with short whitish brows. *Crown, nape, and upperparts sandy grey-brown to olive-brown, crown and nape spotted whitish to buff*, with narrow cinnamon band across base of nape; scapulars and wing coverts spotted cinnamon to whitish. *Tail brown with 3–4 slightly broken, white to pale buff bars on upper-*

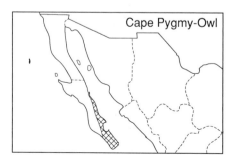

Cape Pygmy-Owl

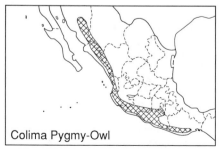

Colima Pygmy-Owl

side (5–6 bars in all, upper 2 covered by tail coverts), upper 1–2 bars may be hidden by wings; 2–3 pale bars visible on underside. Sides of chest cinnamon to buffy brown, spotted buff, underparts whitish with cinnamon to buffy-brown streaking. **Juv:** *unspotted grey crown and nape contrast with brown upperparts,* forehead may be flecked whitish to buff, false eyespots sootier, pale cinnamon nape band lacking or indistinct, tail bars may be pale buffy cinnamon.

Voice. Hollow hooting; usually shorter phrases than Ferruginous Pygmy-Owl, with variable, often successively increasing, number of notes, *hoo-hoo, hoo-hoo-hoo, hoo-hoo-hoo-hoo, hoo-hoo-hoo-hoo-hoo,* ... etc., up to 24 notes (or more?) in series; typically hollower and slightly slower than Ferruginous Pygmy-Owl; 10/3–4 s. Also gives quavering rolls which may precede hooting series.

Habs. Thorn forest, semideciduous forest, coffee plantations, locally pine–oak forest. See genus note.

SS: See Mountain Pygmy-Owl. Ferruginous Pygmy-Owl larger and longer-tailed, mainly in tropical lowlands, often fairly bright rufous overall, crown has fine pale streaks, whitish scapular spots usually more prominent, tail has 6–7 narrow pale bars and usually barred rufous and brown.

SD/RA: F to C resident (near SL–1500 m) on Pacific Slope from cen Son (SNGH, SW) to Oax and in interior along Balsas drainage to Mor.

NB: Traditionally considered a single species, Least Pygmy-Owl *G. minutissimum,* together with Tamaulipas and Central American pygmy-owls but differs in morphology, voice, habitat, etc.

TAMAULIPAS [LEAST] PYGMY-OWL
Glaucidium sanchezi   Plate 25
Tecolotito Tamaulipeco

ID: 5.5–6 in (14–15 cm). Endemic to NE Mexico. **Descr. Adult:** facial disks brownish, flecked white to buff, with short white brows. *Crown, nape, and upperparts rufous-brown* (♀) *to rich olive-brown with greyer crown and nape* (♂); *forecrown finely spotted pale cinnamon to whitish,* spots may extend along sides of crown to nape; wing coverts spotted cinnamon to white, scapulars usually show indistinct pale spots. *Tail brown with 3–4*

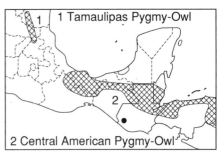

1 Tamaulipas Pygmy-Owl

2 Central American Pygmy-Owl

broken whitish to pale buff bars on upperside (5 bars in all, upper bar covered by tail coverts), 2–3 pale bars visible on undertail. Sides of chest rufous-brown, spotted buff, underparts whitish with rufous-brown streaking. **Juv:** *unspotted grey crown and nape contrast with brown upperparts,* forehead may be flecked whitish to buff, false eyespots sootier, tail bars pale cinnamon.

Voice. Song 1–3 slow-paced, hollow, ringing hoots, typically 2, repeated after a pause; also a quavering roll which may precede hoots.

Habs. Humid evergreen and pine–evergreen forest, typically cloud forest. See genus note.

SS: See Mountain Pygmy-Owl. Ferruginous Pygmy-Owl allopatric in tropical lowlands. Central American Pygmy-Owl (allopatric?) has relatively shorter tail which shows 2–3 pale bars above, voice and habitat distinct.

SD/RA: U to F but local resident on Atlantic Slope (1500–2100 m) in S Tamps and SE SLP; should be sought in cloud forest S at least to cen Ver.

NB: Traditionally considered a single species, Least Pygmy-Owl *G. minutissimum,* together with Colima and Central American pygmy-owls.

CENTRAL AMERICAN [LEAST] PYGMY-OWL
Glaucidium griseiceps   Plate 25
Tecolotito Centroamericano See Tamaulipas
                                   Pygmy-Owl for map

ID: 5.5–6 in (14–15 cm). SE Mexico and C.A. **Descr. Adult.** Facial disks grey-brown, flecked whitish, with short white brows. *Crown and nape brownish grey to grey-brown, typically contrasting slightly with rich brown upperparts; forecrown finely spotted buff* to whitish, spots may extend back to nape. Wing coverts spotted cinnamon to white, scapulars usually show indistinct pale spots. *Tail brown with 2–4 broken whitish to pale buff bars on upperside* (4 bars in all, upper 1–2 bars covered by tail coverts), 2 pale bars visible on undertail. Sides of chest rufous-brown, spotted buff, underparts whitish with rufous-brown streaking. **Juv:** *unspotted grey crown and nape contrast with brown upperparts,* forehead may be flecked whitish to buff, false eyespots sootier, tail bars may be pale cinnamon.

Voice. Hollow, ringing hoots in short series, starting with 2–4 hoots and series typically ranging from 6–9 (but at times to 18) in steady or varied series, *huu-huu, huu-huu-huu, huu-huu,* ..., etc.; 10/2–3.5 s. Also quavering trills which may precede hooting series.

Habs. Humid evergreen forest (rain forest). See genus note.

SS: Ferruginous Pygmy-Owl larger and longer-tailed, not in forest, often fairly bright rufous overall, crown with fine pale streaks, whitish scapular spots usually prominent, tail usually barred rufous and brown, with 6–7 narrow pale bars. Tamaulipas and Mountain pygmy-owls allopatric.

SD: U to F but local resident (near SL–1200 m) on Atlantic Slope from N Oax and SE Ver (SNGH, HGS, tape) to Honduras; and (1300 m) on Pacific Slope in Guatemala (BW, tape) and probably also Chis (?).

RA: SE Mexico to Brazil.

NB: Traditionally considered a single species, Least Pygmy-Owl *G. minutissimum,* together with Colima and Tamaulipas pygmy-owls.

FERRUGINOUS PYGMY-OWL
Glaucidium brasilianum*   Plate 25
Tecolotito Común

ID: 6.5–7.5 in (16.5–19 cm). Common in tropical lowlands. Polymorphic. **Descr. Adult:** facial disks brown to rufous, flecked whitish, white brows often fairly bold. Crown, nape, and *upperparts grey-brown to rufous, crown and nape streaked whitish to buff,* outer scapulars broadly tipped whitish. *Uppertail varies* from washed out rufous, through barred rufous and dark brown, to barred dark brown and whitish, *but consistently 5–7 narrow paler bars;* 3–5 pale bars visible on underside. Under-

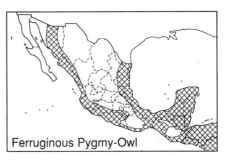

Ferruginous Pygmy-Owl

parts whitish with coarse dark brown to rufous streaking heaviest at sides of chest where may be flecked whitish. **Juv:** crown and nape often greyer, fine pale streaks restricted to forecrown, false eyespots sootier.

Voice. Hollow, often prolonged whistled hooting, easily imitated, *hoo-hoo-hoo...,* 10/3–3.5 s, at times faster and more insistent, *whi'whi'whi...,* which may break into bursts of high, yelping twitters; at times gives irregular short series suggesting Colima Pygmy-Owl.

Habs. Semiopen areas with hedges and scattered forest patches, open forest and edge, semiopen thorn forest, plantations, etc. See genus note.

SS: See Colima, Central American, and Mountain pygmy-owls; Elf Owl (nocturnal) small with short tail and long wings, buff-spotted crown and upperparts, grey bill, no dark eyespots on sides of neck, blurry underpart streaking.

SD: C to F resident (SL–1400 m, locally to 1900 m in C.A.) on both slopes from Son and Tamps, and locally in interior from Balsas drainage, to El Salvador and Honduras.

RA: Mexico (locally in SW USA) to N Argentina.

ELF OWL
Micrathene whitneyi*   Not illustrated
Tecolotito Enano

ID: 5.5–6 in (14–15 cm). *A tiny migratory owl of N deserts.* Wings fairly long, tail short, lacks ear tufts, toes bristled. **Descr.** Ages differ slightly. **Adult:** eyes yellow, bill greyish with horn tip. Facial disks cinnamon to buff, white brows narrow. *Crown, nape, and upperparts grey-brown, spotted buff to cinnamon, with 2 bold white rows of scapular and wing-covert spots,* narrow

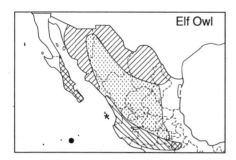

Elf Owl

whitish band across base of nape. Wings and tail grey-brown, barred whitish to pale cinnamon, tail with 3–4 narrow paler bars. *Underparts whitish, heavily and blurrily mottled cinnamon and grey.* **Juv:** greyish crown unspotted, pale spots indistinct on upperparts, underparts mottled grey and white.

Voice. Song (Mar–Jul) a yelping, accelerating to bouncing-ball roll of up to *c.*20 notes, may be repeated rapidly, every 1–2 s, *kyeu-kyeu* ... Calls include sharp, barking yelps, *kyew*! or *kyeuh*!, a rougher *kyeh*, and a plaintive, drawn-out, often soft and ventriloquial *hew* or *hee-ew*.

Habs. Arid to semiarid wooded canyons, thorn forest, semiopen areas with scrub and scattered trees. Hunts for insects at low to mid-levels, from perches and also in flight when may hover; at times migrates in flocks (Ligon 1968*b*); nests, and often roosts, in tree cavities, especially old woodpecker holes; 2–3 eggs.

SS: See Ferruginous Pygmy-Owl.

SD: F to C but local breeder (Mar–Aug) on Pacific Slope from Son to N Sin, in interior from Chih to NL (AMS), possibly elsewhere S over Plateau; disjunctly in upper Balsas drainage in S Pue. Migrates late Feb–early Apr, Aug–Sep; winters (Sep–Mar) mainly in Balsas drainage from Mich to Pue; R (?) to Gro (MVZ spec.) and on Pacific Slope N to Sin. F to C resident in BCS (mostly S of 24° N) and (at least formerly) on Isla Socorro. Reports of breeding in Gto (AOU 1983) refer probably to migrants (Ligon 1968*b*).

RA: SW USA to cen Mexico; winters Mexico.

BURROWING OWL
*Athene cunicularia**   Figure 27
Búho Llanero

ID: 9–10 in, W 19.5–22.5 in (23–25.5 cm,

Fig. 27 Burrowing Owl

W 50–57 cm). Widespread *terrestrial owl with long legs*, toes bristled. **Descr.** Ages differ. **Adult:** eyes yellow, bill greyish horn. *Facial disks pale brownish with contrasting thick white brows* and broad white forecollar. *Crown, nape, and upperparts sandy brown, boldly spotted whitish* to pale buff, wings and tail barred pale buff, tail with 3–4 pale bars. *Underparts whitish to pale buff, coarsely scalloped and barred dark brown*, heaviest on chest. **Juv:** crown unspotted, upperparts with sparser spotting, upper chest brownish, rest of underparts unmarked pale buff.

Voice. Song a hollow, slightly quavering, monotonously repeated *hui-poor* or *coo-cooh*. Calls include a rough, yelping chatter, and screaming yelps and chatters.

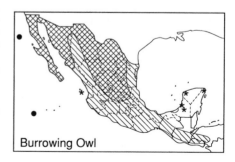

Burrowing Owl

Habs. Arid to semiarid, open and semi-open country. Terrestrial, seen often during the day, roosts on ground, under bushes, in crevices, burrows, etc. Hunts from perches such as posts and wires, often hovers; flight undulating, long legs trail behind. Nests in burrows, often colonially, 7–9 eggs.

SD: Poorly known in much of region, complicated by presence of N migrants and vagrants. C to F resident (near SL–2000 m) in Baja, on Pacific Slope to N Sin, and on Plateau; also Islas Guadalupe and Clarión. Irregularly U to F transient and winter visitor (Sep–mid-Mar) S to Honduras (may breed locally in winter range), mainly in interior and on Pacific Slope, vagrant (Sep–Mar) to Yuc Pen (Storer 1961; G and JC, BM).

RA: W N.A. to cen Mexcio and Caribbean; winters S to C.A.; resident locally in S.A.

GENUS STRIX

These are large to medium-large nocturnal owls of forest and woodland, lack ear tufts, eyes dark brown (at least in the region), 'eyeshine' at night amber to red. Roost low to high in shady trees, understory. Neotropical species (including Mottled and Black-and-white) have traditionally been placed in the genus *Ciccaba*.

MOTTLED OWL
Strix virgata * Plate 26
Búho Café Figure 26

ID: 13–15 in (33–38 cm). Common tropical owl. Dimorphic. **Descr.** See genus note. **Adult. Light morph** (mostly drier areas): bill horn to yellowish, toes greyish to yellow. *Facial disks brownish with bold white brows and whiskers.* Crown, nape, and upperparts dark brown, flecked and sparsely

barred whitish to pale buff, outer scapulars boldly tipped whitish; remiges dark brown, barred pale grey to whitish, uppertail dark brown with 3–4 narrow whitish bars. *Underparts whitish to pale buff, mottled dark brown at sides of chest, rest of underparts streaked dark brown,* thighs and undertail coverts buff. **Dark morph** (mostly humid SE): darker. Whitish to pale brown brows less striking, upperparts sparsely vermiculated greyish buff; uppertail blackish with 3–4 narrow grey to whitish bars; underparts buff to ochraceous buff, sides of chest heavily mottled dark brown, rest of underparts coarsely streaked dark brown. **Juv:** bill pinkish. Pale buff to ochraceous buff overall with whitish facial disks.

Voice. Deep, gruff to resonant hoots: a single *wh-owh'* and *wooh'*, and longer series, typically 3–10 hoots, often accelerating and becoming stronger, before fading with last 1–2 notes, *wo-oh' wo-oh' wo-oH' wo-oH' wo-oh*; also a rapid, bouncing-ball series of about 20 hoots suggesting Spectacled Owl or possibly an *Otus*, *wup wup wup-wup-wup-wup ...*, and wailing screams *eeihrr-rr-rr*, or *wheeahrr*, etc. Dueting (?) birds may give excited-sounding but fairly soft series of resonant hoots, *wuuh-wuuh-wuuh*, *wuUH, wuuh-wuuh-wuuh-wuuh wuuh-wuuh-wuuh*, etc.

Habs. Wide variety of wooded and forest habitats from thorn forest and plantations to humid evergreen forest and pine–oak. Hunts along forest edges and in semiopen areas. Nests in tree cavities, old nests of other birds; 2 eggs.

SS: Barred, Spotted, and Fulvous owls larger, not in lowlands, facial disks paler, upperparts boldly spotted and barred; note voice.

SD: C to F resident (SL–2500 m) on both slopes from S Son and cen NL (JCA), and in interior from Balsas drainage, to El Salvador and Honduras. A report from Gto (Friedmann *et al.* 1950) is probably erroneous.

RA: Mexico to Ecuador and N Argentina.

BLACK-AND-WHITE OWL
Strix nigrolineata Plate 26
Búho Blanquinegro Figure 26

ID: 15–16 in (38–40.5 cm). SE humid forest and edge. **Descr.** See genus note. **Adult:** *bill*

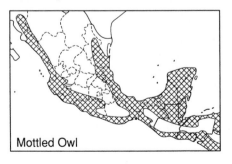

Mottled Owl

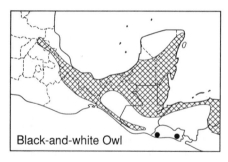

Black-and-white Owl

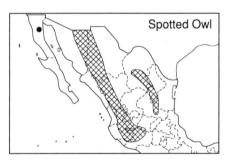

Spotted Owl

and feet orange-yellow. *Facial disks black with white-speckled brows. Crown, nape, and upperparts blackish*, tail with 4 narrow whitish bars. *Underparts white, narrowly barred blackish*, black-and-white barring extends around sides of neck as partial collar. **Juv:** whitish overall, upperparts narrowly barred dark brown, underparts washed buff, narrowly barred blackish.

Voice. Deep, gruff, barking hoots. An emphatic *hu whoOOo*, 1st note often inaudible at distance, also series of 3–5 or more notes, last or penultimate note emphatic, *woh-woh-WHOW' woh* or *ha-ha-ha HAH*, etc. Also fairly rapid series of 8–10 notes, *whuk-whuk-whuk* ..., suggesting Spectacled or Mottled owls, a harder *heh-heh-heh* ... up to 10×, loud, wailing screams, *wheeahew* or *meeeowh*, much like Mottled Owl, and a harder *rreah'k*.

Habs. Humid to semihumid evergreen and semideciduous forest and edge, plantations. May be conspicuous at forest edge in fairly open trees, at times hunts bats under artificial lights. Nests in crotches of large bromeliads; 1 egg.

SS: Plumage unmistakable; voice suggests Mottled Owl but less resonant and usually more emphatic.

SD: U to F resident (SL–1200 m) on Atlantic Slope from SE SLP and Ver to Honduras, on Pacific Slope from E Oax to El Salvador.

RA: Mexico to Peru and Venezuela.

SPOTTED OWL
Strix occidentalis* Not illustrated
Búho Manchado

ID: 17–19 in (43–48.5 cm). N and W mountains. **Descr.** See genus note. **Adult:** *bill greenish yellow*, toes dull yellowish. Facial disks grey-brown, rimmed dark brown, with short paler brows. Crown, nape, and upperparts dark brown, crown and nape

spotted white, upperparts with short, thick white bars and spots. Uppertail dark brown with 4–6 whitish to buff bars. *Underparts pale buff to whitish with coarse dark brown bars and scallops producing a boldly spotted effect*, undertail coverts barred dark brown. **Juv:** pale buff to greyish buff overall, barred darker, facial disks brown.

Voice. In N.A., a loud, gruff, barking *whoo, whoo-whoo, whoOAH*, and a rising, slightly screeching whistle, *eeeeeeahr*. Also shorter to longer varied series of barking hoots, *whoo, whoo-whoo, whoo-whoo*, or *hoo-hoo, hoo*, or *whoo-whoo-whoo whoo, whoo-whoo, whoo-whoo*, and strengthening series of 6–10 barking hoots, *hoh-hoh-hoh-hoh-hoh-hoh* ..., etc.

Habs. Conifer and pine–oak forests. Hunts in open forest and along edges. Nests in old nests of other birds and in tree cavities; 2–3 eggs.

SS: Barred Owl larger and greyer, facial disks pale grey, bill brighter yellow, upperparts with bolder whitish spots and bars, upper chest barred, rest of underparts streaked. See Mottled Owl.

SD: U to F and apparently local resident (1200–2500 m) in interior and on adjacent Pacific Slope from Son to Mich, and from Coah to E SLP (LSU spec.); also (900–2500 m) in BCN (Grinnell 1928; Wilbur 1987).

RA: W N.A. to cen Mexico.

BARRED OWL
Strix varia sartorii Not illustrated
Búho Barrado

ID: 19–21 in (48.5–53.5 cm). Cen and W mountains. **Descr.** See genus note. **Adult:** *bill yellow*, toes dull ochre-yellow. Facial disks pale grey, rimmed dark brown, with short paler brows. Crown, nape, and upperparts dark grey-brown, with coarse but

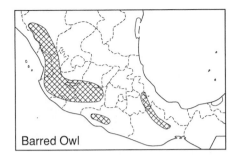

Barred Owl

relatively sparse whitish bars, scapulars and wing coverts boldly spotted whitish. Uppertail dark brown with 4–5 whitish bars. *Underparts whitish, upper chest coarsely barred dark grey-brown, rest of underparts coarsely streaked dark grey-brown,* undertail coverts may be unmarked. **Juv:** greyish buff to pale buff overall, barred darker, facial disks brown.

Voice. In N.A., deep, barking hoots, typically 2 series of 4 notes, ending with an emphatic, at times disyllabic note, *whoo whoo whoo-whoo', whoo whoo whoo-WHOO'AH.*

Habs. Humid to semiarid pine–oak, fir, and pine–evergreen forest. Habits much like Spotted Owl.

SS: See Spotted and Mottled owls.

SD: Probably U to F resident (1500–2500 m) on Pacific Slope and in adjacent interior from Dgo to Mich and Gro, and on Atlantic Slope and in adjacent interior from Ver to N Oax; reported infrequently.

RA: N.A. to Oax.

FULVOUS OWL

Strix fulvescens   Plate 26
Búho Fulvo   Figure 26

ID: 15–17 in (38–48.5 cm). Endemic to mountains S of Isthmus. **Descr.** See genus note. **Adult:** bill and toes yellow. Facial

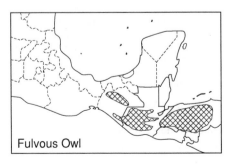

Fulvous Owl

disks pale grey-brown, rimmed dark brown, with short paler brows. Crown, nape, and upperparts dark brown, *crown and nape coarsely scalloped tawny to ochraceous buff,* upperparts with coarse but relatively sparse buff to ochraceous buff bars, scapulars and wing coverts boldly spotted, and remiges sparsely barred, buff to pale buff. Uppertail dark brown with 3–4 pale cinnamon to pale buff bars. *Underparts buff to ochraceous buff, upper chest coarsely barred dark brown, rest of underparts coarsely streaked dark brown,* undertail coverts may be unmarked. **Juv:** ochraceous buff overall, faintly barred darker on upperparts, facial disks brownish.

Voice. Loud, barking hoots, often with rhythm suggesting Morse code, *a'hoo a'hoo-hoo a'hoo, hoo,* or *a'hoo a'hoo-hoo a'hoo hoo-hoo,* etc.

Habs. Humid evergreen and pine–oak forest. Habits much like Spotted Owl. Nest and eggs undescribed (?).

SS: See Mottled Owl.

SD/RA: F resident (1200–3000 m), mostly in interior and on adjacent Pacific Slope, from Chis to El Salvador and Honduras. Reports from Oax are erroneous (Binford 1989).

GENUS ASIO

These are large, relatively long-winged, 'eared' owls of temperate origin, inhabiting forests and open country. Toes feathered to bristled, ear tufts usually noticeable. Bills dark grey, plumage overall cryptically patterned in browns, buffs, etc.; downy juvs often show dark facial disks.

LONG-EARED OWL

Asio otus wilsonianus   Not illustrated
Búho-cornudo Caricafé

ID: 14–16 in, W 34–37.5 in (35.5–40.5 cm, W 86–95 cm). N and cen Mexico. Long ear tufts, toes feathered. **Descr.** See genus note. **Adult:** eyes yellow. *Facial disks cinnamon, eyes set in vertical dark patches.* Crown, ear tufts, nape, and upperparts dark brown, patterned with buff and whitish, outer scapulars boldly tipped whitish, *broad tawny bars on base of primaries form conspicuous flash in flight;* underwings whitish with bold dark carpal mark and dark barring on tips of outer primaries.

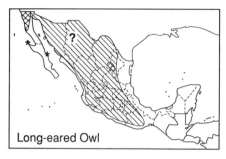

Long-eared Owl

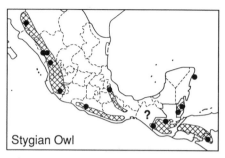

Stygian Owl

Uppertail dark brown with 5–7 narrow paler greyish bars, undertail buff with 5–7 narrow dark bars. Underparts whitish, coarsely streaked and barred dark brown. **Juv:** pale grey overall, barred dark brown. **Voice.** Song (N.A.) a single low hoot, *whoo* or *woop*, every 2–3 s.

Habs. Semideciduous riparian woods, pine and pine–oak forest, more widespread in winter, including open and semiopen country with thickets. Nocturnal, roosts in trees. Hunts in open areas and forest, flight fairly buoyant and erratic. Uses old nests of other birds; 4–5 eggs.

SS: Short-eared Owl (winter visitor to open areas) similar but paler and buffier overall, especially on underparts where heavily marked only on upper chest, black eye patches horizontal on pale grey facial disks, ear tufts short and inconspicuous, more active by day. Stygian Owl (mountain forests) darker overall, facial disks dark grey; note whitish patch on forehead.

SD: U to F resident (50–1000 m?) in NW BCN (at least formerly, now R?); locally resident in NL (Hubbard and Crossin 1974), probably also Chih (MVZ specs, Jun and Aug). Irregularly U to F transient and winter visitor (Nov–Mar; near SL–2500 m) to BCS (LACM spec.), on Pacific Slope in Son, and in interior and on adjacent slopes S to cen volcanic belt; R to Gro (MVZ spec.) and Oax (Binford 1989).

RA: Holarctic breeder; N populations migratory; breeds in New World S to N Mexico; winters to S Mexico.

STYGIAN OWL

Asio stygius *   Plate 26
Búho-cornudo Oscuro

ID: 15–17 in, W 36–40.5 in (38–43 cm, W 91.5–103 cm). *Pine forests. Long ear tufts,* toes bristled. **Descr.** See genus note. **Adult:**

eyes yellow. *Whitish forehead contrasts with blackish facial disks.* Crown, ear tufts, nape, and *upperparts blackish brown,* sparsely barred and spotted buff, outer scapulars coarsely tipped whitish to buff, primaries sparsely barred buff to cinnamon; underwings appear dark overall. Uppertail blackish brown with 3–4 narrow buff bars. *Underparts dirty buff, heavily streaked and barred dark brown.* **Juv:** pale buff overall, barred dark brown. **Voice.** In NW Mexico, a single deep, emphatic *woof* or *wupf,* at a distance may sound like *whu,* repeated at 6–10 s intervals. In Belize, a softer, slightly higher, and less emphatic *whuh,* at 4–5 s intervals, can be deceptively soft at close range; also a short, screamed *rre-ehhr* or *mehrr* (♀ calling to ♂). In display, ♂ claps wingtips together below body in short flight, wing claps usually not in quick succession.

Habs. Humid to semiarid pine and pine–oak forest. Nocturnal. Roosts at and calls from mid- to upper levels of trees. Nests in trees in old nests of other birds, and (Caribbean at least) at times on ground; 2 eggs (Bond 1985).

SS: See Long-eared and Great Horned owls.

SD: Poorly known. F to C resident (1500–2000 m) in W Dgo and E Sin, locally (1500–3000 m) elsewhere in Pacific Slope mountains (Chih, Jal, Gro, Chis); on Atlantic Slope known from cen Ver and Guatemala. U to F resident (near SL–800 m) in Belize; visitor (Feb–Apr, resident?) to Isla Cozumel (SNGH, DRP, DEW).

RA: Mexico and Caribbean locally to N Argentina.

STRIPED OWL

Asio clamator forbesi   Plate 26
Búho-cornudo Cariblanco

ID: 13–15 in; W 30–37 in (33–38 cm; W

76–94 cm). *S open country. Ear tufts*, toes feathered. **Descr.** See genus note. **Adult:** *eyes brownish. Whitish facial disks boldly rimmed blackish.* Crown, ear tufts, nape, and upperparts dark brown, coarsely streaked and mottled buff, *broad buff to tawny bars on base of primaries form conspicuous flash in flight.* Underwings pale buff with bold dark carpal mark and dark barring on tips of outer primaries. Uppertail dark brown with 5–7 narrow buff bars, undertail buff with 5–7 narrow dark bars. *Underparts pale buff, coarsely streaked dark brown.* **Juv:** buff overall, upperparts barred dark brown.

Voice. A single, nasal hoot lasting about 1 s, loudest and highest in middle: *hooOOOoh* or *hnnNNNnh*, hoot of ♀ higher. Also 7–8 sharp, dog-like barks: *hu-how! how! how!* etc., sometimes given by mates in chorus (Stiles and Skutch 1989). Imm has a slightly shrill, whistled scream, *heeeen*(!) or *hreeeeh*, repeated.

Habs. Savannas, open and semiopen areas with scattered bushes, thickets, hedges, etc. Nocturnal. Often hunts from roadside wires, posts, etc. Flies with fairly shallow, rapid wingbeats. Nests on ground, often at base of tree or bush; 2–3 eggs.

SS: Short-eared Owl (winter visitor) darker and buffier overall, lacks bright, whitish, black-rimmed face, yellow eyes set in black patches, ear tufts short, underparts less boldly striped. See Long-eared Owl, Barn Owl.

SD: U to F resident (SL–900 m) on Atlantic Slope from S Ver to cen Chis and Tab, also in S Belize (Howell *et al.* 1992*b*), and Sula Valley, Honduras; on Pacific Slope from Chis to Nicaragua.

RA: Mexico to N Argentina.

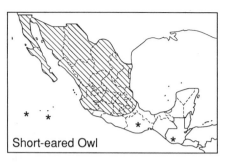

Short-eared Owl

SHORT-EARED OWL

Asio f. flammeus　　　　Not illustrated
Búho Orejicorto

ID: 15–17 in, W 35.5–40 in (38–43 cm, W 90.5–102 cm). *Winter visitor. Short ear tufts*, toes feathered. **Descr.** See genus note. Eyes yellow. *Facial disks pale buff, eyes set in horizontal dark patches.* Crown, ear tufts, nape, and upperparts dark brown, heavily streaked and mottled buff, *broad buff to tawny bars on base of primaries form conspicuous flash in flight.* Underwings whitish with bold dark carpal mark and coarse dark barring on tips of outer primaries. *Uppertail dark brown with 3–4 broad buff bars, undertail buff with 3–4 dark bars. Underparts buff, coarsely streaked dark brown on chest,* thighs and undertail coverts unstreaked, *rest of underparts finely streaked dark brown,*

Habs. Open country in general, especially grasslands and marshes. May be active by day but seen mainly at dusk, hunting with buoyant floppy flight; perches on fence posts and ground. Roosts on ground. Usually silent in winter, rarely uttering short barks.

SS: See Striped and Long-eared owls, Barn Owl.

SD: Irregularly U to R transient and winter visitor (Oct–Mar; SL–3000 m) S to cen, rarely S, Mexico; only reaches NW Mexico most years. Vagrant to Islas Revillagigedo and Guatemala.

RA: Holarctic breeder; disjunct populations in Hawaii, Galapagos, and Caribbean; winters S in New World to Mexico.

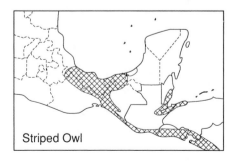

Striped Owl

GENUS AEGOLIUS

These are small owls of highland forests; large, squarish heads lack ear tufts, toes feathered; nocturnal. Nest in tree cavities, 5–6 eggs (in Northern Saw-Whet, N.A.).

NORTHERN SAW-WHET OWL

Aegolius a. acadicus   Not illustrated
Tecolote-abetero Norteño

ID: 8–8.5 in (20.5–21.5 cm). Highlands N of Isthmus. **Descr.** Ages differ, sexes similar. Adult: eyes yellow, bill dark grey. *Facial disks pale buff to brownish with contrasting broad white brows.* Crown, nape, and upperparts dark brown, crown and nape streaked white, outer scapulars, wing coverts, and primaries spotted white. Uppertail dark brown with 2 broken bars of white spots. *Underparts whitish, coarsely streaked cinnamon-brown.* Juv: *head and upper chest dark brown with broad white brows*; upperparts dark brown; *underparts ochraceous buff.*
Voice. Song a steadily to hesitantly repeated whistle, *whu whu whu* ... or *hoo hoo hoo* ..., 10/3–6 s; less regular than Mountain Pygmy-Owl; also may give surprisingly loud, yelping screams, *meeeah*, etc. ♀ has a high, shrill *siirr*; begging young give an insistent rough hissing *shhii* or *sssii.*
Habs. Humid to semihumid pine, fir, and pine–oak forests. Most often detected by voice.

SD: F to U resident (1800–3500 m) on both slopes and in interior from Chih and NL to cen volcanic belt, thence to cen Oaxaca.
RA: N.A. to Mexico.

UNSPOTTED SAW-WHET OWL

Aegolius ridgway* or *A. acadicus* (in part)
Tecolote-abetero Sureño   Plate 25

ID: 8–8.5 in (20.5–21.5 cm). Highlands S of Isthmus. **Descr.** Ages/sexes similar (?). Eyes yellow, bill dark grey. *Facial disks brown with paler central area and narrow outer rim,* broad white brows. Upper chest brown, rest of *underparts ochraceous buff.* Upperparts brown, primaries and tail with trace of sparse white spots. Juv undescribed (?).
Voice: A whistled *hoo hoo hoo* ..., etc., 10/3–5 s, much like allopatric Northern Saw-Whet Owl but slightly lower. ♀ gives a high, slightly hissing *ssirr.* Also a surprisingly loud shriek and a shrill short chipper.
Habs. Humid pine–oak and oak forest. Most often detected by voice. Nest and eggs undescribed (?).
SD: F to U resident (1650–3000 m), mostly in interior and on adjacent Pacific Slope, from Chis to Guatemala and NW El Salvador.
RA: Chis locally to W Panama.

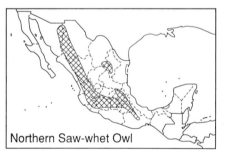

Northern Saw-whet Owl

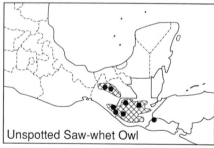

Unspotted Saw-whet Owl

NIGHTHAWKS AND NIGHTJARS: CAPRIMULGIDAE (14)

A poorly known, cosmopolitan family of crepuscular and nocturnal birds, many of which are migratory. In general, nighthawks hunt their prey in flight, while nightjars hunt either from perches, often from the ground, or in flight; both roost in trees and on ground. Wings fairly long and pointed in nighthawks, relatively rounded in nightjars; tails typically long, rounded to slightly cleft. Small, slender, and slightly decurved bills open to reveal very broad gapes, rimmed with bristle-like feathers to aid in capturing prey. Legs short, feet small and weak, central toes pectinate. Eyes dark brown but reflected 'eyeshine' at night bright orange to red. Bills and feet greyish to flesh. Ages differ; young altricial and downy; juvs typically have paler upperparts than adults, but soon resemble adults (for example, before arrival of northern migrants in the region) after the partial 1st prebasic molt. Molts poorly known but nightjars retain juv flight feathers through first year. Sexes often differ in wing and/or tail patterns, but similar in poorwills. All caprimulgids are cryptically patterned in browns, greys, black, and white; their plumage is notably soft and fluffy; most species have a white throat collar. Color may vary with breeding environment; several species, such as *Chordeiles* nighthawks, are darker in more humid areas, paler in more arid areas; some species are dimorphic, ♀♀ more often exhibiting a rufous morph.

Seen at night, nightjars often appear oddly puffy and large-headed and may show a shaggy chest apron, quite different from the sleeker shapes shown by roosting birds. The main plumage characters to note are head, throat, and tail patterns; scapular and wing covert patterns can also be useful. Voice is often the key to identification and detection of caprimulgids. Warm moonlit nights are best for hearing nightjars, especially if not at the height of the breeding season. Songs and calls include whistles, clucks, and trills. Most species are vocal only for a few months in the breeding season, and it still is unknown whether some species are resident or migratory in the region.

Almost wholly insectivorous, nighthawks and nightjars prey mostly on moths and beetles. Eggs (usually 2) are laid on the ground (except *Lurocalis?*) in a shallow scrape.

SHORT-TAILED NIGHTHAWK
Lurocalis semitorquatus ssp?   Plate 27
Chotacabras Colicorta   Figure 28

ID: 8–8.3 in, W 22–24.4 in, T 2.7 in (20–21 cm, W 56–62 cm, T 7 cm). SE rain forest. *Tail short and squared to slightly cleft, wings long and bluntly pointed; at rest wingtips project well beyond tail.* **Descr.** Sexes similar. **Adult:** *head and upperparts blackish,* spotted cinnamon-rufous, *typically with silvery mottling on scapulars* and tertials; *no white in wings.* Tail blackish with narrow greyish-cinnamon bars, small white tip to inner web of next-to-central rectrix. *Throat and chest sooty blackish brown, spotted paler, with white throat chevron; belly and undertail coverts cinnamon, barred blackish brown.* Juv undescribed (?).

Voice. Often silent. Calls in flight (only in breeding season?), a sharp, slightly liquid *g'wik* or *gweek,* and *whik whik whik* or *gwik wik wik,* a longer *whik whik whik-*

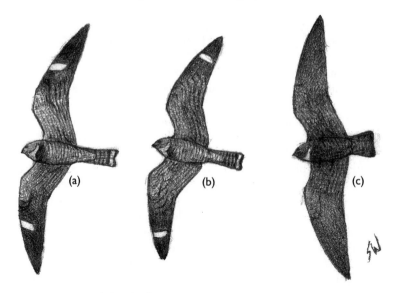

Fig. 28 (a) Common Nighthawk; (b) Lesser Nighthawk; (c) Short-tailed Nighthawk

whik whik, and a rougher *gwirrk*, etc. (Chis, Apr 1992).

Habs. Humid evergreen forest, adjacent plantations. Usually seen at dusk or dawn, flight erratic and typically low over canopy, especially (?) along rivers. Roosts along tree branches. *L. (s?) nattereri* nests on large horizontal branches, high in trees; it lays 2 whitish eggs, spotted with browns and greys (Straneck *et al.* 1987).

SS: Fairly large size and short-tailed appearance distinctive; note lack of white or pale bars in wings or tail.

SD: Apparently U to F resident (SL–500 m) on Atlantic Slope from E Chis to Honduras; overlooked until recently. First noted Feb 1989 in Chis (Howell 1989c); subsequent records (Nov 1991, Apr 1992) elsewhere in E Chis (SNGH, JSa, SW),

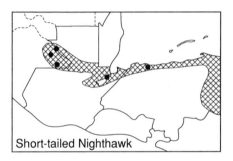

Short-tailed Nighthawk

and in NE Guatemala (Feb 1991, Howell and Webb 1992b). Also found recently in N Honduras (Jun 1988, Howell and Webb 1992b).

RA: SE Mexico to S.A.

NB: The Amazonian (*rufiventris*) and Andean (*nattereri*) forms may be specifically distinct (AOU (1983) incorrectly switched the names/distributions).

GENUS CHORDEILES

These nighthawks are widespread and conspicuous nightjars with long, pointed wings and fairly long, cleft tails. Ages/sexes differ. The two species found in the region are notoriously similar and care must be taken separating them in the field; the shape and position of the white bar on the primaries is an important character. Eggs: 2, cream to dusky cinnamon with fine dark markings (WFVZ).

LESSER NIGHTHAWK
Chordeiles acutipennis *   Plate 27
Chotacabras Menor   Figure 28

ID: 8–9 in, W 20–23.5 in, T 3–3.5 in (20.5–23 cm, W 51–60 cm, T 7.5–9 cm). Widespread and common. *At rest, wingtips usually close to tail tip,* wingtips relatively

rounded (next-to-outermost primary longest). **Descr.** ♂: crown, nape, and upperparts cryptically patterned pale grey, blackish, and buff to cinnamon, with pale supercilium. Remiges dark brown with cinnamon spotting on bases of outer primaries, *white bar across outer 4 primaries, if visible at rest, usually about even with or beyond tips of tertials* (also some distance from primary coverts but this often hard to see); *note that distal border of white band is not strongly staggered and is very rarely sharply 'toothed'.* Tail dark brown, barred paler, with white subterminal band. *Underparts greyish buff to pale buffy cinnamon*, with white throat chevron, chest mottled dark brown and grey, rest of underparts barred dark brown. *In flight, white bar on primaries falls about 60–70% out along leading edge of outer wing.* ♀: *primary bar buff to pale cinnamon, often indistinct, note cinnamon notching and barring on bases of outer primaries*; white throat chevron less extensive, often washed buff, lacks white tail bar. **Juv:** upperparts overall paler, sandier, and more uniform, primaries broadly tipped cinnamon to buff (soon fading to whitish), buff throat chevron often indistinct, whitish bar on primaries and tail band of ♂ often smaller and less distinct. 1st prebasic molt partial.
Voice. Song (Apr–Aug) given from on or near ground, a low, churring, toad-like trill, *urrrr ...*, typically in bursts of 7–13 s in duration (at times to 26 s) which often run into series lasting to 3+ minutes; the churrs swell quickly and fade abruptly with a 1–3 s pause between churrs in long series. Also a bleating *whik* given in flight.
Habs. Open and semiopen areas with scattered trees, towns, forest and edge, mangroves, etc. Aerial, flight erratic with fairly quick, flicking wingbeats interspersed with faster fluttering and short

glides with wings held in dihedral; flies low to high, though usually fairly low on breeding grounds, flocks often high overhead in migration. Also hunts from ground, perching on quiet roads, etc. Post-roosting flights of hundreds not uncommon in migration and winter. Roosts in trees and on ground.
SS: Common Nighthawk has longer wings and longer tail, wings more pointed, flies with slower, often deeper, and more rangy wingbeats interspersed with erratic fluttering and glides with wings in dihedral; both sexes have white band on primaries about 50–60% out along leading edge of outer wing; at rest, wingtips often project beyond tail. Common often darker overall, white on primaries often strongly staggered, with sharply toothed distal edge; note voice.
SD: C to F but often local breeder (SL–2500 m) throughout most of region, exact distribution poorly known; mostly absent from Atlantic Slope S of Ver except N Yuc Pen, and mainly in interior from Isthmus S. C to F transient and winter visitor (Aug–early May) from BCS, S Son and Ver, and in interior from cen Mexico, to El Salvador and Honduras.
RA: SW USA to Mexico, locally to S.A.; N populations migratory.
NB: Sometimes known as Trilling Nighthawk.

COMMON NIGHTHAWK
Chordeiles minor * Plate 27
Chotacabras Mayor Figure 28

ID: 9–9.5 in, W 21.5–26.2 in, T 3.5–4 in (23–24 cm, W 55–67 cm, T 9–10 cm). *Summer resident*, widespread but local. *At rest, wingtips usually longer than tail tip*, wingtips relatively pointed (outermost primary longest). **Descr.** Sexes differ slightly. ♂: crown, nape, and upperparts

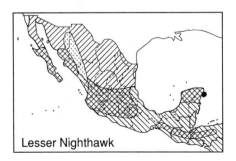

Lesser Nighthawk

Common Nighthawk

cryptically patterned blackish, pale grey, and buff to cinnamon, with pale supercilium. Remiges blackish brown, *white bar across outer 5 primaries, if visible at rest, usually falls short of or is even with tips of tertials* (also close to primary coverts but this often hard to see); *note that distal border of white band is often strongly staggered and is sharply 'toothed'.* Tail dark brown, barred paler, with white subterminal band. *Underparts pale greyish to greyish buff*, with white throat chevron, chest mottled dark brown and grey, rest of underparts barred dark brown. *In flight, white bar on primaries about 50–60% out along leading edge of outer wing.* Flight feathers molted on winter grounds. ♀: white bar on primaries narrower and less distinct (mostly on inner webs so often hard to see at rest), white throat chevron less extensive, often washed buff, lacks white tail bar. Rare rufous morph ♀ has ground color ruddier overall than either Lesser or Antillean nighthawks. **Juv:** resembles ♀ but upperparts overall paler, greyer and buffier, and more uniform, primaries tipped cinnamon-buff to whitish, buff throat chevron often indistinct, whitish bar on primaries of ♂ usually bolder than adult ♀ bar. Attains adult-like tail in 1st winter (1st prebasic possibly complete?).

Voice. Usually silent except during breeding season, call a sharp, nasal *beenk* or *peehn*; in display flight, ♂ stoops with wings making a low rushing 'boom' at bottom of dive. May give a clucking *gwek* when flushed.

Habs. Open and semiopen areas, typically in fairly arid situations; also humid forest and edge, scrub, etc., in migration. Roosts in trees and on ground. Flight similar to Lesser Nighthawk but wingbeats more rangy and less fluttery, flies low to high; ♂♂ on breeding grounds often very high and hard to see though audible. In flocks during migration when not known to associate with Lesser.

SS: See Lesser Nighthawk.

SD: C to F but local summer resident (Apr–Aug; near SL–2000 m) in interior and on adjacent Pacific Slope from Son and Chih at least to Zac (SNGH, SW), and from N Mich and SW Mex (SNGH, SW) locally to Honduras; possibly also in Coah (Urban 1959); on Atlantic Slope from Tamps locally to Isthmus, and in pine savannas of Belize and the Mosquitia. F to C transient (Mar–May, Aug–Nov), commonest on Atlantic Slope, few records from Pacific Slope. Winter reports (as in Belize Christmas Counts) require verification. Vagrant to Clipperton.

RA: Breeds N.A. to Panama; winters in S.A.

NB1: Sometimes known as Booming Nighthawk. **NB2: Antillean Nighthawk** *C. gundlachii* (8–8.5 in, W 20–23.2 in, T 3.5 in; 20.5–21.5 cm, W 51–59 cm, T 9 cm), formerly considered conspecific with Common Nighthawk, may occur as transient or vagrant (Apr, Sep–Oct) on Caribbean islands in region (Apr spec. from Swan Islands). Much like Common Nighthawk but smaller, shorter-winged (wingtips often fall short of tail on perched birds), penultimate primary tip spacing usually shorter than adjacent inner primary tip spacing at rest (longer in Common). Underbody buff to pale cinnamon, barred dark brown; usually not safely told by plumage. Flight more fluttery than Common Nighthawk, less rangy. May call on migration, species-specific call a sharp nasal *pi-dik* or *pi-di-dik* with slightly buzzy or metallic quality. Breeds in Caribbean, probably winters in S.A.

PAURAQUE
Nyctidromus albicollis *   Plate 27
Tapacaminos Picuyo

ID: 11–12 in, T 5–6 in (28–30.5 cm, T 12.5–15 cm). Common in tropical lowlands. *Tail very long and rounded, projects well beyond wingtips at rest.* Dimorphic. **Descr.** Sexes differ. **Grey morph ♂:** crown greyish with bold black streaks, *auriculars rufous*; upperparts cryptically patterned pale grey, black, and buff, *wing coverts tipped buff, forming neat rows of spots,*

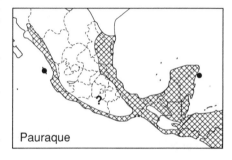

Pauraque

scapulars marked with bold black triangles or diamonds neatly edged buff to cinnamon. *Primaries blackish with bold white bar across outer 5, striking in flight.* Inner rectrices greyish with indistinct narrow wavy dark bars, *white inner webs of outer rectrices (but outermost all dark) form striking oblique white flashes on spread tail.* Underparts buff, with *white throat side chevrons,* chest mottled grey and brown, rest of underparts narrowly barred dark brown. ♀: bar on primaries narrower, pale buff to cinnamon, white in tail reduced to tips of outer rectrices; often looks plain in flight. **Rufous morph:** ground color cinnamon to cinnamon-brown. **Juv:** upperparts overall paler, sandy buff, with black spots, brighter wing covert edgings. Throat buff, underparts pale cinnamon-buff to pale buff, barred and scalloped dark brown. Wings and tail resemble ♀ but ♂ has more white in tail (though less than adult ♂), white bar on primaries often brighter than ♀.

Voice. Loud whistles, given year round: a loud, slightly explosive *p'weeEER,* often preceded by one or more, often hesitant, clucks, *puc, puc, puc, puc p'weeEER;* also (mostly Mar–Aug), an incessantly repeated *p'weeer, p'weeer ... or whee'oo, whee'oo ...,* 10/19–32 s, rarely a longer *poo wee-ee-oo'oo* or *pu pree-o-ree-u,* possibly two birds calling together. Perched birds often utter low clucks while bobbing nervously.

Habs. Forest, second growth, scrub, edges, and adjacent open and semiopen areas. Hunts from ground and low perches, often seen on quiet roads whence ♂♂ flush with a blur of white in wings and tail. Eggs: 2, pinkish to pale cinnamon, marked with browns (WFVZ).

SS: Very long tail, voice, and white bars on primaries of ♂ distinguish Pauraque from other nightjars; note also scapular pattern.

SD: C to F resident (SL–1800 m) on both slopes from S Son and Tamps to El Salvador and Honduras.

RA: Mexico (and S Texas) to Peru and N Argentina.

NB: Formerly known as Common Pauraque.

COMMON POORWILL
Phalaenoptilus nuttallii ∗   Plate 27
Pachacua Norteña

ID: 7–7.5 in, T 2.7–3 in (18–19 cm, T 7–7.5

cm). N Mexico. *Tail relatively short.* **Descr.** Sexes similar. **Adult:** crown, nape, and upperparts grey with cryptic black markings, broad paler sides of crown contrast with darker median crown and *blackish auriculars;* note also marbled pale grey wing coverts, black arrowhead marks on scapulars. Primaries barred dark brown and rufous. Inner rectrices grey with fine black bars, *outer 3 rectrices* blackish distally, *boldly tipped white;* narrower in ♀ and may be tinged buffy cinnamon. *White forecollar contrasts with sooty throat and upper chest,* underparts pale grey (at times buff on belly), barred blackish. **Juv:** upperparts paler and more uniform, silvery grey, with sparse black dots, cinnamon mottling, forecollar pale cinnamon-buff.

Voice. Song (Mar–Aug) a mellow 3-syllable *pur-whi-u'* or *poor-will-up',* 10/9–18 s; last note often inaudible at a distance. Also quiet clucks like most nightjars.

Habs. Arid to semiarid, open and semi-open country, scrub. Terrestrial, hunts from ground with short fluttering jumps into the air; often seen on quiet roads. Eggs: 2, creamy to pale pinkish, rarely flecked dusky (WFVZ).

SS: Small size, relatively short tail, and voice distinguish Common Poorwill from most nightjars in range; Eared Poorwill (W Mexico) more arboreal, longer-tailed, browner, lacks contrasting dark auriculars; note voice.

SD: C to F resident (near SL–2500 m) in Baja, on Pacific Slope S to Son, on Plateau and adjacent slopes S to N Jal and SLP, and in Tamps. More widespread (Sep–Apr) when N migrants present. S limit of winter range poorly known.

RA: W N.A. to cen Mexico.

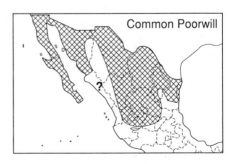

Common Poorwill

GENUS NYCTIPHRYNUS

These poorwills are small, poorly known neotropical nightjars. Sexes similar but ♀ commonly exhibits rufous morph. Chest feathers often distended into an 'apron'.

EARED POORWILL

Nyctiphrynus mcleodii   Plate 27
Pachacua Prío

ID: 8–8.5 in, T 3.7 in (20.5–21.5 cm, T 9.5 cm). *Endemic to W Mexico.* Name derives from erectile elongated postocular feathers. **Descr.** Sexes similar. **Grey morph:** crown and nape sandy grey with narrow buff hindcollar. Upperparts grey-brown, *scapulars and wing coverts with chestnut blotches, wing coverts spotted white;* primaries barred dark brown and rufous. Tail sandy grey-brown, outer 4 rectrices blackish distally, boldly tipped white. Throat and underparts finely patterned sandy brown and greyish, with white forecollar, belly flecked white. **Rufous morph** (mainly ♀): ground color vinaceous cinnamon to cinnamon-brown. Juv undescribed (?).
Voice. Song (mostly Feb–Jun, irregularly Dec–Jan) a sharp, clipped, resonant *ree-oo'* or *preeOO* or *wheeOO*, ends emphatically, 10/21–30 s; may have slight burry quality. Also a slightly sharp *gwik* or *wuik.*
Habs. Arid to semihumid oak and pine-oak forest and woodland on rocky slopes; also (at least seasonally, Feb–Mar) overgrown fields with scattered trees and wooded gullies. Inconspicuous unless calling; hunts mostly from perches at mid- to upper levels in trees, less often along edges of quiet roads. Eggs: 2, white, unmarked (WFVZ).
SS: Appears fairly pale and uniform overall, note distinct whitish tip to tail (obvious below), clean white forecollar. See Common Poorwill. Buff-collared Nightjar larger and greyer, cinnamon hindcollar usually obvious; note tail pattern and voice. Whip-poor-wills larger and darker, less uniform, often show pale brows and pale braces, ♂♂ have more white in tail, ♀♀ have buffy-cinnamon tail tips; note voice.
SD/RA: F to C presumed resident (1200–2500 m) on Pacific Slope locally from SW Chih to Oax.
NB: Formerly placed in the genus *Otophanes.*

YUCATAN POORWILL

Nyctiphrynus yucatanicus   Plate 27
Pachacua Yucateca

ID: 8–8.5 in, T 3.7 in (20.5–21.5 cm, T 9.5 cm). *Endemic to Yuc Pen.* **Descr.** Sexes similar. **Brown morph:** crown, nape, and upperparts dark rich grey-brown, cryptically patterned black and brown, narrow buffy-cinnamon hindcollar sometimes noticeable, scapulars coarsely marked black, wing coverts spotted white. Primaries barred dark brown and rufous. Tail dark rich grey-brown, *outer 4 rectrices blackish distally, boldly tipped white.* Throat and underparts mottled grey-brown and brown, forecollar white, belly mottled whitish. **Rufous morph** (mainly ♀): ground color pinkish cinnamon. Juv undescribed (?).
Voice. Song (mostly Feb–Oct?) a loud, slightly resonant *whirr'* or *whirrrr*, 10/12–16 s, rarely 10/25 s; also a slightly liquid, at times accelerating clucking *puk-puk …*, and a sharp *week week week'* in flight (alarm call?). Song traditionally attributed to Yucatan Nightjar *Caprimulgus badius* until the 1980s (Pierson 1986; Howell 1990a).
Habs. Humid to semihumid forest, scrub

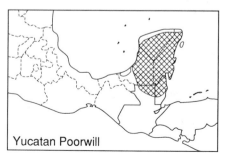

Eared Poorwill

Yucatan Poorwill

and edge. Calls and hunts from trees, may also hunt from ground (?). Eggs buff, speckled brown (Van Tyne 1935).

SS: Yucatan Nightjar and Northern Whip-poor-will (migrant, not recorded in Yuc Pen) larger with longer wings and tail, different tail patterns; note voice.

SD/RA: F to C presumed resident (SL–250 m) in Yuc Pen, S to N Guatemala and S Belize (Howell *et al.* 1992*b*).

NB: Formerly placed in the genus *Otophanes*.

GENUS CAPRIMULGUS

These are often considered typical nightjars of wooded and forested habitats. Sexes differ, most noticeably in tail pattern. Overall patterned in rich browns, grey-brown, and black, with contrasting white to buff forecollar, and also often a hindcollar.

CHUCK-WILL'S-WIDOW
Caprimulgus carolinensis   Plate 27
Tapacaminos Carolinense

ID: 11.5–12.5 in, T 4–4.5 in (29–33 cm, T 10–11.5 cm). *Winter visitor. A large dark brown nightjar.* **Descr.** ♂: *dark rich brown overall*, crown streaked blackish, upperparts cryptically patterned black and cinnamon, primaries blackish brown, barred rufous. Throat rufous with *whitish to pale buff forecollar*, chest mottled blackish, cinnamon, and buff, belly cinnamon, barred dark brown. Inner rectrices brown with fairly bold, broken black bars, *inner webs of outer 3 rectrices white (buff from below), flashed in flight if tail spread.* ♀/imm: forecollar pale buff, tail barred black and cinnamon-brown, lacks white.
Voice. Rarely heard on spring migration; song a rapid, clipped *tchk wee-u' wee-u'*.
Habs. Humid to semiarid forest and woodland, adjacent open and semiopen

areas at night. Roosts in trees, rarely on ground, hunts from perches, also from ground (?).

SS: Other nightjars smaller; Tawny-collared and Yucatan nightjars have tawny-cinnamon hindcollars; note tail patterns. See Pauraque.

SD: F to C transient (mid-Feb–early May, late Aug–mid-Nov) on Atlantic Slope from Tamps and Yuc Pen S. Probably (?) U to F winter visitor (Sep–Apr, SL–1500 m) on Atlantic Slope from S Tamps to Honduras (except Yuc Pen), and on Pacific Slope S from Isthmus; most records, at least in Mexico, are during migration periods.

RA: Breeds E N.A.; winters from S USA to N C.A.

TAWNY-COLLARED NIGHTJAR
Caprimulgus salvini   Plate 27
Tapacaminos Ti-cuer

ID: 9.5–10 in, T 4.3–5 in (24–25.5 cm, T 11–12.5 cm). Endemic; E Mexico and N C.A. **Descr.** ♂: crown, nape, and upperparts dark rich grey-brown, cryptically patterned black, brown, and cinnamon, *tawny-cinnamon hindcollar conspicuous*, scapulars mottled whitish to pale buff. Primaries blackish brown, barred rufous. Inner rectrices barred blackish and brown, *outer 3 rectrices blackish, broadly tipped white* to buffy white (mostly on inner webs). Throat and underparts mottled blackish, cinnamon-brown, and creamy, *forecollar buff to pale cinnamon*, belly coarsely spotted whitish. ♀: outer rectrices narrowly tipped *pale cinnamon to cinnamon-buff.* Juv undescribed (?).
Voice. Song (mostly Mar–Jul) an abrupt, clipped *chi-wihw'* or *tchi-wheeu*, repeated rapidly, 10/8–11 s; also a shorter, faster *chi-weeu*, 10/8–9 s.
Habs. In NE Mexico, arid to semihumid,

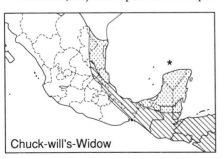

Chuck-will's-Widow

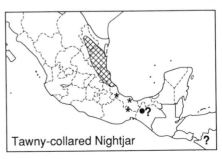

Tawny-collared Nightjar

brushy woodland, thorn forest, and dense scrub, not in overgrazed or more humid, wooded and forested areas. Also humid to semihumid forest edge and clearings in S Mexico (at least in winter). Calls and hunts from perches in trees and bushes, often well hidden when calling; may also hunt from ground (?). Eggs: 2, whitish, marked with browns and greys (MLZ).

SS: See Chuck-will's-widow, Pauraque. Northern Whip-poor-will (migrant) paler overall, lacks bold hindcollar; note tail pattern.

SD/RA: C to F breeder (Apr–Aug; SL–500 m) on Atlantic Slope from cen NL to N Ver; some apparently withdraw in winter when recorded (Dec–Feb) in S Ver (Winker *et al.* 1992c) and N Oax (Binford 1989). Possibly local resident (separate ssp?) in SE Mexico and N C.A., with Jun spec. from N Chis (Alvarez del Toro 1954), and Apr spec. from Nicaragua (AMNH). Reports of calling birds in S Ver in Jan–Mar (Winker *et al.* 1992c) require verification.

NB1: Traditionally considered conspecific with Yucatan Nightjar C. *badius*, the Nicaragua specimen sometimes incorrectly attributed to that form (Sibley and Monroe 1991). NB2: Also known as Chip-Willow.

YUCATAN [TAWNY-COLLARED] NIGHTJAR

Caprimulgus badius   Plate 27
Tapacaminos Yucateco

ID: 9.5–10 in, T 4.3–5 in (24–25.5 cm, T 11–12.5 cm). *Endemic to Yuc Pen.* **Descr.** ♂: crown, nape, and upperparts dark rich grey-brown, cryptically patterned black, grey, and cinnamon, *tawny-cinnamon hindcollar conspicuous*, scapulars mottled silvery white. Primaries blackish brown, barred rufous. Tail appears blackish overall with *outer 3 rectrices very broadly tipped white*. Throat dark, *forecollar white,*

underparts mottled blackish, brown, and white. ♀: outer rectrices tipped buff to pale buff. **Imm** ♂ (?): outer rectrices broadly tipped buffy white to buff. Juv undescribed (?).

Voice. Song (mostly Feb–Aug) a loud, clear *puc ree-u-reeeu'* or *pc weeu weeweeeu,* 10/19–20 s, may suggest Chuck-will's-widow; the first note quiet and not always audible at a distance; also hard, hollow clucking, *k-lok k-lok … or p-tok …* Song traditionally confused with Yucatan Poorwill.

Habs. Scrub and brushy woodland, forest edge. Habits much as Tawny-collared Nightjar. Eggs: 2, white, with sparse dark brown and mauve-grey flecks and blotches (MCZ).

SS: See Chuck-will's-widow, Yucatan Poorwill, Pauraque. Northern Whip-poor-will (migrant) paler overall, lacks bold hindcollar and white mottling below.

SD/RA: C to F resident (around SL) in Yuc Pen (including Isla Cozumel) S to S Camp (SNGH) and cen QR. Some apparently withdraw S in winter (Dec–Feb), when recorded in Belize and N Honduras (KB photo), including an apparent migrant on Half Moon Cay; old reports from Guatemala appear to be erroneous (Van Rossem 1934).

NB: See NB1 under Tawny-collared Nightjar.

BUFF-COLLARED NIGHTJAR

*Caprimulgus ridgwayi**   Plate 27
Tapacaminos Préstame-tu-cuchillo

ID: 8.5–9 in, T 4–4.3 in (21.5–23 cm, T 10–11 cm). W Mexico and interior CA. **Descr.** Sexes differ. ♂: crown, nape, and *upperparts brownish grey,* cryptically patterned black and pale grey, *cinnamon hindcollar conspicuous*; primaries blackish brown, barred cinnamon-rufous. Tail brownish

Yucatan Nightjar

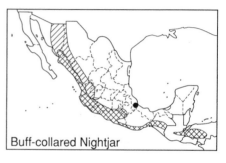

Buff-collared Nightjar

grey, *outer 3 rectrices darker, broadly tipped white*, mostly on inner webs hence not always conspicuous. Throat dusky, *forecollar buff to buffy white*, underparts greyish buff, chest mottled dark brown, rest of underparts barred dark brown. ♀: *outer 3 rectrices tipped pale cinnamon to whitish.* Juv undescribed (?).

Voice. Song (mostly Mar–Aug) a striking, rapid, accelerating series of hollow clucks ending with a sharp note, *ku-ku-kukukukukuku-a-chia'*, or *koo-koo-kookookookoo-oo-chee'a*, 10/28–38 s; at times shorter and faster series, less often slower, 1/7 s. Calls include a low, clucking *chuuk* and *kruk kruk, kruk . . .*, and harder, clucking chatters.

Habs. Arid to semiarid brushy woodland, scrub, thorn forest, and edge. Calls and hunts from or near ground; sits on quiet roads, at times within a few m of Pauraques. Eggs: 2, buff, marked with browns (WFVZ).

SS: See Pauraque, Common and Eared poorwills. Whip-poor-wills darker, lack striking hindcollar, forecollar often brighter white in ♂♂; note voice.

SD/RA: F to C resident (SL–1800 m) on Pacific Slope from S Son to Oax, in interior (150–1600 m) in Balsas drainage, and locally from Chis to Honduras; in summer (May–Aug?) F to C N to N Son (also S Arizona) and interior NW Mexico. F to C resident on Atlantic Slope in cen Ver (Coffey and Coffey 1989; Howell and Webb 1990*b*).

NB: Sometimes known as Cookacheea or Tucuchillo.

NORTHERN WHIP-POOR-WILL

Caprimulgus vociferus   Fig. 29
Tapacaminos Cuerprihuiu

ID: 9–9.5 in, T 3.7–4.2 in (23–24 cm, T 9.5–10.5 cm). Winter visitor. **Descr.** Sexes

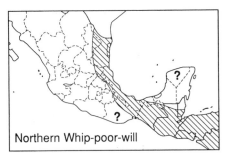

Northern Whip-poor-will

Fig. 29 Typical tail patterns of male Northern Whip-poor-will (a) and Mexican Whip-poor-will (b)

differ. ♂: crown, nape, and *upperparts brownish grey*, cryptically patterned black and cinnamon, *narrow buff hindcollar* indistinct, scapulars may appear as contrastingly pale braces; primaries blackish brown, barred cinnamon-rufous. Inner rectrices brownish grey, *outer rectrices blackish with broad white tips* (Fig. 29, and see Mexican Whip-poor-will) which may be tinged cinnamon distally. Throat dark, forecollar whitish to buff, underparts greyish buff to dirty cinnamon, mottled dark brown. ♀: forecollar buff, outer 3 rectrices tipped pale cinnamon to buffy white.

Voice. Unlikely to be heard in region, except perhaps in spring migration when may give a slightly burry *whir-pr-iweeu*, distinctly clearer and typically faster-paced than Mexican Whip-poor-will, 10/10–12 s. Calls include hollow clucks.

Habs. Forest and woodland in general, including pine–oak; may occur in coastal scrub during migration.

SS: See Buff-collared, Tawny-collared, and Yucatan nightjars, Chuck-will's-widow. Mexican Whip-poor-will rarely distinguishable except by voice.

SD: Probably an U to F transient and winter visitor (Sep–Apr; near SL–1800 m) on Atlantic Slope from Tamps to Honduras, and on Pacific Slope from Isthmus S, but few winter records; unrecorded Yuc Pen but should occur there as transient (?).

RA: Breeds E N.A.; winters from S USA to W Panama.

MEXICAN WHIP-POOR-WILL

*Caprimulgus arizonae** or *C. vociferus*
(in part)   Plate 27
Tapacaminos Cuerporruin   Figure 29

ID: 9–9.5 in, T 3.7–4.2 in (23–24 cm, T

9.5–10.5 cm). Highlands throughout. **Descr.** Much like Northern Whip-poor-will. Plumage differences between Northern and Mexican whip-poor-wills are rarely absolute, and most characters appear to overlap when large series are examined. Birds from Oax S, however, average darker and redder than Northern Whip-poor-wills, with pale braces often reduced to absent. One feature which appears to be useful in separating extremes of the two forms is the amount of white on the outer rectrices of ♂♂, which is noticeably more extensive in Northern (Fig. 29).

In Northern Whip-poor-will ($n=61$), white area (measured along shaft) on the outermost rectrix averages 35.3 mm (26–45 mm), and along the next-to-outermost rectrix, 50.5 mm (42–60 mm). In Mexican Whip-poor-will ($n=70$), the white area averages 25.1 mm (11–34 mm), and 41.1 mm (32–48 mm) on outermost and next-to-outermost rectrices respectively. (Two specimens labeled Mexican, from the range of Mexican Whip-poor-will (Oax and Honduras), were excluded from this analysis. These showed white of 36–38 mm on outermost rectrix, 50–52 mm on next-to-outermost); both were collected in Nov and likely represent misidentified migrant Northerns.) ♀: tail tips may average narrower (but not useful in field) and also darker, buff to cinnamon-buff, in Mexican. **Juv:** upperparts paler with more warm buff and cinnamon spotting and marbling, pale buff throat bordered by dark malar stripes. Underparts paler, cinnamon to pale buff, coarsely barred and scalloped dark brown.

Voice. Song (mostly Feb–Aug) a burry *pwurr-p'wiun* or *whirr'p'wiirr* (notably slower and lower than Northern Whip-poor-will), 10/9.5–15 s, averages slower southward. Calls include hollow clucks.

Habs. Humid to semiarid pine and pine–oak forest, probably to adjacent woodland and forest in foothills (and lowlands?) in winter. Calls mostly from trees; hunts from trees or ground. Eggs: 2, whitish, unmarked or with fine dusky markings (WFVZ).

SS: See Buff-collared, Tawny-collared, and Yucatan nightjars, Chuck-will's-widow, and Northern Whip-poor-will.

SD/RA: C to F resident (1400–3000 m, locally to 500 m in winter) in interior and on adjacent slopes from Chih and Coah (and SW USA in summer) to Honduras; also (summer resident at least) in BCS (Banks 1967a; Howell and Webb 1992f). Probably withdraws (?), at least in part, from NE and NW Mexico in winter, and possibly altitudinal migrant elsewhere.

SPOT-TAILED NIGHTJAR
Caprimulgus (?) maculicaudus Plate 27
Tapacaminos Colimanchado

ID: 8–8.5 in, T 3.2 in (20.5–21.5 cm, T 8 cm). S savannas. **Descr.** Sexes differ. ♂: face brown with pale buff supercilium and wedge-shaped blackish malar stripe bordered buff; blackish crown flecked rufous, *cinnamon hindcollar usually distinct.* Upperparts grey-brown, cryptically vermiculated dark brown and buff, with *prominent buff scapular V, wing coverts and secondaries tipped buff,* forming distinct rows of spots; primaries blackish brown, barred cinnamon-rufous basally. Inner rectrices greyish, barred black, *outer rectrices* blackish with cinnamon bars on outer webs, *tipped white;* 2–3 pairs of oval white spots on underside of tail rarely visible in the field. Throat pale buff, ruffled chest coarsely mottled cinnamon to buffy white, rest of underparts pale cinnamon with sparse dark barring, unbarred on belly. ♀: outer rectrices barred grey-brown and cin-

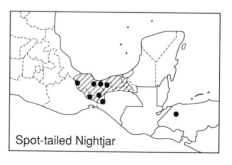

Mexican Whip-poor-will

Spot-tailed Nightjar

namon with narrow pale tips, no white spots below. Juv undescribed (?).

Voice. Song (late Mar–Jun) a high, passerine-like *t-seet'* or *t-seeit'*, at times varied to *t'tsuwee* or *t't'-swee*; carries surprisingly well but not always repeated incessantly as in other *Caprimulgus*, 10/22–30 s; also a more leisurely *pt swee-i'*, 1/5 s. Calls include a rapid *t-seet t-seet t-seet t-seet'*, an accelerating *t-seet seet-seet*, and a shriller, slightly wailing *seeeu* or *see-ee-eeii* in flight, which may be accompanied by a triple wing flutter, *futt-a-futt* or *flut-flut-flut*.

Habs. Savannas, open short grassy areas with scattered bushes. Calls from ground or low perch, also in flight. Flies low; hunts from ground and in flight. Begins calling before dark, earlier than most nightjars. Eggs: 2, creamy pink with fine dark markings (SNGH, SW).

SS: Small size, habitat, buff braces, and voice distinctive.

SD: C to F but local summer resident (SL–500 m; late Mar–Jul at least) on Atlantic Slope from S Ver to Tab, and in the Mosquitia; probably elsewhere in savannas of N C.A. (for example, near Lake Yojoa, Honduras, Monroe 1968). Wintering grounds unknown, suspected to be in S.A.; occasional winter reports from region require verification. Feb reports from Oax (Blake 1949a) are based on Avilés specimens (Binford 1989) and thus are not credible.

RA: SE Mexico, locally in C.A., and S.A. from Colombia to S Brazil.

NB1: Sometimes known as Pit-Sweet. **NB2:** Appears sufficiently different from typical *Caprimulgus* to warrant separate generic treatment.

POTOOS: NYCTIBIIDAE (2)

A neotropical family of large nocturnal birds that superficially resemble nightjars. Potoos hunt from perches in a manner recalling pewees, often returning to the same perch after a sally and using favored perches regularly. Roost by day in trees where cryptic coloration and nearly motionless upright posture make them appear much like branches; in alarm, potoos assume a very attenuated posture with the head and bill pointing upwards. Wings and tails long and rounded, flight silhouette may suggest large raptor. Bills relatively small, slender, and decurved, but gapes enormous, rimmed with bristles to aid in capturing prey. Legs short, feet fairly strong with well-developed claws. Ages differ, sexes similar though some tendency to dimorphism, ♀♀ often more rufescent. Young altricial; molts poorly known; juv plumage mostly lost within a few weeks. Voices guttural, heard mainly on moonlit nights.

Potoos feed mostly on insects. A single egg (whitish, flecked with browns and black; WFVZ) is laid in a hollow along a branch or at the end of a stub.

GREAT POTOO

Nyctibius grandis guatemalensis   Plate 26
Biemparado Grande

ID: 19–23 in, W 44–49 in (48.5–58.5 cm, W 112–125 cm). *SE humid forest. Very large, overall pale appearance.* **Descr. Adult:** *eyes brown,* nocturnal 'eyeshine' bright reddish orange. *Finely and cryptically patterned overall* with pale cinnamon, greys, browns, black, and white; *crown and nape finely vermiculated, rarely with distinct dark streaks.* Underparts more coarsely vermiculated and barred, chest usually with coarse blackish mottling; flight feathers barred. **Juv:** more whitish overall.
Voice. Deep, guttural, strangled cries, memorable if not frightening, a drawn-out *awhrrr,* a groaning *rroh-rr,* and *aahrrrr.* Also a slightly owl-like, deep *woh', woh'.*

Habs. Humid evergreen forest and edge. Roosts high in trees and rarely found by day. At night usually perches on exposed high snags (more than 8–10 m above ground), along edges, rivers, etc. Flight powerful with fairly slow, deep wing-beats.
SS: Common Potoo markedly smaller and proportionately smaller-headed, darker and browner overall, plumage more coarsely marked with distinct streaking on crown and underparts; head rarely appears pale (beware juvs), and often shows a dark gape line. At night, Common inhabits more open areas, often perches lower. Nightjars much smaller.
SD: F to U resident (SL–500 m) on Atlantic Slope in E Chis (Rangel-S. and Vega-R. 1989; SNGH, SW) and Guatemala; reports from Belize (Wood *et al.* 1986) are erroneous, those from Honduras (Monroe 1968) require verification.
RA: SE Mexico to S Brazil.

NORTHERN [COMMON] POTOO

Nyctibius jamaicensis*   Plate 26
Biemparado Norteño

ID: 15–17 in, W 36–40 in (38–43 cm, W 91–101 cm). *Tropical lowlands. Large, brownish overall.* **Descr. Adult:** *eyes yellow,* nocturnal 'eyeshine' bright reddish orange. *Crown and nape heavily streaked blackish and brown, head typically shows dark gape line.* Upperparts cryptically

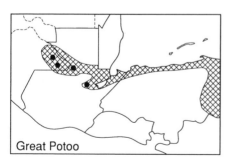

Great Potoo

patterned with browns, greys, black, and white; flight feathers barred. *Throat and underparts* grey-brown to pale greyish cinnamon, *with dark streaking*, throat often paler, chest with variable blackish mottling. **Juv:** *paler, pale grey overall*, cryptically patterned with greys, browns, and black, *usually has dark gape line.*

Voice. A deep, guttural, slightly eerie call, usually followed by 1–2 short notes, *wahhrrr wah-wah* or *bwaahhhr, ah-ah*; at a distance, often only the first part audible.

Habs. Open and semiopen areas with scattered trees and hedges, forest and edge. Roosts fairly high to high in trees. Hunts from exposed perches such as fence posts, corn stalks, usually fairly low but also in forest canopy; flight strong with deep wingbeats.

SS: See Great Potoo. Nightjars smaller.

SD: F (but may be local) resident (SL–1500 m) on both slopes from S Sin and S Tamps to El Salvador and Honduras; possibly extirpated from much of El Salvador (Thurber *et al.* 1987).

RA: Mexico and West Indies to cen Costa Rica.

NB: Traditionally considered conspecific with *N. griseus*, Grey (Common) Potoo, of S C.A. and S.A. See Davis (1978).

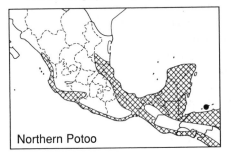

Northern Potoo

SWIFTS: APODIDAE (10)

Swifts are supreme aerialists with long, pointed wings; although they resemble swallows superficially, the two families are quite different. Tails vary from relatively short and squared with 'spines' (projecting stiff shaft tips) to long and deeply forked. Small, short bills open into very wide gapes which take in prey during flight. Small feet have well-developed claws for perching on vertical faces. Bills blackish, dusky to pink feet rarely visible in the field. Ages/sexes similar or slightly different. Plumage blackish overall, often with contrasting paler or white areas, especially on throat, chest, and rump. Molts poorly known. Voices consist of chattering and screaming calls; solitary birds rarely vocalize; calls usually made by chasing pairs or flocks.

Swifts feed on insects and other invertebrates caught in flight. Most species nest in caves, crevices, tree cavities, behind waterfalls, etc., singly or colonially. Nests typically are cups of saliva and small sticks (*Chaetura*), or mud and plant matter. Eggs: 1–6, white (WFVZ).

Note: Swifts are rarely seen other than in flight, often very high overhead, and species identification can be a real problem. Shape and flight manner are the two best characters; in general, smaller species flap faster and more often than larger species; all swifts glide on bowed wings, larger species soar. Plumage differences are often not visible (or barely exist anyway) given typical views. First, become familiar with common and widespread species such as Vaux's and White-collared. Frequenting ridge tops, passes, or other good lookouts, especially on cloudy or rainy days, may be rewarded with prolonged close views of swifts, at times of several species together. After a while, shape differences, such as those between *Cypseloides* and *Chaetura*, will become apparent. In many cases, however, one will still have to settle for *Chaetura* sp. or *Cypseloides* sp.

GENUS CYPSELOIDES

These are medium-sized swifts with long, pointed, but relatively broadly based wings, usually swept back (scimitar-like); tails squared to slightly cleft. *Cypseloides* are among the highest fliers and are faster and more streamlined than the stubbier, more fluttery *Chaetura*. White scalloping or mottling on the belly has been traditionally attributed to ♀♀ but appears to be age-related (MM, personal communication); 1–2 years required to attain adult plumage but plumage sequences are not well known. Buzzy to hard chipping calls. Nest singly or in small groups, in caves or on cliff ledges, often near water such as behind waterfalls; 1–2 eggs.

BLACK SWIFT
Cypseloides niger*   Plate 28
Vencejo Negro

ID: 6–7 in, W 14.3–16.5 in, T 2–2.3 in (15–18 cm, W 36.5–42 cm, T 5–6 cm). Widespread summer visitor; N.A. migrants larger (W 15–16.5 in, 38–42 cm). **Descr.** Ages/sexes differ slightly. Relatively long-tailed (especially ♂). ♂: *Tail cleft to slightly cleft. Blackish brown overall* (looks black), *whitish scaling on forehead and supraloral area may be noticeable as a frosty brow on flying birds*; throat may also appear pale at times. ♀: *tail squared and shorter than* ♂. **Juv:** head and body extensively tipped whitish. Imms (1st years) may have whitish scalloping on belly (hard to see in field), more so in ♀♀ (?).

Voice. Heard infrequently but distinctive, fairly hard, dry chipping, *chi-chi-chi-chit* or *shi-shi-shi-shi-shi*, etc., not strikingly hard or buzzy.

Habs. Aerial and wide-ranging, nests in highlands, notably behind waterfalls, wanders to and migrates through lowlands. Wingbeats often relatively slow and rangy but can be quick and dashing, often glides and soars without flapping, associates with other swifts.

SS: Chestnut-collared Swift smaller, wingbeats faster, tail squared in both sexes, appears shorter-winged and relatively *Chaetura*-like, chestnut collar distinctive when present. White-chinned Swift proportionately shorter-winged, shorter tail squared in both sexes, flight often more fluttery (MM). White-collared and White-naped swifts much larger, wingbeats slow and powerful, soar with wings in paddle-like bulge.

SD: F to C but local summer resident (Mar–Aug; 1200–3000 m, ranging locally to lowlands) in interior and on adjacent Pacific Slope from W cen Chih (SNGH, SW) to Oax, on Atlantic Slope from Hgo (Howell and Webb 1992e) to N Oax. Probably F to U but local summer resident in interior and on adjacent Pacific Slope from Chis to Honduras but few records (none from El Salvador). N.A. breeders occur as U to F transients (Apr–May, Aug–Sep) on Pacific Slope and offshore from Baja to at least Guatemala. Winter status of birds breeding in region unclear; most (all?) withdraw (Oct–Feb?) from N of Isthmus.

RA: W N.A. to Middle America and Caribbean; N populations (at least) migratory, probably wintering in S.A.

WHITE-CHINNED SWIFT
*Cypseloides cryptus**   Plate 28
Vencejo Barbiblanco

ID: 5.5–6 in, W 13–14 in, T 1.5–1.8 in (14–15 cm, W 33–35.5 cm, T 4–4.5 cm). Two ssp: *storeri* (SW Mexico) and *cryptus* (C.A.). Status poorly known. Descr. Ages differ slightly, *relatively short-tailed*. Adult. *Cryptus*: tail squared. Blackish brown overall (looks black), whitish scaling on chin and sides of forehead rarely visible in the field. *Storeri* ('White-fronted Swift'): whitish frosting extends across forehead. Juv (unknown in *storeri*): extensive white scaling on belly and undertail coverts hard to see in field. Imms (1st years) may have whitish scalloping on belly (hard to see in field), more so in ♀♀ (?).

Voice. Birds thought to be White-chinned Swifts in Belize gave hard, buzzy calls and chatters unlike Black or Chestnut-collared swifts (SNGH).

Habs. Aerial and wide-ranging. Flight much as Black Swift but higher wing-loading results in more fluttery wingbeats, less gliding; may not soar (MM).

SS: See Black Swift. Chestnut-collared Swift slightly smaller with longer tail, more *Chaetura*-like flight, buzzy crackling call, often shows chestnut collar.

SD: Apparently R summer resident (?) in SW Mexico (recorded Jul–Sep, 1500–2500 m, in Jal, Mich, and Gro, Navarro *et al.* 1992b; per ATP), and locally (Jun–Aug at least) from S Belize to N Nicaragua; breeding not documented in region. Mar–Jun sight reports of unidentified *Cypseloides* from Ver to Chis and S Belize (SNGH, SW, RGW) may pertain to this species, in-hand identification desirable.

RA: N and NW S.A., N to Costa Rica, locally N to SW Mexico (resident?).

NB: *Storeri* was described as a full species by

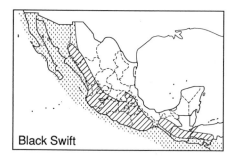

Black Swift

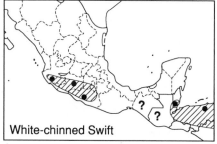

White-chinned Swift

Navarro *et al.* (1992*b*). Little unequivocal evidence supports this evaluation, however, and no data indicate that *storeri* is resident or even breeds in Mexico. Its specific 'distinctness' rests on more extensive whitish frosting in the face and on a subjective description of facial shape (Navarro *et al.* 1992*b*). Critical studies may show *storeri* to be a good species, endemic to SW Mexico, but based on present information we consider its recognition as a species premature.

CHESTNUT-COLLARED SWIFT
*Cypseloides rutilus**   Plate 28
Vencejo Cuellicastaño

ID: 5–5.5 in, W 12–13 in, T 1.5–1.8 in (12.5–14 cm, W 30.5–33 cm, T 4–4.5 cm). Widespread. Shape resembles *Chaetura*, tail squared (may appear slightly cleft when closed) with short spines ('invisible' in the field). **Descr.** Ages/sexes differ. **Adult:** blackish brown overall (looks black) with broad chestnut collar, broadest across upper chest where often mottled with grey-brown. Some adult ♀♀ may lack collar. **Juv:** sooty brown overall, may show partial or rarely full chestnut collar (fuller in juv ♂♂?).
Voice. Buzzy, crackling chatters, *zzchi zzchi-zzchi-zzchi-zzchi*, etc., quality may recall electricity crackling in pylon wires, occasional screechy notes thrown in; also a single, hard, buzzy *zrrrt*.
Habs. Aerial and wide-ranging, breeds in highlands, ranges to lowlands. At times in flocks up to a few hundred birds, associates with other swifts. Flight *Chaetura*-like but glides more often and spreads tail widely, wings more scimitar-like.
SS: See Black and White-chinned swifts. *Chaetura* swifts similar but throat, chest, and rump contrastingly paler, flight more

fluttery with shorter wings less swept back, shorter tails less often spread.
SD: F to C resident (1500–3000 m, ranging to SL) in interior and on adjacent Pacific Slope from Dgo and cen volcanic belt to Honduras, on Atlantic Slope from Hgo to Honduras. May occur N locally to W cen Chih (Jul 1991, SNGH, SW). Withdraws locally (Oct–Feb at least) from higher and colder interior in winter (Wilson and Ceballos-L. 1986). Some birds also may withdraw from N Mexico in winter (Oct–Mar?) as apparent migrating flocks seen (mid-Mar–May) in W Mexico (Mich, Col, Jal, SNGH, SW), but F to U in winter N to Sin and Nay (SNGH, PP), and in interior and on Atlantic Slope at least to Oax (SNGH).
RA: Mexico to E Peru and W Bolivia.
NB: Has been placed in the genus *Chaetura*; some authors suggest placement with *Streptoprocne*.

GENUS STREPTOPROCNE

These are very large swifts (White-naped being the largest swift in the world) with powerful flight and deep, slow wingbeats; spend much time soaring without flapping, wings spread into distinctive paddle-shaped bulges. Often in large screaming flocks whose wings make a loud rushing sound when fairly close. In comparison, they truly dwarf other swifts. Loud, screeching calls. Nest colonially in caves, sinkholes, etc., often in limestone regions near waterfalls; eggs may be laid on ledges with no nest structure built; 2 eggs.

WHITE-COLLARED SWIFT
*Streptoprocne zonaris**   Plate 28
Vencejo Cuelliblanco

ID: 8–8.5 in, W 19–21 in (20–21.5 cm, W 48–53 cm). S and E. *Tail slightly forked*

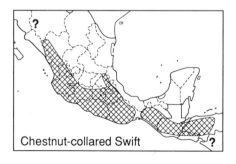

Chestnut-collared Swift

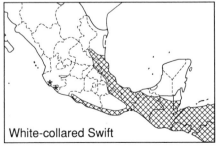

White-collared Swift

and with short spines. **Descr.** Ages differ, sexes similar. **Adult:** blackish overall with *bold white collar, broadest across upper chest.* **Juv:** *white usually restricted to broad white hindcollar, chest blackish* with narrow whitish scaling (rarely visible in the field). Some birds may attain a complete white collar at end of 1st year, showing coarse white scallops on chest in 1st winter and spring. Plumage variation and sequences need study.

Voice. Loud, screeching, and chattering cries, may recall parakeets.

Habs. Aerial and wide-ranging; nests in humid foothills and highlands but often ranges to coastal lowlands. Flight fast and powerful, soars for long periods with no flapping. Associates with other swifts.

SS: White-naped Swift larger and thicker-set with squared to slightly cleft tail, lacks white across chest (as may juv White-collared). See Bat Falcon.

SD: C to F resident (SL–2500 m) on both slopes from Gro and S Tamps, and in interior from Chis, to El Salvador and Honduras; absent from Yuc Pen. Irregularly (?) U visitor (Dec–Mar) N on Pacific Slope to Col and Jal (CB, SNGH), perhaps the S C.A. ssp *albicincta,* which has been collected (Nov–Dec) N to Chis.

RA: Mexico and Caribbean to N Argentina.

WHITE-NAPED SWIFT

Streptoprocne semicollaris   Plate 28
Vencejo Nuquiblanco

ID: 8.7–9 in, W 22–23.2 in (22–23 cm, W 56–59 cm). Endemic to W Mexico. *Tail squared to slightly cleft,* with short spines. **Descr.** Ages/sexes similar. *Blackish overall with broad white hindcollar,* white rarely visible from below but can be striking if seen slightly in profile.

Voice. Loud, burry screeching and chat-

tering, deeper and less shrieky than White-collared Swift.

Habs. Aerial and wide-ranging, typically associated with arid highlands and interior but ranges to humid coastal lowlands, at least in NW Mexico. Flight and habits much like White-collared Swift.

SS: See White-collared Swift, Black Swift.

SD/RA: F to C resident (1500–3000 m, ranging to SL in NW Mexico) on Pacific Slope and in adjacent interior from S Chih to Nay, and in interior and on adjacent Pacific Slope from Gro to Mex and Mor; range break may simply reflect absence of suitable nesting/roosting habitat. Some leave higher and colder areas in winter (absent DF Aug–Mar, Wilson and Ceballos-L. 1986). Vagrant (?) to Pacific Slope of Chis (3 Mar 1942, WFVZ, spec.). Sight reports from Oax, Chis, and Honduras (Monroe 1968) unconfirmed; a report from Hgo (AOU 1983) may be erroneous.

GENUS CHAETURA

These are small, relatively stubby swifts with very rapid wingbeats, squared tails (with spines at tips) may appear rounded when spread. Rarely glide for long periods without bursts of quick, almost fluttery flapping. Wings appear less swept-back than *Cypseloides.* Throat and chest contrastingly paler if seen well. High, twittering and chipping calls.

CHIMNEY SWIFT

Chaetura pelagica   Plate 28
Vencejo de Chimenea

ID: 4.7–5 in, W 11.8–12.8 in, T 1.5 in (12–12.5 cm, W 30–32.5 cm, T 4 cm). *Transient. Wings relatively long and swept back for a Chaetura.* **Descr.** Ages/sexes similar.

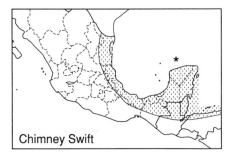

White-naped Swift   Chimney Swift

Sooty brown overall (looks blackish) with *pale grey throat and upper chest, rump and uppertail coverts paler sooty grey-brown, not contrasting strongly.*

Voice. Rarely heard in migration; high, chipping and twittering calls, fuller and mellower than Vaux's Swift.

Habs. Aerial, usually in flocks. May associate with Vaux's Swift when feeding but migrating flocks seem to be single-species. In late March and April, loose flocks of up to several hundred birds stream north day after day.

SS: Vaux's Swift smaller (M.A. breeding populations smallest), can be told with experience by shorter, stubbier wings which create different jizz; N.A. migrant Vaux's paler below than Chimney but M.A. populations of Vaux's similar below to Chimney; Vaux's typically shows paler and more contrasting greyish rump. Calls of Vaux's are higher and shriller. See Chestnut-collared Swift.

SD: C to F spring transient (mid-Mar–mid-May, mainly Apr; SL–1500 m) on Atlantic Slope from Honduras through E Mexico; U to F autumn transient (late Aug–early Nov) through N and E Yuc Pen and Atlantic Slope of C.A.; R on Gulf Slope of Mexico.

RA: Breeds E N.A.; winters in S.A.

VAUX'S SWIFT
Chaetura vauxi*   Plate 28
Vencejo de Vaux

ID: 4–4.5 in, W 9.7–11.5 in, T 1–1.3 in (10–11.5 cm, W 24.5–29 cm, T 2.5–3.5 cm). *Common and widespread; gaumeri of Yuc Pen smallest* (W 9.7–10.7 in, 24.5–27 cm). *Wings and tail relatively short and stubby.* Descr. Ages/sexes similar. Sooty brown overall (looks blackish) with *pale grey throat and upper chest* (pale down to belly on N.A. *vauxi*), *rump and uppertail*

coverts *contrastingly paler greyish to brownish grey.*

Voice. High, thin, chipping and twittering calls, given often; shriller in N.A. *vauxi*.

Habs. Aerial, mainly in foothills and highlands (except *gaumeri* of Yuc Pen), ranges to lowlands mainly in migration and winter. Usually in flocks (up to a few hundred birds at times) flying quickly and often fairly low, rarely high like *Cypseloides*. Flight fast with rapid, fluttery wingbeats, glides for short periods, at times spreads tail. Nests singly or in loose colonies in hollow trees and buildings. Eggs: 3–6 (in N.A.).

SS: See Chimney and Chestnut-collared swifts.

SD: C to F resident (SL–3000 m), breeding locally, on both slopes from Sin and S Tamps, and in interior from cen volcanic belt to El Salvador and Honduras; disjunctly in Yuc Pen. Migrant *vauxi* from W N.A. occur mid-Sep–May, wintering from cen Mexico to W Honduras, and are F to C transients (Apr–May, mid-Sep–Oct) in NW Mexico.

RA: Breeds W N.A. to N S.A.; N populations winter from Mexico to Honduras.

WHITE-THROATED SWIFT
Aeronautes saxatalis*   Plate 28
Vencejo Gorjiblanco

ID: 5.5–6 in, W 12.5–14 in, T 2.3 in (14–15 cm, W 32–36 cm, T 5.5 cm). Widespread in arid country. *Wings long and narrow, tail long and cleft.* Descr. Ages/sexes similar. Blackish brown overall with *white throat and chest, white continues as narrow stripe to belly, white hind-flank patches visible from above and below,* trailing edge to secondaries white.

Voice. Distinctive, high, liquid to slightly screechy twittering and trilling calls.

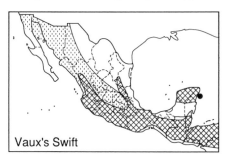

Vaux's Swift

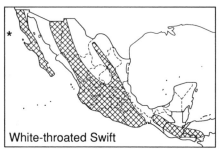

White-throated Swift

Habs. Aerial, mostly in fairly arid areas, deserts and highlands, canyons, about cliff faces, etc. Flight fast with fairly quick wingbeats, soars and glides readily. Often in small groups, rarely flocks up to 100+ birds, associates loosely with other swifts. Nests colonially in rock crevices, often on inaccessible cliffs. Eggs: 3–5.

SS: Swallow-tailed swifts have complete white collars, black underparts, long tails usually held closed. In poor light, White-throated told from *Cypseloides* by long, narrower-looking wings and longer, narrow cleft tail.

SD: F to C resident (SL–2000 m) in Baja and (1000–2500 m, locally lower in winter) in interior and on adjacent slopes from Son and Coah to cen volcanic belt; U to F (commoner Nov–Feb) to El Salvador and Honduras. Reports from Belize (such as Russell 1964) are unconfirmed.

RA: W N.A. (winters from SW USA) to El Salvador and Honduras.

GENUS PANYPTILA

These are spectacular swifts with very long, narrow, swept-back wings and long, deeply forked tails, usually held closed in a point. The two species are virtual replicas of one another except in size and should not be confused with any other species. Calls reedy and screeching. Nests are elongated, downward-tapering tubes (up to 1 m long) of plant material and saliva, built against or hanging from tree, cliff, or wall, entrance at the bottom.

LESSER SWALLOW-TAILED SWIFT

Panyptila cayennensis　　　　　Plate 28
Vencejo-tijereta Menor

ID: 5–5.5 in, W 11.5–12.3 in, T 2.3 in (12.5–14 cm, W 29–31 cm, T 5.5 cm).

Small. SE humid lowlands. See genus note. **Descr.** Ages/sexes similar. *Blackish overall, white throat continues back as hindcollar isolating black cap*, white flank patches often visible from below. At close range, note whitish preocular spot and narrow whitish trailing edge to secondaries and inner primaries.

Voice. High, shrill twitters and reedy chipping, *pssi ssississi* …, etc., and a distinctive, high, thin, slightly wheezy, abruptly fading *speez* or *psee-ih*.

Habs. Humid evergreen forest and adjacent areas. Aerial, usually fairly high to very high, associates loosely with other swifts. Flight fast with rapid, at times almost fluttery wingbeats, glides but usually not for long without flapping. Rarely in groups of more than 5–10 birds. Eggs: 2–3.

SS: Great Swallow-tailed Swift larger, wingbeats slower and glides for long periods without flapping, tail more deeply forked, white trailing edge to wings bolder.

SD: F to U resident (near SL–1000 m) on Atlantic Slope from S Ver to Honduras, on Pacific Slope in Chis (SNGH, PP) and Guatemala (Land 1970).

RA: S Mexico to Brazil.

GREAT SWALLOW-TAILED SWIFT

Panyptila sanctihieronymi　　　　Plate 28
Vencejo-tijereta Mayor

ID: 7.5–8 in, W 17.7–18.7 in, T 3.3 in (19–20.5 cm, W 45–47.5 cm, T 8 cm). *Large. W and S highlands. See genus note.* **Descr.** Ages/sexes similar. Blackish overall, *white throat continues back as hindcollar isolating black cap*, white flank patches often visible from below. At close range, note whitish preocular spot and whitish trailing edge to secondaries and inner primaries.

Voice. Reedy, screechy chippering, including a distinctive, accelerating series,

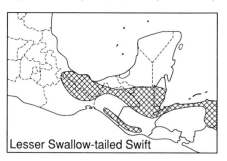

Lesser Swallow-tailed Swift

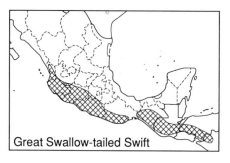

Great Swallow-tailed Swift

ending emphatically, may recall Pileated Flycatcher song: *kree kri-kri-kri-kri kree-kreeh!*

Habs. Aerial, generally in arid to semi-arid highlands and interior valleys, wandering locally to humid lowlands. The most spectacular swift in the region; flight fast and graceful, often long periods without flapping, scythe-like wings cut the air as birds arc and sweep over ridges, changing direction effortlessly; wingbeats deep and slow, almost floppy at times. Eggs undescribed (?).

SS: See Lesser Swallow-tailed Swift.

SD/RA: U to F resident (600–2000 m) in interior and on adjacent Pacific Slope from Nay (PA, SNGH) and Balsas drainage to Honduras, ranging (Dec–Jan) to near SL in NW Mexico (SNGH). R visitor (Sep–Nov, Mar–Apr) S to Nicaragua and Costa Rica.

HUMMINGBIRDS: TROCHILIDAE (65)

A very distinctive New World family. Up to 30 genera (yet only about 65 species) are recognized in the region. Although hummingbird taxonomy is in need of a thorough revision, there is remarkable diversity in the family. It is still unclear, however, what constitutes a hummingbird species and the allocation of several allopatric populations remains little more than educated guesswork.

Hummingbirds are best known for their small size, remarkable flight powers, and brilliant colors. Their slender, pointed, and proportionately long bills vary from straight to strongly arched and are used to probe flowers for nectar and insects; the tubular tongues have brush-like tips to aid in nectar gathering. Hummingbirds are unique in being able to reverse their primaries while hovering (in effect, rotating the wings through 180°), which enables them to fly backwards; the wings are moved so quickly in flight as to appear blurred. Ages usually differ, young altricial and naked. Hummingbird molts are poorly known, confused by protracted breeding seasons which vary locally in timing; adults typically have a complete molt after breeding but this may not necessarily be an annual molt (Wagner 1957). Juvs show variable traces of adult plumage and most species appear to attain adult plumage by a protracted complete 1st prebasic molt through their first year; imm in the species accounts refers to commonly seen juv/imm plumages. Juvenal plumage is characterized in almost all hummingbirds by pale buff to rufous tipping of the upperparts and sometimes underparts. Sexes similar or different. Most species have metallic-looking plumage (at least dorsally), most commonly various greens. Many species show iridescence, usually on the throat, or throat and upper chest, which may appear as a contrasting *gorget*. Iridescent colors are referred to as 'glittering' in the accounts and, as they are produced by refracted light rather than pigmentation, often appear blackish unless seen in the 'right light'. In the species accounts, all colors are metallic except black, white, greys, and browns (including buff, cinnamon, and rufous). Most hummingbirds have blackish-brown to dark grey remiges and this is assumed unless stated otherwise. Songs vary from essentially non-existent (in species where displays are important) to complex and startlingly loud; calls are generally chips and trills. Several species, mainly those breeding in North America, are migratory and many others have little-known local and seasonal movements related to the abundance and distribution of flowers.

Hummingbirds feed mostly on nectar and insects. Their nests are small cups of lichen, plant fiber, spider silk, hair, and other fine material, usually saddled on a branch or attached to a more vertical twig or stem by fibers and cobwebs; hermits characteristically hang their cone-shaped nests under the tips of *Heliconia* or banana leaves. All hummingbirds, as far as is known, lay two whitish eggs, and the ♂ typically plays no part in nesting. Nesting seasons often do not coincide with those of most other species in the region and an indication of nesting dates (i.e. nests with eggs and/or young) is given for each species, when available; in some cases, nesting seasons have been deduced from birds in breeding condition (b.c.). If no locations are given, the nesting season appears to be similar throughout a species' range. The months given for many species represent at best a skeleton of the full breeding season(s).

Note: Field identification of hummingbirds can be difficult at best. Geographic (subspecific) variation often has to be taken into account and some species are only 'safely' identified by range. For an observer new to the Neotropics, the variety of species, combined with their fast flight, doesn't help matters. Relative bill length and bill shape are important characters to note; bill color should also be noted. Tail shape and tail patterns are frequently important; in the species accounts, *inner rectrices* refer to the inner 2–3 pairs of rectrices, *central rectrices* to the innermost pair, *outer rectrices* to the outer 2–3 pairs (hummingbirds have 5 pairs of rectrices). Relative wing to tail projections of perched birds can be useful. A good thing to learn is which flowers hummingbirds like. Waiting patiently by a mass of suitable flowers is often the best way to see hummingbirds well. With experience, calls are important identification aids.

GENERA THRENETES, PHAETHORNIS, and PYGMORNIS: Hermits

Hermits are humid-forest hummingbirds with long, arched bills, striped faces, and long, graduated tails which are wagged persistently while singing. Feed low and often 'trap-line', making predictable circuits through the forest to single flowers or small groups of flowers. Some species form singing leks.

BAND-TAILED BARBTHROAT
Threnetes ruckeri ventosus   Plate 70
Ermitaño Barbudo

ID: 4.3–4.8 in (11–12 cm). C.A. Bill 2× head, tail fairly long and graduated, wings < tail. **Descr.** Ages differ, sexes similar. **Adult:** bill black above, yellowish below with dark tip. *Whitish moustache borders*

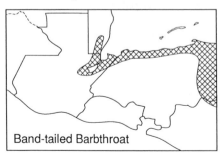

Band-tailed Barbthroat

blackish upper throat, lower throat and upper chest cinnamon-rufous; rest of underparts dusky grey, washed pale cinnamon on lower belly. White postocular stripe contrasts with dull green crown and blackish auriculars. Nape and upperparts dull green, uppertail coverts tipped buff. *Central rectrices green, tipped white, rest of tail white with broad black median band.* **Imm:** throat and chest dusky, usually with only a small area of cinnamon-rufous.

Voice. A high, sharp *sqzzk*, lower and buzzier than hermit calls. Song high, 1st part buzzy, 2nd part squeaky, *squee-i-zzk ssi-ssi-u ii-ii-ii*, etc., repeated.

Habs. Humid evergreen forest and edge, *Heliconia* thickets. See genus note. Sings from low perch in forest understory. Nesting: May (Honduras).

SS: Black throat and broad black band on rounded tail distinguishes Barbthroat from other hermits.

SD: U to F resident (SL–500 m) on Atlantic Slope from S Belize and NE Guatemala to Honduras.

RA: C.A. to N Ecuador.

LONG-TAILED HERMIT
Phaethornis superciliosus *   Plate 29
Ermitaño Colilargo

ID: 6.3–6.8 in (16–17 cm). SE humid forest. *Bill 2.5× head, tail long and graduated with very long central rectrices,* wings < tail. **Descr.** Ages/sexes similar. Bill black above, flesh to yellowish below with dusky

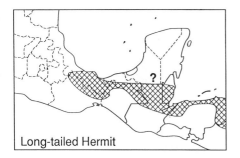

Long-tailed Hermit

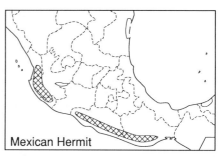

Mexican Hermit

tip. *Blackish auriculars bordered by whitish postocular stripe and moustache,* throat sides dusky grey, median throat whitish to buff; underparts pale buffy grey, washed cinnamon on belly and undertail coverts. Crown, nape, and upperparts bronzy green, rump and uppertail coverts broadly edged cinnamon. Tail black (dusky green basally), narrowly tipped pale cinnamon to white, *projecting tips of central rectrices white.*
Voice. A high, sharp, slightly explosive *sweik!* Song an insistently repeated, buzzy *zzreih* or *zzziiu,* 10/5–6 s, rarely as fast as 10/3 s.
Habs. Humid evergreen forest and edge, *Heliconia* thickets. ♂♂ form leks, singing from low perches in forest understory. Hermits are noted for appearing suddenly in front of an observer, hovering for a few moments (at times within inches of one's face!), then whipping away with a call and flash of white from the tail. Nesting: Apr–May.
SS: Little Hermit much smaller, cinnamon below, lacks elongated white central rectrices. Mexican Hermit allopatric. See Band-tailed Barbthroat (C.A.).
SD: F to C resident (SL–1500 m) on Atlantic Slope from S Ver to Honduras.
RA: SE Mexico to Brazil.
NB: Mexican Hermit traditionally considered conspecific with Long-tailed Hermit; distinguished by larger size, longer straighter bill, longer tail, duskier plumage.

MEXICAN [LONG-TAILED] HERMIT

*Phaethornis mexicanus** or *P. superciliosus* (in part)   Plate 29
Ermitaño Mexicano

ID: 7–7.5 in (18–19 cm). *Endemic to W Mexico. Bill 3× head, tail long and graduated with very long central rectrices,* wings

< tail. **Descr.** Ages/sexes similar. Bill black above, orangish below with dusky tip. *Blackish auriculars bordered by pale cinnamon postocular stripe and moustache,* sides of throat dusky grey, median throat pale buff; underparts dusky brownish grey, washed cinnamon on belly and undertail coverts. Crown dusky, nape and upperparts dark bronzy green, rump and uppertail coverts broadly edged cinnamon. Tail black (dusky green basally), broadly tipped white, *projecting tips of central rectrices white. P. m. griseoventer* of NW Mexico slightly paler with whitish to pale buff throat stripe, crown dull greenish, central rectrices white from point of projection beyond rest of tail.
Voice. A sharp, slightly explosive *sweik!,* much like Long-tailed Hermit. Song undescribed (?).
Habs. Humid evergreen forest and edge, *Heliconia* thickets, shady semideciduous arroyos. Habits much as Long-tailed Hermit but unknown if Mexican Hermit forms leks (?). Nesting: May–Jul.
SS: Long-tailed Hermit allopatric.
SD/RA: U to F but often local resident (100–2000 m) on Pacific Slope from Nay to Col; F to U from Gro to Oax.
NB: See Long-tailed Hermit.

LITTLE HERMIT

*Pygmornis longuemareus** Plate 29
Ermitaño Chico

ID: 3.5–4 in (9–10 cm). SE humid forest. Bill 1.5× head, tail long and graduated, wings < tail. **Descr.** Ages/sexes similar. Bill blackish above, yellowish below with dark tip. *Blackish auriculars contrast with pale cinnamon postocular stripe and moustache. Throat and underparts cinnamon,* washed greyish on median throat. Crown, nape, and upperparts bronzy green, *uppertail coverts rufous.* Tail dusky green,

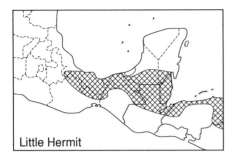

Little Hermit

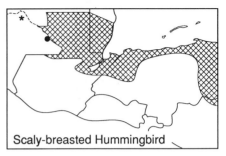

Scaly-breasted Hummingbird

tipped pale cinnamon, central rectrices tipped whitish.

Voice. A high, sharp *siip!*, weaker than Long-tailed Hermit. Variable song a high, squeaky, jerky, ventriloquial *see squee-ee-eek si-eek sik sik sik*, or *sik sik see seesee sik sik see sik . . .*, etc.

Habs. Humid evergreen forest and edge. Sings from low perch, often in fairly dense understory near the ground. Feeds low to high, rarely in canopy, at times in fairly horizonal posture with tail cocked. In display, ♂ (?), hovers, tail cocked, in front of perched ♀ (?), moving back and forth, and the two birds may hover together and alternate perches. Nesting: May–Jul.

SS: See Long-tailed Hermit, Band-tailed Barbthroat (C.A.).

SD: F to U resident (SL–1500 m) on Atlantic Slope from S Ver to Honduras.

RA: SE Mexico to Ecuador and Brazil.

NB1: *Pygmornis* often merged with *Phaethornis*. **NB2:** A recent re-evaluation of the complexities of hermit taxonomy and nomenclature indicates that this species may be called more correctly *Pygmornis striigularis* (RLZ).

GENERA PHAEOCHROA and CAMPYLOPTERUS: Sabrewings

Sabrewings are large hummingbirds of humid forest and edge. Bills black, straightish to slightly arched, tails fairly long and broad, wings shorter than tail at rest. Songs strong and varied, ♂♂ may form leks. Shafts of outer primaries broad and flattened, most strongly in adult ♂♂.

SCALY-BREASTED HUMMINGBIRD

Phaeochroa cuvierii roberti　　　　Plate 29
Fandangero Pechiescamoso

ID: 4.5–5 in (11.5–12.5 cm). *SE humid*

forest edge. Bill straightish, 1.25× head, tail broad and squared. **Descr.** Ages/sexes similar. *Face, throat, and chest heavily mottled green; white postocular spot; belly dusky cinnamon,* flanks mottled green; undertail coverts dusky green, edged whitish. Crown, nape, and upperparts greyish green. Inner rectrices green, *outer rectrices blackish with broad white tips* visible from below and from above when tail spread.

Voice. A sharp, warbler-like *chip*, and more metallic *tchik*. Song strong and jerky, a varied squeaky warble with distinctive sharp chips thrown in, *chi ssi-i-ee'ur ssi ssiur Chik Chik see chi-iu Chik . . .* etc., may recall Olive-backed Euphonia but more prolonged and varied.

Habs. Humid evergreen forest and edge, clearings with scattered trees, gardens, plantations. Perches and feeds low to high, often fairly high on exposed twigs when singing. Nesting: Feb (Chis).

SS: '♀-plumaged' White-necked Jacobin has bolder dark-scaled whitish throat and chest, belly whitish, outer rectrices more narrowly tipped white, usually silent.

SD: F to C resident (SL–500 m) on Atlantic Slope from N Guatemala and Belize to Honduras. Status in Mexico unclear, two records from NE Chis: ♂ singing (Jan 1976; Feltner 1976); 3–4 adults, one feeding two nestlings (Feb 1989; Howell 1989b). May be regular (seasonal?) in NE Chis and S Camp.

RA: C.A. to N Colombia.

WEDGE-TAILED SABREWING

*Campylopterus curvipennis**　　　　Plate 29
Fandangero Colicuña

ID: 4.7–5.3 in (12–13.5 cm). *Endemic; E and SE forest.* Bill straightish, 1.25–1.5× head. **Descr.** Ages/sexes differ slightly. ♂: *tail long and graduated. Face, throat, and* *underparts pale grey*, auriculars duskier with white postocular spot; undertail coverts washed pale cinnamon. *Crown glittering violet to violet-blue*, nape and upperparts green to blue-green. Tail blue-green to green, outer rectrices blackish distally, outermost rectrices may be mottled pale grey on distal outer webs. ♀: tail shorter and less graduated, outer rectrices broadly tipped whitish. **Imm:** resembles adult but crown duller, tipped buff, underparts washed pale cinnamon, tips of outer rectrices washed buff.

Voice. A steady sharp chipping, *chip chip chip chip-ip' chip* ..., and a nasal *peek*, often given in flight. Song a loud, prolonged, gurgling warble interspersed with squeaky chipping, starting typically with hesitant, nasal, reedy, insect-like chippering which may go on for minutes before breaking into full song.

Habs. Humid to semiarid, evergreen and semideciduous forest and edge, second growth, and edge with flowers. Perches and feeds low to high, sings from bare twigs, usually at mid- to upper levels, in understory. Nesting: Mar–Jul.

SS: Long-tailed Sabrewing allopatric (?). ♂ has longer tail, ♀ tail similar in length to ♂ Wedge-tailed but note extensive whitish tips to outer rectrices, less flattened primary shafts; imms of the two may be indistinguishable in field.

SD/RA: F to C resident (SL–1400 m, in foothills N of Isthmus) on Atlantic Slope from S Tamps to N Oax, possibly to Isthmus (SNGH), and from NE Chis to Yuc Pen, S to N Guatemala and Belize; disjunctly in E Honduras.

NB1: Some authors consider birds S of the Isthmus specifically distinct, *C. pampa*, distinguished by their slightly shorter average culmen length. Behavior and voice of both forms are similar and we prefer to

treat them as allopatric ssp. **NB2:** See Long-tailed Sabrewing.

LONG-TAILED [WEDGE-TAILED] SABREWING.
Campylopterus (curvipennis?) excellens
Fandangero Colilargo   Plate 29

ID/Descr: 5.2–5.5 in (13.5–14 cm). *Endemic to SE Mexico. Resembles Wedge-tailed Sabrewing but larger;* ♂ *has markedly longer tail*, underparts average paler.

Voice. Much like Wedge-tailed Sabrewing. Song perhaps not distinguishable but seems louder; also persistent chipping and nasal *peek* calls.

Habs. Humid evergreen forest and edge. Habits much as Wedge-tailed Sabrewing. Nest and eggs undescribed (?).

SS: See Wedge-tailed Sabrewing (allopatric?).

SD/RA: F to C resident (near SL–1200 m) in Los Tuxtlas and in Isthmus from SE Ver to E Chis (AMNH spec. 824955).

NB: Perhaps simply a large ssp of Wedge-tailed Sabrewing. Field work should be concentrated in the Isthmus to determine whether *curvipennis* and *excellens* breed sympatrically and/or intergrade.

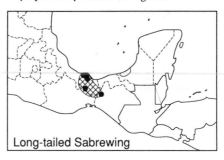
Long-tailed Sabrewing

RUFOUS SABREWING
Campylopterus rufus   Plate 29
Fandangero Rufo

ID: 5–5.5 in (12.5–14 cm). *Endemic to highlands S of Isthmus.* Bill straightish, 1.33–1.5× head, tail broad and squared to slightly rounded. **Descr.** Ages/sexes similar. *Face, throat, and underparts cinnamon,* auriculars duskier with white postocular spot. Crown, nape, and upperparts green to bluish green. Inner rectrices bronzy green, *outer rectrices pale cinnamon with broad black subterminal band*, outermost web without black.

Voice. A sharp, nasal *squihk'*, a slightly

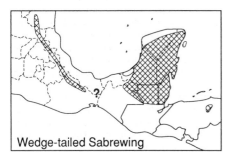
Wedge-tailed Sabrewing

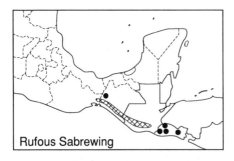

Rufous Sabrewing

more metallic *pli'ik*, and a hard, chipping *chi'i'rr chik-chik-chik-chik*, etc. Song varied, strong, squeaky chipping and chattering, also short, rich, warbled phrases.

Habs. Humid evergreen forest and edge, plantations. Sings and feeds mostly fairly low, often in dense understory where ♂♂ form small leks; also feeds in open situations at edges, at times at mid- to upper levels. Nesting: Apr–May (Oax).

SD/RA: F to C resident (900–2000 m, locally to 50 m in El Salvador) on Pacific Slope and in adjacent interior from E Oax to El Salvador.

VIOLET SABREWING
Campylopterus h. hemileucurus   Plate 29
Fandangero Morado

ID: 5.5–6 in (14–15 cm). *S humid forest. Bill arched, 1.5× head, tail broad and squared to slightly rounded.* **Descr.** Ages/sexes differ. ♂: *often looks blackish with large white tailflashes. Throat and underparts deep violet to bluish violet,* undertail coverts dull green. White postocular spot contrasts with dark greenish crown and violet auriculars, nape and upperparts violet becoming blue-green to green on uppertail coverts; upperwing coverts green. Tail violet-black (inner rectrices sometimes with green sheen), *outer rectrices broadly*

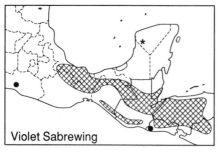

Violet Sabrewing

tipped white. ♀: *gorget bluish violet to blue,* bordered by pale grey moustache, white postocular spot contrasts with dull green crown and dusky auriculars. Nape and upperparts green to blue-green. Underparts pale grey, sides and flanks slightly mottled green, undertail coverts green. Inner rectrices blackish green, *outer rectrices blackish, broadly tipped white.* **Imm** ♂: upperparts resemble ♀, tail as ♂, *underparts blackish green, often with some violet in throat.* **Imm** ♀: resembles ♀ but throat pale grey with a few violet spots.

Voice. High, sharp chippering and prolonged, hard chipping, at times run into a rattle, single sharp chips given in flight. Song varied, loud, sharp chipping and warbles, often punctuated with fairly shrill, slightly explosive notes.

Habs. Humid evergreen forest and edge, plantations. Habits much as Rufous Sabrewing. Nesting: Apr–Aug.

SS: Blue-throated Hummingbird (unlikely in same areas) has long straightish bill, white postocular stripe, and blackish tail with bold white corners.

SD: F to C resident (near SL–2000 m, mostly above 500 m, locally to 2500 m in C.A.) on Pacific Slope in Gro (Navarro 1986; SNGH, SW) and from Chis to El Salvador, on Atlantic Slope from cen Ver to Honduras and N cen Nicaragua. Some apparently withdraw S from Atlantic Slope N of Isthmus in winter (Nov–Feb at least). Vagrant to Yuc (Dec 1980, G and JC).

RA: Mexico to W Panama.

WHITE-NECKED JACOBIN
Florisuga m. mellivora   Plate 29
Jacobino Nuquiblanco

ID: 4.3–4.8 in (11–12 cm). *SE humid forest and edge.* Bill straight or slightly decurved at tip, 1.25× head, tail squared to slightly cleft, wings = tail. *Elongated uppertail coverts cover most of closed tail.* ♀/imm plumages complex and need study. **Descr.** Ages/sexes differ. ♂ (**some** ♀♀): bill black. *Head, throat, and chest violet-blue, white nape band may be hard to see.* Upperparts blue-green. Sides of chest green, rest of underparts white. *Tail white, rectrices narrowly edged black.* ♀: polymorphic, some resemble ♂ (above). Two other plumages are common: (i) *throat and chest dusky green with broad white scalloping* and

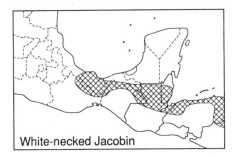

White-necked Jacobin

sometimes a few violet feathers, *rest of underparts whitish*, mottled bronzy green on chest and sides, undertail coverts blackish, tipped white. Crown, nape, and upperparts green, some birds with trace of white nape band. *Tail blue-green to green, blackish distally, outer rectrices tipped white*; (ii) resembles ♂ but duller. Crown and nape greenish blue, malar stripe cinnamon, white nape band often less distinct. *Central rectrices green, becoming black distally, bases may be white but covered by long uppertail coverts; rest of tail white, tipped blue-black with narrow black edges to outer webs except all-white outermost rectrices.* Imm ♂: *resembles ♀* (ii) but often *more extensive cinnamon in face*, extending to sides, back also may be mottled cinnamon, central rectrices white basally.

Voice. Mostly silent; at times utters fairly hard, slightly wiry *tsit*ing and trilling, a high, sharp *sit*, and a rolled *tssirrr* or *siiiirrr*, etc.

Habs. Humid evergreen forest and edge, second growth with scattered tall trees, plantations. Usually at mid- to upper levels in fairly open canopy, 'σσ' in particular perch conspicuously on bare twigs; also at mid-levels in forest interior. Often feeds with fairly horizontal posture, tail cocked and occasionally flashed open. B.c.: Apr–Jun

SS: See Scaly-breasted Hummingbird.

SD: Locally/seasonally U to C resident (SL–500 m) on Atlantic Slope from S Ver to Honduras; possibly some withdrawal from Atlantic Slope of Mexico in winter (Nov–Feb?).

RA: S Mexico to Ecuador and Brazil.

GREEN-BREASTED MANGO
Anthracothorax prevostii*   Plate 29
Mango Pechiverde

ID: 4.5–4.8 in (11.5–12 cm). *E and SE low-*

lands. Bill slightly arched, 1.33–1.5× head, tail broad and squared to slightly cleft, wings = tail. **Descr.** Ages/sexes differ. ♂ (some ♀♀): bill black. Often looks all dark. *Black throat bordered by glittering blue-green to bluish moustache and chest* (black may extend onto chest), rest of underparts blue-green medially, green on sides and flanks, undertail coverts dusky purplish. Crown, nape, auriculars, and upperparts deep green. Central rectrices dark bronzy green, rest of *tail purple to rufous-purple*, rectrices edged blackish. ♀: *throat and median underparts whitish with broad black stripe on throat becoming glittering blue-green on underbody*, chin may be white; sides and flanks green, undertail coverts dusky, broadly edged white. Crown, auriculars, nape, and upperparts green. Central rectrices dark bronzy green, *rest of tail blue-black, tipped white, often with broad purple band basally.* **Imm:** resembles ♀ but chin and upper throat white (may be spotted black), *white border to throat and chest mixed with cinnamon-brown*, outer rectrices of ♂ (at least) with noticeable purple base.

Voice. Fairly hard chipping, at times repeated steadily, *chik chik chik . . .*, a high, sharp *sip* or *sik*, and high, shrill, slightly tinny twittering in chases. Song a high, thin *tsi si-si-si si-si-si*. Often silent.

Habs. Semiopen areas with scattered tall trees and hedges, forest edge, clearings. Typically perches and nests conspicuously on tall bare branches and twigs. Feeds high to low, flashes tail open while feeding. Nesting: Mar–Jun (Atlantic Slope), Oct–Feb (E El Salvador).

SS: Starthroats also have dark throats and favor conspicuous perches but note very long, straight bills, white back patches;

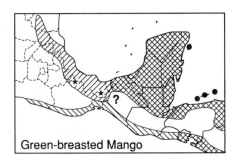

Green-breasted Mango

dark on throat of ♀ Mango extends onto chest.

SD: Locally/seasonally F to C resident (SL–1000 m) on Atlantic Slope from N Chis to Honduras, most common on Caribbean coast and islands; on Pacific Slope from E El Salvador S. F to C summer resident (late Feb–Sep) on Atlantic Slope N to S Tamps, R to absent late Sep–Feb; also U to R in Yuc Pen (Sep–Feb). F to C winter visitor (Oct–Feb) on Pacific Slope from Isthmus to cen El Salvador, R N to S Gro (Nov 1987, RB).

RA: Mexico (migratory in N) to Peru and N Venezuela.

NB1: Sometimes considered conspecific with *A. nigricollis* of Panama and S.A. **NB2:** Taxonomy of the forms of Green-breasted Mango from Panama to S.A. poorly understood.

BROWN VIOLET-EAR
Colibri delphinae   Plate 70
Orejavioleta Café

ID: 4–4.5 in (10–11.5 cm). *C.A. Bill straightish, 1× head*, tail broad and squared to slightly cleft, wings ≤ tail. **Descr.** Ages differ, sexes similar. **Adult:** bill black. *Whitish postocular streak contrasts with grey-brown crown and nape and glittering violet auriculars; whitish lores and broad moustache* border dusky brown chin and glittering greenish gorget, lower gorget with a few blue spots. Upperparts grey-brown, *lower rump and uppertail coverts broadly edged cinnamon*. Underparts pale brownish grey, mottled darker on chest, undertail coverts broadly edged pale cinnamon. *Tail dull bronzy green with broad blackish subterminal band* and pale tip. **Imm:** crown, nape, and upperparts broadly tipped cinnamon, tail tip washed cinnamon, little or no violet in face.

Voice. A hard, smacking *chik* and a dry, rattling chatter. Song a series of 3–8 strong, rough, slightly shrill to nasal chips, *shrih shrih...*, repeated after a short pause; also a fairly strong, scratchy, slightly squeaky warble with occasional liquid phrases thrown in.

Habs. Humid evergreen forest and edge, plantations. Feeds and sings at mid- to upper levels, often in fairly open canopy. B.c.: Feb–Mar (Guatemala).

SD: Locally/seasonally U to F resident (near SL–1000 m) on Atlantic Slope from S Belize and NE Guatemala to Honduras.

RA: C.A. to Ecuador and Brazil.

GREEN VIOLET-EAR
Colibri t. thalassinus   Plate 31
Orejavioleta Verde

ID: 4.2–4.5 in (11–11.5 cm). *Common in highlands.* Bill straightish, 1.25–1.33× head, tail broad and squared, wings = tail. **Descr.** Ages differ, sexes similar. **Adult:** bill black. *Deep green overall*, more bluish green below, with *glittering violet auriculars* and, usually, bluish-violet patch on chest (mainly ♂♂?), undertail coverts edged pale cinnamon. *Tail bronzy blue-green with broad blackish subterminal band.* **Imm:** duller, *throat and underparts dusky green*, often with patches of adult color.

Voice. A hard, dry rattle, often given when feeding. Song a rhythmic, jerky, metallic chipping, *t'issik-t'issik, t'issik-t'issik...*, punctuated irregularly with *tik* and *tssi* notes, or *ch-it chi-i-it chi-chi-it chi-it chi-i-it...*, etc., often prolonged.

Habs. Humid to semiarid pine–oak, oak, and evergreen forest and edge, clearings with flowers. Feeds low to fairly high; perches and sings, often tirelessly, on exposed bare twigs. Nesting: Mar, Jul–Sep (N of Isthmus), Aug–Jan (Guatemala).

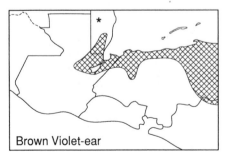
Brown Violet-ear

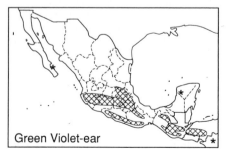
Green Violet-ear

SD: Locally/seasonally C to F resident (1000–3000 m) in interior and on adjacent slopes from Jal and S SLP to El Salvador and Honduras; N of Isthmus (except Pacific Slope?) most ♀♀ and imms, and some ♂♂, withdraw S (Nov/Jan to Jun/Jul) after breeding, presumably to N C.A. where ♀♀ and imms greatly outnumber ♂♂ at that time (Dickey and Van Rossem 1938; Wagner 1945b; Wilson and Ceballos-L. 1986); R to N cen Nicaragua (Dec 1985; Martinez-S. 1989). Vagrant (Jan–Feb) to Yuc (G and JC) and BCS (RS).

RA: Mexico to Ecuador and N Bolivia.

EMERALD-CHINNED HUMMINGBIRD
*Abeillia abeillei**   Plate 32
Colibri Barbiesmeralda

ID: 2.7–3 in (7–7.5 cm). *Endemic to SE cloud forest. Bill short* and straight, 0.5–0.75× head, tail fairly broad and squared to slightly cleft, wings = tail. **Descr.** Ages/sexes differ. ♂: bill black. *Bold white postocular spot contrasts with deep green crown and auriculars, upper gorget glittering green, lower throat blackish* (throat often looks grey); underparts dusky grey, mottled green on chest and flanks. Nape and upperparts deep blue-green. Inner rectrices blue-green, outer rectrices blue-black, tipped pale grey. ♀: *throat and underparts pale grey, auriculars dusky with distinct white postocular spot*, sides and flanks mottled green, outer rectrices tipped whitish. **Imm** ♂: resembles ♀ but throat with some emerald green spots.
Voice. A liquid, rattling trill, much like Stripe-tailed Hummingbird (and thus suggesting Costa's Hummingbird), at times only single notes given, *puip, puip* ..., often when feeding, and a sharp *sii'ing* when perched. Song a high, thin, slightly

squeaky chipping, *tsin, tsin-tsin tsin-tsin tsin-tsin*, or *tsi tsi tsi-si, tsi-tsi tsi tsi-tsi* ..., etc.
Habs. Humid evergreen and pine–evergreen forest and edge. Usually keeps fairly low, rarely in open situations, flight fast. Sings from low to mid-levels of fairly open forest understory. Nest is a fairly deep cup bound to a vertical shoot by cobwebs etc., low in forest understory (SNGH, LJP). Nesting: Feb–Mar (N Chis).

SS: Tiny size, very short bill, and bold white postocular spot distinctive. Imm ♂ Canivet's and Salvin's emeralds may appear similar (small, green throat) but note white postocular stripe, forked tail, voice, habitat. See Violet-headed Hummingbird (E Honduras, Appendix E).

SD/RA: F to C resident (1000–2200 m) on Atlantic Slope from N Oax (SNGH) to Honduras, on Pacific Slope and in interior from E Oax to Honduras and N cen Nicaragua; U N to cen Ver (no recent records?). Reports from Yuc Pen (Edwards 1978) and Belize (Wood *et al.* 1986) are not credible.

SHORT-CRESTED [RUFOUS-CRESTED] COQUETTE
Lophornis brachylopha   Plate 32
Coqueta Cresticorta

ID: 2.7–3 in (7–7.5 cm). *Endemic to Gro.* Bill straight, 0.5–0.66× head. **Descr.** Ages differ, sexes differ. ♂: tail double-rounded, wings ≤ tail. *Bill black. Forehead and filamentous crest cinnamon* (longest crest feathers tipped green), *gorget glittering emerald green with* green-flecked *cinnamon side-tufts.* Auriculars dusky, nape and upperparts deep emerald green with pale *buffy-cinnamon to whitish rump band*, lower rump bronzy purple, uppertail

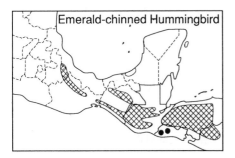

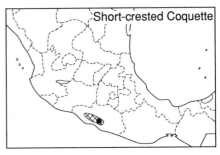

coverts green. Underparts dirty pale cinnamon with broad whitish band below gorget, sides mottled green. Central rectrices green, tipped darker, rest of *tail cinnamon-rufous, broadly tipped black,* outer rectrices edged green. ♀: tail rounded, wings ≤ tail. *Forehead dull cinnamon (inconspicuous),* auriculars dusky, crown, nape, and upperparts green with *pale buffy-cinnamon to whitish rump band,* uppertail coverts duller, tipped cinnamon. *Throat whitish,* bordered below by white chest band, underparts washed pale cinnamon. Central rectrices green, tipped blackish, rest of tail cinnamon with a broad blackish subterminal band, and pale buff tips to outer rectrices. **Imm** ♂: resembles ♀ but with spiky crest of slightly shorter and thicker feathers than ♂, crest more broadly tipped green; throat pale buff, finely flecked dusky.

Voice. Mostly silent; feeding birds sometimes give high, sharp chips, *siip* or *tsiip, tsiip tsiip . . .*, and quiet, dry chips *chi chichi . . .*, etc.

Habs. Humid to semihumid evergreen, pine–evergreen, and semideciduous forest and edge, plantations. Feeds on flowers including *Inga* and *Cecropia,* where usually dominated by other hummingbirds. Feeds low to high, with tail cocked or held down and wagged. Often perches on prominent twigs. Nest and eggs undescribed (?).

SS: See ♀ Sparkling-tailed and Bumblebee hummingbirds, the only potentially similar small hummingbirds in this coquette's range.

SD/RA: Locally/seasonally F to U resident (900–1800 m) in Sierra Madre del Sur of Gro. First described from 2♂ collected May 1947 (Moore 1949), rediscovered when 1♂ and 2♀ collected Jan–May 1986 (Ornelas 1987). Subsequent observations suggest breeding at higher elevations (above 1500 m?; Nov–Feb?) with downslope movement (Mar–Aug at least) (Howell 1992a; RGW); San Vicente de Benitez, where the 1947 birds were taken, is at about 3000 ft, not 1500 ft as given by Moore (1947), an error perpetuated by Banks (1990a).

NB: Originally described as a ssp of Rufous-crested Coquette, *L. delattrei,* of Panama to S.A.

BLACK-CRESTED COQUETTE
Lophornis helenae	Plate 32
Coqueta Crestinegra

ID: 2.5–2.8 in (6.5–7 cm). *SE cloud forest.* Bill straight, 0.75× head. **Descr.** Ages/sexes differ. ♂: tail cleft. Bill red with black tip. *Forehead and long filamentous crest dark green (often looking black),* longest crest feathers black, upper gorget glittering green, *lower throat black with broad buff streaks at sides.* Underparts whitish, boldly spotted greenish bronze, undertail coverts cinnamon. Auriculars, nape, and upperparts green with *bold whitish rump band,* uppertail coverts blackish. Central rectrices dark green, rest of tail rufous, rectrices edged dusky green. ♀: tail rounded. Bill blackish above, red below with dark tip. *Blackish auriculars contrast with dark green crown and dark-flecked whitish throat.* Nape and upperparts green with *bold whitish rump band,* uppertail coverts bronzy. Chest bronzy green, *belly whitish, boldly spotted greenish bronze,* undertail coverts cinnamon. Tail dusky green with broad cinnamon tip, cinnamon on base of outer rectrices visible on spread tail. **Imm** ♂: resembles ♀ but shows vestige of crest, throat whitish with slight blackish apron on upper chest, older imms have ♂-like tail.

Voice. Mostly silent, song (?) a clear, upslurred *tsuwee,* repeated; also thin, high twittering when fighting and a quiet, slightly metallic *teek* when feeding. Wings make a low insect-like humming.

Habs. Humid evergreen forest and edge, clearings with flowers. Feeds low to high; flight often slow and bee-like, tail cocked; can be confused, more than momentarily, with a sphinx moth. Often perches on fairly prominent bare twigs. B.c.: Jan (Guatemala).

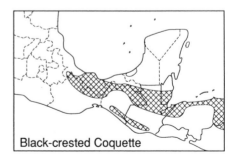

Black-crested Coquette

PLATES

Species are portrayed to scale on a given plate or a line divides the plate into two scales; flying birds are clearly to different scales than perched birds on the same plate. The plates and facing page notes will often be sufficient to identify a species but still we recommend a thorough reading of the species accounts. Generally a sense of taxonomic sequence (AOU 1983 Checklist) has been followed through the plates.

Mean lengths (in inches and centimeters) are given for each species. These should serve only as a rough guide; check species accounts for size variation.

Abbreviations and format (see list of abbreviations, p. xv)

The style of scientific names conveys a variety of information.

Spizastur melanoleucus: a binomial indicates that a species is monotypic.

Burhinus b. bistriatus: a trinomial without an asterisk indicates that only one subspecies occurs in the region (the nominate in this case; the specific name is abbreviated to its first letter when the subspecific name is the same).

Tinamus major robustus*: a trinomial with an asterisk indicates that more than one subspecies occurs in the region but that the illustration is of the subspecies named (in this case, *robustus*).

Crypturellus cinnamomeus*: a binomial with an asterisk indicates that more than one subspecies occurs in the region, and more than one is illustrated. Subspecies illustrated (and their ranges) are given in the species caption.

Space is often not available on facing pages to give alternative common names or full taxonomic options (see Taxonomy, pp. 61–4). Thus, names are the first options indicated in the species account headings.

Endemic indicates *endemic to the region covered by this guide* which, from a birder's point of view, generally means a species not readily seen in the USA or Costa Rica.

PLATE 1: HERONS AND LIMPKIN

1 **Boat-billed Heron** *Cochlearius cochlearius phillipsi** 19.5in (50cm) **p. 144**
Tropical lowlands, nocturnal. Distinctive; note huge broad bill. Flight shape more compact than 2 and 3; note bill shape, black wing-linings of adult.

2 **Black-crowned Night-Heron** *Nycticorax nycticorax hoactli* 23.5in (60cm) **p. 143**
Widespread. From 1 in flight by pointed bill, *wok!* call; from 3 by more compact shape, shorter legs.

3 **Yellow-crowned Night-Heron** *N. v. violaceus** 21.5in (55cm) **p. 144**
Widespread, mostly lowlands. In flight, note rangier shape, longer legs than 1 and 2.

4 **Limpkin** *Aramus guarauna dolosus* 24in (61cm) **p. 248**
S and E marshes. Distinctive; note long, rail-like bill, exensive white streaking, broad wings, and crane-like flight.

5 **American Bittern** *Botaurus lentiginosus* 22in (56cm) **p. 136**
Widespread migrant. From larger 6 by striped neck, less contrastingly marked upperparts, black malar stripe.

6 **Pinnated Bittern** *B. pinnatus caribaeus* 27.5in (70cm) **p. 135**
SE lowlands, local. From smaller 5 by barred neck, more boldly patterned upperparts; heavier flight suggests tiger-heron. Lacks bold bars on upperparts of 8, 9.

7 **Agami Heron** *Agamia agami* 28in (71cm) **p. 142**
SE swamps and forest, rarely seen. Note long, slender bill and relatively short legs. Adult beautiful and unmistakable; juv distinctive, note shape.

8 **Bare-throated Tiger-Heron** *Tigrisoma m. mexicanum** 30in (76cm) **p. 137**
Tropical lowlands. Note naked yellow throat, dark greyish-olive legs. Adult distinctive; note grey face, black cap. Juv from imm 9 (hypothetical in region) by naked throat, dark legs, less rufescent tones. Imm darker and more narrowly barred.

9 **Rufescent Tiger-Heron** *T. l. lineatum* 27in (69cm) **p. 138**
(See NB under 8 in main text.) Hypothetical in region, known from E Honduras (see Appendix B). Note feathered white throat, yellowish legs. Adult unmistakable. Juv distinctive; note rufescent neck with dark chevrons, boldly barred upperparts. Imm similar to juv 8; note feathered white throat, colors of bare parts.

1

juv

1

2

3

4

5

6

7 juv

8 imm

8 juv

9

9 imm

9 juv

SW 12-91

PLATE 2: KITES, ACCIPITERS, FOREST-FALCONS

1 **Hook-billed Kite** *Chondrohierax u. uncinatus** 16.5in (42cm) **p. 179**
Widespread, mostly tropical lowland forest. Strongly hooked bill with greenish base, pale eyes (adult), highly variable plumage (see text). Light morph adult has 1–2 pale tail bands; dark morph adult (1(a)) slaty to brownish black with 1–2 white tail bands. Juv has 3 pale tail bands, underparts barred to unbarred. See Plates 8, 10, 12.

2 **Mississippi Kite** *Ictinia mississippiensis* 14.3in (36cm) **p. 183**
Transient in E and S. Juv from juv 3 by longer tail, rufous streaking below, less blackish upperparts. Adult distinctive (see text). For imm see text and Plate 13.

3 **Plumbeous Kite** *I. plumbea* 13.8in (35cm) **p. 184**
Summer resident (Feb–Aug) in SE. From 2 by shorter tail. Adult darker without pale panel on secondaries of 2; juv darker above (primaries not contrastingly darker than back), underparts streaked grey. For imm see text and Plate 13.

4 **Collared Forest-Falcon** *Micrastur semitorquatus naso* ♂ 21.5in (55cm), ♀ 24.5in (62cm) **p. 214**
Tropical lowland forest; does not soar, rarely seen. Large, lanky; long, graduated tail. From 7 by pale collar, dark auricular crescent, longer legs, strongly graduated tail. Adult face, collar, and underparts white to buff, or all black (4(a)), or black with white-barred belly. Juv underparts scalloped, collar and auricular crescent less distinct; dark juv browner than adult. Hollow, far-carrying *cowh* call.

5 **Barred Forest-Falcon** *M. ruficollis guerilla** 14in (36cm) **p. 214**
S humid forest; does not soar, rarely seen. Note long legs, graduated tail. Adult distinctive; note bright facial skin, barred underparts. Juv much smaller than 4, underparts variably barred; compare with unbarred 6, 7. Sharp, yapping bark call.

6 **White-breasted Hawk** *Accipiter chionogaster* 12in (30cm) **p. 187**
Endemic, highlands S of Isthmus. Slaty upperparts, white underparts, 3 broad pale tail bands. Face and underparts of juv finely streaked. See Plate 13.

7 **Bicolored Hawk** *A. b. bicolor** ♂ 14in (36cm), ♀ 18in (46cm) **p. 187**
SE rain forest, rarely seen. Adult grey, paler below with rufous thighs. Juv variable, rarely with pale collar; underparts creamy to cinnamon, 3 narrow greyish tail bands.

8 **Double-toothed Kite** *Harpagus bidentatus fasciatus* 13.3in (34cm) **p. 183**
SE rain forest. May suggest accipiter, but note longer primary projection, smaller legs and feet, 3 narrow whitish tail bars. Adult barred below, juv variably streaked or can be unmarked below. See Plates 8, 13.

2

1

1a

2
juv

3

juvs

♂

♀

juv

4

juv

5

juvs

4a

6

juv

juv

7

juvs

8

juv

SW 12-88

PLATE 3: BUTEOS AND HARRIS' HAWK

1 **Grey Hawk** *Buteo nitidus plagiatus** 17in (43cm) **p. 196**
Widespread, especially arid lowlands. Adult distinctive. Juv from juv 2 by shorter primaries, longer tail, and tail pattern (progressively wider wavy bands distally); note boldly striped face, barred thighs, longer legs, more strikingly white uppertail coverts. See text and Plate 9.

2 **Broad-winged Hawk** *B. p. platypterus* 16in (41cm) **p. 199**
Widespread migrant. Juv from juv 1 by longer primaries, shorter tail, etc. See text and Plate 9.

3 **Roadside Hawk** *B. magnirostris griseocauda** 14.5in (37cm) **p. 197**
Tropical lowlands, especially humid areas. Note short primaries, long tail with evenly broad bands, streaked chest, and barred belly; also whitish eyes of adult. See Plates 9, 69.

4 **Zone-tailed Hawk** *B. albonotatus* 19.5in (50cm) **p. 202**
Widespread, mostly migrant in S. Distinctive; note long wings and tail; juv flecked whitish below. Rarely seen perched. Compare with dark 5. See Plate 8.

5 **Short-tailed Hawk** *B. brachyurus fuliginosus* 17in (43cm) **p. 199**
Widespread, rarely seen perched. Dark morph all dark (5(a)); note whitish lores and tail pattern; compare with larger and blacker 4. Light morph distinctive; note long wingtips close to tail tip, juv face streaked. Dark juv flecked whitish below. See Plates 8, 9.

6 **Harris' Hawk** *Parabuteo unicinctus harrisi** 20.5in (52cm) **p. 195**
Deserts of N and cen Mexico, local in S Mexico, C.A. Note long, graduated tail with bold white tip. Adult unmistakable, often looks all dark; juv pattern suggests adult. See Plate 11.

7 **White-tailed Hawk** *Buteo albicaudatus hypospodius* 21in (53cm) **p. 201**
Widespread but local, savannas. Long primaries usually project to or slightly beyond tail tip. Adult unmistakable. Variable juv told from most species by long wings; compare with Swainson's Hawk (see text). See Plates 8, 9.

juv

1

juv

2

juv

3

juv

3

1

2

3

4

5a

5

juv

6

juv

7

juv

SW 9-88

PLATE 4: NEAR BUTEOS AND ALLIES

1 **Snail Kite** *Rostrhamus sociabilis major* 19in (48cm) **p. 182**
S marshes. Distinctive; note slender hooked bill, long wingtips project just beyond tail tip, broad white band at base of tail. See Plates 10, 11.

2 **Black-collared Hawk** *Busarellus n. nigricollis* 19in (49cm) **p. 190**
Tropical lowland marshes. Adult unmistakable. Juv distinctive; note structure (especially wingtips equal with short tail), flesh legs, hint of adult head and chest pattern. See Plate 11.

3 **Crane Hawk** *Geranospiza caerulescens nigra** 19.5in (50cm) **p. 189**
Tropical lowlands. Lanky and small-headed; told by small bill and grey lores, long red legs, long tail with 2 white bands. See Plates 8, 10.

4 **Common Black Hawk** *Buteogallus a. anthracinus** 19.5in (50cm) **p. 191**
Widespread. Adult: from 5 by longer primary projection accentuating shorter tail, shorter legs; also note bright orange-yellow lores, single white tail band. From much larger 6 by blacker plumage, more extensively bright yellow loral area and base of bill, less massive bill and legs. Juv: from juv 5 by structure (especially longer primary projection, shorter tail), bold dark malar stripe, lack of whitish-headed look, and fewer, bolder tail bands (tail of juv Common variable, with 4–7 white to pale brownish bands above). Tail of juv Common can be similar to imm 5; note structure and plumage differences under juv. Much larger 6 has relatively uniform tail, solidly dark thighs, etc. See text and Plates 8, 10, 11.

5 **Great Black Hawk** *B. urubitinga ridgwayi* 22in (56cm) **p. 193**
Widespread. See 4 for differences; note especially structure (wing–tail differences, leg length). Adult from more massive 6 by wing–tail projection, blacker plumage, greyish lores, 2 white tail bands. Juv and imm from 6 by structure, tail pattern, lack of dark malar stripe, etc. 1st basic much like juv but note tail pattern (5(b)) more like 4, distal dark band proportionately wider; 2nd basic tail (5(a)) variable, may be shown by birds in mostly 1st basic plumage, also by adult-like birds. See text and Plates 8, 10, 11.

6 **Solitary Eagle** *Harpyhaliaetus solitarius sheffleri** 28in (71cm) **p. 194**
Rare and local, foothills and highlands. Massive bulk with stout bill and legs, wingtips of adult project to or beyond tail tip. See 4, 5 for differences. 1st basic resembles juv but note tail pattern (6(a)). See text and Plates 8, 10, 11.

4

1 juv

1 ♂

♀

juv

2

3

juv

4

5

6

6a

5 juv

juv

4

5a

5b

juv

J W 9-89

PLATE 5: LARGE FOREST RAPTORS

1 **Grey-headed Kite** *Leptodon cayanensis* 19.5in (50cm) **p. 178**
S. Adult distinctive; note soft grey head, blue-grey lores and feet. Juv variable. 1(a)
Dark morph: dark head, black streaking on chest; 1(b) light morph: lacks black
lores, orange cere, feathered tarsi, and white shoulder of larger and crested 2; also
note tail pattern. See Plates 8, 12.

Hawk-Eagles are large crested eagles of tropical forests, 1–2 years required to attain
adult plumage. Note feathered tarsi.

2 **Black-and-white Hawk-Eagle** *Spizastur melanoleucus* 22.5in (57cm) **p. 208**
S (local W Mexico). Striking; note black lores and orange cere, white shoulder,
short crest. Lacks long crest, barred thighs of juv 3. Juv similar but upperparts
blackish brown. See Plates 8, 12.

3 **Ornate Hawk-Eagle** *Spizaetus ornatus vicarius* 24.5in (62cm) **p. 210**
E and S (local W Mexico). Adult unmistakable. Juv from 2 by longer crest, barred
thighs and flanks, pale lores, and yellowish cere, lack of white shoulder. See
Plates 8, 12.

4 **Black Hawk-Eagle** *S. tyrannus serus* 26.5in (67cm) **p. 209**
S (local SW Mexico). Adult unmistakable; note bushy, white-based crest, barred
thighs. Juv distinctive; note dark cheeks. See Plates 8, 12.

5 **White Hawk** *Leucopternis albicollis ghiesbreghti** 20.5in (52cm) **p. 191**
S rain forest. Unmistakable; juv has blackish barring on wing coverts, more black
in tail. Birds of S Belize, NE Guatemala, and Pacific Slope from Chis S appear
intermediate toward *costaricensis* (E of Sula Valley) which has extensively black
wing coverts, and black secondaries broadly tipped white in adult. See Plate 11.

5

juvs

1b

2

1a

3

juv

juv

5

4

juv

juv

SW 1-89

PLATE 6: FALCONS,
HARPY AND CRESTED EAGLES

1 **Aplomado Falcon** *Falco femoralis septentrionalis** 16.5in (42cm) **p. 216**
SE savannas, local. Unmistakable; note long wings and tail, broad whitish
supercilium, blackish chest band, pale cinnamon thighs. See Plate 13.

2 **Bat Falcon** *F. rufigularis albigularis** 10in (25cm) **p. 217**
Common in E and S, local on Pacific Slope. Small and compact, overall dark with
white throat, rufous thighs. Juv chest and sides of neck more strongly washed
cinnamon, undertail coverts barred black. See Plate 13.

3 **Orange-breasted Falcon** *F. deiroleucus* 14.8in (38cm) **p. 218**
Rare, SE rain forest. From 2 by blacker upperparts (blackish, edged blue-grey
versus slaty grey with dark centers), larger bill and feet, rufous chest (beware Bat
Falcons with cinnamon-washed chest). Juv thighs spotted black. See Plate 13.

4 **Crested Eagle** *Morphnus guianensis* 31in (79cm) **p. 206**
Rare, SE rain forest (not recorded Mexico). Smaller than 5 with less massive legs
and feet. Adult dimorphic. 4(a) Dark morph: underparts barred black; 4(b) light
morph: from 5 by single-pointed crest, lack of black chest band, pale mottling on
wing coverts. Juv and imm plumages variable, note tail patterns (1st basic shown,
including 1st basic tail variation). Attains adult plumage in about 3 years. Does
not soar. See text and Fig. 20.

5 **Harpy Eagle** *Harpia harpyja* 38in (97cm) **p. 207**
Rare, SE rain forest. Huge and powerful, legs and feet massive. Adult
unmistakable (compare with 4). Juv and imm plumages variable, note tail
patterns. 2nd basic shown; 1st basic tail (shown) can be much like 1st basic tail
variation of 4. Attains adult plumage in about 4 years. Does not soar. See text and
Fig. 20.

6

1

juv

2

juv

3

juv

juv

4

4a

4b

imm

juv

5

5

imm

juv

imm

SW 7-89

PLATE 7: VULTURES, CARACARA, LAUGHING FALCON

1 **Laughing Falcon** *Herpetotheres c. cachinnans* 20in (51cm) **p. 213**
Tropical lowlands. Unmistakable; does not soar.

2 **Crested Caracara** *Caracara plancus audubonii** 21in (53cm) **p. 213**
Widespread. Unmistakable; note striking flight pattern.

3 **Turkey Vulture** *Cathartes a. aura** 28in W 68.5in (71cm W 174cm) **p. 174**
Widespread. From 4 by shorter wing projection beyond tail; adult has browner
plumage, dark red head with pale bill, flesh legs, brownish-white primary shafts
on upperwings. Juv (shown with neck ruff raised) head greyish at first. In flight
from 5 by two-tone wings, long tail, strong dihedral.

4 **Lesser Yellow-headed Vulture** *C. b. burrovianus* 23in W 61.5in (58cm W 156cm)
p. 175
S savannas and marshes. Much like 3; note blacker plumage, whiter primary
shafts on upperwings, multicolored head of adult. Juv undescribed (?) and
separation criteria from juv 3 poorly known; note wing–tail projection on perched
birds, head probably greyish at first (?), whitish legs as adult (?), may have paler,
browner edgings on upperparts (?).

5 **Black Vulture** *Coragyps atratus* 24in W 58.5in (61cm W 150cm) **p. 174**
Widespread, distinctive. From 3 (and 4) in flight by broader wings, short tail,
white panel in primaries, flatter wings. See juv 6.

6 **King Vulture** *Sarcoramphus papa* 30in W 72.5in (76cm W 184cm) **p. 176**
S forest, widely extirpated. Adult unmistakable. Juv from smaller 5 by lack of
white panel in primaries, usually shows some white mottling on underwing
coverts.

imm

3

5

6

juv 3

4

juv

1

2

juv

3

4

juv

5

6

juv

|| - 9|

PLATE 8: TROPICAL FOREST RAPTOR SHAPES/ DARK BUTEOS

Tropical forest raptor shapes not to scale

1 **Grey-headed Kite** *Leptodon cayanensis* W 39in (99cm) **p. 178**
   Broad, rounded wings with checkered panel in primaries, tail usually held closed.
   Pale body contrasts with dark underwings (adult). See Plates 5, 12.

2 **Double-toothed Kite** *Harpagus bidentatus* W 27in (69cm) **p. 183**
   Broad, tapered wings with checkered panel in primaries, puffy white undertail
   coverts, tail rarely spread. See Plates 2, 13.

3 **Hook-billed Kite** *Chondrohierax uncinatus* W 34.5in (88cm) **p. 179**
   Broad, rounded wings with base pinched in at rear, broad-fingered 'hands' with
   checkered panel in primaries, tail often spread. See Plates 2, 10, 12.

4 **Black Hawk-Eagle** *Spizaetus tyrannus* W 55.5in (141cm) **p. 209**
   Broad wings and long, broad tail, primaries checkered. See Plates 5, 12.

5 **Ornate Hawk-Eagle** *S. ornatus* 51in (130cm) **p. 210**
   Broad wings and long, broad tail, body darker than underwings (adult), panel in
   primaries pale but not boldly barred. See Plates 5, 12.

6 **Black-and-white Hawk-Eagle** *Spizastur melanoleucus* W 51in (130cm) **p. 208**
   Long, broad, tapered wings, relatively short tail; overall bright white below. See
   Plates 5, 12.

7 **Common Black Hawk** *Buteogallus anthracinus* W 46.5in (118cm) **p. 191**
   Adult: broad wings with very broad secondaries, short tail often spread. Juv:
   wings more broadly based and tail shorter than 9; often appears dark with pale
   tail and panel in primaries. See Plates 4, 10, 11.

8 **Solitary Eagle** *Harpyhaliaetus solitarius* W 67in (170cm) **p. 194**
   Adult: long, broad wings relatively less broadly based than 7, short tail. Shape
   somewhat intermediate between 7 and 9. See Plates 4, 10, 11.

9 **Great Black Hawk** *Buteogallus urubitinga* W 50.5in (128cm) **p. 193**
   Adult: long, broad wings relatively less broadly based than 7, tail longer than 7
   and often less widely spread. Juv: from 7 by longer tail, longer and less broadly
   based wings; often appears dark with pale tail and panel in primaries. See Plates 4,
   10, 11.

10 **Crane Hawk** *Geranospiza caerulescens* W 38.5in (98cm) **p. 189**
   Broad wings, long tail often held closed (2 white bands). See Plates 4, 10.

11 **White-tailed Hawk** *Buteo albicaudatus hypospodius* W 51.5in (131cm) **p. 201**
   Widespread but local; savannas. Juv from 12 by more tapered wings, less
   distinctly and more finely barred flight feathers, whitish undertail coverts, lack of
   broad pale panel in primaries; from 14 by broader based wings, bolder whitish
   mottling on underparts, pale undertail coverts. 13 much smaller, undertail coverts
   dark. See text and Plates 3, 9.

12 **Red-tailed Hawk** *B. jamaicensis calurus** W 50in (127cm) **p. 203**
   Widespread. Juv dark morph: note coarsely and relatively strongly barred flight
   feathers, broad pale panel in primaries, thickset shape. Compare with 11, 13, 14.
   See text and Plate 9.

13 **Short-tailed Hawk** *B. brachyurus fuliginosus* W 37in (94cm) **p. 199**
   Widespread. Dark morph. Note small size, medium-length tail, fairly plain, pale
   flight feathers (dark bars indistinct). Compare with 11, 12, 14. See text and
   Plates 3, 9.

14 **Zone-tailed Hawk** *B. albonotatus* W 51in (130cm) **p. 202**
   Widespread. Note long, narrow wings and tail; easily passed off in flight as a
   Turkey Vulture by flight manner. Adult shows 1 bold white tail band. Juv:
   stronger two-tone underwings, plain tail. Compare with 11, 12, 13. See Plate 3.

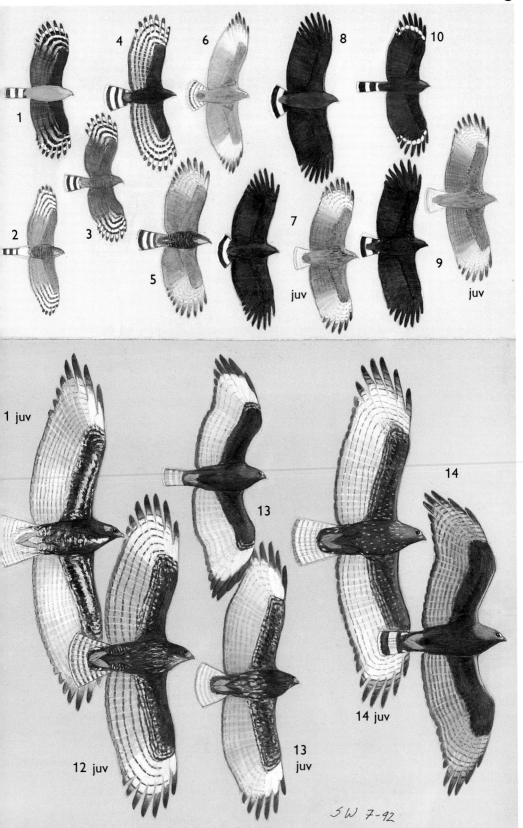

8

PLATE 9: BUTEOS

1 **Roadside Hawk** *Buteo magnirostris griseocauda** W 29in (74 cm) **p. 197**
Tropical lowlands. Relatively short, rounded wings, fairly long tail usually held closed. Rarely soars except in noisy display. Note 2–3 evenly broad pale bands on tail, rufous panel in primaries. See Plate 3.

2 **Broad-winged Hawk** *B. p. platypterus* W 34in (86 cm) **p. 199**
Widespread migrant. From 3 by longer, tapered wings with bolder dark trailing edge, shorter tail. Adult: 1 white tail band, rufous-barred underparts. Juv: note more distinct tail bars, broad pale panel in primaries often visible; compare with 3. See Plate 3.

3 **Grey Hawk** *B. nitidus plagiatus** W 34.5in (88 cm) **p. 196**
Widespread, especially arid lowlands. From 2 by less tapered wings with secondary bulge pinched in at base, longer tail. Note less distinct dark trailing edge to wings, pale body, and 2 white tail bands of adult; juv notably plain and pale below, flight feathers translucent when backlit. See Plate 3.

4 **Red-shouldered Hawk** *B. l. lineatus** W 38in (97 cm) **p. 197**
Resident in Baja, migrant in E Mexico. Note boldly banded tail, checkered remiges; wings have narrow pale panel in primaries.

5 **Red-tailed Hawk** *B. jamaicensis calurus** W 50in (127 cm) **p. 203**
Widespread, variable. Juv: from 6 by less tapered wings, coarser and more distinct flight feather barring, broad pale panel in primaries, dark lesser coverts. See text and Plate 8.

6 **White-tailed Hawk** *B. albicaudatus hypospodius* W 51.5in (131 cm) **p. 201**
Widespread but local. Note broadly based, long, tapered wings. Adult: short white tail with single black band, duskier inner primaries. Juv: compare with 5. See text and Plates 3, 8.

7 **Swainson's Hawk** *B. swainsoni* W 49.5in (126 cm) **p. 200**
Widespread transient, breeds N Mexico. Note long, fairly narrow, tapered wings with contrasting dark remiges. Light morph adult has dark head and chest, white bib. Plumage variable (see text).

8 **Short-tailed Hawk** *B. brachyurus fuliginosus* W 37in (94 cm) **p. 199**
Widespread. Light morph. Note dark hood, fairly plain, white underparts, secondaries often form dusky panel (especially juv). See Plates 3, 8.

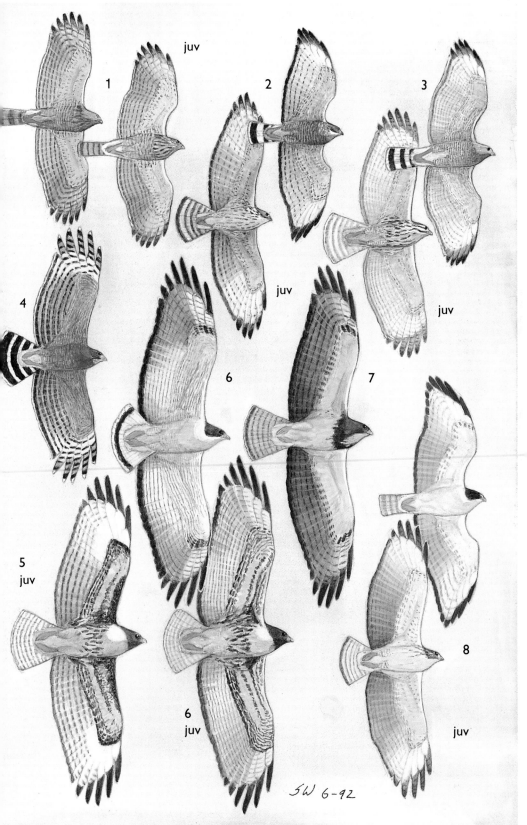

9

1

juv

2

3

4

6

7

juv

juv

5
juv

6
juv

8

juv

SW 6-92

PLATE 10: BLACK HAWKS

1 **Hook-billed Kite** *Chondrohierax u. uncinatus** W 34.5in (88cm) **p. 179**
Widespread. Note broad wings pinched in at base, broad-fingered 'hands', longish
tail. From 3 by shape, lack of single bold white band in primaries, short legs,
greenish facial skin, hooked bill. Plumage variable, adult may have 2 white tail
bands, pale bars on primaries. See text and Plates 2, 8, 12.

2 **Snail Kite** *Rostrhamus sociabilis major* W 44in (112cm) **p. 182**
S marshes. Note long, broad wings (held arched), fairly long tail with broad white
base. Underwings may be all dark or with more extensive pale panel. See Plate 11
for ♀ and juv; also see Plate 4.

3 **Crane Hawk** *Geranospiza caerulescens nigra** W 38.5in (98cm) **p. 189**
Tropical lowlands. Note broad wings with white band in primaries, long tail with
2 white bands, long red legs. Compare with 1. See Plates 4, 8.

4 **Solitary Eagle** *Harpyhaliaetus solitarius sheffleri** W 67in (170cm) **p. 194**
Rare and local. Adult: huge, all dark with 1 white tail band. From smaller 5 by
relatively longer outer primary, relatively longer but less broadly based wings, grey
plumage (often looks all dark until light catches head and neck). From smaller 6
by single white tail band, grey plumage. See text and Plates 4, 8, 11.

5 **Common Black Hawk** *Buteogallus a. anthracinus** W 46.5 in (118cm) **p. 191**
Widespread. Adult: often looks huge due to very broad wings with broad
secondaries, slow flapping wingbeats. Plumage variable. 5(a) (dark type, common
in S Mexico and C.A.): all dark with 1 white tail band; note shape, especially
relatively short outer primary. Compare with 4, 6. 5(b) (paler type, common in N
Mexico): note paler remiges with broad black trailing edge, white flash at base of
outer primaries. See text and Plates 4, 8, 11.

6 **Great Black Hawk** *B. urubitinga ridgwayi* W 50.5in (128cm) **p. 193**
Widespread. Note relatively longer, less broadly based wings than 5, and slightly
longer tail, longer legs. Adult: from 5 by shape, 2nd white tail band often visible
at base of tail, feet usually project farther into white tail band; underwings all
dark with small whitish flash at base of outer primaries, can also show small
distal pale window at tips of inner primaries. 2nd basic: note pale, barred panel
on inner primaries, 2–4 white tail bands. See text and Plates 4, 8, 11.

juv
1
2
♂
juv
3
4
5a
5b
6
imm

SW 6-92

PLATE 11: NEAR BUTEOS AND ALLIES

1 **Snail Kite** *Rostrhamus sociabilis major* W 44in (112cm) **p. 182**
S marshes. Distinctive; note long, broad wings (held arched), fairly long tail with broad white base, primaries often appear as contrasting pale panel when backlit. ♂ on Plate 10; also see Plate 4.

2 **White Hawk** *Leucopternis albicollis ghiesbreghti** W 48.5in (123cm) **p. 191**
SE rain forest. Unmistakable; wings of birds in S Belize and NE Guatemala (2(a)) often more opaque when backlit, with broad translucent trailing edge. See Plate 5.

3 **Solitary Eagle** *Harpyhaliaetus solitarius sheffleri** W 67in (170cm) **p. 194**
Rare and local. Huge; long, broad wings, fairly short tail. Juv: from smaller 4 and 5 by dusky secondaries and tail with no distinct barring, all-dark thighs. 1st basic (3(a)) has single broad pale band on tail, plumage otherwise much like juv. See text and Plates 4, 8, 10.

4 **Common Black Hawk** *Buteogallus a. anthracinus** W 46.5in (118cm) **p. 191**
Widespread. Juv: from juv 5 by relatively shorter wings with secondaries bulging slightly, shorter tail, shorter legs, fewer tail bars, darker malar stripe; note similarity of juv tail to 1st basic 5(a). Plumage variable, pale panel in primaries may be barred or unbarred. Compare with 3. See text and Plates 4, 8, 10.

5 **Great Black Hawk** *B. urubitinga ridgwayi* W 50.5in (128cm) **p. 193**
Widespread. From 4 by relatively longer and more evenly broad wings, longer tail. Juv flight feathers (most obvious on tail) more narrowly barred than 4; tail may have very broad distal band (see Plate 4); 1st basic (5(a)) similar to 4. Compare with 3. See text and Plates 4, 8, 10.

6 **Black-collared Hawk** *Busarellus n. nigricollis* W 49in (124cm) **p. 190**
Tropical lowland marshes, rare on Pacific Slope. Note very broad, long wings and short tail; in silhouette from 4 by less broad secondaries; from both 4 and 5 by shorter and smaller tail. Adult plumage unmistakable; juv shows hint of adult pattern. See Plate 4.

7 **Harris' Hawk** *Parabuteo unicinctus harrisi** W 43.5in (110cm) **p. 195**
Deserts of N and cen Mexico, local S Mexico and C.A. Note long graduated tail, broad secondaries pinched in at base. Adult unmistakable, juv distinctive. From 1 (S marshes) by shape, dark throat, longer legs, etc. See Plate 3.

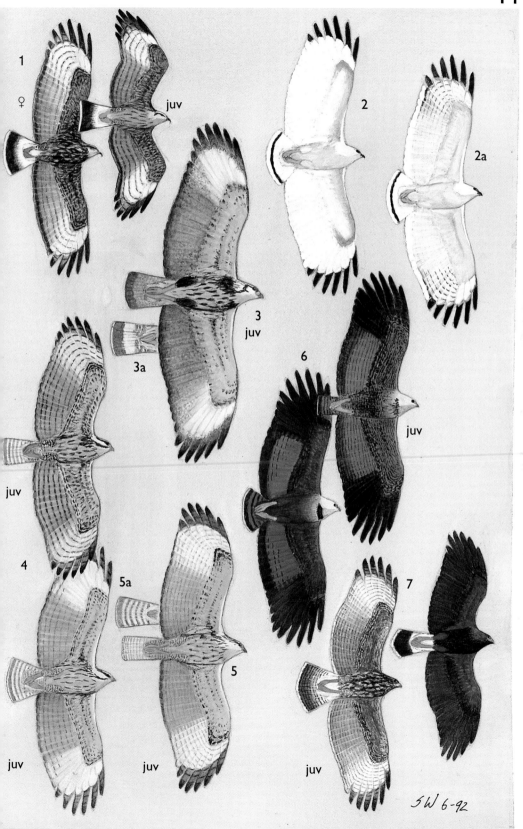

11

1
♀
juv
2
2a
3
juv
3a
4
juv
5a
5
juv
6
juv
7
juv
juv

SW 6-92

PLATE 12: HAWK-EAGLES AND LARGE FOREST KITES

1 **Grey-headed Kite** *Leptodon cayanensis* W 39in (99cm) **p. 178**
SE lowlands. Note broad wings with checkered outer primaries, tail rarely widely spread. Adult distinctive; pale body contrasts with dark underwings. Juv dimorphic. 1(a) Dark morph: pale underwings with checkered barring on outer primaries, dark-streaked body; 1(b) light morph: from larger 2 by shape (especially longer tail, shorter and blunter wings), bolder barring on outer primaries, tail pattern, lack of black mask, etc. From juv 3 by shape, tail pattern, less distinct barring on secondaries. See text and Plates 5, 8.

2 **Black-and-white Hawk-Eagle** *Spizastur melanoleucus* W 51in (130cm) **p. 208**
SE rain forest. Long, broad wings and relatively short tail. Bright white below with dark wingtips, tail often appears dark; note black mask and, if seen head-on or from above, distinctive bold white leading edge to inner wings. From juv 5 by shape (especially short tail), lack of dark barring below, etc. Compare with 1(b). See Plates 5, 8.

3 **Hook-billed Kite** *Chondrohierax u. uncinatus** W 34.5in (88cm) **p. 179**
Widespread. Note shape with broad wings pinched in at base, broad-fingered 'hands'; strongly hooked bill and green facial skin distinctive at close range. Floppy flight often suggests a larger bird. Adult has 1–2 pale tail bands, juv has 2–3 narrower pale tail bands. Plumage highly variable (see text). Remiges more boldly barred than 1(b). See Plates 2, 8, 10.

4 **Black Hawk-Eagle** *Spizaetus tyrannus serus* W 55.5in (141cm) **p. 209**
SE rain forest. Long, broad wings and tail. Often calls while soaring, far-carrying whistles, *wh wh wheeoo*, etc. Adult distinctive; black wing-linings contrast with checkered remiges. Juv from adult 5 by tail pattern, bolder dark bars on remiges, etc. See Plates 5, 8.

5 **Ornate Hawk-Eagle** *S. ornatus vicarius* W 51in (130cm) **p. 210**
SE rain forest. Shape rarely distinguishable from 4 but may appear relatively shorter and broader winged with relatively longer, more rounded tail. Often calls while soaring, far-carrying whistles, *whi whee-whee-wheep*, etc. Adult: note dark body contrasting with pale underwings, no bold dark bars on wings, primaries often appear as translucent panel when backlit. Juv: note shape (especially long tail; compare with 2), black barring on underparts. See Plates 5, 8.

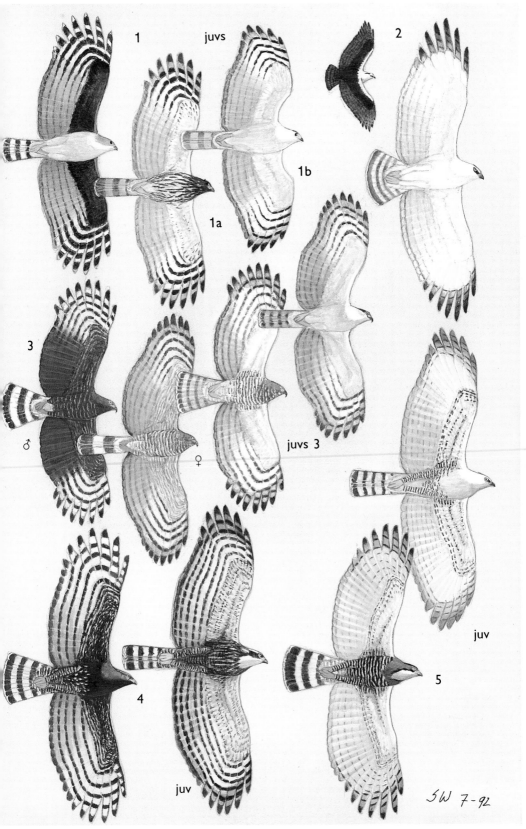

12

1 juvs 2

1b

1a

3

♂ ♀ juvs 3

juv

4 juv 5

juv

SW 7-92

PLATE 13: KITES, ACCIPITERS, FALCONS

1 **Double-toothed Kite** *Harpagus bidentatus fasciatus* W 27in (69cm) **p. 183**
SE rain forest. Accipiter-like shape but wings relatively longer, more tapered, tail rarely spread; note puffy white undertail coverts. Flies with wings distinctively bowed. Adult: note dark body contrasting with whitish underwings, outer primaries checkered black. Juv: note shape, 3 narrow whitish tail bands; underparts may be unstreaked or heavily streaked dark. See Plates 2, 8.

2 **Mississippi Kite** *Ictinia mississippiensis* W 35in (89cm) **p. 183**
Transient in E and S. From 3 by longer tail, longer outer primary, lack of rufous in primaries. Imm: similar to imm 3 (not shown) but paler overall; note duller tail bars, features noted above. See text for adult and juv plumages; also see Plate 2.

3 **Plumbeous Kite** *I. plumbea* W 35in (89cm) **p. 184**
Summer resident (Feb–Aug) in SE lowlands. From 2 by shorter tail, shorter outer primary, rufous flash in primaries. See Plate 2.

4 **Sharp-shinned Hawk** *Accipiter striatus suttoni** W 22.5in (57cm) **p. 186**
Widespread. Juv from juv 5 by heavy dark streaking below. See Cooper's Hawk (text).

5 **White-breasted Hawk** *A. chionogaster* W 22.5in (57cm) **p. 187**
Endemic, highlands S of Isthmus. Size and shape like Sharp-shinned Hawk; note bright white underparts, finely streaked and flecked dusky in juv. See Plate 2.

6 **American Kestrel** *Falco s. sparverius** W 25in (64cm) **p. 215**
Widespread. Relatively long-tailed and slender-winged with floppy flight; hovers. Underparts overall pale; compare with more compact 7, 8.

7 **Merlin** *F. c. columbarius** W 26.3in (67cm) **p. 216**
Widespread migrant. Small and compact; note streaked underparts; compare with darker 8. Juv 11 larger and proportionately shorter-tailed.

8 **Bat Falcon** *F. rufigularis albigularis** W 26.5in (67cm) **p. 217**
Tropical lowlands, rare on Pacific Slope. Small, dark, compact falcon with white throat. From larger and rarer 9 by less thickset shape, more extensively dark underbody. See Plate 6.

9 **Orange-breasted Falcon** *F. deiroleucus* W 33in (84cm) **p. 218**
Rare, SE rain forest. Thickset shape suggests 11; note rufous chest. Compare with 8. See Plate 6.

10 **Aplomado Falcon** *F. femoralis septentrionalis** W 35in (89cm) **p. 216**
SE savannas, local. Long, narrow wings and long tail with numerous distinct, narrow bars; note pale cinnamon thighs, bold white supercilium. See Plate 6.

11 **Peregrine Falcon** *F. peregrinus tundrius* W 42in (107cm) **p. 218**
Widespread. Note thickset shape with broadly based wings and relatively short tail, bold dark moustache. Juv: dark underwings, streaked underparts. Adult has pale underwings, finely barred underparts.

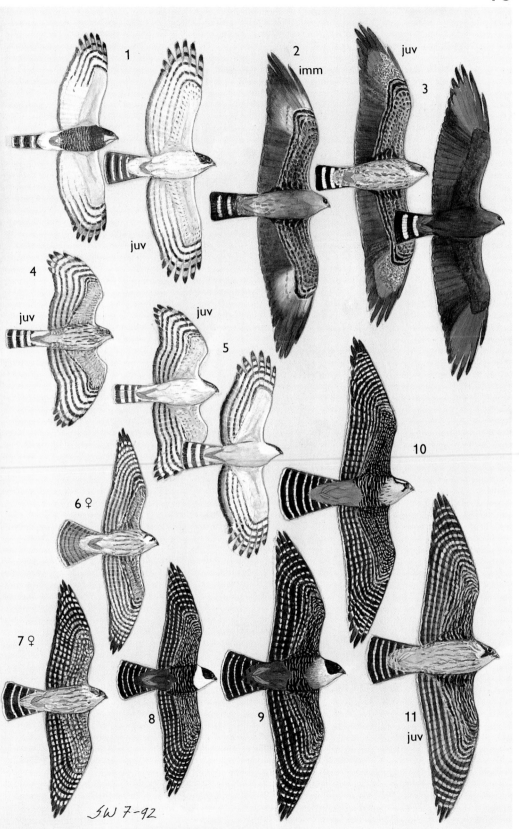

PLATE 14: CRACIDS AND TURKEY

All of these species, except the noisy and conspicuous chachalacas, have been decimated or locally extirpated by overhunting combined with habitat loss. They are rarely seen other than in remote and/or protected areas.

1 **Rufous-bellied Chachalaca** *Ortalis (poliocephala?) wagleri* 25in (64cm) **p. 221**
   Endemic to NW Mexico. Chestnut belly and tail tip. Intergrades abruptly with 2 in NW Jal.

2 **West Mexican Chachalaca** *O. p. poliocephala** 25in (64cm) **p. 221**
   Endemic to W Mexico. Whitish belly, creamy-buff tail tip; from smaller 4 by broader, creamier tail tip, cinnamon undertail coverts, reddish orbital skin. 3 allopatric (?).

3 **Plain Chachalaca** *O. v. vetula** 19in (48cm) **p. 220**
   Atlantic Slope. Buffy-brown belly, buff tail tip. Belly and tail tip whitish in *pallidiventris* of Yuc Pen.

4 **White-bellied Chachalaca** *O. leucogastra* 20.5in (52cm) **p. 222**
   Endemic, Pacific Slope S from Chis. From larger 2 by narrower, clean white tail tip, whitish undertail coverts, grey orbital skin.

5 **Crested Guan** *Penelope p. purpurascens* 34in (86cm) **p. 222**
   Both slopes, local. Blackish brown with red wattle, white streaking may be hard to see. 5(a) Crest raised.

6 **Horned Guan** *Oreophasis derbianus* 33in (84cm) **p. 223**
   Endemic: very local in cloud forests of Chis and Guatemala, endangered. Unmistakable; horn develops through 1st year.

7 **Highland Guan** *Penelopina nigra* 24.5in (62cm) **p. 222**
   Endemic; cloud forests S of Isthmus. ♂ black with red bill, bright red wattle. ♀ barred brown and black. Imm ♂ blackish brown, barring indistinct, wattle small. 5 larger, browner, with dark bill, mostly in lowlands.

8 **Great Curassow** *Crax r. rubra** 33in (84cm) **p. 223**
   SE rain forest, local. Unmistakable; note curly crest. ♀ polymorphic: 8(a) barred morph; 8(b) dark morph.

9 **Ocellated Turkey** *Meleagris ocellata* ♂ 38in (97cm), ♀ 30in (76cm) **p. 226**
   Endemic to Yuc Pen. Unmistakable. (Domestic turkeys common in Ocellated's range.)

14

1

2

3

4

5

5a

imm

6

imm ♂

7

♂

♀a

imm ♂

7 ♀

♀b

9

SW 10-88

PLATE 15: TINAMOUS, WOOD-PARTRIDGES, FOREST QUAIL

1 **Thicket Tinamou** *Crypturellus cinnamomeus** 10.8in (27cm) **p. 90**
Widespread. 1(a) *occidentalis* (W Mexico): only tinamou in range, ♂ fairly plain greyish, ♀ barred above; 1(b) *goldmani* (Yuc Pen): cinnamon chest, upperparts boldly barred. Loud, hollow *hoo-oo* or *hoo o' oop* call.

2 **Little Tinamou** *C. soui meserythrus* 9in (23cm) **p. 89**
Humid SE. Small and plain; note olive legs, grey head, cinnamon belly. Tremulous whistles often in series.

3 **Slaty-breasted Tinamou** *C. boucardi** 10.5in (27cm) **p. 90**
SE humid forest. 3(a) *boucardi* (W of Sula Valley): neck and chest slaty grey, ♀ barred above; 3(b) *costaricensis* (E of Sula Valley): greyish-cinnamon underbody, ♀ boldly barred above. Low, mournful, drawn-out *ooo-o-oooo* call.

4 **Great Tinamou** *Tinamus major robustus** 16.5in (42cm) **p. 89**
SE rain forest. Large, plain; note greyish legs. Tremulous, eerie whistles often paired.

5 **Singing Quail** *Dactylortyx thoracicus** 8.5in (22cm) **p. 228**
Endemic; widespread, mostly humid montane forest. ♂ cinnamon face and throat, grey underparts. ♀ pale grey face and throat, cinnamon underparts. 5(a) *thoracicus* (E Mexico); 5(b) *sharpei* (Yuc Pen): paler overall. Striking 'song' unmistakable.

6 **Spotted Wood-Quail** *Odontophorus guttatus* 9.5in (24cm) **p. 228**
SE humid forest. Face and throat blackish, underparts spotted white, erectile crest (orange-based in ♂) often obvious. 6(a) Rufous morph; 6(b) brown morph. Loud, rollicking choruses often prolonged.

Dendrortyx Wood-Partridges are long-tailed, chicken-like quail of highland forests. The 3 following species are mostly (entirely?) allopatric; note face patterns. Voices are loud and rollicking, often involving duets.

7 **Bearded Wood-Partridge** *D. barbatus* 13.5in (34cm) **p. 227**
Endemic to E Mexico, endangered. Rich brown and cinnamon overall with pale blue-grey throat and upper chest.

8 **Long-tailed Wood-Partridge** *D. macroura** 13.5in (34cm) **p. 226**
Endemic to cen Mexico. Grey-brown overall with boldly white-striped black head. 8(a) *striatus* (W Mexico); 8(b) *oaxacae* (N Oax) shown with crest down, white face stripes less distinct.

9 **Buffy-crowned Wood-Partridge** *D. l. leucophrys* 12.5in (32cm) **p. 227**
SE Chis to C.A. Whitish forecrown and throat, dark bill.

15

1a ♂ ♀

2 ♀ ♂

1b ♂ ♀

3 ♂ ♀

4

5a ♀

5b ♂

6a ♀

6b ♂

7

8a

8b

9

SW 8-89

PLATE 16: BRUSH QUAIL

Most species are readily identified in their range and habitat.

1 **Elegant Quail** *Callipepla d. douglasii** 9.5in (24cm) **p. 233**
Endemic to NW Mexico. Note tawny crest (♂) or dark brown crest (♀), white-spotted belly.

2 **Scaled Quail** *C. s. squamata** 9.5in (24cm) **p. 232**
Plateau. Tufted whitish crest, scaly grey chest.

3 **Banded Quail** *Philortyx fasciatus* 8in (20cm) **p. 232**
Endemic to Balsas drainage. Pointed crest, boldly barred plumage; imm has blackish face.

4 **Montezuma Quail** *Cyrtonyx montezumae** 8.5in (22cm) **p. 229**
Pine–oak highlands N of Isthmus. ♂ harlequin face pattern, ♀ face pattern suggests ♂; note bushy crest. 4(a) *mearnsi* (N Mexico); 4(b) *sallei* (SW Mexico).

5 **Ocellated Quail** *C. (m.?) ocellatus* 8.5in (22cm) **p. 230**
Endemic to pine–oak highlands S of Isthmus. Distinctive in range; ♀ like 4.

6 **Northern Bobwhite** *Colinus virginianus** 8.2in (21cm) **p. 231**
Widespread. Marked geographic variation in ♂ plumages (see text). 6(a) *texanus* (NE Mexico); 6(b) *ridgwayi* (Son, endangered); 6(c) *graysoni* (cen Mexico); 6(d) *coyolcos* (Pacific Slope, S Mexico). ♀ buff supercilium and throat. Other bobwhites allopatric.

7 **Yucatan Bobwhite** *C. n. nigrogularis** 7.8in (20cm) **p. 231**
Endemic to Yuc Pen and Mosquitia. ♂ unmistakable; ♀ much like ♀ Northern Bobwhite. Other bobwhites allopatric.

Spot-bellied Bobwhite *C. leucopogon* C.A. See Plate 70.

16

1 ♀ ♂

2

imm 3

6a ♂ 6b ♂

♀

6c ♂

4a ♂ ♀

6d ♂

4b ♂

♀

5 ♂

♀ 7 ♂

SW 2-91

PLATE 17: CRAKES, RAILS, JACANA

1 **Sora** *Porzana carolina* 8.3in (21cm) **p. 242**
Widespread migrant. Adult distinctive; note black mask. Juv (plumage kept till Dec) may suggest smaller 2; note heavier yellow bill, plainer face, olive legs.

2 **Yellow-breasted Crake** *P. flaviventer woodi* 5.3in (13cm) **p. 243**
Local in Tropics, rarely seen. Tiny crake with buff face and chest, dark lores and cap, straw-yellow legs; compare with juv 1. Juv neck and chest barred dusky.

3 **Ruddy Crake** *Laterallus ruber* 5.8in (15cm) **p. 237**
S lowlands, common. Overall ruddy with slaty-grey hood, olive-grey legs; ♀ has rufous-brown rump and uppertail coverts. Juv overall sooty grey.

4 **Grey-breasted Crake** *L. exilis* 5.8in (15cm) **p. 237**
C.A., local. Distinctive; note bright lime-green base of bill, rufous nape, yellowish legs. Compare with Black Rail (see text). Juv similar to 3.

5 **Northern Jacana** *Jacana s. spinosa* 9in (23cm) **p. 263**
Tropical lowlands. Unmistakable; note strikingly different appearance of imm.

6 **Virginia Rail** *Rallus l. limicola** 9in (23cm) **p. 239**
Local. From much larger 8 and 9 by blue-grey auriculars, redder legs; in flight, note rufous upperwing coverts. Juv (not shown) extensively blackish, flank bars indistinct.

7 **Spotted Rail** *R. maculatus insolitus* 10.5in (27cm) **p. 240**
Local in Tropics. Adult unmistakable; note neon-lime-green bill with reddish spot at base. Juv variable, dark brown with sparse white spots; 'barred morph' similar to adult but browner, spots and bars less contrasting.

8 **Clapper Rail** *Rallus longirostris** 13.5in (34cm) **p. 238**
Salt marshes and mangroves, NW Mexico, Tamps, and Yuc Pen. From 9 by habitat (also see text). Compare with smaller 6. 8(a) *beldingi* (Baja); 8(b) *pallidus* (Yuc Pen).

9 **King Rail** *R. (l?) elegans tenuirostris** 14.8in (38cm) **p. 239**
Freshwater marshes, cen Mexico. From 8 by habitat. Compare with smaller 6.

10 **Uniform Crake** *Amaurolimnas concolor* 8.5in (22cm) **p. 242**
SE humid lowlands. Plain; note greenish-yellow bill, bright pinkish-red legs. Juv darker and greyer; from larger and greyer juv 11 (mangroves) by shorter bill, grey undertail coverts.

11 **Rufous-necked Wood-Rail** *Aramides axillaris* 12in (30cm) **p. 241**
Mangroves, local. Adult distinctive; note rufous neck, blue-grey lower hindneck patch. Juv duller with greyish head, neck, and underparts; compare with juv 10.

12 **Grey-necked Wood-Rail** *A. cajanea mexicana** 16in (41cm) **p. 241**
S lowlands. Adult unmistakable; juv duller with pale-mottled grey belly.

17

PLATE 18: SUNBITTERN, SUNGREBE, PLOVER, THICK-KNEE/PIGEONS

1 **Sunbittern** *Eurypyga helias major* 18in (46cm) **p. 247**
SE rain forest streams, not in Mexico (?). Unmistakable. Usually wary, runs and flies, does not swim.

2 **Sungrebe** *Heliornis fulica* 11in (28cm) **p. 246**
E and S lowlands; slow-moving rivers, lagoons. Unmistakable. Usually seen swimming near cover, does not dive. ♀ has tawny cheeks.

3 **Collared Plover** *Charadrius collaris* 5.8in (15cm) **p. 254**
Widespread. Slender and elegant, neat black and cinnamon head markings, narrow black chest band, flesh legs, lacks white hindcollar. Juv head plain, blackish chest side patches. See also Fig. 22.

4 **Double-striped Thick-knee** *Burhinus b. bistriatus* 19in (48cm) **p. 251**
S savannas, local. Unmistakable, mostly nocturnal.

Columba Large, mostly arboreal pigeons, widespread in the region; often hunted for food and frequently wary. Voices distinctive.

5 **Band-tailed Pigeon** *C. f. fasciata** 14in (36cm) **p. 323**
Pine–oak highlands. Orange-yellow bill and feet, white collar, broad pale tail tip (indistinct in *vioscae* of BCS). Low *wh'hoo wh'hoo* ...

6 **Pale-vented Pigeon** *C. cayennensis pallidicrissa* 12.5in (32cm) **p. 321**
SE savannas and riparian forest. From 8 by grey head, black bill, pale tail, whitish belly. *Whoooo, who'koo-koo who'koo-koo* ...

7 **White-crowned Pigeon** *C. leucocephala* 14in (36cm) **p. 322**
Caribbean islands and coasts. Dark slaty grey with snow-white cap, juv sootier with whitish forehead. *Whu, cu-cu-cuu cu-cu-cuu* ...

8 **Red-billed Pigeon** *C. f. flavirostris** 13in (33cm) **p. 322**
Widespread. Note pale horn bill (red base hard to see), vinaceous head, slaty blue-grey belly and blackish tail. Compare with 6, 10. *Whoooo, oo'koo-koo-koo oo'koo-koo-koo* ...

9 **Short-billed Pigeon** *C. nigrirostris* 11.5in (29cm) **p. 323**
SE rain forest. Dark overall; head and underparts dark vinaceous, bill black. Fairly bright *hoo'koo-koo-koo.*

10 **Scaled Pigeon** *C. speciosa* 13in (33cm) **p. 321**
SE rain forest. Note red bill, scaly neck and chest, whitish belly contrasting with black tail; compare with 6, 8. ♂ upperparts chestnut, ♀ upperparts brownish, juv duller, scaling indistinct. Low *hoo, oo'hoo oo'hoo* ...

18

1

♀ ♂

2

juv 3

4

juv

juv 5

6

7

juv

8 9 10

♀ ♂

juv

5 6 7

8 9 10

JW 10-88

PLATE 19: GROUND-DOVES

1 **Inca Dove** *Columbina inca* 8.5in (22cm) **p. 325**
Widespread, except Yuc Pen. Long white-sided tail, scaly overall. *Who'pu* call, 10/18–27 s.

2 **Common Ground-Dove** *C. passerina pallescens** 6.5in (17cm) **p. 326**
Widespread, except humid SE. Orangish-based bill, scaly chest. *Whuu* or *h'woo* call, 10/10–12 s.

3 **Ruddy Ground-Dove** *C. talpacoti rufipennis** 6.8in (17cm) **p. 327**
Widespread, mostly humid tropical lowlands. ♂ distinctive. ♀ very similar to smaller ♀ 4 but tends to be warmer-colored and browner, wing covert marks blackish (versus violet in 4), rufous in wings brighter. *Per-woop* or *h-woop* call, 10/6–8 s.

4 **Plain-breasted Ground-Dove** *C. minuta interrupta* 5.8in (15cm) **p. 326**
Local in lowland savannas. Small, plain, ♂ colors suggest 2 but lacks orangish bill and scaly chest. ♀ very like ♀ 3 but smaller and greyer, wing spots violet, rufous in wings duller. *Wuup* or *w-up* call, 10/6.5–7.5 s.

5 **Maroon-chested Ground-Dove** *Claravis mondetoura salvini* 8in (20cm) **p. 328**
Very local in humid montane forests (especially bamboo?). Distinctive. ♂ often appears all dark, white sides of tail may flash in flight. ♀ from lowland ♀ 6 by bolder wingbars, pale-tipped brownish tail, brownish wing-linings. *Woop* call, 10/12–13 s, typically faster and more prolonged than 6.

6 **Blue Ground-Dove** *C. pretiosa* 8.3in (21cm) **p. 327**
E and S. ♂ unmistakable. ♀ note rufous rump and tail, dark rufous wingbars. *Boop* call, usually 1–6×, 6/7–9 s.

19

SW 3-89

PLATE 20: DOVES AND QUAIL-DOVES

1 **White-winged Dove** *Zenaida asiatica mearnsi** 11.8in (30cm) **p. 324**
Widespread. Bold white wingbars distinctive. Burry *hoo-koo'koo-koo* call.

2 **Mourning Dove** *Z. macroura marginella** 11in (28cm) **p. 324**
Widespread. Long, tapered tail, rapid flight. Mournful *whoo'hoo hu, hu hu* call.

3 **Zenaida Dove** *Z. aurita salvadorii* 10.8in (27cm) **p. 324**
N coastal Yuc Pen. Pinkish brown overall (♀ duller), shortish squared tail, white trailing edge to secondaries. Mournful *hoo'ooo-oo oo-ooo* call.

Leptotila Retiring terrestrial doves of forest and woodland. All have yellowish eyes, blackish bills, dark reddish orbital rings, white-tipped tails. Low, mournful voices distinctive.

4 **White-tipped Dove** *L. verreauxi fulviventris** 11.5in (29cm) **p. 329**
Widespread. Vinaceous head and chest. *Wh' whoo-ooo* call, 1/9–10 s.

5 **Caribbean Dove** *L. jamaicensis gaumeri* 11.5in (29cm) **p. 330**
Yuc Pen and Caribbean islands. Whitish face, blue-grey nape, rosy hindneck sides. More extensively rufous underwings than 4. *Hoo coo'hoo-ooo* call, 1/7–9 s.

6 **Grey-headed Dove** *L. p. plumbeiceps* 10.5in (27cm) **p. 329**
SE rain forest and edge. Blue-grey nape and hindneck, pale vinaceous face. Short *whoo'* call, 10/21–28 s.

7 **Grey-chested Dove** *L. cassini cerviniventris* 10.5in (27cm) **p. 330**
SE rain forest and edge. Dusky overall, crown and nape warm brown, chest greyish. Drawn-out *whoooo* call, 10/48–65 s (compare with 8).

Geotrygon Quail-Doves are stocky, retiring doves of shady forest understory, easily overlooked unless calling; lack white tail tips of *Leptotila*.

8 **Ruddy Quail-Dove** *G. m. montana* 9.5in (24cm) **p. 332**
Both slopes. ♂ distinctive, bright ruddy overall. ♀ red bill and orbital ring, pale face with dusky cheek stripe. Low, mournful *whoooo*, 10/25–98 s (compare with 7; see text), fades away (versus more abrupt ending of 9).

9 **White-faced Quail-Dove** *G. albifacies rubida** 12in (30cm) **p. 331**
Endemic; E and S cloud forest. Appears ruddy overall with whitish head. Low, mournful *whoooo*, 10/25–45 s (compare with 8).

10 **Purplish-backed Quail-Dove** *G. lawrencii carrikeri* 12in (30cm) **p. 331**
Endemic to Sierra de Los Tuxtlas, S Ver. Whitish face with black cheek stripe, grey hindneck and chest. Low, slightly burry or twangy *hu' w-wohw*, 10/27–33 s.

Socorro Dove *Zenaida graysoni* Island endemic. See Plate 69.

20

SW 7-89

PLATE 21: PARROTS

For parrots in flight see Plate 23.

1 **Lilac-crowned Parrot** *Amazona f. finschi** 12.8in (33cm) **p. 342**
Endemic to W Mexico. Longish tail, dark reddish forehead contrasts with pale bill, crown lilac. Juv similar.

2 **White-fronted Parrot** *A. albifrons nana** 10.8in (27cm) **p. 341**
W and S. From 4 by red lores, lack of dark cheek patch and red shoulder; ♀ duller without red in wing.

3 **Red-crowned Parrot** *A. viridigenalis* 12.5in (32cm) **p. 342**
Endemic to NE Mexico. Red crown (less extensive in ♀), bright green cheeks, pale bill. Note voice, compare with 6.

4 **Yucatan Parrot** *A. xantholora* 10.5in (27cm) **p. 341**
Endemic to Yuc Pen. From 2 by yellow lores, dark cheek patch, red on shoulder of ♂, ♀ face often fairly plain.

5 **White-crowned Parrot** *Pionus senilis* 9.5in (24cm) **p. 340**
E and S. Head and chest bluish with white forecrown, shoulders bronzy. Juv head and chest greenish with smaller whitish forehead patch.

6 **Red-lored Parrot** *Amazona a. autumnalis* 13.8in (35cm) **p. 342**
E and S. From 3 by yellow cheeks (reduced in juv), darker bill, voice.

7 **Brown-hooded Parrot** *Pionopsitta h. haematotis* 8.5in (22cm) **p. 340**
SE rain forest. Small; brown hood, pale bill and orbital ring; hood paler in juv.

8 **Yellow-naped Parrot** *A. a. auropalliata** 14.5in (37cm) **p. 345**
Pacific Slope S from Isthmus. Also Honduras Bay Islands and Mosquitia (see Plate 70). Yellow nape, grey bill. Juv head has bluish cast with little or no yellow.

9 **Yellow-headed Parrot** *A. o. oratrix** 14.5in (37cm) **p. 343**
Local, both slopes, endangered. Mexican adult has full yellow head; in juv yellow restricted to forecrown; imm intermediate; bill pale. See Plates 69 and 70.

10 **Mealy Parrot** *A. farinosa guatemalae** 16in (41cm) **p. 343**
SE rain forest. Large, green, bold pale orbital ring, blue crown can be hard to see; crown greenish in *virenticeps* (E of Sula Valley).

21

1

2 ♀ ♂

3 ♀ ♂

4 ♀ ♂

5 juv

6 juv

7 juv

8 juv

9 imm 9 juv

10 juv

SW 10-88

PLATE 22: MACAWS, PARAKEETS, AND ALLIES

For flying birds see Plates 23 and 24.

1 **Scarlet Macaw** *Ara macao* 35in (89cm) **p. 337**
   SE rain forest, endangered, very local. Unmistakable.
2 **Military Macaw** *A. militaris* ssp.* 28.5in (72cm) **p. 337**
   Local in NW and NE Mexico. Huge, long-tailed, overall green and blue. Bill all
   dark in W Mexico.
3 **Thick-billed Parrot** *Rhynchopsitta pachyrhyncha* 16in (41cm) **p. 337**
   Endemic to NW Mexico. Highland pines. Large; long tail. Red forehead and
   shoulder, black bill.
4 **Maroon-fronted Parrot** *R. terrisi* 17in (43cm) **p. 338**
   Endemic to NE Mexico. Highland pines. Distinctive in range; note dark red
   forehead.

5 **Mexican Parrotlet** *Forpus c. cyanopygius** 5.3in (13cm) **p. 339**
   Endemic to NW Mexico. Tiny, social. ♂ has turquoise rump and wing patches.
6 **Aztec Parakeet** *Aratinga astec** 9in (23cm) **p. 336**
   E Mexico and C.A. Bold whitish orbital ring, blue in wings, dusky olive throat
   and chest. 6(a) *vicinialis* (NE Mexico): throat and chest paler, greenish olive;
   6(b) *astec* (SE Mexico and C.A.).
7 **Orange-fronted Parakeet** *A. canicularis eburnirostrum** 9.5in (24cm) **p. 336**
   W Mexico and C.A. Orange forehead, blue forecrown, blue in wings.
8 **Orange-chinned Parakeet** *Brotogeris j. jugularis* 7.3in (19cm) **p. 339**
   Pacific Slope S from Isthmus. Compact shape, bronzy shoulders, orange chin often
   hard to see.
9 **Pacific Parakeet** *Aratinga (holochlora?) strenua* 12.5in (32cm) **p. 334**
   Endemic to Pacific Slope S from Isthmus. Large, bright green overall, bill stout,
   may show orange flecks in throat. See text.
10 **Green Parakeet** *A. h. holochlora** 11.5in (29cm) **p. 333**
   Endemic to Mexico. From 9 by smaller, less deep bill, more often has orange
   flecks in throat. See text.
11 **Barred Parakeet** *Bolborhynchus l. lineola* 6.8in (17cm) **p. 338**
   SE cloud forest. Tiny; long, pointed wings and tail. Black barring may be hard to
   see (duller in ♀).

Red-throated Parakeet *Aratinga (holochlora?) rubritorques* C.A. See Plate 70.
Socorro Parakeet *A. brevipes* Island endemic. See Plate 69.

22

SW 9-91

PLATE 23: PARAKEETS, PARROTS, AND ALLIES

Note: Flying parrots can be difficult to identify; voice is useful with experience. For perched parrots see Plate 21; for parakeets and allies see Plate 22.

1 **Mexican Parrotlet** *Forpus c. cyanopygius** 5.3in (13cm) **p. 339**
Endemic to NW Mexico. Tiny; short, squared tail, ♀ lacks turquoise. Flight bounding, often in twittering flocks.

2 **Orange-chinned Parakeet** *Brotogeris j. jugularis* 7.3in (19cm) **p. 339**
Pacific Slope S from Isthmus. Compact; short, pointed tail, yellow wing-linings. Bounding flight.

3 **Orange-fronted Parakeet** *Aratinga canicularis eburnirostrum** 9.5in (24cm) **p. 336**
W Mexico and C.A. Long, pointed tail, remiges grey below; note head pattern, fast direct flight.

4 **Aztec Parakeet** *A. a. astec** 9in (23cm) **p. 336**
E Mexico and C.A. Long, pointed tail, remiges grey below; note head pattern (especially bold eyering), fast direct flight.

5 **Barred Parakeet** *Bolborhynchus l. lineola* 6.8in (17cm) **p. 338**
SE cloud forest. Tiny; long, pointed wings and tail. Flight rapid, often high.

6 **Green Parakeet** *A. h. holochlora** 11.5in (29cm) **p. 333**
Endemic to Mexico. Long, pointed tail, remiges yellowish below. Wingbeats less hurried, flight often higher than 3 and 4. Pacific Parakeet *A. (h.?) strenua* 12.5in (32cm) similar (see text and Plate 22).

7 **White-fronted Parrot** *Amazona albifrons nana** 10.8in (27cm) **p. 341**
Widespread. Hurried, shallow wingbeats, red on forewing of ♂ only (more extensive than 8); note duller rump and tail with contrasting broad bright green tip, face pattern.

8 **Yucatan Parrot** *Amazona xantholora* 10.5in (27cm) **p. 341**
Endemic to Yuc Pen. Much like 7; note extensively bright green rump and tail, red restricted to primary coverts of ♂, face pattern (especially dark cheek patch).

Large Amazons All have red patch on secondaries above, shallow wingbeats slower than 7 and 8. Voice, range, and habitat important for identification. Distinctive head patterns often hard to see.

9 **Red-crowned Parrot** *A. viridigenalis* 12.5in (32cm) **p. 342**
Endemic to NE Mexico. From 10 by voice, pale bill, head pattern.

10 **Red-lored Parrot** *A. a. autumnalis* 13.8in (35cm) **p. 342**
E Mexico and C.A. Note voice, head pattern (yellow cheeks may be lacking in juv).

11 **Mealy Parrot** *A. farinosa guatemalae** 16in (41cm) **p. 343**
SE rain forest. Large, plain; note voice, bold pale orbital ring. Sympatric with 10.

12 **Lilac-crowned Parrot** *A. f. finschi** 12.8in (33cm) **p. 342**
Endemic to W Mexico. Long-tailed, sympatric with 7 and 14.

13 **Yellow-naped Parrot** *A. a. auropalliata** 14.5in (37cm) **p. 345**
Pacific Slope from Isthmus S where sympatric with 7, sympatric with 10 in Honduras Bay Islands. Yellow nape absent in juv.

14 **Yellow-headed Parrot** *A. o. oratrix** 14.5in (37cm) **p. 343**
Local, both slopes. Yellow head reduced in juv, imm, and C.A. forms (see Plates 21, 69, 70); note voice.

15 **White-crowned Parrot** *Pionus senilis* 9.5in (24cm) **p. 340**
E Mexico and C.A. Deep wingbeats quite different from *Amazona*. Bluish overall, note white forecrown, reddish undertail coverts, voice.

16 **Brown-hooded Parrot** *Pionopsitta h. haematotis* 8.5in (22cm) **p. 340**
SE rain forest. Small; rapid flight with deep wingbeats. Note bright red axillars, brownish head, voice.

23

SW 10-91

For perched parrots see Plate 22.

1 **Military Macaw** *Ara militaris mexicana** 28.5in (72cm) **p. 337**
Local in NW and NE Mexico. Huge; long tail, green and blue overall, underside of flight feathers metallic yellowish.

2 **Thick-billed Parrot** *Rhynchopsitta pachyrhyncha* 16in (41cm) **p. 337**
Endemic to NW Mexico. Highland pines. Distinctive in range; note yellow underwing stripe.

3 **Maroon-fronted Parrot** *R. terrisi* 17in (43cm) **p. 338**
Endemic to NE Mexico. Highland pines. Distinctive in range.

4 **Scarlet Macaw** *Ara macao* 35in (89cm) **p. 337**
SE rain forest, endangered, very local. Unmistakable.

5 **Greater Roadrunner** *Geococcyx californianus* 22in (56cm) **p. 350**
N Mexico and Plateau. From slightly smaller 6 by heavily striped throat and chest, bigger bill.

6 **Lesser Roadrunner** *G. velox melanchima** 18in (46cm) **p. 350**
Endemic, Pacific Slope, S Interior, N Yuc Pen. From larger 5 by unstreaked buff throat and median chest, smaller bill. 6(a) Crest flattened.

7 **Smooth-billed Ani** *Crotophaga ani* 13.8in (35cm) **p. 351**
Caribbean islands. From smaller 8 by arched culmen, smooth bill, voice.

8 **Groove-billed Ani** *C. s. sulcirostris** 12.8in (33cm) **p. 351**
Widespread. Unmistakable in most of range; see 7.

9 **Mangrove Cuckoo** *Coccyzus minor palloris** 12.5in (32cm) **p. 347**
Both slopes. Dark mask, buffy underparts, large white tail-spots.

10 **Squirrel Cuckoo** *Piaya cayana** 17.8in (45cm) **p. 348**
Both slopes. Unmistakable, mostly bright rufous. 10(a) *mexicana* (W Mexico): paler, tail rufous below; 10(b) *thermophila* (E and S): darker, tail blackish below.

11 **Pheasant Cuckoo** *Dromococcyx phasianellus rufigularis* 14.5in (37cm) **p. 349**
S and E. Long, broad tail, small head with dark eyestripe, chestnut crest. Hard to see, mostly heard: far-carrying, whistled *whee-whee wheerr*, last note quavering. Juv (brood parasite) has throat and chest uniform rich buff.

12 **Striped Cuckoo** *Tapera naevia excellens* 11.5in (29cm) **p. 348**
SE. Buff upperparts striped blackish, erectile crest rises in song: clear, whistled *whee whee*, 2nd note higher, carries well. Juv (brood parasite) has cinnamon-spotted upperparts, dusky barred chest.

13 **Lesser Ground-Cuckoo** *Morococcyx e. erythropygus** 10.5in (27cm) **p. 349**
Pacific Slope and S interior. Terrestrial, often heard but hard to see: loud, rolled referee whistles in speeding then slowing series. Note pale yellow and blue orbital ring, cinnamon underparts; flight feathers have greenish sheen, noticeable on flushed birds.

24

5W 7-89

PLATE 25: SMALL OWLS

Glaucidium **Pygmy-Owls** Tiny, long-tailed owls, all have 'false eyes' on their nape sides. Taxonomy poorly understood (see text).

'Northern Pygmy-Owl' complex Highland forests. Long tail has 5–6 narrow pale bars on upperside, crown and nape finely spotted, upperparts usually spotted. Juvs: unspotted grey crown and nape, upperparts mostly unspotted, thus much like 'Least Pygmy-Owl' group; note longer tail with 5–6 pale bars.

1 **Cape Pygmy-Owl** *G. hoskinsi* 6.3in (16cm) **p. 361**
Endemic to mountains of BCS. Other species allopatric. Strikingly slow 'double-hoots', 5 double notes/30–40 s.

2 **Mountain Pygmy-Owl** *G. gnoma** 6.3in (16cm) **p. 360**
Mountains S to Honduras. Brownish grey to rufous overall. From 3 and 4 (5 allopatric) by longer tail, tail pattern, spotted upperparts, voice. 2(a) *gnoma* (N of Isthmus): double-hoots, 5 double notes/4.5–6.5 s; 2(b) *cobanense* (S of Isthmus): voice undescribed.

'Least Pygmy-Owl' complex Relatively short tail with 2–4 slightly broken pale bars on upperside, spotted crown, mostly unspotted upperparts. Juvs: compare with 2.

3 **Colima Pygmy-Owl** *G. p. palmarum** 5.8in (15cm) **p. 361**
Endemic to W Mexico. Thorn forest. 4 and 5 allopatric. Compare with 2. From 6 by shorter tail with fewer bars, spotted crown, lack of bold pale scapular spots, voice. Hollow hoots, 2–24 in series, 10/3–4 s.

4 **Tamaulipas Pygmy-Owl** *G. sanchezi* 5.8in (15cm) **p. 362**
Endemic to NE Mexico. Cloud forest. 3, 5, and 6 allopatric. Compare with 2. ♂ much like 5, ♀ has rufescent morph (shown). 1–3 slow-paced ringing hoots.

5 **Central American Pygmy-Owl** *G. griseiceps* 5.8in (15cm) **p. 362**
SE Mexico and C.A. Rain forest. From 6 (more open areas) by shorter tail with fewer bars, spotted crown, lack of bold pale scapular spots, voice. Ringing hoots, 2–18 in series, 10/2–3.5 s.

6 **Ferruginous Pygmy-Owl** *G. brasilianum ridgwayi** 7 in (18cm) **p. 363**
Tropical lowlands. Grey-brown to rufous overall; note long tail with 5–7 narrow pale bars, streaked crown and nape, voice. Juv head greyer, streaks restricted to forehead. Hollow hooting, often prolonged, 10/3–3.5 s.

Otus **Screech-Owls** Small to medium-sized owls. Cryptic plumage, ear tufts often held flat. Voice, range, and habitat important for identification.

7 **Balsas Screech-Owl** *O. seductus* 10in (25cm) **p. 356**
Endemic to SW Mexico. Arid scrub. Large; note brown eyes. Gruff bouncing-ball hoots.

8 **Pacific Screech-Owl** *O. c. cooperi** 9.5in (24cm) **p. 357**
Pacific lowlands S from Oaxaca. Larger and paler than 9 with bristled toes, gruff bouncing-ball hoots.

9 **Vermiculated Screech-Owl** *O. g. guatemalae** 8.5in (22cm) **p. 358**
Humid tropical forest and dense woodland. Note naked flesh toes, voice. Toad-like purring trill, 6–15 s long.

10 **Bearded Screech-Owl** *O. barbarus* 7in (18cm) **p. 358**
Endemic to mountains of Chis and Guatemala. From 11 by bolder whitish brows, naked pink toes, less coarsely streaked and more scalloped underparts, voice. Cricket-like trill, 3–5 s long.

11 **Whiskered Screech-Owl** *O. t. trichopsis** 7in (18cm) **p. 357**
Pine–oak highlands. Note coarsely streaked underparts, voice. Bouncing-ball and Morse-code hoots. Compare with 10 and Western Screech-Owl (see text). 11(a) Rufous morph (mainly cen Mexico S).

12 **Unspotted Saw-whet Owl** *Aegolius ridgwayi rostratus* 8.3in (21cm) **p. 370**
Mountains of Chis and Guatemala. Unmistakable in range. Hollow hooting, often persistent but not steady, 10/3–5 s.

25

SW 9-92

PLATE 26: LARGE OWLS AND POTOOS

1 **Great Horned Owl** *Bubo virginianus pallescens* * 20.5in (52cm) **p. 359**
Widespread, mostly in temperate areas. Thick, bushy ear tufts at sides of head, underparts barred. Varied low hooting series, *hoo hoo-hoo, hoo, hoo*, etc.

2 **Stygian Owl** *Asio stygius robustus* * 16in (41cm) **p. 368**
Widespread but local, mostly pine and pine–oak. Dark facial disks with whitish forehead blaze, ear tufts central, underparts streaked. Single deep *woof*, repeated.

3 **Mottled Owl** *Strix virgata* * 14in (36cm) **p. 365**
Widespread, common. Lacks ear tufts, eyes brown. Dimorphic. Note bold whitish brows and whiskers. 3(a) *squamulata* (W Mexico); 3(b) *centralis* (SE Mexico and C.A.). Voice varied, typically series of resonant barking hoots. See Fig. 26.

4 **Striped Owl** *Asio clamator forbesi* 14in (36cm) **p. 368**
S savannas. Whitish facial disks rimmed black, central ear tufts, pale buff underparts streaked blackish. Single hoot, repeated.

5 **Fulvous Owl** *Strix fulvescens* 16in (41cm) **p. 367**
Endemic to mountains S of Isthmus. Lacks ear tufts, eyes brown in pale facial disks, chest barred, rest of underparts streaked. Varied series of barking hoots. See Fig. 26.

6 **Black-and-white Owl** *S. nigrolineata* 15.5in (39cm) **p. 365**
SE rain forest. Unmistakable, black and white. Voice suggests Mottled Owl but with strongly emphatic hoots. See Fig. 26.

7 **Crested Owl** *Lophostrix cristata stricklandi* 16in (41cm) **p. 358**
S humid forest. Unmistakable, spectacular white ear tufts (may be held flattened). Throaty growl, repeated. See Fig. 26.

8 **Spectacled Owl** *Pulsatrix perspicillata saturata* 18in (46cm) **p. 359**
SE rain forest. Unmistakable, large. Accelerating series of resonant hoots. See Fig. 26.

9 **Northern Potoo** *Nyctibius jamaicensis mexicanus* * 16in (41cm) **p. 382**
Tropical lowlands. Recalls large, long-tailed, and long-winged nightjar. Sits vertically, at roost blends with branches. Large head and big yellow eyes (nocturnal 'eyeshine' amber), overall streaked cryptic plumage. Juv paler overall. Deep, guttural *bawrr ah-ah* call on moonlit nights.

10 **Great Potoo** *N. grandis guatemalensis* 21in (53cm) **p. 382**
SE rain forest and edge. Large; huge pale head with deep brown eyes (nocturnal 'eyeshine' amber) suggests seal; overall vermiculated cryptic plumage. Juv paler. Deep, guttural, strangled cries, *awhrrr*, etc.

26

2

3a

3b

4

5

6

7

8

9

juv

10

SW 12-88

PLATE 27: NIGHTJARS

Silent nightjars can be very difficult to identify: note patterns on head, throat, scapulars, and tail; range and habitat useful. See text for all species.

1 **Common Nighthawk** *Chordeiles minor neotropicalis** 9.3in (24cm) **p. 373**
Summer resident. Long, fairly pointed wingtips often project beyond tail at rest. ♂ white bar on primaries (reduced, often not visible in ♀) strongly staggered, closer to primary coverts than 2. Sharp, nasal *beenk* in flight. See Fig. 28.

2 **Lesser Nighthawk** *C. acutipennis texensis** 8.5in (22cm) **p. 372**
Widespread. Long, fairly blunt wingtips close to tail tip at rest, primaries barred rufous basally. ♂ white bar on primaries (buffy, rarely visible in ♀) not strongly staggered, farther out on wing than 1. Toad-like churring from ground. See Fig. 28.

3 **Short-tailed Nighthawk** *Lurocalis semitorquatus* ssp? 8.2in (21cm) **p. 371**
SE rain forest. Dark, wingtips well beyond tail. No white in wings. Mostly silent. See Fig. 28.

4 **Common Poorwill** *Phalaenoptilus n. nuttallii** 7.3in (19cm) **p. 375**
N deserts. Stumpy; black mask, marbled grey above, white tail corners. Mellow *pur-whi-u'*.

5 **Chuck-will's-widow** *Caprimulgus carolinensis* 12in (30cm) **p. 377**
E and S, winter. Large, dark brown. ♀ outer rectrices tipped cinnamon. ♂ undertail pale cinnamon (outer webs barred), white visible on uppertail when spread. Rarely vocal.

6 **Pauraque** *Nyctidromus albicollis yucatanensis** 11.5in (29cm) **p. 374**
Tropical lowlands. Long tail extends well beyond wingtips. ♂ bold white primary bar and tail flashes. ♀ buff primary bar and white tail tips inconspicuous. 6(a) Rufous morph. Varied loud whistles, *puc, puc, p'weeEER*, or *whee'oo*, etc.

7 **Mexican Whip-poor-will** *Caprimulgus arizonae chiapensis** 9.3in (24cm) **p. 379**
Highland pine–oak. Dark brown overall, often shows pale braces and slight cinnamon hindcollar. S birds generally darker and redder. ♂ bold white tail corners. ♀ buff tail corners. Northern Whip-poor-will *C. vociferus* (migrant) ♂ has more white in tail on average. Burry *whirr'-p'wiirr*. See Fig. 29.

8 **Buff-collared Nightjar** *C. r. ridgwayi** 8.8in (22cm) **p. 378**
Pacific Slope and Interior, thorn forest. Greyish overall with bold cinnamon hindcollar. ♂ bold white tail corners (mostly inner webs). ♀ buff tail corners. Rapid clucking *koo-koo-kookookookoo-oo-chee'a*, ends emphatically.

9 **Tawny-collared Nightjar** *C. salvini* 9.8in (25cm) **p. 377**
Endemic to NE Mexico; thorn forest, woodland. Dark overall with tawny-cinnamon hindcollar, buff forecollar, pale braces. ♂ bold white tail corners (mostly on inner webs, hidden at rest). ♀ buff tail corners. Abrupt clipped *chi-wihw'*.

10 **Yucatan Nightjar** *C. badius* 9.8in (25cm) **p. 378**
Endemic to Yuc Pen. Thorn forest, woodland. Dark overall with tawny-cinnamon hindcollar, whitish forecollar, pale braces. ♂ very bold white tail corners (tail looks all white below). ♀ bold buffy-white tail corners. Loud, clear *pc ree-u-reeeu'*.

11 **Eared Poorwill** *Nyctiphrynus mcleodii* 8.3in (21cm) **p. 376**
Endemic to W Mexico; oak woods. Fairly pale and uniform, clean white forecollar, white wing covert spots, white tail tip. 11(a) Rufous morph, shown with 'ears' erect. Clipped, resonant *preeOO*.

12 **Yucatan Poorwill** *N. yucatanicus* 8.3in (21cm) **p. 376**
Endemic to Yuc Pen; humid forest. Dark overall with clean white forecollar, white tail tip. 12(a) Rufous morph. Resonant *whirrr*.

13 **Spot-tailed Nightjar** *Caprimulgus maculicaudus* 8.3in (21cm) **p. 380**
SE savannas. Buff braces and wing covert spots, cinnamon hindcollar. ♂ white tail corners (spots below hard to see). ♀ narrow pale tail corners. High, passerine-like *t-seet'*.

27

SW 10-91

Chaetura Small, relatively stubby swifts with squared tails, pale throats and rumps. Fly with rapid wingbeats and little gliding, tails spread infrequently. Often fairly low, rarely high. High, twittering calls.

1 **Vaux's Swift** *C. vauxi** W 10.8in (27cm) **p. 388**
   Common and widespread; smallest and stubbiest swift. From very similar 2 by relatively shorter, less swept-back wings. From larger 4 by more hurried flight with less gliding, tail rarely spread, pale throat and rump, voice. 1(a) *vauxi* (N.A. migrant): larger, pale below extends to belly; 1(b) *richmondi* (widespread resident): smaller, pale below more restricted, like 2.

2 **Chimney Swift** *C. pelagica* W 12.3in (31cm) **p. 387**
   Atlantic Slope transient. Larger and longer winged than 1 (especially resident ssp). From 4 by pale throat and rump, voice.

3 **White-throated Swift** *Aeronautes s. saxatalis** W 13.3in (34cm) **p. 388**
   NW deserts, mountains S to Honduras. Distinctive; note white median stripe below, long cleft tail. In silhouette, from *Cypseloides* by longer narrow tail, narrower-looking wings; note voice.

Cypseloides Medium-sized swifts with long, swept-back wings, squared to cleft (♂ Black) tails. Fly fast and often very high, fast wingbeats interspersed with glides, tails often spread. Buzzy to hard chipping calls.

4 **Chestnut-collared Swift** *C. rutilus griseifrons** W 12.5in (32cm) **p. 386**
   Widespread. From larger 5 and 6 by relatively *Chaetura*-like flight; chestnut collar distinctive. Note shorter tail than 5; longer tail than 6. Juv uniform sooty; compare with 1, 2.

5 **Black Swift** *C. niger costaricensis** W 15in (38cm) **p. 384**
   Widespread summer resident. Larger and longer-tailed than 4 and 6, with rangier wingbeats, frequent gliding; note cleft tail of ♂. Frosty whitish brows often visible. N.A. transients larger.

6 **White-chinned Swift** *C. cryptus** W 13.5in (34cm) **p. 385**
   Local, SW Mexico and C.A. Recorded in region Jun–Sep. From 5 by relatively shorter wings, shorter tail, flight heavier with little gliding. Compare with 4. 6(a) *cryptus* (C.A.): frosty whitish chin and sides of forehead hard to see; 6(b) *storeri* 'White-fronted Swift' (SW Mexico): whitish frosting extends across forehead.

Panyptila Spectacular, unmistakable swifts with long, narrow wings, long, deeply forked tails usually closed in a point. Calls reedy and screeching.

7 **Lesser Swallow-tailed Swift** *P. cayennensis* W 12in (30cm) **p. 389**
   SE humid lowlands. Small, fast flight; compare with much larger 8.

8 **Great Swallow-tailed Swift** *P. sanctihieronymi* W 18.3in (46cm) **p. 389**
   Endemic to Pacific Slope and Interior. Large, often glides and soars; compare with much smaller 7.

Streptoprocne Very large swifts with powerful flight, spend much time soaring, wings spread in paddle-shaped bulge. Loud, screeching calls.

9 **White-naped Swift** *S. semicollaris* W 22.5in (57cm) **p. 387**
   Endemic to W Mexico. Larger and thicker-set than 10, with squared to slightly cleft tail.

10 **White-collared Swift** *S. zonaris mexicanus** W 20in (51cm) **p. 386**
    S and E. From larger 9 by slightly forked tail, white collar often striking (can be indistinct or lacking in juv).

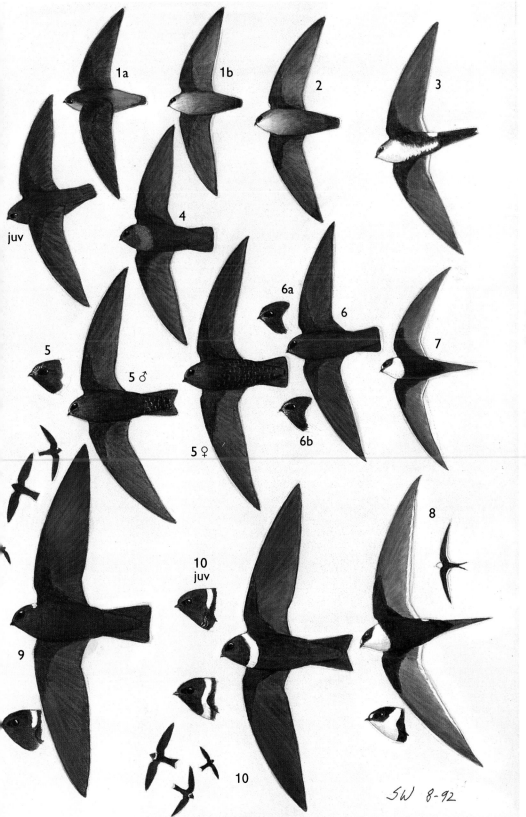

28

1a

1b

2

3

juv

4

5

5 ♂

6a

6

7

5 ♀

6b

8

10
juv

9

10

SW 8-92

PLATE 29: HERMITS, SABREWINGS, STARTHROATS, etc.

1 **Mexican Hermit** *Phaethornis (superciliosus?) m. mexicanus** 7.3in (19cm) **p. 393**
Endemic to W Mexico. Unmistakable in range; note very long, arched bill, long, white-tipped central rectrices. 2 allopatric.

2 **Long-tailed Hermit** *P. superciliosus longirostris** 6.5in (17cm) **p. 392**
SE rain forest. Unmistakable in range; note long, arched bill, long, white-tipped central rectrices. 1 allopatric.

3 **Little Hermit** *Pygmornis longuemareus adolphi** 3.7in (9cm) **p. 393**
SE rain forest. Distinctive. Note long, arched bill, striped face, rufous rump, graduated tail with white-tipped central rectrices.

4 **Wedge-tailed Sabrewing** *Campylopterus c. curvipennis** 5in (13cm) **p. 394**
Endemic to E Mexico and N C.A. From larger 5 (allopatric?) by shorter tail, dingier underparts. ♀ outer rectrices tipped whitish.

5 **Long-tailed Sabrewing** *C. (c.?) excellens* 5.3in (13cm) **p. 395**
Endemic to SE Mexico. Compare with smaller and shorter-tailed 4 (allopatric?).

6 **Plain-capped Starthroat** *Heliomaster constantii leocadiae** 4.8in (12cm) **p. 420**
Pacific Slope and interior valleys. From 7 by grey-green crown, whitish postocular stripe, greyish sides, black spot on white tail corners (variable); pale grey base to undertail often hard to see. Imm throat dark.

7 **Long-billed Starthroat** *H. longirostris pallidiceps** 4.7in (12cm) **p. 419**
S lowlands. From duller 6 by white postocular spot, greenish sides. Adult crown turquoise (green in some ♀♀), lower throat pinkish. Juv crown greenish, throat sooty.

8 **Green-breasted Mango** *Anthracothorax p. prevostii** 4.7in (12cm) **p. 397**
E and S lowlands. Distinctive. ♂ (some ♀♀) deep green overall with purple tail. ♀ blackish-green underpart stripe bordered white. Imm like ♀; note cinnamon sides of neck.

9 **White-necked Jacobin** *Florisuga m. mellivora* 4.5in (11cm) **p. 396**
SE rain forest and edge. ♂ (some ♀♀) unmistakable; note deep blue head, white tail. ♀ underparts boldly scalloped, white tail corners narrow, compare with 10. 9(a) Imm ♂ (some ♀♀?) like dull ♂ with cinnamon moustache, less white in tail.

10 **Scaly-breasted Hummingbird** *Phaeochroa cuvierii roberti* 4.8in (12cm) **p. 394**
SE rain forest. Dull green overall with white postocular spot, dusky cinnamon belly, bold white tail corners. Scaly chest often hard to see.

11 **Rufous Sabrewing** *Campylopterus rufus* 5.3in (13cm) **p. 395**
Endemic; Pacific Slope, SE Oax to El Salvador. Unmistakable; note large size, broad tail, etc.

12 **Violet Sabrewing** *C. h. hemileucurus* 5.7in (14cm) **p. 396**
S humid forest. ♂ unmistakable, often looks black with bold white tail flashes. ♀ dusky-grey below with purple throat, white postocular spot, white tail corners. Imm ♂ appears intermediate between ♂ and ♀.

13 **Purple-crowned Fairy** *Heliothryx barroti* 4.8in (12cm) **p. 419**
SE rain forest. Unmistakable, white sides of tail constantly flash in active, darting flight. ♂ purple crown and cheek spot. ♀ longer-tailed, lacks purple. Imm crown (and initially upperparts) edged cinnamon, chest flecked dusky.

Band-tailed Barbthroat *Threnetes ruckeri* C.A. See Plate 70.

29

1

2

3

imm ♀
4

4 ♂

5 ♂

imm
6

imm
7

8

9a

9

9 ♀

10

♀

imm

12

imm ♀

12

13

♂

♀

13

11

imm ♂
♂

♀

13
imm ♀

SW 7-90

PLATE 30: AMAZILIA, CYNANTHUS, EUPHERUSA, etc.

1 **Golden-crowned Emerald** *Chlorostilbon auriceps* ♂ 3.7in (9cm), ♀ 3.2in (8cm) **p. 403**
Endemic to W Mexico. ♂ unmistakable. ♀ small, forked tail. See Plate 70 for
allopatric Canivet's Emerald (*C. canivetii*) of E Mexico.

2 **Doubleday's Hummingbird** *Cynanthus (latirostris?) doubledayi* 3.4in (9cm) **p. 405**
Endemic to Pacific lowlands of Gro and Oax. ♂ from larger 3 by glittering blue
crown, bluish underparts, blue-black undertail coverts, more deeply cleft tail. ♀
like 3 (see text).

3 **Broad-billed Hummingbird** *C. l. latirostris** 3.7in (9cm) **p. 404**
N and cen Mexico. ♂ distinctive, but compare with 2 (allopatric?). ♀ from mostly
allopatric 4 by brighter emerald-green upperparts and central tail, more
contrasting tail pattern. See Plate 69.

4 **Dusky Hummingbird** *C. sordidus* 3.7in (9cm) **p. 404**
Endemic to SW Mexico. ♀ from 3 by duller back and central tail (see text). ♂
duskier, lacks white tail corners.

5 **Xantus' Hummingbird** *Basilinna xantusii* 3.5in (9cm) **p. 408**
Endemic to BCS. Unmistakable in range.

6 **Mexican Woodnymph** *Thalurania ridgwayi* 3.7in (9cm) **p. 406**
Endemic to W Mexico. ♂ often looks all dark; note green gorget, violet-blue
crown, blue-black tail. ♀: no similar species in range and habitat; note voice.

7 **Blue-throated Sapphire** *Hylocharis eliciae* 3.4in (9cm) **p. 407**
S humid lowland forest, local. Distinctive; note red bill (dark above in ♀), bluish-
violet gorget, gold uppertail coverts, green-gold tail.

8 **Stripe-tailed Hummingbird** *Eupherusa e. eximia** 3.8in (10cm) **p. 414**
E and SE humid forest. ♂ distinctive; note rufous wing panel, spread tail flashes
white. ♀ plain face, pale grey underparts, white in tail. 14 and 15 allopatric.

9 **Blue-tailed Hummingbird** *Amazilia cyanura guatemalae** 3.8in (10cm) **p. 411**
Endemic, Pacific Slope S of Isthmus. Deep blue tail distinctive, rufous in wings
may be hard to see. ♀ belly buffy grey. See Plate 70.

10 **Berylline Hummingbird** *A. beryllina viola** 3.8in (10cm) **p. 410**
Endemic, mostly Pacific Slope and Interior. From 12 and 13 by rufous wing panel
(usually obvious), dark bill, voice. See Plate 70.

11 **Cinnamon Hummingbird** *A. r. rutila** 4.3in (11cm) **p. 412**
Pacific Slope, Yuc Pen. Unmistakable; note cinnamon underparts. See Plate 69.

12 **Rufous-tailed Hummingbird** *A. t. tzacatl* 4.2in (11cm) **p. 411**
SE rain forest. From 13 by squared to cleft tail, mostly rufous central rectrices,
dark cinnamon undertail coverts contrast with dusky-grey to greyish-cinnamon
belly. ♀ bill blackish above. Lacks rufous wing panel of 10.

13 **Buff-bellied Hummingbird** *A. yucatanensis** 4in (10cm) **p. 412**
E and SE drier woodland. From 12 by slightly forked tail with mostly green
central rectrices, undertail coverts same color as belly: greyish buff in *chalconota*
(13 (a); NE Mexico) to cinnamon in *yucatanensis* (13 (b); Yuc Pen). ♀ bill
blackish above. Lacks rufous wing panel of 10.

14 **White-tailed Hummingbird** *Eupherusa poliocerca* 4.1in (10cm) **p. 415**
Endemic to Sierra Madre del Sur of Gro and W Oax. From allopatric 15 only by
green crown of ♂. Note mostly white tail (looks all white below), rufous wing panel.

15 **Blue-capped Hummingbird** *E. (p. ?) cyanophrys* 4.1in (10cm) **p. 415**
Endemic to Sierra Madre del Sur, E Oax. Note violet-blue crown of ♂; compare
with allopatric 14.

Salvin's Emerald *Chlorostilbon salvini* C.A. See Plate 70.
Crowned Woodnymph *Thalurania colombica* C.A. See Plate 70.
Honduran Emerald *Amazilia luciae* C.A. See Plate 70.
Cozumel Emerald *Chlorostilbon forficatus* Island endemic. See Plate 69.

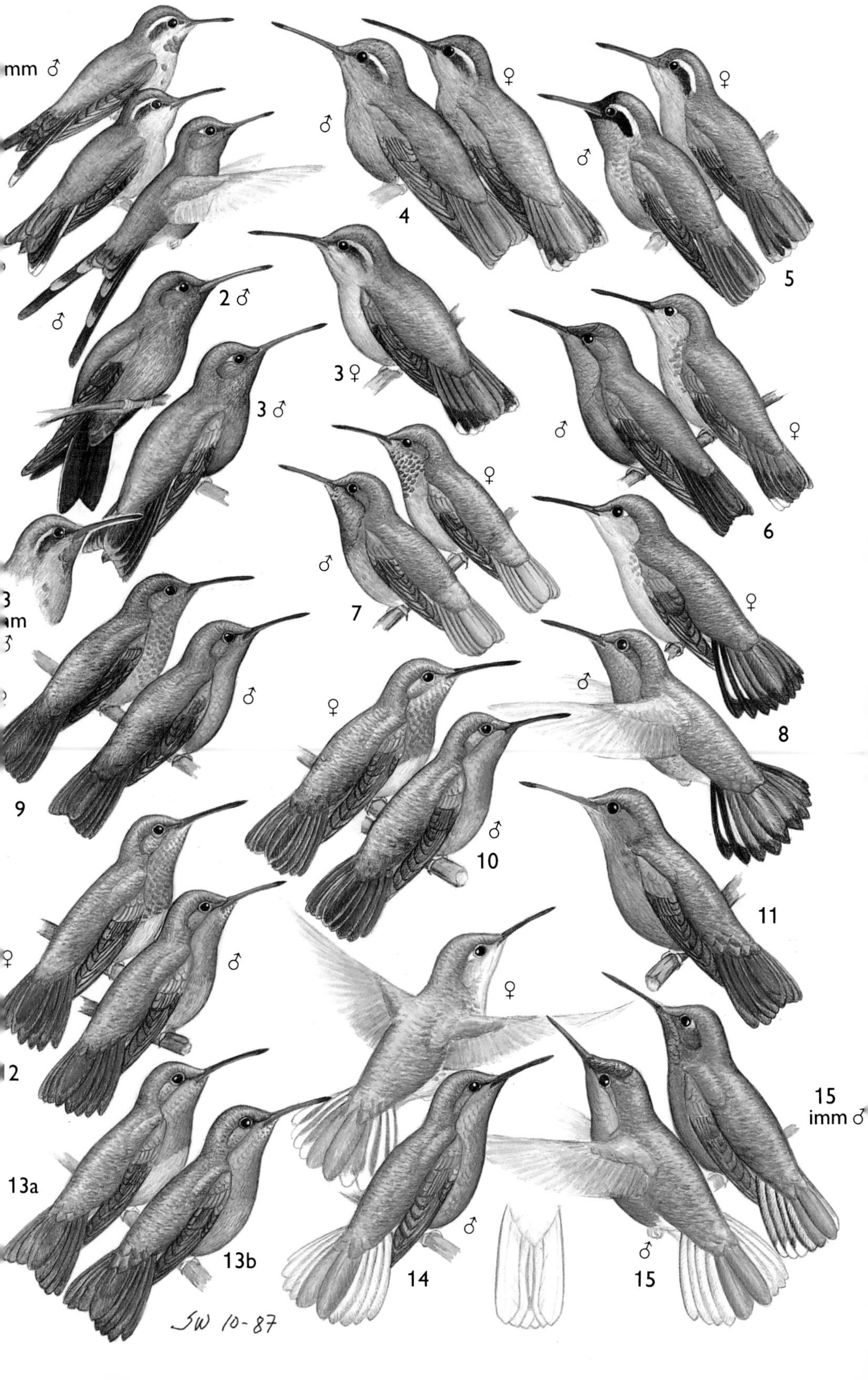

imm ♂

♂

2 ♂

3 ♂

3 ♀

3
imm
♂

9

♂

♀

4

♀

♂

5

7

♂

♀

♂

♀

6

♂

8

♀

10

♂

11

♀

12

13a

♀

♂

13b

14

♂

15
imm ♂

♂

15

SW 10-87

1 **Violet-crowned Hummingbird** *Amazilia violiceps** 4.3in (11cm) **p. 413**
From 3 by violet crown, dull upperparts, bolder postocular spot (2 allopatric).
Imm crown duller, bill black above. 1(a) *ellioti* (W Mexico): tail dull grey-green
to brownish; 1(b) *violiceps* (SW Mexico): tail bronzy. Throat often stained yellow
by pollen. Compare with 3.

2 **Cinnamon-sided Hummingbird** *A. (viridifrons?) wagneri* 4.3in (11cm) **p. 414**
Endemic to S Oaxaca. From 3 by cinnamon sides of neck, dull tawny wing patch
of ♂. ♂ tail burnished rufous, edged bronzy green, ♀ tail broadly edged bronzy-
green (central rectrices mostly greenish). Bill dark above in imm, throat often
stained yellow, compare with 3.

3 **Green-fronted Hummingbird** *A. v. viridifrons** 4.3in (11cm) **p. 413**
Endemic to S Mexico. From 1 by dark green crown, green upperparts. ♂ tail
coppery, ♀ tail bronzy; compare with 2. Lacks cinnamon sides of neck (*rowleyi* of
interior Oax has extensively vinaceous-cinnamon sides, see text). Bill dark above
in imm, throat often stained yellow by pollen.

4 **Azure-crowned Hummingbird** *A. c. cyanocephala** 4.3in (11cm) **p. 409**
Endemic, E and S. Larger than 5 with blue crown and face, dusky sides, duller
rump, uniform tail (tail bronzy in C.A.). Bill appears dark. See Plate 70.

5 **White-bellied Emerald** *A. c. candida** 3.6in (9cm) **p. 408**
E and S humid forest. Smaller than 4, green above, white below; note tail pattern.

6 **White-eared Hummingbird** *Basilinna l. leucotis* 3.7in (9cm) **p. 407**
Endemic, mountains. Unmistakable, ♂ head often looks black with bold white
stripe (6(a)) unless colors catch light. ♀ may suggest ♀ Broad-billed Hummingbird
(Plate 30) but white postocular stripe and black mask bolder, crown duller,
underparts spotted green.

7 **Green Violet-ear** *Colibri t. thalassinus* 4.3in (11cm) **p. 398**
Mountains. Distinctive, dark green overall; note dark band on broad bluish tail.
Imm duller.

8 **Amethyst-throated Hummingbird** *Lampornis amethystinus** 4.7in (12cm) **p. 417**
Endemic, E and S humid montane forest. From 9 by lack of bold white tail
corners, voice. 8(a) *amethystinus* (widespread): gorget pink; 8(b) *margaritae* (SW
Mexico): gorget bluish purple. ♀ and imm throat buff to grey.

9 **Blue-throated Hummingbird** *L. c. clemenciae** 5in (13cm) **p. 417**
Mountains N of Isthmus. From 8 and 10 by bold white tail corners; from 10 also
by shorter bill, whitish face stripes, ♂ throat blue.

10 **Magnificent Hummingbird** *Eugenes f. fulgens** 5in (13cm) **p. 418**
Highlands. Note long straight bill (may suggest starthroats, Plate 29), white
postocular spot. ♂ usually looks all dark, ♀ has mostly greenish tail with small
white corners. Compare with 9.

11 **Garnet-throated Hummingbird** *Lamprolaima r. rhami** 5in (13cm) **p. 418**
Endemic, E and S cloud forest. Unmistakable; note bright rufous wings, short
straight bill, white postocular spot, deep purplish tail.

12 **Green-throated Mountain-gem** *Lampornis viridipallens ovandensis** 4.5in (11cm)
p. 416
Endemic, cloud forest S of Isthmus. No similar species in range. Note white throat
(green spots of ♂ often hard to see), pale grey tail sides often striking. See Plate 70.

Brown Violet-ear *Colibri delphinae* C.A. See Plate 70.
Green-breasted Mountain-gem *Lampornis (viridipallens?) sybillae* C.A. See Plate 70.

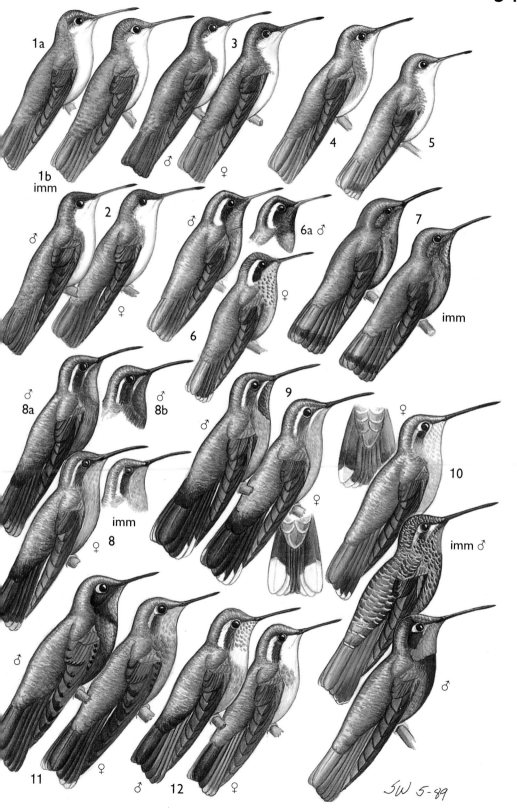

31

PLATE 32: SMALL GORGETED HUMMINGBIRDS

1 **Rufous Hummingbird** *Selasphorus rufus* 3.5in (9cm) **p. 428**
Migrant, W Mexico. ♂ unmistakable (see text). Imm ♀ (and Allen's Hummingbird, see text) from 2 by rufous-edged uppertail coverts, more rufous in tail. From smaller 3 by bigger bill, wingtips shorter than tail, more rufous in tail. ♀ and imm ♂ have red spots on throat, ♂ tail mostly rufous.

2 **Broad-tailed Hummingbird** *S. p. platycercus** 3.7in (9cm) **p. 427**
Highlands. ♂ from 5 and 6 by rose-pink gorget, paler auriculars. ♀ lacks rufous uppertail covert fringes of 1, larger than 3 with bigger bill, wingtips shorter than tail.

3 **Calliope Hummingbird** *Archilochus calliope* 3.1in (8cm) **p. 426**
Migrant, W Mexico. ♂ unmistakable (see text). ♀ from 1, 2, and 7 (8 allopatric) by shorter tail (wingtips longer than tail), also from 7 by longer bill, fainter throat marks, paler sides, less rufous in tail.

4 **Costa's Hummingbird** *A. costae* 3.2in (8cm) **p. 426**
NW deserts. ♂ unmistakable (see Anna's Hummingbird in text). ♀ from 5 and 6 by shorter tail (wingtips often slightly longer than tail), generally dingier; note pale postocular stripe, lack of blackish lores; may show whitish sides of neck, compare with 5, 6.

5 **Ruby-throated Hummingbird** *A. colubris* 3.5in (9cm) **p. 424**
Widespread migrant. ♂ (not shown) from 2 and 6 by ruby-red gorget. ♀ hard to tell from 6; note shorter bill, pointed (versus truncate) tips of primaries, adult has green crown (often dull in imm, like 6). Imms may be inseparable in field. Compare with 4.

6 **Black-chinned Hummingbird** *A. alexandri* 3.6in (9cm) **p. 425**
N and cen Mexico. ♂ gorget distinctive, compare with ♂ 5. ♀ much like 5. Adult ♀ has tapered tips of rectrices (rounded in imm, like 5), buff often restricted to spot on flanks (more extensive, like 5, in imm). Compare with 4.

7 **Bumblebee Hummingbird** *Selasphorus h. heloisa** 2.8in (7cm) **p. 429**
Endemic to mountains N of Isthmus. ♂ distinctive (8 allopatric). ♀ from 3 by longer tail, heavily spotted throat, darker sides, etc. From larger 1 and 2 by shorter bill; note darker sides than 2, green uppertail coverts (compare with 1).

8 **Wine-throated Hummingbird** *S. e. ellioti** 2.7in (7cm) **p. 429**
Endemic to mountains S of Isthmus. ♂ distinctive (7 allopatric). ♀ from 2 by small size, short bill, more rufous in tail, darker sides.

9 **Sparkling-tailed Woodstar** *Philodice dupontii* ♂ 3.7in (9cm), ♀ 2.7in (7cm) **p. 420**
Endemic, cen Mexico to C.A. Unmistakable; note bold white rump squares.

10 **Slender Sheartail** *Calothorax enicura* ♂ 4.6in (12cm), ♀ 3.4in (9cm) **p. 422**
Endemic, highlands S of Isthmus. ♂ unmistakable. ♀ from (allopatric?) 11 and 12 by darker underparts, tail more deeply forked than 11 (similar to imm ♂ 12), longer bill more arched than 12.

11 **Lucifer Hummingbird** *C. lucifer* 3.7in (9cm) **p. 422**
N and cen Mexico. Rarely separable from 12 but longer bill more arched, ♂ outer rectrices attenuated. Compare with 10.

12 **Beautiful Hummingbird** *C. pulcher* 3.5in (9cm) **p. 423**
Endemic to SW Mexico. Rarely separable from 11. Compare with 10.

13 **Mexican Sheartail** *C. eliza* ♂ 3.8in (10cm), ♀ 3.4in (9cm) **p. 421**
Endemic to N Yuc and cen Ver. Unmistakable in range; note long, arched bill, ♀ underparts whitish.

14 **Short-crested Coquette** *Lophornis brachylopha* 2.8in (7cm) **p. 399**
Endemic to Sierra Madre del Sur of Gro. Unmistakable; note white to pale buff rump band.

15 **Black-crested Coquette** *L. helenae* 2.7in (7cm) **p. 400**
S humid forest. Unmistakable; note whitish rump band. Younger imm ♂ much like ♀, including tail.

16 **Emerald-chinned Hummingbird** *Abeillia a. abeillei** 2.8in (7cm) **p. 399**
Endemic, S cloud forest. Distinctive; note small size, very short bill, bold white postocular spot, broad tail.

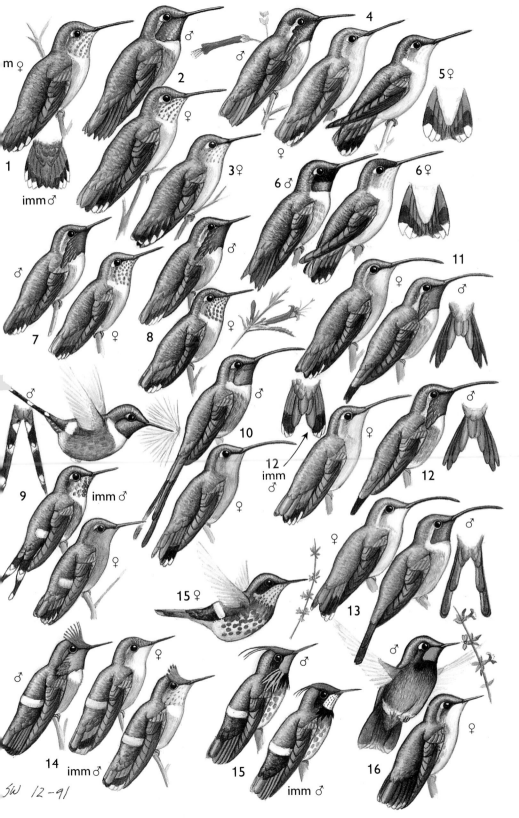

m ♀

1

imm ♂

2

♂

♀

3 ♀

♀

4

5 ♀

6 ♂

6 ♀

7

♀

♂

8

♀

11

♀

♂

♂

9

imm ♂

♀

10

♂

12
imm ♂

♀

12

♀

13

♂

15 ♀

♂

14

imm ♂

15

imm ♂

16

♀

SW 12-91

PLATE 33: QUETZALS AND TROGONS

Songs of all species are distinctive and birds are heard more often than they are seen. See text for song descriptions.

1 **Eared Quetzal** *Euptilotus neoxenus* 13.5in (34cm) **p. 435**
Endemic to NW Mexico. Unmistakable; note small head, grey bill, white undertail; voice unlike *Trogon* trogons. Sexes similar but adult ♀ head and chest slaty, red on belly less extensive.

2 **Slaty-tailed Trogon** *Trogon m. massena* 13.5in (34cm) **p. 435**
SE rain forest. No white chest band, undertail slaty grey (outer webs barred whitish in imm). ♂ stout orange bill, ♀/imm bill dark above.

3 **Resplendent Quetzal** *Pharomachrus m. mocinno* 15.5in (39cm) + plumes. **p. 436**
Cloud forest S of Isthmus. ♂ unmistakable, ♀ distinctive.

4 **Elegant Trogon** *Trogon elegans ambiguus** 11.5in (29cm) **p. 433**
Woodland N of Isthmus; also C.A. (Plate 70). ♂ from 5 and 6 by broadly black-tipped uppertail (coppery N of Isthmus), paler wing panel; note also whitish edges of primaries; compare with 5. ♀ from 5 and 6 by white postocular stripe, broader black tail tip above. Note undertail patterns (note similarity of imm ♂ 6), voice.

5 **Mountain Trogon** *T. m. mexicanus** 12in (30cm) **p. 433**
Endemic to mountain forests. ♂ above from 4 and 6 by vermiculated wing panel, dark primaries; note narrower black tail tip than 4. ♀ above similar to 6 but longer-tailed; note brownish chest band; compare with ♀ 4. Note undertail patterns, voice.

6 **Collared Trogon** *T. collaris puella** 11in (28cm) **p. 434**
Humid E and S forest. ♂ above from 4 and 5 by coarsely peppered wing panel, from 4 by narrow black tail tip, from 5 by whitish edges of primaries. ♀ very similar above to 5 but smaller, more compact, note wholly red belly; compare with ♀ 4. Note undertail patterns (imm ♂ may be mistaken for 4, see Plate 70), voice.

7 **Citreoline Trogon** *T. citreolus sumichrasti** 10.8in (27cm) **p. 432**
Endemic to W Mexico, thorn forest. Only yellow-bellied trogon in most of range. From overall darker 8 (sympatric in W Chis?) by yellow eyes, solidly white undertail of ♀.

8 **Black-headed Trogon** *T. m. melanocephalus* 10.8in (27cm) **p. 431**
E and S. From smaller and more compact 9 by blue orbital ring, dark wings, mostly white undertail, ♂ has bluish-violet rump; note voice. Compare with 7.

9 **Violaceous Trogon** *T. violaceus braccatus* 9.7in (25cm) **p. 432**
SE humid forest. ♂ above from 8 by yellow orbital ring, pale wing panel, green rump; ♀ from 8 by white eye-crescents, white-barred wing panel. Note undertail patterns, voice.

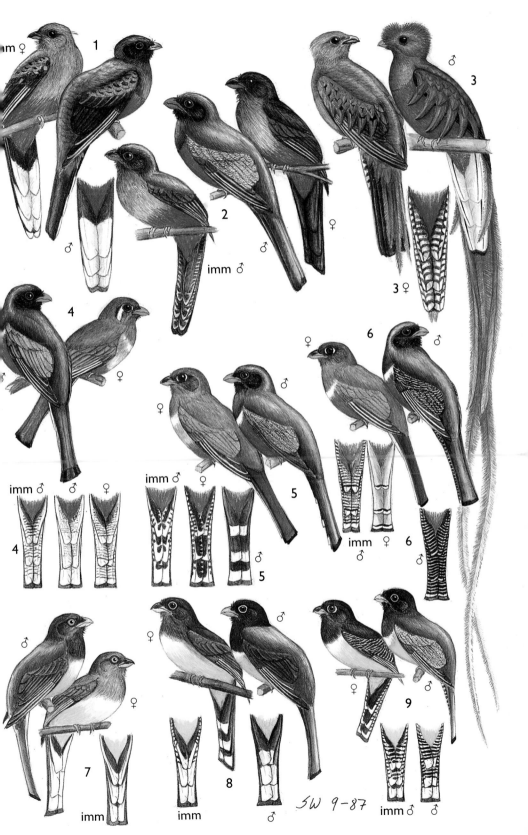

imm ♀ 1 ♂

imm ♂

2 ♀ ♂

♂ 3 3 ♀

4 ♀

imm ♂ ♂ ♀
4

imm ♂ ♀ ♂
5

♀ 6 ♂ 5

imm ♀ 6
♂ ♂

♂ 7 ♀

imm imm ♂

♀ ♂ ♂ ♀

7 8 9

imm ♂ imm ♂ ♂

SW 9-87

PLATE 34: TOUCANS, MOTMOTS, KINGFISHERS, etc.

1 **Keel-billed Toucan** *Ramphastos s. sulfuratus* 21.5in (55cm) **p. 448**
SE rain forest and edge. Unmistakable.

2 **Collared Araçari** *Pteroglossus t. torquatus* 16in (41cm) **p. 447**
S rain forest and edge. Unmistakable.

3 **Emerald Toucanet** *Aulacorhynchus prasinus** 13.5in (34cm) **p. 447**
E and S humid forest, mainly cloud forest. Unmistakable. 3(a) *prasinus* (E and
SE); 3(b) *wagleri* (Sierra Madre del Sur of Gro and Oax).

4 **Turquoise-browed Motmot** *Eumomota s. superciliosa** 14in (36cm) **p. 439**
S lowlands. Distinctive; from larger 5 by long bare tail shafts with larger racket
tips, bright turquoise brow, rufous back, etc. Nasal, ringing *wohh* call.

5 **Blue-crowned Motmot** *Momotus momota** 16in (41cm) **p. 438**
E and S lowlands. From 6 (rare in Mexico) by slender bill, head pattern, voice.
Low, owl-like double-hoot, *oot oot.* 5(a) *coeruliceps* (NE Mexico); 5(b) *lessonii*
(SE Mexico and C.A.).

6 **Keel-billed Motmot** *Electron carinatum* 12.5in (32cm) **p. 439**
SE rain forest, rare in Mexico. From 5 by broad bill, brighter turquoise brow,
rufous forehead, larger black chest spots. Nasal, ringing *ohnng*, suggests 4.

7 **Blue-throated Motmot** *Aspatha gularis* 10.5in (27cm) **p. 437**
Endemic to mountains S of Isthmus. Unmistakable. Far-carrying low *hoot.*

8 **Russet-crowned Motmot** *Momotus m. mexicanus** 13in (33cm) **p. 438**
Endemic, Pacific Slope and Interior valleys. Distinctive; note rufous crown,
glaucous-green underparts. Burry *krrup* call.

9 **Tody Motmot** *Hylomanes m. momotula** 6.7in (17cm) **p. 437**
SE rain forest. Unmistakable; small, wags tail. Hollow, ringing *woh woh woh...*
call.

10 **White-necked Puffbird** *Notharchus macrorhynchos hyperrhynchus* 9.7in (25cm)
p. 444
SE humid lowland forest and edge. Unmistakable; mostly silent.

11 **White-whiskered Puffbird** *Malacoptila panamensis inornata* 7.7in (20cm) **p. 444**
SE rain forest. Unmistakable; mostly silent.

12 **Rufous-tailed Jacamar** *Galbula ruficauda melanogenia* 9in (23cm) **p. 446**
SE rain forest and edge. Unmistakable; often noisy.

13 **Ringed Kingfisher** *Ceryle t. torquata* 15.5in (39cm) **p. 441**
Tropical lowlands. Huge; note massive bill, rufous underparts, compare with 14.

14 **Belted Kingfisher** *Ceryle a. alcyon** 12.5in (32cm) **p. 441**
Widespread migrant. Relatively small bill, blue-grey upperparts, compare with
13, 15.

15 **Amazon Kingfisher** *Chloroceryle amazona mexicana* 11.2in (28cm) **p. 442**
Tropical lowlands, rare W Mexico. From small 17 by more massive bill, tufted
crest, lacks white barring in wings, ♀ 17 has 2 chest bands.

16 **Pygmy Kingfisher** *Chloroceryle aenea stictoptera* 5.3in (13cm) **p. 443**
S lowlands. From larger 17 by duller and narrower collar, rufous underparts, lack
of bold white spotting on wings.

17 **Green Kingfisher** *Chloroceryle americana septentrionalis** 7.7in (20cm) **p. 442**
Widespread. Small, compare with 15. Note bold broad white collar, compare with
16.

34

SW 5-87

1 **Gila Woodpecker** *Centurus uropygialis sulfuriventer** 9in (23cm) **p. 453**
NW Mexico. Sympatric with 2(a), 3. Note plain head, ♂ has red cap.

2 **Golden-fronted Woodpecker** *C. aurifrons** 9.8in (25cm) **p. 452**
Widespread. Complex variation. 2(a) *aurifrons* (N Mexico): sympatric with 1, 3.
Note wide barring on upperparts, unbarred tail, ♂ red crown patch separate from
golden nape, ♀ head like 2(b). 2(b) *polygrammus* (Pacific Slope, S Mexico):
narrow barring on upperparts, ♂ red crown patch merges with golden nape;
santacruzi (much of Guatemala and Honduras) similar to *polygrammus* but yellow
areas of head and belly often mixed with red. 2(c) *dubius* (Yuc Pen): very narrow
barring on upperparts, belly red. From smaller 6 by large bill, red nasal tufts,
darker upperparts, voice. 2(d) *leei* (Isla Cozumel): like *dubius* but head and
underparts duskier. *C. a. grateloupensis* (E Mexico) shows variably intermediate
characters of *dubius, aurifrons*, and *polygrammus*.

3 **Golden-cheeked Woodpecker** *C. c. chrysogenys** 8in (20cm) **p. 451**
Endemic to W Mexico. Black eye-patch, golden-yellow cheeks and nape, ♂ crown
red. Sympatric with 1, 2(a), 4.

4 **Grey-breasted Woodpecker** *C. hypopolius* 7.7in (20cm) **p. 451**
Endemic to interior SW Mexico. Dusky grey head and underparts with whitish
forehead, white eye-crescents, dull red cheek patch; ♂ has red crown spot.

5 **Black-cheeked Woodpecker** *C. pucherani perileucus* 7.3in (19cm) **p. 450**
SE rain forest. Black mask, white flash behind eye, barred belly, ♂ has red crown.

6 **Yucatan Woodpecker** *C. pygmaeus** 6.7in (17cm) **p. 452**
Endemic to Yuc Pen and Isla Guanaja (Plate 71). Small. From 2(c) and 2(d) by
small bill, yellow nasal tufts (encircling bill), silvery back (wider white barring),
voice. 6(a) *rubricomus* (Yuc Pen); 6(b) *pygmaeus* (Isla Cozumel): head and
underparts duskier.

7 **Grey-crowned Woodpecker** *Piculus a. auricularis** 8in (20cm) **p. 459**
Endemic to W Mexico. Bronzy-green overall, barred below, grey crown. ♂ red
moustache. 8 and 9 allopatric.

8 **Golden-olive Woodpecker** *Piculus rubiginosus yucatanensis** 8.5in (22cm) **p. 459**
E and S. Bronzy-green overall, barred below. From 9 by shorter tail, more
narrowly barred underparts, ♂ red nape patch extends forward to bill, ♀ nape
patch evenly broad. Note voice.

9 **Bronze-winged Woodpecker** *P. (r.?) aeruginosus* 9.3in (24cm) **p. 458**
Endemic to NE Mexico. From 8 by longer tail, coarsely scalloped underparts, ♂
red nape patch ends behind eye, ♀ red nape patch U-shaped. Note voice.

10 **Smoky-brown Woodpecker** *Veniliornis fumigatus sanguinolentus** 6.2in (16cm) **p.458**
E and S, local W Mexico. Brown overall with pale face, ♂ crown red.

11 **Chestnut-colored Woodpecker** *Celeus castaneus* 9in (23cm) **p. 460**
SE humid forest. Unmistakable; chestnut with crested blond head, ♂ moustache
red.

Olivaceous Piculet *Picumnus olivaceus* C.A. See Plate 71.
Hoffmann's Woodpecker *Centurus hoffmannii* C.A. See Plate 71.

35

1 ♀ ♂

2b ♀
2a ♂
2d ♀
2b ♂
2c ♂

3 ♀ ♂

4 ♀ ♂

6 ♀ ♂

5 ♀ ♂

7 ♂ ♀

9 ♂ ♀

8 ♂ ♀

10 ♂ ♀

11 ♀ ♂

SW 7-87

PLATE 36: WOODPECKERS II

1 **Arizona Woodpecker** *Picoides arizonae fraterculus** 7.5in (19cm) **p. 457**
   NW Mexico. Oak and pine–oak. Brown back, white face with dark cheek spot, spotted underparts.

2 **Strickland's Woodpecker** *P. stricklandi aztecus** 7in (18cm) **p. 458**
   Endemic to cen Mexico. Pine woods. Dark brown back with barred white patch, white face with dark cheek spot, streaked and spotted underparts, bill small; compare with sympatric populations of 4.

3 **Ladder-backed Woodpecker** *P. scalaris** 7in (18cm) **p. 456**
   Widespread (local in C.A.). Black-and-white striped face, ladder-barred back. Compare with Nuttall's Woodpecker of NW BCN (see text). 3(a) *azelus* (S cen Mexico): small bill typical of S ssp; 3(b) *scalaris* (NE Mexico).

4 **Hairy Woodpecker** *P. villosus** 8in (20cm) **p. 457**
   Highlands. Black-and-white-striped face, blackish upperparts with white back patch. Underparts whitish (NW Mexico) to brownish in smaller S ssp. 4(a) *sanctorum* (Chis and Guatemala); 4(b) *icastus* (NW Mexico).

5 **Acorn Woodpecker** *Melanerpes formicivorus** 8.8in (22cm) **p. 450**
   Widespread, social, and noisy; oak and pine–oak. Unmistakable; note clown-like face; forehead black in ♀. 5(a) *lineatus* (N C.A.); 5(b) *formicivorus* (N of Isthmus).

6 **Northern Flicker** *Colaptes auratus** 11in (28cm) **p. 460**
   Widespread, large and distinctive, but variable (see text). Red to yellow flight feather shafts and white rump striking in flight. 6(a) *mexicanus* (Red-shafted Flicker; highlands N of Isthmus); 6(b) *mexicanoides* (Guatemalan Flicker; S of Isthmus): bold narrow barring above, crown and nape cinnamon, red shafts. Gilded Flicker (*C. a. chrysoides* group) of NW deserts has yellow shafts.

7 **Lineated Woodpecker** *Dryocopus lineatus** 13in (33cm) **p. 461**
   Both slopes. From larger 8 by more tufted crest, black face, open white V on back, voice. E and S populations have white cheek stripe (for example, 7(a) *similis*). White stripe reduced or lacking in *scapularis* of W Mexico (7(b)).

8 **Pale-billed Woodpecker** *Campephilus g. guatemalensis** 14.5in (37cm) **p. 461**
   Both slopes. From smaller 7 by squarer crest, all-red head of ♂ (forehead and throat black on ♀), closed white V on back, voice. Resonant double-rap drum distinctive.

36

1

2

3a ♀

3b ♂

4a ♀

5a ♀

5b ♂

6a ♂

4b ♂

6b ♀
♀

8 ♀

7a ♀

7b ♀

♂

♂

SW 8-89

PLATE 37: WOODCREEPERS

Voices are often useful for detecting and identifying these species. See text for voice descriptions.

1 **White-striped Woodcreeper** *Lepidocolaptes l. leucogaster** 8.8in (22cm) **p. 472**
Endemic to mountains of W Mexico. Distinctive; note white face and throat, slender decurved bill, white-striped underparts.

2 **Spot-crowned Woodcreeper** *L. a. affinis** 8.2in (21cm) **p. 472**
Highlands. From 3 and 4 (both rarely sympatric) by spotted crown, voice; often has dark malar stripe; compare with 4. Note stout, straighter bill and more boldly streaked back of 3. 5 has bicolored bill, spotted back and chest.

3 **Ivory-billed Woodcreeper** *Xiphorhynchus flavigaster eburneirostris** 9.5in (24cm) **p. 470**
Tropical lowlands, to foothills in W Mexico. Often confused with smaller 4; note stouter, straight bill, more boldly streaked back, voice; dusky malar stripe can be lacking. Birds in W Mexico and N Yuc Pen paler and greyer overall. Compare with 2. Buff-throated Woodcreeper (C.A., Plate 71) has bicolored bill, note voice (see text).

4 **Streak-headed Woodcreeper** *Lepidocolaptes souleyetii insignis** 7.7in (20cm) **p. 472**
S lowlands. Note finely streaked crown, slender decurved bill, voice, lack of dark malar stripe; compare with 2, 3.

5 **Spotted Woodcreeper** *Xiphorhynchus e. erythropygius** 9.3in (24cm) **p. 471**
Cloud forest. Note bicolored bill, distinctly spotted plumage (back spots indistinct to lacking in *parvus*, E of Sula Valley), voice. Compare with 2.

6 **Strong-billed Woodcreeper** *Xiphocolaptes promeropirhynchus omiltemensis** 12.2in (31cm) **p. 469**
S pine—oak and evergreen forest, mostly highlands. Distinctive; very large.

7 **Wedge-billed Woodcreeper** *Glyphorynchus spirurus pectoralis* 5.8in (15cm) **p. 468**
SE rain forest. Small, distinctive. Note wedge-shaped bill, buff-spotted chest.

8 **Olivaceous Woodcreeper** *Sittasomus griseicapillus sylvioides** 6.3in (16cm) **p. 468**
Both slopes (local on Pacific). Small, distinctive. Note slender bill, unstreaked plumage, plain grey head.

9 **Black-banded Woodcreeper** *Dendrocolaptes picumnus puncticollis* 10.5in (27cm) **p. 470**
Highlands S of Isthmus, rarely seen. Note dark bill, streaked throat, barred belly. 2, 5, and 6 are the only sympatric species.

10 **Barred Woodcreeper** *D. certhia** 10.5in (27cm) **p. 469**
S humid tropical forest, often with ant swarms. Overall barred plumage distinctive (often looks fairly uniform). 10(a) *sanctithomae* (humid SE); 10(b) *sheffleri* (Sierra Madre del Sur of Gro and Oax).

11 **Tawny-winged Woodcreeper** *Dendrocincla a. anabatina** 7.3in (19cm) **p. 467**
SE humid lowland forest, often with 10 at ant swarms. Distinctive; note pale supercilium, plain upperparts with two-tone wings.

12 **Ruddy Woodcreeper** *D. h. homochroa* 7.7in (20cm) **p. 468**
SE humid lowland forest, often with 11 at ant swarms. Distinctive; bright ruddy overall with bushy face.

Buff-throated Woodcreeper *Xiphorhynchus guttatus* C.A. See Plate 71.

37

SW 6-87

PLATE 38: OVENBIRDS, ANTBIRDS, GNATWREN

1 **Rufous-breasted Spinetail** *Synallaxis erythrothorax furtiva** 6.2in (16cm) **p. 463**
Endemic to SE lowland second growth; skulking. Distinctive; note long tail,
rufous chest and wings.

2 **Spectacled Foliage-gleaner** *Anabacerthia v. variegaticeps* 6.5in (17cm) **p. 463**
S cloud forest; arboreal. Prominent buff spectacles, head greyer than back.

3 **Buff-throated Foliage-gleaner** *Automolus ochrolaemus cervinigularis** 7.7in
(20cm) **p. 464**
SE rain forest. Broad buff supercilium, throat and chest rich buff.

4 **Ruddy Foliage-gleaner** *A. rubiginosus guerrerensis** 8.2in (21cm) **p. 464**
S cloud forest; skulking. Deep ruddy-brown overall, throat tawny-rufous. From
terrestrial 7 by stouter bill, longer rufous tail, voice.

5 **Plain Xenops** *Xenops minutus mexicanus* 4.7in (12cm) **p. 465**
SE rain forest; arboreal and conspicuous. Distinctive; note white moustache,
patterned wings and tail.

6 **Scaly-throated Leaftosser** *Sclerurus g. guatemalensis* 6.9in (18cm) **p. 466**
SE rain forest; terrestrial. Long, straightish bill, pale throat checkers hard to see.

7 **Tawny-throated Leaftosser** *S. m. mexicanus* 6.4in (16cm) **p. 465**
E and S cloud forest; terrestrial. Long, slender bill, tawny-rufous throat and chest.
Compare with 4.

8 **Plain Antvireo** *Dysithamnus mentalis septentrionalis* 4.5in (11cm) **p. 476**
SE rain forest; arboreal. Distinctive; note heavy bill, shortish tail, indistinct wing
bars. ♀ may suggest Tawny-crowned Greenlet (Plate 53) but note stout bill, dark
eye, and whitish eyering.

9 **Long-billed Gnatwren** *Ramphocaenus melanurus rufiventris** 4.9 in (12cm) **p. 577**
SE humid forest and edge, vine tangles. Distinctive; note very long, slender bill,
cocked tail, cinnamon face and underparts.

10 **Russet Antshrike** *Thamnistes a. anabatinus* 6in (15cm) **p. 475**
SE rain forest; arboreal. Distinctive; note heavy bill, dark eyestripe and buff
supercilium.

11 **Slaty Antwren** *Myrmotherula s. schisticolor* 4.2in (11cm) **p. 476**
SE humid forest. ♂ distinctive; note black bib. ♀ plain; note pale throat and eyering.

12 **Dot-winged Antwren** *Microrhopias quixensis boucardi** 4.3in (11cm) **p. 477**
SE rain forest; active and conspicuous. Unmistakable; note bold wing and tail spots.

13 **Great Antshrike** *Taraba major melanocrissa* 7.8in (20cm) **p. 474**
SE humid second growth; skulking. Unmistakable; note very stout bill, red eyes.

14 **Barred Antshrike** *Thamnophilus doliatus intermedius** 6.7in (17cm) **p. 474**
E and S second growth. ♂ unmistakable. ♀ distinctive; note stout bill, pale eyes.

15 **Dusky Antbird** *Cercomacra tyrannina crepera* 5.2in (13cm) **p. 477**
SE humid second growth; skulking. ♂ slaty overall, indistinct wingbars. ♀ note
blank tawny face and underparts.

16 **Mexican Antthrush** *Formicarius m. moniliger** 7.3in (19cm) **p. 478**
Endemic to SE humid forest; terrestrial. Distinctive; note rufous-bordered black
face and throat, pale blue orbital ring.

17 **Scaled Antpitta** *Grallaria guatimalensis ochraceiventris** 7.3in (19cm) **p. 479**
Humid forest, mostly foothills and highlands; terrestrial. Unmistakable; a hopping
tail-less ball on long legs; underparts pale dusky-cinnamon (cen Mexico) to bright
cinnamon (S of Isthmus); ssp shown (SW Mexico) intermediate in brightness of
underparts.

Slaty Antshrike *Thamnophilus punctatus* C.A. See Plate 71.
Bare-crowned Antbird *Gymnocichla nudiceps* C.A. See Plate 71.

38

PLATE 39: TROPICAL FLYCATCHERS

Many tropical flycatchers are best detected by voice (see text for descriptions) and may otherwise be overlooked easily.

1 **Northern Beardless Tyrannulet** *Camptostoma i. imberbe** 4in (10cm) **p. 482**
Widespread in drier areas. Bushy crest, small bill bright orange below, wags tail.

2 **Paltry Tyrannulet** *Zimmerius v. vilissimus* 4.5in (11cm) **p. 480**
Chis and C.A. Broad whitish supercilium, small bill, cocked tail.

3 **Yellow-bellied Tyrannulet** *Ornithion semiflavum* 3.4in (9cm) **p. 482**
SE rain forest canopy. Tiny, short-tailed, white supercilium, yellow underparts.

4 **Northern Bentbill** *Oncostoma cinereigulare* 4.0in (10cm) **p. 486**
Humid SE, forest understory and thickets. 'Bent' bill, pale eyes.

5 **Common Tody-Flycatcher** *Todirostrum cinereum finitimum** 3.8in (10cm) **p. 487**
SE, open country, scrub. Black face, whitish eyes, yellow underparts, cocked tail.

6 **Slate-headed Tody-Flycatcher** *T. sylvia schistaceiceps* 3.7in (9cm) **p. 486**
Humid SE, second growth and thickets. Grey head (olive in juv), white supraloral stripe, pale eyes, ochre-yellow wingbars.

7 **Ochre-bellied Flycatcher** *Mionectes oleaginus assimilis* 5.3in (13cm) **p. 485**
SE humid forest understory. Plain, slender bill, ochre belly.

8 **Stub-tailed Spadebill** *Platyrinchus cancrominus* 3.6in (9cm) **p. 488**
SE humid forest understory. Stub-tailed; note face pattern, broad bill.

9 **Yellow-bellied Elaenia** *Elaenia flavogaster subpagana* 6.3in (16cm) **p. 484**
SE, open country, scrub. From 10 by dull flesh base to bill below, brighter yellowish belly, voice. Spiky white-based crest usually raised, rarely flat (9(a)).

10 **Caribbean Elaenia** *E. martinica remota* 5.7in (14cm) **p. 483**
E QR and Belize, coasts and islands. From 9 by bright orange lower mandible, overall duller and more uniform greyish appearance, voice. Bushy white-based crest rarely raised (10(a)).

11 **Sepia-capped Flycatcher** *Leptopogon amaurocephalus pileatus* 5.3in (13cm) **p. 485**
SE rain forest, mid-levels. Dark brown cap, pale face with dark cheek mark, slender bill.

12 **Greenish Elaenia** *Myiopagis viridicata placens** 5.4in (14cm) **p. 483**
Both slopes, woodland and forest. Upright stance, long tail, flattish head, slender dark bill. Narrow whitish supercilium, no wingbars. Compare with 13.

13 **Yellow-olive Flycatcher** *Tolmomyias sulphurescens cinereiceps* 5.3in (13cm) **p. 488**
SE forest and edge. Grey head, pale eyes (dark in juv), broad bill pale below. Compare with 12.

14 **Eye-ringed Flatbill** *Rhynchocyclus b. brevirostris** 6.5in (17cm) **p. 487**
S humid forest, mid-levels. Overall olive-green, face greyer with distinct white eyering, broad bill.

15 **Ruddy-tailed Flycatcher** *Terenotriccus erythrurus fulvigularis* 3.8in (10cm) **p. 489**
SE rain forest interior. Small, cinnamon underparts, ruddy wings and tail.

16 **Sulphur-rumped Flycatcher** *Myiobius s. sulphureipygius* 5.0in (13cm) **p. 490**
SE rain forest interior. Bold pale lemon rump and black fanned tail, ochre chest.

17 **Royal Flycatcher** *Onychorhynchus coronatus mexicanus* 6.7in (17cm) **p. 489**
SE rain forest interior. Hammerhead crest and long bill, pale tawny rump merges with cinnamon tail. Crest (rarely raised) red in ♂, orange in ♀.

Mountain Elaenia *Elaenia frantzii* C.A. See Plate 71.

39

SW 2-89

1(a) **Pacific Slope Flycatcher** *Empidonax difficilis cineritius** 5.3in (13cm) **p. 500**
Breeds Baja; winters Pacific Slope N of Isthmus. Rarely distinguishable from 1(b) in field (see text). From 4 by shorter primary projection, wider bill, voice. 3 allopatric. High *si* and *pseep* calls.

1(b) **Cordilleran Flycatcher** *E. (d. ?) o. occidentalis** 5.3in (13cm) **p. 501**
Breeds highlands N of Isthmus, altitudinal migrant in winter. Compare with 1(a). From 4 by shorter primary projection, wider bill, voice. 3 allopatric. High *si* and *w-seen* calls.

2 **Buff-breasted Flycatcher** *E. fulvifrons** 4.6in (12cm) **p. 502**
Local in highlands. Distinctive *Empidonax*. 2(a) *pygmaeus* (NW Mexico): paler and duller; 2(b) *inexpectatus* (N C.A.): brighter, darker cap. Mellow to sharp *pic* call.

3 **Yellowish Flycatcher** *E. flavescens dwighti** 5.3in (13cm) **p. 502**
Cloud forest S of Isthmus. Distinctive in range; note teardrop-shaped eyering, bright plumage, compare with migrant Yellow-bellied Flycatcher (see text). 1 allopatric. Brighter and greener than 4(b). High *sii* and *w-seen* calls.

4 **Pine Flycatcher** *E. affinis** 5.5in (14cm) **p. 500**
Endemic to pine–oak in mountains of Mexico and Guatemala. From 1 by longer primary projection, narrower bill, voice; note also paler wingbars. Compare with migrant Hammond's Flycatcher (see text). 4(a) *pulverius* (NW Mexico); 4(b) *trepidus* (S of Isthmus). Mellow *pwip* call.

5 **White-throated Flycatcher** *E. a. albigularis** 5.3in (13cm) **p. 497**
Local, breeds highlands, winters lowlands. From other *Empidonax* by combination of ochre flanks, buffy-cinnamon wingbars, indistinct eyering, voice. Compare with Willow Flycatcher (see text). Worn summer birds (5(a)) from 8 by short primary projection, less peaked nape, habits. Burry *rreeah* call.

6 **Greater Pewee** *Contopus p. pertinax** 7.2in (18cm) **p. 492**
Highlands. Distinctive; note large size, spiky tufted crest, bright orange lower mandible, voice, *C. p. pallidiventris* of NW Mexico paler and greyer overall.

7 **Tropical Pewee** *C. cinereus brachytarsus** 5.5in (14cm) **p. 494**
SE lowlands. From 8 and Eastern Pewee by shorter primary projection, voice (see text).

8 **Western Pewee** *C. sordidulus richardsonii** 5.6in (14cm) **p. 493**
Widespread transient, local summer resident (breeds temperate areas). From 7 (rarely sympatric) by long primary projection, voice, bill often dark-tipped below. Compare with Eastern Pewee (see text).

9 **Tufted Flycatcher** *Mitrephanes p. phaeocercus** 5in (13cm) **p. 491**
Breeds highland forests, altitudinal migrant. Unmistakable; conspicuous.

10 **Belted Flycatcher** *Xenotriccus callizonus* 4.7in (12cm) **p. 490**
Endemic to interior S of Isthmus, local. Unmistakable; skulking.

11 **Pileated Flycatcher** *X. mexicanus* 5.5in (14cm) **p. 490**
Endemic to interior SW Mexico. Distinctive. Note spiky crest (may be held flat), short primary projection, long tail, bright pinkish-orange lower mandible. 11(a) Worn plumage. Juv wing edgings pale cinnamon. Compare with migrant Dusky Flycatcher (see text).

40

SW 1-92

PLATE 41: MYIARCHUS, STREAKED FLYCATCHERS, AND ALLIES

Myiarchus A remarkably uniform-looking genus whose members are best identified by voice; note also face patterns, bill size, tail patterns.

1 **Ash-throated Flycatcher** *M. c. cinerascens** 7.7in (20cm) **p. 507**
Pacific Slope and interior Mexico, migratory. From 4 by voice, greyer face, more peaked nape, rufous primary edgings contrast with whitish secondary edgings, dark shaft usually spreads across tips of outer rectrices, mouth flesh. From 5 by voice, smaller size, and smaller bill, tail pattern. Short *pip* call.

2 **Dusky-capped Flycatcher** *M. tuberculifer** 6.8in (17cm) **p. 507**
Widespread. From all except 3 by small bill, dull wingbars, reduced rufous in tail. 2(a) *olivascens* (NW Mexico): dull, basically no rufous in tail; 2(b) *lawrencei* (SE Mexico): from 3 by contrastingly darker head with peaked nape, dull rufous wing-bars, pale cinnamon tertial edgings, voice. Plaintive *wheeeeu* call.

3 **Yucatan Flycatcher** *M. yucatanensis** 7.3in (19cm) **p. 506**
Endemic to Yuc Pen. From 2 by less peaked nape, dull grey-brown wingbars, whitish tertial edgings, greyish lores, and often a broad pale eyering, voice. 3(a) *yucatanensis*; 3(b) *lanyoni* (Isla Cozumel): notably darker. Plaintive, drawn-out *whee-ee-eeu*.

4 **Nutting's Flycatcher** *M. nuttingi inquietus** 7.3in (19cm) **p. 508**
Mainly Pacific Slope, resident. Very similar to 1, told by voice, browner face, less peaked nape, rufous primary edgings blend with cinnamon to pale lemon secondary edgings, dark shaft usually does not spread across tips of outer rectrices, mouth orange. From 5 by voice, smaller bill, browner face, tail pattern. Sharp *wheek!*

5 **Brown-crested Flycatcher** *M. tyrannulus** 8.7in (22cm) **p. 509**
Widespread, partly migratory. Larger and bulkier than 1 with larger bill; note tail pattern and voice. Paler overall than 6. 5(a) *magister* (W Mexico): very large bill; 5(b) *cooperi* (E Mexico). Disyllabic *h-whik*.

6 **Great Crested Flycatcher** *M. crinitus* 8.5in (22cm) **p. 509**
Migrant to E and S. Distinctive; little contrast between grey face and olive crown, strong contrast between grey chest and bright lemon belly, bill often extensively pale below at base; note voice. Rising *wheep*.

7 **Flammulated Flycatcher** *Deltarhynchus flammulatus* 6.3in (16cm) **p. 510**
Endemic to W Mexico. Thorn forest. Suggests *Myiarchus* but bill relatively broader, usually with obvious flesh below at base, head not strongly crested. Note pale spectacles and dusky loral stripe, cinnamon wing and tail edgings, voice. Duller and greyer when worn (7(a)).

8 **Piratic Flycatcher** *Legatus leucophaius variegatus* 6.5in (17cm) **p. 513**
Summer resident in humid SE. Distinctive; note small bill, head pattern, blurred dusky streaking on chest. Perches prominently atop bare trees.

9 **Sulphur-bellied Flycatcher** *Myiodynastes l. luteiventris** 8.0in (20cm) **p. 512**
Summer resident, both slopes. From less common 10 by smaller black bill, bold black malar stripes and chin (variable, 9(a)), whitish supercilium and moustache, voice. Juv has rufous wing edgings.

10 **Streaked Flycatcher** *M. maculatus insolens* 8.5in (22cm) **p. 512**
Summer resident in humid SE. From commoner 9 by larger, pale-based bill, whitish chin, narrower malar stripes, pale lemon supercilium and moustache, voice.

11 **Bright-rumped Attila** *Attila spadiceus** 8.0in (20cm) **p. 505**
Widespread. Distinctive; staring face, stout, hooked bill, bright rump, streaked chest, flicks tail often. Voice strident. 11(a) *mexicanus* (W Mexico); 11(b) *gaumeri* (Yuc Pen).

41

Manakins   Small stocky birds of tropical forest. ♂♂ stunning, ♀♀ olive overall.

1   **Long-tailed Manakin** *Chiroxiphia linearis* 4.5in (11cm) + tail streamers. **p. 526**
Pacific Slope S from Isthmus. ♂ unmistakable. ♀ olive with orange legs, tail streamers. Imm ♂ like ♀ with crimson cap; in 2nd-year tail streamers long, face and chest mottled black. Other manakins allopatric.

2   **Red-capped Manakin** *Pipra m. mentalis* 4.2in (11cm) **p. 527**
SE rain forest. ♂ unmistakable. ♀ lacks yellow belly of 3, bill dull flesh, legs dark.

3   **White-collared Manakin** *Manacus candei* 4.7in (12cm) **p. 526**
SE rain forest edge. ♂ unmistakable. ♀ from 2 by yellow belly, black bill, orange legs. Imm ♂ throat pale grey.

4   **Boat-billed Flycatcher** *Megarynchus pitangua mexicanus** 9.3in (24cm) **p. 511**
Widespread. From 5 by heavier and broader bill, olive upperparts, voice. 6 much smaller with small bill; note voice.

5   **Great Kiskadee** *Pitangus sulphuratus derbianus** 9.5in (24cm) **p. 510**
Widespread. From 4 and smaller 6 by extensively rufous wings and tail, bill, voice.

6   **Social Flycatcher** *Myiozetetes similis texensis** 7in (18cm) **p. 511**
Widespread. Distinctive; note small stubby bill, voice. Juv has rufous-edged wings and tail.

Tyrannus: **Kingbirds**   Large conspicuous flycatchers of open and semiopen country, often on roadside wires, fences, etc.

7   **Thick-billed Kingbird** *T. c. crassirostris** 9in (23cm) **p. 515**
Pacific Slope and Interior. Distinctive; note dark head, thick bill, voice.

8   **Cassin's Kingbird** *T. v. vociferans** 8.5in (22cm) **p. 515**
Pacific Slope and Interior. Head and chest dusky grey with cottony-white chin, squared blackish tail.

9   **Western Kingbird** *T. verticalis* 8in (20cm) **p. 516**
Pacific Slope and Interior. Pale grey head and chest, small bill (compare with 10, 11), white sides of tail.

10   **Couch's Kingbird** *T. couchii* 8.8in (22cm) **p. 514**
Atlantic Slope. From 8 and 9 by large bill, forked tail, from 11 safely told only by voice; note also bill often relatively broader and shorter, primary tips (adults only) more evenly spaced.

11   **Tropical Kingbird** *T. melancholicus chloronotus** 8.5in (22cm) **p. 514**
Both slopes. From 8, 9, by large bill, forked tail; from 10 safely told only by voice, bill often longer, note less regular primary tip spacing (adults only).

12   **Fork-tailed Flycatcher** *T. savana monachus* 13in (33cm) (juv 7.7in (20cm)). **p. 517**
SE savannas. Adult unmistakable; note black head and tail. Juv crown dusky, tail shorter.

13   **Scissor-tailed Flycatcher** *T. forficatus* 12in (30cm) **p. 517**
Mostly E and S in winter. Adult unmistakable; note pale grey head, pink sides, mostly white undertail; juv tail shorter.

Grey-headed Piprites *Piprites griseiceps* C.A. See Plate 71.

42

SW 8-90

1 **Grey-collared Becard** *Pachyramphus major** 5.8in (15cm) **p. 521**
Endemic; widespread but uncommon. Distinctive (compare with mostly allopatric White-winged Becard, Plate 71). ♀ from 2 by graduated tail, pale supraloral stripe, patterned wings and tail. Imm ♂ variable (see text). 1(a) *major* (E Mexico); 1(b) *uropygialis* (W Mexico): chestnut cap bordered black.

2 **Rose-throated Becard** *P. aglaiae** 6.7in (17cm) **p. 521**
Widespread. ♂ distinctive: paler in N, darker in humid SE where rose on throat may be absent. ♀ in N greyish above with buff hindcollar, compare with 1. ♀ in humid SE from 3 by bushy blackish cap without pale supraloral stripe, ungraduated tail, voice. Imm ♂ variable (see text). 2(a) *albiventris* (NW Mexico); 2(b) *sumichrasti* (SE Mexico); 2(c) *aglaiae* (SW Mexico).

3 **Cinnamon Becard** *P. cinnamomeus fulvidior* 5.7in (14cm) **p. 520**
Humid SE. Sexes similar. From ♀ 2(b) by graduated tail, dark rufous cap with pale supraloral stripe, voice.

4 **Masked Tityra** *Tityra semifasciata** 9in (23cm) **p. 522**
Both slopes. Distinctive; note bright bare pinkish-red face and bill base. 4(a) *griseiceps* (W Mexico); 4(b) *personata* (E and S).

5 **Black-crowned Tityra** *T. inquisitor fraserii* 7.7in (20cm) **p. 523**
Humid SE. From 4 by stout blackish bill, lack of red in face; note black cap of ♂, chestnut face of ♀.

6 **Lovely Cotinga** *Cotinga amabilis* 7.3in (19cm) **p. 524**
Humid SE. ♂ unmistakable. ♀ distinctive; note shape, spotted underparts, scalloped upperparts.

7 **Speckled Mourner** *Laniocera r. rufescens* 8.3in (21cm) **p. 519**
SE rain forest. From 8 and 9 by wing-covert pattern, faint dusky scallops on chest, voice; yellow tufts often concealed. Uncommon.

8 **Rufous Mourner** *Rhytipterna h. holerythra* 8.0in (20cm) **p. 506**
SE rain forest. From larger 9 by voice, uniformly rufous throat and chest. Compare with 7.

9 **Rufous Piha** *Lipaugus u. unirufus* 9.8in (25cm) **p. 523**
SE rain forest. From smaller 8 by voice; note paler throat, grey eyering (hard to see), bushier face. Compare with 7.

10 **Thrushlike Mourner** *Schiffornis turdinus veraepacis* 6.7in (17cm) **p. 519**
SE rain forest. Distinctive; best located by haunting song. Brown overall, note paler eyering accentuating big eye.

White-winged Becard *Pachyramphus polychopterus* C.A. See Plate 71.

43

SW 3-89

PLATE 44: SWALLOWS AND MARTINS

Flying

1 **Mangrove Swallow** *Tachycineta a. albilinea* 4.5in (11cm) **p. 531**
Mangroves, lowlands. Distinctive; note white rump and tertial edgings.

2 **Violet-green Swallow** *T. t. thalassina** 4.7in (12cm) **p. 532**
NW deserts, elsewhere highlands. Note white sides of rump (often appears white-rumped in NW Mexico, compare with 1). ♀ face less well marked.

3 **Blue-and-white Swallow** *Notiochelidon cyanoleuca patagonica* 5.2in (13cm) **p. 532**
Austral migrant (May–Jul) S of Isthmus. Note strongly forked tail, flight resembles 4, undertail coverts black.

4 **Black-capped Swallow** *N. pileata* 5in (13cm) **p. 533**
Endemic to mountains S of Isthmus. Dusky below with large, oval white belly patch; note long, forked tail, 'twinkling' flight.

5 **Cave Swallow** *Hirundo fulva citata** 5.3in (13cm) **p. 535**
Local resident, nomadic. From 6 by less contrasting throat–chest border, darker rump. Overhead from larger and longer-winged 9 and 10 by squarer tail, dusky undertail coverts.

6 **Cliff Swallow** *H. pyrrhonota swainsoni** 5.3in (13cm) **p. 535**
Widespread transient, summer resident N of Isthmus. Note contrasting dark throat, compare with 5.

7 **Sinaloa Martin** *Progne sinaloae* 7in (18cm) **p. 530**
Endemic summer resident, NW Mexico. ♂ blue-black with white belly; ♀ shown perched. Note more deeply forked tail than 8.

8 **Grey-breasted Martin** *P. c. chalybea** 6.8in (17cm) **p. 530**
Tropical lowlands (withdraws from E Mexico in winter). Overhead from smaller 9 and 10 by more deeply cleft tail. See ♀ Sinaloa and Purple martins (shown perched).

9 **Ridgway's Rough-winged Swallow** *Stelgidopteryx ridgwayi stuarti** 5.3in (13cm) **p. 534**
Endemic to SE (mainly Yuc Pen). Darker overall than 10 with black distal undertail coverts, more strongly cleft tail. Compare with 5, 8.

10 **Northern Rough-winged Swallow** *S. serripennis fulvipennis** 5.3in (13cm) **p. 533**
Widespread. Compare with 5, 8, 9.

Perched

1 **Mangrove Swallow** *Tachycineta a. albilinea* 4.5in (11cm) **p. 531**
Distinctive; also shown in flight.

2 **Black-capped Swallow** *Notiochelidon pileata* 5in (11cm) **p. 533**
Note long forked tail, compare with 3, 4. Also shown in flight.

3 **Ridgway's Rough-winged Swallow** *Stelgidopteryx ridgwayi stuarti** 5.3in (13cm) **p. 534**
Note pale lore spots, compare with 4. Also shown in flight.

4 **Northern Rough-winged Swallow** *S. serripennis fulvipennis** 5.3in (13cm) **p. 533**
Compare with 3. 8 larger with forked tail. Also shown in flight.

5 **Cave Swallow** *Hirundo fulva** 5.3in (13cm) **p. 535**
Note less contrasting throat–chest border and darker rump than 6. 5(a) *pallida* (NE Mexico); 5(b) *citata* (Yuc Pen). Also shown in flight.

6 **Cliff Swallow** *H. pyrrhonota** 5.3in (13cm) **p. 535**
Note contrasting dark throat, compare with 5. 6(a) *tachina* (breeds NW Mexico); 6(b) *swainsoni* (breeds S to Oax). Also shown in flight.

7 **Purple Martin** *Progne s. subis** 7.7in (20cm) **p. 529**
Widespread transient (summer resident N and cen Mexico). From smaller 8 and 9 by paler forehead, pale collar, mottled dusky throat and chest; imms less mottled below with narrow dark streaks on undertail coverts. ♂ all blue-black.

8 **Grey-breasted Martin** *P. c. chalybea** 6.8in (17cm) **p. 530**
From ♀ 9 by paler throat and chest, less strongly contrasting white belly, less deeply forked tail. Upperparts mostly blue-black in adult ♀. Compare with 3, 4, and 7. Also shown in flight.

9 **Sinaloa Martin** *P. sinaloae* 7in (18cm) **p. 530**
Compare with 7, 8. Upperparts may be mottled dusky. ♂ shown in flight.

44

JW 2-92

PLATE 45: JAYS

1 **Black-throated Magpie-Jay** *Calocitta (formosa?) colliei* 26.5in (67cm) **p. 538**
Endemic to NW Mexico. Unmistakable, but intergrades with 2 in Jal and Col.
Some Black-throateds may be mostly white-throated (1(a)) well to N of White-throated range.

2 **White-throated Magpie-Jay** *C. f. formosa** 19.5in (50cm) **p. 538**
Pacific Slope from cen Mexico S. Unmistakable, but intergrades with 1 in Jal and Col.

3 **Green Jay** *Cyanocorax yncas** 11.3in (29cm) **p. 539**
Both slopes, local W Mexico. Unmistakable. Eyes dark in NE and W Mexico, yellow in S; underparts greenish to yellow. 3(a) *luxuosa* (NE Mexico); 3(b) *maya* (Yuc Pen).

4 **Tufted Jay** *C. dickeyi* 14.5in (37cm) **p. 539**
Endemic to mountains of NW Mexico (local). Unmistakable and striking.

5 **Brown Jay** *C. morio** 16in (41cm) **p. 539**
Atlantic Slope lowlands. Unmistakable; large, noisy, obnoxious. Dark morphs occur in E Mexico, light morphs in SE Mexico and C.A., both common from S Ver to N Chis. 5(a) *morio* (NE Mexico); 5(b) *fuliginosus* (SE Mexico). Yellow or mostly yellow bill of imm retained for a year or so.

Cissilopha Now considered a subgenus of *Cyanocorax*, four black-and-blue jays with mostly allopatric distributions. Imms have yellow bill for a year or so.

6 **Purplish-backed Jay** *C. beecheii* 14.5in (37cm) **p. 540**
Endemic to NW Mexico. From smaller 7 (sympatric around San Blas, Nay) by deeper purplish-blue back, brighter yellowish legs, voice.

7 **San Blas Jay** *C. s. sanblasianus** 13in (33cm) **p. 540**
Endemic to W Mexico. From larger 6 by brighter, contrasting blue back, duller legs, voice; juv and imm crest usually obvious.

8 **Yucatan Jay** *C. y. yucatanicus** 13in (33cm) **p. 540**
Endemic to Yuc Pen. No similar species in range. Striking juv plumage kept only a short period (Jul–Sep).

Bushy-crested Jay *C. melanocyaneus* C.A. See Plate 70.

45

juv

1

1a

2

3a

3b

imm

4

juv

5a

5b

imm

6

imm

7

8

juv

imm

SW 11-88

PLATE 46: JAYS AND CROWS

1 **Steller's Jay** *Cyanocitta stelleri** 11.5in (29cm) **p. 537**
Highlands. Distinctive; note tufted crest. 1(a) *diademata* (N and cen Mexico):
eye-crescents may be lacking in shorter-crested juv; 1(b) *coronata* (cen Mexico to
C.A.): shorter crest shown flattened, note white eye-crescents; compare with 4.

2 **Scrub Jay** *Aphelocoma coerulescens** 11.5in (29cm) **p. 543**
N of Isthmus. Variable. From larger and chunkier 3 by white supercilium,
blackish mask, dusky necklace, voice. 2(a) *hypoleuca* (BCS); 2(b) *cyanotis* (cen
Mexico). Juv sooty with trace of adult head pattern.

3 **Grey-breasted Jay** *A. ultramarina wollweberi** 12.2in (31cm) **p. 544**
From slimmer, longer-tailed 2 by plain bluish head, pale pinkish-grey underparts,
voice. Juv bill initially yellow.

4 **Unicolored Jay** *A. u. unicolor** 13.5in (34cm) **p. 544**
Endemic; E and S humid montane forest. Uniformly deep blue, compare with 1(b).
Juv sooty, bill initially yellow.

Cyanolyca Small jays of cloud forest and pine–evergreen forest, mostly allopatric.

5 **White-throated Jay** *C. m. mirabilis** 9.5in (24cm) **p. 543**
Endemic to Sierra Madre del Sur of Gro and Oax. Unmistakable; note striking
face pattern, duller in juv.

6 **Dwarf Jay** *C. nana* 9.2in (23cm) **p. 542**
Endemic to E cen Mexico. Distinctive; note pale blue throat, black mask. Juv
duller.

7 **Black-throated Jay** *C. p. pumilo** 10in (25cm) **p. 542**
Endemic, Chis and N C.A. Black face and throat with narrow pale supercilium. Juv
duller, lacks supercilium.

8 **Azure-hooded Jay** *C. cucullata guatemalae** 11.5in (29cm) **p. 542**
From smaller 7 by sky blue hood (hard to see from below), duller and without
white border in juv.

9 **Sinaloa Crow** *Corvus sinaloae* 14.5in (37cm) **p. 546**
Endemic to NW Mexico. Small, glossy, tail squared; note voice. 12 allopatric.

10 **Northern Raven** *C. corax sinuatus** 23.5in (60cm) **p. 547**
Widespread. From slightly smaller 11 by longer bill with nasal bristles less than
half bill length, shaggier throat, voice useful.

11 **Chihuahuan Raven** *C. cryptoleucus* 19in (48cm) **p. 546**
N and cen Mexico, grasslands. From very similar 10 by shorter bill, long nasal
bristles accentuate sawn-off bill profile, voice useful. From smaller and slighter 12
by relatively shorter, graduated tail, voice.

12 **Tamaulipas Crow** *C. imparatus* 14.8in (38cm) **p. 546**
Endemic to NE Mexico, counterpart of allopatric 9. Voice distinctive.

46

SW 9-91

PLATE 47: GNATCATCHERS, TITMICE, AND ALLIES

1 **Mexican Chickadee** *Parus s. sclateri** 4.8in (12cm) **p. 548**
Pine and pine–oak N of Isthmus. No similar species in range; note extensive black bib.

2 **Bridled Titmouse** *Parus w. wollweberi** 5.2in (13cm) **p. 549**
Oak and pine–oak N of Isthmus. Unmistakable; note striking face pattern.

3 **Bushtit** *Psaltriparus minimus** 4.3in (11cm) **p. 552**
Temperate areas S to Guatemala. Small stubby bill, long tail. Complex transition from plain to black-eared forms (see text). 3(a) *personatus* (cen Mexico); 3(b) *melanurus* (NW BCN).

4 **Verdin** *Auriparus flaviceps ornatus** 4.1in (10cm) **p. 551**
N Mexico. Deserts. Adult distinctive; juv suggests 3 but note pointed bill (initially orange).

Polioptila: Gnatcatchers   Small active birds with long graduated tails often cocked and swung about. Juvs resemble ♀♀.

5 **California Gnatcatcher** *Polioptila c. californica** 4.1in (10cm) **p. 578**
Baja. Undertail mostly black with narrow white tips (becoming wider in S). Alt ♂ black cap, dusky overall (shown; NW Baja chaparral) to pale grey and whitish (deserts), much like 6 but note tail pattern, indistinct eyering, voice. Basic ♂ cap reduced to lacking. ♀ brownish above, head greyer, dusky to whitish below. From 6 in NE Baja by narrower white tail tips, indistinct eyering of alt ♂, voice.

6 **Black-tailed Gnatcatcher** *Polioptila melanura lucida** 4.1in (10cm) **p. 578**
N Mexico including NE Baja (where sympatric with 5). Deserts. Undertail mostly black with wide white tips. Pale grey and whitish overall. Alt ♂ black cap with bright white eyering. Basic ♂ cap reduced or lacking. ♀ grey head contrasts with brown back. Note voice. Compare with 5.

7 **Blue-grey Gnatcatcher** *Polioptila caerulea deppei** 4.3in (11cm) **p. 577**
Widespread. Undertail mostly white. From 8 by voice, shorter bill, less graduated tail, wings often darker, bill often flesh below at base. Alt ♂ black brows. ♀ (and basic ♂) complete white eyering, no black on head.

8 **Black-capped Gnatcatcher** *Polioptila n. nigriceps** 4.2in (11cm) **p. 579**
Endemic to NW Mexico. Thorn forest. Undertail mostly white. From 7 by voice, longer blackish bill, strongly graduated tail. Alt ♂ black cap, eyering indistinct. Basic ♂ often retains a few black flecks. ♀ eyering often less clean-cut than 7. 9 allopatric (?).

9 **White-lored Gnatcatcher** *Polioptila albiloris vanrossemi** 4.3in (11cm) **p. 579**
SW Mexico, N Yuc, C.A. Thorn forest. Alt ♂ solid black cap. Basic ♂ narrow white supercilium. ♀ narrow white supercilium year-round. 8 and 10 allopatric.

10 **Tropical Gnatcatcher** *Polioptila plumbea brodkorbi* 4.1in (10cm) **p. 580**
SE rain forest canopy. Tail relatively short. Blank-faced look, broad white supercilium year-round. ♂ black cap. ♀ grey cap slightly darker than back.

Long-billed Gnatwren *Ramphocaenus melanurus*. See Plate 38.

47

1

3a ♂

3b ♂

3a ♀

juv

4

2

5

6

♀

♀

6 ♂ basic

♂ alt

♂ alt

7

8

♀

♀

♂ alt

8 ♂ basic

♂ alt

10

9

♀

♀

9

♂ basic

♂

9 ♂ alt

SW 10-86

PLATE 48: LARGE WRENS

Campylorhynchus: **Cactus Wrens** Large, conspicuous, noisy, and usually social wrens of open and wooded habitats. Identification problems are few and most species are allopatric, particularly when habitats are considered. Gruff, chattering calls and rollicking songs (often with duets) attract attention.

1 **Cactus Wren** *C. brunneicapillus** 7.3in (19cm) **p. 560**
NW Mexico and Plateau. Deserts. Potentially sympatric only with smaller 2 (oak woods); note heavy black spotting on chest; smaller and fainter spots on juv more even on underparts, flanks lack barring. 1(a) *affinis* (BCS); 1(b) *couesi* (N Plateau). Juvs of both forms similar.

2 **Spotted Wren** *C. gularis* 6.5in (17cm) **p. 558**
Endemic to NE and W Mexico. Oak woods. Distinctive; note band of spots across chest, black malar stripe. Juv plain below. From larger 3 by shorter bill and tail, less extensively spotted underparts (2 and 3 probably allopatric, sympatry possible in N Gro).

3 **Boucard's Wren** *C. jocosus* 6.6in (17cm) **p. 559**
Endemic to interior SW Mexico. Arid scrub. No similar species known to be sympatric; compare with 2.

4 **Grey-barred Wren** *C. megalopterus nelsoni** 7.3in (19cm) **p. 557**
Endemic to montane forests of cen Mexico. Arboreal. Distinctive in range. Possibly sympatric locally with 5 in W cen Ver; note juv paler below than 5 with trace of barring on flanks, usually in groups with adults.

5 **Band-backed Wren** *C. zonatus restrictus** 7.7in (20cm) **p. 556**
SE lowlands and S highlands. Arboreal. Distinctive in range, compare with 4.

6 **Rufous-naped Wren** *C. rufinucha** 6.8in (17cm) **p. 557**
Marked geographic variation (see text). 6(a) *humilus* (W Mexico): intergrades abruptly in W Chis with 6(b) *nigricaudatus* ('Rufous-backed Wren', Pacific Slope in Chis and W Guatemala). 6(b) from larger 7 by barred wings. 6(c) *rufinucha* (cen Ver): underparts spotted. Juv similar to adult.

7 **Giant Wren** *C. chiapensis* 8.5in (22cm) **p. 557**
Endemic to Pacific Slope of Chis. Large, distinctive, compare with 6(b). Juv similar to adult.

8 **Yucatan Wren** *C. yucatanicus* 7in (18cm) **p. 559**
Endemic to N coastal Yuc. Unmistakable in range.

48

1a

1b

juv

2

juv

3

juv

4

juv

5

juv

6a

6b

6c

7

8

juv

BW 12-87

PLATE 49: SMALL WRENS

1 **Rock Wren** *Salpinctes o. obsoletus** 5.5in (14cm) **p. 560**
Open rocky areas. Distinctive; note pale flecking above, cinnamon tail corners. See also Plate 71.

2 **Canyon Wren** *Catherpes m. mexicanus** 5.3in (13cm) **p. 561**
Canyons, towns, etc. Distinctive; note bright white bib.

3 **Carolina Wren** *Thryothorus ludovicianus berlanderi** 5.3in (13cm) **p. 565**
NE Mexico. White supercilium, cinnamon underparts.

4 **White-browed Wren** *T. (l. ?) a. albinucha** 5in (13cm) **p. 565**
Endemic; SE Mexico, local N C.A. White supercilium, dull whitish underparts, compare with 8.

5 **Bewick's Wren** *Thryomanes bewickii murinus** 5in (13cm) **p. 566**
N Mexico and interior S to Oaxaca. Long, white-tipped tail often cocked, white supercilium.

6(a) **Northern House Wren** *Troglodytes aedon parkmani** 4.6in (12cm) **p. 567**
Widespread in winter N of Isthmus. Plain, grey-brown and pale grey (see text).

6(b) **Brown-throated Wren** *T. (a. ?) brunneicollis compositus** 4.6in (12cm) **p. 568**
Mountains N of Isthmus. Buff supercilium and underparts, barred flanks (see text).

6(c) **Southern House Wren** *T. (a. ?) musculus intermedius** 4.6in (12cm) **p. 568**
Resident S of Isthmus. Plain, warm brown and pale buff (see text).

7 **Rufous-browed Wren** *T. r. rufociliatus** 4.3in (11cm) **p. 570**
Endemic; mountains S of Isthmus. Winter Wren jizz. Cinnamon supercilium, throat, and chest.

8 **White-bellied Wren** *Uropsila leucogastra brachyura** 4in (10cm) **p. 566**
Endemic; both slopes, often local. Pale lores accentuate blank-faced expression, compare with 4.

9 **White-breasted Wood-Wren** *Henicorhina leucosticta prostheleuca** 4.3in (11cm)
p. 572 SE rain forest. Chunky, stub-tailed, white throat and chest (pale grey in juv), cinnamon-brown flanks.

10 **Grey-breasted Wood-Wren** *H. leucophrys mexicana** 4.3in (11cm) **p. 572**
Humid montane forest. Chunky, stub-tailed, grey throat and chest, rufous-brown flanks.

11 **Sinaloa Wren** *Thryothorus s. sinaloa** 5.3in (13cm) **p. 564**
Endemic to W Mexico. Plain face with neck side stripes, pale grey chest. Note voice.

12 **Spot-breasted Wren** *T. maculipectus umbrinus** 5.3in (13cm) **p. 562**
E and S. Spotted chest, striped face. Juv duller, chest indistinctly marked.

13 **Happy Wren** *T. felix pallidus** 5.3in (13cm) **p. 563**
Endemic to W Mexico. Striped face, buff underparts. Note voice.

14 **Plain Wren** *T. m. modestus** 5.3in (13cm) **p. 565**
S of Isthmus. Plain face and pale grey chest, cinnamon flanks, rufous rump, black-barred tail.

15 **Banded Wren** *T. pleurostictus oaxacae** 5.7in (14cm). **p. 564**
SW Mexico to C.A. Black-banded sides and flanks.

16 **Rufous-and-white Wren** *T. r. rufalbus** 6.1in (15cm) **p. 563**
S Chis and C.A. Rufous upperparts, whitish underparts. Juv duller.

17 **Nightingale Wren** *Microcerculus philomela* 4.3in (11cm) **p. 573**
SE rain forest. Long bill, stub-tailed. Great song.

18 **Sumichrast's Wren** *Hylorchilus sumichrasti* 6.3in (16cm) **p. 561**
Endemic to cen Ver and N Oax. Forested limestone outcrops, local. Distinctive; note bill, tawny chest. 19 allopatric.

19 **Nava's Wren** *H. navai* 6,3in (16cm) **p. 562**
Endemic to SE Ver and NW Chis. Forested limestone outcrops, local. Distinctive; note bill, whitish throat and chest. 18 allopatric.

Cozumel Wren *Troglodytes beani* Island endemic. See Plate 69.
Clarion Wren *T. tanneri* Island endemic. See Plate 69.
Socorro Wren *T. sissonii* Island endemic. See Plate 69.

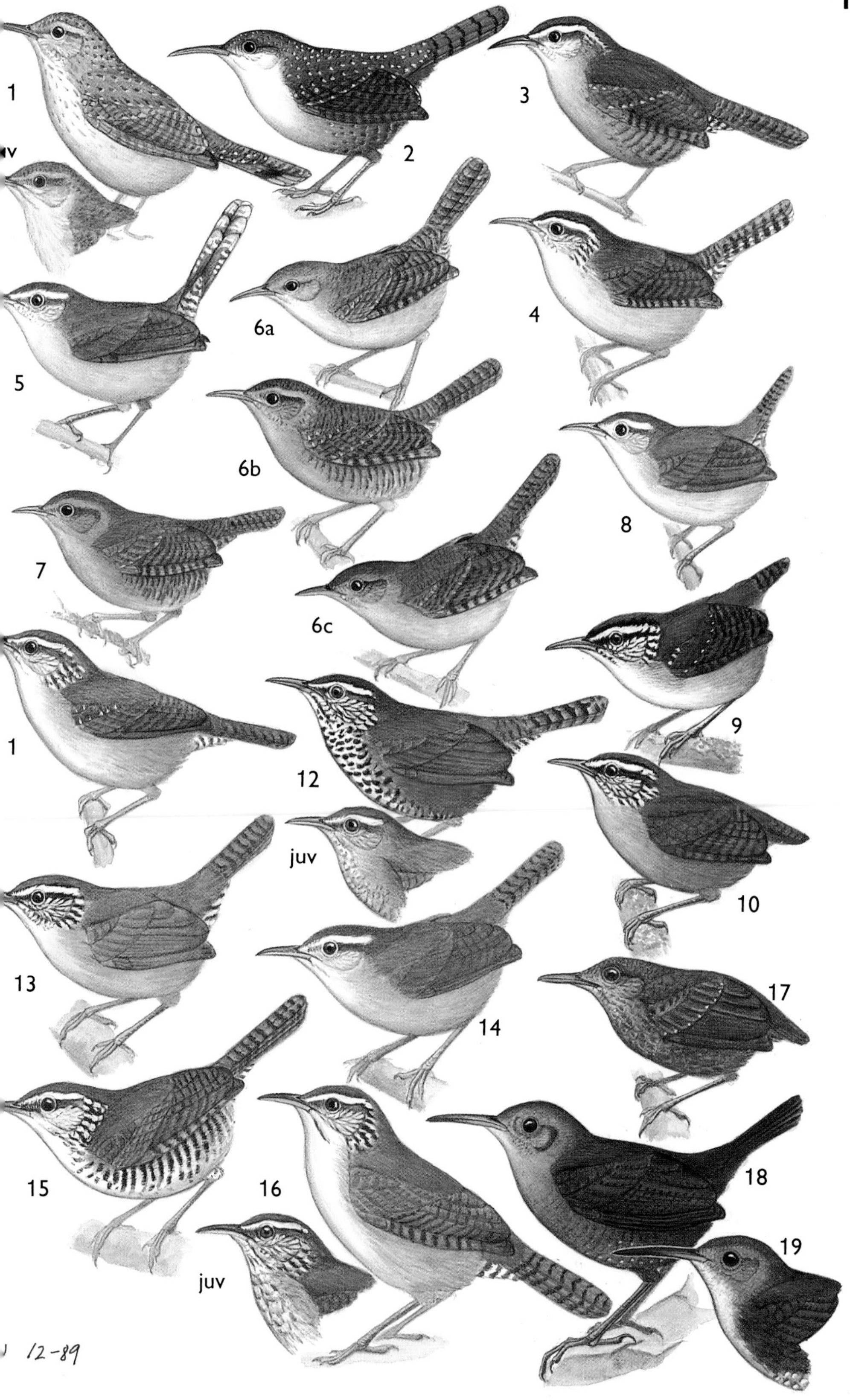
1
2
3
juv
6a
4
5
6b
8
7
6c
9
11
12
juv
10
13
14
17
15
16
18
juv
19
12-89

49

1 **Western Bluebird** *Sialia m. mexicana** 6.7in (17cm) **p. 582**
N and cen Mexico. ♂ deep blue overall with chestnut on chest and flanks. ♀ throat dusky.

2 **Eastern Bluebird** *S. sialis fulva** 6.7in (17cm) **p. 581**
Widespread. ♂ throat, chest, and flanks rufous, belly whitish. ♀ throat pale cinnamon.

3 **Slate-colored Solitaire** *Myadestes u. unicolor** 7.8in (20cm) **p. 584**
Endemic; SE cloud forest. Slaty overall, thick white eye-crescents. Ethereal song.

4 **Brown-backed Solitaire** *M. o. occidentalis** 8.3in (21cm) **p. 583**
Endemic; montane forests. Upperparts brownish, head and underparts grey with white eye-crescents. Striking song.

5 **Orange-billed Nightingale-Thrush** *Catharus aurantiirostris melpomene** 6.4in (16cm) **p. 584**
Widespread. Bright orange bill (often with dark culmen, 5(a)), orange orbital ring, and bright yellow-orange legs. Note voice.

6 **Russet Nightingale-Thrush** *C. occidentalis** 6.5in (17cm) **p. 585**
Endemic to mountains N of Isthmus. From 7 by voice, bill flesh below with dark tip, grey underparts often tinged buff on lower throat and mottled dusky on chest, buff stripe on underwing across bases of remiges. 6(a) *olivascens* (NW Mexico); 6(b) *occidentalis* (Oax). Compare with 5.

7 **Ruddy-capped Nightingale-Thrush** *C. frantzii** 6.5in (17cm) **p. 585**
Mountains from cen Mexico S. From 6 by voice, bill yellow-orange below, throat and underparts dusky grey, underwings smoky grey. 7(a) *confusus* (E Mexico); 7(b) *alticola* (S Chis to El Salvador). Compare with 5.

8 **Black-headed Nightingale-Thrush** *C. m. mexicanus** 6.3in (16cm) **p. 586**
E and S cloud forest. Distinctive; note bright orange orbital ring and bill, blackish cap. ♀ browner on head and chest.

9 **Spotted Nightingale-Thrush** *C. dryas ovandensis** 7.2in (18cm) **p. 586**
SE cloud forest. Distinctive; black head with bright orange orbital ring and bill, lemon underparts spotted olive on chest, upperparts olive.

10 **Aztec Thrush** *Zoothera p. pinicola** 9in (23cm) **p. 593**
Endemic to mountains N of Isthmus. Unmistakable; note striking wing pattern, duller on ♀ and juv.

♀

♂

♂

50

1

2

♀

3

4

juv

5

5a

6a

juv

6b

7a

7b

♀

8

♂

10

♀

♂

9

GW 2-90

PLATE 51: LARGE THRUSHES (ROBINS)

1(a) **American Robin** *Turdus migratorius phillipsi** 9.5in (24cm) **p. 592**
Mountains N of Isthmus. Distinctive. Black head (grey in some ♀♀), brick-orange underparts.

1(b) **San Lucas Robin** *T. (m.?) confinis* 9.5in (24cm) **p. 592**
Endemic to mountains of S BCS. Washed-out form of 1(a). Underparts buff.

2 **Clay-colored Thrush** *T. grayi** 9.8in (25cm) **p. 590**
E and S. Variable, brown overall, paler below. Note greenish-yellow bill, pinkish legs. Lacks white forecollar of 4(a). From 5, 6, and 7 by color of bare parts. 2(a) *tamaulipensis* (NE Mexico); 2(b) *grayi* (S Chis and Guatemala).

3(a) **Rufous-backed Thrush** *T. r. rufopalliatus** 9in (23cm) **p. 591**
Endemic to W Mexico. Grey head and rump, rufous back distinctive.

3(b) **Grayson's Thrush** *T. (r.?) graysoni* 9.7in (25cm) **p. 591**
Endemic to Nay (mostly Islas Tres Marías). Resembles washed-out 3(a); note greyish olive-brown back, pale cinnamon wash to sides.

4 **White-throated Thrush** *T. assimilis** 9.5in (24cm) **p. 590**
Variable. Notably dark with bright bare parts in humid SE, may thus suggest ♂ 6 from behind. White forecollar distinctive. 4(a) *lygrus* (W Mexico); 4(b) *leucauchen* (humid SE).

5 **Rufous-collared Thrush** *T. rufitorques* 9.5in (24cm) **p. 592**
Endemic to mountains S of Isthmus. ♂ distinctive, ♀ pattern similar but less striking. Some imm ♀♀ dark overall, recalling imm ♂ 6 but lack yellow orbital ring, larger and bulkier, white-streaked undertail coverts distinctive.

6 **Black Thrush** *T. infuscatus* 8.7in (22cm) **p. 589**
Endemic; E and S cloud forest. ♂ distinctive. ♀ from all except 7 by black bill. From 7 by yellow legs, warmer brown plumage, streaked throat, uniform undertail coverts. Imm ♂ greyer than ♀ with yellow bill and orbital ring, often has some black patches in 1st summer.

7 **Mountain Thrush** *T. plebejus differens** 9.5in (24cm) **p. 589**
Cloud forest S of Isthmus. From all except ♀ 6 by black bill; note dark legs, pale-edged undertail coverts.

51

1a

1b

2a

2b

3b

3a

4a

4b

5

♂

♀

imm ♀ 5

♂
6

♀
6

imm ♂
6

7

SW 2-89

PLATE 52: SILKIES AND MIMIDS

1 **Phainopepla** *Phainopepla n. nitens** 7.3in (19cm) **p. 610**
   N and cen Mexico. ♂ unmistakable; ♀ distinctive; note crest.
2 **Grey Silky** *Ptilogonys c. cinereus** 7.7in (20cm) **p. 610**
   Endemic, mountains S to Guatemala. Unmistakable; note bushy crest, white eyering, long tail with white basal patches.

Toxostoma: **Thrashers** Often skulking but sing their usually loud and varied songs from prominent perches, especially early and late in the day.
   3 **Crissal Thrasher** *T. c. crissale** 11.5in (29cm) **p. 603**
      N and cen Mexico. From California Thrasher (not shown, see text) of N BCN by chestnut undertail coverts, greyer overall.
   4 **Grey Thrasher** *T. c. cinereum** 10in (25cm) **p. 601**
      Endemic to Baja. Distinctive in range. Migrant Sage Thrasher (not shown) has short, thrush-like bill.
   5 **Long-billed Thrasher** *T. l. longirostre** 11in (28cm) **p. 600**
      NE Mexico. Brown Thrasher (winter vagrant, see text) has slightly shorter bill with pale flesh base below.
   6 **Ocellated Thrasher** *T. ocellatum* 11in (28cm) **p. 602**
      Endemic to E cen and SW Mexico. Distinctive; note pale supercilium, dark eyes, bold black chest spots.
   7 **Curve-billed Thrasher** *T. curvirostre** 10.5in (27cm) **p. 602**
      N of Isthmus. Distinctive; note orange eyes. Dusky mottling and spotting on chest less distinct on Pacific Slope, more distinct in interior. 7(a) *palmeri* (N Son); 7(b) *curvirostre* (cen Plateau).

8 **Black Catbird** *Dumetella g. glabrirostris** 7.7in (20cm) **p. 596**
   Endemic to Yuc Pen, including Isla Cozumel. Unmistakable; size and shape as Grey Catbird but glossy black. From other all-black birds by slender bill.

9 **Tropical Mockingbird** *Mimus gilvus gracilis** 9.5in (24cm) **p. 598**
   S of Isthmus. From 10 by wing and tail pattern; note lack of large white wing flashes.
10 **Northern Mockingbird** *M. polyglottus leucopterus* 9.5in (24cm) **p. 598**
   N of Isthmus. From 9 by wing and tail pattern.
11 **Blue Mockingbird** *Melanotis c. caerulescens** 10in (25cm) **p. 597**
   Endemic N of Isthmus, mostly highlands (also lowlands in W Mexico). Distinctive; dull blue with black mask, red eyes often hard to see.
12 **Blue-and-white Mockingbird** *M. hypoleucus* 10.5in (27cm) **p. 597**
   Endemic, counterpart of 11 S of Isthmus. Unmistakable.

Socorro Mockingbird *Mimodes graysoni* Island endemic. See Plate 69.
Cozumel Thrasher *Toxostoma guttatum* Island endemic. See Plate 69.

52

SW 10-89

PLATE 53: VIREOS

1(a) **Plumbeous Vireo** *Vireo (solitarius?) pinicolus** 5.5in (14cm) **p. 619**
Widespread. Overall grey and white N of Isthmus, residents S of Isthmus much like
1(b) in color (see text).

1(b) **Cassin's Vireo** *V. (s.?) cassini** 5.2in (13cm) **p. 619**
Widespread migrant (breeds Baja). White spectacles broken in front of eye by dark
lores. Head greyish, upperparts olive, flanks washed yellow-olive (see text).

2 **Bell's Vireo** *V. bellii medius** 4.5in (11cm) **p. 616**
Whitish supraloral stripe and dark lores, bill often flesh below, upper wingbar often
indistinct. Baja breeders (*pusillus*) greyer above.

3 **Hutton's Vireo** *V. huttoni mexicanus** 4.7in (12cm) **p. 620**
Pale lores and whitish eyering broken above eye. Head and upperparts olive,
underparts paler.

4 **Dwarf Vireo** *V. nelsoni* 4.2in (11cm) **p. 616**
Endemic to SW Mexico. Kinglet-like. Whitish lores and eyering broken over eye. From
♀ 5 by less contrast between head and back, duller lores, whiter wingbars. Duller when
worn (4(a)).

5 **Black-capped Vireo** *V. atricapillus* 4.2in (11cm) **p. 616**
Breeds NE Mexico, winters W Mexico. Bold white lores and eyering broken over eye.
♂ unmistakable; ♀ (head color of imm ♀ very like ♂ 4, see text) from 4 by grey head
contrasting more with back, creamy to pale lemon wingbars.

6 **Golden Vireo** *V. h. hypochryseus** 5in (13cm) **p. 620**
Endemic to W Mexico. Broad yellow supercilium, olive-green upperparts, yellow
underparts.

7 **Mangrove Vireo** *V. pallens semiflavus** 4.5in (11cm) **p. 615**
Pacific coast and Yuc Pen. Distinctive; note pale lemon lores. Some birds greyer above,
paler below than shown. See Plate 71.

8 **White-eyed Vireo** *V. griseus perquisitor** 4.5in (11cm) **p. 614**
Yellow spectacles, whitish eyes. Most ssp have contrasting greyer head, pale grey
throat, whitish underparts, yellow sides.

9 **Slaty Vireo** *V. brevipennis* 4.6in (12cm) **p. 614**
Endemic to SW Mexico. Unmistakable and striking. Imm may have dull eyes, head and
upperparts olive.

10 **Warbling Vireo** *V. gilvus swainsonii** 5in (13cm) **p. 621**
Widespread. From 11 by greyish crown and upperparts. No wingbars. E migrants in
particular can have greyish-olive back, lemon sides and undertail coverts; compare
with 12.

11 **Brown-capped Vireo** *V. leucophrys amauranotus** 5in (13cm) **p. 622**
E and S cloud forest. From 10 by brownish crown and upperparts. No wingbars.

12 **Yellow-green Vireo** *V. f. flavoviridis** 5.7in (14cm) **p. 622**
Summer resident, both slopes. Pale grey supercilium with narrow dark edging above,
sides and undertail coverts lemon.

13 **Yucatan Vireo** *V. m. magister* 6in (15cm) **p. 624**
E QR and Caribbean islands. Broad whitish supercilium and bold dark eyestripe, large bill.

14 **Rufous-browed Peppershrike** *Cyclarhis gujanensis flaviventris** 6.3in (16cm) **p. 626**
E and S. Distinctive; note very stout bill. See Plate 69.

15 **Green Shrike-Vireo** *Vireolanius p. pulchellus** 5.5in (14cm) **p. 626**
SE rain forest canopy. Note yellow throat, stout grey bill. Compare with Blue-crowned
Chlorophonia (Plate 56).

16 **Chestnut-sided Shrike-Vireo** *V. m. melitophrys** 6.7in (17cm) **p. 625**
Endemic, mountains from cen Mexico to Guatemala. Unmistakable; ♀ duller.

17 **Tawny-crowned Greenlet** *Hylophilus o. ochraceiceps* 4.8in (12cm) **p. 624**
SE rain forest understory. Distinctive; bushy tawny crown, greyish face and throat,
ochre chest, pale eyes.

18 **Lesser Greenlet** *H. d. decurtatus* 4.2in (11cm) **p. 625**
SE humid lowland forest. Grey head with whitish eyering, greenish-olive upperparts.

Cozumel Vireo *Vireo bairdi* Island endemic. See Plate 69.

53

1a

1b

2

♂

3

4a

♀

5

4 ♂

8

6

7

imm

10

14

9

11

15

12

juv

13

♀

16

♂

17

18

SW 3-90

PLATE 54: YELLOW WARBLERS

1 **Mangrove [Yellow] Warbler** *Dendroica (petechia?) erithachorides bryanti** 4.7in (12cm) **p. 633**
Mangroves. ♂ chestnut head. ♀ greyish nape, chestnut-washed crown. 1st basic ♀ can be dull and washed-out.

***Geothlypis*: Yellowthroats** Attain their greatest diversity in Mexico. Species-level taxonomy still is unclear. Note head patterns, range.

2 **Altamira Yellowthroat** *G. (trichas?) flavovelata* 4.8in (12cm) **p. 650**
Endemic to NE Mexico. ♂ broad yellow supercilium, all-yellow underparts. ♀ bright yellow face and underparts; compare with 4(a).

3 **Belding's Yellowthroat** *G. beldingi** 5.5in (14cm) **p. 651**
Endemic to BCS. Larger than 4. 3(a) *beldingi* (S BCS): ♂ yellow supercilium, golden-yellow underparts. ♀ buffy-yellow supercilium. 3(b) *goldmani* (cen BCS); ♂ supercilium pale lemon to whitish, underparts paler.

4 **Common Yellowthroat** *G. trichas** 4.8in (12cm) **p. 649**
Widespread, variable. 4(a) *occidentalis* (widespread in winter): ♂ whitish supercilium, yellow throat and chest. ♀ yellow mostly on throat, chest, and undertail coverts; imm ♀ can be very washed-out with little yellow.
4(b) *chapalensis* (W cen volcanic belt): ♂ broad pale lemon supercilium, mostly yellow underparts. 4(c) *melanops* (cen and E cen volcanic belt): ♂ broad white supercilium, mostly yellow underparts. ♀ mostly yellow underparts. Imm ♂ (all ssp) mask mottled blackish, supercilium duller.

5 **Black-polled Yellowthroat** *G. s. speciosa** 5.2in (13cm) **p. 652**
Endemic to marshes of cen volcanic belt. ♂ head extensively black, legs dark. ♀ duskier overall than 4, legs dark; note voice. Imm ♂ black on head less extensive than ♂.

6 **Hooded Yellowthroat** *G. n. nelsoni** 5in (13cm) **p. 651**
Endemic to highlands of E and S Mexico, montane scrub. Duskier overall than 4. ♂ grey supercilium, yellow underparts with olive flanks. ♀ indistinct greyish supercilium, dusky olive flanks. Note voice.

7 **Grey-crowned Yellowthroat** *Chamaethlypis p. poliocephala** 5.5in (14cm) **p. 652**
Both slopes, grassy areas. Stout bill flesh with dark culmen. Grey crown, black lores, white eye-crescents (reduced or lacking E of Sula Valley). ♂ generally brighter than ♀. Imm duller, head washed olive. Note voice.

Basileuterus A genus of neotropical warblers. Sexes similar.

8 **Rufous-capped Warbler** *B. rufifrons** 5in (13cm) **p. 657**
Endemic; both slopes. 8(a) *jouyi* (E Mexico; W Mexico similar): rufous cap, white supercilium, yellow bib, greyish belly, long cocked tail; 8(b) *salvini* (humid SE): underparts mostly yellow.

9 **Chestnut-capped Warbler** *B. (r.?) d. delattrii* 5in (13cm) **p. 658**
SE Chis and C.A. Head rufous with dark lores, white supercilium, underparts yellow.

10 **Golden-crowned Warbler** *B. c. culicivorus** 5in (13cm) **p. 657**
Both slopes, humid forest. Indistinctly striped face with blackish lateral crown stripes; greyish upperparts.

11 **Golden-browed Warbler** *B. b. belli** 5in (13cm) **p. 658**
Endemic; humid montane forests. Head rufous with yellow supercilium.

12 **Fan-tailed Warbler** *B. lachrymosa* 6in (15cm) **p. 656**
Endemic; both slopes, local in rocky, wooded areas. Distinctive. Note long, expressive, white-tipped tail, white face spots, orange wash on chest.

54

1 ♂

imm
1

♀
imm

♂

imm ♂

2 ♂

♀

♀ 4a

3a ♂

3b ♂

4b ♂

♂ 4a

imm ♂

4c

♀ 4c

5 ♀

♂

♂

7

♀
imm

♀ 6

♂

imm

♂

♂

8a

8b

10

9

12

11

SW 9-90

PLATE 55: WARBLERS, BANANAQUIT, FLOWERPIERCER

1 **Olive Warbler** *Peucedramus taeniatus giraudi** 5in (13cm) **p. 660**
Highland pines. ♂ orange head with black mask. ♀ head and chest yellowish with dusky mask and darker cheek patch. Note forked tail.

2 **Grace's Warbler** *Dendroica graciae remota** 4.7in (12cm) **p. 640**
Widespread; pines. Short yellow supercilium and subocular crescent, yellow throat and chest, blue-grey upperparts. ♀ duller.

3 **Colima Warbler** *Vermivora crissalis* 5in (13cm) **p. 629**
Highlands; breeds chaparral of NE Mexico, winters W Mexico. Grey-brown overall with white eyering, orchre-yellow undertail coverts.

4 **Tropical Parula** *Parula pitiayumi nigrilora** 4.2in (11cm) **p. 632**
Both slopes, often local. Blue-grey above with olive back patch, white wingbars; underparts yellow. ♂ has black mask, orange chest. Lacks white eye-crescents of Northern Parula. See Plates 69, 71.

5 **Crescent-chested Warbler** *Vermivora superciliosa mexicana** 4.5in (11cm) **p. 631**
Endemic; highland oak and pine–oak. White supercilium, chest spot, no wingbars.

6 **Painted Redstart** *Myioborus p. pictus** 5.2in (13cm) **p. 655**
Highland oak and pine–oak. Unmistakable; note bold white wing panel.

7 **Slate-throated Redstart** *M. m. miniatus** 5.3in (13cm) **p. 655**
Humid highland forest. Slaty grey upperparts, red belly, expressive tail. See Plate 71.

8 **Red-faced Warbler** *Cardellina rubrifrons* 5in (13cm) **p. 654**
Highland pine–oak. Unmistakable; note face pattern.

9 **Red Warbler** *Ergaticus ruber** 5.2in (13cm) **p. 654**
Endemic to humid montane forests N of Isthmus. Unmistakable. 9(a) *ruber* (cen and SW Mexico); 9(b) *melanauris* (NW Mexico).

10 **Pink-headed Warbler** *E. versicolor* 5.2in (13cm) **p. 655**
Endemic to humid montane forests of Chis and Guatemala. Unmistakable.

11 **Cinnamon-bellied Flowerpiercer** *Diglossa baritula montana** 4.5in (11cm) **p. 706**
Endemic; highland flower banks. Note hooked bill. ♂ unmistakable. ♀ greyish olive above, dusky cinnamon below. Juv greyish olive overall, paler below, dull cinnamon wingbar.

12 **Bananaquit** *Coereba flaveola** 4in (10cm) **p. 662**
Humid SE lowlands. Distinctive; note short, decurved bill, white supercilium and wing check. 12(a) *mexicana* (SE); 12(b) *caboti* (NE QR and Isla Cozumel).

Granatellus Two allopatric species; note stout bills, expressive white-tipped tails often fanned.

13 **Red-breasted Chat** *G. v. venustus** 6in (15cm) **p. 659**
Endemic to W Mexico, thorn forest. ♂ unmistakable. ♀ broad buff supercilium on plain face, flashy tail. Imm ♂ head and chest have traces of ♂ pattern.

14 **Grey-throated Chat** *G. sallaei boucardi** 5.2in (13cm) **p. 660**
Endemic to Yuc Pen and adjacent SE lowlands. ♂ unmistakable. ♀ broad buff supercilium on plain face. Imm like adult but wings browner.

55

SW 2-91

PLATE 56: EUPHONIAS, HONEYCREEPERS, TANGARAS

Euphonias Small, stubby tanagers, usually in pairs or small groups; species associate readily with one another at fruiting trees and shrubs.

1 **Scrub Euphonia** *E. affinis** 4in (10cm) **p. 666**
Both slopes. 1(a) *godmani* (W Mexico): note white undertail coverts; 1(b) *affinis* (E Mexico and C.A.): ♂ from 2 by black throat, from smaller 4 by yellow undertail coverts; yellow chest less orangish than 2 or 4. ♀ fairly uniform, dull yellow-olive underparts. Juv like ♀ but ♂ has blue-black remiges.

2 **Yellow-throated Euphonia** *E. h. hirundinacea** 4.3in (11cm) **p. 667**
E and S. ♂ distinctive; note yellow throat. ♀ pale grey throat and median underparts, yellow flanks and undertail coverts. Imm ♂ upperparts olive.

3 **Blue-hooded Euphonia** *E. e. elegantissima** 4.3in (11cm) **p. 667**
Widespread (mostly foothills, highlands), usually near mistletoe. ♂ unmistakable. ♀ turquoise hood. Juv blue hood restricted to crown (♂ has cinnamon chin, blue-black flight feathers).

4 **White-vented Euphonia** *E. minuta humilis* 3.8in (10cm) **p. 668**
SE humid lowlands, not Mexico (?). From larger 1(b) by white undertail coverts, orange-yellow chest. ♀ has pale grey throat, yellow chest and flanks, white undertail coverts. 1(a) allopatric.

5 **Olive-backed Euphonia** *E. g. gouldi* 4.2in (11cm) **p. 667**
SE rain forest, edge. Distinctive. ♂ yellow forehead, chestnut belly to undertail coverts. ♀ forehead and undertail coverts chestnut.

6 **Blue-crowned Chlorophonia** *Chlorophonia occipitalis* 5in (13cm) **p. 665**
Endemic; SE cloud forest. Note stocky euphonia shape, stubby bill. ♂ yellow underparts below dark chest bar. ♀ yellow belly and undertail coverts. Ventriloquial, cooing *hoo* call. Compare with Green Shrike-Vireo (Plate 53).

Honeycreepers Colorful tanagers with slender, decurved bills.

7 **Green Honeycreeper** *Chlorophanes spiza guatemalensis* 5.5in (14cm) **p. 664**
SE humid forest. ♂ unmistakable; note unique blue-green color. ♀ plain green, bill pale yellow below. Compare with ♀ 8.

8 **Red-legged Honeycreeper** *Cyanerpes cyaneus carneipes* 4.5in (11cm) **p. 665**
E and S lowlands. Tail longer, bill stouter and less decurved than 9. Alt ♂ from 9 by bright red legs, turquoise crown, black back, lacks black bib. ♀ more uniform than 9, legs reddish, lacks blue malar stripe and blue-streaked necklace. Basic ♂ and imm ♂ like ♀ but wings and tail black. Molting ♂ (8(a)) blotchy appearance; compare with imm ♂ 9.

9 **Shining Honeycreeper** *C. l. lucidus* 4.3in (11cm) **p. 664**
SE humid forest. From 8 by shorter tail, finer, more decurved bill, waxy yellow legs, etc.

10 **Golden-hooded Tanager** *Tangara l. larvata** 5.2in (13cm) **p. 663**
SE rain forest and edge. Adult unmistakable. Juv has hint of adult colors.

11 **Cabanis' Tanager** *T. cabanisi* 5.8in (15cm) **p. 663**
Endemic to humid evergreen forest (1000–1700 m) on Pacific Slope of Chis and Guatemala, local. Unmistakable.

56

1a ♀
♂
1b ♀
♂
♀
2 ♀
♂
8a
♀
8
♂ alt

♀
3 juv ♀
♂
♀ 4
♂
juv
imm ♂

imm ♂
9 ♀
♂

7 ♀
♂

♀ 5
♂
♀ 6
♂
juv
10
11

SW 4-99

PLATE 57: TROPICAL TANAGERS

1 **Common Bush-Tanager** *Chlorospingus ophthalmicus** 5.5in (14cm) **p. 678**
E and S cloud forest. Distinctive but variable; note tanager bill, white postocular
mark. 1(a) *albifrons* (Sierra Madre del Sur of Gro and Oax); 1(b) *postocularis* (S
Chis and Guatemala); 1(c) *ophthalmicus* (E Mexico).

2 **Blue-grey Tanager** *Thraupis episcopus cana* 6.5in (17cm) **p. 668**
SE lowlands. Open country, towns. Unmistakable.

3 **Yellow-winged Tanager** *T. abbas* 7.2in (18cm) **p. 669**
SE humid lowlands. Distinctive; note yellow wing check.

4 **Red-crowned Ant-Tanager** *Habia rubica** 7.4in (19cm) **p. 671**
Both slopes. Very similar to 5, often most easily separated by higher, squeaky,
chattering calls. ♂ from 5 by lack of dark lores, bill averages smaller, red in crown
often more conspicuous and bordered black. ♀ more olive-brown than 5 without
strongly contrasting buff throat, smaller bill pale below. 4(a) *rubicoides* (SE); 4(b)
rosea (W Mexico).

5 **Red-throated Ant-Tanager** *H. fuscicauda** 7.8in (20cm) **p. 671**
E and S lowlands. Very similar to 4 but note lower, rasping, scolding calls. ♂ from
4 by dark lores, larger bill. ♀ more rufous-brown than 4 with contrasting buff
throat, black bill. 5(a) *salvini* (SE Mexico to C.A.); 5(b) *peninsularis* (N Yuc Pen).

6 **Black-throated Shrike-Tanager** *Lanio aurantius* 8in (20cm) **p. 670**
Endemic; SE rain forest, arboreal. ♂ unmistakable, from orioles by habits, stout
hooked bill. ♀ distinctive; note tawny-brown upperparts brightest on rump, grey
head, stout bill; compare with 7.

7 **Grey-headed Tanager** *Eucometis penicillata pallida* 6.7in (17cm) **p. 670**
SE rain forest understory. Distinctive; note clean-cut, bushy grey head, bright
olive-green upperparts, bright yellow underparts. Compare with ♀ 6.

8 **Scarlet-rumped Tanager** *Ramphocelus p. passerinii* 7in (18cm) **p. 677**
SE humid second growth. ♂ unmistakable. ♀ distinctive; note bluish-white bill,
bright tawny-ochre rump, greyish head, ochre chest.

9 **Crimson-collared Tanager** *Phlogothraupis s. sanguinolenta* 7.7in (20cm) **p. 676**
SE humid forest edge. Unmistakable.

10 **Rosy Thrush-Tanager** *Rhodinocichla rosea schistacea* 8.2in (21cm) **p. 677**
W Mexico, thorn forest. Unmistakable; skulking, mimid-like behaviour and song.

Stripe-headed Tanager *Spindalis zena* Island endemic. See Plate 69.

57

SW 1-90

PLATE 58: TEMPERATE TANAGERS AND YELLOW GROSBEAK

1 **Western Tanager** *Piranga ludoviciana* 6.8in (17cm) **p. 674**
Basic ♂ yellow head and rump, red-tinged face, lemon and white wingbars (alt ♂
brighter red face). ♀ greyish to olive above, dirty pale lemon to yellow below, two
whitish wingbars.

2 **Summer Tanager** *P. rubra cooperi** 7.2in (18cm) **p. 673**
♂ bright red overall, bill horn. ♀ yellow-olive overall, underparts paler and
yellower with ochre wash strongest on undertail coverts. Imm ♂ variably mottled
red.

3 **Flame-colored Tanager** *P. bidentata** 7.3in (18cm) **p. 675**
Note grey bill, bold white wingbars, tertial tips, and tail corners. ♀ suggests larger
8 but wingbars bolder, lacks gross bill. 3(a) *bidentata* (W Mexico); 3(b)
sanguinolenta (E Mexico and C.A.).

4 **Hepatic Tanager** *P. flava** 7.7in (20cm) **p. 673**
Note stout dark bill. 4(a) *albifacies* (N C.A.): ♂ dull red, ♀ olive overall; 4(b)
hepatica (W Mexico): ♂ dusky reddish, brightest on forecrown and throat. ♀
dusky cheeks, brighter yellowish forehead and throat, from smaller 5 by stout bill,
voice.

5 **Red-headed Tanager** *Spermagra erythrocephala* 5.5in (14cm) **p. 676**
Endemic to mountains of W Mexico. ♂ unmistakable. ♀ from larger 4(b) by
stubbier blackish bill, paler belly.

6 **White-winged Tanager** *S. l. leucoptera* 5.5in (14cm) **p. 675**
Distinctive (imm ♂ flame-red); note dark wings with neat white wingbars.

7 **Rose-throated Tanager** *Piranga r. roseogularis** 6.3in (16cm) **p. 672**
Endemic to Yuc Pen. Distinctive; grey overall with broken white eyering. ♂ rosy
throat, dull reddish crown, wings, and tail. ♀ pink and red of ♂ replaced by dull
yellowish.

8 **Yellow Grosbeak** *Pheucticus chrysopeplus** 9in (23cm) **p. 682**
Note massive bill. 8(a) *chrysopeplus* (W Mexico): ♂ head and underparts yellow,
black wings and tail boldly marked white. ♀ crown and upperparts olive, streaked
darker, auriculars dusky, wings and tail duller; compare with 3. Imm ♂ appears
intermediate between ♂ and ♀. 8(b) *aurantiacus* (Chis and Guatemala): ♂ head
and underparts golden orange.

♂ basic

1

♀

2

imm ♂

♀

♂

4a

3a

♀

♂

3b

♂

imm ♂

4b

♂

5

♀

♀

♂

8a

♀

6

♂

♂

8b

♀

7

♂

SW 9-90

PLATE 59: FINCHES AND GROSBEAKS

1 **Lesser Goldfinch** *Carduelis psaltria colombianus** 4.3in (11cm) **p. 760**
Widespread. ♂ unmistakable (in NW Mexico has black cap, olive upperparts).
♀ note white wing check.

2 **Black-headed Siskin** *C. notata forreri** 4.4in (11cm). **p. 760**
Endemic; highland oak and pine–oak. Adult unmistakable (♀ duller); note black
hood, bold yellow wing patch. Imm: note pointed bill, yellow wing patch.

3 **Black-capped Siskin** *C. atriceps* 4.6in (12cm) **p. 759**
Endemic to highlands of Chis and Guatemala. Adult distinctive (♀ duller); note
pointed pink bill. Juv recalls 4; note dark cap, bill.

4 **Pine Siskin** *C. pinus perplexus** 4.6in (12cm) **p. 758**
Highlands S to Guatemala. N of Isthmus overall streaky (distinct dark streaking
below; see text). S of Isthmus (shown) much duller, diffuse dusky streaking below.

5 **Grassland Yellow-Finch** *Sicalis luteola chrysops** 4.3in (11cm) **p. 707**
Savannas, local. ♂ distinctive, bright yellow rump obvious in flight. ♀ and juv
duller; note stubby bill, face pattern, streaked back.

6 **Greyish Saltator** *Saltator coerulescens** 9in (23cm) **p. 679**
Widespread. Distinctive; note greyish head and upperparts; paler overall in W
Mexico. Imm washed olive overall, supercilium lemon. 6(a) *vigorsii* (NW
Mexico); 6(b) *grandis* (E Mexico to C.A.).

7 **Buff-throated Saltator** *S. maximus** 8.5in (22cm) **p. 680**
SE humid lowlands. Smaller and quieter than 8; note buff throat. 7(a) *magnoides*
(S from Isthmus); 7(b) *gigantoides* (N from Isthmus).

8 **Black-headed Saltator** *S. atriceps** 10.5in (27cm) **p. 680**
E and S. Large and noisy, also from 7 by darker head, white throat in most
populations. 8(a) *atriceps* (E Mexico to C.A.); 8(b) *suffuscus* (Los Tuxtlas).

9 **Crimson-collared Grosbeak** *Rhodothraupis celaeno* 8.5in (22cm) **p. 681**
Endemic to NE Mexico. Unmistakable; note stubby, thick bill. Black on head of
♀/imm varies with age/sex (see text).

10 **Hooded Grosbeak** *Coccothraustes a. abeillei** 7in (18cm) **p. 762**
Endemic; highlands S to Guatemala. ♂ unmistakable, ♀ from Evening Grosbeak
(see text) by black cap, larger pale wing panel.

11 **Black-faced Grosbeak** *Caryothraustes p. poliogaster* 7.2in (18cm) **p. 680**
SE rain forest canopy. Distinctive; olive-yellow and blue-grey with black face,
often in flocks.

59

1 ♂
2 ♂
3 ♂
juv
♂ 5
♀
imm
3 juv
4
♀

6a
imm ♀
b
mm
♂
9
♀
10
7b
♂
8a
♀
8b
11

SW 2-90

PLATE 60: BUNTINGS AND BAMBOO SEEDEATERS

1 **Orange-breasted Bunting** *Passerina l. leclancherii** 4.7in (12cm) **p. 688**
Endemic to SW Mexico. Most plumages unmistakable. Dull imm from 2 by
yellowish lores, less uniformly greenish plumage (underparts yellower).

2 **Painted Bunting** *P. ciris pallidior** 5in (13cm) **p. 689**
♂ unmistakable (see text). ♀ and imm green, paler below, with pale eyering,
compare with 1.

3 **Rosita's Bunting** *P. rositae* 5.5in (14cm) **p. 686**
Endemic to Pacific Slope of Isthmus. ♂ unmistakable; other plumages distinctive;
note white eye-crescents.

4 **Varied Bunting** *P. versicolor** 4.7in (12cm) **p. 688**
Widespread N of Isthmus, also interior valleys of Chis and Guatemala.
♂ distinctive (often looks dark), ♀ variably dull. From 5 by smaller bill and paler,
often bluish, wing and tail edgings, call. From 7 and 8 by dusky-flesh lower
mandible (but bill may be blackish in 1st year ♂), paler wing edgings (often
bluish) and undertail coverts, call. 4(a) *versicolor* (E and cen Mexico); 4(b)
dickeyae (NW Mexico). Compare with ♀/imm Indigo Bunting (see text).

5 **Blue Bunting** *Cyanocompsa parellina** 5.3in (13cm) **p. 685**
Endemic. ♂ distinctive, from 7 and 8 by bright bluish areas in face, shoulder,
rump, heavier bill. ♀ from 7 and 8 by heavier bill (dusky flesh below in W
Mexico), call. 5(a) *indigotica* (W Mexico); 5(b) *parellina* (SE Mexico).

6 **Blue-black Grosbeak** *C. cyanoides concreta* 7in (18cm) **p. 684**
Humid SE lowland forest edge. Distinctive; note very stout blackish bill.
♂ uniform blue-black, ♀ uniform rich brown.

7 **Slate-blue Seedeater** *Amaurospiza (concolor?) relicta* 5.3in (13cm) **p. 704**
Endemic to SW Mexico. Bamboo. 8 allopatric. ♂ uniform dull slaty blue.
♀ uniform cinnamon-brown, from 4 by dark grey bill, more uniform plumage,
call. From more thickset 5 by smaller dark grey bill, call. Imm ♂ like ♀ but often
has dark slaty face and throat in 1st summer.

8 **Blue Seedeater** *A. c. concolor* 5in (13cm) **p. 704**
SE; mostly foothills, highlands. Bamboo. 7 allopatric. ♂ uniform dull dark bluish,
compare with 5. ♀ uniform cinnamon-brown, from 4 by dark grey bill, more
uniform plumage, call. From more thickset 5 by smaller dark grey bill, call.

9 **Slaty Finch** *Haplospiza rustica uniformis** 5.3in (13cm) **p. 706**
E and S highlands; bamboo. Rare, little known. Note longish, pointed bill with
straight culmen. ♂ uniformly slaty. ♀ brownish, throat and chest paler with dark
streaking. Imm ♂ intermediate in appearance between ♂ and ♀.

imm ♀

1

♀

imm ♂

imm ♂

3

♂

♀

2 ♀

4a

♀

4b

♀

♂

3 imm ♀

7

♂

5b

♀

5a ♀

♂

♀

8 ♂

♀ 8

6

♂

♂

imm ♂

9

♀

SW 10-90

PLATE 61: TOWHEES, BRUSHFINCHES, etc.

1 **California Towhee** *Pipilo crissalis** 8.2in (21cm) **p. 699**
Baja. Distinctive in range (2 allopatric). 1(a) *senicola* (NW BCN); 1(b) *albigula* (BCS). Compare with Abert's Towhee (NE BCN, see text).

2 **Canyon Towhee** *P. fuscus** 8.2in (21cm) **p. 699**
NW Mexico and Plateau (1 allopatric). From 3 (sympatric in NW Oax) by pale brownish throat, dark chest spot, lack of white wingbars. 2(a) *mesoleucus* (N Plateau); 2(b) *fuscus* (cen Mexico).

3 **White-throated Towhee** *P. a. albicollis** 8in (20cm) **p. 700**
Endemic to interior SW Mexico. Distinctive: note white throat with cinnamon band, narrow whitish wingbars, compare with 2.

4 **Rufous-sided Towhee** *P. erythrophthalmus macronyx** 8.3in (21cm) **p. 698**
Marked geographic variation (see text). All ssp have dark head and chest, rufous sides. Birds in cen volcanic belt have olive upperparts, elsewhere upperparts blackish to dark olive-grey with whitish markings. Interbreeds locally with 5.

5 **Collared Towhee** *P. ocai alticola** 8.4in (21cm) **p. 697**
Endemic to highlands of cen Mexico. From 10 by pale grey supercilium, broad black chest band, brighter upperparts, voice. Interbreeds locally with 4, hybrids variable.

6 **Olive Sparrow** *Arremonops rufivirgatus** 6in (15cm) **p. 694**
Both slopes. No similar species in most of range. Note brownish head stripes, dull olive back. From 7 (sympatric in Yuc Pen) by dark brown head stripes (often flecked black in front of eyes), duller back, paler and buffier undertail coverts, song. 6(a) *sumichrasti* (W Mexico); 6(b) *verticalis* (Yuc Pen, compare with 7); 6(c) *rufivirgatus* (NE Mexico); 6(d) *crassirostriis* (SE Mexico).

7 **Green-backed Sparrow.** *A. c. chloronotus** 6.2in (16cm) **p. 695**
Endemic; humid SE forest. From 6 by black head stripes (flecked brown in imm), brighter olive-green back, brighter buffy-lemon undertail coverts, song.

8 **Orange-billed Sparrow** *Arremon aurantiirostris* 6.3in (16cm) **p. 693**
Humid SE forest. Unmistakable; neon-orange bill 'glows' in shady understory.

9 **Green-striped Brushfinch** *Atlapetes v. virenticeps** 7.5in (19cm) **p. 693**
Endemic to mountains of NW and cen Mexico. Black head with green stripes, bright white throat. 10 allopatric.

10 **Chestnut-capped Brushfinch** *A. brunneinucha** 7.3in (19cm) **p. 692**
E and S mountains. Compare with 5. 9 allopatric. 10(a) *apertus* (Los Tuxtlas, 'Plain-breasted Brushfinch'); 10(b) *brunneinucha* (E Mexico).

11 **Rufous-capped Brushfinch** *A. p. pileatus** 6.2in (16cm) **p. 692**
Endemic to mountains N of Isthmus. Distinctive; rufous cap absent in juv.

12 **Yellow-throated Brushfinch** *A. gutteralis griseipectus** 7.3in (19cm) **p. 692**
Mountains of S Chis and C.A. Distinctive. 13 allopatric.

13 **White-naped Brushfinch** *A. albinucha* 7.5in (19cm) **p. 691**
Endemic to mountains of E Mexico, S to N Chis. Distinctive. 12 allopatric.

14 **Rusty-crowned Ground-Sparrow** *Melozone k. kieneri** 6.5in (17cm) **p. 695**
Endemic to W Mexico. Distinctive; note white spectacles, rufous crown.

15 **Prevost's Ground-Sparrow** *M. biarcuatum* 6.5in (17cm) **p. 696**
Endemic; S of Isthmus. Distinctive; note white face, dark 'sideburns'.

16 **White-eared Ground-Sparrow** *M. leucotis occipitalis** 7.3in (19cm) **p. 696**
Pacific Slope S of Isthmus. Distinctive; note striking face pattern. See Plate 71.

61

SW 1-90

PLATE 62: GRASSQUITS AND SEEDEATERS

1   **Blue-black Grassquit** *Volatinia jacarina splendens* 4.2in (11cm) **p. 700**
Both slopes. Greyish pointed bill with straight culmen. Alt ♂ blue-black, white
shoulder tufts concealed. Basic ♂ mottled brown and blue-black. ♀ brown, paler
below, dark-streaked chest. Imm ♂ like ♀ but wings blue-black, edged brown.

2   **Yellow-faced Grassquit** *Tiaris olivacea pusilla** 4.2in (11cm) **p. 705**
E and S, often local. Greyish bill with straight culmen. ♂ distinctive; note face
pattern. ♀ olive; note face pattern. Imm ♀ duller and greyer, face pattern
indistinct.

3   **Ruddy-breasted Seedeater** *Sporophila minuta parva* 3.8in (10cm) **p. 703**
Pacific Slope. Alt ♂ distinctive. Basic ♂ like ♀ but note dull chestnut rump, dusky
cinnamon smudges below. ♀ plain buffy brown (greyer when worn), similar to
4(a); note orangish bill, more contrasting wing edgings, especially tertials. Imm ♂
like ♀; note white wing check.

4   **White-collared Seedeater** *S. torqueola** 4.3in (11cm) **p. 702**
Both slopes, variable. 4(a) *torqueola* (W Mexico): alt ♂ distinctive (back may be
black), ♀ much like 3 (see text). Basic ♂ like ♀ but with white wing check and pale
cinnamon rump. 4(b) *sharpei* (NE Mexico): alt ♂ distinctive but variable. Basic ♂
similar to imm ♂ (fresh plumage shown). ♀ similar to 4(c). Intergrades with 4(c)
from SE Mexico to N C.A. 4(c) *moreletti* (S Mexico and C.A.): alt ♂ distinctive,
rump and flanks often buffy. ♀ (worn plumage shown, browner and buffier when
fresh, like imm ♂ 4(b)); note white wingbars.

5   **Thick-billed Seedfinch** *Oryzoborus funereus* 4.7in (12cm) **p. 704**
Humid SE lowlands. ♂ from 6 by stout, grosbeak-like bill. ♀ distinctive, rich
brown overall; note bill.

6   **Variable Seedeater** *Sporophila aurita corvina* 4.3in (11cm) **p. 702**
Humid SE lowlands. ♂ from 5 by stubby, seedeater bill. ♀ distinctive, olive
overall; note stubby bill.

7   **Slate-colored Seedeater** *S. schistacea* ssp. 4.3in (11cm) **p. 701**
Humid SE lowlands; bamboo. ♂ distinctive; note striking bill. ♀ suggests allopatric
4(a). Note olive-brown upperparts, dirty buff underparts, no wingbars.

♂ basic

imm ♂

♀

♀ imm ♀

alt ♂

1

♂ 2

alt ♂

basic ♂

♂

♂

imm ♂

3

♀ 3

4a

4b

imm ♂

♀

♀

♀

♂

4c

5

♀

♀

♂

6

♂

♀

7

♂

SW 6-89

PLATE 63: SPARROWS I

Juv sparrows are duller than adults and typically have streaked chests; note that all show subdued face pattern suggesting adult. Most quickly attain adult-like plumage, and usually are attended by adults while in juv plumage.

1 **Five-striped Sparrow** *Aimophila quinquestriata septentrionalis** 6in (15cm) **p. 708**
   NW Mexico. Thorn forest. Distinctive, unstreaked. Note white face stripes, black malar stripe, black chest spot.

2 **Rufous-collared Sparrow** *Zonotrichia capensis septentrionalis* 5.7in (14cm) **p. 727**
   Mountains S of Isthmus. Unmistakable in range, often in towns and villages.

3 **Black-chested Sparrow** *Aimophila humeralis* 6.3in (16cm) **p. 708**
   Endemic to SW Mexico. Thorn forest. Unmistakable and striking.

4 **Black-throated Sparrow** *Amphispiza bilineata grisea** 5.5in (14cm) **p. 714**
   N and cen Mexico. Deserts. Distinctive, smooth grey with white supercilium, white-edged black bib.

5 **Stripe-headed Sparrow** *Aimophila ruficauda** 6.8in (17cm) **p. 709**
   Pacific Slope and interior valleys. Thorn forest, brush. Unmistakable in range and habitat; note bold head pattern. 5(a) *acuminata* (W Mexico); 5(b) *lawrencei* (SW Mexico).

6 **Bridled Sparrow** *A. mystacalis* 6.3in (16cm) **p. 707**
   Endemic to interior SW Mexico. Arid scrub. Unmistakable in range; note bold white malar stripe, black throat, rufous shoulders and rump.

7 **Black-chinned Sparrow** *Spizella a. atrogularis** 5.7in (14cm) **p. 718**
   N of Isthmus. Pink bill, grey head and underparts, streaked back. Alt ♂: black chin (reduced in ♀). Basic ♂ lacks black chin.

8 **Baird's Junco** *Junco bairdi* 5.8in (14cm) **p. 730**
   Endemic to mountains of S BCS. Unmistakable in range; note golden eyes.

9 **Yellow-eyed Junco** *J. phaeonotus** 6.2in (16cm) **p. 731**
   Highlands S to Guatemala. Distinctive; note golden-yellow eyes (dusky in juv) offset by dark lores. 9(a) *phaeonotus* ('Mexican Junco', N of Isthmus); 9(b) *alticola* ('Guatemalan Junco', SE Chis and Guatemala). *J. p. fulvescens* ('Chiapas Junco', N Chis) is intermediate in appearance between 9(a) and 9(b), with buffy-brown flanks.

10 **Striped Sparrow** *Oriturus s. superciliosus** 6.8in (17cm) **p. 714**
    Endemic to highlands of NW and cen Mexico. Bunch grass, open pine woods. Large and fat, distinctive in range. Note bold white supercilium, blackish mask, scaly back.

Guadalupe Junco *Junco insularis* Island endemic. See Plate 69.

63

1

juv

2

juv

3

juv

4

juv

5a

juv

5b

6

juv

♂ alt

8

7

basic

juv

9a

9b

10

juv

SW 1-88

PLATE 64: SPARROWS II

See comments about juvenile sparrows on Plate 63.

1 **Cassin's Sparrow** *Aimophila cassinii* 5.7in (14cm) **p. 711**
N and cen Mexico. Desert grassland. From 4 by straightish culmen, plainer face, upperparts with dark scallops, outer rectrices tipped whitish, streaked flanks often covered by wings. Note song. 1(a) Rufous morph. Juv back scalier, chest streaked.

2 **Worthen's Sparrow** *Spizella w. wortheni** 5.3in (13cm) **p. 718**
Endemic to Mexican Plateau; local, little known. Grasslands with scattered shrubs. Distinctive; note pink bill, grey forehead, rufous cap, white eyering, grey rump. 2(a) Worn plumage (summer): face greyer, wingbars paler.

3 **Rufous-winged Sparrow** *Aimophila c. carpalis** 5.6in (14cm) **p. 712**
NW Mexico. Thorn forest. Bicolored bill, grey face with rufous crown; also note rufous shoulders, whitish wingbars. Juv lacks rufous.

4 **Botteri's Sparrow** *A. botterii** 5.7in (14cm) **p. 710**
Widespread. N birds (such as 4(a), *texanus*) much like 1; note decurved culmen, streaked upperparts, unstreaked flanks, song. 4(b) *vulcanica* (N C.A.): darker and greyer. Juv face washed buffy lemon.

5 **Sierra Madre Sparrow** *Ammodramus baileyi sierrae** 4.8in (12cm) **p. 724**
Endemic to mountains of NW and cen Mexico; very local, endangered. Bunch grass. Note grey supercilium, dark eyestripe and malar stripe, extensively rufous upperparts, whitish underparts streaked black, often forming chest spot. Compare with Song Sparrow (see text).

6 **Oaxaca Sparrow** *Aimophila notosticta* 6.3in (16cm) **p. 713**
Endemic to interior cen Oaxaca. Arid scrub. Suggests 9 but allopatric (?); note smaller black bill, white eyering boldly offset by black lores, sharp black back streaks, brown tail. Compare with sympatric 7(a) (smaller, duller face, greyish bill, etc.).

7 **Rufous-crowned Sparrow** *A. ruficeps** 5.7in (14cm) **p. 712**
N of Isthmus. Scrub on rocky slopes. Suggests small plain 6 or 9; note small greyish bill, fairly plain face with dark lores, whitish eyering. 7(a) *australis* (SW interior); 7(b) *sororia* (S Baja).

8 **Sumichrast's Sparrow** *A. sumichrasti* 6.2in (16cm) **p. 710**
Endemic to Pacific Slope in Isthmus. Thorn forest. No similar species in limited range; note face pattern.

9 **Rusty Sparrow** *A. rufescens** 6.8in (17cm) **p. 713**
Widespread, variable. Large and bulky with proportionately large bill, black above, grey below. 9(a) *pyrgatoides* (SE Mexico and C.A.): from allopatric (?) 6 by large bill, duller face pattern, rufous tail; 9(b) *mcleodii* (NW Mexico): from smaller 7 by large bill, duller back markings, brighter tertial edgings. Juv face washed buffy lemon.

64

SW 3-91

PLATE 65: ICTERIDS

1 **Red-winged Blackbird** *Agelaius phoeniceus gubernator** 9in (23cm) **p. 735**
Widespread, marshes. ♀ 'Bicolored Blackbird' (cen Mexico): uniform dark brown, throat flecked paler. Other ssp have obvious pale streaking. ♂ black with red shoulder; red bordered creamy yellow in most ssp.

2 **Shiny Cowbird** *Molothrus bonariensis* ssp. 7in (18cm) **p. 742**
(see NB under Bronzed Cowbird). Unrecorded in region but rapidly expanding through Caribbean, should be looked for in Yuc Pen. Bill more slender and pointed than 3 or 4. ♂ eyes dark, lacks ruff. ♀ from 4 by bill shape.

3 **Bronzed Cowbird** *M. aeneus** 8in (20cm) **p. 741**
Widespread. Forehead slopes into stout bill (compare with smaller 4), eyes red. ♂ much smaller than 7 with shorter bill etc. 3(a) ♀ *aeneus* (E Mexico and C.A.): duller and browner than ♂. 3(b) ♀ *assimilis* (W Mexico): from smaller 4 by stouter bill, plainer face, red eyes.

4 **Brown-headed Cowbird** *M. ater obscurus** 6.8in (17cm) **p. 742**
Widespread N of Isthmus. Smaller bill than 3; compare with 2. ♂ glossy black with brown head.

5 **Yellow-billed Cacique** *Amblycercus h. holosericeus** 9.2in (23cm) **p. 753**
E and S tropical lowlands; skulking. Distinctive, heard far more often than seen (see text). Note pale eyes and bill.

6 **Melodious Blackbird** *Dives dives* 10.3in (26cm) **p. 738**
E and S; conspicuous and noisy. All black with dark eyes, stout pointed bill; compare with 3, 7.

7 **Giant Cowbird** *Scaphidura oryzivora impacifica* ♂ 13in (33cm), ♀ 11.5in (29cm) **p. 742**
SE lowlands. Large; note stout, pointed bill, long primary projection, red eyes. ♂ neck ruff accentuates small head; compare with 3, 6. In flap-flap-flap-glide flight, from 8 by broader-based, more pointed wings, shorter squared tail.

8 **Great-tailed Grackle** *Quiscalus m. mexicanus** ♂ 16in (41cm) ♀ 11.5in (29cm) **p. 740**
Widespread, common and conspicuous. Note long, keel-shaped tail, fairly slender, pointed bill, pale eyes (dusky in juv). ♀ in NW Mexico smaller and paler below. Flies with fairly steady flapping; compare with 7.

9 **Yellow-winged Cacique** *Cacicus melanicterus* 12.5in (32cm) **p. 753**
Endemic to Pacific Slope. Unmistakable, noisy and conspicuous.

10 **Montezuma Oropendola** *Psarocolius montezuma* ♂ 19in (48cm), ♀ 15.5in (39cm) **p. 754**
SE lowlands. Large, ♂ much larger than ♀. Unmistakable. Shape and flight may suggest Brown Jay.

11 **Chestnut-headed Oropendola** *P. w. wagleri* ♂ 13.5in (34cm), ♀ 10.2in (26cm) **p. 754**
SE lowlands. Distinctive, ♂ much larger than ♀. Flight hurried relative to 10, with deep wingbeats; note pale bill, narrower wings with more emarginated primaries.

65

SW 4-92

PLATE 66: YELLOW ORIOLES

Icterus **New World Orioles** (Plates 66–68) Attractive, brightly colored icterids. Identification problems are common and are compounded by marked age/sex and geographic variation in many species. Bill shape and size often are important for species identification. See text for further details on age/sex and ssp differences not shown on the plates.

1 **Black-vented Oriole** *I. w. wagleri** 8.5in (22cm) **p. 745**
Endemic; Interior and Pacific Slope. Bill straight and pointed, fairly slender. Adult distinctive (4 allopatric in SE humid lowlands); note lack of wingbars. Imm: note black face or face and bib, dull pale wing edgings, dusky back streaks. Molting imm has blacker wings and tail; note dark-mottled back. Scott's Oriole (Plate 68) has bold white wingbars.

2 **Yellow-tailed Oriole** *I. m. mesomelas* 8.6in (22cm) **p. 747**
SE humid lowlands, often fairly skulking. Bill stout with slightly decurved culmen. Adult flashy and unmistakable. Imm: note bill shape, long yellow wingbar.

3 **Bar-winged Oriole** *I. maculialatus* 8.5in (22cm) **p. 743**
Endemic to Interior and adjacent Pacific Slope, SE Oax to El Salvador. Bill small, slender, and slightly decurved. ♂ unmistakable; note single bold white wingbar. ♀ suggests ♀ or imm 4 (allopatric), wings and tail less dark, usually with pale lower wingbar. Imm ♀ black restricted to lores and bib; note bill shape, single pale wingbar, plain back, compare with imm 1 and 5 (both sympatric).

4 **Black-cowled Oriole** *I. dominicensis prosthemelas* 7.8in (20cm) **p. 743**
SE humid lowlands. Bill small, slender, and slightly decurved. ♂ unmistakable (1 allopatric). ♀ variable, plumage becomes increasingly like ♂ to S: that is, more black on head and back, thus some like imm shown but wings and tail black; sexes become similar in S C.A. Imm: black on head and chest variable, usually like ♀ but wings and tail dark brown, others have more black on head. Often mistaken for larger 5 and Audubon's Oriole (Plate 68, rarely sympatric); note bill shape, more black in face than 5, voice.

5 **Yellow-backed Oriole** *I. chrysater** 8.8in (22cm) **p. 746**
S from Isthmus. Note straight, fairly stout bill. Adult unmistakable; note bright yellow crown and back; Audubon's Oriole (Plate 68) has full black hood. Imm crown and back washed olive, wings dark brown; from imm 4 by bill shape, less black in face. 5(a) *chrysater* (Isthmus to C.A.): larger; 5(b) *mayensis* (Yuc Pen): smaller.

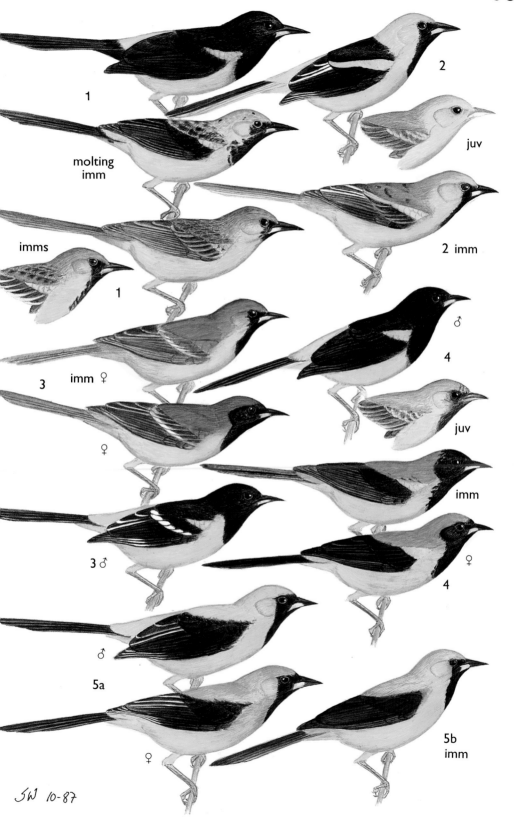

66

1

2

juv

molting
imm

imms

1

2 imm

3

imm ♀

♂

4

♀

juv

imm

3 ♂

♀

4

♂

5a

♀

5b
imm

JW 10-87

PLATE 67: ORANGE ORIOLES

See note on Plate 66.

1 Streak-backed Oriole *Icterus pustulatus** 8.2in (21cm) **p. 748**
Pacific Slope and interior valleys. Marked geographic variation. Note straight, fairly stout bill. 1(a) *microstictus* (NW Mexico, winters to Gro): ♂ distinctive; note flame-colored head, streaked back, bold white wing edgings. ♀ duller; note bill shape and dusky back streaks, compare with 2. Imm ♀ dull, from imm Bullock's Oriole (Plate 68) by all-black upper mandible, less distinct wingbars, belly washed yellow. 1(b) *formosus* (S Mexico): back heavily streaked black, sexes similar (♂ generally brighter). 1(c) *alticola* (Rio Motagua Valley, Guatemala): large, heavily marked back can appear solidly black, from 3 by less stout bill with larger blue-grey area at base, voice. See text to compare imm with imm 3 and 4.

2 Hooded Oriole *I. cucullatus** 7.5in (19cm) **p. 745**
Breeds NW and NE Mexico, resident E Mexico to Yuc Pen, winters W Mexico S to Isthmus. Note slender, decurved bill. 2(a) *nelsoni* (NW Mexico): ♂ tawny-yellow, back broadly edged yellow-olive when fresh, appears scalloped, not streaked; compare with 1(a) (note bill shapes). ♀ and imm from smaller Orchard Oriole (Plate 68) by relatively longer tail, longer bill, narrower lower wingbar, brighter orange-yellow undertail coverts. 2(b) *igneus* (Yuc Pen): ♂ orange to flame-orange, from much larger 3 by slender, decurved bill, bold white upper wingbar, etc. Fresh plumage has broad pale back edges. ♀ distinctive, lacks black lores and bib. Imm ♂ from 5 by slender decurved bill, head and back duller, black often more extensive in face.

3 Altamira Oriole *I. gularis** 9.5in (24cm) **p. 750**
E and S lowlands, interior valleys S from Isthmus. Large; note heavy, deep-based bill with straight culmen, small blue-grey area below at base. From 4 by bill shape, wing pattern, lack of black chest spots. Compare with 1(c) and 2(b). 3(a) *yucatanensis* (Yuc Pen); 3(b) *flavescens* (Gro).

4 Spot-breasted Oriole *I. pectoralis** 9in (23cm) **p. 749**
SW Mexico to C.A., Pacific Slope and interior valleys. Note stout bill with decurved culmen, triangular white panel on tertial and secondary edgings. Adult distinctive; compare with 3. Imm can be confused with 3 and 1(c) (see text). 4(a) *gutullatus* (Chis to C.A.); 4(b) *carolynae* (SW Mexico).

5 Orange Oriole *I. auratus* 7.3in (19cm) **p. 749**
Endemic to Yuc Pen. Note straight culmen; compare with 2(b). ♂ unmistakable; ♀ and imm from imm ♂ 2(b) by bill shape, brighter head and back, less extensive black in face, worn wings show bolder white edgings, especially flash at primary bases.

67

SW 7-87

PLATE 68: NORTHERN ORIOLES

See note on Plate 66.

1 **Scott's Oriole** *Icterus parisorum* 7.8in (20cm) **p. 752**
   NW deserts and Interior N of Isthmus. Bill long and straight, relatively slender.
   ♂ unmistakable. ♀ and imm plumage variable; note 2 bold white wingbars, yellow
   underparts. Compare with imm Black-vented Oriole (Plate 66).

2 **Audubon's Oriole** *I. graduacauda** 9in (23cm) **p. 747**
   Both slopes N of Isthmus. Note straight, fairly stout bill. 2(a) *graduacauda* (E
   Mexico): distinctive; note black hood and yellow-olive back, white wing edgings
   often reduced in S birds. 2(b) *dickeyae* (SW Mexico): striking, black and golden
   yellow (♀ duller). Imm: black head may be reduced, back washed olive. Compare
   with imm Black-cowled Oriole (Plate 66).

3 **Orchard Oriole** *I. spurius** 6.4in (16cm) **p. 744**
   Breeds N and cen Mexico, widespread migrant. ♂ unmistakable (3(a) *phillipsi*
   breeds N and cen Mexico; 3(b) *fuertesi* breeds E Mexico). ♀ and imm from larger
   Hooded Oriole (Plate 67) by shorter bill, relatively shorter tail, bolder lower
   wingbar, olive-yellow undertail coverts.

4 **Bullock's Oriole** *I. (galbula?) bullockii** 7.5in (19cm) **p. 752**
   4(a) *bullockii* (breeds N and cen Mexico; winters to Guatemala): ♂ distinctive
   (interbreeds with 4(b) and 5). ♀ and imm much like 4(b) but show trace of ♂ face
   pattern. From 5 by paler face, pale grey belly, but not always safely told (and
   intergrades occur). 4(b) *abeillei* (breeds cen Mexico; winters to S Mexico): ♂
   distinctive (but interbreeds with 4(a)). ♀ and imm show trace of ♂ face pattern;
   compare with 4(a). See imm ♀ Streak-backed Oriole (Plate 67, 1(a)).

5 **Baltimore Oriole** *I. galbula* 7.5in (19cm) **p. 751**
   Widespread migrant. ♂ and ♀ distinctive (but interbreeds with 4(a)). Imm not
   always safely told from 4 (see text).

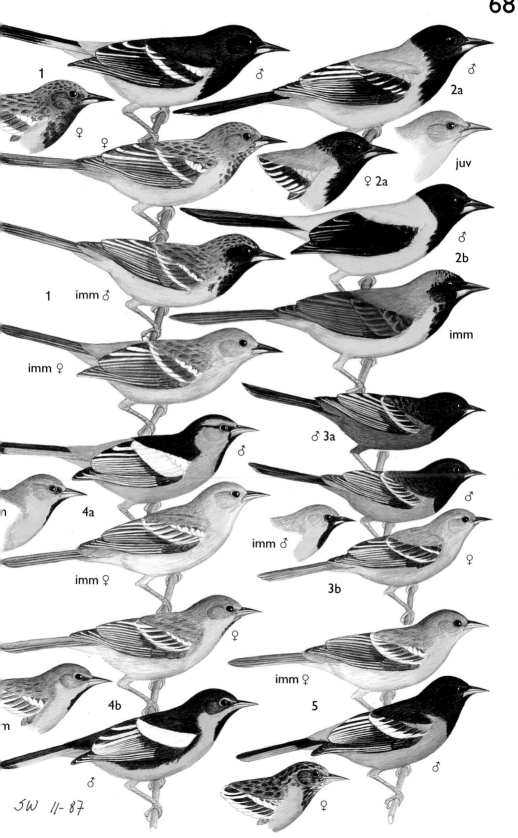

68

1

♀

♀

♂

2a

♂ 2a

juv

♂
2b

1 imm ♂

imm

imm ♀

♂ 3a

4a

♂

imm ♂

♀

imm ♀

3b

♀

imm ♀

5

4b

♂

♀

JW 11-87

PLATE 69: ISLAND ENDEMICS

Isla Guadalupe

1 **Guadalupe Junco** *Junco insularis* 5.8in (15cm) **p. 730**
Endangered. Distinctive; pointed bluish-grey bill, dusky grey head and chest, vinaceous sides.

Islas Revillagigedo

10 **Tropical Parula** *Parula pitiayumi graysoni* 4.2in (11cm) **p. 632**
Socorro. Both sexes resemble dull ♀ Tropical Parula (*nigrilora* group), lack white tail-spots.

11 **Socorro Wren** *Troglodytes sissonii* 4.5in (11cm) **p. 569**
Socorro. Only wren on island.

12 **Clarion Wren** *T. tanneri* 5.3in (13cm) **p. 569**
Clarion. Only wren on island.

13 **Socorro Mockingbird** *Mimodes graysoni* 10in (25cm) **p. 599**
Endangered. Socorro. Upperparts brownish, underparts dirty whitish with dusky flank streaks, no white wing or tail flashes.

14 **Socorro Parakeet** *Aratinga brevipes* 12.5in (32cm) **p. 335**
Socorro. Only parrot in Islas Revillagigedo.

15 **Socorro Dove** *Zenaida graysoni* 11.3in (29cm) **p. 765**
Extinct in the wild? Socorro. Like thickset Mourning Dove, face and underparts deep cinnamon. Note voice.

Islas Tres Marías

2 **Lawrence's [Broad-billed] Hummingbird** *Cynanthus latirostris lawrencei* 3.7in (9cm) **p. 404**
♂ throat bluish green. ♀ much like mainland birds (see Plate 30).

3 **Cinnamon Hummingbird** *Amazilia rutila graysoni* 4.7in (12cm) **p. 412**
Larger and duskier than mainland Cinnamon Hummingbirds.

18 **Yellow-headed Parrot** *Amazona oratrix tresmariae* 14.5in (37cm) **p. 343**
Head and chest generally more extensively yellow than mainland birds.

19 **Streak-backed Oriole** *Icterus pustulatus graysonii* 8in (20cm) **p. 748**
♂ bright yellow-orange with few black back streaks. ♀ duller: crown, nape, and upperparts olive, face and chest yellow-orange.

Isla Cozumel

4 **Cozumel Emerald** *Chlorostilbon forficatus* ♂ 3.7in (9cm), ♀ 3.3in (8cm) **p. 403**
♂ emerald green with long, deeply forked tail. For ♀, see text and Fig. 30 (no similar species on island).

5 **Cozumel Vireo** *Vireo bairdi* 4.9in (12cm) **p. 615**
Unmistakable; tawny-olive upperparts with whitish lores and eyering, sides cinnamon.

6 **Cozumel Wren** *Troglodytes beani* 4.8in (12cm) **p. 568**
Only wren on island.

7 **Rufous-browed Peppershrike** *Cyclarhis gujanensis insularis* 6.3in (16cm) **p. 626**
Dirty greyish-olive upperparts, dirty pale lemon throat and chest, buffy-white belly.

8 **Golden [Yellow] Warbler** *Dendroica petechia rufivertex* 4.8in (12cm) **p. 633**
♂ has distinct rufous cap, ♀ like Yellow Warbler.

9 **Stripe-headed Tanager** *Spindalis zena benedicti* 6in (15cm) **p. 669**
♂ unmistakable. ♀ greyish olive overall with paler belly, hint of ♂ wing pattern.

16 **Roadside Hawk** *Buteo magnirostris gracilis* 13.5in (34cm) **p. 197**
Smaller than mainland birds. Juv has blank pale face, sparsely streaked underparts. Adult similar to mainland birds.

17 **Cozumel Thrasher** *Toxostoma guttatum* 9in (23cm) **p. 601**
Long, slightly decurved black bill, eyes amber to orange. Only thrasher on Cozumel.

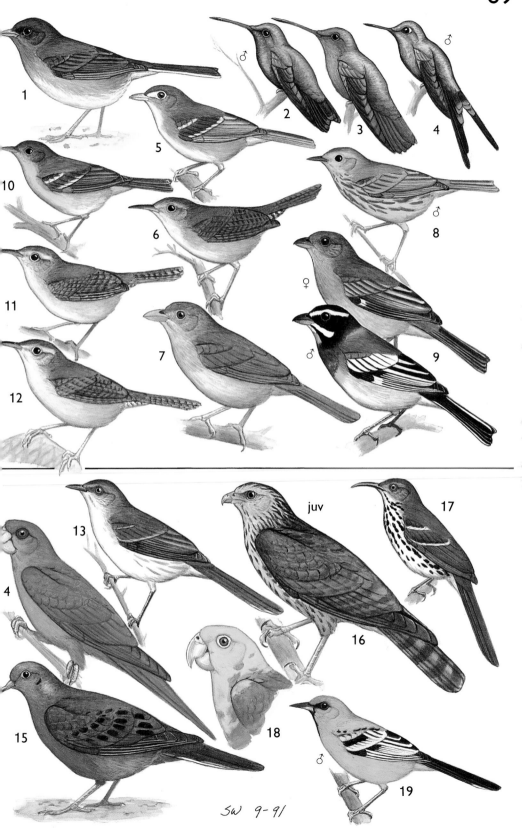

69

1

2

3

4 ♂

5

10

6

8 ♂

11

9 ♀

7

9 ♂

12

13

4

16 juv

17

15

18

19 ♂

SW 9-91

1 **Canivet's Emerald** *Chlorostilbon canivetii* 3.3in (8cm) **p. 401**
Endemic; E Mexico to Yuc Pen and Honduras Bay Islands. From 2 (allopatric?)
by longer, more deeply forked tail; ♂ tail tips paler. See text and Fig. 30.

2 **Salvin's Emerald** *C. (c.?) salvini osberti* 3.2in (8cm) **p. 403**
SE Chis to C.A. Compare with 1. See text and Fig. 30.

3 **Crowned Woodnymph** *Thalurania colombica townsendi* 4in (10cm) **p. 406**
N to S Belize. Rain forest. ♂ distinctive. ♀ plain; note black bill, green-mottled
flanks, white-tipped blue-black tail.

4 **Azure-crowned Hummingbird** *Amazilia cyanocephala chlorostephana** 4.3in
(11cm) **p. 409**
Endemic; ssp to Mosquitia. Note glittering green crown. See Plate 31.

5 **Honduran Emerald** *A. luciae* 3.7in (9cm) **p. 409**
Honduran endemic. Arid interior valleys, local. Distinctive. Note glittering
turquoise throat and chest, pale grey belly, cleft bronzy tail with dark subterminal
band.

6 **Berylline Hummingbird** *A. beryllina devillei** 3.8in (10cm) **p. 410**
Endemic; ssp from S Chis to C.A. Tail bronzy-purple, small rufous wing panel, ♂
underparts green. See Plate 30.

7 **Blue-tailed Hummingbird** *A. c. cyanura** 3.8in (10cm) **p. 411**
Endemic; ssp from E El Salvador to W Nicaragua. Rufous wing panel more
extensive than *guatemalae* (Plate 30). Note blue tail.

8 **Green-throated Mountain-gem** *Lampornis v. viridipallens** 4.5in (11cm) **p. 416**
Endemic; ssp from E Guatemala to W Honduras. Throat and chest more solidly
green than N ssp, separated by white forecollar (see Plate 31). 9 allopatric.

9 **Green-breasted Mountain-gem** *L. (v.?) sybillae* 4.5in (11cm) **p. 416**
Endemic to mountains of E Honduras and N cen Nicaragua. Distinctive. ♂ throat
and chest mottled 'solidly' green. ♀ throat pale buff. Whitish sides of tail flash in
flight. 8 allopatric.

10 **Band-tailed Barbthroat** *Threnetes ruckeri ventosus* 4.5in (11cm) **p. 392**
N to S Belize. Rain forest. Distinctive; note rufous chest, banded tail.

11 **Brown Violet-ear** *Colibri delphinae* 4.3in (11cm) **p. 398**
N to S Belize. Rain forest. Distinctive; brown overall with short bill, broad pale
moustache.

12 **Bushy-crested Jay** *Cyanocorax m. melanocyaneus** 12.5in (32cm) **p. 540**
Endemic to N C.A. No similar species in range.

13 **Elegant Trogon** *Trogon elegans lubricus** 11.5in (29cm) **p. 433**
Undertail more strongly barred than Mexican birds (Plate 33), ♂ uppertail bronzy
green.

14 **Red-throated Parakeet** *Aratinga (holochlora?) rubritorques* 11in (28cm) **p. 335**
Endemic to N C.A. Note orange-red throat (reduced or lacking in juv); Pacific
Parakeet (Plate 22) larger.

15 **Spot-bellied Bobwhite** *Colinus (cristatus?) leucopogon** 8in (20cm) **p. 230**
N to E Guatemala. Variable, other bobwhites allopatric. 15(a) *hypoleucus* (E
Guatemala to cen El Salvador); 15(b) *leucopogon* (Honduras to W Nicaragua).

16 **Yellow-naped Parrot** *Amazona auropalliata caribaea** 14.5in (37cm) **p. 345**
Bay Islands and Mosquitia (see also Plate 21). No similar species in range.

17 **Yellow-headed Parrot** *A. oratrix** 14.5in (37cm) **p. 343**
17(a) *belizensis* (Belize to NW Honduras where complex transition occurs to
yellow-crowned and yellow-naped birds, all characterized by pale bills); 17(b)
'Yellow-crowned Parrot' (NW Honduras) occurs with yellow-naped birds which
appear identical except for large yellow nape patch. See text.

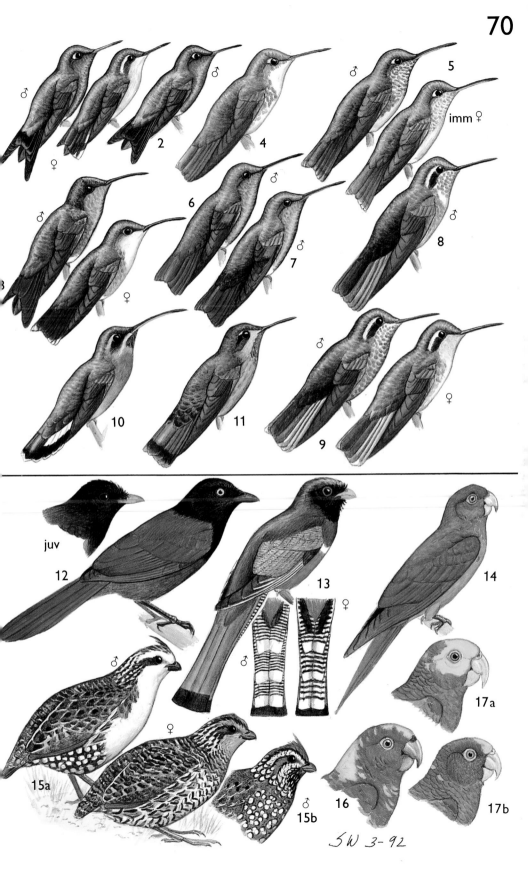

70

SW 3-92

1 **White-winged Becard** *Pachyramphus polychopterus cinereiventris* 5.7in (14cm) **p. 520**
N to N Guatemala (possibly E Chis?). ♂ darker than Grey-collared Becard (Plate 43) with blackish hindneck, lacks pale supraloral stripe. ♀ distinctive; note pale spectacles, olive upperparts, cinnamon wing edgings.

2 **Grey-headed Piprites** *Piprites griseiceps* 4.3in (11cm) **p. 523**
One record from NE Guatemala, presumably resident (rare?) in rain forest. Note stocky, manakin-like shape, grey head, whitish eyering.

3 **Olivaceous Piculet** *Picumnus olivaceus dimotus* 3.6in (9cm) **p. 449**
N to NE Guatemala. Unmistakable; tiny. Active low to high in rain forest and edge.

4 **Slaty Antshrike** *Thamnophilus punctatus atrinucha* 5.7in (14cm) **p. 475**
N to S Belize. Distinctive. ♂ note black cap, white wingbars. ♀ note pale buff wingbars, white-tipped tail.

5 **Bare-crowned Antbird** *Gymnocichla nudiceps chiroleuca* 6in (15cm) **p. 478**
N to S Belize. ♂ unmistakable. ♀ note bright blue orbital ring.

6 **Buff-throated Woodcreeper** *Xiphorhynchus guttatus confinis** 8.8in (22cm) **p. 470**
N to NE Guatemala. From Ivory-billed Woodcreeper (Plate 37) by bicolored bill, voice.

7 **Slate-throated Redstart** *Myiborus miniatus connectens** 5.3in (13cm) **p. 655**
Differs from Mexican birds (Plate 55) in orange-red underparts.

8 **Tropical Parula** *Parula pitiayumi speciosa** 4.2in (11cm) **p. 632**
N to Honduras. Deeper blue above, more extensively yellow below than N ssp. See Plate 55.

9 **Mountain Elaenia** *Elaenia frantzii ultima* 5.7in (14cm) **p. 484**
Mountains N to E Guatemala. Distinctive; lacks crest. Note pale eyering, bold whitish tertial edgings.

10 **Yucatan Woodpecker** *Centurus pygmaeus tysoni** 6.7in (17cm) **p. 452**
Isla Guanaja (Honduras Bay Islands); no similar species on island. Head pattern differs from Yuc Pen birds (Plate 35).

11 **Hoffmann's Woodpecker** *C. hoffmannii* 7.7in (20cm) **p. 452**
Pacific Slope S from Honduras. From larger Golden-fronted Woodpecker (Plate 35) by yellow nape.

12 **Mangrove Vireo** *Vireo p. pallens** 4.5in (11cm) **p. 615**
Pacific Slope S from Honduras; mangroves. Washed-out olive-grey above, dirty buffy white below; compare with Mexican birds (Plate 53).

13 **Rock Wren** *Salpinctes obsoletus fasciatus** 5.5in (14cm) **p. 560**
El Salvador to Nicaragua. Strikingly different from Mexican birds (Plate 49); note boldly spotted and barred underparts.

14 **White-eared Ground-Sparrow** *Melozone leucotis nigrior** 7.2in (18cm) **p. 696**
N cen highlands of Nicaragua. Unmistakable; skulking. See Plate 61.

♂

71

1

imm ♂

♀

2

♀

3

♂

4

♂

5

♀

6

7

♂

8

9

10

♂

♀

11

♂

♀

12

13

14

SW 10-92

SD: Locally/seasonally F to U resident (near SL–1500 m) on Atlantic Slope from S Ver to Honduras and N cen Nicaragua, on Pacific Slope in Chis and Guatemala (MAT; Land 1970).
RA: S Mexico to Costa Rica.

GENUS CHLOROSTILBON: Emeralds

Emeralds are small hummingbirds named for the overall brilliant emerald green color of the ♂♂. Bills straight, 1–1.25× head. Tails cleft to deeply forked. Ages/sexes differ. Bills red with black tip in ♂♂, black above and red to pinkish below with dark tip in ♀♀ and imms. Songs and displays undescribed(?)

To consider the four forms found in the guide region as species is consistent with treatment of this genus in other areas such as the Caribbean or S.A.; *auriceps* and *forficatus* in particular are far more distinct from other *Chlorostilbon* than are most other accepted species in this genus (Howell 1993*a*). Figure 30 and Table 2 illustrate differences among the four forms of *Chlorostilbon* in the guide region.

CANIVET'S [FORK-TAILED] EMERALD
Chlorostilbon canivetii   Plate 70
Esmeralda de Canivet   Figure 30

ID: ♂ 3.3–3.6 in (8.5–9 cm); ♀ 3–3.3 in (7.5–8 cm). *Endemic; E Mexico to N C.A.* Descr. See genus note. ♂: *tail moderately forked.* Head and body emerald green, glittering on crown and underparts. *Tail blue-black, inner rectrices tipped grey.* ♀: tail slightly forked, outermost rectrices shorter than next pair. *Whitish postocular stripe contrasts with green crown and blackish auriculars.* Nape and upperparts golden green. Throat and underparts pale

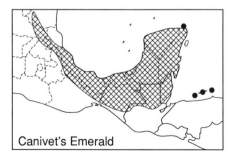

Canivet's Emerald

grey. *Inner rectrices bluish green, outer rectrices blackish, tipped white, outer 2 rectrices with whitish median band.* Imm ♂: resembles ♀ but blue-black tail longer, inner 3 rectrices tipped green, outer 2 rectrices tipped pale grey to whitish, underparts with green patches, at times forming a green gorget. Older imm ♂♂ have adult-like tail with inner 3 rectrices tipped dull grey, washed green, outer 2 rectrices blue-black to tips, underparts mostly green or with pale grey throat.

Voice. Dry, rattling and chattering calls, slightly harder than Broad-billed Hummingbird.

Habs. Brushy woodland and scrub, overgrown clearings, forest edge, mainly in arid to semihumid areas. Feeds at low to mid-levels, often near ground, tail wagged and held open so fork noticeable. Nesting: Feb–May.

SS: Salvin's Emerald possibly sympatric in N Guatemala; ♂ has shorter, less forked tail with duller grey tips (tail looks blue-black), ♀ has shorter, less deeply forked tail which usually has less distinct whitish median band on outer rectrices. ♀ and imm ♂ Broad-billed Hummingbird larger, tail squarer, imm ♂ often has blue on throat. Golden-crowned Emerald allopatric.

SD/RA: F to C resident (SL–1600 m) on Atlantic Slope from S Tamps to Yuc Pen, S to N Guatemala and Belize; also on Honduras Bay Islands. Reports of *canivetii* from DF (Villada 1875, and subsequent authors) are questionable (Wilson and Ceballos-L. 1986; Howell 1993*a*). Reports of *canivetii* and of intergrade *canivetii* × *auriceps* from Mor (Friedmann *et al.* 1950) are in error: the former refers to Avilés specs. (MLZ spec. 10170, 10189, 10190, 10210), the latter to an imm ♂ *auriceps* (MLZ spec. 14072).

NB1: *C. canivetii*, *C. auriceps*, and *C. forficatus* are allopatric (see SD/RA above), are quite distinct morphologically (Fig. 30, Table 2), and are best treated as separate species. NB2: *C. canivetii* and *C. salvini osberti* have been collected within about 15 km of one another in the vicinity of the Río Cajabón valley, N Guatemala (Griscom 1932). Field work should be concentrated in that area to document whether intergradation occurs (none is evident from specimens, *contra* Griscom 1932).

Fig. 30 Tail patterns of *Chlorostilbon* emeralds. (a) Golden-crowned Emerald; (b) Cozumel Emerald; (c) Canivet's Emerald; (d) Salvin's Emerald

Table 2. Wing chord, tail, culmen, and tail fork measurements (in mm) of *Chlorostilbon* in Mexico and northern Central America. *n*: number of birds examined; RL–R1: longest rectrix minus central rectrix. Mean values in parentheses.

Species	n	Wing chord	Tail length	RL–R1	Culmen
auriceps	34♂	41.9–45.8 (43.7)	40.6–46.5 (43.1)	24.6–32.5 (28.7)	11.7–13.7 (13.0)
	23♀	41.9–44.5 (43.4)	29.5–35.6 (33.1)	12.5–16.1 (14.9)	12.9–15.2 (13.9)
canivettii	34♂	42.0–48.0 (46.2)	30.5–38.0 (34.7)	16.0–20.0 (17.1)	13.0–15.0 (14.3)
	30♀	42.5–48.8 (45.0)	27–32.0 (29.7)	6.5–9.0 (7.8)	13.9–16.0 (15.0)
salvini	48♂	43.0–47.0 (45.7)	27.5–33.5 (30.8)	9.0–14.0 (11.7)	13.0–15.5 (14.4)
	40♀	42.5–47.0 (45.3)	24.5–30.0 (26.4)	3.0–6.0 (4.4)	14.0–16.5 (15.1)
forficatus	18♂	47.0–50.5 (48.4)	39.0–45.8 (43.1)	23.0–30.0 (26.3)	13.9–15.5 (14.8)
	11♀	46.0–49.0 (47.1)	30.5–35.0 (32.2)	9.5–11.5 (10.3)	14.6–16.5 (15.5)

SALVIN'S [FORK-TAILED] EMERALD
Chlorostilbon (canivetii?) salvini osberti
Esmeralda de Salvin
Plate 70
Figure 30

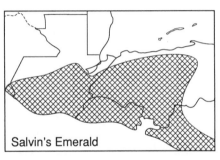

Salvin's Emerald

ID: ♂ 3.5 in (9 cm); ♀ 3–3.3 in (7.5–8 cm). **C.A. Descr.** See genus note. ♂: tail slightly forked. Head and body emerald green, glittering on crown and underparts. *Tail blue-black, inner rectrices with dull grey tips* (tail usually looks blue-black). ♀: tail slightly cleft, outermost rectrices shorter than next pair. *Whitish postocular stripe contrasts with green crown and blackish auriculars.* Nape and upperparts green. Underparts pale grey. Inner rectrices blue-green, outer rectrices blackish, tipped white, *outer 2 rectrices with indistinct whitish median band.* **Imm ♂:** plumages apparently analogous to Canivet's Emerald.

Voice. Dry, rattling and chattering calls, much like Canivet's Emerald.

Habs. Brushy woodland and scrub, overgrown clearings, forest edge, mainly in arid to semihumid areas. Habits much as Canivet's Emerald. Nest and eggs undescribed (?). B.c.: Dec–Feb (El Salvador).

SS: See Canivet's Emerald. Blue-tailed Hummingbird larger, rump bronzy purplish, tail blue. Broad-billed and Doubleday's hummingbirds allopatric.

SD: F to C resident (SL–1900 m) on Pacific Slope from SE Chis (IBUNAM spec.; SNGH) and in interior from W Guatemala, to Honduras and W Nicaragua, also locally on Atlantic Slope from E Guatemala to Honduras.

RA: SE Chis and Guatemala to S C.A.

NB: The form *assimilis* of S C.A. may be specifically distinct, as treated by AOU (1983), but see Stiles and Skutch (1989).

GOLDEN-CROWNED [FORK-TAILED] EMERALD
Chlorostilbon auriceps
Esmeralda Mexicana
Plate 30
Figure 30

ID: ♂ 3.7 in (9.5 cm); ♀ 3–3.3 in (7.5–8 cm). *Endemic to W Mexico.* **Descr.** See genus note. ♂: *tail very long and deeply forked. Head and body emerald green, glittering on crown and underparts. Tail blue-black, inner rectrices tipped grey.* ♀: *tail forked. Whitish postocular stripe contrasts with green crown and blackish auriculars.* Nape and upperparts golden green. Underparts pale grey. Inner rectrices golden green,

outer rectrices blackish, outer 2 pairs tipped white, outermost rectrices whitish basally on outer webs. **Imm ♂:** resembles ♀ but tail longer, blue-black, inner 3 rectrices tipped pale grey, washed green, outer 2 rectrices tipped pale grey to dull grey, underparts with green patches, at times forming a green gorget.

Voice. Dry, rattling and chattering calls, slightly harder than Broad-billed Hummingbird.

Habs. Brushy woodland and scrub, overgrown clearings, forest edge, mainly in arid to semihumid areas. Habits much as Canivet's Emerald. Nesting: Feb–Jul.

SS: ♀ and imm ♂ may recall Broad-billed or Doubleday's hummingbirds but smaller, tail more deeply forked, lack blue on throat.

SD/RA: F to C resident (SL–1800 m) on Pacific Slope from Sin to Oax, and in interior along Balsas drainage to S Mor.

NB: See NB1 under Canivet's Emerald.

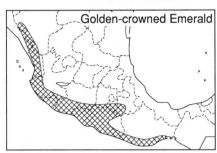

Golden-crowned Emerald

COZUMEL [FORK-TAILED] EMERALD
Chloristilbon forficatus
Esmeralda de Cozumel
Plate 69
Figure 30

ID: ♂ 3.6–3.8 in (9–9.5 cm), ♀ 3.2–3.5 in (8–9 cm). *Endemic to Isla Cozumel.* **Descr.** See genus note. ♂: *tail very long and deeply forked. Head and body emerald green,*

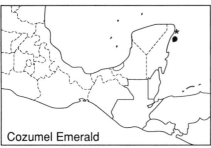

Cozumel Emerald

glittering on crown and underparts. *Tail blue-black*, inner rectrices tipped brownish grey. ♀: tail forked. *Whitish postocular stripe contrasts with green crown and blackish auriculars.* Nape and upperparts green. Underparts pale grey. Inner rectrices deep blue-green, outer rectrices blackish, outer 3 pairs tipped white, outer 2 pairs whitish basally on outer webs. **Imm** ♂: undescribed (?).

Voice. Dry, rattling and chattering calls, much like other *Chlorostilbon*.

Habs. Brushy woodland and scrub, second growth, etc. Habits much as Canivet's Emerald. Nest and eggs undescribed (?).

SS: None on Cozumel.

SD/RA: F to C resident (around SL) on Isla Cozumel, R visitor (?) to Isla Mujeres (CM spec.).

NB: See NB1 under Canivet's Emerald.

DUSKY HUMMINGBIRD

Cynanthus sordidus   Plate 30
Colibri Prieto

ID: 3.5–4 in (9–10 cm). *Endemic to SW Mexico.* Bill straightish, 1.25–1.33× head, tail slightly cleft, wings = tail. **Descr.** Ages/ sexes differ slightly. ♂: *bill bright red with black tip.* Whitish postocular stripe (to elongated spot) contrasts with dull green

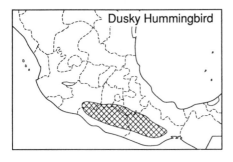

Dusky Hummingbird

crown and blackish auriculars. *Nape and upperparts golden green. Throat and underparts dusky grey*, often merging with blackish mask, throat rarely flecked blue-green. *Tail greyish green.* ♀: bill blackish above, reddish below with dark tip. *Underparts paler than* ♂, contrasting more with mask, throat rarely buffy grey. *Outer rectrices with blackish subterminal band, tipped whitish.* **Imm:** resembles adult of respective sex but outer rectrices narrowly tipped pale buff.

Voice. Chattering calls suggest Broad-billed Hummingbird but softer and more liquid or spluttering. Also dry, slightly buzzy chips and a quiet, dry, chippering warble (song?).

Habs. Arid scrub, semiopen and open areas with scattered trees. Usually feeds at mid- to upper levels in trees; tail persistently wagged and spread when feeding. Nesting: Mar–May, Aug, Nov–Dec.

SS: ♀ Broad-billed Hummingbird (locally sympatric in Balsas Basin) brighter, more emerald green above (including tail), blue-black subterminal tail band and white tail corners more contrasting, the black area often broader than on Dusky (rarely most of tail blue-black). Some individuals, however, may not be safely identified, even in the hand. Doubleday's Hummingbird smaller, more *Chlorostilbon*-like, upperparts bright emerald green, call drier.

SD/RA: C to F resident (900–2200 m) in interior from E Mich and Mor to Oax. A report from DF (Binford 1985) is based on an Avilés spec. and is thus questionable; reports from Jal and Hgo (AOU 1983) are probably also in error.

NB: Reported hybrid Dusky × Broad-billed hummingbirds apparently pertain to variations of adult ♂ Dusky Hummingbird (Binford 1985).

BROAD-BILLED HUMMINGBIRD

*Cynanthus latirostris**   Plates 30, 69
Colibri Piquiancho

ID: 3.5–4 in (9–10 cm). N and cen Mexico. Bill straightish, 1.25–1.33× head, tail slightly cleft, wings = tail. Two groups: *latirostris* group (widespread) and *lawrencei* (Islas Tres Marías). **Descr.** Ages/ sexes differ. *Latirostris.* ♂: bill bright red with black tip. *Throat glittering blue to violet-blue, underparts glittering blue-green, undertail coverts white*, often with dusky centers, to dusky with whitish edgings. Crown, nape, and upperparts emerald green. Tail blue-black, inner rectrices tipped greyish. ♀: bill blackish above, red below with dark tip. White postocular stripe contrasts with dull green crown and blackish auriculars. *Nape and upperparts emerald green* to golden green. Throat and underparts pale grey to dusky grey. *Tail variable, typically emerald green to blue-green* (rarely blue-black), *outer rectrices with fairly broad blue-black subterminal band, tipped white.* Imm ♂: resembles ♀ but tail blue-black, outer rectrices tipped white, inner rectrices tipped greyish, *underparts with blue and green patches. Lawrencei:* glittering turquoise-green throat in ♂, more deeply cleft tail.

Voice. Dry, chattering calls suggest Ruby-crowned Kinglet.

Habs. Arid to semiarid brushy woodland, scrub, semiopen areas with scattered trees. Feeds low to high, tail persistently wagged and spread when feeding. Nesting: Jan–Jul, Sep.

SS: Dusky and Doubleday's hummingbirds also tail-wag persistently. ♂ Doubleday's (allopatric?) has turquoise forecrown, mostly blue underparts, blackish undertail coverts, ♀ Doubleday's perhaps not safely told in field but smaller, more *Chlorostilbon*-like. ♂ Mexican Woodnymph has black bill, green gorget, different habitat and call. ♀ White-eared has green-spotted underparts, hard chipping call. See Golden-crowned Emerald.

SD: Locally/seasonally C to F resident (SL–2500 m) on Pacific Slope from Son to Col (possibly to NW Gro), in interior from Chih and NL to cen volcanic belt, locally on Atlantic Slope (mostly above 900 m) from S Tamps to cen Ver; also on Islas Tres Marías (*lawrencei*), and in interior in lower Balsas drainage (*toroi*). Most withdraw (Oct–Feb) from N (N Son to NL). N migrants winter S to N Gro; reports from Oax have been questioned (Binford 1989). Vagrant (Oct–Jan) to BCS (DFD, Howell and Webb 1992*f*).

RA: SW USA to cen Mexico; N populations migratory.

NB1: *Lawrencei* sometimes considered specifically distinct, Lawrence's Hummingbird.
NB2: Broad-billed and Doubleday's hummingbirds often considered conspecific, with *toroi* suggested as an intermediate form although it is clearly closer to Broad-billed. Broad-billed (migrants?) and Doubleday's were found within 20–30 km of one another in coastal W Gro in Apr 1988 and ♂♂ showed no signs of intergradation in plumage characters (SNGH, SW); another problem in need of study.

DOUBLEDAY'S [BROAD-BILLED] HUMMINGBIRD
Cynanthus doubledayi or *C. latirostris* (in part)   Plate 30
Colibri de Doubleday

ID: 3.2–3.5 in (8–9 cm). *Endemic to SW Mexico.* Bill straightish, 1.25–1.33× head, *tail cleft*, wings ≤ tail. Descr. Ages/sexes differ. ♂: resembles Broad-billed Hummingbird but *forecrown glittering turquoise-blue, throat deeper glittering violet-blue, underbody glittering greenish blue, undertail coverts blue-black* with narrow pale grey edgings rarely visible in the field. ♀: resembles Broad-billed, perhaps not safely distinguished in field, but smaller, tail may appear duller, grey-green, contrasting slightly with emerald green back. Imm ♂: resembles Broad-billed but can be told if blue visible on forecrown and underparts.

Voice. Dry chattering much like Broad-billed Hummingbird. Also a buzzy chipping while perched, *zzchik zzchik zzchik*

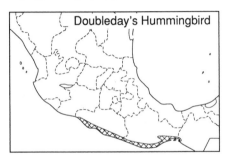

zzchik zzchik, accented on 1st and last notes.

Habs. Arid to semiarid brushy woodland, scrub, semiopen areas with scattered trees. Habits much as Broad-billed Hummingbird. Nesting (?).

SS: See Broad-billed (allopatric?) and Dusky hummingbirds, Golden-crowned Emerald.

SD/RA: C to F resident (SL–900 m) on Pacific Slope from W Gro to W Chis.

NB: See NB2 under Broad-billed Hummingbird.

MEXICAN [CROWNED] WOODNYMPH
Thalurania ridgwayi   Plate 30
Ninfa Mexicana

ID: 3.5–4 in (9–10 cm). *Endemic to W Mexico.* Bill straightish, 1.25× head, *tail fairly broad and slightly cleft*, wings = tail. **Descr.** Ages/sexes differ. ♂: bill black. *Forehead and forecrown glittering violet-blue,* hindcrown blue-green, nape and upperparts green. *Throat and chest glittering emerald green, belly and undertail coverts dusky green; often looks blackish overall. Tail blue-black.* ♀: Face, throat, and underparts pale grey, with slight whitish postocular spot, neck and sides of chest flecked green, Crown, nape, and upperparts green. *Tail blue-black, inner rectrices with green sheen, outer rectrices tipped white.* **Imm** ♂: undescribed (?).

Voice. Liquid, rattled notes, recalling *Eupherusa* hummingbirds, may suggest Costa's Hummingbird; also prolonged, rolled, liquid rattles or trills.

Habs. Humid semideciduous woodland, coffee plantations, especially in shady barrancas. Feeds and perches low to high in forest, often along streams, rarely at edges. Nest and eggs undescribed (?). B.c: Feb–Mar

SS: See Broad-billed Hummingbird, Berylline Hummingbird has rufous tail (looks black when closed) and wing patch, pale belly, hard, buzzy call.

SD/RA: F to U but local resident (250–1200 m) on Pacific Slope from S Nay to Col.

NB: Sometimes considered conspecific with Crowned (or Common) Woodnymph (*T. colombica*), and has been called Ridgway's Woodnymph.

CROWNED WOODNYMPH
Thalurania colombica townsendi   Plate 70
Ninfa Coronada

ID: ♂ 4–4.3 in (10–11 cm); ♀ 3.75–4 in (9.5–10 cm). C.A. Bill straightish, 1.33× head. **Descr.** Ages/sexes differ. ♂: *tail fairly broad, long, and forked,* wings < tail. Bill black. *Forehead and forecrown glittering violet,* hindcrown and nape blackish green, *upper mantle violet,* rest of upperparts dark blue-green. *Throat and underparts glittering green, mixed with bluish violet on sides,* undertail coverts blue-black. *Tail violet-black.* ♀: *tail slightly cleft,* wings = tail. *Face, throat, and underparts pale grey,* neck and sides of chest flecked green, flanks mottled green, undertail coverts whitish. Crown, nape, and upperparts blue-green. Tail blue-black, central rectrices with blue-green sheen, outer rectrices tipped white. **Imm** ♂: resembles ♀ but underparts dusky grey mixed with emerald green (mainly on throat), undertail coverts blue-black. Tail resembles ♂ but less deeply forked.

Voice. Dry, fairly hard, rattling chips, at times run into chatters.

Habs. Humid evergreen forest and edge, often near streams. Feeds low to high, perches mostly at mid- to upper levels, often quite conspicuously on bare twigs. B.c: Jan–Feb.

SD: F to C resident (SL–750 m) from S Belize and NE Guatemala to Honduras.

RA: Guatemala to N S.A.

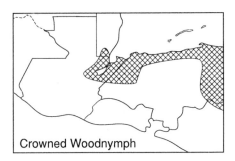

Mexican Woodnymph

Crowned Woodnymph

NB1: Crowned Woodnymph is sometimes treated as two species (the C.A. form being Purple-crowned Woodnymph, *T. colombica*), or sometimes regarded as conspecific with the S.A. Common Woodnymph *T. furcata*. See also Mexican Woodnymph.

BLUE-THROATED SAPPHIRE (GOLDENTAIL)
Hylocharis eliciae   Plate 30
Zafiro Gorjiazul

ID: 3.2–3.5 in (8–9 cm). *S Mexico and C.A.* Bill straightish, 1.25–1.33× head, *tail slightly cleft* and *fairly short*, wings = tail. **Descr.** Ages/sexes differ. ♂: *bill bright red with black tip. Gorget glittering bluish violet to violet*, chin spotted buff, *underparts dirty buff, mottled green on sides of chest.* Crown, nape, and upperparts green, becoming *burnished gold on uppertail coverts. Tail glittering greenish gold* (often looks greyish). ♀: bill may be blackish above, red below with dark tip. *Gorget mixed with greyish buff*, underparts paler. **Imm:** resembles ♀ but throat greyish buff, sparsely spotted bluish violet; bill blackish above, red below with dark tip.
  Voice. Song (Costa Rica) a lisping call followed by a chipping chatter, *tssir, chi-ti-ti-ti-tit*, or *tssir chi-chi chi-chi chi-chi-ch-it-it*, etc. Also high, squeaky chippering, especially when fighting.
  Habs. Humid to semihumid forest, second growth woodland, gallery forest, plantations. Prefers shady but fairly open forest interior and edge where feeds mainly at mid- to upper levels. Sings from mid-level perches on fairly conspicuous bare twigs. Nesting (?).
SS: Rufous-tailed Hummingbird larger with relatively longer rufous tail, throat and chest green, bill rarely so intensely red, calls harder.
SD: Locally/seasonally U to F resident (SL–

1000 m) on Atlantic Slope from S Ver to Honduras, on Pacific Slope from E Oax to El Salvador.
RA: S Mexico to N Colombia.

WHITE-EARED HUMMINGBIRD
*Basilinna leucotis**   Plate 31
Colibri Orejiblanco

ID: 3.5–4 in (9–10 cm). Endemic; *common in highlands.* Bill straightish, 1.25–1.33× head, tail squared to slightly cleft, wings = tail. **Descr.** Ages/sexes differ. ♂: bill bright red with black tip. *White postocular stripe contrasts with dark head.* Forehead and chin glittering violet, lower gorget glittering turquoise-green, auriculars black. Crown, nape, and upperparts green, uppertail coverts tipped rufous. *Underparts whitish, heavily mottled green on chest and flanks*, undertail coverts with dusky centers. *Inner rectrices green, outer rectrices blackish, often tipped green.* ♀: bill blackish above, red below with dark tip. *White postocular stripe contrasts with greenish brown crown and blackish auriculars, white throat spotted green.* Nape and upperparts green, uppertail coverts tipped rufous. Underparts whitish, mottled green on chest and flanks. Inner rectrices green, outer rectrices blackish, tipped white. **Imm** ♂: resembles ♀ but only outermost rectrices conspicuously tipped white, throat and forehead often show some ♂ colors.
  Voice. Song a tedious, metallic chipping, *chi'tink chi'tink ch'tink ...*, or *chi'dit chi'dit chi'dit ...*, or simply *tink tink ...*, etc.; lacks hesitant, jerky rhythm typical of Green Violet-ear. Fairly hard, dry chips, at times repeated steadily, may break into short, quiet gurgles.
  Habs. Pine–oak, oak, and pine–evergreen forest, clearings with flowers. Feeds and perches at low to mid-levels, often

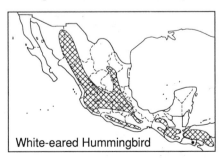

Blue-throated Sapphire   White-eared Hummingbird

abundant along low banks of flowers. Nesting: Mar–Aug (N and cen Mexico), May–Aug and Nov–Feb (S Mexico), Aug–Dec (C.A.).

SS: See ♀ and imm ♂ Broad-billed Hummingbird. Green-throated and Green-breasted mountain-gems larger with black bills, different tail patterns; note voice.

SD/RA: C to F resident (1200–3500 m) in interior and on adjacent slopes from Son and S Coah to Honduras and N cen Nicaragua. Most withdraw from N interior (Chih to NL at least) in winter (Oct–Mar).

NB: *Basilinna* has been merged with *Hylocharis*.

XANTUS' HUMMINGBIRD

Basilinna xantusii   Plate 30
Colibri de Xantus

ID: 3.3–3.8 in (8–9 cm). *Endemic to S Baja.* Bill straightish, 1.25–1.33× head, tail squared, wings = tail. **Descr.** Ages/sexes differ. ♂: bill bright red with black tip. *White postocular stripe contrasts with black forehead and auriculars* and dull greenish crown; chin black, lower gorget glittering green. Nape and upperparts green, uppertail coverts tipped rufous. *Underparts cinnamon,* mottled greenish at sides of chest. *Tail rufous-chestnut,* central rectrices edged green. ♀: bill blackish above, red below with dark tip. *Whitish postocular stripe contrasts with dull greenish crown and dark auriculars.* Nape and upperparts green, uppertail coverts tipped rufous. *Throat and underparts pale cinnamon. Tail dark rufous with green central rectrices,* outer rectrices with paler tips. **Imm** ♂: resembles ♀ but shows some ♂ colors on head and gorget.

Voice. A dry, chattering rattle suggesting Broad-billed Hummingbird, a high, metallic *chi-ti,* or *ti-tink,* or *chi-tiik,* which may be given 2–3× and suggests White-

eared Hummingbird, a single *tiik,* rapid, chippering series, *ssi ti-ti-ti-ti-ti* or *chi-ti ti-ti ti-ti,* etc., and a high, sharp *siik* in chases. Song (?) a varied, fairly soft, buzzy, scratchy to crackling (rarely squeaky) twitter or warble.

Habs. Arid to semiarid scrub, open brushy woodland, gardens. Feeds and perches at low to mid-levels. Nesting: Feb–Apr (N), Jul–Sep (S).

SD/RA: C to F resident (SL–1500 m) in BCS, N to Sierra de la Giganta (26° 30′ N), U to F and local (Dec–Apr) N to 28° N. Reports N to 31° N (Wilbur 1987) appear to be in error. Vagrant (late Dec–Mar) to S California (Hainebach 1992).

NB: See NB under White-eared Hummingbird.

GENUS AMAZILIA

One or more members of this genus of 'typical' hummingbirds are common in most lowland and foothill habitats. Bills straightish and red basally, at least on lower mandible, tails squared to cleft, wings ≤ tail. Ages/sexes differ slightly or similar. Heads lack striking patterns; several similar species cause identification problems.

WHITE-BELLIED EMERALD

Amazilia candida*   Plate 31
Esmeralda Vientre-blanco

ID: 3.5–3.7 in (9–9.5 cm). E and S. Bill 1–1.25× head, tail squared, wings = tail. *Small, plain.* **Descr.** Ages/sexes similar. *Bill black above, reddish below with dark tip. Crown, nape, and upperparts emerald green,* duller on rump and uppertail coverts, auriculars green with small white postocular spot. *Throat and underparts white,* neck and sides of chest flecked green. *Tail greyish green to bronzy,* outer

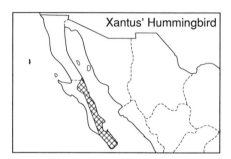

Xantus' Hummingbird

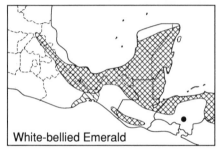

White-bellied Emerald

rectrices with dark subterminal band, tipped pale grey.

Voice. Song varied, high, thin, slightly shrill chipping, *tsi'si-sit' tsi-tsin'*, 2nd phrase repeated up to 4×, or *syiik syik-syik syiik syik*, repeated; also a monotonous, high, squeaky *tssi-ip tssi-ip ...* or *tsip tsip ...*, etc. Calls mostly rolled or trilled chips, *tsirr* and *ti-ti* or *tsi-tsir*, and longer *drii-i-i-it* and *tsi si-si-si-sit*, etc., softer and not as hard or rattling as Rufous-tailed Hummingbird calls.

Habs. Humid evergreen to semideciduous forest and edge. Feeds and sings low to high, typically feeds fairly low, often at same flowers as Rufous-tailed hummingbirds which usually dominate emeralds. Nesting: Feb–May.

SS: Azure-crowned Hummingbird larger and longer-billed, dusky sides of chest contrast with white throat, sides of face bluish with bolder white postocular spot (turquoise forecrown may be hard to see), tail lacks dark subterminal band; note voice.

SD/RA: C to F resident (SL–1500 m) on Atlantic Slope from SE SLP to Honduras and Nicaragua, locally on Pacific Slope from E Chis (SNGH, HGS) to Guatemala; mostly in foothills N of Isthmus. U winter visitor (late Nov–Feb) on Pacific Slope in Isthmus of Oax and W Chis (RAB).

HONDURAN EMERALD
Amazilia luciae   Plate 70
Esmeralda Hondureña

ID: 3.5–4 in (9–10 cm). *Honduran endemic.* Bill 1.33× head, *tail cleft to slightly forked*, wings slightly < tail. **Descr.** Ages/sexes differ slightly. ♂: bill black above, reddish below with dark tip. *Throat and upper chest glittering turquoise to blue-green* (often looks pale grey, mottled dusky), *underparts pale grey*, mottled green on flanks, undertail coverts dusky, broadly

edged white. Crown and auriculars green with *white postocular spot and smaller preocular spot*, nape and upperparts emerald-green, more bronzy on uppertail coverts. Tail bronzy green, outer rectrices with blackish subterminal band. ♀: turquoise gorget less intense, not extending on to upper chest. **Imm:** resembles ♀ but throat greyish, spotted turquoise, outer rectrices tipped pale buff.

Voice. A hard, slightly metallic ticking, may be repeated steadily, *chik, chik-chik, chik chik ...*; also hard, slightly buzzy chatters in flight, *zzchi-zzchi-zzchi-zzchi* and *chik chi zzhi-zzchi-zzchi-zzchi*, recalling Chestnut-collared Swift, and a high, sharp *siik* in flight chases, similar to other *Amazilia*; ♂♂ also give a dry, quiet, gruff warble (song?).

Habs. Arid thorn forest, scrubby woodland and edge, semiopen areas with scattered trees and scrub. Feeds and perches low to high, perches mainly at mid-levels. Makes prolonged insect-catching flights. Nest and eggs undescribed (?).

SS: None in habitat or range.

SD/RA: Until recently known only from 11 specimens, the most recent taken in 1950. Rediscovered in Jun 1988 in upper Río Aguán Valley, where common (Howell and Webb 1989c, 1992b). Probably F to C but local resident (150–300 m) in arid interior valleys on Atlantic Slope of Honduras.

AZURE-CROWNED HUMMINGBIRD
*Amazilia cyanocephala**   Plates 31, 70
Colibrí Coroniazul

ID: 4–4.5 in (10–11.5 cm). *Endemic; E and S, mostly foothills and highlands.* Bill 1.25–1.33× head, tail slightly cleft to squared, wings = tail. Two groups: *cyanocephala* group (widespread) and *chlorostephana* (Mosquitia). **Descr.** Ages differ

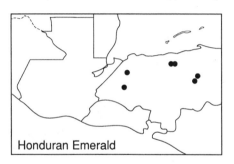

Honduran Emerald

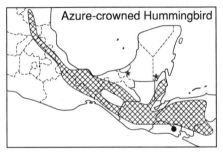

Azure-crowned Hummingbird

slightly, sexes similar. *Cyanocephala.* **Adult:** bill blackish above, pinkish red below with black tip. *Forehead and crown glittering turquoise to violet-blue,* auriculars emerald green to blue-green with white postocular spot. Nape and mantle green becoming dull brownish on back and rump. *White throat and upper chest contrast with dusky green sides,* undertail coverts dusky, edged white. *Tail greyish green* (bronzy green in *guatemalensis* from Guatemala S). **Imm:** *forehead and crown duller* (blue can be hard to see), throat and chest off-white or washed buff, outer rectrices pale grey distally. *Chlorostephana:* crown glittering emerald green to turquoise-green; tail bronzy green.

Voice. A low, fairly hard, buzzy *dzzzrt* in flight and perched, may be repeated steadily from perch; also fairly mellow, strong chipping, at times run into a trill or rattle. Voice of *chlorostephana* undescribed (?).

Habs. Oak and pine–oak woodland, scrub, second growth, humid evergreen forest edge, pine savannas. Feeds and perches low to high, at times abundant around flowering *Inga* trees. Nesting: Mar–Aug (E and S Mexico), Jan–Jul (Belize); b.c.: Jul–Sep (Guatemala).

SS: See White-bellied Emerald. Green-fronted Hummingbird bright white below with dark crown, black-tipped red bill.

SD/RA: *Cyanocephala* group: C to F resident (near SL–2400 m, in Mexico mainly above 600 m) on Atlantic Slope from S Tamps to Honduras, on Pacific Slope and in interior from E Oax and Chis to W Nicaragua. Apparently R visitor (Mar, Jul; around SL) to base of Yuc Pen (Peters 1913; SNGH). *Chlorostephana*: F to C resident in the Mosquitia.

NB: '*A. microrhyncha*' of Honduras, known from a 'unique' type specimen (AMNH 38481), is a juv Azure-crowned Hummingbird (not 'aberrant or possibly juvenile' (AOU 1983, p. 343)), see Phillips (1971*b*).

BERYLLINE HUMMINGBIRD
*Amazilia beryllina**   Plates 30, 70
Colibri de Berilo

ID: 3.7–4 in (9.5–10 cm). *Endemic; W and S foothills and highlands.* Bill 1.33× head, tail squared to slightly cleft, wings = tail. Two groups: *beryllina* group (N of Isthmus) and *devillei* group (S of Isthmus).

Descr. Ages/sexes differ. *Beryllina.* ♂: *bill black,* reddish below at base. Throat and chest glittering green, *belly greyish cinnamon, undertail coverts cinnamon, edged whitish. Conspicuous rufous flash across base of secondaries and primaries.* Crown, nape, and back green, becoming brownish on rump, uppertail coverts purplish. *Tail deep rufous, central rectrices darker purplish (tail often looks blackish when closed).* ♀: throat less intense green, *belly pale greyish.* **Imm** ♂: resembles ♀ but throat and chest buffy cinnamon, mottled green at sides, glittering green medially, belly greyish cinnamon. *Devillei: tail bronzy purple to violet-purple, ♂ has entire underparts glittering green,* rufous in wing often restricted to secondaries, ♀ belly mottled green.

Voice. A dry, fairly hard, buzzy *dzzzzir* or *drrzzzt,* and a more liquid, rolled, buzzy *dzzzzrrt,* etc., which may be repeated steadily. Song varied, slightly gruff, high twitters, usually with 1–2 lisping introductory notes, *ssi kirr-i-rr kirr-i-rr,* or *sirrr, ki-ti ki-dik,* or *sssi-ir sssiir chit-chit chit-chit chit-chit,* etc.; also (*devillei* at least) a squeaky *jer'eek 'r-eek,* repeated 2–3×.

Habs. Woodland and edge, scrub, especially with oaks, clearings, plantations. Feeds and perches low to high, often common and dominant over most species at flowering trees and flower banks. Nesting: Jun–Oct (W and S Mexico).

SS: Rufous wing patch distinguishes Berylline from Buff-bellied and Rufous-tailed hummingbirds in the few areas of overlap, ♂♂ of latter two species also have black-tipped red bills, raspy call of Berylline distinctive. Blue-tailed Hummingbird has deep blue tail. Perched ♂ *Eupherusa* can look surprisingly like Berylline from behind (rufous wing patch, dark tail), but all green below,

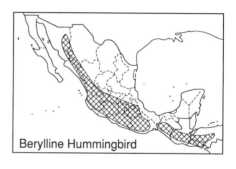

Berylline Hummingbird

white in tail flashes in flight. See Mexican Woodnymph.

SD/RA: C to F resident (600–2000 m, locally to 50 m, at least post-breeding) on Pacific Slope and in interior from Son and S Chih to El Salvador and Honduras, and on Atlantic Slope locally from cen Ver to Honduras.

NB1: The two groups are as distinct as many recognized hummingbird species and warrant further taxonomic investigation. **NB2:** Reported hybridization with Blue-tailed Hummingbird in cen Guatemala and El Salvador (AOU 1983) requires verification.

BLUE-TAILED HUMMINGBIRD

Amazilia cyanura*   Plates 30, 70
Colibri Coliazul

ID: 3.7–4 in (9.5–10 cm). *Endemic to SE Chis and C.A.* Bill 1.25–1.33× head, tail slightly cleft, wings = tail. Two distinct, allopatric ssp: *guatemalae* and *cyanura* (see SD/RA below). **Descr.** Ages/sexes differ. *Guatemalae.* ♂: *bill black*, reddish below at base. Throat and underparts glittering green, white leg puffs often striking, undertail coverts dark violet-blue. *Rufous flash across base of outer secondaries and inner primaries.* Crown, nape, and back green, becoming purplish on rump, uppertail coverts and *tail deep blue.* ♀: throat and chest less intense green, becoming mostly buffy grey on belly, *rufous in wings reduced to absent.* **Imm** ♂: resembles adult but throat and underparts dull green, mottled glittering green, tail slightly paler. *Cyanura: rufous flash across base of outer secondaries and inner primaries more extensive and conspicuous,* distinct in ♀.

Voice. A hard, raspy *bzzzrt*, similar to Berylline Hummingbird, hard chips, and a high, sharp *siik!* in flight. Song undescribed (?). Voice of *cyanura* undescribed (?).

Habs. Woodland and edge, plantations, scrub, clearings. Feeds and perches low to high. Nest and eggs undescribed (?).

SS: Deep blue tail distinctive.

SD/RA: *Guatemalae:* F to C resident (100–1800 m) on Pacific Slope from Chis to Guatemala. *Cyanura:* F to C resident (near SL–1200 m) on Pacific Slope and in interior from E El Salvador and Honduras to W Nicaragua. Vagrant (?) to Costa Rica.

NBI: The two ssp are as distinct as many recognized hummingbird species and warrant further taxonomic investigation. **NB2:** Replaced in cen Nicaragua by similar but allopatric **Steely-vented Hummingbird** *A. saucerrottei* which lacks rufous in wings. **NB3:** See NB2 under Berylline Hummingbird.

RUFOUS-TAILED HUMMINGBIRD

Amazilia t. tzacatl   Plate 30
Colibri Colirrufo

ID: 4–4.3 in (10–11 cm). *SE lowland forest.* Bill 1.33× head, *tail squared,* wings = tail. **Descr.** Ages/sexes differ. ♂: bill bright red with black tip. Throat and chest glittering green, *belly dusky brownish grey to greyish cinnamon,* flanks mottled green, *undertail coverts dark cinnamon-rufous, contrasting with belly and vent.* Crown, nape, and upperparts green, distal uppertail coverts rufous. *Tail deep rufous, rectrices tipped bronzy green.* ♀: bill blackish above, red below with dark tip. Throat and chest less intense green, *belly paler greyish.* **Imm** ♂: resembles ♀ but throat and chest duller green, mottled glittering green, belly darker.

Voice. A fairly hard, smacking *tchik-tchik* ... or *tchi tchi ...,* at times repeated insistently, and dry, hard chips often run into a rattling *chirr-rr-rr-rr-rr,* etc. Song varied, high, thin, squeaky chipping, *tsi, tsi-tsi-*

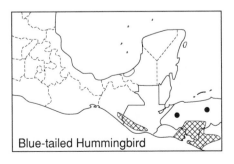

Blue-tailed Hummingbird

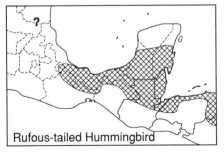

Rufous-tailed Hummingbird

tsit tsi-tsi-tsi tsi-si-si, a shorter *t'sin t'sit t'chin* and *tsi-sink si-sik tsi-sink*, etc.

Habs. Humid evergreen forest and edge, plantations, clearings with flowers. Feeds and perches low to high, often abundant at patches of roadside flowers. Nesting: Dec–Sep.

SS: Buff-bellied Hummingbird has greyish cinnamon to cinnamon belly (same color as undertail coverts in s ssp), cleft tail with mostly green central rectrices. See Berylline Hummingbird.

SD: C to F resident (SL–1200 m) on Atlantic Slope from S Ver to Honduras, possibly N to SE SLP (SNGH) and S Tamps (Zimmerman 1957c). A report from the Pacific Slope of Guatemala (Griscom 1932) is probably erroneous (spec. [AMNH 394014] appears to be lost).

RA: SE Mexico to Ecuador.

BUFF-BELLIED HUMMINGBIRD
Amazilia yucatanensis*   Plate 30
Colibri Vientre-canelo

ID: 3.8–4.3 in (10–11 cm). *E and SE drier woodland.* Bill 1.33× head, *tail cleft to slightly forked, wings ≤ tail. Noticeable geographic variation.* **Descr.** Ages differ, sexes similar. **Adult:** bill bright red with black tip (♂) to blackish above, red below with dark tip (♀). Throat and chest glittering green (less intense in ♀?), *belly pale greyish buff* (*chalconota* of NE Mexico) *to cinnamon* (*yucatanensis* of Yuc Pen); *undertail coverts buff to cinnamon, not contrasting strongly with belly.* Crown, nape, and upperparts golden green, uppertail coverts tipped rufous. *Tail rufous, central rectrices mostly to all green, outer rectrices edged bronzy green.* **Imm:** throat and chest greyish buff to pale cinnamon, mottled green at sides, glittering green medially.

Voice. Fairly hard chips, at times run into

a rattle (not as hard as Rufous-tailed Hummingbird), and a high, sharp *siik* given by chasing birds. In apparent display makes fast erratic flights over an area, repeating a sharp *siik*. Song (?) hard chips run into a short series of wheezy, squeaky notes (JKi tape).

Habs. Arid to semihumid forest and edge, brushy scrub, clearings with flowers. Feeds and perches low to high, mainly at low to mid-levels. Nesting: Apr–Jul (E Mexico), Feb–Apr (Yuc Pen).

SS: See Berylline and Rufous-tailed hummingbirds.

SD/RA: C to F resident (SL–1200 m) on Atlantic Slope from Tamps (and S Texas) to Yuc Pen. A report from Honduras (Monroe 1968) is in error, being based on MLZ spec. 13887 which appears to be a hybrid Rufous-tailed × Cinnamon hummingbird.

NB: Has been called Fawn-breasted Hummingbird.

CINNAMON HUMMINGBIRD
Amazilia rutila*   Plates 30, 69
Colibri Canelo

ID: 4–4.5 in (10–11.5 cm). *Mostly Pacific Slope and Yuc Pen.* Bill 1.33–1.5× head, tail squared to slightly cleft, wings = tail. **Descr.** Ages/sexes similar. Bill bright red with black tip (♂) to mostly black above (♀/imm). *Throat and underparts cinnamon.* Crown, auriculars, nape, and upperparts green, distal uppertail coverts tipped rufous. *Tail deep rufous, rectrices edged bronzy green. A. r. graysoni* of Islas Tres Marías larger (4.5–4.8 in; 11.5–12 cm), darker and more bronzy above and duskier below than mainland birds.

Voice. A hard to sharp *chik* which may be run into hard rattles. Song varied, high, thin, slightly squeaky chips, *si ch chi-chit* or *tsi si si-si-sit*, or *chi chi-chi chi chi*, etc.

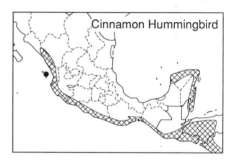

Buff-bellied Hummingbird

Cinnamon Hummingbird

Habs. Woodland and edge, thorn forest, plantations, typically in fairly arid areas. Feeds and perches low to high. Nesting: Nov–Feb, Jun–Jul, Sep (W Mexico), Jul–Oct (El Salvador), Feb–Apr, Aug–Nov (Yuc Pen).

SD: C to F resident (SL–1600 m) on Pacific Slope from cen Sin, and in interior from E Guatemala, to El Salvador and Honduras, and on Atlantic Slope from N Yuc to Belize, locally in NE Guatemala and Honduras.

RA: Mexico to Costa Rica.

VIOLET-CROWNED HUMMINGBIRD

Amazilia violiceps*   Plate 31
Colibri Corona-violeta

ID: 4–4.5 in (10–11.5 cm). *Endemic to W Mexico.* Bill 1.33–1.5× head, tail squared to slightly cleft, wings = tail. **Descr.** Ages differ slightly, sexes similar. **Adult:** *bill bright red with black tip. Crown bluish violet* (rarely turquoise-blue), nape and auriculars dull green, often mixed with blue to violet, small *white postocular spot* usually present. *Upperparts dull greenish brown to greenish. Throat and underparts white,* sides and flanks dusky greenish (usually covered by wings at rest), throat may be stained yellow by pollen, undertail coverts rarely with slight dusky centers. *Tail dull greyish green to greenish brown* (NW Mexico), *burnished bronzy copper in* violiceps *of SW Mexico.* **Imm:** bill blackish above. *Forehead and crown dark bluish green, tipped cinnamon* (often looks dark), a few violet feathers soon apparent.

Voice. Dry, hard chips, often run into a rattle. Song (?) a single plaintive chip, repeated, *chieu chieu chieu …*

Habs. Arid to semiarid scrub, thorn forest, riparian woodland, semiopen areas with hedges and scattered trees. Feeds and perches low to high, usually at mid- to upper levels. Nesting: Apr–Jul.

SS: Green-fronted Hummingbird has dark green crown, brighter green upperparts; note tail patterns, voice.

SD/RA: F to C resident (near SL–2400 m) on Pacific Slope from Son (U to extreme SW USA), and in interior from S Dgo and Hgo, to Balsas drainage. Most withdraw in winter (Sep–Mar) from N and interior areas when commoner and more widespread on Pacific Slope. Recorded in interior S to NW Oax (Mar–Oct; Binford 1989; IBUNAM specs) where possibly resident. A report from interior E Guatemala (Wendelken and Martin 1989) is not credible.

NB: Has been considered conspecific with Green-fronted Hummingbird *A. viridifrons.*

GREEN-FRONTED HUMMINGBIRD

Amazilia viridifrons*   Plate 31
Colibri Corona-verde

ID: 4–4.5 in (10–11.5 cm). *Endemic to SW Mexico.* Bill 1.33–1.5× head, tail squared to slightly cleft, wings = tail. **Descr.** Ages/sexes differ slightly. ♂: *bill bright red with black tip. Crown blackish green,* often with oily *blue-green sheen,* nape and auriculars golden green, rarely with faint whitish postocular spot. *Upperparts green to bronzy green. Throat and underparts white* (throat can be stained yellow by pollen), sides of chest and flanks spotted bronzy to green on pale dusky to dusky vinaceous wash; basal undertail coverts rarely with subterminal cinnamon marks. *Tail burnished bronzy copper to purplish copper* with narrow bronzy green edgings. ♀: crown dark green, rarely with blue-green sheen, *tail greenish gold to copper.* **Imm:** bill blackish above, forehead and crown duller, tipped cinnamon, tail mostly

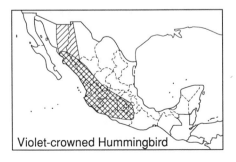

Violet-crowned Hummingbird

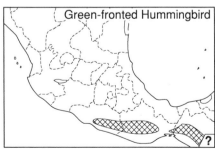

Green-fronted Hummingbird

deep purplish to purplish copper, tipped cinnamon. *A. v. rowleyi* of interior cen Oaxaca has more extensively vinaceous cinnamon sides and flanks (suggesting Cinnamon-sided Hummingbird), duller upperparts, ♀ may have more bronzy purple tail, ♂ has concealed cinnamon on base of secondaries.

Voice. Dry chattering, reminiscent of Broad-billed Hummingbird but slightly harder; softer than calls of Violet-crowned Hummingbird.

Habs. Arid to semiarid scrub, thorn forest, riparian woodland, semiopen areas with hedges and scattered trees. Feeds and perches low to high, usually at mid- to upper levels. Nesting: Dec–Feb, Apr–Jun.

SS: See Violet-crowned and Azure-crowned hummingbirds. Cinnamon-sided Hummingbird (sympatric in E Oax, and in upper Río Grande drainage) has bright cinnamon neck and chest sides, burnished rufous tail edged bronzy green, ♂ has dull rufous wing panel.

SD/RA: Locally/seasonally F to C resident (900–1400 m) in interior from cen Gro to cen Oax and, disjunctly (60–1300 m) from Pacific Slope of E Oax through Central Valley of Chis; ranges at least seasonally to Pacific Slope in Gro (Apr 1988, SNGH, SW). Probably occurs in interior NW Guatemala.

NB: See Violet-crowned Hummingbird.

CINNAMON-SIDED [GREEN-FRONTED] HUMMINGBIRD

Amazilia (viridifrons?) wagneri   Plate 31
Colibri Flanquicanelo

ID: 4–4.5 in (10–11.5 cm). *Endemic to S Oaxaca.* **Descr.** Ages/sexes differ slightly. ♂: *bill bright red with black tip. Crown blackish*, often with oily green sheen, nape and auriculars bronzy green. *Upperparts bronzy green* to bronzy. *Dull rufous patch*

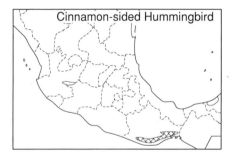
Cinnamon-sided Hummingbird

across secondaries and, less noticeably, on base of central primaries. *Throat and underparts white, lower auriculars and sides of neck mottled cinnamon, sides and flanks bright cinnmon* to cinnamon-rufous (often covered by wings at rest); undertail coverts often with subterminal cinnamon marks and shaft streaks. *Tail rufous-chestnut, edged bronzy green.* ♀: crown dark green, *cinnamon on face and sides paler, rufous in wing* restricted to tips of tertials and inner webs of secondaries (*mostly concealed*), inner rectrices bronzy to bronzy gold, *outer rectrices rufous, narrowly edged bronzy.* **Imm:** forehead and crown duller, blackish green in ♂, tipped cinnamon when fresh, wing panel dull cinnamon and restricted to base of secondaries (concealed in ♀), tail tipped cinnamon, central rectrices purplish copper basally in ♂.

Voice. Dry chattering, much like Green-fronted Hummingbird; also a quiet, hard crackling *zzzzrr'k chiuk*, repeated from perch.

Habs. Arid to semihumid scrub, thorn forest, riparian woodland. Feeds and perches low to high, usually at mid- to upper levels. Nesting: Jan–Feb, May, Aug–Oct.

SS: See Green-fronted Hummingbird.

SD/RA: F to C resident (250–900 m) in Oax on Pacific Slope from Sierra de Miahuatlán to Isthmus, and inland along lower Río Grande drainage.

GENUS EUPHERUSA

These are medium-sized hummingbirds typical of humid evergreen forest, especially cloud forest. Ages/sexes differ, bills black and straightish, 1.25–1.33× head, tails squared, wings = tail. Wings have conspicuous rufous patches, tails flash white when spread.

STRIPE-TAILED HUMMINGBIRD

*Eupherusa eximia**   Plate 30
Colibri Colirrayado

ID: 3.7–4 in (9.5–10 cm). *SE humid forest.* **Descr.** ♂: emerald green overall, glittering below, with buff lower belly, white undertail coverts. *Rufous secondaries form conspicuous panel on wing.* Inner rectrices green, *white flash on inner webs of blackish outer rectrices visible from above if tail spread.* ♀: throat and underparts pale grey,

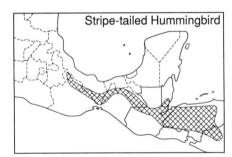

Stripe-tailed Hummingbird

auriculars dusky with small white postocular spot, white in tail extends to outer webs. **Imm** ♂: resembles ♀ but underparts dull green, usually with some glittering green on median throat and chest, tail as ♂.
Voice. A liquid, rattling trill, often given in flight, much like call of Emerald-chinned Hummingbird, and may suggest Costa's Hummingbird; also short, rolled, liquid chips. Song a rapid, high, liquid, chipping warble.
Habs. Humid evergreen forest and edge. Feeds and perches low to high, flashes tail open while feeding. Nesting: Apr–Aug.
SS: See Berylline Hummingbird.
SD: F to C resident (600–1800 m, to near SL in winter) on Atlantic Slope from S Ver, and in interior from Honduras, to N cen Nicaragua; U N of Isthmus. U to F on Pacific Slope in E Oax (Binford 1989). Reports from Pue (AOU 1983) require verification.
RA: SE Mexico to W Panama.
NB: Has been considered conspecific with White-tailed and Blue-capped hummingbirds.

WHITE-TAILED [STRIPE-TAILED] HUMMINGBIRD
Eupherusa poliocerca Plate 30
Colibri Guerrerense

ID: 4–4.3 in (10–11 cm). *Endemic to SW*

Mexico. **Descr.** ♂: emerald green overall, glittering below, undertail coverts white. *Rufous secondaries form conspicuous panel on wing*, rufous on outer webs of inner primaries duller. *Central rectrices greyish green (may look blackish), rest of tail whitish.* ♀: *throat and underparts pale grey*, auriculars dusky with small white postocular spot, less rufous in wings, inner rectrices green, *outer rectrices white with dusky green outer edges* (from below closed tail looks all white). **Imm** ♂: resembles ♀ but underparts dull green, mixed with greyish cinnamon on lower belly, usually some patches of glittering green on median throat and chest, tail as ♂ but outer rectrices noticeably tipped white.
Voice. A liquid to slightly buzzy, rolled chip, often run into rattled trills, much like Stripe-tailed Hummingbird. Song a high, rapid, slightly liquid to squeaky warbling, 2–8 s in duration, faster and more warbled than Blue-capped Hummingbird.
Habs. Humid evergreen forest and edge, plantations, ranges into semideciduous and pine–evergreen forest. Feeds and perches low to high, flashes tail open while feeding. Nest and eggs undescribed (?).
SS: See Berylline Hummingbird. Not known to be sympatric with Blue-capped Hummingbird and only ♂♂ can be told apart; ♂ Blue-capped has violet-blue crown.
SD/RA: F to C resident (800–2300 m) on Pacific Slope of Gro and W Oax.
NB: See Stripe-tailed Hummingbird.

BLUE-CAPPED [STRIPE-TAILED] HUMMINGBIRD
Eupherusa poliocerca (in part) or
E. cyanophrys Plate 30
Colibri Oaxaqueño

ID: 4–4.3 in (10–11 cm). *Endemic to Pacific Slope of Oaxaca.* **Desc.** ♂: *resembles ♂*

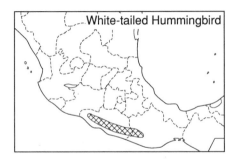

White-tailed Hummingbird

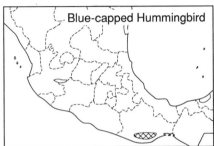

Blue-capped Hummingbird

White-tailed Hummingbird but forecrown violet-blue, hindcrown turquoise. ♀/imm: apparently indistinguishable from ♀/imm White-tailed Hummingbird, even in the hand (?).

Voice. Calls and song similar to White-tailed Hummingbird, possibly not distinguishable in the field (?) but song jerkier, less hurried.

Habs. Humid evergreen forest and edge, ranging to semideciduous and pine-evergreen forest. Habits much as White-tailed Hummingbird. Nesting: Sep–Nov, May.

SS: See Berylline and White-tailed hummingbirds.

SD/RA: F to C resident (700–1800 m, U to 2500 m) on Pacific Slope of Sierra de Miahuatlán, Oax.

NB: See White-tailed Hummingbird.

GENUS LAMPORNIS

Mountain-gems are fairly large hummingbirds characteristic of highland forests, especially cloud forest. Ages/sexes differ, bills black and straight, tails fairly long and broad, squared to slightly cleft, wings = tail. Faces patterned with whitish postocular stripes and dark auriculars; ♂♂ have glittering gorgets.

GREEN-THROATED MOUNTAIN-GEM
Lampornis viridipallens*   Plates 31, 70
Colibrí-serrano Gorjiverde

ID: 4.3–4.7 in (11–12 cm). *Endemic to cloud forest S of Isthmus.* Bill 1.33× head. **Descr.** ♂: *white postocular stripe contrasts with green crown and dark auriculars, white throat spotted glittering blue-green to green (often looks whitish, flecked dusky).* Nape and mantle green becoming purplish bronzy on rump, uppertail coverts

blue-black. *Underparts white, mottled green on sides* (heaviest in *viridipallens* of E Guatemala and Honduras in which green may meet across chest), flanks and undertail coverts dusky. *Inner rectrices blackish, outer rectrices pale grey, conspicuous when tail spread.* ♀: *lacks green throat spotting,* upperparts entirely emerald-green, *paler outer rectrices* may be tipped white. Imm ♂: resembles ♀ but throat washed pale buff.

Voice. A hard, buzzy *zzrrt*, at times repeated rapidly, and high, thin chips; also a quiet, gurgling warble, up to several s in duration.

Habs. Humid evergreen and pine-evergreen forest and edge. Feeds and perches low to high, flashes tail open while feeding. Nesting: Mar–Apr, Jun–Jul.

SS: See ♀ White-eared Hummingbird.

SD/RA: F to C resident (900–2700 m) on both slopes from SE Oax and Chis, and in interior from Guatemala, to W Honduras, W of the Sula Valley.

NB: The allopatric Green-breasted Mountain-Gem may be specifically distinct from Green-throated. Monroe (1963*b*, p. 3) claimed that 'the display patterns of the two ... differ considerably' but in the same paragraph said 'the display patterns ... are thought to be isolating mechanisms among closely related ... species ...' and '... would probably act as such in the *L. viridipallens* complex.' The displays (if they exist) in the *L. viridipallens* complex remain unknown.

GREEN-BREASTED [GREEN-THROATED] MOUNTAIN-GEM
Lampornis viridipallens (in part) or *L. sybillae*   Plate 70
Colibrí-serrano Pechiverde

ID: 4.3–4.7 in (11–12 cm). *Endemic to N C.A.* Bill 1.33× head. *Counterpart of*

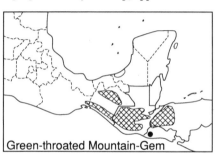

Green-throated Mountain-Gem

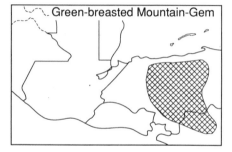

Green-breasted Mountain-Gem

Green-throated Mountain-Gem E of Sula Valley. **Descr.** ♂: *resembles Green-throated Mountain-Gem but throat and chest glittering blue-green to green* (can look pale greyish), *rump green, whitish outer rectrices contrast more strongly with dark inner rectrices,* striking when tail spread. ♀: *resembles ♀ Green-throated but throat usually washed buff, outer rectrices more contrastingly whitish.* **Imm** ♂: resembles ♀ but throat whitish, only faintly washed buff.

Voice. A hard, raspy buzz, *zzrrt,* much like Green-throated Mountain-Gem. Song undescribed (?).

Habs. Humid evergreen and pine—evergreen forest and edge. Habits much as Green-throated Mountain-Gem. Nest and eggs undescribed (?).

SS: See ♀ White-eared Hummingbird.

SD/RA: F to C resident (1400–2200 m) in interior from E Honduras to N cen Nicaragua.

NB: See Green-throated Mountain-Gem.

AMETHYST-THROATED HUMMINGBIRD
Lampornis amethystinus*   Plate 31
Colibri-serrano Gorjiamatisto

ID: 4.5–5 in (11.5–12.5 cm). Endemic, highlands from cen Mexico S. Bill 1.33× head. Two groups: *amethystinus* (widespread) and *margaritae* (SW Mexico). **Descr.** *Amethystinus.* ♂: *white postocular stripe contrasts with dark green crown and dusky auriculars, gorget glittering rose-pink.* Nape and mantle dark green becoming bronzy on rump, distal uppertail coverts blackish. *Underparts dusky grey,* washed dull green, undertail coverts edged pale buff. *Tail blue-black, outer rectrices narrowly tipped pale grey.* ♀: *throat dusky pale cinnamon.* **Imm:** resembles ♀ but throat dusky, usually with faint cinnamon

edgings, ♂ often has a few pink throat feathers. *Margaritae:* ♂: gorget glittering bluish violet. ♀ and imm like *amethystinus* but imm ♂ shows violet throat feathers.

Voice. Song (?) a high, sharp *tsip tsip ...* or *tsii tsii ...,* etc. (Jal, Chis, *amethystinus*) and a harder, sharp chipping *chik chik ...* or *chiup chiup ...* (Gro, *margaritae*), often repeated insistently when perched. Also a rasping, buzzy *tzzzzir,* given mostly while feeding, a rapid, fairly hard, clicking *ki-ki-ki-ki-kik ki-kik ...,* a high, slightly shrill *siik* in flight, not as thin as Blue-throated Hummingbird, and a quiet, dry, gurgling chatter.

Habs. Humid to semihumid evergreen and pine—evergreen forest and edge. Usually feeds and perches at low to mid-levels, often at banks of flowers. Nesting: Oct–Dec, May–Jul (W Mexico to Guatemala). B.c.: Sep (Hgo).

SS: Blue-throated Hummingbird has large white tail tips, ♂ has blue throat.

SD/RA: *Amethystinus:* F to C resident (900–3000 m) on both slopes from S Nay and S Tamps, and in interior from Mex and Mor, to El Salvador and cen Honduras; absent S Mich to SW Oax. *Margaritae:* F to C resident in Pacific Slope mountains from Mich to Oax; possibly sympatric with *amethystinus* in S Oax. See Phillips (1966b) and Binford (1989) for discussion of this confusing situation. If specific status is warranted for *margaritae,* we suggest the English name Violet-throated Hummingbird.

BLUE-THROATED HUMMINGBIRD
Lampornis clemenciae*   Plate 31
Colibri-serrano Gorjiazul

ID: 4.8–5.3 in (12–13.5 cm). Highlands N of Isthmus. Bill 1.25–1.33× head. **Descr.** ♂: *white postocular stripe contrasts with green crown and dusky auriculars, gorget*

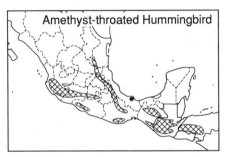
Amethyst-throated Hummingbird

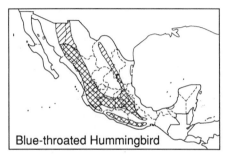
Blue-throated Hummingbird

glittering blue, usually bordered by white moustache. Nape and upperparts green, distal uppertail coverts blackish. Underparts dusky grey, mottled green at sides of chest, undertail coverts edged whitish. *Tail blue-black with broad white tips to outer rectrices.* ♀: *throat dusky grey.* **Imm** ♂: resembles ♀ but with some blue in throat.

Voice. A high, thin, squeaky *tsii*, often given in flight, a fuller *tsiuk*, and a high *seek* or *siik* repeated steadily by perched ♂ (song?).

Habs. Pine–oak and oak woodland and edge, clearings with flowers. Feeds and perches at low to mid-levels, ♂♂ may be higher when singing (?); often at banks of flowers along streams. Nest usually on slender twigs along overhanging wet bank or rock face. Nesting: Mar–Jul (NW Mexico), Jul–Sep (Oax).

SS: ♀ Magnificent Hummingbird has longer bill, white postocular spot, smaller white tail tips, often perches high and conspicuously. See Amethyst-throated Hummingbird, ♀ Violet Sabrewing.

SD/RA: F to U resident (1800–3000 m) on both slopes and in interior from Son and Coah (also SW USA) to Oax, most withdraw (Oct–Mar?) from N parts of range (N Son to NL, at least).

GARNET-THROATED HUMMINGBIRD
Lamprolaima rhami*   Plate 31
Colibri Alicastaño

ID: 4.8–5 in (12–12.5 cm). *Endemic to E and S highlands.* Bill straight and sharp, 1× head; tail broad and cleft, wings = tail. **Descr.** Ages/sexes differ. ♂: bill black. *Often looks all dark with rufous wings. Blackish face with white postocular spot borders narrow, glittering rose-pink gorget, chest glittering violet,* rest of underparts sooty grey, mottled green on flanks.

Remiges *rufous*, tipped brown. Crown, nape, and upperparts deep green; tail dark purplish. ♀: *throat and underparts dusky grey* (usually with a few pink throat feathers), sides and flanks mottled green, outer rectrices narrowly tipped whitish. **Imm** ♂: resembles ♀ but darker below with fine cinnamon fringes to blue-mottled chest, rufous in wings less extensive than ♂.

Voice. A nasal *nyik* and *choiw* and high, sharp chips. Song (?) a quiet, gruff, dry, crackling warble with nasal, gurgling notes thrown in; also a fairly sharp, slightly buzzy *tis-i-tyu'tyu'*, repeated 2–3×.

Habs. Humid evergreen and pine-evergreen forest, edge, and adjacent shrubby thickets. Feeds and perches low to high, makes conspicuous high flycatching sallies with distinct jerky movements. Nesting: Dec–Mar (Pacific Slope, Oax), Apr–May (Atlantic Slope, Oax).

SD/RA: Locally/seasonally F to U resident (1200–3000 m) on both slopes from Gro and Pue to El Salvador and Honduras, in interior from Guatemala S.

MAGNIFICENT (RIVOLI'S) HUMMINGBIRD
Eugenes fulgens*   Plate 31
Colibri Magnífico

ID: 4.7–5.3 in (12–13.5 cm). *Highlands.* Bill straight, 1.33–1.5× head, tail broad and squared, wings = tail. **Descr.** Ages/sexes differ. ♂: bill black. *Often looks all dark with long bill, white postocular spot.* Forehead and crown glittering violet, gorget glittering turquoise-green, auriculars dusky with distinct white postocular spot. Nape and upperparts green. Underparts blackish, sides and flanks mottled green, undertail coverts dusky with white edges. *Tail green to bronzy green.* ♀: *white post-*

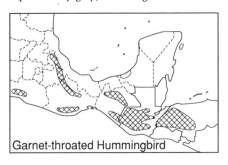

Garnet-throated Hummingbird

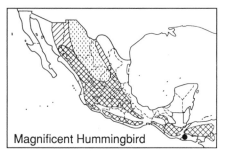

Magnificent Hummingbird

ocular spot contrasts with green crown and dusky auriculars. Nape and upperparts green. Throat and underparts dusky grey (darker in imm with neat whitish to buff scaly edgings), undertail coverts edged whitish. *Tail green, outer rectrices with broad black subterminal band* (indistinct in imm), *tipped white.* **Imm** ♂: green upperparts with neat, scaly, pale edgings, throat and underparts dusky green with neat, pale scaly edgings; often shows traces of violet forehead and green gorget; outer rectrices lack well-defined dark band, whitish tips narrower.

Voice. A fairly loud, sharp *tchik* or *tcheep*, often repeated insistently from perch; also a soft, slightly buzzy, gurgling warble.

Habs. Pine–oak and pine–evergreen forest, edge, clearings with flowers, also desert grasslands and open farmland with flowering agaves, etc., in migration. Feeds and perches low to high, often fairly high on exposed perches whence makes flycatching sallies like a starthroat. Nesting: May–Aug (N and cen Mexico). B.c.: Jul–Aug (Guatemala).

SS: See ♀ Blue-throated Hummingbird. Starthroats have longer bills, white back patches, blackish throats, Plain-capped has whitish postocular stripe.

SD: F to C resident (1000–3000 m, mainly above 1500 m) on both slopes and in interior from Son and Coah to Honduras and N cen Nicaragua, most withdraw (Oct–Mar?) from N of range (N Son to NL, at least).

RA: SW USA to W Panama.

NB: Populations in Costa Rica and Panama sometimes considered specifically distinct, *E. spectabilis.*

PURPLE-CROWNED FAIRY

Heliothryx barroti   Plate 29
Hada Coronimorada

ID: 4.5–5.2 in (11.5–13 cm). *SE humid lowland forest. Bill straight and sharply pointed, 1× head, tail long and strongly graduated,* wings < tail. **Descr.** Ages/sexes differ. ♂: bill black. *Forecrown glittering violet, black mask* contrasts with green hindcrown and green moustache; glittering violet droplet at rear of mask. Nape and upperparts green. *Throat and underparts white.* Inner rectrices blue-black, *outer rectrices white.* ♀: tail markedly longer than

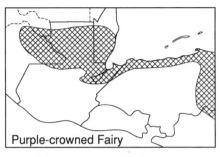

Purple-crowned Fairy

♂, crown green, moustachial area white. **Imm:** crown, nape, and upperparts broadly tipped cinnamon, throat and chest with sparse fine dusky spotting, ♂ lacks violet on head.

Voice. Usually silent; at times gives a high, thin, slightly metallic *sssit*, which may be run into longer rapid series.

Habs. Humid evergreen forest and edge. Feeds and perches at mid- to upper levels, often in canopy. Often hunts insects, flashing tail open and moving quickly and erratically along edges or in openings; also feeds by piercing flower bases. Splash-bathes in quiet puddles. Nesting: Mar.

SD: U to F resident (SL–500 m, rarely higher) on Atlantic Slope from E Chis and E Tab to Honduras.

RA: SE Mexico to Ecuador.

LONG-BILLED STARTHROAT

*Heliomaster longirostris**   Plate 29
Picolargo Coroniazul

ID: 4.5–4.8 in (11.5–12.5 cm). S Mexico to C.A. *Bill very long and straight, 2–2.33× head, tail squared to slightly rounded,* wings ≤ tail. **Descr.** Ages/sexes differ slightly. ♂: bill black. *Crown glittering turquoise-blue to turquoise-green, white postocular spot* contrasts with dusky auriculars, *broad white moustache borders*

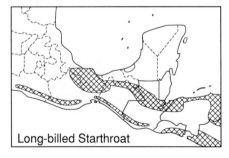

Long-billed Starthroat

blackish chin and glittering rose-pink lower gorget; postocular stripe may be pale but rarely striking. Underparts pale grey, *mottled green on sides and flanks*, belly whitish, undertail coverts dusky with broad white edges. Nape and upperparts emerald-green to golden green with *white flash on lower back*; white flank tufts sometimes exposed. Inner rectrices green (often tipped black), outer rectrices blackish, tipped white. ♀: crown green to turquoise-green with little or no blue, gorget narrower with more black, less red, sides and flanks less heavily mottled green. ♀ *masculinus* (Pacific Slope N of Isthmus) much like ♂, with bluish crown, more red in throat. **Imm:** resembles ♀ but gorget sooty, ♂ usually with a few pink feathers; blue crown apparently last feature attained in complete 1st prebasic molt.

Voice. A sharp *pik* or *peek*, softer than Plain-capped Starthroat.

Habs. Humid to semihumid forest edge, clearings, semiopen areas with hedges and scattered trees. Feeds low to high, often perches high on exposed twig or wire whence hawks insects. Nesting: Nov–Mar (Pacific Slope). B.c.: Aug (Ver).

SS: Plain-capped Starthroat duller overall, sides and flanks greyish, crown greyish green with whitish postocular stripe; in good views look for orange-red lower gorget, paler greyish base to outer rectrices below, and (usually) dark spot on outer web of white tips of outer rectrices. See Magnificent Hummingbird (rarely sympatric).

SD: U to F resident (near SL–1600 m) on both slopes from Gro (Navarro 1986; SNGH, SW) and S Ver to El Salvador and Honduras.

RA: S Mexico to Peru and Brazil.

PLAIN-CAPPED STARTHROAT
*Heliomaster constantii**   Plate 29
Picolargo Coronioscuro

ID: 4.7–5 in (12–13 cm). *Pacific Slope and interior valleys. Bill very long and straight, 1.75–2× head, tail squared, wings = tail.* **Descr.** Ages differ slightly, sexes similar. **Adult:** bill black. *White postocular stripe* contrasts with dull green crown and dusky auriculars, broad *white moustache borders sooty grey upper throat and glittering pinkish red to orangish red lower gorget.* Nape and upperparts greyish green to gol-

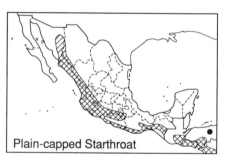

Plain-capped Starthroat

den green with *white flash on lower back*; white flank tufts sometimes exposed. Underparts pale grey with slight green mottling on sides and flanks, belly white, undertail coverts dusky with broad white edges. Inner rectrices green (often tipped black), outer rectrices blackish (pale grey basally), tipped white, *usually with black spot at tip of outer webs.* **Imm:** throat sooty grey with little or no red.

Voice. A sharp, fairly loud *peek*, recalls Black Phoebe. Song (?) a series of sharp chips interspersed with varied chips, *chip chip chip chip pi-chip chip chip ...*, or *chi chi chi chi whit-it chi ...*, etc. (JKi tape).

Habs. Arid to semiarid forest edge, thorn forest, semiopen areas with scattered trees and hedges. Habits much as Long-billed Starthroat. Nesting: Jan–Jun (W Mexico). B.c.: Aug (Gro), Oct–Nov (El Salvador).

SS: See Long-billed Starthroat, Magnificent Hummingbird.

SD: F to C resident (SL–1500 m) on Pacific Slope from S Son to Isthmus and from E Guatemala to El Salvador and Honduras, and in interior from Balsas drainage to Honduras; U (Jun–Sep?) to N Son.

RA: W Mexico to NW Costa Rica.

SPARKLING-TAILED WOODSTAR (HUMMINGBIRD)
Philodice dupontii   Plate 32
Colibri Colipinto

ID: ♂ 3.5–3.8 in (9–10 cm); ♀ 2.5–3 in (6.5–7.5 cm). *Endemic; foothills and highlands from cen Mexico S.* Bill straightish, 1× head. **Descr.** Ages/sexes differ. ♂: *tail very long and deeply forked, usually held closed in a point,* wings < tail. Bill black. *Gorget glittering blue to violet-blue,* crown and auriculars green with white postocular spot. Nape and upperparts green with *squarish white patches on sides of rump.* White chest forms contrasting band below gorget, rest of underbody dusky green. Tail blackish, outer rectrices tipped white and outer 2 pairs with white median band bordered rufous; *closed tail looks black with 3 white bands and white tip.* ♀: *tail cleft* and double-rounded, wings = tail. *Face, throat, and underparts cinnamon* to pale cinnamon with short dark eyestripe, pale postocular spot. Crown, nape, and upperparts green with *whitish patches on sides of rump.* Inner rectrices green, outer rectrices black, tipped whitish to pale cinnamon. **Imm** ♂: some resemble ♀ but with green mottling on underparts, outer rectrices tipped white. Others resemble ♀ but 1st basic (?) tail longer with white median band when closed; throat and underparts may be mottled blue and green.

Voice. Usually silent, rarely gives high, sharp, twittering chips. ♂ sings from an exposed perch, 'a very high, thin, but musical squeaking ... lasting for many secs at a time' (Rosenberg and Rosenberg 1979).

Habs. Humid to semiarid forest edge, brushy second growth, scrubby woodland, often with oaks. Feeding flight slow and bee-like, tail cocked, wings make low, insect-like humming. Feeds and perches low to high, makes prolonged insect-catching flights from exposed perches. Nest and eggs undescribed (?). B.c.: Aug (Mich).

SS: White squares on sides of rump distinctive; note shortish straight bill, compared with *Calothorax*.

SD/RA: Locally/seasonally F to U resident (750–2500 m, locally/seasonally to near SL) on both slopes from cen Sin and cen Ver, and in interior from cen Mexico, to Honduras and N cen Nicaragua.

NB: Sometimes known as Dupont's Hummingbird; traditionally placed in the monotypic genus *Tilmatura*.

GENUS CALOTHORAX

Sheartails are small hummingbirds of semi-open scrubby areas, black bills long and arched, ♂ tails long and forked, wings < tail. Ages/sexes differ, ♂♂ with glittering gorgets, ♀♀ pale buff to whitish below, with distinctive dusky auricular stripe. ♂ displays (as far as known) consist of rocking pendulum flights, high climbs, and steep dives. Field identification can be practically impossible, not helped by the fact that ♀♀ and imms typically outnumber ♂♂ by up to 10 to one.

MEXICAN SHEARTAIL
Calothorax eliza   Plate 32
Tijereta Yucateca

ID: ♂ 3.7–4 in (9.5–10 cm); ♀ 3.3–3.5 in (8.5–9 cm). *Endemic to SE Mexico.* Bill 1.5× head. **Descr.** ♂: *tail very long and deeply forked, usually held closed in a point. Gorget glittering rose-pink,* whitish postocular stripe between gorget and dull green crown. Nape and upperparts bronzy green. Median underparts whitish, sides and flanks mottled bronzy green, undertail coverts washed cinnamon. Inner rectrices green, *outer rectrices blackish with pale cinnamon edging to inner webs* (tail looks black when closed). ♀: *tail forked and double-rounded.* Face whitish with dusky auricular stripe. Throat and underparts whitish (tinged buff in imm), sides, flanks, and undertail coverts washed buffy cinnamon. Crown, nape, and upperparts bronzy green. *Inner rectrices green, outer rectrices rufous with broad black subterminal band and white tips.* **Imm** ♂: resembles ♀ but

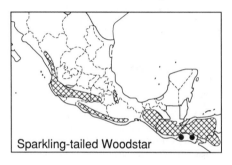

Sparkling-tailed Woodstar

Mexican Sheartail

throat flecked pink, tail with broader black subterminal band.

Voice. Fairly hard to slightly liquid, rapid chipping, often slightly rolled. ♂ makes a dry chattering at the bottom of his display dive, and a rapid, dry, buzzy *zz-zz-zzt* as he flies back and forth in front of ♀; both sounds possibly made with his wings (?).
Habs. Arid scrub and semiopen mangrove edge, gardens. Feeds and perches at low to mid-levels, often feeds near ground; nest may be fairly high in palms though usually low in scrubby bush or tree. ♂ feeds with tail held closed and nearly vertical, ♀ feeds with rapid tail-wagging. Wings make low, insect-like humming. In display, ♂ buzzes more loudly, back and forth in front of perched ♀, with his tail spread to 60–100° and twisted to almost 90° from his body plane, then he climbs steeply, tail closed, to 30+ m before plummeting down to swoop up and perch near ♀. Nesting: Aug–Apr (Yuc). B.c.: May (Ver).
SS: ♀ Lucifer Hummingbird (allopatric) buffier below with shorter, less strongly arched bill, closed tail tapered (square-tipped in Mexican Sheartail).
SD/RA: C to F resident (around SL) in coastal Yuc Pen from NW Camp to N QR, vagrant to NE QR (Apr 1977, BM); F but local resident (near SL–300 m) in cen Ver.
NB: Traditionally placed in the genus *Doricha*.

SLENDER SHEARTAIL
Calothorax enicura Plate 32
Tijereta Centroamericana

ID: ♂ 4.3–5 in (11–12.5 cm); ♀ 3.2–3.5 in (8–9 cm). *Endemic to highlands S of Isthmus.* Bill 1.25–1.5× head. **Descr.** ♂: *tail extremely long and deeply forked, usually held closed in a point.* Chin blackish, lower *gorget glittering pinkish violet,* crown and auriculars green with white postocular

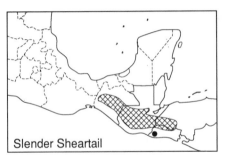

Slender Sheartail

spot. Nape and upperparts green. Whitish chest forms contrasting band below gorget, rest of underparts dull green, central belly whitish. Inner rectrices green, *outer rectrices blackish.* ♀: *tail forked and double-rounded.* Face pale buff with dusky auricular stripe, whitish postocular spot. *Throat and underparts buffy cinnamon* to buff, throat paler, belly whitish. Crown, nape, and upperparts green. *Inner rectrices green, outer rectrices rufous with broad black subterminal band, tipped white* (rectrix 3 tipped white to cinnamon, rarely all dark). **Imm** ♂: resembles ♀ but rectrices narrower, tail slightly longer and more deeply forked (?); outer rectrices more narrowly tipped white, often shows violet spots on throat. 1st basic tail intermediate between ♂ and ♀ in length, outer rectrices tipped cinnamon.
Voice. Fairly hard, rapid chips, often repeated steadily or slightly trilled, given while perched or feeding.
Habs. Humid to semiarid scrub, brushy second growth, forest edge. Feeds and perches mainly at low to mid-levels, often feeds near ground. ♂ feeds with tail held closed and nearly vertical, ♀ feeds with rapid tail-wagging and often flashes tail open. Wings make low, insect-like humming. Nesting Oct (Chis); display undescribed (?).
SS: ♀ and imm ♂ Beautiful Hummingbird (allopatric?) have straighter bill, underparts not so uniformly buffy cinnamon (but similar to pale Slender Sheartail) and often dirty pale grey on throat, ♀ has shorter tail but imm ♂ tail similar to ♀ Slender Sheartail. In hand, note narrower and slightly falcate tip to outer primary of Beautiful (and Lucifer) versus blunter and straighter tip in Slender Sheartail.
SD/RA: Locally/seasonally F to C resident (1000–2200 m), mostly in interior and on adjacent Pacific Slope, from Chis to W Honduras and N El Salvador.
NB: Traditionally placed in the genus *Doricha*.

LUCIFER HUMMINGBIRD
Calothorax lucifer Plate 32
Tijereta Norteña

ID: 3.5–4 in (9–10 cm). *Arid scrub in N and W Mexico.* Bill arched, 1.5× head. **Descr.** ♂: *tail fairly long and deeply forked, outermost rectrices sharply pointed. Gorget* *glittering rose-pink with violet-blue highlights*, whitish postocular stripe between gorget and green crown. Nape and upperparts green. Whitish chest contrasts with dark gorget and dusky green underbody, flanks washed buff, undertail coverts whitish. Inner rectrices blue-green, outer rectrices blackish. ♀: tail cleft and double-rounded. *Face whitish to pale buff with dusky auricular stripe*, whitish postocular spot. Crown, nape, and upperparts green. Throat and underparts pale buffy cinnamon, washed greyish on throat, belly whitish. Tail green, outer rectrices with broad black subterminal band, tipped white, rufous at base. **Imm** ♂: resembles ♀ but rectrices narrower, usually shows a few pink throat feathers.

Voice. High, twittering chips.

Habs. Arid to semiarid scrub, brushy woodland, semiopen areas with scattered trees and bushes. Feeds and perches low to high, mainly fairly low. In display, ♂ buzzes loudly back and forth in pendulum motion, gorget flared out, less often climbing to 20+ m and stooping steeply; when ♀ present, ♂ goes into pendulum-like rocking at bottom of dive (Wagner 1946*b*). Nesting: late Apr–Oct.

SS: In winter, Lucifer reaches range of Beautiful Humminbird which has straighter, slightly shorter bill, ♂ has rounded tips to outermost rectrices, ♀ generally paler and greyer below; the two species rarely separable in field.

SD/RA: F to C breeder (Mar–Oct, 1000–2500 m) in interior from NE Son and Coah (also SW USA) to cen Mexico, at least to DF (Wilson and Ceballos-L. 1986); winters (Aug–Apr, 600–2500 m) on Pacific Slope and in interior from Nay and Jal to Balsas drainage, R to U N to Sin (DMNH spec.) and (Oct–Dec at least) S to Oax (Binford 1989).

BEAUTIFUL HUMMINGBIRD

Calothorax pulcher   Plate 32
Tijereta Oaxaqueña

ID: 3.2–3.7 in (8–9 cm). *Endemic to SW Mexico. Bill slightly arched, 1.33× head.* **Descr.** ♂: *tail fairly long and deeply forked, outer pair of rectrices bluntly pointed. Gorget glittering rose-pink with violet highlights*, whitish postocular stripe between gorget and green crown. Nape and upperparts green. Whitish chest contrasts with dark gorget and dusky green underbody, flanks washed buff, undertail coverts whitish. Inner rectrices blue-green, outer rectrices blackish. ♀: tail cleft and double-rounded. *Face whitish to pale buff with dusky auricular stripe*, whitish postocular spot. Crown, nape, and upperparts green. *Throat and underparts pale buffy cinnamon*, washed greyish on throat, belly whitish. *Tail green, outer rectrices with broad black subterminal band, tipped white, cinnamon-rufous at base* (rectrix 3 rarely dark to tip). **Imm** ♂: resembles ♀ but rectrices narrower, tail slightly longer and more deeply forked, usually shows a few pink throat feathers.

Voice. High, twittering chips much like Lucifer Hummingbird. A very high, thin to slightly squeaky *t-swiik t-swiik* . . ., or *w-siiu w-siiu* . . ., to 4×, in rocking display.

Habs. Arid to semiarid scrub, thorn forest, semiopen areas with scattered trees and bushes. Feeds and perches low to high, mainly fairly low. In display, ♂ rocks back and forth, bill tip to bill tip with perched ♀, his tail fanned and twisted almost to 90° from his body plane, then he climbs steeply to 10+ m and dives down to swoop over ♀. Nesting: Nov, May, Jul.

SS: See Lucifer Hummingbird, ♀ Slender Sheartail (allopatric?).

SD/RA: Incompletely known due to con-

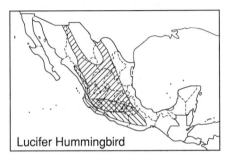

Lucifer Hummingbird

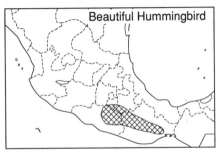

Beautiful Hummingbird

fusion with Lucifer Hummingbird. Locally/ seasonally F to C resident (1000–2200 m) in interior from cen Gro and S Pue to Oax; seasonally to near SL in Isthmus (Dec– Apr, SNGH, SW). Reports from DF and Mor (AOU 1983) are based on MLZ specs of equivocal identity, although culmen length suggests Beautiful. Perpetuated reports from Chis (AOU 1983) doubtful; based on Avilés spec. at MLZ (which may even be Slender Sheartail because of shape of outer primary and tail length).

GENUS ARCHILOCHUS

These are small migratory hummingbirds mainly of N and W Mexico. Black bills straightish, tails forked in ♂♂, slightly rounded in ♀♀. Ages/sexes differ, ♂♂ with glittering gorgets, ♀♀ and imms often hard to identify. Some authors unite this genus with *Selasphorus*, Broad-tailed Hummingbird being intermediate between the two genera.

RUBY-THROATED HUMMINGBIRD

Archilochus colubris   Plate 32
Colibri Gorjirrubi

ID: 3.2–3.7 in (8–9 cm). Widespread migrant. *Bill 1–1.33× head. Wings ≤ tail.* **Descr. ♂:** chin blackish, *gorget glittering ruby-red*, underparts dusky, mottled green on sides, with broad white forecollar below gorget. *Crown, nape, and upperparts emerald green*, small white postocular spot. Inner rectrices green, outer rectrices blackish. **♀:** *White postocular spot* contrasts with *green crown* (crown dusky in some imms) and dusky auriculars, throat and underparts whitish with *broad whitish forecollar extending as squarish patches on sides of neck, pale cinnamon wash on sides. Nape and upperparts emerald-green to golden green. Tail green,* outer rectrices with broad black subtermi-

nal band, outer 3 tipped white. **Imm ♂:** resembles ♀ but throat with lines of dusky flecks and usually one or more red feathers; by late winter (Mar) resembles ♂ but gorget patchy, often with large white central area.
  Voice. High, twittering chips, *tchi tchi tchi-chit, tchi* ... etc..
  Habs. Forest and edge, plantations, brushy woodland, semiopen areas with forest patches and scattered trees. Feeds and perches low to high, usually at low to mid-levels, often common at banks of flowers with swarms of other hummingbirds.
  SS: Black-chinned Hummingbird has longer bill (especially noticeable in ♀), ♂ gorget colors distinctive but often hard to see, ♀/ imm upperparts duller golden green to grey-green, forecrown dusky, tail typically bluish green, underparts often slightly dirtier with less extensive cinnamon wash on flanks. More tapered tips of primaries of Ruby-throated may be visible in good view (versus blunt wingtip of Black-chinned); in hand, note also notched tips of outer rectrices of adult ♀ Black-chinned (but rounded in imm). ♀/imm Costa's Hummingbird (deserts) shorter-tailed, dingier underparts typically lack bold whitish squares on sides of neck, upperparts dingier greyish green without blackish lores but typically with whitish postocular stripe (though postocular spot often brighter), calls distinctly liquid. ♂ Broad-tailed Hummingbird has rose-pink gorget and paler auriculars, longer tail, ♀/imm sides strongly washed cinnamon, throat with lines of dusky flecks. ♀/imm Anna's Hummingbird (unlikely to overlap) larger and bulkier, typically with a splash of reddish pink in throat, underparts pale grey.
  SD: F to C transient and winter visitor (Sep– Apr, SL–3000 m) on Pacific Slope from S Sin, and in interior from Oax, to El Salvador and Honduras. U to F in winter on Atlantic Slope from S Ver to Honduras. C to F transient (mid-Feb–mid-May, Aug– Oct) on Atlantic Slope and in E interior N of winter range. Vagrant to BCS (Dec 1983, Howell and Pyle 1988).
  RA: Breeds E N.A.; winters Mexico to Panama.

BLACK-CHINNED HUMMINGBIRD

Archilochus alexandri   Plate 32
Colibri Barbinegro

ID: 3.3–3.8 in (8.5–9.5 cm). N and W Mexico. *Bill 1.25–1.5× head.* Wings ≤

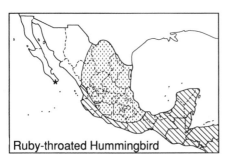
Ruby-throated Hummingbird

tail. **Descr.** ♂: *throat blackish above broad band of glittering bluish violet,* underparts dusky, mottled green on sides, with broad white forecollar below gorget. Crown, nape, and upperparts green, small white postocular spot. Inner rectrices green, outer rectrices blackish. ♀: *white postocular spot contrasts with dusky green crown and dusky auriculars,* throat and underparts whitish (throat may show lines of dusky flecks) with *broad whitish forecollar extending as squarish patches on sides of neck,* sides pale greyish, *with only slight pale cinnamon wash in adult often restricted to hind flanks;* but sides and flanks of imm often washed buffy cinnamon. *Nape and upperparts greyish green to golden green. Tail typically slightly bluish green,* outer rectrices with broad black subterminal band, outer 3 tipped white. **Imm** ♂: resembles ♀ but throat with lines of dusky flecks, usually one or more black and bluish violet feathers. As ♂ by spring.
Voice. High twittering chips, much like Ruby-throated Hummingbird.
Habs. Breeds in arid to semiarid scrub, riparian woodland, gardens. In winter also in humid areas. Feeds and perches low to high, mainly at low to mid-levels. ♂'s wings make dry buzz in display flight as he flies back and forth with pendulum motion. Nesting: Apr–Aug.
SS: See Ruby-throated Hummingbird. ♀/imm Costa's Hummingbird shorter-tailed, often shorter-billed, underparts lack bold whitish squares on sides of neck, upperparts dingier greyish green without blackish lores but typically with whitish postocular stripe (though postocular spot often brighter), imm ♂ usually has patch of purple on throat, calls distinctly liquid. ♀/imm Anna's Hummingbird larger and bulkier, proportionately shorter-billed, typically has a splash of reddish pink in throat, underparts pale grey.

SD: F to C breeder (Mar–Sep, near SL–2000 m) in N BCN and in interior from NE Son to Coah and NL. F to C transient and winter visitor (Aug–Apr; SL–1500 m) on Pacific Slope from S Son to Mich and in Balsas drainage E to Mor, R in BCS, in N interior Mexico, and to W Ver.
RA: Breeds W NA to N Mexico; winters W Mexico.

ANNA'S HUMMINGBIRD
Archilochus anna   Not illustrated
Colibri de Anna

ID: 3.5–4 in (9–10 cm). NW Mexico. Bill 1.25–1.33× head. Wings < tail. **Descr.** ♂: *crown and gorget glittering rose-pink to orangish red,* underparts dusky greyish, mottled green. Nape and upperparts green, white postocular spot. Inner rectrices green, outer rectrices blackish. ♀: whitish postocular spot contrasts with green crown and dusky auriculars, nape and upperparts green. *Throat and underparts pale greyish, throat with lines of dusky spots,* typically *with glittering rose-pink to orange-red patch* (reduced or absent in imm). Tail green, outer rectrices with broad black subterminal band, and outer 2–3 rectrices tipped white. **Imm** ♂: resembles ♀ but often some ♂ color on crown, outer 1–2 rectrices tipped white.
Voice. Song often prolonged, a high, squeaky, and wiry warble. Calls include sharp to buzzy chips often run into chatters; calls notably more liquid and spluttering in Isla Guadalupe birds.
Habs. Semiarid to arid scrub, semiopen areas with bushes and trees, riparian woodland. Feeds and perches at low to mid-levels, ♂ often perches on conspicuous bare twigs when singing, and makes steep dives in display flight. Nesting: Dec–Apr.
SS: See Black-chinned and Ruby-throated

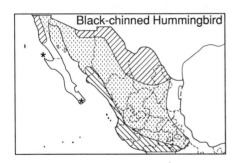

Black-chinned Hummingbird

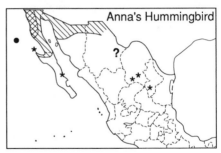

Anna's Hummingbird

hummingbirds. Costa's Hummingbird smaller and shorter-tailed, ♀ paler below, dingier above, throat lacks distinct lines of dusky streaks and reddish patch. ♂ Broad-tailed has green crown, ♀ has cinnamon wash on sides.

SD: F to C resident (SL–2500 m), commonest Dec–Jul, in N BCN (including Isla Guadalupe); U to F winter visitor (Sep–Mar) to N BCS (Howell and Webb 1992*f*) and in N Mexico from N Son to N Chih, R to NL (Feb 1983, AMS).

RA: W N.A. to N Baja; winters to N Mexico.

NB: Traditionally placed in the genus *Calypte*.

COSTA'S HUMMINGBIRD
Archilochus costae   Plate 32
Colibri de Costa

ID: 3–3.4 in (7.5–8.5 cm). *NW Mexico, deserts.* Bill 1.25–1.33× head, *wings ≥ tail*. **Descr. ♂:** *crown and elongated gorget glittering purplish,* nape and upperparts grey-green to golden green. Underparts whitish, mottled greenish on sides, often with broad whitish band below gorget. Inner rectrices green, outer rectrices blackish. **♀:** *whitish postocular stripe (often with whiter postocular spot)* between green crown and dusky auriculars, *throat and underparts dirty whitish, often tinged greyish to greyish buff,* rarely with whitish patches at sides of neck; may show a few violet flecks on throat. Nape and upperparts dingy grey-green to golden green. *Tail typically slightly bluish green,* outer rectrices with broad black subterminal band, outer 3 tipped white. **Imm ♂:** resembles ♀ but tail dotted with lines of dusky flecks and *usually one or more purple feathers forming a small central chevron.*

Voice. In display flight, ♂ gives a very high, thin, whining whistle as he makes high oval loops. Calls include distinctly liquid, twittering chips.

Habs. Arid to semiarid scrub, open areas with scattered bushes and trees. Feeds and perches at low to mid-levels, ♀ feeds with much tail-wagging. Nesting: Jan–Jul.

SS: See Black-chinned, Anna's, and Ruby-throated hummingbirds.

SD: C to F resident (SL–2200 m) in Baja and on Pacific Slope in Son, possibly N Sin; U in winter (Aug–Feb) to cen Sin, R to Nay (Oct–Dec at least, Baltosser 1989).

RA: SW USA and NW Mexico.

NB: Traditionally placed in the genus *Calypte*.

CALLIOPE HUMMINGBIRD
Archilochus calliope   Plate 32
Colibri de Caliope

ID: 3–3.2 in (7.5–8 cm). *W Mexico.* Bill 1× head. *Wings slightly > tail.* **Descr. ♂:** *gorget formed by glittering rose-pink streaks* (may look solidly pink), white postocular spot contrasts with green crown and dusky auriculars. Underparts white, sides mottled greenish, buff wash on hind flanks. Nape and upperparts green. *Tail blackish,* central rectrices finely edged rufous basally. **♀:** throat whitish with lines of dusky spots (fainter in imm), whitish postocular spot contrasts with green crown and dusky auriculars. *Underparts whitish, washed pale cinnamon on flanks* and undertail coverts. Nape and upperparts green. *Tail dull greenish basally, blackish distally, the outer 3 rectrices tipped white* (outer rectrices with inconspicuous rufous edge at base of outer webs). **Imm ♂:** resembles ♀ but with some pink on throat.

Voice. High, fairly shrill, sharp chips, *si tsi-tsi, tsi ...,* etc. ♂ makes a high, thin *zzing* in display.

Habs. Breeds in open conifer forest with meadows. In winter, also humid to semiarid pine–oak, brushy scrub and edge. Feeds and perches at low to mid-levels, often at banks of flowers where usually

Costa's Hummingbird

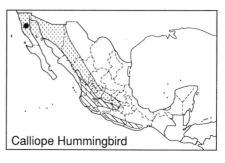

Calliope Hummingbird

dominated by other hummingbirds. Quivers but rarely wags tail while feeding. In display, ♂ flies back and forth in shallow swoops. Nesting: May–Jul.

SS: ♀ Bumblebee Hummingbird darker overall, more heavily marked throat and darker cinnamon flanks separated by contrasting white forecollar, rufous at base of tail may be hard to see, wings < tail at rest, often feeds horizontally with longer tail cocked. ♀ Rufous, Allen's, and Broad-tailed hummingbirds larger with longer, heavier bills, conspicuous rufous in tails, wings < tail at rest.

SD: U to R breeder (May–Aug, 2200–3000 m) in Sierra San Pedro Martír, BCN (perhaps formerly F, see Grinnell 1928). F to U transient and winter visitor to Pacific Slope and adjacent interior of W Mexico, R to S Oax (Binford 1989); in autumn birds mainly move S, inland of mountains, and in spring move N on Pacific Slope (Phillips 1975).

RA: Breeds W N.A. to BCN; winters W Mexico.

NB1: Traditionally placed in the monotypic genus *Stellula*. NB2: *A. c. lowei*, described from Gro (Griscom 1934), was based on winter specimens; there is no evidence that this species breeds in S Mexico.

GENUS SELASPHORUS

These are small to very small hummingbirds nesting in temperate to subtropical habitats, N species migratory. Bills straightish and black, tails graduated to slightly double-rounded, wings < tail. Ages/sexes differ, ♂♂ with glittering gorgets, ♀♀ and imms often hard to identify (at times, possible only in the hand). Closely related to, and sometimes merged with, *Archilochus*.

BROAD-TAILED HUMMINGBIRD

*Selasphorus platycercus**   Plate 32
Zumbador Coliancho

ID: 3.5–4 in (9–10 cm). Highlands. Bill 1.25–1.33× head. *Tail double-rounded.* Descr. ♂: *gorget glittering rose-pink,* white postocular spot contrasts with green crown and *pale greyish auriculars,* nape and upperparts green. Underparts whitish, mottled green, with broad whitish band below gorget. Inner rectrices green, *outer rectrices blackish with fine rufous edges to*

bases of outer webs. ♀: whitish postocular spot contrasts with green crown and dusky auriculars, nape and *upperparts green to blue-green.* Throat and underparts whitish, throat with lines of dusky flecks (rarely one to several pink throat feathers), *sides and undertail coverts washed cinnamon.* Tail blue-green, central rectrices rarely with darker tips, outer rectrices with broad blackish subterminal band and white tips, *rufous edging at base of outer webs often hard to see.* Imm ♂: resembles ♀ but often with some glittering pink throat feathers, rufous in tail reduced and inconspicuous. Older imm resembles ♂ but throat whitish, heavily flecked dusky.

Voice. Fairly sharp but not hard chips, *chi chi-chip, chip, chip ...,* etc. In display a high, thin, slightly buzzy, slurred *szzzzeuu.* ♂'s wings make high, 'singing' buzz in flight.

Habs. Arid to semihumid open pine and pine–oak woodland, clearings with flowers, also grasslands with scattered trees and bushes (at least in migration). Feeds and perches at low to mid-levels, tail held closed while feeding, rarely flashed open or wagged. ♂ makes high, steep loops in display flight. Nesting: Apr–Jul (N and cen Mexico).

SS: See Calliope and Ruby-throated hummingbirds. ♀ and imm Rufous and Allen's hummingbirds less blue-green above, uppertail coverts edged rufous, sides deeper cinnamon, rufous in tail more conspicuous, central rectrices typically with dark tips, throat often with glittering orange-red spots.

SD: F to C breeder (Mar–Sep, 1800–3500 m) in interior and on adjacent slopes from Son and Coah to cen volcanic belt, withdraws in winter from N of range (N Son to NL, at least), winters mostly in cen volcanic belt, U to F (Oct–Mar) S to Oax;

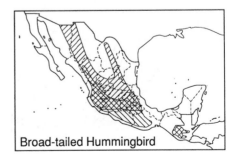

Broad-tailed Hummingbird

also F to C resident (2700–3500 m) in SE Chis and W Guatemala.

RA: W USA to W Guatemala; winters from N Mexico S.

NB: Might be better placed in the genus *Archilochus*.

RUFOUS HUMMINGBIRD

Selasphorus rufus   Plate 32
Zumbador Rufo

ID: 3.2–3.7 in (8–9 cm). Migrant to W Mexico. Bill 1.25–1.33× head. Rectrices broader than Allen's Hummingbird, outer primary of ♂ attenuated. **Descr.** ♂: tail graduated. *Gorget glittering orange-red,* forehead green, crown, auriculars, and *upperparts rufous,* white postocular spot, back may have a few green feathers. Broad whitish forecollar below gorget, rest of underparts rufous. Tail rufous, tipped dark green. ♀: tail slightly rounded. Whitish postocular spot contrasts with green crown and dusky auriculars, nape and upperparts green, *uppertail coverts tipped rufous.* *Throat and underparts whitish, throat with lines of dark spots and a few glittering orange-red spots* (slight or lacking in imm). Central rectrices green (to blue-green in imm), edged rufous and usually dark-tipped, *outer rectrices rufous with blackish subterminal band, tipped white.* **Imm** ♂: resembles ♀ but uppertail coverts mostly rufous, central rectrices mostly rufous to base.

Voice. Fairly sharp chips, *chi-chik, chi chi chi-chik* ..., etc.

Habs. Forest edge, scrub, clearings with flowers. Feeds and perches at low to midlevels, ♂'s wings usually make high, trilling buzz in flight.

SS: Allen's Hummingbird often not safely distinguished in field, adult ♂ has green upperparts, ♀ and imm may be told from Rufous, with experience and good views,

by narrower rectrices. See Broad-tailed and Calliope hummingbirds. ♀ Bumblebee Hummingbird smaller with shorter, finer bill, brighter white forecollar.

SD: C to F transient and winter visitor to W and interior Mexico. In autumn (Aug–Oct) moves S in interior, from Son and Coah to DF; by winter (Dec–Jan) most of population in Pacific Slope mountains from Jal to Oax; in spring (Feb–Apr) moves N along Pacific Slope and through BCN (Phillips 1975); vagrant to S Ver (Sep 1984, SNGH).

RA: Breeds W N.A.; winters W Mexico.

ALLEN'S HUMMINGBIRD

Selasphorus sasin   Not illustrated
Zumbador de Allen

ID: 3.2–3.5 in (8–9 cm). Migrant to W Mexico. Bill 1.25–1.33× head. Rectrices narrower than Rufous Hummingbird, outer primary of ♂ finely attenuated. **Descr.** ♂: tail graduated. *Gorget glittering orange-red,* auriculars rufous with white postocular spot, crown, *nape, and upperparts green.* Broad whitish forecollar below gorget, rest of underparts rufous. Tail rufous, tipped dark green. ♀: tail slightly rounded. Whitish postocular spot contrasts with green crown and dusky auriculars, nape and upperparts green, *uppertail coverts tipped rufous.* *Throat and underparts whitish, throat with lines of dark spots and a few glittering orange-red spots* (slight or lacking in imm). Central rectrices green (to blue-green in imm), edged rufous and usually dark-tipped, *outer rectrices rufous with blackish subterminal band, tipped white.* **Imm** ♂: resembles ♀ but uppertail coverts mostly rufous, central rectrices mostly rufous to base.

Voice. Sharp chips much like Rufous Hummingbird.

Habs. Forest edge, scrub, clearings with

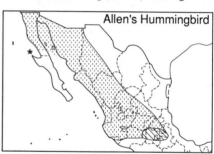

Rufous Hummingbird

Allen's Hummingbird

flowers. Habits much as Rufous Hummingbird, including ♂'s wing buzz.

SS: See Rufous, ♀ Broad-tailed, and ♀ Calliope hummingbirds. ♀ Bumblebee Hummingbird smaller with shorter, finer bill, brighter white forecollar.

SD: F to C transient and winter visitor to NW and interior Mexico. In autumn (Jun–Sep) moves S through BCN and then inland of Sierra Madre Occidental to cen Mexico; winters (Aug–Dec) mainly in cen Mexico (Mex–Pue); spring migration (Dec–Mar) up Pacific Slope and through BCN (Phillips 1975).

RA: Breeds W USA; winters W Mexico.

BUMBLEBEE HUMMINGBIRD
Selasphorus heloisa* Plate 32
Zumbador Mexicano

ID: 2.7–3 in (7–7.5 cm). *Endemic to highlands N of Isthmus. Bill 0.75–1× head, outer primary of ♂ attenuated at tip.* **Descr.** ♂: *tail rounded. Gorget glittering rose-pink with violet-blue highlights,* whitish postocular stripe between green crown and dusky auriculars. Nape and upperparts green. Underparts whitish, *flanks washed greyish cinnamon and slightly mottled green. Central rectrices green* or with some rufous edging basally, *rest of tail bright rufous with broad black subterminal band* (often green basally) *and white tips to outer rectrices.* ♀: *tail double-rounded. Throat whitish with bold lines of dusky spots,* whitish postocular spot contrasts with green crown and dusky auriculars. Nape and upperparts green. Median underparts whitish with *cinnamon flanks* and undertail coverts, *upper chest appears as contrasting white forecollar. Tail as ♂* but with slightly more black and less rufous (rufous may be hard to see unless tail spread), tips of outer rectrices may be

washed buff. **Imm** ♂: resembles ♀ but with some pink on throat.

Voice. ♂ song a high, thin, whining *sss sssssssiu* or *seeuuuuu,* drawn out and fading at end, given from perch and often hard to locate. Calls include high chips much like other *Selasphorus.*

Habs. Humid to semihumid evergreen and pine–evergreen forest edge, adjacent shrubby growth and clearings with flowers. In display, ♂ hovers in horizontal posture with tail cocked nearly vertically, gorget catching the light. Feeds and perches at low to mid-levels, feeds often in horizontal posture, tail cocked; ♂'s wings make low, insect-like buzz, louder in display. Nest and eggs undescribed (Kiff and Hough 1985). B.c.: May–Jun (Tamps), Apr (N Oax), Jul (Jal, Col, Ver), Dec (Gro, N Oax).

SS: See ♀ Calliope, Rufous, Allen's, and Broad-tailed hummingbirds. Wine-throated Hummingbird allopatric.

SD/RA: Locally/seasonally F to C resident (1500–3000 m) on both slopes from Jal and S Tamps to Oax, locally in interior from cen volcanic belt to Oax; U N to S Chih.

NB: Traditionally placed in the genus *Atthis.*

WINE-THROATED HUMMINGBIRD
Selasphorus ellioti* Plate 32
Zumbador Centroamericano

ID: 2.5–2.8 in (6.5–7 cm). *Endemic to highlands S of Isthmus. Bill 0.75× head.* **Descr.** ♂: *tail rounded. Gorget glittering rose-pink with slight violet highlights,* whitish postocular stripe between green crown and dusky auriculars. Nape and upperparts green. Underparts whitish, *flanks washed cinnamon and slightly mottled with green. Central rectrices green* or with some rufous edging basally, *rest of tail bright rufous with broad black subterminal band* (often

Bumblebee Hummingbird

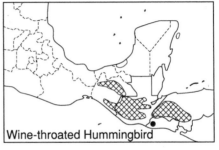

Wine-throated Hummingbird

green basally), *white tips to outer rectrices*
may be washed cinnamon. ♀: tail double-
rounded. *Throat whitish with heavy lines
of dusky spots*, whitish postocular spot
contrasts with green crown and dusky
auriculars. Nape and upperparts green.
*Median underparts whitish, with deep cin-
namon flanks and undertail coverts*, upper
chest appears as contrasting white forecollar. *Tail as* ♂ but with slightly more
black and less rufous, tips of outer rectrices
often washed cinnamon. **Imm** ♂: resembles
♀ but with some pink on throat.

Voice. Song a series of high, slightly buzzy,
squeaky chips that break into a warble.
Calls include high, thin, sharp chips, often
repeated steadily, *sip-sip* ... or *cheup
cheup* ..., etc.

Habs. Humid to semihumid evergreen
and pine–evergreen forest edge, adjacent
shrubby growth and clearings with flowers.
Habits much as Bumblebee Hummingbird,
but calls steadily in display flight when ♂
moves back and forth in front of ♀. ♂'s
wings make low, insect-like buzz, louder
in display. Nesting: Aug–Oct (C.A.). B.c.:
Feb (Chis).

SS: See ♀ Broad-tailed Hummingbird;
Bumblebee Hummingbird allopatric.

SD/RA: Locally/seasonally F to C resident
(1500–3500 m) in interior and on adjacent
slopes from Chis to El Salvador and Honduras.

NB: Traditionally placed in the genus *Atthis*.

TROGONS: TROGONIDAE (9)

A pantropical family of brightly colored forest birds which spend much time perched motionless in the forest and thus may be overlooked easily. Bills short and stout, tomia serrated. Small weak feet heterodactylous, generally greyish in color and rarely visible in the field. Tails long, broad, and squared. Ages/sexes differ; young altricial, 1st prebasic molt partial. ♂♂ have metallic golden-green to blue-green upperparts and chest, and brilliant red or yellow belly. ♀♀ patterned similarly but most species have grey or brown upperparts and chest. Upperwing coverts, tertials, and outer webs of secondaries often barred or vermiculated black and white, forming a distinct panel (= wing panel in species accounts). Juvs of most species mottled brown overall, with buff and whitish spotting, barring, and vermiculations on wing coverts, tertials, and secondaries; belly may be whitish, often with brownish mottling or only a trace of adult color. First basic quickly assumed and, except in some larger species, resembles adult of respective sex. Juv tails, however, may be patterned quite differently from those of adults and can cause identification problems; note that juv rectrices are markedly more tapered than the truncate rectrices of adults. Adult plumage attained by 2nd prebasic molt when about one year old. Black-and-white patterns on the undertail, formed by markings on the strongly graduated outer three pairs of rectrices, are important specific characters.

Trogons usually are solitary or in pairs, but several birds may gather at fruiting trees. Flight undulating and slightly rocking but usually only species of montane forests fly more than short distances. Songs typically hollow or plaintive hoots, repeated; calls include clucks and chatters; most trogons are heard far more often than they are seen.

Food mostly fruit and insects plucked from outer branches and foliage in short sallying flights. Nests are in natural tree cavities or may be excavated out of rotten stumps, termitaries, or wasp nests. Eggs: 2–3, white to bluish white in trogons, pale blue in quetzals (WFVZ).

BLACK-HEADED [CITREOLINE] TROGON

Trogon m. melanocephalus   Plate 33
Trogon Cabecinegro

ID: 10.5–11 in (26.5–28 cm). *Yellow-bellied; SE forest.* **Descr.** ♂: bill pale blue-grey, *orbital ring blue.* Head and chest blackish, *back blue-green becoming violet on rump and uppertail coverts.* Wings blackish, primaries white on outer webs. Uppertail blue-green, tipped black. *Yellow belly and undertail coverts* separated from dark chest by white band. *Undertail black, white-tipped outer rectrices often form solid white block distally.* ♀: bill may be dark above. Head, chest, and upperparts dark grey. *Undertail black, white tips of outer rectrices form 3 pairs of bold white blocks.* **Imm:** *outer webs of undertail white with black bars basally.*

Voice. Song an accelerating series of nasal hoots that slows and ends fairly abruptly, *kow-kow-kow-kow-kowkowkow ...* or *nyah-nyah-nyah ...,* recalls Great Antshrike. Also a hollow, rolling chatter *kow kohkohkohkohkoh* or *kuk-kuk-kukkuk ...,* ending abruptly, hollow thrush-like clucks, a rolled chatter, *k-week k-weh-k-rrek,* and a rolled *kwarr-rrak* in flight.

Habs. Forest edge, semiopen areas with scrub and scattered trees, plantations,

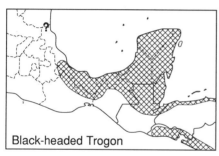

Black-headed Trogon

mangroves. ♂♂ may form noisy leks of up to 8–10 birds, often attended by several ♀♀. Nests in termitaries.

SS: Citreoline Trogon (locally sympatric in Chis?), paler overall, eyes yellow, more white on undertail; Violaceous Trogon smaller and shorter-tailed, ♂ has yellow orbital ring, pale grey wing panel, greenish rump, ♀ has white eye-crescents, barred wing panel, note undertail and voice. See ♂ Black-throated Trogon (E Honduras, Appendix E).

SD: C to F resident (SL–1000 m) on Atlantic Slope from S Ver (formerly (?) S Tamps, Richmond 1895) to Honduras, on Pacific Slope in El Salvador and Honduras.

RA: SE Mexico to N Costa Rica.

NB: Has been considered conspecific with Citreoline Trogon.

CITREOLINE TROGON

*Trogon citreolus**   Plate 33
Trogon Citrino

ID: 10.5–11 in (26.5–28 cm). *Yellow-bellied; endemic to W Mexico.* **Descr.** ♂: *eyes yellow,* bill pale blue-grey. Head and chest dark grey, mantle golden green to blue-green becoming violet-blue on rump and uppertail coverts. Wings blackish, primaries white on outer webs. Uppertail blue-green, tipped black. *Yellow belly and*

undertail coverts separated from dark chest by broad white band. *Undertail white overall.* ♀: chest paler grey, belly often paler yellow, undertail black at base and may show black bars basally on outer webs; upperparts dark grey. **Imm:** eyes can be dark into early winter, usually more extensive black on undertail, including black bars basally on outer webs.

Voice. Much like Black-headed Trogon. Song and accelerating series of rolled hoots run into a chatter, *kyow-kyow-kyow-kyow-kyow-kyowkowkow* . . ., often with nasal quality; calls include hollow, thrush-like clucks.

Habs. Arid to semiarid woodland, thorn forest, plantations, mangroves. Forms noisy leks much like Black-headed Trogon. Nests in termitaries.

SS: See Black-headed Trogon.

SD/RA: C to F resident (SL–1000 m) on Pacific Slope from S Sin to W Chis (vicinity of Ocozocuatla where may be sympatric with Black-headed Trogon).

NB: See Black-headed Trogon.

VIOLACEOUS TROGON

Trogon violaceus braccatus   Plate 33
Trogon Violáceo

ID: 9.5–10 in (24–25.5 cm). *Yellow-bellied; small and fairly short-tailed. E lowlands.* **Descr.** ♂: bill blue-grey, *orbital ring yellow.* Head blackish with violet nuchal collar, *upperbody green. Wing panel finely vermiculated black and white (looks pale grey overall),* primaries black with white outer webs. Uppertail blue-green with black tip. Violet-blue chest separated from yellow belly and undertail coverts by white band. *Undertail barred black and white,* outer rectrices broadly tipped white. ♀/juv: bill may be dark above. Head, chest, and upperparts blackish, white crescents in front of and behind eye. *Blackish wing-*

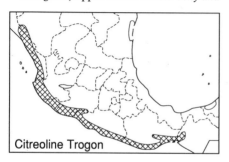

Citreoline Trogon

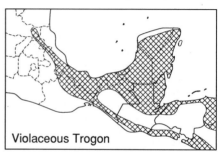

Violaceous Trogon

covert panel narrowly barred white. Outer webs of undertail coarsely barred black and white, inner webs black, tipped white. **Imm** ♂: wing panel browner than ♂, undertail with more black on inner webs.

Voice. A fairly rapid, persistent, hollow hooting, *kyow-kyow-kyow* ... or *ku-ku-ku* ..., 10/2.5–3 s, may suggest Ferruginous Pygmy-Owl; calls include a rolled chatter, *krr-rr-rah-ah* or *kwahrr-rr-rr-rr*, not as hard or hollow as Black-headed Trogon.

Habs. Humid to semiarid forest and forest edge, plantations, mangroves. At mid- to upper levels. Nests in wasp nests, termitaries, rotten trunks.

SS: See Black-headed Trogon, also Black-throated Trogon (E Honduras, Appendix E).

SD: C to F resident (SL–1800 m) on Atlantic Slope from S Ver to Honduras, on Pacific Slope from E Oax to El Salvador; U to F N to SE SLP.

RA: E Mexico to Ecuador and Brazil.

MOUNTAIN TROGON

Trogon mexicanus *   Plate 33
Trogon Mexicano

ID: 11.5–12.5 in (29–31.5 cm). *Red-bellied; endemic to highland forests.* **Descr.** ♂: bill *bright yellow*, orbital ring orange-red. Face and throat blackish, green chest separated from red belly and undertail coverts by white band. Crown, nape, and upperparts green; *wing panel* coarsely vermiculated black and white (*looks grey*), blackish primaries lack white outer webs. Uppertail blue-green to green, tipped black. *Undertail black with white-tipped outer rectrices forming 3 pairs of bold white blocks.* ♀: bill *dark above*, no bright orbital ring. Head, chest, and *upperparts warm brown*, bold white crescent behind eye, small whitish crescent in front of eye. Pale brown

wing panel vermiculated dusky, blackish primaries white on outer webs. Rufous-brown uppertail tipped black. *Brown lower chest and red belly below white chest band. Outer webs of undertail barred black and white, inner webs black with broad white tips.* **Imm** ♂: browner wing panel than adult, often some brown mottling on upper belly. *Undertail resembles ♀ but usually shows more white.*

Voice. Song a series of plaintive paired whistles, usually 3 pairs per series but at times up to 15+ pairs, *kyow-kyow kyow-kyow kyow-kyow* ..., also a softer, higher, bouncing-ball series of 10–14 notes, *hu-hu-hu* ... or *ha-ha* ..., which may follow song. Calls include hollow, thrush-like clucks, *chook* and *cho-owk*, and a rapid, cackling chatter *kwa-ahr-ra-rak*, mostly in flight.

Habs. Pine–oak and pine–evergreen forest. At mid- to upper levels, may associate loosely with mixed-species flocks. Nests in tree cavities and rotten stumps.

SS: Undertail pattern, grey wing panel, blackish primaries, and voice of ♂ and undertail pattern of ♀ distinguish Mountain from Collared and Elegant trogons; note also yellow bill and white postocular stripe of ♀ Elegant, mostly red belly of ♀ Collared. Eared Quetzal larger with grey bill, bluish upperparts, and white undertail, lacks white chest band; note voice.

SD/RA: C to F resident (1200–3500 m, rarely lower) in interior and on adjacent slopes from S Chih and S Tamps to El Salvador (Thurber *et al.* 1987) and Honduras.

NB: Sometimes known as Mexican Trogon.

ELEGANT TROGON

Trogon elegans *   Plates 33, 70
Trogon Elegante

ID: 11–12 in (28–30.5 cm). *Red-bellied.* Two groups: *ambiguus* group (Mexico)

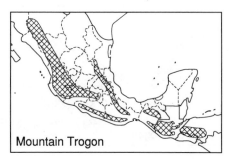

Mountain Trogon

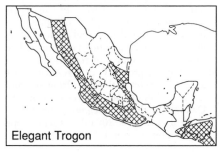

Elegant Trogon

and *elegans* group (C.A.). **Descr. *Ambiguus*.** ♂: bill bright yellow, orbital ring red. Face and throat black, green chest separated from red belly and undertail coverts by white band. Crown, nape, and upperparts green to blue-green; *wing panel* finely vermiculated black and white (*looks pale grey*), blackish primaries white on outer webs. *Uppertail burnished gold to greenish copper, broadly tipped black. Undertail white overall, outer rectrices vermiculated dusky basally.* ♀: head, chest, and *upperparts grey-brown*, with *broad white stripe behind eye*, whitish crescent in front of eye. Rufous-brown uppertail broadly tipped black. *Brown lower chest and red belly below white chest band, and usually a 2nd whitish area above red. Undertail white, barred dusky, with broad white tips.* **Imm** ♂: wing panel browner than adult, chest and belly often mixed with brown. *Undertail resembles ♀ but dark barring coarser*, shows more white. ***Elegans*:** differs mainly in undertail pattern. ♂: *undertail barred black and white, outer rectrices broadly tipped white; upperparts blue-green, uppertail golden green.* ♀/**imm:** undertail more coarsely barred than ♂, much like tail of imm *ambiguus* but dark bars bolder.

Voice. Song (mainly Mar–Aug) a distinctly gruff or hoarse, disyllabic *kwa'h* or *krow'h*, usually repeated rapidly 4–6×, also more prolonged series; unlike songs of other trogons. Calls include a rolled chattering *wehrr-rr-rr rr-rr-rr* or *kwerr-rrrerrrr*, a fairly quiet, steady, gruff *ahrr ahrr ...*, and a dry, churring *churrrrrrrr*.

Habs. Arid to semiarid woodland, thorn forest, deciduous riparian woods in pine zones. Nests in tree cavities.

SS: See Mountain Trogon. Collared Trogon prefers more humid forest; note darker wing panel, green uppertail, undertail pattern, and voice of ♂, warmer brown upperparts, mostly red belly, lack of white postocular stripe, and undertail pattern of ♀; imm ♂ Collared undertail much like Elegant and a source of confusion.

SD: *Ambiguus* group: C to F resident (near SL–2400 m) on both slopes from Son and E NL, and in interior from cen Mexico, to Oax (N of Isthmus). *Elegans* group: C to F resident (near SL–1800 m) on Pacific Slope and in interior from E Guatemala to Hon-duras and W Nicaragua. A report from Chis (Parker *et al.* 1976) is erroneous.

RA: Mexico (also S Arizona) to NW Costa Rica.

NB: The two groups have been considered separate species.

COLLARED TROGON
Trogon collaris *   Plate 33
Trogon Collarejo

ID: 10.5–11.5 in (26.5–29 cm). *Red-bellied; S humid forests.* **Descr.** ♂: bill bright yellow, orbital ring orange-red. Face and throat blackish, green chest separated from red belly and undertail coverts by white band. Crown, nape, and upperparts green; *black wing panel narrowly barred white, black primaries white on outer webs.* Uppertail metallic golden green to blue-green, tipped black. *Blackish undertail narrowly barred white.* ♀: bill dark above, lacks bright orbital ring. Head, chest, and upperparts warm brown, white crescent in front of and behind eye. Pale brown wing panel vermiculated dusky, blackish primaries white on outer webs. Deep rufous-brown uppertail tipped black. *Red underparts separated from brown chest by whitish band. Undertail appears silvery grey, whiter on outer webs, outer rectrices with black subterminal bars, tipped white.* **Imm** ♂: browner wing panel less distinctly barred; *undertail barring and white tail tips wider.* **Imm** ♀: inner webs of undertail coarsely freckled blackish and white.

Voice. Song a plaintive 2–3-noted *kyow'-kyow* or *caow'caow*, and a faster *kyow kyow-kyow*, occasionally run into longer laughing, or bouncing-ball series of 5–8 notes. Calls include a rolling growl, *ahrrrr* or *irrrrrr*, often repeated steadily, and a single *kyow*.

Habs. Humid to semihumid evergreen

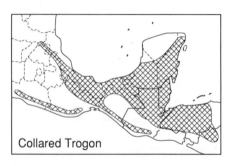

Collared Trogon

and semideciduous forest, pine–evergreen forest. Nests in rotten stumps and snags.

SS: See Mountain and Elegant trogons. ♂ Slaty-tailed Trogon larger; note orange bill, lack of white chest band, pale grey wing panel, slaty undertail. In E Honduras note voice of Black-throated Trogon, very similar to Collared.

SD: C to F resident (near SL–2400 m) on both slopes from Gro (Navarro 1986; SNGH, SW, tape) and SE SLP to El Salvador and Honduras.

RA: Mexico to Ecuador and Brazil.

NB: May be conspecific with *T. aurantiiventris* of S C.A.

SLATY-TAILED TROGON

Trogon m. massena   Plate 33
Trogon Colioscuro

ID: 13–14 in (33–35.5 cm). Red-bellied; SE rain forest. *Lacks white chest band, bill very stout.* **Descr.** ♂: *bill and orbital ring orange.* Face and throat blackish, crown, nape, chest, and upperparts green; *belly and undertail coverts red. Wing panel* vermiculated black and white (*looks pale grey*), blackish primaries white on outer webs. Uppertail green, tipped black. *Undertail blackish grey.* ♀: bill blackish above, orbital ring dull. *Head, chest, upper belly, and upperparts dark grey, lower belly and undertail coverts red.* Wing panel faintly vermiculated whitish (looks dark grey). Uppertail blackish, *undertail as* ♂. **Imm** ♂: bill mostly blackish above, orbital ring dull. Wing panel more coarsely vermiculated (looks grey), underparts resemble ♀ but with some green on chest. *Undertail blackish overall, outer webs narrowly barred and tipped white.* **Imm** ♀: undertail as imm ♂.

Voice. Song a fairly hard, steady *koh-koh-koh* ... or *ka-ka-ka* ..., 10/2.5–4 s, rarely a rapid clucking *kohkohkoh* ..., 10/1.5 s. Calls include a quiet clucking *huh-huh-huh-huh* and a hard, chuckling chatter.

Habs. Humid evergreen forest. Nests in termitaries and rotten trunks.

SS: See Collared Trogon.

SD: F to C resident (SL–600 m) on Atlantic Slope from S Ver to Honduras.

RA: SE Mexico to N S.A.

EARED QUETZAL (TROGON)

Euptilotis neoxenus   Plate 33
Quetzal Mexicano

ID: 13–14 in (33–35.5 cm). Red-bellied; *endemic to NW Mexico. Small-headed with relatively small bill, tail broad and squared.* Named for inconspicuous, filamentous postocular plumes. *Lacks white chest band.* **Descr.** ♂: bill grey, often with dark tip and culmen, orbital ring grey. Face and throat blackish, crown, nape, chest, and upperparts blue-green; belly and undertail coverts red. *Wings blackish*, white on outer webs of outer primaries. *Uppertail deep blue, undertail white overall.* ♀: head slaty grey, chest and upper belly grey-brown with slight green sheen at sides of chest. **Imm:** black at base of undertail more extensive, inner rectrices tipped white.

Voice. Quite unlike *Trogon* trogons. Song a reedy, slightly shrill, quavering, whistled series, beginning softly and quickly swelling strongly, carries well. Initially 1–2-syllable whistles becoming 3-syllable whistles which, at a distance, may sound hollow and eerie: *whee whee wheerr-i wheerr-i-hi wheerr-i-hi* ..., or *whii whii whii whii-ii whii-ii-iir whii-ii-iir* ..., etc., usually 8–15 whistles in series, at times to 24+. Calls include an ascending, grackle-like squeal that ends abruptly with a low cluck which is often inaudible at a distance, *huweeeh chk* or *wheeeuh*

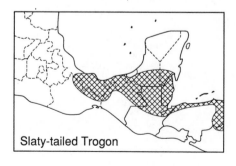

Slaty-tailed Trogon

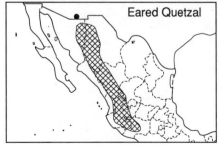

Eared Quetzal

chk, and a hard clucking or cackling *ka-kak* or *k-tk*, and *ka-ka-kak* or *k-t-tk*, etc., usually given when disturbed or in flight. **Habs.** Pine, pine–oak, and pine–evergreen forest, especially (?) in canyons. At upper and mid-levels of trees, more mobile and wide-ranging than trogons, pairs and groups moving readily through open woods, across canyons, etc. Nests in tree cavities (KK); breeds Jun–Oct, at least in NW Mexico. Eggs pale blue (MVZ); eggs attributed to this species at USNM are probably mislabeled.

SD/RA: Poorly known, may be declining due to loss of large trees for nesting (?). U to F but local resident (1800–3000 m) in interior and on adjacent Pacific Slope from Chih to W Mich; R visitor to S Arizona.

RESPLENDENT QUETZAL

Pharomachrus m. mocinno   Plate 33
Quetzal Centroamericano

ID: 15–16+ in (38–40.5+ cm). SE cloud forests. Considered by many the most spectacular New World bird. *Head slightly crested, upperwing coverts elongated;* ♂ *has extremely elongated filmy uppertail covert plumes* that extend up to 24 in (61 cm) beyond tail tip. **Descr.** ♂: *bill yellow. Intense green and blue-green overall* (colors change with the light from golden green to violet-blue) with red belly and undertail coverts. Blackish remiges usually covered by plumes, *undertail white.* ♀: *bill blackish. Head greyish,* chest and upperparts green, plumes extend to or just beyond tail tip. *Belly grey, vent and undertail coverts red. Undertail broadly barred black and white.* **Imm** ♂: resembles ♀ but bill yellow, *undertail more extensively white,* outer rectrices barred black only at base.

Voice. Song a plaintive, fairly leisurely, steadily repeated *k'yoiw-k'yow k'yoiw-k'yow* ..., or *whee'o-whee'u whee'o-whee'u,* etc. Calls include a fairly hard *k'wah,* a steady *ka-ka-ka* ... suggesting Barred Forest-Falcon, and a rolled chatter in flight, *kwar-ahrr-rrak,* etc.

Habs. Humid evergreen forest. Spends much time high in trees and overlooked easily. Nests in rotten tree stumps.

SD: F to C resident (1400–3000 m) in undisturbed cloud forest in interior and on both slopes from E Oax and Chis to Honduras and N cen Nicaragua. U to R in much of range due to forest destruction and hunting for plumes.

RA: S Mexico to W Panama.

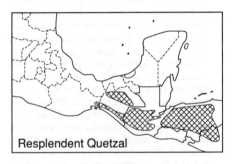

Resplendent Quetzal

MOTMOTS: MOMOTIDAE (6)

A neotropical family of colorful birds whose center of diversity lies in Middle America. Motmots are large-headed, long-tailed inhabitants of forest and edge. Bills fairly long and stout with decurved culmen. Feet small and weak, often not visible in field; bills black and feet greyish to dull flesh in most species. Wings short and rounded; barbs near tips of central rectrices often weakly attached, soon wearing off to reveal distinct 'racket-tips'. Ages differ slightly, sexes similar; young altricial. Juvs duller but quickly attain adult-like plumage with completion of 1st prebasic molt. Plumage colored overall in greens and browns, typically with a black auricular mask, some bright blue on head, and two black spots on central chest. Songs are low hoots or honks, given mainly early and late in the day, often before dawn, when birds are most active. Motmots spend much time perched quietly, with a fairly vertical stance, in shady spots and are overlooked easily, although most species often twitch their tails sharply from side to side.

Feed on invertebrates, small vertebrates, and fruit; prey often caught by sallies to the ground where birds pose momentarily with tail cocked, or in canopy with adept, quick flights. Nest in burrows excavated in banks or root masses of fallen trees, etc., often near the ground. Eggs: 2–5, white (WFVZ).

TODY MOTMOT
Hylomanes momotula *   Plate 34
Momoto Enano

ID: 6.5–7 in (16.5–18 cm). *SE humid forests. Tail lacks racket tips.* **Descr. Adult:** bill blackish, often paler below at base. *Crown and nape rufous, short supercilium turquoise,* blackish mask contrasts with pale greyish auriculars and *bushy whitish lores and moustache. Broad dusky stripe borders whitish throat,* underparts dull greenish olive becoming whitish on belly. Upperparts green, flight feathers darker. **Juv:** duller overall, crown and nape grey-brown.
Voice. A steady, slightly ringing *woh woh*

woh … or *wah-wah-wah* …, often prolonged, 10/10–12 s, at times faster series to 10/3–6 s, may suggest Violaceous Trogon but rougher, more nasal, and usually slower; rarely a quavering *Otus*-like *hoo-oo-oo-oo-oor.*
Habs. Humid evergreen forest. Perches quietly, low in shady understory, often not seen until flushed when flies a short distance with low whirr of wings. Wags tail slowly while perched, often quite confiding. Nest and eggs undescribed (?).
SD: F to C resident (SL–1500 m) on Atlantic Slope from S Ver to Honduras; U to F on Pacific Slope locally from E Oax (Binford 1989) to El Salvador (Thurber *et al.* 1987).
RA: S Mexico to W Colombia.

BLUE-THROATED MOTMOT
Aspatha gularis   Plate 34
Momoto Gorjiazul

ID: 10–11 in (25.5–28 cm). *Endemic to highlands S of Isthmus. Tail lacks racket tips.* **Descr.** Bill blackish, often pale horn below at base, feet straw. *Green overall,* tail bluer, *with turquoise-blue throat, ochre face, bold black auricular spot,* 2

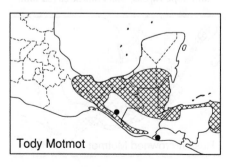

Tody Motmot

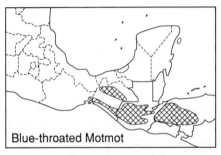

Blue-throated Motmot

black chest spots, and turquoise wash to belly. **Juv:** head and upperparts duller, washed olive.

Voice. A single, far-carrying, low *hoot* or *huuk*, usually repeated every few s. Also a rapid yodeling series of (usually) 10–20 hoots, *hoodloodloodloodl* ... or *uut-a-uudauudauuda* ..., may involve dueting and at times has slightly burry quality.

Habs. Humid to semihumid evergreen and pine–evergreen forest. Often shy and elusive; perches low and quietly, wings make a distinctive low whirr in flight; jerks tail from side to side. Calls from high in trees, often on an exposed branch but can be hard to locate.

SD/RA: F to U resident (1500–3000 m) in interior and on adjacent slopes from E Oax to El Salvador and Honduras.

BLUE-CROWNED MOTMOT

Momotus momota *   Plate 34
Momoto Coroniazul

ID: 15–17 in (38–43 cm). *E and S lowlands. Tail racket-tipped, bill narrow.* **Descr. Adult:** eyes red. *Black crown broadly encircled by turquoise-blue* (crown all blue in *coeruliceps* of NE Mexico), *black mask edged turquoise.* Nape and upperparts green, flight feathers bluer, blue rackets tipped black. *Throat and underparts greenish olive to greenish cinnamon,* chin

pale turquoise, 2 black chest spots. **Juv:** eyes brownish, black chest spots reduced or absent.

Voice. A distinctive, low double-hoot, *oot-oot* or *hoop-hoop*, often heard before dawn, suggesting an owl; at times only single hoots, or several hoots in series if 2–3 birds together; also a hard, hollow clucking, *kluk-kluk-kluk* ... or *klok klok* ... in alarm, and a slightly bouncing-ball-like *wuuh wuh-wuh-wuh-wuh-wuh-wuh-wuh*, suggesting Spectacled Owl.

Habs. Humid to semiarid open woodland, forest and edge, clearings with scattered trees, plantations, gardens. Mostly at low to mid-levels but often calls from fairly open perches in canopy; jerks tail from side to side.

SS: Turquoise-browed Motmot smaller with very long bare tail shafts and larger racket tips, bright turquoise-blue flight feathers, shaggy black throat stripe, rufous back patch; note voice. Keel-billed Motmot usually in mid- to upper levels of forest interior, bill very broad; note turquoise supercilium, rufous forehead, large black chest spots, less developed racket tips, and voice.

SD: C to F resident (SL–1600 m) on Atlantic Slope from E NL and S Tamps to Honduras, on Pacific Slope and in interior from Chis to El Salvador and Honduras.

RA: Mexico to Peru and N Argentina.

NB: Taxonomy of forms in S M.A. and S.A. unclear, possibly more than one species involved.

RUSSET-CROWNED MOTMOT

Momotus mexicanus *   Plate 34
Momoto Coronicafé

ID: 12–14 in (30.5–35.5 cm). *Endemic; arid Pacific Slope and interior.* Tail racket-tipped. **Descr. Adult:** eyes reddish. *Crown and nape rufous, black mask edged violet.*

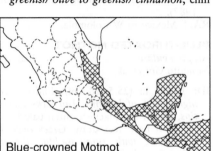

Blue-crowned Motmot

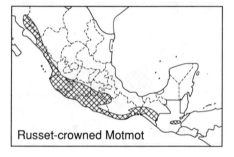

Russet-crowned Motmot

Upperparts green, flight feathers bluer, rackets tipped black. *Throat and underparts pale glaucous-green* with 2 black chest spots. **Juv:** duller, eyes brownish.

Voice. A low, rolled *krrrp* or *krrup*, and *kru krr-up*, also rhythmic dueting *krrru-uu krrru-uu krrru-uu krrru-uu*, etc.; rarely a single hollow *ook*, suggesting Blue-crowned Motmot but with slight burry quality.

Habs. Arid to semiarid woodland and edge, thorn forest, semiopen areas with hedges and trees. Habits much as Blue-crowned Motmot.

SD/RA: C to F resident (SL–1800 m) on Pacific Slope from S Son to W Chis, in interior locally from cen Mexico to Guatemala.

KEEL-BILLED MOTMOT
Electron carinatum   Plate 34
Momoto Piquianillado

ID: 12–13 in (30.5–33 cm). *SE humid forest.* Tail racket-tipped but often not strongly so, *bill very broad.* **Descr. Adult:** eyes amber; *turquoise-blue supercilium and rufous forehead*, black mask. Crown, nape, and upperparts green, flight feathers bluer, rackets tipped black. *Throat and underparts greenish to greenish cinnamon*, chin pale turquoise, 2 *large black chest spots.* Juv undescribed (?).

Voice. A far-carrying, nasal, ringing *ohnng* or *oyhng*, lower and more resonant than Turquoise-browed Motmot, much like Broad-billed Motmot, 1/3.5–5.5 s; 2♂♂ courting a ♀ gave a prolonged, rapid, rhythmic, nasal clucking, *ohn k-k-owhng k-k-k-owhng k-k-owhng ...*; also other clucks similar to Turquoise-browed and Broad-billed motmots.

Habs. Humid evergreen forest. Perches at mid- to upper levels, often in forest canopy where overlooked easily if silent.

Feeds by agile sallies for prey, moving from perch to perch. Jerks tail from side to side. Nest and eggs undescribed (?).

SS: Broad bill (obvious from below), turquoise brow, rufous forehead, and bold black chest spots distinctive. See Blue-crowned Motmot. Broad-billed Motmot (E Honduras, see Appendix E) shares broad bill but adult has cinnamon head and chest with bold black mask.

SD: Poorly known. Traditionally considered a R resident (near SL–1500 m) on Atlantic Slope from SE Ver to Honduras, but recently found to be locally F to C in foothill forest in S Belize, E Guatemala, and N Honduras (Howell and Webb 1992*b*).

RA: E Mexico to Costa Rica.

NB: A mixed pair of Keel-billed and Broad-billed motmots has been observed in Costa Rica (1985 and 1986), including apparent courtship feeding of Keel-billed by Broad-billed (SNGH, M. and P. Fogden). This, in addition to apparent plumage similarity of juv Broad-billed (Stiles and Skutch 1989) and adult Keel-billed, and similar vocalizations of the two species raises a question about their taxonomic status.

TURQUOISE-BROWED MOTMOT
Eumomota superciliosa *   Plate 34
Momoto Cejiturquesa

ID: 13–15 in (33–38 cm). *S and E lowlands. Very long, bare shafts of central rectrices accentuate large racket tips.* Bill narrow. **Descr. Adult:** *pale turquoise supercilium*, black mask. Rest of head and upperparts green with *rufous mantle patch, flight feathers turquoise-blue*, remiges and rackets tipped black. *Throat and chest olive-green with shaggy black median stripe edged turquoise, belly cinnamon.* **Juv:** black throat stripe and rufous back patch reduced or lacking, turquoise brow duller and shorter.

Voice. A hollow, slightly nasal, ringing

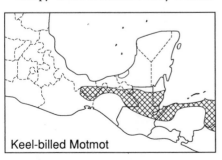
Keel-billed Motmot

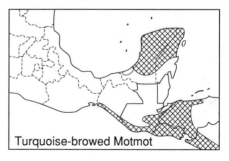
Turquoise-browed Motmot

wohh or *owhh*, 1/3.5–4.5 s, similar to *Electron* motmots but higher and less resonant; also varied hollow, slightly hoarse, clucking calls, including a rhythmic *k'wok k'wok* ... and a longer (dueting?) *k'wok t'k'wok t'k'wok t'k'wok*, etc.

Habs. Forest and woodland edge, plantations, semiopen areas with hedges and scattered trees, gardens. Nests singly or in colonies up to 100+ pairs. Perches conspicuously on roadside wires, especially when nesting. Known in Yuc Pen as 'pájaro reloj' (clock bird) for pendulum-like twitching of its tail.

SS: See Blue-crowned and Keel-billed motmots.

SD: C to F resident (SL–1400 m) on Pacific Slope from Chis, and in interior from Guatemala, to El Salvador and Honduras, on Atlantic Slope in Yuc Pen and locally in Honduras. Also old records from Isthmus (Oax, and possibly SE Ver) where no longer occurs (?).

RA: S Mexico to Costa Rica.

KINGFISHERS: ALCEDINIDAE (5)

A cosmopolitan family poorly represented in the New World. Typically associated with water, kingfishers spend much time sitting quietly, overlooking rivers and pools. Long, fairly heavy and pointed bills used for catching fish. Small weak feet with two forward-pointing toes fused together, greyish to pale flesh, inconspicuous in the field. Heads large and often crested, wings rounded, tails medium length and fairly narrow when closed, often held cocked. Ages/sexes differ, mainly in pattern and color of underparts; young altricial; adult plumage attained in first year (molts poorly known). Plumage blue-grey or dark green above, whitish to rufous below, throat color extends around hindneck in a collar. Voices unmusical, dry rattles, clicks, and rasping buzzes. Flight usually low and fast over water; larger species also fly high and may hunt by hovering.

Food mostly fish but also invertebrates and small reptiles caught by splash-diving or in flight. Nest burrows excavated in banks, usually near water. Eggs: 3–6, white (WFVZ).

RINGED KINGFISHER

Ceryle t. torquata   Plate 34
Martin-pescador Collarejo

ID: 15–16 in (38–40.5 cm). *Very large. Bushy erectile crest usually obvious, bill very stout.* **Descr.** Ages/sexes differ. ♂: blackish bill usually paler, grey to horn below at base. *Head and upperparts blue-grey* with broad white collar, white lore spot, wing coverts sparsely flecked white; flight feathers blackish, secondaries and rectrices broadly edged blue-grey, rectrices barred and spotted white. *Underparts rufous, undertail coverts white.* ♀: *upper chest blue-grey with narrow white band below, undertail coverts rufous.* **Juv:** resembles ♀ but chest band darker greyish, mottled cinnamon, upperparts more extensively flecked white.
    Voice. A loud, chattering rattle, deeper

and often shorter than Belted Kingfisher; also a gruff, deep *rrruk* in flight, and loud, excited, rough, buzzy, and screechy chatters.
Habs. Rivers, lakes, marshes, etc. Often perches on roadside wires, bobs and flicks tail when wary. Flight strong and often high, with sweeping floppy wingbeats.
SS: Belted Kingfisher markedly smaller with proportionately smaller bill, underparts mostly white.
SD: U to R resident (SL–1500 m) on Pacific Slope from S Sin to Isthmus; F to C on Atlantic Slope from NL and Tamps, and on Pacific Slope and locally in interior from Isthmus, to El Salvador and Honduras.
RA: Mexico (also S Texas) to Tierra del Fuego.

BELTED KINGFISHER

*Ceryle alcyon**   Plate 34
Martin-pescador Norteño

ID: 12–13 in (30.5–33 cm). *Widespread winter visitor. Bushy erectile crest usually obvious.* **Descr.** Ages/sexes differ. ♂: blackish bill often paler at base. *Head and upperparts blue-grey* with broad white collar, white lore spot, wing coverts sparsely flecked white; flight feathers blackish, secondaries and rectrices broadly edged blue-grey, rectrices barred and

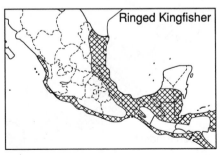
Ringed Kingfisher

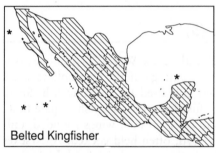

Belted Kingfisher

spotted white. *Underparts white with blue-grey band across upper chest.* ♀: *also has a rufous band across lower chest*, rufous flanks. **Juv:** resembles adult but upper chest band darker greyish, mottled cinnamon, ♂ may show slight cinnamon mottling at sides of chest, upperparts more extensively flecked white.

Voice. A loud, machine-gun-like rattle of variable length.

Habs. Rivers, lakes, marshes, also sea coasts and estuaries. Habits much as Ringed Kingfisher.

SS: See Ringed Kingfisher. In silhouette, Amazon Kingfisher has much larger and heavier bill, tufted rather than bushy crest.

SD: C to F transient and winter visitor (Aug–Apr, U to R to mid-May; SL–3000 m) throughout region; U in interior N Mexico, commonest in lowlands.

RA: Breeds N.A.; winters in N S.A.

AMAZON KINGFISHER
Chloroceryle amazona mexicana	Plate 34
Martin-pescador Amazona

ID: 11–11.5 in (28–29 cm). Tropical lowlands. *Tufted crest, bill very long and heavy.* **Descr.** Ages/sexes differ. ♂: bill black. *Head and upperparts dark glossy green* with broad white collar, *outer rectrices sparsely spotted white.* Underparts white with broad rufous chest band, flanks

mottled dark green. ♀: underparts white with *dark green patch on sides of chest, rarely meeting across center.* **Juv:** resembles ♀ but chest patches finely flecked buff, upper chest may be washed cinnamon in ♂.

Voice. A low, slightly rasping *krrrik*, and a hard, buzzy *zzzzrt* in flight, at times run into a short, rattling chatter.

Habs. Rivers, lakes, mangroves, marshes, etc. Habits much as Ringed Kingfisher but hovers rarely, flight usually low over water.

SS: Green Kingfisher smaller, head flatter, wings conspicuously spotted white, tail flashes white when spread, ♀ has 2 chest bands. See Belted Kingfisher.

SD: R to U resident (SL–1200 m) on Pacific Slope from S Sin (Miller *et al.* 1957) to Isthmus; F to C resident on Atlantic Slope from S Tamps, and on Pacific Slope and locally in interior from Isthmus, to El Salvador and Honduras. A report from Mor (Davis and Russell 1953) is erroneous.

RA: Mexico to Argentina and Uruguay.

GREEN KINGFISHER
*Chloroceryle americana**					Plate 34
Martin-pescador Verde

ID: 7.5–8.2 in (19–21 cm). Widespread, *small. Nape peaked, without obvious crest.* **Descr.** Ages/sexes differ. ♂: bill black. *Head and upperparts dark glossy green with broad white collar, wings spotted and barred white, outer rectrices extensively white.* Underparts white with broad rufous chest band, flanks mottled dark green. ♀: underparts white, often washed creamy buff, with *2 dark green mottled chest bands, the lower one often incomplete.* **Juv:** resembles ♀ but ♂ has upper chest washed cinnamon.

Voice. Hard, dry, clicking notes, often run into a short rattle, a chattering, slightly

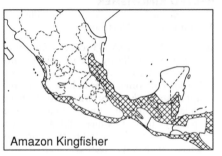

Amazon Kingfisher

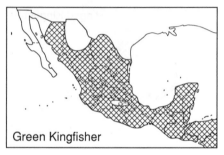

Green Kingfisher

buzzy *sree-ssee-ssee-ee srri-srri-srri*, a rough rolled *zchrrk*, and excited chattering. **Habs.** Rivers, lakes, mangroves, marshes, etc. Habits much as Amazon Kingfisher but usually perches low; does not hover.

SS: See Amazon Kingfisher. Pygmy Kingfisher smaller, throat and collar pale cinnamon, underparts mostly rufous, wings less boldly spotted.

SD: C to F resident (SL–2100 m) on both slopes from Son and Tamps, and in interior from S Plateau, to El Salvador and Honduras; F to U in Yuc Pen and locally on N Plateau.

RA: Mexico (also SW USA) to N Chile and Argentina.

PYGMY KINGFISHER

Chloroceryle aenea stictoptera   Plate 34
Martin-pescador Enano

ID: 5–5.5 in (12.5–14 cm). *Tropical lowlands*. Nape peaked without obvious crest. **Descr.** Ages similar, sexes differ. ♂: bill black. Head and upperparts dark glossy green with *pale cinnamon to whitish collar*, supraloral stripe cinnamon; wings spotted and barred whitish to pale cinnamon, outer rectrices extensively white. *Throat pale cinnamon, chest and flanks rufous, belly and undertail coverts white.* ♀: *narrow dark green band across upper chest.* **Juv:** resembles adult but ♂ may have partial green chest band.

Voice. Hard, dry clicking, much like Green Kingfisher but quieter; also a shrill *chrreik* and a trilled *chrreeiiiiir*.

Habs. Mangroves, forest streams and pools, rarely in open situations. Small and easily overlooked. Typically keeps low, flight darting and sprite-like. At times catches insects like a flycatcher.

SS: See Green Kingfisher.

SD: F to C resident (SL–750 m, most below 350 m) on Atlantic Slope from S Ver, on Pacific Slope from Chis, to El Salvador and Honduras; reports from SE SLP (Davis 1952) require verification.

RA: S Mexico to Ecuador and Brazil.

NB: Also known as American Pygmy Kingfisher.

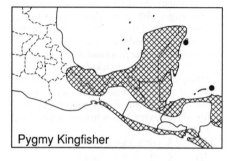

Pygmy Kingfisher

PUFFBIRDS: BUCCONIDAE (2)

A neotropical family of forest and forest-edge birds, most diverse in South America. Chunky with large bulky heads, short, rounded wings, and medium-length tails usually held closed. Bills stout and hooked, feet small and weak, inconspicuous. Ages/sexes differ slightly; young altricial, molts poorly known. Like many other tropical forest birds, puffbirds spend much time perched quietly, fairly upright, and are overlooked easily.

Food mostly insects and small vertebrates caught with infrequent sallying flights. Nest in burrows, eggs white.

WHITE-NECKED PUFFBIRD

*Notharchus macrorhynchos**   Plate 34
Buco Collarejo

ID: 9.5–10 in (24–25.5 cm). *SE humid forest and edge. Bill very heavy and broad.* **Descr.** Ages differ slightly, sexes similar. **Adult:** eyes reddish, bill black. *Black crown with white forehead and broad white collar*; upperparts blackish, wing coverts, rump, and uppertail coverts finely barred whitish. *Underparts white with broad black chest band,* flanks barred blackish. **Juv:** duller overall, eyes brown, white areas washed dirty buff.
  Voice. A fairly rapid, full, bubbling trill, typically of 3–6 s duration, at times breaking into bright, upward-inflected, nasal phrases, *wik wii-ii-ii-ii-ii ... rrr, h-wik' h-wik' i wi-di-ik' wi-di-dik' wi-di-dik',* etc., the soft introductory note optional and often hard to hear.
  Habs. Humid to semiarid forest edge, second growth woods, plantations. Perches at mid- to upper levels on exposed bare branches, making infrequent short flights after prey or to another perch; wings make low whirr in flight. Nests in termitaries; eggs undescribed (?).

SD: U to F resident (SL–750 m) on Atlantic Slope from N Oax, and on Pacific Slope from SE Oax, to El Salvador and Honduras.
RA: S Mexico to Ecuador and N Argentina.

WHITE-WHISKERED PUFFBIRD

Malacoptila panamensis inornata   Plate 34
Buco Barbón

ID: 7.5–8 in (19–20 cm). SE humid forest. **Descr.** Ages/sexes differ. ♂: eyes red, bill black above, grey below. *Cinnamon-brown overall with bushy whitish lores and moustache, upperparts spotted pale cinnamon.* Chest streaked dark brown, belly whitish to pale cinnamon, streaked darker. ♀: ground color grey-brown overall with more contrasting buff markings. **Juv:** resembles ♀ but upperparts more scaled (less spotted), throat scaled brownish, moustache smaller.
  Voice. A high, thin, slightly reedy, drawn-out *tsssiiiw* or *tssssiiir,* fading away, suggests Red-capped Manakin but fuller and reedier; other thin, often sibilant calls, *seeeeu* or *tsssii,* etc.
  Habs. Humid evergreen forest and edges

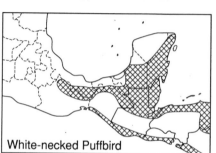
White-necked Puffbird

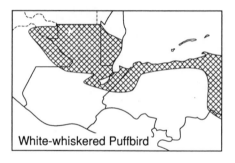
White-whiskered Puffbird

of shady clearings. At low to mid-levels, rarely in canopy, sometimes joins mixed-species flocks. Nest burrow in level or sloping ground, 2–3 eggs.

SD: F to U resident (SL–750 m) on Atlantic Slope from N Chis and E Tab to Honduras.

RA: SE Mexico to Ecuador.

JACAMARS: GALBULIDAE (I)

One member of this neotropical family reaches the region; most are restricted to South America. Bill long, slender, and straight, feet weak and inconspicuous. Wings short and rounded, tail long and strongly graduated. Ages similar, sexes differ; young altricial, molts poorly known. Metallic green and cinnamon overall. Perches fairly upright and still for long periods, sallies out for prey in dashing flights. Voice loud and noisy, shrieks and trills.

Food mostly insects caught in flight. Nests in burrows, often near the ground, in banks, etc. Eggs: 2–4, glossy white (WFVZ).

RUFOUS-TAILED JACAMAR
Galbula ruficauda melanogenia Plate 34
Jacamar Colirrufo

ID: 8.7–9.2 in (22–23.5 cm). *Humid SE forest and edge. See family account.* **Descr.** ♂: *bill black, feet straw. Head, chest, and upperparts metallic golden green* with *white throat; belly and undertail coverts cinnamon. Inner rectrices metallic green, outer*

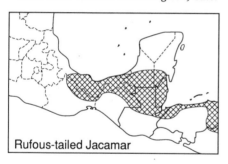

Rufous-tailed Jacamar

rectrices cinnamon. ♀: *throat* pale cinnamon to buff.

Voice. Varied loud, sharp, explosive screams; song a series of loud notes run into a rapid, excited trill, *whee twee twee twee twee-twee ti-illllll*; calls include a sharp, upward-inflected *whee'k* or *peeeu'k*, suggesting Royal or Great Crested flycatchers, and varied series, *whi' whEEh whEEh whee whee*, etc.

Habs. Humid evergreen forest and edge, plantations. See family account. Mostly detected by voice, at times conspicuous on swinging vines at forest edge or dust-bathing in quiet dirt roads.

SD: C to F resident (SL–750 m) on Atlantic Slope from E Chis to Honduras; U, N to S Ver and Yuc Pen.

RA: SE Mexico to Ecuador and N Argentina.

NB: M.A. populations sometimes regarded as specifically distinct, Black-chinned Jacamar, *G. melanogenia*.

TOUCANS: RAMPHASTIDAE (3)

A New World family of spectacular frugivorous birds with huge, laterally compressed bills; culmens decurved, tomia serrated. The bill is made of plates and is quite light despite its size. Legs and feet fairly strong and grasping. Fairly slender overall with rounded wings and long, often graduated tails. Ages differ slightly, sexes similar; young altricial. Molts poorly known; typically juvs are duller with duller-patterned bills. Colors and patterns varied, bright, and striking, bills and bare skin around eye often brightly colored. Voices include low croaks, squeaky and reedy calls.

Food mostly fruit, also invertebrates, and small vertebrates, including nestling birds. Nest in unlined tree cavities. Eggs: 2–4, white (WFVZ).

EMERALD TOUCANET

Aulacorhynchus prasinus*   Plate 34
Tucaneta Verde

ID: 12.5–14.5 in (32–37 cm). E and S forest. *Green overall.* Two groups: *prasinus* group (E and S) and *wagleri* (SW Mexico). **Descr. Prasinus:** orbital skin dark grey-brown to rufous-brown, *bill mostly black below, yellow above*, with black stripe at base of culmen, legs greyish. *Emerald-green overall* with whitish throat and chestnut undertail coverts, outer rectrices tipped chestnut. *Wagleri:* bill black with broad yellow stripe on culmen; crown washed yellowish, lower throat and chest paler, tinged bluish.
> **Voice.** Low, frog-like notes, repeated steadily, *rronk-rronk-rronk* ... or *wup-wup-wup* ..., etc., 10/3.5–6 s, often with more emphatic notes thrown in to break up an otherwise monotonous repetition.
> **Habs.** Humid to semihumid forest and edge, plantations, clearings. Usually in groups of 3–10 birds at mid- to upper levels, feeding quietly. Flight direct with rapid, whirring wingbeats.

SD: *Prasinus:* C to F resident (near SL–3000 m) on both slopes from SE SLP and SE Oax, and in interior from Chis, to El Salvador and Honduras; U at N edge of range in base of Yuc Pen. *Wagleri:* C to F resident (250–2700 m) in Sierra Madre del Sur of Gro and Oax.
RA: Mexico to Peru.
NB: *Wagleri* sometimes regarded as specifically distinct, as is *A. p. caeruleogularis* of Costa Rica and Panama.

COLLARED ARACARI

Pteroglossus torquatus*   Plate 34
Tucancillo Collarejo

ID: 15–17 in (38–43 cm). S humid lowlands. **Descr. Adult:** eyes pale yellow, orbital skin black and red, *bill black below, pale greyish horn above with dark tip and purplish flush at base*, legs blue-grey. Head, upper chest, and upperparts black with narrow maroon hindcollar. Rest of underparts yellow, mottled red on chest, with black chest spot and belly band, thighs chestnut-brown, underside of tail greenish.

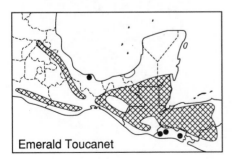
Emerald Toucanet

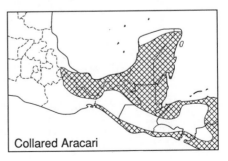
Collared Aracari

Juv: bill duller, eyes bluish, orbital skin pale grey.

Voice. A high, sharp, squeaky *pi-cheet* or *squi-zeek*, not usually repeated steadily, may recall Groove-billed Ani. Also a rough, rattling chatter given by chasing birds.

Habs. Humid to semihumid forest and edge, plantations, second growth woods. Habits much as Emerald Toucanet.

SD: C to F resident (SL–1300 cm) on both slopes from S Ver and E Oax to El Salvador and Honduras.

RA: S Mexico to N S.A.

NB1: Sometimes considered conspecific with *P. frantzii* of Costa Rica and Panama. **NB2:** Taxonomy of Collared Aracaris in N S.A. poorly understood, more than one species may be involved.

KEEL-BILLED TOUCAN
*Ramphastos sulfuratus**   Plate 34
Tucán Pico-multicolor

ID: 20–23 in (51–58.5 cm). SE humid forest. *Huge rainbow-colored bill.* **Descr.** Orbital skin green, *bill pale green with maroon tip, a patch of orange above and a flash of turquoise-blue below,* legs bluish. Face, throat, and chest yellow, narrowly bordered red on chest; otherwise black overall with maroon cast to nape, white uppertail coverts, and red undertail coverts.

Voice. A low, throaty, frog-like note, repeated, *rrrk-rrrk-rrrk* ... or *rruk-rruk-rruk* ..., etc. 10/4.5–7 s.

Habs. Humid evergreen forest and edge, semiopen areas with forest patches. Flight undulating and often high, rapid flaps interspersed with sailing glides, wingbeats make loud rushing sound at close range. At mid- to upper levels, often on prominent bare snags; in pairs or small groups.

SS: In E Honduras, see Chestnut-mandibled Toucan (Appendix E).

SD: F to C resident (SL–1400 m) on Atlantic Slope from S Ver to Honduras; U N to SE SLP.

RA: E Mexico to N S.A.

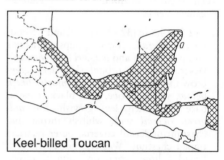

Keel-billed Toucan

PICULETS AND WOODPECKERS: PICIDAE (27*)

A nearly cosmopolitan family of birds (absent from Australasia) adept at climbing trees. Two distinct subfamilies occur in the region: Picumninae, the tiny piculets, and Picinae, the typical woodpeckers. Bills stout, pointed, and chisellike, often long; legs short, feet strong and zygodactylous. Tongues of typical woodpeckers are remarkably long and tubular with barb-like tips to aid in penetrating small holes to capture prey. Central rectrices of typical woodpeckers stiffened to act as props on branches and trunks. Ages/sexes usually differ, mainly in head patterns. Young altricial; 1st prebasic molt incomplete (to complete?); some species require a few months to attain adult head and body plumage. Piculets brownish to olive overall. Woodpeckers vary from mostly brown, through greens, to strikingly black and white, usually with some red on the head; several species strikingly barred above or below. Most species have black or greyish bills, greyish feet, and dark (brown to reddish eyes); only if different are these mentioned in the species accounts. Mainly solitary but some species social, at times joining mixed-species flocks. Flight undulating. Unmusical voices vary from sharp chips and chucks to rattles. Most woodpeckers (both sexes) 'drum', i.e. hammer rapidly a resounding branch with their bill to produce a carrying rattle or drumming sound used to proclaim territories. Specific drums are distinguishable with experience, but the volume and quality of a drum depend on the nature of the surface used.

Food mostly insects (particularly larvae) and fruit; some species drink sap. Nest in holes drilled out of trees, large cacti, etc., cavities unlined. Eggs: 2–6 (mostly 2–4), white (WFVZ).

Woodpecker taxonomy has undergone a recent trend of overenthusiastic generic lumping and inconsistent specific splitting (see Short 1982; AOU 1983). A more uniform treatment of the family is desirable.

OLIVACEOUS PICULET
Picumnus olivaceus dimotus   Plate 71
Carpinterito Oliváceo

ID: 3.5–3.7 in (9–10 cm). C.A. **Tiny,** bill fairly short and pointed. **Descr.** Ages/sexes differ. ♂: *crown and nape blackish, crown*

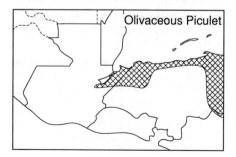

Olivaceous Piculet

spotted red, nape flecked white. Face and throat buff with *dark auricular mark,* chest brownish olive, rest of underparts streaked dusky and buff. Upperparts olive, flight feathers darker; secondaries and tertials edged yellow-green, *inner webs of central rectrices pale lemon, outer rectrices with oblique pale lemon stripes.* ♀: crown and nape blackish brown, flecked white. **Juv:** resembles ♀ but crown and nape lack white flecks.

Voice. A high, thin, slightly twittering, insect-like trill, suggests Yellow-faced Grassquit; also a high *ssip-ssip.* Does not drum.

Habs. Humid evergreen forest and edge, semiopen woodland, plantations. Feeds low to high, active and restless, foraging may resemble a nuthatch or Plain Xenops.

Often puzzling when first encountered.

SD: F to U resident (SL–500 m) on Atlantic Slope from NE Guatemala to Honduras.

RA: E Guatemala to Ecuador.

LEWIS' WOODPECKER

Melanerpes lewis   Not illustrated
Carpintero de Lewis

ID: 9.2–10.2 in (23.5–26 cm). *NW Mexico in winter.* **Descr.** Ages differ, sexes similar. **Adult:** *head and upperparts glossy greenish black with* dark red face and *silvery white hindcollar. Chest silvery white becoming pinkish on belly,* thighs and undertail coverts blackish. **Imm:** *hindcollar indistinct or lacking,* head and upperparts duller, throat and underparts dirty whitish with red wash on belly.

Voice. Generally silent, calls include sharp chips and short churrs.

Habs. Open and semiopen woodland, especially with oaks. Singly or in small groups, spends much time flycatching with soaring flight.

SD: Irregularly U to R transient and winter visitor (late Sep–Apr; SL–2000 m) to N Mexico, from BCN to N Chih; vagrant to N Coah (Dec 1981, JRG, PRG).

RA: Breeds W N.A.; winters to NW Mexico.

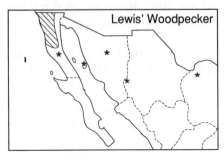
Lewis' Woodpecker

ACORN WOODPECKER

Melanerpes formicivorus*   Plate 36
Carpintero Arlequín

ID: 8.3–9.3 in (21–23.5 cm). *Widespread in oak woodland,* social and noisy. **Descr.** Ages/sexes differ. ♂: *eyes whitish* (dark in *angustifrons* of BCS?). *Cream forehead and lower throat joined by band across lores, crown red, rest of head and upperparts black; broad white patch across base of primaries striking in flight.* Chest blackish, flecked white (heavily streaked white in *lineatus* of N C.A.), rest of underparts white, streaked black on flanks. ♀: fore-

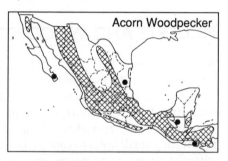
Acorn Woodpecker

crown black. **Juv:** resembles ♂ but eyes dusky.

Voice. Nasal, burry chattering and rhythmic laughing, *ya'ka ya'ka ya'ka,* and *aah'aah* or *eh'eh ...,* etc., a nasal *rrah* and *ahh,* and a longer *ahrr-ah-ah,* etc.

Habs. Oak and pine–oak woodland. Typically social and noisy, often in groups up to 12–15 birds.

SD: C to F resident (500–3000 m) in Baja, and on both slopes and in interior from Son and Coah to El Salvador and Nicaragua. Also (near SL–900 m) locally on Atlantic Slope from Tamps to the Mosquitia.

RA: W N.A. to N Colombia.

GENUS CENTURUS

These woodpeckers are common and conspicuous throughout most of the region. Ages similar, though juvs usually have dusky streaked underparts, and juv ♀♀ may have a trace of red on their crown; sexes differ. Heads and underparts greyish, ♂♂ with red and/or yellow head markings, reduced or absent in ♀♀; vents washed red or yellow in most species. Upperparts barred black and white. Calls mostly inflected nasal chucks and chatters. Feed mainly at mid- to upper levels, singly or in pairs. Some authors (AOU 1983) merge this genus with *Melanerpes.*

BLACK-CHEEKED WOODPECKER

Centurus pucherani perileucus   Plate 35
Carpintero Cachetinegro

ID: 7–7.5 in (18–19 cm). *SE humid forest.* **Descr.** ♂: *black mask with white postocular flash,* forehead yellow, crown and nape red. Upperparts barred black and white, rump and uppertail coverts white; lacks noticeable white flash in primaries in flight. Tail black, central and outermost rectrices with sparse white bars. Throat

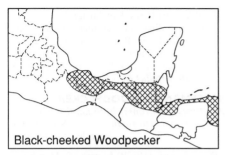

Black-cheeked Woodpecker

and underparts greyish, belly and undertail coverts barred blackish, vent red. ♀: crown greyish with black band posteriorly.

Voice. A rolled *chuh-uh-uh* or *chu-uh-uht*, less nasal than Golden-fronted Woodpecker, a sharp, slightly nasal, rolled *kweh-eh-eh* much like Golden-fronted, a rolled, chuttering *kwehrr-err-errr*, a drier *ahrr-rr-rr*, etc.

Habs. Humid evergreen forest and edge, clearings, plantations. See genus note.

SS: Golden-fronted Woodpecker larger, lacks black mask and white postocular flash, belly less extensively barred.

SD: C to F resident (SL–750 m) on Atlantic Slope from S Ver to Honduras. Reports from Pue (AOU 1983) may be in error.

RA: SE Mexico to Ecuador.

GOLDEN-CHEEKED WOODPECKER
*Centurus chrysogenys** Plate 35
Carpintero Cachetidorado

ID: 7.5–8.5 in (19–21.5 cm). *Endemic to W Mexico.* **Descr.** ♂: *face and chin yellowish with black eye patch*, forehead whitish, red crown becomes yellow-orange on nape. Upperparts barred black and white; white flash across base of primaries conspicuous in flight. Throat and underparts dusky grey, hind flanks and undertail coverts barred blackish, vent yellow. ♀: crown greyish with black band posteriorly.

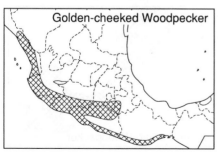

Golden-cheeked Woodpecker

Voice. A sharp, nasal *ki-di-dik*, sharper than Gila Woodpecker, a sneezy, fairly hard, nasal *keh-eh-eh-ehk* or *keh-i-heh-hek*, a softer *keh-i-heh* or *kuh-uh-uh*, and *churr-i-huh*, a sneezy *ch'dik ch'dik*, and longer, nasal, bickering series that may include short, rolled churrs, etc.

Habs. Arid to semihumid forest and edge, plantations, semiopen areas with scattered trees and forest patches. See genus note. Eggs undescribed (?).

SS: Golden-fronted Woodpecker locally sympatric, lacks black eye patch, underparts paler grey. Gila Woodpecker has paler grey head and underparts, head plain, ♂ has red cap. Grey-breasted Woodpecker duskier with white eye-crescents on grey face, gruffer voice.

SD/RA: C to F resident (SL–1500 m) on Pacific Slope from S Sin to Oax, in interior along Balsas drainage.

GREY-BREASTED WOODPECKER
Centurus hypopolius Plate 35
Carpintero Pechigris

ID: 7.5–8 in (19–20.5 cm). *Endemic to SW Mexico.* **Descr.** ♂: *head and underparts dusky grey with whitish forehead, white eye-crescents*, black around eye, *red crown patch*, and dull red wash below eye; hind flanks and undertail coverts barred blackish. Upperparts barred black and white, white flash across base of primaries most obvious in flight from below; tail black, central and outermost rectrices barred white. ♀: lacks red crown patch.

Voice. A rhythmic, nasal *yek'a yek'a* and a dry, chattering *chi-i-i-ir chi-i-i-ir*, etc.; lower, drier, and gruffer than Golden-fronted Woodpecker.

Habs. Arid open and semiopen areas with scattered trees and bushes, organpipe cacti. See habits under genus note; more often in small groups than other *Centurus*.

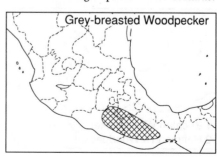

Grey-breasted Woodpecker

SS: Golden-fronted Woodpecker larger, head and underparts paler, nape yellow-orange, crown red in ♂. See Golden-cheeked Woodpecker.

SD/RA: C to F resident (900–1800 m) in interior from N Gro and Mor to Oax.

YUCATAN (RED-VENTED) WOODPECKER

*Centurus pygmaeus**   Plates 35, 71
Carpintero Yucateco

ID: 6.5–7 in (16.5–18 cm). *Endemic; mostly Yuc Pen. Bill short.* Descr. ♂: head and underparts pale greyish (duskier in island ssp) *with yellow nasal tufts (yellow usually encircles base of bill)*, red crown and nape; hind flanks and undertail coverts barred blackish, vent red. Upperparts barred narrowly with black and white, rump and uppertail coverts white; white flash across base of primaries indistinct. Tail black, outermost rectrices with narrow white bars. ♀: *red on head restricted to nape patch. C. p. tysoni* of Isla Guanaja has red of crown slightly separate from orange-red nape in ♂; ♀ has narrow red nape patch.

Voice. A mellow, rolled, chuttering *pwurr-r-r pwurr-r-r*, and a fairly rapid, dry, nasal, laughing *chuh-uh-uh-uh-uh-uh* or *cheh-cheh-cheh cheh-eh-eh*, etc., typically softer and less nasal than Golden-fronted Woodpecker. *C. p. tysoni*, at least, has a nasal *keh heh heh-heh heh-heh heh-heh*, more like Golden-fronted in quality.

Habs. Forest and edge, scrubby woodland, clearings. See genus note; often forages lower than other *Centurus*. Eggs undescribed (?).

SS: Golden-fronted Woodpecker larger with obviously larger bill, reddish nasal tufts, ♀ with larger red nape patch.

SD/RA: F to C resident (around SL) in Yuc Pen, including Isla Cozumel, S to cen Campeche and cen Belize; also on Guanaja Island, Honduras.

NB: Sometimes considered conspecific with *C. rubricapillus* of S C.A. and N S.A.

HOFFMANN'S WOODPECKER

Centurus (aurifrons?) hoffmannii   Plate 71
Carpintero de Hoffmann   See Yucatan Woodpecker for map

ID: 7.5–8 in (19–20.5 cm). *S Honduras.* Descr. ♂: head and underparts pale greyish, nasal tufts and *nape yellow, crown red*; hind flanks and undertail coverts barred blackish, vent yellow. *Upperparts barred black and white*, rump and uppertail coverts white; white flash across base of primaries indistinct. Tail black, central and outermost rectrices barred black and white. ♀: crown pale grey.

Voice. Much like Golden-fronted Woodpecker, also a hard rattle.

Habs. Arid to semihumid open and semi-open areas with scattered trees, forest and edge. See genus note.

SS: Golden-fronted Woodpecker larger, crown and nape orange-red to red, upperparts and wings more narrowly barred.

SD: F to C resident (SL–200 m) on Pacific Slope of Honduras.

RA: S Honduras to Costa Rica.

NB: Interbreeds with Golden-fronted Woodpecker in S Honduras (Monroe 1968, pp. 215–16) and with Red-crowned Woodpecker in Costa Rica (Stiles and Skutch 1989).

GOLDEN-FRONTED WOODPECKER

*Centurus aurifrons**   Plate 35
Carpintero Frentidorado

ID: 8.5–10 in (21.5–25.5 cm). *E and S. Complex geographic variation.* Three

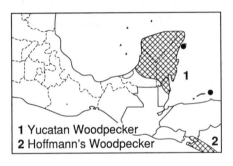

1 Yucatan Woodpecker
2 Hoffmann's Woodpecker

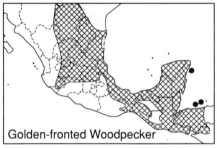

Golden-fronted Woodpecker

groups, all with linking, variably intermediate populations: *aurifrons* (N and cen Mexico), *dubius* group (Atlantic Slope, S Mexico to Honduras), and *polygrammus* (Pacific Slope of Isthmus and interior of Chis). **Descr. Aurifrons.** ♂: head and underparts pale greyish, *nasal tufts and nape yellow to orange-yellow, crown patch red*; hind flanks and undertail coverts barred blackish, *vent yellow. Upperparts broadly barred black and white*, rump and uppertail coverts white; white patch across base of primaries striking in flight. Tail black, outermost rectrices with slight white barring. ♀: crown pale greyish. Intergrades from S Tamps to Isthmus with *Dubius*: resembles *aurifrons* but *nasal tufts, crown, and nape red, vent red, upperparts narrowly barred*, white flash across base of primaries indistinct; nasal tufts and belly become yellow and nape more orange-red in interior Guatemala and Honduras. *Polygrammus*: resembles *aurifrons* but *red crown joins yellow-orange nape*, upperparts narrowly barred, *central rectrices barred white*. Intergrades with *dubius* group along Pacific Slope of Chis.

Voice. A gruff, rhythmic *chow'ka chow'ka* or *ch'krr ch'krr*, a rolled, nasal *ka-du-duk ka-da-duk*, and *ka-a-h ka-a-ah*, a rolled *pwurr-rr-rr*, etc.

Habs. Open and semiopen areas with scattered trees, hedges, forest edge, scrubby woodland. See genus note.

SS: See Golden-cheeked, Grey-breasted, Yucatan, and Hoffmann's woodpeckers. Gila Woodpecker told by plain head, ♂ has small red cap.

SD: C to F resident (SL–2500 m) in interior from Chih and NL to Jal and N Mich, and from Isthmus to N cen Nicaragua; on Atlantic Slope from NL and Tamps to Honduras; on Pacific Slope from Isthmus to El Salvador and Honduras.

RA: Mexico and SW USA to W Nicaragua.

NB1: Red-bellied Woodpecker *C. carolinus* is an irregular rare winter visitor to lower Rio Grande Valley, Texas, and could reach NE Mexico. Resembles Golden-fronted (*aurifrons*) but ♂ has nasal tufts, crown, and nape red (crown pale grey in ♀), vent reddish, central rectrices barred white.

NB2: Golden-fronted interbreeds locally with Gila Woodpecker in NW Mexico (Selander and Giller 1963) and with Red-

bellied Woodpecker *C. carolinus* in Texas (Smith 1987). See NB under Hoffmann's Woodpecker.

GILA WOODPECKER

Centurus uropygialis*   Plate 35
Carpintero de Gila

ID: 8.5–9.5 in (21.5–24 cm). NW Mexico. **Descr.** ♂: *head and underparts pale greyish with red crown patch*, hind flanks and undertail coverts barred blackish, vent yellow. Upperparts broadly barred black and white, white rump and uppertail coverts with sparse black bars; white patch across base of primaries striking in flight. Tail black, central and outermost rectrices barred white. ♀: crown pale greyish.

Voice. A rolled *wurr uh-rr wurr uh-rr ...*, and *wurr-rr* or *churrr*, an excited, intensifying, slightly ringing to screaming series, *kyuh-kyuh-kyuh ...* or *rruhk-rruhk ...*, and a nasal, slightly mewing screech, *woihk' ...*, etc.

Habs. Arid to semihumid open and semi-open areas with scattered trees, organpipe cacti, plantations. See genus note.

SS: See Golde.·-cheeked and Golden-fronted woodpeckers.

SD: C to F resident (SL–1600 m) in Baja (except NW), and on Pacific Slope and in adjacent interior from Son and Chih to Jal and Ags.

RA: SW USA to NW Mexico.

NB: See NB2 under Golden-fronted Woodpecker.

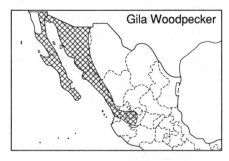

Gila Woodpecker

GENUS SPHYRAPICUS

Sapsuckers are a distinctive group of woodpeckers that characteristically feed on sap obtained by drilling small holes, often in neat rows, in trees. Ages/sexes similar or different, most show a distinctive white stripe on the median and inner greater wing coverts, referred to below as wing-covert stripe. Relatively sluggish, quiet, and overlooked easily, feed mainly at mid- to upper levels. All are migratory; one species breeds in the region.

YELLOW-BELLIED SAPSUCKER
Sphyrapicus varius   Figure 31
Chupasavia Vientre-amarillo

ID: 7.5–8 in (19–20.5 cm). *Widespread winter visitor.* **Descr.** Ages/sexes differ. ♂: *red crown bordered black,* broad white postocular stripe and moustache and broad black eyestripe all continue down sides of neck; *red throat bordered by black malar stripe and broad black chest patch;* some birds show red on nape. Underparts whitish, streaked blackish on sides, belly washed yellow. *Upperparts barred and mottled black and whitish to creamy buff with bold white wing-covert stripe,* rump and uppertail coverts mostly white. Tail black, central and outermost rectrices barred white. ♀: *throat white,* crown may be partly or wholly black. **Imm:** resembles adult but *head brownish with indistinct pattern, usually shows whitish moustache and red mottling* on crown; *chest and sides dusky brown, scalloped darker;* upperparts washed brownish. Attains adult plumage through 1st winter, most resemble adults by Mar–Apr, black chest band is last adult feature to appear. Intergrades with Red-naped Sapsucker may occur in region.
   Voice. Mainly silent; calls include a nasal mewing *meeah* recalling Grey Catbird, and quiet chuks.
   Habs. Forest and edge, rarely in pure conifers. See genus note.
SS: Red-naped Sapsucker has less extensive pale mottling on upperparts, both sexes

Fig. 31 Immature Yellow-bellied Sapsucker

have red on throat, adult has distinct red nape patch (rarely lacking), imm resembles adult by early winter. ♀ Williamson's Sapsucker told from imm Yellow-bellied by plain brownish head with no red, lack of white wing-covert stripe, neat barring overall.

SD: F to C transient and winter visitor (late Sep–mid-Apr; SL–3500 m) on both slopes and in interior from S Sin and Coah to El Salvador and W Nicaragua. Vagrant (Dec–Feb) to BCN (Patten *et al.* 1993), Isla Socorro (SNGH), and Son (DS).
RA: Breeds N.A.; winters to Panama.

RED-NAPED [YELLOW-BELLIED] SAPSUCKER
Sphyrapicus (varius?) nuchalis
Chupasavia Nuquirroja   Not illustrated

ID: 7.5–8.2 in (19–21 cm). *NW Mexico in winter.* **Descr.** Ages/sexes differ slightly. **Adult:** *similar to ♂ Yellow-bellied Sapsucker but* typically with *obvious red nape patch* (rarely indistinct or almost absent), red of throat extends into malar region of

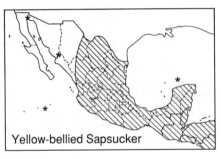

Yellow-bellied Sapsucker

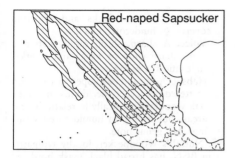

Red-naped Sapsucker

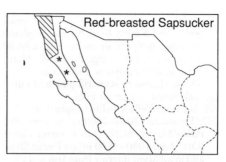

Red-breasted Sapsucker

♂, obscuring black malar stripe; less extensive pale mottling of upperparts tends to form two broad stripes down sides of back; ♀ usually has chin and upper throat whitish. **Imm:** *mostly like adult upon arrival in region* but *red nape patch may be indistinct or lacking,* face may be washed brownish, *chest and sides dusky brown, scalloped darker;* as adult by Jan, black chest patch last to appear. Intergrades with Yellow-bellied Sapsucker may occur in region.

Voice. A mewing *meeah*, like Yellow-bellied Sapsucker.

Habs. Forest and edge. See genus note.

SS: See Yellow-bellied Sapsucker.

SD: F transient and winter visitor (late Sep–mid-Apr; SL–2500 m) in Baja, and on Pacific Slope and in interior from Son and NL to Jal; possibly R and irregular (at least formerly) to Guatemala and Honduras (Salvin and Godman 1895; Monroe 1968).

RA: Breeds W N.A.; winters to NW Mexico.

NB: Has hybridized with Williamson's Sapsucker (Short 1982).

RED-BREASTED [YELLOW-BELLIED] SAPSUCKER

Sphyrapicus (varius?) ruber daggetti
Chupasavia Pechirroja   Not illustrated

ID: 8–8.7 in (20.5–22 cm). *N Baja in winter.* Descr. Ages/sexes similar. *Head and chest dull red* with whitish nasal tufts and short moustache, blackish eye-patch. Underparts whitish, streaked blackish on sides, belly washed yellow. *Upperparts black with bold white wing-covert stripe,* whitish mottling down back, white bars on primaries, rump and uppertail coverts mostly white. Tail black, central rectrices barred white.

Voice. A mewing *meeah* and quiet *puc*

and *pwuc* calls, much as Yellow-bellied Sapsucker.

Habs. Forest and edge, often in conifers. See genus note.

SD: U to R winter visitor (Oct–Mar; 500–3000 m) to NW BCN, vagrant S to about 29 °N (Nov 1987, TEW).

RA: Breeds W N.A.; winters to N Baja.

WILLIAMSON'S SAPSUCKER

*Sphyrapicus thyroideus**   Not illustrated
Chupasavia de Williamson

ID: 8.5–9 in (21.5–23 cm). NW Mexico, mainly winter. Descr. Ages/sexes differ. ♂: *head, chest, and upperparts glossy black with white postocular stripe and moustache,* red median throat, *bold white wing-covert stripe,* and white rump and uppertail coverts; primaries and central rectrices sparsely spotted white. *Underbody yellow,* sides and undertail coverts scalloped black and white. ♀: *head brownish,* often with darker malar stripe; *body and flight feathers barred blackish and pale brown overall,* usually with black chest patch, rump and uppertail coverts white, *median belly yellow.* **Juv:** resembles adult, ♂ duller, sooty black with white median throat, back with sparse white spots; ♀ barred less contrastingly, lacks black chest patch. Attains adult plumage before migration.

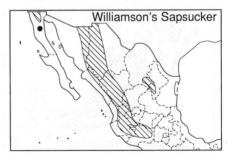

Williamson's Sapsucker

Voice. A quiet, slightly purring growl *ahrrr*, and a fairly loud, hoarse, emphatic mewing, typically in series of 4, *wheeha wheeha ...*, may suggest a hawk. Hard, rattling drum given in 4–8 bursts.

Habs. Humid to semiarid pine and pine–oak forest. See genus note.

SS: See imm Yellow-bellied Sapsucker.

SD: F resident (2000–3000 m) in Sierra San Pedro Martír, BCN; U to F winter visitor (Oct–Apr; 1000–3500 m) on Pacific Slope and in adjacent interior from Son and Chih to Jal and N Mich, R to U E to NL (DEW).

RA: Breeds W N.A. to BCN; winters to W Mexico.

NB: See Red-naped Sapsucker.

GENUS PICOIDES

These strikingly patterned, black- or brown-and-white woodpeckers are of temperate origin. Ages/sexes differ, ♂ has red patch on hindcrown, juv has red crown patch. Fairly active and conspicuous, feed low to high. Calls are sharp *chik* notes and rapid chatters. Most species formerly placed in the genus *Dendrocopos*.

LADDER-BACKED WOODPECKER
Picoides scalaris*   Plate 36
Carpintero Listado

ID: 6–8 in (15–20 cm). *Widespread in dry areas.* **Descr.** ♂: crown and nape black with large red hindcrown patch, white and red flecks on forecrown often wear off; white postocular stripe continues down sides of neck, *broad white moustache surrounded by black eyestripe and malar stripe. Upperparts barred black and white,* lower rump to tail black, outer rectrices barred white. *Throat and underparts dirty whitish to dusky buff* (darkest from cen Mexico S), spotted blackish on sides, often

forming bars on flanks and undertail coverts. ♀: hindcrown black.

Voice. A sharp, shrill, rapid chatter, a high, sharp *chiik* or *peek*, and a jerky, rhythmic *kee'ki-krr*, etc.

Habs. Open and semiopen areas with scattered trees, cacti, deciduous and pine–oak woodland, typically in relatively arid areas, but locally in humid forest clearings. See genus note.

SS: Nuttall's Woodpecker, locally sympatric in BCN, has broad black mask bordered by narrow white supercilium and moustache, upper mantle lacks white bars, nasal tufts and underparts typically whiter, ♂'s red hindcrown contrasts more sharply with black forecrown.

SD: C to F resident (SL–3000 m) in Baja, on Pacific Slope to Sin and locally to Oax, in interior S to Chis, and on Atlantic Slope from NL and Tamps locally to Yuc Pen, N Belize, and the Mosquitia; also (SL–1800 m) locally in Guatemala (Howell and Webb 1992*b*; AD) and Honduras. Reports from Isla Cozumel (AOU 1983) based on Gaumer specimens are erroneous (see Parkes 1970).

RA: W N.A. to N Nicaragua.

NB: Hybridizes with Nuttall's Woodpecker in BCN (Short 1971).

NUTTALL'S WOODPECKER
Picoides nuttallii   Not illustrated
Carpintero de Nuttall

ID: 6.7–7.5 in (17–19 cm). *N Baja.* **Descr.** ♂: crown and nape black with large red hindcrown patch, white flecks on forecrown often wear off; *broad black mask bordered by narrow white postocular stripe and narrow white moustache,* postocular stripe continues down sides of neck. Upperparts barred black and white, except *upper mantle which forms black band above barred area;* lower rump to tail

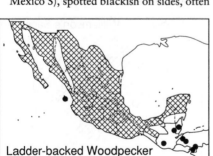
Ladder-backed Woodpecker

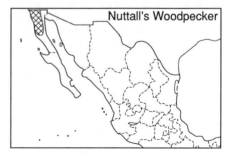
Nuttall's Woodpecker

black, outer rectrices barred white. *Throat and underparts whitish*, spotted blackish on sides, often forming bars on flanks and undertail coverts. ♀: hindcrown black.

Voice. A sharp, disyllabic *pi-dik* or *pr-dik*, may be repeated in series and run into a rattling chatter.

Habs. Arid to semiarid riparian woods, semiopen areas with scattered trees, scrubby woodland. See genus note.

SS: See Ladder-backed Woodpecker.

SD: F to C resident (SL–2000 m, mostly below 1250 m) in NW BCN, N of 30° N.

RA: California and N Baja.

NB1: See Ladder-backed Woodpecker. **NB2:** Has hybridized with Downy Woodpecker in S California (Short 1982).

HAIRY WOODPECKER
Picoides villosus*   Plate 36
Carpintero-velloso Mayor

ID: 7–9 in (18–23 cm). N>S. *Widespread in highlands. Bill fairly long.* Two groups: *villosus* group (NW Mexico; including *icastus*, shown) and *jardinii* group (NE and cen Mexico to C.A.; including *sanctorum*, shown). **Descr. *Villosus*.** ♂: crown and nape black with red hindcrown patch, *face broadly striped* with white supercilium, black auriculars, fan-shaped white moustache, and black malar stripe. *Upperparts black with broad white stripe down back*, primaries spotted white. *Tail black with white outer rectrices.* Throat and underparts whitish. ♀: hindcrown black. *Jardinii*: smaller and shorter-billed, *throat and underparts dusky brownish* to sooty brown.

Voice. A sharp *chiik!* or *chriek* and a rapid, chattering rattle, harder than calls of Ladder-backed and Strickland's woodpeckers; also a slightly nasal, ringing *cree'chree chree'chree* and *chree'ka chree'ka ...*, etc.

Habs. Pine–oak and oak forest. See genus note.

SS: Unbarred white stripe down back distinctive, bill longer than Strickland's Woodpecker.

SD: C to F resident (1200–3500 m) in BCN, and on both slopes and in interior from Son and S Coah to El Salvador and N cen Nicaragua.

RA: N.A. to Panama.

NB: Downy Woodpecker *Picoides pubescens* (6–6.5 in, 15–16.5 cm) may occur as R visitor to NW BCN (see Appendix B). Resembles Hairy Woodpecker (*villosus* group) but much smaller with short bill, remiges more distinctly spotted white, white outer rectrices with 2 broken blackish bars distally. Sharp, shrill, rapid chatter and sharp *chik*, softer than Hairy Woodpecker.

ARIZONA [STRICKLAND'S] WOODPECKER
Picoides arizonae* or *P. stricklandi* (in part)
Carpintero de Arizona   Plate 36

ID: 7–8 in (18–20.5 cm). *NW Mexico, highlands.* **Descr.** ♂: crown and nape dark brown with red hindcrown patch, *face white with bold dark brown auricular patch* and malar stripe. *Upperparts brown* (may show trace of white bars on back), primaries sparsely spotted white. Tail blackish, outer rectrices barred white. Throat and underparts whitish, *underparts heavily spotted dark brown*, often forming bars on flanks. ♀: hindcrown dark brown.

Voice. A sharp, slightly shrill *chik* or *peek*, a nasal *chriek'a ...* usually 2–5×, and a rapid, slightly rough, screeching chatter, gruffer than Ladder-backed Woodpecker.

Habs. Arid to semihumid oak and pine–oak forest. See genus note.

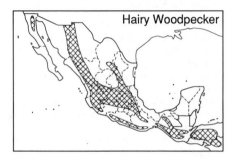

Hairy Woodpecker

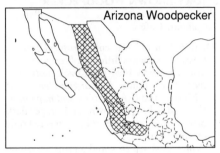

Arizona Woodpecker

SS: See Hairy Woodpecker. Strickland's Woodpecker allopatric (?).

SD/RA: C to F resident (1200–2400 m) on Pacific Slope and in adjacent interior from Son (also S Arizona) to Jal and Mich.

STRICKLAND'S (BROWN-BARRED) WOODPECKER

Picoides stricklandi *   Plate 36
Carpintero de Strickland

ID: 6.7–7.2 in (17–18.5 cm). *Endemic to cen Mexico.* Descr. Resembles Arizona Woodpecker but bill shorter, *upperparts and wings dark brown, back with thick white bars (often appearing as a solid whitish stripe on back)*, white outer rectrices with slight dark bars, chest more distinctly striped dark brown.
     Voice. A sharp *chiik* and a rattling chatter, etc.
     Habs. Open pine woodland. See genus note. Eggs undescribed (?).
SS: See Hairy Woodpecker; Arizona Woodpecker allopatric (?).
SD/RA: F to C resident (2500–4000 m) in cen volcanic belt from Mex (and E Mich?) to cen Ver.

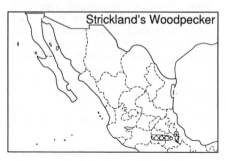
Strickland's Woodpecker

SMOKY-BROWN WOODPECKER

Veniliornis fumigatus *   Plate 35
Carpintero Café

ID: 6–6.3 in (15–16 cm). *Tropical forests.* Descr. Ages/sexes differ. ♂: *brown overall*, brighter tawny-brown on back, *with pale buff face* (duller in *sanguinolentus* of S Mexico and C.A.), *crown and nape mottled reddish.* ♀: crown and nape dark brown. Juv: resembles ♂ but ♀ has red only on forecrown.
     Voice. A sharp *chik* and a gruff, chortling

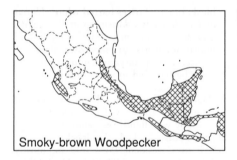
Smoky-brown Woodpecker

rattle, recalling *Picoides*; also a squeaky *chik'a chik'a* or *choi-it choi-it* ..., and a quiet, gruff *chick*.
     Habs. Humid to semihumid evergreen and semideciduous forest, edge, plantations. Feeds mostly at low to mid-levels, at times in dense tangles near ground.
SS: In shady forest, larger Grey-crowned Woodpecker (W Mexico) may recall Smoky-brown but note barred underparts, brighter bronzy wings, grey crown, ♂ has red moustache.
SD: F to C resident (near SL–1500 m) on both slopes from S Nay and S Tamps to El Salvador and Honduras.
RA: Mexico to Peru and N Argentina.

GENUS PICULUS

These are green woodpeckers of humid tropical forest. Wings contrastingly bronzy, underparts barred. Ages/sexes differ, ♂♂ have red moustaches. Common but relatively inactive and overlooked easily, heard far more often than seen. Feed low to high. Calls are sharp screams and rattles.

BRONZE-WINGED [GOLDEN-OLIVE] WOODPECKER

Piculus (rubiginosus?) aeruginosus   Plate 35
Carpintero Alibronceado

ID: 9–9.5 in (23–24 cm). *Endemic to NE Mexico.* Descr./SS: *Resembles Golden-olive Woodpecker but* larger and longer-tailed, upperparts greener, *underparts coarsely scalloped*; red nape band of ♂ narrower with *red on sides of crown typically extending forward only to eye*; red on nape of ♀ typically in a narrower, more U-shaped band.
     Voice. A sharp, nasal, squirrel-like *kyow'n* or *chey-eh*, at times repeated in

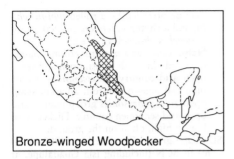

Bronze-winged Woodpecker

short series, and a steady series of sharp *weeyk!* or *wheeir* notes, 10/4–6 s, suggesting Squirrel Cuckoo song; also a low, short, guttural chatter, audible at close range. Voice thus quite different from Golden-olive Woodpecker.

Habs. Humid to semiarid forest and edge, plantations. See genus note. Eggs undescribed (?).

SD/RA: C to F resident (near SL–2100 m) on Atlantic Slope from cen NL and Tamps to cen Ver.

NB: Distinct vocalizations and plumage differences suggest specific status for Bronzewinged and Golden-olive woodpeckers. Field studies are needed to investigate the extent of intergradation (if any?) in cen Ver.

GOLDEN-OLIVE WOODPECKER
Piculus rubiginosus*   Plate 35
Carpintero Oliváceo

ID: 8.3–8.8 in (21–22.5 cm). *E and S.* **Descr.** ♂: *grey crown and red nape*, red extends forward narrowly along sides of crown to bill; red moustache borders greyish buff face. *Upperparts olive-green, wings contrastingly bronzy*; tail dark brown, paler basally. Throat mottled dark and whitish, *underparts barred* dark olive and cream to pale lemon. ♀: moustachial area mottled

dark and whitish, red restricted to nape. **Juv:** resembles adult but ♂'s moustache mixed with grey, underpart barring less distinct.

Voice. A sharp, slightly explosive *kee'ah* or *k'yaah*, recalling a flicker, and a rapid, shrill, churring rattle.

Habs. Humid to semihumid forest and edge, plantations. See genus note.

SS: See Bronze-winged Woodpecker.

SD: C to F resident (SL–2100 m) on Atlantic Slope from cen Ver to Honduras; on Pacific Slope and in interior from Isthmus to El Salvador and N cen Nicaragua.

RA: Mexico to Peru and N Argentina.

NB: See Bronze-winged Woodpecker.

GREY-CROWNED WOODPECKER
Piculus auricularis or *P. rubiginosus* (in part)
Carpintero Coronigris   Plate 35

ID: 7.7–8.2 in (19.5–21 cm). *Endemic to W Mexico.* **Descr.** ♂: *crown and nape grey*, often with red flecks along sides; red moustache borders greyish buff face. *Upperparts olive-green, wings contrastingly bronzy*; tail dark brown, paler basally. Throat mottled dark and whitish, *underparts barred* dark olive and cream to pale lemon. ♀: moustachial area mottled dark and whitish, no red on sides of crown. **Juv:** resembles adult but ♂'s moustache mixed with grey, underpart barring less distinct.

Voice. A sharp, slightly explosive *kee'ah*, a gruff mewing *growh*, and a rapid, shrill, churring rattle much like Golden-olive Woodpecker.

Habs. Humid to semihumid forest and edge, locally in pine–oak forest. See genus note. Eggs undescribed (?).

SS: See Smoky-brown Woodpecker.

SD/RA: F to C resident (near SL–2400 m) on Pacific Slope from S Son to Oax.

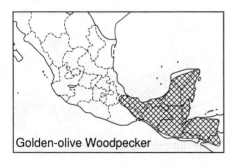

Golden-olive Woodpecker

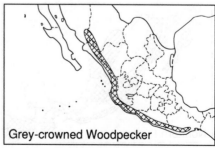

Grey-crowned Woodpecker

NORTHERN (COMMON) FLICKER

Colaptes auratus*   Plate 36
Carpintero Collarejo

ID: 10–12 in (25.5–30.5 cm). A large wood-pecker of temperate regions. Four groups: *cafer* group (Red-shafted Flicker, N of Isthmus; including *mexicanus*, shown), *mexicanoides* (Guatemalan Flicker, S of Isthmus), *chrysoides* group (Gilded Flicker, Baja and NW Mexico), and *auratus* group (Yellow-shafted Flicker, N Mexico in winter). **Descr.** Ages similar, sexes differ. **Cafer.** ♂: *crown and nape grey-brown* (cinnamon-brown in *rufipileus* of Isla Guadalupe), *face and throat grey with red moustache.* Upperparts grey-brown, barred black; *rump white*, upptail coverts barred black and white; tail black. *Underside of flight feathers and feather shafts red, striking in flight.* Broad black chest band, underparts whitish to buffy white, boldly spotted black. ♀: lacks red mous-tache but may show brownish moustache. *Mexicanoides*: resembles *cafer* group but *crown and nape cinnamon, upperparts narrowly barred black and pale cinnamon,* ♀ has cinnamon moustache. *Chrysoides*: resembles *cafer* group but *crown and nape pale cinnamon-brown, underside of flight feathers and feather shafts yellow.* *Auratus*: resembles *chrysoides* group but *crown and nape greyish with red nape patch, face and throat vinaceous,* ♂ has *black moustache.* Cafer and *auratus* groups intergrade widely in N.A. and *cafer* and *chrysoides* groups intergrade in SW USA, producing a variety of intermediate plumages which may occur in N Mexico in winter.

Voice. A screaming, slightly explosive *kyeea!*, a varied, rapid, laughing rattle, *puh-puh-puh* ... or *hih-hih-hih* ..., which may be preceded by a rolled *whi-hi whi-hi*, and a rhythmic *tch-wik tch-wik* ... or *ch-week'a ch-week'a* ..., etc.

Habs. *Cafer* and *auratus* groups mainly pine and pine–oak forest, open and semi-open arid country; *mexicanoides* mainly pine and pine–oak forest, adjacent semi-open areas; *chrysoides* group mainly arid open and semiopen country. Flickers feed low to high, often on the ground.

SD: *Cafer* group: C to F resident (900–4000 m) in BCN (including Isla Guadalupe, at least formerly), and in interior and on adjacent slopes from Son and Coah to Oax; *mexicanoides*: C to F resident (750–3300 m) in interior and on adjacent slopes from N Chis to El Salvador and N cen Nicaragua; *chrysoides* group: C to F resi-dent (SL–1800 m), in Baja (except NW) and on Pacific Slope from Son to N Sin; *auratus* group: R (to U?) winter visitor (Oct–Mar) to N Mexico from Son (ARP) to Tamps (JCA).

RA: N.A. to W Nicaragua.

NB1: *Mexicanoides* and *chrysoides* some-times considered specifically distinct, as is *chrysocaulosus* of Cuba and Grand Cay-man. **NB2:** *Cafer, auratus,* and *chrysoides* groups formerly considered separate species.

CHESTNUT-COLORED WOODPECKER

Celeus castaneus   Plate 35
Carpintero Castaño

ID: 8.5–9.5 in (21.5–24 cm). SE humid forest. *Prominent shaggy crest.* **Descr.** Ages/sexes differ. ♂: *bill pale greenish yellow.* Head and crest ochre with red patch below eye, rest of plumage chestnut-brown, coarsely scalloped and barred black except on flight feathers; primaries and tail dark distally. ♀: lacks red on face. **Juv:** duller, malar area mottled dusky.

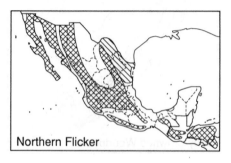

Northern Flicker

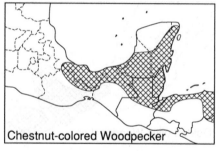

Chestnut-colored Woodpecker

Voice. A sharp, slightly explosive, hollow *whehoo!* or *kyow*, at times followed by a laughing series of 2–10 notes, *kyow! heh-heh-heh* ...; also a loud, sharp, slightly squeaky *pi'chk!* or *ki'chi*.
Habs. Humid evergreen and semideciduous forest. Fairly active, mostly at mid- to upper levels.
SD: F to C resident (SL–1000 m) on Atlantic Slope from S Ver to Honduras.
RA: SE Mexico to W Panama.

LINEATED WOODPECKER
Dryocopus lineatus*   Plate 36
Carpintero Lineado

ID: 12.5–13.5 in (31.5–34 cm). *Tufted crest striking.* **Descr.** Ages/sexes differ. ♂: eyes pale yellow, bill pale horn. *Crown and crest red, face black with red moustache and white stripe across lower auriculars* (white stripe less distinct or lacking in *scapularis* of W Mexico) that continues down sides of neck. Throat mottled black and white, neck, chest, and upperparts black with broken white V on scapulars. Underbody barred black and whitish. ♀: forecrown and moustachial area black. **Juv:** eyes dusky, ♂'s moustache mostly blackish, underpart barring less distinct.
Voice. A fairly rapid, often intensifying, loud, laughing series of 3–6 s in duration, usually rising and falling slightly, *yih-yih* ... or *nyeh-nyeh* ... Also a sharp, slightly ringing *chik* or *choik*, often followed by a gruff growl *ahrrrrr* or *urrrrr*, i.e. *chik, chik, chik, chik-urrrrr*, suggesting Squirrel Cuckoo. Drum hard and rapid, 2–3 s in duration.
Habs. Forest and edge, plantations, mangroves, semiopen areas with scattered trees and forest patches. Feeds low to high, at times in loose groups up to 5–6

birds. Associates readily with Pale-billed Woodpecker, may even nest in the same tree.
SS: Pale-billed Woodpecker's head usually looks squarer, white V joined on upperparts, ♂ has red head, ♀ head red with black forecrown and throat.
SD: F to C resident (SL–1600 m) on both slopes from S Son and E NL, and locally in interior from Isthmus, to El Salvador and Honduras.
RA: Mexico to Peru and N Argentina.
NB: Pileated Woodpecker *D. pileatus* (16–18 in; 40.5–45.5 cm) of N.A. may reach N Mexico in winter. Resembles Lineated but larger, bill dark greyish, underparts blackish, lacks white V on back; in flight, note white flash across bases of primaries on upperwing. See Appendix B.

PALE-BILLED WOODPECKER
Campephilus guatemalensis*   Plate 36
Carpintero Piquiclaro

ID: 14–15 in (35.5–38 cm). *Bushy crest prominent.* **Descr.** Ages/sexes differ. ♂: eyes pale yellow, bill pale horn. *Head red*, neck, chest, and upperparts black, white line down sides of neck runs into *closed white scapular V*. Underbody barred black and whitish. ♀: forecrown and throat black. **Juv:** resembles ♀ but eyes dusky, sides of head mixed with blackish, underpart barring less distinct.
Voice. Sharp, nasal, squirrel-like clucks, *kuh kuh kuh-uh-uh* or *chuk-chuk-chuk* ..., etc. Drum a distinctive, loud, rapid double-rap, less often up to 7 distinct raps in rapid, resonant series.
Habs. Humid to semiarid forest and edge, plantations, mangroves. Habits much as Lineated Woodpecker.
SS: See Lineated Woodpecker.
SD: F to C resident (SL–2000 m) on both

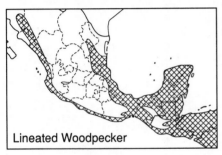

Lineated Woodpecker

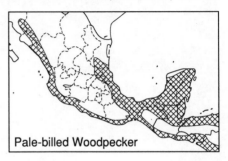

Pale-billed Woodpecker

slopes from S Son and S Tamps to El Salvador and Honduras.
RA: Mexico to W Panama.

IMPERIAL WOODPECKER
Campephilus imperialis
Presumed extinct. See Appendix A.

OVENBIRDS: FURNARIIDAE (7)

A diverse neotropical family. Most are forest dwellers but some inhabit second growth and scrub. Ages similar or slightly different, sexes similar; young altricial. Juv plumage usually lost within a few weeks and 1st basic resembles adult. Plumage predominantly brown, often with some buff and rufous. Most are skulking, hard to observe well, and live on or near the forest floor. Relatively few species occur in the region and identification problems are slight; note bill length and shape, head and chest color and pattern, behavior, and habitat. Often best detected by voice.

Primarily insectivores. Most species in the region nest in burrows or holes. The family name derives from the large domed (oven-like) nests built by many S.A. species.

RUFOUS-BREASTED SPINETAIL

Synallaxis erythrothorax*   Plate 38
Guitio Pechirrufo

ID: 5.8–6.5 in (14.5–16.5 cm). Endemic. A skulking, long-tailed, and somewhat wren-like bird. **Descr.** Ages differ. **Adult:** eyes reddish, bill blackish, legs grey. Head and back brownish grey, *wings rufous*, rump and tail brown, *tail tips usually abraded and spiny-looking*. Whitish throat marked with lines of black spots that merge into a broad black forecollar. *Rufous chest and sides blend to form a bright band with wings*, belly grey, flanks and undertail coverts brown. *S. e. pacifica* (Pacific Slope from Chis to El Salvador) paler overall, throat spotting and forecollar indistinct or lacking. **Juv:** eyes brown, bill yellow with dark culmen and tip; chest dusky brown, throat markings and forecollar less distinct.

Voice. A sneezy, nasal *wytch'err* or *whit'chew*, 1st part often repeated 2–4×, *whi whi'chew*, etc.; also a hard *cheurr* and a nasal, slightly barking *kyow* or *kyeh*.

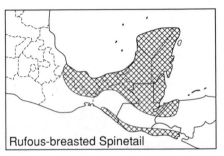
Rufous-breasted Spinetail

Song an excited, rapid, rough to yodeling rolled chatter, usually with 1–3 gruff introductory notes: *chree ree rreu-rreu-rreu* ... or *br, briih briih bri-bri-bri* ..., etc., may suggest Lineated Woodpecker or a screaming small falcon.

Habs. Second growth, scrub, overgrown clearings, forest edge. Usually keeps low and skulks, hard to see well, mostly detected by voice. Nest a bulky globular stick structure, entered by a long covered passage, low in bush or small tree. Eggs: 2–4, pale greenish blue, unmarked (MVZ).

SS: See Slaty Spinetail (E Honduras, Appendix E).

SD/RA: C to F resident (SL–750 m) on Atlantic Slope from S Ver to NW Honduras, on Pacific Slope from Chis to El Salvador.

SPECTACLED (SCALY-THROATED) FOLIAGE-GLEANER

Anabacerthia v. variegaticeps   Plate 38
Breñero Cejudo

ID: 6.3–6.8 in (16–17.5 cm). Highlands; arboreal. **Descr.** Ages similar. Bill grey above, pale horn below, legs grey. Greyish head with *ochraceous buff eyering and postocular stripe ('spectacles')*, contrasts slightly with brown upperparts; *tail rufous*. Throat and chest creamy with indistinct dusky scaling (often hard to see), rest of underparts olive-brown.

Voice. A sharp, explosive *squeer* or *squeezk!*; song a series of sharp, high *squeek* or *see* notes run into a rapid series

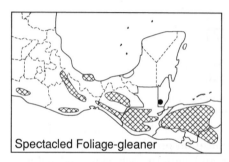

Spectacled Foliage-gleaner

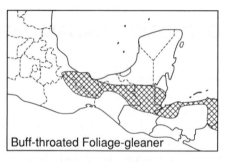

Buff-throated Foliage-gleaner

of several s in duration, suggesting a shrill Downy Woodpecker rattle or, in early stages, sharp hummingbird chips. Song often preceded by a buzzy sneeze, like a softened call note.

Habs. Humid evergreen forest. Forages low to high around epiphytes and mossy tangles, in an acrobatic manner recalling a titmouse. Often in groups of 2–6 birds, regularly joins mixed-species flocks. Nest undescribed (?). Eggs: 2, white (MLZ).

SS: Buff-throated Foliage-Gleaner (lower elevation) larger, usually on or near ground; note bright buff face and throat; Plain Xenops (lower elevation) smaller with black and rufous patterned wings and tail, white moustache.

SD: F to C resident (400–2500 m) on both slopes from Gro and S Ver, and in interior from Chis, to El Salvador and N cen Nicaragua.

RA: S Mexico to Ecuador.

NB1: S.A. populations sometimes considered specifically distinct, *A. temporalis.* **NB2:** *A. striaticollis* of S.A. sometimes considered conspecific with the *A. variegaticeps* group.

BUFF-THROATED FOLIAGE-GLEANER
*Automolus ochrolaemus**   Plate 38
Breñero Gorjipalido

ID: 7.5–8 in (19–20.5 cm). SE humid lowlands. *Bill fairly long and stout.* **Descr.** Ages smiliar. Bill dark grey above, pale horn below, legs greyish. *Face, throat, and chest ochraceous buff with broad dark eyestripe*; upper chest indistinctly scaled dusky, rest of underparts dusky buff, fading to pale buff on undertail coverts. Lores, crown, nape, and upperparts rich brown becoming rufous on uppertail coverts and tail.

Voice. A low, gruff *chuk,* a sharp, nasal

pe-duk, and a hard *tchehrr.* Song a descending chortling rattle of 1–1.5 s duration, often repeated steadily, mostly early and late in the day.

Habs. Humid evergreen forest. Forages in shady understory tangles and dead leaf clusters, also on the forest floor, digging in litter and flicking leaves aside with its bill; less often at mid-, rarely upper, levels, associates loosely with mixed-species flocks. Sings from perches at low to mid-levels. Nests in burrows in banks. Eggs: 2–3, white (WFVZ).

SS: See Spectacled Foliage-Gleaner. Leaf-tossers shorter-tailed with more slender bills, lack striking head patterns. Juv Ruddy Foliage-Gleaner (higher elevation) lacks buff supercilium.

SD: F to C resident (SL–1000 m) on Atlantic Slope from S Ver to Honduras.

RA: SE Mexico to NW Ecuador and Bolivia.

NB: Reports from the Pacific highlands of Guatemala (Griscom 1932) pertain to misidentified juv Ruddy Foliage-Gleaner (AMNH spec.).

RUDDY FOLIAGE-GLEANER
*Automolus rubiginosus**   Plate 38
Breñero Rojizo

ID: 7.8–8.5 in (20–21.5 cm). Highlands. *Bill fairly long and stout.* **Descr.** Ages differ. **Adult:** bill dark grey above, pale horn

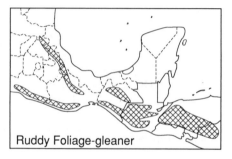

Ruddy Foliage-gleaner

below, legs dusky flesh. Head and upperparts rufous-brown with narrow pale orbital ring creating distinct facial expression; uppertail coverts and tail rufous. *Throat and chest bright tawny-rufous* grading to duller tawny underparts, flanks washed olive. **Juv:** throat and chest ochraceous buff with slight dusky scaling on chest.

Voice. Song a disyllabic, nasal mewing *yeh'nk' yeh'nk'* or *yeh-enk' yeh-enk'*, given 2–3×; suggests Grey Catbird or Azure-hooded Jay. Calls include a hard, dry chatter and a slowly repeated *chak*.

Habs. Humid evergreen and pine–evergreen forest. Skulks low in forest understory, dense second growth along banks and cuttings, at times on the forest floor. Nests in burrows in banks. Eggs: 2, white (WFVZ).

SS: Tawny-throated Leaftosser smaller, shorter-tailed, terrestrial, bill slender, only rump and uppertail coverts ruddy, tail dark.

SD: F to C resident (500–2500 m) on both slopes from Gro and SE SLP, and in interior from Chis, to El Salvador and N cen Nicaragua.

RA: S Mexico to Bolivia.

NB: Populations from E Panama to W Ecuador (*A. r. nigricauda*) sometimes considered specifically distinct.

PLAIN XENOPS

Xenops minutus mexicanus　　　Plate 38
Picolezna Sencillo

ID: 4.5–5 in (11–12.5 cm). Lowlands. Small, arboreal. *Bill short and wedge-shaped.* **Descr.** Ages similar. Bill grey, paler below at base, legs grey. Head brown with narrow pale buff supercilium and *bright white moustache.* Upperparts brown, tertials rufous, primaries and secondaries blackish with broad rufous band. *Tail rufous with 2*

oblique black stripes. Creamy throat blends into dusky underparts.

Voice. A thin *tseep* and a lisping hiss, *psssi.* Song a high, rapid, silvery to liquid trill suggesting Olivaceous Woodcreeper, begins typically with 1–4 *tsiip* or *pip* notes.

Habs. Humid evergreen and semideciduous forest. Feeds conspicuously at mid-levels, in vine tangles, along branches and trunks, etc., notably agile, often hangs upside down like a chickadee. Singly or in pairs, joins mixed-species flocks. Carves nest hole in soft wood at mid-levels. Eggs: 2, white.

SS: See Spectacled Foliage-Gleaner. Wedge-billed Woodcreeper lacks striking face pattern, feeds on trunks like Brown Creeper.

SD: F to C resident (SL to 1000 m) on Atlantic Slope from S Ver to Honduras.

RA: SE Mexico to Paraguay and NE Argentina.

GENUS *SCLERURUS*

Leaftossers, formerly known as leafscrapers, are shortish-tailed terrestrial birds of humid forest interiors. Ages similar. Bills long and slender. Usually found singly, hopping on shady forest floor and tossing leaves aside with their bill. If flushed, they fly off low with a sharp call, and often land on a low trunk or branch and bob warily in plain view for several s. Nest in burrows in banks. Eggs: 2, white (WFVZ).

TAWNY-THROATED LEAFTOSSER

Sclerurus m. mexicanus　　　Plate 38
Hojarasquero Gorjirrufo

ID: 6.3–6.5 in (16–17 cm). *Bill slightly decurved.* **Descr.** Bill dark above, pale horn below, legs dusky flesh. *Dark rich brown*

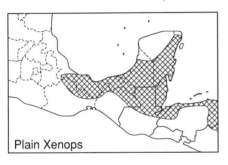

Plain Xenops

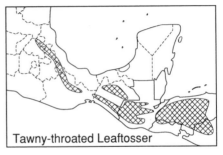

Tawny-throated Leaftosser

overall, throat and chest tawny-rufous, rump and uppertail coverts chestnut.

Voice. A sharp, explosive *sweek!* or *skweek.* Song a descending, slowing series of 3–9 or more rich, plaintive notes, *squee squee . . .,* etc.

Habs. Humid evergreen forest. Favors shady areas such as tangled gullies and overgrown banks. See genus note.

SS: See Ruddy Foliage-Gleaner. Scaly-throated Leaftosser (lowlands) dark brown overall without contrasting ruddy coloration, bill straight.

SD: F to C resident (500–2000 m) on Atlantic Slope from Hgo (Bjelland and Ray 1977), and in interior from Chis, to Honduras and N cen Nicaragua, on Pacific Slope in Chis and Guatemala.

RA: E Mexico to Bolivia.

SCALY-THROATED LEAFTOSSER

Sclerurus g. guatemalensis   Plate 38
Hojarasquero Oscuro

ID: 6.8–7 in (17–18 cm). *Bill straight,* lower mandible tapers toward tip. **Descr.** Bill blackish, pale horn below at base, legs dusky flesh. *Dark brown overall, throat checkered whitish* (often hard to see), coarser pale brown mottling on upper chest; may show narrow pale supercilium.

Voice. A sharp, explosive *squeek!* or *sweeik* much like Tawny-throated Leaftosser, and a low gruff *chuk.* Song a slightly laughing, rippling series of 7–12 or more sharp, slightly liquid *week* or *swee* notes, accelerating and then slowing at the end, at times repeated over and over.

Habs. Humid evergreen forest. Favors fairly open but shady forest floor. See genus note.

SS: See Tawny-throated Leaftosser. Nightingale Wren much smaller with wren-like actions, bobs persistently.

SD: F to U resident (SL–750 m) on Atlantic Slope from SE Ver to Honduras.

RA: SE Mexico to Colombia.

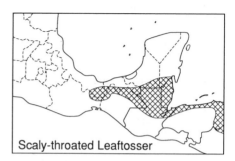

Scaly-throated Leaftosser

WOODCREEPERS: DENDROCOLAPTIDAE (13)

A neotropical family of rufous-brown birds that climb on trees like woodpeckers. Shape suggests creepers (*Certhia*) of temperate regions but the two families are not closely related. Bills vary from short and slender to long and stout, feet strong with well-spread toes to aid in gripping trunks and branches; legs and feet greyish. Tails fairly long with stiffened shafts which act as props, wings fairly short and rounded. Ages similar or slightly different, sexes similar; young altricial. Molts poorly known. Plumage overall colored and patterned in browns, wings and tail brighter rufous; head, back, and underparts often with contrasting buff streaks or spots. One or more species common in most wooded and forested habitats. Woodcreepers often join mixed-species flocks when up to five or more species may occur together. Some species (especially *Dendrocincla* and Barred Woodcreeper) often attend army-ant swarms to prey upon insects and spiders flushed by the ants. Several larger species forage commonly in bromeliads. Voices are varied whistles and trills.

Food mostly invertebrates. Nest in tree cavities, hollow trunks. Eggs: 2–3, white, laid on chips of bark (Stiles and Skutch 1989; WFVZ).

Despite the fact that relatively few species of woodcreepers occur in the region, field identification can be difficult since birds are commonly seen against trunks, often high and in poor light. Note relative size, shape, and color of bills, any patterns visible, voice.

GENUS DENDROCINCLA

These are medium-sized, chunky woodcreepers with erectile bushy crests, stout, medium-length straightish bills, and medium-length broad tails. Plumage lacks prominent pale streaking or spotting. Generally at low to mid-levels, sluggish and overlooked easily, often at army-ant swarms.

TAWNY-WINGED WOODCREEPER

Dendrocincla anabatina *   Plate 37
Trepatroncos Alileonado

ID: 7–7.5 in (18–19 cm). SE humid forest.
 Descr. Eyes pale, bill greyish. *Buff supercilium and pale buff throat with dark moustache create distinctive facial expression,* median throat often dark. Otherwise brown overall, *remiges contrastingly tawny* with dark tips, tail rufous-chestnut; underparts paler grey-brown, chest flecked buff.
 Voice. A plaintive to sharp, whistled *tchee-u* or *tcheu* which may be repeated steadily; song an insistent, intensifying, rapid, churring rattle *chri-chri* ..., up to 70–80 s in duration, at times ends with *cheeu 'cheeu.*
 Habs. Humid evergreen and semideciduous forest. See genus note. Usually more numerous, or conspicuous, than Ruddy Woodcreeper although both may be common at ant swarms.
SS: See Plain-brown Woodcreeper (E Honduras, Appendix E).
SD: F to C resident (SL–1200 m, mostly below 500 m) on Atlantic Slope from S Ver to Honduras.
RA: SE Mexico to W Panama.

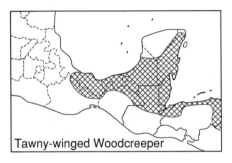
Tawny-winged Woodcreeper

RUDDY WOODCREEPER
Dendrocincla h. homochroa Plate 37
Trepatroncos Rojizo

ID: 7.5–8 in (19–20 cm). SE humid forest. *Head fairly squared with bushy forehead.* **Descr.** Bill greyish. *Broad pale greyish eye-ring creates distinctive facial expression; bright rufous overall,* paler on throat, wings deeper rufous, tail rufous-chestnut. **Voice.** Mostly silent. Calls include a thin, reedy, slightly plaintive *sreeah* or *tcheeu* and a more drawn-out *tleeeoo*. Song a churring, slightly slurred rattle of 2.5–4 s in duration, slowing slightly at the end, notes not distinct as is often the case with Tawny-winged Woodcreeper's more pro-longed rattle. **Habs.** Humid to semihumid evergreen and semideciduous forest, mangroves. See genus note. Rarely seen other than at army-ant swarms, usually low and at times on forest floor. **SD:** F to C resident (SL–1600 m, mainly below 1000 m) on Atlantic Slope from N Oax to Honduras, on Pacific Slope from E Oax to El Salvador (Thurber *et al.* 1987). **RA:** SE Mexico to N S.A. **NB:** Reports from Islas Cozumel and Mujeres (AOU 1983) are probably erroneous.

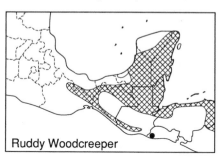
Ruddy Woodcreeper

OLIVACEOUS WOODCREEPER
Sittasomus griseicapillus * Plate 37
Trepatroncos Olivaceo

ID: 6–6.5 in (15–16.5 cm). *Plain; bill small and slender.* **Descr.** Bill greyish. *Head and underparts grey,* back olive-brown becoming rufous on rump and uppertail coverts, tertials and tail rufous. Broad pale tawny stripe across base of remiges often striking in flight. **Voice.** A rapid, liquid trill recalling Plain

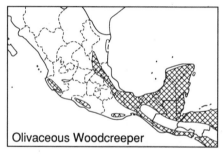
Olivaceous Woodcreeper

Xenops but fuller, lacking introductory chips; also prolonged, quiet, churring trills which may begin hesitantly, last up to 2–3 min, and a short, dry, rattling trill suggesting Long-billed Gnatwren. **Habs.** Humid to semihumid forest and edge, plantations. Feeds low to high, works its way up a tree, often in spirals, then drops to low on a nearby tree and climbs up again. **SS:** Unmistakable if seen well; Wedge-billed Woodcreeper has stouter wedge-shaped bill, buff-spotted chest. **SD:** F to C resident (near SL–2000 m) locally on Pacific Slope from Jal to El Salvador; C to F (SL–1500 m) on Atlantic Slope from S Tamps to Honduras. **RA:** Mexico to Peru and N Argentina.

WEDGE-BILLED WOODCREEPER
Glyphorynchus spirurus pectoralis Plate 37
Trepatroncos Piquicuña

ID: 5.5–6 in (14–15 cm). SE humid forest. *Bill wedge-shaped.* **Descr.** Bill dark above, greyish below. *Face brown, flecked buff,* with indistinct buff supercilium; *crown and nape dark brown.* Back olive-brown becoming rufous on rump and uppertail coverts; tertials and tail rufous, broad tawny-buff stripe across base of remiges often striking in flight. Underparts grey-brown, *chest coarsely spotted buff.*

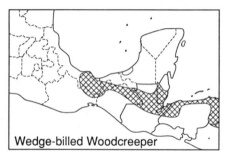
Wedge-billed Woodcreeper

Voice. A sharp chip, *chrrik*, often doubled *chrik-chrik*, and a longer series of 6–8 notes, slightly squirrel-like. Song a rapid, intensifying, slightly rippling series of 10–15 rich, sharp chips ending abruptly, *wiih wihh-wihh-wihh-wihhwihhwihhwihhwihh.*

Habs. Humid evergreen forest. Habits much as Olivaceous Woodcreeper.

SS: See Olivaceous Woodcreeper.

SD: F to C resident (SL–1200 m, mostly below 500 m) on Atlantic Slope from S Ver (Winker *et al.* 1992*b*) and E Oax to Honduras.

RA: SE Mexico to Peru and Brazil.

STRONG-BILLED WOODCREEPER
Xiphocolaptes promeropirhynchus* Plate 37
Trepatroncos Gigante

ID: 11.8–12.5 in (30–31.5 cm). *Very large (flicker-sized), bill long and stout.* **Descr.** *Bill greyish.* Crown, nape, and auriculars dark brown, streaked buff, *lores and supercilium pale buff,* whitish throat bordered by dark malar stripe. Back dull tawny-brown with *sparse fine buff streaks on upper mantle,* becoming rufous on rump and uppertail coverts; remiges rufous, tail dark rufous. *Underparts tawny-brown,* brighter on flanks, *chest finely streaked buff.*

Voice. Song an overall descending, slightly slowing series of 6–10 rich, disyllabic whistles with slightly screechy, metallic quality, the 1st part slurred, the 2nd shorter and fairly abrupt: *chu'ik chu'ik . . .* or *tchooh'ih tchooh'ih . . .*, given mostly at or before dawn. Calls include a muffled, drawn-out snarl slurred into a short, emphatic note, *rrieh-chk!* or *ryehhhr ehk!.*

Habs. Humid to semiarid pine and pine-oak forest, humid evergreen forest. Feeds low to high, often at bromeliads, at times on the ground.

SS: Black-banded Woodcreeper smaller with more slender blackish bill, whitish throat streaked dark, mantle more distinctly streaked buff, belly and flanks barred blackish.

SD: F to U resident (near SL–3500 m, above 1500 m in Mexico) on Atlantic Slope from SE SLP, and in interior from Chis, to Honduras and N cen Nicaragua, on Pacific Slope in Gro and El Salvador.

RA: Mexico to Bolivia.

BARRED WOODCREEPER
Dendrocolaptes certhia* Plate 37
Trepatroncos Barrado

ID: 10–11 in (25.5–28 cm). S humid forests. Barred overall. Bill stout and straightish. Two ssp: *sanctithomae* (SE Mexico to C.A.) and *sheffleri* (SW Mexico). **Descr.** *Sanctithomae:* bill blackish, flesh below at base. Often looks brown with dark lores. Head, back, and underparts tawny-brown, barred blackish, barring less distinct on back; rump and uppertail coverts rufous. Remiges rufous, tail dark rufous. *Sheffleri:* bill mostly flesh, upperparts less distinctly barred, throat and upper chest with heavier, mottled blackish barring.

Voice. Song (both ssp) a varied series of loud, rich, inflected whistles, accelerating and slowing, *whee'pa whee'pa . . .* or *dwoi'ik' dwoi'ik' . . .* or *chew'ee chew'ee . . .*, etc., may recall Bright-rumped Attila and may run into agitated-sounding, rapid, chuckling clucks; also a quiet *wh-whee* while foraging.

Habs. Humid evergreen and semideciduous forest. Most often seen at ant swarms, notably sluggish and overlooked easily, usually at low to mid-levels.

SS: Overall brown appearance, large size, and dark barring distinctive.

SD: *Sanctithomae:* F to U resident (SL–1500 m) on Atlantic Slope from S Ver to

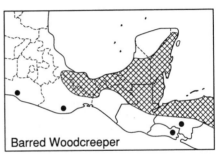

Strong-billed Woodcreeper

Barred Woodcreeper

Honduras, U to R (70–1100 m) on Pacific Slope in El Salvador. *Sheffleri*: U resident (250–1600 m) in Sierra Madre del Sur of Gro (SNGH, SW, tape) and Oax.
RA: S Mexico to Ecuador and Brazil.

BLACK-BANDED WOODCREEPER

Dendrocolaptes picumnus puncticollis
Trepatroncos Vientre-barrado   Plate 37

ID: 10–11 in (25.5–28 cm). *Highlands S of Isthmus. Little known in the region.* Bill fairly stout and straightish. **Descr.** *Bill blackish*, paler at tip. Crown, nape, and auriculars dark brown, streaked buff, supercilium pale buff, *throat whitish, streaked dark brown.* Back dull tawny-brown, brighter on rump, with *fine buff streaks on upper mantle*, uppertail coverts rufous; remiges rufous, tail dark rufous. *Underparts grey-brown, chest finely streaked buff, belly and flanks barred blackish.*
Voice. Undescribed in region (?). In Ecuador, an accelerating, rapid, bubbling series of 30+ whistles fading out fairly abruptly, *whi-whi-whiwhiwhi* ... (Hardy *et al.* 1991, tape).
Habs. Humid evergreen and pine–oak forest. Mostly silent, sluggish and overlooked easily, attends ant swarms in S C.A. and S.A.
SS: See Barred and Strong-billed woodcreepers.
SD: U to F resident (1500–3000 m, locally to 750 m in winter) on Atlantic Slope and in adjacent interior from Chis to Honduras.
RA: S Mexico to N Argentina.

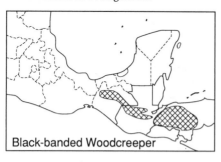

Black-banded Woodcreeper

GENUS XIPHORHYNCHUS

These are fairly large forest woodcreepers with *fairly stout, straightish bills.* Plumage with conspicuous buff streaking and spotting. Feed low to high, often at bromeliads.

BUFF-THROATED WOODCREEPER

Xiphorhynchus guttatus *   Plate 71
Trepatroncos Gorjipálido

ID: 8.5–9 in (21.5–23 cm). *C.A. Bill long and fairly stout.* **Descr.** *Bill dark above, pale horn-flesh below.* Face brown, streaked buff, with buff eyering and supercilium; buff throat often has narrow dusky malar stripe. Crown and nape dark brown, spotted buff. Back tawny-brown, *mantle streaked buff (often inconspicuous),* rump, uppertail coverts, and flight feathers rufous. Underparts grey-brown, *chest coarsely streaked buff.*
Voice. A slightly laughing series of 10–20 rich notes, typically steady or rising and/or falling only slightly, *weet-weet* ... or *whee-whee* ... Calls include a fairly rich, rolled *chirrr* or *pri-i-i-iu*, much like Boat-billed Flycatcher, in descending series 1–5×, often 3×, and a nasal *chey-oo*.
Habs. Humid evergreen forest and edge, plantations. See genus note.
SS: Ivory-billed Woodcreeper has a pale bill, back streaks typically bolder, pale streaking on chest bolder, extending to belly, note voice; juv more similar to Buff-throated but bill less distinctly bicolored. Streak-headed Woodcreeper often at edges and in clearings, smaller with slender pale bill. See Long-tailed Woodcreeper (E Honduras, Appendix E).
SD: F to C resident (SL–500 m) on Atlantic Slope from NE Guatemala to Honduras.
RA: E Guatemala to Brazil.

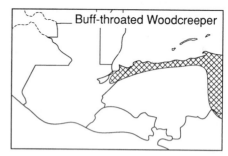

Buff-throated Woodcreeper

IVORY-BILLED WOODCREEPER

Xiphorhynchus flavigaster *   Plate 37
Trepatroncos Piquiclaro

ID: 8.5–10.5 in (21.5–26.5 cm). E > W

Mexico. *Widespread and common. Bill long and stout.* **Descr.** *Bill pale horn to pale flesh*, may be dusky on culmen (especially in W Mexico). Face brown, streaked buff, with buff eyering and supercilium; *buff throat usually has narrow dusky malar stripe.* Crown and nape dark brown with buff droplets and streaks running into *bold lines of buff streaks on tawny-brown to grey-brown back*; rump brighter, upper-tail coverts and flight feathers rufous. *Underparts grey-brown, coarsely streaked buff, boldest on chest, fading out on belly and flanks.* Birds in W Mexico (N of Isthmus) and N Yuc Pen paler and greyer overall. **Juv:** bill may be dusky above, back streaks less distinct.

Voice. Varied, loud, laughing series, rising and falling, less often steady, *whee-whee ...* or *wheetweetwee ...*, may suggest Canyon Wren, and often ending with an upward-inflected *whee whee-wheep'* which may be given as a separate call; lower churring and chortling laughs in W Mexico. Calls include a rich, sharp, slightly explosive *tchee-oo'* or *cheeoo'*; often a less sharp *chreeu* and *chehr* in W Mexico.
Habs. Forest and edge, plantations, mangroves. See genus note.

SS: Streak-headed Woodcreeper smaller (often hard to judge), slender bill slightly decurved, crown with fine buff streaks, back less boldly streaked buff, lacks dark malar stripe, prefers more open areas (Ivory-billed can easily be misidentified as Streak-headed, but the other way around is unlikely). Spot-crowned Woodcreeper (rarely sympatric) has slender bill, crown finely spotted, back streaks inconspicuous. Spotted Woodcreeper (rarely sympatric) has bill dark above, crown, nape, back, and underparts spotted buff. See Buff-throated Woodcreeper (C.A.).

SD: C to F resident (SL−1500 m, to 2400 m in W Mexico) on both slopes from S Son and S Tamps, and in interior from Isthmus, to El Salvador and Honduras.
RA: Mexico to NW Costa Rica.

SPOTTED WOODCREEPER
Xiphorhynchus erythropygius*   Plate 37
Trepatroncos Manchado

ID: 9−9.5 in (23−24 cm). *Highland forests. Bill straightish and fairly stout.* **Descr.** *Bill dark above, pale horn to grey below.* Face brown, streaked buff, with distinct buff eyering and postocular stripe creating *spectacled look*; lores brown, pale buff throat flecked dark brown. Crown and nape dark brown, spotted buff, *back tawny-brown with bold buff spots* (indistinct or lacking in *parvus*, E of Sula Valley), rump and uppertail coverts rufous; flight feathers dark rufous. *Underparts grey-brown, boldly scalloped and spotted pale buff.*

Voice. A plaintive, descending and slowing series of 2−4 (usually 4) rich, slurred whistles, *wheeeoo, wheeeoo, wheeeoo, wheeeoo,* or *kee-ow ...,* etc. Less often a single *wheeoo* which may suggest a muted Rufous Piha.
Habs. Humid evergreen forest. See genus note.

SS: Spot-crowned Woodcreeper has slender pale bill, back lacks distinct markings, underparts coarsely streaked pale buff, often shows dark malar stripe; note voice. See Ivory-billed Woodcreeper.

SD: F to C resident (600−2200 m) on Atlantic Slope from SE SLP, and in interior from Isthmus, to Honduras and N cen Nicaragua, locally in S Belize (Miller and Miller 1992b), and on Pacific Slope in Gro and from E Oax to Guatemala.
RA: Mexico to Ecuador.

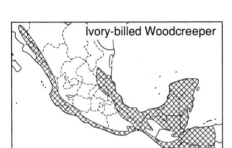
Ivory-billed Woodcreeper

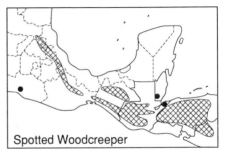
Spotted Woodcreeper

GENUS LEPIDOCOLAPTES

These are medium-sized woodcreepers of open forests. *Bills slender and slightly decurved.* Plumage with conspicuous buff to whitish streaking and spotting. Feed low to high, fairly active; often spiral up one tree, then drop to the base of a nearby tree and start again.

WHITE-STRIPED WOODCREEPER
Lepidocolaptes leucogaster *   Plate 37
Trepatroncos Blanquirrayado

ID: 8.5–9 in (21.5–23 cm). Endemic to highlands N of Isthmus. Bill long, slender, and decurved. **Descr.** Bill dark above, pale horn-flesh below. *Face and throat whitish with dark eyestripe* and rarely a trace of a dark malar stripe; *whitish chest boldly scalloped blackish, rest of underparts dusky with coarse white striping.* Crown and nape dark brown, boldly spotted whitish, back bright tawny-brown, upper mantle with inconspicuous fine whitish streaks; rump, uppertail coverts, and flight feathers rufous.
   Voice. A rapid trill, at first relatively dry then slower, descending and liquid, 1st part may recall Chipping Sparrow song. Call a fairly rough, short trill *ssirrr.*
   Habs. Humid to semiarid pine and pine-oak woodland. See genus note. Eggs undescribed (?).
SS: White face and throat and boldly scalloped chest unmistakable; note also long slender bill.
SD/RA: F to C resident (900–3500 m) on Pacific Slope and in interior from S Son to Mor and Oax; R to U on Atlantic Slope and in adjacent interior from W cen Ver to Oax. A report from SLP (AOU 1983) requires verification.

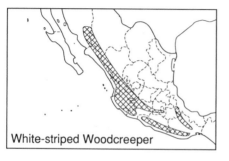
White-striped Woodcreeper

STREAK-HEADED WOODCREEPER
Lepidocolaptes souleyetii *   Plate 37
Trepatroncos Corona-rayada

ID: 7.5–8 in (19–20.5 cm). *S and E lowlands. Bill slender and slightly decurved.* **Descr.** *Bill pale horn to horn-flesh,* culmen rarely slightly dusky. Face brown, flecked pale buff, with pale buff eyering and supercilium; throat pale buff. *Crown dark brown, finely streaked buff, back cinnamon-brown, streaked buff,* rump, uppertail coverts, and flight feathers rufous. *Underparts grey-brown, coarsely streaked buff, streaks bold on chest, only slightly less distinct on belly and flanks.*
   Voice. A rapid, rolled trill or rattle suggesting Olivaceous Woodcreeper but fuller and woodier, less liquid; also short, rolled trills.
   Habs. Humid to semiarid forest and edge, semiopen areas with scattered trees and hedges, plantations. See genus note.
SS: See Ivory-billed and Buff-throated woodcreepers. Spot-crowned Woodcreeper (rarely sympatric) has finely spotted crown, fine buff streaks on back hard to see, often has dark malar stripe.
SD: F to C resident (SL–1500 m) on both slopes from Gro and S Ver to El Salvador and Honduras.
RA: S Mexico to Peru and Brazil.

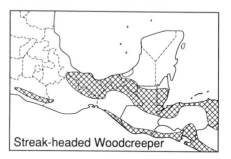
Streak-headed Woodcreeper

SPOT-CROWNED WOODCREEPER
Lepidocolaptes affinis *   Plate 37
Trepatroncos Corona-punteada

ID: 8–8.5 in (20.5–21.5 cm). *Highland forests. Bill slender and slightly decurved.* **Descr.** *Bill pale greyish flesh, dusky at base.* Face brown, streaked buff, with buff eyering and supercilium, *buff throat often has narrow dark malar stripe.* Crown and nape dark brown, *neatly spotted buff,* back tawny-brown, upper mantle with fine buff streaks (often hard to see), rump, uppertail coverts, and flight feathers rufous. *Underparts grey-brown, coarsely and boldly streaked buff.*

Voice. A plaintive, thin, squeaky *see-yih'* or *syeih*, a longer, plaintive, reedy *tw see'i tchew*, suggesting Rose-throated Becard, and longer series of similar quality, *jeeer dee dee deet*, etc. Song a reedy note followed by a rapid laugh, *syeehr see-see-see-see-see-see-see-syn*, or *rreeer hee-hee-hee-hee-hee-hee-hee*, etc.

Habs. Humid to semiarid pine and pine–oak woodland, evergreen forest. See genus note.

SS: See Streak-headed, Ivory-billed, and Spotted woodcreepers.

SD: F to C resident (1000–3500 m) on both slopes from Gro and S Tamps, and in interior from Chis, to El Salvador and N cen Nicaragua.

RA: Mexico to N Bolivia.

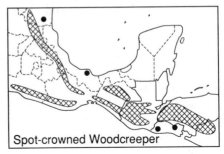

Spot-crowned Woodcreeper

ANTBIRDS: FORMICARIIDAE (II)

A large neotropical family whose members have diversified to fill many niches. Mainly forest, forest edge, and second growth inhabitants. Ages/sexes usually differ; young altricial. Juv plumage typically soft, loose, and lost within a few days, often much of it while still in the nest. Imms resemble adults in most species. Plumage typically colored in greys, browns, black, and white; some species have concealed white or light-colored back patches displayed in courtship or aggression. Many are skulking and best located by voice; songs of most species consist of rich piping notes and hollow whistles.

Antbirds eat mostly insects and other arthropods. The family name derives from many species' habit (mainly in C.A. and S.A.) of attending army-ant swarms to prey upon fleeing insects. Nests typically are open cups, often suspended, vireo-like, by the rims; one species nests in cavities.

The present family may actually be two families: the Thamnophilidae (typical antbirds) comprising antshrikes, antvireos, antwrens, antbirds, etc.; and the Formicariidae (ground antbirds) comprising antthrushes, antpittas, etc.

GREAT ANTSHRIKE
Taraba major melanocrissa   Plate 38
Batará Mayor

ID: 7.5–8.2 in (19–21 cm). Large, bicolored. *Bill very stout and hooked.* Erectile crest often raised. **Descr.** Ages/sexes differ. ♂: *eyes red,* bill blackish, legs blue-grey. *Head and upperparts black* with white wingbars; concealed white back patch rarely visible. *Underparts white* with dusky grey flanks and undertail coverts. ♀: *head and upperparts rufous* without wingbars, *underparts white* with buffy brown flanks and undertail coverts. **Imm:** resembles adult of respective sex but with dusky scalloping at sides of chest, ♂ may retain cinnamon lower wingbar for a month or so. **Juv ♂:** resembles adult but eyes brown, upperparts sooty grey, underparts softly scalloped dusky, heaviest on chest, wing coverts tipped cinnamon. **Juv ♀:** resembles adult but head and upperparts barred blackish, underparts scalloped dusky as juv ♂.

Voice. Song suggests Black-headed Trogon, a bouncing-ball series of 12–30 hollow *cow* or *hoh* notes, accelerating toward the end and fading quickly; often ends with a quiet snarl, *hehh,* hard to hear at a distance. Calls include a gruff rattle with a rough introductory note, *ahrr, kuhkuhkuh ...,* and varied, hard *chak* notes and growls.

Habs. Humid second growth and forest edge, rarely in forest. Often in pairs, typically low and skulking. Nest is a bulky cup of vines, leaves, etc., slung low in thickets. Eggs: 2, whitish, with brown and mauve scrawls (WFVZ).

SS: Unlikely to be confused with other species; juv (rarely seen) may recall Barred Antshrike but much less contrasting, eyes dark.

SD: U to F resident (SL–750 m) on Atlantic Slope from S Ver to Honduras.

RA: SE Mexico to Uruguay.

BARRED ANTSHRIKE
*Thamnophilus doliatus**   Plate 38
Batará Barrada

ID: 6.5–7 in (16.5–18 cm). Commonest and most familiar Mexican antbird. *Bill stout and hooked, erectile crest typically striking.*

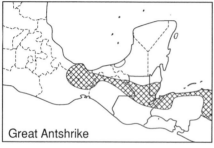

Great Antshrike

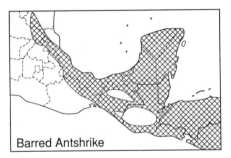

Barred Antshrike

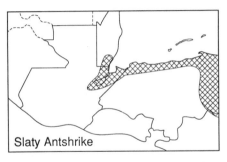

Slaty Antshrike

Descr. Ages/sexes differ. ♂: *eyes whitish,* bill blackish, paler below, legs blue-grey. *Head striped black and white,* crest with black tips and white base; *body, wings, and tail strikingly barred black and white.* ♀: *auriculars and nape striped creamy and blackish,* crown and upperparts rufous. *Underparts ochraceous buff,* throat paler. **Imm** ♂: flanks and belly washed buff. **Imm** ♀: dusky bars and spots on chest. **Juv:** imm head feathers probably appear while still in nest. ♂ resembles adult but patterned blackish brown and buff, tail like adult. ♀ resembles adult but head and upperparts barred blackish, underparts scalloped dusky.
Voice. A crow-like cawing *arrrr,* a slow hooting *hoo-hoo* ... reminiscent of a pygmy-owl, and a series of 6–8 nasal *kyow* notes, rising then falling. Song an accelerating series of about 25–30 *cah* or *cow* notes ending with a gruff *ahr.*
Habs. Humid to semiarid second growth, forest edge, thickets, rarely in forest interior. Fairly skulking but seen often, usually in pairs, at low to mid-levels. Nest is a deep cup of vines, rootlets, etc., at low to mid-levels in thickets. Eggs: 2–3, whitish, with purple and grey scrawls (WFVZ).
SS: ♂ unmistakable, ♀ may recall certain wrens but note crest, stout bill, and unbarred wings and tail. See Fasciated Antshrike (E Honduras; Appendix F).
SD: C to F resident (SL–1500 m) on Atlantic Slope from S Tamps, and on Pacific Slope and in interior from E Oax, to El Salvador and Honduras.
RA: E Mexico to N Argentina.

SLATY ANTSHRIKE
Thamnophilus punctatus atrinucha Plate 71
Batará Apizarrada

ID: 5.5–6 in (14–15 cm). C.A. Bill stout and

hooked. Both sexes have concealed white back patch. **Descr.** Ages/sexes differ. ♂: eyes brown, bill and legs grey. *Slaty grey overall with black crown,* back darker and mottled black; *wings and tail black with white wingbars,* outer rectrices tipped white. ♀: *head and upperparts olive-brown with tawny-brown crown; wings blackish with pale buff wingbars,* tail rufous, outer rectrices tipped white. *Throat and underparts buffy brown,* greyer on throat. **Imm** ♂: brownish remiges contrast with black wing coverts. **Juv:** head and upperparts greyish brown, edged paler, with indistinct buff wingbars. Throat and underparts buffy brown, greyer on throat.
Voice. Song a rapid series of 25–30 *cah* or *aah* notes accelerating toward the end and run into an emphatic *whek!,* suggests Barred Antshrike but steadier, accelerates less strongly, and ends emphatically rather than with a snarl (RAB tape, Panama). Calls include a growling *ah'rrrrrrrr.*
Habs. Humid to semihumid evergreen and semideciduous forest and edge, plantations, second growth. At low to mid-levels, often in fairly open understory, joins mixed-species flocks, often in pairs. Nest cup of dark fibers slung in shallow fork at low to mid-levels. Eggs: 2, whitish, marked with browns and greys (WFVZ).
SS: ♀ Plain Antvireo smaller, wingbars less distinct, tail lacks white tips, throat whitish, belly lemon.
SD: U to F resident (SL–500 m) on Atlantic Slope from S Belize and NE Guatemala to Honduras. A report from Petén (Beavers *et al.* 1991) is questionable.
RA: E Guatemala to Ecuador and Brazil.

RUSSET ANTSHRIKE
Thamnistes a. anabatinus Plate 38
Batará Café

ID: 5.7–6.2 in (14.5–16 cm). Arboreal,

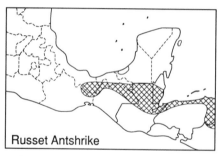

Russet Antshrike

plain. *Bill stout and hooked.* **Descr.** Ages/sexes similar. Bill dark grey, paler below, legs grey. *Dusky eyestripe and broad pale buff supercilium give face a distinctive expression suggesting Warbling Vireo*; auriculars dusky buff, throat and underparts buff, washed dusky on flanks. Upperparts brown, wings and tail rufous; ♂ has concealed tawny patch on mantle, rarely visible.

Voice. Quiet and missed easily. Song (?) a high, thin, slightly sibilant *tssip-i-tssip* or *tseep-it-seet*, dropping on middle syllable, carries fairly well; also an up-slurred *sweek*.

Habs. Humid evergreen forest. Arboreal, forages at mid- to upper levels in a sluggish, vireo-like manner; often in pairs, joins mixed-species flocks. Nest is a deep cup of rootlets, fine leaves, etc., slung at mid- to upper levels in tree tangles and mossy branches. Eggs: 2, whitish, speckled with browns.

SD: U to F resident (SL–400 m) on Atlantic Slope from E Oax to Honduras.

RA: SE Mexico to Bolivia.

PLAIN ANTVIREO
Dysithamnus mentalis septentrionalis
Batarito Sencillo   Plate 38

ID: 4.3–4.7 in (11–12.5 cm). Small, vireo-like. *Bill heavy and slightly hooked, tail*

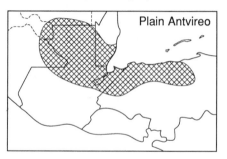

Plain Antvireo

fairly short. **Descr.** Ages/sexes differ. ♂: bill black above, grey below, legs grey. *Head greyish with paler supercilium*, upperparts greyish olive with 2 narrow white wingbars. Throat whitish, chest and flanks greyish, belly and undertail coverts pale lemon. ♀: *crown rufous, face grey with whitish eyering*, upperparts olive-brown with pale buff wingbars. Throat whitish, chest and flanks brownish, belly and undertail coverts pale lemon. **Juv** (nestling): dark brown overall, quickly attains adult-like plumage.

Voice. A low nasal *nyeu-nyeut*, a quiet, nasal *rroynk'* and *dyerr-k'*, a nasal barking *keh keh ... or kyeh kyeh ...*, 10/8–9 s (suggests quiet Barred Forest-Falcon), and a quiet, fairly rapid, and usually accelerating and slowing series of notes, often coupled, *bee-bee bee bee-bee ...*, etc. Song an accelerating, laughing series of hollow, slightly nasal *how* notes, usually 25–28 notes, slowing at the end into a few harder notes, *how-how-how-how ... heh-heh-heh*; reminiscent of Barred Antshrike but more laughing, lacks snarl at end.

Habs. Humid evergreen forest. At low to mid-levels, often in pairs, joins mixed-species flocks. Actions sluggish and vireo-like. Nest is a deep cup of rootlets, etc., slung low in forest understory. Eggs: 2, whitish, marked with purplish brown.

SS: ♀ may recall Tawny-crowned Greenlet, a slighter bird with longer tail, smaller bill, and no wingbars or eyering. See ♀ Slaty Antshrike (C.A.). Streak-crowned Antvireo (E Honduras, see Appendix E) has boldly streaked crown and chest.

SD: F to C resident (near SL–1200 m) on Atlantic Slope from NE Chis and S Camp to N Honduras.

RA: SE Mexico to N Argentina.

SLATY ANTWREN
Myrmotherula s. schisticolor   Plate 38
Hormiguerito Apizarrado

ID: 4–4.3 in (10–11 cm). Small, unobtrusive. *Tail fairly short.* **Descr.** Ages/sexes differ. ♂: bill blackish, legs grey. *Slaty grey overall with contrasting black throat and chest*, wing coverts blackish with white wingbars, outer rectrices narrowly tipped white. White shoulder tufts usually concealed. ♀: bill pale flesh below. *Head and upperparts olive-brown with pale eyering. Throat pale buff, underparts tawny-*

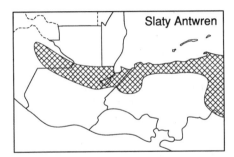

Slaty Antwren

brown. **Imm** ♂: resembles ♀ but duller,
head and upperparts greyer without pale
eyering, face and throat often with black
patches. **Juv** (nestling): dusky brown overall, quickly attains imm plumage.

Voice. A nasal, up-slurred *reeah* or *nyieh*;
song a fairly quiet, whistled *weep*, repeated 4–8×.

Habs. Humid evergreen forest. Active but
inconspicuous at low to mid-levels, often
in pairs, joins mixed-species flocks. Nest
cup of dark rootlets, etc., slung low in
bush or small tree in understory. Eggs: 2,
whitish, marked with purplish browns
(WFVZ).

SS: ♂ Dot-winged Antwren (lowlands) blacker
overall, wingbars bolder, tail longer and
white-tipped. Dusky Antbird (lowlands)
larger, skulks low in dense second growth,
♂ uniform grey, ♀ tawny below.

SD: U to F resident (near SL–1500 m, above
500 m in Mexico) on Atlantic Slope from E
Chis to Honduras.

RA: SE Mexico to S Peru.

DOT-WINGED ANTWREN
Microrhopias quixensis*   Plate 38
Hormiguerito Alipunteado

ID: 4.2–4.5 in (11–12 cm). Small, attractive,
and rather wren-like. *Tail fairly long and
strongly graduated.* Both sexes have concealed white back patch. **Descr.** Ages/sexes

differ. ♂: bill and legs blackish. *Velvety
black overall with sooty grey flanks, solid
white bar on greater coverts, and neat white
spotting on median and lesser coverts, all
but central rectrices broadly tipped white.*
♀: upperparts slaty grey, *underparts
rufous-chestnut, wings and tail as* ♂. **Juv**
(nestling): sooty blackish above, sooty
chestnut-brown below. Quickly attains
adult-like plumage.

Voice. Varied chipping and piping notes
include a loud, clear *tchip teeoo* and
tchip-teeu, teeu-teeu, a sharp, liquid *tew,*
and a rich, sharp *tchoo-choot;* song
begins with 3–6 high, thin, piping *pee* or
pi notes and runs into an accelerating
series of similar notes, suggests Dusky
Antbird song but thinner, less rich.

Habs. Humid evergreen forest. Forages
actively at low to mid-levels, in pairs or
small groups up to 6–10 birds, joins
mixed-species flocks. Nest is a deep cup of
dead leaves, fibers, etc., slung at low to
mid-levels in small trees. Eggs: 2, white,
spotted with brown.

SS: See Slaty Antwren.

SD: F to C resident (SL–750 m) from N Oax
to Honduras. Reports from Los Tuxtlas
(Estrada and Estrada 1985) require verification.

RA: SE Mexico to N Bolivia.

DUSKY ANTBIRD
Cercomacra tyrannina crepera   Plate 38
Hormiguero Negruzco

ID: 5–5.5 in (12.5–14 cm) Skulking, plain,
and wren-like. **Descr.** Ages/sexes differ. ♂:
bill blackish, legs dark grey. *Dusky grey
overall with inconspicuous whitish wingbars;* concealed white back patch rarely
visible. ♀: bill horn below. Head and
upperparts olive-brown with *tawny lores
and narrow eyering.* Throat and underparts tawny. **Imm** ♂: resembles ♀ but (Jan–

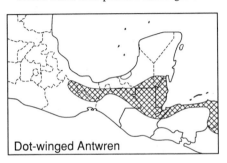

Dot-winged Antwren

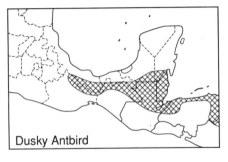

Dusky Antbird

Jun) with patches of adult plumage. **Juv** (nestling): darker overall than ♀.

Voice. A series of 6–8 rich, piping whistles, 1st 2 separate followed by a rapid series, *peu peu peupeupeupeupeupeu* or *piu piu piupiu* ...; calls include a hard, churring *keh-rrr* and a harsh, rasping *brreeea*.

Habs. Humid second growth and forest edge, rarely in forest understory. Often heard but hard to see. Skulks low in dense second growth, often in pairs, rarely joins mixed-species flocks. Nest is a deep cup of dead leaves, fibers, etc., slung mostly low in second growth. Eggs: 2, whitish, spotted and scrawled with purplish brown (WFVZ).

SS: ♀ may suggest a wren but lacks barring, blank-faced expression distinctive; see also more arboreal Slaty Antwren, a foothill bird. ♀ Bare-crowned Antbird (C.A.) has bold blue orbital ring, paler wingbars.

SD: F to C resident (SL–750 m, rarely to 1100 m) on Atlantic Slope from N Oax to Honduras. Reports from Los Tuxtlas (Winker *et al.* 1992*b*) require verification.

RA: SE Mexico to Brazil.

BARE-CROWNED ANTBIRD
Gymnocichla nudiceps chiroleuca   Plate 71
Hormiguero Calvo

ID: 5.8–6.2 in (15–16 cm). C.A. Both sexes have concealed white back patch. **Descr.** Ages/sexes differ. ♂: *naked forehead and crown bright blue*, bill and legs grey-blue. *Black overall with narrow white wingbars.* ♀: *broad blue orbital ring.* Head and upperparts grey-brown, flight feathers darker, wing coverts edged cinnamon-rufous; throat and underparts cinnamon-rufous. **Imm** ♂: crown initially feathered, greater coverts tipped cinnamon-brown. **Juv** (nestling): dark sooty brown overall.

Voice. A loud, shrill, slightly squeaky *k-leey* or *keeyp!*, suggesting a leaftosser.

Song a steady series of 8–12 loud, rich, fairly sharp notes, at times accelerating slightly at the end, *tchee-tchee* ... or *cheeu-cheeu* ..., etc. (RAB, tape, Panama), may suggest Red-crowned Ant-Tanager but faster, steadier, often longer.

Habs. Humid evergreen forest and edge, second growth thickets. Skulks and rarely seen, keeps low in understory. Regularly at army-ant swarms, often in pairs. Nest and eggs undescribed (?).

SS: See ♀ Dusky Antbird.

SD: U (to F?) resident (SL–250 m) on Atlantic Slope from S Belize and NE Guatemala to Honduras. A second-hand report from Petén (Taibel 1955) is probably erroneous.

RA: E Guatemala to N Colombia.

MEXICAN [BLACK-FACED] ANTTHRUSH
Formicarius moniliger*   Plate 38
Hormiguero-gallito Mexicano

ID: 7–7.5 in (18–19.5 cm). *Endemic*, humid *S forest. Terrestrial, rather rail-like* with fairly long legs. **Descr.** Ages differ, sexes similar. **Adult:** bill black, legs dusky flesh. *Face and throat black, orbital ring pale blue*, lore spot whitish, *sides of neck and narrow foreneck collar rufous-chestnut.* Upperparts rich dark brown, crown blacker, uppertail coverts chestnut. Tail blackish. Underbody sooty grey, washed olive on flanks, undertail coverts pale cinnamon. **Juv:** throat sooty brown, chest washed olive-brown.

Voice. Song distinctive and far-carrying, a series of 9–13 rich, sharp whistles, a single note followed by a rapid series *piu, piupiu* ..., series often accelerating and at times slurring up at the end. Call a clipped *p-tuk*, may suggest Summer Tanager and may be run into rapid, clucking series, *pituk ptuk-ptukptukptuk* ...

Habs. Humid evergreen and semi-

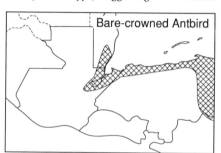
Bare-crowned Antbird

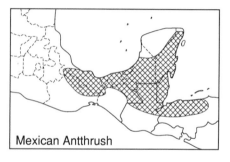
Mexican Antthrush

deciduous forest. Terrestrial. Walks with distinctive rail-like gait on shady forest floor, tail cocked. If flushed, usually calls and flies off low with whirring wings. Nest mat of vegetation in tree cavities within 3 m of ground. Eggs undescribed (?) but white in closely related *F. analis*.

SD/RA: F to C resident (SL–1800 m) on Atlantic Slope from SE Pue and cen Ver to N Honduras.

RA: SE Mexico to Bolivia.

NB: Traditionally considered conspecific with birds of S C.A. and S.A. which differ in morphology, voice, and habitat (Howell 1994). See Appendix E.

SCALED ANTPITTA

Grallaria guatimalensis* Plate 38
Hormiguero-cholino Escamoso

ID: 7–7.5 in (18–19.5 cm). *Terrestrial; rotund and almost tail-less. Bill stout and fairly long, legs long and stick-like.* Descr. Ages differ, sexes similar. Adult: bill dusky horn, legs dusky flesh. *Crown and nape grey, scaled black, lores pale buff, auricu-*lars olive-brown, *bushy moustachial stripe creamy. Upperparts greyish olive, scaled black*, wings brown. Throat dusky, streaked buff, *underparts pale greyish cinnamon* to dusky cinnamon (N of Isthmus) *to deep cinnamon (guatimalensis*, S from Isthmus) with creamy foreneck collar and black scalloping on chest. Juv: bill blackish. Head, mantle, and upper chest sooty blackish, streaked cream; lower chest and belly creamy buff, scalloped black on chest and flanks.

Voice. Song heard infrequently (mostly at dawn and late in day), a low resonant series of notes that starts as a trill, rises in pitch, and slows to distinct, pulsating notes, stops abruptly; also a low, pig-like grunt or croak in alarm (Stiles and Skutch 1989). Song of birds in highlands N of Isthmus undescribed (?).

Habs. Humid evergreen, pine–oak, and fir forest. Terrestrial and solitary, keeps to shady and usually dense forest understory where retiring and hard to see except by luck. Moves by kangaroo-style hopping, often surprisingly quickly. Flushes with a whirr of wings but without calling. Nest is a bulky cup or platform low in dense growth on a stump, fallen log, tree fork. Eggs: 2, unmarked bluish (WFVZ).

SD: U to F resident (50–3500 m, above 1000 m in N), on both slopes from Jal and Hgo (Howell and Webb 1992e), and in interior from cen volcanic belt, to El Salvador and Honduras; in lower foothills only on Atlantic Slope (mainly in vicinity of Isthmus).

RA: Cen Mexico to S Peru.

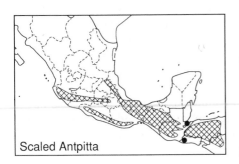

Scaled Antpitta

TYRANT-FLYCATCHERS: TYRANNIDAE (67)

A large New World family that attains its greatest diversity in the Neotropics. Widespread and common in all habitats throughout the region, tyrant-flycatchers generally possess somewhat flattened and often slightly hooked bills, and rictal bristles. Most perch fairly upright and make aerial sallies for prey. Beyond this, they exhibit great diversity in external appearance and habits. Ages/sexes usually similar or only slightly different; young altricial; typically imms molt quickly into adult-like plumage. For field purposes three groups may be recognized, corresponding roughly to presently recognized subfamilies.

Tropical flycatchers (Elaeniinae and some Fluvicolinae) comprise 18 species. The most diverse grouping, and that least familiar to the majority of observers. Very small to medium-sized flycatchers of humid forest, edge, second growth, and scrub. Colored mainly in browns and greens, facial expressions often subtly distinctive. Although usually common, most are overlooked easily and best detected by voice. Mostly sedentary. Food mainly insects, some also eat fruit. Most build cup nests but several make elaborate hanging nests which are conspicuous features of tropical forests (Fig. 32).

Empidonax and allies (most Fluvicolinae) comprise 21 species. Small to medium-sized flycatchers of forest, woodland, scrub, and edge. Relatively uniform in appearance, most have peaked to crested napes, olive to greyish upperparts with pale wingbars, and pale underparts which are often washed dusky on the chest. Most breed in temperate habitats and are at least partly migratory; seven occur entirely as migrants in the region. Food mostly insects. Nests are cup structures, some saddled on branches.

Kingbirds and *Myiarchus* and allies (Tyranninae, some Fluvicolinae) comprise 28 species. Medium-sized to large flycatchers of open and wooded habitats; most are conspicuous and noisy. Often strikingly patterned and brightly colored. Several species are migratory. Food mostly insects and fruit. Most build cup nests but *Myiarchus* and allies are obligate cavity nesters.

AOU (1983) united tyrant-flycatchers with cotingas and manakins in the superfamily Tyrannoidea. Relationships within this group are still far from clear, as they have been traditionally clouded by great diversity in external appearance.

PALTRY TYRANNULET

Zimmerius v. vilissimus   Plate 39
Mosquerito Cejiblanco

ID: 4.3–4.7 in (11–12 cm). *S of Isthmus, arboreal.* Bill small and stubby. **Descr.** Ages differ slightly, sexes similar. **Adult:** eyes pale, bill and legs blackish. *White supercilium, fading out behind eye, contrasts with dark grey crown* and brownish-olive auriculars. Upperparts olive, wings and tail dark grey with *bright sulphur edgings to greater coverts, tertials, and secondaries. Throat and underparts pale grey,* washed dusky on chest, lemon on belly and undertail coverts. **Juv:** eyes darker, olive-washed head contrasts less with back.

Voice. A clear, slightly ringing to plaintive *peeu* or *pyeu,* at times repeated steadily in series, and *pee-peeu'* or *pyi-pyeu',* at times *pyi pyi pyeu,* etc.; calls lack the strident or

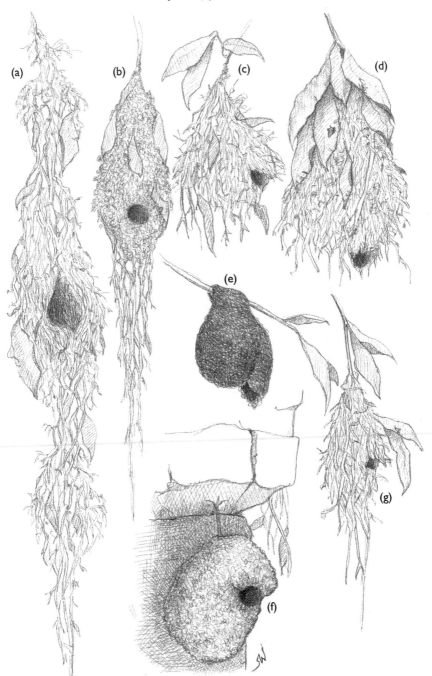

Fig. 32 Pendulous tropical flycatcher nests. (a) Royal Flycatcher; (b) Ochre-bellied Flycatcher;
(c) Sulphur-rumped Flycatcher (Ruddy-tailed Flycatcher similar); (d) Eye-ringed Flatbill;
(e) Yellow-olive Flycatcher; (f) Sepia-capped Flycatcher; (g) Common Tody-Flycatcher

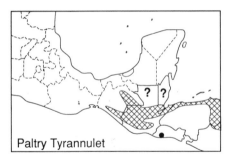

Paltry Tyrannulet

emphatic quality of Yellow-bellied Tyrannulet. Dawn song undescribed (?).

Habs. Humid evergreen forest and edge, plantations, semiopen pastures with large trees, shrubby bushes, etc. Arboreal, from mid- to upper levels in large trees to fairly low in second growth; can be hard to find when calling, overlooked easily. Forages fairly actively, tail held cocked. Nest is a globular mass of rootlets, mosses, etc., with side entrance, set in a mossy tangle, epiphyte, etc., at mid- to upper levels. Eggs: 2, whitish, speckled rusty (WFVZ).

SS: No other species in region similar.

SD: F to C resident (500–2500 m, lower in winter) on both slopes and in interior of Chis and Guatemala; also (near 600 m) in Sierra de Balsamo, El Salvador, and U to F resident (near SL–300 m) on Atlantic Slope from NE Guatemala to Honduras. Reports from Petén (Smithe 1966; Beavers *et al.* 1991) and Belize (Wood *et al.* 1986) require verification.

RA: S Mexico to N Venezuela.

NB: Formerly placed in the genus *Tyranniscus*.

YELLOW-BELLIED TYRANNULET
Ornithion semiflavum　　　　Plate 39
Mosquerito Vientre-amarillo

ID: 3.3–3.5 in (8.5–9 cm). *Tiny and stubtailed*, bill small; *arboreal*. **Descr.** Ages dif-

fer slightly, sexes similar. **Adult:** bill and legs blackish. *White supercilium contrasts with slate-grey crown* and brownish-olive auriculars. Upperparts olive, wings and tail darker with narrow olive edgings. *Throat and underparts yellow.* **Juv:** crown and upperparts tinged brownish, underparts paler.

Voice. A strong, clear, slightly plaintive *dee* or *pee* and *peer*, more strident than Paltry Tyrannulet; also given in series, usually 5–8× with the main part of the series descending, *wheeeu pee-pee-pee ...* or *bee! bee! bee, bee-bee-bee-bee*, etc., may recall Northern Beardless Tyrannulet or Cinnamon Becard. Also a strong, fairly hard, nasal *biihk-biihk ...*, which may be repeated persistently with *pee* notes thrown in. Dawn song *dee dee whi'chu*, repeated incessantly, or an excited, rapid, slightly nasal chortling *t-chee-ee-ee eee-eee-eet* or *t-chi-ee pee-pee-pee t-chi-ee t-chi-ee*, etc.

Habs. Humid evergreen forest and edge. Arboreal, at mid- to upper levels, less often low in second growth. Fairly active but overlooked easily. Nest (globular?) at mid-levels in a curled-up dead *Cecropia* leaf still hanging in the tree (Ver, SNGH) or in clump of small bromeliads (Belize, SNGH). Eggs undescribed (?).

SD: F to U resident (near SL–1500 m) on Atlantic Slope from S Ver to Honduras.

RA: SE Mexico to Costa Rica.

NORTHERN BEARDLESS TYRANNULET
Camptostoma imberbe*　　　　Plate 39
Mosquerito Lampino Norteño

ID: 3.8–4.2 in (9.5–10.5 cm). Widespread. *Erectile crest usually raised.* Bill small. **Descr.** Ages differ slightly, sexes similar. **Adult:** *bill pinkish orange with dark culmen*, legs blackish. Head and upperparts

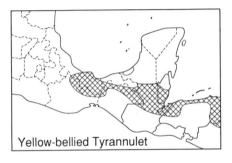

Yellow-bellied Tyrannulet

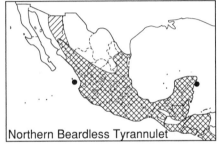

Northern Beardless Tyrannulet

greyish olive with *short whitish super-cilium and narrow dark eyestripe*; wings and tail darker with 2 pale buff wingbars and buffy-lemon panel on secondaries. Throat and underparts pale grey, washed dusky on chest, pale lemon on belly and undertail coverts. **Juv:** brighter wingbars buffy cinnamon, panel on secondaries buff.

Voice. A slightly nasal *peeert* or *peeeet* and a descending series of 3–5 or more short, plaintive notes *dee dee dee ...* or *twee twee ...*; also a short, burry *peerrr*, a more plaintive, slurred *peeeeeu* and *hoo-eet*, and a shorter call followed by a rapid, liquid trill, *peeeut, di-il-il-il*. Dawn song a varied series of piping, reedy whistles typically including 1–2 stronger notes, *cheet chee chee chee chee chee chEE-cheel' it* or *dee dee dee DEE dee dee dee dee'*; also shorter series, *si pee-peeur pee-peeu* or *pi pee pee PEE PEE*, etc.

Habs. Arid to semihumid woodland, second growth, semiopen areas with scrub and scattered trees. Forages low to high; posture fairly upright, tail often wagged. Overlooked easily. Nest is a globular structure of fine grasses, fibers, etc., with side entrance, set in clump of dead seeds or leaves, epiphyte, etc., at low to midlevels. Eggs: 2, whitish, speckled rusty and grey (WFVZ).

SS: Fairly plain but distinctive; note crest, tail-wagging, voice.

SD: C to F resident (SL–2100 m, mostly below 750 m in C.A.) on both slopes from Son (withdraws from N, Oct–Feb) and Tamps, and in interior from S Plateau (where U to F), to El Salvador and Honduras.

RA: Mexico (locally in SW USA) to Costa Rica.

GREENISH ELAENIA
*Myiopagis viridicata**   Plate 39
Elenia Verdosa

ID: 5–5.7 in (12.5–14.5 cm). Both slopes. *Head flattish* (slight crest rarely raised), *bill fairly small and slender, tail long.* **Descr.** Ages differ, sexes similar. **Adult:** bill blackish, often flesh below at base, legs blackish. Head and upperparts olive (head may look greyer), *short whitish supercilium and dark lore stripe create facial expression recalling Orange-crowned Warbler,* yellow crown patch rarely visible. Wings and tail darker with olive edgings *and sulphur panel on secondaries.* Throat and chest

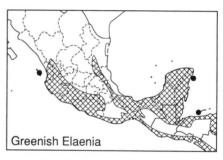
Greenish Elaenia

greyish with whitish streaking heaviest on throat, belly and undertail coverts lemon; *note streaky border between chest and belly.* **Juv:** head brownish, upperparts brownish olive, lacks crown patch.

Voice. A high, thin, slightly reedy to slurred, descending *seei-seeir* or *teez eez,* varied to a clearer *tee-eeu* or *tsee-chu,* and a plaintive *cheee'eu,* suggesting a soft Rose-throated Becard, also a single *teeez* or *tzeeeu.* Dawn song a slightly plaintive *chew-ee-ee'u* or *chewee-eeu',* repeated over and over, 10/30–31 s, may be run into an excited *ts teeu-i tew-eh'*, etc.

Habs. Forest, woodland, adjacent scrub, plantations. At mid- to upper levels. Stance fairly upright, sits quietly for long periods and overlooked easily. Nest is a shallow cup of rootlets, spider silk, etc., in a fork at mid- to upper levels. Eggs: 2, whitish, marked with purplish and greys (WFVZ).

SS: Yellow-olive Flycatcher has broad bill (pale below), shorter tail, often less upright stance, whitish spectacles, more contrasting grey head, brighter wing covert edgings; note voice.

SD: F to C resident (SL–1800 m, R to 2500 m) on both slopes from S Dgo and S Tamps, and in interior from cen Mexico, to El Salvador and Honduras.

RA: Mexico to Ecuador and N Argentina

CARIBBEAN ELAENIA
Elaenia martinica remota   Plate 39
Elenia Caribeña

ID: 5.5–6 in (14–15 cm). Caribbean islands. Fairly chunky-bodied, *erectile crest usually held flat.* **Descr.** Ages differ slightly, sexes similar. **Adult:** *bill dark above, pinkish orange below,* legs blackish. Head and upperparts greyish olive with *short pale grey supercilium and dusky lore stripe,* base of crest whitish. Wings and tail darker with whitish to lemon-buff wingbars and

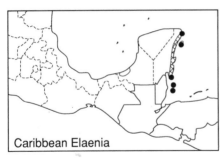

Caribbean Elaenia

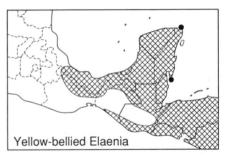

Yellow-bellied Elaenia

panel on secondaries, tertials broadly edged whitish. Throat and underparts pale grey, chest washed greyish olive, belly and undertail coverts usually washed pale lemon. **Juv:** crest shorter, lacks crown patch, wingbars washed cinnamon.

Voice. A clear *whee-u* or *whee'u* and *peeu*; also a short, trilled *pirrrr*, and a hoarse *brreeeah* suggesting Yellow-bellied Elaenia. Song 2 plaintive, clipped phrases, repeated after a pause, *s-peeu, si-pit* or *s-weeu', see-bit*, etc.

Habs. Humid to semihumid woodland and edge, brushy second growth. At mid- to upper levels; perches fairly conspicuously on low trees and bushes, or may be hard to see in forest canopy; feeds at fruiting trees. Nest is a roughly built cup in shrub or tree. Eggs (West Indies): 2–3, whitish, spotted with browns and greys (WFVZ).

SS: Yellow-bellied Elaenia occurs in open and semiopen habitats, spiky crest usually prominent, bill dull below, without orange color, belly brighter yellow.

SD: F to C resident (around SL) on islands off QR (Mujeres, Cozumel); also recorded (status unknown?) from Banco Chinchorro, QR, and Lighthouse Reef and Glover's Reef, Belize. U to F visitor (Sep–early Apr, local resident?) on mainland of E QR (Lopez O. *et al.* 1989) and Ambergris Cay, Belize (Howell *et al* 1992*b*).

RA: Caribbean islands, adjacent QR.

YELLOW-BELLIED ELAENIA

Elaenia flavogaster subpagana   Plate 39
Elenia Viente-amarillo

ID: 6–6.5 in (15–16.5 cm). E and S. Fairly chunky-bodied, *erectile crest usually raised.* **Descr.** Ages differ, sexes similar. **Adult:** *bill blackish, pale flesh below at base,* legs blackish. Head and upperparts brownish olive with *pale lores and narrow*

pale eyering, base of crest whitish. Wings and tail darker with whitish to buffy-lemon wingbars, whitish tertial edgings, and pale lemon panel on secondaries. Throat pale grey, chest washed greyish olive, belly and undertail coverts lemon. **Juv:** crest shorter, crown patch indistinct or lacking, wingbars buff.

Voice. A burry, drawled *brreuhh* or *wheeeuuh,* a harsher *rreear,* and a hoarse, nasal, rhythmic bickering *rreeahr-ch'reer* ..., repeated quickly 3–5×, etc. Dawn song a burry *fri-di-yu* or *prri di-di-eu,* repeated over and over.

Habs. Semiopen and open areas with scattered trees, hedges, scrub, etc. At mid-levels in trees and on tops of bushes, feeds at fruiting trees. Overlooked easily if not calling. Nest cup of fine vegetation and mosses at low to mid-levels in bush or tree, slung in fork or saddled on horizontal branch. Eggs: 2, whitish, spotted with browns and greys (WFVZ).

SS: See Caribbean Elaenia. Mountain Elaenia (highlands) has rounded head, whitish eyering, mostly pale bill.

SD: C to F resident (SL–1600 m) on both slopes from S Ver and Chis, and in interior from E Guatemala, to El Salvador and Honduras; most birds withdraw in winter (Sep–Feb?) from Atlantic Slope N of Isthmus.

RA: SE Mexico to Peru and N Argentina.

MOUNTAIN ELAENIA

Elaenia frantzii ultima   Plate 71
Elenia Serrana

ID: 5.5–6 in (14–15 cm). *C.A. highlands.* Fairly chunky-bodied, *round-headed.* **Descr.** Ages differ slightly, sexes similar. **Adult:** *bill flesh with dusky culmen and tip,* legs blackish. Head and upperparts brownish olive with *pale lemon eyering;* wings and tail darker with pale lemon wingbars

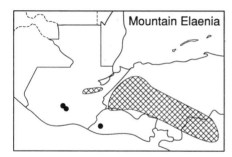

Mountain Elaenia

and panel on secondaries, *bold white tertial edgings.* Throat and chest greyish olive, belly and undertail coverts pale lemon. **Juv:** wingbars brighter, buff.

Voice. A burry, drawn-out, down-slurred *peeeer* or *peeeah*, or a clearer whistled *peer* or *pee-er*; dawn song a rather buzzy *d'weet d'weet* or *ch'weet ch'weet*, at times interspersed with gurgling short syllables (Stiles and Skutch 1989).

Habs. Humid to semihumid forest edge, second growth, and semiopen areas with trees and shrubs. Forages low to high, often at fruiting trees. Nest cup of root-lets, mosses, etc., at low to mid-levels in bush or tree. Eggs: 2, whitish, spotted with browns and greys (WFVZ).

SS: See Yellow-bellied Elaenia.

SD: U to F resident (1000–2200 m) in the interior of cen Guatemala and from Honduras to N cen Nicaragua.

RA: Guatemala to Venezuela.

OCHRE-BELLIED FLYCATCHER

Mionectes oleaginus assimilis　　　Plate 39
Mosquero Vientre-ocre

ID: 5–5.5 in (12.5–14 cm). *Humid forest. Plain*; head rounded, *bill relatively long and narrow.* **Descr.** Ages/sexes similar. Bill blackish, pale flesh below at base, legs greyish. *Head and upperparts dark olive,* slightly paler lores enhance *big-eyed ex-*

pression; wings and tail darker with olive to tawny-olive edgings. Throat greyish, chest greyish olive, *belly and undertail coverts ochre.*

Voice. Song a nasal *s-yuk' s-yuk* ... or *whiuk'-whiu'* ..., long series often varied with series of sharper *kyeh-kyeh* ... or *ke-ke* ..., repeated by ♂♂ in loose leks in open understory; calls include a sharp *pee-k* or *p-lik*, and a plaintive *cheu.*

Habs. Humid evergreen and semi-deciduous forest and edge, adjacent second growth. At low to mid-levels in shady understory, at times in canopy, often eats berries. Jerks head slowly back and forth and quickly lifts one wing at a time. Joins mixed-species flocks. Nest is a matted pyriform structure of mosses, fine fibers, etc., with side entrance, suspended from tip of hanging vine or epiphytic root, usually with hanging tail of dead vegetation up to 0.5–0.8 m (Fig. 32). Eggs: 2–3, white (WFVZ).

SD: C to F resident (SL–1600 m) on both slopes from S Ver and Chis to El Salvador and Honduras, commoner on Atlantic Slope; U N to SE SLP (Lowery and Newman 1951).

RA: SE Mexico to Ecuador and Brazil.

SEPIA-CAPPED FLYCATCHER

Leptopogon amaurocephalus pileatus
Mosquero Gorripardo　　　Plate 39

ID: 5–5.5 in (12.5–14 cm). *SE humid lowland forest. Distinct face pattern,* nape slightly peaked, bill narrow. **Descr.** Ages/sexes similar. *Eyes pale,* bill blackish, often pale flesh below at base, legs greyish. *Crown and nape dark rich brown, face pale greyish with dark auricular crescent, dusky smudge below eye.* Upperparts olive, wings and tail darker with *ochre wingbars* (buffier in juv) and tawny-yellow panel on secondaries. Throat and

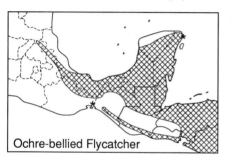

Ochre-bellied Flycatcher

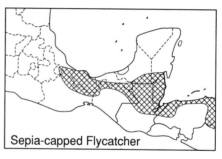

Sepia-capped Flycatcher

chest greyish olive, belly and undertail coverts pale lemon.

Voice. A rolled, slightly frog-like trill, suggests Northern Bentbill but longer, more emphatic, at times with 1–2 introductory notes, *purrrrrrrrrr* or *whik, whik prrrrrrrrrru*, and a longer, spluttering trill *prri' prriiiiiiew eup'*.

Habs. Humid evergreen and semideciduous forest and edge. Low to high, mainly at mid-levels in fairly open understory. Sits quietly for long periods and overlooked easily. Joins mixed-species flocks. Nest is a matted globular structure of mosses, fine fibers, etc., with side entrance, suspended from shady recess or overhang (Fig. 32). Eggs: 2, white (MLZ).

SD: F to C resident (near SL–1300 m) on Atlantic Slope from S Ver to Honduras.

RA: SE Mexico to N Argentina.

NORTHERN BENTBILL

Oncostoma cinereigulare   Plate 39
Picocurvo Norteño

ID: 3.8–4.2 in (9.5–10.5 cm). SE humid forest. *Bill broad, with broken nose profile.* **Descr.** Ages differ slightly, sexes similar. **Adult:** *eyes pale lemon*, blackish bill pale flesh below at base, legs flesh. *Greyish olive head often contrasts with olive upperparts*, lores paler; wings darker with narrow greenish lemon edgings. Throat and underparts pale grey, chest streaked whitish, belly and undertail coverts washed pale lemon. **Juv:** eyes dusky, olive-washed head contrasts less with back, wing edgings buffier.

Voice. Frog-like: quiet, low churrs, most often *urrrrrr* or *rrrrrr*, also a shorter *krrrrp* and *rrr-ip'*, rarely *pic ahrrrrr*, and short chips, *pip-pip-pip-pip* or *puip-puip ...*, etc.; song (?) *pirrrip pirrrip p-p-prrrrr*, 1st note given 1–3×.

Habs. Humid to semihumid forest and woodland. At low to mid-levels in understory, often in thickets and overlooked easily. Nest a matted pyriform structure of fine fibers, moss, etc., with side entrance, hung from slender branch low in edge of thicket (Fig. 32). Eggs: 2, white, faintly flecked pale brown.

SS: Slate-headed Tody-Flycatcher more often in second growth and edge; note thick white supraloral stripe and narrow eye-ring, bolder ochre-lemon wingbars, more contrast between grey head and olive back, and longer and straight bill.

SD: F to C resident (SL–1500 m) on both slopes from S Ver and Chis, and locally in interior from Chis, to El Salvador and Honduras.

RA: SE Mexico to W Panama; possibly also NW Colombia.

NB: Sometimes considered conspecific with *O. olivaceum* of Panama and N Colombia.

SLATE-HEADED TODY-FLYCATCHER

Todirostrum sylvia schistaceiceps   Plate 39
Espatulilla Cabecigris

ID: 3.5–3.8 in (9–9.5 cm). *Skulks* in SE humid second growth. *Bill relatively long and broad.* **Descr.** Ages differ, sexes similar. **Adult:** eyes dull pale yellowish, black bill pale flesh below at base, legs dull flesh. *Slate grey head contrasts with olive upperparts, bold white supraloral stripe and narrow white eye-crescents create vireo-like expression.* Wings darker with *bright ochre-lemon wingbars* and panel on secondaries. Throat and chest pale grey, streaked whitish, flanks and undertail coverts washed lemon. **Juv:** eyes brownish, head olive like back, wingbars brighter.

Voice. A short, rolled, frog-like *prrrrr* or *chrrrr*, often preceded by clipped, hollow, frog-like clucks, *puc, puc prrrrr,* or *tchk chrrr,* etc.

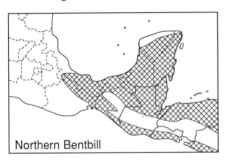

Northern Bentbill

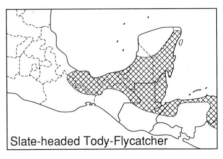

Slate-headed Tody-Flycatcher

Habs. Humid to semihumid second growth, forest edge, thickets. Skulking and hard to see, low in dense vegetation. Nest is a matted pensile structure of fine fibers, moss, etc., with visor-like side entrance, suspended from tip of slender branch at low to mid-levels in thicket. Eggs: 2, white, speckled brown (WFVZ).
SS: See Northern Bentbill.
SD: F to C resident (SL–1000 m, mainly below 500 m) on Atlantic Slope from S Ver to Honduras.
RA: SE Mexico to Brazil.

COMMON TODY-FLYCATCHER

Todirostrum cinereum *   Plate 39
Espatulilla Común

ID: 3.7–4 in (9.5–10 cm). *Open second growth in SE. Tail strongly graduated, bill relatively long and broad.* **Descr.** Ages/sexes differ slightly. **Adult:** *eyes pale lemon,* bill black above, below flesh (♀) or blackish with flesh base (♂), legs greyish. *Blackish face blends into dark grey crown* and upper mantle. Upperparts greyish olive, wings darker with greenish-lemon edgings; *tail black, outer rectrices broadly tipped white. Throat and underparts yellow.* **Juv:** eyes dusky, face greyer, underparts paler.
Voice. A sharp, warbler-like, smacking *tchik* or *chik* and a short, rolled, liquid trill, usually repeated 2–4×, *tril-il-il-it tril-il-il-it ...,* or *ssi-ssi ssi-ssih ssi-i-i ssit,* etc., suggests Tropical Kingbird.
Habs. Open and semiopen areas with trees, hedges, orchards, thickets, mangroves, and edge. At low to mid-levels in trees and bushes, tail held cocked. Often overlooked although fairly conspicuous. Nest is a matted pensile structure of fine fibers, moss, etc., with visor-like side entrance, suspended from tip of slender branch, low to high in bush or tree (Fig. 32). Eggs: 2, white, may be faintly speckled brown (WFVZ).
SD: F to C resident (SL–1000 m) on both slopes from S Ver and Chis to El Salvador and Honduras. A record from Isla Mujeres (AOU 1983) is based on a Gaumer spec. and is thus questionable.
RA: SE Mexico to Peru and Brazil.

EYE-RINGED FLATBILL

Rhynchocyclus brevirostris *   Plate 39
Picoplano de Anteojos

ID: 6.3–6.8 in (16–17 cm). Humid S forests. *Large headed, long tailed, bill very broad.* **Descr.** Ages/sexes similar. Bill blackish above, flesh below, legs greyish. *Head and upperparts olive-green with bold whitish eyering,* face paler greyish with dusky auricular mark and faint smudge below eye. Wings and tail darker with yellow-olive edgings. Throat and chest dusky olive, rest of underparts lemon, diffusely streaked olive on upper belly.
Voice. Insect-like: a rising, high, squeaky *zweeip* or *sweik,* a slightly rasping, cicada-like *zzrrip* or *rrrip,* and a shrill, lisping *siiir* or *ssssi,* all of which may be given 1–5×; last call much like Yellow-olive Flycatcher but less lisping. Also a high, lisping, and spluttering *syiip-yiip.*
Habs. Humid evergreen and semideciduous forest. At mid-levels in shady understory where generally quiet and overlooked easily. Joins mixed-species flocks. Nest is a matted, bulky pyriform structure of fine fibers, dead leaves, etc., with sloping chute entrance below, suspended from tip of hanging vine at mid-levels in open understory (Fig. 32). Eggs: 2–3, vinaceous, spotted dark brown (WFVZ).
SD: F to C resident (SL–1800 m) on both

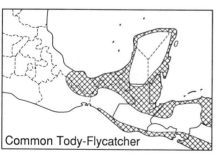
Common Tody-Flycatcher

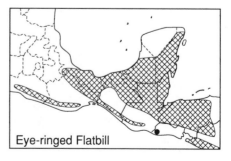
Eye-ringed Flatbill

slopes from Gro (Navarro 1986; SNGH) and S Ver to El Salvador and Honduras; in interior from Chis S.

RA: S Mexico to Ecuador.

YELLOW-OLIVE FLYCATCHER
Tolmomyias sulphurescens cinereiceps
Picoplano Ojiblanco　　　　　　　Plate 39

ID: 5–5.5 in (12.5–14 cm). SE lowland forest. *Bill broad.* **Descr.** Ages differ slightly, sexes similar. **Adult:** eyes whitish, *bill blackish above, greyish flesh below,* legs dusky flesh. *Grey head contrasts with olive-green upperparts, whitish supraloral stripe and eye-crescents create slight spectacled expression.* Wings and tail darker with bright yellow-olive edgings. Throat and chest pale grey, washed dusky on chest, belly and undertail coverts pale lemon. **Juv:** eyes brownish, head washed olive, underparts paler.

Voice. A lisping, strengthening hiss, usually 2–5×, with increasing intensity in series, *sssii sssii sssii,* or *sssih sssih …,* etc., at times harder and buzzier, *sssiir …* or *zzzieh …,* much like Eye-ringed Flatbill; also a quieter single *sssi.*

Habs. Humid to semiarid forest and edge, clearings, plantations. Low to high, mainly at mid-levels; overlooked easily if not calling. Nest is a matted pyriform structure of fine dark fibers with sloping chute entrance below, slung on slender branch or vine at mid-levels in bush or tree at edge or fairly open situation (Fig. 32); a side entrance often develops later, used for feeding young. Eggs: 2–3, creamy, speckled reddish brown (WFVZ).

SS: See Greenish Elaenia.

SD: C to F resident (SL–1200 m) on both slopes from S Ver and E Oax, and locally

in interior from Chis, to El Salvador and Honduras.

RA: SE Mexico to Peru and Brazil.

STUB-TAILED [WHITE-THROATED] SPADEBILL
Platyrinchus cancrominus　　　　　Plate 39
Picochato Rabón

ID: 3.5–3.7 in (9–9.5 cm). SE humid forest. *Small, stub-tailed; bill very broad.* **Descr.** Ages/sexes differ slightly. Bill blackish above, pale flesh below, legs flesh pink. *Face pale greyish with broad dark auricular mark and dark smudge below eye setting off pale eyering.* Crown, nape, and upperparts brown; ♂ has yellow crown patch, usually concealed. Wings and tail darker, edged cinnamon-brown. *Throat white,* chest and sides pale tawny-brown, belly and undertail coverts whitish, washed lemon. **Juv:** face pattern less distinct, ♂ lacks crown patch.

Voice. A bright, nasal *chi-dih* or *ch-dee,* and *chi-di-dit* or *ki-dih-dih,* etc., also a nasal, rolled *pirririrr.* Dawn song an excited, nasal, rapid, rolled trill alternated with sharp nasal calls, *ki-di-di-di-rril, ki-di-di-di-drri-l-l sy-iik' …,* etc.

Habs. Humid evergreen and semideciduous forest. At low to mid-levels in shady understory. Sits quietly for long periods, flights short and quick, overlooked easily. Nest is a cone-shaped cup of fine grass, fibers, etc., in fork of low understory shrub. Eggs: 2, buff, marked with reddish brown (WFVZ).

SD: C to F resident (SL–1500 m) on both slopes from S Ver and Chis to El Salvador and Honduras.

RA: SE Mexico to NW Costa Rica.

NB: Formerly considered conspecific with White-throated Spadebill (*P. mystaceus*) of Costa Rica to S.A.

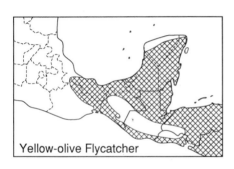
Yellow-olive Flycatcher

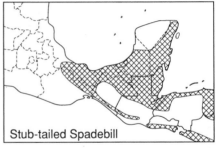
Stub-tailed Spadebill

ROYAL FLYCATCHER
Onychorhynchus coronatus mexicanus
Mosquero Real   Plate 39

ID: 6.5–7 in (16.5–18 cm). SE humid lowland forest. *Erectile crest usually held flat which, with longish bill, creates hammerhead profile.* **Descr.** Ages/sexes differ. ♂: eyes pale brown, bill blackish above, orange-yellow below, legs orange-yellow. *Crest fiery red, tipped with black and metallic bluish-purple eyespots (some red often visible when held flat). Head and upperparts brown, becoming tawny on uppertail coverts; wing coverts spotted buff,* tertials edged pale cinnamon. *Tail cinnamon, dusky distally.* Throat whitish, *underparts pale cinnamon, washed and mottled brown on chest.* ♀: crest yellow-orange. **Juv:** upperparts and chest scalloped dusky, wing coverts, tertials, and rectrices with dusky subterminal bars, tipped pale cinnamon; ♂ crest fiery orange.
Voice. A squeaky to hollow, plaintive *whee-uk'* or *see-yuk'*, suggesting a muffled Jacamar. Song a descending, slowing series of plaintive whistles, usually 5–8, following a shorter introductory note, *whi' peeu peeu peeu peeu peeu ...*, or *wh' wheeu wheeu ...*
Habs. Humid evergreen and semideciduous forest, often along streams. At mid- to low levels in shady understory, fairly active flycatching flights often attract attention. Joins mixed-species flocks. Nest is an ovate matted structure with side entrance, held in scraggly pensile mass of rootlets, dead leaves, etc., with a tail up to 1 m long (Fig. 32); suspended from slender branch or vine at mid-levels, typically along forest stream. Eggs: 2, dark vinaceous brown, marked with darker browns (WFVZ).

SS: Bright-rumped Attila has darker tail, lacks hammerhead crest, stouter bill mostly pale, face and chest streaked dusky.
SD: F to U resident (SL–1200 m) on both slopes from S Ver and E Oax to El Salvador and Honduras, possibly extirpated from El Salvador by habitat loss (Thurber *et al.* 1987).
RA: SE Mexico to Peru and Brazil.
NB: *Coronatus* group of S.A. sometimes considered specifically distinct.

RUDDY-TAILED FLYCATCHER
Terenotriccus erythrurus fulvigularis Plate 39
Mosquerito Colirrufo

ID: 3.8–4 in (9.5–10 cm). SE humid lowland forest. **Descr.** Ages/sexes similar. Bill blackish above, flesh below, legs flesh. *Head and upperparts greyish olive becoming cinnamon on uppertail coverts; wings darker with cinnamon edgings. Tail cinnamon-rufous. Throat and underparts cinnamon,* palest on throat.
Voice. A quiet, high, slightly reedy *seeeu siirr* or *psee sirrr*, the 1st part up-slurred, the 2nd slightly trilled, less often a single *speeu* or *pseee*, and *psee seeip*, etc.
Habs. Humid evergreen forest. At midlevels in shady understory, quiet and overlooked easily. Nest is a matted pyriform structure of fine fibers, dead leaf fragments, etc., with visor-like side entrance, suspended from vine or slender twig at mid-levels in forest understory (see Fig. 32). Eggs: 2, white, marked with browns.
SD: U to F resident (SL–500 m) on Atlantic Slope from NE Chis and S Camp (MB, AMS, PW) to Honduras.
RA: SE Mexico to Brazil.
NB: *Terenotriccus* is sometimes merged with *Myiobius* (Lanyon 1988b).

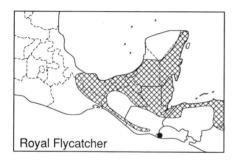

Royal Flycatcher

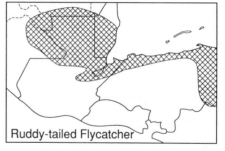

Ruddy-tailed Flycatcher

SULPHUR-RUMPED FLYCATCHER

Myiobius s. sulphureipygius　　　Plate 39
Mosquerito Rabadilla-amarilla

ID: 4.7–5.2 in (12–13.5 cm). SE humid low-land forest. Very long rictal bristles often obvious. **Descr.** Ages/sexes differ slightly. Bill blackish above, flesh below with dusky tip, legs greyish. *Large dark eyes emphasized by pale face with dark auricular mark.* Crown, nape, and upperparts dark brownish olive with *striking pale lemon rump*; yellow crown patch (reduced or absent in ♀ and juv) usually concealed. Dark wings edged brownish, tail blackish. Throat buffy white, *chest and sides ochre*, belly pale lemon.

Voice. A wet, fairly sharp *plik* or *pic*; also a rising and falling series of 4–6 quiet notes, *tchew tchew tchew tchew tchew tchew*, slightly suggesting Dot-winged Antwren. Dawn song a rapidly repeated *chu wee-da-wiit'* or *chu wee-da-ti-wit*.

Habs. Humid evergreen and semideciduous forest. At low to mid-levels in shady understory, often near water. Fairly active, usually holds wings drooped and tail spread to show off bright rump. Joins mixed-species flocks. Nest is a matted pyriform structure of fine brownish fibers, with overhung side entrance, suspended from vine or slender twig at mid-levels in open forest understory (Fig. 32). Eggs: 2, whitish to buff, marked with browns and greys (WFVZ).

SD: F to C resident (SL–1000 m) on Atlantic Slope from S Ver to Honduras.

RA: SE Mexico to Ecuador.

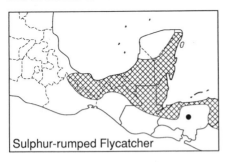

Sulphur-rumped Flycatcher

BELTED FLYCATCHER

Xenotriccus callizonus　　　Plate 40
Mosquero Fajado

ID: 4.5–5 in (11.5–12.5 cm). *Endemic to in-*

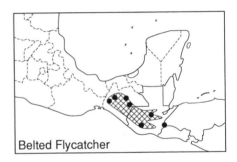

Belted Flycatcher

terior S of Isthmus. Erectile spiky crest usually obvious. **Descr.** Ages/sexes similar (?). Bill blackish above, yellow-orange below, legs dark grey. Head greyish olive with *teardrop-shaped pale lemon eyering*. Upperparts brownish olive, wings darker with *cinnamon wingbars*, pale cinnamon tertial edges and panel on secondaries; tail dark grey-brown, outer webs of outer rectrices pale grey. Throat and *underparts pale lemon*, throat paler, *with broad cinnamon band across upper chest*.

Voice. A fairly hard, buzzy, nasal *rreah* or *choi-ih*, a sharper *cheuh'*, and a burry *pic'weehr* or *pi'kweehr*. Song a bright, short, accelerating, slightly liquid to nasal *pic pi-pi-pi-pi-chi-i-weer* or *chi-chi-chi-chi-i-ir*, often repeated quickly a few times.

Habs. Semiarid to semihumid brushy and scrubby woodland, especially with oaks. Keeps well hidden (even when singing), low in dense understory, detected rarely except by voice. Nest cup of fine grasses and fibers placed in a fork, low in brushy scrub. Eggs: 3, whitish, marked with browns.

SD/RA: F to U (and apparently local) resident (1200–2000 m) in interior and on adjacent Pacific Slope from Chis to NW El Salvador; C at El Sumidero, Chis, poorly known away from there.

PILEATED FLYCATCHER

Xenotriccus mexicanus　　　Plate 40
Mosquero del Balsas

ID: 5.2–5.7 in (13.5–14.5 cm). *Endemic to SW Mexico. Erectile spiky crest usually raised.* **Descr.** Ages differ slightly, sexes similar. **Adult:** *bill blackish above, pinkish orange below*, legs dark grey. Head and upperparts brownish olive to greyish olive with *whitish lores and slightly teardrop-*

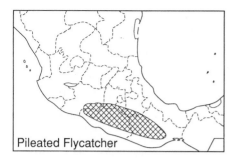

Pileated Flycatcher

shaped *eyering*; wings and tail darker with 2 whitish wing bars, tertial edgings, and panel on secondaries. Throat whitish, chest washed dusky, belly washed lemon. **Juv:** crest shorter, wing edgings pale cinnamon. Attains adult-like plumage quickly.

Voice. A rich, slightly nasal *bree* and *b-rree*, a rich, explosive *cheeoo'* or *cheeup'*, and a scolding (alarm?) *whee whee-eu' whee-eu' whee-eu*. Song typically a bright introductory note followed by a spluttering, intensifying series suggesting a referee's whistle, and clearing for a final upward-inflected whistle, *p'iweer pi-pi-pirrr-i-it' cheeu*, or *wheeu whirrr-rr-rr-rr whee-u'*.

Habs. Arid scrub and thorn forest, less often in semiopen overgrazed areas. At low to mid-levels, sits quietly for long periods and overlooked easily in dense scrub; best detected by voice and often sings from conspicuous perches. Flight heavy and direct. Jerks tail up in manner similar to Dusky Flycatcher. Nest cup of grasses, fibers, etc., suspended from a fork by spider silk, often with trailing grass streamers; placed low in thorny thickets. Eggs: 2–3, buff, speckled with reddish brown and greys (WFVZ).

SS: Superficially *Empidonax*-like but more heavily built, relatively long-tailed, bill stouter and pinkish orange below; with crest flattened might be confused with Dusky Flycatcher; note structure and bill, crest projects at back of crown, teardrop-shaped eyering.

SD/RA: F resident (900–2000 m) in interior from cen Mich and Mor to Oax. May withdraw from Oaxaca Valley (and other higher areas?) in winter, possibly descending into Balsas drainage; C at Monte Alban, Oax, in summer, R or absent there in winter.

TUFTED FLYCATCHER
Mitrephanes phaeocercus*   Plate 40
Mosquero Penachudo

ID: 4.7–5.2 in (12–13.5 cm). Highlands throughout. *Tufted crest obvious.* **Descr.** Ages differ slightly, sexes similar. Bill blackish above, orange-yellow below, legs blackish. **Adult:** *face, throat, and underparts cinnamon* with pale eyering. Crown, narrow nape stripe, and upperparts brownish olive, wings and tail darker with 2 dull cinnamon wingbars, whitish to lemon tertial edgings and panel on secondaries. **Juv:** crown, nape, and upperparts tipped pale cinnamon, wingbars broader and brighter. Attains adult-like plumage quickly.

Voice. A burry, bright, rolled *tchwee-tchwee* or *turree-turree*, less often a single *tchwee* or longer series, a sharp, usually fairly quiet *pic* or *beek* suggesting Hammond's Flycatcher, and a bright, slightly emphatic to penetrating *seeeu* or *pseeeu*, repeated every few s and given mainly in breeding season.

Habs. Humid to semihumid pine–oak and evergreen to semideciduous forest and edge, clearings, plantations, etc.; to arid semiopen areas, riparian groves, etc., in winter. Habits pewee-like: perches conspicuously in open situations, flycatches actively at mid- to upper levels, often returns to same perch and quivers its tail on landing. Nest cup of rootlets, moss, etc., saddled on branch at mid- to upper levels of trees. Eggs (Costa Rica): 2, whitish, marked with browns.

SD: C to F resident (1200–3500 m) on both slopes from Son and S Tamps, and in interior from cen Mexico, to El Salvador and N cen Nicaragua. Locally descends to near SL in NW Mexico (Sin to Jal) in winter;

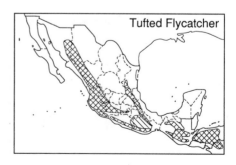

Tufted Flycatcher

altitudinal migrant elsewhere, at least N of Isthmus.

RA: Mexico to Bolivia.

NB: *M. olivaceus* of E Andes in Peru and Bolivia sometimes considered conspecific.

GENUS CONTOPUS

Pewees have peaked to crested napes, relatively long and broad bills, and long wings. Head and upperparts relatively dark, no obvious pale eyering, pale wingbars poorly contrasting. Ages differ slightly, sexes similar; quickly attain adult-like plumage by first prebasic molt. Typically perch prominently and return to same perch after aerial sallies for prey. Nests are cups of dead grasses, fibers, moss, etc., at mid- to upper levels in tree or bush, typically saddled on horizontal branch, less often in vertical fork. Eggs: 2–4, white to buffy white, marked with browns and greys (WFVZ).

OLIVE-SIDED FLYCATCHER
Contopus borealis   Not illustrated
Pibí Boreal

ID: 6.8–7.3 in (17–18.5 cm). Mainly a transient. *Nape peaked, tail fairly short.* **Descr. Adult:** *bill* blackish above, *flesh below with dark tip,* rarely all dark, legs blackish. Head and upperparts dark greyish olive to grey-brown, head slightly darker, whitish rump side tufts often covered by wings; wings and tail darker with dull paler wingbars, whitish to brownish-white tertial and secondary edgings. *Whitish throat continues as stripe down median chest to belly and contrasts strongly with dark greyish sides, creating vested appearance.* **Juv:** wingbars and tertial edgings washed buff. **Voice.** A fairly sharp, slightly nasal to ringing *bik-bik-bik* or *beehk-beehk-beehk,* may be repeated steadily, much like Greater Pewee. Song a loud emphatic *whit! whee-pew,* often repeated over and over at dawn.

Habs. Breeds in conifer forests; migrates and winters mainly in pine–oak, evergreen, and semideciduous forest and edge. See genus note; perches conspicuously atop dead snags.

SS: Greater Pewee longer-tailed with tufted crest, more uniform greyish overall, bill bright orange below; other pewees smaller, relatively longer-tailed, lack strongly vested underparts. White rump side tufts of Olive-sided distinctive.

SD: F to C breeder (May–Aug; 2000–2800 m) in BCN; F to C transient (Apr–early Jun, Aug–early Nov, mainly in foothills and highlands) on Pacific Slope and in interior throughout; U to R on Atlantic Slope, including S Yuc Pen (PW). U winter visitor (Sep–Apr; 1500–2000 m) on Pacific Slope from Jal (SNGH, PP, SW), on Atlantic Slope from S Ver, and in interior from Chis, to Guatemala and Belize, probably elsewhere in C.A.

RA: Breeds W and N N.A. to BCN; winters S Mexico to Peru.

NB: Formerly placed in the monotypic genus *Nuttallornis.*

GREATER PEWEE
Contopus pertinax*   Plate 40
Pibí Mayor

ID: 6.8–7.5 in (17–19 cm). Widespread in highlands. *Tufted crest obvious.* **Descr. Adult:** *bill* blackish above, *bright yellow-orange below,* legs blackish. Head and upperparts greyish to greyish olive, lores faintly paler with dark lore stripe, wings and tail darker with paler wingbars, whitish to brownish-white tertial and secondary edgings. *Throat and underparts pale greyish,* throat often paler, *belly paler and washed lemon. C. p. pallidiventris* of N

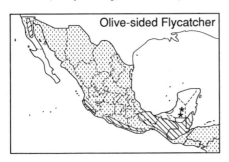

Olive-sided Flycatcher

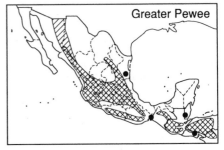

Greater Pewee

and cen Mexico (winters S to Oax) paler and greyer; resident S ssp darker with upperparts washed olive, belly washed ochre. **Juv:** wingbars buffy cinnamon, tertial and secondary edgings washed lemon-buff.

Voice. A sharp to mellow *beek beek beek* or *puip-puip-puip*, at times repeated steadily, much like Olive-sided Flycatcher. Song a bright to slightly plaintive *wheee tee whee-wheu'* (or *Jo-sé Ma-ría*), varied to *whee'tee-tee whee-wee'* or *wheeu tew t-tee*, etc., and followed by a short, rolled *whee di-irit* or *chewdl-it*, in full song, this last phrase may also be repeated over and over.

Habs. Pine–oak, evergreen, and semi-deciduous forest and edge, clearings. Habits much as Olive-sided Flycatcher.

SS: See Olive-sided Flycatcher. Other pewees smaller, lack tufted crests; note voice.

SD/RA: C to F resident (750–3500 m), locally to 250 m, at least in winter) on both slopes and in interior from Son and NL (in summer to SW USA) to El Salvador and N cen Nicaragua. Most withdraw from colder parts of range in winter (Nov–Feb), at least from Chin to NL, altitudinal migrant elsewhere.

NB1: Also known as Coues' Flycatcher. **NB2:** Sometimes considered conspecific with *C. lugubris* of S C.A.

WESTERN PEWEE (WOOD-PEWEE)
Contopus sordidulus* Plate 40
Pibí Occidental

ID: 5.3–6 in (13.5–15 cm). *Highlands throughout in summer.* Nape slightly peaked. **Descr. Adult:** *bill blackish above, pale orangish below with dark tip or all dark,* legs blackish. *Head and upperparts dark olive-grey,* lores may be slightly paler; *wings and tail darker with paler wingbars. Pale grey throat contrasts with dusky grey*

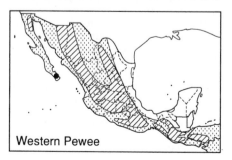

Western Pewee

chest and sides, belly and undertail coverts whitish, often washed lemon. **Juv:** pale buff tips to head and upperparts quickly lost, wingbars brighter buff.

Voice. A burry, drawled *brreeu* or *bzzeiu,* less often a plaintive, slurred *peeeu* to slightly up-slurred, at times disyllabic *pee-eeur,* suggesting Eastern Pewee. Song a *brreeu*-type call followed by a rapid, rolled *chu-i-lit* which may also be given separately.

Habs. Breeds in pine and pine–oak forest and edge, clearings, semiopen areas with trees, hedges, etc.; widely in woodland and open areas with hedges, fences, etc., in migration. See genus note; often quivers wings upon landing but does not wag tail like *Empidonax.*

SS: See Olive-sided Flycatcher, Greater Pewee. Eastern Pewee more olive above, less dusky below, bill often all pale below (rarely so in Western); note voice; usually not safely identified unless heard. *Empidonax* behave differently, do not habitually return to prominent perches and quiver their wings, all have shorter wings (which do not reach well down the tail), smaller bills, most (except Willow and Alder) have obvious pale eyerings and wingbars; note voice. Tropical Pewee best told by shorter primary projection, voice.

SD: F to U summer resident (Apr–Sep; 1000–3000 m) in Baja and in interior and on adjacent slopes from Son and Coah to Honduras; often U and local S of Isthmus. C to F transient (mid-Mar–early Jun, Aug–early Nov; near SL–3000 m) on Pacific Slope and in interior throughout; U to F on Atlantic Slope (mainly above 500 m). A report from Belize (Russell 1964) is based on a Gaumer spec. and thus questionable.

RA: Breeds W N.A. to Honduras; winters in S.A.

EASTERN PEWEE (WOOD-PEWEE)
Contopus virens Not illustrated
Pibí Oriental

ID: 5.3–6 in (13.5–15 cm). *Transient in E and S.* Nape slightly peaked. **Descr.** *Bill blackish above, pale orangish below, often with dark tip. Head and upperparts greyish olive,* lores may be slightly paler; *wings and tail darker with paler wingbars* (brighter and buffer in imm). *Whitish throat contrasts with greyish-olive chest,*

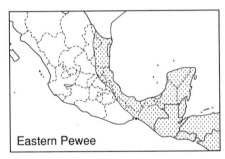

Eastern Pewee

belly and undertail coverts pale lemon to whitish.

Voice. A plaintive, slurred *d-weee* or *du-weee*, and a longer *whee-u-hwee*, at times followed by a rapid, rolled *hew-di-lit.*

Habs. Forest and edge, clearings, open areas with hedges, fences. Habits much as Western Pewee.

SS: Tropical Pewee best told by shorter primary projection; browner above, may appear more crested with contrastingly darker crown; note voice. See Western Pewee and distinctions from *Empidonax* under Western Pewee SS.

SD: C to F transient (late Mar–early Jun, Aug–early Nov; SL–2700 m, mainly below 1000 m) on Atlantic Slope from NL and Tamps to Honduras, also on Pacific Slope and in interior from Isthmus S. AMNH specs from Mor and Chih, if correctly identified, indicate a more westerly migration route, at least in spring.

RA: Breeds E N.A.; winters in S.A.

TROPICAL PEWEE

Contopus cinereus *   Plate 40
Pibí Tropical

ID: 5.2–5.7 in (13.5–14.5 cm). *E and S lowlands.* Nape peaked to slightly crested, *primary projection relatively short.* **Descr.** **Adult:** *bill* blackish above, *pale orangish below*, rarely with dusky tip, legs blackish.

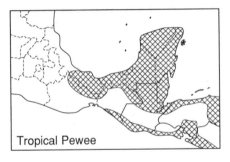

Tropical Pewee

Head and upperparts brownish olive, usually with pale lores, *crown contrastingly darker*; wings and tail darker with paler wingbars. Whitish throat contrasts with brownish-olive chest, belly and undertail coverts pale lemon. **Juv:** pale buff tips to head and upperparts quickly lost, wingbars brighter, buff to pale cinnamon.

Voice. A short, liquid trill, *ti-i-i-i-il* or *tree-ee-ee-eet*, and a high, up-slurred *psee-eep* or *pee-eep* which may suggest Western Flycatcher; also a fuller *w-heep*, and a quiet *bzzzir*, softer and higher than Western Pewee call. Dawn song (?) a simple, high, clear, up-slurred *s-iep* or *w-iep*, repeated, 10/20–27 s.

Habs. Humid to semihumid forest edge, clearings with trees and bushes, semiopen areas with hedges, fences, scattered trees. Habits much as Western Pewee.

SS: See Eastern and Western pewees, and distinctions from *Empidonax* under Western Pewee SS.

SD: F to C resident (SL–1200 m) on Atlantic Slope from Yuc Pen to Honduras, on Pacific Slope from Chis S; F to U summer resident on Atlantic Slope from Chis to S Ver whence most withdraw S in winter (Sep–Feb).

RA: SE Mexico to Peru and N Argentina.

GENUS EMPIDONAX

This is a remarkably uniform-looking genus of smallish flycatchers; typically brownish to olive above, whitish to pale grey below with a variable lemon wash, especially on the belly, and a dusky wash across the chest. All have paler wingbars, tertial edgings, and secondary panels, and most have distinct pale eyerings. Ages differ slightly (usually in the color or brightness of wingbars), sexes similar; juv plumage mostly lost by migrants before arrival in region.

Much has been written about *Empidonax* identification but most of it reflects a state of confusion and virtually nothing has been directed at winter identification. Field differences among *Empidonax* are usually *relative* but, like many identification problems, familiarity is the key. Voice is usually the most important starting point, combined with structure and habits. Habitat and known ranges are also useful (see below). Correctly judging bill width and the degree to which primaries project on closed wings is important. Many other traits often quoted as field marks are of little value without considerable experience: all species flick their wings and tail depending on mood (though the way in which the tail is flicked may be useful); all species have paler lores; almost all show variably paler outer webs to the outermost rectrices; head shape varies with mood; and tail shape varies with wear, molt, and mood. We consider that an overall impression of the bird, combined with factors such as voice, habits, and habitat, as well as plumage, is more likely to be useful than attempting to rely on one or two specific field marks. The relative features found under SS sections, combined with Habs, provide most identification clues. All species vary in appearance due to plumage wear and/or age and/or subspecies; the following descriptions are, by necessity, simplified.

All migrant North American *Empidonax* occur in the region, though Acadian and Alder occur only as transients. Observers should be aware of broad *known* distribution patterns. The following applies to most N.A. migrants in *winter*: in the Atlantic Slope lowlands one finds Least and Yellow-bellied (the latter also on Pacific Slope and in interior S of Isthmus); in the Pacific Slope lowlands, Least, Willow and, N of the Isthmus, Pacific Slope; in the interior and highlands N of the Isthmus, Grey and Dusky; and in highlands throughout, Hammond's. All have preferred habitats within these broad regions.

YELLOW-BELLIED FLYCATCHER
Empidonax flaviventris   Not illustrated
Mosquero Vientre-amarillo

ID: 5–5.5 in (12.5–14 cm). *Winter visitor to E and S. Bill relatively broad.* **Descr.** Bill blackish above, orange-flesh below, legs grey. *Head and upperparts olive with pale lemon eyering often extending into slight*

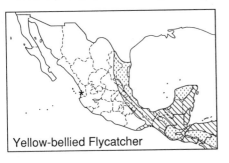

Yellow-bellied Flycatcher

teardrop behind eye. Wings and tail *contrastingly darker* (blackish brown) with pale lemon wingbars (buff in imm), tertial edges, and panel on secondaries. *Throat and underparts pale lemon,* palest on throat, with dusky wash across chest. Worn adults in autumn can appear whitish below with only slight lemon wash.

Voice. A sharp, slightly explosive *peeik!* or *spee-ip;* also a more leisurely *peeyp* or *pee-eep,* and less often a slurred, plaintive *tch-wee* suggesting Eastern Pewee but harder and perhaps faster.

Habs. Humid to semihumid evergreen and semideciduous forest, plantations, and edge. At low to upper levels, usually in shady understory, rarely in open situations except during migration.

SS: Cordilleran Flycatcher locally sympatric in Atlantic Slope foothills N of Isthmus, typically shows more strongly teardrop-shaped eyering, wings and wingbars less contrasting, upperparts more brownish olive; note voice. Yellowish Flycatcher (highlands S of Isthmus) rarely sympatric, brighter overall with brighter eyering, less contrasting wings, and ochraceous-buff wingbars; note voice. Acadian Flycatcher (transient) has whitish throat, wingbars usually washed buff; note longer primary projection.

SD: C to F transient and winter visitor (Aug–May; SL–1200 m), on both slopes from S Tamps (U in winter) and Oax, and locally in interior from Chis, to El Salvador and Honduras, U to R in Yuc Pen (SNGH, PP); C to F transient (Aug–Sep, Apr–May: to 3000 m) on Atlantic Slope throughout. Vagrant to Nay (Dec 1982, GHR *et al.,* tape).

RA: Breeds E N.A.; winters E Mexico to Panama.

ACADIAN FLYCATCHER
Empidonax virescens   Not illustrated
Mosquero Verdoso

ID: 5.2–5.7 in (13–14.5 cm). *Transient in E and S. Bill relatively long and broad, primary projection relatively long.* **Descr.** Bill blackish above, orange-flesh below (often with slight dusky tip), legs grey. *Head and upperparts olive with pale lemon eyering. Wings* and tail *contrastingly darker* (blackish brown) with *pale buff to lemon-buff wingbars* (may be pale lemon when worn), and pale buff to lemon tertial edges and

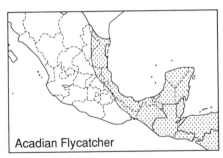

Acadian Flycatcher

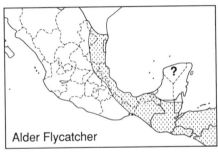

Alder Flycatcher

panel on secondaries. *Throat whitish* (rarely tinged lemon in autumn), underparts whitish with dusky chest, belly and flanks usually washed pale lemon.

Voice. A sharp, slightly explosive *sweeit!* or *psee'ip*, much like Yellow-bellied Flycatcher.

Habs. Similar to Yellow-bellied Flycatcher.

SS: See Yellow-bellied Flycatcher. Willow and Alder flycatchers have indistinct eyerings, less contrasting wings with dull wingbars, browner upperparts, often in relatively open habitats; note calls. Least Flycatcher has smaller bill, upperparts more brownish olive, wingbars and eyering often bright whitish; note shorter primary projection and call.

SD: F to C transient (mid-Mar–May; SL–2500 m) on Atlantic Slope from Tamps, and in interior from Isthmus, to Honduras, and (Aug–mid-Oct) from Yuc Pen E. Winter reports from Mexico and Belize (e.g. Rogers *et al.* 1986; Whitney and Kaufmann 1986) are probably in error.

RA: Breeds E N.A.; winters E Nicaragua to Ecuador.

ALDER FLYCATCHER
Empidonax alnorum   Not illustrated
Mosquero Ailero

ID: 5.3–6 in (13–15 cm). *Transient in E and S. Bill relatively long and broad.* **Descr.** Bill blackish above, orange-flesh below, legs blackish. *Head and upperparts brownish olive to olive with indistinct narrow pale eyering and pale lores*; wings and tail darker with buff to pale lemon (worn) wingbars (broader and brighter buff in imm) and pale lemon tertial edgings and panel on secondaries. Throat whitish, chest washed dusky, belly pale lemon.

Voice. A sharp *peek* or *piic*, recalling Hammond's Flycatcher.

Habs. Humid to semiarid scrubby fields with hedges, fences, woodland and edge, plantations, etc. At low to mid-levels, often in open situations.

SS: Willow Flycatcher safely told only by voice, often distinctly browner above than Alder. See Acadian Flycatcher. White-throated Flycatcher browner above, wingbars buffy cinnamon, flanks washed ochraceous; note voice. Other *Empidonax* have smaller bills, usually show distinct wingbars and eyerings. See pewees under Western Pewee SS.

SD: C to F transient (Apr–early Jun, Aug–Sep; SL–2500 m) on Atlantic Slope from Tamps and Ver, and on Pacific Slope and in interior from Isthmus (SNGH) to El Salvador and Honduras (SNGH); no verified records from Yuc Pen (?). More data needed to clarify status and timing of migration.

RA: Breeds N N.A.; winters S.A.

NB: Formerly considered conspecific with Willow Flycatcher.

WILLOW FLYCATCHER
*Empidonax traillii**   Not illustrated
Mosquero Saucero

ID: 5–5.8 in (12.5–14.5 cm). *Winters on Pacific Slope, transient in E. Bill relatively long and broad.* **Descr. Adult:** bill blackish above, orange-flesh below, legs blackish.

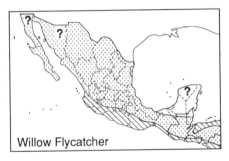

Willow Flycatcher

Head and upperparts brownish to brownish olive with indistinct narrow pale eyering and pale lores; wings and tail darker with buff to pale lemon (worn) wingbars and pale lemon tertial edgings and panel on secondaries. Throat whitish, chest washed dusky, belly pale lemon. **Juv:** wingbars broader and brighter buff.

Voice. A mellow *whit* or *whuit* and a slightly sharper *whik*, much like Least Flycatcher. Dawn song a clipped, nasal, buzzy *spit'chew*.

Habs. Much as Alder Flycatcher. Nest cup of grasses, fibers, etc., placed low in bush or small tree. Eggs: 2–4, whitish, usually speckled with reddish browns (WFVZ).

SS: See Alder and Acadian flycatchers. White-throated Flycatcher warmer brown above, wingbars buffy cinnamon, flanks washed ochre; note voice. Other *Empidonax* have smaller bills, usually show distinct wingbars and eyerings. See pewees under Western Pewee SS.

SD: R breeder (May–Aug, at least formerly) in BCN and N Son (Unitt 1987; SNGH). F to C transient and winter visitor (Sep–May; SL–1500 m) on Pacific Slope from Nay (SNGH, PP) to El Salvador and Honduras, U in interior from Balsas drainage S (SNGH, PP); U to F in winter on Atlantic Slope from Honduras S, possibly from Ver (AOU 1983). F to C transient (Apr–early Jun, Aug–Oct; SL–2500 m) on Pacific Slope; U to F in interior and on Atlantic Slope but no verified records from Yuc Pen (?).

RA: Breeds N.A.; winters W Mexico to Panama.

NB: See Alder Flycatcher.

WHITE-THROATED FLYCATCHER
Empidonax albigularis*   Plate 40
Mosquero Gorjiblanco

ID: 5–5.5 in (12.5–14 cm). *Breeds in highlands, winters mainly in marshes. Bill relatively long and broad.* **Descr. Adult:** bill blackish above, orange-flesh below, legs blackish. *Head and upperparts brown* to grey-brown with indistinct narrow pale eyering and *pale lores, rump and uppertail coverts brighter tawny-brown.* Wings and tail darker with buffy-cinnamon to whitish (worn) wingbars and pale cinnamon to pale lemon tertial edgings and panel on secondaries. *Whitish from throat often ex-*

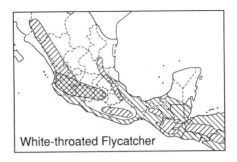

White-throated Flycatcher

tends back under auriculars, chest dusky, belly and undertail coverts pale lemon, becoming *ochraceous on flanks.* **Juv:** wingbars broader and brighter cinnamon.

Voice. A rough, burry to nasal *neeark* or *rreeah*, quite different from other *Empidonax*; song (?) a burry *bree-uh*, repeated.

Habs. Breeds in open and semiopen areas with hedges, shrubby growth, etc., usually near water, in damp meadows, along streams, irrigation ditches, etc. Winters mainly in marshes, especially with tall rushes, reeds, and scrubby edges. Habits much as Alder Flycatcher but can be hard to see in dense rushes or reedbeds. Nest cup of grasses, fibers, etc., placed low in bush or small tree. Eggs: 2–3, creamy, spotted with reddish browns (WFVZ).

SS: See Willow and Alder flycatchers. See pewees under Western Pewee.

SD: F to U but local breeder (Apr–Sep; 1200–3000 m) on Pacific Slope and in interior from SW Chih to N cen Nicaragua, on Atlantic Slope from S Tamps to Honduras. F to C but local transient and winter visitor (Aug–early May; SL–1800 m) on Pacific Slope from Nay (SNGH, PP) to El Salvador; U to F on Atlantic Slope from cen Ver through base of Yuc Pen (SNGH, PP) to N Honduras; U and local in interior from Lerma drainage S.

RA: Mexico to W Panama.

LEAST FLYCATCHER
Empidonax minimus*   Not illustrated
Mosquero Mínimo

ID: 4.8–5.2 in (12–13 cm). *Widespread winter visitor in lowlands. Bill relatively broad.* **Descr.** Bill blackish above, orange-flesh to flesh below and often tipped dusky, legs blackish. Head and upperparts brownish olive to greyish olive, with *bold whitish eyering.* Wings and tail darker with whitish to pale fawn wingbars and whitish to pale lemon tertial edges and panel on secondaries. Throat whitish, chest dusky, belly and undertail coverts pale lemon. May appear whitish below with dark chest side patches in worn plumage (mainly Sep–Dec).

Voice. A mellow to fairly sharp, liquid *whit* or *schwic*, much like Willow or Dusky flycatchers but often wetter and more insistent.

Habs. Humid to semiarid woodland and edge, open and semiopen weedy fields and scrub with hedges, fences, rarely in forest although, at least in Yuc Pen where essentially the only wintering *Empidonax*, sometimes in open forest canopy. Usually at low to mid-levels, often fairly active; holds tail slightly below body plane; after tail flicks, tail wobbles slightly, unlike most *Empidonax* where tail usually stops after a discrete flick.

SS: Dusky Flycatcher mainly in arid scrub of interior; bill slightly longer and relatively narrow, tail longer, often shows contrasting pale lores but eyering narrow; typically sits more vertically than Least, tail held in same plane as body, gives discrete high tail jerks. Hammond's Flycatcher mainly in highland forests; bill small and narrow, eyering most distinct behind eye in slight teardrop, tail held in same plane as body or often slightly cocked, relatively long primary projection accentuates relatively short tail; note voice. See Acadian Flycatcher.

SD: C to F transient and winter visitor (Aug–May; SL–2500 m, most below 1500 m) on both slopes from Sin and E NL (SNGH), and in interior from Balsas drainage, to El Salvador and Honduras; R N to S Son and Chih (SNGH). U to F transient (mid-Jul–Oct, Apr–May; to 3000 m) in interior almost throughout region; C on Atlantic

Slope. Vagrant (Oct) to BCN (Wilbur 1987; Howell and Pyle 1993).

RA: Breeds N.A.; winters Mexico to Costa Rica.

HAMMOND'S FLYCATCHER
Empidonax hammondii   Not illustrated
Mosquero de Hammond

ID: 5–5.5 in (12.5–14 cm). *Winters in highland forests. Bill relatively small and narrow; relatively long primary projection accentuates shortish tail.* **Descr.** Bill blackish above, orange-flesh to dusky below, legs blackish. Head and upperparts bright olive to brownish grey (worn) with whitish eyering broadest behind eye where usually forms slight teardrop. Wings and tail darker with whitish to pale fawn wingbars and whitish to pale lemon tertial edges and panel on secondaries. Throat pale grey to whitish, chest dusky, belly pale lemon (fairly bright lemon in fresh basic).

Voice. A high, sharp *peek* or *pic*, quite different from Dusky, Pine, or Least flycatchers.

Habs. Humid to semiarid pine, pine–oak, and evergreen forest and edge. Usually at mid- to upper levels, forages actively in canopy. Stance relatively horizontal with tail in same plane as body or slightly cocked, often flicks tail nervously several times in succession.

SS: See Least Flycatcher. Pine Flycatcher has longer bill and longer tail but actions and jizz often much like Hammond's, lower mandible of Pine orange-flesh; note voice. Dusky Flycatcher, usually in more open areas and forest edge, has longer bill, shorter primary projection accentuates longer tail, sits more vertically and usually gives discrete high tail jerks; note more prominent pale lores, less distinct teardrop eyering, and voice.

SD: C to F transient and winter visitor (Aug–

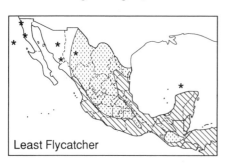

Least Flycatcher

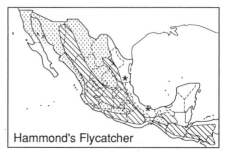

Hammond's Flycatcher

May; 1000–3500 m, locally to 300 m in W Mexico) in interior and on adjacent slopes from Son and S Coah to El Salvador and N cen Nicaragua. C to F transient (Aug–Oct, Apr–May) in NW Mexico but U in BCN.
RA: Breeds W N.A.; winters Mexico to W Nicaragua.

DUSKY FLYCATCHER
Empidonax oberholseri   Not illustrated
Mosquero Oscuro

ID: 5.2–5.7 in (13–14.5 cm). *Winter visitor to arid interior and highlands N of Isthmus. Bill relatively long and narrow. Relatively short primary projection accentuates longish tail.* **Descr.** Ages similar. Bill blackish above, flesh to dusky below, legs blackish. Head and upperparts brownish grey to greyish olive with *relatively contrasting pale lores and narrow whitish eyering.* Wings and tail darker with whitish to pale fawn wingbars and whitish to pale lemon tertial edges and panel on secondaries. Throat pale grey to whitish, chest dusky, belly pale lemon. Juv similar with pale buffy-cinnamon wingbars and tertial edgings.
Voice. A mellow *whuit* or *whit*, much like Grey, Least, and Willow Flycatchers, also a sharp *s-pik* or *s-pit* given in summer. Song of bright and burry notes which may vary in sequence, *si-pit p-rieh p'see*, etc.
Habs. Arid to semiarid scrub, semiopen areas and clearings with scattered trees and scrub, forest edge. Nests in aspen-conifer woodland. At low to mid-levels, usually sits fairly vertically, tail in same plane as body and gives single high tail jerks. Nest cup of plant fibers, grasses, etc., usually low in deciduous tree or bush. Eggs: 2–4, white, rarely with brownish flecks (WFVZ).
SS: See Least and Hammond's flycatchers.

Pine Flycatcher usually in pine–oak forest and clearings, primary projection longer, jizz including nervous tail flicks more like Hammond's Flycatcher, eyering brighter and teardrop-shaped behind eye, bill orange-pink below; note voice. Grey Flycatcher prefers more open desert scrub, dips tail down like a phoebe, bill longer and typically orange-flesh below with dark tip, outer webs of outer rectrices brighter whitish, call very like Dusky.
SD: Apparently U and local breeder (May–Aug?; 2700 m) in Sierra San Pedro Martír, BCN (Howell and Webb 1992*f*). C to F transient and winter visitor (Aug–May; 1000–3000 m) in interior from cen Mexico to Oax; U to F in winter in interior and locally on both slopes N to Son and NL, possibly R to NW Guatemala (AOU 1983). F to C transient (Apr–May, Aug–Sep; near SL–3000 m) in N Mexico, apparently R in BCN.
RA: Breeds W N.A.; winters Mexico.
NB: Formerly known as Wright's Flycatcher (*E. wrightii*); much confusion still surrounds many records of Dusky and Grey flycatchers prior to 1957.

GREY FLYCATCHER
Empidonax wrightii   Not illustrated
Mosquero Gris

ID: 5.5–6 in (14–15 cm). *Winters in N deserts. Bill relatively long and narrow.* **Descr.** Head and upperparts greyish olive to brownish grey (worn) with narrow whitish eyering, pale lores. Wings and tail darker with whitish to pale fawn wingbars and whitish to pale lemon tertial edges and panel on secondaries; outer webs of outermost rectrices contrastingly whitish, more so than other *Empidonax*. Throat and underparts pale grey, chest duskier, belly washed lemon.

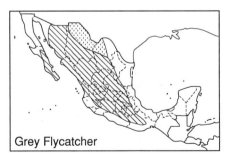

Dusky Flycatcher   Grey Flycatcher

Voice. A fairly sharp to slightly liquid *whit* or *pwit*, much like Dusky Flycatcher.

Habs. Arid open and semiopen areas with scrub and scattered trees. Unlike other *Empidonax*, does not flick tail up but dips it down like a phoebe. At low to mid-levels, often fairly conspicuous on tops of bushes.

SS: See Dusky Flycatcher.

SD: F to C transient and winter visitor (Aug–Apr; near SL–2500 m) in BCS, on Pacific Slope from Son to Nay, and in interior from Chih and W Tamps to Mich and Mor, U to R to cen Oax (RAB, SNGH). U to F transient (Apr–early May, Aug–Sep) on Pacific Slope and in interior of N Mexico.

RA: Breeds W N.A.; winters N and cen Mexico.

PINE FLYCATCHER
Empidonax affinis* Plate 40
Mosquero Pinero

ID: 5.2–5.7 in (13–14.5 cm). *Endemic; highlands S to Guatemala. Bill relatively narrow, primary projection and tail relatively long.* **Descr. Adult:** bill blackish above, pale flesh-orange below (rarely with fine dusky tip), legs blackish. Head and upperparts olive to greyish olive (greyest in *pulverius* of NW Mexico), with bright whitish eyering broader behind eye where usually it forms a teardrop. Wings and tail darker with whitish to pale fawn wingbars and pale lemon tertial edges and panel on secondaries. Throat and underparts pale lemon to greyish lemon, belly and sometimes throat paler (rarely whitish in fresh *pulverius*) with dusky to dusky-olive wash on chest. **Juv:** wingbars buff.

Voice. A mellow, fairly sharp *whip* or *pwip* or *whiup*, distinct from other *Empidonax* 'whit' calls. Song in N and cen

Mexico typically 2–4 hesitant phrases, often including nasal House Finch-like notes and a slightly stammering phrase suggesting Pileated Flycatcher, *prrp, p-wiet, chit p-p-p-reer'* or *p-rip, p-rip p'rr-ree*; in Chis a leisurely, irregular series of plaintive, burry, and clipped notes, *chri-k whee-u', chik-wheer,* or *cheenk, cheenk t-weeree,* etc. Also (N Mexico) a rapid, excited, slightly liquid trill in alarm, *dri-i-i-rr,* and a *chu-wik* call.

Habs. Semiarid to humid pine–oak woodland and edge, clearings, semiopen areas with scrub and scattered trees in pine–oak forest. Low to high, habits much as Hammond's Flycatcher. Nest cup in fork at mid-levels of trees. Eggs undescribed (Kiff and Hough 1985).

SS: See Hammond's and Dusky flycatchers. Cordilleran (and Pacific Slope) Flycatcher chunkier with shorter primary projection and tail, broader bill, wings and wingbars less contrasting, thicker eyering pale lemon with strong teardrop shape; note voice.

SD/RA: C to F but often local resident (1600–3500 m) in interior and on adjacent slopes from Chih and S Coah to Guatemala. Much confusion has surrounded this species' SD due to differing subspecific determinations of Guatemala winter specimens. This species is common, however, in the N of its range in winter (SNGH, PP, SW) and territorial birds have been noted in Chis in late May (SNGH). We consider this species resident, though local wandering may be expected in autumn and winter.

NB: The apparent song differences N and S of the Isthmus suggest that more than one species may be involved.

PACIFIC SLOPE [WESTERN] FLYCATCHER
Empidonax difficilis* Plate 40
Mosquero Occidental

ID: 5.2–5.5 in (13–14 cm). *Breeds in Baja, winters Pacific Slope. Bill relatively broad.* **Descr.** Bill blackish above, orange-flesh below, legs grey. *Head and upperparts olive (greyer when worn) with teardrop-shaped pale lemon eyering.* Wings and tail brownish with lemon-buff (to whitish when worn in summer) wingbars, lemon-buff to pale-lemon tertial edges and panel on secondaries. Throat and underparts lemon to pale

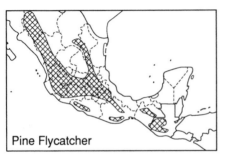

Pine Flycatcher

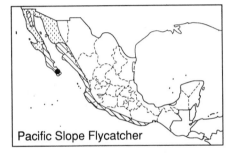

Pacific Slope Flycatcher

lemon, washed dusky to brownish on chest. **Juv:** upperparts washed brownish, wingbars brighter (cinnamon-buff in N migrant imms).

Voice. A high, thin, insect-like *si* or *tsi*, and (♂ only) a high, rising *pseep* or *wseep*, rarely strongly disyllabic. Dawn song *si ti-swee pi-tik* or *si p-seep p-tup*, the 1st part soft and often hard to hear, the 2nd (much like ♂ call) strong and usually attracting attention, and the 3rd clipped and typically rising. The 2nd note may be repeated and varied several times, *whi see-it si-it si-ip p-tik*, etc.

Habs. Breeds in semiarid pine–oak forest, often in shady arroyos. Winters in forest and woodland. At low to upper levels, usually in shady understory, rarely in open situations except during migration. Nest cup of moss and fibers at low to mid-levels, typically placed against a flat surface such as a bank, cliff, trunk, etc., or may be saddled in vertical fork; often near water. Eggs: 3–4, whitish, speckled with reddish browns (WFVZ).

SS: Cordilleran Flycatcher rarely identified safely in field but ♂ call more strongly disyllabic *wi-seet* or *w-seen*. See Yellow-bellied and Pine flycatchers (rarely sympatric).

SD: C to F breeder (Mar–Sep; 1000–3500 m) in Baja. C to F transient and winter visitor (SL–1500 m; Aug–Apr) in BCS and on Pacific Slope from S Son to Oax. Winter distributions of Pacific Slope and Cordilleran flycatchers not well understood. Reports of Western Flycatcher from Guatemala (such as Vannini 1989*b*) are erroneous.

RA: Breeds W N.A. to Baja; winters W Mexico.

NB: Yellowish Flycatcher has been considered conspecific with the *E. difficilis* complex.

CORDILLERAN [WESTERN] FLYCATCHER

*Empidonax (difficilis?) occidentalis** Plate 40
Mosquero Barranqueño

ID: 5.2–5.5 in (13–14 cm). *Highlands N of Isthmus.* **Descr.** Bill blackish above, orange-flesh below, legs grey. *Head and upperparts olive to brownish olive with teardrop-shaped pale lemon eyering.* Wings and tail dark brownish with ochraceous-buff wingbars (*occidentalis,* most of Mexico) to lemon-buff wingbars (*hellmayri* of N Mexico), lemon-buff to pale-lemon tertial edges and panel on secondaries. Throat and underparts lemon to pale lemon, washed dusky to brownish on chest. **Juv:** upperparts washed brownish, wingbars brighter, buffy.

Voice. A high, thin, insect-like *si* and *ssii*, and (♂ only) a high, bright, rising, disyllabic *wi-seet* or *w-seen*. Dawn song repeated over and over, a fairly strong *sii swee'in pi-luk* or *si wee-ee pi-dik*, analogous to Pacific Slope song with 2nd part much like ♂ call, 3rd part typically dropping; 2nd note may be repeated and varied several times, as in Pacific Slope Flycatcher.

Habs. Breeds in humid to semiarid evergreen and pine-oak forest, often in shady arroyos; most leave conifer zones in winter. Winters in forest and woodland. Habits, nest, and eggs (WFVZ) as Pacific Slope Flycatcher.

SS: See Pacific Slope, Yellow-bellied, and Pine flycatchers. Yellowish Flycatcher allopatric.

SD: C to F breeder (Mar–Sep; 1000–3500 m) in interior and on adjacent slopes from Son and Coah to Oax. Most move to foothills (and Pacific lowlands?) in winter; on Atlantic Slope winters (600–1500 m) from S NL to Oax. Winter ranges of

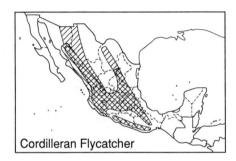

Cordilleran Flycatcher

Cordilleran and Pacific Slope flycatchers not well known.

RA: Breeds W N.A. to Mexico; winters Mexico.

NB: See NB under Pacific Slope Flycatcher.

YELLOWISH FLYCATCHER
Empidonax flavescens* Plate 40
Mosquero Amarillento

ID: 5.2–5.5 in (13–14 cm). *Humid montane forest S of Isthmus. Bill relatively broad.* **Descr. Adult:** bill blackish above, orange-flesh below, legs grey. *Head and upperparts olive-green with contrasting pale lemon teardrop-shaped eyering.* Wings and tail darker, brownish, with ochraceous-buff wingbars, ochraceous-buff to ochraceous-lemon tertial edges and panel on secondaries. Throat and underparts lemon-yellow, palest on throat, washed olive to ochraceous olive on chest. **Juv:** browner above, paler below, with cinnamon wingbars.

> **Voice.** A high *w-seen* or *tsee'n*, and *see-ip'*, and a quiet, high *ssi*, all much like Cordilleran Flycatcher. Dawn song repeated over and over, a high thin, slightly clear *si-iin ch-ik w-seein*, or often simply *si-iin w-seein*, repeated such that sequence is hard to determine, the low, clipped *chik* note omitted, the last note much like ♂ call; quality suggests Cordilleran Flycatcher (NKJ tape, N cen Nicaragua).
> **Habs.** Humid evergreen to pine–evergreen forest. At mid- to upper levels in forest where can be hard to find unless calling. Nest and eggs (2–3) much like Pacific Slope Flycatcher (WFVZ).

SS: See Yellow-bellied Flycatcher.

SD: C to F resident (900–3000 m) in S Ver (Los Tuxtlas, where U down to 100 m in winter), on Pacific Slope from E Oax, and in interior and on Atlantic Slope from Chis, to El Salvador and N cen Nicaragua.

RA: S Mexico to W Panama.

NB: See NB under Pacific Slope Flycatcher.

BUFF-BREASTED FLYCATCHER
Empidonax fulvifrons* Plate 40
Mosquero Pechicanelo

ID: 4.5–4.7 in (11.5–12 cm). *Highlands almost throughout. A petit and unmistakable* Empidonax; plumage varies markedly with wear and ssp. **Descr. Adult:** bill blackish above, orange-flesh below, legs blackish. *Face, sides of neck, throat, and underparts cinnamon* to pale buff (worn) with whitish eyering; throat usually paler. Crown, nape, and upperparts grey-brown to brown (crown contrastingly darker S of Isthmus). Wings darker with pale cinnamon to whitish (worn) wingbars and whitish to pale lemon tertial edges and panel on secondaries; outer webs of outer rectrices whitish. **Juv:** wingbars broader and brighter cinnamon.

> **Voice.** A sharp to mellow *pic* or *whic*, unlike more leisurely *whit* of Dusky Flycatcher. Song a jerky, hesitant series of short, sharp phrases, *p-teek!, pit p-teek! pi-tik, peek, pi-chu' . . .*, or shorter phrases repeated, *sipit siu, sipit piu*, etc.
> **Habs.** Grassy and scrubby clearings, fields, pastures with hedges, in pine and oak woodland. Fairly active and conspicuous at low to mid-levels, at times high in pines, often perches atop low bushes. Nest cup of fine leaves, fibers, etc., saddled in a sloping fork at mid-levels in tree. Eggs: 2–4, whitish, unmarked (WFVZ).

SD/RA: F to C breeder (mid-Mar–mid-Sep; 1000–3500 m) in interior and on adjacent slopes from Son (locally in SW USA) and S Coah to cen volcanic belt; winters (Sep–Apr; near SL–2500 m, mostly above 600 m) from S Son and SLP to cen Mexico, R to Oax. U to F resident in Sierra Madre del

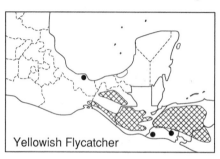

Yellowish Flycatcher

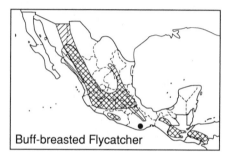

Buff-breasted Flycatcher

Sur of Gro and Oax, and in N Chis; F to C resident (600–2600 m) in interior from Guatemala to cen Honduras; recorded Nov and Mar in N El Salvador.

GENUS SAYORNIS

Phoebes are conspicuous, medium-sized flycatchers of open and semiopen areas. Bills relatively narrow. Ages differ slightly, sexes similar; quickly attain adult-like plumage though juv wing coverts may be retained for a few months. Bills and legs blackish. Phoebes perch at low to mid-levels on bushes and trees, often along fences and hedges, and frequently dip their tails. Nest cups of grasses, bound with mud in Black Phoebe, placed under bridges, against rock faces, in crevices. Eggs: 3–7, white, may be speckled brown (WFVZ).

BLACK PHOEBE

Sayornis nigricans* Figure 33
Mosquero Negro

ID: 6–7 in (15–17.5 cm). Widespread near water. **Descr. Adult:** *black overall with* paler wing edgings, *white belly and under-tail coverts; S. n. aquatica* (S of Isthmus) has white restricted to median belly and vent. **Juv:** pale cinnamon wing edgings and fringing on upperparts.
  Voice. A sharp *peek!* or *seek*. Song a bright *pidl-eee'* or *pi-ts-lee*, often repeated over and over.
  Habs. Open and semiopen areas near water, typically along streams, beside pools and lakes. See genus note.
SD: C to F resident (SL–3000 m) in interior and on adjacent slopes throughout region except Yuc Pen. F to U transient and winter visitor (Sep–Mar) on Pacific Slope S to Nay.
RA: W USA to N Argentina.

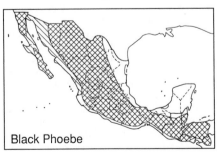
Black Phoebe

EASTERN PHOEBE

Sayornis phoebe Not illustrated
Mosquero Fibí

ID: 6.3–6.8 in (16–17 cm). Winters in N Mexico. **Descr.** *Head and upperparts dark greyish olive, head contrastingly darker;* wings darker with paler edgings and whitish panel on secondaries. *Throat and under-parts whitish with dusky sides of chest, belly and undertail coverts washed lemon.*
  Voice. Call a sharp *chik* or *piik*, softer than Black Phoebe.
  Habs. Open and semiopen areas with hedges, fences, scattered trees, often near water. See genus note.
SS: Contrasting dark head, tail-dipping habit, and lack of pale eyering and wingbars distinctive.
SD: C to F transient and winter visitor (late Sep–Apr; near SL–2500 cm) on Atlantic Slope and in interior from Coah and Tamps to cen Ver, U to F W to Chih and Jal, S to S Ver and Oax, R to Yuc Pen. Vagrant (Nov–Mar) to NW Mexico. Reports from Belize (Wood *et al.* 1986) and Arrecife Alacrán (Boswall 1978*b*) require verification.
RA: Breeds E N.A.; winters S USA to cen Mexico.

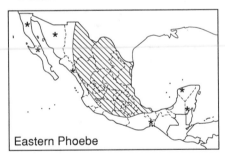
Eastern Phoebe

SAY'S PHOEBE

Sayornis saya* Figure 33
Mosquero Llanero

ID: 7–7.5 in (17.5–19 cm). *Dry country N of Isthmus.* **Descr. Adult:** head and upper-parts grey-brown, darker on head; wings darker with paler edgings. *Tail contrast-ingly blackish.* Throat and chest greyish, *belly and undertail coverts cinnamon.* **Juv:** wingbars broader and brighter cinnamon.
  Voice. Call a plaintive, often slurred *peeu* or *p-eeu* and *pu-weet*. Dawn song a plain-tive *p'dew* or *pi'di-hew*, repeated rapidly

Fig. 33 (a) Vermilion Flycatcher; (b) Black Phoebe; (c) Say's Phoebe

over and over, interspersed with rolled, burry *pi-rrep* or *prriip* and *pi-di-reep* phrases, may be given in flight.

Habs. Arid to semiarid open country with scrub, fences, scattered trees. See genus note.

SS: ♀ Vermilion Flycatcher usually has pale supercilium, throat and chest whitish, streaked dusky, lacks contrasting black tail and cinnamon belly. Cassin's and Western kingbirds can appear similar in bright light, but both have dark mask, yellow belly, olive upperparts, relatively heavier bill; Cassin's has contrasting whitish throat, Western has white sides of tail.

SD: C to F breeder (Mar–Sep; near SL–2500

m) in interior from NE Son and Coah to cen Mexico, possibly to NW Oax (Binford 1989); U to F in BCN. U to F in winter (Nov–Mar) on N Plateau; F to C transient and winter visitor (Oct–Mar) in Baja, on Pacific Slope in Son, and in interior from cen Mexico to Oax, R to Chis (Feb 1985, SNGH, SW); U on Atlantic Slope to Tamps, R to Ver.

RA: Breeds W N.A. to Mexico; winters SW USA to Mexico.

VERMILION FLYCATCHER
Pyrocephalus rubinus*   Figure 33
Mosquero Cardenal

ID: 5–5.7 in (12.5–14.5 cm). Widespread in open country. *Phoebe-like structure and habits.* **Descr.** Ages/sexes differ. ♂: bill and legs blackish. *Head and underparts bright red with blackish-brown mask.* Upperparts blackish brown, wings with narrow paler edgings. ♀: head and upperparts grey-brown with paler, often whitish, supercilium. Wings and tail darker with pale wing edgings. *Throat and underparts whitish, chest streaked dusky, lower belly and undertail coverts washed pink* (pale

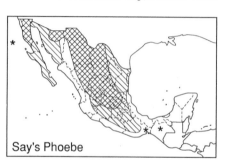

Say's Phoebe

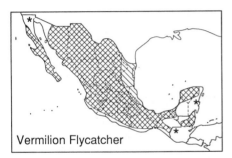

Vermilion Flycatcher

yellowish in imm). **Imm ♂:** resembles ♀ but crown and underparts with variable red mottling, may be mostly red in 1st summer. Attains adult plumage when about 1 year old. **Juv:** resembles ♀ but wing edgings broader and brighter, upperparts appear mottled and streaked, lower belly and undertail coverts whitish.

Voice. Call a sharp, thin *pseeup* or *pseep*. Song (may be given at night) a rapid, accelerating, bubbling trill, *pi pi-li-li-li-zing* or *ti ti-li-li-li-liu*, repeated rapidly in display flight.

Habs. Open and semiopen areas with hedges, scattered trees and bushes, often near water. Perches at low to mid-levels, dips tail frequently like phoebes. In display flight, ♂ climbs with exaggerated slow 'butterflying' wingbeats, then drops back to perch. Nest is a shallow cup of fine twigs, grasses, fibers, placed low to high in horizontal fork of tree or bush. Eggs: 2–3, whitish, marked with dark browns and greys (WFVZ).

SS: See Say's Phoebe.

SD: F to C but often local resident (SL–2500 m) on Pacific Slope S to Nay, in interior and on adjacent slopes S to Oax, in interior Chis, on Atlantic Slope from S Ver to W Camp, disjunctly in N Yuc Pen and pine savannas from Petén to the Mosquitia; U to F and local in cen and S Baja. N populations migratory: U to F (Oct–Mar) on N Plateau, F to C on Pacific Slope S to Oax, on Atlantic Slope in Tamps and Ver, R in interior to Guatemala.

RA: SW USA to Nicaragua, locally in S.A. and the Galapagos.

BRIGHT-RUMPED ATTILA
*Attila spadiceus**   Plate 41
Atila Rabadilla-brillante

ID: 7.8–8.5 in (19.5–21.5 cm). Tropical forest. Staring face, *stout hooked bill.*

Descr. Ages differ slightly, sexes similar. **Adult:** *eyes red to yellow-orange, bill flesh with dark subterminal band,* or culmen and rarely whole bill dark, legs blue-grey. *Head* brownish, *streaked blackish and* often washed grey or olive, *with short pale grey supercilium and dark eyestripe.* Upperparts warm dark brown to tawny-brown with *contrasting tawny to golden-yellow rump and uppertail coverts;* wings dark with dull cinnamon to olive-brown edgings. Tail brown (E Mexico and C.A.) to cinnamon (*mexicanus* of W Mexico). *Throat and underparts* whitish, *streaked dusky* on throat and chest, usually down to upper belly in E Mexico and C.A., *with variable tawny to lemon wash* on chest, flanks, and undertail coverts. **Juv:** eyes brown, wing edgings brighter.

Voice. Varied, loud, rich, far-carrying whistles, *wh whee-du whee-du whee-du whee-du wheeu,* or *whee-deet whee-deet whee-deet whee-deet whew,* etc., and *wh wee-ba wee-ba wee-ba wee-ba,* and a more hurried *dweet weet weedweedweed-weedweedwee-a',* etc. Also a loud, sharp *pi-dik'* and *ki-di-dik',* like an overgrown Stub-tailed Spadebill, quiet growls *prrr* or *pi-rrrr,* and soft, excited, and often prolonged series, *whi-di-di di-di deer di-di-deer . . .* interspersed with a sharp *whee'k* and a slightly plaintive *whee dee,* the 1st note rising.

Habs. Humid to semiarid forest and edge, plantations, scrubby woodland. Low to high, may join mixed-species flocks and attend army ants; at other times sits quietly for long periods and overlooked easily, hard to locate when singing. Often sits slightly hunched, wings drooped to show off rump, and flicks tail up. Nest is a bulky cup of rootlets, fine leaves, etc., low to high in niche amid epiphytes, in tree cavity (SNGH), bank, etc. Eggs (Costa

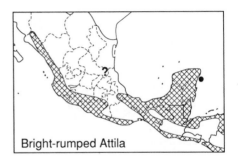

Bright-rumped Attila

Rica): 3–4, pinkish white to pale buff, marked with browns and greys (WFVZ).

SS: See Royal Flycatcher.

SD: C to F resident (SL–1600 m, U to 2100 m) on both slopes from S Son and Ver to El Salvador and Honduras, also in interior in Balsas drainage and Chis. Reports N to SLP (AOU 1983) require verification.

RA: Mexico to Ecuador and Brazil.

NB: Sometimes known as Polymorphic Attila.

RUFOUS MOURNER

Rhytipterna h. holerythra   Plate 43
Papamoscas Alazán

ID: 7.7–8.2 in (19.5–21 cm). *SE humid lowland forest.* **Descr.** Ages/sexes similar. Bill blackish, flesh below at base, legs greyish. Overall rufous-brown, paler below, palest on belly and undertail coverts.

　　Voice. A leisurely, mournful 'wolf-whistle' *wheeeu peeeu* or *whi heeeu*, rising then falling, also a mournful, descending *wheeeu* or *peeeu*. Song a steady, rich, slightly plaintive *whee-u' whee-u' whee-u'* ... to 15× or more, often with an introductory *whi-heeeu*; family groups may give quiet, reedy to plaintive calls, *wh'peeu* and *druk*, sometimes combined, *druk ruk pi-i-pee*, etc.

　　Habs. Humid evergreen forest. At mid- to upper levels, rarely low at edges. Sits quietly for long periods and overlooked easily. Nest apparently in tree cavities such as old woodpecker hole (Honduras, SNGH). Eggs undescribed (?).

SS: Rufous Piha larger but often hard to distinguish in the field; note stouter bill and paler throat, bushy lores and narrow pale orbital ring often give the Piha a facial expression recalling Squirrel Cuckoo; note voice. Speckled Mourner slightly heavier set with stouter bill, wing coverts mottled dark and rufous, chest indistinctly scal-

loped dusky, yellow flashes at sides of chest usually covered by wings; note voice.

SD: F to C resident (SL–750 m) on Atlantic Slope from N Oax to Honduras.

RA: SE Mexico to Ecuador.

GENUS MYIARCHUS

This is a remarkably uniform-looking genus of fairly large to large flycatchers, one or more of which occur in most of the region. This genus is a major identification problem, perhaps more so than the dreaded *Empidonax*. All are fairly slender and large-headed with peaked to slightly crested napes. Ages differ slightly, sexes similar. Juvs have extensive cinnamon-rufous edgings to wings and tails and often are not identifiable in the field (or hand!). Generally, they quickly undergo a complete molt into adult plumage. Bills blackish, often with some (usually slight) flesh below at base, legs blackish. At mid- to upper levels in trees, or at low to mid-levels in scrub, often in canopy where commonly at fruiting trees. Vocal mostly in early morning, often silent for long periods later in day and overlooked easily. Nest in tree cavities, hollow posts, old woodpecker holes, etc. Eggs: 3–6, creamy to pinkish buff, heavily scrawled with browns and greys (WFVZ).

　　Up to four species are sympatric in some areas and identifications are best based on diagnostic call notes (all species have bickering and chattering calls), with due attention paid to relative bill size (♂♂ have larger bills than ♀♀) and tail patterns if visible.

YUCATAN FLYCATCHER

Myiarchus yucatanensis*   Plate 41
Copetón Yucateco

ID: 7–7.5 in (17.5–19 cm). *Endemic to Yuc Pen.* Bill slightly heavier than Dusky-capped Flycatcher. **Descr.** Bill blackish, flesh below at base; mouth orange. **Adult:**

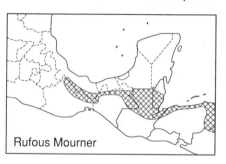

Rufous Mourner

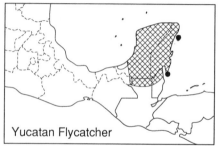

Yucatan Flycatcher

head brown with paler greyish lores and often a subtle but noticeable paler and greyer oval eyering creating distinct facial expression; upperparts greyish olive-brown. Wings and tail dark brown, *coverts edged paler grey-brown, tertials edged whitish, remises edged rufous, inner webs of inner rectrices edged cinnamon-rufous and not very striking.* Throat and chest pale greyish, belly and undertail coverts lemon. *M. y. lanyoni* of Isla Cozumel darker than mainland birds.

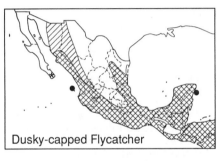

Dusky-capped Flycatcher

Voice. A plaintive, drawn-out, slurred *whee-eee-eu* or *hoooooo-eu*, rising then falling, also a shorter slurred *hooo weeu*, a long, chattering *whee bee-bee-bee ...*, and a drawn-out, plaintive whistle, often followed by a bright, upward-inflected whistle, *wheeeeur hweep*. Dawn song a fairly loud, clear *hoor-eeep* or *hoow'eeep*, repeated monotonously, 10/35–38 s, the second part up-slurred strongly.

Habs. Humid to semiarid forest, scrubby woodland and edge. See genus note. Eggs undescribed (?).

SS: Face pattern subtle but distinctive. Dusky-capped Flycatcher has contrastingly darker head with brownish lores, tertials edged pale cinnamon, nape often peaked, bill slightly more slender; note voice. Brown-crested Flycatcher larger with heavier bill, greyish face contrasts slightly with brown crown, wingbars brighter whitish, tail extensively rufous; note voice.

SD/RA: C to F resident (SL–250 m) in Yuc Pen (including Isla Cozumel), S to E Tab, N Guatemala, and N Belize.

DUSKY-CAPPED (OLIVACEOUS) FLYCATCHER

*Myiarchus tuberculifer**   Plate 41
Copetón Triste

ID: 6.3–7.3 in (16–18.5 cm). *Smallest and most widespread* Myiarchus. *Two groups: olivascens group (NW Mexico) and lawrencei group (rest of the region). Bill relatively slender.* **Descr. Olivascens. Adult:** bill blackish, flesh below at base; mouth orange. Head and upperparts greyish olive to grey-brown, head not contrastingly darker. Wings dark brown, coverts edged paler, grey-brown to rufous-brown, tertials edged whitish becoming lemon on secondaries, rufous on primaries. *Tail dark brown, rectrices finely edged ochre.* Throat and chest pale grey, duskier on chest, belly

and undertail coverts pale lemon. *Lawrencei*. **Adult:** head and upperparts brownish olive to olive, *head contrastingly darker brownish. Wing coverts edged dull cinnamon to grey-brown*, tertials edged pale cinnamon (rarely whitish), *flight feathers edged rufous.* Belly and undertail coverts lemon. Juv may migrate before molting.

Voice. A plaintive, drawn-out *wheeeeu* or *peeeu*, may be varied to a slightly burry *wheeer* or a shorter, raptor-like *reeeu*, repeated steadily; also a shorter note quickly followed by a rolled whistle, *whee peeerrr-rr* or *whee' wheerrrrrrr*, and a fairly sharp *ki-dee ew*, repeated 2–3×. Dawn song a varied repetition of short, often plaintive phrases, *wheeu reehr-peu* or *whee'bew, wheeeu* or *whit' whee'peu, whee-beu*, etc.

Habs. Forest and woodland (including pine–oak), semiopen areas with hedges, scrub, etc. See genus note.

SS: See Yucatan Flycatcher. Ash-throated and Nutting's flycatchers slightly larger with stouter bills, paler overall without darker head, wing coverts edged whitish, tails more extensively rufous; note voice.

SD: C to F resident (SL–2800 m) on both slopes from Son and E NL, and in interior from cen Mexico, to El Salvador and Nicaragua. Withdraws slightly from N Mexico (*olivascens* winters S to Isthmus) and higher and colder areas in winter (Oct–Mar?); also absent from Isla Cozumel in winter. Vagrant (Jun 1896) to BCS.

RA: Mexico (SW USA in summer) to Peru and N Argentina.

ASH-THROATED FLYCATCHER

*Myiarchus cinerascens**   Plate 41
Copetón Gorjiceniizo

ID: 7.5–8 in (19–20.5 cm). *N and W deserts, migratory. Bill moderately stout. Nape*

often looks peaked or crested. **Descr.**
Adult: bill black, rarely flesh below at
base; *mouth flesh.* Head and upperparts
greyish olive to grey-brown, crown often
slightly browner than greyer face which
may extend back as indistinct greyish hindcollar. Wings dark brown, coverts and
tertials edged whitish to pale fawn, *secondaries edged pale lemon* to pale fawn,
primaries edged rufous. Tail dark brown,
*inner webs of all but central rectrices mostly
cinnamon-rufous, dark on outer webs
usually spreads across top of inner web.*
Throat and chest pale grey, duskier on
chest, belly and undertail coverts pale
lemon; in fresh plumage often shows
whitish area between grey chest and lemon
belly. Juv may migrate before molting.
   Voice. A wet to slightly sharp *pip* or *puip*
or *pic,* also a rolled *prreeer* suggesting a
referee's whistle, and a nasal *ke-bek' ke-
bek,* etc. Dawn song a varied arrangement
of calls, *ha ke-wher prri ki-di-wheer ki-di-
wheer,* or *k'brik, k-br-di-deer,* etc.
   Habs. Arid to semiarid scrub, riparian
woods. See genus note.
SS: Nutting's Flycatcher hard to tell in the
field unless heard but often appears less
crested, smaller-billed, upperparts browner
with brown of crown suffusing down into
face (versus grey of throat coming up into
face on Ash-throated), secondaries edged
cinnamon (thus less contrast between
primary and secondary edgings) and wingbars and tertial edgings duller (especially in
S), dark on webs of outer rectrices rarely
expands across tip of inner web, mouth
orange. Brown-crested Flycatcher similar
in plumage, but larger with larger and
heavier bill (especially in W Mexico), tail
more extensively rufous; note voice. See
Dusky-capped Flycatcher.
SD: C to F breeder (Apr–Sep; SL–2500 m)
in Baja, on Pacific Slope in Son, and in

interior from Chih and NL to N Mich,
Gto, and SLP. C to F winter visitor (Aug–
May; SL–2000 m) in Baja (mostly BCS),
on Pacific Slope from Son, and in interior
from Balsas drainage, to Isthmus, thence
F to U, mainly in interior, to El Salvador
and Honduras; R in winter (Oct–Mar) on
Plateau; U to F on Atlantic Slope S to Ver.
Reports from Yuc Pen and N Guatemala
are in error (Lanyon 1965b).
RA: Breeds W USA and N Mexico; winters
SW USA and Mexico to Nicaragua, rarely
Costa Rica.

NUTTING'S FLYCATCHER
*Myiarchus nuttingi**   Plate 41
Copetón de Nutting

ID: 7–7.5 in (18–19 cm). *W and S thorn
forest, resident. Bill moderately stout.
Nape typically bushy but rarely crested.*
Descr. Adult: bill blackish, often flesh
below at base; *mouth orange.* Head and
upperparts brownish olive to grey-brown,
face may look slightly paler and greyer in
contrast to brownish-olive crown. Wings
dark brown, *coverts and tertials edged pale
fawn* to whitish (worn), *secondaries edged
cinnamon,* primaries edged rufous. Tail
dark brown, *inner webs of all but central
rectrices mostly cinnamon-rufous, dark on
outer webs rarely spreads across tip of
inner web.* Throat and chest pale grey,
duskier on chest, belly and undertail
coverts lemon to pale lemon (brightest in
flavidior of Pacific Slope S of Isthmus).
   Voice. A sharp *wheek!* or *wheep,* suggesting Great Crested Flycatcher but shorter,
more emphatic, also varied to *whee'u,* and
sometimes a sharp doubled *kwee-week!*
and *wh-ik whi-ik.* Dawn song a varied
series of 1–3 or more sharp calls followed
by a short, often accelerating, rolled chatter, *wheek! ki-di di-di-drr* or *wheek,*

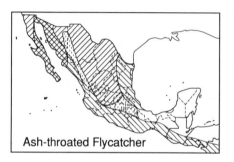

Ash-throated Flycatcher

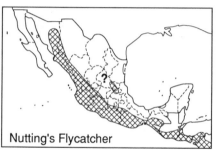

Nutting's Flycatcher

wheep, ki di-di-dir, etc., may be repeated monotonously.

Habs. Arid to semiarid scrubby woodland, thorn forest, semiopen areas with scrub and small trees. See genus note.

SS: See Ash-throated Flycatcher. Brown-crested Flycatcher larger, bill larger and heavier (especially in W Mexico), often looks large-headed with peaked or crested nape, tail more extensively and brighter rufous, mouth flesh; note voice. See Dusky-capped Flycatcher.

SD: C to F resident (SL–1800 m) on Pacific Slope from Son, and in interior from Jal, to El Salvador and Nicaragua; disjunctly on Atlantic Slope from S SLP to N Hgo.

RA: W Mexico to NW Costa Rica.

GREAT CRESTED FLYCATCHER
Myiarchus crinitus   Plate 41
Copetón Viajero

ID: 8–8.8 in (20.5–22 cm). *Transient and winter visitor in E and S. Bill stout and fairly heavy.* **Descr.** *Bill* blackish, *extensively flesh below at base*; mouth yellow-orange. *Face, throat, and chest grey, with little tonal contrast between face and olive crown. Lemon belly and undertail coverts contrast strongly with grey chest. Upperparts olive.* Wings dark brown, coverts and tertials edged whitish to pale lemon (duller and buffier in some autumn imms), secondaries edged pale lemon, primaries edged rufous. Tail dark brown, *inner webs of all but central rectrices mostly cinnamon-rufous.*

Voice. A rising, slightly drawn-out *wheep*, often repeated steadily, most like Nutting's Flycatcher but less emphatic. Dawn song 2 bright phrases, *wheerreep whee-uh ... or wheerrep wheeruh ...*, alternated and repeated.

Habs. Humid to semiarid forest and edge, wooded habitats in general during migration; nests in deciduous woodland, often along streams. See genus note.

SS: Large size and heavy bill, lack of contrast between olive crown and grey face and throat, strong contrast between grey chest and lemon belly, and extensively flesh lower mandible distinctive; note voice. Rarely in same habitat as Ash-throated and Nutting's flycatchers.

SD: F to C transient and winter visitor (Sep–Apr; SL–1000 m) on both slopes from S Ver and E Oax S, commoner on Atlantic Slope, F in interior Chis. F to C transient on Atlantic Slope of E Mexico. Probably breeds locally (May–Aug) in NE Coah (Urban 1959).

RA: Breeds E N.A.; winters S Mexico to N S.A.

BROWN-CRESTED FLYCATCHER
Myiarchus tyrannulus*   Plate 41
Copetón Tirano

ID: 8–9.2 in (20.5–23.5 cm). W > E. *Largest Myiarchus, nape usually peaked or crested. Bill long and heavy,* especially in *magister* of W Mexico. **Descr. Adult:** bill black, often flesh below at base; mouth yellowish flesh. *Pale grey face usually contrasts slightly with brownish crown, throat whitish, chest pale grey, belly and undertail coverts pale lemon. Upperparts greyish olive to grey-brown.* Wings dark brown, coverts and tertials edged whitish to pale fawn, secondaries edged pale lemon, primaries edged rufous. Tail dark brown, *inner webs of all but central rectrices mostly cinnamon-rufous, dark on outer webs usually spreads onto inner webs, especially in Mexican populations. Cozumelae* (Isla Cozumel) and *insularum* (Honduras Bay Islands) darker and browner above than mainland birds.

Voice. A sharp, slightly to strongly di-syllabic *hwuik'* or *h-whik'* or *ha-wik'*, sharply inflected upwards, and a longer

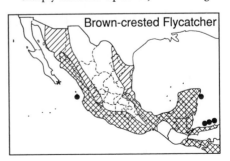

whi kir-ir-ik, etc. Dawn song a varied repetition of calls and burry rolled phrases, *wheeu', prri-di-whew* or *pi-d-whee pi-di-birr* or *whik' pri-di whee pi-di-we-eu*, etc.

Habs. Arid to semihumid forest and edge, scrubby woodland, humid to semihumid open areas with hedges, scattered trees, forest clearings. See genus note.

SS: See other *Myiarchus*.

SD: C to F resident (SL–2000 m) on Pacific Slope from Son, and in interior from Balsas drainage, to Isthmus, and on Pacific Slope and in adjacent interior (SL–1500 m) from El Salvador to Nicaragua. C to F breeder (Apr–Aug; SL–1000 m) on Atlantic Slope (including Isla Cozumel) from NE Coah and Tamps to Honduras; N populations migratory (mostly resident from Guatemala S), wintering mainly on Pacific Slope and in interior from Isthmus to El Salvador; U to R in winter (Oct–Mar) on Atlantic Slope N to SE SLP and Yuc Pen.

RA: Mexico (SW USA in summer) to NW Costa Rica; also Colombia to N Argentina.

FLAMMULATED FLYCATCHER
Deltarhynchus flammulatus   Plate 41
Copetón Piquiplano

ID: 6–6.5 in (15–16.5 cm). *Endemic to W Mexico. Resembles a chunky* Myiarchus, *nape bushy. Bill relatively broader than* Myiarchus. **Descr.** Ages differ slightly, sexes similar. **Adult:** *bill blackish, flesh below at base,* legs dark grey; mouth orange. Head and upperparts olive to grey-brown (worn), *whitish supraloral stripe and subocular crescent and dusky lores create facial expression recalling Greenish Elaenia.* Wings dark brown, *coverts and remiges edged pale cinnamon.* Tail dark brown, narrowly edged pale cinnamon.

Throat whitish, chest pale grey with dusky streaking (often inconspicuous), belly and undertail coverts pale lemon. **Juv:** tail broadly edged pale cinnamon.

Voice. Song (mostly Apr–Aug) a plaintive whistle followed by short, quick roll, *chuu wi'ti-li-liu*, less often by distinct notes, *tew, wee-dee-dee-dee.* Calls include a plaintive, slurred *teeuu* or *chew*, often 3–5× in descending series, and variations, *tew 'i-chu-chu*, or *tew 'i-chu*; also a squeaky, spluttering chatter.

Habs. Arid to semiarid thorn forest and scrubby woodland. Relatively sluggish, usually keeps well hidden (even when singing) and overlooked easily. Nests in tree cavities. Eggs: 3, identical to *Myiarchus* eggs in color and pattern (Lanyon 1982*a*).

SS: Distinctive but rarely seen. Differs from *Myiarchus* in more skulking habits, face pattern, relatively broader bill, extensively pale cinnamon (not rufous) wing and tail edgings, dusky streaking on chest, and voice.

SD/RA: F resident (SL–1400 m) on Pacific Slope from Sin to W Chis; should be sought in interior Chis E to Guatemala border, where habitat remains.

GREAT KISKADEE
*Pitangus sulphuratus**   Plate 42
Luis Grande

ID: 9–10 in (23–25.5 cm). *Bill stout and relatively long.* **Descr.** Ages differ, sexes similar. **Adult:** bill and legs blackish. White forehead and supercilium contrast with black crown and mask, yellow crown patch usually concealed. Throat white, underparts yellow. Upperparts olive-brown, wings and tail darker, broadly edged rufous, *appearing rufous in flight.* **Juv:** wing coverts more extensively edged rufous, lacks crown patch.

Voice. Calls loud, raucous, and scream-

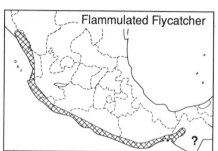

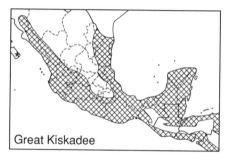

ing, including a loud *k-reah!* or *kih-kerrr*, and *ki ke-reeh* or *kis-k-dee*, etc. Dawn song (?) a raucous *kyah k-yah zzk-zzik ky-ar*, and longer variations, repeated.

Habs. Open and semiopen areas with hedges, scattered trees, second growth, forest edge, often near water. Noisy and conspicuous. At times catches fish like a kingfisher. Nest is a bulky roofed structure of grasses, weeds, etc., with side entrance, on tree support, in acacias, on telegraph poles, etc. Eggs: 3–4, creamy to pale buff, sparsely marked with browns and greys (WFVZ).

SS: Boat-billed Flycatcher mainly a forest bird, bill stouter and very broad, wings and tail brownish; note voice. Social Flycatcher smaller with small bill, wings and tail brownish (edged rufous in juv); note voice.

SD: C to F resident (SL–1800 m) on both slopes from S Son and Tamps, and in interior from cen Mexico, to El Salvador and Honduras. Vagrant to BCS (Jan 1987, Collins *et al.* 1990).

RA: Mexico (and S Texas) to cen Argentina.

BOAT-BILLED FLYCATCHER
*Megarynchus pitangua** Plate 42
Luis Piquigrueso

ID: 9–9.5 in (23–24 cm). *Bill very stout, broad, and fairly long.* Descr. Ages differ slightly, sexes similar. Adult: bill and legs blackish. White supercilium contrasts with blackish-brown crown and black mask, yellow to orange crown patch usually concealed. Throat white, underparts yellow. *Upperparts olive, wings and tail brown, narrowly and inconspicuously edged pale cinnamon, appearing brownish in flight.* Juv: crown and upperparts tipped cinnamon, cinnamon wing and tail edgings broader, lacks crown patch.

Voice. Varied harsh and slightly squeal-

ing gruff chatters, most often a grating, complaining *eeihrrr* or *krrrah* and a querulous, gruff *quee-zika quee-zika* and *eehr, eehr ki-di-rrik*, etc. Dawn song a simple, monotonously repeated, rolled *pprrri-iu* or *chirr-rr*, and a shorter *prriu* or *chrr-ee*, 10/11–16 s.

Habs. Humid to semihumid forest and edge, plantations. At mid- to upper levels, often in canopy. Nest is a shallow cup of twigs, fibers, etc., on tree branch. Eggs: 2–3, whitish, marked with browns and greys (WFVZ).

SS: See Great Kiskadee. Social Flycatcher smaller with much smaller bill.

SD: F to C resident (SL–1500 m) on Atlantic Slope from S Tamps, and on Pacific Slope and in interior from Isthmus, to El Salvador and Honduras; F to U and local on Pacific Slope N of Isthmus to Sin.

RA: Mexico to Peru and N Argentina.

SOCIAL FLYCATCHER
*Myiozetetes similis** Plate 42
Luis Gregario

ID: 6.7–7.2 in (17–18.5 cm). *Bill small and stubby.* Descr. Ages differ, sexes similar. Adult: bill and legs blackish. *White supercilium contrasts with dark grey crown and blackish mask,* flame-colored crown patch usually concealed. Throat white, underparts yellow. *Upperparts olive, wings and tail brown,* edged olive. Juv: wings and tail edged rufous, lacks crown patch.

Voice. A shrill, piercing, screamed *seea'* or *tcheea* or *see-yh!*, and varied, excited, high to nasal or slightly shrill, rapid twittering and bickering phrases, *t-cheer-cheer chee-tiqueer* or *chiir t-chiir t-chiir*, etc. Dawn song (?) a shrill note, repeated one or more times, may be followed by a longer series, *seeu, seeu, see-u-chu'* or *sree, sree, sree si-si-chuhr*, or simply *chirrrr* or *cheirrrr*, repeated over and over.

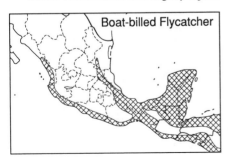

Boat-billed Flycatcher

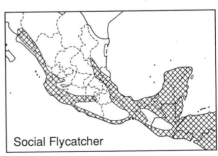

Social Flycatcher

Habs. Humid to semiarid semiopen areas with hedges, scattered trees, second growth, forest and edge, plantations. Noisy and conspicuous, often in small groups. Nest is a bulky roofed structure of grasses, weeds, etc., with side entrance, at mid- to upper levels in tree or bush. Eggs: 2–4, whitish, sparsely marked with browns and greys (WFVZ).

SS: See Great Kiskadee, Boat-billed Flycatcher.

SD: C to F resident (SL–1800 m) on both slopes from cen Sin and S Tamps, and in interior from Balsas drainage, to El Salvador and Honduras, U N to S Son.

RA: Mexico to Peru and N Argentina.

STREAKED FLYCATCHER

Myiodynastes maculatus insolens   Plate 41
Papamoscas Rayado

ID: 8–9 in (20.5–23 cm). *Summer visitor to Atlantic Slope. Bill stout and fairly broad.* **Descr.** Ages differ slightly, sexes similar. **Adult:** bill blackish, *extensively flesh below at base,* legs blackish. *Blackish mask bordered by narrow pale lemon supercilium and pale lemon moustache, throat whitish, dusky malar stripe rarely meets base of bill.* Underparts pale lemon, finely streaked dusky on chest and sides. Crown, nape, and upperparts brownish olive, streaked and mottled darker, yellow crown patch usually concealed, uppertail coverts mostly rufous. Wings dark brown, outer coverts and primaries edged cinnamon, inner coverts, tertials, and secondaries edged whitish to pale lemon. Tail cinnamon-rufous with dark shaft streaks. **Juv:** wings more extensively edged cinnamon, lacks crown patch.

 Voice. Calls include a sharp, nasal, slightly woodpecker-like *behk* or *pehk!*, often repeated steadily, and a slightly liquid *w-see'u* or *t-cheu*, at times repeated; also a

fairly hard, smacking chatter *chi w-chi' ki-chi w-chi ki-chi* ... Dawn song (Costa Rica) a persistently repeated, soft, clear, liquid *kawee-teedly-wink* or *whee-cheerily-chee* (Stiles and Skutch 1989).

Habs. Humid evergreen to semideciduous forest, edge and clearings, plantations. At upper to mid-levels where missed easily if not calling, often nests in dead trees where perches conspicuously on guard, at times in trees adjacent to nesting Sulphur-bellied Flycatchers. Nests in tree cavities at mid- to upper levels, often in clearings or at forest edge. Eggs (Colombia): 2–3, whitish to buffy white, marked with purplish browns and greys (AMNH).

SS: Sulphur-bellied Flycatcher has smaller bill with little or no flesh below at base, and different head pattern: face of Sulphur-bellied appears to have broad white moustache bordered by black mask and thick black malar stripe, chin black, supercilium whitish; underpart streaks coarser in Sulphur-bellied and belly often brighter yellowish, wing edgings of adult mostly whitish.

SD: U to F summer resident (late Mar–Sep; SL–1500 m) on Atlantic Slope from S Tamps to W Honduras.

RA: Summer resident from E Mexico to N C.A.; winters S.A. Resident from Costa Rica to Peru and N Argentina.

SULPHUR-BELLIED FLYCATCHER

Myiodynastes luteiventris*   Plate 41
Papamoscas Vientre-amarillo

ID: 7.5–8.5 in (19–21.5 cm). *Summer visitor to both slopes. Bill fairly stout and broad.* **Descr.** Ages differ slightly, sexes differ. **Adult:** *bill blackish,* often pale flesh below at base, legs blackish. *Broad whitish moustache bordered by black mask and thick black malar stripe, supercilium and throat*

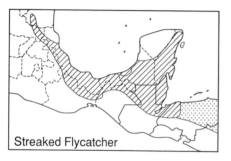

Streaked Flycatcher

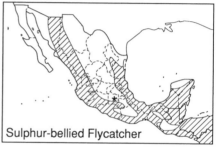

Sulphur-bellied Flycatcher

whitish, chin blackish; underparts pale lemon, streaked dusky on chest and sides. Crown, nape, and upperparts greyish olive, streaked and mottled darker, yellow crown patch usually concealed, uppertail coverts mostly rufous. Wings dark brown, edged whitish, outer covert edges often washed pale cinnamon. Tail cinnamon-rufous with dark shaft streaks. **Juv:** wings more extensively edged cinnamon, lacks crown patch.

Voice. An excited, piercing to shrill, 'squeezy-toy'-like *wee'iz-uh* and *weez-ih*, often doubled, may be preceded by short, gruff notes, *pek, pek pek kweez-i-zik kweez-i-zik*, or run into longer series *whee whee whee-whee-whee whee-i-eezk'* or *si-chu' w-chee w-chee w-chee*, etc. Dawn song a bright, slightly slurred phrase followed by a clipped, slightly liquid phrase, *chee-a-leet s-lik* or *chew-ee ti-lit* or *doo-ee ti-chu'*, etc., repeated over and over, often before dawn, 10/19–20 s.

Habs. Humid to semiarid forest, edge and clearings, gallery woodland, plantations. Habits, nest, and eggs (3–4; WFVZ) much like Streaked Flycatcher.

SS: See Streaked Flycatcher.

SD: C to F summer resident (Mar–Sep; SL–1800 m) on both slopes from Son, E NL, and Tamps, and in interior from Balsas drainage, to El Salvador and Honduras.

RA: Breeds Mexico to Costa Rica; winters S.A.

PIRATIC FLYCATCHER

Legatus leucophaius variegatus   Plate 41
Papamoscas Pirata

ID: 6.3–6.7 in (16–17 cm). *Summer visitor to S. Bill small and stubby.* **Descr.** Ages differ slightly, sexes similar. **Adult:** bill and legs blackish. *Whitish supercilium contrasts with blackish crown and mask,* yellow crown patch usually concealed.

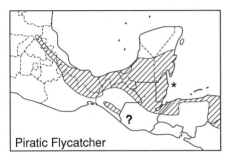

Piratic Flycatcher

Throat whitish with indistinct dusky malar stripe, underparts pale lemon with blurry dusky streaking on chest and sides. Upperparts dark olive-brown, wings and tail darker with narrow whitish wingbars and tertial edges, pale lemon panel on secondaries, tail narrowly fringed cinnamon-lemon. **Juv:** crown feathers tipped cinnamon, wing coverts edged pale cinnamon, lacks crown patch.

Voice. A bright *whee-ee* or *swee-u'*, often followed by a rolled *ji-ji-jit* or *whee di-weet*, suggesting an overgrown Yellow-throated Euphonia. Also a longer, at times prolonged, series of excited piping whistles, rising and falling, *whii-whii-whii ...*

Habs. Humid evergreen forest edge and clearings, open areas with scattered trees and hedges, plantations. Perches conspicuously at upper to mid-levels, often atop bare trees. Steals already-built closed nests from other species (mainly oropendolas and becards) by persistent and aggressive diving attacks on the rightful owners until they abandon the site. Eggs: 2–3, grey-brown to brown, marked with browns (WFVZ).

SD: F to C summer resident (Mar–Aug; SL–1000 m) on Atlantic Slope from S Ver to Honduras; U N to SE SLP and on Pacific Slope and in adjacent interior of Chis (SNGH, SW).

RA: Breeds E Mexico to Ecuador and N Argentina; winters S.A.

GENUS TYRANNUS

Kingbirds are large conspicuous flycatchers of open and semiopen country, often seen on roadside wires and fences. Some species are migratory, moving by day, often in spectacular waves; many species roost communally in winter. Ages differ slightly, sexes similar. Juvs lack concealed brightly colored crown patches of adults, but most species molt quickly into adult-like plumage. Bills and legs blackish. Note bill size, color of throat and underparts, tail shape and pattern, and voice. Songs typically given before dawn, and repeated over and over. Nests are frail to bulky cups of twigs, grasses, fibers, etc., low to high in bushes and trees. Eggs: 2–5, white to buff, marked with dark browns and greys (WFVZ).

TROPICAL KINGBIRD
Tyrannus melancholicus*   Plate 42
Tirano Tropical

ID: 7.8–9.3 in (19.5–23.5 cm). Common in lowlands. *Bill relatively long and heavy. Tail relatively long and forked.* **Descr. Adult:** head grey with darker mask, flame-colored crown patch usually concealed; upperparts olive. Wings and *tail blackish brown*, wings edged paler. *Throat whitish, underparts yellow, chest washed dusky.* **Juv:** wing coverts and tail edged pale cinnamon, remiges edged pale lemon, uppertail coverts edged cinnamon.
> **Voice.** Bright, slightly liquid trills, quite different from Couch's Kingbird, *tree-ee-eer* and *tril-il-il-iil-l* or *trri tri-li-lit i-il*, etc., drier and tinnier in W Mexico; also a bright *si-seep*. Songs include a loud, sweet series, the 2nd part accelerating brightly, *weet weet weet-weet-weet*, a stuttering series of call-like trills, and a short, bright note followed by a liquid trill *sweeit tri-i-i-i-il*, etc.
> **Habs.** Open and semiopen areas with scattered trees, fences, hedges, forest clearings, beach scrub. In Yuc Pen, C in arid beach scrub, U in scrubby woodland of interior. See genus note.

SS: Couch's Kingbird often has smaller bill but there is some overlap; best told by voice; irregular spacing of outermost primaries in Tropical differs from more regular spacing in Couch's but this can be hard to see, and is only reliable on adults. Cassin's Kingbird has small bill, shorter squared black tail tipped whitish, head and chest dusky with contrasting whitish chin; note voice. Western Kingbird has small bill, squared black tail with white sides; note voice.

SD: C to F resident (SL–1800 m) on both slopes from Son and cen Tamps (SNGH, SW), and in interior from cen Mexico, to El Salvador and Honduras. Partly migratory in N and winters more widely, for example U in Baja (Sep–Mar; Wilbur 1987; SNGH, SW) and U to R (to 2200 m) in DF (Wilson and Ceballos-L. 1986).

RA: Mexico (SW USA in summer) to Peru and cen Argentina.

NB: Has been considered conspecific with Couch's Kingbird.

COUCH'S [TROPICAL] KINGBIRD
Tyrannus couchii   Plate 42
Tirano de Couch

ID: 8–9.5 in (20.5–24 cm). *Atlantic Slope.* **Descr.** *Rarely distinguishable in the field from Tropical Kingbird except by voice, but Couch's bill averages smaller and this can be a useful field mark.* Plumage differences are at best subtle and not practical for field purposes. Couch's has less deeply forked tail than ♂ Tropical but similar to ♀ Tropical.
> **Voice.** A nasal *bihk* or *pik* and a buzzy, nasal *brrrear* or *brreeah*, often combined, *pik pik pik brreeah*, etc. Dawn song a series of bright, intensifying whistles with sneezy inflections, ending abruptly, *s'wee' s'wee s'wee s'wee-i-chu'* or *k-leep' k-leep k-leep k-leep k-lee-i-chu'*, burry *brrreeu* calls sometimes thrown in; also varied to a shorter *prreeu wi-wit wee wi-chu'*, etc.
> **Habs.** Scrubby woodland, forest and edge, plantations, open areas with scattered trees, hedges, fences, etc. In Yuc Pen, C in scrubby woodland of interior, U in arid beach scrub. See genus note.

SS: See Tropical Kingbird. See Western and Cassin's kingbirds (rare in Couch's range) under Tropical Kingbird SS.

SD/RA: C to F resident from E NL and Tamps (also S Texas) to Yuc Pen (S to N Guatemala and Belize); partly migratory (*contra* Traylor 1979), with some withdrawal from Texas (Oberholser 1974;

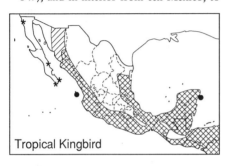

Tropical Kingbird

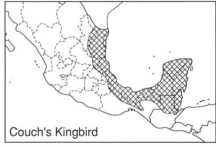

Couch's Kingbird

DeSante and Pyle 1986) and probably NE Mexico in winter, when commoner in S Mexico, including interior Chis where R or absent in summer (SNGH). C in N and F in S Belize (SNGH, SW) (although omitted by Wood *et al.* (1986)). More work is needed to clarify the seasonal status of this species in S Mexico.

NB: See Tropical Kingbird.

CASSIN'S KINGBIRD
Tyrannus vociferans* Plate 42
Tirano de Cassin

ID: 8–9 in (20.5–23 cm). *Interior and W Mexico. Bill relatively short, tail relatively short and squared to slightly cleft.* **Descr.** **Adult:** *head and chest dusky grey with slightly darker mask and contrasting white chin;* flame-colored crown patch usually concealed; underbody yellow. Upperparts greyish olive. Wings dark brown, edged paler; *tail black, tipped whitish.* **Juv:** wing coverts edged pale cinnamon.

Voice. A loud, burry *ch-beehr* or *k-beehrr*, a burry, nasal *breeahr* or *beeah*, repeated and followed irregularly by a rapid, sharp, nasal *ki-dih ki-dih ki-dih ...,* and nasal, bickering chatters. Dawn song typically phrases of rough, querulous notes which run into a shrieky, slightly emphatic ending, may suggest Kiskadee: *rruh rruh rruh-rruh rreahr, rruh ree ree-uhr,* etc.

Habs. Arid to semihumid open areas with scattered trees, hedges, fences, scrub, etc. See genus note.

SS: Western Kingbird has paler throat and chest, white-sided tail; note voice. See Tropical and Couch's kingbirds, Say's Phoebe.

SD: C to F breeder (mid-Mar–Aug; near SL-3000 m) in BCN and in interior and on adjacent slopes from Son and Coah to Oax. C to F transient and winter visitor (Sep–Apr; SL–2500 m) in BCS (irregularly

F to C in BCN), on Pacific Slope from S Son to Nay, and in interior and on adjacent slopes from S Plateau to Oax, U to F in interior to Chis, R (to U?) to cen Honduras. Vagrant to SW Camp (Nov 1992, RAB). A report from Belize (Wood *et al.* 1986) requires verification.

RA: Breeds W USA to S Mexico; winters Mexico to N C.A.

THICK-BILLED KINGBIRD
Tyrannus crassirostris* Plate 42
Tirano Piquigrueso

ID: 8.5–9.5 in (21.5–24 cm). *W Mexico. Bill very stout and broad,* tail squared to slightly cleft. **Descr.** **Adult:** *dark grey head with darker mask contrasts with greyish-olive upperparts,* yellow crown patch usually concealed. Wings and tail dark brown, coverts and tail narrowly edged pale cinnamon. *Throat and underparts whitish with slight dirty wash on chest, belly washed lemon* (can be bright in fresh plumage). **Juv:** greyish-olive head, with darker mask, does not contrast with back, wings and tail edged cinnamon.

Voice. A bright, nasal *di-diweek* or *kidi-wik*, at times *di-i-diweek* or *ki-di wi-eu*, a burry to buzzy *chweeer* or *bzzzeiu*, repeated over and over, at times interspersed with clipped *chk ch-weer* phrases, and varied, nasal, bickering calls. Song a stuttering, nasal, bickering series clearing into a somewhat emphatic ending, *ki-di di-di di-di-dee-yew*, or *pri-di di-di di-di ree-hew*, etc.

Habs. Woodland and edge, plantations, semiopen areas with trees, hedges, scrub, etc., mostly in arid to semiarid areas. Often perches higher and more conspicuously than Tropical Kingbird. See genus note.

SD/RA: C to F resident (near SL–2000 m) on Pacific Slope from S Son (in summer N to

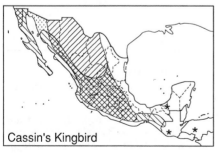

Cassin's Kingbird

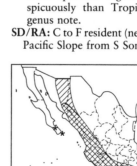

Thick-billed Kingbird

SE Arizona) to Oax; F to U in interior from cen Mexico to Oax. U winter visitor (Sep–Apr) on Pacific Slope to W Chis, R to Guatemala.

WESTERN KINGBIRD

Tyrannus verticalis   Plate 42
Tirano Occidental

ID: 7.8–8.3 in (19.5–21 cm). *N and W. Bill relatively short, tail squared.* **Descr. Adult:** head grey with darker mask, upperparts greyish olive, flame-colored crown patch usually concealed. Wings dark brown, edged paler; *tail black, outer webs of outermost rectrices white. Whitish throat blends into pale grey chest,* underbody yellow. **Juv:** usually greyer and more faded than adult in autumn.

    Voice. A sharp, clipped *bec* or *bek* and querulous, bickering chatters. Song excited, rapid, bickering clucks accelerating into a short, squeaky, and slightly emphatic chatter, *pc, pc, pc, pc pc, pc-pc pc, pree pree pr-prrr,* or *pic, pic, pic-pic-pic pic bree bree bruh-brr,* etc.

    Habs. Breeds in open areas with hedges, scattered trees, fences, ranches, open pine woods, etc. In winter also in canopy of open woods, thorn forest, plantations. Migrant flocks often at fruiting trees. See genus note.

SS: See Cassin's, Tropical, and Couch's kingbirds, Say's Phoebe.

SD: F to U breeder (Apr–Aug; near SL–1800 m) in BCN, Son, and N Chih; C to F transient (Apr–May, Aug–early Dec; SL–2500 m) on Pacific Slope and through interior to S Mexico; C to F in winter (Sep–Apr; SL–1800 m) on Pacific Slope and in interior from Gro to El Salvador and Nicaragua, U and local N to S Jal (SNGH). R transient (Apr–early May) on Atlantic Slope N of Isthmus.

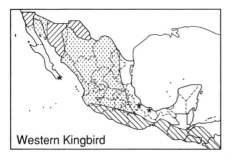

Western Kingbird

RA: Breeds W N.A. to NW Mexico; winters SW Mexico to Costa Rica.

EASTERN KINGBIRD

Tyrannus tyrannus   Not illustrated
Tirano Viajero

ID: 7.7–8.2 in (19.5–21 cm). *Transient in E and S. Bill small and short, tail squared.* **Descr. Adult:** *blackish head contrasts with dark grey upperparts,* flame-colored crown patch usually concealed. Wings blackish brown, edged whitish. *Tail black, broadly tipped white. Throat and underparts whitish,* washed grey on chest. **Juv:** told in autumn by fresh, white-tipped primaries.

    Voice. A high, slightly burry twittering, usually given by flocks.

    Habs. Forested and open country in general. Impressive flocks, up to thousands, can occur in Apr–May and Sep, often at fruiting trees. See genus note.

SS: Grey Kingbird has heavier bill, longer forked tail lacks white tip, head grey with dark mask.

SD: C to F transient (late Mar–May, late Aug–Oct; SL–2000 m, mostly in lowlands) on Atlantic Slope from NL and Tamps to Honduras; F in spring and R to U in autumn on Pacific Slope S of Isthmus; R W to DF (Wilson and Ceballos-L. 1986) and E Chih (Vuillemier and Williams 1964). Vagrant (Aug–Sep) to BCS (Lamb 1925*b*; CAS spec.). Winter reports from N C.A. (Thurber *et al.* 1987) require verification.

RA: Breeds E N.A., winters S.A.

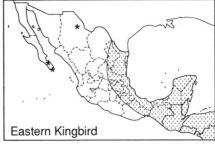

Eastern Kingbird

GREY KINGBIRD

Tyrannus d. dominicensis   Not illustrated
Tirano Gris

ID: 8.2–9 in (21–23 cm). *Transient on Caribbean islands. Bill relatively long and heavy, tail relatively long and cleft to*

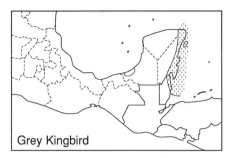

Grey Kingbird

slightly forked. **Descr.** *Head and upperparts grey* with darker mask, yelloworange crown patch usually concealed, uppertail coverts edged cinnamon. Wings dark brown, edged whitish. *Tail blackish brown. Throat and underparts whitish,* washed grey on chest. **Juv:** wing covert edgings washed cinnamon.
Voice. Trilled calls much like Tropical Kingbird, *trii-il-il-it*, etc., but harder and slightly rougher.
Habs. Open and semiopen areas with scrub, scattered trees, etc. See genus note.
SS: See Eastern Kingbird.
SD: U to R transient (late Mar–Apr, Sep–early Nov) on Caribbean islands (R on adjacent coasts) from QR to Belize (Howell *et al.* 1992*b*), probably also in N Honduras.
RA: Breeds Caribbean and adjacent coasts of SE USA; winters Caribbean to N S.A.
NB: We consider the report of Giant Kingbird *Tyrannus cubensis* on Isla Mujeres (Salvin 1889) hypothetical (see Appendix B).

SCISSOR-TAILED FLYCATCHER
Tyrannus forficatus   Plate 42
Tirano-tijereta Rosado

ID: 7.5–14 in (19–35.5 cm). *Winters in E and S. Tail very long and forked* (♂ > ♀) to fairly long and forked (imm). **Descr. Adult:**

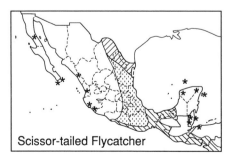

Scissor-tailed Flycatcher

head and upperparts pale grey, red crown patch (reduced or absent in ♀) usually concealed; wings dark brown, edged whitish. *Outer rectrices white, tipped black, inner rectrices black.* Throat and *underparts* white, *washed pink on sides and undertail coverts*, brighter and redder at bend of wing (often concealed when perched). **Juv:** sides and undertail coverts washed paler, orange-pink. Attains adult-like plumage through 1st winter but tail often still relatively short in spring.
Voice. A sharp *pik* or *bik*, a more nasal *bik err*, and dry, buzzy, bickering chatters.
Habs. Breeds in arid to semiarid open country with brushy scrub, scattered trees, etc.; more humid areas in winter. Often in spectacular flocks in migration and winter. See genus note.
SS: Fork-tailed Flycatcher has black head, lacks pink wash on underparts, tail appears black in flight.
SD: F to U summer resident (Apr–Aug; near SL–1500 m) on Atlantic Slope from N Coah to N Tamps. C to F transient (Mar–May, Aug–Nov; to 2500 m) on Atlantic Slope N of Isthmus and on Pacific Slope from Isthmus S; U spring transient in interior N of Isthmus W to Mor (SNGH, RGW) and DF (Wilson and Ceballos-L. 1986). U to F winter visitor (Sep–Apr) on Atlantic Slope from S Ver to W Camp, and in interior Chis; C to F on Pacific Slope from Gro to El Salvador and Honduras. U to R vagrant (mainly Oct–Apr) to W Mexico, N to Sin (RJC) and Baja, and (Oct–May) to Yuc Pen (SNGH, PP, SW) and Belize.
RA: Breeds S cen USA and NE Mexico; winters S Mexico to W Panama.

FORK-TAILED FLYCATCHER
Tyrannus savana monachus   Plate 42
Tirano-tijereta Sabanero

ID: 7.5–16 in (19–40.5 cm). *S savannas. Tail extremely long and deeply forked* (♂ > ♀) to fairly long and forked (imm). **Descr. Adult:** *black head contrasts with pale grey upperparts*, yellow crown patch usually concealed; uppertail coverts blackish. Wings dark brown, edged whitish. *Tail black*, outer web of outermost rectrices white. Throat and *underparts whitish.* **Juv:** head sooty grey, often with contrasting black mask, wing coverts edged pale cinnamon.

Voice. A sharp, slightly liquid *sik* or *plik*, and sharp, dry, crackling to buzzy twitters and chatters.

Habs. Savannas and other open areas with scattered bushes. See genus note; mainly on low perches, often on or near ground; at times in flocks of 50+ birds.

SS: See Scissor-tailed Flycatcher.

SD: F to C but local and somewhat nomadic resident (SL–1500 m) on Atlantic Slope from S Ver to SW Camp, in Petén, Belize (and extreme S QR?), and the Mosquitia. U to F and local (nomadic?) resident in interior valleys from Chis to Honduras; local movements need study, perhaps confused by occasional vagrants from S.A. (which could occur Mar–Oct). Vagrant to cen QR (Feb 1984, Lopez O. *et al.* 1989).

The basis for 'winters irregularly through Middle America' (AOU 1983) is unclear; no true migration is documented for N M.A. birds but nomadic wandering occurs.

RA: S Mexico to cen Argentina.

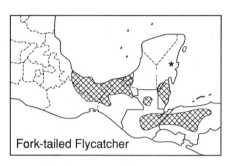

Fork-tailed Flycatcher

COTINGAS: COTINGIDAE (12)

A diverse neotropical family whose taxonomy, like that of the closely related Manakins and Tyrant-Flycatchers, is far from clear (all are placed in the Superfamily Tyrannoidea by AOU (1983)). We place several problematic genera, including the *Schiffornis* group, with the Cotingidae following the suggestions of Prum and Lanyon (1989) and R. Prum (personal communication), but acknowledge that this treatment remains tentative. It is impossible to characterize concisely even the few cotingas in the region. Classic cotingas are strongly sexually dimorphic and frugivorous. Ages/sexes differ or are similar. Molts and breeding biology poorly known. Some species have loud and striking vocalizations, others are virtually mute.

THRUSHLIKE MOURNER (MANAKIN)
Schiffornis turdinus veraepacis　　　Plate 43
Llorón Café

ID: 6.5–7 in (16.5–18 cm). *SE humid lowland forest*. Descr. Ages/sexes similar. Bill blackish, paler below at base, legs greyish. *Overall dark brownish olive with slightly paler eyering*, wings and tail warmer and browner, belly and undertail coverts duller and greyer.
　Voice. Haunting and unmistakable. A rich, whistled phrase, the 1st note slurred, the 2nd short, the last sharply up-slurred, *djeeeeu whee-chee'* or *djoooo di wi-chu'*, etc., at times a two-note *dweeeer weet*; also a short, hard, rattling chatter.
　Habs. Humid evergreen forest. Singly at low levels in shady, often dense, understory, often within a meter of the ground; rarely seen unless a singing bird is tracked down. Nest is a bulky cup of dead leaves, rootlets, etc., low in dense understory. Eggs (Costa Rica): 2, pale buff, marked with black, dark brown, and pale lilac.
SS: Unmistakable but relatively nondescript; ♀ ant-tanagers larger and stouter-billed, richer colored overall, more conspicuous, usually noisy.
SD: F to C resident (SL–750 m) on Atlantic Slope from E Chis to Honduras, U N to S Ver and NE Yuc Pen.
RA: SE Mexico to Ecuador and Brazil.
NB: An enigmatic bird of uncertain taxonomic relationships; certainly not a true manakin; may include more than one species.

SPECKLED MOURNER
Laniocera r. rufescens　　　Plate 43
Llorón Moteado

ID: 8–8.5 in (20.5–21.5 cm). SE humid forest. Descr. Ages/sexes similar. Bill blackish, flesh below at base, legs greyish. *Rufous-brown overall*, paler below. *Dark wing coverts tipped rufous* (appearing mottled), *underparts indistinctly scalloped dusky, boldest on chest, yellow flash at sides of chest* often covered by wings.
　Voice. A rich, plaintive, whistled *chee-oo-ee* or *tew i-dee*, repeated 2–6×, occasionally with a suffix to the 1st phrase of a series, *tew i-dee tcheu-u, tew i-dee tew*

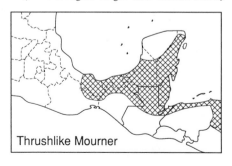
Thrushlike Mourner

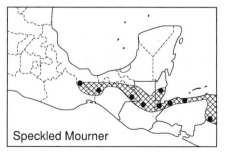
Speckled Mourner

i-dee ...; vaguely suggests White-breasted Wood-Wren but more slurred, quality much like Thrushlike Mourner. Also a plaintive, drawn-out, slightly mewing *peeeeeeu* or *wheeeeeeu,* ending fairly abruptly.

Habs. Humid evergreen forest. Singly, at mid- to upper levels in forest where perches quietly for long periods, overlooked easily. Nest and eggs undescribed (?).

SS: See Rufous Mourner. Rufous Piha larger, lacks mottled wing coverts, dusky scalloping, and yellow side flashes; Piha's bushy lores and narrow pale orbital ring give it a Squirrel Cuckoo-like facial expression. Note voice.

SD: Apparently U to R resident (near SL–750 m) on Atlantic Slope from N Oax (Binford 1989) to Honduras.

RA: SE Mexico to Ecuador.

NB: AOU (1983) placed Speckled and Rufous mourners with *Myiarchine* flycatchers; we consider them not closely related and retain the former with cotingas.

GENUS PACHYRAMPHUS

Becards are large-headed, slightly crested birds of open woodland and edge, plantations, etc. Adult ♂♂ have a short and attenuated 9th primary. Smaller species have strongly graduated tails, Rose-throated has a squared tail. Ages/sexes differ or (Cinnamon) similar, juvs resemble ♀♀ and quickly attain imm plumage; adult plumage attained when about 1 year old. Bills blackish, often paler greyish to greyish flesh below, legs dark grey. Nests are bulky globular masses of fine dead leaves, fibers, moss, etc., with entrance near the bottom, slung from or wedged among outer branches at mid- to upper levels.

CINNAMON BECARD
Pachyramphus cinnamomeus fulvidior
Cabezón Canelo   Plate 43

ID: 5.5–6 in (14–15 cm). *SE humid forest edge.* **Descr.** Crown, nape, and *upperparts rufous, crown darker with pale supraloral stripe and dusky lores. Throat and underparts ochraceous tawny,* palest on throat and belly.

Voice. Song a slightly plaintive, reedy *tee dee-dee-dee ...* or *cheei dee-dee ...,* the 2nd part of 5–9 notes rising or falling; calls are shorter, reedy series, *swee-dwee-*

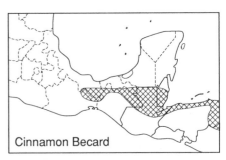

Cinnamon Becard

dwee or *seer eeer eeeur,* and a plaintive *seeeeiu,* etc.

Habs. Humid evergreen forest edge, clearings, plantations. Singly or in pairs, at mid- to upper levels, easily missed if silent. May join mixed-species flocks. Eggs: 3–4, whitish to greyish buff, marked with browns (WFVZ). See genus note.

SS: ♀ Rose-throated Becard larger and more heavily built, tail not strongly graduated, deeper cap blackish, often has paler hindcollar.

SD: F to U resident (SL–500 m) on Atlantic Slope from SE Ver to Honduras; may be partly migratory in Mexico, with some withdrawal from edge of range in winter (Sep–Feb?).

RA: SE Mexico to Ecuador.

WHITE-WINGED BECARD
Pachyramphus polychopterus cinereiventris
Cabezón Aliblanco   Plate 71

ID: 5.5–6 in (14–15 cm). C.A. **Descr.** ♂: *crown, nape, and upperparts glossy black,* with white scapular stripe, grey uppertail coverts. Wings black, *coverts, tertials, and secondaries edged white.* Tail black, outer rectrices broadly tipped white. *Throat and underparts grey.* ♀/juv: *crown, nape, and upperparts brownish olive; pale supraloral stripe, eyering, and dark lores create vireo-like facial expression.* Wings black-

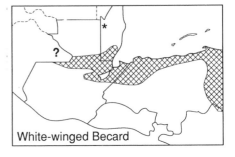

White-winged Becard

ish, *coverts, tertials, and secondaries edged cinnamon*. Central rectrices olive-brown, black distally in juv ♂, rest of tail blackish, broadly tipped pale cinnamon. *Throat and underparts pale lemon*, whiter on throat, duskier on chest. **Imm ♂**: resembles ♀ but often has grey throat, black flecks on head and back; central rectrices blackish distally. After 1st prealternate molt (Feb–May), resembles ♂ but usually retains *olive rump, lemon belly, most flight feathers, and greater wing coverts.*

Voice. Song a varied arrangement of rich, mellow whistles *chu chu chu wee'* or *chu chuwee chuwee*, an accelerating *chew chu-chu-chu-chu-chu*, etc.

Habs. Humid to semiarid forest edge, open areas with scattered trees, plantations. Habits much as Cinnamon Becard. Eggs (Colombia): 2–3, buffy white, blotched with reddish brown and grey (AMNH). See genus note.

SS: ♂ Grey-collared Becard (rarely sympatric) paler overall with grey hindcollar, pale supraloral stripe. ♀ may suggest a *Myiarchus* flycatcher; note face pattern, graduated tail, often with ♂.

SD: U to F resident (SL–500 m) on Atlantic Slope from S Belize and Guatemala to Honduras. Reports from Chis (e.g. Sada 1989*b*) require verification.

RA: Guatemala to N Argentina.

GREY-COLLARED BECARD
Pachyramphus major*   Plate 43
Cabezón Cuelligris

ID: 5.5–6.2 in (14–15.5 cm). Endemic; widespread but uncommon. Two groups told readily in ♀ plumage: *uropygialis* (W Mexico) and *major* group (E Mexico and C.A.). **Descr. ♂**: *glossy black crown contrasts with pale grey hindcollar, pale supraloral stripe; throat and underparts pale grey*. Back black or mottled grey and black

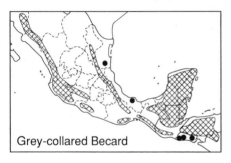

Grey-collared Becard

(mostly grey in Yuc Pen), with white scapular stripe, rump and uppertail coverts grey. *Wings black, coverts, tertials, and secondaries edged white*. Tail black, outer rectrices broadly tipped white. ♀/juv. *Major*: blackish-brown to blackish crown contrasts with pale cinnamon to tawny-buff hindcollar, pale supraloral stripe; throat and underparts buff to cinnamon-buff (buffy lemon in *matudai* of Pacific Slope of Chis and W Guatemala). *Upperparts cinnamon-brown, wings blackish, coverts, tertials, and secondaries edged cinnamon to cinnamon-brown*. Central rectrices cinnamon-brown, often with black subterminal mark, rest of *tail blackish, broadly tipped cinnamon.* **Uropygialis**: *cinnamon-rufous crown contrasts with broad black eyestripe, underparts and hindcollar pale lemon*. **Imm ♂**: resembles ♀ but back may be mottled black, wing edgings lemon-buff, grey-brown central rectrices blackish distally, outer rectrices tipped whitish. Often much like ♂ following 1st pre-alternate molt (Feb–May), but *rump cinnamon-brown.*

Voice. Song typically a rich, whistled *hoo wee-deet'* ... or *hu whi-diit'* repeated steadily, usually 4–6×, 10/10 s; also a plaintive to clear *peeu peeu* and longer series *beeh beeh beeh* ..., up to 6–7×.

Habs. Humid to semiarid forest, especially oak and pine–oak, forest edge, plantations. Habits much as Cinnamon Becard, often with mixed-species flocks. Nest and eggs undescribed (?). See genus note.

SS: ♀ Rose-throated Becard larger and often bushy-headed, lacks strongly graduated tail, upperparts uniform, crown grey and underparts and collar buff in W Mexico; see White-winged Becard (C.A.).

SD/RA: U to F but local resident (SL–2500 m, mostly lower in winter) on both slopes from S Son and S NL, and in interior from cen Mexico, to El Salvador and N cen Nicaragua. Apparently an altitudinal migrant, at least N of Isthmus.

ROSE-THROATED BECARD
Pachyramphus aglaiae*   Plate 43
Cabezón Degollado

ID: 6.5–7 in (16.5–18 cm). Common and widespread. Two groups: *aglaiae* group (widespread except humid SE) and *hypophaeus* group (humid SE, incl *sumichrasti* shown). **Descr. Aglaiae. ♂**: *blackish crown* contrasts with grey to dark grey *upperparts; underparts pale grey to grey with rose-pink patch on lower throat/upper chest.* ♀/juv: *grey to dark grey crown contrasts with ochraceous-buff hindcollar,* lores paler; throat and underparts pale buff to ochraceous buff. *Upperparts greyish to grey-brown or, in gravis of NE Mexico, cinnamon-brown; wings and tail browner.* **Hypophaeus** ♂: *head and upperparts blackish; underparts dark grey, smaller rose-pink patch may be lacking.* ♀/juv: *sooty blackish crown contrasts with cinnamon sides of neck,* lores paler; throat and underparts ochraceous buff to ochraceous tawny. *Upperparts deep rufous.* **Imm** ♂: *resembles* ♀ *but* head and upperparts *variably like* ♂, usually mottled blackish and grey, usually some rose-pink on throat.

Voice. Common call a plaintive, downslurred *t-sseu* or *tzeeeu*, and a shorter *sseeu* or *teew*, often run into a reedy, rolled, spluttering chatter or trill, or a chatter may slur into a plaintive *tcheu* or *tew*. In alarm near nest, a quiet *pik* and *pii-dik*. Rarely heard dawn song a slightly reedy, plaintive *si-tchew wii-chew*, or *si-tseeu wii-tzeeu*, repeated, at times *si-tseeu wii-tzee, si tseeeu*, etc.

Habs. Forest edge, including pine–oak, clearings, open areas with scattered trees and hedges, riparian groves. Singly or in pairs, fairly conspicuous, often joins mixed-species flocks. Eggs: 3–4, pale brownish to whitish, marked with browns (WFVZ). See genus note.

SS: See Grey-collared and Cinnamon becards.

SD: C to F resident (SL–2700 m) on both slopes from Son and Tamps, and in interior from cen Mexico, to El Salvador and Nicaragua. Withdraws from colder N areas (N Son to NL) in winter (Sep–Feb?).

RA: Mexico (SW USA in summer) to W Panama.

NB: Often placed in the genus *Platypsaris*.

MASKED TITYRA
*Tityra semifasciata**   Plate 43
Titira Enmascarada

ID: 8.5–9.5 in (21.5–24 cm). Common and widespread. Two groups readily told in ♀ plumage: *griseiceps* (W Mexico) and *personata* group (E Mexico and C.A.). **Descr.** Ages/sexes differ. ♂: eyes reddish, *bill, bare lores, and eyering pinkish red, bill tipped black,* legs blackish. *Pale silvery grey overall, paler below, with black face, wings,* and broad subterminal tail band. ♀. **Griseiceps:** *head and upperparts pale grey,* back washed grey-brown, paler and greyer on innerwing coverts and tertials, *flight feathers and outerwing coverts black,* tail broadly tipped whitish. *Throat and underparts pale grey.* **Personata:** head and upperparts grey-brown, darker on head. **Juv:** resembles ♀ but whitish tail tip narrower, paler upperparts washed brown. Quickly attains imm plumage. **Imm** ♂: resembles ♀ but upperparts paler and greyer, attains adult plumage Mar–Aug.

Voice. Distinctive buzzy or fart-like calls, *zzzu rrk* or *zzr zzzrt*, and *rr-rr-rrk*, etc.; also a quiet *rruk rruk*.

Habs. Humid to semiarid forest edge, woodland, second growth, semiopen areas with scattered trees, plantations. Usually in pairs or small groups at mid- to upper levels, often at fruiting trees. Perches conspicuously on bare branches. Slightly undulating flight suggests a woodpecker. Nests in tree cavities, often old woodpecker holes, at mid- to upper levels. Eggs: 2–3, dark vinaceous buff, marked with browns (WFVZ).

SS: Black-crowned Tityra smaller with

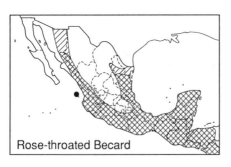

Rose-throated Becard

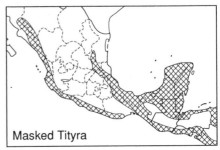

Masked Tityra

heavier black bill, lacks red face, ♂ has black crown, ♀ black crown and chestnut-brown face.

SD: C to F resident (SL–2500 m, mainly below 1500 m) on both slopes from S Son and S Tamps, and locally in interior from Isthmus S, to Honduras and W Nicaragua.
RA: Mexico to Ecuador and Brazil.

BLACK-CROWNED TITYRA

Tityra inquisitor fraserii   Plate 43
Titira Piquinegra

ID: 7.5–8 in (19–20.5 cm). SE lowlands. **Descr.** Ages/sexes differ. ♂: *bill and legs black. Pale silvery grey overall, paler below, with black crown, wings,* and broad subterminal tail band; underwings with white panel across base of primaries. ♀: *black crown and chestnut-brown face,* upperparts grey-brown, paler and greyer on innerwing coverts and tertials, flight feathers and outerwing coverts black, tail broadly tipped whitish. Throat and underparts pale grey. **Juv:** resembles ♀ but crown mottled black and cinnamon, upperparts mottled whitish and grey. Quickly attains imm plumage. **Imm:** resembles ♀ but ♂ has pale grey upperparts, crown may be mottled cinnamon-brown; attains adult plumage May–Aug (?).
   Voice. A gruff, slightly rasping *sheh-shehk,* and variations, suggests Band-backed Wren.
   Habs. Humid to semihumid forest edge, clearings, plantations, etc. Habits and nest much as Masked Tityra with which it often occurs. Eggs undescribed (Kiff and Hough 1985).
SS: See Masked Tityra.
SD: F resident (SL–1200 m) on Atlantic Slope from S Ver to Honduras, U N to SE SLP and in N Yuc Pen.
RA: E Mexico to Ecuador and N Argentina.

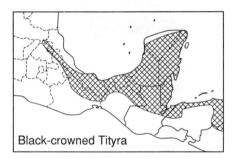
Black-crowned Tityra

GREY-HEADED PIPRITES (MANAKIN)

Piprites griseiceps   Plate 71
Saltarín Cabecigris

ID: 4–4.5 in (10–11.5 cm). C.A. **Descr.** Ages differ slightly, sexes similar. **Adult:** bill blackish, paler below, legs grey. *Slate grey head with white eyering contrasts with olive upperparts. Throat and underparts yellow-olive, paler yellowish on throat* and belly. **Juv:** head olive with whitish eyering.
   Voice. A soft, liquid, rolling *purrr* or *wurrr;* song an elaborate, structured medley of staccato and rolling notes: *pik pi'k prrrity pee-kit prrrity peer,* etc. (Stiles and Skutch 1989).
   Habs. Humid evergreen forest. At mid-levels, often with mixed-species flocks (Stiles and Skutch 1989). Nest and eggs undescribed (?).
SS: Plumage and behavior may suggest a flycatcher but stocky shape, large head, and stout, short bill distinguish this little-known bird.
SD: Probably an U resident (near SL–500 m) on Atlantic Slope from NE Guatemala to Honduras, but only one record in region: one collected in Izabál, Mar 1959 (Land 1963).
RA: Guatemala to Costa Rica.
NB: Taxonomic affinities uncertain; traditionally considered a manakin.

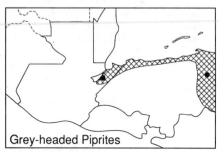
Grey-headed Piprites

RUFOUS PIHA

Lipaugus u. unirufus   Plate 43
Piha Rufa

ID: 9–10.5 in (23–26.5 cm). SE humid lowland forest. **Descr.** Ages/sexes similar. Bill blackish, flesh below at base, legs greyish. *Overall rufous-brown, paler below, palest on throat,* belly, and undertail coverts.
   Voice. Loud, rich, explosive whistles given irregularly, *p-wee-e-loo!* and *whee-oo! p-wee'oo!,* often simply *p-weeOO!* or

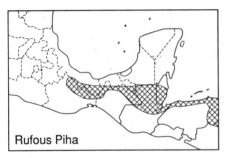

Rufous Piha

whee-ooo!, suggesting Pauraque; also short bursts of hard, squirrel-like chattering *chuh-uh-uh-uh*, etc.

Habs. Humid evergreen forest. Singly, at mid- to upper levels in forest where perches for long periods without moving, overlooked easily. Nest is a small flimsy platform of fine twings at mid-levels. Eggs: 1, greyish, heavily mottled brown.

SS: See Rufous and Speckled mourners.

SD: F to C resident (SL−900 m) on Atlantic Slope from N Oax to Honduras.

RA: SE Mexico to Ecuador.

LOVELY COTINGA

Cotinga amabilis   Plate 43
Cotinga Azuleja

ID: 7−7.5 in (18−19 cm). SE humid lowlands. **Descr.** Ages/sexes differ. ♂: bill and legs dark grey. *'Electric' turquoise-blue overall with blackish-purple throat patch and deep purple patch on lower chest*; wings and tail black, edged turquoise-blue. ♀: *head and upperparts grey-brown, flecked and scalloped whitish*; wings and tail darker, edged brown. *Throat and underparts whitish, with dark spotting and mottling on chest and sides.* **Juv:** resembles ♀, attains ♂ plumage over 1st spring and summer.

Voice. Mostly silent; in flight dry, fluttering rattles may be made by both sexes.

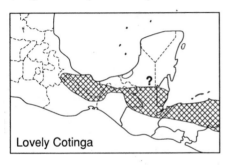

Lovely Cotinga

Habs. Humid evergreen forest and edge, open areas with scattered trees and forest patches. Perches conspicuously atop bare branches, singly or in pairs, in small groups at fruiting trees. Nest is a shallow cup in the canopy, eggs undescribed (?).

SD: U to F resident (near SL−1500 m) on Atlantic Slope from S Ver to Honduras. May be an altitudinal migrant locally (?).

RA: SE Mexico to Costa Rica.

THREE-WATTLED BELLBIRD

Procnias tricarunculata   Figure 34
Campanero Tricarunculado

ID: ♂ 12−13 in (30.5−33 cm); ♀ 10 in (25.5 cm). *C.A.* ♂ has 3 string-like black wattles hanging over bill. **Descr.** Ages/sexes differ. ♂: bill and legs blackish. *Chestnut overall with white head*, neck, and upper chest. ♀/ **juv:** *olive overall, throat and underparts streaked lemon.* In Feb−Apr, ♂ begins protracted molt into **1st basic:** shows patches of chestnut on underparts. By following Feb−Apr, resembles adult but extensively mottled olive, wings and tail mixed olive and chestnut, wattles often shorter. Appears to attain adult plumage by following spring but molts need study.

Voice. ♂♂ (Mar−Jun) give a remarkable, far-carrying, resonant, wooden *boi-nng* or *bohng*, usually preceded, at irregular intervals, by one or more piercing whistles and twangs. Silent most of year.

Fig. 34 Male Three-wattled Bellbird

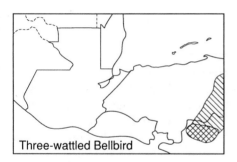

Three-wattled Bellbird

Habs. Humid evergreen forest. Singly or, in migration, in small groups. At mid- to upper levels, overlooked easily if not calling. ♂ sings from exposed perch in canopy; feeds at fruiting trees. Flight strong and fairly direct; a noted altitudinal migrant. Nest and eggs undescribed (?).

SD: F breeder (Mar–Sep; 1000–1800 m) in N cen Nicaragua. U to F winter visitor (Oct–Jan; 300–700 m) on Atlantic Slope of E Honduras and probably N Nicaragua.

RA: Breeds Nicaragua to Panama, to E Honduras in winter.

MANAKINS: PIPRIDAE (3)

A neotropical family of forest birds achieving its greatest diversity in S.A. Generally small and stocky with large heads and short bills; tails short or with elongated central rectrices, wings short and rounded. Ages/sexes similar or different; ♂♂ strikingly patterned in bright colors, ♀♀ dull olive. Young altricial; juvs resemble ♀♀ and quickly attain imm plumage. Molts poorly known. Voices vary, *Pipra* and *Manacus* noted for making loud wing snaps and buzzes at leks. Manakins spend much time sitting quietly and are overlooked easily. They move with abrupt jumps and short flights, their wings often making loud buzzes.

Food mostly fruit, also insects obtained by short sallying flights. Nests are shallow cups of fine dead leaves, rootlets, fibers, etc., suspended in horizontal forks at low to mid-levels. Eggs: 1–2.

WHITE-COLLARED MANAKIN
Manacus candei Plate 42
Saltarín Cuelliblanco

ID: 4.5–4.8 in (11.5–12 cm). SE humid forest. **Descr.** Ages/sexes differ. ♂: *bill blackish, legs yellow-orange. Black crown contrasts with white throat, chest, and broad hindcollar; belly and undertail coverts yellow.* Back and innerwing coverts whitish, outerwing coverts, remiges, upper rumps, and tail black, lower rump and uppertail coverts olive. ♀/**juv:** *olive overall, belly contrastingly yellowish.* **Imm** ♂: resembles ♀ but throat and median upper chest pale grey. May attain adult plumage when about 1 year old.
Voice. A nasal, slightly buzzy *rreu* or *pee-eer,* a rough *ehrr,* and a burry *chrru* and *chi-rr.* In display, ♂♂ make remarkably loud 'firecracker' wing snaps alternated with buzzy calls.
Habs. Humid evergreen forest edge and thickets, especially *Heliconia* patches, plantations. Low to mid-levels in shady understory, at times high in fruiting trees.

Often detected by strong buzz made by wings in flight. ♂♂ form leks low in shady understory: each clears a circle of forest floor and makes several displays centred about two slender saplings on either side of the cleared area. Eggs: 2, whitish, marked with browns (WFVZ).
SS: ♀ Red-capped Manakin uniformly olive, bill dull flesh, legs greyish.
SD: F to C resident (SL–500 m) on Atlantic Slope from N Oax to Honduras.
RA: SE Mexico to W Panama.
NB: Sometimes considered conspecific with the *M. vitellinus* complex of S C.A. and S.A.

LONG-TAILED MANAKIN
Chiroxiphia linearis Plate 42
Saltarín Colilargo

ID: 4.2–4.5+ in (11–11.5+ cm). *Pacific Slope S of Isthmus. Elongated central rectrices add 4–5 in (10–12.5 cm) to ♂ length, 1–1.5 in (2.5–4 cm) to ♀.* **Descr.** Ages/sexes differ. ♂: bill blackish, *legs orange. Black overall with crimson hind-*

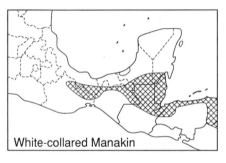

White-collared Manakin

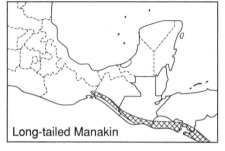

Long-tailed Manakin

crown patch and sky blue back. ♀/juv: *overall olive, slightly paler and greyer below.* 1st basic ♂: resembles ♀ but with red hindcrown patch. Successive molts to adult ♂ occur mainly Jun–Nov. 2nd basic ♂: head and underparts variably mottled black, central rectrices elongated as adult. 3rd basic ♂: more extensively blackish, back with variable amount of blue. Attains adult plumage by completion of 4th prebasic molt when about 3 years old.

Voice. A hollow to plaintive, ringing *te dee oh*, a richer, slightly explosive *weehoo'* or *whee-oo'*, and *clee-oo'*, a quieter *wheu*, and a low nasal *rreh* or *reeah*.

Habs. Humid to semihumid evergreen and semideciduous forest, gallery woodland. At low to mid-levels in shady understory. In display, 2 ♂♂ leap-frog one another rapidly along a display perch. Eggs: 2, buff, marked with browns (WFVZ).

SD: C to F resident (near SL–1500 m) on Pacific Slope from E Oax to El Salvador and Honduras.

RA: S Mexico to Costa Rica.

NB: Sometimes considered conspecific with *C. lanceolata* and *C. pareola* of Panama and S.A.

RED-CAPPED MANAKIN

Pipra m. mentalis   Plate 42
Saltarín Cabecirrojo

ID: 4–4.3 in (10–11 cm). SE humid forest.
Descr. Ages/sexes differ. ♂: eyes whitish, *bill dull flesh with dusky tip*, culmen may be dusky, *legs greyish. Black overall with flaming red head*, yellow chin and thighs.

♀/juv/imm ♂: eyes brown or flecked brown and white. *Overall dull olive*, darker above, palest on throat and belly. Some ♀♀ have red flecks on head. ♂ may attain adult plumage when about 1 year old.

Voice. Common call a high, thin, drawn-out lisp, typically followed by a sharp chip, *tsi sssssiu chik!* or *sssssiu plik*, sharp chips also given separately. In display, a hard, buzzy *zzzzrk*, a rapid snapping *kak-kak-kak-kak-kak-kak-kak-kak*, and a sharp, slightly liquid *lik-lik* and *lik-lik-lik'*, etc.

Habs. Humid evergreen and semideciduous forest. Low to high, mainly at mid-levels, often detected by strong buzz made by wings in flight. ♂♂ form leks in mid-level understory, each making several display movements, accompanied by loud wing snaps and buzzes. Eggs: 2, dark greyish buff, mottled with browns.

SS: See ♀ White-collared Manakin.

SD: C to F resident (SL–750 m) on Atlantic Slope from S Ver to Honduras.

RA: SE Mexico to Ecuador.

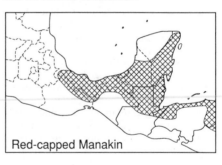

Red-capped Manakin

LARKS: ALAUDIDAE (I)

A primarily Old World family of terrestrial birds with one representative in the Americas. Bill fairly stout but pointed, hind claws long. Larks inhabit open, typically arid, temperate habitats. Ages/sexes differ, young altricial; juv attains adult plumage by complete 1st prebasic molt.

Food mostly seeds and insects. Nest is a slight hollow in open ground, lined with grasses. Eggs: 2–5, greyish white, densely speckled with browns (WFVZ).

HORNED LARK
Eremophila alpestris*   Figure 35
Alondra Cornuda

ID: 6.2–7.2 in (16–18.5 cm). *Open arid country N of Isthmus.* **Descr.** ♂: bill and legs blackish. *Face and throat washed lemon, with black mask and band across upper chest;* black forecrown extends back into slight 'horns' at sides. *Hindcrown* and upperparts *pinkish cinnamon* to pinkish brown, indistinctly streaked dusky; wings and central rectrices browner, rest of *tail blackish, contrasting in flight,* outer web of outermost rectrices white. ♀: black head and chest markings less distinct, hindcrown and upperparts with distinct dark streaking. **Juv:** crown and upperparts brown, spotted whitish, face with hint of adult pattern, wings edged buff to pale cinnamon. Underparts whitish, throat washed lemon-buff, chest spotted dusky.

Voice. Song 2 or more high, thin chips, run into a high, tinkling and trilled warble; also a more prolonged chipping and tinkling warble which may be given in flight, and a quiet, drier, prolonged chip-ping warble. Varied calls include a high, sweet *t-sleep* or *tsi-leep,* a buzzy *tzzzip* or *tzz-zzik,* and a slightly buzzy, rolled *trri-it.*

Habs. Arid to semiarid open country, grassland, ploughed fields, etc. Often in large flocks in autumn and winter when may associate loosely with American Pipits and longspurs. Runs and walks on ground, flight slightly undulating.

SD: C to F resident (SL–3500 m) in N and cen Baja, on Pacific Slope in Son, in interior from Chih and NL to cen Mexico, on Atlantic Slope in Tamps; locally F to U in interior Oax and on Pacific coast in Isthmus. More widespread in N Mexico in winter when N migrants present.

RA: Holarctic breeder, S in New World to Mexico, disjunctly in Colombia; N populations migratory.

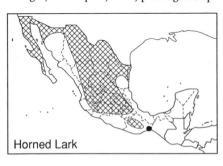
Horned Lark

Fig. 35 Horned Lark

SWALLOWS: HIRUNDINIDAE (14)

A cosmopolitan family of aerially feeding birds that resemble swifts superficially. Martins, at least in the New World, are simply large swallows. Bills small but gapes wide, feet small and weak. Wings long and pointed but fairly broad, tails fairly broad and often somewhat forked. Ages/sexes similar or different; young altricial. First prebasic molt usually complete but may occur over several months and be completed on the winter grounds. Swallows usually are dark, often glossy, above and pale below, with distinct head patterns; flight feathers blackish to dark grey, only described in the species accounts if different. Bills blackish, feet blackish to flesh, rarely visible in the field. Most species in the region are at least partial migrants and wintering areas are often not well defined, with birds being somewhat nomadic. Swallows are notably gregarious and several species join in feeding or roosting flocks. Drink by dipping into rivers, ponds, etc., while flying. Voices typically high twitters and buzzes.

Food mostly flying insects caught in the air. Some species attach their mud nests to buildings, rock faces, etc., others nest in cavities in trees, walls, banks, etc. Several species nest colonially. Eggs: usually 3–7, white, or finely marked with browns.

GENUS PROGNE

Martins are large swallows with cleft to forked tails. Taxonomy of the genus is problematic as virtually all species are allopatric breeders. Ages/sexes differ. Identification in the field may not always be possible and any martin suspected to be out of its range should be studied carefully. Nests are in cavities in dead trees, cliffs, large cacti, buildings, etc. Eggs: 2–5, white (WFVZ).

PURPLE MARTIN
Progne subis*　　　　　　Plate 44
Martín Azul

ID: 7.3–8 in (18.5–20.5 cm). *Large. Widespread transient, breeds N Mexico. Descr.* ♂: *glossy blue-black overall.* ♀: face dark grey-brown, crown blue-black, *forehead grey and hindcollar pale grey.* Upperparts mottled blue-black and grey. *Throat dusky pale grey with fine dark streaks, chest and flanks darker, dusky-brown with pale edgings creating mottled appearance; rest of underparts dingy whitish, streaked dusky, undertail coverts often show bold dark centers.* Underwings dark. **Juv/imm:** duller than ♀, head and upperparts dark

grey-brown with paler forehead, *pale grey collar,* blue-black sheen on crown and back (stronger in ♂). *Chest and flanks paler and more uniform overall than* ♀, with indistinct fine dark spots and streaks, rest of underparts whitish, washed dusky buff when fresh, with fine dark streaks. **1st alt** ♂: resembles ♀ but underparts often have some blue-black feathers, undertail coverts have narrow dark shaft streaks.

Voice. Rolled, burry, and twangy chirps much like Grey-breasted Martin; song a rich, often surprisingly loud, bubbling warble with buzzes and crackles thrown in.

Habs. Breeds in pine–oak woodland, open and semiopen arid areas, towns;

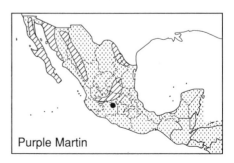
Purple Martin

widespread in migration. In large flocks during migration on Atlantic Slope.

SS: Other martins smaller, only ♀ Purple has contrasting pale forehead and collar; on other species, whiter belly and undertail coverts contrast more with chest; note cleft (versus forked) tail of Grey-breasted.

SD: F but local summer resident (Apr–Aug; SL–3000 m) in Baja, on Pacific Slope in Son, and in interior from Chih and NL to Mich and Gto. C to F transient (late Jan–May, Jul–Oct) on Atlantic Slope from NL and Tamps to Honduras; U to F on Pacific Slope from Son to Honduras. Vagrant (Aug–Nov) to Clipperton, and in winter to Nay (Dec 1983–Jan 1984, SNGH, PP, JS).

RA: Breeds N.A. to N Mexico; winters S.A.

NB: Cuban Martin *P. cryptoleuca* (7–7.7 in; 18–19.5 cm) breeds on Cuba and Isle of Pines, may occur in migration on Caribbean coast and islands of the region. ♂ told in hand from ♂ Purple Martin by concealed, extensively white abdomen. ♀ resembles Grey-breasted Martin but throat, chest, and sides duskier, belly contrastingly whiter (with a few indistinct dusky shaft streaks), undertail coverts have large dusky centers in adult; imm ♂ resembles ♀ but often has blue feathers below, mainly on chest, narrow dark shaft streaks on undertail coverts. Probably safely told only in the hand.

SINALOA MARTIN

Progne sinaloae　　　　　　　Plate 44
Martín Sinaloense

ID: 6.8–7.3 in (17–18.5 cm). *Endemic breeder; NW Mexico in summer.* **Descr.** ♂: *glossy blue-black overall with clean-cut white belly and undertail coverts,* dusky shaft streaks on undertail coverts inconspicuous. Underwings dark. ♀: *face, throat, chest, and flanks dusky brownish, median throat typically slightly paler* (hard

to see), *whitish belly and undertail coverts contrast strongly.* Crown, nape, and upperparts glossy blue-black, may be mottled grey. **Imm** ♂: resembles ♀ but upperparts usually solidly glossy blue-black, has a few blue feathers on chest. **Juv:** undescribed (?).

Voice. Undescribed (?); probably much as other *Progne*.

Habs. Pine and pine–oak forest and open woods, widely in migration. Breeds in small colonies, seen about tall dead snags or coming to drink at lakes and rivers. Eggs undescribed (?).

SS: See ♀ Purple Martin. Grey-breasted Martin has cleft (versus forked) tail, less contrast between white belly and dark chest; belly has fine dusky streaks visible at close range, throat paler and chest less sooty than Sinaloa.

SD/RA: Poorly known. U to F but local summer resident (Mar–Aug) on Pacific Slope and in adjacent interior from Son to Mich. Migration route (Mar–Apr and Aug) and wintering ground unknown (?); a Mar spec. from Petén, Guatemala, and sight reports from El Sumidero, Chis (GHR), and of several Sinaloa/Caribbean martins heading E out to sea over Belize City with a heavy movement of Purple Martins (late Aug 1962, BLM) suggest a migration route to/from the north coast of South America, perhaps crossing to/from the Pacific Slope mountains in the Isthmus. A report from Bermuda (Phillips 1986) refers to *P. dominicensis* (AMNH spec.).

NB: Carribbean Martin *P. dominicensis* (7–7.5 in; 18–19 cm), sometimes considered conspecific with Sinaloa Martin, has been reported from Isla Cozumel (Phillips 1986) but we consider the record hypothetical (see Appendix B). Much like Sinaloa Martin and probably safely identified only in the hand; ♂ has slightly narrower white median belly patch and dark spots on undertail coverts, ♀ has more extensively pale throat.

GREY-BREASTED MARTIN

*Progne chalybea**　　　　　　Plate 44
Martín Pechigris

ID: 6.3–7 in (16–18 cm). *Widespread in tropical lowlands.* **Descr.** ♂: *face, throat, chest, and flanks grey-brown,* throat often slightly paler, chest mottled darker; *belly and undertail coverts contrastingly whit-*

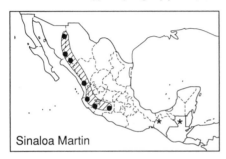

Sinaloa Martin

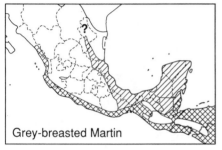

Grey-breasted Martin

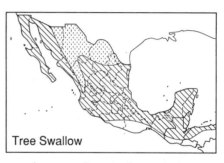

Tree Swallow

ish, *upper belly with fine dusky shaft streaks* (hard to see in field), lower belly to undertail coverts also may show fine dusky streaks. Crown, nape, and upperparts glossy blue-black. Underwings dark. ♀/imm: resemble ♂ but upperparts mottled dusky, throat often paler, chest more uniformly dusky, less mottled. **Juv:** resembles ♀ but upperparts dark brownish, paler throat and chest contrast less with belly.

Voice. Calls twangy and rolled, *chrru* or *chrrreu* and *chew*, a more burry *chiurr chi-chrr*, etc. Song often prolonged, a chirpy, slightly jerky warble.

Habs. Towns, forest and edge, clearings with dead trees, etc. Breeds singly or colonially, nests in cavities in trees, buildings, rocky islets.

SS: See ♀ Purple and Sinaloa martins. Rough-winged swallows smaller with less cleft tails, brown upperparts.

SD: C to F breeder (Feb–Aug; SL–1500 m) on both slopes from S Sin and S Tamps (N locally to NL) to El Salvador and Honduras; on Pacific Slope in winter, U to F N of Isthmus to Nay, F to C from Isthmus S; withdraws in winter from Atlantic Slope N of Isthmus, U to R (Nov–mid-Jan) from Belize S. C to F transient (mid-Jan–Apr, Aug–Sep) on both slopes. See Phillips (1986) for clarification of this species' M.A. range.

RA: Mexico to Peru and N Argentina; N populations partly migratory.

TREE SWALLOW

Tachycineta bicolor　　Not illustrated
Golondrina Arbolera

ID: 5.2–5.7 in (13.5–14.5 cm). *Widespread winter visitor*. Tail forked to cleft. Descr. Ages/sexes differ but many ♀♀ may resemble ♂. **Adult:** *head and upperparts metallic greenish blue to blue, throat and*

underparts white. Underwings dark. ♀/imm: *head and upperparts brown or mottled with metallic greenish blue.* Throat and underparts white, *chest often washed dusky in imm* and may form band suggesting Bank Swallow.

Voice. A chirping *chrip* and *chi-ip*.

Habs. Open and semiopen areas, often near lagoons and marshes. Winter movements poorly known; at times in huge flocks.

SS: Violet-green Swallow (mainly highlands) has white sides of rump, white to whitish face, mainly green upperparts. Mangrove Swallow smaller with white rump. Rough-winged swallows have dusky throat and chest, squarer-tipped tails. Bank Swallow smaller with cleaner-cut brown chest band.

SD: Irregularly F to C transient and winter visitor (Aug–May in NW Mexico, Oct–Apr in SE Mexico; SL–2500 m) throughout most of region N of Isthmus (U to R in winter N in interior to S Chih and S Coah, SNGH) and on Atlantic Slope to Belize; U to F on Atlantic Slope to Honduras and in interior from Isthmus to Honduras. Main winter concentrations are on Atlantic Slope in marshes from S Ver to Belize.

RA: Breeds N.A.; winters S USA to M.A.

MANGROVE SWALLOW

Tachycineta a. albilinea　　Plate 44
Golondrina Manglera

ID: 4.3–4.7 in (11–12 cm). *Widespread in coastal lowlands.* Tail cleft to slightly cleft. **Descr.** Ages differ, sexes similar. **Adult:** *head and upperparts metallic blue-green to greenish blue with bold white rump,* also narrow white V on forehead, tertials edged white. *Throat and underparts white;* underwing coverts white. **Juv:** bluish-green areas of adult are grey-brown, attains adult-like plumage by early winter.

Voice. Varied chirping calls, *chiri-chrit*

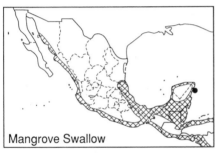

Mangrove Swallow

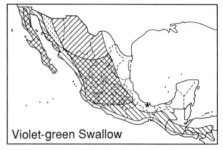

Violet-green Swallow

and *chrit* or *chriet*, also buzzier chirps. Song a varied series of chirps and burry chips.

Habs. Mangroves, rivers, coastal lagoons, ponds, and marshes in farmland. Singly or in small groups, at times in hundreds near roosts. Typically flies low over water and perches at low to medium heights. Nests singly or colonially in cavities in trees, pitted rocks, walls, etc. Eggs: 3–5, white, often with faint cinnamon markings (WFVZ).

SS: Violet-Green Swallow has white face, green upperparts, white only on sides of rump but often looks white-rumped, especially in NW Mexico. See Tree Swallow. White chevron on forehead of Mangrove distinctive if seen.

SD: C to F resident (SL–600 m) on both slopes from cen Son and S Tamps to El Salvador and Honduras, though R in El Salvador due to pesticide pollution and habitat loss (Thurber *et al.* 1987). Mainly along coast but also inland along rivers and at lakes, on Pacific Slope locally along Balsas drainage in Mich, on Atlantic Slope inland from SE SLP to Honduras. Large flocks may wander in winter, at least in NW Mexico (Sin and Nay), depending on local conditions. Seasonal movements need study.

RA: Mexico to Panama, disjunctly in Peru.

NB: *T. (a.?) stolzmanni* of Peru sometimes considered specifically distinct.

VIOLET-GREEN SWALLOW
*Tachycineta thalassina**   Plate 44
Golondrina Cariblanca

ID: 4.5–5 in (11.5–12.5 cm). *Widespread, mainly highlands.* Tail cleft. **Descr.** Ages/ sexes differ. ♂: *crown, nape, and upperparts glossy dark green*, becoming violet on uppertail coverts, narrow violet hindcollar rarely visible; some birds with purplish back and dark green uppertail coverts; tertials broadly edged white when

fresh. *Face, throat, underparts, and sides of rump white* (can appear white-rumped). Underwings dark. ♀: duller above, face washed dusky. **Juv:** resembles ♀ but crown, nape, and upperparts brown, underparts may be dirty whitish.

Voice. A slightly buzzy or burry *chi-chit* and clipped *ji-jit*; song a varied, twangy, chipping series with short burry warbles thrown in.

Habs. Open and semiopen areas, mainly in temperate regions; in Baja and around Gulf of California in deserts, elsewhere mainly pine and pine–oak woodland. In flocks, often high overhead, perches at mid- to upper levels. Nests in cavities in trees, cacti, etc. Eggs: 4–5, white, often with faint cinnamon markings (WFVZ).

SS: See Tree and Mangrove swallows.

SD: C to F breeder (Feb–Aug; SL–3000 m) in Baja, on Pacific Slope in Son, and in interior and on adjacent slopes from Chih and NL to cen volcanic belt, possibly in interior to Oax (Binford 1989). Irregularly C to F winter visitor on Pacific Slope from S Son to Nay and in interior and on adjacent slopes from S Plateau to Oax, U to F (Sep–Mar; mainly 500–3000 m) to El Salvador and N cen Nicaragua. Migrates Jan–May, Jul–Oct.

RA: Breeds W N.A. to Mexico; winters Mexico to C.A.

BLUE-AND-WHITE SWALLOW
Notiochelidon cyanoleuca patagonica
Golondrina Azuliblanca   Plate 44

ID: 5–5.3 in (12.5–13.5 cm). *Rare austral migrant.* Tail fairly deeply forked. **Descr.** Ages differ, sexes similar. **Adult:** *head and upperparts metallic blue, throat and underparts white with blackish undertail coverts.* Underwings dusky. **Imm:** head and upperparts mottled brown, chest

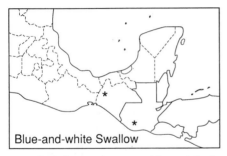

Blue-and-white Swallow

washed brownish. Resembles adult by Jun–Jul.

Voice. Flight calls a buzzy *dzzzhir* and *zzhie* (Chile).

Habs. Aerial, likely to be found with other swallows, especially Black-capped which it resembles in shape, flight, jizz, etc.

SS: Black-capped Swallow has more deeply forked tail, black cap contrasts with brown upperparts, throat and chest flecked dusky. Tree Swallow unlikely in S Mexico at same season, tail less forked, flight manner different, undertail coverts white.

SD: R (irregularly U?) non-breeding visitor (May–Jul; 900–1500 m) S of Isthmus; recorded Chis (Paynter and Alvarez del Toro 1956) and Guatemala (PH). Probably more regular, should be looked for.

RA: Costa Rica to Tierra del Fuego; S populations winter to M.A.

NB: Traditionally placed in the genus *Pygochelidon*. Blue-and-white and Black-capped swallows sometimes placed in the genus *Atticora*.

BLACK-CAPPED SWALLOW
Notiochelidon pileata Plate 44
Golondrina Gorrinegra

ID: 5–5.2 in (12.5–13.5 cm). *Endemic to highlands S of Isthmus. Tail deeply forked.* **Descr.** Ages differ slightly, sexes similar. **Adult:** *black head contrasts with brown*

upperparts. Throat and chest whitish, flecked brown, white belly forms contrasting oval patch set off by brown flanks and blackish vent and undertail coverts. Underwings dark. **Juv:** throat and chest lack dark flecking and often washed pale cinnamon; upperparts and tertials tipped pale cinnamon.

Voice. An up-slurred nasal *sreet* or *zrieh*, a buzzy *zrrieh*, a liquid trilled *tri-i-it*; song a few dry chips run into a buzzy crackling chatter.

Habs. Pine and pine–oak woodland, towns, forest edge, clearings. Flight fast and 'twinkling', often in flocks high overhead. Nests in holes excavated in banks, road cuts, etc., possibly in buildings, often in colonies. Eggs: 4, white.

SS: Flight and deeply forked tail unlike other swallows occurring regularly in region. See Blue-and-white Swallow.

SD/RA: Locally/seasonally C to F resident (1000–3000 m) in interior and on adjacent slopes from Chis to El Salvador and W Honduras; wanders in winter, at least in Chis (SNGH), movements need study.

NB: See Blue-and-white Swallow.

NORTHERN ROUGH-WINGED SWALLOW
Stelgidopteryx serripennis * Plate 44
Golondrina-aliserraada Norteña

ID: 5–5.5 in (12.5–14 cm). *Widespread. Tail slightly cleft.* **Descr.** Ages differ slightly, sexes similar. **Adult:** *head and upperparts brown. Throat, chest, and flanks dusky, belly and undertail coverts whitish.* Underwings dark. **Juv:** wing coverts and tertials edged cinnamon, throat and chest washed pale cinnamon.

Voice. A buzzy *zzrt* or *zzzzr*, and *zzz-zzz-zzit*, etc.

Habs. Open and semiopen country in general. Singly or in flocks, at times in

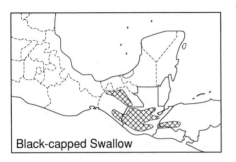

Black-capped Swallow

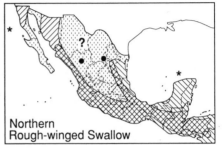

Northern Rough-winged Swallow

hundreds. Associates readily with other swallows, especially during migration. Nests in holes in banks, road cuts, etc., often colonially. Eggs: 4–6, white, often with faint cinnamon markings (WFVZ).

SS: Ridgway's Rough-winged Swallow darker overall, told by pale forehead spots, blackish distal undertail coverts, more deeply cleft tail. Bank Swallow smaller with white throat and brown chest band. Cave Swallow smaller with squarer tail, cinnamon to chestnut rump, face and throat cinnamon, undertail-coverts dusky. See Grey-breasted Martin, Southern Rough-winged Swallow (E Honduras, Appendix E).

SD: C to F but often local breeder (Mar–Aug, near SL–2500 m; commonest in foothills) in BCN, on Pacific Slope from Son to Oax, in interior from cen Mexico to N cen Nicaragua, and on Atlantic Slope from S NL to Honduras; U to F and local on Plateau (Dgo and S Coah at least). More widespread and common in winter and during migration (Aug–Apr), winters mainly from S Son and Tamps S.

RA: Breeds N.A. to Costa Rica; winters S USA and Mexico to Panama.

NB: Sometimes considered conspecific with Ridgway's Rough-winged Swallow and Southern Rough-winged Swallow *S. ruficollis* (S C.A. and S.A.), but see Phillips (1986) and Stiles (1981*b*), respectively.

RIDGWAY'S [NORTHERN] ROUGH-WINGED SWALLOW
Stelgidopteryx ridgwayi * Plate 44
Golondrina-aliserrada Yucateca

ID: 5–5.5 in (12.5–14 cm). Endemic; mainly Yuc Pen. Tail cleft. **Descr.** Ages differ, sexes similar. **Adult:** *head and upperparts dark brown with whitish supraloral spots. Throat, chest, and flanks dusky* (more extensive in *stuarti* of S Ver to Belize), *belly*

and undertail coverts whitish, longest undertail coverts broadly tipped blackish. Underwings dark. **Juv:** wing coverts and tertials edged cinnamon, throat and chest washed pale cinnamon.

Voice. A hard, buzzy *zzzrrt* or *zzzzr*, harder than Northern Rough-winged Swallow, and a spluttering, liquid warble *splik spli spli-tit* . . ., etc.

Habs. Forest edge, clearings, open and semiopen areas. Habits much as Northern Rough-winged Swallow, associates loosely with other swallows. Nests in cavities in caves, limestone sinks, ruins. Eggs undescribed (?).

SS: See Northern Rough-winged Swallow, Grey-breasted Martin. Cave Swallow smaller with squarer tail, chestnut rump, cinnamon face and throat. Bank Swallow smaller with white throat and brown chest band.

SD/RA: C to F resident (SL–500 m) from Yuc Pen to S Ver, N Guatemala, and Belize. Range limits still need to be determined.

NB: See Northern Rough-winged Swallow.

BANK SWALLOW
Riparia r. riparia Not illustrated
Golondrina Ribereña

ID: 4.7–5.2 in; 12–13 cm. *Widespread transient. Tail cleft.* **Descr.** Ages differ slightly, sexes similar. **Adult:** *head and upperparts brown. Throat and underparts whitish with contrasting brown chest band.* Underwings dark. **Juv:** wing coverts and tertials edged cinnamon, throat often washed pale cinnamon.

Voice. A rolled, gravelly *zzzr* and *dzzz zzzrt* or *zzzr zzzt*, etc., and a higher, slightly reedy *zrieh* in alarm. Song a more prolonged, rolled, gravelly twittering.

Habs. Widely in migration, mainly near water, often with other swallows. Nests

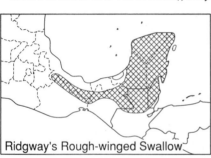

Ridgway's Rough-winged Swallow

Bank Swallow

colonially in holes in banks, typically near water. Eggs: 4–5, white, often with faint cinnamon markings (WFVZ).

SS: See Rough-winged swallows.

SD: U and local summer resident (Apr–Aug) from N Coah to N Tamps (Phillips 1986; SNGH, SW). F to C transient (mid-Mar–May, Aug–Oct; SL–3000 m) throughout most of region. Irregularly U to F winter visitor (Nov–Feb) on Pacific Slope from BCS (SNGH, SW) and S Sin (Phillips 1986) to Gro, probably also to N C.A.

RA: Breeds N.A. to N Mexico, also Eurasia; winters in New World mainly in S.A., irregularly to W Mexico.

CLIFF SWALLOW
Hirundo pyrrhonota* Plate 44
Golondrina Risquera

ID: 5–5.5 in (12.5–14 cm). *Summer visitor and transient. Tail squared to slightly cleft.* **Descr.** Ages differ, sexes similar. **Adult:** forehead chestnut (pale buff in NW breeders and N.A. migrants), chestnut from face continues back as narrow hindcollar behind blue-black crown; *chestnut throat with blackish central patch contrasts strongly with whitish underparts,* chest and flanks washed greyish cinnamon, undertail coverts dusky cinnamon with dark centers. *Upperparts glossy blue-black,* streaked whitish on mantle, *with contrasting pale cinnamon rump* (buff in many N migrants). Underwings dusky. **Juv:** head and upperparts duller and browner, throat and chest paler with hint of adult pattern, forehead dull cinnamon to dusky (all populations), tertials edged dull cinnamon.

Voice. A dry *chrri-chrri* and more nasal *chreh,* a slightly burry, nasal *rrih* and lower *rreh,* a burry *rrieh,* and a nasal *jeh,* often doubled. Song a crackling, spluttering warble.

Habs. Open and semiopen areas, towns, and villages, widely in migration. Nests colonially, builds gourd-shaped mud nests on buildings, cliffs, etc. Eggs: 3–5, white, may be flecked with browns (WFVZ).

SS: Cave Swallow has cinnamon throat not contrasting strongly with whitish underparts, rump less contrasting (especially Yuc Pen), could be confused with juv Cliff Swallow.

SD: C to F summer resident (Feb–Aug; near SL–2700 m) in BCN, on Pacific Slope S to N Sin, and in interior S to cen Oax (local S of cen volcanic belt). N.A. breeders are C to F transients (Feb–May, mid-Aug–Oct) throughout, commonest in coastal lowlands.

RA: Breeds N.A. to Mexico; winters Brazil to Argentina.

NB: Formerly placed in the genus *Petrochelidon.*

CAVE SWALLOW
Hirundo fulva* Plate 44
Golondrina Pueblera

ID: 5–5.7 in (12.5–14.5 cm). *Local in E and S Mexico. Tail squared to slightly cleft.* **Descr.** Ages differ slightly, sexes similar. **Adult:** forehead cinnamon-rufous (chestnut in *citata* of Yuc Pen), cinnamon from face continues back as narrow hindcollar behind blue-black crown; *cinnamon throat and upper chest do not contrast strongly with whitish underparts,* flanks washed greyish cinnamon, undertail coverts dusky cinnamon with dark centers. *Upperparts glossy blue-black,* streaked whitish on mantle, rump cinnamon (chestnut and poorly contrasting in *citata*). Underwings dusky. **Juv:** head and upperparts duller and browner, throat paler, chest washed dusky, forehead duller, tertials edged dull cinnamon.

Voice. In Yuc Pen, a soft, nasal *sweik* or

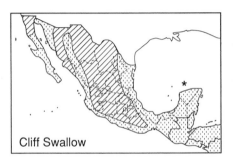

Cliff Swallow

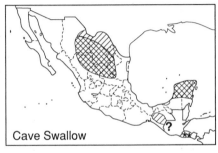

Cave Swallow

whiet; song a dry, nasal, crackling warble. In NE Mexico calls richer, *zweih* and *chrieh*, suggesting Barn Swallow; song a dry, slightly buzzy, chattering warble, not as nasal or crackling as Yuc Pen birds.

Habs. Towns, cities, open and semiopen areas. Nests colonially, builds deep mud cup nests on buildings, in caves, etc. Eggs (NE Mexico): 3–5, much like Cliff Swallow (WFVZ).

SS: See Cliff Swallow; also rough-winged swallows in Yuc Pen.

SD: C to F breeder (Feb–Oct; 500–2100 m) in interior from Chih and W Tamps to N Zac and SLP; in interior Chis, and (around SL) in Yuc Pen S to cen Camp and QR. Irregularly (?) F to C in winter (Nov–Jan) in interior at least from Chih and Dgo to NL (SNGH, SW) though wintering in such numbers may be a recent trend; expanding its breeding range in SW USA but comparable data in Mexico lacking. Nomadic during winter in Chis (absent Nov–Jan?) and Yuc Pen; movements need study. Recently found to be locally F (Dec 1992–Jan 1993) in El Salvador (OK), perhaps wintering birds from Chis.

RA: SW USA, Mexico, and Caribbean. Migratory status unclear.

NB1: Formerly placed in the genus *Petrochelidon*. **NB2:** *H. rufocollaris* of Ecuador and Peru sometimes considered conspecific with Cave Swallow.

BARN SWALLOW

Hirundo rustica erythrogaster Not illustrated
Golondrina Ranchera

ID: 5–5.7+ in (12.5–14.5+ cm). *Widespread migrant, breeds in Mexico. Tail deeply forked; adults have long outer tail streamers* adding 1–1.5 in (2.5–3.5 cm) to length. **Descr.** Ages differ, sexes similar. **Adult:** *head and upperparts glossy blue-black with chestnut forehead; all but cen-*tral *rectrices have bold white central spot on inner webs. Throat and upper chest cinnamon with dark mark at sides of neck, underparts pale vinaceous cinnamon to buffy cinnamon.* Underwings dusky. **Juv:** throat and underparts paler, upperparts duller, chest often washed dusky.

Voice. A clipped *chi-dit* or *ch-jit* and an up-slurred *zwieh*, usually near nesting areas, also a mellow, rolled *cheht* and *cheh-cheht* or *chi-chi*. Song a rich, often prolonged, burbling warble with scratchy notes thrown in; ♂ has a wet buzz often incorporated into song.

Habs. Open and semiopen areas, typically near human habitation. Widely in migration. Builds mud cup nests on buildings, bridges, etc. Eggs: 4–6, whitish, speckled with browns and greys (WFVZ).

SD: C to F breeder (Feb–Oct; near SL–3000 m), mainly in interior, S to cen Mexico, U in NW BCN. Irregularly U to C in winter (Oct–Mar) in BCS (SNGH, PP, SW), on both slopes from Nay and S Ver, and in interior from cen Mexico, through region, though R to U in Yuc Pen. N.A. breeders are C to F transients (Mar–May, Aug–Nov) throughout region. Winter movements need study.

RA: Breeds N.A. to Mexico, also Eurasia; winters in New World from Mexico to Tierra del Fuego.

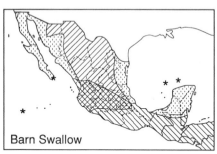

Barn Swallow

JAYS AND CROWS: CORVIDAE (24)

A cosmopolitan family of social, intelligent, and often noisy birds, most of which occur in wooded and forested habitats. Jays are brightly colored and patterned, mainly in blues, black, and white; several have crests. Crows are black. Bills and legs fairly stout and strong, feet strong and grasping. Ages usually differ in jays, similar in crows; sexes similar; young altricial. Juvs quickly attain adult-like plumage by partial 1st prebasic molt. Bills and legs usually blackish, but young of several jays have yellow bills.

Varied diet includes insects, fruit, seeds, nestling birds, reptiles, amphibians, and carrion. Nests are bulky cups of sticks and twigs, with finer lining of rootlets, hairs, etc., usually at mid- to upper levels in trees. Some species breed cooperatively with several birds helping to build and attend a single nest; other birds of the same species may nest as separate pairs.

STELLER'S JAY

Cyanocitta stelleri*   Plate 46
Chara de Steller

ID: 11–12 in (28–30.5 cm). Highland forests. Two ssp groups: *diademata* group (N and cen Mexico) and *coronata* group (E and SW Mexico to C.A.); *erectile crest often conspicuous* (longer in *diademata* group). **Descr.** Ages differ, *Diademata*. **Adult:** bill and legs blackish. *Head and crest blackish with white eye-crescents* and whitish streaking on chin and forehead. Back smoky grey, becoming pale blue on rump and uppertail coverts. Wings and tail blue, greater coverts and tertials barred black, tail indistinctly barred blackish. Upper chest blackish grey, becoming blue on underbody. **Juv:** head and body sooty grey, crown and crest blacker, white eye-crescents indistinct or lacking, greater coverts unbarred. *Coronata*. **Adult:** *deep blue overall, face blacker with white eye-crescents*, whitish streaking on chin and

forehead. Greater upperwing coverts and tertials barred black, tail with faint dark barring. **Juv:** head and body sooty blue, white eye-crescents indistinct or lacking, greater coverts unbarred.

Voice. Varied, including a rasping *sshhehk-sshhehk*, a ringing *cheh-cheh-cheh-chenk* or *sheh-eh-eh-ehnk*, a rasping, nasal *kyeh-kyeh* or *cheh-cheh* ..., a shrill, slightly metallic *bik-bik bik-bik*, a hard, dry rattle, and a hawk-like scream *reeeah*.

Habs. Humid to semihumid conifer, pine–oak, and oak forest and edge. In pairs or groups, often joins mixed-species flocks. Eggs: 2–5, pale bluish, speckled with browns and greys (WFVZ).

SS: Unicolored Jay lacks crest, overall more uniform without white eye-crescents.

SD: *Diademata*: C to F resident (1000–3500 m) in interior and on adjacent slopes from Son and S Coah to cen volcanic belt from E Mich to W Ver. Vagrant (May 1885) to BCN. *Coronata*: C to F resident (1000–3500 m) from W Mich and SE SLP to N cen Nicaragua.

RA: W N.A. to N cen Nicaragua.

NB1: The plumages of the two ssp groups, including juvs, are quite distinct, although the darker head of *purpurea* (W Mich) suggests introgression between the two groups; studies should be carried out in potential areas of contact (cen Mich or W cen Ver?) to investigate their taxonomic status. **NB2: Blue Jay** *Cyanocitta cristata* (10–11 in; 25.5–28 cm) probably occurs as irregular winter visitor to NE Mexico;

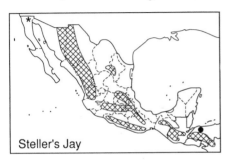

Steller's Jay

has blue crown and crest, whitish face and throat bordered black, underparts pale greyish. Upperparts bluish, wings and tail barred black, greater coverts, tertials, and tail boldly tipped white. Raucous *keeah-keeah* call. See Appendix B.

BLACK-THROATED MAGPIE-JAY
Calocitta (formosa?) colliei   Plate 45
Urraca-hermosa Carinegra

ID: 23–30 in (58.5–76.5 cm). *Endemic to NW Mexico. Tail very long and graduated; crest conspicuous.* **Descr.** Ages differ slightly. **Adult:** bill and legs black. *Head, crest, and chest black* with pale blue eye-crescents and subocular patch. *Nape and upperparts blue*, brightest on tail, *outer rectrices broadly tipped white.* Underparts white. May show partly or mostly white throat and chest, N at least to Sin. **Juv:** crest tipped white, subocular patch smaller and darker blue. Apparently interbreeds with White-throated Magpie-Jay in Jal and Col; birds with intermediate characters may be locally common.
  Voice. Varied, loud, raucous cries may suggest large parrots, a rolling *rrrik* or *krrrrup*, and *rroik*, a loud, clear, whistled *wheeoo*, a hard, hollow ringing *kyooh*, a raucous, rolled *krrowh* and *rrow*, a nasal *rriihk*, etc.
  Habs. Arid to semihumid woodland, semiopen areas with scattered trees and forest patches. In pairs or small groups, flight slow and graceful, long tail sweeping out behind. Cooperative breeder (?). Eggs: 3–7, whitish, spotted with browns and greys (WFVZ).
SS: White-throated Magpie-Jay locally sympatric in Jal and NW Col, shorter tailed, paler overall with mostly white face and throat, narrow black chest band.
SD/RA: C to F resident (SL–1800 m) on

Pacific Slope from S Son to Jal and NW Col (HG).
NB: Individual Black-throateds often show some white in throat and chest N at least to Sin; White-throateds N of Isthmus often show variable black in face from Jal to Gro. The eggs and juv plumages of the two forms are quite different, at least where allopatric; studies in areas of sympatry are needed.

WHITE-THROATED MAGPIE-JAY
*Calocitta formosa**   Plate 45
Urraca-hermosa Cariblanca

ID: 17–22 in (43–56 cm). *Pacific Slope. Tail long and graduated; crest conspicuous,* often recurved. **Descr.** Ages similar. Bill and legs black. *Crown and crest black, white face* margined black; crown mostly blue from Isthmus S. *Throat and underparts white with narrow black chest band; C. f. formosa* (N of Isthmus) often has a partial black malar stripe (slight or lacking S of Isthmus) and some birds (mainly juvs?) have mostly black face. Upperparts grey-blue, bluer on tail, outer rectrices broadly tipped white. Apparently interbreeds with Black-throated Magpie-Jay in Jal and Col.
  Voice. Varied, loud, and noisy. Calls range from harsh and raucous to mellow and squeaky, and often suggest large parrots: including a harsh *rrah* and *rrah rrah*, a sharp *whuit'*, a mellow, whistled *who-whee*, a loud *whee-wheep'*, etc.
  Habs. Arid to semihumid woodland, semiopen areas with scattered trees and forest patches. Habits much as Black-throated Magpie-Jay. Cooperative breeder. Eggs: 2–6, pale grey, densely speckled brown (WFVZ).
SS: See Black-throated Magpie-Jay.
SD: C to F resident (SL–1400 m) on Pacific Slope from S Jal, and in interior from

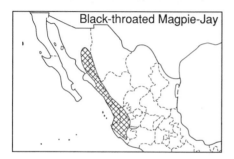

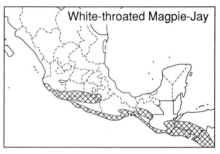

Balsas drainage, to El Salvador and Honduras.

RA: W Mexico to NW Costa Rica.
NB1: See Black-throated Magpie-Jay. **NB2:** Has hybridized with Brown Jay (Pitelka *et al.* 1956).

TUFTED JAY
Cyanocorax dickeyi   Plate 45
Chara Pinta

ID: 14–15 in (35.5–38 cm). Endemic to NW Mexico. *Spectacular, with stiff bushy crest,* shorter in juv. **Descr.** Ages differ. **Adult:** *eyes yellow,* bill and legs blackish. *Crest, face, throat,* and upper chest *black with white supraocular spot and subocular patch. Nape,* upper mantle, and underparts *white.* Upperparts blackish blue, *tail blackish blue at base, white distally* (looks all white from below). **Imm:** lacks supraocular spot, subocular spot smaller and bluish. **Juv:** resembles imm but upperparts sooty greyish.
  Voice. Varied, including a ringing *cheh-cheh chen-chen-chenk,* a dry, rattling chatter, a hard, dry *chah-chak* in series up to 3×, a fairly soft, nasal *beehn beehn beehn* ..., and a slightly ringing *chuh-chuh chuh-chuk,* etc.
  Habs. Humid to semihumid semideciduous and pine–oak forest. In groups up to 12–15 birds, feeds low to high, at times associates with Steller's Jay. Cooperative breeder. Eggs: 2–4, greyish white, densely speckled and mottled with browns and greys (WFVZ).
SD/RA: F to C resident (1500–2100 m) on Pacific Slope from Sin and Dgo to N Nay.
NB: Sometimes known as Dickey Jay.

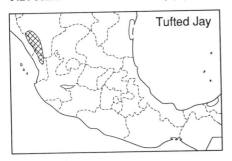
Tufted Jay

GREEN JAY
*Cyanocorax yncas**   Plate 45
Chara Verde

ID: 10.5–12 in (26.5–30.5 cm); N > S.

Small and flashy. Noticeable geographic variation. **Descr.** Ages similar. Eyes yellow (brown in NE and W Mexico, and in juvs), bill and legs black. *Crown and nape blue,* face, throat, and chest black with blue supraocular spot and subocular patch. *Upperparts green,* tail slightly bluer with *yellow outer rectrices striking when tail spread.* Underparts greenish to greenish yellow.
  Voice. Varied calls include a shrill, almost parakeet-like chatter, *cha-cha* ..., a loud, ringing *chenk chenk chenk* or *jink-jink-jink,* a drier scolding *cheh-cheh* ..., a buzzier, nasal scolding *jehr jihjihjih* ..., and a throaty, frog-like croaking, *ahrrrrrrr.*
  Habs. Forest and edge, plantations, scrubby woodland. In pairs or small groups, usually at low to mid-levels. Cooperative breeder; nest may be at low to mid-levels in thickets. Eggs: 3–5, greyish white, speckled and marked with browns and greys (WFVZ).
SD: C to F resident (SL–2000 m) on Pacific Slope from Gro to Guatemala, on Atlantic Slope from E NL and Tamps to W Honduras; U to F and local on Pacific Slope N to Nay.
RA: Mexico (also S Texas) to Honduras, disjunctly in S.A. from Colombia to Bolivia.

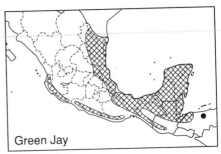
Green Jay

BROWN JAY
*Cyanocorax morio**   Plate 45
Chara Papán

ID: 15–17 in (38–43 cm). *Atlantic Slope; large and noisy.* Dimorphic. **Descr.** Ages differ in color of bare parts: adult bill and legs black; juv bill and orbital ring yellow, becoming black by 2nd winter. **Dark morph** (E Mexico): *overall dark brown with paler belly* and undertail coverts. **Light morph** (SE Mexico and C.A.): *overall dark brown with contrasting whitish*

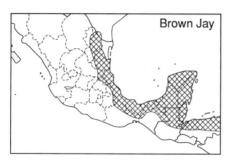

Brown Jay

belly and undertail coverts, *outer rectrices broadly tipped white.* The two morphs interbreed readily from Ver to N Chis.

Voice. Monotonous and obnoxious, an all-too-soon familiar loud screaming *kyeeah!* or *k'yaah!* and *kaah! kaah!*, etc., often repeated mercilessly; also a mewing, slightly drawn-out *reyaah* which may be repeated steadily.

Habs. Humid to semiarid forest and edge, second growth, plantations semiopen areas with scattered trees and hedges. In groups up to 5–15 birds, feeds low to high, at times on ground in open fields. Cooperative breeder. Eggs: 2–7, pale grey, densely speckled brown (WFVZ); larger clutches laid by more than 1 ♀.

SS: When flashing quickly across the road, may suggest Plain Chachalaca which has longer neck, small head.

SD: C to F resident (SL–1500 m, mainly below 1000 m) on Atlantic Slope from NL and Tamps to Honduras.

RA: E Mexico (also S Texas) to W Panama.

NB1: Often placed in the genus *Psilorhinus*.

NB2: The two morphs were once considered separate species. **NB3:** See NB2 under White-throated Magpie-Jay.

SUBGENUS CISSILOPHA

The following four species of 'black-and-blue' jays constitute the subgenus *Cissilopha*, sometimes considered a full genus, endemic to northern M.A. Ages differ, mostly in color of bare parts. All are cooperative breeders, usually found in small groups, sometimes in flocks of 15–20 birds. Forage low to high, often on or near ground, and attend army-ant swarms. Eggs: 3–6, pale pinkish buff (Yucatan and San Blas) to vinaceous (Purplish-backed and Bushy-crested), marked with reddish browns (WFVZ).

BUSHY-CRESTED JAY

Cyanocorax melanocyaneus*   Plate 70
Chara Centroamericana

ID: 12–13 in (30.5–33 cm). *Endemic to N C.A.* Erectile bushy crest usually held flat. **Descr. Adult:** eyes yellow, bill black, legs blackish. *Head and chest black, belly and undertail coverts dull blue. Upperparts turquoise-blue,* tail deeper blue. **Juv:** head and chest blackish grey, rest of underparts smoky grey, upperparts dull bluish, wings and tail blue. *Eyes brown,* yellow by 2nd winter, bill yellowish, mostly black by 1st summer.

Voice. A nasal scolding *kyah kyah ...,* a more raucous *kyeh-kyeh ...,* and a drier *cheh-cheh-cheh ...* which may suggest a small parrot.

Habs. Arid to semihumid forest and edge, second growth, plantations. See genus note.

SS: None in range.

SD/RA: C to F resident (600–2400 m) in interior and on Pacific Slope from Guatemala to Honduras and N cen Nicaragua.

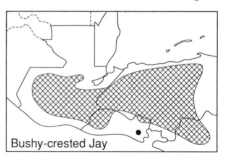

Bushy-crested Jay

PURPLISH-BACKED JAY

Cyanocorax beecheii   Plate 45
Chara de Beechy

ID: 14–15 in (35.5–40.5 cm). *Endemic to NW Mexico.* Erectile bushy crest usually held flat. **Descr. Adult:** eyes yellow, bill black, legs flesh-yellow. *Head and underparts black, upperparts deep purplish.* **Juv:** head and chest blackish becoming smoky grey on rest of underparts. Upperparts dull bluish, wings and tail deep purplish. Eyes brown, yellow by 3rd winter, bill yellowish, black by 3rd winter.

Voice. A loud, raucous, metallic to slightly rasping *ahhr* or *ehrrr* and *tchaah,* repeated, a rough, slightly bell-like *ownk,*

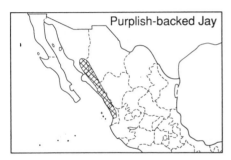

Purplish-backed Jay

usually doubled, a raspier *ehrr ehrr*, and a throaty *rowhr*, etc.

Habs. Arid to semiarid forest and edge, mangroves. See genus note.

SS: See San Blas Jay.

SD/RA: F to C resident (SL–600 m) on Pacific Slope from S Son to Nay (vicinity of San Blas).

SAN BLAS JAY

Cyanocorax sanblasianus ⁎   Plate 45
Chara de San Blas

ID: 12.5–13.5 in (32–34.5 cm). *Endemic to W Mexico.* Erectile frontal crest may be held flattened. **Descr. Adult:** eyes yellowish, bill black, *legs dull yellowish. Head and underparts black,* undertail coverts dark bluish; *upperparts blue,* deepest on tail. **Juv:** head blackish, becoming smoky grey on underbody, upperparts dull bluish, wings and tail blue. Frontal crest longer, like adult after 2nd prebasic molt. *Eyes brown,* like adult in 1–3 years, bill yellowish horn, black by 2nd winter, legs dull brownish in juv, like adult by 1st summer.

Voice. Varied, including a hard, dry chattering, a clipped, dry *cheh-det cheh-det,* a ringing *chook,* a nasal, fairly rough *chah-ichah,* and a nasal, buzzy, often insistent scolding *zherr zherr* ...

Habs. Arid to semihumid forest and

edge, plantations. May associate with Purplish-backed Jay. See genus note.

SS: Purplish-backed Jay larger, upperparts deep purplish, not strongly contrasting with black, legs brighter yellow.

SD/RA: C to F resident (SL–1200 m) on Pacific Slope from Nay (vicinity of San Blas) to Gro.

YUCATAN JAY

Cyanocorax yucatanicus ⁎   Plate 45
Chara Yucateca

ID: 12.5–13.5 in (32–34.5 cm). *Endemic to Yuc Pen.* **Descr. Adult:** eyes dark brown, bill black, legs yellow. *Head and underparts black, upperparts turquoise-blue,* deepest on tail. **Juv** (Jul–Sep): *bill and orbital ring yellow. Head and underparts white,* upperparts as adult but outer rectrices tipped white. **Imm:** told from adult by white-tipped rectrices and color of bare parts: bill and orbital ring yellow, becoming black by 3rd winter.

Voice. Varied calls include a hard, often rapid, dry chattering or scolding *cha-cha* ..., a more rattled *chahah-urr-rr,* a rolled *chirr-rrr-rrt,* a loud, ringing nasal *ch'ik chi-chik,* a ringing, slightly metallic *chi'diu* and *jihnk jihnk,* and a hollow *cah* suggesting a soft Collared Forest-Falcon.

Habs. Arid to semihumid forest and edge, second growth, plantations. See genus note.

SS: None in range.

SD/RA: C to F resident (SL–250 m) in Yuc Pen, S to W Tab, N Guatemala, and N Belize.

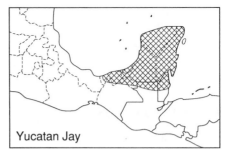

Yucatan Jay

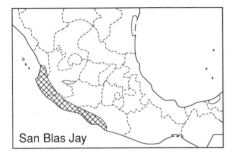

San Blas Jay

GENUS CYANOLYCA

These are fairly small jays of humid montane forests. Ages differ slightly, sexes similar. Bills and legs blackish. Often relatively skulking, low to high, in shady forest and edge. In pairs or small groups, may join mixed-species flocks, best detected by voice. Apparently not cooperative breeders.

AZURE-HOODED JAY
*Cyanolyca cucullata** Plate 46
Chara Gorriazul

ID: 11–12 in (28–30.5 cm). *E and S cloud forests.* **Descr. Adult:** overall dark blue with blackish head and upper chest, and *sky blue hind crown and nape,* edged white. **Juv:** duller overall, duller blue hood lacks white edges.
> **Voice.** A loud, bright *oink-oink ...* or *eihnk-eihnk ...,* usually 4–5×, a nasal *ehr-ehn* or *eh'enk,* usually 2×, and a low, hard, gruff *cheh-r.*
> **Habs.** Humid evergreen and pine– evergreen forest. See genus note. Joins mixed-species flocks including Unicolored Jays, Emerald Toucanets, etc. Eggs un- described (?).
SS: Azure hood often hard to see from below. Black-throated Jay smaller with narrow whitish supercilium; note voice. Uni- colored Jay larger, throat and chest deep blue; note voice.
SD: F to C resident (1000–2000 m) on Atlan- tic Slope and in adjacent interior from SE SLP to Honduras.
RA: E Mexico to Panama.

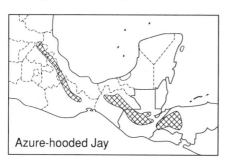
Azure-hooded Jay

BLACK-THROATED JAY
*Cyanolyca pumilo** Plate 46
Chara de Niebla

ID: 9.3–10.5 in (23.5–26.5 cm). *Endemic to highlands S of Isthmus.* **Descr. Adult:** *over- all deep blue with black face and throat, narrow white supercilia meet across fore- head.* **Juv:** duller, face and throat sooty grey, lacks white supercilium.
> **Voice.** A dry, nasal, slightly rasping

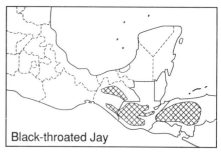
Black-throated Jay

tchehv eh cheh and *sheeva-sheeva,* a slightly nasal scolding *sheh sheh sheh,* or a more whining *shrieh shrieh,* and a quiet *whiew* and *rrah riew* when foraging.
> **Habs.** Humid evergreen and pine– evergreen forest. See genus note. At times in flocks with Azure-hooded Jays. Nest built of twigs in outer branches of tree canopy, Chis (SW). Eggs undescribed (?).
SS: Unicolored Jay larger, face and throat not contrasting black; note voice. See Azure- hooded Jay.
SD/RA: F to C resident (1500–3000 m) in interior and on adjacent slopes from Chis to El Salvador and Honduras.

DWARF JAY
Cyanolyca nana Plate 46
Chara Enana

ID: 8.7–9.7 in (22–24.5 cm). *Endemic to mountains of E Mexico.* **Descr. Adult:** over- all dull blue with *black mask and pale blue throat,* forehead and narrow supercilium pale blue. **Juv:** overall sooty bluish, face pattern less distinct with duller throat, bluish supercilium.
> **Voice.** A nasal *shiev'a shiev'a,* given in- frequently.
> **Habs.** Humid pine–evergreen and pine– oak forest. See genus note. In winter, flocks with Steller's Jays, Band-backed

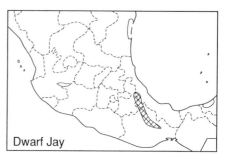
Dwarf Jay

Wrens, orioles, etc. Nest largely built of mosses. Eggs: 2–3, pale greenish blue, marked with olive (Hardy 1971).

SD/RA: F to C but local resident (1600–3000 m) on Atlantic Slope and in adjacent interior from cen Ver (few recent reports) to N Oax.

WHITE-THROATED (OMILTEMI) JAY
Cyanolyca mirabilis*   Plate 46
Chara de Omiltemi

ID: 9–10 in (23–25.5 cm). *Endemic to Sierra Madre del Sur of Gro and Oax.* Descr. Adult: *broad black mask bordered by white throat and white stripe from forehead around auriculars.* Crown, nape, and upper chest blackish, rest of plumage dull blue. Juv: face duller with less distinct white borders, whitish throat ill-defined, underparts sooty greyish, upperparts dull bluish.

Voice. A slightly raspy, nasal to shrill *sheev-idee sheev-idee* or *shiev-a shiev-a*, and a slightly buzzy *sheir sheir*.

Habs. Humid evergreen and pine–evergreen forest. See genus note. In nonbreeding season, flocks with Unicolored Jay, Emerald Toucanet, etc. Nest and eggs undescribed (?).

SD/RA: F to C resident (1800–3000 m) on Pacific Slope and in adjacent interior of Gro and Oax.

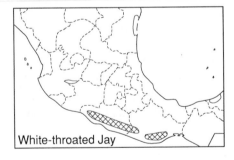

White-throated Jay

GENUS APHELOCOMA

These are medium-sized jays similar in shape to *Cyanolyca*; some authors merge the two genera. Ages differ. Bills and legs blackish in adults. In pairs or groups up to 10–15 birds, at mid- to upper levels, often feed on ground.

SCRUB JAY
Aphelocoma coerulescens*   Plate 46
Chara Azuleja

ID: 10.5–12.5 in (26.5–32 cm). Widespread N of Isthmus. Descr. Adult: *head blue with blackish mask bordered by narrow white supercilium* (distinct in Baja, indistinct or lacking in much of Mexico). *Dusky greyish back typically forms contrasting patch* (indistinct in *cyanotis* of E and cen Mexico) *on otherwise blue upperparts.* Throat and upper chest whitish with *dusky streaks* that *often appear to form a chest band* above pale grey to pinkish-grey underparts. Juv: head and upperparts sooty grey-brown with hint of adult face pattern, throat and underparts duller, often with more distinct dusky chest band.

Voice. A rasping *sheir sheir sheir* ... or *rrieh'k rrie'k* ..., and *ehrk ehrk* or *shehk shehk*, notably higher and shriller in S Mexico, *shee'nk shee'nk* etc.; also a scolding *chen-chen-chen-chen* ..., etc.

Habs. Arid to semiarid scrub, open areas with hedges and patches of scrubby woodland, open pine–oak and juniper woodland, etc. See genus note. Often in pairs, perches prominently, shrike-like, on wires and bushes. Eggs: 3–6, blue-green to bluish white, marked with browns and greens (WFVZ).

SS: Grey-breasted Jay larger and more heavily built with relatively longer, broader wings and shorter tail, prefers pine–oak woodland, rarely in open scrubby areas; overall more uniform bluish above, pale pinkish grey below; note voice.

SD: C to F resident (SL–2100 m) in Baja and (1000–2700 m) in interior and on adjacent slopes locally from Son and Coah to Oax.

RA: W USA to S Mexico; disjunctly in Florida.

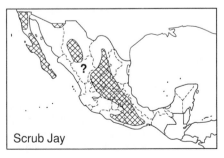

Scrub Jay

NB: Birds in California Channel Islands and Florida sometimes considered specifically distinct.

GREY-BREASTED (MEXICAN) JAY
Apheloxoma ultramarina*   Plate 46
Chara Pechigris

ID: 11.5–13 in (29–33 cm). Interior N and cen Mexico. **Descr. Adult:** *head and upperparts dull blue,* lores and auriculars darker, back typically washed greyish. *Throat and underparts pale greyish to pinkish grey,* becoming white on vent and undertail coverts. **Juv:** bill yellow (quickly becomes black in NE and W-cen populations). Head and upperparts greyish, wings and tail duller than adult. Underparts pale greyish, chest washed dusky. Bill becomes black in 1–2 years in NW and S populations.
   Voice. Common call a slightly shrieking *sheenk sheenk* or *schweenk schweenk,* often repeated, also varied to *jehnk-jehnk* or *rrienk rrienk,* etc.
   Habs. Arid to semiarid pine–oak, oak, and juniper woodland and edge. See genus note. Often in flocks at mid- to upper levels. Cooperative breeder. Eggs: 4–7, unmarked blue to greenish blue, or marked with greens (WFVZ).
SS: See Scrub Jay.
SD/RA: C to F resident (1000–3500 m) in interior and on adjacent slopes from NE Son and Coah (also SW USA) to cen volcanic belt.

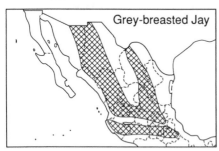
Grey-breasted Jay

UNICOLORED JAY
Apheloxoma unicolor*   Plate 46
Chara Unicolor

ID: 13–14 in (33–35.5 cm). *Endemic; E Mexico and C.A.* **Descr. Adult:** *overall deep blue* to purplish blue, lores and auriculars darker. **Juv:** yellow bill quickly

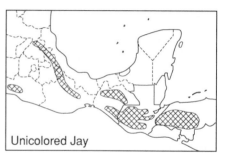
Unicolored Jay

becomes black. Head and body sooty grey, wings and tail duller than adult.
   Voice. A loud, slightly squeaky to nasal *chwenk* or *chweenk* or *reenk,* etc., often doubled or in series, a more ringing *shee-ik' shee-ik'* or *shiek ...,* and a gruff *ssha ssha,* etc.
   Habs. Humid evergreen and pine–oak forest and edge. See genus note. Usually in small flocks at mid- to upper levels, joins mixed-species flocks. Cooperative breeder (?); eggs undescribed (?).
SS: See Azure-hooded and Black-throated jays.
SD/RA: C to F resident (1500–3000 m) on both slopes locally from Gro and Hgo (Bjelland and Ray 1977), and in interior from Chis, to El Salvador and Honduras. A report from Mex (Miller *et al.* 1957) is probably in error.

PINYON JAY
Gymnorhinus cyanocephalus Not illustrated
Chara Piñonera

ID: 10–11 in (25.5–28 cm). *Mountains of BCN. Bill relatively long and pointed, tail relatively short.* **Descr.** Ages differ. **Adult:** bill and legs black. *Overall greyish blue,* throat streaked whitish. **Juv:** overall sooty

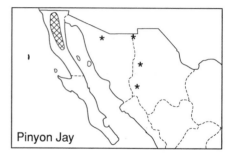
Pinyon Jay

grey-brown, wings and tail tinged blue.
Voice. A nasal ringing *whee-uh* or *waah aah*, and nasal, slightly laughing calls, *nyeh* and *nyeh-eh* and *aah aah-aah aah-aah aah-aah* ..., etc.
Habs. Arid to semiarid pine (especially pinyon) and pine–oak forest. In flocks, at times in hundreds, moving through forest low to high, often on ground. Breeds colonially. Eggs: 3–5, bluish to greyish white, densely speckled with browns and greys (WFVZ).
SD: C resident (1500–3000 m) in Sierra Juarez and Sierra San Pedro Martír, BCN. Irregular R visitor (Oct–Dec, May–Jun) to N Son and N Chih (Phillips 1986).
RA: W USA to N Baja.

CLARK'S NUTCRACKER

Nucifraga columbiana   Not illustrated
Cascanueces Americano

ID: 10.5–11.5 in (26.5–29 cm). *Local in N Mexico. Bill relatively long, stout and pointed.* **Descr.** Ages differ slightly. **Adult:** bill and legs black. *Head and body pale grey* with white undertail coverts. Wings black with *broad white tips to secondaries, striking in flight. Central rectrices black, rest of tail white, striking in flight.* **Juv:** duller overall, head and body brownish grey.
Voice. A harsh rasping *rrrahh* or *krraaah*.
Habs. Arid to semiarid pine forest and clearings around timberline. Singly, in pairs, or in small groups, perches on dead snags, often feeds on ground. Eggs: 2–5, pale greenish to greenish white, speckled with browns and greys (WFVZ).
SD: U to R presumed resident (2500–3700 m) in Sierra San Pedro Martír, BCN, and on Cerro Potosí, NL. Irregular R visitor (Oct–Nov, May–Jun; to near SL) to NW

Mexico, S to BCN and Son, possibly to Dgo (Evenden *et al.* 1965).
RA: W N.A. to N Mexico.

GENUS CORVUS

Crows are all-black birds found in temperate areas of the region. (Ravens are simply large crows.) Ages differ slightly in that juvs are sootier, dull blackish versus glossy blackish of adults. Bills and legs blackish. Feed mostly on ground. Size and shape, particularly of the bill and tail, together with voice, are the main identification marks; wingtips of all crows in the region are noticeably shorter than tails at rest, but in ravens wingtips are about equal with tails.

AMERICAN CROW

Corvus brachyrhynchos hesperis
Cuervo Americano   Not illustrated

ID: 16–17 in (40.5–45.5 cm). *Extreme NW Mexico. Bill moderately stout and heavy. Tail squared* to slightly rounded. **Descr.** Overall black, upperparts with slight purplish gloss.
Voice. A hard cawing *ahh* or *ahhr*, often in series, and a higher *aahr aahr*, etc.
Habs. Open and semiopen areas with trees, hedges, open woodland. In pairs or small groups. Eggs: 4–6, pale greenish, speckled to mottled with greys and browns (WFVZ).
SS: Northern Raven larger with heavier, stouter bill and wedge-shaped tail.
SD: U resident (near SL–500 m) in NW BCN; formerly C (?) (see Grinnell 1928). Irregular R winter visitor (Dec–Jan?) to Colorado Delta in extreme NW Son. Reports from Chih (Hubbard 1987) require verification.
RA: N.A. to NW Mexico.

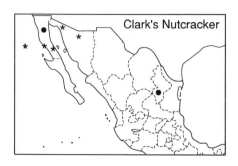

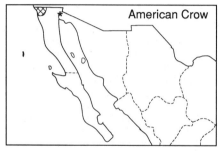

TAMAULIPAS [MEXICAN] CROW
Corvus imparatus   Plate 46
Cuervo Tamaulipeco

ID: 14–15.5 in (35.5–39.5 cm). *Endemic to NE Mexico.* Tail squared. **Descr.** *Overall glossy black*, head and upperparts with strong purplish sheen, underparts with bluish to blue-green gloss.
   Voice. Short, low, guttural to hoarse, frog-like calls, *ahrrr* or *arrhk*, a doubled *rrah-rrahk* or *ahwr ahwr*, and a buzzy *zzr*, etc.
   Habs. Open and semiopen areas with trees and hedges, towns, rubbish dumps, etc. Often in large flocks, also in pairs and small groups. Eggs: 4–5, pale grey to blue-grey, marked with browns (WFVZ).
SS: ♂ Great-tailed Grackle has whitish eyes, longer keel-shaped tail; ravens much larger and more heavily built with heavy stout bills, wedge-shaped tails.
SD/RA: C to F resident (near SL–800 m) on Atlantic Slope from Tamps (also S Texas) and E NL to N Ver.
NB: Traditionally considered conspecific with Sinaloa Crow.

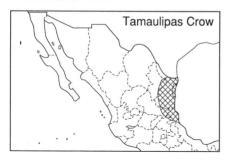
Tamaulipas Crow

SINALOA [MEXICAN] CROW
Corvus sinaloae   Plate 46
Cuervo Sinaloense

ID: 14–15 in (35.5–38 cm). *Endemic to NW Mexico.* Tail squared. **Descr.** *Overall glossy black*, head and upperparts with strong purplish sheen, underparts with bluish to bluish-purple gloss. Juv undescribed (?).
   Voice. A cawing *caahr* or *rraah*, often doubled, a disyllabic *cah-ow*, a short *caah!*, and a nasal cawing *aah* or *raah*, etc.
   Habs. Open and semiopen areas with trees and hedges, towns, rubbish dumps, etc. Often in flocks, also in pairs and small groups. Eggs: 4–5, pale blue to blue-grey, marked with browns (WFVZ).

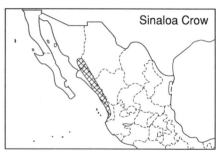
Sinaloa Crow

SS: ♂ Great-tailed Grackle has whitish eyes, longer keel-shaped tail; ravens much larger and more heavily built with heavy stout bills, wedge-shaped tails.
SD/RA: C to F resident (SL–1000 m) on Pacific Slope from S Son to Nay. Reports from Col (AOU 1983) are in error (see Schaldach 1963), and a report from the Islas Tres Marías (Miller *et al.* 1957) is probably also erroneous.
NB: See Tamaulipas Crow.

CHIHUAHUAN (WHITE-NECKED) RAVEN
Corvus cryptoleucus   Plate 46
Cuervo Llanero

ID: 18–20 in (45.5–51 cm). *N and cen Mexico. Bill stout and heavy, relatively blunt-tipped, rictal bristles extend more than half way along bill. Tail wedge-shaped.* **Descr.** Overall glossy black, upperparts with slight purplish sheen, *bases of neck feathers white*, at times visible when feathers blown by wind or when preening.
   Voice. A throaty cawing *krrah* and *rrr-ahk* or *rrahk*, typically higher than Northern Raven.
   Habs. Arid to semihumid open and semiopen country, especially arable fields, with hedges, tall trees, etc. Often in flocks, up to hundreds locally. Nests often conspicuous on telegraph poles, pylons, etc.

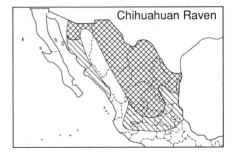
Chihuahuan Raven

Eggs: 3–7, pale greenish white, speckled to scrawled with browns and greys (WFVZ).

SS: Northern Raven larger, longer and heavier bill looks less blunt-tipped, rictal bristles less than half bill length, tail longer, bases of neck feathers pale grey; note voice. See crows.

SD: C to F resident (near SL–2400 m) in interior from NE Son and Coah to NE Jal, Gto, and NW Hgo, on Atlantic Slope in E NL and N Tamps. In winter, recorded S on Pacific Slope to Sin and in interior to N Mich, Mex, and N Ver (Phillips 1986) but winter status confused owing to similarity with Northern Raven.

RA: S cen USA to cen Mexico; N populations migratory.

NORTHERN (COMMON) RAVEN

*Corvus corax**   Plate 46
Cuervo Grande

ID: 22–25 in (56–63.5 cm). *Temperate areas throughout. Bill very stout and heavy, rictal bristles extend less than half way along bill. Tail wedge-shaped.* **Descr.** Overall glossy black, upperparts with slight purplish sheen, bases of neck feathers pale grey, at times visible when feathers blown by wind.

Voice. A deep, throaty *rrok* and *rronk* or *rrrahk*, higher in juvs.

Habs. Arid to semihumid open and semi-open country, mainly deserts, rocky mountainous areas, also farmland. Singly, in pairs or flocks, rarely of more than 50 birds. Eggs: 3–7, pale bluish to greenish, speckled to scrawled with browns and greys (WFVZ).

SS: See Chihuahuan Raven, American Crow.

SD: C to F resident (SL–3500+ m) in Baja, on Isla Clarión (and formerly San Benedicto), on Pacific Slope from Son to Nay, and in interior and on adjacent slopes from Son and Coah to N cen Nicaragua, mostly above 1000 m from cen Mexico S; absent from humid SE.

RA: Holarctic, in New World S to N cen Nicaragua.

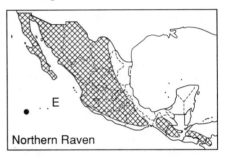

Northern Raven

CHICKADEES AND TITMICE: PARIDAE (5)

Small and agile arboreal birds typical of temperate regions in the Northern Hemisphere. Titmice have prominent crests, at least in the New World. Small bills stout and pointed, feet strong and grasping. Wings short and rounded, tails fairly long and squared. Ages/sexes typically similar, but juvs notably fluffier, often with paler wingbars, quickly attain adult-like plumage; young altricial. Plumage patterned in greys, black, and white; chickadees have dark caps and bibs. Paler edged tertials and secondaries often form contrasting wing panels. Bills and legs blackish. Varied calls often high and buzzy, songs mostly slightly plaintive whistles.

Food mostly insects, also seeds and fruit. Nest in tree cavities. Eggs: 5–8, white, usually flecked with reddish browns (WFVZ).

MEXICAN CHICKADEE
Parus sclateri* Plate 47
Paro Mexicano

ID: 4.7–5 in (12–12.5 cm). *Highlands N of Isthmus.* **Descr.** *Black cap and broad bib contrast with white auriculars.* Upperparts greyish, underparts paler greyish with whitish band below bib continuing as stripe down belly.
Voice. Varied buzzy and excited twittering calls include a buzzy *dzi-ssi-ssi* or *tssir-irr-rr* and *dzshi-chiik-i-dit*, a shorter buzzy *chi-pi-tit*, an often loud, nasal scolding *ssheher-sshehr* or *cheh-cheh*, a nasal *jih-jih*, a slightly rolled *brri-i-i-it*, etc. Song a short, clear to slightly burry warble *ss chirr-i-chu* or *chis'i chu-wur*, and a rich *chee-dlee* or *cheelee*, repeated rapidly 4–6× etc.
Habs. Arid to semihumid conifer and pine–oak forest. In pairs or small groups, often joins mixed-species flocks.
SD/RA: C to F resident (1500–3500 m) in interior and on adjacent slopes from Son

(also SW USA) and S Coah to Oax.
NB: Sometimes called Grey-sided Chickadee.

MOUNTAIN CHICKADEE
Parus gambeli* Not illustrated
Paro Cejiblanco

ID: 5–5.3 in (12.5–13.5 cm). *Mountains of BCN.* **Descr.** *Black cap with whitish lores and narrow (often indistinct) whitish supercilium*; auriculars white, bib black. Upperparts greyish, underparts paler greyish.
Voice. A buzzy *chi-chi-chit* and *si-si-si chi-chi-chi* and *si-chi chi-chi-chit*, and a bubbling, liquid, rolled warble, etc. Song a plaintive *swee chee chee* or *dee dee dee*, and longer *tee chee chee-chee-cheu*, etc.
Habs. Arid to semiarid pine and pine–oak forest and woodland. Habits much as Mexican Chickadee.
SD: C to F resident (1500–3000 m, at times lower in winter) in Sierra Juarez and Sierra San Pedro Martír, BCN.
RA: W N.A. to N Baja.

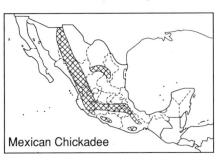

Mexican Chickadee

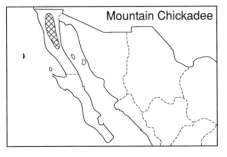

Mountain Chickadee

BRIDLED TITMOUSE
Parus wollweberi *   Plate 47
Paro Embridado

ID: 5–5.3 in (12.5–13.5 cm). *Highlands N of Isthmus.* **Descr.** *Grey crown, crest, and nape edged black, face white with black eyestripe continuing around auriculars to black throat.* Upperparts greyish, underparts pale grey.
  Voice. A high, slightly buzzy *sir-r-r si-si-sirr-rr-rr*, a more twittering *spi-spi-spi* ... and *pi-i-i-i-r* suggesting Bushtit, a similar phrase run into nasal notes, *tsi-tsi-si si-si serr-eh-ehr*, and dry, wren-like churring. Song a rich *cheewee* ..., 4–8×, or a rapid *weet-weet* ..., up to 10–12×.
  Habs. Arid to semihumid oak and pine–oak woodland. Habits much as Mexican Chickadee.
SD/RA: C to F resident (1000–3000 m) in interior and on adjacent slopes from Son (also SW USA) and NL to Oax.

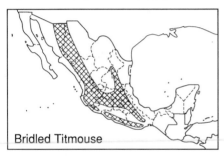

Bridled Titmouse

PLAIN TITMOUSE
Parus inornatus *   Not illustrated
Paro Sencillo

ID: 5.2–5.5 in (13.5–14 cm). *NW Mexico.* **Descr.** *Overall plain greyish,* browner

above in Baja, face and underparts paler.
  Voice. A high *si-si-pit*, and a rougher, buzzy *sin chih-chih chih-chih* or *si chih-chih*, etc. Song a bright *pee-chee pee-chee pee-chee.*
  Habs. Arid to semiarid oak woodland, locally in pine–oak. Habits much as Mexican Chickadee.
SD: C to F resident (500–2400 m) in Baja, N of 30° N and in mountains of Cape District; also presumed resident in extreme N Son.
RA: W USA to NW Mexico.

BLACK-CRESTED [TUFTED] TITMOUSE
Parus (bicolor?) atricristatus *   Figure 36
Paro Crestinegro

ID: 5–5.5 in (12.5–14 cm). *NE Mexico.* **Descr.** Ages differ slightly. *Face, throat, and underparts whitish with black crown and crest, cinnamon wash on flanks.* Upperparts greyish. **Juv:** crest greyish, flanks duller.
  Voice. A fairly hard, nasal scold, often doubled, *ssi ssi, t-chay t-chay* ... or a buzzier *ssi-cheh ssi-cheh* or *ssi cheh-cheh-cheh*, and a nasal *neh-neh* ... or *nyeh-nyeh* ..., etc. Song a plaintive *chew chew chew chew* or *chee-u chee-u chee-u*, at times in longer series.
  Habs. Arid to semihumid oak and semideciduous woodland, open areas with hedges, scattered trees, scrubby woodland. Habits much as Mexican Chickadee.
SD: C to F resident (near SL–2300 m) on Atlantic Slope from Coah and NL to N Ver, U disjunctly in S Ver (Andrle 1967b); possibly also in E Chih (Phillips 1986).
RA: E N.A. to NE Mexico.

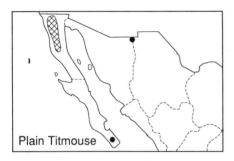

Plain Titmouse

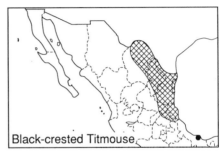

Black-crested Titmouse

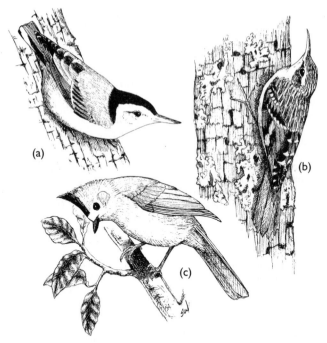

Fig. 36 (a) White-breasted Nuthatch; (b) Brown-Creeper; (c) Black-crested (Tufted) Titmouse

VERDINS: REMIZIDAE (1)

The Verdin is a chickadee-like bird of open deserts. Bill pointed, culmen straight. Ages differ, sexes similar; young altricial. Juvs quickly attain adult-like plumage.

Food mostly insects, also fruits. Verdins build bulky globular nests of twigs at low to mid-levels in bushes. Eggs: 3–6, pale bluish, flecked with reddish browns (WFVZ).

Formerly placed with the Paridae, current opinion places the Verdin with the Old World Remizidae; relationships are still in need of study.

VERDIN
Auriparus flaviceps*   Plate 47
Baloncillo

ID: 4–4.2 in (10–11 cm). N deserts. **Descr.**
Adult: bill and legs blackish. *Greyish over-all,* paler below, *with orange-yellow face and throat,* dark lores; reddish-chestnut lesser wing coverts often covered by scapulars. **Juv:** bill bright orange-pink, quickly

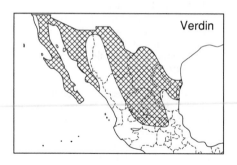

Verdin

becoming black. Overall grey-brown, paler greyish below.
Voice. A hard, sharp chipping, at times repeated insistently, *chik ...* or *chip ...,* a sharp, slightly ringing *cheen,* a sharp chipping *tchiep,* and a buzzy *chrru-chrru ...,* etc. Song a bright, slightly plaintive to clear *tee ch-weet* or *chee ti-cheu,* and variations, *tsee tweet* and *chee chu-i-cheeh,* etc.
Habs. Arid open and semiopen country with scrub, bushes, trees, etc. Singly, in pairs, or family groups, quick and active. See family account.
SS: Juv told from Bushtit by shorter tail and deeper-based, pointed orangish bill with straight culmen; usually with adults (unlikely to be with Bushtits).
SD: C to F resident (SL–2000 m) in Baja, on Pacific Slope from Son to N Sin, in interior from Son and NL to N Jal and NW Hgo, and on Atlantic Slope in Tamps.
RA: SW USA to cen Mexico.

BUSHTITS: AEGITHALIDAE (1)

The Bushtit is a tiny social bird recalling a small chickadee; it has a small stubby bill and long squared tail. Ages/sexes vary in appearance, with a complex transition between two forms in N Mexico; young altricial. Juv quickly attains adult plumage by complete 1st prebasic molt.

Food mostly insects and arthropods. Bushtits build large oval nests of mosses, lichens, cobwebs, etc., placed at low to mid-levels in bushes. Eggs: 5–13, white (WFVZ).

Formerly placed with the Paridae, current opinion places the Bushtit with the Old World Aegithalidae; relationships are still in need of study.

BUSHTIT
Psaltriparus minimus* Plate 47
Sastrecillo

ID: 4.3–4.5 in (11–11.5 cm). See family account. *Marked geographic variation*: Plain Bushtit (NW Mexico; including *melanurus*, shown) and Black-eared Bushtit (cen Mexico to Guatemala; including *personatus*, shown). **Descr.** *Eyes pale yellowish (♀) to dark brown (♂ and juv).* **Plain Bushtit** (sexes similar): bill and legs blackish. *Overall greyish, paler below, with brownish crown (Baja) or grey crown and brownish auriculars (NW Mexico).* **Black-eared Bushtit.** ♂: *black mask contrasts with grey crown,* upperparts grey-brown. Throat and underparts whitish, washed cinnamon on flanks. ♀: *auriculars brownish with black spot at rear and, usually, black stripe above eye.* See NB below.

Voice. A high, slightly buzzy, excited twittering given by flocks; also a thin, trilled *sir-r-rrrrr*, and a high, rolled *spih* or *spik*, etc.

Habs. Arid to semihumid scrub, brushy woodland, oak and pine–oak forest. Active and restless; usually in flocks of 10–30 birds at low to mid-levels, often with mixed-species flocks. See family account.

SS: See juv Verdin.

SD: C to F resident (near SL–2500 m) in NW and S Baja; and (1500–3500 m) in interior and on adjacent slopes from Son and Coah to Guatemala.

RA: W N.A. to Guatemala.

NB: The two forms have been regarded as separate species but Raitt (1967) documented a complex transition zone in SW USA and N Mexico, whereby black-eared morphs become increasingly common southward, at first appearing in juv ♂♂ but never in adult ♀♀. By cen Mexico, the Black-eared form has apparently stabilized, as described above.

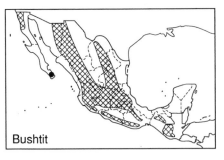

Bushtit

NUTHATCHES: SITTIDAE (3)

A mainly Old World family of temperate origin. Small tree-climbing birds which often travel head-down on trunks. Bills straight and pointed, lower mandible tapers to chisel-like tip; legs short and feet strong. Tails short and squared. Ages/sexes similar or slightly different; young altricial; 1st prebasic molt partial. Bills blackish, whitish below at base, legs blackish. Overall grey-blue above, light-colored below, with distinctive head markings. Voices are nasal or chipping notes.

Food mostly insects, some seeds. Nest in tree cavities which they may excavate. Eggs: 4–8, whitish, heavily speckled with reddish browns and greys (WFVZ).

RED-BREASTED NUTHATCH

Sitta canadensis   Not illustrated
Saltapalos Canadiense

ID: 4.3–4.5 in (11–11.5 cm). *N Mexico in winter.* **Descr.** Ages/sexes differ. ♂: *white supercilium contrasts with* glossy *black crown and nape and broad eyestripe;* lower auriculars white, *throat and underparts cinnamon.* Upperparts grey-blue with white tail-spots. ♀: crown and nape grey. **Juv:** crown duller than adult.
    Voice. A nasal, often prolonged *ehk-ehk-ehk-ehk* ... or *yeh-yeh-yeh* ..., etc., rarely single notes.
    Habs. Conifer and pine–oak woodland. Singly or in pairs, low to high on trunks and branches.
SD: U (extirpated?) resident (around 1200 m) on Isla Guadalupe, BCN, last recorded 1971 (Jehl and Everett 1985). Irregular R (to U?) winter visitor (Sep–Apr; 1000–2700 m) in BCN (Ruiz-C. and Quintana-B. 1991; TEW) and in interior and on adjacent Pacific Slope S to Sin (RAB) and

NL (AMS); vagrant to Isla Socorro (Mar 1957, UBC spec.).
RA: W N.A.; winters irregularly to N Mexico.

WHITE-BREASTED NUTHATCH

Sitta carolinensis *   Figure 36
Saltapalos Pechiblanco

ID: 5.2–5.5 in (13.5–14 cm). Mountains N of Isthmus. **Descr.** Ages/sexes similar. *White face with* slightly duskier auriculars *contrasts with* glossy *black* to dark grey *crown and nape.* Throat and underparts whitish to pale grey, washed cinnamon on flanks. Upperparts grey-blue with white tail-spots.
    Voice. Nasal, often slightly laughing phrases, may suggest Acorn Woodpecker but less varied, *kyeh-kyeh-kyeh* or *neh-neh-neh-neh* ..., etc.; song a rapidly repeated nasal series *eh-eh-eh-eh-eh-eh* ...
    Habs. Arid to semihumid pine and pine–oak woodland. Habits much like Red-

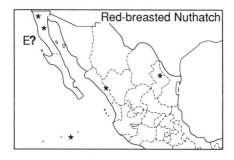
Red-breasted Nuthatch

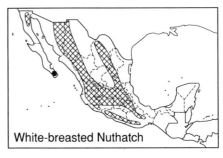
White-breasted Nuthatch

breasted Nuthatch; joins mixed-species flocks.

SD: F to C resident (1500–3500 m) in Baja and in interior and on adjacent slopes from Son and Coah to Oax.

RA: N.A. to S Mexico.

PYGMY NUTHATCH
Sitta pygmaea*   Not illustrated
Saltapalos Enano

ID: 4.3–4.5 in (11–11.5 cm). Mountains of N and cen Mexico. **Descr.** Ages/sexes similar. *Head grey-brown with blackish auricular mask and whitish spot on nape.* Lower auriculars, throat, and underparts whitish, washed pale pinkish buff on underbody. Upperparts grey-blue with white tail-spots.

Voice. A high, excited, often insistent *beep, bee-beep bee-bee.bee-bee-beep*, etc., also a quieter *tweek-i-tweek*, quiet, sócial twittering, *kip, kip-kip* …, a single sharp *beek!*, and rolled, beeping series.

Habs. Arid to semiarid pine forest and woodland. In pairs or small groups, actively at mid- to upper levels, often with mixed-species flocks.

SD: F to C resident (2000–3500 m) in BCN and in interior and on adjacent slopes from Son and Coah to cen volcanic belt.

RA: W N.A. to cen Mexico.

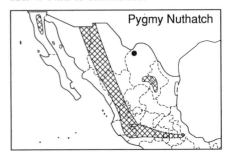

TREECREEPERS: CERTHIIDAE (I)

One species of this temperate Old World family is currently recognized in the Americas. Resembles a tiny woodcreeper in appearance and behaviour. Bill slender and decurved. Tail long and spiky, used as a prop when climbing. Ages/sexes similar; young altricial; rectrices replaced in incomplete 1st prebasic molt.

Food mostly insects. Nests behind loose bark or in tree crevices, where makes a hammock of fine vegetable matter. Eggs: 4–6, white, heavily speckled with reddish browns and greys (WFVZ).

BROWN CREEPER

Certhia americana*   Figure 36
Trepador Americano

ID: 4.7–5.2 in (12–13.5 cm). Highlands almost throughout. **Descr.** See family account. Bill blackish above, horn below, legs flesh. *Head and upperparts dark brown, streaked white, with white supercilium* and rufous rump and uppertail coverts. Wings dark brown with white covert and tertial edgings and *zig-zag buff bars on remiges; broad pale buff stripe across base of remiges striking in flight.* Tail brown. *Throat and underparts whitish* to pale grey, often washed pale cinnamon on flanks.

Voice. A high, thin *siin* or *tsii*, and *tsii-siin*, etc. Song a series of high, thin notes, often ending with a burrier trill, *tssir tsi tsir-ti-it* or *tssi ssi-ssi ssi-ssir*, etc.

Habs. Conifer and pine–oak forest and woodland. Singly or in pairs, creeps up in spirals, then drops to base of nearby tree and climbs again. May join mixed-species flocks.

SD: F to C resident (1500–3500 m, lower locally in C.A.) in interior and on adjacent slopes from Son and S Coah to N cen Nicaragua. Irregular U to R winter visitor (Nov–Apr; near SL–3500 m) to BCN and from Son to Tamps (Webster, in Phillips 1986).

RA: N.A. to N cen Nicaragua; N populations migratory.

NB: Formerly considered conspecific with *C. familiaris* of Old World.

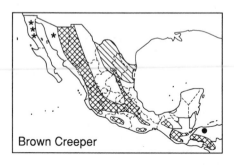

Brown Creeper

WRENS: TROGLODYTIDAE (32)

A mainly New World family that achieves its greatest diversity in Middle America. Small to medium-sized birds with short rounded wings and short to longish tails. Bills slender and usually fairly long. Ages similar or different, sexes similar; young altricial. First prebasic molt typically partial and quite rapid in most smaller species; often more extensive and prolonged in *Campylorhynchus*. Plumage colored in browns, black, and white, often with striking patterns and barred flight feathers. Marked plumage variation, combined with numerous allopatric populations, means that the taxonomic relationships of several species are still somewhat uncertain. Although common, the smaller wrens can often be hard to see, keeping to dense tangles and understory. Usually found in pairs or singly, though several *Campylorhynchus* occur in small groups. Some wrens are among the best singers in the region and their loud songs may be given throughout the year; several species duet, with ♂ and ♀ singing together, at times contributing different parts of the song. Calls include scolding chatters and rattles.

Food mostly insects. Nests typically are ball-shaped masses of sticks and other vegetation; several species nest in crevices or holes in rocks and trees. Eggs: 2–8.

GENUS CAMPYLORHYNCHUS: Cactus Wrens

Cactus Wrens are large, conspicuous, and social wrens of open and wooded habitats. There are few identification problems, each species occurring in a distinct habitat and most being allopatric. Often located by their loud, gruff calls and rollicking, chortling songs. Eyes reddish to amber in adults, brown in juvs. Bills blackish above, paler below, at least basally; legs greyish to greyish flesh. Nests are conspicuous bulky masses of sticks placed low in cacti, bushes, etc., or (Band-backed and Grey-barred wrens) at mid- to upper levels in palms, conifers, or other trees. Eggs: 3–6.

BAND-BACKED WREN
Campylorhynchus zonatus * Plate 48
Matraca-barrada Tropical

ID: 7.3–8 in (18.5–20.5 cm). *E and S, forest and edge.* **Descr.** Ages differ. **Adult:** face streaked greyish and white with white supercilium and dark eyestripe; crown greyish, spotted dark brown; nape streaked blackish and white. *Upperparts blackish brown, barred whitish to pale cinnamon. Throat and chest white, boldly spotted blackish, belly and undertail coverts cinnamon.* **Juv:** face dusky with whitish to cinnamon supercilium and dark eyestripe; crown dark brown. Nape and upperparts steaked dark brown and pale cinnamon, wings and tail as adult. Throat and chest pale grey, chest indistinctly mottled dusky, belly washed cinnamon.

Voice. Gruff, rasping and rollicking chatters, *sheh-eh eh-eh-eh*, or *cheh-cheh* ..., and a gruff *shehk-shehk* ..., at times a rhythmic *cheh-eh eh'eh'eh* and *shehr k-rreh shehr k-rreh*, etc., repeated.

Habs. Evergreen to pine–oak woodland and edge, plantations, gardens, semiopen areas with trees and hedges. In groups of 4–10 birds mainly at mid- to upper levels where clambers with agility; often joins

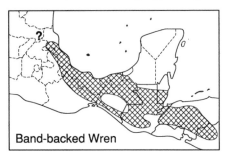

Band-backed Wren

mixed-species flocks. Eggs whitish, unmarked or faintly speckled brown (WFVZ).

SS: Grey-barred Wren probably allopatric (may overlap in W Ver?), lacks cinnamon belly and cinnamon wash to upperparts, flanks boldly barred; juv paler below with trace of barring on flanks, usually with adults.

SD: C to F resident (SL–3000 m) on Atlantic Slope from Ver to Chis, thence locally F to Honduras; F to C in interior and on adjacent slopes from E Oax to N cen Nicaragua. Reports from SLP (Miller *et al.* 1957) require verification.

RA: E Mexico to N Ecuador.

GREY-BARRED WREN
Campylorhynchus megalopterus * Plate 48
Matraca-barrada Serrana

ID: 6.8–7.8 in (17–19.5 cm). *Endemic to highland forests of cen Mexico.* **Descr.** Ages differ. **Adult:** face streaked greyish and white with white supercilium and dark eyestripe; crown greyish, spotted dark brown; nape streaked blackish and white. *Upperparts barred dark brown and whitish. Throat and underparts white, spotted dark brown, becoming barred on flanks.* **Juv:** face whitish to pale buff with pale cinnamon supercilium and dark eyestripe; crown dark brown. Nape and upperparts streaked dark brown and pale cinnamon, wings and tail like adult but washed cinnamon. Throat and underparts whitish, often tinged cinnamon, flanks indistinctly barred dusky.
Voice. Gruff, dry, chattering calls suggest Band-backed Wren, *chah ah-ra-ra-rah*, and *cheh-cheh . . .*, and a more varied and rollicking, *ke wook'a ke wook'a ke wook'a*, or *kurr-e te-rek kurr-e te-rek*, etc.
Habs. Humid to semiarid conifer and pine–oak forest and edge. Habits much as

Band-backed Wren. Eggs undescribed (?).
SS: See Band-backed Wren.
SD/RA: C to F resident (2100–3000 m, U to 1500 m in Ver) in cen volcanic belt from Jal (and possibly adjacent Col) to W Pue; disjunctly from W Ver to N Oax.

GIANT WREN
Campylorhynchus chiapensis Plate 48
Matraca Chiapaneca

ID: 8–8.8 in (20.5–22 cm). *Endemic to Pacific Slope of Chis.* **Descr.** Ages similar. White supercilium contrasts with blackish crown, nape, and eyestripe. *Upperparts dark rufous,* tail darker, outer rectrices blackish with white subterminal band. *Throat and underparts whitish, washed cinnamon on flanks* and undertail coverts.
Voice. Bizarre, hollow, and gurgling rhythmic chortling and rollicking phrases, repeated, *kar-ah-u'too*, or *kar-rah-du-ow*, etc. Calls include a gruff rasp and a dry, soft, rattled *chirrr, chirrr. . . .*
Habs. Humid to semihumid open country with hedges, fences, scattered trees, gardens, orchards, etc. In pairs or small groups at low to mid-levels, often on ground. Eggs pale buff, heavily mottled brown.
SS: Rufous-naped Wren (*C. rufinucha nigricaudatus*) smaller with barred wings, blackish tail.
SD/RA: C to F resident (SL–300 m) on Pacific Slope of Chis, from vicinity of Tonalá to Puerto Madero. Apparently absent from neighboring Guatemala.
NB: Sometimes considered conspecific with *C. griseus* of S.A.

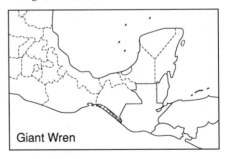

Giant Wren

RUFOUS-NAPED WREN
Campylorhynchus rufinucha * Plate 48
Matraca Nuquirrufa

ID: 6–7.5 in (15–19 cm). Complex geographic variation, with Pacific Slope and

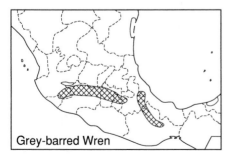

Grey-barred Wren

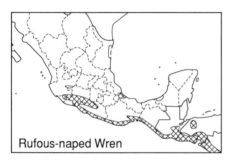

Rufous-naped Wren

Atlantic Slope—Interior populations both comprising several distinct ssp. Ages similar. **Pacific Slope.** *Humilus* (W Mexico): small. White supercilium contrasts with *rufous-brown crown* and dark eyestripe. *Nape and upper mantle rufous, rest of upperparts streaked and spotted dark brown and whitish to pale cinnamon; tail barred blackish brown and pale grey.* Throat and underparts whitish, washed pinkish on sides and flanks, *blackish malar stripe usually distinct,* undertail coverts barred black. In N of range (interior Col) duller overall, sides of chest may be spotted black, flanks with dark bars. Intergrades abruptly in W Chis (vicinity of Tonalá) with **nigricaudatus** ('Rufous-backed Wren' of Chis and W Guatemala): large. *Crown and eyestripe blackish. Nape and upperparts plain rufous to chestnut,* uppertail coverts barred dark brown. *Wings barred dark brown and cinnamon* to whitish. Tail blackish, outer rectrices barred pale grey and with broad subterminal white band. Lacks dark malar stripe. Intergrades in Guatemala with **capistratus** (E Guatemala to Honduras): rufous back usually has some dark and whitish bars and streaks, tail distinctly barred, may show slight dark malar stripe. **Atlantic Slope—Interior forms.** *Rufinucha* (cen Ver): small. Suggests *humilus* but *crown and eyestripe blackish, rufous band restricted to nape,* back more boldly marked, *underparts finely spotted dusky.* **Xerophilum** (Motagua Valley) and **nicaraguae** (interior W Nicaragua): similar to *rufinucha* but rarely show hint of dark malar stripe, *underparts unspotted* and undertail coverts rarely with dark spots or bars. **Castaneus** (Sula Valley): *back plain rufous-chestnut,* atypical of Atlantic Slope—Interior group.
   Voice. Varies geographically, but all (?) ssp give rich, rollicking phrases, typically repeated 3–5×, often in chortling duet, *ch-ree* or *ri chee'choo* ..., or *whi-cheeu ch-wik* ..., a rapidly repeated *chi-koh* ..., a longer *ch ree-choo-reek* ... and *wheer who'to-yu* ..., etc.; also a rapid *whi-chi-whi-chi* ..., and dry, gruff, slightly oriole-like chattering to slightly nasal clucking *cheh-cheh* ..., and a gruff *wh'rrr* or *gu'rrrr.*
   Habs. Arid to semiarid scrub, thorn forest, open and semiopen areas with bushes, cacti, scattered trees, similar habitat on humid Pacific Slope of Chis and W Guatemala. In pairs or small groups at low to mid-levels. Eggs whitish, speckled to spotted with browns and greys (WFVZ).
   SS: See Giant Wren (Chis only).
   SD: See Descr. for ssp ranges. C to F resident (SL–1200 m) on Pacific Slope from S Mich to Honduras, in interior along Balsas drainage and N to Col; also (SL–300 m) on Atlantic Slope in cen Ver, upper Motagua Valley, Sula Valley, and interior of W Nicaragua.
   RA: Mexico to NW Costa Rica.

SPOTTED WREN
Campylorhynchus gularis   Plate 48
Matraca Manchada

ID: 6.2–7 in (16–18 cm). *Endemic to N and cen Mexico.* **Descr.** Ages differ. **Adult:** face whitish with dark eyestripe and white supercilium; crown brown. Nape and upperparts brown, spotted and streaked white and black. Wings barred dark brown and buff to whitish, tail grey-brown, barred darker. *Throat and underparts whitish with dark malar stripe, chest and flanks boldly spotted black,* flanks washed cinnamon. **Juv:** face and supercilium washed buff, crown dark brown. Nape and upperparts streaked dark brown and pale cinnamon, wings and tail similar to adult.

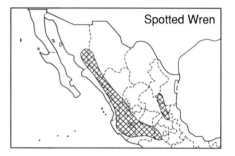

Spotted Wren

Throat and underparts whitish, washed creamy buff, with dark malar stripe.

Voice. A gruff, dry, and fairly quiet *cheh-cheh-cheht* and longer scolding and chattering series. Song varied, gruff to slightly gurgling and rollicking phrases, repeated 2–4×, *whir-i-i-whirrih*, or *chorr-i-choik*, or *wichoo wit-choo'wit*, etc.

Habs. Arid to semihumid oak woodland and scrub, open areas with hedges, stone walls, scattered trees. In small groups at low to mid-levels, often on ground. Eggs undescribed (?).

SS: Cactus Wren may overlap locally but prefers deserts, throat and chest heavily spotted black, face streaked blackish. Boucard's Wren apparently allopatric, larger with longer bill and tail, browner upperparts, more extensive black spotting below.

SD/RA: C to F resident (800–2500 m) on Pacific Slope and in adjacent interior from S Son to N Mich and W Mex, possibly to N Gro (Navarro *et al.* 1991); on Atlantic Slope from S Tamps to Hgo.

BOUCARD'S WREN

Campylorhynchus jocosus　　　Plate 48
Matraca del Balsas

ID: 6.3–7 in (16–18 cm). *Endemic to interior SW Mexico.* **Descr.** Ages differ slightly. **Adult:** face whitish with dark eyestripe and white supercilium; crown brown. Nape streaked blackish and white, upperparts brown, spotted and streaked whitish and black. Wings barred dark brown and pale cinnamon to whitish, tail grey-brown, barred darker. *Throat and underparts whitish, with dark malar stripe, bold black spots on underparts form bars on flanks.* **Juv:** upperparts less contrasting, underparts with sparser, less distinct dusky spots and bars.

Voice. A gruff *kyah-kyah* ..., a rapid, dry *chi-chi* ..., and dry, oriole-like chatters, etc. Song varied short, rollicking phrases *ka-yow' ka-yow* ..., or *ch'wi-it* ..., etc.

Habs. Arid to semiarid areas with scrub, scattered bushes, cacti. In small groups or pairs at low to mid-levels, at times relatively skulking in dense scrub. Eggs whitish, densely speckled with browns and greys (WFVZ).

SS: Spotted Wren apparently allopatric.

SD/RA: C to F resident (800–2500 m) in interior from Gro and Mor to Oax.

YUCATAN WREN

Campylorhynchus yucatanicus　　Plate 48
Matraca Yucateca

ID: 6.7–7.2 in (17–18.5 cm). *Endemic to coastal Yuc.* **Descr.** Ages differ slightly. **Adult:** face whitish with dark eyestripe and white supercilium; crown grey-brown, spotted darker. Nape and *upperparts streaked dark brown, blackish, and white*, wings barred dark brown and whitish. Tail barred blackish and pale grey, outer rectrices tipped whitish. *Throat and underparts whitish with dark malar stripe, underparts sparsely spotted blackish becoming barred on flanks.* **Juv:** upperparts less contrasting, underparts with sparser, less distinct dusky spots and bars.

Voice. Gruffer and huskier than other *Campylorhynchus*, a husky *krrohr*, a gruff, rasping *shorr-shl-shl-shlohr-shrr*, a rhythmic, chortling *ch-hor k-k wohrk*, repeated, etc.

Habs. Arid scrub with cacti, gardens, plantations. In pairs or small groups, usually low or on ground. Eggs undescribed (Kiff and Hough 1985).

SD/RA: C to F resident (around SL) in coastal Yuc Pen from extreme NW Camp (near Celestún) to E Yuc.

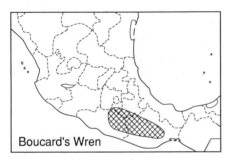

Boucard's Wren

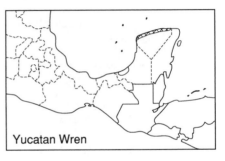

Yucatan Wren

CACTUS WREN
Campylorhynchus brunneicapillus *   Plate 48
Matraca Desértica

ID: 7–7.5 in (18–19 cm). *Deserts of N and cen Mexico.* **Descr.** Ages differ. Two groups: *affinis* group (Baja, except extreme NW) and *brunneicapillus* group (N and cen Mexico, and NW Baja; including *couesi*, shown). **Affinis. Adult:** face streaked brown and whitish with white supercilium and dark eyestripe, *crown rufous-brown. Nape and upperparts brown, streaked whitish.* Wings and tail dark brown, barred and spotted whitish and grey-brown. *Throat and underparts whitish, streaked black on throat and chest, spotted black on belly.* **Juv:** crown brown, underparts with sparser, less distinct dusky spots, flanks washed buff. **Brunneicapillus. Adult:** differs mainly in shorter, *broken white streaks on greyer-brown upperparts, heavy black spotting on lower throat and upper chest, and stronger cinnamon wash on flanks and belly.* **Juv:** underparts with sparse fine black spots, faint cinnamon wash on flanks.
Voice. A rough rasping *rrehrrk* or *rrehr*, a hard *chak* or *chek* which may run into chattering, a gruff *chuh* or *chowk*, a hard, dry *cha-cha* ..., suggesting somewhat a mockingbird, and a hard *chak-ak-ak* ..., etc. Song varied, low, hard, dry notes repeated in rapid chattering or rattling series: a low, gruff *grru-grru-grru-grru* ..., a buzzier, more rasping *chri-chri* ...
Habs. Arid to semiarid open areas with cacti, scrub, etc. In pairs or small groups, usually low or on ground. Eggs pale pinkish to creamy, densely speckled with reddish brown and greys (WFVZ).
SS: See Spotted Wren.
SD: C to F resident (SL–2000 m) in Baja, on Pacific Slope from Son to cen Sin, in interior from Son and NL to N Mich and Hgo, and on Atlantic Slope in Tamps.
RA: SW USA to cen Mexico.

ROCK WREN
Salpinctes obsoletus *   Plates 49, 71
Saltapared Roquero

ID: 5.2–5.7 in (13.5–14.5 cm). Dry rocky areas. Two groups: *obsoletus* group (Mexico to Guatemala) and *fasciatus* (Nicaragua). **Descr.** Ages differ. **Obsoletus. Adult:** bill and legs dark grey. *Head and upperparts grey-brown with whitish supercilium and fine dark and white flecking, rump cinnamon.* Wings and tail grey-brown, barred darker, *outer rectrices with black subterminal band and pale cinnamon tips.* Throat and underparts whitish, washed cinnamon on flanks; *throat and chest flecked dusky* (rest of underparts sparsely spotted dusky in *neglectus* of Guatemala). **Juv:** bill flesh below. Head and upperparts grey-brown with pale buff supercilium. Throat and underparts pale buff (*neglectus* shows faint dusky spotting). **Fasciatus:** Head and upperparts darker than *obsoletus*, underparts whitish with coarse and dense dark brown barring and spotting. **Juv:** underparts less distinctly marked. *S. o. guttatus* of El Salvador and Honduras intermediate between *obsoletus* and *fasciatus*.
Voice. A bright, springy, slightly trilled *ch-reer* or *b-reehrr*, and a scolding *jehr-jeh-je-je-je*, etc. Song a varied series of buzzes, trills, chatters, and whistles, usually repeated 2–6×, often 3×, suggesting a mockingbird, *ch-wee ch-wee ch-wee, jeer-r-r-r-r, ch-reeoo ch-reeoo, jeh-jeh-jeh, jirr-rr jirr-rr* ..., etc.
Habs: Arid to semiarid open rocky country with sparse vegetation, especially

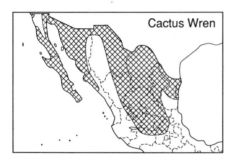

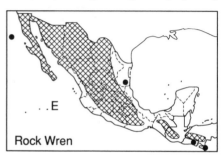

broken ground and loose rock slopes. Bobs conspicuously on rocks and often quite confiding. Nest is a shallow cup of grasses in crevices and cavities among rocks. Eggs: 4–8, white, with sparse dark speckling (WFVZ).

SS: Juv *obsoletus* may suggest House Wren but usually attended by adults.

SD: C to F resident (SL–3000 m) in Baja, on Isla Guadalupe (and formerly San Benedicto), on Pacific Slope S to N Sin, and in interior and on adjacent slopes from Son and Coah to cen volcanic belt, thence (1000–3500 m) locally to El Salvador and W Nicaragua.

CANYON WREN

Catherpes mexicanus *   Plate 49
Saltapared Barranquero

ID: 5–5.5 in (12.5–14 cm). *Canyons and arid country S to Chis.* **Descr.** Ages differ slightly. **Adult:** bill and legs dark grey. Crown and auriculars brownish, flecked white, upperparts rufous-brown, flecked white, becoming rufous on uppertail coverts; wings barred blackish. *Tail bright rufous, barred black.* Throat and chest white, belly and undertail coverts dark rufous, flecked white. **Juv:** lacks white flecks on upperparts and belly, underparts duller.

Voice. A bright, springy, nasal *beeihr* or *bzeeihr*, at times longer rollicking series, *beei-di-di beei-beeir* or *beeih-di-beih ...*, etc. Song a descending series of rich whistles, usually with quieter introductory notes and often ending with one or more buzzy mews, audible at close range, *ss ssi see-see-see-syee-syee-syee-syee-syee-syee-syoo-syoo-syoo-syoo, jirr jirr,* etc.; typically 12–20 or more notes.

Habs. Canyons, cliffs, buildings, ruins, etc., mostly in fairly arid and open country. Moves easily over rock surfaces, poking in and out of crevices and caves;

often bobs like Rock Wren, tail held slightly cocked. Nest cup of twigs and other vegetation in rock crevices. Eggs: 4–6, white, speckled with reddish brown and greys (WFVZ).

SD: C to F resident (SL–3000 m) in Baja, on Pacific Slope in Son, and (mainly above 1000 m) in interior and on adjacent slopes from Son and NL to cen volcanic belt, locally to W Chis.

RA: W N.A. to S Mexico.

SUMICHRAST'S [SLENDER-BILLED] WREN

Hylorchilus sumichrasti   Plate 49
Cuevero de Sumichrast

ID: 6–6.5 in (15–16.5 cm). *Endemic to E Mexico. Bill long and straightish.* **Descr.** Ages differ slightly. **Adult:** bill blackish, pale orange-yellow below at base, legs dark grey. *Head and upperparts dark rich brown, face paler greyish with dark auricular mark. Throat pale buff becoming tawny-brown on chest and dark brown on belly and undertail coverts,* belly flecked white. **Juv:** throat washed dirty buff and faintly scaled dusky, underparts darker sooty brown with paler mottling on chest, sparse whitish flecks on belly.

Voice. A loud, rich, explosive *cheeoo!* or *wheeoo!,* a slightly nasal, squirrel-like *chowh* or *chowk,* often repeated, and a rapid, gruff scolding *keh-keh-keh ...* Song a varied descending series of loud rich whistles, suggesting Canyon Wren. Full song typically 10–18 notes slowing toward the end, the last note distinct: *swee twee twee twee ... tew.* Also gives shorter series, usually of 3–5 notes, the last note rising: *whee whee hew',* etc.

Habs. Limestone (karst) outcrops in humid evergreen to semideciduous forest and coffee plantations. Hops easily over

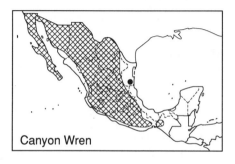

Canyon Wren

Sumichrast's Wren

rocks, in and out of crevices and caves; often bobs like a dipper, tail usually cocked, often nearly vertical. Can be elusive but also often curious, appearing suddenly at very close range and then disappearing. Singing posture usually (?) fairly horizontal, tail cocked. Nest cup of grass, rootlets, moss, etc., in caves and crevices. Eggs: 3, white (Bangs and Peters 1927).

SS: Tawny-throated Leaftosser (higher elevation) has tawny-rufous throat and chest, chestnut rump and uppertail coverts, prefers humid forest floor. Nava's Wren allopatric.

SD/RA: F to C but local resident (75–1000 m) on Atlantic Slope in cen Ver and N Oax.

NB1: Sometimes known as Slender-billed Wren. **NB2:** Some authors (such as Hardy and Delaney 1987) have suggested merging *Hylorchilus* with *Catherpes*, but see Atkinson *et al.* (1993).

NAVA'S [SLENDER-BILLED] WREN
Hylorchilus navai   Plate 49
Cuevero de Nava

ID: 6–6.5 in (15–16.5 cm). *Endemic to SE Mexico. Bill long and straightish.* **Descr.** Ages differ slightly (?). **Adult:** bill blackish, pale orange-yellow below at base, legs dark grey. *Head and upperparts dark rich brown, face tawny-brown with slightly greyer lores;* remiges faintly barred black. *Throat and upper chest whitish* becoming pale grey, faintly scalloped dusky on lower chest, belly with coarse whitish scalloped spots offset by subterminal dark crescents; sides and flanks dark sooty brown, undertail coverts dusky grey-brown. **Juv:** undescribed (?).

Voice. A tinny, nasal *iihn* or *eiihn* to *ihnk*, often repeated steadily or may be doubled, *iihn-iihn;* suggests Magnolia Warbler call but tinnier. Song a varied, often slightly

jerky warble of rich whistles, may suggest Blue-black Grosbeak: *ch tweeu tew-i-weeeu* and *chwee chu-e-wee-u* and longer series, *ch wee-u ch-wee chu-ee-choo,* etc. Also may be introduced by a few soft, slightly accelerating notes and end with a strongly up-slurred note, *che ch-ch-che-oo, che ch-re-oo weeu wee-oo reeeuh* (AK tape).

Habs. Limestone (karst) outcrops in humid evergreen forest. Habits much as Sumichrast's Wren; singing posture usually (?) fairly vertical, tail held straight down. Nest and eggs undescribed (?).

SS: None in range. Sumichrast's Wren allopatric.

SD/RA: F to C but local resident (75–800 m) on Atlantic Slope in SE Ver (KC, SNGH tape) and N Chis; probably also adjacent Oax.

NB: Originally described as a ssp of Sumichrast's Wren (Crossin and Ely 1973). The strikingly different vocalizations of *navai,* however, together with its distinct plumage, indicate that it should be considered specifically distinct. See Atkinson *et al.* (1993).

GENUS THRYOTHORUS

This is a mostly tropical genus of small wrens that inhabit the understory of forest and scruby woodland. Rarely are more than 2 species sympatric. Songs loud and varied, some species duet. Bills blackish above, paler, often greyish flesh, below; legs grey to greyish flesh. Singly or in pairs in shady, often tangled understory and edge where hard to see well. Heard far more often than seen. Nests (except Carolina Wren) are distinctive globular or retort-shaped structures of rootlets, grass stems, plant fibers, etc., usually with a downward-pointing entrance chute at one end, placed at low to mid-levels in bushes or thickets, often in acacias Eggs: 3–5.

SPOT-BREASTED WREN
Thryothorus maculipectus *   Plate 49
Saltapared Pechimanchado

ID: 5–5.5 in (12.5–14 cm). *E and S lowlands.* **Descr.** Ages differ. **Adult:** face streaked black and white with contrasting dark eye-stripe and white supercilium. *Throat and underparts whitish, spotted black, flanks*

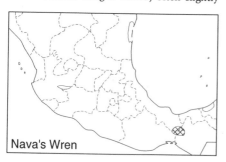

Nava's Wren

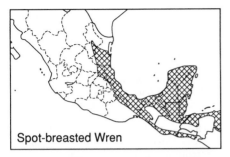

Spot-breasted Wren

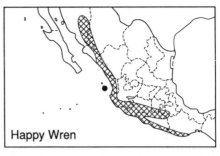

Happy Wren

cinnamon-brown, undertail coverts barred black. Crown, nape, and upperparts tawny-brown to rufous-brown, remiges indistinctly barred dusky. Tail brown, barred black. **Juv.** paler above, face and underparts greyish to grey-brown, with hint of adult pattern.

Voice. Song varied, bright, rollicking series, *swee hu-a wee-a-hew*, or *sweeu swee-i sui-sweery-ew*, or *chu tee chi-wee weery-chuh*, etc. Calls include a springy, rising trill, *purr-r-r-r-r-r-r-r-rip*, like running one's fingernail along a comb, and a chattering, nasal scold.

Habs. Forest and edge, second growth, plantations, mainly in humid areas. See genus note. Nest usually with little or no entrance chute. Eggs whitish, heavily marked with reddish browns (WFVZ).

SS: Spotted chest distinctive. Song longer and more varied than White-breasted Wood-Wren, much like allopatric Happy Wren.

SD: C to F resident (SL–1300 m) on Atlantic Slope from E NL and cen Tamps to Honduras, on Pacific Slope from Chis to El Salvador.

RA: Mexico to N Costa Rica.

NB: Sometimes considered conspecific with *T. rutilus* of Costa Rica to S.A., and with *T. sclateri* of NW S.A.

HAPPY WREN
Thryothorus felix *   Plate 49
Saltapared Feliz

ID: 5–5.5 in (12.5–14 cm). *Endemic to W Mexico.* **Descr.** Ages differ slightly. **Adult:** *face striped black and white* (less extensively striped and whiter-faced in Islas Tres Marías) with contrasting dark eyestripe and white supercilium. *Throat whitish, underparts ochraceous buff* to cinnamon (rarely pale buff), undertail coverts barred black and white. Crown, nape, and upperparts brown, tail barred black. **Juv:** face

less distinctly marked, underparts washed greyish buff.

Voice. Suggests Spot-breasted Wren. Song varied, rich, bright phrases, repeated, *si wee puree pee*, or *sweer di weeree weet weet*, or *s wee-see chuh-uhrrrr*, etc. Calls include a dry, gruff chattering and a springy, rising trill; note voice.

Habs. Forest and edge, second growth, plantations. See genus note. Eggs bluish white, unmarked (WFVZ).

SS: Sinaloa Wren lacks boldly striped face, underparts mainly whitish; note voice.

SD/RA: C to F resident (SL–2000 m) on Pacific Slope from Son to Oax, and in interior along Balsas drainage.

RUFOUS-AND-WHITE WREN
Thryothorus r. rufalbus   Plate 49
Saltapared Rufiblanco

ID: 5.8–6.5 in (14.5–16.5 cm). *Pacific Slope S of Isthmus.* **Descr.** Ages differ. **Adult:** *face striped black and white*, with dark eyestripe and white supercilium. *Throat and underparts whitish*, washed greyish on sides, dusky brown on flanks, undertail coverts barred black. *Crown, nape, and upperparts rufous*, wings and tail barred black. **Juv:** duller above, face indistinctly marked, underparts often scalloped dusky, undertail coverts brownish, barred black.

Voice. Song haunting, typically a rapid

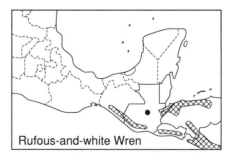

Rufous-and-white Wren

series of hollow, quavering hoots, beginning and/or ending with one or more whistled notes, *whu whew hu-hu-hu-hu-hu-hu-hu-hu hoo*, or *whoo-whoo hee whu-hu-hu-hu-hu-hu*, etc. Calls include a hard, slightly hollow scolding *chek* or *chehk*; also a persistent, fairly hard ticking or clicking *ti ti ti* ... (HG tape).

Habs. Humid to semiarid forest and edge, plantations. See genus note. Eggs greenish blue, unmarked (WFVZ).

SS: Plain Wren smaller, lacks boldly striped face, upperparts grey-brown, flanks washed cinnamon; note voice.

SD: C to F resident (near SL–1800 m, mainly 300–1500 m) on Pacific Slope from Chis to Honduras, locally U in interior Guatemala and Honduras, and on Atlantic Slope in N Honduras.

RA: S Mexico to N Colombia.

SINALOA (BAR-VENTED) WREN
Thryothorus sinaloa*   Plate 49
Saltapared Sinaloense

ID: 5–5.5 in (12.5–14 cm). *Endemic to W Mexico.* **Descr.** Ages differ slightly. **Adult:** *face whitish, indistinctly streaked dusky, with contrasting dark eyestripe and white supercilium*, sides of neck striped black and white. *Throat and underparts whitish, washed greyish cinnamon on flanks*, undertail coverts barred black. Crown, nape, and upperparts brown, wings and tail barred black. **Juv:** face less distinctly marked, sides washed dusky, undertail coverts dusky pale cinnamon.

Voice. Song of varied loud and rich phrases, often with rapid repetition of notes, *whi-whi-whi-whi wheet wheet wheet whurrrrrr*, or *hoo hoo wi-chu ho whi-chooa-chooa-chooa*, or *whi whi-whi chu-i weet-weet-weet-weet*, etc.; recalls Banded Wren and quite distinct from Happy Wren. Calls include a rough, buzzy to

slightly mewing rasp, *dzzzshrr* or *rreihrr*, a rough scolding *rahr* or *rahrr*, and *rreh-rreh* ..., and a hard, dry chatter.

Habs. Forest and edge, second growth, plantations. See genus note. Eggs white to bluish white, unmarked (WFVZ).

SS: Banded Wren has sides and flanks boldly barred black, but song similar. See Happy Wren.

SD/RA: C to F resident (SL–2000 m) on Pacific Slope from S Son to W Oax, and in interior along Balsas drainage.

BANDED WREN
Thryothorus pleurostictus*   Plate 49
Saltapared Vientre-barrado

ID: 5.5–6 in (14–15 cm). *Pacific Slope from SW Mexico S.* **Descr.** Ages differ. **Adult:** face whitish, indistinctly streaked dusky, with dark eyestripe and white supercilium sides of neck striped black and white. *Throat and underparts white, boldly barred black on sides, flanks, and undertail coverts*, flanks washed cinnamon. Crown, nape, and upperparts brown, wings and tail barred black. **Juv:** face and underparts duller, indistinctly streaked and barred dusky.

Voice. Varied songs striking, loud, rich, and resonant, typically including rapid series of rich whistles, *whie-whie-whie hoo-hoo hoo-hoo whee whee whi-whi-whi-whi-whi*, or *choorrrrrrrr whu hu-hu-hu hu-hu*, or *hew, hoo-oo-oo-oo-oo-oo-oo-oo wheet wheet wheet wheet*, etc., recall Sinaloa Wren. Calls include a nasal scolding *cheh-cheh-cheh* or *jeh-jeh-jeh*, and a springy, slightly liquid, rolled trill, *prrirrr*.

Habs. Arid to semihumid forest and edge. See genus note. Eggs pale bluish to white, unmarked (WFVZ).

SS: See song of Sinaloa Wren (W Mexico).

SD: C to F resident (SL–1500 m) in upper

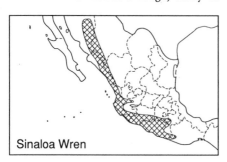

Sinaloa Wren

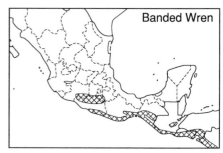

Banded Wren

Balsas drainage, on Pacific Slope from cen Gro to Honduras, and locally in interior from Chis to Honduras.
RA: Mexico to Costa Rica.

CAROLINA WREN
Thryothorus ludovicianus *   Plate 49
Saltapared de Carolina

ID: 5–5.5 in (12.5–14 cm). *NE Mexico.* **Descr.** Ages similar. White supercilium contrasts with brown crown and broad dark eyestripe; *face and throat whitish, underparts cinnamon,* often barred dusky on flanks (especially in S), undertail coverts barred black. Upperparts brown with white-tipped greater coverts, wings and tail indistinctly barred blackish.
Voice. Song varied, loud, rich phrases repeated rapidly 6–12×, *choo-wee choo-wee ...,* or *shee wee-wee ...,* etc. Calls include a bright, slightly explosive *beenk* or *beehn,* often repeated, a bright, slightly metallic *pli-dik,* and a low rasping *t-shihrr,* sometimes repeated in scolding.
Habs. Humid to semiarid forest and edge. See genus note. Nests in tree cavities, in tangles near ground, etc. Eggs (N.A.) whitish, marked with reddish browns and greys (WFVZ).
SD: C to F resident (SL–2000 m) on Atlantic Slope from E Coah and Tamps to E SLP.
RA: E N.A. to NE Mexico.

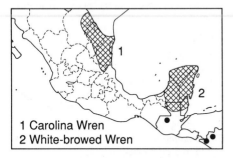

1 Carolina Wren
2 White-browed Wren

WHITE-BROWED [CAROLINA] WREN
Thryothorus ludovicianus (in part) or
T. albinucha   Plate 49
Saltapared Yucateco   See Carolina Wren
for map

ID: 4.5–5.2 in (11.5–13.5 cm). *Endemic to Yuc Pen and N C.A.* **Descr.** Ages similar(?). *White supercilium contrasts with brown crown and broad dark eyestripe.* Face, throat, and underparts whitish,

washed buffy brown on flanks, sides of neck slightly striped black and whitish, undertail coverts barred black. Upperparts brown, *wings and grey-brown tail indistinctly barred blackish.* In disjunct S populations, underparts washed buff to buffy brown, flanks indistinctly barred dusky. **Juv** undescribed (?).
Voice. Song of rich, rollicking phrases much like Carolina Wren, *chwee chwee ... or ch-ree ch-ree ...,* or *w'chee w'chee ...,* etc. Calls include a slightly explosive, buzzy *bzzeeu* or *bzzz-iu',* a quieter, burry *bzzeirr,* a rapidly repeated, buzzy scold *zzhi-zzhi ...,* and a scolding *chih-chih ... or jeh-jeh ...*
Habs. Forest and edge. See genus note. Nest and eggs undescribed (?).
SS: White-bellied Wren smaller, whitish lores give distinctive bare-faced expression. Plain Wren (possibly sympatric in N C.A.) has rufous tail sharply barred black, sides of chest washed greyish, flanks cinnamon; note voice.
SD/RA: C to F resident (SL–300 m) on Atlantic Slope in Yuc Pen and locally (to 1200 m) in the upper Rio Negro Valley, Guatemala, S Belize (Howell *et al.* 1992*b*) and W Nicaragua.

PLAIN WREN
Thryothorus m. modestus   Plate 49
Saltapared Sencillo

ID: 5–5.5 in (12.5–14 cm). *S of Isthmus.* **Descr.** Ages similar. *Face whitish with dark eyestripe and white supercilium. Throat and underparts whitish, washed grey on sides of chest, cinnamon on flanks and undertail coverts.* Crown, nape, and upperparts brown, *washed grey on nape, cinnamon on rump*; wings and tail brown, wings indistinctly barred dusky, tail barred black.
Voice. Relatively short, bright, rich

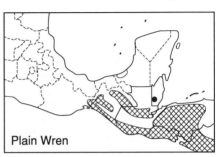

Plain Wren

phrases, usually with a lisping introductory note and often repeated in quick, rollicking succession, *ss ti-been-ti-been ss ti-been-ti-been*, or *ssi ti whee-twee-tweet*, and thinner phrases, *tsii sii sii-wii*, repeated, etc. Calls include a dry *ch-ch-cht* and a rippling, tinkling *chi-chi-chik* and *chi-i-i.*

Habs. Humid to semiarid second growth, forest edge, scrubby woodland. See genus note. Eggs bluish white, unmarked (WFVZ).

SS: See White-browed Wren (possibly sympatric).

SD: C to F resident (near SL–1800 m) on both slopes and in interior from E Oax and Chis to W Nicaragua; disjunctly (450–900 m) in Mountain Pine Ridge, Belize.

RA: S Mexico to Panama.

Fig. 37 White-bellied Wren nest (*Thryothorus* wren nests similar)

WHITE-BELLIED WREN

Uropsila leucogastra *   Plate 49
Saltapared Vientre-blanco   Figure 37

ID: 3.8–4 in (9.5–10 cm). Endemic; local on both slopes. **Descr.** Ages/sexes similar. Bill blackish above, pale below, legs greyish. *Whitish lores and supercilium contrast with brown crown and postocular stripe to create blank-faced expression.* Upperparts brown, flight feathers barred dusky; tail distinctly barred blackish in Yuc Pen. Throat and underparts whitish, washed buff to grey-brown on flanks, undertail coverts barred dark brown.

Voice. Song varied, a 2–3× repetition of short, simple, tinkling or bubbling phrases, *chid-l-er*, or *chidl-il-i-chu*, or *whi didl-oo*, etc.; drier, not bubbly, in W Mexico, for example a pleasant *tew dee-oo*, etc. Calls include a hard, dry, crackling rattle, a gruff, scolding chatter, and a mellow, fairly low *chek* or *whek.*

Habs. Semideciduous to deciduous woodland and forest, thorn forest, locally humid evergreen forest. Singly or in pairs at low to mid-levels, joins mixed-species flocks. Nest is a retort-shaped mass of grass, fibers, moss, etc. (Fig. 37), much like *Thryothorus* nests, placed at low to mid-levels in bush or tree, often in acacias. Eggs: 4, pale blue, unmarked (Sutton 1948a).

SS: See White-browed Wren (Yuc Pen).

SD/RA: C to F but local resident (SL–500 m) on Pacific Slope from Jal to Gro, on Atlantic Slope from S Tamps to Yuc Pen; disjunctly in N Honduras (Aguán Valley).

BEWICK'S WREN

Thryomanes bewickii *   Plate 49
Saltapared de Bewick

ID: 4.7–5.2 in (12–13.5 cm). Temperate regions N of Isthmus. *Tail relatively long.* **Descr.** Ages differ. Bill blackish above, pale below, legs greyish. **Adult:** *white supercilium contrasts with brown crown and broad eyestripe.* Throat and underparts whitish, washed grey on sides and flanks, undertail coverts barred blackish. Upperparts grey-brown, rump usually richer and browner; wings and tail indistinctly barred blackish, *outer rectrices broadly tipped*

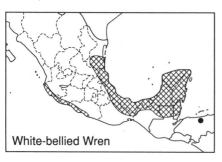

White-bellied Wren

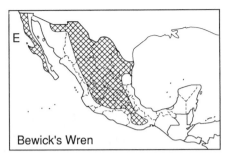

Bewick's Wren

Eggs: 2–7, whitish to pinkish white, marked with reddish browns (WFVZ).

NORTHERN HOUSE [HOUSE] WREN
Troglodytes aedon *   Plate 49
Saltapared-continental Norteño

ID: 4.5–4.7 in (11.5–12.5 cm). Widespread in winter N of Isthmus. **Descr.** Ages differ slightly. **Adult:** *head and upperparts grey-brown with narrow whitish eyering* and often an indistinct pale supercilium. Wings and tail brown, barred blackish. *Throat and underparts pale grey*, tinged brown on chest and flanks; flanks (often covered by wings) and undertail coverts barred dark brown. **Juv:** indistinctly scalloped dusky on chest.

Voice. A rough scolding *cheh-cheh* ..., often run into a rapid chatter, a dry, nasal *chehr*, a dry *chi-chi* ..., and a longer nasal mewing *meeeah*. Song varied, a scratchy, chortling, warbling and trilled series, usually beginning with a few separate gruff notes.

Habs. Open and semiopen areas, scrubby woodland and edge, gardens. Skulking like most small wrens but seen often; sings from prominent perches.

SS: Brown-throated Wren has more distinct supercilium, underparts washed buff (especially S populations); Southern House Wren warmer brown above, underparts pale buff. Winter Wren smaller, darker and browner, with shorter cocked tail, strongly barred belly and flanks; note voice.

SD: C resident (near SL–2500 m) in NW BCN; C to F transient and winter visitor (mid-Sep–mid-May; SL–3000 m) in Baja and on both slopes and in interior S to Oax (mostly in interior from cen Mexico S), possibly to W Tab (Phillips 1986). A May spec. from Chis (Webster 1984) requires verification.

whitish to pale grey. **Juv:** underparts flecked dusky, often forming scallops.

Voice. Song notably varied, typically one or more notes followed by a trill, *sseu iiiiiiiiiiiiirr* or *tss-iu si-i-i-i-r*, etc., but also more complex, at times suggesting Rufous-sided Towhee, *jer weeh zw-i zw-i zw-i zw-i*, etc. Calls include a sharp, springy, slightly nasal *whee-ik*, a wetter *schwenk*, often repeated, a rasping, buzzy *shirrrr*, an excited *whiut whiut* ..., a gruff scolding *jerr jerr* ..., etc.

Habs. Arid to semiarid scrub, open areas with scattered trees and bushes, brushy thickets, gardens. Usually low, in or near cover, but sings from atop bushes; tail often cocked high. Nests in cavities in trees, buildings, etc. Eggs: 5–8, whitish, marked with reddish browns and greys (WFVZ).

SS: Bold white supercilium and long, white-tipped tail distinctive.

SD: C to F resident (SL–3000 m) in Baja (N of Cape district), in interior and on adjacent slopes (mainly above 1000 m) from NE Son and NL to Oax, and on Atlantic Slope in Tamps.

RA: N.A. to S Mexico.

GENUS TROGLODYTES

These small brownish wrens lack the striking patterns shown by many *Thryothorus*. Inhabit open and forested areas, nest in cavities in trees, buildings, etc. Wings and tails barred blackish to dark brown, outer webs of outer primaries notched buff to whitish in several species. Bills dark grey above, orange-flesh to horn below, legs dusky flesh to greyish. Songs are varied chortled and trilled arrangements. Nests are often bulky, domed masses of grasses and other vegetation at low to midlevels in tree cavities, burrows in banks, etc.

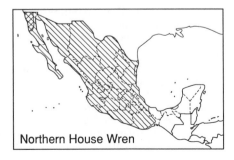

Northern House Wren

RA: Breeds N.A. to NW Baja; winters S USA to S Mexico.

NB1: Populations in the SW USA and N Mexico appear intermediate to various degrees between Northern House Wren and Brown-throated Wren.

BROWN-THROATED [HOUSE] WREN

Troglodytes (aedon?) brunneicollis*
Saltapared-continental Gorjicafé   Plate 49

ID: 4.5–4.7 in (11.5–12.5 cm). *Highlands N of Isthmus.* **Descr.** Ages differ. **Adult:** *head and upperparts brown with distinct buff supercilium*; wings and tail barred blackish. *Throat and underparts greyish buff* (*cahooni* of N Mexico) *to ochraceous buff*, flanks and undertail coverts barred dark brown. **Juv:** underparts boldly scalloped dusky.
   Voice. Much like Northern House Wren but also a bright springy trill, *tseeeurr* or *ssreeuur*, suggesting Rock Wren.
   Habs. Pine–oak and oak woodland, clearings. Habits much as Northern House Wren.
SS: See Northern House Wren. Southern House Wren allopatric.
SD/RA: C to F resident (1600–3000 m) in interior and on adjacent slopes from Son (also SE Arizona) and Coah to Oax.
NB1: See Northern House Wren.

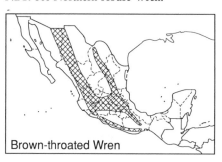

Brown-throated Wren

SOUTHERN HOUSE [HOUSE] WREN

Troglodytes aedon (in part) or *T. musculus**
Saltapared-continental Sureño   Plate 49

ID: 4.5–4.7 in (11.5–12.5 cm). *Widespread S of Isthmus.* **Descr.** Ages differ slightly. **Adult:** *head and upperparts brown with narrow pale buff eyering and paler supercilium*; wings and tail barred blackish. *Throat and underparts pale buff*, undertail coverts barred dark brown, flanks may show slight dark barring. **Juv:** underparts

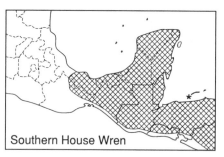

Southern House Wren

indistinctly flecked dusky, often forming scallops.
   Voice. Much like Northern House Wren, probably not safely told by voice.
   Habs. Open and semiopen areas from towns to clearings in humid forest. Habits much as Northern House Wren.
SS: See Northern House Wren. Rufous-browed Wren (highland forest) has bright cinnamon face with dark eyestripe, dark-barred flanks, Winter Wren-like jizz.
SD: C to F resident (SL–3000 m) on both slopes and in interior from S Ver (Andrle 1967b; WJS) and E Oax to Honduras and W Nicaragua, though absent from most of Pacific lowlands.
RA: S Mexico to Tierra del Fuego.

COZUMEL [HOUSE] WREN

Troglodytes beani   Plate 69
Saltapared de Cozumel

ID: 4.7–5 in (12–12.5 cm). *Endemic to Isla Cozumel.* **Descr.** Ages differ slightly. **Adult:** *head and upperparts brown, becoming rufous-brown on rump and uppertail coverts, with narrow pale buff eyering and slight buff supercilium.* Wings and tail rufous-brown, barred darker. *Throat and underparts whitish, flanks washed buffy brown*, undertail coverts washed cinnamon, barred dark brown. **Juv:** underparts finely and inconspicuously flecked

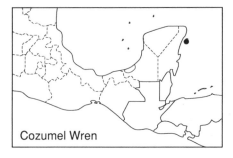

Cozumel Wren

dusky, rarely hinting at dusky scallops on chest.

Voice. A scolding *sheh-sheh* ... and a rolled *jirr-rr-rr-rr-rr*, etc. Song recalls Northern House Wren but fuller and richer, without trills, typically begins with a slight scolding *chih-chih* ... or *sh sh*, 2–5×, breaks into a short, rich warble which ends with a bright *wheet wheet wheet wheet* or *reeh reeh ree-ree-ree*, or with longer series of bright notes which may suggest White-browed Wren; also longer, rambling, scratchy warbles, etc.

Habs. Scrubby woodland and edge, semiopen areas. Habits much as Northern House Wren. Nest and eggs undescribed (?).

SD/RA: C to F resident (around SL) on Isla Cozumel.

NB: Sometimes considered conspecific with Southern House Wren, but appears closer to the Caribbean forms (species?) *musicus* of St Vincent and *mesoleucus* of St Lucia. Comprehensive studies of house wrens outside the USA are still needed.

CLARION WREN

Troglodytes tanneri Plate 69
Saltapared de Clarión

ID: 5–5.5 in (12.5–14 cm). *Endemic to Isla Clarión.* **Descr.** Ages differ slightly. **Adult:** head and upperparts grey-brown with narrow pale buff eyering and buff supercilium; wings and tail brown, barred darker. *Throat and underparts pale buff*, undertail coverts barred dark brown. **Juv:** underparts flecked dusky, forming faint scallops.

Voice. A rather insect-like, dry, churring rattle *tchiirr-rr-rr* ... or *djurr-rr-rr* ..., up to one minute or longer, a gruff *tchehk* ..., usually in short series up to 4×, and a more rapid, gruff, chattering scold. Song resembles Northern House Wren, typically

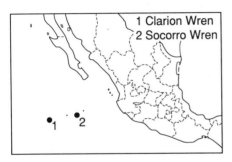

1 Clarion Wren
2 Socorro Wren

begins with 2 or more gruff *chut* notes then runs into warbling, lower and with more gruff chattering than Socorro Wren song. Also a prolonged scratchy warbling (post-breeding) which may be preceded by a dry, gruff *cheh-eh-eh-ehr*.

Habs. Arid scrub, open and semiopen areas over most of the island. Habits much as Northern House Wren. Fledging (1st brood?) occurred Feb–Mar 1988 (SNGH, SW). Nests in bushes, cacti, buildings.

SD/RA: C resident (SL–250 m) on Isla Clarión; 175–200 pairs in Feb 1988 (Howell and Webb 1989*b*).

NB: Has been considered conspecific with mainland house wrens but differs in morphology, plumage, and vocalizations. Clarion Wrens showed no more than brief curiosity in response to playback of Socorro Wren songs but strong aggression and much singing in response to songs of their own species (SNGH, SW).

SOCORRO WREN

Troglodytes sissonii Plate 69
Saltapared de Socorro See Clarion Wren
for map

ID: 4.3–4.8 in (11–12 cm). *Endemic to Isla Socorro.* **Descr.** Ages differ. **Adult:** head and upperparts brown with narrow pale buff eyering and buff supercilium; wings and tail barred blackish. *Throat and underparts whitish, washed buff* (mainly on chest), undertail coverts barred dark brown. **Juv:** underparts boldly scalloped dusky.

Voice. A fairly quiet, dry scolding *chuk-chuk chuk-chuk* ..., a fairly rapid, harder *shehk-shehk* ... or *chak-chak* ..., and a rapid scolding *jehr-jehr* ... recalling Northern House Wren. Song recalls Northern House Wren, typically begins with 2 or more low, gruff *chuk* notes, breaks into short, rich, slightly scratchy warble, and often ends with a rich chortle or clear *sweet-weet-weet-weet*; ♀ may countersing with ♂, contributing gruff chatters. Post-breeding song longer and more varied.

Habs. Semiopen and wooded habitats over most of the island. Habits much as Northern House Wren. Apparently breeds earlier than Clarion Wren, most young (1st brood?) had completed 1st prebasic

molt in Feb 1988 (SNGH, SW). Nest and
eggs undescribed (?).

SD/RA: C resident (SL−1000 m) on Isla
Socorro.

NB1: Traditionally placed in the genus
Thryomanes, this species is quite obviously
a House Wren in its manners, song, plumage, etc. **NB2:** See NB under Clarion
Wren.

RUFOUS-BROWED WREN

Troglodytes rufociliatus*   Plate 49
Saltapared Cejirrufo

ID: 4−4.5 in (10−11.5 cm). *Highlands S of
Isthmus.* Winter Wren-like jizz. **Descr.**
Ages differ. **Adult:** *face and throat cinnamon with paler eyering and dark eyestripe.* Underparts cinnamon, paler on
belly; flanks and undertail coverts barred
dark brown. Crown, nape, and upperparts
brown, wings and tail barred blackish,
wing coverts flecked white. **Juv:** underparts heavily scalloped dusky, crown and
upperparts barred dusky.
  Voice. A loud, sharp, scolding, buzzy
rasp, persistent when excited, *zzrrit* ... or
zzzrei, and a longer *jhrrrirr* ... Song a
varied, scratchy warble run into a tinkling
trill, recalls Winter Wren.
  Habs. Humid to semihumid pine, pine−
oak, and evergreen forest, clearings. In
pine and pine−oak, often found at low to
mid-levels, for example in brush piles; in
humid evergreen forest, also creeps along
branches and among epiphytes at mid- to
upper levels.
SS: See Southern House Wren.
SD/RA: C to F resident (1750−3500 m) in
interior and on adjacent slopes from Chis
to Honduras; U (?) in N cen Nicaragua
(1250 m, Martinez-S. 1989).
NB1: Relationship to *T. ochraceous* of S C.A.
and similar forms in S.A. in need of study.

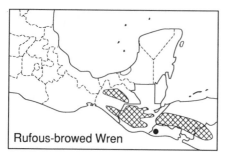
Rufous-browed Wren

WINTER WREN

Troglodytes troglodytes hiemalis
Saltapared Invernal   Not illustrated

ID: 3.5−4 in (9−10 cm). *Rare winter visitor
to N Mexico. Tail relatively short, often
cocked.* **Descr.** Head and upperparts
brown with *pale buff supercilium*, rump
and uppertail coverts brighter rufous-
brown. Wings and tail rufous-brown, barred blackish, wing coverts flecked whitish.
Throat whitish, *underparts pale buffy
brown, barred and vermiculated blackish on
belly, flanks, and undertail coverts.*
  Voice. A dry, fairly hard *tik* and *tik-tik* or
tk-tk, sometimes repeated steadily, suggests Wilson's Warbler; may be run into a
ticking rattle.
  Habs. Brushy understory of forest and
woodland, often near streams. Usually
skulking and hard to see, tail often cocked.
SS: See Northern House Wren, Brown-
throated Wren.
SD: R to U winter visitor (Nov−Mar; near
SL−2000 m) to Atlantic Slope from Coah
to Tamps (Phillips 1986; JCA, RR).
RA: Holarctic breeder; in New World winters S to S USA, rarely N Mexico.

Winter Wren

SEDGE WREN

Cistothorus platensis*   Not illustrated
Saltapared Sabanero

ID: 4−4.5 in (10−11.5 cm). Widespread but
local. **Descr.** Ages differ slightly. **Adult:** *bill
yellow-orange to flesh with dark culmen,*
legs flesh. *Head and upperparts buffy
brown, streaked darker,* with whitish
supercilium, white-streaked, triangular,
black back patch, and *cinnamon rump.*
Wings and tail barred dark brown and buff
to pale cinnamon. Throat and underparts
whitish, washed buff on flanks, chest and
undertail coverts. Crown and rump strongly

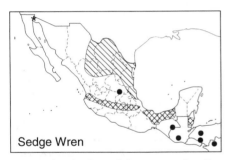

Sedge Wren

grants) so back patch less contrasting. **Juv**: white back streaks reduced or absent.

Voice. Song varied, short, buzzy chatters, often beginning with one or more sharp chips or hard, low clucks; also prolonged series of short rattled and churring phrases, repeated with short pauses and no introductory chips. Call a sharp *chik* or *chk*.

Habs. Grassy marshes, especially with sedges, locally in dry bunch grass. Skulking and overlooked easily unless singing. If flushed, flies a short distance and drops into cover whence it may be impossible to flush again. Nest is a globular mass of grasses with side entrance, near ground in tussock of grass. Eggs: 4–7, white (WFVZ).

SS: Marsh Wren larger with longer bill, crown dark, overall darker and browner (less buffy) above, rump dark rufous, note voice; N migrants have contrasting whitish supercilium.

SD: F to U but local resident (600–3000 m) in interior and on adjacent slopes from Nay and SE SLP to El Salvador (Thurber *et al.* 1987) and W Nicaragua, on Atlantic Slope (SL–1000 m) from S Ver to the Mosquitia. N.A. migrants are F to U in winter (Oct–Apr) in N Mexico S to Dgo (Phillips 1986) and N Ver (SNGH, SW); vagrant to NW Son (Dec 1983, SNGH, PP).

RA: N.A., locally in M.A. and S.A.

NB1: S.A. populations sometimes regarded as a separate species, Grass Wren C. *platensis*, the N populations then being *C. stellaris*. **NB2**: Formerly known as Shortbilled Marsh-Wren.

MARSH WREN

*Cistothorus palustris** Not illustrated
Saltapared Pantanero

ID: 4.5–5 in (11.5–12.5 cm). Widespread winter visitor, local in summer. Bill relatively long. Two groups: *tolucensis* (cen

volcanic belt) and *paludicola* group (breeds Colorado delta, widespread in winter). **Descr.** Ages differ slightly. **Adult.** *Tolucensis*: bill blackish, yellow-orange to flesh below at base, legs flesh. *Crown dark brown, face plain pale buffy with poorly contrasting whitish supercilium becoming pale grey in front of eye*, dusky postocular stripe. White-streaked triangular black back patch contrasts with cinnamon-brown scapulars and *dark rufous rump* and uppertail coverts. Wings and tail barred blackish and brown. Throat and chest greyish white, flanks dusky rich tawny-brown, belly whitish, undertail coverts whitish, barred buffy brown. *Paludicola*: *sides of crown blackish, face dusky brownish with contrasting whitish supercilium*. Underparts dirty whitish, washed dusky buff on chest, sides and flanks dusky cinnamon-buff, undertail coverts washed cinnamon and distinctly barred dark, uppertail coverts barred blackish. **Juv** (both groups): duller overall, head darker with indistinct whitish supercilium (*paludicola* group), white back streaks reduced or absent.

Voice. A dry *cheh* or *chi* often run into chatters, a gruff dry *cheh-cheht*, and a low, fairly sharp *chk*, etc. Song often prolonged, a varied arrangement of churring, buzzy, and dry rattling trills. *Tolucensis* song a fairly simple medley of dry buzzes and chatters.

Habs. Freshwater and brackish marshes, typically with tall reeds. Fairly skulking but more readily seen than Sedge Wren. Nest is a globular mass of reeds and grass with side entrance, low in reeds or bush. Eggs: 3–6, dull brownish, marked with darker browns (WFVZ).

SS: See Sedge Wren.

SD: *Tolucensis*: F to C resident (1000–2500 m) in cen volcanic belt from Mex and Hgo

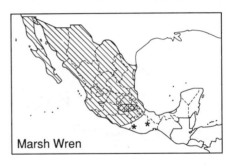

Marsh Wren

to Pue. *Paludicola*: F to U breeder (Apr–Aug; around SL) in Colorado Delta. C to F transient and winter visitor (Sep–mid-May; SL–2500 m) in Baja and on both slopes and in interior from Son and Tamps to N Oax and cen Ver, R and irregular to cen Oax (SNGH).

RA: Breeds N.A., disjunctly in cen Mexico; winters S USA to Mexico.

NB1: Formerly known as Long-billed Marsh-Wren. **NB2:** Studies in N.A. suggest that western and eastern populations may be specifically distinct (Kroodsma 1988); the distinctive-looking allopatric cen Mexican population remains unstudied. E N.A. populations winter to S Texas and may also occur in NE Mexico. In the field, the ssp most likely to occur might be distinguishable from W birds by its brighter whitish underparts with only a slight buff wash on chest, richer dusky tawny-brown sides and flanks, slight to distinct dusky bars on undertail coverts, and usually unbarred uppertail coverts.

GENUS HENICORHINA: Wood-Wrens

Wood-Wrens are small but chunky stub-tailed wrens of humid forest interior. Although common, they stay well hidden in dense understory close to the forest floor and are difficult to see. Bills black, legs dark greyish. Songs rich and rollicking. Nests are globular structures of rootlets, moss, fibers, etc., with a downward-facing or visored side entrance, on or near ground, in banks, etc. Eggs: 2–3, white (WFVZ).

WHITE-BREASTED WOOD-WREN
*Henicorhina leucosticta** Plate 49
Saltapared-selvatico Pechiblanco

ID: 4–4.5 in (10–11.5 cm). *Tropical lowland*

forest. **Descr.** Ages differ. **Adult:** *face boldly striped black and white*, with broad black eyestripe and white supercilium. Crown olive-brown, becoming rufous-brown on upperparts; wings and tail barred black, wing coverts flecked white. *Throat and median underparts whitish with grey sides*, cinnamon hind flanks and undertail coverts. **Juv:** throat and median underparts pale greyish, face pattern less distinct.

Voice. Song varied, rich to slightly plaintive phrases repeated, *ss chee ree-eu*, or *ss cheree-choo*, or *hoo-ee hoo'ee-ee*, etc., longer and more complex in duets, *ss weery tee ti wee-chi-wee*, etc. Calls include a ventriloquial, bright, metallic *peenk* or *biink*, which grades through a *cheep* and *chup* to a scolding *chek-chek*; also a scolding, dry chatter, *cherr-errrt* or *cherr-rrrt-rrrt*, and a hard, dry, rattling chatter.

Habs. Humid evergreen forest. See genus note.

SS: Grey-breasted Wood-Wren overlaps locally, whitish throat contrasts with grey chest; but beware juv wood-wrens (usually attended by parents).

SD: C to F resident (SL–1300 m) on Atlantic Slope from S Ver to Honduras, on Pacific Slope in Chis and Guatemala; U to F N to SE SLP.

RA: E Mexico to N Peru.

NB: Sometimes known as Lowland Wood-Wren.

GREY-BREASTED WOOD-WREN
*Henicorhina leucophrys** Plate 49
Saltapared-selvatico Pechigris

ID: 4–4.5 in (10–11.5 cm). *Humid montane forest.* **Descr.** Ages differ. **Adult:** *face boldly striped black and white*, with broad black eyestripe and white supercilium. Crown olive-brown, becoming rufous-brown on upperparts; wings and tail barred black, wing coverts flecked white. *Throat*

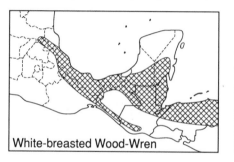

White-breasted Wood-Wren

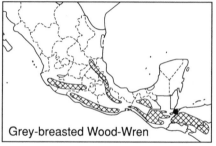

Grey-breasted Wood-Wren

whitish, underparts grey, becoming rufous-brown on hind flanks and undertail coverts. **Juv:** throat greyish, face pattern less distinct.

Voice. Analogous to White-breasted Wood-Wren. Songs of loud, bright, rich phrases involve dueting more often and are richer, fuller, and often more varied and longer than White-breasted: *wheer heer chee-dee-hu-weedee*, or *cheer de-choo*, or *cheedl-eet ch-weet'chee*, etc. Calls include a hollow, ringing *chuuh*, less bright and metallic than White-breasted, which grades into a low, gruff *chuk* or *chuh-chuk*; also a hard, scolding chatter and a hard, dry, clicking or rattling chatter, hollower and harder, more clicking than White-breasted.

Habs. Humid evergreen and pine–evergreen forest. See genus note.

SS: See White-breasted Wood-Wren.

SD: C to F resident (600–3000 m, mostly above 900 m) on both slopes from Jal and E SLP, and in interior from W Mexico, to El Salvador and N cen Nicaragua.

RA: Mexico to Ecuador and N Bolivia.

NB: Sometimes known as Highland Wood-Wren.

NIGHTINGALE WREN

Microcerculus philomela   Plate 49
Saltapared Ruiseñor

ID: 4–4.5 in (10–11.5 cm). *Humid foothills S of Isthmus.* Stub-tailed and longish-billed. **Descr.** Ages differ. **Adult:** bill and legs blackish. *Overall dark brown, throat and chest paler and greyer.* At close range note grey scalloping on chest, white to pale buff spots on tips of greater wing coverts. **Juv:** crown and upperparts with pale grey scaling, underparts sparsely flecked whitish.

Voice. Song unmistakable and haunting, a seemingly random, confident to hesitant, rising and falling series of whistles interspersed with thin lisps, often prolonged, *hee hoo, hee hoo, hoo hoo hee hoo, ss hoo hee ...* or *tee tee-tee-tee ssi tee tee-tee-tee tee-tee tee tee-tee ssi ...*, on and on. Calls include a sharp *chek*.

Habs. Humid evergreen forest. Keeps to dense forest understory, often on forest floor, where hard to find unless singing; sings from low perches. Walks with almost constant bobbing motion. Nest and eggs undescribed (?).

SD: F to U resident (near SL–1400 m) on Atlantic Slope from E Chis to Honduras; R to U (?) to NW Chis (SWC spec.).

RA: SE Mexico to Costa Rica.

NB1: Birds from S Costa Rica to S.A. formerly considered conspecific. **NB2:** Sometimes known as Northern Nightingale-Wren.

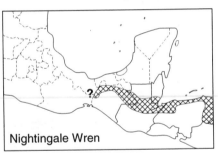

Nightingale Wren

DIPPERS: CINCLIDAE (1)

Dippers are plump, semiaquatic birds of temperate regions. Wings short and rounded, tails fairly short. Legs fairly long and sturdy. Ages differ, sexes similar; young altricial. Imms quickly resemble adults after partial 1st prebasic molt. Plumage dense and waterproof. Dippers swim and dive readily, seeking food from the bottom of streams and pools.

Food mostly insects. Nest is a large, bulky, oven-shaped mass of vegetation lodged in a bank or under a waterfall. Eggs: 3–6, white (WFVZ).

AMERICAN DIPPER
Cinclus mexicanus *   Figure 38
Mirlo-acuático Americano

ID: 6–7 in (15–17.5 cm). *Local along mountain streams.* **Descr. Adult:** bill dark grey, legs greyish flesh. *Slaty grey overall,* head and chest browner, wings and tail slightly

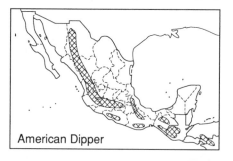

American Dipper

darker. **Juv:** bill pinkish. Overall grey, throat paler, underparts mottled pale grey.
Voice. Strong, rasping to shrill, slightly metallic calls carry over the sound of rushing water: *zzeip* or *zzreip* and *rreip,* 2–3 note phrases, especially in flight, *zzir-rit* and *kirri-di-rit* and *rrei-rreirrip,* etc. Song, given year-round, an often prolonged series of rich to metallic warbles, trills, buzzes, etc., with mimid-like repetition.
Habs. Fast-flowing rocky streams. See family account. Perches on rocks and bobs almost constantly; flight fast and direct, usually low.
SD: U to C but local resident (1000–3000 m) in interior and on adjacent slopes from Chih and Pue to Honduras and (at least formerly) N cen Nicaragua.
RA: W N.A. to W Panama.

Fig. 38 American Dipper

OLD WORLD WARBLERS, KINGLETS, AND GNATCATCHERS: SYLVIIDAE (II)

A primarily Old World family of small, mostly delicately built, insectivorous birds. New World members are characterized by the following features. Bills fine and slender, legs slender. Ages/sexes usually different; seasonal variation in some gnatcatchers; young altricial. Imms quickly resemble adults after partial 1st prebasic molt. Head patterns often important for age/sex and species identification. Forage actively in trees and bushes, often with short hovering flights to pick food from under leaves. Old World Warblers are vagrants in the region.

The Sylviidae has been merged with the Muscicapidae, including thrushes and allies (AOU 1983). Gnatwrens have been placed with the Formicariidae.

DUSKY WARBLER
Phylloscopus fuscatus   Not illustrated
Reinita Fusca

ID: 4.5–4.8 in (11.5–12 cm). Vagrant. *Bill slender and pointed.* Descr. Ages/sexes similar. Bill dark, flesh below at base, legs dull flesh. Crown, nape, and *upperparts dusky brown, face buffy brown with dirty pale buff supercilium* and dark eyestripe. *Underparts dirty pale buff to brownish buff* becoming whitish on throat and belly, undertail coverts buff.
  Voice. Call an emphatic *tack tack*, often repeated 4–6× (Madge 1987).
  Habs. Typically skulks low in thickets and brush.
  SS: No N.A. species similar.
  SD: Vagrant, one record: near Maneadero, BCN, on 15 Oct 1991 (TEW).
  RA: Breeds NE Asia; winters SE Asia.

ARCTIC WARBLER
Phylloscopus b. borealis   Not illustrated
Reinita Artica   See Dusky Warbler for map

ID: 5–5.5 in (13–14 cm). Vagrant. *Bill fairly stout* and pointed. Descr. Ages/sexes similar. *Bill dark above, bright orangish below with dark tip*, legs dull flesh. Crown, nape, and upperparts olive, wings and tail slightly darker, edged yellow-olive; *whitish lower wingbar distinct*, may show hint of short pale upper wingbar. *Long whitish supercilium* tinged lemon, especially in front of eye, broad dark eyestripe, and dirty pale lemon auriculars, streaked dusky. Throat and underparts whitish with hint of lemon wash, dusky sides and flanks usually concealed by wings but often shows slight dusky chest side streaks.
  Voice. Call a hard metallic *dzik* or *chick* (Harris *et al.* 1989).
  Habs. Forages low to high in low trees and bushes, may associate loosely with New World wood-warblers.
  SS: No N.A. species similar.
  SD: Vagrant; one record: Punta Eugenia, BCS, on 12 Oct 1991 (Pyle and Howell 1993).
  RA: Breeds N Eurasia to W Alaska; winters SE Asia to Indonesia.

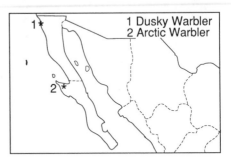

1 Dusky Warbler
2 Arctic Warbler

GENUS REGULUS: Kinglets

Kinglets have overall olive upperparts and whitish underparts, and well-cleft tails. Ages/sexes differ. Bills and legs blackish, feet with conspicuous yellow soles. Nest is a deep pouch of mosses, lichens, plant fibers, open at top and slung from drooping fork at mid- to upper levels in conifers. Eggs (N.A.): 5–10, whitish, faintly and finely marked wih browns (WFVZ).

GOLDEN-CROWNED KINGLET
*Regulus satrapa** Not illustrated
Reyezuelo Corona-dorada

ID: 3.8–4 in (9.5–10 cm). *Highland fir forests.* **Descr.** Adult: *white supercilium contrasts with black sides of crown and dark eyestripe, median crown yellow* (♂ has orange center to crown, often concealed). Nape greyish, upperparts olive. Wings and tail blackish with 2 whitish wingbars, thick black bar across base of secondaries, and yellowish flight feather edgings. Throat and underparts pale grey, washed buff on flanks. **Juv:** face pattern indistinct, median crown greyish olive.
  Voice. A high, thin *ssi-ssi-ssi*, and *ssii*, etc. Song a high, thin series accelerating into a buzzy trill, *tsi-tsi-tsi-tsi-si-sisisisisizzzi,* etc.
  Habs. Humid fir forest; N migrants may occur in mixed woodland. At mid- to upper levels, often hard to see among dense branches.
SS: Ruby-crowned Kinglet lacks striped face; note voice.
SD: C to F resident (2000–3500+ m) in cen volcanic belt from Jal (AMNH spec.) to Pue and S Hgo, in cen Gro (MLZ specs), Oax (formerly?) and S Guatemala. Irregular U to R winter visitor (Nov–early Apr) to N Mexico, from BCN to Tamps. Reports of residents in Sierra Madre Occidental S to Jal (Phillips 1991) appear to be in error.
RA: N.A. to Guatemala.

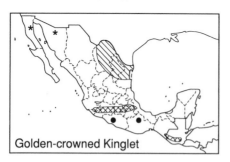

Golden-crowned Kinglet

RUBY-CROWNED KINGLET
*Regulus calendula** Not illustrated
Reyezuelo Sencillo

ID: 4–4.2 in (10–11 cm). *Winter visitor,*

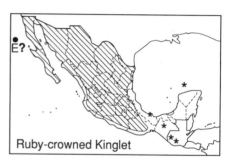

Ruby-crowned Kinglet

common N of Isthmus. **Descr.** Adult: *head and upperparts greyish olive* (♂ has red crown patch, usually concealed), *with bold whitish eyering broken above and below eye.* Wings and tail blackish with 2 *whitish wingbars, thick black bar across base of secondaries,* and yellowish flight feather edgings. Throat and underparts pale grey, washed buff on flanks. **Juv:** upperparts washed brown, wingbars washed buff, ♂ lacks crown patch.
  Voice. Call a dry chattering *dr-dr-drrt* or *ch-ch-chut,* often prolonged when scolding. Song, also heard late in winter, typically 1–4 soft, thin introductory notes, then mellow chips run into a short warble, *si si hew hew hew hewi wi-i wi* or *ssi, chu chu chu chu chu-i-chir-ichi,* etc.
  Habs. Arid to semihumid forest and woodland, mainly in temperate regions. Feeds low to high, often with mixed-species flocks.
SS: Hutton's (highland forests) and Dwarf (arid scrub) vireos are more heavily built and larger-headed, with stout bills, stout blue-grey legs and feet, and squared to notched tails; their whitish lores create different facial expressions and both have contrasting black greater coverts between the wingbars (versus a contrasting black bar across base of secondaries). See Golden-crowned Kinglet.
SD: F breeder (1000–1200 m; nesting last confirmed 1953) on Isla Guadalupe, BCN. C to F transient and winter visitor (Sep–early May; SL–3500 m) almost throughout N of Isthmus, but R to absent in humid SE and from lowlands S of N Mexico; U to R (2000–3000 m) to Chis and Guatemala. Vagrant to Yuc Pen (Howell 1989a).
RA: Breeds N.A.; winters S USA to Mexico, rarely Guatemala.

LONG-BILLED GNATWREN
*Ramphocaenus melanurus** Plate 38
Soterillo Picudo

ID: 4.8–5 in (12–12.5 cm). S lowlands. *Bill very long, straight, and slender, mosquito-like.* Tail fairly long and graduated. **Descr.** Ages/sexes similar. Bill flesh with dark culmen, legs grey. *Face, chest, and flanks cinnamon* with indistinct dusky eyestripe; whitish throat indistinctly streaked dusky, belly and undertail coverts whitish. Crown, nape, and upperparts grey-brown, tail darker, *outer rectrices tipped white.* **Juv:** duller overall.
> **Voice.** Common song a high, rising and falling trill (quality recalls trill of Tropical Pewee), often preceded by a quiet chortle; also lower and gruffer trills, often preceded by quiet chatters; wren-like calls include a nasal scold *cheut-cheut*, repeated rapidly, a slower *chir-r-r chir-r*, and a dry, crackling chatter reminiscent of White-bellied Wren.
> **Habs.** Humid to semihumid evergreen and semideciduous forest and edge. Active and restless, often in pairs, in vine tangles at low to mid-levels. Tail typically held cocked and swung loosely around; joins mixed-species flocks. Nest is a deep cup of fine plant material in dense growth near ground. Eggs: 2, white, speckled brown (WFVZ).
SD: F to C resident (SL–750 m, locally to 1000 m) on Atlantic Slope from S Ver to Honduras, on Pacific Slope from Chis to El Salvador.
RA: SE Mexico to NE Peru and SE Brazil.

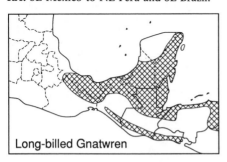
Long-billed Gnatwren

GENUS POLIOPTILA: Gnatcatchers

Gnatcatchers have long graduated tails which are frequently cocked and swung from side to side. Ages/sexes differ, juvs resemble ♀♀. Plumage grey above, whitish below; wings darker, whitish-edged tertials and secondaries may form contrasting panels, at least in fresh plumage; tails blackish, outer rectrices patterned with white. Bills and legs blackish. Nests are neat cups of plant fibers, lichens, etc., at low to mid-levels in tree or bush. Eggs: 2–6, whitish to pale bluish, speckled brown to red-brown (WFVZ).

BLUE-GREY GNATCATCHER
*Polioptila caerulea** Plate 47
Perlita Grisilla

ID: 4–4.5 in (10–11.5 cm). *Widespread. Tail slightly graduated.* **Descr.** Bill *usually flesh below at base.* Head and upperparts blue-grey with *white eyering (alt ♂ has black supraloral stripe).* Remiges dark brown, edged whitish. Tail black with *mostly white outer rectrices (closed tail looks white below).* Throat and underparts whitish (dusky in *cozumelae* of Isla Cozumel).
> **Voice.** A mewing, slightly nasal *meear* and *jeehrt*, and a chattering *chi-chi-chi-chi-chi-chi-chir; cozumelae* has a mewing, buzzy *dzeeehr* or *jehhr*, often repeated 2–3×, and a harder *meeihr*, distinct from mainland Blue-greys. Song varied, a high, thin, scratchy and squeaky warble with nasal mews thrown in; undescribed in *cozumelae* (?).
> **Habs.** Woodland and edge, especially with oaks and scrub, thorn forest, semi-open areas with scrub, scattered trees, clearings in humid forest etc. Forages

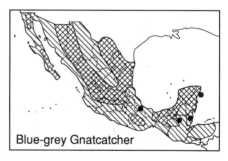
Blue-grey Gnatcatcher

actively low to high, often joins mixed-species flocks.

SS: Black-capped Gnatcatcher (NW Mexico) has more strongly graduated tail, bill usually notably longer, grey below at base, wings often browner (can be similar in worn Blue-greys), basic ♂ often shows flecks of black on face; note voice. Black-tailed and California gnatcatchers smaller with slighter bills, mainly black inner webs of outer rectrices; note voice. White-lored and Tropical gnatcatchers told by face patterns and voice.

SD: F to C but local resident (SL–2500 m) in Baja, in interior and on adjacent slopes from Son and Coah to cen Mexico, thence locally in interior to Chis (and adjacent NW Guatemala?), and on Atlantic Slope from Tamps to Yuc Pen and Belize. C to F transient and winter visitor (Sep–Apr; SL–2500 m) throughout region to Guatemala; U to F to Honduras.

RA: Breeds N.A. to Mexico and Belize; winters S USA to Honduras.

BLACK-TAILED GNATCATCHER
Polioptila melanura*   Plate 47
Perlita Colinegra

ID: 4–4.2 in (10–11 cm). *N Mexico, deserts.* **Descr.** ♂: head and upperparts blue-grey (greater coverts brownish when worn), *crown black in alt* (Mar–Aug) *with contrasting white eyering*, often some black around eye in basic. Remiges dark brown, edged whitish. Tail black, *outer rectrices with white outer webs and broad white tips (closed tail looks mostly black below).* Throat and underparts whitish, washed buff on flanks in fresh plumage. ♀: resembles basic ♂ but with pale lores and indistinct eyering, no black on face, back washed brownish.

Voice. A low, nasal, slightly buzzy *jehrr* or *jeehrr*, often doubled, a thinner *jiihh*, a

sharp *chip chip chip*, and a low, rasping scold, *ssheh ssheh ... or cheh-cheh ...* that may suggest House Wren. Song a rapidly repeated, nasal *jeh-jeh ...*

Habs. Arid to semiarid, open and semi-open areas with scattered bushes and scrub, riparian brushy wooodland. Forages actively at low levels, often sings from atop bushes.

SS: California Gnatcatcher, where sympatric, is duskier below, undertail more extensively black with only narrow white tips to outer rectrices, ♂ eyering indistinct; note voice. Black-capped Gnatcatcher prefers thorn forest, has mostly white undertail, longer bill, alt ♂ has less extensive black cap, indistinct eyering. See Blue-grey Gnatcatcher.

SD: C to F resident (SL–2000 m) in NE Baja, on Pacific Slope in Son, and on Plateau from Chih and NL to Gto and SLP.

RA: SW USA to N Mexico.

CALIFORNIA [BLACK-TAILED] GNATCATCHER
Polioptila californica*   Plate 47
Perlita Californiana

ID: 4–4.2 in (10–11 cm). *Baja.* Two groups: *californica* (NW chaparral) and *margaritae* group (deserts). **Descr.** *Californica.* ♂: head and *upperparts dark blue-grey* (greater coverts browner when worn) with whitish eyering, *crown black in alt* (Mar–Aug), often some black around eye in basic. Remiges dark brown, edged whitish. Tail black, *outer rectrices with white-edged outer webs and narrow white tips (closed tail looks mostly black below). Throat and underparts dusky*, washed buff on flanks. ♀: resembles basic ♂ but with pale lores and indistinct eyering, no black on face, back washed brownish. *Margaritae: paler overall, much like Black-tailed Gnatcatcher* but alt ♂ has dull eyering.

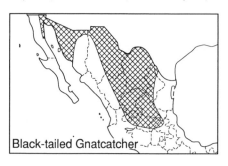

Black-tailed Gnatcatcher

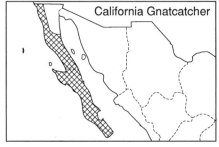

California Gnatcatcher

Voice. Distinctive call a drawn-out, mewing *meeeeah* or *meeeahr*, also a buzzier *nyeh* and *nyeh nyeh*, and a soft *cheh*. Song (?) a buzzy, mewing *zzihr zzihr zzihr* or *chih-chih-chih-chih*.

Habs. Arid to semiarid, open and semi-open areas with scattered bushes and scrub, riparian brushy woodland. Habits much as Black-tailed Gnatcatcher.

SS: See Black-tailed and Blue-grey gnatcatchers.

SD/RA: C to F resident (SL–1500 m) in Baja (also SW California).

NB: See Black-tailed Gnatcatcher.

BLACK-CAPPED GNATCATCHER
Polioptila nigriceps *   Plate 47
Perlita Gorrinegra

ID: 4–4.5 in (10–11.5 cm). *Endemic to NW Mexico; thorn forest. Tail strongly graduated, bill long.* **Descr.** Bill black with grey base below. Head and upperparts blue-grey (greater coverts brownish when worn) with *paler face and whitish eyering, crown black in alt* ♂ (Mar–Aug), usually a few *black flecks over eye of basic* ♂. Remiges dark brown, edged whitish. Tail black with *mostly white outer rectrices (closed tail looks white from below).* Throat and underparts whitish, tinged smoky grey on chest. **Juv:** upperparts washed brownish, brighter white eyering, may show diffuse dark brow over eye but not extending forward to bill.

Voice. A mewing *meeihr* or *reeihr*, rougher and more drawn-out than Blue-grey Gnatcatcher, a rougher *meiyhrr* and shorter *eihr*, a thinner, nasal mewing *pseuih*, a quiet, rough scolding *cheh-cheh*, a soft, dry chattering *chehchehcheh* ... when agitated, and a quiet, metallic *tink* or *dink* suggesting Lucy's Warbler. Song a short, jerky, scratchy warble, lacking the plaintive or mewing quality of Blue-grey.

Habs. Arid to semiarid scrub, thorn forest. Often forages in dense scrub and can be overlooked easily, best detected by call.

SS: See Blue-grey and Black-tailed gnatcatchers. White-lored Gnatcatcher allopatric (?), ♀ and basic ♂ told by narrow white supercilium, alt ♂ has more extensive black cap, probably all plumages have bolder white tertial edgings than Black-capped.

SD/RA: C to F resident (SL–1200 m) on Pacific Slope from Son to Col, should be looked for in W Mich. Has bred SE Arizona.

WHITE-LORED GNATCATCHER
Polioptila albiloris *   Plate 47
Perlita Cejiblanca

ID: 4–4.7 in (10–12 cm). *S Mexico and C.A., arid scrub.* **Descr.** ♂: *crown black, in basic (Aug–Mar?) has narrow white supercilium.* Upperparts blue-grey. Remiges dark brown, edged paler, *tertials blackish with bold white edgings forming panel.* Tail black with mostly white outer rectrices (closed tail looks white below). Throat and underparts whitish, throat often contrastingly white. ♀: resembles basic ♂ but crown and eyestripe grey.

Voice. A nasal, low, buzzy mew, *jehrr* or *zshirr*, a longer *bzzzzeir*, and a higher, drawn-out *meeeur*, all quite different from Blue-grey Gnatcatcher; also a sharp, warbler-like *chik* often preceded by a bright nasal *oo-eik'*, given one or more times, and a short, high trill, *tli-li-li-li*. Song a nasal, squeaky, jerky chatter.

Habs. Arid to semiarid scrub, thorn forest. Relatively conspicuous in its habitat, often in pairs.

SS: See Blue-grey and Black-capped gnatcatchers. Tropical Gnatcatcher occurs in humid forest canopy; note face pattern.

SD: C to F resident (SL–1800 m) on Pacific

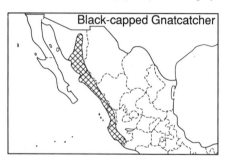

Black-capped Gnatcatcher

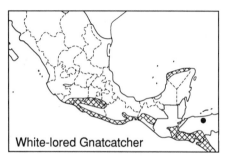

White-lored Gnatcatcher

Slope from Mich to Isthmus and from E Guatemala to Honduras, in interior locally from Balsas drainage S; disjunctly in N Yuc and N Honduras (Aguán Valley).

RA: Mexico to NW Costa Rica.

TROPICAL GNATCATCHER

Polioptila plumbea brodkorbi Plate 47
Perlita Tropical

ID: 4–4.2 in (10–11 cm). *SE humid forest. Relatively short-tailed.* Descr. ♂: *white lores and broad supercilium contrast with*

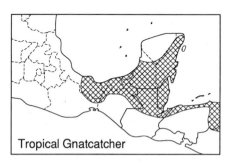

Tropical Gnatcatcher

black crown and postocular stripe, creating bare-faced expression, upperparts blue-grey. Remiges blackish, edged white. Tail black with mostly white outer rectrices (closed tail looks white below). Throat and underparts whitish. ♀: crown and postocular stripe grey, slightly darker than upperparts.

Voice. A nasal, mewing *meeah* or *myieh,* often doubled, *meeah-meah,* etc., shorter and buzzier than Blue-grey Gnatcatcher call, suggests Red-legged Honeycreeper. Song a high, rapid, silvery trill, often rippling and descending, 2–3 s in duration; also a quiet, scratchy warble.

Habs. Humid evergreen to semideciduous forest. Singly or in pairs in forest canopy, rarely lower, often joins mixed-species flocks.

SS: See Blue-grey and White-lored gnatcatchers.

SD: F to C resident (SL–500 m) on Atlantic Slope from E Chis to Honduras, U to S Ver (Los Tuxtlas, Winker *et al.* 1992*b*) and E Oax (Binford 1989).

RA: SE Mexico to Peru and Brazil.

THRUSHES AND ALLIES: TURDIDAE (26)

A cosmopolitan family with its greatest diversity in temperate areas, especially highland forests. One or more members are found throughout the region. Bills slender, legs fairly long. Wings usually long and pointed, tails moderately long and squared. Ages differ, sexes similar or different; young altricial. Juvs quickly lose their spotted and mottled plumage in their partial 1st prebasic molt; most species then resemble adults. Plumages mainly sombrely colored in browns, black, and white, but patterns often attractive. Songs among the finest of all birds.

Food mostly insects and fruit. Nests are cups of vegetation, reinforced with mud in some *Turdus*, in bushes or trees, on the ground, or in cavities. Eggs vary from unmarked blue to whitish mottled with browns.

NORTHERN WHEATEAR
Oenanthe oenanthe ssp.   Not illustrated
Collalba Norteña

ID: 6–6.5 in (15–16.5 cm). *Vagrant*. **Descr.** Ages/sexes differ. Bill and legs blackish. *In all plumages, white uppertail coverts and base of outer rectrices striking in flight*, rest of tail black. **Basic (autumn):** head and upperparts grey-brown with *creamy supercilium and narrow dusky mask*; wings blackish brown, edged cinnamon. *Throat and underparts buff.* **Alt ♂ (Mar–Aug):** crown and upperparts blue-grey with white supercilium and black mask, wings black. Throat and underparts pale buff. **♀:** resembles basic but wing edgings narrower.
Voice. A dry, fairly hard *tchek* or *chehk* (Britain).
Habs. Open areas with scattered bushes, most likely to be along or near coast. Runs quickly, stands fairly upright, and perches prominently.
SD: Vagrant (mid–late Nov) to Yuc Pen, two records: near Telchac Puerto, Yuc (1973,

G and JC, photo) and Cancún, QR (1983, BM photo).
RA: Breeds NE Canada, N Eurasia to Alaska; winters Africa and S Asia.

GENUS SIALIA: Bluebirds

Bluebirds are colorful thrushes. Ages/sexes differ. ♂♂ extensively bright blue, ♀♀ duller but with hint of ♂ pattern. Juvs have grey-brown head and upperparts, streaked buff; underparts whitish, mottled brown, wings and tail as adult. Bills and legs blackish. Nest in tree cavities. Eggs: 4–6, pale blue, rarely white in Eastern (WFVZ).

EASTERN BLUEBIRD
Sialia sialis *   Plate 50
Azulejo Gorjicanelo

ID: 6.5–7 in (16.5–18 cm). *Widespread.* **Descr.** ♂: *head and upperparts blue*, back washed rufous in fresh plumage. *Throat, chest, and flanks cinnamon* (darker

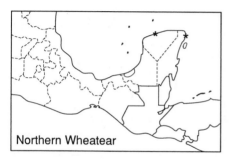

Northern Wheatear

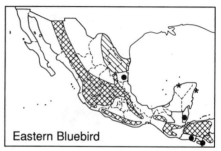

Eastern Bluebird

cinnamon-rufous in *sialis*, E Mexico in winter) with white belly and undertail coverts. ♀: head and upperparts grey-brown with whitish eyering, rump and uppertail coverts blue; wings and tail darker, edged blue. *Throat, chest, and flanks pale cinnamon* with whitish belly and undertail coverts. Juv: see genus note.

Voice. A slightly nasal *cheu* or *tcheu*, and *chew-it* or *tew-ee*. Song varied, short, rich to slightly burry warbles.

Habs. Arid to semihumid oak and pine–oak woodland, open areas with scattered trees and bushes. Often in pairs or small groups, perches conspicuously on bushes, trees, roadside wires. Flies from perch to ground for prey or hovers in open country.

SS: Western Bluebird ♂ deeper blue, throat and underparts dull bluish with dark rufous mainly on sides of chest, ♀ has trace of ♂ pattern, throat greyish. ♀ Mountain Bluebird paler and greyer overall with greyish throat and underparts, longer primary projection.

SD: C to F resident (600–2700 m, lower in winter) in interior and on adjacent slopes from Son and S Tamps to cen volcanic belt, thence mostly in interior to Nicaragua, and (SL–1000 m) locally on Atlantic Slope from S Tamps to the Mosquitia. Irregular R winter visitor (Nov–Mar; near SL–1500 m) to Atlantic Slope in NL and Tamps; vagrant (Feb–Mar) to N Yuc Pen (G and JC, JCo).

RA: E N.A. to W Nicaragua.

WESTERN BLUEBIRD
Sialia mexicana*   Plate 50
Azulejo Gorjiazul

ID: 6.5–7 in (16.5–18 cm). *N and cen Mexico*. **Descr.** ♂: *head and upperparts*

deep blue (back often chestnut in NW Mexico). *Throat and underparts dusky blue, dark rufous patches at sides of chest* sometimes meet across center. ♀: head and upperparts grey-brown with whitish eyering, rump and uppertail coverts blue; wings and tail darker, edged blue. *Throat and underparts pale greyish with dull rufous chest and flanks*. Juv: see genus note.

Voice. A nasal twangy *chew* or *choy*, sometimes doubled, given in flight, and a quiet, dry, rattling chatter. Dawn song short, rich, twangy warbles or caroling phrases, often with soft, dry rattles thrown in.

Habs. Arid to semihumid pine woodland with grassy clearings, open areas with scattered bushes and trees. Habits as Eastern Bluebird.

SS: See Eastern Bluebird. ♀ Mountain Bluebird paler and greyer overall with greyish throat and underparts, longer primary projection.

SD: C to F resident (1500–3500 m) in N Baja, and in interior and on adjacent slopes from Son and Coah to cen volcanic belt; U to R to near SL (Nov–Apr) in NW Mexico.

RA: W N.A. to cen Mexico.

MOUNTAIN BLUEBIRD
Sialia currucoides   Not illustrated
Azulejo Pálido

ID: 6.5–7 in (16.5–18 cm). *N Mexico in winter*. **Descr.** ♂: *sky blue overall*, deeper-colored above, washed greyish in fresh plumage. ♀: head and upperparts grey-brown with whitish eyering, rump and uppertail coverts blue; wings and tail darker, edged blue. *Throat, chest, and flanks pale grey-brown* with whitish belly and undertail coverts.

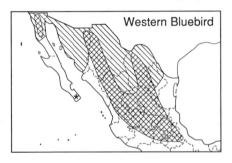

Western Bluebird

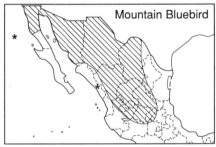

Mountain Bluebird

Voice. A nasal *peu* or *peur*.

Habs. Open country, especially grassland, with scattered bushes and trees. Usually in flocks, at times 50+ birds. Perches on ground and hovers more than other bluebirds.

SS: See ♀ Eastern and Western bluebirds.

SD: Irregularly C to F winter visitor (Nov–Mar; 1000–2500 m, to SL in NW Mexico) in N Baja, on Pacific Slope in Son (R to S Sin), and in interior from Chih and Coah to Zac and SLP; R and irregular to N Mich and Mex.

RA: Breeds W N.A.; winters to N Mexico.

Singly, or in small groups in winter, perches fairly upright and often still for long periods, overlooked easily.

SS: Brown-backed Solitaire has brownish-olive upperparts, rufous-edged wings, white eye-crescents, note voice.

SD: U to F resident (1800–3500 m) in interior and on adjacent Pacific Slope from Chih to Zac, possibly to Jal (Webster 1984). Irregular F to R winter visitor (Oct–Apr; 1500–3000 m) to BCN and in interior and on adjacent slopes to N Jal, Coah, and NL (SNGH, SW).

RA: W N.A. to NW Mexico; winters to NE Mexico.

GENUS MYADESTES: Solitaires

Solitaires are slender, relatively plain, long-tailed thrushes of highland forests. Sexes similar. Juvs have head and body whitish to pale buff, heavily scalloped dark brown, wings and tail as adult. Songs striking. Bills blackish. Nest cups of moss, grass, plant fibers, etc., placed on or near the ground at base of tree or in banks. Eggs: 2–5, whitish to bluish white, heavily speckled and marked with reddish browns and greys (WFVZ).

TOWNSEND'S SOLITAIRE

Myadestes townsendi*   Not illustrated
Clarín Norteño

ID: 8–8.5 in (20.5–21.5 cm). N Mexico. **Descr.** Legs blackish. **Adult:** grey overall with dark lores and white eyering. Wings and tail darker with buff stripe across base of remiges, mostly white outer rectrices. **Juv:** see genus note.

Voice. A plaintive, whistled *hew* or *whee*. Song a pleasant, rich warble with slightly mechanical quality.

Habs. Arid to semiarid conifer and pine–oak forest, also oak woodland in winter.

BROWN-BACKED SOLITAIRE

Myadestes occidentalis*   Plate 50
Clarín Jilguero

ID: 8–8.5 in (20.5–21.5 cm). *Endemic, highlands almost throughout.* **Descr.** Legs greyish to flesh. **Adult:** head and underparts grey with dark lores and *white eye-crescents. Upperparts brownish olive, wings darker, edged rufous,* with buff stripe across base of remiges. Central rectrices greyish, rest of tail blackish with mostly *white outer rectrices.* **Juv:** see genus note.

Voice. Unique song heard year-round, a characteristic sound of highland forests: an accelerating, squeaky, metallic, jangling series, beginning hesitantly before running into a jumbled crescendo; introductory notes may be repeated several times, hesitantly, without breaking into full song. Calls a metallic, slightly whining, and up-slurred *wheeu* or *yeeh*, and a nasal rasping *shiehh* in alarm.

Habs. Humid to semiarid evergreen, semideciduous, and pine–oak forest, often along streams. Habits much as Townsend's Solitaire.

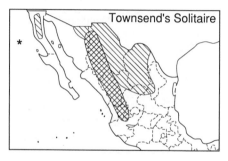

Townsend's Solitaire

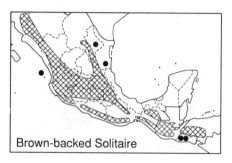

Brown-backed Solitaire

SS: See Townsend's Solitaire. Slate-colored Solitaire smaller, darker slaty grey overall, with bolder white eye-crescents, flesh legs, note voice.

SD/RA: C to F resident (600–3500 m, lower locally in winter in W Mexico) on both slopes from S Son and NL (and SE Coah?), and in interior from cen Mexico, to N El Salvador and cen Honduras. A report from Belize (Wood and Leberman 1987) requires substantiation.

NB: Formerly known as *Myadestes obscurus*.

SLATE-COLORED SOLITAIRE

*Myadestes unicolor**   Plate 50
Clarín Unicolor

ID: 7.5–8 in (19–20.5 cm). *Endemic to E and S cloud forest.* **Descr.** Legs flesh. **Adult:** *slaty grey overall* with dark lores and *thick white eye-crescents*; wings and tail darker with pale buff stripe across base of remiges, mostly *white outer rectrices.* **Juv:** see genus note.

Voice. Ethereal, haunting song characteristic of E and S cloud forests: begins, often hesitantly, with a few poor notes before breaking into a varied series of clear to quavering, fluty whistles, often ending with a loose trill. Calls a hard, nasal *rrank* or *rran*, and a buzzier *zzrink*.

Habs. Humid evergreen and pine–evergreen forest. Habits much as Townsend's Solitaire.

SS: See Brown-backed Solitaire.

SD/RA: C to F resident (1000–2700 m, in winter to near SL in Los Tuxtlas and Belize, SNGH) on Atlantic Slope from Hgo to Honduras, on Pacific Slope and locally in interior from E Oax to El Salvador and N cen Nicaragua. Reduced (extirpated?) locally in E Mexico due to capture for cagebird trade.

NB: Also known as Slaty Solitaire.

GENUS CATHARUS

These are small thrushes mainly of temperate regions, many with beautiful songs. M.A. species are usually known as nightingale-thrushes while N.A. species are simply thrushes. Sexes similar. Terrestrial and retiring. Unless singing (mainly Apr–Aug), the M.A. species may seem rare or even absent in areas where they are common. Early and late in the day, all species may be found at mid- to upper levels in fruiting trees or on open trails but after dawn most 'melt' into the forest. Nest cups of plant fibers, moss, grass, etc., placed at low to mid-levels in bushes, thickets, trees, on banks, at times on ground. Eggs: 2–3.

ORANGE-BILLED NIGHTINGALE-THRUSH

*Catharus aurantiirostris**   Plate 50
Zorzalito Piquinaranja

ID: 6–6.8 in (15.5–17 cm). *Widespread.* **Descr. Adult:** *orbital ring orange, bill bright orange*, often with dark culmen, *legs orangish. Crown, nape, and upperparts rich brown*, brighter rufous-brown on uppertail coverts, wings, and tail. Face, throat, and underparts pale grey, becoming whitish on throat, belly, and undertail coverts. **Juv:** upperparts duller, spotted pale cinnamon; throat and underparts whitish, heavily mottled dark brown on chest and flanks.

Voice. Song a varied, short, scratchy warble with a tinny to fairly melodic quality: *chiviree che-oo'*, or *chik'r ssir-irr-it*, etc., jerkier and scratchier than other nightingale-thrushes. Call a nasal, mewing *meehr* or *meeeahr*, much like a Blue Mockingbird call.

Habs. Brushy to fairly open understory

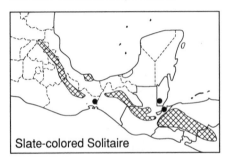

Slate-colored Solitaire

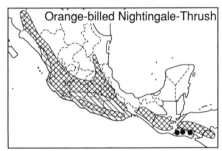

Orange-billed Nightingale-Thrush

ranging from arid thorn forest to stream sides in humid pine–oak forest. Keeps to shady understory but seen relatively often, sings fairly conspicuously at low to mid-levels. Eggs pale bluish, heavily speckled with reddish browns (WFVZ).

SS: Russet and Ruddy-capped nightingale-thrushes lack bright orbital rings and bright legs, crown often contrastingly brighter than back; Russet has dark bill pale below at base, and buff wash on throat and chest; Ruddy-capped darker overall, bill dark above, yellow-orange below; note voices.

SD: C to F resident (600–2500 m) on both slopes from S Chih and S Tamps, and in interior from cen Mexico, to Honduras and N cen Nicaragua. F to C winter visitor (Oct–Apr; to SL) on Pacific Slope from S Son to Mich (SNGH), and may withdraw (Nov–Mar) from northernmost parts of range (Phillips 1991). Withdrawal and altitudinal migrations poorly known.

RA: Mexico to N S.A.

RUSSET NIGHTINGALE-THRUSH
Catharus occidentalis*   Plate 50
Zorzalito Piquipardo

ID: 6–7 in (15.5–18 cm). NW > S. *Mexican endemic; highlands N of Isthmus.* **Descr. Adult:** *bill blackish above, flesh below with dusky tip,* legs flesh. Crown, nape, and upperparts rich brown (paler, greyish olive-brown in *olivascens* of NW Mexico), crown often brighter rufous-brown. In flight, note buff stripe on underwing across base of remiges. Face greyish, *throat and underparts pale grey, washed buff on throat and chest, mottled dusky on chest,* flanks washed dusky. **Juv:** upperparts flecked cinnamon-buff; throat and underparts whitish, mottled dark brown on chest and flanks.

Voice. Song varied, thin phrases with tinny quality, not particularly musical *she-vee-ee-i-lu* or *chee ti-vee,* etc., sometimes repeated in fairly rapid succession. Calls a quiet, low, slightly gruff *chuk* or *chruh,* and a nasal, mewing *reear.*

Habs. Humid to semiarid pine–oak and oak forest. Seen fairly often in the relatively open understory it prefers. Eggs pale blue, unmarked (WFVZ).

SS: Where sympatric, Ruddy-capped Nightingale-Thrush prefers denser, more humid understory (and is seen less often), orange-yellow lower mandible lacks dusky tip, throat and chest greyer, greyish underwings lack buff stripe; note voice. Migrant N.A. *Catharus* have more distinctly spotted chests, pale eyerings; Russet could be mistaken for Veery. See Orange-billed Nightingale-Thrush.

SD/RA: C to F resident (1500–3500 m) in interior and on adjacent slopes from Chih and S Coah to Oax; some may withdraw from northernmost parts of range in winter (Phillips 1991).

NB1: *Olivascens* sometimes regarded as specifically distinct although songs similarly short and squeaky, unmusical. **NB2:** Has sometimes been considered conspecific with Ruddy-capped Nightingale-Thrush, an error most recently corrected by Rowley and Orr (1964*b*) and Phillips (1969).

RUDDY-CAPPED NIGHTINGALE-THRUSH
Catharus frantzii*   Plate 50
Zorzalito de Frantzius

ID: 6–7 in (15.5–18 cm). S > N. *Cen Mexico to C.A., highlands.* **Descr. Adult:** *bill blackish above, orange-yellow below,* legs flesh. Crown, nape, and upperparts rich brown to dark brown, crown usually brighter rufous-brown. *Face, throat, and underparts greyish,* with paler streaked throat, whitish vent and undertail coverts;

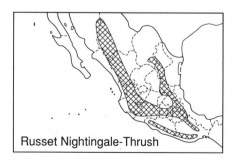

Russet Nightingale-Thrush

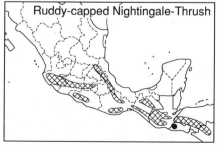

Ruddy-capped Nightingale-Thrush

chest washed brown in E Mexico. **Juv:** bill tipped dusky below. Head and upperparts dark rich brown, throat and underparts whitish, heavily mottled dark brown on chest and flanks.

Voice. Song of varied short phrases, richer and more musical than Russet Nightingale-Thrush: *shee-vee-li-ee-ree*, or *shee-vee-shee-oo*, etc.; best distinguished from Russet by solitaire-like quality. Call a quavering *whierrr* to burry *wheeer*.

Habs. Humid to semihumid conifer, pine–oak, and evergreen forest. Keeps to dense humid understory, seen infrequently. Eggs pale bluish, heavily marked with brownish red (WFVZ).

SS: See Russet and Orange-billed nightingale-thrushes.

SD: C to F resident (1500–3500 m) in interior and on adjacent slopes from Jal and SE SLP to Honduras and N cen Nicaragua.

RA: Cen Mexico to W Panama.

NB1: Also known as Highland or Frantzius' Nightingale-Thrush. **NB2:** See NB2 under Russet Nightingale-Thrush.

BLACK-HEADED NIGHTINGALE-THRUSH
Catharus mexicanus*   Plate 50
Zorzalito Coroninegro

ID: 6–6.5 in (15–16.5 cm). *Cloud forest in E and S.* **Descr. Adult:** bill bright orange with dark culmen, orbital ring and legs bright orange. *Black crown contrasts with greyish face and brownish-olive upperparts.* Throat and underparts pale grey, washed dusky on chest and flanks. ♀ has duller head pattern, crown blackish brown, face and chest washed brownish. **Juv:** head and upperparts dark brownish, flecked cinnamon-buff; throat and underparts whitish, mottled brown and buff on chest and flanks.

Voice. Song a slightly tinny, jumbled warble, pleasant though slightly scratchy, at times run into a warbled trill. Call a buzzy to fairly harsh complaining mew, *rreahr* or *meahh*.

Habs. Humid evergreen forest and adjacent second growth. Keeps to dense undergrowth, seen infrequently unless singing. Eggs bluish white, speckled with reddish browns (WFVZ).

SD: Mostly C to F resident (750–1800 m; in winter to near SL locally in Los Tuxtlas) on Atlantic Slope from S Tamps to Honduras and N cen Nicaragua but may withdraw (Oct–Mar?) from NE Mexico (Phillips 1991). A report from Mex (AOU 1983) is erroneous.

RA: E Mexico to W Panama.

SPOTTED NIGHTINGALE-THRUSH
Catharus dryas*   Plate 50
Zorzalito Pechiamarillo

ID: 7–7.5 in (18–19 cm). *Cloud forest S of Isthmus.* **Descr. Adult:** bill, orbital ring, and legs bright orange. *Black head contrasts with olive upperparts,* wings and tail darker and greyer. *Throat and underparts lemon, spotted dusky on chest,* washed grey on sides and flanks. **Juv:** undescribed (?).

Voice. Song of rich fluty phrases suggests Slate-colored Solitaire in quality: *clee-oo-leew clee-oo-low*, or *whee-i-lee, wee-i-lou*, etc. Calls a nasal, complaining *rrehr* and *rreh'hu* or *reh'chew*.

Habs. Humid evergreen forest. Keeps to dense undergrowth, seen infrequently unless singing. Eggs bluish white, speckled with rusty and greys (WFVZ).

SD: C to F presumed resident (1200–3000 m) in interior and on adjacent slopes from E Oax and Chis to Honduras.

RA: S Mexico to Honduras, also Colombia to N Argentina.

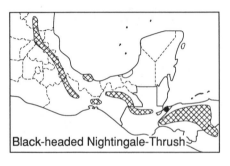

Black-headed Nightingale-Thrush

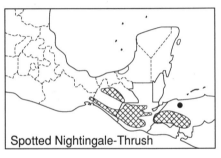

Spotted Nightingale-Thrush

VEERY
Catharus fuscescens   Not illustrated
Zorzalito Rojizo

ID: 6.8–7.5 in (17–19 cm). *Spring and autumn transient.* **Descr.** Bill blackish above, flesh below with dusky tip, legs flesh. *Head and upperparts rufous-brown* to brown *with grey face and indistinct whitish eyering.* Throat and underparts whitish, *washed buff on throat and chest with poorly defined dark spotting on upper chest,* flanks washed greyish. Buff stripe across underside of remiges visible in flight.
    Voice. A mellow, burry *breeuh*, rarely heard in migration.
    Habs. Forest and woodland understory where mostly retiring and terrestrial.
SS: Lightly spotted chest and narrow whitish eyering distinctive, as are rufous-brown upperparts of most birds.
SD: F to C transient (Apr–mid-May, Sep–Oct, mainly near SL) on Atlantic Slope from S Ver to Honduras, possibly also through NE Mexico.
RA: Breeds N.A.; winters Colombia to Brazil.

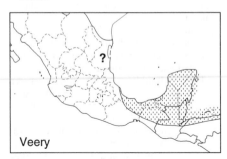

Veery

GREY-CHEEKED THRUSH
Catharus minimus   Not illustrated
Zorzalito Carigris

ID: 6.5–7.2 in (16.5–18.5 cm). *Spring and autumn transient.* **Descr.** Bill blackish above, flesh below with dusky tip, legs flesh. *Head and upperparts grey-brown* to olive-brown *with grey face and narrow whitish eyering most distinct behind eye.* Throat and underparts whitish, *washed buff on chest, throat sides and chest spotted dark brown,* flanks washed greyish. Buff stripe across underside of remiges visible in flight.

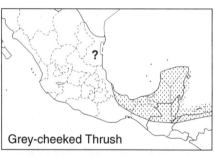

Grey-cheeked Thrush

    Voice. A thin, slightly burry *phreu*, rarely heard in migration.
    Habs. Much as Veery.
SS: Swainson's Thrush has bold buff spectacles, brighter and more extensive buff wash on throat and chest; Hermit Thrush has rufous tail, bolder dark chest spots; see Veery.
SD: U to F transient (mid-Apr–mid-May, late Sep–Nov; mainly near SL) on Atlantic Slope from S Ver to Honduras, possibly also through NE Mexico, at least in spring. Vagrant to Clipperton.
RA: Breeds N.A.; winters Colombia to Brazil.

SWAINSON'S THRUSH
Catharus ustulatus   Not illustrated
Zorzalito de Swainson

ID: 6.5–7.5 in (16.5–19 cm). Transient and winter visitor. **Descr.** Bill blackish above, flesh below with dusky tip, legs flesh. *Head and upperparts rich olive-brown* to brown (grey-brown in *swainsoni*, a spring and autumn transient), *broad buff eyering and supraloral stripe create spectacles.* Throat and underparts whitish, washed buff on throat and chest, throat sides and chest spotted dark brown, flanks washed brownish. Buff stripe across underside of remiges visible in flight.
    Voice. A mellow *whiut* or *whiuh*, and *hwee*, and a burry *rrehrr*. Spring migrants

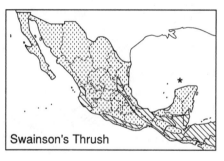

Swainson's Thrush

often sing: a fluty upward-spiraling *wh'*
wheedl-oo wheedl-oo . . ., the main phrase
given 2–4×.

Habs. Humid to semihumid evergreen
and semideciduous forest; more widely in
migration. Habits much as Veery.

SS: Bold buff spectacles distinctive; see other
N.A. migrant *Catharus*.

SD: C to F transient (Apr–May, Sep–Oct;
SL–3000 m) throughout most of region
but U (?) on N Plateau and U to F in Yuc
Pen; F to U winter visitor (Sep–May; SL–
2000 m, mostly above 300 m) on Pacific
Slope from Nay to El Salvador; U to F on
Atlantic Slope from S Tamps to Honduras
and in interior from Isthmus S.

RA: Breeds N.A.; winters Mexico to N
Argentina.

HERMIT THRUSH
*Catharus guttatus**   Not illustrated
Zorzalito Colirrufo

ID: 6–7 in (15.5–18 cm). Mainly a winter
visitor N of Isthmus. **Descr. Adult:** bill
blackish above, flesh below with dusky tip,
legs flesh. *Head and upperparts grey-
brown to brown* with narrow whitish eyering and pale lores, *uppertail coverts and
tail brighter rufous-brown to rufous.*
Throat and underparts whitish, often with
slight buff wash on chest, *throat sides
and chest boldly spotted dark brown*,
flanks washed greyish. Buff stripe across
underside of remiges visible in flight. **Juv:**
head and upperparts with buff streaks,
wing coverts with buff terminal spots,
chest with coarse, blurry, dark spots on
buff wash.

Voice. A low *chup* or *tchup*, recalls Hepatic Tanager but often 2–3× in rapid series,
tchup chup-chup, etc. Song a rich, fluty
warble, *whee ti-heedle-i-heu* or *wheeu
heedle-i-eeu*, etc., with quality of Slate-
colored Solitaire.

Habs. Pine–oak and oak forest, more
widely in migration, nests in conifer-aspen.
Habits much as Veery. Eggs pale blue,
unmarked (WFVZ).

SS: Contrasting rufous tail distinctive; see
other N.A. migrant *Catharus*.

SD: F but local breeder (May–Aug; 2700 m)
in BCN (Howell and Webb 1992*f*). C to F
transient and winter visitor (mid-Sep–
May; near SL–3500 m, mainly above
1000 m from cen Mexico S) in Baja, on
both slopes, and in interior from Son and
Tamps to cen volcanic belt, U to F to Oax,
U to R to Guatemala and NW El Salvador.

RA: Breeds N.A.; winters to N C.A.

WOOD THRUSH
Catharus mustelinus   Not illustrated
Zorzalito Maculado

ID: 7–8 in (18–20.5 cm). Winter visitor to E
and S. **Descr.** Bill blackish above, flesh
below with dusky tip, legs flesh-pink. *Head
and upperparts bright rufous-brown*, duller on rump and tail, with narrow whitish
eyering, face streaked whitish. Throat and
underparts white, *boldly spotted blackish
on throat sides, chest, and flanks.* Buff
stripe across underside of remiges visible in
flight.

Voice. A rapid, mellow, liquid *whiu-
whiu-whiut* or *whi-whi-whi-whuit*, and a
lower clucking *whe-whe-whe-wheh*, often
heard early and late in the day.

Habs. Humid to semihumid evergreen
and semideciduous forest. Habits much
as Veery, but often more conspicuous
than other *Catharus.*

SS: Bright rufous upperparts and boldly spotted underparts distinctive.

SD: C to F transient and winter visitor (late
Aug–Apr; SL–1500 m) on Atlantic Slope
from S Ver to Honduras, U to R N to SE
SLP and on Pacific Slope from Isthmus to
El Salvador. F to C transient (late Aug–

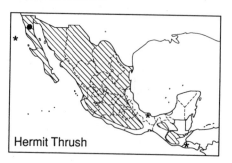

Hermit Thrush

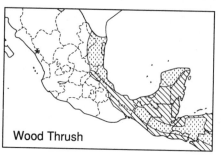

Wood Thrush

Oct, Apr–mid-May) on Atlantic Slope N of Isthmus. Vagrant to Sin (Jan 1964, PA).
RA: Breeds E N.A.; winters SE Mexico to N Colombia.
NB: Traditionally placed in the monotypic genus *Hylocichla*.

GENUS TURDUS

These are large thrushes of semiopen to forested habitats, traditionally called robins in N.A. Sexes similar or different; some species have an imm plumage kept through 1st year. Nests are bulky cups of moss, grass, rootlets, etc., at times reinforced with mud, low to high (usually at mid- to upper levels) in tree or bush. Eggs: 2–3 (rarely 4).

BLACK THRUSH (ROBIN)
Turdus infuscatus   Plate 51
Zorzal Negro

ID: 8.5–9 in (21.5–24 cm). *Endemic, cloud forest.* **Descr.** Sexes differ. ♂: *bill, orbital ring and legs bright yellow. Plumage entirely black.* ♀: *bill black, legs yellowish. Brown overall, slightly paler below with buffy brown throat streaked dusky.* **Imm** ♂: *resembles* ♀ *but head and upperparts colder grey-brown, underparts greyish*, often with a few black feathers in 1st summer; *bill, orbital ring, and legs yellow by 1st summer.* Attains adult plumage by 2nd prebasic molt when about 1 year old. **Juv:** bill dark, resembles ♀ but warmer brown overall, upperparts flecked buff, chest mottled brown.
    Voice. A dry clucking *chuh-chuh-chuh-chuh-chuh*, a harder *chehk-chehk ...*, and a high, thin *ssi*, mostly in flight. Song a rich warbling with frequent 2–3× repetition, overall hesitant with irregular pauses (versus the continuous, less rich song of White-throated Thrush).

Habs. Humid evergreen to pine–evergreen forest and edge. Mainly at mid- to upper levels, at times on the ground in clearings. In small flocks in winter which may associate with other thrushes at fruiting trees. Eggs blue, unmarked (WFVZ).
SS: Mountain Thrush has dark legs, grey-brown overall without contrasting paler throat, undertail coverts edged pale buff. Clay-colored Thrush has yellowish bill, dull flesh legs, warmer brown overall, usually in more open areas, clearings, etc. From behind, White-throated Thrush (*leucauchen* group) may resemble ♂ Black Thrush but note underparts. Dark imm Rufous-collared Thrush told from imm ♂ Black by lack of yellow eyering and no black patches in plumage, also somewhat larger and bulkier.
SD/RA: F to C resident (1200–3500 m, lower locally in winter) on both slopes from S Tamps and Gro, and in interior from Chis, to El Salvador and Honduras. Vagrant to Mor (Mar 1982, Wilson and Ceballos-L. 1986).
NB: Sometimes considered conspecific with *T. serranus* of S.A.

MOUNTAIN THRUSH (ROBIN)
Turdus plebejus*   Plate 51
Zorzal Serrano

ID: 9–10 in (23–25.5 cm). *Highlands S of Isthmus.* **Descr.** Sexes similar. **Adult:** *bill blackish, legs greyish brown. Grey-brown overall* (browner in *differens* of Chis and Guatemala), paler below, *undertail coverts broadly edged pale buff.* **Juv:** upperparts flecked cinnamon, throat and underparts cinnamon-brown, mottled darker, undertail coverts less distinctly patterned.
    Voice. A clucking *kwek-kwek-kwek-kwek-kwek* or *chowk-chowk ...*, and a high, thin *sip* or *siip*, mostly in flight. Song monotonous, a slightly jerky

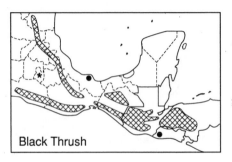
Black Thrush

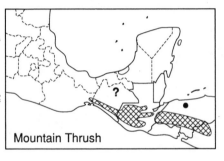
Mountain Thrush

warbling, on and on, relatively poor and unmusical for a thrush, *cheechuchree-chichuchee* ..., etc.

Habs. Humid evergreen forest and edge. Habits much as Black Thrush. Eggs greenish blue, unmarked (WFVZ).

SS: See ♀ and imm ♂ Black Thrush. Clay-colored Thrush warmer brown overall with pale throat streaked dusky, yellowish bill. Dark imm Rufous-collared Thrush darker overall with yellow bill.

SD: F to C resident (1800–3500 m, lower locally in C.A.) on Pacific Slope from SE Oax, and in interior and on adjacent Atlantic Slope from Guatemala, to N cen Nicaragua. Reports from N Chis (Alvarez del Toro 1964; Bubb 1991) require verification.

RA: S Mexico to W Panama.

CLAY-COLORED THRUSH (ROBIN)
Turdus grayi*   Plate 51
Zorzal Pardo

ID: 9–10.5 in (23–26.5 cm). *E and S.* Noticeable geographic variation, paler overall in drier areas (as in Yuc Pen), darker in humid areas (as on Pacific Slope of Chis). **Descr.** Sexes similar. *Bill yellowish, legs dull flesh.* **Adult:** *head and upperparts grey-brown to dark tawny-brown, throat and underparts buff to dusky tawny, throat paler and streaked dusky.* **Juv:** upperparts flecked cinnamon, underparts spotted and mottled brown.

Voice. A rich, slurred *reeeur-ee* or *hoouree* and *clee-ee-eu*, a clucking *kyuh-kyuh* ... or *kluh-kluh* ..., and a high, thin *ssi* or *ssip*, mostly in flight. Song of varied rich phrases with little or irregular repetition.

Habs. Open and semiopen areas with hedges, scattered trees, forest edge, gardens. On ground or at low to mid-levels in trees and bushes, at times in canopy of fruiting trees. Singly or in pairs, in small flocks at fruiting trees. Eggs pale bluish, heavily flecked and spotted with reddish browns and greys (WFVZ).

SS: See ♀ Black Thrush, Mountain Thrush. Pale imm ♀ Rufous-collared Thrush (highlands) greyer above, usually with at least a hint of paler hindcollar. Clay-colored does not occur in W Mexico where occasional reports pertain presumably to White-throated Thrush, *assimilis* group (colder brown with contrasting white forecollar), Grayson's Thrush (vent and undertail coverts whitish), or (rarely) escaped cage birds.

SD: C to F resident (SL–2100 m) on Atlantic Slope from NL and Tamps to Honduras, on Pacific Slope and locally in interior from Chis to El Salvador and W Nicaragua. A population around Oaxaca City probably derived from escaped cage birds (Howell 1990*b*). Reports from Gro and Mex (AOU 1983) are erroneous; one from Jal (Thompson 1962) requires verification.

RA: E Mexico to N Colombia.

WHITE-THROATED THRUSH (ROBIN)
Turdus assimilis*   Plate 51
Zorzal Gorjiblanco

ID: 9–10 in (23–25.5 cm). *Widespread. Two groups:* assimilis *group (NE Mexico and Pacific Slope; including* lygrus, *shown) and* leucauchen *group (humid SE).* **Descr.** Sexes similar. **Adult.** *Assimilis: bill greyish to dull yellowish, orbital ring and legs greyish* (W Mexico) *to yellowish* (mostly S of Isthmus). *Head and upperparts grey-brown to brown.* Throat whitish, heavily streaked dark brown with *contrasting white forecollar.* Underparts pale greyish (NW Mexico) *to grey-brown* (C.A.), becoming whitish on vent and undertail coverts. *Leucauchen: bill, orbital ring, and legs yellow to orange-yellow. Head and upperparts dark slaty grey to blackish.*

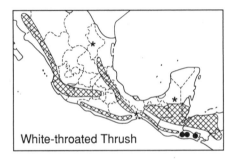

Clay-colored Thrush   White-throated Thrush

Throat whitish, heavily streaked dark brown, with *contrasting white forecollar.* Underparts dusky greyish becoming whitish on vent and undertail coverts. **Juv (both groups):** head and upperparts brown, streaked pale cinnamon, underparts whitish to pale buff, spotted and mottled dark brown.

Voice. A loud, slightly nasal to gruff *rreeuh* or *rreuh*, often doubled, a clucking *kyow* or *ch-uhk*, rapid clucking that may run into a rich yodel *wheeljeeujeeujeeu...*, and a high, thin *ssi* or *ssee*, mostly in flight. Song a rich, continuous warbling with frequent mimid-like 2–3× repetition of phrases.

Habs. Forest and edge, from conifers to humid evergreen and arid deciduous (at least seasonally) forest, plantations, etc. Feeds low to high, rarely on ground; often flocks in winter and mixes readily with other species. Eggs whitish to pale blue, densely speckled with reddish browns and greys (WFVZ).

SS: Contrasting white forecollar distinctive in adults of all ssp. See Clay-colored and Black thrushes.

SD: C to F resident (near SL–3000 m, rarely above 2000 m S of Isthmus, lower elevations mainly in winter) on both slopes from S Son and S Tamps, and in interior from cen Mexico, to El Salvador and N cen Nicaragua. Winter wandering occurs but poorly known; vagrant to NL (Feb 1984, DAS).

RA: Mexico to Ecuador.

NB1: Sometimes considered conspecific with *T. albicollis* of S.A. **NB2:** Populations of *T. assimilis* from E Panama to S.A. sometimes considered specifically distinct.

RUFOUS-BACKED THRUSH (ROBIN)
Turdus rufopalliatus *   Plate 51
Zorzal Dorsirrufo

ID: 8.5–9.5 in (21.5–24 cm). Endemic to W Mexico. **Descr.** Sexes similar (female averages duller). **Adult:** bill and orbital ring yellow, legs flesh. *Greyish head and nape contrast with rufous* to olive-rufous *back,* rump and uppertail coverts grey; wings and tail dark greyish. Throat white, heavily streaked blackish, *chest and flanks rufous,* belly and undertail coverts white. **Juv:** head and upperparts duller with whitish to buff flecks, rump and uppertail coverts brownish. Underparts whitish to cinna-

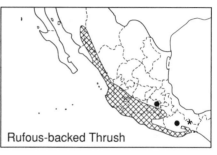

Rufous-backed Thrush

mon, spotted dark brown, undertail coverts white.

Voice. A plaintive, mellow, drawn-out whistle, *cheeoo* or *teeeuu,* a fairly hard clucking *chuk chuk chuk ...* or *chok ...,* and a high, thin *ssi* or *ssit,* mostly in flight. Song a leisurely rich warbling, includes 2–3× repetition of some phrases.

Habs. Arid to semihumid deciduous and semideciduous forest and edge, plantations, gardens. On ground and low to high in trees and bushes; often in flocks in winter. Eggs whitish, heavily marked with reddish browns (WFVZ).

SS: Grayson's Thrush resembles washed out Rufous-backed, back only slightly brighter than head and nape, underparts greyish with slight cinnamon wash.

SD/RA: C to F resident (SL–1500 m) on Pacific Slope from S to Oax, in interior along Balsas drainage. Populations in DF (2200–2500 m), apparently established in the last 50 years (Wilson and Ceballos-L. 1986), and Oaxaca City (Rowley 1984; Howell 1990b) probably derived from escaped cage birds.

GRAYSON'S [RUFOUS-BACKED] THRUSH
Turdus (rufopalliatus?) graysoni   Plate 51
Zorzal de Grayson

ID: 9.5–10 in (24–25.5 cm). *Endemic to NW Mexico.* Descr. Sexes similar. **Adult:** bill (and orbital ring?) yellowish, legs flesh. *Head and upperparts greyish with slightly contrasting greyish-rufous* to olive-brown *back;* wings and tail grey. *Throat whitish, streaked dark brown; underparts pale greyish, washed cinnamon on chest and flanks,* becoming *whitish on vent and undertail coverts.* **Juv:** undescribed (?).

Voice. Undescribed (?).

Habs. Arid to semiarid forest and edge, plantations. Habits probably much as

Grayson's Thrush

Rufous-backed Thrush. Nest and eggs undescribed (?).

SS: See Rufous-backed Thrush, White-throated Thrush; Clay-colored Thrush allopatric.

SD/RA: C resident (SL–600 m) on Islas Tres Marías, Nay. U to R visitor (Dec–Jun, mainly Dec–Apr; around SL) to adjacent Nay (vicinity of San Blas) where may be resident. See Phillips (1981). Status needs further study.

RUFOUS-COLLARED THRUSH (ROBIN)
Turdus rufitorques   Plate 51
Zorzal Cuellirrufo

ID: 9–10 in (23–25.5 cm). *Endemic, highlands S of Isthmus*, plumage variable. **Descr.** Sexes differ. *Undertail coverts streaked white in all plumages.* ♂: *bill and legs yellow* to yellowish. *Black overall with rufous chest and broad rufous hindcollar*, throat often washed rufous and streaked blackish, belly rarely washed rufous. ♀: *head and upperparts greyish* to greybrown *with greyish-cinnamon hindcollar*. Throat dusky, streaked blackish, chest dull cinnamon-rufous to greyish cinnamon, belly and undertail coverts dusky grey. **Imm** ♂: *resembles* ♀ *but darker overall* with brighter cinnamon-rufous chest and hindcollar, belly often washed cinnamon. **Imm** ♀: *resembles* ♀ but duller, hindcollar less contrasting, underparts paler and belly

washed cinnamon, or *underparts dusky greyish* with little or any trace of cinnamon. Probably attains adult plumage by 2nd prebasic molt when about 1 year old. **Juv:** bill mostly dark. Head and upperparts dark brown, spotted cinnamon-buff, with buff eye-crescents and supraloral stripe; throat whitish, underparts cinnamon-buff, spotted dark brown.

Voice. Much like American Robin. A clucking *cheik-chuk-chuk* ... and *kweh-kweh-kweh* ..., and a sharper *wheuk*, a high *ssir* which may be doubled, etc. Song a hesitant, rich, repetitive caroling, *cheer chrri-chrri chree-ip chree-ip chee cheer* ..., etc.

Habs. Pine and pine–oak woodland and edge, grassy clearings, villages. Often on the ground in open situations, also low to high in trees; sings from high prominent perches. Eggs blue, unmarked.

SS: Imm ♀♀ can be confused with other species, see Mountain, imm ♂ Black, and Clay-colored thrushes. Bright imm ♀ paler on head and upperparts than American Robin, usually with at least a hint of a paler hindcollar.

SD/RA: C to F resident (1500–3500 m), mainly in interior and on adjacent Pacific Slope, from Chis to W El Salvador; vagrant (?) to Honduras.

AMERICAN ROBIN
*Turdus migratorius**   Plate 51
Zorzal Petirrojo

ID: 9–10 in (23–25.5 cm). *Highlands N of Isthmus.* Two ssp groups: *migratorius* group (widespread) and *confinis* ('San Lucas Robin', BCS). **Descr.** Sexes similar. *Migratorius.* **Adult:** bill yellow, legs flesh. *Head and upperparts slaty grey*, head often blacker, *with white eye-crescents* and supraloral stripe; wings darker, tail black. N.A. migrants have broad white tips to outer rectrices. *Throat white, heavily streaked*

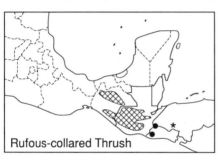

Rufous-collared Thrush

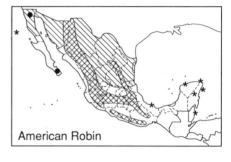

American Robin

black, *underparts rufous* with whitish vent and undertail coverts; in fresh plumage, underparts broadly tipped whitish. **Juv:** bill mostly dark. Head and upperparts dark brown, spotted cinnamon-buff, with whitish eye-crescents and supraloral stripe; throat whitish, underparts cinnamon-buff, spotted dark brown. *Confinis.* **Adult:** *head and upperparts grey*, head often slightly darker, with short whitish supercilium and white subocular crescent; wings and tail darker. Throat white, streaked dark brown, *underparts buff to ochraceous buff*, becoming whitish on vent and undertail coverts. **Juv:** head and upperparts grey-brown, flecked pale buff, with distinct pale cinnamon supercilium. Throat and underparts whitish, washed cinnamon on dark brown mottled chest and sides.

Voice. Song a rich, at times slightly hesitant, caroling warble with frequent repetition, suggests Black-headed Grosbeak; notably burrier and less strident in *confinis* and unlikely to be mistaken for a grosbeak. Calls (similar in both forms) include a clucking *cluk cluk* ..., often preceded by one or more *sheek* notes, a high, thin *ssir* or *ssip*, mostly in flight, and other shrieking and clucking calls.

Habs. Arid to semihumid pine and pine–oak woodland and edge, grassy clearings; to oak and semideciduous woods in winter. Habits much as Rufous-collared Thrush. Eggs blue, unmarked (WFVZ).

SS: See imm ♀ Rufous-collared Thrush.

SD: C to F resident (1500–3500 m, lower locally in winter) in interior and on adjacent slopes from Chih and S Coah to Oax; U breeder (recent colonist?) in BCN (Howell and Webb 1992*f*). U to R winter visitor (mid-Oct–Apr; SL–3000 m) in N Mexico from BCN to Tamps, and S to Yuc Pen, vagrant to Belize (Jan 1981, per BWM). Reports from Guatemala (Land 1970) require verification. *Confinis* is a C resident (1000–2000 m, lower locally in winter) in Sierra Victoria, BCS.

RA: N.A. to Mexico.

NB: *Confinis* has been regarded as a distinct species.

VARIED THRUSH

Zoothera naevia meruloides　Not illustrated
Zorzal Pechicinchado

ID: 8.5–9.5 in (21.5–24 cm). *BCN in winter.*

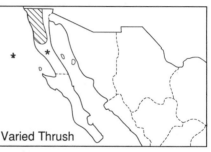

Varied Thrush

Descr. Sexes differ. ♂: bill black, straw below at base, legs flesh. *Head and upperparts dark blue-grey with cinnamon postocular stripe*; wings darker with 2 *cinnamon wingbars, remiges patterned with cinnamon.* Tail dark grey, outer rectrices tipped white. Throat and *underparts cinnamon-rufous with broad blackish chest band*, flanks and undertail coverts mottled whitish and grey. ♀: head and upperparts brownish, underparts paler with *less distinct, dusky chest band.*

Voice. A quavering to slightly eerie *wheeeirr*, a more trilled *whirrr*, and a low *chuk.*

Habs. Conifer forests, mixed woodland. Feeds on the ground, often in forest, and in trees and bushes at edges; overlooked easily.

SD: Irregular U to R winter visitor (Nov–Mar; SL–2500 m) to BCN, possibly a vagrant to Son (Phillips 1991).

RA: W N.A.; winters to BCN.

NB: Traditionally placed in the monotypic genus *Ixoreus.*

AZTEC THRUSH

Zoothera pinicola *　　　　　　Plate 50
Zorzal Azteca

ID: 8.5–9.5 in (21.5–24 cm). *Mexican endemic; highlands N of Isthmus.* **Descr.** Sexes differ. ♂: bill black, legs flesh. *Head, upper chest, and upperparts blackish brown*, streaked buff on head and mantle, *with white uppertail coverts. Wings blackish, greater coverts with creamy edgings, remiges with complex pattern of white edgings and tips*; in flight, note broad white stripe across base of remiges. *Black tail broadly tipped white*, outer webs of outermost rectrices edged white. Rest of underparts whitish, mottled dark brown on flanks. ♀: head, upper chest, and upperparts brown, with more pronounced dark

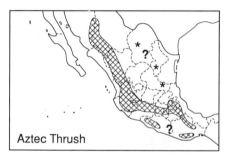

Aztec Thrush

mask and paler supercilium than ♂, *chest streaked and mottled buff*, tips to secondaries and inner primaries duller, pale greyish, not forming bold silvery-white panel. **Juv:** head and upperparts dark brown, coarsely streaked buff, uppertail coverts tipped pale buff; wing coverts and tertials edged ochraceous buff. Remiges patterned like ♀, retained through 1st year. Throat and underparts buff, heavily scalloped blackish on chest and flanks, becoming whitish on belly and undertail coverts.

Voice. A quavering, slightly burry, whining *wheeerr* or *whieeer*, a slightly metallic, whining *whein* or *wheen*, and a slightly nasal to clear *sweee-uh* in alarm. Song (?) a louder variation of call, repeated steadily by ♂, *wheeerr* ... or *dweir* ..., etc.

Habs. Pine, pine–oak, and pine-evergreen forest. Feeds mainly in bushes and trees, also on ground. Sits still for long periods and overlooked easily; best detected by voice. In pairs or small groups, rarely up to 50 birds, which join mixed-species flocks with *Turdus* thrushes, Grey Silkies, etc. Nest cup of grasses, moss, etc., at mid- to upper levels of trees. Eggs: 2, pale blue, unmarked (WFVZ).

SD/RA: U to F resident (1800–3500 m) in interior and on adjacent Pacific Slope from Chih to Oax; may withdraw (Oct–Feb?) from NW Mexico (Phillips 1991), occurs in winter N at least to Sin (SNGH, PP). Probably local resident in NE Mexico (Coah to E SLP) but all records Nov–Jan.

NB: Traditionally placed in the monotypic genus *Ridgwayia*.

BABBLERS: TIMALIIDAE (1)

The Wrentit is thought to be the sole New World representative of the diverse Old World babblers. It was formerly placed in its own family, the Chamaeidae, and sometimes the Timaliidae is merged with the Muscicapidae (AOU 1983). The Wrentit has a short bill, short rounded wings, and a long graduated tail. Ages/sexes similar; young altricial; 1st prebasic molt complete.

Food mostly insects and berries. Nest is a neat cup of plant materials, low in bush. Eggs: 3–5, bluish, unmarked (WFVZ).

WRENTIT
Chamaea fasciata canicauda Not illustrated
Camea

ID: 6–6.5 in (15.5–16.5 cm). *NW Baja.*
  Descr. *Eyes pale lemon,* bill and legs blackish. *Head and upperparts greyish, throat and underparts pale pinkish, indistinctly streaked dusky.*
  Voice. Song, given year-round, a series of 3–10 or more bright, sharp, whistled notes *beehk beehk beehk* ... or *peeh-peeh-peeh* ... which, in ♂, runs into a rolled, bouncing-ball churr or trill, *beehk* (usually 3–5×) *bee birr-rrrrrrrrrr,* etc. Calls include a hard, dry, scolding rattle or churr.
  Habs. Semiarid scrub, especially chaparral. In pairs, low in dense scrub and bushes. Usually skulking and hard to see; best detected by voice.

SD: C to F resident (SL–2500 m) in NW Baja, N of 30° N.
RA: W USA to N Baja.

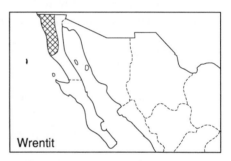
Wrentit

MOCKINGBIRDS, THRASHERS, AND ALLIES: MIMIDAE (18)

Mimids, or mocking-thrushes, are noted for their loud and varied songs. Typically they are slender birds with fairly short, rounded wings and long graduated tails. Slender bills long and often decurved in typical thrashers, used for digging for food, or shorter and straighter in catbirds and mockingbirds; legs fairly long. Ages similar or different, sexes similar; young altricial. Imms quickly resemble adult after 1st prebasic molt which varies from partial to complete. Plumage overall colored in browns, greys, black, and white, sometimes in striking patterns. Generally skulking and often hard to see well, although most sing from exposed perches, especially early and late in the day. Songs typically loud and varied, including repetition of phrases and mimicry in many species; calls include whistles, mews, and chacks.

Food mostly insects and fruit. Nest cups built at low to mid-levels in bushes. Eggs: 3–5, pale green or bluish, speckled brown in most species.

GREY CATBIRD
Dumetella carolinensis*   Not illustrated
Pájaro-gato Gris

ID: 8–8.5 in (20.5–21.5 cm). Winter visitor to S and E. **Descr.** Ages similar. Bill and legs blackish. *Slaty grey overall with black crown, blackish tail, and chestnut under-tail coverts.*
> **Voice.** A complaining nasal mew, *meeah* or *myeh* and a harder, slightly harsher *nyaah*; less often, a hard, dry rattle.
> **Habs.** Humid to semihumid evergreen and semideciduous forest and edge, second growth. Fairly skulking at low to mid-levels, but may feed higher in fruiting bushes and trees.

SD: C to F transient and winter visitor (late Sep–mid-May; SL–2000 m) on Atlantic Slope from Tamps to Honduras; U on Pacific Slope from E Oax to Guatemala and in interior from Chis S. Vagrant to BCN (May 1991, Palacios and Alfaro 1992*b*) and Nay (Phillips 1986).

RA: Breeds N.A.; winters SE USA to Panama.

BLACK CATBIRD
Dumetella glabrirostris*   Plate 52
Pájaro-gato Negro

ID: 7.5–8 in (19–20.5 cm). *Endemic to Yuc Pen and islands.* Shape much like Grey Catbird. **Descr.** Ages differ slightly (?). Eyes dark reddish, bill and legs blackish. *Plumage entirely glossy black.* Juv duller (?).
> **Voice.** Varied. A nasal, slightly gruff *chehr*, a harsh *rrriah*, and a slightly metallic, rasping *tcheeu*, etc. Song a squeaky to sweet to scratchy warble, often repeated over and over, *klee tu-who-wiik'*, or *ti che-wee rr cher-wer*, etc., often with

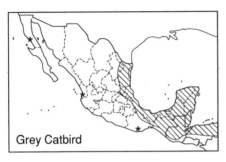
Grey Catbird

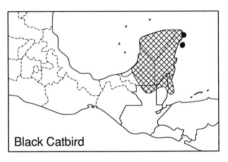
Black Catbird

metallic clicking buzzes thrown in; less often prolonged warbling.

Habs. Humid to semiarid scrubby woodland, forest edge. Habits much as Grey Catbird; sings from fairly prominent perches. Nest is a bulky cup of grass, twigs, etc., at low to mid-levels in thicket, low tree, etc. Eggs: 2, greenish blue, unmarked (AMNH).

SS: Other all-black birds (such as Melodious Blackbird, Bronzed Cowbird) are larger and more heavily built with stout bills, usually less skulking.

SD/RA: Status unclear in much of Yuc Pen. C to F resident (around SL) from E QR to Ambergris Cay, Belize; also on Isla Cozumel, and C to U on Cay Corker, Lighthouse Reef, and Glover's Reef, Belize. U to R (mainly Nov–May, seasonal?) in much of Yuc Pen, W to Camp, S to N Guatemala (Van Tyne 1935) and N Belize. No records from Honduras since the type was purportedly collected at Omoa in 1855 or 1856. Reports from Isla Holbox (Salvin 1888) are based on Gaumer specs and require verification.

NB: Traditionally placed in the monotypic genus *Melanoptila*.

BLUE MOCKINGBIRD
*Melanotis caerulescens**   Plate 52
Mulato Azul

ID: 9.5–10.5 in (24–26.5 cm). Endemic to Mexico, N of Isthmus. **Descr.** Ages differ. **Adult:** eyes reddish, bill and legs blackish. *Slaty blue overall with black mask;* crown, throat, and chest streaked paler blue. **Imm:** duller, without paler streaking. **Juv:** eyes dull, dull slaty grey overall, wings and tail bluer. Attains adult plumage when about 1 year old.

Voice. Extremely varied. Song a varied arrangement of rich phrases, often with repetition, can be hard to distinguish from Ocellated or Curve-billed thrashers. Common calls include a loud, rich *choo* or *cheeoo*, a nasal mewing *mejhr* or *meeahr* much like Orange-billed Nightingale-Thrush, a loud, rich *wee-cheep* or *wheeep*, a low *chruk* or *chuk*, a loud, rich *choo'l-eep*, a sharp *pli-tik*, and any combination of these and many other noises.

Habs. Scrubby woodland and understory from pine forest to arid oak scrub and thorn forest. Usually skulking, low or on ground but often sings from conspicuous perches. Nest cup of twigs, rootlets, etc., at low to mid-levels in dense bush or tree. Eggs: 2, blue, unmarked (WFVZ).

SD/RA: C to F resident (SL–3000 m, only near SL on Pacific Slope) on both slopes from S Son and S Tamps (mainly in interior from cen Mexico S) to Oax.

NB: Sometimes considered conspecific with Blue-and-White Mockingbird.

BLUE-AND-WHITE MOCKINGBIRD
Melanotis hypoleucus   Plate 52
Mulato Pechiblanco

ID: 10–11 in (25.5–28 cm). Endemic, S of Isthmus. **Descr.** Ages differ. **Adult:** eyes reddish, bill and legs blackish. *Head and upperparts slaty blue with black mask. Throat and underparts white,* becoming dull slaty blue on flanks and undertail coverts. **Juv:** eyes duller, dull slaty grey overall, wings and tail bluer, throat and underparts slightly mottled white. Apparently lacks imm plumage and soon resembles adult.

Voice. Much like Blue Mockingbird. Song varied rich phrases, often with 2–3× repetition. Common calls include a loud, rich *tcheeoo* or *whee-oo*, and *chweet*, *chrr*, a low hollow *chuk*, a gruff growl followed by hollow clucking, *ahrrr took-took-took-took*, etc.

Habs. Humid to semiarid scrub and

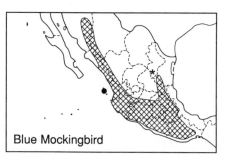

Blue Mockingbird

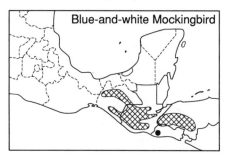

Blue-and-white Mockingbird

forest edge, second growth. Habits, nest, and eggs (WFVZ) much like Blue Mockingbird.

SD/RA: C to F resident (1000–3000 m) in interior and on Pacific Slope from Chis to Honduras and El Salvador.

NB: See Blue Mockingbird.

NORTHERN MOCKINGBIRD

Mimus polyglottos leucopterus   Plate 52
Cenzontle Norteño

ID: 9–10 in (23–25.5 cm). *N of Isthmus.*
Descr. Ages differ. **Adult:** eyes pale yellowish, bill and legs blackish. Face, throat, and underparts whitish to dusky pale grey, with dusky lores. Crown, nape, and upperparts grey. Wings blackish with 2 white wingbars, white primary coverts, and whitish tertial and secondary edgings; *broad white patch across bases of primaries striking in flight.* Tail blackish, *mostly white outer rectrices appear as white sides on spread tail.* **Juv:** eyes dusky, upperparts browner, chest and flanks spotted dusky.
Voice. Song a loud, varied series of rich to gruff phrases, each commonly repeated 3–4×. Calls include a fairly full, gruff *chek* or *tchek.*
Habs. Arid to semihumid, open and semiopen areas with scrub, scattered bushes, and trees. Often on or near ground where runs easily, tail often cocked. Perches atop bushes, on roadside wires, etc. Nest is a bulky cup of grass, twigs, etc., at low to mid-levels in bush or tree. Eggs: 3–5, pale bluish to greenish blue, heavily marked with reddish browns and greys (WFVZ).
SS: Tropical Mockingbird lacks white patch across bases of primaries, spread tail appears broadly white-tipped; apparent hybrid Northern × Tropical may occur rarely in Isthmus region. Loggerhead Shrike

has black mask, stout hooked bill, perches more vertically.
SD: F to C resident (SL–3000 m) in Baja, on Isla Socorro (colonized in 1970s), on Pacific Slope S to Sin, and in interior and on adjacent slopes S to cen volcanic belt; U to F in interior to Oax; possibly R resident on Atlantic Slope to S Ver (formerly?). More widespread in winter (Sep–Apr), when F to C on much of Pacific Slope and U to R on Atlantic Slope S to W Tab. Vagrant (Jan–Mar) to Islas Guadalupe, and Clarión (Swarth 1933).
RA: N.A. to Mexico; N populations migratory.
NB: Sometimes considered conspecific with Tropical Mockingbird; apparent hybrids occur but the role of escaped cage birds in this situation is unknown.

TROPICAL MOCKINGBIRD

Mimus gilvus*   Plate 52
Cenzontle Sureño

ID: 9–10 in (23–25.5 cm). *S of Isthmus.*
Descr. Ages differ. **Adult:** eyes pale yellowish, bill and legs blackish. Face, throat, and underparts whitish with dusky lores. Crown, nape, and upperparts grey; wings blackish with 2 white wingbars and whitish tertial and secondary edgings. Tail blackish, *outer rectrices broadly tipped white.* **Juv:** eyes dusky, upperparts browner, chest and flanks spotted dusky.
Voice. Probably not safely told from Northern Mockingbird by voice. Varied song of rich to gruff phrases often features 3–4× repetition, calls include a hard *chek* or *shahk*, etc.
Habs. Much as Northern Mockingbird, including nest and eggs (WFVZ).
SS: See Northern Mockingbird.
SD: C to F resident (SL–500 m) on Atlantic Slope from N Chis to Yuc Pen and N Belize, including Isla Cozumel and N and

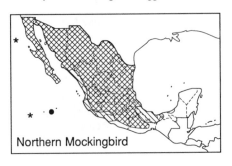

Northern Mockingbird

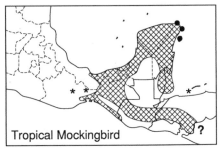

Tropical Mockingbird

outer Belize cays; U and local to SE Ver. C to F in interior (500–2400 m) from Chis to Guatemala, U to F to Honduras. U to F and local (near SL–500 m) on Pacific Slope from E Oax to W Guatemala, and in Honduras; should be looked for in W Nicaragua. Records from interior cen Oax (such as 5–6 birds in Jan 1987, SNGH, SW) and Isla Utila (Mar 1991, Howell and Webb 1992*b*) may pertain to escaped cage birds.

RA: S Mexico to Honduras, also in NE S.A. and S Caribbean.

NB: See Northern Mockingbird.

SOCORRO MOCKINGBIRD

Mimodes graysoni Plate 69
Cenzontle de Socorro

ID: 9.5–10.5 in (24–26.5 cm). *Endemic to Isla Socorro, endangered.* **Descr.** Ages differ. **Adult:** eyes amber, bill and legs blackish. *Head and upperparts brown* with dusky lores and short pale supercilium. Wings and tail slightly darker with *2 narrow whitish wingbars,* outer rectrices tipped paler. *Throat and underparts whitish, flanks streaked brown.* **Juv:** eyes duller, wingbars pale cinnamon, throat and chest spotted dusky.
  Voice. Traditional song a varied, scratchy to rich warble. Calls include a rich *whichoo* and *whi-chee-oo.* Northern Mockingbirds can incorporate such calls into their songs but usually repeat them 2–3× in succession, unlike Socorro Mockingbird. However, either species can mimic the song of the other (LFB)!
  Habs. Formerly scrub, woodland, and edge over the entire island; now mostly found in wooded canyons and chaparral scrub at middle and higher elevations (300–900 m; Castellanos and Rodriguez-E. 1993; SNGH, SW). Habits recall Northern Mockingbird but often more skulking. Nest and eggs undescribed (?).

SS: Northern Mockingbird grey above with bold white wing flashes and tail sides.

SD/RA: Formerly an abundant resident on Isla Socorro but, since human settlement (in 1957) and its associated problems, has declined drastically. Causes of decline discussed by Castellanos and Rodriguez-E. (1993) who estimated a recent population of 50–60 pairs of Socorro Mockingbirds. Critical conservation measures are needed to ensure this species' survival.

RA: Formerly known as Socorro Thrasher.

SAGE THRASHER

Oreoscoptes montanus Not illustrated
Cuitlacoche de Artemesia

ID: 8–9 in (20.5–23 cm). *Winter visitor to N Mexico. Bill relatively short and straight, thrush-like.* **Descr.** *Eyes yellow* to orange-yellow, bill greyish, paler below at base, legs dark greyish flesh. *Head and upperparts brownish grey,* indistinctly streaked darker, with whitish supercilium; wings browner with 2 narrow whitish wingbars and pale tertial edgings. Tail browner, *outer rectrices boldly tipped white,* outer web of outermost rectrices edged white. Throat and underparts whitish to pale buff (in fresh basic, flanks pale cinnamon), with indistinct dusky malar stripe, *underparts marked with elongated dark brown spots,* except unmarked tail coverts.
  Voice. A quiet, low, slightly gruff *tchuh* and a slightly harder *chreh.*
  Habs. Arid to semiarid, open and semi-open country with scrub, scattered bushes, sage brush. Singly or in small groups, on or near ground, usually skulking; at times perches conspicuously on posts, fences, atop bushes. Flight strong and thrush-like.

SS: Grey Thrasher (Baja) larger with longer, slightly decurved bill, longer tail, plainer face, golden-yellow eyes.

Socorro Mockingbird

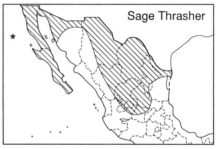

Sage Thrasher

SD: F to C transient and winter visitor (late Sep–Apr; SL–2400 m) in BCN, on Pacific Slope in N Son, and in interior from Son and NL to Dgo, U to R to BCS, N Gto, and on Atlantic Slope in N Tamps.

RA: Breeds W N.A.; winters SW USA to N Mexico.

BROWN THRASHER

Toxostoma rufum longicauda Not illustrated
Cuitlacoche Rojizo

ID: 10–11 in (25.5–28 cm). *Vagrant. Bill medium-long and straightish.* **Descr.** Ages differ slightly. *Eyes yellow to yellow-orange, bill blackish, flesh below at base, legs greyish flesh. Head and upperparts rufous-brown,* face greyer; 2 whitish wing-bars (buff in imm), outer rectrices narrowly tipped paler. Throat and underparts whitish to pale buff with dark malar stripe, chest and sides coarsely streaked blackish to dark brown, undertail coverts buff.

Voice. A hard smacking *tchuk* or *tsuk*, and a slightly rasping, nasal *tchah.*

Habs. Arid to semihumid brushy scrub, woodland understory. Typically skulking, on or near ground.

SS: Long-billed Thrasher duller above (but beware nominate *longirostre,* from SE SLP and N Ver S, which is redder than Texas Long-billeds), head greyer, bill longer and slightly decurved with paler but not flesh base, face often bordered below by dark moustache.

SD: Vagrant (Dec–mid-Apr, probably Oct–Apr) on Pacific Slope from Son to N Nay (Russell and Lamm 1978; Phillips 1986), and in Tamps (Phillips 1986) where may be R but regular in winter.

RA: Breeds E N.A; winters SE USA, vagrant to N Mexico.

LONG-BILLED THRASHER

Toxostoma longisrostre*   Plate 52
Cuitlacoche Piquilargo

ID: 10.5–11.5 in (26.5–29 cm). *NE Mexico. Bill fairly long and slightly decurved.* **Descr.** Ages differ slightly. **Adult:** *eyes orange to orange-yellow,* bill blackish, paler below at base, legs greyish flesh. *Head grey-brown with greyer face often bordered below by dark moustache. Upperparts rich brown to cinnamon-brown* (duller and greyer in *sennetti* from Tamps to N Ver), *brightest on rump and tail;* 2 whitish wing-bars, outer rectrices narrowly tipped paler. Throat and underparts whitish with dark malar stripe, chest and sides coarsely streaked blackish, *dusky centers to undertail coverts usually hidden.* **Juv:** face and underparts less distinctly marked, wing-bars pale buff.

Voice. A loud, rich, whistled *cleeooeep;* also may have gruff call like Brown Thrasher (?). Song a loud, rich to slightly scratchy warbling, phrases often repeated 2–4×.

Habs. Arid to semihumid brushy woodland and edge, scrubby thickets, hedges, etc. Typically skulking, on or near ground, but often sings from conspicuous perches. Nest is a bulky cup of twigs, grass, etc., at low to mid-levels in dense undergrowth, bush, or tree. Eggs: 2–5, bluish white, densely speckled with reddish browns and greys (WFVZ).

SS: See Brown Thrasher.

SD/RA: C to F resident (SL–1500 m) on Atlantic Slope from N Coah and Tamps (also S Texas) to cen Ver. A report from DF (Wilson and Ceballos-L. 1986) probably pertains to an escaped cage bird; a report from S Ver (Winker *et al.* 1992b) is unsatisfactory.

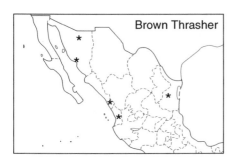
Brown Thrasher

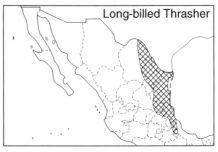
Long-billed Thrasher

COZUMEL THRASHER
Toxostoma guttatum Plate 69
Cuitlacoche de Cozumel

ID: 8.5–9.5 in (21.5–24 cm). *Endemic to Isla Cozumel.* Bill fairly long and slightly decurved. **Descr.** Ages differ slightly (?). **Adult:** eyes amber, bill blackish, legs greyish. *Head and upperparts rich brown to cinnamon-brown,* face slightly paler with pale supercilium; 2 whitish wingbars, outer rectrices narrowly tipped paler. *Throat and underparts whitish,* often with dark malar stripe, *chest and sides coarsely streaked blackish.* **Juv:** undescribed (?).
Voice. Song a rich, varied warbling, slightly scratchy and with little repetition. Calls undescribed (?).
Habs. Scrubby woodland and edge. Habits much as Long-billed Thrasher. Nest and eggs undescribed (?).
SD/RA: Apparently U resident (around SL) on Isla Cozumel, QR; F to C prior to Sep 1988 hurricane.

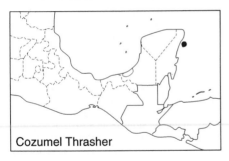
Cozumel Thrasher

GREY THRASHER
*Toxostoma cinereum** Plate 52
Cuitlacoche Peninsular

ID: 9.5–10.5 in (24–26.5 cm). *Endemic to Baja.* Bill medium length, slightly decurved. **Descr.** Ages differ slightly. **Adult:** eyes golden yellow, bill and legs greyish.

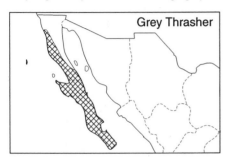
Grey Thrasher

Head and upperparts brownish grey to brownish, becoming *brighter cinnamon-brown on rump and uppertail coverts;* 2 whitish to pale buff wingbars, tertials edged whitish when fresh, outer rectrices tipped whitish. Throat and underparts whitish, flanks washed buff to cinnamon, *underparts marked with elongated blackish spots* (heaviest in *mearnsi,* N of 28° N), undertail coverts unmarked. **Juv:** eyes duller, underparts less distinctly marked, wingbars and tertial edgings pale cinnamon.
Voice. A rolled, rippling to rough *whirr-rr-rr* or *chirr-rri-rrit,* and a gruff *chrek.* Song a loud, fairly scratchy warbling with frequent 2–3× repetition of phrases and occasional rich notes thrown in.
Habs. Arid to semiarid, open and semi-open areas with cacti, scrub, scattered bushes, trees, grassy areas, etc. Usually low and skulking but often sings from high exposed perches. Nest is a bulky cup of twigs, grass, etc., at low to mid-levels in bush, cactus, or tree. Eggs: 2–4, bluish white to blue-green, heavily speckled and mottled with reddish browns and greys (WFVZ).
SS: See Sage Thrasher.
SD/RA: C to F resident (SL–1500 m) in Baja, S of about 31° N on W side, S of 28° N on E side.

BENDIRE'S THRASHER
Toxostoma bendirei Not illustrated
Cuitlacoche de Bendire

ID: 9.5–10 in (24–25.5 cm). *NW Mexico. Bill medium length and straightish.* **Descr.** Ages differ slightly. **Adult:** eyes yellow to yellow-orange, *bill dark grey, paler to flesh below at base,* legs greyish. Head and upperparts grey-brown with 2 indistinct whitish wingbars, outer rectrices tipped whitish. *Throat and underparts pale buff* to dirty whitish, *chest and sides finely marked with dusky triangular spots* (may

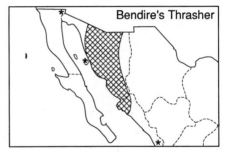
Bendire's Thrasher

appear as dusky streaks and often hard to see when worn). **Juv:** wingbars washed cinnamon, underparts more finely and neatly marked with short dusky streaks.

Voice. A low, dry, slightly gruff *chek* or *shek*, at times a doubled *chuk-chek*. Song a mellow warbling with frequent repetition of phrases, often not loud or strident.

Habs. Arid to semiarid, open and semiopen country and grassy areas, with scattered scrub, cacti, hedges, etc. Habits much as Grey Thrasher. Nest cup of twigs, grass, etc., at low to mid-levels in bush or cactus. Eggs: 3–4, bluish white to blue-green, heavily speckled and mottled with reddish browns and greys (WFVZ).

SS: Curve-billed Thrasher larger and bulkier with longer, slightly decurved bill (lower mandible decurved versus straightish in Bendire's), underparts spotted and coarsely mottled dusky, eyes often brighter orange; note call.

SD: F to U resident (SL–750 m) on Pacific Slope in Son (limits of breeding range uncertain); U transient and winter visitor (Sep–Feb) to N Sin, R to S Sin. Vagrant to NE BCN (Jan 1989, Patten *et al.* 1993); other reports from Baja (Wilbur 1987) require confirmation.

RA: SW USA to NW Mexico; N populations migratory.

CURVE-BILLED THRASHER

Toxostoma curvirostre*   Plate 52
Cuitlacoche Piquicurvo

ID: 10–11 in (25.5–28 cm). *Widespread N of Isthmus. Bill medium length and slightly decurved.* Two groups: *palmeri* group (Pacific Slope) and *curvirostre* group (interior). **Descr.** Ages differ slightly. *Palmeri.* **Adult:** *eyes orange* to yellow-orange, bill blackish, legs greyish. Head and upperparts grey-brown with 2 narrow whitish wingbars, outer rectrices tipped whitish

(narrowly tipped buff and poorly contrasting in imm) to pale buff (*occidentale* in S). Throat and underparts whitish to pale buff with dusky malar, chest and flanks spotted and mottled dusky. **Juv:** eyes pale greyish, bill shorter and straighter (soon attains full length), outer rectrices lack contrasting pale tips, wingbars pale cinnamon. *Curvirostre.* **Adult:** wingbars often more distinct, underparts more coarsely and often more boldly spotted and mottled dusky, white tips of outer rectrices average broader. **Juv:** whitish tips to outer rectrices may be tinged buff, underparts more finely marked, wingbars cinnamon. Both ssp groups attain adult plumage when about 1 year old.

Voice. A bright, emphatic *whit whuit* or *quidi quit*, and *whit whuit whuit*, etc., less often a gruff *chuk* and *chuh-uh-uh-uh.* Song a loud, varied, rich to slightly scratchy warbling, often with 2–3× repetition of phrases.

Habs. Arid to semiarid, open and semiopen areas with scattered bushes, trees, fences, cacti, etc. Habits much as Grey Thrasher but more often in open situations. Nest is a bulky cup of twigs, grass, etc., at low to mid-levels in bush, cactus, etc. Eggs: 2–4, pale bluish green, densely speckled with reddish browns and greys (WFVZ).

SS: See Bendire's Thrasher.

SD: C to F resident (*palmeri* group; SL–1500 m) on Pacific Slope from Son to N Nay and (*curvirostre* group; 1000–3000 m) in interior from NE Son and NL to Oax, locally on Pacific Slope in S Nay and Jal, also (to near SL) in N Tamps.

RA: SW USA to S Mexico.

OCELLATED THRASHER

Toxostoma ocellatum   Plate 52
Cuitlacoche Manchado

ID: 10.5–11.5 in (26.5–29.5 cm). *Endemic*

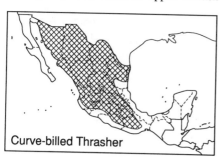

Curve-billed Thrasher

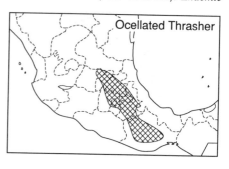

Ocellated Thrasher

to cen Mexico. Bill longish and slightly decurved. **Descr.** Ages differ slightly. **Adult:** *eyes deep reddish to amber-brown,* bill blackish, legs greyish. Head and upperparts grey-brown to brown with pale buff supercilium; 2 whitish wingbars, outer rectrices tipped paler. *Throat and underparts whitish* with dark malar stripe, *chest and flanks boldly spotted blackish.* **Juv:** eyes brownish (?), face and underparts less distinctly marked, underparts with blurry blackish spots, wingbars and tips of outer rectrices washed buff.

Voice. A hard, slightly smacking to gruff *chehk* or *tchehk*. Song a varied, rich warbling with frequent 2–3× repetition of phrases, not always safely told from Curve-billed Thrasher song.

Habs. Arid to semiarid scrub (especially oaks), brushy understory of oak woodland. Skulking, usually on or near ground in dense brush, but sings conspicuously from tops of bushes and trees. Nest is a bulky cup of twigs, grass, etc., at low to mid-levels in bush or tree. Eggs: 2, greenish blue, heavily speckled with reddish browns and greys (WFVZ).

SD/RA: F to C resident (1400–3000 m) in interior from NE Gto and cen Hgo to Oax.

CALIFORNIA THRASHER

Toxostoma r. redivivum   Not illustrated
Cuitlacoche Californiano

ID: 11–12 in (28–30.5 cm). *N Baja. Bill long and decurved.* **Descr.** Ages similar. Eyes brown, bill and legs blackish. *Head and upperparts grey-brown to brown with pale supercilium, dark moustache blends with face;* wings and tail darker with 2 indistinct paler wingbars. Throat whitish, chest dusky, rest of *underparts pale cinnamon, deeper on undertail coverts.*

Voice. A rolled *chreip* or *chur-eip*, a fairly strong, slightly nasal *hwuit* or *hreit* suggesting House Finch, a gruff, clipped *chuk!* or *chruk*, and harder *chep* or *tchep!* Song a rich, slightly jerky warbling with frequent repetition of phrases.

Habs. Chaparral, riparian thickets, semiopen areas with bushes. Usually low or on ground, skulking and hard to find unless singing from atop bushes. Nest is a bulky cup of twigs, grass, etc., low in bush or scrubby tree. Eggs: 2–4, pale bluish, usually heavily speckled reddish brown (WFVZ).

SS: Crissal Thrasher (NE Baja) greyer, dark malar stripe stands out on throat sides, undertail coverts chestnut.

SD: C to F resident (near SL–2000 m) in NW Baja.

RA: W California to NW Baja.

CRISSAL THRASHER

Toxostoma crissale*   Plate 52
Cuitlacoche Crisal

ID: 11–12 in (28–30.5 cm). *N Mexico. Bill long and decurved.* **Descr.** Ages similar. Eyes pale brown, bill and legs blackish. *Brownish grey overall,* paler below, wings and tail slightly darker; throat whitish with dusky malar stripe, *undertail coverts chestnut.*

Voice. A rich, warbled *cheeoo-ree-eep* or *choochereep*, and a doubled *ree-o-ree ree-o-reep*. Song a rich, often slightly hesitant warbling with frequent 2–3× repetition of phrases.

Habs. Arid to semiarid, semiopen areas with scrub, scattered bushes, riparian thickets. Habits much as California Thrasher. Nest cup of twigs, grass, etc., at low to mid-levels in bush or tree. Eggs: 2–4, pale blue, unmarked (WFVZ).

SS: See California Thrasher.

SD: F to C resident (near SL–2500 m) in NE

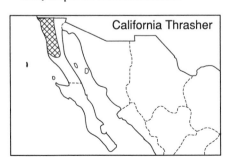
California Thrasher

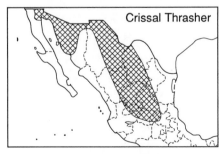
Crissal Thrasher

Baja, on Pacific Slope in Son, and on Plateau from NE Son and Coah to Zac and Hgo.

RA: SW USA to cen Mexico.

LE CONTE'S THRASHER
*Toxostoma lecontei**   Not illustrated
Cuitlacoche Pálido

ID: 10.5–11.5 in (26.5–29 cm). *NW Mexico. Bill long and decurved.* **Descr.** Ages similar.

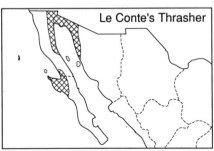
Le Conte's Thrasher

Eyes brown, bill and legs blackish. *Pallid grey-brown overall,* paler below, wings slightly darker, *tail contrastingly blackish* to dark brown; throat whitish with indistinct dusky malar stripe, vent and *undertail coverts ochre* (paler in juv).

Voice. A bright, up-slurred, slightly plaintive *chreep* or *hreep* to *ch-reeip* or *hu-reep*, a shorter *reeih* or *reeip*, and a slightly mewing *rreea*; shorter calls may suggest House Finch. Song a rich, robin-like warbling with infrequent repetition.

Habs. Arid, open, sandy desert with scattered bushes and scrub. Mostly terrestrial and elusive but sings from conspicuous perches. Runs quickly with tail cocked high. Nest cup of twigs, grass, etc., low in bush or cactus. Eggs: 2–4, pale bluish to greenish, speckled reddish brown (WFVZ).

SD: F to C but local resident (around SL) in NE and cen W Baja and NW Son.

RA: SW USA and NW Mexico.

WAGTAILS AND PIPITS: MOTACILLIDAE (4)

This mainly Old World family is represented in the Americas by several species of pipits; wagtails occur only as vagrants in the region. Both wagtails and pipits are fairly small, slender, mostly terrestrial birds with fairly long to long tails that are often wagged. Very long tertials cover most or all of the primaries at rest. Bills slender and pointed, legs fairly long with long hind claws. Ages/sexes usually similar in pipits, different in wagtails, several species have distinct alt plumages; young altricial. First prebasic molt partial. Pipits are colored mainly in browns and whitish, plumage variably streaked; wagtails are strikingly patterned in black, white, and greys; outer rectrices mostly white.

Food mostly insects picked off or near the ground.

WHITE WAGTAIL
Motacilla alba ocularis   Not illustrated
Lavandera Blanca

ID: 7–7.5 in (18–19 cm). *Vagrant.* **Descr.** Ages/sexes differ. **Basic** (Sep–Mar): bill and legs blackish. *Forhead and face whitish* with dark *eyestripe*; forehead often dusky in Oct–Nov imms. Crown grey (imm and some adult ♀) to all black (adult ♂); nape and upperparts grey (duskier in imm, brighter blue-grey in adult). *Lower rump slightly darker grey*, uppertail coverts black. Wings darker with *2 whitish wingbars (imm)* to *solid white panel (adult ♂)*, whitish tertial and secondary edgings. Tail black, mostly white outer rectrices appear as *white sides to tail*. *Throat and underparts whitish with black forecollar from lower malar area across chest*, flanks washed pale grey. Begins molt to 1st alt in mid-winter. Attains adult plumage by 2nd prebasic molt when about 1 year old. **Alt** (Feb–Sep): throat and chest solidly black, back cleaner blue-grey.
Voice. A bright to fairly sharp *tchissik!* and *tchi-ssi-sik* in flight (N.A.).

Habs. Open areas, typically near water. Walks quickly with almost constant tail-wagging. Flight strong and undulating.
SS: Black-backed Wagtail (see NB) not always safely told in 1st basic but typically lower rump mostly blackish, shaft of 4th rectrix often white, wings show bold white wingbars tending towards a panel (not the distinct, relatively narrow wingbars of imm White). Adult Black-backeds have mostly white remiges and, in alt, white chin and upper throat; alt ♂ has mostly black back, alt ♀ has usually at least some black mottling on back; 1st alt ♀ Black-backed, however, may have black throat, grey back with very little dark mottling; note duskier grey back, more extensive black on rump, though some may not be safely told from alt White Wagtail. Intergrades may occur which are not identifiable as either form.
SD: Vagrant (Oct–Feb); two records; an unsexed specimen (identified as *M. alba ocularis*, when *lugens* was considered a synonym) was collected at La Paz, BCS, 9 Jan 1882 (Ridgway 1883a). It was deposited at the USNM but has since been lost. Ridgway (1883a) gave its culmen length as 0.42 in (10.7 mm) which falls in the range of White Wagtail (10.2–13.2 mm) and outside that of Black-backed (11.4–14.7 mm; Howell 1990c). An adult ♂ White Wagtail wintered at San José del Cabo, BCS (Oct 1984–Feb 1985, JCA photo). Also four reports (Nov–Apr) of White or Black-backed wagtails: three from BCS (Wilbur 1987) and one from Son (Morlan 1981); see Appendix B.

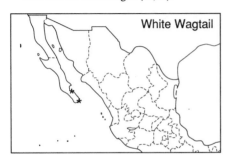

White Wagtail

RA: Breeds N Eurasia to W Alaska; winters S Eurasia and Africa.

NB: Black-backed Wagtail *M. (a.?) lugens* (7–7.5 in, 18–19 cm) breeds NE Asia, winters E Asia, has occurred as vagrant to W USA (see Appendix B). Habits, voice, etc., much like White Wagtail. Field identification criteria poorly known for 1st-year birds (see SS and SD above).

RED-THROATED PIPIT

Anthus cervinus   Not illustrated
Bisbita Gorjirrufa

ID: 5.5–5.7 in (14–14.5 cm). *Vagrant.* **Descr.** Ages differ. **Basic** (Sep–Mar): bill horn with dusky culmen, *legs flesh-pink.* Head and upperparts brown with pale supercilium, *crown and upperparts heavily streaked blackish,* paler stripes on each side of mantle form contrasting braces; wings dark brown with 2 whitish wingbars and pale edgings. Tail dark brown, outermost rectrices mostly white. *Throat and underparts pale buff* (faintly pinkish buff on some adults) *with dark malar stripe, chest and flanks boldly streaked blackish.* **Alt** (Feb–Aug): *face, throat, and chest pinkish cinnamon,* auriculars browner, dark malar stripe and streaking on chest reduced or lacking.
  Voice. A thin, high, exhaled, hissing to slightly wheezy *tsssi* or *teez*, also doubled to *pseeu-seeu.*
  Habs. Open country, most likely near the coast. Walks quickly, wags tail often. Flight strong and slightly undulating.
SS: American Pipit lacks bold dark streaking and pale braces above, legs dusky; note voice. Sprague's Pipit (mainly interior) has plain face, fine dark streaks below mainly on chest; note voice. Suspected Red-throated Pipits should be thoroughly checked to eliminate other Asian species, several of which are notoriously similar.

SD: Vagrant (Oct–Apr) to W Mexico: BCN (Oct 1991, Howell and Pyle 1993), BCS (Jan 1883, Ridgway 1883b), Col (Mar 1992, SNGH), and Mich (Apr 1988, Howell and Webb 1989a).
RA: Breeds N Eurasia to W Alaska; winters S Eurasia and N Africa.

AMERICAN [WATER] PIPIT

Anthus rubescens*   Figure 39
Bisbita Americana

ID: 6–6.2 in (15–16 cm). *Widespread winter visitor N of Isthmus.* **Descr.** Ages similar. **Basic** (Sep–Mar): bill blackish, flesh below at base, *legs dusky. Head and upperparts grey-brown to olive-brown with* pale lores and *whitish supercilium,* often a whitish subocular crescent; crown and back indistinctly streaked darker. Wings dark brown with 2 whitish to pale buff wingbars, tertials edged buff. Tail blackish, outer rectrices mostly white. Throat and underparts whitish to pale buff with dark malar stripe, dark streaking on chest and flanks. **Alt** (Mar–Aug): *head and upperparts greyer, underparts washed pinkish buff to pale cinnamon,* palest on throat, with dark streaking reduced or absent. **Juv:** resembles basic but with blurry dark spots on chest and back.
  Voice. A high, distinctive *sip-it* and *sip,* at times run into *si-si-si-sif* when flushed. Song given in descending display flight, a rich, fairly rapid, chipping *chiir-chiir* ... or *chwee-chwee* ...
  Habs. Open areas, often near water. Walks quickly, wags tail often; flight strong and slightly undulating. At times in flocks of 100+ birds. Nest cup of grasses in recess or under rock, grass clump, etc. Eggs: 4–6, pale grey, densely speckled with browns (WFVZ).

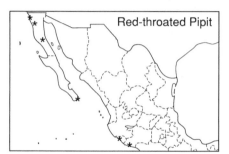
Red-throated Pipit

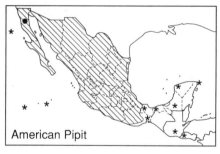
American Pipit

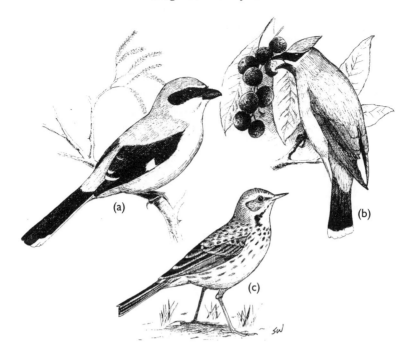

Fig. 39 (a) Loggerhead Shrike; (b) Cedar Waxwing; (c) American Pipit

SS: Sprague's Pipit occurs in grassy areas where rarely seen unless flushed; upperparts streaked buff and blackish, face plain with pale eyering, underparts whitish with neat band of fine dark streaks across chest, bill flesh with dusky tip, legs flesh-pink; note call.

SD: C to F transient and winter visitor (late Sep–mid-May; SL–3000 m) in Baja, on Pacific Slope in Son and locally to Col, in interior from NE Son and NL to cen volcanic belt, and on Atlantic Slope from Tamps to cen Ver; U to F to Oax, mainly in interior, R and irregular to Guatemala and NW El Salvador; U to R (late Oct–mid-Feb) in Yuc Pen (G and JC, BM). Vagrant (Nov–Feb) to Islas Revillagigedo and Clipperton. May breed in Sierra San Pedro Martír, BCN (Howell and Webb 1992*f*).

RA: Breeds N and W N.A. and NE Asia; winters in New World from S N.A. to Mexico, rarely to N M.A.

NB: Sometimes considered conspecific with the Rock/Water Pipit complex of Old World.

SPRAGUE'S PIPIT
Anthus spragueii　　　　Not illustrated
Bisbita de Sprague

ID: 5.8–6 in (14.5–15 cm). Winter visitor to N and cen Mexico. **Descr.** Ages/sexes similar. *Bill flesh with dusky culmen and tip, legs flesh-pink. Face plain buff, pale eyering accentuates big-eyed expression; crown, nape, and upperparts streaked buff and blackish.* Wings dark brown, edged buff. Tail dark brown, *outer rectrices mostly white. Throat and underparts*

Sprague's Pipit

whitish, finely streaked blackish across chest, chest and flanks often washed buff.

Voice. A slightly explosive, squeaky *speep* or *sweep*, often doubled, at times *speep beep beep* when flushed.

Habs. Open grassland. Usually not seen until flushed; if followed and relocated, may be watched at close range in fairly open short grass where runs quickly, at times freezes and hard to see.

SS: See American Pipit.

SD: F to C transient and winter visitor (Oct–early Apr; 1000–2500 m) in interior from NE Son and NL to Zac and SLP, U to cen volcanic belt; F to C (to near SL) on Atlantic Slope from Tamps to cen Ver, R to W Tab. Vagrant to S Gro (Mar 1990, Howell and Wilson 1990).

RA: Breeds interior W N.A.; winters S USA to cen Mexico.

WAXWINGS: BOMBYCILLIDAE (1)

One member of this small family of northern forest birds reaches the region in winter. Bill small and stubby, legs short. Wings fairly long and pointed, tail squared. Pointed crest conspicuous. Ages differ, sexes similar. First prebasic molt partial.

Food mostly fruit, also insects.

CEDAR WAXWING

Bombycilla cedrorum   Figure 39
Ampelis Americano

ID: 6–7 in (15–17.5 cm). Widespread winter visitor. **Descr. Adult:** bill and legs blackish. *Head and crest vinaceous cinnamon with black loral mask*, upperparts grey-brown, becoming grey on rump and uppertail coverts. Wings and tail darker, secondaries often with waxy red tips, *tail tipped yel-*

Cedar Waxwing

low. Chin blackish, throat and chest dusky vinaceous, belly lemon, undertail coverts whitish. **Imm:** duller, whitish underparts streaked dusky; resembles adult by mid-winter.

Voice. A high, sibilant, at times slightly trilled *sssir* or *ssssirr*, in flight and when perched.

Habs. Open woodland, forest edge, semi-open areas with scattered trees and bushes, parks, gardens. Often in flocks, at times to several hundred birds, which fly in compact formation recalling European Starlings. Usually seen at mid- to upper levels in fruiting trees.

SD: Irregularly C to F transient and winter visitor (Oct–early Jun; SL–3000 m) throughout most of region S to Honduras and N cen Nicaragua, but irregularly U to R in Yuc Pen (late Dec–early Mar) and in coastal lowlands S of N Mexico.

RA: Breeds N.A.; winters to Panama.

SILKIES: PTILOGONATIDAE (2)

Silkies, or Silky-flycatchers, are slender long-tailed birds with crests and small bills. This small Middle American family is closely related to waxwings and sometimes merged with that family. Ages/sexes differ; young altricial; 1st prebasic molt variable. Bills and legs blackish. Plumage soft and sleek ('silky') like that of waxwings; patterned mainly in greys, black, and white. Voices varied chatters, warbles, and mellow whistles.

Food fruit and insects. Nests are cups of fine plant materials and lichens saddled low to high in bush or tree.

GREY SILKY (SILKY-FLYCATCHER)
*Ptilogonys cinereus** Plate 52
Capulinero Gris

ID: 7.2–8.2 in (18.5–21 cm). Endemic to highland forests. *Short crest.* **Descr.** ♂: *head, crest, and upperparts blue-grey with dark lores, white eye-crescents* and supraloral stripe. Flight feathers black with *broad white band across base of outer rectrices* (basal half of tail looks white below). Throat and chest grey, flanks ochre-yellow becoming *bright yellow on undertail coverts,* belly whitish. ♀: *head and crest greyish, upperparts dusky vinaceous,* becoming grey on uppertail coverts. *Throat and underparts dull vinaceous with yellow undertail coverts,* whitish belly. **Juv:** resembles ♀ but belly washed yellow, quickly attains adult-like plumage.

Voice. Varied chattering and nasal, bickering calls, flight calls may suggest House Finch. A fairly dry *chi-che-rup che-chep,* a clipped, slightly nasal *k-lik* or *ch-pik* and *k-li-lik,* a clear, fairly sharp *chureet* or *chu-leep,* and *ch-tuk,* etc. Song a pleasant, fairly soft, warbled series of clucks with quiet, plaintive whistles thrown in.

Habs. Humid to semiarid pine–oak and pine–evergreen forest and edge, wanders to adjacent habitats in winter. In pairs or flocks, rarely to a few hundred birds, at mid- to upper levels. Often perches conspicuously atop tall trees, flies high and in loose flocks. Eggs: 2, bluish white, speckled and spotted with dark browns and greys (WFVZ).

SD/RA: C to F resident (1000–3500 m, lower locally in W Mexico, at least in winter) in interior and on adjacent slopes from Sin and S Coah to Guatemala, U N to Son (per GM). May withdraw from NW Mexico (N of cen Sin?) in winter (Phillips 1991).

PHAINOPEPLA
*Phainopepla nitens** Plate 52
Capulinero Negro

ID: 6.7–7.7 in (17–19.5 cm). Arid country of N Mexico. *Spiky crest conspicuous.* **Descr.** ♂: eyes reddish. *Glossy blue-black overall, broad white patch across base of primaries striking in flight.* ♀/**juv:** *slaty grey overall,* undertail coverts edged whitish. Wings and tail dark grey to blackish, edged whitish (edgings duller in juv), paler flash across base of primaries indistinct; eyes dull in

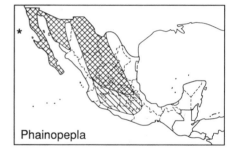

Grey Silky   Phainopepla

juv. **Imm ♂:** *intermediate between ♂ and ♀ in appearance*, wings and tail also variably mixed black and greyish. Attains adult plumage when about 1 year old.

Voice. A mellow *wheu* or *whiut*, also a slightly gruff *rreh*. Song a mellow, slightly gurgled, short warble, *tlee-oo-ee* or *teedl-ee*.

Habs. Arid to semiarid, open and semi-open country with scattered trees and bushes, scrub, riparian washes. Singly or in loose groups, often perches conspicuously atop bushes. Eggs: 2–4, whit-ish, densely speckled dark brown and grey (WFVZ).

SD: F to C breeder (Apr–Aug; SL–2400 m) in Baja, on Pacific Slope in Son and in interior from NE Son and NL S at least to N Gto and Hgo. Migratory and nomadic in some regions, winters/wanders (Sep–Apr?) to cen Mexico (Jal to W Ver, S to NW Oax). Some extralimital reports may refer to escaped cage birds and the status of this species in much of the region needs further study.

RA: Breeds SW USA to N Mexico; winters to cen Mexico.

SHRIKES: LANIIDAE (1)

Only one member of this mainly Old World family occurs in the region. Bill stout and hooked, legs fairly strong. Tail fairly long and strongly graduated. Ages differ, sexes similar; young altricial. First prebasic molt variable and often includes tail and tertials.

Varied diet includes mice, insects, and small birds. Nest is a bulky cup of twigs at low to mid-levels in bush or tree. Eggs: 2–6, greyish white, marked with browns and greys (WFVZ).

LOGGERHEAD SHRIKE
Lanius ludovicianus*   Figure 39
Lanio Americano

ID: 8.2–9 in (21–23 cm). Widespread N of Isthmus. **Descr. Adult:** bill and legs blackish. *Head and upperparts grey with broad black mask* narrowly bordered whitish above, uppertail coverts and outer scapulars often contrasting whitish. *Wings black with white-tipped tertials and secondaries, white band across base of primaries striking in flight.* Tail black, outer rectrices broadly tipped and edged white, appearing as white sides to spread tail. Throat and underparts whitish. **Juv:** head and upperparts washed brown, underparts finely scalloped dusky. Resembles adult by early winter.

Voice. A rasping, buzzy *jehrr* or *zzchrr*, often in series, a springy *bzzcheir* which may precede rasping series, a rough, mewing *meeahhr*, etc. Song a varied series of buzzy gurgles, rasps, and mews.

Habs. Open and semiopen country with scattered trees, bushes, etc. Singly or in pairs, perches prominently on wires or atop bushes on lookout for prey which is caught on the ground or in the air. Flight strong and fairly direct.

SS: See Northern Mockingbird.

SD: C to F resident (SL–3000 m) in Baja, on Pacific Slope from Son to N Sin and in interior and on adjacent slopes from Son and NL to Oax. Reports of breeding on Atlantic Slope of NE Mexico lack confirmation (Phillips 1986). F to C transient and winter visitor (late Aug–early Apr) to coastal lowlands from Sin to Nay, on Atlantic Slope from Tamps to S Ver, R and irregular (Nov–Feb) to Chis (SNGH photo, KK) and NW Guatemala (Ericsson 1981).

RA: Breeds N.A. to Mexico.

NB: Northern (Great Grey) Shrike *L. excubitor* (9.5–10.5 in, 24–26.5 cm) may occur as irregular R winter visitor (Nov–Feb) to extreme N Chih. Distinctive imms most likely to occur. Larger and usually paler than Loggerhead Shrike, longer and stouter bill pale at base, at least below. Lores pale, narrower mask (dusky, not black, in imm) sets off bold white eye-crescents, underparts scalloped dusky; upperparts often brownish, and flanks washed vinaceous brown in imm.

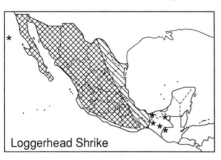

Loggerhead Shrike

STARLINGS: STURNIDAE (1)

One species of this Old World family was introduced into New York, USA, in 1890 and has spread across most of N.A. Why it has not (yet?) spread through Mexico is unknown. Stocky overall with pointed wings and shortish squared tail. Bill medium length and pointed. Ages differ, sexes similar; young altricial; 1st prebasic molt complete.

Food varied, mainly insects, also small invertebrates, fruits, seeds. Nests in cavities in buildings, trees, etc. Eggs: 2–8, pale blue (WFVZ).

EUROPEAN STARLING

Sturnus v. vulgaris   Not illustrated
Estornino Europeo

ID: 8–9 in (20.5–23 cm). N Mexico. **Descr. Adult:** bill dusky (autumn) to yellow (spring/summer), when ♂ has bluish base, ♀ pinkish base; legs dusky flesh. *Black overall, glossed purple on head, green on body, extensively spotted tawny-brown above, whitish below* (autumn/winter); wings and tail edged tawny-brown. Through winter, spots wear off to produce sleek glossy 'breeding' plumage by spring. **Juv:** grey-brown overall, throat whitish, belly indistinctly streaked whitish. Resembles adult by early winter.

Voice. Varied: often prolonged series of high, thin whines, warbles, and chattering gurgles may include mimicry of numerous other species. Calls include a slightly wet, buzzy to slightly rasping *shirr* or *shehrr*.

Habs. Open and semiopen areas, towns, and cities. Singly or in flocks which fly in compact formation. Feeds mainly on ground.

SS: Great-tailed Grackle and perhaps other icterids may drop their rectrices simultaneously in autumn molt and can be mistaken for Starlings but larger with longer heavier bills, different flight.

SD: F to C but local resident (SL–2500 m) in Mexico/USA border areas from BCN to Tamps, first recorded 1939 (N Tamps), reached BCN in early 1960s (Edwards and Morton 1963), S to N BCS and cen Sin in 1991 (Howell and Webb 1992*f*); also in Mexico City (first recorded 1983, Wilson and Ceballos-L. 1986). Irregular U to R winter visitor (Nov–Mar) in interior S to Ags (SNGH) and on Atlantic Slope to S Ver (Phillips 1991) and N Yuc Pen (SNGH, KK, ARP, PP). Early reports from Gto (Donagho 1965) and Ver (Coffey 1959) require verification. This species is offered for sale in cages in Oaxaca City (Dec 1989, SNGH) so odd extralimital reports could pertain to escapes.

RA: Eurasia; introduced to N.A.; N populations migratory.

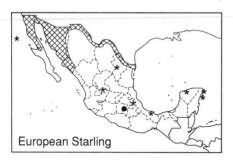

European Starling

VIREOS: VIREONIDAE (24)

This New World family consists mainly of fairly small birds that resemble heavily built warblers. Bills stout and hooked, legs stout. Ages similar or different, sexes usually similar; young altricial, most species soon resemble adults. Bills blackish and legs blue-grey in most typical vireos (genus *Vireo*). Plumages mainly colored in browns, greens, yellows, black, and white. Songs vary from rich warbles to repetition of simple phrases, calls typically include scolding chatters and mews.

Food mostly insects in summer, fruit in winter. As far as is known, nests are cups of plant material suspended in shallow forks of branches. Eggs of typical vireos: 2–5, white with sparse fine brown to reddish-brown flecking (unmarked white in Black-capped and Dwarf vireos).

SLATY VIREO

Vireo brevipennis　　　　　Plate 53
Vireo Pizarra

ID: 4.5–4.7 in (11.5–12 cm). Endemic to cen and SW Mexico. **Descr.** Ages differ. **Adult:** *eyes staring white*, bill black, legs dark grey. Head and body *slaty grey overall with yellow-green crown, white chin*, and white belly and undertail coverts. *Wings and tail* darker, broadly edged *yellow-green*. **1st basic:** head and upperparts olive, chest and flanks paler grey. **Juv:** eyes dull. Head and upperparts olive with trace of pale lemon spectacles, 2 buffy-lemon wingbars. Throat and underparts whitish, dusky at sides of chest.

Voice. A gruff, scolding *chichichi...* or *shehshehsheh ...* recalling House Wren. Song varied, rich, slightly burry phrases often beginning or ending with an emphatic note, *chik wi-di-weuw*, or *wheer chi-i-wik*, etc.; may suggest White-eyed Vireo.

Habs. Arid to semihumid brushy scrub, tangled thickets, forest edge. Skulking, moves sluggishly low through dense cover and overlooked easily, but sings from conspicuous perches; tail often cocked and twitched from side to side while foraging. Nests at low to mid-levels in tree or bush.

SD/RA: F to U resident (1200–3000 m) in interior, and locally on adjacent slopes, from Jal and W Ver to Oax.

NB: Formerly placed in the monotypic genus *Neochloe*.

WHITE-EYED VIREO

*Vireo griseus** 　　　　　Plate 53
Vireo Ojiblanco

ID: 4.3–4.7 in (11–12 cm). *E and S, mainly in winter*. Noticeable geographic variation. **Descr.** Ages differ. **Adult:** *eyes whitish*, bill blackish, legs blue-grey. *Head greyish to greyish olive with yellow spectacles* and dusky lores. Upperparts olive; wings and tail blackish brown with 2 *whitish wingbars* and whitish tertial edgings, flight feathers edged yellow-olive. *Throat and underparts whitish, washed grey on throat, yellow on sides*. **Juv:** eyes dusky, head olive with less distinct face pattern, wingbars

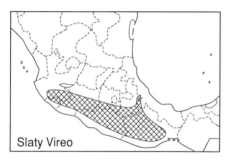

Slaty Vireo

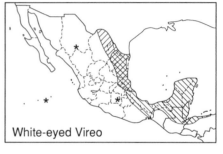

White-eyed Vireo

washed buff, underparts washed lemon. *V. g. perquisitor* (breeding in E Mexico) has *head and upperparts olive, underparts washed lemon;* wingbars and tertial edges may be washed lemon.

Voice. A nasal, scolding *sheh-sheh* ... often repeated insistently. Song, given all year, a varied short warble often beginning and/or ending with a sharp chip, *tchk iweedle-iwee chik,* or *chur di-wer chik,* etc.

Habs. Humid to semiarid brushy woodland, forest edge, second growth, semiopen areas with scrub and thickets. At low to mid-levels, less often in low canopy, moves sluggishly. Often joins mixed-species flocks. Nests at low to mid-levels in tree or bush.

SD: F to C breeder (Mar–Sep; near SL–1200 m) on Atlantic Slope from E Coah to SE SLP, *perquisitor* (near SL–1000 m, resident?) in N Ver and NE Pue. C to F transient and winter visitor (late Sep–early May; SL–1500 m) on Atlantic Slope from NL and Tamps to Yuc Pen; U to Honduras and on Pacific Slope in Isthmus. R transient W to DF (Wilson and Ceballos-L. 1986). Vagrant to Chih and to Isla Socorro (Feb 1988, Howell and Webb 1992g).

RA: Breeds E USA to NE Mexico; winters SE USA to N Nicaragua.

MANGROVE VIREO
Vireo pallens*   Plates 53, 71
Vireo Manglero

ID: 4.3–4.7 in (11–12 cm). *Pacific coast, Yuc Pen, and Caribbean islands.* Noticeable geographic variation. **Descr.** Ages differ slightly. **Adult:** eyes pale brownish to amber (rarely whitish), bill greyish, legs blue-grey. *Head and upperparts olive* to greyish olive *with yellow lores* and narrow supraorbital crescent. Yuc Pen birds (yellowish morphs at least, see below) may

also show narrow pale subocular crescent. Wings and tail dark brown with *2 whitish to pale lemon wingbars* and tertial edgings, flight feathers edged yellow-olive. *Throat and underparts pale lemon,* washed olive on sides of chest. *V. p. pallens* (Pacific coast of Honduras and Nicaragua) *olivegrey* to greyish olive *above* with pale lemon to buffy lemon lores and postocular spot, wingbars whitish; *throat and underparts dirty buffy white,* flanks and belly may be washed pale lemon. Atlantic Slope birds have color morphs (yellowish olive versus greyish olive) apparently absent in Pacific coast birds (Parkes 1991). **Juv:** facial pattern less distinct than adult.

Voice. Notably varied. A nasal to buzzy scolding, often rapid and persistent, *chi-chi-chi-chi* ... or *ji-ji* ... a slightly emphatic *tckrrr,* a drawn-out buzzy *tchrrrirrr,* and a drawn-out, mewing *j-weehr.* Song varied, a fairly rapid, nasal, twanging series of 3–12 (typically 4–6) slightly disyllabic notes, *j-wee'* ... or *bwie* ... or *chr-ei* ..., etc., at times with a single, often different introductory note, *jih j'iu-j'iu* ..., etc. Also, at least in Honduras Bay Islands, a complex series of buzzes, chips, and warbles, at times somewhat suggesting House Wren.

Habs. Along Pacific coast, restricted to mangroves. On Atlantic Slope, scrubby woodland, overgrown brushy fields, forest edge, second growth, mangroves. At low to mid-levels, singly or in pairs, associates loosely with mixed-species flocks. Nest and eggs undescribed (?).

SS: See White-eyed Vireo.

SD: F to C resident on Pacific coast from S Son to Nay and from Oax (SNGH, SW) to Honduras; apparent range break may simply reflect lack of observation. C to F resident (SL–250 m) on Atlantic Slope from Yuc Pen (including Isla Holbox) to S Belize (including Turneffe Islands) and NE Guatemala, on Honduras Bay Islands (Utila and Roatan), and in the Mosquitia.

RA: Mexico to NW Costa Rica.

COZUMEL VIREO
Vireo bairdi   Plate 69
Vireo de Cozumel

ID: 4.8–5 in (12–12.5 cm). *Endemic to Isla Cozumel.* **Descr.** Ages differ slightly. **Adult:** eyes pale brownish, *bill flesh with dusky tip,* legs blue-grey to flesh. *Head and*

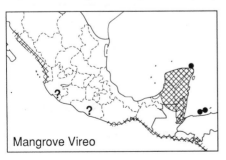

Mangrove Vireo

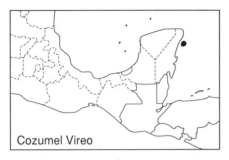

Cozumel Vireo

upperparts tawny-olive, greyer on head, with *whitish lores and broad, teardrop-shaped eyering*. Wings and tail blackish brown with 2 *whitish wingbars* and whitish tertial edgings, flight feathers edged yellow-olive. *Throat and underparts whitish, with cinnamon sides* of neck and chest. **Juv:** eyes dark; paler overall, lacks bright cinnamon sides.

Voice. A rapid, wren-like scolding *cheh-cheh* ... and a nasal *breeah* or *rreeew*. Song varied, a twangy, nasal *rreeu-rreeu* ... or *tchew* ... or *twee* ... 3–9×, recalls Mangrove Vireo, at times with an introductory note, *whee twee-twee* ..., etc.

Habs. Scrubby woodland and edge, second growth. Habits much as Mangrove Vireo. Nest and eggs undescribed (?).

SD/RA: C to F resident (around SL) on Isla Cozumel.

BELL'S VIREO
*Vireo bellii**   Plate 53
Vireo de Bell

ID: 4.3–4.7 in (11–12 cm). *Widespread, lacks bold spectacles.* **Descr.** Ages differ slightly. **Adult:** eyes brown, bill greyish to greyish flesh with dark culmen, legs blue-grey. *Head and upperparts greyish to olive* (greyer in W breeders) with *narrow whitish supraloral stripe and subocular*

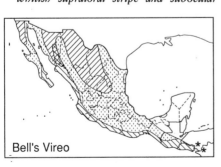

Bell's Vireo

crescent. Wings and tail blackish brown with *whitish lower wingbar*, less distinct upper wingbar may be lacking; tertials edged whitish, flight feathers edged yellow-olive. Throat and underparts whitish, washed lemon on underparts. **Juv:** head and upperparts grey-brown, throat and underparts whitish, wingbars washed buff.

Voice. A nasal scolding, often fairly rapid and insistent *sheh-sheh* ... or *chih-chih* ... Song a fairly rapid, jumbled, scratchy warble, often with mellow phrases thrown in.

Habs. Arid to semihumid scrub, brushy woodland, semiopen areas with bushes, scrub, hedges. At low to mid-levels, joins mixed-species flocks. Often holds tail cocked like a Gnatcatcher. Nests at low to mid-levels in tree or bush.

SS: Grey Vireo has short thick bill, complete white eyering.

SD: F to C breeder (late Mar–Aug; SL–1500 m) in NW Baja S to 30° N and (formerly?) in Rio Colorado delta, on Pacific Slope in Son, and in interior from Chih and Coah to N Zac and S NL; formerly on Atlantic Slope in N Tamps. F to C transient and winter visitor (late Aug–Apr; SL–1500 m) in BCS and on Pacific Slope from cen Son to Isthmus, U to F on Pacific Slope and in interior from Isthmus to El Salvador, R (to U?) to Nicaragua. F to U transient (Apr–mid-May, Sep) on Atlantic Slope N of Isthmus and in interior. May be R winter visitor on Atlantic Slope (AOU 1983).

RA: Breeds USA to N Mexico; winters Mexico to Nicaragua.

BLACK-CAPPED VIREO
Vireo atricapillus   Plate 53
Vireo Gorrinegro

ID: 4–4.4 in (10–11 cm). Breeds N Mexico,

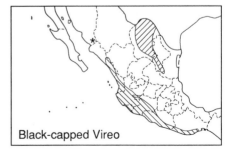

Black-capped Vireo

winters Pacific Slope. **Descr.** Ages/sexes differ. ♂: *eyes reddish*, bill blackish, legs blue-grey. *Head glossy black with white lores and eyering* broken over eye. Upperparts olive-green. Wings and tail blackish with 2 *lemon wingbars* and pale lemon tertial edgings, flight feathers edged yellow-olive. Throat and underparts whitish, sides and flanks washed pale olive-lemon. ♀/**imm:** eyes duller, amber in imm, *head slaty-grey* to bluish grey, upperparts olive, wingbars pale lemon. **1st alt ♂** (**Mar–Aug**): head mixed black and grey. **Juv:** resembles ♀ but eyes dull, head and upperparts greyish olive.

Voice. A dry chattering *dr-drt* or *dr-dr-drit*, much like Ruby-crowned Kinglet but drier, more staccato, repeated less persistently, and a low, gruff scolding *sherr sherr ...*, etc. Song varied: rich to slightly scratchy, hurried warbled phrases, often prolonged, and short, scratchy, warbled phrases, all much like Dwarf Vireo.

Habs. Arid to semiarid scrub, especially with oaks, in winter also humid brushy second growth and forest edge. Usually low, in tangled understory where overlooked easily. Nests at low to mid-levels in bush or scrubby tree.

SS: Most Dwarf Vireos have less contrasting greyish-olive head; note also whiter wingbars (not tinged lemon) offset by blacker greater coverts, duller loral area. Fresh basic adult ♂ Dwarf Vireo (at least N of Balsas drainage) has bluish-grey head much like dull ♀ Black-capped but cap suffused olive, contrasting less with back; possibly not always safely distinguished, but note that imm ♀ Black-capped can have amber eyes at least through Mar (reddish in adult ♂ Dwarf). Cassin's and Blue-headed vireos larger with dark line through lores, heavier bills.

SD: F but local breeder (Apr–Aug; 1000–2000 m) in interior from Coah to S NL, probably to W Tamps (JCA). U to F transient and winter visitor (late Aug–Apr; SL–1600 m) on Pacific Slope and locally in adjacent interior from Sin to Oax, R N to S Son (Phillips 1991). Presumably transient in intervening areas but few records.

RA: Breeds SW USA and N Mexico; winters W Mexico.

NB: Closely related to, and perhaps conspecific with, Dwarf Vireo.

DWARF VIREO
Vireo nelsoni Plate 53
Vireo Enano

ID: 4–4.4 in (10–11 cm). *Endemic to cen and SW Mexico.* **Descr.** Ages/sexes differ slightly. Eyes reddish (dull in juv), bill blackish, legs blue-grey. *Head and upperparts greyish olive* to olive, head often slightly greyer, *with pale grey lores and whitish eyering broken over eye.* Adult ♂ (at least in cen Mexico, N of Balsas drainage) has bluish-grey head suffused olive, most evident in fresh plumage. Head and upperparts in worn plumage dull greyish olive overall. Wings and tail blackish brown, 2 *whitish to creamy wingbars contrast with blackish greater coverts*, tertials edged whitish, flight feathers edged yellow-olive. *Throat and underparts whitish*, washed pale ochraceous lemon on sides and flanks.

Voice. Apparently not distinguishable from Black-capped Vireo. A short, dry *dri-dri-it* or *chi-chi chi-chi-chi*, etc., like Ruby-crowned Kinglet, a gnatcatcher-like mewing *meearr-meear*, and a scolding *cheh-cheh ...* or *jeh-jeh-jehr*. Song varied: rich to slightly scratchy, hurried warbled phrases, often prolonged, and short, scratchy, warbled phrases, *whee chi-a-wee wee-chi*, or *wee ch'wee wee chir'-awee*, or *wi chee'r ch wit*, etc.

Habs. Arid to semiarid scrub, often with oaks. Usually low and fairly skulking, singly or in pairs, but often sings from conspicuous perches. Nests at low to mid-levels in tree or bush.

SS: See ♀ Black-capped Vireo (perhaps not always safely told from ♂ Dwarf Vireo), Ruby-crowned Kinglet. Chunkier Hutton's Vireo (highland forests) has dingier underparts; note voice. Cassin's Vireo larger with different facial expression, larger bill.

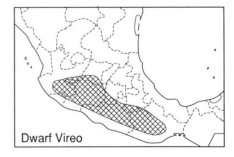

Dwarf Vireo

SD/RA: F to U but local resident (1000–2500 m) in interior from Jal and S Gto to Oax. N breeders may move to Pacific Slope in winter(?).

NB: See Black-capped Vireo.

GREY VIREO
Vireo vicinior   Not illustrated
Vireo Gris

ID: 5–5.5 in (12.5–14 cm). *N Mexico. Bill notably short and thick.* **Descr.** Ages differ slightly. **Adult:** bill blackish, legs blue-grey. *Head and upperparts grey with whitish lores and complete white eyering.* Wings and tail darker with whitish lower wingbar and less distinct upper wingbar, flight feathers edged whitish. Throat and underparts whitish, washed grey on sides. **Juv:** wingbars and tertial edgings washed buff.

Voice. A low, rasping, scolding *jerr* and *jerr-jerr-jerr-jerr*, and a harder, rough *cherr cherr* ... or *chierr chierr* ..., etc. Song hesitant and slightly jerky, recalls Solitary Vireo but faster, *ch-ree' ch-ruh chee-r ch-ree* ... or *ch-ree chew, ch-weet chrew* ..., etc., often repeated tirelessly.

Habs. Breeds in chaparral, brushy scrub in pinyon–juniper woodland; in winter also in semiopen areas with scattered scrub, semiopen brushy woodland. At low to mid-levels, sluggish. Often cocks tail and swings it loosely like a gnatcatcher. Nests at low to mid-levels in bush or tree.

SS: Plumbeous Vireo more heavily built, shorter tail not cocked, bold white spectacles distinctive, also 2 bold wingbars. See Bell's Vireo.

SD: F to C but local breeder (late Mar–Aug; 1000–1600 m) in NW Baja, and in Sierra del Carmen, Coah. F to U transient and winter visitor (Sep–Apr; near SL–1500 m) in BCS and Son, U and local on N Plateau (SNGH, SW).

RA: Breeds SW USA to N Mexico; winters mainly in N Mexico.

BLUE-HEADED [SOLITARY] VIREO
Vireo s. solitarius   Not illustrated
Vireo Solitario

ID: 5–5.7 in (12.5–14.5 cm). Winters in E and S. **Descr.** Ages similar. Bill blackish, legs blue-grey. *Bold white spectacles, eyering broken in front of eye* by dusky lores. *Blue-grey head* (washed olive when fresh) contrasts with olive to greenish-olive upperparts; wings and tail blackish brown with *2 broad white* to pale lemon *wingbars* and narrower tertial edgings, secondaries edged olive. Throat and underparts whitish, washed olive on sides, lemon on flanks and vent.

Voice. An accelerating, scolding chatter, *sheh, cheh-cheh-cheh-cheh-cheh* ... or *shih, ch-chi-ch-chi-ch-chi-ch-chi*, etc., much like Yellow-throated Vireo, less often a single *shehr* or *sheihr*, and short series *shiehr cheh-cheh*, etc.

Habs. Woodland and forest in general, but usually avoids SE humid forest. At mid- to upper levels, less often low; sluggish like most vireos. Often joins mixed-species flocks.

SS: Cassin's Vireo lacks contrasting blue-grey head, bill more slender. See Black-capped and Dwarf vireos.

SD: F to C transient and winter visitor (late Sep–mid-May; SL–2500 m) on both slopes from Gro and E Coah, and in interior from Oax, to N cen Nicaragua, R to Mich and Mor and (Nov–Mar) in Yuc Pen.

RA: Breeds N and E N.A.; winters SE USA to C.A.

NB: Some authors consider Blue-headed, Cassins, and Plumbeous vireos as separate species. Studies including all M.A. populations are needed.

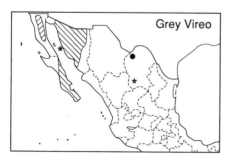

Grey Vireo

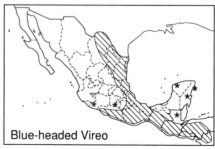

Blue-headed Vireo

CASSIN'S [SOLITARY] VIREO
Vireo (solitarius?) cassini*
Vireo de Cassin   Plate 53

ID: 5–5.5 in (12.5–14 cm). Breeds Baja, winters W Mexico. **Descr.** Ages similar. Bill blackish, legs blue-grey. *Bold white spectacles, eyering broken in front of eye* by dusky lores. *Head greyish*, washed olive when fresh, *upperparts greyish olive to olive*. Wings and tail blackish brown with *2 broad white* to pale lemon *wingbars* and narrower tertial edgings, secondaries edged greyish olive. Central rectrices edged olive, outer rectrices edged whitish. Throat and underparts dingy whitish, washed olive on sides, lemon to yellow-olive on flanks.

Voice. Song varied: a jerky series of rich to slightly burry phrases, *chreu ch'ree choo'reet* or *ch-ree ch-ri'chi-roo*, or *ch-ree ree-e-eu, ree-u-yuh, ree-uh* ..., often repeated tirelessly; less often, phrases may be repeated steadily, *reer chu, reer chu, reer chu* ..., or *ree-uh ree-uh* ..., etc. Calls like Blue-headed Vireo.

Habs. Breeds in pine–oak forest, also thorn forest in BCS. Winters in most woodland and forest habitats. Habits as Blue-headed Vireo. Nests low to high in tree or bush.

SS: See Blue-headed, Black-capped, and Dwarf vireos. Plumbeous Vireo has thicker bill, N birds grey and whitish overall. Hutton's Vireo smaller, face pattern different.

SD: F breeder (Apr–Aug; 1000–2400 m) in BCN, resident (breeds 400–2500 m) in Cape District of BCS. F to C transient and winter visitor (late Aug–May; SL–3000 m) on Pacific Slope from Son and in interior from Dgo and S NL to Isthmus.

RA: Breeds W N.A. to Baja; winters SW USA to Mexico.

NB: See NB under Blue-headed Vireo.

PLUMBEOUS [SOLITARY] VIREO
Vireo solitarius (in part) or *V. plumbeus**
Vireo Plomizo   Plate 53

ID: 4.8–5.8 in (12–14.5 cm). Widespread. Two groups: *plumbeus* group (N of Isthmus) and *notius* group (S of Isthmus). **Descr.** Ages similar. *Plumbeus*: larger. Bill blackish, legs blue-grey. *Bold white spectacles, eyering broken in front of eye* by dusky lores. *Head and upperparts grey* becoming olive-grey on rump; wings and tail blackish brown with *2 broad white wingbars* and narrower tertial edgings, secondaries edged olive-grey; rectrices edged whitish. *Throat and underparts whitish*, washed dusky on flanks where may show olive-lemon wash. *Notius*: smaller. Plumage much like Cassin's Vireo, not grey and white like N birds.

Voice. Song varied: hesitant, scratchy, and nasal phrases much like Cassin's Vireo, *chureeh, ch-ireet', ch-reeh ch-ireet*, etc., often repeated tirelessly. Calls much like Blue-headed Vireo.

Habs. Pine–oak forest, more widely in winter to other wooded habitats. Habits, nest, and eggs as Cassin's Vireo.

SS: See Cassin's and Grey vireos.

SD: F to U but local breeder (Apr–Sep; 1500–3000 m) in interior and on adjacent Pacific Slope from Son to Nay, resident (?) locally in interior and on both slopes from Mich and Hgo to El Salvador and Honduras; disjunctly in S Belize. Also F to U transient and winter visitor (late Aug–Apr; SL–2500 m) in Baja (winters in S, SNGH) and on Pacific Slope from S Son, and in interior from cen Mexico, to Oax.

RA: Breeds W USA to N C.A.; winters Mexico to N C.A.

NB: See NB under Blue-headed Vireo.

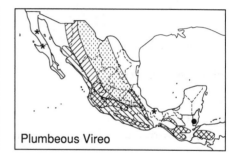

Cassin's Vireo

Plumbeous Vireo

YELLOW-THROATED VIREO

Vireo flavifrons Not illustrated
Vireo Gorjiamarillo

ID: 5–5.5 in (13–14 cm). *Winter visitor to E and S*. **Descr.** Ages similar. Bill blue-grey with dark culmen, legs blue-grey. *Head and upperparts olive-green, with bold yellow spectacles, grey rump and uppertail coverts.* Wings and tail blackish brown with *2 broad white wingbars* and whitish tertial edgings, flight feathers edged pale grey. *Throat and chest yellow,* washed olive at sides of chest, *belly and undertail coverts white,* flanks washed grey.
Voice. A nasal, often accelerating, scolding series, *sheh cheh-cheh-cheh ... or shi chi-chi-chi-chi-chi,* recalling Blue-headed Vireo, and a buzzy, rasping *kzzchik,* repeated. Song of leisurely, rich, often slightly hesitant, varied phrases, *ch-i-ree chr-eu, chi-wi, chu-ee-u chi chu-ee-u-ree ...,* etc.
Habs. Woodland, forest, plantations. Habits much as Blue-headed Vireo.
SS: Pine Warbler (R in NE Mexico in winter) has slender pointed bill, forked tail, lacks bold spectacles; note white tail-spots.
SD: F to U transient and winter visitor (late Aug–Apr; SL–1800 m) on both slopes, and locally in interior, from S Ver and E Oax to Honduras, R to U N to S Tamps; F transient (Apr, late Aug–Oct) on Atlantic Slope N of Isthmus. Vagrant (Nov–Jan) to Nay (Clow 1976; PP). Increasing in S Texas and may breed in NE Coah (Van Hoose 1955).
RA: Breeds E N.A.; winters S Mexico to N S.A.

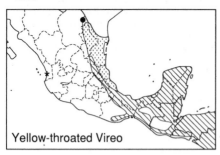

Yellow-throated Vireo

HUTTON'S VIREO

Vireo huttoni* Plate 53
Vireo de Hutton

ID: 4.5–5 in (11.5–12.5 cm). *Highland forests*. **Descr.** Ages differ slightly. **Adult:**

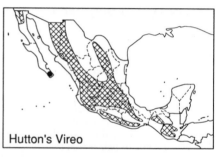

Hutton's Vireo

bill greyish, legs blue-grey. Head and upperparts greyish olive with *pale lores and whitish eyering broken above eye.* Wings and tail blackish brown with 2 whitish wingbars and whitish tertial edgings, flight feathers edged yellow-olive. *Throat and underparts dusky olive,* paler buffy lemon on belly and undertail coverts. *V. h. cognatus* of BCS paler overall, greyish above, whitish below, washed dusky on throat and chest. **Juv:** head and upperparts washed brownish.
Voice. A scolding *jehr or rreah,* often followed by a rapid, descending nasal laugh, *rrehr, heh-heh-heh-heh;* also a scolding *rreeah,* often 2–3× in descending series. Song a monotonously repeated *ch'ree* or *ch'weesu* or *siree,* etc., 10/8–12.5 s.
Habs. Pine–oak and oak forest and woodland. At mid- to upper levels, relatively active. Often joins mixed-species flocks. Nests low to high in trees.
SS: See Ruby-crowned Kinglet, Dwarf and Cassin's vireos.
SD: C to F resident (1200–3500 m, locally to near SL in NW Mexico in winter) in Baja, and in interior and on adjacent slopes of Son and Coah to Guatemala. A report from Belize (Russell 1964) is in error (see Phillips 1991).
RA: W N.A. to Guatemala.

GOLDEN VIREO

Vireo hypochryseus* Plate 53
Vireo Dorado

ID: 4.7–5.2 in (12–13 cm). Endemic to W Mexico. **Descr.** Ages differ slightly. **Adult:** bill flesh-grey, legs blue-grey. *Head and upperparts olive-green to olive with broad yellow supercilium. Throat and underparts yellow.* **Juv:** crown and upperparts washed brown, underparts paler.
Voice. An accelerating, nasal scolding

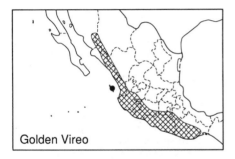

Golden Vireo

chih cheh-chehchehchehcheh, or *chih chih-chih chehchehchehcheh*, etc., a drawn-out scolding *shehh*, often repeated steadily, a dry *chk or chik*, recalling Wilson's Warbler. Song a rapid series of bright notes, the last usually sharply inflected upwards, *whee-whee-whee-wheet* or *jujujujujeet*, at times preceded and followed by a sharp chip, *tsik, deeu-deeu-deeu-deeu-deeu-deeu tik*, and locally varied to a descending series suggesting Ivory-billed Woodcreeper; also a quiet, scratchy warble.

Habs. Arid to semihumid scrub, thorn forest, woodland, forest edge, plantations. Low to high, joins mixed-species flocks. Tail often cocked, raises crown when agitated. Nests at mid-levels in trees.

SS: Wilson's Warbler smaller and slimmer, with slender bill and flesh legs, large beady eye, lacks such a distinct broad supercilium, most have some black on crown.

SD/RA: C to F resident (near SL–1900 m) on Pacific Slope from S Son to Oax, and in interior from Jal to Mor, U to Oax.

WARBLING VIREO
*Vireo gilvus** Plate 53
Vireo Gorjeador

ID: 4.7–5.2 in (12–13 cm). *Mainly Pacific Slope and interior.* **Descr.** Ages differ

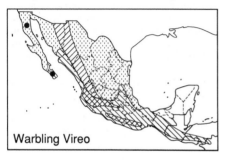

Warbling Vireo

slightly. **Adult:** bill greyish with dark culmen, legs blue-grey. *Broad whitish supercilium contrasts with greyish crown and dusky auriculars,* paler subocular crescent. *Upperparts greyish olive. Throat and underparts whitish,* washed lemon on flanks. Many winter birds in S Mexico and N C.A. (migrants from E N.A.?) are notably brighter than most W Mexican winter birds, with back washed olive, flanks and undertail coverts lemon. **Juv:** duller, with pale cinnamon wingbars.

Voice. A harsh, nasal, up-slurred scolding *rreih* or *nyeeah*, often repeated insistently, and a dry, clipped *ch* which may be run into short chatters, *ch-ch-ch* … Song a rich, slightly scratchy warble of about 1.5–2.5 s duration.

Habs. Humid to semiarid mixed woodland, in winter widely in woodland, scrub, plantations. Feeds low to high, a prominent member of mixed-species flocks in W Mexico. May occur in loose flocks during migration. Nests at mid- to upper levels in trees.

SS: Philadelphia Vireo has dark eyestripe, yellow wash on throat and chest. Yellow-green Vireo (compare with bright winter Warbling) has larger bill, whitish supercilium with indistinct dark border above. Tennessee Warbler has slender pointed bill, upperparts brighter olive, throat and chest often washed yellow. Brown-capped Vireo (cloud forest) has brown crown, darker olive-brown upperparts.

SD: F to C breeder (Apr–Aug; 900–2500 m) in Baja, on Pacific Slope from Son to Oax, and in interior E to Mor. C to F transient and winter visitor (mid-Aug–mid-May; SL–3000 m) on Pacific Slope from S Son, and in interior from cen volcanic belt, to Guatemala and El Salvador, U to Honduras, R to W Nicaragua; U to R in winter on Atlantic Slope from SE SLP to Chis. C to F transient (Aug–Oct, Mar–mid-May) in NW Mexico, U to F transient (Apr–early May, Sep–Oct) on Atlantic Slope N of Isthmus. Reports from N Guatemala (Beavers 1992) and of breeding in NE Mexico (Phillips 1991) require verification.

RA: Breeds N.A. to Mexico; winters W Mexico to Nicaragua.

NB1: Brown-capped Vireo sometimes considered conspecific with Warbling Vireo, while other authors suggest that two species of Warbling Vireo breed in N.A. (Sibley and Monroe 1990).

BROWN-CAPPED [WARBLING] VIREO
Vireo leucophrys* or *V. gilvus* (in part)
Vireo Gorripardo   Plate 53

ID: 4.5–4.7 in (11.5–12 cm). *Cloud forest.* **Descr.** Ages differ slightly. **Adult:** bill greyish with dark culmen, legs blue-grey. *Broad whitish supercilium contrasts with brown crown and dusky auriculars with paler subocular crescent. Upperparts brownish olive. Throat and underparts whitish,* washed lemon on flanks. **Juv:** duller, with pale cinnamon wingbars.
Voice. Rarely distinguishable from Warbling Vireo. A nasal up-slurred *meihk* or *rreih,* often given 2–5× in fairly rapid succession. Song a rich to slightly scratchy warble of about 1.5–2.5 s duration.
Habs. Humid evergreen forest and adjacent woodland. At mid- to upper levels, often joins mixed-species flocks. Nest and eggs undescribed (?), probably like Warbling Vireo.
SS: See Warbling Vireo. Philadelphia Vireo has dark eyestripe, yellow wash on throat and chest, greyish crown. Tennessee Warbler has slender pointed bill, upperparts brighter olive, throat and chest often washed yellow, crown olive to grey.
SD: C to F resident (1200–2000 m) on Atlantic Slope from S Tamps and E SLP to Chis, thence in interior and on adjacent slopes to Honduras.
RA: E Mexico to NW Bolivia.
NB: See Warbling Vireo.

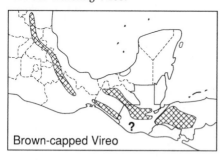

Brown-capped Vireo

PHILADELPHIA VIREO
Vireo philadelphicus   Not illustrated
Vireo de Filadelfia

ID: 4.5–4.7 in (11.5–12.5 cm). *Migrant in E and S.* **Descr.** Ages similar. Bill greyish with dark culmen, legs blue-grey. *Broad whitish supercilium contrasts with grey crown and dark eyestripe,* auriculars dusky. Upperparts olive. *Throat and underparts pale lemon,* becoming whitish on belly, or lemon dull and mostly restricted to throat and chest.
Voice. Rarely gives a low nasal *rreh* or *reh,* at times doubled.
Habs. Humid to semiarid forest and edge, plantations. At mid- to upper levels, joins mixed-species flocks.
SS: See Warbling Vireo. Tennessee Warbler has slender pointed bill, upperparts brighter olive. Yellow-green Vireo (summer visitor) larger with heavier bill, throat and median underparts whitish.
SD: U to F transient and winter visitor (Sep–mid-May: SL–2000 m) on Atlantic Slope from S Ver (not casual, as stated by AOU 1983) to Honduras, on Pacific Slope from Chis to El Salvador. U to F transient (Apr–mid-May, Sep–Oct) on Atlantic Slope N of Isthmus, R in spring W to Gro. Vagrant (Oct–Dec) to BCN (Howell and Pyle 1993) and Nay (Clow 1976).
RA: Breeds E N.A.; winters S Mexico to N Colombia.

Philadelphia Vireo

YELLOW-GREEN [RED-EYED] VIREO
Vireo flavoviridis*   Plate 53
Vireo Amarillo-verdoso

ID: 5.5–6 in (14–15 cm). *Summer visitor to tropical lowlands.* **Descr.** Ages differ slightly. **Adult:** eyes reddish, bill greyish with dark culmen and pale base below, legs blue-grey. Pale grey to *whitish supercilium (edged darker above) contrasts with grey crown and pale olive auriculars,* lores dusky. Upperparts olive. *Throat and underparts whitish, washed lemon on sides and undertail coverts.* **Juv:** eyes dark, duller overall, upperparts tinged brownish, face pattern less distinct.
Voice. A soft, dry chatter and a rough

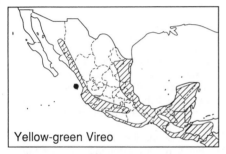

Yellow-green Vireo

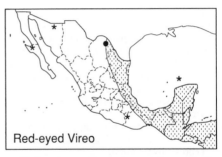

Red-eyed Vireo

mewing *rreh* or *rrieh*, repeated steadily at times. Song of varied rich to nasal chirps given in jerky, hesitant manner, *ch-ree chree swi ch-ree chree chree* ..., often repeated tirelessly, suggests House Sparrow.
Habs. Forest and woodland, including riparian woodland, plantations, scrubby forest edge, not breeding in heavy forest. At mid- to upper levels, fairly sluggish. Sometimes cocks tail and raises crown when agitated. In flocks during migration. Nests at mid-levels in trees.
SS: Red-eyed Vireo has blackish lores and black border above whiter supercilium, giving different facial expression, underparts whitish. Black-whiskered Vireo lacks obvious dark border above whitish supercilium, dark eyestripe more contrasting, underparts mainly whitish; note dark malar stripe. See Philadelphia and Warbling vireos.
SD: C to F summer resident (late Mar–mid-Oct, mainly Apr–Sep; SL–1500 m) on both slopes from Son and Tamps, and in interior from Balsas drainage, to Honduras and W Nicaragua. Occasional winter reports (Nov–Feb), mostly from Yuc Pen and Belize, require verification.
RA: Breeds Mexico (also S Texas) to Panama; winters W Amazonia.

RED-EYED VIREO
Vireo olivaceus   Not illustrated
Vireo Ojirrojo

ID: 5.5–6 in (14–15 cm). *Transient in E and S.* **Descr.** Ages similar. Eyes red (dull in imm), bill grey with dark culmen, legs blue-grey. *Whitish supercilium (edged black above) contrasts with grey crown and pale olive-brown auriculars, lores blackish.* Upperparts olive. *Throat and underparts whitish,* washed lemon on undertail coverts, flanks may be washed lemon in autumn.

Voice. A rough mewing *rrieh,* much like Yellow-green Vireo; mostly silent during migration. Song of varied mellow phrases given in fairly continuous flow, *ch'ree ch-uh-ree ch-t-ree ch-u-ree tch-ree* ..., suggests American Robin.
Habs. Forest, woodland and edge, plantations. Habits much as Yellow-green Vireo.
SS: See Yellow-green Vireo. Black-whiskered Vireo has larger bill, lacks obvious dark border above whitish supercilium, often shows broad dusky eyestripe, note dark malar stripe.
SD: C to F transient (late Mar–May, mid-Aug–early Nov; SL–1500 m) on Atlantic Slope from NL and Tamps to Honduras; U in interior and on Pacific Slope from Oax S. Apparently a local summer resident (May–Aug?) in NE Coah (Urban 1959). Vagrant to BCS (Howell and Webb 1992*f*).
RA: Breeds N.A.; winters Amazonia.

BLACK-WHISKERED VIREO
Vireo altiloquus   Not illustrated
Vireo Bigotudo

ID: 5.5–6 in (14–15 cm). *Transient on Caribbean coast and islands.* Bill large and heavy. **Descr.** Ages similar. Eyes amber (dull in imm), bill greyish with dark culmen, legs blue-grey. *Whitish supercilium*

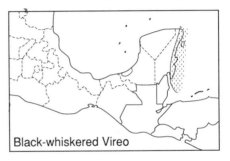

Black-whiskered Vireo

(with faint darker edge above) contrasts with grey crown and broad dusky eye-stripe, auriculars pale brownish, bordered below by *narrow dark malar stripe*. Upperparts olive-brown. Throat and underparts whitish, washed dusky olive on sides, lemon on undertail coverts.

Voice. Mewing call much like Red-eyed Vireo (DAS).

Habs. Woodland and edge, scrub, mangroves. Low to high, habits much as Yellow-Green Vireo.

SS: See Red-eyed and Yellow-green vireos. Yucatan Vireo has heavier bill, brownish eyes, lacks dark malar stripe, broad creamy supercilium and dark eyestripe more contrasting, upperparts greyer.

SD: Probably R to U transient (mid-Mar–Apr?, Sep–Oct?) on offshore Caribbean islands and adjacent coast. Three records to date: NE QR (Apr and Sep 1978, BM), and Half Moon Cay, Belize (Mar 1926, Russell 1964).

RA: Breeds Caribbean; winters Amazonia.

YUCATAN VIREO

Vireo m. magister Plate 53
Vireo Yucateco

ID: 5.7–6.2 in (14.5–15.5 cm). *Caribbean coast and islands. Bill large and heavy.* **Descr.** Ages similar (?). Eyes brownish, bill grey, paler below at base, legs blue-grey to blue. *Broad whitish supercilium contrasts with greyish crown and broad dark eyestripe. Upperparts greyish olive*, wings and tail edged yellow-olive. *Throat and underparts whitish, often tinged buff*, with dusky auriculars, sides, and flanks. **Juv:** undescribed (?).

Voice. A soft, dry chatter, *shi tchi-chi-chi-chi* ..., similar to Yellow-Green Vireo, and a sharp, slightly nasal *beenk* or *peek*, and *peenk, peenk*, etc. Song of varied rich phrases given in jerky, hesitant manner,

phrases often repeated 2×, *chu-ree' chu-i-chu ch-weet ch'ee chu chu ch'oo choo-choo* ..., etc. More leisurely and mellower than Yellow-Green Vireo, may suggest a soft, leisurely mockingbird.

Habs. Humid to semihumid scrubby woodland and edge, mangroves. Low to high, habits much as Yellow-green Vireo. Nests at low to mid-levels in tree or bush.

SS: See Black-whiskered and Red-eyed vireos.

SD: F to C resident (around SL, mostly along coast) from E QR S to Ambergris Cay, on Islas Cozumel and Mujeres, on Belize Cays (Cay Corker, Turneffe Islands, Lighthouse and Glover's reefs), and on Honduras Bay Islands; U on inshore Belize Cays and along coast S to cen Belize.

RA: Caribbean coast and islands (see above), also Grand Cayman.

TAWNY-CROWNED GREENLET

Hylophilus o. ochraceiceps Plate 53
Verdillo Corona-leonada

ID: 4.6–5 in (11.5–12.5 cm). *SE humid forest.* **Descr.** Ages differ. **Adult:** *eyes pale grey, bill greyish, paler below, legs flesh. Tawny crown contrasts with greyish face* and olive-brown upperparts; wings and tail edged cinnamon-brown. Throat greyish, chest ochraceous, belly pale lemon, flanks washed dusky. **Juv:** eyes dull, head and upperparts vinaceous brown, throat white, underparts ochre, quickly attains adult-like plumage.

Voice. A nasal *dwoi-dwoi-dwoi-dwoi* or *dwoi-h' dwoi-h'* ..., a clear, plaintive *tew-hee* or *wee-u'*, and a more drawn-out *teeeeu*, etc. Song a high, plaintive, insect-like whining *wheee* or *swee*, repeated, 10/20–25 s, often only given 2–3×, may be preceded by a short, liquid trill or chortling chatter.

Habs. Humid evergreen and semi-

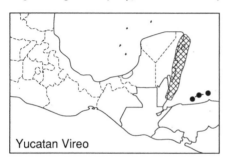

Yucatan Vireo

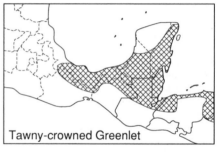

Tawny-crowned Greenlet

deciduous forest. At low to mid-levels in forest, moves actively in pairs or small groups, often with mixed-species flocks, especially with Golden-crowned Warbler and Red-crowned Ant-Tanager. Crown typically raised and bushy. Nests at low to mid-levels in trees. Eggs: 2, white, marked with purplish and greys (WFVZ).

SS: See ♀ Plain Antvireo.

SD: C to F resident (SL–1200 m) on Atlantic Slope from S Ver to Honduras, including S and E Yuc Pen (*contra* AOU 1983).

RA: SE Mexico to Ecuador and Brazil.

LESSER GREENLET
Hylophilus d. decurtatus   Plate 53
Verdillo Menor

ID: 4–4.4 in (10.5–11.5 cm). *Humid forests of E and S.* Descr. Ages differ. **Adult:** bill grey above, flesh below with dark tip, legs greyish. *Head grey with whitish lores and white eyering. Upperparts olive-green. Throat and underparts whitish*, washed yellow-olive on flanks, lemon on undertail coverts. **Juv:** head vinaceous-grey-brown, quickly resembles adult.

Voice. A nasal, often persistent scolding *chih-chih ...* or *yiih yiih ...*, and a quiet, dry, rolled *chri-i-ih* or *chri chi-chi.* Song a repetition of 2–3-note phrases, repeated for a while then often changed to another phrase, *wi-chi-wee*, or *si-chi-vee*, or *chree-chee*, or *chir-i-chee*, etc., rarely run into a more prolonged warbling.

Habs. Humid evergreen and semideciduous forest. At mid- to upper levels, typically in canopy, in pairs or small groups, often joins mixed-species flocks. Nest at mid-levels in trees. Eggs: 2, white, marked with pale brown.

SS: Nashville Warbler has slender pointed bill, throat and chest yellowish.

SD: C to F resident (SL–1500 m) on Atlantic Slope from S Ver to Honduras, U N to SE SLP; F to U resident on Pacific Slope from E Oax to Honduras

RA: E Mexico to Ecuador.

NB: M.A. forms formerly considered specifically distinct, Grey-headed Greenlet.

CHESTNUT-SIDED SHRIKE-VIREO
*Vireolanius melitophrys**   Plate 53
Vireón Pechicastaño

ID: 6.5–7 in (16.5–18 cm). *Endemic; highlands from cen Mexico to Guatemala.* Descr. Ages/sexes differ. ♂: eyes whitish (greyish in imm), bill blackish, legs flesh. *Head boldly marked with yellow supercilium, broad black eyestripe, white auriculars, and blackish malar stripe.* Crown and nape grey, upperparts olive-green, tail tipped whitish. *Throat and underparts whitish with chestnut band across chest extending down flanks.* ♀: face duller (yellow supercilium paler, eyestripe slaty grey), chest band generally narrower and sides less extensively chestnut (often with only a faint wash). **Juv:** duller, crown washed brownish, face pattern less distinct, throat whitish, underparts washed pale vinaceous cinnamon without distinct chest band or stronger wash on sides.

Voice. A mewing scream, *rreeeah*, which may run into a scolding chatter, *rreah cha-cha-cha-cha*, etc. Song a drawn-out, burry to slightly plaintive, far-carrying mew, *wheeeu* or *wheeeurr*, 10/20–25 s, suggests a hawk.

Habs. Humid to semiarid pine–oak and oak woodland, evergreen forest. Forages high to low, mainly in canopy; joins mixed-species flocks. Sluggish and overlooked easily. Nests at mid-levels in trees. Eggs undescribed (?).

SD/RA: F to U resident (1200–3500 m) on both slopes and in interior from Jal and E

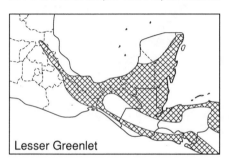

Lesser Greenlet

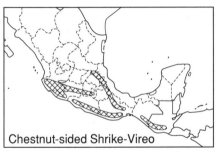

Chestnut-sided Shrike-Vireo

SLP to Oax, and on Pacific Slope in Chis and Guatemala.

NB: Sometimes known as Highland Shrike-Vireo.

GREEN SHRIKE-VIREO
Vireolanius pulchellus*   Plate 53
Vireón Esmeraldo

ID: 5.3–5.8 in (13.5–14.5 cm). Humid SE forest. **Descr.** Ages/sexes differ. ♂: bill black above, grey below, legs grey. Bright green overall with turquoise-blue crown, yellow throat, underparts washed yellowish. ♀: duller, yellow throat less contrasting. *V. p. verticalis* of W Nicaragua has blue on head restricted to forehead and nape. **Juv:** head and upperparts olive-green with dirty pale lemon supercilium and moustache, indistinct dirty pale lemon wingbars; throat and underparts dull yellowish.

Voice. Commonly heard song a monotonously repeated, chanting *chew chew chew chew* or *chewy chewy chewy*, either phrase repeated 3–5×; may suggest a Titmouse. Calls include a hard rasping scold, *djehr djehr . . .*

Habs. Humid evergreen forest and edge. Typically in canopy where overlooked easily and hard to see when singing. Joins mixed-species flocks ranging down to understory. Nests at mid-levels in trees. Eggs undescribed (?).

SS: Blue-crowned Chlorophonia chunkier, stub-tailed, with shorter bill, ♂ has black chest band and yellow belly, ♀ has greenish throat.

SD: C to F resident (SL–1800 m) on Atlantic Slope from S Ver to Honduras, R to cen QR (JKi tape). Local resident (500–1000 m) on Pacific Slope from Chis (and SE

Oax?) to El Salvador (Land 1970; Thurber *et al.* 1987; Phillips 1991).

RA: S Mexico to Panama.

NB: Often considered conspecific with *V. eximius* of E Panama and N S.A.

RUFOUS-BROWED PEPPERSHRIKE
Cyclarhis gujanensis*   Plates 53, 69
Vireón Cejirrufo

ID: 6–6.5 in (15–16.5 cm). E and S tropical woods. Two groups: *flaviventris* group (E Mexico to C.A.) and *insularis* (Isla Cozumel). **Descr.** Ages differ slightly. **Adult.** *Flaviventris: eyes amber to red, bill greyish horn*, legs flesh. *Head grey with broad rufous supercilium*, crown often tinged brownish. Upperparts olive. Throat and underparts yellow. *Insularis:* throat and chest dusky-yellowish, belly and undertail coverts buffy whitish, upperparts greyish olive. **Juv:** eyes brown, crown brownish, paler supercilium less distinct.

Voice. A loud, rich, sad, descending series *ree treeu treeu treeu . . .,* or *djreeu djreeu . . .,* less often gives 1–2 quieter notes; also a loud, screamed *creeah rrah.* Song varied, bright, rich, warbled phrases, *chikee wheer peeripee pee-oo,* or *ch'weecheewee ch-weern,* or *chee wee cheery-choo,* etc.

Habs. Forest, woodland and edge, scrubby thickets, plantations. Nests at mid-levels in trees. Eggs: 2–3, pinkish white, spotted and speckled brown (WFVZ).

SD: F to C resident (SL–2000 m) on Atlantic Slope from S Tamps to Isthmus and in Yuc Pen, in interior, on adjacent Atlantic Slope, and on Pacific Slope from Chis to Honduras; apparently decreased, for unknown reasons, in E Mexico (except Yuc Pen) in 1960s and 1970s (JCA).

RA: E Mexico to cen Argentina.

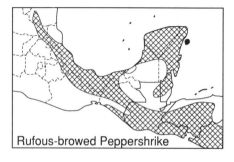

Green Shrike-Vireo

Rufous-browed Peppershrike

WOOD-WARBLERS: PARULINAE (66)

Formerly considered a family in their own right, the New World wood-warblers are now considered a subfamily of the Emberizidae. For ease of reference we treat subfamilies of the now vast emberizid family separately. Typical wood-warblers are small, active, often colorful birds with slender, usually pointed bills and slender legs. Like other emberizids they have only nine primaries. The family Parulidae has traditionally acted as a clearing house for warbler-like birds of uncertain taxonomic affinities, and some species included may be better placed elsewhere. Age/sex/seasonal variation differs between and within genera, and is typically complex in *Dendroica*, the largest genus of parulines. Young altricial, 1st prebasic molt partial, rarely complete. Juv plumage of some species is remarkably ephemeral and lost in less than a week, much of it while still in the nest (and hence rarely seen); in other species it may be kept for a few weeks. Several species, particularly *Dendroica*, have contrasting wingbars and tail-spots. Wintering as well as breeding habitats are often well defined, but in migration any species may be out of its preferred habitat, particularly in arid coastal scrub or semiopen country. Individuals of several migratory species defend winter and probably migration territories. Wood-warblers are often a dominant feature of mixed-species winter flocks in many areas. Songs of most N.A. breeding species are rarely if ever heard in the region. Calls include *chip* notes, most of which can be distinguished fairly readily, and high-pitched contact or *tsit* notes often given in flight, which require considerable experience to distinguish.

Wood-warblers are primarily insectivorous but some also eat fruit and *Vermivora* in particular take nectar. Nests of most warblers breeding in the region are cup-shaped or domed structures of grasses and other vegetation on or near the ground. Eggs: 2–6, typically whitish, speckled reddish brown.

GENUS VERMIVORA

Bills sharply pointed, tail-spots usually lacking. In general, adult ♂♂ brighter, imm ♀♀ duller. Seasonal variation slight or lacking. Bills blackish, paler below, to horn-flesh with dark culmen, legs greyish to dull flesh. The genus *Parula* should probably be merged with *Vermivora*.

BLUE-WINGED WARBLER

Vermivora pinus   Not illustrated
Chipe Aliazul

ID: 4.5–4.7 in (11.5–12 cm). Winters in E and S. **Descr.** Ages/sexes differ slightly. *Head and underparts yellow with black lores*, crown washed olive in all except adult ♂, undertail coverts whitish. Hind-neck and *upperparts olive, wings and tail blue-grey with 2 whitish wingbars*, white tail-spots. Interbreeds with Golden-winged Warbler, producing two main plumage types: 'Brewster's Warbler' and 'Lawrence's

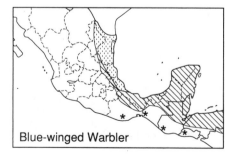

Blue-winged Warbler

Warbler' (rarer). Brewster's is colored like Golden-winged (grey and white) but with face pattern of Blue-winged, variable wingbars, and yellow wash on chest. Lawrence's is colored like Blue-winged (olive and yellow) but with face pattern and dark throat of Golden-winged, variable wing pattern.

Voice. A high, thin, slightly buzzy *tssi* or *tzzii*, often doubled.

Habs. Humid to semihumid evergreen and semideciduous forest and edge. Singly at low to mid-levels, joins mixed-species flocks.

SD: U to F transient and winter visitor (Sep–Apr; SL–2000 m) on Atlantic Slope from SE SLP to Honduras, R to S Tamps (JCA); U to R on Pacific Slope from Oax to El Salvador. U to F transient (Mar–Apr, Sep–Oct) in NE Mexico.

RA: Breeds E N.A.; winters E Mexico to cen Panama.

NB: Interbreeds extensively with Golden-winged Warbler (*V. chrysoptera*) and perhaps best regarded as conspecific with that form, despite striking plumage differences.

GOLDEN-WINGED WARBLER
Vermivora chrysoptera or *V. pinus* (in part)
Chipe Alidorado   Not illustrated

ID: 4.5–4.7 in (11.5–12 cm). Winters in S forests. **Descr.** Ages similar, sexes differ. ♂: head strikingly patterned with *yellow crown, white supercilium and moustache, and black auricular mask and throat*; underparts pale grey. Hindneck and *upperparts grey*, flight feathers darker with *yellow wing panel*, olive secondary edgings, and white tail-spots. ♀: *face pattern duller*, auricular mask and throat dusky grey, underparts duller, wing panel reduced. See Blue-winged Warbler for intergrade plumages.

Voice. A high, slightly buzzy *ssi* or *tssi*, often doubled, much like Blue-winged Warbler.

Habs. Humid evergreen and semideciduous forest and edge. At mid- to upper levels, joins mixed-species flocks.

SD: U to F transient and winter visitor (Sep–mid-May; SL–1800 m) on Atlantic Slope from Chis to Honduras, R to E Yuc Pen and S Ver (SNGH) and on Pacific Slope from Oax (ARP) to Honduras. U transient (Apr–mid-May, Sep–Oct) elsewhere in E Mexico. Vagrant (Oct–Dec) to NW Mexico (Schaldach 1969; Clow 1976; Howell and Pyle 1993) and Clipperton.

RA: Breeds E N.A.; winters S Mexico to N S.A.

NB: See NB under Blue-winged Warbler.

TENNESSEE WARBLER
Vermivora peregrina   Not illustrated
Chipe Peregrino

ID: 4.2–4.7 in (11–12 cm). Winters in E and S. **Descr.** Seasonal/sex variation. **Basic:** *narrow pale lemon to whitish supercilium contrasts with dark eyestripe.* Crown and upperparts olive to yellow-olive, flight feathers slightly darker with indistinct pale tail-spots. *Throat and underparts whitish to dirty pale lemon, undertail coverts white,* rarely tinged lemon. **Alt:** supercilium whitish, crown and nape blue-grey (♂) to olive-grey, throat and underparts whitish (♀ often washed lemon, mainly on chest). Birds feeding at flowering trees often stain their throat pink to orange, forming a neat and initially confusing 'gorget' which may extend to the whole face.

Voice. A high, thin *tsik* or *tsii*, often doubled, and a sharp *tchip*.

Habs. Humid to semihumid forest and edge, plantations, gardens. Often in small flocks; joins mixed-species flocks, often at flowering trees.

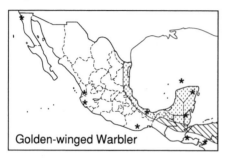

Golden-winged Warbler

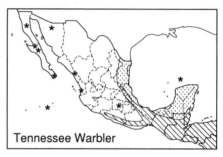

Tennessee Warbler

SS: See Philadelphia and Warbling vireos. Orange-crowned Warbler usually duller, especially on underparts where often streaked dusky on throat and chest, face pattern less distinct, undertail coverts lemon. ♀ Black-throated Blue Warbler has broken supercilium, whitish subocular crescent, and white flash at bases of primaries; note call and habits.

SD: C to F transient and winter visitor (late Sep–mid-May; near SL–3000 m, mostly above 100 m) on both slopes and in interior from Chis to Honduras and W Nicaragua, U to F to SE SLP and Gro (SNGH), R to Sin (PP) and BCS (SNGH, RS). F to C transient (Apr–May, mid-Sep–early Nov) through E Mexico, R to NW Mexico. Vagrant (Oct–Nov) to Clipperton and (Apr) to Isla Socorro.

RA: Breeds N N.A.; winters S Mexico to N S.A.

Habs. Arid to semihumid second growth, scrub, woodland, overgrown fields. Often in flocks at flowering trees. Nest cup of moss, rootlets, etc., low in bushes. Eggs: 2–4, white, speckled reddish brown (WFVZ).

SS: See Tennessee Warbler, Yellow Warbler has blank facial expression with pale eyering, less pointed bill, yellow tail-spots.

SD: U to F summer resident (Mar–Aug; SL–1500 m) in NW Baja, including Islas Los Coronados and Todos Santos. C to F transient and winter visitor (Aug–May, mainly Sep–Apr; SL–3000 m) throughout N of Isthmus, U (mainly in interior) to Guatemala, possibly to El Salvador (Thurber *et al* 1987), R (Nov–Mar) in Yuc Pen (Parkes 1970; BM) and possibly Belize.

RA: Breeds N and W N.A. to NW Mexico; winters S USA to N C.A.

ORANGE-CROWNED WARBLER
Vermivora celata   Not illustrated
Chipe Corona-naranja

ID: 4.6–5 in (11.5–12.5 cm). *Widespread in winter* N of Isthmus. **Descr.** Ages/sexes similar. Orange crown patch of adult usually concealed. Head and upperparts greyish olive to yellow-olive with *narrow pale lemon supercilium and eye-crescents, narrow dusky eyestripe*; head may be contrastingly greyer. *Throat and underparts dirty pale lemon to yellowish, indistinctly streaked dusky on throat and chest.* **Juv:** upperparts brownish olive with 2 pale cinnamon wingbars, face pattern indistinct.
  Voice. A sharp *tsik* or *sik*. Song a rapid trilled series of chips, steady or rising then falling abruptly at end, 1.5–2 s in duration, at times slows into separate chips or a soft, short warble; on Islas Los Coronados also a steady, accelerating series of chips *chi-chi-chi* ...

COLIMA WARBLER
Vermivora crissalis   Plate 55
Chipe Colimense

ID: 4.7–5.2 in (12–13 cm). Mexico, breeds in N, winters in W. **Descr.** Ages differ, sexes similar. **Adult:** *face grey with white eyering*, pale rufous crown patch usually concealed. *Crown and upperparts grey-brown, tinged olive, with bright olive-yellow rump* and uppertail coverts. *Throat and underparts grey, washed buffy brown on flanks, undertail coverts ochre-yellow*; chest washed buff in fresh plumage. **Juv:** duller and more brownish overall with 2 buff wingbars, eyering tinged lemon, lacks crown patch.
  Voice. A metallic *tchiu* or *chink*, fuller than Virginia's Warbler call. Song suggests Orange-crowned Warbler, a rising, rapid, trilled series of chips, 1–2 s in duration, ending fairly abruptly with soft, warbled notes, *sisisisisisisiiriu*, at times ends

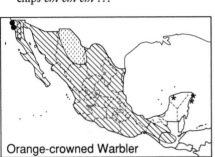

Orange-crowned Warbler

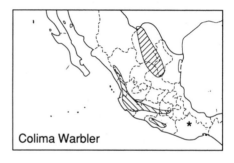

Colima Warbler

with a few separate chips, *siiiiiiiiiir swee-twee-twee-twee*, etc.

Habs. Breeds in chaparral with scattered oaks and conifers, oak scrub, and pinyon–oak woodland; winters in brushy understory and flower banks of humid to semihumid montane forests. Fairly skulking, especially in winter, but sings from prominent perches atop trees or bushes; often wags tail while hopping about. Nest of fine grasses, moss, etc., embedded in bank. Eggs: 4, white, speckled brown (Texas).

SS: Virginia's Warbler (winters in arid scrub) smaller, paler and greyer overall, lacks brown tones.

SD/RA: F to C but local breeder (mid-Mar–Aug; 1500–2500 m) in interior from Coah and NL to NE Zac and N SLP (Lanning *et al.* 1990), possibly to N Hgo (RR); also SW Texas. U to F winter visitor (Sep–Apr; 1500–3500 m) on Pacific Slope from S Sin to Gro, in interior to Mor (SNGH, RGW) R to N Oax (SNGH); possibly to DF where recorded late Feb–Apr (RGW).

NASHVILLE WARBLER

Vermivora ruficapilla*   Not illustrated
Chipe de Nashville

ID: 4.3–4.7 in (11–12 cm). Widespread in winter. **Descr.** Ages/sexes similar. Chestnut crown patch usually concealed. *Head grey with complete white eyering*, upperparts greyish olive to greyish, rump brighter yellow-olive. *Throat and underparts yellow*, becoming white on vent. Imm may be washed-out below, yellow mainly on chest, with pale lemon throat and flanks.

Voice. A fairly sharp, tinny *tink* or *chik*.
Habs. Woodland and edge, plantations, scrub. Often in flocks with other species at flowering trees, wags tail while hopping about.

SS: *Oporornis* warblers superficially similar but larger and more heavily built with larger bills, mostly skulking and terrestrial, grey extends across chest. Virginia's Warbler greyer overall, yellow on underparts restricted to chest and undertail coverts.

SD: C to F transient and winter visitor (late Aug–May; SL–3000 m), on both slopes and in interior from S Son, Dgo, S NL, and Tamps to Isthmus; F to U (mainly in interior) to Guatemala and NW El Salvador; R in BCS (Howell and Webb 1992*f*) and Honduras (Marcus 1983). Vagrant (Oct–Feb) to Yuc Pen (Howell 1989*a*; Lopez O. *et al.* 1989). More widespread transient, C to F in N Mexico, U in BCN.

RA: Breeds N and W N.A.; winters SW USA to N C.A.

VIRGINIA'S WARBLER

Vermivora virginiae   Not illustrated
Chipe de Virginia

ID: 4.3–4.7 in (11–12 cm). Winters SW Mexico. **Descr.** Ages/sexes similar. Chestnut crown patch usually concealed. *Head and upperparts grey with white eyering and olive-yellow rump* and uppertail coverts. *Throat and underparts whitish* to pale grey with *yellow wash across chest* (brightest in ♂), *and yellow undertail coverts*, flanks often washed buff. Imm may have throat, chest, and flanks washed buff, with little or no yellow on chest.

Voice. A sharp, metallic *tink*, stronger than Nashville Warbler call.
Habs. Arid to semiarid scrub, in migration also scrub in open pine woods. Singly, rarely in small groups at flowering trees. Often wags tail while hopping about.

SS: Lucy's Warbler smaller, plain face lacks distinct eyering, uppertail coverts chestnut, undertail coverts white. See Nashville and Colima warblers.

SD: C to F transient and winter visitor (Sep–Apr: 500–2000 m) in interior, and locally

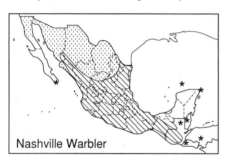

Nashville Warbler

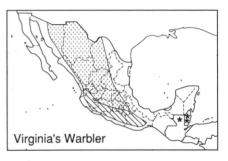

Virginia's Warbler

on adjacent Pacific Slope, from Nay (SNGH, PP) and Balsas drainage to Oax. F to C transient (late Aug–Sep, late Mar–mid-May) in N Mexico from Son to Coah and W SLP. Vagrant (mid-Dec–mid-Mar) to N Guatemala (Beavers *et al.* 1991) and Belize (Howell *et al.* 1992b).

RA: Breeds SW USA; winters W Mexico.

LUCY'S WARBLER
Vermivora luciae   Not illustrated
Chipe de Lucy

ID: 3.8–4.2 in (9.5–10.5 cm). Winters W Mexico. **Descr.** Ages differ, sexes similar. **Adult:** chestnut crown patch usually concealed. *Face and underparts whitish to pale grey*, with duskier auriculars. Crown, nape, and upperparts grey, with *chestnut lower rump and uppertail coverts*. **Juv:** lacks crown patch, lower rump and uppertail coverts dull cinnamon, 2 dirty buff wingbars.
> **Voice.** A metallic *chik* or *tink*. Song full and fairly loud, a varied series of accelerating slurred chips ending fairly abruptly, *chi-chi-chi-chi-chi shi-shi-shi-shi*, or *si-si-si-si-si-si chi-chi chii chi-chi-wii*, etc.
> **Habs.** Arid scrub, brushy riparian thickets, especially with mesquite. Often in small groups in winter, visits flowering trees. Nest cup of fine plant material placed in tree cavities, also in old Verdin nests. Eggs: 3–6, white, speckled reddish brown (WFVZ).

SS: See Virginia's Warbler.
SD: F to C summer resident (Mar–Aug; SL–1500 m) in NE BCN and N Son. C to F transient and winter visitor (Sep–Mar; SL–1800 m) on Pacific Slope and in adjacent interior from cen Sin (RAB, PP) to Gro, U to R S to Oax (SNGH, SJ, PP), N to S Son (per GM). C to F transient (Mar, Aug–Sep) in Son and Sin. Vagrant (Oct–Apr) to Baja (Wilbur 1987).

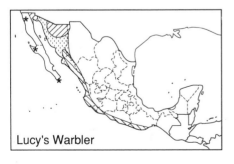
Lucy's Warbler

RA: Breeds SW USA, NW Mexico; winters W Mexico.

CRESCENT-CHESTED WARBLER
*Vermivora superciliosa**   Plate 55
Chipe Cejiblanco

ID: 4.2–4.7 in (11–12 cm). Endemic; highland forests. **Descr.** Ages differ, sexes similar. **Adult:** bill greyish, legs dusky flesh. *Head blue-grey head with white supercilium* and subocular crescent. *Throat and chest yellow with small chestnut crescent on upper chest* (faint or lacking in some ♀♀), belly and undertail coverts whitish, washed dusky on flanks. *Upperparts greenish olive* with *blue-grey uppertail coverts, wings, and tail*. **Juv:** head and upperparts duller with 2 pale cinnamon wingbars, throat and underparts dirty buff becoming whitish on belly.
> **Voice.** A high, thin, sharp *tsit* or *sik*. Song a steady, dry, buzzy trill, *dzzzzzzr* or *zzzirrr*, 0.5–1 s in duration.
> **Habs.** Humid to semiarid pine–oak, oak, and pine–evergreen forest. At mid- to upper levels, singly or in pairs; joins mixed-species flocks. Nest cup of grass, moss, pine needles, etc., placed on or near ground, usually in bank or clump of grass. Eggs: 3, white, unmarked (WFVZ).

SS: Tropical Parula lacks white supercilium and chest spot, song (in NE Mexico) a longer and harder buzzy trill.
SD/RA: C to F resident (1500–3500 m) on both slopes from S Chih and NE NL (JCA), and in interior from cen Mexico, to N cen Nicaragua.
NB: Has been placed in the genus *Parula*.

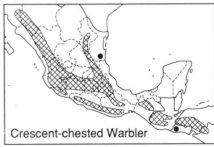
Crescent-chested Warbler

NORTHERN PARULA
Parula americana   Not illustrated
Parula Norteña

ID: 4–4.3 in (10–11 cm). Winters in E and S.

Northern Parula

Descr. Ages/sexes differ. ♂: bill blackish above, yellowish flesh below, legs yellowish flesh. *Head and upperparts grey-blue with black lores and white eye-crescents, olive back patch,* 2 white wingbars, white tail-spots. *Throat and chest yellow with blackish-slate and rufous band across upper chest,* rest of underparts whitish. ♀: *duller* (especially imm), underparts lack chest bands or show at most a trace of orange wash.

Voice. A sharp *chik* or *tchik*, often surprisingly loud.

Habs. Humid to semiarid forest and edge, plantations, mangroves. Singly at mid- to upper levels, joins mixed-species flocks. Fairly active, tail often held cocked.

SS: Tropical Parula lacks white eye-crescents, yellow more extensive on underparts, often has dark mask.

SD: F to C transient and winter visitor (late Aug–Apr; SL–1000 m) on Atlantic Slope from S Ver to N Honduras, U to R to S Tamps and on Pacific Slope from Oax to El Salvador. F to U transient (Mar–early May, mid-Aug–Oct; SL–2500 m) on Atlantic Slope W to Coah and DF. Vagrant (Dec–Jan) to Nay (Clow 1976; SNGH).

RA: Breeds E N.A; winters E Mexico and Caribbean to N C.A.

NB: Northern and Tropical parulas are sometimes considered conspecific.

TROPICAL PARULA
Parula pitiayumi *   Plates 55, 69, 71
Parula Tropical

ID: 4–4.3 in (10–11 cm). Widespread but often local. Four groups: *nigrilora* (N of Isthmus), *graysoni* (Isla Socorro), *inornata* (Chis and Guatemala), and *speciosa* (Honduras and S). **Descr.** Ages/sexes differ. *Nigrilora*. ♂: bill blackish above, yellowish flesh below, legs yellowish flesh. *Blackish lores blend into dark mask* (bolder in E

Mexico). *Head and upperparts greyish blue with olive patch on back,* 2 white wingbars (broader in W Mexico), white tail-spots. *Throat and underparts yellow, washed orange on chest,* becoming white on belly and undertail coverts, flanks often washed cinnamon (darkest in *insularis* of Islas Tres Marías). ♀: duller, lacks black mask though lores may be dark, head often washed olive. *Graysoni*: resembles dull ♀ *nigrilora* but underparts without orange wash, lacks white tail-spots. *Inornata*: resembles *nigrilora* but duller, mask less contrasting, *wingbars reduced* to narrow white tips of outer greater coverts, *underparts lack strong orange wash*. *Speciosa*: ♂ and ♀ *have black lores,* ♂ usually has slight mask. *Head and upperparts deep bluish with poorly contrasting olive back patch, indistinct whitish lower wingbar. Throat and underparts orange-yellow,* brightest on chest, becoming white on undertail coverts. **Juv** (all ssp): head and upperparts greyish, tinged olive on back, throat and underparts whitish, tinged lemon on chest.

Voice. A sharp *chik,* much like Northern Parula. Song variable. In W Mexico (including Socorro), a varied arrangement of high, thin buzzes and chips, at times accelerating into a rapid trill or ending abruptly with an upward-inflected note, *syi-syi-syi-syi-syi sisisi siiiiiirr,* or *ssi ssi ssi ssi swi-si-si-it,* etc.; in NE Mexico, a buzzy trill *zzzzzzzirr,* and varied wiry buzzes *zzi zzi zzir zzzi zzi zzi zzzii,* etc.; in Honduras, high thin chips run into a trill, *si-si-si-sisisisisi si-i-il,* etc.

Habs. Humid to semiarid woodland and forest, riparian groves with large trees and Spanish moss, also arid thorn forest and mangroves in W Mexico. Habits much as Northern Parula; on Socorro commonly in groups of up to 20 birds. Nest tucked into or hollowed out from hanging moss

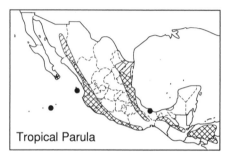

Tropical Parula

at mid-levels (undescribed on Socorro?). Eggs: 3–4, whitish, speckled reddish brown (WFVZ).

SS: See Northern Parula, Crescent-chested Warbler.

SD: F to C but local resident (SL–1800 m) on Pacific Slope from Sin to Oax, and on Islas Tres Marías and Isla Socorro. C to F but local resident on Atlantic Slope (mainly above 500 m S of N Ver) from S Tamps to Honduras, and in interior from Chis to N cen Nicaragua. F but local breeder (Mar–Sep) N to cen Son and NL. Altitudinal migrant locally (e.g. Los Tuxtlas). Vagrant (Jul–Feb) to BCS. Reports from Belize (Wood *et al.* 1986) and Petén (Land 1970) require verification.

RA: Mexico (and S Texas) to Peru and N Argentina.

NB1: *Graysoni* sometimes considered specifically distinct, Socorro Parula. **NB2:** See NB under Northern Parula.

GENUS DENDROICA

This is a widespread genus of mainly temperate-breeding warblers (only four species breed in the region, 16 others occur as migrants). Plumages typically complex with much age/sex and seasonal variation. In general, adult ♂♂ are brightest, adult ♀♀ and imm ♂♂ slightly duller, and imm ♀♀ dullest; basic adults resemble imms in several species. For more detail of age/sex variation, see Pyle *et al.* (1987). All *Dendroica* have tail-spots. Alt usually worn from Mar through summer, basic attained on breeding grounds prior to migration; the months for alt given below (usually Mar–May) reflect only the period when one sees this plumage in the region, i.e. during spring migration. Bills usually black or dark grey, especially in alt, often paler horn or flesh below in imms, legs flesh to blackish; rarely of use in identification except when noted. Call notes of many species are classic 'chip' notes; songs of migrants may be heard in spring migration.

YELLOW WARBLER
Dendroica petechia ⃰   Plates 54, 69
Chipe Amarillo

ID: 4.5–5 in (11.5–12.5 cm). Widespread. Three groups: *aestiva* group (widespread); *bryanti* group ('Mangrove Warbler', mangroves), and *rufivertex* ('Golden Warbler',

Isla Cozumel). **Descr.** Ages/sexes differ. *Aestiva.* ♂: *head and underparts bright yellow, streaked chestnut on chest and flanks.* Nape and upperparts olive to yellow-olive, wings and tail dark, broadly edged olive-yellow, including *yellow tail-spots.* ♀/imm ♂: *duller,* olive wash extends onto crown, *blank-looking face with paler eyering,* underparts with fewer or no chestnut streaks. Imm ♀: resembles ♀ but duller and paler. Crown and upperparts greyish olive to greyish. Face, throat, and underparts dirty pale lemon, at times whitish to pale grey but usually with pale lemon undertail coverts; *note yellowish tail-spots, facial expression. Bryanti.* ♂: *head chestnut, chest finely streaked chestnut* (streaks indistinct in *castaneiceps* of BCS). ♀: typically has *chestnut wash on crown, greyish nape.* Face and throat may be mottled chestnut. Imm ♂: *head mottled chestnut,* crown and upperparts greyish olive, greyest on nape. Imm ♀: *often very plain and washed-out, lacks chestnut on head, underparts often with lemon restricted to undertail coverts. Rufivertex.* ♂: chestnut crown, bolder underpart streaking than *aestiva.* ♀: like *aestiva* group.

Voice. Call a fairly sharp *tchik* or *chip,* often slightly fuller in *bryanti* group. Song varied, an intensifying series of sweet

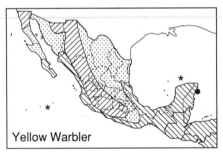

Yellow Warbler

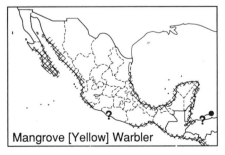

Mangrove [Yellow] Warbler

chips ending fairly abruptly, *swee chee-chee chee-chee chee-chi-choo'*, or *swee swee-swee-swee-sweeziweet*, etc.

Habs. *Aestiva* breeds in deciduous woodland, especially riparian groves, winters in brushy woodland, gardens, scrub, wooded marshes, mangroves; *bryanti* resident in mangroves; *rufivertex* resident in scrubby woodland, mangrove edge, etc. Nest cup of fine plant fibers at low to mid-levels in tree or bush (mangroves in *bryanti*). Eggs: 2–5, whitish to greenish white, speckled and spotted with browns and greys (WFVZ).

SS: Wilson's Warbler often cocks longer darker tail; note larger eye, more uniform upperparts, lack of yellow tail-spots. See Orange-crowned Warbler.

SD: *Aestiva*: F to C breeder (May–Aug; SL–2000 m) in BCN, in interior and on adjacent Pacific Slope from Son and Chih to N Gro and Pue. C to F transient and winter visitor (late Jul–May; SL–2000 m, mainly below 1000 m) in BCS and on both slopes from cen Sin and Ver to Honduras and El Salvador, U to F in interior from Isthmus S; U to R in winter N to S Son and S Tamps. Vagrant (mid-Aug–mid-Apr) to Isla Socorro and Clipperton. *Bryanti*: C to F resident along coasts from BCS, cen Son, and S Tamps to El Salvador and Honduras. A report from Isla Socorro is equivocal. *Rufivertex*: F to C resident (around SL) on Isla Cozumel.

RA: Breeds N.A. to Peru and Venezuela; winters S USA and Mexico to Peru and Brazil.

Basic (Sep-Mar): *face and underparts pale grey with whitish eyering*, undertail coverts white; ♂ may have chestnut patches on sides. *Crown, nape, and upperparts yellow-olive*, rump and blue-grey uppertail coverts with black centers. Wings and tail blackish with *2 pale lemon* to whitish *wing-bars*, white tail-spots. **Alt** (Mar–May): *face white with broad black eyestripe, black moustache, and yellow crown*. Throat and underparts white with *chestnut sides and flanks*. Nape and upperparts greyish olive, heavily streaked black. ♀ duller than ♂, with less distinct face pattern, less chestnut on sides.

Voice. A full *chik* or *chiuk*.

Habs. Humid to semihumid evergreen forest and edge, plantations. At mid- to upper levels, tail often cocked like a gnatcatcher. Joins mixed-species flocks.

SS: Basic Chestnut-sided often unfamiliar to N.A. observers (may suggest a vireo) but unmistakable when learned.

SD: F to C transient, U to F in winter (Sep–May; SL–1200 m) on Atlantic Slope from S Ver to Honduras. U to F transient (Sep–Oct, Apr–May; SL–2000 m) in Yuc Pen and on Atlantic Slope N of Isthmus, R W to DF (Wilson and Ceballos-L. 1986). Vagrant (May, Oct) on Pacific Slope in BCN (Keith and Stejskal 1987), Son (GHR), and El Salvador (Thurber *et al.* 1987), wintering (Nov–Mar) locally from Nay (Clow 1976) to Gro (Howell and Wilson 1990).

RA: Breeds E N.A.; winters S Mexico to N S.A.

CHESTNUT-SIDED WARBLER
Dendroica pensylvanica   Not illustrated
Chipe Flanquicastaño

ID: 4.5–5 in (11.5–12.5 cm). Winters in S forests. **Descr.** Age/sex/seasonal variation.

MAGNOLIA WARBLER
Dendroica magnolia   Not illustrated
Chipe de Magnolia

ID: 4.5–5 in (11.5–12.5 cm). Winters in E and S lowlands. **Descr.** Age/sex/seasonal

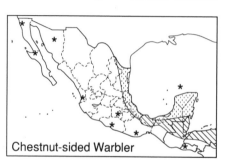

Chestnut-sided Warbler

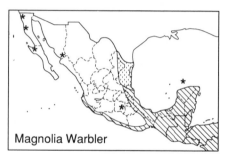

Magnolia Warbler

variation. **Basic/imm** (Aug–Apr): *head and nape greyish with whitish eyering,* often a faint whitish supercilium. *Throat and underparts yellow with white undertail coverts, sides and flanks streaked dusky to blackish,* chest often washed dusky. Back olive (often streaked darker) with yellow rump patch, uppertail coverts grey with black centers. Wings and tail blackish with 2 white wingbars, *basal white tail-spots appear as distinctive white basal half on black undertail.* **Alt** ♂ (Apr–May): *white postocular stripe contrasts with blue-grey crown and black auricular mask.* Throat and underparts yellow with *band of bold black streaks across chest* and down flanks, undertail coverts white. Upperparts black with yellowish rump. Wingbars often wider than basic, forming panel on inner coverts. ♀: auriculars greyish, upperparts olive with black centers, black streaks below less distinct, wingbars narrower.

Voice. A distinctly nasal, slightly squeaky *iihh* or *iihh* and *shiih.*

Habs. Humid to semiarid forest and edge, second growth, plantations. Feeds low to high, tail often slightly spread. Joins mixed-species flocks.

SS: From below may suggest Canada Warbler (transient migrant); note dark legs, distinctive tail pattern, wingbars, and call.

SD: C to F transient and winter visitor (Sep–May; SL–1500 m) on Atlantic Slope from SE SLP to Honduras; U to F on Pacific Slope from Nay to El Salvador, and in interior from Isthmus S. C to F transient (Apr–May, Sep–Oct) in NE Mexico, R W to DF. Vagrant (Oct–early Dec) to NW Mexico (Howell and Pyle 1993; per GM) and Clipperton.

RA: Breeds N N.A.; winters Mexico and Caribbean to Panama.

CAPE MAY WARBLER

Dendroica tigrina   Not illustrated
Chipe Atigrado

ID: 4.5–5 in (11.5–12.5 cm). Winters Caribbean islands and coasts. Bill sharply pointed. **Descr.** Age/sex/seasonal variation. **Basic** ♂ (Aug–Mar): *face and sides of neck yellow with chestnut auricular patch* (at times indistinct). *Throat and underparts yellow* becoming white on vent and undertail coverts, *heavily streaked black*

on *lower throat, chest, and flanks.* Crown, nape, and upperparts olive with black centers, yellowish rump. Wings and tail blackish, edged olive, with *whitish wing panel* and white tail-spots. **Alt** ♂ (Mar–May): brighter, crown blackish, chestnut auricular patch bolder. ♀: face dusky with *indistinct pale lemon supercilium, yellow wash on sides of neck.* Throat and underparts whitish, often washed lemon on chest, with *fine dusky streaking on lower throat, chest, and flanks.* Crown, nape, and upperparts greyish olive to greyish with *yellow-olive rump.* Wings and tail darker with 2 whitish to dirty buff wingbars, white tail-spots. **Imm** ♂ (Aug–Feb): resembles ♀ but brighter, yellower below, white wingbars often form panel.

Voice. A high, thin, slightly wiry *tsii* or *siik.*

Habs. Humid to semiarid scrub and woodland, gardens. Low to high, often at flowering trees.

SS: Basic Myrtle Warbler larger, browner above, with coarse dusky streaks below, lacks yellowish sides of neck but shows yellow patch at sides of chest; note voice. Palm Warbler wags tail, has distinct pale supercilium, yellow undertail coverts, usually near ground.

SD: U transient and winter visitor (Sep–early May; SL–500 m) on Atlantic Slope in S Ver (WJS, DEW) and from N Yuc to N Honduras, mainly on Caribbean islands (not casual, as stated by AOU 1983). Vagrant (Nov–Apr) on Pacific Slope from Son (DS) to Nay (Clow 1976) and Isla Socorro (Howell and Webb 1992g), on Pacific Slope and in interior from Oax (SNGH, PP) to Guatemala and El Salvador (Mason 1976), and on Atlantic Slope N to SLP (RAB).

RA: Breeds NE N.A.; winters Caribbean and adjacent coasts of M.A.

Cape May Warbler

BLACK-THROATED BLUE WARBLER

Dendroica caerulescens   Not illustrated
Chipe Azuloso

ID: 4.7–5.2 in (12–13 cm). Winters on Caribbean coasts and islands. **Descr.** Ages similar, sexes differ. ♂: *black face and throat contrast with bluish crown and upperparts.* Wings and tail darker with *white flash on base of primaries*, white tailspots. Underparts white with black mottling on sides. Imm (Aug–Apr) often told by narrow olive edgings to upperparts and primary coverts. ♀: *head and upperparts olive to greyish olive, typically with narrow whitish supercilium* (often broken in front of eye) *and subocular crescent. White check on base of primaries* may be absent, pale tail-spots indistinct. *Throat and underparts dirty buffy lemon*, often washed dusky on throat and flanks, undertail coverts palest.

Voice. A fairly hard, smacking *tsik* or *tchik*.

Habs. Humid to semiarid woodland and edge. At low to mid-levels; rarely joins mixed-species flocks.

SS: See Tennessee Warbler.

SD: U to R transient and winter visitor (late Aug–Apr; around SL) on Atlantic Slope from E Yuc Pen to N Honduras, mainly on Caribbean islands (not casual, as stated by AOU 1983), R to S Ver (PP, WJS) and (Nov–Dec) to N Chis (Hunn 1971; SNGH) and Guatemala. Vagrant (Oct–Nov) to NW Mexico.

RA: Breeds NE N.A.; winters Caribbean and adjacent mainland coasts.

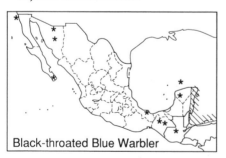

Black-throated Blue Warbler

YELLOW-RUMPED WARBLER

Dendroica coronata*   Figure 40
Chipe Rabadilla-amarilla

ID: 5–5.5 in (13–14 cm). Widespread, mainly in winter. Three groups (see SD below):

Fig. 40 Male Goldman's [Yellow-rumped] Warbler

coronata group ('Myrtle Warbler'), *auduboni* group ('Audubon's Warbler'), and *goldmani* ('Goldman's Warbler'). **Descr.** Age/sex/seasonal variation. *Coronata.* Basic/imm (Aug–Mar): *head brownish grey with narrow* whitish to *pale buff supercilium and subocular crescent.* Upperparts brownish to grey-brown, streaked darker, with *yellow rump.* Wings and tail dark with 2 buff to whitish wingbars, white tail-spots. Throat and underparts whitish with *yellow flash on sides of chest*, dusky to black streaking on chest and flanks; throat and chest often tinged buff.

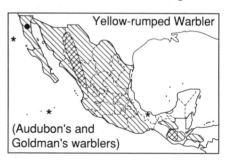

Yellow-rumped Warbler

(Audubon's and Goldman's warblers)

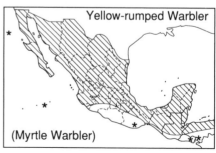

Yellow-rumped Warbler

(Myrtle Warbler)

Alt ♂ (Mar–May): *throat white, white supercilium* (broken in front of eye) *and subocular crescent contrast with black auricular mask* and blue-grey crown; yellow crown patch often visible. Upperparts blue-grey, heavily streaked black, with yellow rump; wingbars whiter than basic. White underparts heavily mottled black on chest and flanks, with yellow flash on sides of chest. ♀: auriculars greyish, crown and upperparts washed grey-brown, chest and flanks less heavily mottled black. Interbreeds with *auduboni* group; many migrants cannot be placed readily with either form. **Auduboni. Basic/ imm:** head and upperparts greyer, *face grey with white to whitish eyering or eye-crescents; throat yellow to dirty buff, tinged yellow;* tail-spots more extensive relative to age and sex. Alt ♂: *head dark blue-grey with black lores, face pattern restricted to white eye-crescents, throat yellow; black mottling on underparts more extensive,* extending to belly in *nigrifrons* (breeding in NW Mexico); white wingbars typically form a panel. ♀: resembles ♀ Myrtle but *throat yellow, note Auduboni's face pattern.* **Goldmani.** ♂: *head and upperparts black with yellow crown patch, yellow rump,* and white wing panel. *Throat yellow with white lower border;* chest, upper belly, and flanks black with yellow chest side flashes, lower belly to undertail coverts white (apparently no seasonal variation, Hubbard 1970). ♀/ imm/basic: resemble *auduboni* but darker overall, especially on head and upperparts. **Juv** (all ssp?): head and upperparts grey-brown, heavily streaked black, rump whitish with dark streaking. Throat and underparts whitish, streaked black. Quickly attains adult-like plumage.

Voice. *Coronata:* a slightly wet to sharp *chwik. Auduboni:* a slightly liquid *dwep* or *chip.* Song a varied series of rich notes, *si-wi chi-wee chi-wee chi-wee* or *si-wee si-wee si-wee-chit,* etc. *Goldmani:* undescribed (?).
Habs. Breeds in pine–oak woodland, winters in deciduous forest and edge, fields and open areas with adjacent hedges and fences, etc. *Coronata* often in small flocks, feeds on ground and in trees. *Auduboni* often with mixed-species flocks in highland forests. Nest cup of fine twigs, rootlets, etc., at mid- to upper levels in conifers. Eggs: 3–4, whitish, flecked and spotted with reddish browns and greys (WFVZ).

SS: *Coronata:* Palm Warbler has yellow undertail coverts, wags tail; see ♀ Cape May Warbler.
SD: *Coronata:* F to C transient and winter visitor (Oct–mid-May; SL–1500 m) on Atlantic Slope from Tamps to Honduras; U to R (to 3000 m) in interior and on Pacific Slope from Baja (where locally F) to Honduras, including Isla Guadalupe and Islas Revillagigedo. *Auduboni:* U to F breeder (May–Aug; 1500–3500 m) in BCN, and in interior from Chih to Dgo. C to F transient and winter visitor (late Sep–May; SL–3500 m) in Baja and on Pacific Slope and in interior from Son and Coah to Chis, F to U (late Oct–Apr) to W Honduras. F to U (late Sep–May) on Atlantic Slope N of Isthmus. *Goldmani:* U to F resident (1800–3500 m) in W Guatemala (and SE Chis?).
RA: Breeds N.A. to Guatemala; winters USA to C.A. and Caribbean.
NB: Myrtle and Audubon's warblers have been considered specifically distinct but interbreed extensively in N.A.

BLACK-THROATED GREY WARBLER
Dendroica nigrescens   Not illustrated
Chipe Negrigris

ID: 4.5–5 in (11.5–12.5 cm). Winters N of Isthmus. **Descr.** Ages/sexes differ. ♂: *broad white postocular stripe and malar stripe contrast with blackish crown, black auricular mask, and black throat;* note *yellow lore spot.* Underparts white, streaked black on sides and flanks. *Upperparts grey* (washed brown in autumn) with black centers. Wings and tail blackish, edged pale grey, with 2 white wingbars, white tail-spots. ♀: *duller,* crown and auriculars mottled grey and black; *throat mottled black and white,* often white with black collar on sides of throat. **Imm** (Aug–Mar): resembles ♀ but back washed brownish, throat and underparts tinged pinkish buff, ♀'s *throat has little or no*

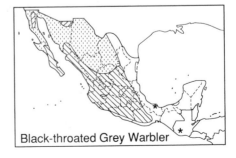

Black-throated Grey Warbler

black. **Juv** (Jun–Aug): resembles imm but throat, chest, and flanks mottled smoky grey.

Voice. Call a low *chip* or *chep*, duller than 'black-throated green' warbler group chips. Song a varied, high, buzzy series, speeding up at the end, *dzee dzee dzee dzee dzee-zzi-ju'* or *zee zee zee zee-zee*, etc.

Habs. Breeds in semiarid pine–oak and juniper woodland. Winters in arid to semi-humid oak and pine–oak forest and scrub. Often joins mixed-species flocks. Nest cup of plant fibers, grass, etc., low to high in tree or bush. Eggs: 3–5, whitish, speckled with reddish browns and greys (WFVZ).

SS: Note yellow lore spot. Black-and-white Warbler has white median crown stripe, black and white striped back, creeps on trees. Cerulean Warbler greenish blue on crown and upperparts; note range and season.

SD: C to F breeder (Apr–Aug; 1500–3000 m) in BCN and NE Son. C to F transient and winter visitor (mid-Aug–Apr; near SL–3000 m) in BCS and, mainly on Pacific Slope and in interior, from S Son and Coah to Oax. C to F transient (Apr–May, Sep–Oct) in NW Mexico. Vagrant to cen Guatemala (Wetmore 1941*b*).

RA: Breeds W N.A. to NW Mexico; winters SW USA to S Mexico.

TOWNSEND'S WARBLER
Dendroica townsendi　　　Not illustrated
Chipe de Townsend

ID: 4.5–5 in (11.5–12.5 cm). Winters in high-lands. **Descr.** Ages/sexes differ. ♂: *black auricular mask* with yellow subocular cres-cent *bordered by broad yellow supercilium, sides of neck, and malar stripe. Throat black, chest yellow,* belly and undertail coverts white; sides and flanks streaked blackish. Crown blackish, *olive upperparts and blue-grey uppertail coverts with dark*

centers. Wings and tail darker with 2 white wingbars, white tail-spots. ♀/**imm** ♂: *duller, throat and chest yellow with dark mottling on sides of chest, crown and auricu-lars mottled olive.* **Imm** ♀: resembles ♀ but duller, auriculars olive, only slight dark mot-tling at sides of chest, flanks with dusky streaks, upperparts with indistinct dark cen-ters. Rare hybrids with Hermit Warbler, usually resemble Hermit but have dark streaks on sides.

Voice. A dry *chip* or *chik*, brighter than Black-throated Grey Warbler.

Habs. Pine–oak, oak, and conifer forests, also scrub in migration. At mid- to upper levels; a common member of mixed-species flocks.

SS: Black-throated Green Warbler has yellow face with less contrasting dusky olive auricu-lars, chest whitish, upperparts greener with-out dark streaks. Golden-cheeked Warbler has bright yellow face with narrow dark eye-stripe, crown and upperparts darker, chest whitish. Imm Hermit Warbler (autumn) has dusky auriculars; note unstreaked sides.

SD: C to F transient and winter visitor (Sep–May; 500–3000 m) mainly in interior and on Pacific Slope, from Son and S Coah (SNGH, SW) to Guatemala, F to U to Hon-duras and W Nicaragua. U to R in winter in BCN. C to F transient (Apr–May, Sep–Oct) in NW Mexico.

RA: Breeds NW N.A.; winters SW USA to Costa Rica.

HERMIT WARBLER
Dendroica occidentalis　　　Not illustrated
Chipe Cabeciamarillo

ID: 4.7–5.2 in (12–13 cm). Winters in high-lands. **Descr.** Ages/sexes differ. ♂: *head yellow with black throat,* blackish mot-tling on nape. *Underparts whitish, un-streaked.* Upperparts greyish (washed olive in autumn) with dark centers; wings and

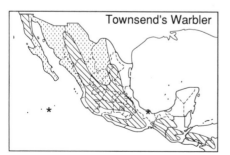

Townsend's Warbler

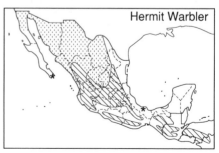

Hermit Warbler

tail darker with 2 white wingbars, white tail-spots. ♀/imm: *duller, crown duskier to olive, throat mottled black and white.* Autumn imms often have dusky olive auriculars, upperparts washed olive with indistinct dark centers, sides and flanks washed buff.

Voice. A *chip*, much like Townsend's Warbler but slightly fuller.

Habs. Pine–oak, oak, and conifer forests. Habits much as Townsend's Warbler with which it often occurs.

SS: Golden-cheeked Warbler has narrow dark eyestripe, dark olive to black crown and upperparts, dark streaks on sides and flanks. Black-throated Green Warbler has olive crown, dark streaking on sides. See Hermit × Townsend's Warbler hybrid under Townsend's description.

SD: C to F transient and winter visitor (Aug–May; 500–3500 m), mainly in interior and on Pacific Slope, from Dgo and S NL to Guatemala, F to U to Honduras and N cen Nicaragua. C to F transient (Apr–May, Aug–Oct) in NW Mexico.

RA: Breeds W USA; winters coastal California and Mexico to Nicaragua.

BLACK-THROATED GREEN WARBLER
Dendroica virens   Not illustrated
Chipe Dorsiverde

ID: 4.5–5 in (11.5–12.5 cm). Winters in E and S. **Descr.** Ages/sexes differ. ♂: *face yellow with broad olive eyestripe. Throat and sides of upper chest black*, whitish underparts streaked black on flanks, sides of vent washed lemon. *Crown, nape, and upperparts greenish olive* with indistinct dark centers, uppertail coverts blue-grey with black centers. Wings and tail darker with 2 white wingbars, white tail-spots. ♀/imm ♂: *auriculars washed olive, throat and sides of chest whitish, mottled black*, throat may be tinged lemon; upperparts

appear unstreaked. **Imm ♀:** resembles ♀ but throat and underparts whitish, tinged lemon, with slight dusky mottling on sides of chest, dusky streaks on flanks.

Voice. A *chip* much like Townsend's Warbler.

Habs. Humid to semihumid evergreen, semideciduous, and pine–oak forest and edge. At mid- to upper levels; often with mixed-species flocks.

SS: Golden-cheeked Warbler has brighter yellow face with distinct dark eyestripe, crown and upperparts dark olive to black. See Townsend's and Hermit warblers.

SD: C to F transient and winter visitor (Sep–May; SL–2000 m) on Atlantic Slope from S NL, and in interior from Isthmus, to Honduras. U to F on Pacific Slope from Nay (SNGH, ARP, PP) to El Salvador. Vagrant (Nov–Dec) to NW Mexico (Wilbur 1987; Rising 1988), and Clipperton.

RA: Breeds E N.A.; winters E Mexico to cen Panama.

GOLDEN-CHEEKED WARBLER
Dendroica chrysoparia   Not illustrated
Chipe Caridorado

ID: 4.7–5.2 in (12–13 cm). Winters S of Isthmus. **Descr.** Ages/sexes differ. ♂: *golden-yellow face with narrow black eyestripe contrasts with black crown, throat, and upper chest.* White underparts streaked black on sides and flanks. *Upperparts black* (narrowly edged olive in autumn) with 2 white wingbars, white tail-spots. ♀: *face yellow with narrow dark eyestripe, crown and upperparts dark olive to olive-grey, streaked blackish.* Chin lemon, throat and upper chest whitish to pale lemon, mottled blackish mainly on upper chest. **Imm ♂:** resembles ♀ but crown and upperparts blackish, edged olive, throat and upper chest black, veiled

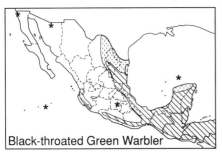

Black-throated Green Warbler

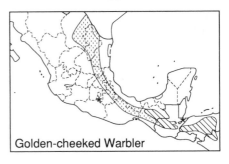

Golden-cheeked Warbler

with pale lemon. **Imm ♀:** resembles ♀ but crown and upperparts with narrower black streaks, black mottling often restricted to sides of upper chest.

Voice. A *chip* or *chik*, much like Black-throated Green warbler but perhaps slightly fuller. Songs a high, buzzy *zzwi zzui-zwi-zwi zzziu-zzi* and a higher, thin *zi-zi-zi-zi zzziu zz-i'*, etc. (Texas).

Habs. Winters in humid to semiarid pine–oak, pine–evergreen, and (rarely?) evergreen forest. At mid- to upper levels; joins mixed-species flocks.

SS: See Black-throated Green, Townsend's, and Hermit warblers, all of which occur in flocks with Golden-cheeked Warbler.

SD: U to F winter visitor (mid-Aug–Feb; 1500–3000 m) from Chis (SNGH, PP) to N cen Nicaragua, mainly in interior; vagrant to Mor (Nov 1986, RGW). U to F transient (late Jun–Sep, mid-Feb–Mar; 1200–3000 m) through mountains of E Mexico, from Coah to Chis. Should be sought breeding in mountains of NL and Coah, in association with mature stands of Ashe Juniper (*Juniperus ashei*). A report from Belize (Miller and Miller 1992a) requires verification.

RA: Breeds cen Texas; winters S Mexico to Nicaragua.

BLACKBURNIAN WARBLER
Dendroica fusca   Not illustrated
Chipe Gorjinaranja

ID: 4.5–5 in (11.5–12.5 cm). Transient in E and S. **Descr.** Ages/sexes differ. ♂: *black auricular mask bordered by flaming orange supercilium, neck sides, throat,* and upper chest. Whitish underparts streaked black on sides and flanks. Crown and upperparts black with orange median stripe on forecrown, *whitish braces.* Wings and tail blackish with *white wing panel,* white tail-spots. ♀/imm ♂: duller, yellow-

orange to ochre-yellow on face, throat, and chest; auricular mask, crown, and upperparts greyish (mottled black in imm ♂), 2 white wingbars. **Imm ♀:** resembles ♀ but duller, dirty pale lemon on face, throat, and chest; crown and upperparts greyish olive. Note face pattern, pale braces.

Voice. A bright to fairly full *chip* or *tchip*.

Habs. Humid to semihumid evergreen and semideciduous forest and edge. At mid- to upper levels; joins mixed-species flocks.

SS: Imm ♀ Townsend's Warbler has yellow throat and chest, more distinct dark streaks on sides, olive crown and upperparts, lacks braces. Yellow-throated Warbler has white supercilium and sides of neck, blue-grey upperparts without braces.

SD: C to F transient (late Mar–May, mid-Aug–early Nov; SL–2000 m) on Atlantic Slope from Tamps to Honduras; U in Yuc Pen, in interior from DF, and on Pacific Slope from Oax S. Vagrant to BCN (Sep 1980, Clark and Kendall 1986).

RA: Breeds NE N.A.; winters Costa Rica to Venezuela and Bolivia.

YELLOW-THROATED WARBLER
*Dendroica dominica**   Not illustrated
Chipe Gorjiamarillo

ID: 4.7–5.2 in (12–13 cm). Winters in E and S. Bill relatively long and stout, pointed. **Descr.** Ages/sexes differ slightly. *Yellow throat and chest bordered by black auricular mask* which extends down into black streaking on white underparts. *White supercilium,* tinged yellow on lores (strongly yellow in *dominica*), *white subocular crescent, and patch on sides of neck complete striking head pattern.* Crown, nape, and upperparts blue-grey, washed brown in imm, forecrown mottled black (more extensively in ♂). Wings and tail

Blackburnian Warbler

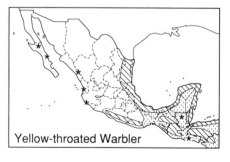

Yellow-throated Warbler

blackish with 2 white wingbars, white tail-spots.

Voice. A strong, sharp *chip*, similar to Yellow Warbler, but often louder.

Habs. Palm groves, pine woods, gardens, woodland and edge. Singly at mid- to upper levels, often in dense clusters, for example among coconuts or pine needles, or creeping on branches.

SS: Grace's Warbler has short yellow supercilium, lacks bold black auricular mask. See Blackburnian Warbler.

SD: F to C transient and winter visitor (Jul–Apr; SL–1500 m) on Atlantic Slope, mainly near coast, from Tamps to Belize; F to U but local in interior (500–2000 m) from Chis S; U on Pacific Slope from Gro to Guatemala. Vagrant elsewhere (Dec–Mar) on Pacific Slope from BCS (KK) to El Salvador, and (Aug) to Clipperton.

RA: Breeds E USA; winters SE USA and Caribbean to M.A.

GRACE'S WARBLER

Dendroica graciae*   Plate 55
Chipe de Grace

ID: 4.5–5 in (11.5–12.5 cm). Widespread in pines. **Descr.** Ages/sexes differ slightly. *Short yellow supercilium and subocular crescent constrast with blue-grey crown and auriculars,* lores and forecrown mottled black in ♂. *Throat and chest yellow,* rest of underparts white, streaked blackish on sides and flanks; flanks washed buff in imm. Upperparts blue-grey, often with fairly distinct black streaks on lower back, back washed brown in imm. Wings and tail blackish with 2 white wingbars, white tail-spots. **Juv:** head and upperparts brownish with yellowish supraloral stripe, throat and underparts dirty whitish, streaked dusky brown, wings and tail as adult.

Voice. A fairly hard *chik* or *tchik*. Song a

rapid series of chips, often run into a faster ending, *chi-chi chi-chi-chit chi-chi-chi-chi-chi-chu* or *chu-chu-chu-chu chichichichichichichi,* etc.

Habs. Pine and pine–oak forest. At mid- to upper levels in pines, singly or in pairs; joins mixed-species flocks. Nest cup of fine plant materials at mid-to upper levels in pines. Eggs: 3–4, whitish, speckled with reddish browns and greys (WFVZ).

SS: See Yellow-throated Warbler.

SD: C to U resident on Pacific Slope and in adjacent interior (600–3000 m, mainly above 1000 m) from Son to Isthmus, in interior and on adjacent slopes to N cen Nicaragua, and (SL–900 m) in pine savannas from Belize to the Mosquitia; withdraws from NW in winter (GM) and R in winter (Oct–Mar) to W cen Ver and NW Oax.

RA: SW USA to Nicaragua; N birds withdraw S in winter.

PINE WARBLER

Dendroica pinus   Not illustrated
Chipe Pinero

ID: 4.7–5.2 in (12–13 cm). Rare winter visitor to NE Mexico. Bill relatively stout. **Descr.** Ages/sexes differ. ♂: *head and upperparts yellow-olive with short yellow supercilium* and subocular crescent; wings and tail darker with 2 white wingbars, white tail-spots. *Throat and underparts yellow, becoming white on vent and undertail coverts, blurrily streaked dusky olive on sides of chest.* ♀: duller, head and upperparts olive, throat and chest dull yellowish, mottled and streaked olive on sides of chest. **Imm:** *head and upperparts brownish to brownish olive with short buff to whitish supercilium and subocular crescent.* Wings and tail dark with 2 pale buff to whitish wingbars, whitish tail-spots. *Throat and underparts dirty pale buff,*

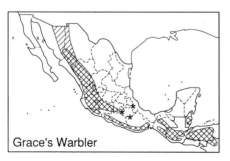

Grace's Warbler

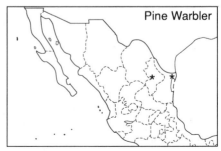

Pine Warbler

often *tinged lemon on throat and chest,* flanks washed buffy brown.

Voice. A sharp, fairly hard, smacking *tsik* or *tchik.*

Habs. Pine–oak and deciduous woodland. Joins mixed-species flocks.

SS: Yellow-throated Vireo has heavier bill, yellow spectacles. Basic Bay-breasted and Blackpoll warblers have long undertail coverts and appear shorter-tailed; unlikely to overlap seasonally.

SD: U to R (irregular?) winter visitor (Dec–Feb) on Atlantic Slope in NL (JCA) and Tamps, probably also NE Coah.

RA: Breeds E N.A.; winters SE USA, rarely to NE Mexico.

PRAIRIE WARBLER

Dendroica discolor   Not illustrated
Chipe Pradeño

ID: 4.2–4.7 in (11–12 cm). Winters on Caribbean islands. **Descr.** Ages/sexes differ. ♂: *short yellow supercilium and thick yellow subocular crescent set off by short black eyestripe and curving black moustache create distinctive facial expression.* Crown, auriculars, nape, and upperparts olive-green, chestnut back centers often visible. Wings and tail dark with 2 dull lemon wingbars, white tail-spots. Throat and underparts yellow, streaked black on sides and flanks. ♀: *duller,* eyestripe and moustache dark olive, back with little or no chestnut. **Imm:** duller than adult, auriculars blue-grey.

Voice. A fairly hard, slightly smacking *tchik* or *chik.*

Habs. Open scrubby woodland and edge, overgrown fields, open areas. At low to mid-levels, frequently wags its tail like a Palm Warbler.

SS: See imm ♀ Magnolia Warbler.

SD: U to R transient and winter visitor (late Aug–Apr; around SL) from QR to N

Honduras, mainly on Caribbean islands. Vagrant (Sep–Mar) to S Ver (WJS), on Pacific Slope from Nay (Clow 1976) to El Salvador, to Chih (Jun 1959, Thompson 1962), and (Nov) to Clipperton.

RA: Breeds E N.A.; winters Caribbean and adjacent mainland coasts of Florida and M.A.

PALM WARBLER

Dendroica palmarum*   Not illustrated
Chipe Playero

ID: 4.5–5 in (11.5–12.5 cm). Winters on Caribbean islands and coast. **Descr.** Ages/sexes similar, seasonal variation. **Basic:** *face dusky with pale buff to whitish supercilium and darker eyestripe.* Crown, nape, and upperparts brownish to grey-brown, rump and uppertail coverts brighter, yellow-olive. Wings and tail dark, edged paler brownish, with distal white tail-spots. *Throat and underparts dirty whitish to pale buff with dusky streaking on chest and flanks, undertail coverts yellow.* **Alt:** crown chestnut, throat, chest, and supercilium yellow; underparts yellow in *D. p. paludicola* (one record from QR, Apr 1987, SJ).

Voice. A sharp, slightly metallic *tchip* or *shik.*

Habs. Open and semiopen areas and adjacent woodland. Usually on or near ground, often in small flocks; wags its tail constantly.

SS: See basic Myrtle Warbler, ♀ Cape May Warbler.

SD: F to C transient and winter visitor (late Sep–Apr; around SL) from N Camp to N Honduras, mainly Caribbean islands and coast, U to R to S Ver (SNGH, WJS). Vagrant (mid-Oct–early Apr) on Pacific Slope from Baja and Son to Oax, Islas Revillagigedo (Howell and Webb 1992g), and Clipperton.

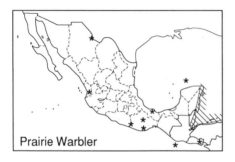

Prairie Warbler

Palm Warbler

RA: Breeds N N.A.; winters SE USA, Caribbean, and adjacent coasts of N M.A.

BAY-BREASTED WARBLER

Dendroica castanea   Not illustrated
Chipe Pechicastaño

ID: 4.7–5.2 in (12–13 cm). Transient in E and S. **Descr.** Age/sex/seasonal variation. **Basic/imm** (Sep–Nov): *narrow pale lemon supercilium and brighter narrow eye-crescents*; face, throat, and *underparts dirty pale buff* to pale lemon, often with *pinkish-cinnamon wash on flanks* (also chestnut mottling on flanks of many adults). Crown, nape, and upperparts olive with dusky streaking on back (blacker in adults, also on crown and nape of ♂); rump grey in adult. Wings and tail dark with 2 white wingbars, distal white tailspots. **Alt** ♂ (Apr–Jun): *face black with chestnut crown and throat, pale buff sides of neck. Underparts dirty pale buff, chestnut on sides and flanks.* Upperparts greyish to brownish grey, streaked black. **Alt** ♀: face dusky with pale buff eye-crescents, pale buff sides of neck often indistinct. Throat and underparts dirty whitish to pale buff with dull chestnut mottling on throat and sides of chest. Crown and upperparts olive-grey, streaked black, some chestnut mottling on crown.
  Voice. A fairly full, slightly sharp *chik* or *chip*.
  Habs. Forest and edge. At mid- to upper levels; joins mixed-species flocks.
SS: Basic Blackpoll Warbler has yellowish-flesh (not grey) legs and feet (sometimes restricted to soles), pale lemon underparts with indistinct dusky streaks on throat and chest, more distinct supercilium and eye-stripe. See Pine Warbler.
SD: F to U transient (mid-Apr–mid-May, late Sep–early Nov; SL–1000 m) on Atlantic Slope from Tamps to Honduras, mainly

from Yuc Pen E in autumn. Vagrant (late May–mid-Jun) to Son (GHR) and Baja and (late Oct–Nov) to Isla San Benedicto, Oax (PDV), and Clipperton.
RA: Breeds N N.A.; winters Panama to Venezuela.

BLACKPOLL WARBLER

Dendroica striata   Not illustrated
Chipe Gorrinegro

ID: 4.7–5.2 in (12–13 cm). Rare migrant. **Descr.** Age/sex/seasonal variation. **Basic/imm:** *narrow pale lemon supercilium and dusky eyestripe*; face, throat, and *underparts dirty pale lemon* (whiter in adult), with *indistinct dusky streaking on chest.* Crown, nape, and upperparts olive with dusky streaking (blacker in adults, also on crown and nape of ♂). Wings and tail dark with 2 white wingbars, distal white tailspots. **Alt** ♂: *black cap contrasts with white auriculars; throat and underparts white with black malar stripe running into black streaks* on sides and flanks. Nape and upperparts greyish, streaked black. **Alt** ♀: face dusky with narrow paler supercilium and subocular crescent. *Throat and underparts whitish to pale lemon, with hint of black malar stripe, narrow black streaking on sides and flanks.* Crown, nape, and upperparts greyish olive, streaked black.
  Voice. A sharp *chik* or *tchip*.
  Habs. Forest and edge, groves of trees near coast. At mid- to upper levels; may join mixed-species flocks.
SS: See Bay-breasted and Pine warblers.
SD: R transient (late Apr–May, early–mid-Nov) in NE Yuc Pen (Zimmerman 1969; Paulson 1986; G and JC, KK, SJ). Vagrant (mid-Oct–mid-Nov) to Baja (Keith and Stejskal 1987; Howell and Pyle 1993), Oax, and Clipperton, and (Jun) to Chih.

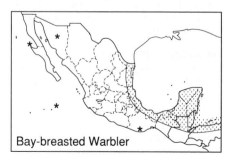

Bay-breasted Warbler

Blackpoll Warbler

RA: Breeds N N.A.; winters Colombia to E
Peru and N Argentina.

CERULEAN WARBLER
Dendroica cerulea   Not illustrated
Chipe Cerúleo

ID: 4.2–4.7 in (11–12 cm). Transient in E.
Relatively short-tailed. Descr. Ages/sexes
differ. ♂: *head and upperparts bluish*,
sometimes with trace of whitish supercilium, back streaked blackish. Wings and
tail blackish with 2 white wingbars, white
tail-spots. *Throat and underparts white
with broken to complete grey-blue chest
band*, dark streaks on sides and flanks. ♀:
*Head and upperparts blue-green with
whitish supercilium*. Whitish underparts
often washed buffy lemon, especially on
chest, sides and flanks streaked dusky.
Imm: resembles ♀ but ♂ often bluer above,
with darker streaks on sides and flanks.
   Voice. A full *tchik* or *chik*.
   Habs. Forest and edge. At mid- to upper
levels; joins mixed-species flocks.
SS: See imm Black-throated Grey Warbler.
SD: U to R transient (late Mar–early May,
mid-Aug–mid-Oct; SL–750 m) on Atlan-
tic Slope from Ver (not casual, as stated by
AOU 1983) to Honduras. Vagrant to BCN
and (Apr) off Guatemala (Jehl 1974). A
Dec report from BCS (Wilbur 1987) re-
quires verification.
RA: Breeds E N.A.; winters Colombia to
Bolivia.

Cerulean Warbler

BLACK-AND-WHITE WARBLER
Mniotilta varia   Not illustrated
Chipe Trepador

ID: 4.5–5 in (11.5–12.5 cm). Widespread in
winter. Descr. Age/sex/seasonal variation.
Basic ♂: *crown black with broad white
median stripe*, blackish auriculars contrast

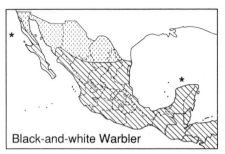

Black-and-white Warbler

with white supercilium, subocular crescent, and throat. White underparts
streaked black on lower throat, sides, and
flanks; undertail coverts with dark centers.
*White streaking on black upperparts forms
neat braces. Wings and tail black, edged
white*, with 2 white wingbars, white tail-
spots. Alt ♂: auriculars and throat black.
♀: resembles basic ♂ but *auriculars dusky
whitish, throat whitish*, streaks on sides
less distinct. Imm: resembles ♀ but underparts often tinged buff, streaks on sides
indistinct in ♀.
   Voice. A sharp, slightly liquid *spik* or
chik, often run into a rapid, spluttering
series when excited; also a high, thin
ssit.
   Habs. Humid to semiarid woodland and
edge. Forages on trunks and branches like
a creeper; joins mixed-species flocks.
SS: See Black-throated Grey Warbler.
SD: C to F transient and winter visitor (Jul–
May; SL–2500 m) on both slopes and in
interior from Nay and Tamps to Honduras
and W Nicaragua; U to R to Baja and in N
Mexico.
RA: Breeds E and N N.A.; winters S USA to
N S.A.
NB: Traditionally placed in the genus
Mniotilta, this species should be merged
with *Dendroica*, in which case the overall
genus name should become *Mniotilta*. The
main barrier to this seems to be the
bureaucracy of sorting out the consequently
jumbled nomenclature.

AMERICAN REDSTART
Setophaga ruticilla   Not illustrated
Pavito Migratorio

ID: 4.7–5.2 in (12–13 cm). Widespread in
winter, mainly lowlands. Descr. Ages/
sexes differ. ♂: *head, chest, and upperparts
black with flaming orange flash on sides of
chest, broad flaming orange band across*

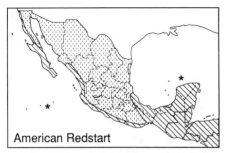

American Redstart

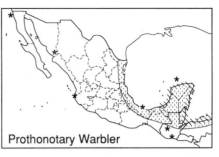

Prothonotary Warbler

base of remiges and outer rectrices. Belly and undertail coverts whitish. ♀: *head* and *upperparts greyish to greyish olive with indistinct whitish spectacles, yellow band across base of remiges (can be indistinct) and outer rectrices. Throat and underparts whitish with yellow to orange-yellow flash on sides of chest.* Imm ♂: resembles ♀ but chest side flashes yellow-orange to orange, usually shows a few black flecks on head, back, and chest from mid-winter on.

Voice. A sharp *tsik* or *tchik*, suggests Yellow Warbler but sweeter and more slurred.

Habs. Forest and edge, scrubby woodland. Joins mixed-species flocks. Feeds actively with much flycatching, tail fanned and wings often drooped.

SD: C to F transient and winter visitor (Aug–May; SL–1500 m, to 2500 m in migration) on Atlantic Slope from Ver to Honduras; U to C on Pacific Slope from Baja and Sin to Isthmus, F to C on Pacific Slope and in interior S of Isthmus. U transient (Aug–Oct, Apr–May) in interior N of Isthmus, C to F in NE Mexico. Vagrant (Aug–May) to Islas Revillagigedo and Clipperton.

RA: Breeds N and E N.A.; winters Mexico and Caribbean to Ecuador and Brazil.

NB: *Setophaga* should probably also be merged with *Dendroica/Mniotilta*.

PROTHONOTARY WARBLER

Protonotaria citrea   Not illustrated
Chipe Protonotario

ID: 5–5.5 in (13–14 cm). Migrant in E. *Bill long, stout, and pointed.* Descr. Ages/sexes similar. *Head and underparts bright golden yellow* to yellow (crown washed olive in ♀), with white undertail coverts. *Upperparts olive* with blue-grey rump and uppertail coverts. *Wings and tail blue-grey with white basal tail-spots.*

Voice. A metallic *tchiin* or *chiik*, suggests Northern Waterthrush and a high, slightly shrill *sssip*.

Habs. Swampy and scrubby woodland, forest understory, mangroves. Feeds low; does not join mixed-species flocks.

SD: F to C transient (late Jul–mid-Oct, Mar–Apr; around SL) on Atlantic Slope from Ver to Honduras, mainly from Yuc Pen E; R (to 2100 m) in interior Chis and Guatemala. R to U winter visitor (Oct–Mar) from S Ver to Honduras. Vagrant (late Apr–May) to Jal, Son (Russell and Lamm 1978), and BCN (Jehl 1977), and (late Oct–Dec) to Clipperton and S Guatemala (Behrstock 1977).

RA: Breeds SE N.A.; winters Caribbean coast of M.A. to N S.A.

WORM-EATING WARBLER

Helmitheros vermivorus   Not illustrated
Chipe Gusanero

ID: 4.8–5.3 in (12–13 cm). Winters in SE. Bill stout and pointed. Descr. Ages/sexes similar. *Head striped with broad buff median crown stripe, dark brown lateral crown stripe, buff supercilium, and dark brown eyestripe. Auriculars, throat and underparts buff,* palest on throat and vent, undertail coverts with dusky centers. *Upperparts greyish olive.*

Voice. A high, lisping *tssi* or *tzip* often doubled, and a sharp *tchik*.

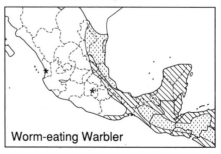

Worm-eating Warbler

Habs. Humid to semiarid forest and edge, second growth woodland and thickets. At low to mid-levels, forages singly in tangles, associates loosely with mixed-species flocks.

SD: F to C transient and winter visitor (Aug–Apr; SL–1500 m) on Atlantic Slope from S Ver to Honduras; U to R to S Tamps and on Pacific Slope from Nay (PA, SNGH, PP) to El Salvador. F to U transient (Aug–Oct, Apr) in NE Mexico, R W to DF (Wilson and Ceballos-L. 1986).

RA: Breeds E USA; winters SE Mexico and Caribbean to Panama.

SWAINSON'S WARBLER

Helmitheros swainsonii   Not illustrated
Chipe de Swainson

ID: *5–5.5* in (13–14 cm). Winters Caribbean coast and islands. *Bill stout and pointed.* **Descr.** Ages/sexes similar. *Creamy supercilium contrasts with rufous-brown crown and dark eyestripe.* Auriculars, throat and underparts dirty whitish, tinged lemon on throat and belly, slightly duskier on flanks. Upperparts warm olive-brown.
   Voice. A full, sharp *cheik* or *chink*.
   Habs. Humid evergreen and semideciduous forest, swampy woodland. Singly, on or near forest floor where forages in leaf litter and tangles.

SD: U transient and winter visitor (Sep–Apr; SL–500 m) on Atlantic Slope from Yuc Pen to Honduras; R in winter to S Ver; U to R transient (late Aug–Oct, Apr) on Atlantic Slope N to Tamps.

RA: Breeds SE USA; winters Caribbean and adjacent coasts of N M.A.

NB: Traditionally placed in the monotypic genus *Limnothlypis*.

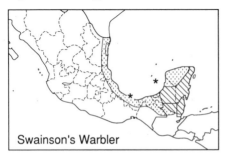

Swainson's Warbler

GENUS SEIURUS

These are distinctive terrestrial warblers that walk rather than hop. Bills fairly stout, flesh-pink legs stout and long. Ages/sexes similar. Do not associate with mixed-species flocks.

OVENBIRD

Seiurus aurocapillus   Not illustrated
Chipe-suelero Coronado

ID: 5.3–5.8 in (13.5–14.5 cm). Widespread in winter, mainly lowlands. **Descr.** *Head greyish olive with white eyering and black-edged, pale rufous median crown stripe.* Throat and *underparts white with* black malar stripe, and *bold black spotting* on chest running into streaks on flanks. Upperparts olive.
   Voice. A sharp, strong *tchip* or *tchk*, often repeated insistently.
   Habs. Forest and edge, scrubby woodland and second growth. On or near forest floor where walks with tail cocked.

SS: See *Catharus* thrushes.

SD: F to C transient and winter visitor (late Aug–mid-May; SL–2200 m) on both slopes from S Sin and S Tamps to Honduras and El Salvador; U to F in interior from cen Mexico S. U to F transient (late Aug–Oct, Apr–mid-May) in NE Mexico. Vagrant (late May–Jun) to NW Mexico and (Oct–Nov) to Clipperton.

RA: Breeds E N.A.; winters Mexico and Caribbean to N S.A.

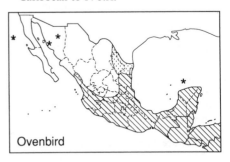

Ovenbird

NORTHERN WATERTHRUSH

*Seiurus noveboracensis**   Not illustrated
Chipe-suelero Charquero

ID: 5.3–5.8 in (13.5–14.5 cm). Widespread in winter, *mainly lowlands.* **Descr.** Head and upperparts brown, with *narrow pale lemon to whitish supercilium.* Throat and underparts pale lemon to whitish, *throat*

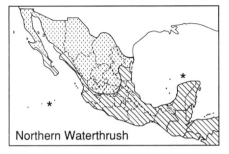

Northern Waterthrush

often flecked dark brown, chest and flanks neatly streaked dark brown.

Voice. A bright, metallic *chink* or *tchink.*
Habs. Mangroves, swampy woodland, margins of lakes, rivers, marshes, etc. Walks with constant tail-wagging.
SS: Louisiana Waterthrush (prefers running water) has stouter bill, white supercilium broadens behind eye, fewer and coarser streaks on underparts; note contrasting buff flanks, duller call.
SD: C to F transient and winter visitor (late Aug–May; SL–2500 m) on both slopes from Sin and S Tamps to Honduras and El Salvador, F to U in interior from Lerma drainage S; U to R to BCS and cen Son (GM). U to F transient (late Aug–Oct, Apr–May) in N Mexico. Vagrant (late Oct–Feb) to Isla Socorro (Howell and Webb 1992g) and Clipperton.
RA: Breeds N N.A.; winters Mexico and Caribbean to Ecuador and Peru.

LOUISIANA WATERTHRUSH

Seiurus motacilla   Not illustrated
Chipe-suelero Arroyero

ID: 5.5–6 in (14–15 cm). Widespread in winter, *mainly foothills. Bill relatively stout.* **Descr.** Head and upperparts brown, *white supercilium broadens behind eye, washed buff in front of eye. Throat and underparts white,* streaked dark brown on

chest and flanks, *note buffy-cinnamon wash on flanks.*
Voice. A metallic *chik* or *chink,* less bright than Northern Waterthrush.
Habs. Running streams, margins of lakes, rivers, marshes, prefers running water. Walks with more leisurely tail-wagging than Northern Waterthrush, often the whole rear end swinging from side to side.
SS: See Northern Waterthrush.
SD: F to U transient and winter visitor (Jul–Apr; near SL–3000 m) on both slopes from Son and S NL, and in interior from Isthmus, to Honduras and W Nicaragua; R to BCS. U transient (Jul–Sep, Mar–Apr) in Yuc Pen, and presumably in interior N of Isthmus (no records for N Plateau). Vagrant (late Apr) to BCN.
RA: Breeds E N.A.; winters Mexico and Caribbean to N S.A.

GENUS OPORORNIS

These are skulking warblers of forest understory and scrub. Overall olive above, yellow below, with distinctive face patterns. Bills flesh to dusky with dark culmen, legs flesh.

KENTUCKY WARBLER
Oporornis formosus   Not illustrated
Chipe de Kentucky

ID: 5–5.2 in (12.5–13.5 cm). Winters in SE.
Descr. Ages similar, sexes differ slightly. ♂: *yellow supraloral stripe curves around behind eye and contrasts with black crown and auricular mask which extends slightly onto sides of neck.* Upperparts olive, throat and underparts yellow. ♀: crown and auricular mask mottled olive.
Voice. A full, fairly deep *chup* or *tchk.*
Habs. Humid to semihumid evergreen and semideciduous forest. Hops about on forest floor in shady understory.

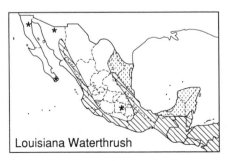

Louisiana Waterthrush

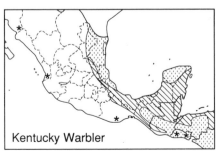

Kentucky Warbler

SS: Imm ♂ Common Yellowthroat lacks neat yellow spectacles, black does not extend onto sides of neck.

SD: C to F transient and winter visitor (Aug–Apr: SL–1500 m) on Atlantic Slope from S Ver to Honduras; U to F on Pacific Slope from Nay (Clow 1976) to El Salvador; R N to S Tamps (Arvin 1990). F transient (Aug–Oct, Apr) in NE Mexico. Vagrant (mid-Jun) to Son.

RA: Breeds E USA; winters SE Mexico to N S.A.

CONNECTICUT WARBLER
Oporornis agilis   Not illustrated
Chipe de Connecticut

ID: 5–5.3 in (13–14 cm). Spring transient through Caribbean islands. *Relatively heavily built, long wings and long undertail coverts accentuate short-tailed look.* **Descr.** Ages/sexes differ. ♂: *head and chest grey with complete white eyering, rest of underparts yellow, flanks washed olive.* Upperparts olive. ♀: *head and chest washed brownish, throat often paler, buffy, underparts duller yellow.* **Imm** (autumn): resembles ♀ but *head and chest brownish, crown and nape washed olive, throat dirty pale buff, eyering whitish to pale buff.*

Habs. Forest and scrubby woodland. Walks on ground, rarely hops, skulking and overlooked easily. Usually silent.

SS: Mourning and MacGillivray's warblers more slender, longer-tailed, hop rather than walk, ♀♀/imms brighter yellow below. Eyering of Mourning (when present) narrow and often broken in front of eye, imm lacks well-defined hood, usually shows trace of pale lemon spectacles. MacGillivray's has distinct white eye-crescents. Imm ♀ Common Yellowthroat has longer tail, eyering less distinct, yellow throat lacks hooded effect, underparts paler; note call. See Nashville Warbler.

SD: Probably R (to U?) transient (early–mid-May) off and along Caribbean coasts but only one record to date: Half Moon Cay, Belize (7 May 1958, Russell 1964). Vagrant to Clipperton, 1 Nov 1987 (Howell *et al.* 1993).

RA: Breeds N N.A.; winters N Colombia to Amazonian Brazil.

MOURNING WARBLER
Oporornis philadelphia   Not illustrated
Chipe Llorón

ID: 4.7–5.2 in (12–13 cm). *Transient* in E and S. **Descr.** Ages/sexes differ. ♂: *Head and chest grey with black lores, heavy black mottling on throat and chest; rest of underparts yellow, washed olive on flanks.* Upperparts olive. ♀: *head and chest grey, often with narrow, broken whitish to pale lemon eyering, throat pale grey to pale lemon.* **Imm** (autumn): resembles ♀ but head olive, *throat and chest pale lemon with greyish sides of chest creating hint of a hood, eyering pale lemon, often has pale lemon supraloral stripe creating vaguely spectacled look.*

Voice. A sharp, slightly liquid *chik* or *whic*.

Habs. Humid to semihumid, evergreen and deciduous forest and edge, second growth. Hops about in undergrowth, skulking.

SS: MacGillivray's Warbler has bold white eye-crescents, imm has dusky chest and whitish to pale buff (rarely dull lemon) throat. Common Yellowthroat has pale grey to dirty buff belly and flanks, tail often held cocked; note voice. See Connecticut Warbler.

SD: F to C transient (mid-Apr–May, Sep–mid-Oct; near SL–2000 m) on Atlantic Slope from Tamps to Honduras (but R in Yuc Pen), R W to DF (Wilson and Ceballos-L. 1986); F to U on Pacific Slope

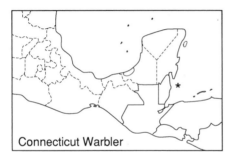
Connecticut Warbler

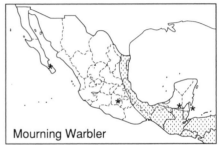
Mourning Warbler

and in interior from Isthmus S. Vagrant to BCS (Jun 1980, Helbig 1983), possibly also BCN (Wilbur 1987).

RA: Breeds NE N.A.; winters S C.A. to Ecuador and Venezuela.

MacGILLIVRAY'S WARBLER

Oporornis tolmiei   Not illustrated
Chipe de Tolmie

ID: 4.7–5.2 in (12–13 cm). *Widespread in winter* except SE. **Descr.** Ages/sexes differ. *♂: head and chest grey with black lores, bold white eye-crescents*, heavy black mottling on throat and chest; rest of underparts yellow, washed olive on flanks. Upperparts olive. *♀: head and chest grey with dusky lores and white eye-crescents*, throat pale grey. **Imm:** *resembles ♀ but head washed olive, throat whitish to pale buff, chest often washed brownish*, may have pale buff supraloral stripe. **Juv:** head and upperparts brownish olive with 2 pale cinnamon wingbars; throat and chest dusky tawny-brownish becoming pale lemon on belly and undertail coverts. Quickly attains adult-like plumage.

  Voice. A fairly hard to slightly liquid *tchik* or *drik*, rarely slightly metallic. Song an accelerating series of chips run into a rolled trill or short warble.

  Habs. Humid to semiarid forest and edge, second growth. Breeds in scrubby chaparral with scattered bushes. Hops about in undergrowth, mostly skulking but sings from conspicuous perches. Nest cup of grasses, twigs, etc., with slight roof of grasses or other material, low in bush or shrub. Eggs: 3–5, whitish, speckled and spotted with reddish browns, greys, and black (WFVZ).

SS: See Mourning and Connecticut warblers.

SD: F to C but local breeder (May–Aug; 1800–2700 m) in SE Coah (Ely 1962) and S NL. C to F transient and winter visitor

(mid-Aug–May; SL–3000 m) in BCS, on both slopes from S Son and S NL, and in interior from cen Mexico, to Honduras and N cen Nicaragua; absent from Yuc Pen. C to F transient (mid-Aug–Oct, Apr–May) in N Mexico.

RA: Breeds W N.A.; winters Mexico to W Panama.

GENUS GEOTHLYPIS: Yellowthroats

Yellowthroats are a fairly uniform-looking group of warblers that inhabit dense brushy vegetation, typically in association with water. Fairly skulking, tails often held cocked. ♂♂ have broad black masks from their forehead across their auriculars, ♀♀ of several species are similar. Species limits within the genus unclear.

COMMON YELLOWTHROAT

*Geothlypis trichas**   Plate 54
Mascarita Común

ID: 4.5–5.2 in (11.5–13.5 cm). *Widespread,* mainly in marshes. Marked geographic variation, three groups: *trichas* group (mainly N.A. migrants, widespread in winter; including *occidentalis*, shown), *modesta* group (NW Mexico), and *melanops* group (cen Mexico). **Descr.** Ages/sexes differ. *Trichas. ♂: bill blackish, often flesh below,* legs flesh to brownish flesh. *Whitish to bluish-white border above black mask.* Crown and upperparts olive. *Throat and chest yellow, becoming whitish to pale lemon on belly,* flanks brownish, undertail coverts yellow. *♀: head and upperparts greyish olive to olive, with pale eyering and usually a pale supercilium. Throat and chest yellow, becoming whitish to pale buff on belly,* flanks brownish,

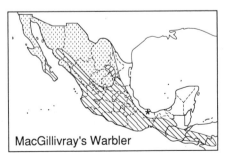

MacGillivray's Warbler

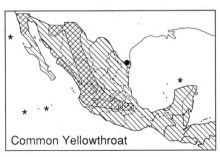

Common Yellowthroat

undertail coverts yellow. **Imm** ♂: resembles ♀ but *face mottled blackish*, resembles ♂ by spring but may show whitish eye-crescents. **Imm** ♀: resembles ♀ but duller, throat and chest may be buff, *yellow only on undertail coverts*. **Modesta**: *underparts mostly or entirely yellow*, washed brownish on flanks, *white border above* ♂*'s mask may be washed lemon*. **Melanops**: *underparts mostly or entirely yellow*, washed brownish to dusky-ochre on flanks. ♂♂ *in cen volcanic belt from Lago Cuitzeo W have pale lemon border above mask (chapalensis* group), *elsewhere border above mask is broader and bright whitish*. ♀♀ *have distinct pale buffy-brown supercilium*. **Juv** (all groups): head and upperparts brownish olive with 2 pale cinnamon wingbars, throat and underparts dusky lemon to dirty buff.

Voice. A low gruff *tchek* or *chrek*, and a dry rattle when agitated. Song a bright, varied, chanting series of rich notes, typically *w whee-ch-too whee-ch-too whee-ch-too* or *si-chi-tee si-chi-tee si-chi-tee*, or *wh wheechi wh wheechi wh wheech*, etc.

Habs. Freshwater and brackish marshes with reeds, grassy vegetation, etc. In migration and winter also wet fields, scrubby second growth, not always near water. Skulks, usually low in dense vegetation, but ♂♂ often sing from conspicuous perches. Nest cup of grasses, fine leaves, etc., low in clump of grass, reeds, etc. Eggs: 2–5, whitish, speckled with reddish browns, greys, and black (WFVZ).

SS: Hooded Yellowthroat (brushy highlands) usually in different habitat, darker and duller overall with olive flanks same colour as upperparts (paler in Common), ♂ has grey border above black mask, ♀ has indistinct greyish forehead and short supercilium; note voice. Belding's Yellowthroat (BCS) larger with heavier, all-dark bill, S populations richer and brighter than migrant Commons; ♂ has yellowish border above mask (as could some migrant Commons), ♀ brighter overall with typically more distinct face pattern than migrant Commons. ♀ Altamira Yellowthroat brighter overall with mostly yellow face, yellow underparts. ♀ Black-polled Yellowthroat has dark legs, darker overall, underparts rich yellow, washed ochre on flanks; note voice.

SD: C to F but local resident (SL–2500 m) on Pacific Slope and in interior S to cen volcanic belt; also N Tamps. C to F transient and winter visitor (mid-Aug–May; SL–3000 m) throughout. Vagrant (Nov–Feb) to Islas Revillagigedo (Howell and Webb 1992g) and Clipperton.

RA: Breeds N.A. to Mexico; winters S USA to N S.A.

NB1: *Chapalensis* has been considered a separate species, Chapala Yellowthroat. **NB2:** Altamira and Belding's yellowthroats may be conspecific with Common Yellowthroat.

ALTAMIRA (YELLOW-CROWNED) YELLOWTHROAT
Geothlypis trichas (in part) or *G. flavovelata*
Mascarita de Altamira   Plate 54

ID: 4.5–5 in (11.5–12.5 cm). *Endemic to NE Mexico.* Bill relatively long and pointed. **Descr.** Ages/sexes differ. ♂: bill black, legs flesh. *Broad yellow border above black mask, hindcrown and upperparts yellow-olive.* Throat and underparts rich yellow, washed olive on flanks. ♀: *face and underparts bright yellow with dusky-olive auriculars*, flanks washed olive. Crown, nape, and *upperparts yellow-olive*. **Imm** ♂/juv: as corresponding plumages of Common Yellowthroat (?), bill may be pale below. Birds in cen Tamps, appear duller than birds to the S; ♂♂ have a pale lemon forehead and brow tending toward the appearance of Common Yellowthroat.

Voice. Much like Common Yellowthroat. A gruff *cheh* or *tcheh*. Song varied, a rich to slightly scratchy *si-wee-chee si-wee-chee si-weet*, or *wi wi-chee-tee wi-chee-tee wi-chee-tee*, etc.

Habs. Freshwater marshes, irrigation ditches, etc., with reedbeds. Habits much

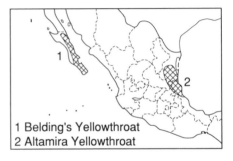

1 Belding's Yellowthroat
2 Altamira Yellowthroat

as Common Yellowthroat. Nest and eggs undescribed (?).

SS: See Common Yellowthroat.

SD/RA: F to C resident (SL–500 m) on Atlantic Slope from cen Tamps (SNGH, SW) to E SLP and N Ver.

BELDING'S (PENINSULAR) YELLOWTHROAT

*Geothlypis beldingi** or *G. trichas* (in part)
Mascarita de Belding   Plate 54
See Altamira Yellowthroat for map

ID: 5.3–5.8 in (13.5–14.5 cm). *Endemic to BCS.* Bill relatively large and heavy. Two ssp (*goldmani* in N, *beldingi* in S). **Descr.** Ages/sexes differ. *Beldingi.* ♂: *bill black-ish, legs flesh. Lemon border above black mask, crown and upperparts yellow-olive. Throat and underparts rich yellow,* washed ochre on flanks. ♀: head and upperparts olive with pale eyering and pale buffy-brown supercilium. Throat and chest yellow, becoming whitish on belly, undertail coverts yellow, flanks washed brown. **Imm ♂:** resembles ♀ but face mottled blackish, resembles ♂ by spring but may show whitish eye-crescents. *Goldmani:* duller overall, upperparts olive, throat and underparts yellow, washed brownish on flanks; *♂ has pale lemon to lemon-and-whitish border above black mask.* **Juv** (both ssp): head and upperparts brownish olive with 2 pale cinnamon wing-bars, throat and underparts dirty greyish buff.

Voice. A gruff *chek* or *chrek*, deeper and fuller than Common Yellowthroat, and a dry rattle when agitated. Song rich and powerful, may even suggest Cardinal by virtue of power, *whi-tee-wee-chu whi-tee-wee-chu whi-tee-wee-chu,* or *whee ti-whi-chu whee ti-whi-chu whee-chi,* or *whee-t-wee-ch-tu whee-t-wee-ch-tu wee-t-weech,* the main phrase usually given 2–3×.

Habs. Freshwater marshes with reedbeds, adjacent marshy growth. Habits much as Common Yellowthroat. Nest cup of reed strips, grasses, etc., low in reeds. Eggs: 2–3, whitish, speckled with browns, greys, and black (WFVZ).

SS: See Common Yellowthroat.

SD/RA: C to F but local resident in BCS, *goldmani* from about 26° to 24° N, *beldingi* in Cape District.

HOODED YELLOWTHROAT

*Geothlypis nelsoni**   Plate 54
Mascarita Matorralera

ID: 4.7–5.2 in (12–13 cm). *Endemic to brushy highlands N of Isthmus.* **Descr.** Ages/sexes differ. ♂: bill black, legs flesh. *Dull greyish to blue-grey border above black mask,* crown and upperparts dull olive. *Throat and underparts yellow, flanks olive.* ♀: head and upperparts olive with indistinctly greyish forehead and short supercilium. Throat and underparts yellow, paler on belly, *flanks olive.* **Imm ♂:** undescribed (?). **Juv:** sooty olive overall, paler below.

Voice. A gruff rolled *tchrrek,* also a softer *tchk* or *tchik,* suggesting MacGillivray's Warbler, and a dry, wren-like rattle *ji-ji-iirrrr* when agitated. Song varied, suggests Common Yellowthroat but typically of 2-syllable phrases (versus 3 syllables in Common) ending with an upward-inflected note: *wi-chee wi wi-chee wi-chee wi-chee wi-chee weet,* or *wee-chu wee-chu wee-chu wee-chu wee-chee weet,* less often *wi wi-chee wi wi-chee ... wi-weet,* etc.

Habs. Arid to semihumid brushy scrub, chaparral, not marshes. Habits much as Common Yellowthroat. Nest cup of grasses, lined with finer grasses, in clump of grass. Eggs: clutch size unknown (?), white, with faint rusty flecks (MLZ).

SS: See Common Yellowthroat.

SD/RA: F to C but local resident (1400–3000 m) in interior and on adjacent Atlantic Slope from S Coah to Oax, also in Sierra Madre del Sur of Oax; should be sought in Gro, and may also occur on Volcanes de Colima (SNGH).

NB: Sometimes known as Brush Yellow-throat.

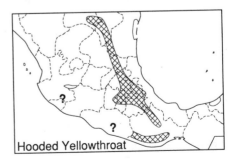

Hooded Yellowthroat

BLACK-POLLED YELLOWTHROAT
*Geothlypis speciosa**   Plate 54
Mascarita Transvolcanica

ID: 5–5.3 in (12.5–13 cm). *Endemic to cen volcanic belt marshes.* **Descr.** Ages/sexes differ. ♂: bill blackish, *legs dark brownish. Head black*, hindcrown and nape olive in fresh plumage, *sometimes washed grey at sides*, upperparts brownish olive. *Throat and underparts rich yellow*, washed orange on chest, ochre on flanks. ♀: *head and upperparts dark olive* with pale buff eyering and indistinct pale supercilium. *Throat and underparts rich yellow, washed ochre on flanks.* **Imm** ♂: resembles ♂ but black restricted mainly to lores and auriculars, forehead often black by 1st spring, crown dull olive, washed grey. **Juv:** head and upperparts brownish olive, underparts paler, dusky olive-lemon.
   Voice. A slightly nasal *chreh* and a more liquid *chwik*, distinct from Common Yellowthroat. Song a fairly rapid to rapid series of rich chips, usually running midway through into descending drier chips, *chree-chree-chree-chree-chree … chi-chi-chi-chi-chi …* or *chree-chree-chree-chree chee-chee-chee-chee*, etc., typically up to 8 chips in each half.
   Habs. Freshwater marshes and lakes with reedbeds. Habits much as Common Yellowthroat. Nest and eggs undescribed (?).
SS: See Common Yellowthroat.
SD/RA: C to F but local resident (1750–2500 m) in cen volcanic belt from E Mich to Mex. Reports from DF, if correct, apparently refer to an extirpated population (Wilson and Ceballos-L. 1986). Threatened by drainage schemes.
NB: Formerly known, confusingly, as Orizaba Yellowthroat. See Dickerman (1970) for clarification of this species' distribution.

GREY-CROWNED YELLOWTHROAT
*Chamaethlypis poliocephala**   Plate 54
Mascarita Piquigruesa

ID: 5.3–5.8 in (13.5–14.5 cm). *Widespread, not in marshes. Bill markedly stouter and deeper than* Geothlypis. **Descr.** Ages differ, sexes similar. **Adult:** bill flesh with blackish culmen, legs flesh. *Head grey with black lores and white eye-crescents* (eye-crescents reduced or lacking E of Sula Valley), crown washed olive in fresh plumage. Upperparts olive. Throat and underparts yellow, palest on belly, flanks washed brownish. **Imm:** resembles adult but head washed olive, lores dark, eye-crescents pale lemon, resembles adult by spring but some ♀♀ relatively dull through 1st year. **Juv:** brownish olive overall, more tawny below, with 2 narrow, pale buffy-lemon wingbars.
   Voice. Call a bright, nasal *chee'dle* or *cheed-l-eet*, rarely a single *chee.* Song a rich to slightly scratchy warble, recalling *Passerina* buntings, sometimes given in flight; also a series of rich, slightly plaintive, rising and falling whistles, *whiu whiu whiu …*, suggesting Canyon Wren.
   Habs. Fields and pastures, often with hedges, fences, scattered bushes and scrub. Usually keeps low, fairly skulking but sings and often calls from prominent perches. Tail often held cocked and swung from side to side. Nest cup of grasses near ground in grass tussock or tangle. Eggs: 2–3, whitish, speckled with reddish browns and black (WFVZ).
SS: Grey head and black lores distinctive; also note stout bill and voice.
SD: C to F resident (SL–1500 m) on both slopes from Sin and S Tamps to El Salvador and Honduras, also in upper Balsas drainage and in interior from Isthmus S.
RA: Mexico to W Panama.
NB1: Also known as Ground-Chat and

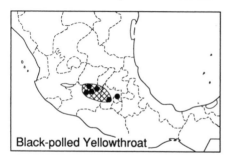

Black-polled Yellowthroat

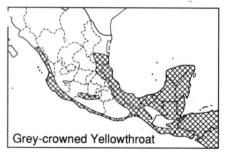

Grey-crowned Yellowthroat

Meadow Warbler. NB2: Sometimes placed in the genus *Geothlypis*.

GENUS WILSONIA

These are active, olive and yellow warblers of low to mid-levels in forest and edge. Bills grey, often paler to flesh below at base, legs flesh. Ages similar, sexes differ or ♀♀ often resemble ♂♂. All are migrants in the region.

HOODED WARBLER
Wilsonia citrina   Not illustrated
Chipe Encapuchado

ID: 4.8–5.3 in (12–13 cm). Winters in S and E. **Descr.** ♂: *yellow face surrounded by black crown*, nape, neck sides, *throat, and upper chest*; rest of underparts yellow. Upperparts yellow-olive with white tail-spots. ♀: *blank yellow face contrasts with olive crown and nape*, lores often dusky. Throat and underparts yellow. Some have trace of a hood, i.e. variable black on crown and neck sides, dusky wash on throat.
Voice. A metallic *tink* or *chink*, much like Blue Bunting.
Habs. Humid to semihumid, evergreen and semideciduous forest and edge. At low to mid-levels in forest understory; often flicks tail to show white tail-spots.
SS: Wilson's Warbler smaller, longer tail lacks white spots and often held cocked; note voice.
SD: C to F transient and winter visitor (mid-Aug–Apr; SL–1500 m) on Atlantic Slope from S Ver to Honduras, U to S Tamps; U to R on Pacific Slope from Isthmus to El Salvador; vagrant (Nov–Mar) to Son (Russell and Lamm 1978) and Nay. F transient (mid-Aug–Sep, early–mid-Apr) in NE Mexico, R W to DF (RGW).
RA: Breeds E USA; winters E Mexico to Panama.

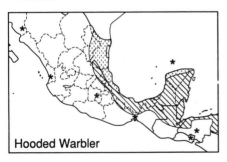
Hooded Warbler

WILSON'S WARBLER
Wilsonia pusilla*   Not illustrated
Chipe de Wilson

ID: 4.3–4.8 in (11–12 cm). Widespread in winter. **Descr.** ♂: *face, throat, and underparts yellow with slightly duskier auriculars* and *glossy black cap*. Nape and upperparts yellow-olive, wings and tail darker. ♀: crown olive to mostly black.
Voice. A dry *chek* or *chik*. Song often heard in spring migration, a series of about 8–15 chips, initially accelerating then fading.
Habs. Humid to semiarid forest and edge, second growth, overgrown fields, etc. At low to mid-levels, often holds tail cocked.
SS: See ♀ Hooded Warbler, Yellow Warbler.
SD: C to F transient and winter visitor (mid-Aug–May; SL–3500 m) in BCS, on both slopes from S Son and Tamps, and in interior from cen Mexico, to Honduras and W Nicaragua; R to U in Yuc Pen (Oct–early May, SNGH, SW, PW). C to F transient (mid-Aug–Oct, Apr–May) in N Mexico.
RA: Breeds N and W N.A.; winters Mexico to W Panama.

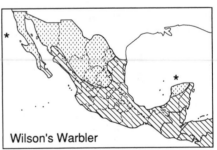
Wilson's Warbler

CANADA WARBLER
Wilsonia canadensis   Not illustrated
Chipe Collarejo

ID: 4.7–5.2 in (12–13 cm). Transient in E and S. **Descr.** ♂: *yellow supraloral stripe and white to pale lemon eyering contrast with black lores and lower auriculars*; forecrown streaked blackish. Rest of head and *upperparts grey*. Throat and underparts yellow with *necklace of black streaks across chest*, undertail coverts white. ♀: forehead greyish to olive, necklace blackish to dusky, often indistinct in imm ♀.
Voice. A slightly smacking *tsik*.

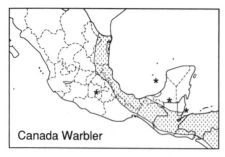

Canada Warbler

Habs. Humid to semihumid forest and edge. At low to mid-levels, tail often cocked and wings flicked.

SS: See Magnolia Warbler.

SD: F to C transient (mid-Apr–May, late Aug–mid-Oct; SL–2500 m) on Atlantic Slope from Tamps to Isthmus, and in interior and on both slopes from Isthmus S; R W to DF (Wilson and Ceballos-L. 1986) and in Yuc Pen (AMS photo, PW). Vagrant to Clipperton (Nov 1987). Winter reports from Mexico (AOU 1983) are based on Avilés spec. (Binford 1989) and thus are not credible.

RA: Breeds N N.A.; winters Colombia to Brazil, rarely to S C.A.

RED-FACED WARBLER

Cardellina rubrifrons   Plate 55
Chipe Carirrojo

ID: 4.8–5.3 in (12–13 cm). Highlands. **Descr.** Ages differ, sexes similar. **Adult:** bill grey, legs dull flesh. *Face, throat, and upper chest red, with black hood over crown to auriculars, whitish nape patch.* Upperparts grey with white rump, 2 indistinct whitish wingbars. Underparts whitish. ♀ averages duller than ♂. **Juv:** sooty brownish overall with whitish rump, belly, and undertail coverts; 2 pale buff wingbars.

Voice. A full *chip* or *tchip*. Song a sweet warbled series, *wi tsi-wi tsi-wi si-wi-si-wichu'*, and variations.

Habs. Arid to semihumid pine–oak and oak woodland, in winter also humid pine–evergreen forest and semideciduous woodland. Singly or in pairs at mid- to upper levels, tail often cocked and swung about loosely like Wilson's Warbler; joins mixed-species flocks. Nest cup of grasses rootlets, etc., concealed on bank or slope. Eggs: 3–4, white, speckled with reddish browns and greys (WFVZ).

SD/RA: F to C breeder (Mar–Aug; 1500–3000 m) on Pacific Slope and in adjacent interior from Son to Dgo (also SW USA). F to U transient and winter visitor (Aug–Apr) on Pacific Slope from Sin, and interior from cen Mexico, to W Honduras and El Salvador.

NB: *Cardellina* is close to *Wilsonia* and perhaps best merged with that genus.

RED WARBLER

Ergaticus ruber*   Plate 55
Chipe Rojo

ID: 5–5.3 in (12.5–13.5 cm). Endemic to highlands N of Isthmus. **Descr.** Ages differ, sexes similar. **Adult:** bill flesh with dark tip, legs flesh. *Red overall with white auricular patch* (slaty grey in *melanauris* of NW Mexico), wings and tail darker, edged pinkish red. **Juv:** overall pinkish cinnamon-brown with whitish (to greyish?) auricular patch, wings and tail darker, edged pinkish cinnamon, with 2 paler wingbars.

Voice. A high, thin *tsiu* or *tsii*. Song a varied series of high, thin chips and occasional short trills, often accelerating and ending with an upward-inflected note.

Habs. Humid to semihumid pine–oak, pine, and fir forest, also oak woodland in winter. Singly or in pairs, sometimes joins

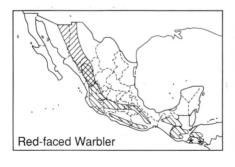

Red-faced Warbler

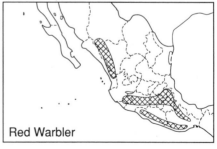

Red Warbler

mixed-species flocks. Nest is an oven-shaped structure of grasses, pine needles, etc., with upward-facing or side opening, on bank or on ground in dense growth. Eggs: 3–4, whitish, speckled with reddish browns and greys (WFVZ).

SD/RA: C to F resident (1800–3500 m) in interior and on adjacent slopes from SW Chih to N Nay (*melanauris*), and from S Jal and S Hgo to Oax.

PINK-HEADED WARBLER
Ergaticus versicolor   Plate 55
Chipe Rosado

ID: 5–5.3 in (12.5–13.5 cm). Endemic to highlands S of Isthmus. Descr. Ages differ, sexes similar. Adult: bill blackish, may be dusky horn below, legs flesh. *Head and chest silvery pink* with reddish forehead and dusky lores, *upperparts dark red*, wings and tail darker, edged dark red. *Underparts pinkish red.* Juv: overall pinkish cinnamon-brown, darker and redder above, wings and tail blackish, edged dark red, with 2 pinkish wingbars.
Voice. A high, slightly tinny *tsiu* or *ssing*. Song a varied series of high chips and short trills, often ending with a slightly emphatic note.
Habs. Humid to semihumid pine–oak, pine–evergreen, and evergreen forest and edge. Habits, nest, and eggs (2–4) much like Red Warbler.
SD/RA: C to F resident (1800–3500 m) in interior and on adjacent slopes of Chis and Guatemala. Populations may fluctuate locally, crashing after volcanic eruptions (when dust blankets out insects) and recovering slowly.

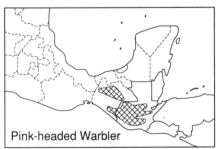
Pink-headed Warbler

PAINTED REDSTART
Myioborus pictus*   Plate 55
Pavito Aliblanco

ID: 5–5.3 in (13–13.5 cm). Highlands

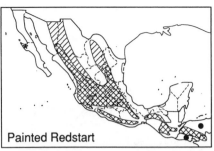
Painted Redstart

throughout. Descr. Ages differ, sexes similar. Adult: bill and legs black. *Head, upper chest, and sides black with contrasting red belly*, white subocular crescent; undertail coverts white with broad black centers. *Upperparts black with bold white wing panel*, white tertial edgings; *outer rectrices mostly white, striking when tail spread.* Juv: sooty black overall, browner below, wings and tail as adult.
Voice. A slightly buzzy *sreeh* or *sreeu*, and a longer *chew'elee*, all suggesting Pine Siskin. Song a varied rich warble, often accelerating initially and ending with an up-slurred note, *wee-chee wee-chee wee-chee wee-chee wee-chee chee-chee-chee* or *chee chee wee-wee-chee chee-chee wee-cheet*, etc.
Habs. Arid to semihumid oak and pine–oak woodland. Singly or in pairs, often with mixed-species flocks. Forages actively, tail fanned and swung from side to side. Nest cup (may be domed?) of grasses, fibers, etc., concealed in bank or on slope. Eggs: 3–4, whitish, speckled with reddish browns and greys (WFVZ).
SS: Slate-throated Redstart has all-dark wings, less white in tail.
SD/RA: C to F resident (1000–3000 m, lower locally in winter) on both slopes from S Son and S NL, and in interior from cen Mexico, to N cen Nicaragua. In summer (Mar–Aug) also N to SW USA. Breeding distribution patchy, for example, only a winter visitor (Aug–Mar) in DF (Wilson and Ceballos-L. 1986). Vagrant to BCN (Oct 1992, Howell and Pyle 1993).

SLATE-THROATED REDSTART
Myioborus miniatus*   Plates 55, 71
Pavito Gorjigris

ID: 5–5.5 in (13–14 cm). *Highlands throughout.* Two groups: *miniatus* group (Mexico to N Guatemala) and *conectens*

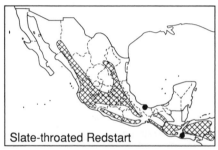

Slate-throated Redstart

group (Pacific mountains of Guatemala to N C.A.). **Descr.** Ages differ, sexes similar. *Miniatus.* **Adult:** bill and legs black. *Face and throat blackish*, crown grey with dull red central patch. *Underparts red with slaty grey flanks*, white undertail coverts with black centers. *Upperparts dark grey*, wings and tail darker, *outer rectrices extensively white*, striking when tail spread. *Connectens.* **Adult:** underparts orange-red, blackish outer rectrices broadly tipped white. **Juv** (both groups): sooty blackish overall, underparts washed pinkish cinnamon.

Voice. A high, sharp *tsi* or *tsit*. Song a varied series of rich chips, often changing in pitch part way through, the 2nd part accelerating and typically ending abruptly or with an up-slurred note, *s-wee s-wee s-wee s-wee s-wee s-chi s-chi s-chi*, or *chi-chi-chi-chi chiree-chiree-chiree-chiree*, etc.
Habs. Humid evergreen and pine–evergreen forest, also semihumid pine–oak and oak woodland in winter. Habits much as Painted Redstart but often lower and at times more skulking. Nest cup of grasses, pine needles, etc., with roofed covering, concealed on ground or bank. Eggs: 3–4, whitish, speckled with reddish browns and greys (WFVZ).
SS: See Painted Redstart.
SD: C to F resident (1000–3000 m, locally to 250 m in winter) on both slopes from S Son and S Tamps, and in interior from cen Mexico, to Honduras and N cen Nicaragua; U and local (Mar–Aug at least) N to S Coah and S NL (SNGH, AMS).
RA: Mexico to N Bolivia and Venezuela.

GENUS BASILEUTERUS

A genus of neotropical warblers. Ages differ, sexes similar, juv plumage ephemeral. Bills

black to horn, legs flesh. Usually at low to mid-levels in forest understory and edge. Nests are oven-like structures of grass, plant fibers, etc., with side opening, placed on bank or under grass tussock, overhanging rock, etc. Eggs: 2–4, whitish, speckled and spotted with reddish browns and greys (WFVZ).

FAN-TAILED WARBLER
Basileuterus lachrymosa　　　Plate 54
Chipe Roquero

ID: 5.8–6.3 in (14.5–16 cm). Endemic; tropical woodland. Tail long and graduated. **Descr.** Adult: *lores and sides of crown black with yellow crown patch, white lore spot, and white eye-crescents.* Rest of head and upperparts grey with *outer rectrices broadly tipped white. Throat and underparts yellow, washed ochre on chest,* undertail coverts whitish. **Juv:** dark sooty grey overall with 2 narrow whitish wingbars, pale lemon vent and undertail coverts.
Voice. A high, thin, slightly tinny *sieu* or *tseein*, and a short *si*. Song a pleasant warbled series up-slurred at the end, *dsiu d-dsiu d-dsiu d-dsiu d-dsiu d-sweet* or *suwee suwee suwee chu*, etc.
Habs. Evergreen to semideciduous forest and edge, typically with open rocky understory. Singly or in pairs, on or near ground, tail held fanned and swung from side to side with downward flipping action.
SD/RA: F to C but often local resident (50–1800 m) on Pacific Slope from Sin, and locally in adjacent interior from Gro, to Honduras and W Nicaragua, U and local (May–Jul) N to Son; F to C and local on Atlantic Slope from S Tamps to Oax, R to S Ver (Winker *et al.* 1992*b*). R (vagrant?) in Mor (Davis and Russell 1953); vagrant

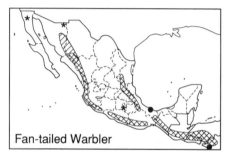

Fan-tailed Warbler

to BCN (Dec 1925, Grinnell and Lamb 1927).

NB: Traditionally placed in the monotypic genus *Euthlypis* but nest, eggs, voice, juv plumage, etc., all typical of *Basileuterus*. See Ridgely and Tudor (1989, pp. 184–5) for a discussion of *Basileuterus*.

GOLDEN-CROWNED WARBLER
*Basileuterus culicivorus**   Plate 54
Chipe Corona-dorada

ID: 4.8–5.3 in (12–13.5 cm). *Humid tropical forest.* **Descr.** *Adult: black lateral crown stripes border dull yellow to tawny-yellow median crown; face dusky olive with paler supercilium and subocular crescent, darker eyestripe. Upperparts olive-grey. Throat and underparts yellow, washed olive on flanks.* **Juv:** brownish overall, becoming buff on belly, with 2 pale cinnamon wingbars.

  Voice. A hard, dry, wren-like *tak* or *chik*, often run into a chatter; also a prolonged, more liquid, churring rattle. Song varied, a simple, slightly plaintive, rich warble ending with a strongly up-slurred note: *s wee-twee wee twee wee-chu* or *chew chew chew si-wi seeu*, etc.

  Habs. Humid to semihumid evergreen and semideciduous forest. In pairs or small groups, active at low to mid-levels in forest understory; joins mixed-species flocks, often with Tawny-crowned Greenlet and Red-crowned Ant-Tanager in SE.

SD: C to F resident (near SL–1600 m) on Atlantic Slope from NL to Honduras; F to C on Pacific Slope (600–2500 m) from Nay to Col (Schaldach 1963), and from Gro (Navarro 1986) to El Salvador; local in interior from Isthmus S.

RA: Mexico to N Argentina.

NB: S.A. forms sometimes regarded as specifically distinct, *B. cabanisi* and *B. auricapillus.*

RUFOUS-CAPPED WARBLER
*Basileuterus rufifrons**   Plate 54
Chipe Gorrirrufo

ID: 4.8–5.3 in (12–13.5 cm). Endemic, two groups: *rufifrons* group (Mexico to W Guatemala including *jouyi*, shown) and *salvini* (Isthmus to Belize). **Descr.** *Rufifrons.* **Adult:** *white supercilium contrasts with rufous-chestnut crown and auriculars, lores and postocular area blackish, lower edge of auriculars whitish.* Upperparts greyish olive to olive, nape greyer. Throat and chest yellow, rest of underparts whitish, washed dusky brown on flanks and undertail coverts. Intergrades with *salvini* from S Ver S. **Salvini.** **Adult:** underparts yellow, washed olive on sides, belly often paler yellow. **Juv** (both groups): brownish overall, becoming buff on belly, with whitish supercilium and 2 pale cinnamon wingbars.

  Voice. A hard *chik* or *chuk*, often run into an excited rapid chipping series; also a quiet, high *tsi* or *tik*. Song a rapid, often accelerating, series of chips which may run into a trill, *chi-chi-chi-chi-chi-ssiu-ssiu-ssiu-ssiu*, and variations, often with longer more trilled series.

  Habs. Scrub, semiopen areas with hedges, scattered bushes, second growth, woodland edge. In pairs or singly, close to ground, tail often held cocked near vertical.

SS: Chestnut-capped Warbler (sympatric in SE Chis and W Guatemala) prefers woodland, relatively shorter-tailed, lacks white stripe below auriculars, nape olive; intergrades may occur but no proven cases.

SD/RA: C to F resident (near SL–3000 m) on both slopes from Son and E NL, and in interior from cen Mexico, to W Guatemala; disjunctly in S Belize.

NB: AOU (1983) stated that Rufous-capped

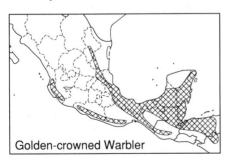

Golden-crowned Warbler

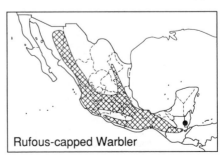

Rufous-capped Warbler

and Chestnut-capped warblers intergrade in Guatemala, El Salvador, and Honduras, but at the same time noted that Rufous-capped does not occur in the last two countries. Differences in plumage, structure, and voice suggest the two may be full species; field studies in area of sympatry are needed.

CHESTNUT-CAPPED [RUFOUS-CAPPED] WARBLER
Basileuterus (rufifrons?) d. delattrii
Chipe Gorricastaño   Plate 54

ID: 4.8–5.3 in (12–13.5 cm). Replaces Rufous-capped Warbler in C.A. **Descr. Adult:** *white supercilium contrasts with chestnut crown and auriculars, lores blackish.* Upperparts olive. *Throat and underparts yellow.* **Juv:** brownish overall, becoming buff on belly, with whitish supercilium and 2 pale cinnamon wingbars.
  Voice. A bright, slightly metallic *plik* or *tink*, suggesting Hooded Warbler (quite distinct from Rufous-capped Warbler), may be combined with a clipped, slightly metallic note, *chi-tink*, etc. Song a rapid, often accelerating series of chips which may run into a trill, *swee swee swee swee si-si-si-si-si-si-si-sit* and longer variations, often ending with an upward-inflected note.
  Habs. Arid to semihumid brushy woodland and edge, thorn forest. Habits much like Rufous-capped Warbler although tail usually less strongly cocked.
SS: See Rufous-capped Warbler. Golden-browed Warbler (montane forests) has yellow supercilium.
SD: C to F resident (near SL–1500 m, mainly foothills) from Pacific Slope of Guatemala to Honduras, also in interior and locally on Atlantic Slope from Honduras to W

Nicaragua; U to F in SE Chis (HG, SNGH tape, PP).
RA: S Mexico to N Venezuela.
NB: See Rufous-capped Warbler.

GOLDEN-BROWED WARBLER
Basileuterus belli*   Plate 54
Chipe Cejidorado

ID: 4.7–5.2 in (12–13.5 cm). Endemic, humid montane forests. **Descr. Adult:** *yellow supercilium, edged black above, contrasts with rufous-chestnut crown and auriculars.* Upperparts olive. Throat and underparts yellow, becoming dusky olive on sides and flanks. **Juv:** head and upperparts olive with trace of pale supercilium, 2 dull cinnamon wingbars. Throat and underparts sooty brownish, washed cinnamon on flanks.
  Voice. A lisping to slightly buzzy *tsssi* or *tssir*, and a disyllabic *ssi-irr*. Song a high, rapid, intensifying series which may slur up at the end, *ssi-ssi ssi-ssi ssi-ssi ssi-ssi-s-s-s-s-s-s syi-sweet* and variations, also a high, thin *ti ti syik, sik*.
  Habs. Humid pine–evergreen, pine–oak, and evergreen forest and edge. In pairs or singly, typically skulks in dense understory.
SS: See Chestnut-capped Warbler.
SD/RA: C to F resident (1200–3500 m) on both slopes from Dgo and S Tamps, and in interior from cen Mexico, to Honduras.

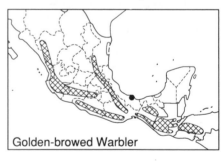
Golden-browed Warbler

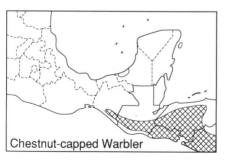
Chestnut-capped Warbler

YELLOW-BREASTED CHAT
Icteria virens*   Not illustrated
Gritón Pechiamarillo

ID: 6.5–7 in (16.5–18 cm). Widespread in winter, breeds N Mexico. Bill stout and heavy, tail long. **Descr.** Ages differ, sexes differ slightly. **Adult:** bill and legs blackish; bill greyish in winter with dark culmen. *Face greyish with black lores* (dark grey in

♀), *white supraloral stripe and eye-crescents.* Crown and upperparts greyish olive. *Bright yellow throat and chest contrast with white belly and undertail coverts.* **Juv:** head and upperparts brownish olive with trace of adult face pattern, throat and underparts pale lemon, washed dusky on chest.

Voice. A hard, dry *chak* or *chk*, often doubled or repeated, *cha-chak ... etc.,* and a dry, but not strong, oriole-like chatter. Song a varied, prolonged, strident series of clucks, rattles, scolds, mews, etc., each note typically repeated 3–8 ×, with single mellow notes often thrown in, overall suggests a mockingbird: *chreeu chreeu chreeu chreeu, huit, hoooh cheh-cheh-cheh cheh-cheh-cheh, joyn-joyn-joyn ...,* etc., sometimes given in parachuting display flight.

Habs. Breeds in deciduous woodland and thickets, especially riparian areas. In winter also in forest edge, second growth, scrub. Singly, skulks low in dense growth, tail usually cocked. Often sings from prominent perches. Nest cup of grasses, dead leaves, etc., low in thicket or dense bush. Eggs: 3–5, whitish, speckled with reddish browns and greys (WFVZ).

SD: F to U breeder (Apr–Aug; near SL–2500 m) in N Baja, on Pacific Slope S to N Sin, and in interior S over Plateau at least to Zac, probably to N Mich(?); perhaps formerly on Atlantic Slope in Tamps. F to C transient and winter visitor (mid-Aug–May; SL–1500 m) on both slopes from Sin and S Tamps, and in interior from Isthmus, to Honduras and W Nicaragua; R in winter along Balsas drainage.

RA: Breeds N.A. to cen Mexico; winters Mexico to Panama.

NB: Taxonomic affinities unclear, sometimes placed with the tanagers.

RED-BREASTED CHAT

Granatellus venustus* Plate 55
Granatelo Mexicano

ID: 5.7–6.2 in (14.5–16 cm). Endemic to W Mexico. **Descr.** Ages/sexes differ. ♂: bill greyish with dark culmen, legs greyish. *White postocular stripe,* edged black above, *contrasts with blue-grey crown and black auricular mask. Black chest band* (often reduced or absent in *francescae* of Islas Tres Marías) *borders white throat, chest and median underparts red, flanks white.* Upperparts blue-grey, tail blackish, *outer rectrices mostly white.* ♀: *buff lores and supercilium contrast with grey crown and dusky auriculars. Throat and underparts buff,* palest on throat and belly, *undertail coverts pink.* Upperparts grey, tail dark grey, *outer rectrices white as* ♂. **Imm** ♂: head and upperparts much like ♂ but with less black, wings brownish. Throat and underparts whitish with variable pink wash from chest to pink undertail coverts. Attains adult plumage by 2nd prebasic molt when about 1 year old. **Imm** ♀: differs from ♀ in brownish wings, paler pinkish to buff undertail coverts. **Juv:** undescribed (?).

Voice. A wet *plek* or *plik,* recalling *Passerina* buntings, at times repeated steadily. Song variable, a fairly sweet warble, at times ending softly, *wheet see-wheet see-wheet see-wheet wheet wheet wheet wheeur,* or *si-wee' si-wee' si-wee' s'weechu weechu,* or simply *s-wee s-wee ...,* up to 7×.

Habs. Arid to semiarid thorn forest, scrubby woodland. Singly or in pairs; at low to mid-levels; moves sluggishly and overlooked easily; tail often cocked high, fanned, and waved about. Associates loosely with mixed-species flocks. Nest and eggs undescribed (?).

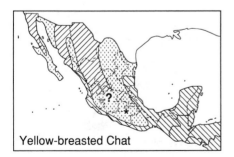
Yellow-breasted Chat

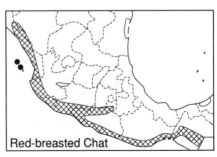
Red-breasted Chat

SD/RA: F resident (SL–1200 m) on Pacific Slope from Sin to Isthmus, and in interior Chis; should be sought in NW Guatemala; R in Balsas drainage to Mor (Rowley 1962)
NB: Taxonomic affinities of *Granatellus* unclear, sometimes placed with the tanagers.

GREY-THROATED CHAT
*Granatellus sallaei**   Plate 55
Granatelo Yucateco

ID: 5–5.3 in (12.5–13.5 cm). Endemic to Yuc Pen and adjacent SE. **Descr.** Ages/sexes differ. ♂: bill greyish with dark culmen, legs greyish. *Head and throat blue-grey to grey with white postocular stripe edged black above. Underparts red with white flanks.* Upperparts blue-grey to grey, wings and tail darker, *outer rectrices narrowly tipped white.* ♀: *pale buff lores and supercilium contrast with grey crown and dusky auriculars. Throat and underparts buff*, palest on throat and belly, undertail coverts rich buff to pale pinkish. Upperparts grey, outer rectrices narrowly tipped whitish. **Juv:** resembles ♀ but face pattern less distinct, quickly attains adult-like plumage but wings and tail browner through first year.
  Voice. A soft *whiut* or *whet*, may recall soft Least Flycatcher call. Song variable, a short, sweet warble, *swee ti-wee ti swee ti-chu*, or *swee-chee' chee-cheu chee-cheu'*, or *chee' ti lee ti-lee ti-eu*, etc.
  Habs. Humid to semiarid scrubby woodland, semideciduous forest and thickets in evergreen forest. Singly or in pairs at low to mid-levels; tail often cocked high, fanned, and swung about. Often joins mixed-species flocks at ant swarms. Nest and eggs undescribed (?).
SD/RA: F to C resident (SL–300 m) in Yuc Pen, S to Petén and N Belize; U to F but local (near SL–1500 m) on Atlantic Slope from S Ver to Guatemala.
NB: See NB under Red-breasted Chat.

OLIVE WARBLER
*Peucedramus taeniatus**   Plate 55
Chipe Ocotero

ID: 4.8–5.3 in (12–13 cm). Highland pines. Bill slender and pointed, *tail forked.* **Descr.** Ages/sexes differ. ♂: bill black, legs grey. *Head and chest orange-ochre with black auricular mask*; belly and undertail coverts whitish, washed dusky to buffy brown on flanks. Upperparts greyish, wings and tail blackish with 2 *white wingbars*, white flash at base of primaries yellow-olive edges to secondaries, *mostly white outer rectrices.* ♀: bill may be flesh below at base. *Head and chest* lemon *yellow* (NW Mexico) to orange-yellow (S Mexico and C.A.) with *dusky auricular mask often most distinct as dark patch on hind auriculars*, crown and nape often washed olive. Underparts dirty whitish, washed buffy brown on flanks. Upperparts greyish, wings and tail resemble ♂ but white slightly less extensive. **Imm** ♂: resembles ♀ but head and chest brighter, tawny-yellow. Some imm ♂♂ in S may resemble adult after 1st prebasic molt but in N, adult plumage attained when about 1 year old. **Imm** ♀: duller than ♀, crown greyish olive. **Juv:** resembles imm ♀ but crown and upperparts brownish olive, face less distinctly marked, throat and chest dirty pale lemon.
  Voice. A mellow, slightly plaintive *teu* or *peu.* Song varied, usually a fairly rapid series of rich chips that may suggest a titmouse, *chroo-chroo-chroo-chroo-chroo-chroo*, or *tree-tree-tree-tree-tree-tree*, and variations, *chi-chi-chu chi-chi-chu*, a more leisurely *chwee chwee chwee chwee*, etc.

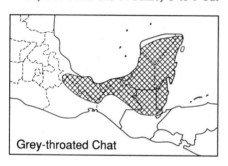
Grey-throated Chat

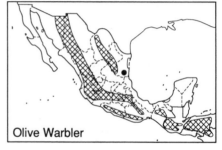
Olive Warbler

Habs. Pine and pine–oak woodland, mainly in arid to semiarid areas. At mid- to upper levels; singly or in small groups; often forages in clumps of pine needles. Joins mixed-species flocks. Nest cup of rootlets, moss, etc., at mid- to upper levels in pines. Eggs: 2–4, blue-grey, speckled with blackish, browns, and greys (WFVZ).

SS: See Hermit Warbler.

SD/RA: C to F resident (1500–3500 m) in interior and on adjacent slopes from Son and Coah (also SW USA in summer) to Honduras and N cen Nicaragua; U to R N of Sin in winter and in N Coah.

NB: Taxonomic position uncertain, has been placed with the Sylviidae.

BANANAQUIT: COEREBINAE (1)

The Bananaquit has been placed in a monotypic family and has also been considered a honeycreeper; current opinion places it in its own subfamily. Ages differ, sexes similar; young altricial; 1st prebasic molt partial (?).

Food mostly nectar, also fruit and insects. Nest is a globular structure of fibers, moss, etc., with opening facing down obliquely, at mid- to upper levels in tree or bush. Eggs: 2, whitish, densely speckled brown (WFVZ).

BANANAQUIT
*Coereba flaveola**   Plate 55
Platanero

ID: 3.7–4.3 in (9.5–10.5 cm). Two ssp: *mexicana* (humid SE) and *caboti* (Isla Cozumel and NE QR). **Descr. Adult.** *Mexicana*: bill black with red gape, legs dark grey. *White supercilium contrasts with black crown and auriculars.* Upperparts slaty grey, washed olive, with *dull yellowish rump* and uppertail coverts; wings and tail blacker with *white flash on base of primaries*, white tips to outer rectrices. *Throat grey, underparts yellow,*

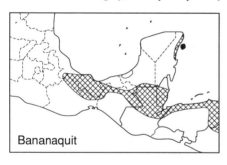

Bananaquit

washed dusky on flanks, undertail coverts pale lemon. *Caboti*: larger, *upperparts blackish with brighter yellow rump, throat and upper chest white.* **Juv** (both ssp): head and upperparts greyish olive with pale lemon supercilium. Throat and underparts dirty pale lemon. Attains adult-like plumage within a few months (?).

Voice. Song of *mexicana* varied, a rapid series of high, sibilant to buzzy notes run into a wiry trill or warble, *wi-si wi-si wi-si...*, or *ssi ssi-iir ssi-ssi-ii-iirr*, etc. Song (?) of *caboti* a high, thin, wiry trill. Calls (both ssp.) mainly high, often lisping, thin to buzzy twitters, also a high, thin, sharp *seiit* or *seeip*.

Habs. Humid evergreen forest and edge, plantations, also (*caboti*) scrubby woodland, second growth, gardens. Singly, in pairs or family groups, often at flowering trees and bushes.

SD: F to C resident (SL–1000 m) on Atlantic Slope from S Ver to Honduras; C resident on Isla Cozumel and locally on adjacent mainland of QR.

RA: Caribbean and Mexico to Peru and N Argentina.

TANAGERS: THRAUPINAE (30)

This mainly neotropical subfamily, formerly considered a family, is now merged with the Emberizidae; honeycreepers were formerly placed in their own family, the Coerebidae. Tanagers are diverse, often brightly colored, mostly finch-like birds which, like other emberizids, have nine primaries. Most are arboreal and some are characteristic members of mixed-species flocks. Bills vary from slender and decurved to stout and hooked, legs from slender to long and stout. Ages usually differ, young altricial. Sexes similar or different; seasonal variation is restricted to a handful of northern migratory species. Molts vary and, in many tropical species, are poorly known. Juv plumage is usually replaced quickly by adult-like plumage, although imm ♂♂ may resemble ♀♀. Songs vary from rich and striking to weak and virtually non-existent; calls are often important for locating and identifying species.

Tanagers are mostly frugivorous, but also eat seeds and insects; honeycreepers in particular take nectar. Nests are usually cups of vegetation, in bushes or trees. Eggs: 2–5.

GENUS TANGARA

Only two species of this mostly South American genus reach the region. Ages differ, sexes mostly similar. Plumage striking and brightly colored, songs and calls unspectacular. Nests are cups of rootlets, leaf strips, moss, etc., at mid- to upper levels in trees or bushes. Eggs: 2–3, whitish to buffy white, speckled with browns and greys (WFVZ).

CABANIS' (AZURE-RUMPED) TANAGER
Tangara cabanisi Plate 56
Tángara de Cabanis

ID: 5.7–6 in (14.5–15 cm). *Endemic to Pacific foothills of Chis and Guatemala.* **Descr. Adult:** bill grey with dark tip, legs greyish. *Face and underparts whitish,*

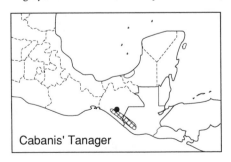

Cabanis' Tanager

washed pale blue, with black lores and black mark on lower auriculars; *chest scalloped black.* Crown mauve, becoming blue on nape, mantle pale greenish mottled darker; rump and uppertail coverts pale blue. *Black scapulars often cover turquoise-blue lesser wing coverts; turquoise-edged remiges contrast with black greater and median coverts,* greater coverts narrowly edged blue. Tail blackish, edged blue. **Juv:** undescribed (?).
Voice. Song (?) a high, thin *wi sseeu* or *ssi wssst,* the 2nd part fading away; calls include a high, thin *ssi* or *ssit* and high, excited twittering.
Habs. Humid evergreen forest and edge. In pairs or small groups in canopy of large trees; often still for long periods, then flies about calling noisily before disappearing into the canopy of another tree. Eggs undescribed (?).
SD/RA: F to U but local resident (1000–1700 m) on Pacific Slope of E Chis and W Guatemala, locally on adjacent Atlantic Slope of Sierra Madre de Chis where first noted Apr 1988 (SNGH, SW).

GOLDEN-HOODED [MASKED] TANAGER
Tangara larvata* Plate 56
Tángara Capucha-dorada

ID: 5–5.3 in (12.5–13 cm). *Humid SE forest.* **Descr. Adult:** bill black, legs grey. *Mauve*

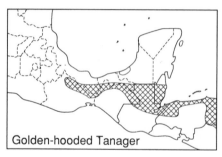

Golden-hooded Tanager

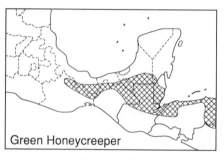

Green Honeycreeper

and blue face with black lores and chin contrasts with golden hood. Upperparts black with turquoise rump; lesser and median coverts mauve to turquoise, greater coverts and flight feathers black, edged blue-green. *Chest black, flanks purple and turquoise, belly and undertail coverts white.* **Juv:** *head greenish with dark lores,* upperparts mottled dark green and black, becoming light green on rump. Wings and tail blackish, edged green. *Throat and chest dusky olive* with dark patches on sides of chest, *rest of underparts pale buff with turquoise flash on flanks.*

Voice. A high trill often preceded by 1–2 sharp chips or low, gruff *cheht* notes, *cheht cheht chi ssiiiiiiiiir.* Also sharp chips *chik* or *chi,* and *tlik* which may be repeated steadily, and a dry chipping trill.

Habs. Humid evergreen forest and edge, plantations. In pairs or small groups in forest canopy; at times with mixed-species flocks.

SD: F to C resident (SL–900 m) on Atlantic Slope from N Oax to Honduras.

RA: SE Mexico to W Ecuador.

NB: Sometimes known as Golden-masked Tanager.

GREEN HONEYCREEPER
Chlorophanes spiza guatemalensis   Plate 56
Mielero Verde

ID: 5.3–5.8 in (13.5–14.5 cm). SE humid forest. *Bill slender and decurved.* **Descr.** Ages/sexes differ. ♂: eyes red, bill horn-yellow with dark culmen, legs grey. *Intense blue-green overall with black hood.* ♀/**juv:** *green overall,* slightly paler below. ♂ attains adult plumage Oct–Jun. Juv eyes red-brown.

   Voice. A sharp chip, *tchi!* or *tchiip!* in alarm, may be repeated insistently; also a

thin, sharp *siip* or *tssip,* given in flight. Rarely heard song a quiet, buzzy twittering interspersed with short trills.

Habs. Humid evergreen forest and edge. Singly or in pairs in the canopy, feeds at flowering trees. At times joins mixed-species flocks. Nest is a shallow cup of rootlets, dry leaves, etc., at mid-levels in tree or bush. Eggs: 2, white, speckled with reddish browns and greys (WFVZ).

SS: See Green Shrike-Vireo, Rufous-winged Tanager (E Honduras, Appendix E). ♀ Blue-crowned Chlorophonia stockier with stout bill, yellowish belly. ♀ Red-legged Honeycreeper smaller and duller with dark bill, reddish legs, pale streaking on underparts.

SD: F to U resident (SL–1500 m) on Atlantic Slope from N Oax (SNGH tape) to Honduras.

RA: SE Mexico to Peru and Brazil.

SHINING HONEYCREEPER
Cyanerpes l. lucidus   Plate 56
Mielero Luciente

ID: 4–4.5 in (10–11 cm). *SE humid foothill forest. Bill slender and decurved.* **Descr.** Ages/sexes differ. ♂: bill black, *legs yellow. Deep glossy blue overall with black lores, throat, wings, and tail.* ♀: *face streaked whitish and green, with dark eyestripe and blue malar stripe; throat buff,* under-

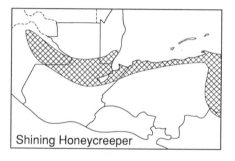

Shining Honeycreeper

parts pale buff with *band of blue streaks across chest*, flanks green. Crown greenish blue, upperparts green. **Juv:** resembles ♀ but streaking on chest green to blue-green, blue malar stripe indistinct or lacking. **Imm ♂** (Aug–Jan): plumage variably intermediate between juv and ♂.

Voice. A high, thin, fairly sharp *chi* or *chit*, and high, thin, at times slightly liquid twittering.

Habs. Humid evergreen forest and edge. In pairs or small groups in forest canopy, often at flowering trees. Nest is a shallow cup of fibers at mid-levels in trees. Eggs: 2, undescribed (?) but dark purplish brown, mottled blackish, in closely related *C. caeruleus* (WFVZ).

SS: Red-legged Honeycreeper has longer tail, stouter and slightly less decurved bill, red legs, yellow underwings, ♂ has blue throat and underparts, ♀ has throat and underparts greenish, indistinctly streaked.

SD: U to F resident (near SL–1500 m, mainly foothills) on Atlantic Slope from E Chis to Honduras.

RA: SE Mexico to NW Colombia.

NB: Sometimes considered conspecific with *C. caeruleus* of S.A.

RED-LEGGED HONEYCREEPER

Cyanerpes cyaneus carneipes   Plate 56
Mielero Patirrojo

ID: 4.2–4.7 in (11–12 cm). E and S humid forest. *Bill slender and slightly decurved.* **Descr.** Ages/sexes differ. **Alt ♂** (Dec–Aug): bill blackish, legs bright red. *Head and underparts deep glossy blue with turquoise crown* and black lores. *Upperparts black with blue scapulars, rump,* and uppertail coverts; *underwing coverts and inner webs of remiges bright yellow, striking in flight.* **Basic ♂** (Aug–Feb): *green overall* with dark lores and pale supraloral stripe, throat and underparts streaked lemon;

wings and tail black, as alt. ♀/**juv:** legs reddish to dull pinkish. *Head and upperparts green* with dark lores, pale supraloral stripe; yellow underwings as ♂. *Throat and chest green, striped lemon*, becoming lemon on belly and undertail coverts, flanks green. ♂ has complete 1st prebasic molt and resembles basic adult by early winter.

Voice. A rough mewing *meeah* or *meeihr* suggesting Tropical Gnatcatcher but rougher and buzzier, a high, rolled, slightly nasal *shihr* or *srrip*, and a high, thin *ssit ssit*, given in flight.

Habs. Humid evergreen and semideciduous forest and edge, plantations, semiopen areas with scattered trees and hedges. In pairs or small groups in canopy, especially at flowering trees. May be in flocks of 50+ birds during migration and winter. Nest cup of rootlets, fibers, etc., at mid-levels in trees. Eggs: 2–3, white, speckled brown.

SS: See Shining Honeycreeper, ♀ Green Honeycreeper.

SD: C to F summer resident (mid-Mar–Aug; SL–1500 m) on both slopes N of Isthmus to Gro (SNGH, SW, RGW) and SE SLP, also to Yuc Pen; C to F resident on both slopes from Isthmus to El Salvador and Honduras, commonest on Pacific Slope in winter when N migrants present. R N of Isthmus in winter (Sep–Feb), at least to S Ver, and U in E QR. Vagrant (Oct–Mar) to Isla Cozumel (SNGH, SJ, DRP photo).

RA: E Mexico to Ecuador and Brazil.

BLUE-CROWNED CHLOROPHONIA

Clorophonia occipitalis   Plate 56
Chlorofonia Coroniazul

ID: 4.8–5.3 in (12–13.5 cm). *S cloud forest. Chunky shape, short tail.* **Descr.** Ages/sexes differ. ♂: bill blackish, grey below at base, legs greyish flesh. Head, chest, and

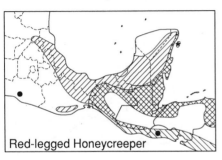

Red-legged Honeycreeper

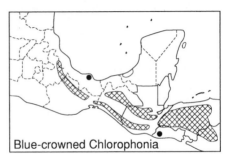

Blue-crowned Chlorophonia

upperparts *brilliant green with turquoise crown patch* and narrow hindcollar. *Narrow dark forecollar above yellow underparts*, flanks green. ♀: *green overall, becoming yellowish on belly and undertail coverts*; turquoise crown patch and narrow hindcollar visible from above. **Juv:** resembles ♀ but duller, lacks turquoise on head, quickly attains ♀-like plumage, ♂ apparently attains adult plumage when about 1 year old.

Voice. A plaintive, ventriloquial cooing *peeu* or *hoo*; also a clucking *kyuk* or *kluh* and *pi-di-luk*, and a soft, slightly liquid *whiut* given in flight.

Habs. Humid evergreen forest and edge. In pairs or small flocks in the canopy where can be difficult to see, even though calling from all around; often at fruiting trees. Nest is a globular structure of rootlets, moss, etc., with side entrance, amid bromeliads at mid-levels in trees. Eggs: 3, white, heavily speckled with reddish browns and greys (WFVZ).

SS: See Green Shrike-Vireo, Green Honeycreeper, Rufous-winged Tanager (E Honduras, Appendix E).

SD/RA: C to F resident (1000–2500 m) on Atlantic Slope from S Ver and N Oax (R (?) N to cen Ver), and on both slopes and in interior from E Oax and Chis, to Honduras and N cen Nicaragua. In winter descends locally to near SL (for example, in Los Tuxtlas).

NB: Sometimes considered conspecific with *C. callophrys* of S C.A.

GENUS EUPHONIA: Euphonias

Euphonias are small stubby tanagers. Usually seen in pairs or small groups at mid- to upper levels, often in forest canopy at fruiting trees; commonly join mixed-species flocks. Ages/sexes usually differ. Bills black, blue-grey below at base, legs blackish to dark grey. Songs and calls notably varied, both sexes appear to sing. Nests are globular structures with a side entrance, placed low to high in hanging masses of vegetation, bromeliads, tree crevices, even on mossy banks. Eggs: 2–5, whitish, speckled reddish brown and brown (WFVZ).

SCRUB EUPHONIA
*Euphonia affinis**   Plate 56
Eufonia Gorjinegro

ID: 3.7–4.2 in (9.5–10.5 cm). *Tropical low-*

lands, two groups: *affinis* (E Mexico to C.A.) and *godmani* (W Mexico). **Descr. Affinis.** ♂: *forehead yellow, head and throat glossy blue-black* with slight purple sheen. Upperparts glossy blue-black; inner webs of outer rectrices mostly white. *Underparts yellow.* ♀: *head greyish* with yellow-olive forehead; upperparts olive, inner webs of outer rectrices with little or no white. *Throat, chest, and flanks dusky greyish olive, belly and undertail coverts yellow.* **Juv:** resembles ♀ but head olive, ♂ has blue-black remiges. Quickly attains adult-like plumage. **Imm ♂:** resembles ♂ but back mottled olive, much like adult by 1st spring. **Godmani.** ♂: *has white undertail coverts.* ♀: *has yellow flanks, whitish belly and undertail coverts,* more white in tail than *affinis.* **Juv:** undescribed (?).

Voice. *Affinis.* Calls include a bright *dwee dwee dwee* to a tinnier *deein deein ...*; a plaintive *syeeu syeeu*, a single *teeu*, and a high, liquid *slip* given in flight. Varied songs include a liquid, twittering warble with a stronger, accelerating series toward the end, a slightly burry warble and a *si chi-chi-chi-chi si chit*, also a prolonged, liquid warble with *been-been* phrases thrown in. *Godmani.* Song a short, high, tinkling warble. Common call a high *dee-dee-dee*, brighter and less plaintive than *affinis.*

Habs. Woodland and forest edge, plantations, scrub, semiopen areas with scattered trees and hedges. See genus note; associates readily with Yellow-throated Euphonia. Eggs of *godmani* undescribed (?).

SS: White-vented Euphonia (C.A.) smaller with small bill, white undertail coverts, ♂ brighter orange-yellow below, ♀ has olive head, pale grey throat contrasts with yellow chest and flanks. ♀ Yellow-throated Euphonia has whitish underparts with yellow sides.

SD: *Affinis:* C to F resident (SL–1500 m,

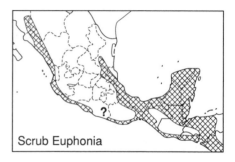

Scrub Euphonia

locally to 1800 m) on both slopes from S Tamps and Oax, and locally in interior from Isthmus, to El Salvador and Honduras. *Godmani*: F to C resident (near SL–1000 cm) on Pacific Slope from S Son to Gro.
RA: Mexico to NW Costa Rica.

YELLOW-THROATED EUPHONIA
Euphonia hirundinacea* Plate 56
Eufonia Gorjiamarillo

ID: 4–4.5 in (10–11 cm). E and S lowlands. **Descr.** ♂: *head and upperparts glossy blue-black with yellow forehead*, inner webs of outer rectrices mostly white. *Throat and underparts rich yellow.* ♀: head and upperparts olive. *Throat and underparts whitish to pale grey with yellow sides* and flanks, undertail coverts lemon. **Juv:** resembles ♀ but throat and underparts tinged yellow. **Imm** ♂: *head resembles* ♂ but crown mottled olive, *upperparts olive.* Throat and underparts yellow, often washed olive on sides.
   Voice. Calls include a bright *chi-chi-chit* or *ji-ji-jit*, a fairly strong *sseeu* or *speeu*, a slightly burry to plaintive *peeur* or *ch-eeu* often doubled, a buzzy, slightly liquid *ssi shi-shi* or *chi-chi-cheeu*, a bright emphatic *bee-bee-beep!*, and a high, thin *ssi* or *si-si* given in flight. Song varied, a liquid to slightly jerky warble, often with sharp *chik* notes thrown in; can be hard to distinguish from Olive-backed Euphonia.
   Habs. Forest and edge, plantations, semi-open areas with hedges, scattered trees. See genus note; associates readily with Scrub Euphonia.
SS: See Scrub Euphonia. White-vented Euphonia (C.A.) smaller with small bill, ♂ has black throat, ♀ has pale grey throat contrasting with yellow chest and flanks, white undertail coverts.
SD: C to F resident (SL–1500 m, locally to 1800 m) on Atlantic Slope from S Tamps

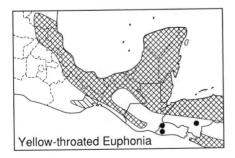
Yellow-throated Euphonia

to Honduras, on Pacific Slope of Chis and Guatemala and locally to Honduras; winter visitor (Oct–early May, resident?) on Pacific Slope of Isthmus in Oax (Binford 1989).
RA: E Mexico to W Panama.

BLUE-HOODED EUPHONIA
Euphonia elegantissima* Plate 56
Eufonia Capucha-azul

ID: 4–4.5 in (10–11 cm). *Highlands and foothills.* **Descr.** ♂: *turquoise-blue hood contrasts with blue-black face and throat,* forehead rufous. *Underparts ochraceous orange.* Upperparts glossy blue-black. ♀: *olive overall, paler below, with turquoise-blue hood,* chestnut forehead, and *cinnamon throat.* **Juv:** *olive overall, paler below,* with dull bluish crown, ♀ *has yellowish forehead,* ♂ *has pale cinnamon chin* and blue-black flight feathers. Quickly attains adult-like plumage.
   Voice. Calls include a slightly nasal *chih chu* or *t-chih*, a nasal *ehnk* or *djeht*, and *teu*, often doubled. Song a prolonged, burbling and tinkling warble.
   Habs. Oak, pine–oak, and evergreen forest, open areas with trees, scrub, especially around clumps of mistletoe. See genus note.
SD: F to C resident (500–3500 m, lower locally in winter) on both slopes from S Son and NL, and in interior from cen Mexico, to Honduras and N cen Nicaragua.
RA: Mexico to W Panama.
NB: Often considered conspecific with *E. musica* of the Caribbean and *E. aureata* of S.A.

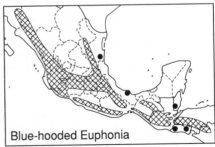
Blue-hooded Euphonia

OLIVE-BACKED EUPHONIA
Euphonia g. gouldi Plate 56
Eufonia Olivácea

ID: 4–4.3 in (10–10.5 cm). *SE humid forest.* **Descr.** Ages similar, sexes differ. ♂: *forehead*

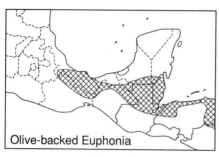

Olive-backed Euphonia

yellow, head and upperparts olive-green with blue sheen. *Throat and underparts olive-green with chestnut belly and undertail coverts.* ♀: *forehead chestnut*, head and upperparts olive-green with blue sheen. *Throat and underparts yellow-olive, undertail coverts chestnut.*

Voice. Song varied, a slightly liquid warble, often slightly jerky and prolonged, with sharp *chik* notes thrown in; short phrases may suggest White-eyed Vireo. Calls include a liquid *dri-dri-drit* or *dri-i-it*, a slightly trilled *chirr-i-rrit* and *sip si-jirr-rr*, a dry, nasal *cheh-leh-let*, a slightly plaintive *hu du-dwit*, and a high *sii* and soft, mellow *pit*, given in flight.

Habs. Humid evergreen forest and edge. See genus note; usually singly or in pairs, associates loosely with other euphonias.

SD: F to C resident (SL–1000 m) on Atlantic Slope from S Ver to Honduras.

RA: SE Mexico to W Panama.

WHITE-VENTED EUPHONIA
Euphonia minuta humilis　　　Plate 56
Eufonia Vientre-blanco

ID: 3.6–4 in (9–10 cm). *Atlantic Slope S of Isthmus.* **Descr.** Ages/sexes differ. ♂: *forehead yellow, head and throat glossy blue-black* with slight purple sheen. Upperparts glossy blue-black, inner webs of outer rec-

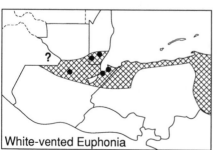

White-vented Euphonia

trices mostly white. *Underparts rich yellow with white undertail coverts.* ♀: *head and upperparts olive*, forehead washed yellowish; inner webs of outer rectrices edged white. *Throat pale grey, chest and flanks yellow, belly and undertail coverts whitish.* **Juv:** resembles ♀ but forehead olive, underparts washed lemon, ♂ has blue-black remiges. Quickly attains adult-like plumage. **Imm ♂:** resembles ♂ but back mottled olive, much like adult by 1st spring.

Voice. Calls include a soft, slightly sharp *s-chik* suggesting a warbler, a wetter *schik* or *s-wik*, short, warbled phrases such as *si-ubble-ubble-ubble* or *t-burble t-burble*, and a high *sip-sip* or *siip-siip* given in flight. Song a varied warble, often with chip notes thrown in.

Habs. Humid evergreen forest and edge, fruiting trees in adjacent pastures and second growth. Associates readily with other euphonias and overlooked easily. See genus note.

SS: See Scrub and Yellow-throated euphonias.

SD: U to F resident (near SL–1000 m, mainly foothills) on Atlantic Slope of Guatemala and S Belize. A spec. reportedly from Mexico (Phillips and Hardy 1965) was collected by Avilés and thus its validity is doubtful; probably occurs, however, in E Chis, and also in Honduras but no documented records to date.

RA: Guatemala (SE Mexico?) to Ecuador and Brazil.

BLUE-GREY TANAGER
Thraupis episcopus cana　　　Plate 57
Tángara Azuligris

ID: 6–7 in (15.5–18 cm). Open country in S lowlands. **Descr.** Ages/sexes similar. Bill black above, blue-grey below with dark tip, legs blackish. *Pale bluish overall,*

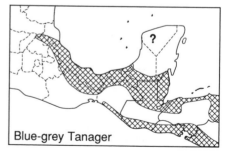

Blue-grey Tanager

darker above, *wings and tail darker, edged turquoise-blue*, bright blue lesser wing coverts usually hidden. **Juv:** overall duller.

Voice. A high, slightly piercing *ssiiu* or *sseu*, and a more nasal, slurred *ssweu* or *sywee*, both of which may be given in flight, a harder *seu*, and a quiet nasal *chu*. Song a varied, squeaky, twittering warble.

Habs. Humid to semihumid, open and semiopen areas with hedges, scattered trees, forest edge, plantations. Singly or in small groups, often at fruiting trees. Joins mixed-species flocks. Nest is a neat cup of grass, fibers, moss, etc., low to high in tree or bush. Eggs: 2, pale blue-grey to grey, speckled and spotted with browns and greys (WFVZ).

SD: C to F resident (SL–1500 m) on both slopes from SE SLP and Chis to El Salvador and Honduras; recent reports from Yuc (per BM) may refer to escaped cage birds. A report from Gto (AOU 1983) is erroneous.

RA: E Mexico to Peru and Brazil.

YELLOW-WINGED TANAGER

Thraupis abbas   Plate 57
Tángara Aliamarilla

ID: 6.8–7.5 in (17.5–19 cm). E and S. **Descr.** Ages differ, sexes similar. **Adult:** bill blackish, legs grey. *Head lilac-blue with black lores; throat and underparts dusky lemon-olive*, washed lilac on throat and chest. Upperparts dusky-bluish, back mottled blackish, rump washed olive. Greater wing coverts dusky lemon-green, flight feathers black with *broad yellow flash across base of remiges*. **Juv:** head and upperparts olive, mottled darker on back, lacks lilac wash on throat and chest. Quickly attains adult-like plumage.

Voice. A high *ssiu* or *sseeru*, and *swee* or *sweek*, both of which may be given in flight. Song a slightly reedy, descending

trill, often with 1–2 introductory notes, *shee iiiiiiiiiiiiirr*, at times preceded by a gruff *shihrr shihrr ...*

Habs. Humid to semihumid, evergreen and semideciduous forest edge, plantations, semiopen areas with scattered trees, hedges. Habits much as Blue-grey Tanager but more often in flocks, at times to 50+ birds. Nest cup of dry leaves, moss, fibers, etc., at mid-levels in trees. Eggs: 3, pale grey, mottled with browns and greys (WFVZ).

SD: C to F resident (SL–1800 m) on Atlantic Slope from SE SLP to Honduras, and on Pacific Slope and in interior from Chis to Honduras. A report from Mex (AOU 1983) is erroneous.

RA: E Mexico to Nicaragua.

STRIPE-HEADED TANAGER

Spindalis zena benedicti   Plate 69
Tángara Cabecirrayada

ID: 5.8–6.3 in (14.5–16 cm). *Isla Cozumel.* **Descr.** Ages/sexes differ. ♂: bill black above, grey below, legs grey. *Head black with striking white supercilium and moustache, broad black malar stripe borders white chin and yellow throat. Upper chest chestnut*, median lower chest yellow, rest of underparts pale grey, duskier on flanks. Upperparts dark brownish olive with chestnut hindcollar, ochraceous-tawny rump, and chestnut uppertail coverts. *Wings and tail black, edged white*, often with solid white panel on greater coverts, outer rectrices mostly white on inner webs. ♀/**juv:** head and upperparts greyish olive with indistinct paler supercilium; wings and tail darker, brownish, *wings narrowly edged pale olive-lemon*. Throat and underparts pale greyish olive, throat paler with indistinct dusky malar stripe, belly and undertail coverts pale lemon. ♂ quickly attains adult-like plumage.

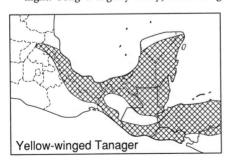

Yellow-winged Tanager

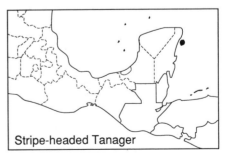

Stripe-headed Tanager

Voice. Song a high, thin, twittering *tsi-si ti-si si-i-tsi-si si-i si-i*, with increasing intensity. Calls a high, thin to slightly shrill *tssi* and *tssi-ssi-ssi*, suggesting a *Turdus* flight call.

Habs. Scrubby forest and edge. In pairs or small groups, often at fruiting trees. Nest cup in bush or tree (Bond 1985). Eggs (Haiti): 2–3, bluish white, densely flecked with dusky browns (WFVZ).

SD: U resident on Isla Cozumel in recent years; F to C prior to hurricane in Sep 1988, when much of the island was defoliated and fruit lost.

RA: Caribbean islands, E to Puerto Rico.

GREY-HEADED TANAGER
Eucometis penicillata pallida   Plate 57
Tángara Cabecigris

ID: 6.5–7 in (16.5–18 cm). *Humid SE forest. Head bushy.* **Descr.** Ages differ, sexes similar. **Adult:** *bill black, legs flesh. Head and throat grey,* lores often appear dusky. Upperparts olive-green, *underparts bright yellow.* **Juv:** head and throat olive, upperparts dull olive, underparts yellow. Quickly attains adult-like plumage.

Voice. A high, sharp *siip* and *ssi-ssip'*, at times run into a slightly lisping twitter, a low, hard, clipped *tk* or *tuk*, and a quiet clucking *chik chik* ... Song varied, a rich, fairly sharp warble, may suggest a euphonia.

Habs. Humid evergreen and semideciduous forest. Singly or in pairs, low in forest understory, overlooked easily; attends ant swarms. Nest is a slight cup at low to mid-levels in spiny palm or bush. Eggs: 2–3, pale blue-grey, mottled brown (WFVZ).

SS: ♀/imm Black-throated Shrike-Tanager larger, arboreal, with heavier bill, tawny rump. ♀ Scarlet-rumped Tanager (second growth and edge) often with striking ♂, bill bluish white, chest tawny-ochre, upperparts tawny-brown.

SD: U to F resident (SL–750 m) on Atlantic Slope from S Ver to Honduras.

RA: SE Mexico to Brazil.

BLACK-THROATED SHRIKE-TANAGER
Lanio aurantius   Plate 57
Tángara-lanio Gorjinegro

ID: 7.5–8.5 in (19–21.5 cm). *Endemic; SE humid forest.* **Descr.** Ages/sexes differ. Bill black, legs grey. ♂: *head and throat black, underparts golden yellow* with brownish-ochre wash on upper chest. *Upperparts yellow,* black on outer scapulars and distal uppertail coverts. *Wings and tail black,* white lesser wing coverts usually covered by scapulars. ♀: *head and throat greyish,* washed olive on crown; *underparts yellow,* washed olive on chest and flanks, ochre on undertail coverts. *Upperparts rich tawny-brown becoming bright tawny on rump* and uppertail coverts, tail edged tawny. **Imm ♂:** resembles ♀ but underparts brighter yellow, upperparts brighter cinnamon-brown. Attains adult plumage when about 1 year old. Juv undescribed (?).

Voice. Common call a sharp, rhythmic *chee'choo* ... or *chi'tuk* ..., repeated, often persistently; may be preceded by 1–2 *shih* notes which also may run into a screechy, squeaky splutter followed by *chee'choo* series. Also a loud *chee-ew* or *teeuw,* and a quiet, varied warble with louder *teeu* notes thrown in.

Habs. Humid evergreen forest. Singly or in pairs at mid- to upper levels in open canopy, often a core species and 'watch-dog' of mixed-species flocks. Perches fairly upright and sits still for long periods, makes fluttering sallies for prey. Nest and eggs undescribed (?); nest of closely related White-throated Shrike-Tanager is a cup in low bush near stream; 2 white eggs, flecked with browns and greys (WFVZ).

SS: ♂ superficially resembles various orioles

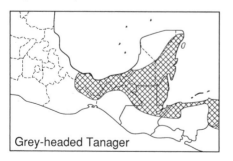

Grey-headed Tanager

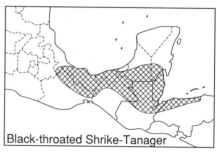

Black-throated Shrike-Tanager

but note habits, stout black bill. See Grey-headed Tanager. Bright-rumped Attila has streaked face and chest, whitish under-parts, brighter and larger rump patch.
SD/RA: F to U resident (SL–1200 m) on Atlantic Slope from S Ver to Honduras.
NB: Sometimes considered conspecific with *L. leucothorax* of S C.A.

RED-CROWNED ANT-TANAGER
Habia rubica*   Plate 57
Tángara-hormiguera Coronirroja

ID: 6.8–8 in (17–20 cm). Humid forest. Marked geographic and individual varia-tion. Two groups: *rosea* group (W Mexico) and *rubicoides* group (E Mexico and C.A.). Descr. Ages/sexes differ. *Rosea*. ♂: bill blackish, paler below, legs flesh to brownish flesh. *Head and upperparts dull pinkish red*; partly concealed red crown patch edged blackish. *Throat and under-parts pale pinkish red*. May show irregular olive patches, including flight feathers. ♀/imm: *bill blackish brown above, dull horn to flesh below. Head and upperparts greyish olive*, tail richer tawny-brown; dull pinkish-tawny crown patch often concealed, re-duced or lacking in imm ♀. *Throat and underparts buff* to pale pinkish cinnamon, palest on throat, washed dusky on flanks. *Rubicoides*. ♂: bill black. Head and upper-parts dull dark reddish; *partly concealed red crown patch edged blackish*. Red throat contrasts slightly with dusky red underparts. ♀/imm: *bill blackish above, dull horn to flesh below. Head and upper-parts olive-brown* to brownish olive, dull tawny crown patch often concealed (indis-tinct or lacking in imm ♀), tail often darker rufous-brown in adult. *Throat and under-parts dusky tawny-brown with paler ochre-buff throat not contrasting strongly*. Some birds (at least in C.A.) appear vari-ably like ♂, with upperparts washed red-dish, wings and tail edged reddish, throat and underparts washed pinkish cinnamon. Attains adult ♂ plumage when about 1 year old. Juv: overall sooty olive-brown, paler below, palest on throat, quickly attains imm plumage.
Voice. Common call a squeaky, splutter-ing chatter, sharper and squeakier in *rosea*; less often rolled chirps, *chriip chriip...*, etc. Song varied but typically a steady repetition of simple phrases, *chee-oo* or *tchee* or *choo* or a slurred *chee-chuwee*, etc., 3–11×, typically 4×; even most 'complex' variations, such as *chee ch-choo chee ch-choo*, lack the jerky qual-ity of Red-throated Ant-Tanager songs.
Habs. Humid to semihumid, evergreen and semideciduous forest, overgrown plantations. In small groups, typically 3–10 birds, low to high in understory, mainly at mid-levels. Often with mixed-species flocks which include Red-throated Ant-Tanager; attends ant swarms. Nest is a shallow cup of rootlets, leaf strips, fibers, etc., at low to mid-levels in bush or tree. Eggs: 2–3, whitish, flecked and spotted with browns and greys (MLZ).
SS: Red-throated Ant-Tanager larger and darker overall (but beware local and indi-vidual variation), bill slightly longer, more often at low levels, note harsh rasping calls; ♂ has distinctive dark lores, red crown patch lacks black edges, ♀ dark rich brown overall with black bill, contrasting buff throat, lacks crown patch. ♂ Hepatic Tanager (mainly pine–oak) more arboreal, habits different, lacks crown patch.
SD: *Rosea* group: F to C resident (near SL–1500 m) on Pacific Slope from Nay to Oax; *rubicoides* group: C to F resident (SL–1500 m) on Atlantic Slope from S Tamps to Honduras, on Pacific Slope from Chis to Nicaragua.
RA: Mexico to N Argentina.

RED-THROATED ANT-TANAGER
Habia fuscicauda*   Plate 57
Tángara-hormiguera Gorjirroja

ID: 7.3–8.5 in (18.5–21.5 cm). E and S. Descr. Ages/sexes differ. ♂: bill black, legs flesh to brownish flesh. Head and upper-parts dull dark reddish with *dark lores*, red crown patch partly concealed. Red throat contrasts slightly with dusky red under-parts. May show irregular olive patches, including flight feathers. ♀/imm: *dark*

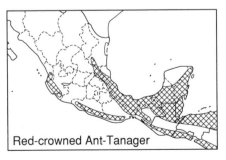
Red-crowned Ant-Tanager

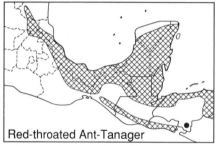

Red-throated Ant-Tanager

rufous-brown overall, paler below, with *contrasting buff to ochraceous-buff throat*, black bill. Some birds have throat and underparts washed pinkish cinnamon. Attains adult ♂ plumage when about 1 year old. Yuc Pen birds paler overall. **Juv:** overall dark sooty cinnamon-brown, quickly attains imm plumage.

Voice. Common call a low, rasping *shehh-shehh-shehh...*, also a harder, dry rattle, *ch-ch-cht*, a liquid, rolled *brrri*, and a harsh *bzzz'iu*. Song varies, typically a jerky, rhythmic phrase, often repeated tirelessly, *chu ree-choo* or *chree-choo-choo chree-choo*, etc., at times a simpler *ch-choo* or *chee'choo*, but jerkier than Red-crowned Ant-Tanager songs.

Habs. Humid to semiarid, evergreen to semideciduous forest and edge, second growth woodland and thickets. Habits much as Red-crowned Ant-Tanager but more often at low levels; attends ant swarms. Nest is a shallow cup of rootlets, leaf strips, fibers, etc., at low to mid-levels in bush or tree. Eggs: 2–3, whitish, unmarked (WFVZ).

SS: See Red-crowned Ant-Tanager. Hepatic Tanager (mainly pine–oak) more arboreal, habits different, lacks crown patch.

SD: C to F resident (SL–1200 m, mainly below 1000 m) on Atlantic Slope from S Tamps to Honduras, on Pacific Slope from Chis to Honduras.

RA: E Mexico to N Colombia.

NB: Sometimes considered conspecific with *H. gutteralis* of S.A.

GENUS PIRANGA

These are large, stout-billed, temperate-breeding tanagers. Ages/sexes differ. Legs dark greyish. As far as is known, juvs resemble washed out ♀♀ with dark brown streaking on head and body. Songs rich and warbling. Nest cups of fine twigs, grasses, pine needles, rootlets, etc., at mid- to upper levels in trees. Eggs: 3–5 in N, 2–3 in S, bluish to greenish, speckled with browns and greys (WFVZ).

ROSE-THROATED TANAGER
Piranga roseogularis＊　　　　Plate 58
Tángara Yucateca

ID: 6–6.5 in (15–16.5 cm). Endemic to Yuc Pen. **Descr.** ♂: bill and legs greyish, culmen darker. *Head and upperparts grey with reddish crown*, white eyering, reddish uppertail coverts; wings darker, broadly edged reddish, tail dull reddish. *Throat rosy, underparts pale grey* with rosy undertail coverts. ♀: head and upperparts grey with *yellow-olive crown*, white eyering, olive uppertail coverts; wings darker, edged yellow-olive. *Throat yellow to pinkish yellow, underparts pale grey* washed yellow on undertail coverts. Some (older birds?) have crown and throat mixed with pinkish, flight feathers edged pinkish. **Imm** ♂: resembles ♀ but often attains dull reddish-edged greater coverts in 1st prealternate molt. **Imm** ♀: duller than ♀, throat washed-out pale lemon. **Juv:** undescribed (?).

Voice. A gruff, rasping, slightly nasal *rreh* or *rrehr*, a nasal mewing *nyieh* or *chehr* suggesting Grey Catbird, and a quiet *chu*. Song a rich, slightly hesitant warble of 3–5 distinct phrases, *whee chee cheer-i-chee cheer-i-cheu chee-i-chu*, or *chee ch'ree ch'ree-u ch'ree-u*, or *chee-u chee chuwee*, etc.

Habs. Humid to semihumid, semideciduous forest and edge. At mid- to upper levels, often in canopy where joins mixed-species flocks; also attends ant swarms. Nest and eggs undescribed (?).

SD/RA: F to C resident (SL–250m) in Yuc Pen, S to N Belize and N Petén.

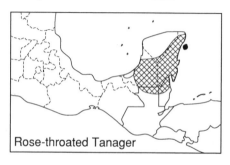

Rose-throated Tanager

HEPATIC TANAGER
Piranga flava* Plate 58
Tángara Encinera

ID: 7.5–8 in (19–20.5 cm). Pine–oak and pine. Two groups: *hepatica* group (Mexico) and *figlina* group (N C.A. including *albifacies*, shown). **Descr. Hepatica. ♂:** bill dark grey, often paler blue-grey below at base. *Crown, throat, and underparts red to orange-red with contrasting dusky auriculars*, flanks usually washed dusky. Upperparts dusky greyish, washed dull red, brighter reddish on uppertail coverts and wing and tail edgings. ♀/**imm:** *forecrown, throat, and underparts yellow to ochraceous orange-yellow with contrasting dusky auriculars*, sides and flanks washed dusky. *Crown, nape, and upperparts dusky greyish olive*, slightly brighter on uppertail coverts and wing and tail edgings. At end of 1st year ♂ attains adult plumage but may show pale yellow patches through 2nd year. **Juv:** resembles ♀ but paler below, with pale buffy wingbars, heavily streaked dark brown overall. **Figlina. ♂:** *head and upperparts dark reddish with dusky, white-flecked auriculars*. Throat and underparts red, slightly duller on chest and flanks. ♀/**imm:** *head and upperparts olive with dusky, white-flecked auriculars*. Throat and underparts yellow to olive-yellow, duskier on chest and flanks.

Voice. A low, full *chuk* or *chuup*, recalling Hermit Thrush, also a quiet nasal *weenk*, usually given in flight. Song a rich warble suggesting Black-headed Grosbeak but often faster.

Habs. Pine–oak and pine woodland, oak scrub. In pairs or small groups, often with mixed-species flocks.

SS: ♀ Red-headed Tanager (W Mexico) smaller; note smaller black bill, pale grey belly. Summer Tanager has paler bill, lacks dusky auriculars, ♂ brighter red overall.

Western and Flame-colored tanagers have bolder pale wingbars than juv Hepatic. Note Hepatic's distinctive call.

SD: *Hepatica*: C to F resident (1000–3000 m, to 150 m in Isthmus) from Son and Coah to N Chis; withdraws in winter (Nov–Mar) from N areas when U to R to near SL in NW Mexico and U (Dec–Feb at least) to NW Guatemala. *Figlina*: C to F resident (600–3000 m) from Guatemala (and S Chis?) to Honduras and N cen Nicaragua, and (near SL–900 m) in pine savannas from Belize to the Mosquitia.

RA: Mexico (SW USA in summer) to Ecuador and N Argentina.

NB: S.A. populations may comprise more than one species.

SUMMER TANAGER
Piranga rubra* Plate 58
Tángara Roja

ID: 6.5–8 in (16.5–20 cm) Widespread. **Descr. ♂:** *bill pale horn to dusky horn. Bright red overall*, upperparts slightly duskier. ♀: *head and upperparts dusky yellow-olive to ochre-olive with pale broken eyering*, rump and uppertail coverts paler. *Throat and underparts yellow to ochre-yellow, typically undertail coverts richer yellow-ochre*; throat and underparts may be washed orangish. **Imm ♂:** resembles richly colored ♀, often with scattered red feathers in winter, becoming *mottled red and yellow* by spring. **Juv:** resembles ♀ but paler below, with pale buffy wingbars, heavily streaked dark brown overall.

Voice. A rolled *pik-u-ruk* or *pt-u-ruk*, similar to but not as hard as Western Tanager call. Song a leisurely rich robin-like warble, typically of 5–7 phrases, *ch-ree ch-oo ch-ree ch-reu ch-reeu ch-reer ch-reeh*, etc.

Habs. Breeds in deciduous woodland, often along watercourses; more widely in

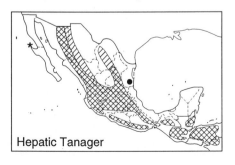

Hepatic Tanager

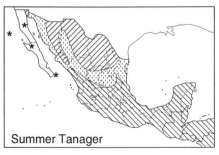

Summer Tanager

migration and winter, from humid ever-green forest to pine–oak woodland. Singly or in pairs, at mid- to upper levels; joins mixed-species flocks.

SS: See Hepatic Tanager. ♀ Scarlet Tanager (transient in E) more greenish-yellow over-all with lemon-yellow undertail coverts. Western Tanager has pale wingbars.

SD: F breeder (late Apr–Aug; near SL–1800 m) in NE BCN (at least formerly), on Pacific Slope S to N Sin, and in interior from Chih and NL to Dgo. F to C transient and winter visitor (late Aug–Apr; SL–2500 m) on both slopes from Sin and Ver, and in interior from cen Mexico, to Honduras and W Nicaragua; U in winter to BCS and S Tamps (Arvin 1990).

RA: Breeds USA to N Mexico; winters Mexico to Ecuador and Brazil.

SCARLET TANAGER
Piranga olivacea   Not illustrated
Tángara Escarlata

ID: 6.5–7 in (16.5–18 cm). Transient in E and S. Bill relatively small. **Descr. Alt** ♂ (Apr–May): bill horn to flesh with dark culmen. *Bright scarlet red with black wings and tail.* **Basic** ♂ (Aug–Nov): *yellow-olive overall,* paler and yellower be-low, *with black wings and tail.* ♀: head and upperparts olive with pale broken eye-ring, wings and tail darker, greyish, edged olive. *Throat and underparts lemon yellowish, undertail coverts brighter lemon-yellow.* **Imm** ♂ (Aug–Nov): re-sembles ♀ but wings and tail darker, greater wing coverts often black. Rarely all plumages may show distinct pale wing-bars (red in alt ♂, yellowish otherwise).

Voice. A low, clipped note often followed by a burry whistle, *chk vrirr.*

Habs. Forest, scrub, and edge. Singly or, in spring at least, in small flocks, at mid-to upper levels; joins mixed-species flocks at fruiting trees.

SS: See ♀ Summer Tanager. ♀ Western Tana-ger typically shows obvious pale wing-bars.

SD: U to F transient (late Mar–early May, Oct–mid-Nov; SL–1200 m) on Atlantic Slope from S Ver to Honduras, mainly from Yuc Pen E. Vagrant (May) to BCN (Wilbur 1987) and (mid-Oct–Nov) to NW Mexico (Keith and Stejskal 1987; Howell and Pyle 1993; ARP) and Clipperton. The source of a report from Jal (Miller *et al.* 1957) is unclear.

RA: Breeds E N.A.; winters Colombia to Bolivia.

WESTERN TANAGER
Piranga ludoviciana   Plate 58
Tángara Occidental

ID: 6.5–7.2 in (16.5–18.5 cm). Winters in W and S. **Descr.** ♂: bill horn to orange-flesh with dark culmen. *Face, crown, and throat red* (heavily veiled yellowish in basic), rest of head and underparts yellow. *Black back edged olive, rump and uppertail coverts yellow.* Wings and tail black with *yellow upper wingbar, whitish lower wingbar,* white tips to tertials and, narrowly, to outer rectrices. ♀: *head and upperparts olive to greyish* with pale broken eyering, rump and uppertail coverts brighter yellow-olive; some show orangish wash on face. Wings and tail dark grey with *pale lemon upper and whitish lower wingbar,* whitish tips to tertials. *Throat and underparts yellow, to pale greyish* with pale lemon only on undertail coverts. **Imm** ♂: re-sembles ♀ but brighter, by spring re-sembles ♂ but note contrasting faded remiges. **Juv:** resembles ♀ but head and body streaked dark brown.

Voice. A rolled *ch-t-ruk* or *piterruk,* or a shorter *ch-dik,* etc. Song 3–4 burry, vireo-

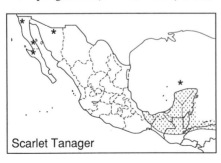

Scarlet Tanager

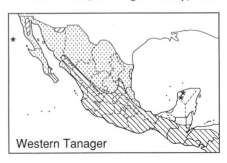

Western Tanager

like phrases, *ch-ree' ch-weeu ch-reeu cheeu*, etc., slightly less hard or burry than Flame-colored Tanager.

Habs. Breeds in pine–oak and deciduous woodland. In winter also to evergreen forest and edge, plantations. Singly or in small flocks at mid- to upper levels; often with mixed-species flocks.

SS: ♀ Flame-colored Tanager has dark bill, dusky auriculars, bolder whitish wing markings, white tail corners, dark-streaked back. See Summer and Hepatic tanagers.

SD: U breeder (Apr–Aug; near SL–2400 m) in N Baja and NE Son. F to C transient and winter visitor (Sep–May; SL–3000 m) in BCS, on Pacific Slope from Son, and in interior from cen Mexico, to Honduras and W Nicaragua; U to F on Atlantic Slope from S Tamps to Guatemala. F to C transient (Apr–May, Sep–Oct) in N Mexico. Vagrant (Dec–Feb) to Yuc Pen (G and JC, PW).

RA: Breeds W N.A. to NW Mexico; winters Mexico to W Panama.

FLAME-COLORED TANAGER

*Piranga bidentata**   Plate 58
Tángara Dorsirrayada

ID: 7–7.5 in (18–19 cm). Oak and pine–oak highlands. Two groups: *bidentata* group (W Mexico) and *sanguinolenta* group (E Mexico and C.A.). **Descr. Bidentata. ♂:** *bill grey. Head and underparts flaming orange* with *dusky auriculars.* Upperparts dusky-orange, often washed olive, *heavily streaked black on back.* Wings and tail blackish, edged orangish, with *2 broad whitish wingbars and bold white spots on tips of tertials and outer rectrices.* ♀: head and underparts yellow with *contrasting dusky auriculars,* crown often washed olive. Upperparts olive, *back streaked dark brown. Wings and tail dark brown,*

patterned as ♂. **Imm ♂:** resembles ♀ but typically with head and underparts mottled orange to orange-red. **Juv:** resembles ♀ but paler yellow underparts streaked dark brown. *Sanguinolenta:* ♂ has head and underparts red to orange-red, wingbars and tertial spots often washed red.

Voice. Much like Western Tanager. A hard, rolled *ch-t-ruk'* or *p-terruk,* or a shorter *ch-duk,* etc. Song 3–5 burry, vireo-like phrases, *chik churree chuwee,* or *churee chiree ch-ree chiwee,* etc.

Habs. Oak, pine–oak, and humid evergreen forest and edge, plantations. Singly or in pairs at mid- to upper levels; joins mixed-species flocks.

SS: See Western and Hepatic tanagers. White-winged Tanager smaller with stubby black bill, narrow white wingbars, ♂ red. Larger ♀ Yellow Grosbeak has massive bill, narrower wingbars.

SD: F to C resident (900–2800 m, lower locally in winter) on both slopes from Son and NL, and in interior locally from cen Mexico, to Honduras and N cen Nicaragua.

RA: Mexico to W Panama.

GENUS SPERMAGRA

These birds recall small *Piranga* (and are sometimes merged with that genus). Bills small and black, may be dull blue-grey below at base, legs dark grey. Unstreaked juvs resemble ♀♀ and quickly resemble adult following 1st prebasic molt. Songs and calls relatively weak and simple. Eggs undescribed (?).

WHITE-WINGED TANAGER

Spermagra l. leucoptera   Plate 58
Tángara Aliblanca

ID: 5.2–5.7 in (13.5–14.5 cm). Humid forest

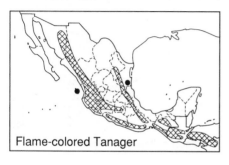

Flame-colored Tanager

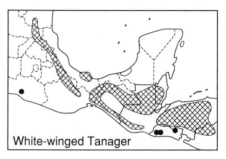

White-winged Tanager

in E and S. **Descr.** ♂: *head and body red with black loral mask.* Scapulars, *wings,* and tail black *with 2 white wingbars.* ♀: head and upperparts olive, face brighter with *dark loral mask.* Wings and tail dark grey with 2 *white wingbars.* Throat and underparts yellow. **Imm** ♂: resembles ♂ but underparts more flame-colored, belly and flanks may be yellow-orange, flight feathers duller, brownish black. **Juv:** resembles ♀ but ♂ has blacker flight feathers.

Voice. Calls high and sharp, a high, clear *swee* or *si-wee,* a more up-slurred *sweet'* or *wi seet',* and a sharp *chiik sii',* etc.; also a high, spluttering call given in flight. Song a high, squeaky, simple, short warble, *si si-see-see chu'* or *ssiu ssiu ssiu si-i-ir,* etc.

Habs. Humid to semihumid evergreen and pine—oak forest and edge, plantations. In pairs or small groups at mid-to upper levels; joins mixed-species flocks. Nest cup at mid-levels in trees.

SS: See Flame-colored Tanager.

SD: F to C resident (100–1800 m, mainly above 600m) on Atlantic Slope from S Tamps, and in interior and on Pacific Slope from E Oax, to Honduras and N cen Nicaragua; also F (900–1400 m) on Pacific Slope in Gro (SNGH, SW, tape).

RA: E Mexico to Ecuador and N Brazil.

RED-HEADED TANAGER
Spermagra erythrocephala   Plate 58
Tángara Cabecirroja

ID: 5.2–5.7 in (13.5–14.5 cm). Endemic to W Mexico. **Descr.** ♂: *head bright red* (averaging more orange-red in imm?) with black lores. *Nape and upperparts olive, underparts yellow.* ♀: *head yellow-olive with dark lores and dusky auriculars,* upperparts olive. *Throat and chest yellow,*

becoming whitish on vent, flanks washed olive, undertail coverts dusky buff. **Juv:** resembles ♀ but brighter.

Voice. Calls include a sharp, high, thin to slightly liquid *spik* or *spi,* often run into high twittering series suggesting overgrown Bushtits. Song a high, thin, fairly rapid, tinkling *tsi-tsi tsee-tsee,* the 1st phrase given 1–3×, or *ts tsi-tsi tsee-tseet',* etc.

Habs. Humid to semihumid pine—oak, evergreen, and semideciduous forest and edge, plantations. In pairs or small groups, often with mixed-species flocks from canopy to low fruiting shrubs. Nest cup of fine twigs and other plant material at mid- to upper levels in outer canopy of tree (Mex, SNGH, SW).

SS: See ♀ Hepatic Tanager.

SD/RA: F to C resident (900–2500 m) on Pacific Slope from S Son to Oax; U in interior from Mich to N Mor; R (2100–3000 m) in N Oax. A report from Gto (AOU 1983) may be in error; a record from Belize (Wood *et al.* 1986) is not credible.

CRIMSON-COLLARED TANAGER
Phlogothraupis s. sanguinolenta   Plate 57
Tángara Cuellirroja

ID: 7.5–8 in (19–20 cm). Humid SE forest and edge. **Descr.** Ages differ, sexes similar. **Adult:** *bill bright bluish-white,* legs bluegrey. *Black overall with crimson hood continuing as collar across chest, tail coverts crimson.* **Juv:** hood dull red, chest mottled black and red. Quickly attains adult-like plumage.

Voice. A piercing *sweeu* or *ssiiew* and a high, thin *ssii-p,* also given in flight. Song a high, squeaky to slightly sweet, jerky series, *ssi ss-ssiip si-si-sip,* etc.

Habs. Humid evergreen forest edge, second

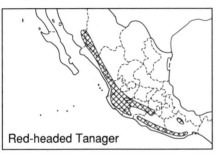

Red-headed Tanager

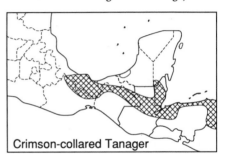

Crimson-collared Tanager

growth. Usually in pairs, at mid- to upper levels along forest edge. Nest cup of rootlets, moss, etc., at mid-levels in trees at forest edge. Eggs: 2, pale blue, spotted blackish (WFVZ).

SD: C to F resident (SL–1200 m) on Atlantic Slope from S Ver to Honduras.
RA: SE Mexico to W Panama.
NB: Sometimes placed in the genus *Ramphocelus*.

SCARLET-RUMPED TANAGER

Ramphocelus p. passerinii Plate 57
Tángara Terciopelo

ID: 6.8–7.3 in (17–18 cm). Humid SE forest edge. **Descr.** Ages/sexes differ. ♂: bill bluish white with fine dark tip, legs grey. *Glossy black overall with flaming red rump and tail coverts.* ♀/juv/imm: head greyish, paler on throat, *upperparts tawny-brown becoming bright tawny-ochre on rump* and uppertail coverts; wings and tail dark brown. *Chest tawny-ochre* becoming dusky, washed ochre, on underparts. Imm ♂ usually has brighter flame-tinged rump, patches of black on head and body by spring.
Voice. Calls include a dry, slightly gruff, nasal *cheht* or *chay*, often repeated as excited nasal chattering, *cheht-cheht, cheht...*, etc., a rough *shih* or *chih*, rough, lisping chatters, and a high, buzzy *ssit* or *zzrit.* Song a buzzy *swit-zhi-chi* or *bzzt chi-chi ih*, etc., up-slurred on the last note.
Habs. Humid evergreen forest edge, second growth, plantations. In pairs or small groups at low to mid-levels. Nest cup of leaf strips, rootlets, etc., at low to mid-levels in bush or tree. Eggs: 2–3, pale bluish, spotted with browns and black (WFVZ).
SS: See Grey-headed Tanager.

SD: C to F resident (SL–750 m) on Atlantic Slope from SE Ver and E Oax (HG, SNGH) to Honduras.
RA: SE Mexico to W Panama.

ROSY THRUSH-TANAGER

Rhodinocichla rosea schistacea Plate 57
Tángara Huitlacoche

ID: 7.8–8.5 in (19.5–21.5 cm). W Mexico. *Thrasher-like,* bill slender, culmen decurved. **Descr.** Ages/sexes differ. ♂: bill grey with blackish culmen, legs grey. *Head and upperparts dark grey with pink supraloral stripe merging into pale pink postocular stripe. Throat and chest bright rose-pink,* belly and flanks sooty grey, undertail coverts rose-pink. ♀: *rose-pink replaced with deep cinnamon*, postocular stripe white. **Juv:** sooty blackish overall, becoming white on belly, whitish postocular stripe indistinct; quickly attains imm plumage. **Imm:** resembles ♀ but throat and chest mottled sooty grey, lores sooty. ♂ may resemble adult after 1st prealternate molt, or underparts retain some dusky and cinnamon feathers.
Voice. Song of varied, loud, rich whistles suggests Blue Mockingbird, *chee-oo* or *p-weeoo*, and *t'chee-uh-oo* or *ch-weeee-oo-woo*, etc., often given in rollicking duets. Calls include a quiet *chowk* and *hu-weep.*
Habs. Arid to semiarid thorn forest, dense understory of brushy woodland. Skulking and rarely seen, best detected by voice. Feeds on ground, tossing litter aside with its bill. Nest is a shallow cup of twigs, rootlets, etc., low in tangled thickets. Eggs: 2, white, with blackish spots and scrawls.
SD: F resident (SL–1000 m) on Pacific Slope from Sin to Mich.
RA: W Mexico, disjunctly in S C.A. and N S.A.

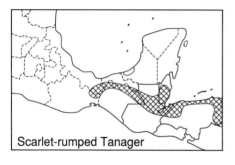

Scarlet-rumped Tanager

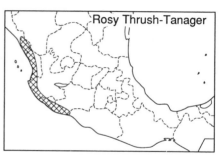

Rosy Thrush-Tanager

COMMON BUSH-TANAGER
Chlorospingus ophthalmicus*   Plate 57
Chinchinero Común

ID: 5.3–5.8 in (13.5–14.5 cm). Common in cloud forest. Three groups: *ophthalmicus* group (NE Mexico), *albifrons* (SW Mexico), and *postocularis* group (S Mexico and C.A.). **Descr.** Ages differ, sexes similar. **Adult.** *Ophthalmicus*: bill black, legs grey. *Head brown with white postocular stripe* and darker brown auriculars, spot at base of bill slightly paler. *Upperparts olive. Throat and underparts whitish, washed lemon on chest, flanks,* and undertail coverts, malar area may be flecked dusky. *Albifrons*: *head brown* with black lores and face, *white postocular spot, and whitish lore spot.* Upperparts olive. *Throat buff* with trace of dark malar stripe, *chest ochraceous lemon*, belly whitish with olive flanks, pale lemon undertail coverts. *Postocularis*: resembles *ophthalmicus* but *head grey with* shorter white postocular stripe and *blackish lateral crown stripe*, pale loral spot indistinct or lacking, chest more strongly washed

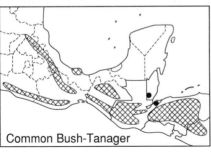

Common Bush-Tanager

lemon. *C. o. honduratius* of Honduras and El Salvador has more uniform grey head, without dark lateral crown stripe; head brownish in *regionalis* of N cen Nicaragua. **Juv** (undescribed in *albifrons?*): head olive, throat and underparts washed to mottled pale lemon. Quickly attains adult-like plumage.

Voice. Song (?) high, thin, hard to slightly liquid, buzzy trills and twitters, up to several s in duration, often preceded by high chip notes; also a varied arrangement of high and buzzy notes, *zzzrp zzzrp zzzrp ssit zzzrp ssit...* Calls are high, thin notes and excited twittering, *tsi* and *tsi-sit*, high, shrill chips that may run into a short, buzzy roll, and longer series *si-si-pit sit sip si-i-i-i-rr*, etc.

Habs. Humid evergreen forest and edge. In pairs or small groups, at times to 40+ birds, moving actively and noisily, low to high. A prominent member of mixed-species flocks. Nest cup of rootlets, grass, moss, etc., in epiphyte at mid-levels in trees, also hidden in undergrowth on banks. Eggs: 2–3, white, flecked with reddish browns and greys (WFVZ).

SD: C to F resident (1000–3500 m, lower locally in winter, for example to near SL in Los Tuxtlas). *Ophthalmicus*: Atlantic Slope from SE SLP to Oax. *Albifrons*: Pacific Slope of Gro and Oax. *Postocularis*: both slopes and in interior from SE Ver and SE Oax to Honduras and N cen Nicaragua.

RA: Mexico to N Argentina.

NB: More than one species may be included in this species.

GROSBEAKS AND BUNTINGS: CARDINALINAE (20)

Grosbeaks and buntings were formerly treated as a separate family but are now merged with other nine primaried passerines in the Emberizidae. Buntings are characterized by their stout seed-eating bills; cardinaline grosbeaks are simply large, stout-billed buntings. Ages differ; young altricial, 1st prebasic molt partial to complete, imm ♂♂ may resemble ♀♀. Sexes differ in temperate-breeding species, but are similar in most tropical species. Seasonal variation restricted to migratory, northern-breeding species. Songs often striking, loud and rich warbled phrases or varied whistled series.

Food mostly seeds and fruit. Nests are cups of vegetation placed in bushes or trees. Eggs: 2–5.

GENUS SALTATOR: Saltators

Saltators are fairly plain tropical grosbeaks of forest edge and second growth. Ages differ, sexes similar. Bills and legs blackish. Nests are bulky cups with stick bases, placed at low to mid-levels in dense shrubbery or thickets. Eggs: 2, pale to fairly bright blue, marked with black scrawls and flecks (WFVZ).

GREYISH SALTATOR
Saltator coerulescens* Plate 59
Saltador Grisáceo

ID: 8.5–9.5 in (21.5–24 cm). Two groups: *vigorsii* group (W Mexico) and *grandis* group (E Mexico and C.A.). **Descr. *Vigorsii.*** **Adult:** *head grey with short white supercilium, dark lores, and blackish malar stripe. Throat white,* underparts dusky pale greyish, washed buff, becoming pale cinnamon on flanks and undertail coverts. *Upperparts pale olive-grey,* wings and tail slightly darker. **Juv:** greyish olive overall with trace of lemon supraloral stripe, belly mottled lemon and pale cinnamon. Quickly attains imm plumage. **Imm** (Aug–Mar): *head and upperparts greyish olive, supercilium pale lemon;* throat whitish, chest washed ochre-lemon, becoming pale cinnamon on undertail coverts. ***Grandis.*** **Adult:** darker, *head and upperparts dark olive-greyish,* (brighter blue-grey in *hesperis* of Pacific Slope from Chis S), *with more extensive black on sides of throat and only a narrow white chin stripe.* Underparts dusky-grey, washed cinnamon, becoming cinnamon on undertail coverts. **Juv:** dark sooty olive overall with trace of lemon supraloral stripe, belly mottled lemon and pale cinnamon. **Imm** (Aug–Mar): head and upperparts dusky-olive with short lemon supercilium, darker lores. Throat may be washed lemon, underparts strongly washed ochre, undertail coverts brighter cinnamon. **Voice.** A rich, up-slurred, drawn-out *ch-wheeet* or *ch'kweeee,* often with a varied, short, jerky, warbling preamble, *choo-choo-chi-choo ch-wheeeet* or *hi'whee chu weeeeh;* a quiet, rolled *wi hu' churrrrr,* suggesting a quiet Pheasant Cuckoo, and a high *tsssi* or *ssit,* given in flight. Song varied, usually a fairly short, slightly jerky, mellow to nasal warble. **Habs.** Forest edge, open plantations, second growth. Singly, in pairs, or small groups, at low to mid-levels; may join mixed-species flocks, including other *Saltator* species; can be skulking. **SS:** Other saltators bright yellow-olive above. **SD:** C to F resident (SL–1500 m) on both slopes from S Sin and S Tamps, and locally in interior from Isthmus, to Honduras and W Nicaragua. **RA:** Mexico to Peru and Brazil.

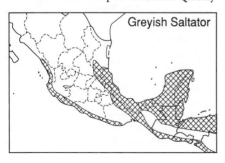

Greyish Saltator

BUFF-THROATED SALTATOR
Saltator maximus *   Plate 59
Saltador Gorjileonado

ID: 8–9 in (20.5–23 cm). Humid SE lowlands. **Descr. Adult:** *head grey with short white supercilium*, crown mostly blackish in *gigantoides* from Isthmus N, mostly olive in *magnoides* S of Isthmus. *White chin and pale cinnamon throat patch contrast with thick black malar stripe and broad black chest band*. Underparts grey, becoming ochre-cinnamon on undertail coverts. *Upperparts bright yellow-olive.* **Juv:** crown washed olive, pale cinnamon throat patch poorly defined against sooty blackish malar stripe and chest band; quickly attains adult-like plumage.
  Voice. A quiet, high, sibilant *ssi* or *tsii*, also given in flight, and a fuller *ssii* or *ssiip*, etc. Song pleasant, a slurred, liquid to slightly tinny warble, may suggest Orange-billed Nightingale-Thrush.
  Habs. Humid to semihumid, evergreen and semideciduous forest and edge, second growth. Habits much as Greyish Saltator.
SS: Black-headed Saltator larger, often in noisy groups, black nape contrasts strongly with upperparts, median throat white except in S Ver.
SD: F to C resident (SL–1500 m, mostly below 1000 m) on Atlantic Slope from S Ver to Honduras; R in SE Yuc Pen. Vagrant to El Salvador (Thurber *et al.* 1987).
RA: SE Mexico to Ecuador and Brazil.

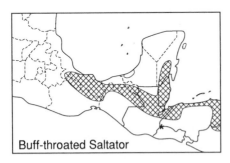
Buff-throated Saltator

BLACK-HEADED SALTATOR
Saltator atriceps *   Plate 59
Saltador Cabecinegro

ID: 10–11 in (25.5–28 cm). Lowlands in E and S. *Large and noisy.* **Descr. Adult:** *head blackish with short white supercilium* and greyish auriculars. *White throat* (cinna-

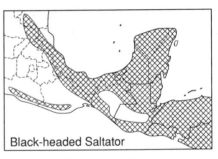
Black-headed Saltator

mon in *suffuscus* of Los Tuxtlas) *contrasts with broad black malar stripe and black chest band* (chest band can be indistinct or lacking, especially S of Isthmus). Underparts grey with cinnamon undertail coverts (pale buffy cinnamon in *flavicrissus* of Pacific Slope in Oax and Gro). *Upperparts bright golden olive.* **Juv:** whitish throat patch (including *suffuscus*?) poorly defined against sooty blackish malar stripe and chest band, quickly attains adult-like plumage.
  Voice. Often loud and noisy. Sharp, smacking barks, *chowk!* or *cheuk!*, etc., a drawn-out *siiip* followed by *chek!*, a sharp *yah!* or *chehh!*, repeated, and sharp, nasal notes run into a gruff, accelerating, rolling laugh or chortling chatter, may suggest Barred Antshrike.
  Habs. Humid to semiarid forest edge, open plantations, second growth. Usually in groups of 3–10 birds, often noisy but can be skulking.
SS: See Buff-throated Saltator.
SD: C to F resident (SL–1800 m) on both slopes from S Tamps and Gro, and locally in interior from Isthmus, to Honduras and W Nicaragua.
RA: Mexico to Panama.

BLACK-FACED GROSBEAK
Caryothraustes p. poliogaster   Plate 59
Picogrueso Carinegro

ID: 7–7.5 in (18–19 cm). Humid SE forest. **Descr.** Ages differ slightly, sexes similar. **Adult:** bill black, blue-grey below at base, legs grey. *Black face and throat contrast with yellowish head and chest, belly and undertail coverts grey.* Upperparts olive, becoming blue-grey on outer scapulars, rump, and uppertail coverts. **Juv:** duller, face pattern less distinct, quickly attains adult-like plumage.

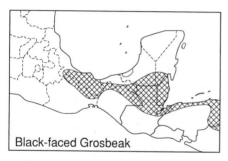

Black-faced Grosbeak

Voice. A buzzy, spluttering *zzzrt* and a bright *tree-dreek* or *ree-ree-reek*, often combined, *zzr-zzzrt tree-treek*, or *zzzrt reek reek*, etc.; also a quiet *pit* given in flight. Song a short, sweet, warbled phrase which may have a quiet, lisping introduction and often ends with an up-slurred note, *tsi tsi-tsi tseu choo*, or *tsi tree-tee choo-choo-choo-choo*, or *si choo-choo chi-chih*, etc.

Habs. Humid evergreen forest and edge. In active flocks up to 50+ birds in forest canopy, often not associating with other species. Nest is a shallow cup at mid-levels in small tree at edge or in clearing. Eggs: 3, white, mottled and spotted with bright brown.

SD: F to C resident (SL–1200 m) on Atlantic Slope from S Ver to Honduras.

RA: S Mexico to cen Panama.

CRIMSON-COLLARED GROSBEAK

Rhodothraupis celaeno   Plate 59
Picogrueso Cuellirrojo

ID: 8.3–8.8 in (21–22 cm). Endemic to NE Mexico. **Descr.** Ages/sexes differ. ♂: bill black, blue-grey below at base, legs grey. *Head and chest black with pinkish-red hindcollar, underparts dusky pinkish red, mottled black on belly*, flanks, and undertail coverts. Upperparts black. ♀: *head and chest black, underparts yellow-olive, upper-*

Crimson-collared Grosbeak

parts olive. **Imm ♂:** 1st basic resembles ♀ but black on head less extensive (hindcrown olive), throat and chest sooty blackish. In 1st alt, head and chest black like adult ♀, upperparts usually with black patches, underparts with red and black patches, often shows some black tertials and rectrices. **Imm ♀:** black restricted to face and throat, chest sooty grey. **Juv:** undescribed (?).

Voice. A high, clear, penetrating slurred *ssseeuu* or *sseeeur*, also a piercing *seeip seeeiyu*. Song a rich to slightly burry warble, often up-slurred at the end.

Habs. Humid to semiarid brushy woodland and edge, second growth. Singly or in pairs, low to high in forest; joins mixed-species flocks. At times skulking, on or near ground. Nest is a bulky cup of twigs and grasses at low to mid-levels in thickets. Eggs: 2–3, pale greyish blue, speckled with browns (Sutton *et al.* 1950).

SD/RA: F to U resident (near SL–1200 m) on Atlantic Slope from cen NL to N Ver. Irregular R visitor (Nov–Apr, mostly Dec–Jan) N to S Tex.

GENUS CARDINALIS: Cardinals

Cardinals are large grosbeaks with conspicuous pointed crests and long squared tails. Ages/sexes differ. Juvs look like ♀♀ but bills grey, little or no red in crest, quickly resemble adults after partial to complete 1st prebasic molt. Nest cups of twigs, grasses, etc., placed at low to mid-levels in bush or small tree. Eggs: 2–5, whitish to bluish white, densely speckled to mottled with browns and greys (WFVZ).

NORTHERN CARDINAL

*Cardinalis cardinalis**   Not illustrated
Cardenal Norteño

ID: 8–9 in (20.5–23 cm). **Descr.** ♂: *bill red, legs grey to dull flesh. Bright red overall*, duller on upperparts, with *black lores and throat*. ♀: *head and upperparts pale brownish grey* to greyish, with dusky lores, red in crest. Wings and tail slightly darker, edged reddish, *red underwing coverts flash in flight*. Throat and underparts buff, paler on throat and vent. **Juv:** see genus note.

Voice. A sharp, slightly liquid to metallic *tik* or *pik*. Song varied, typically a series of 3–6 or more loud rich whistles, *weeoo*

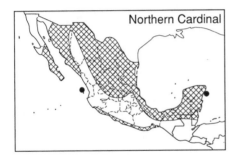

Northern Cardinal

weeoo... or *si-weet si-weet...* or *chip'ee chip'ee...*, etc., at times a more complex *wee choo-choo-choo weet-weetweetweetweetweet*, or *see ch-wee ch-weecha weecha weecha*, etc.

Habs. Semiopen scrub, brushy forest and edge, thickets, overgrown clearings. Singly or in pairs, at low to mid-levels, often on ground; sings from prominent perches.

SS: ♀ Pyrrhuloxia has stubbier yellow-horn bill, greyer overall.

SD: C to F resident (SL–2000 m) in BCS, on Pacific Slope from Son to cen Sin and from Col to Oax, in interior S over Plateau to NE Jal and Hgo, and on Atlantic Slope from Tamps to Yuc Pen, S to N Petén and N Belize.

RA: E N.A. and SW USA to Mexico and N Guatemala.

PYRRHULOXIA

Cardinalis sinuatus*   Not illustrated
Cardenal Desértico

ID: 7.5–8.5 in (19–21.5 cm). N deserts.
  Descr. ♂: *bill yellow to horn*, legs grey to dull flesh. *Grey overall with red face, red mottling from throat to belly*; crest tipped red, undertail coverts have pinkish-red centers. Wings and tail slightly darker, edged red on alula and primaries, underwing coverts red. ♀: *head and upperparts grey* to brownish grey, with *red-tipped*

crest, red flecking on face. Throat and underparts slightly paler greyish, washed buff, often with pink wash on throat and chest. **Juv:** see genus note.
  Voice. A bright metallic *plik* or *tlink*, also a sharp *pik* at times run into a spluttering, slightly liquid chatter. Song varied, loud, rich whistles repeated 3–6× or more, much like Northern Cardinal, *wheet wheet...*, or *chuwee chuwee...* or *whee t-wee t-wee...*, etc.
  Habs. Arid to semiarid scrub, semiopen areas, riparian washes. Habits much as Northern Cardinal, at times in small flocks in winter.
  SS: See ♀ Northern Cardinal.
  SD: U to F resident (SL–2000 m) in BCS; C to F on Pacific Slope S to Nay, in interior S over Plateau to N Mich and Qro, and on Atlantic Slope from Tamps to E SLP.
  RA: SW USA to N Mexico.

GENUS PHEUCTICUS

These are mainly arboreal grosbeaks of temperate forest and woodland. Ages/sexes differ, plumage strikingly patterned. 1st basic ♂ resembles ♀, 1st alt ♂ variably intermediate in appearance between ♂ and ♀. Legs greyish. Nest cups of twigs, grasses, etc., placed at mid-levels in bush or small tree. Eggs: 2–5, pale bluish to greenish, heavily speckled with browns and greys (WFVZ).

YELLOW GROSBEAK

Pheucticus chrysopeplus*   Plate 58
Picogrueso Amarillo

ID: 8.5–9.5 in (21.5–24 cm). *Bill massive.* Two groups: *chrysopeplus* groups (W Mexico) and *aurantiacus* (S of Isthmus). **Descr. Chrysopeplus.** ♂: bill black above, grey below. *Head and underparts yellow*, undertail coverts white. Back black,

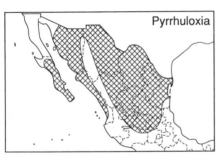

Pyrrhuloxia

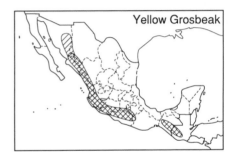

Yellow Grosbeak

mottled yellow down center, rump yellow, uppertail coverts black, boldly tipped white. *Wings black, boldly patterned with 2 white wingbars, white flash at base of primaries,* white tips to tertials and secondaries, and white primary edgings. Tail black, outer rectrices broadly tipped white on inner webs. ♀: bill greyish. *Face and underparts yellow,* with dusky auriculars, undertail coverts white. *Crown, nape and upperparts olive to yellow-olive, streaked blackish on crown and back,* uppertail coverts grey. *Wings and tail dark greyish, patterned much as ♂ but less boldly,* white patch on base of primaries smaller. **Imm ♂:** *resembles ♀ but head yellower, note bold white flash on base of primaries,* rump often has dark centers. Usually attains varying degrees of adult plumage by 1st summer, especially greater wing coverts and tertials, rarely rectrices. **Juv:** resembles ♀ but throat and underparts pale lemon, quickly attains imm plumage. *Aurantiacus:* brighter, ♂ has *brilliant golden-orange head and underparts.*
Voice. A sharp, slightly nasal to metallic *iehk* or *plihk,* and a soft mellow *whoi* or *hu-oi,* given while foraging and in flight. Song a variable, short, rich warble, *chee wee chee-r weer weeuh'* or *ch-reer wi-wi h'ree ree-e-e-eer,* etc., suggesting Black-headed Grosbeak.
Habs. Deciduous to semideciduous forest and edge, scrubby woodland, thorn forest, also (*aurantiacus*) clearings in humid evergreen forest. Singly, in pairs, or small groups, low to high, often at fruiting trees.
SS: See ♀ Flame-colored Tanager. Orioles have slender pointed bills.
SD: *Chrysopeplus:* F to C resident (near SL–2500 m) on Pacific Slope from S Son to Col, thence in interior to SW Pue and NW Oax; F to C in summer (Mar–Sep) N to cen Son. *Aurantiacus:* F to C resident (300–2500 m) on Pacific Slope and in interior of Chis and Guatemala.
RA: Mexico to Guatemala, disjunctly in S.A.
NB: S.A. forms sometimes considered specifically distinct.

ROSE-BREASTED GROSBEAK

Pheucticus ludovicianus   Not illustrated
Picogrueso Pechirrosado

ID: 7–7.7 in (18–20 cm). Winters in E and S. **Descr. Alt ♂** (Mar–Jun): bill pale flesh. *Head and upperparts black with white*

rump, thick white wingbars, broad white flash across base of primaries,* and bold white tips to tertials and outer rectrices; underwing coverts red. *Median chest red, rest of underparts white.* **Basic ♂** (Aug–Feb): bill often dusky above. *Head and upperparts tawny-brown, streaked black, with white supercilium; wings and tail as alt.* Throat, chest, and flanks buffy brown, spotted black, with *rose mottling on chest,* belly and undertail coverts white. ♀: bill often dusky above. *Head and upperparts brown,* streaked darker, *with white supercilium and median crown stripe.* Wings and tail dark brown with 2 white wingbars, narrow white flash on base of primaries, and white tertial spots; *underwing coverts ochre-yellow,* often mixed with rose in adult. *Throat and underparts whitish, coarse dark brown streaks on chest and flanks often on a buff wash,* especially in imm. **Imm ♂:** resembles ♀ with chest and flanks washed buff, but some rose on chest, underwing coverts red, greater secondary coverts black, white wingbars bolder. By spring resembles ♂ but head and upperparts mottled brown; note contrasting brown remiges and usually some rectrices.
Voice. A squeaky to nasal *iihk* or *ihk,* much like Black-headed Grosbeak.
Habs. Humid to semiarid forest and edge, plantations, gardens, semiopen areas with hedges. Singly or in flocks (up to 50+ birds in migration), often with mixed-species flocks. Low to high, often at fruiting trees.
SS: ♀ Black-headed Grosbeak has pale cinnamon wash on chest, finer dark streaking on chest rarely meets across center, underwing coverts lemon yellow.
SD: F to C transient and winter visitor (Oct–May; SL–2500 m) on both slopes from Gro and S Ver, and in interior from Isthmus, to Honduras and W Nicaragua, U

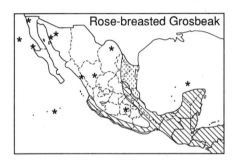

Rose-breasted Grosbeak

to F to Nay (PA, PP) and S Tamps. C to F transient (Oct–Nov, Apr–May; SL–3000 m) on Atlantic Slope throughout; U in interior N of Isthmus W to DF and Coah. Vagrant (Oct–May) to NW Mexico and Isla Socorro (SFB).

RA: Breeds E N.A.; winters Mexico to Peru.

BLACK-HEADED GROSBEAK
Pheucticus melanocephalus * Not illustrated
Picogrueso Tigrillo

ID: 7–8 in (18–20.5 cm). Widespread N of Isthmus. **Descr.** ♂: bill blackish above, blue-grey below. *Head and upperparts black with cinnamon hindcollar* (and often a cinnamon postocular stripe), pale brown back edgings (broader, tawny-brown in basic), and *cinnamon rump.* Wings and tail black with thick white wingbars, broad white flash across base of primaries, and bold white tips to tertials and outer rectrices; underwing coverts yellow. *Underparts cinnamon with lemon median belly.* ♀: bill dark above, greyish to flesh below. *Head and upperparts brown,* streaked darker, *with whitish supercilium and median crown stripe.* Wings and tail dark brown with 2 white wingbars, narrow white flash on base of primaries, and white tertial spots; *underwing coverts lemon yellow. Throat and underparts pale cinnamon to pale buff* (imm), *dark brown streaks on chest and flanks relatively fine, often do not meet across chest,* belly often washed lemon in adult. **Imm** ♂: resembles ♀ but underparts often deeper cinnamon; by spring, plumage is variably intermediate between ♂ and ♀. **Juv:** resembles ♀ but upperparts more brightly streaked tawny-brown, wingbars washed pale cinnamon.

Voice. A sharp, slightly nasal, squeaky to tinny *plik* or *iihk.* Song a rich warble, suggests American Robin but slightly harder, less repetitive.

Habs. Breeds in arid to semihumid oak, pine–oak, and deciduous woodland, often near water. More widespread in winter, rarely to humid evergreen and semideciduous forest and edge. Habits much as Rose-breasted Grosbeak, also in flocks in winter.

SS: See ♀ Rose-breasted Grosbeak.

SD: C to F breeder (Apr–Aug; near SL–2500 m) in BCN, and (1500–3000 m) from Son and Coah to Oax, mainly in interior and on adjacent Pacific Slope. C to F transient and winter visitor (Aug–May; SL–3000 m) in BCS, on Pacific Slope from Son to Mich, and in interior from Zac and S NL S (less common in N); also U to F on Atlantic Slope from Tamps to Oax.

RA: Breeds W N.A. to Mexico; winters Mexico.

GENUS CYANOCOMPSA

Bills deep and stout with curved culmens. Ages/sexes differ, juv resembles ♀. ♂♂ dark blue overall, ♀♀ brown. Nest cups of grasses, tendrils, and rootlets, at low to mid-levels in bush or small tree.

BLUE-BLACK GROSBEAK
Cyanocompsa cyanoides concreta   Plate 60
Picogrueso Negro

ID: 6.8–7.3 in (17–18.5 cm). Humid SE forest and edge. **Descr.** ♂: bill black, blue-grey below at base, legs greyish. *Blue-black overall,* slightly brighter on forehead and in supercilium region. ♀: bill all black. Plumage *deep rich brown overall,* slightly paler on throat. **Juv/imm:** resembles ♀ but slightly darker, underparts streaked dusky, ♂ often has a few blue-black feathers in first summer.

Voice. A sharp, nasal to slightly squeaky *plik* or *chik,* and *pli-dik* or *ch-dik.* Song a

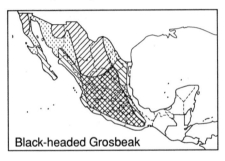
Black-headed Grosbeak

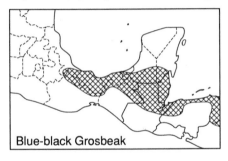
Blue-black Grosbeak

rich, slightly sad, slurred, descending warble often fading into a soft, squeaky ending.

Habs. Humid evergreen forest edge, dense second growth. Singly or in pairs at low to mid-levels, often skulking. Eggs: 2, whitish to bluish white, flecked and spotted with rusty and greys (WFVZ).

SS: Blue Bunting smaller, bill less stout, ♂ has brighter blue face, ♀ paler, less rufous-brown; note voice.

SD: C to F resident (SL–900 m) on Atlantic Slope from S Ver to Honduras.

RA: SE Mexico to Ecuador and Brazil.

BLUE BUNTING
Cyanocompsa parellina *   Plate 60
Colorín Azulinegro

ID: 5–5.5 in (13–14 cm). Endemic, two groups readily separable in ♀ plumage: *parellina* group (E Mexico and C.A.) and *indigotica* (W Mexico). **Descr.** ♂: bill blackish, blue-grey below at base, legs greyish. *Deep blue overall with contrasting bright blue forecrown, supercilium, malar region,* rump, and lesser wing coverts (often covered by scapulars). ♀/juv. *Parellina*: bill blackish. *Warm brown overall, slightly paler cinnamon-brown below.* May show faintly paler lores and narrow eyering. Some ♀♀ show bluish edges to lesser wing coverts. ♂ quickly attains adult plumage. *Indigotica*: bill dusky flesh with dark culmen. Plumage overall paler and greyer.

Voice. A metallic *chik* or *chink*, much like Hooded Warbler. Song varied, a sweet, slightly sad warble often with 1–2 separate notes at the start, and fading away at the end, *swee slee lee swee' t-lee-it* or *seeu, seeu syee su-ee si-si-see*, at times a shorter *seet chee cheer-u*, etc.

Habs. Brushy forest, woodland, and edge, scrubby thickets. In pairs or singly, low in

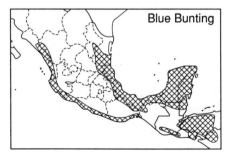

Blue Bunting

forest understory. Eggs: 2, bluish white, unmarked (WFVZ).

SS: Blue and Slate-Blue seedeaters less chunky with smaller bills (culmen straighter), usually in bamboo, ♂♂ duller uniform blue overall, ♀♀ remarkably similar in plumage to Blue Bunting, best told by small, mostly dark grey bills. ♀ Varied Bunting less thickset with smaller bill, paler wing edgings and bluish cast to primaries and tail often noticeable, adult has bluish uppertail coverts. See Blue-Black Grosbeak.

SD/RA: F to C resident (SL–1800 m) on both slopes from Sin and NL, and in interior from Isthmus, to Honduras and N cen Nicaragua. A report from Isla Mujeres (AOU 1983) may be in error.

GENUS PASSERINA

All species in this genus occur in Mexico. Ages/sexes differ, ♂♂ brightly colored, ♀♀ duller, often brownish. Bills blackish above, blue-grey below in alt ♂♂ and some ♀♀, often blue-grey to greyish flesh with dark culmen in other plumages; legs greyish. Songs and calls of all species similar, songs being pleasant warbles mostly of 2–5 s duration, usually given from fairly prominent perches. Molts complex in studied species; for example, Indigo, Lazuli, and Painted buntings attain a supplemental plumage in autumn or early winter (as do other species?). Nest cups of grasses, rootlets, etc., placed at low to mid-levels in bush, viny tangle, small tree, etc. Eggs: 2–5, bluish white, unmarked (Lazuli and Varied) or speckled with reddish brown and grey (Rosita's, Painted, some Varied?) (WFVZ).

BLUE GROSBEAK
Passerina caerulea *   Not illustrated
Picogrueso Azul

ID: 6.5–7 in (16.5–17.5 cm). Widespread. **Descr.** ♂: *blue overall* (extensively tipped brown in fresh basic) with black lores; back mottled darker. Wings and tail blackish with 2 *chestnut wingbars* (the lower paler and narrower), pale edgings to remiges when fresh. ♀: *head and upperparts brown,* becoming grey-brown on rump and uppertail coverts, back streaked darker; crown and rump may be tinged bluish. Wings and tail blackish with 2

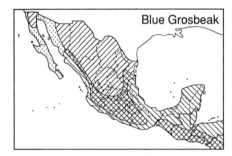

cinnamon wingbars, paler edgings when fresh, lesser wing coverts tinged blue. *Throat and underparts pale cinnamon-brown*, throat paler, flanks often duskier. **Imm:** resembles ♀ but ♂ often deeper brown overall. By spring ♂ has head and upperparts mottled blue. **Juv:** resembles ♀ but with brighter pale cinnamon wingbars, lacks blue.

Voice. A strong, sharp, metallic *chik* or *chink*, and a buzzy *zzir* or *zzzir*. Song a rich, slightly scratchy, full warble, much like *Passerina* buntings but stronger and often scratchier.

Habs. Breeds in deciduous to semideciduous scrub, woodland and edge, semiopen areas with hedges, often near water. More widely in winter, including clearings in humid evergreen forest. In pairs or flocks, at times to 50+ birds, on or near ground.

SS: Imm ♂ may resemble Lazuli Bunting but larger with heavier bill, blue darker and chest duller. Indigo Bunting smaller with slighter bill, ♀ and imm ♀ have brownish wingbars.

SD: F to C breeder (Apr–Aug; near SL–2500 m) in interior and on Pacific Slope from Son and Coah to Honduras and W Nicaragua (local from cen Mexico S), and on Atlantic Slope in Tamps; U in BCN. C to F transient and winter visitor (late Aug–mid-May; SL–3000 m) in BCS, on both slopes from S Son and S Tamps, and in interior from cen Mexico, S through region.

RA: Breeds USA to Costa Rica; winters Mexico to Costa Rica.

NB: Traditionally placed in the monotypic genus *Guiraca*.

ROSITA'S (ROSE-BELLIED) BUNTING
Passerina rositae   Plate 60
Colorín de Rosita

ID: 5.3–5.7 in (13.5–14.5 cm). Endemic to Pacific Slope of Isthmus. **Descr.** ♂: *head*

and upperparts electric blue, brightest on rump, *with black lores and white eye-crescents*. Throat, chest, and flanks electric blue, mottled reddish pink on chest, *belly and undertail coverts pink*. ♀: head and upperparts greyish with *whitish eye-crescents*, becoming *turquoise-blue on rump and uppertail coverts*; some (older birds?) with a few blue flecks on head. Wings and tail darker, remiges edged turquoise-blue, tail washed turquoise-blue. *Throat and underparts vinaceous*, duskier on throat, chest, and flanks. **Imm ♂:** resembles ♀ but underparts brighter in 1st alt, often rose-pink on belly, flecked blue on head, chest, and upperparts. **Imm ♀:** duller than ♀, rump and uppertail coverts only faintly tinged blue, underparts dusky vinaceous-buff, sometimes with dusky streaks on sides and flanks. **Juv:** undescribed (?).

Voice. A wet *plik* or *plek*. Song a sweet, slightly burry warble.

Habs. Arid to semiarid thorn forest, semihumid deciduous gallery woodland and edge, in hilly areas. In pairs or singly, at low to mid-levels in brush.

SS: Smaller Varied Bunting allopatric (?), lacks white eye-crescents.

SD/RA: C to F resident (150–800 m) on Pacific Slope in E Oax and W Chis.

LAZULI BUNTING
Passerina amoena   Not illustrated
Colorín Lazulita

ID: 5–5.5 in (13–14 cm). W Mexico in winter. **Descr.** ♂: *head and upperparts turquoise-blue*, heavily veiled brown in autumn, back mottled darker in spring. Wings and tail black with *2 white wingbars*, pale brown tertial edgings, and blue flight feather edgings. *Chest cinnamon, rest of underparts whitish*, washed cinnamon on flanks. ♀: *head and upperparts* *grey-brown*, paler eyering accentuates blank face, rump often tinged bluish; *wingbars often pale buff. Throat and underparts dirty whitish with warm cinnamon wash across chest. In fresh plumage,* head and upperparts warmer brown, wingbars buff; *throat, chest, and flanks cinnamon-buff* becoming whitish on belly and undertail coverts. **Imm:** resembles autumn ♀. By Nov–Dec, most imm ♂♂ show some turquoise-blue on head and upperparts and by spring, most resemble ♂ but note contrasting brown secondaries and primary coverts. **Juv:** resembles imm but underparts paler, often with indistinct dusky streaks.

Voice. A wet *plik* or *tlik*, and a short wet buzzy *zzzrt*. Song sweet to slightly buzzy, often with paired phrases, *swee-swee suwee-suwee swee swi-si-chu*, etc.

Habs. Breeds in scrub and deciduous woodland, often near water. In winter, also open and semiopen areas such as overgrown fields, thorn forest. Singly or in loose flocks, on or near ground, associating readily with other seed-eating birds.

SS: ♀ Indigo Bunting lacks such distinct pale wingbars, warm brown to pale buffy brown below with dusky streaking on chest. ♀ Varied Bunting can show pale wingbars but more uniform overall; note stubbier bill, lacks warm cinnamon wash across chest and contrasting pale tertial edgings. See imm ♂ Blue Grosbeak.

SD: F breeder (May–Aug; near SL–2400 m) in NW BCN; F to C transient and winter visitor (Aug–mid-May; SL–2000 m) on Pacific Slope from Son to Mich, and in interior along Balsas drainage to Pue and N Gro, R to cen Oax (SNGH). F to C transient in BCS (mid-Aug–Oct) and on W Plateau.

RA: Breeds W N.A. to NW Baja; winters W Mexico.

INDIGO BUNTING

Passerina cyanea   Not illustrated
Colorín Azul

ID: 4.8–5.3 in (12–13 cm). Widespread winter visitor. **Descr. Alt** ♂ (Mar–Jul): *bright blue overall*, lores darker; wings and tail black, edged blue. **Basic** ♂ (Aug–Mar): *head and body brown, mottled blue*, especially on throat and underparts; wings and tail black with 2 brownish wingbars, brown tertial edgings, flight feathers edged blue. ♀: *head and upperparts brown*, often tinged blue on rump. Wings and tail darker, wing coverts edged brown, tertials dark with fairly broad, contrasting rufous-brown edges; flight feathers edged blue. *Throat and underparts whitish, chest and flanks washed buffy brown and streaked dusky*; throat and sides of neck often tinged blue. *In fresh basic, head and upperparts richer, cinnamon-brown, throat and underparts cinnamon-brown*, paler on throat, only faintly streaked dusky on chest and flanks. **Imm:** resembles ♀ but remiges edged brownish, dusky streaking on underparts often more distinct. By Nov–Dec, most imm ♂♂ show patches of blue and by spring, resemble ♂,· but may retain some brown mottling; note contrasting brown secondaries and primary coverts.

Voice. A wet *plik* or *tlik* and a short wet buzzy *zzzrt*, much like Lazuli Bunting. Song also similar to Lazuli, although paired phrases often better defined.

Habs. Open and semiopen areas, brushy forest edge and second growth. In flocks up to 150+ birds, or singly, on or near ground, at times in canopy of fruiting trees; readily joins flocks of other seed-eating birds.

SS: ♀ Varied Bunting has more decurved culmen, lacks streaking below, tertial edges

Lazuli Bunting

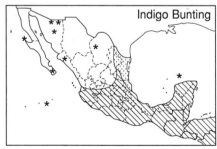

Indigo Bunting

narrower and less contrasting, warm olive-brown to tawny-brown. See ♀ Lazuli Bunting, Blue Grosbeak.

SD: F to C transient and winter visitor (mid-Sep–mid-May; SL–2500 m) on both slopes from Nay and S Ver, and in interior from cen Mexico, to Honduras and W Nicaragua, U N to Tamps. C to F transient (mid-Sep–Nov, Apr–mid-May) on Atlantic Slope N of Isthmus. Vagrant (mainly late May–Jul) to NW Mexico, and Isla Socorro (LFB).

RA: Breeds S N.A.; winters Mexico to Panama.

VARIED BUNTING

Passerina versicolor*　　　　　　Plate 60
Colorín Morado

ID: 4.5–5 in (11.5–12.5 cm). Widespread. **Descr.** ♂: *often looks blackish with some but not all areas of varied colors catching the light.* Black lores and chin contrast with *violet-blue face, nape reddish* with violet-blue hindcollar. Back dark reddish, *rump violet-blue*; wings and tail black, edged bluish. Throat and chest dark reddish becoming deep purple on belly and undertail coverts. Head and body heavily veiled buffy brown in fresh basic. ♀: *notably nondescript; dickeyae* of NW Mexico (breeds S at least to Jal, winters to Gro) notably richer brown than other ssp; grey and brown ssp occur together in S Jal and Col in winter. *Head and upperparts grey-brown to warm brown,* warmer in fresh plumage; paler eyering accentuates blank face, rump often tinged blue-green. Wings and tail darker brown, tertials with narrow, poorly contrasting brown edges, flight feathers edged bluish: some birds show paler wingbars. *Throat and underparts dirty pale buff to warm buffy brown,* throat and belly often slightly paler but *underparts* usually *appear uniform* overall

with *paler vent and undertail coverts.* **Imm:** resembles ♀ but flight feathers edged brownish. ♂ attains adult plumage when about 1 year old. **Juv:** resembles ♀ but generally richer brown with narrow buff wingbars.

Voice. A wet to slightly tinny *plik* or *spik* and *whuit,* also a quiet, slightly rolled *chihr.* Song a rich, slightly scratchy warble.

Habs. Thorn forest, forest edge, scrubby woodland, overgrown clearings, mainly in arid situations but in winter also humid forest edge. Singly or in small groups; often slightly skulking, on or near ground, at times in canopy of fruiting trees.

SS: ♀ Blue and Slate-Blue seedeaters (usually in bamboo) have slightly heavier grey bills (♀ Varied Bunting bill usually dusky flesh below), overall more uniform cinnamon-brown to rufous brown (undertail coverts not paler) without paler wing edgings and bluish tinges. Slate-blue Seedeater slightly longer-tailed. See ♀ Lazuli and Indigo buntings.

SD/RA: F to C resident (near SL–1200 m) in BCS. C to F breeder (Mar–Sep; SL–2000 m) on Pacific Slope from Son to Jal, and F to C on Plateau from Chih and Coah S to N Mich; also locally in SW USA. F to C local resident in interior from Balsas drainage to Guatemala, and on Atlantic Slope in Tamps and cen Ver. N populations mostly withdraw S in winter: F to C in winter (Oct–Mar) on Pacific Slope from S Son to Gro, in interior from cen Mexico S, and on Atlantic Slope S to S Ver (Los Tuxtlas, SNGH).

ORANGE-BREASTED BUNTING

Passerina leclancherii*　　　　　　Plate 60
Colorín Pechinaranja

ID: 4.5–5 in (11.5–12.5 cm). Endemic to W Mexico. **Descr.** ♂: *head and upperparts*

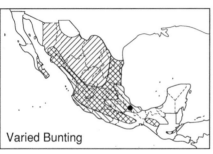

Varied Bunting

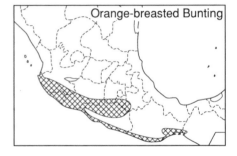

Orange-breasted Bunting

electric turquoise-blue with yellow-green crown, yellow lores and eyering, green wash to back. *Bright yellow throat and underparts suffused golden orange on chest.* ♀: *head and upperparts greenish with yellow lores and eyering,* auriculars often washed turquoise, uppertail coverts dull turquoise. Wings and tail darker, edged dull turquoise-blue. *Throat and underparts yellow.* **Imm** ♂: averages brighter than ♀, auriculars mostly turquoise, uppertail coverts turquoise, chest often suffused orange. **Imm** ♀: head and upperparts duller and greyer than ♀ with little or no turquoise, throat and underparts duller yellow, washed dusky on chest. **Juv:** resembles dull imm ♀ with indistinct dusky streaking on chest.

Voice. A wet to fairly hard *chlik* or *tchik*. Song a sad, sweet warble, often shorter and slower than other *Passerina* buntings.

Habs. Arid to semiarid thorn forest, brushy deciduous woodland and edge, overgrown clearings. In pairs or small groups, on or near ground, often along roadsides. Eggs undescribed (Kiff and Hough 1985).

SS: ♀ Painted Bunting has greenish head with narrow pale eyering but no pale lores, often brighter and greener overall, lacks bluish uppertail coverts.

SD/RA: C to F resident (SL–1200 m) on Pacific Slope from S Nay (EH, PP) to W Chis, and in interior along Balsas drainage to W Pue. Extralimital records may refer to escaped cage birds.

PAINTED BUNTING
*Passerina ciris**   Plate 60
Colorín Sietecolores

ID: 4.8–5.3 in (12–13 cm). Winters in lowlands. **Descr.** ♂: *blue head* with red eyering *clashes with yellow-green back and red throat and underparts; rump and uppertail*

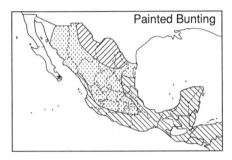

Painted Bunting

coverts red. Wings and tail darker, median coverts edged reddish, greater coverts edged green. ♀: *head and upperparts greenish with narrow pale eyering. Throat and underparts yellowish,* washed olive on chest and flanks. **Imm** ♂: averages brighter than ♀, often with a few blue and red feathers by spring. Attains adult plumage when about 1 year old. **Imm** ♀: duller than ♀, *head and upperparts greyish olive, throat and underparts dirty pale greyish, washed lemon.* **Juv:** resembles ♀ but underparts washed pale buffy brown, often with indistinct dusky streaks, narrow buff lower wingbar.

Voice. A wet *plik*. Song a sweet, rich warble.

Habs. Arid to semihumid, semiopen areas with hedges, thickets, woodland edge; more widely in winter, from arid thorn forest to clearings in humid evergreen forest. Singly or in small flocks, on or near ground; readily joins flocks of other seed-eating birds.

SS: See ♀ Orange-breasted Bunting.

SD: F to U breeder (mid-Apr–Aug; 500–1200 m) in interior and on adjacent Atlantic Slope from Chih to N Tamps. F to C transient and winter visitor (Sep–early May; SL–1800 m) on both slopes from Sin and S Ver, and in interior from Balsas drainage, to Honduras and W Nicaragua, U in winter to S Tamps. F to C transient (late Jul–Oct) in NW Mexico and (Mar–early May, late Jul–Oct) on Plateau and in NE Mexico. Vagrant to BCS (Nov 1968, Wilbur 1987).

RA: Breeds S USA and NE Mexico; winters Mexico to cen Panama.

DICKCISSEL
Spiza americana   Not illustrated
Arrocero Americano

ID: 5.8–6.3 in (14.5–16 cm). Nomadic migrant. **Descr.** Ages/sexes differ. ♂: bill blackish above, blue-grey to greyish flesh below, legs brownish flesh. *Head grey with yellowish supercilium* and whitish subocular crescent; *white chin and moustache border black throat patch,* often has broken black malar stripe. *Chest yellow,* rest of underparts dirty whitish, washed grey on flanks. Upperparts brown, streaked black on back; wings and tail blackish, edged brown, with *chestnut* *shoulder.* ♀: *face pattern duller, lacks black throat patch* but black malar stripe often more distinct; chestnut shoulder paler and less extensive; *flanks* (and sometimes chest) *streaked dark brown.* **Imm:** *resembles* ♀ *but duller; head brownish with indistinct short yellowish-buff supercilium, chest washed ochre to buff with fine dark streaks extending to brownish-washed flanks.* By spring (Mar–Apr) resembles adult.

Voice. A wet, buzzy *zzzrt* or *dzzzrt*, suggesting an emphatic rough-winged swal-

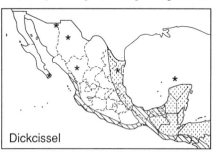

Dickcissel

low call, and a full slightly liquid *fwit* or *plit*; in spring, calls may run into a buzzy *dik-cizz-l*; also a hard, slightly wet *chek.*

Habs. Open and semiopen areas, especially overgrown weedy fields. Singly or in flocks, at times to thousands, which fly in fairly compact, undulating formation.

SS: ♀ House Sparrow has stubbier bill, lacks pale throat with dark malar stripe, flanks unstreaked.

SD: Irregularly F to C transient and winter visitor (late Aug–May; SL–1500 m) on Pacific Slope from Sin (PP) to El Salvador and Honduras; R to U on Atlantic Slope from Tamps (AGL spec.) to SW Camp (SNGH). C to F transient (mid-Aug–Oct, Apr–May) on Atlantic Slope from Tamps to Honduras; U to F in Yuc Pen. Vagrant to NW Mexico (late Sep–Nov) and Clipperton (May 1987, RLP photo).

RA: Breeds cen N.A.; winters nomadically from W Mexico to N S.A.

NB: Taxonomic affinities uncertain, perhaps better placed with the Emberizinae; sometimes placed with the Icteridae.

BRUSHFINCHES, SEEDEATERS, SPARROWS, AND ALLIES: EMBERIZINAE (75)

A large subfamily of nine-primaried seed-eating birds with which the preceding four subfamilies now form the large family Emberizidae. Several species are migratory, and 18 species occur in the region only as migrants or vagrants. Wings typically short and rounded, tails long to fairly short. Most species have conical, deep-based bills. Ages differ; young altricial; 1st prebasic molt usually partial to incomplete, but complete in a few species. Sexes usually similar; when different, basic ♂♂ often resemble ♀♀. Plumage typically dull, predominantly colored with browns, greys, and olive, often streaked. Most species are terrestrial and many are fairly skulking, either in forest understory, brush, or grassy fields, though most species sing from conspicuous perches. Songs weak to quite striking, and often quite varied within and between populations. Calls often include high *tsit* notes, hard to distinguish among species.

Food mostly seeds, but also insects, and flowerpiercers (formerly placed with honeycreepers) take nectar. Nests typically are open cups of grasses and other vegetation. Eggs: 2–5, typically whitish to pale blue, flecked with browns in many species.

GENUS ATLAPETES:
Brushfinches

Brushfinches are large, secretive, towhee-like finches of temperate montane forests. Ages differ, sexes similar, juvs quickly attain adult-like plumage. Bills black, legs flesh to dusky flesh. Nests are bulky cups of grasses, fibers, pine needles, rootlets, etc., placed low in bushes or other dense understory. Eggs: 2–3, bluish white to pale bluish, rarely with fine brown flecks (WFVZ).

WHITE-NAPED BRUSHFINCH
Atlapetes albinucha   Plate 61
Saltón Nuquiblanco

ID: 7–8 in (18–20.5 cm). Endemic to E Mexico. **Descr. Adult:** *head black with broad white stripe down center of crown to nape*; upperparts dark slaty grey. *Throat and underparts yellow*, washed dusky on flanks. **Juv:** head and upperparts dark sooty brown, head blacker with hint of adult pattern. Throat and underparts yellow, brightest on throat, underparts streaked dusky, flanks washed dusky cinnamon.

Voice. Song varied, fairly penetrating series of 2–5 high, slightly slurred notes: *see-seeu* or *seeu seeir*, alternating with longer variations, *seeu seeir seeir si* or *tseeu si-si-sseu*, etc. Call of similar quality, a high, slurred *ti-seeiu* or *t'sseeu*, fading away.

Habs. Humid to semihumid pine–evergreen and evergreen forest edge, clearings, adjacent second growth. Singly or in pairs, usually fairly skulking on or near ground, but sings from more conspicuous perches. Nest and eggs undescribed (?).

SS: Yellow-throated Brushfinch allopatric.

SD/RA: C to F resident (900–2700 m) on Atlantic Slope from SE SLP to E Chis; probably also adjacent NW Guatemala.

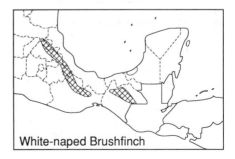

White-naped Brushfinch

YELLOW-THROATED BRUSHFINCH
Atlapetes gutteralis * or *A. albinucha*
(in part)   Plate 61
Saltón Gorjiamarillo

ID: 6.8–7.8 in (17–19.5 cm). Replaces White-Naped Brushfinch in S Chis and C.A. **Descr. Adult:** *head black with broad white stripe down center of crown to nape*; upperparts dark slaty grey. *Yellow throat contrasts with pale grey underparts*, flanks washed brownish. **Juv:** head and upperparts dark sooty brown, head blacker with hint of adult pattern. Yellow upper throat contrasts with dark-streaked, dusky cinnamon-brown underparts.
   Voice. Song of 3–5 varied, high, slightly slurred, whistled phrases, recalls White-naped Brushfinch: *t-see tswee t-si tswee*, or *sseeu chi-chit*, or *sseeu chu-i-ee*, etc. Call a high, thin, drawn-out *ti-siiiu* or *tsiii-iii*, and *siiiu*
   Habs. Humid to semiarid pine-evergreen and pine-oak forest edge, clearings, adjacent second growth, locally in brushy second growth and hedges. Habits much as White-naped Brushfinch.
SS: White-naped Brushfinch allopatric.
SD: C to F resident (1200–3000 m) on Pacific Slope and in interior from S Chis to N cen Nicaragua.
RA: S Mexico to N Colombia.
NB: See White-naped Brushfinch.

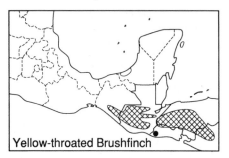

Yellow-throated Brushfinch

RUFOUS-CAPPED BRUSHFINCH
Atlapetes pileatus *   Plate 61
Saltón Gorrirrufo

ID: 5.8–6.5 in (14.5–16.5 cm). Endemic to highlands N of Isthmus. **Descr. Adult:** *rufous crown contrasts with grey face*, dark lores; nape greyish, upperparts olive-brown to olive-grey. *Throat and underparts yellow*, washed olive on chest, dusky on sides and flanks, undertail coverts dusky

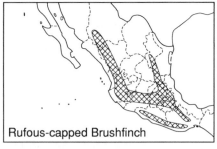

Rufous-capped Brushfinch

pale cinnamon. **Juv:** head and upperparts greyish olive to olive-brown with dark lores, dull pale cinnamon wingbars. Throat and underparts paler.
   Voice. Sharp excited chips, often repeated insistently, *chi chi chi-chi-chi...* or *chik chik...*, etc. Song typically 1–2 high, thin notes followed by varied series of (usually) 4 chips, rarely run into a trill, *ti-sein chi-chi-chi-chi*, or *t'i ssi-ssi-ssi-ssiu*, or *ch chi'in ssiu ssiu ss-wee ss-wee*, etc.
   Habs. Humid to semiarid, brushy and weedy growth, understory, and overgrown clearings in pine–oak, oak, and pine–evergreen forest. In pairs or small groups, relatively conspicuous and noisy. Sings from low to mid-levels. Nests in brush or dense grass near ground.
SD/RA: C to F resident (900–3500 m) in interior and on adjacent slopes from SW Chih and NL to Oax.

CHESTNUT-CAPPED BRUSHFINCH
Atlapetes brunneinucha *   Plate 61
Saltón Gorricastaño

ID: 6.8–7.8 in (17–19.5 cm). E Mexico and C.A. Two groups: *brunneinucha* group (widespread) and *apertus* ('Plain-breasted Brushfinch', Los Tuxtlas). **Descr. Brunneinucha. Adult:** *broad black auricular mask with white loral spot contrasts with white throat and upper chest*; mask bor-

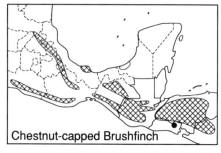

Chestnut-capped Brushfinch

dered above by narrow tawny supercilium, crown rufous-chestnut. Nape and upperparts dark olive. Underparts sooty grey with broken black chest band and whitish belly. **Juv:** bill yellow-orange with dark tip. Head blackish brown with darker mask and hint of paler supercilium; upperparts brownish olive. Throat and underparts olive-brown with trace of whitish malar stripe and white mottling on throat, belly streaked buffy lemon. **Apertus:** *adult lacks tawny supercilium, black chest band lacking or indistinct*; juv: undescribed (?).
Voice. Song a variable series of high, thin, often slightly lisping notes, *tsi-si si-si si-si si-wee si-s-chu*, or *tsi si-si' si-si' si-si' sii wi si t-ssiu*, etc. Calls include a high, thin *ssi* and *sii*.
Habs. Humid evergreen and pine–evergreen forest, and adjacent second growth woodland and plantations. In pairs or singly, on or near shady forest floor. Rarely sings from conspicuous perches. May nest on ground.
SS: Collared Towhee larger, bulkier, and brighter greenish olive above; note its whitish supercilium, broader black chest band, voice.
SD: C to F resident (900–3500 m, *apertus* occurs down to 400 m) on both slopes from SE SLP and Gro, and in interior from Chis, to N cen Nicaragua.
RA: E Mexico to S Peru.

GREEN-STRIPED BRUSHFINCH
Atlapetes virenticeps*   Plate 61
Saltón Verdirrayado

ID: 7–8 in (18–20.5 cm). *Endemic to W and cen Mexico.* **Descr.** **Adult:** *black head with yellow-olive median and lateral crown stripes* and white loral spot *contrasts with white throat and upper chest.* Upperparts dark greenish olive. Underparts dusky grey, mottled white on belly. **Juv:** bill

orange-yellow with dark tip. Head blackish brown with darker mask and indistinct paler head stripes; upperparts brownish olive. Throat and underparts olive-brown with white mottling on throat, belly streaked buffy lemon.
Voice. Song a variable, often prolonged series of high, thin notes with slightly jerky, irregular rhythm, recalling Chestnut-capped Brushfinch: *ssi-ssi ssi ssi-ssi, ssi ssi-ssi ssi ssi-ssi ssi-ssi ssi-ssi-ssi*, etc. Calls include a high, thin, sharp *ssii*.
Habs. Humid to semihumid pine–evergreen and fir forest. Habits similar to Chestnut-capped Brushfinch.
SD/RA: C to F resident (1800–3500 m) on Pacific Slope from cen Sin to N Nay, and in cen volcanic belt from Jal to W Pue.
NB: Sometimes considered conspecific with the *A. torquatus* group of S C.A. and S.A.

ORANGE-BILLED SPARROW
Arremon aurantiirostris   Plate 61
Rascador Piquinaranja

ID: 5.8–6.5 in (14.5–16.5 cm). SE rain forest. **Descr.** Ages differ, sexes similar. **Adult:** *bill neon orange, legs dusky flesh. Black head, with bold white supercilium* and dull grey median crown stripe, *and broad black chest band set off white throat*; chest band averages narrower in ♀, and may be sooty grey (imm only?). Upperparts dark olive with bright yellow bend of wing. Underparts whitish with dusky greyish-olive flanks and undertail coverts. **Juv:** bill dusky. Dark sooty olive overall with paler throat, quickly attains adult-like plumage.
Voice. A fairly hard, smacking *tssik* or *tssk!*, at times doubled, suggests Lincoln's Sparrow; also a high, rapid, wiry, rattling chatter in alarm. Song a high, thin, slightly tinny and jerky series, *sii ti-si tsi ti-sii tsi ti-*

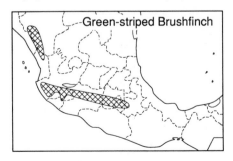
Green-striped Brushfinch

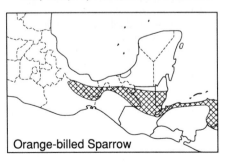
Orange-billed Sparrow

sii tsi-n, or *tsi tsi-si-si si-sin*, etc., may suggest Little Hermit.

Habs. Humid evergreen forest and adjacent dense second growth, *Heliconia* thickets, etc. In pairs or groups of 3–4 birds on shady forest floor, often in fairly dense understory and hard to see; sings from on or near forest floor. Nest is a bulky covered structure of dead leaves, grasses, rootlets, etc., often with fern hanging over side entrance, on slope or bank. Eggs: 2, white, sparsely flecked brown and black (WFVZ).

SS: Superficially similar brushfinches are highland birds.

SD: F to C resident (SL–900 m) on Atlantic Slope from N Oax to Honduras.

RA: SE Mexico to NW Ecuador.

OLIVE SPARROW

Arremonops rufivirgatus *   Plate 61
Gorrión Oliváceo

ID: 5.5–6.5 in (14–16.5 cm). Widespread. Three groups: *sumichrasti* group (W Mexico), *rufivirgatus* group (E Mexico), and *verticalis* (Yuc Pen). **Descr.** Ages differ, sexes similar. **Adult. Sumichrasti:** bill blackish above, flesh-grey to flesh below, legs flesh. *Head greyish with dark eyestripe and dark rufous-brown lateral crown stripe. Upperparts greyish olive.* Throat, chest, and flanks pale dusky buff to greyish buff becoming whitish on belly, undertail coverts buff to lemon-buff. **Rufivirgatus:** bill less bicolored, especially in N where dark brownish horn above, pale dusky flesh below. *Head stripes less distinct*, especially in *crassirostris* (SE Mexico) in which crown washed olive. **Verticalis:** bill black above, blue-grey to flesh-grey below. Resembles *sumichrasti* but *face greyer, head stripes flecked with black* (especially in front of eye), chest and flanks greyish with contrasting whitish throat, flanks may be washed olive, *undertail coverts buff*. **Juv** (nestling *rufivirgatus*): head and upperparts olive, streaked dark brown; throat and underparts pale buff, streaked dark brown except on belly. Quickly attains adult-like plumage.

Voice. A sharp *tsik*, a thin lisping *tssssssir* or *ssssiu*, and a sharp *whee-k' whee-k' whee-k'*; at times calls run into longer series, *tssi-ssi-ssi-ssit-ssit-ssit-ssit*, etc. Song an accelerating series of chips pre-

ceded by one to several quiet notes, *ssit, ssit ssi, chu-chu-chuchuchuchu,* or *tsik, tseeu tseu ssiu-ssiu-ssiu-ssiussiussiu,* or *chee h-wee, whee chi chi chichichichichichi,* etc.

Habs. Arid to semihumid, deciduous and semideciduous forest, woodland, thorn forest, etc., also (mainly *verticalis* group) humid, rarely evergreen, forest and edge, adjacent thickets. Singly or in pairs, skulks on or near ground; sings from low to mid-level, but rarely prominent, perches. Nest is an oven-like mass of grasses, rootlets, etc., with side entrance, low in bush or on ground. Eggs: 2–5, white, unmarked (WFVZ).

SS: Green-backed Sparrow has grey head with black stripes and brighter greenish back, but this can be hard to see in shady understory; note lemon undertail coverts and song.

SD/RA: C to F resident (SL–1800 m) on Pacific Slope from Sin to Oax, and in interior of Chis (*sumichrasti* group); on Atlantic Slope from E Coah and Tamps (also S Texas) to Isthmus (*rufivirgatus* group); and in Yuc Pen and pine savannas of Petén and Belize (*verticalis* group).

NB1: The three ssp groups are sometimes considered specifically distinct: Pacific Sparrow, Olive Sparrow, and Yucatan Sparrow respectively; their songs and calls are all similar. **NB2:** Although Monroe (1963c) thought the affinities of *chiapensis* lay with Atlantic Slope birds, its relatively long tail, plumage, and biogeographic affinities ally it with Pacific Slope birds. **NB3:** AMNH specs 811145 and 811148, from cen QR, appear to be hybrids between Olive and Green-backed sparrows. **NB4:** The status of *A. chloronotus (?) twomeyi* is not fully resolved, and in some ways it appears more similar to Olive Sparrow.

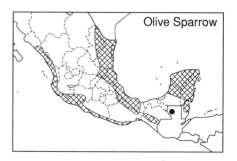
Olive Sparrow

GREEN-BACKED SPARROW
Arremonops chloronotus *   Plate 61
Gorrión Dorsiverde

ID: 6–6.3 in (15–16 cm). Humid SE. **Descr.**
Ages differ, sexes similar. **Adult:** bill blackish above, blue-grey below, legs flesh.
*Head grey with black eyestripe and lateral
crown stripe; upperparts olive-green.*
Chest and flanks greyish with whitish
throat and belly, *undertail coverts ochre-
lemon to buffy lemon,* chest rarely tinged
buff. **Imm:** head duller, black stripes can
be mixed with brown. **Juv:** undescribed
(?), probably similar to Olive Sparrow.
  Voice. A sharp *tsik* or *sik,* and a thin
sssirr, also (*twomeyi* at least) a longer *ssi-
i-ssir,* all much like Olive Sparrow. Song a
variable series of (usually) 3–8 rich chips,
usually with 1–2 introductory notes. The
introduction is typically simpler than
Olive Sparrow and the richer chips, while
sometimes fast, do not accelerate markedly: *chee, chew-chew-chew-chew,* or *ss,
tcheu chi-chi-chi-chi-chi-chi-chi-chi,* or
wheeu wheeu chi-chi-chi, etc. One variation may suggest Green Shrike-Vireo if
introductory notes are not heard. Song of
twomeyi: ee chu-chu-chu-chu-chu-chu,
and 'whisper song' *tik, ssi, seeu-seeu-seeu-
seeu* or *tik, seiu, seeu-seeu-seeu-seeu-seeu,*
etc., not accelerating and thus much like
Green-backed Sparrow.
  Habs. Humid to semihumid evergreen
and semideciduous forest and edge. Also
(*twomeyi*) arid scrub and thorn forest.
Habits much as Olive Sparrow. Nest and
eggs undescribed (?).
SS: See Olive Sparrow (sympatric in Yuc
Pen). Black-striped Sparrow (E Honduras,
Appendix E) allopatric (*contra* Monroe
1963*c*).
SD/RA: C to F resident (SL–750 m) on
Atlantic Slope from N Chis to W Hon-

duras; disjunctly (*A. c. twomeyi*) in
interior in upper Río Aguán Valley and
(status unknown, same habitat?) Agalta
Valley, Honduras.
NB1: See NB3 under Olive Sparrow. **NB2:**
Has been considered conspecific with
Black-striped Sparrow. **NB3:** *A. (c?) twomeyi* was originally described as a ssp of
Green-backed Sparrow (Monroe 1963*c*)
but, atypically inhabits arid thorn forest
and scrub; also, biogeography tends to
ally it with Pacific Slope forms of Olive
Sparrow and it may be a link between the
two species. Field studies are needed to
resolve its taxonomic status.

GENUS MELOZONE

Ground-Sparrows are towhee-like birds endemic to Middle America. Bills black, legs
dusky flesh to flesh. Ages differ, sexes similar.
Note striking head patterns. Nests are bulky
cups of dead leaves and grass, low in bush,
small tree, or (some White-eareds at least) on
ground.

RUSTY-CROWNED GROUND-SPARROW
Melozone kieneri *   Plate 61
Rascador Coronirrufo

ID: 6–7 in (15–17.5 cm). Endemic to W
Mexico. **Descr. Adult:** *face olive-brown
with broken white eyering and white lore
spot,* contrasting *rufous crown and sides of
neck.* Upperparts greyish olive. *Throat and
underparts white with dark central chest
spot,* flanks dusky brown, undertail
coverts pale cinnamon. **Juv:** head and
upperparts rich dusky brown, streaked
darker, with paler lores, darker auriculars,
and dusky malar stripe. Throat and underparts dirty pale lemon, streaked dusky
brown, undertail coverts pale cinnamon.

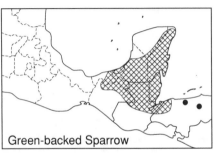
Green-backed Sparrow

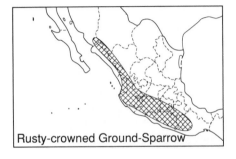
Rusty-crowned Ground-Sparrow

Voice. A high, thin, slightly wiry lisping *dzzzzzzzziu* or *ssssssziu*, a high, thin *ti*, often doubled, and high, sharp chips, *tk* or *tsik* that may be repeated persistently and run into dry, querulous chattering. Song a series of often hesitant, jerky chips, some notes doubled, run into an accelerating series of chips: *tsi tsi ssit s'weet sweet chi-chi-chi-chi-chi*, or often a simpler series without a chipping series at the end, *ssi ssii chi-u chiu chi-chi ssiu*, etc.; suggests Botteri's Sparrow.

Habs. Arid to semihumid brushy woodland and thorn forest, sometimes brushy hedges and thickets in semiopen areas. In pairs or singly, on or near ground, retiring and overlooked easily, but often sings from conspicuous perches. Eggs: 2–3, bluish white, may be finely flecked rusty (WFVZ).

SS: Prevost's Ground-Sparrow allopatric.

SD/RA: F to C resident (SL–2000 m) on Pacific Slope from S Son, and in interior from Jal, to NW Oax.

NB: Has been considered conspecific with Prevost's Ground-Sparrow.

PREVOST'S (WHITE-FACED) GROUND-SPARROW

Melozone biarcuatum   Plate 61
Rascador Patilludo

ID: 6–7 in (15–17.5 cm). Endemic, S of Isthmus. **Descr. Adult:** *white face contrasts with blackish forehead, rufous crown, and black auricular mark.* Upperparts olive-brown. Throat and underparts creamy white, flanks dusky brown, undertail coverts pale cinnamon. **Juv:** head and upperparts rich dusky brown, streaked darker, with paler lores and darker auriculars. Throat and underparts dirty pale lemon, streaked dusky brown, undertail coverts pale cinnamon.

Voice. A slightly nasal, metallic *chiih* or *tchih* which may run into a sharp, chipping, bickering chatter; also a high, thin *ti* or *tsi*. Song a short, pleasant, chipping series ending with a rolled note, *chu dididii durrr* or *choo didi-ir djerrr* (GWL tape).

Habs. Brushy woodland and edge, coffee plantations, overgrown clearings, often in more humid situations than Rusty-crowned Ground-Sparrow. Habits similar to Rusty-crowned. Nest and eggs undescribed (?).

SS: Rusty-crowned Ground-Sparrow allopatric.

SD/RA: C to F resident (100–3000 m, mainly 250–1800 m) in interior and on adjacent slopes from Chis to W Honduras.

NB1: See Rusty-crowned Ground-Sparrow.

NB2: *M. cabanisi* (Cabanis' Ground-Sparrow) of cen Costa Rica often considered conspecific with *M. biarcuatum* but morphologically and vocally quite distinct.

WHITE-EARED GROUND-SPARROW

Melozone leucotis*   Plates 61, 71
Rascador Orejiblanco

ID: 7–7.5 in (18–19 cm). Pacific Slope S of Isthmus. Two distinct ssp: *occipitalis* (Chis to El Salvador) and *nigrior* (N cen Nicaragua). **Descr. *Occipitalis.* Adult:** *face and throat black with whitish lores and broken eyering, white auricular spot, and yellow postocular stripe which curves down to sides of neck;* crown black with grey median stripe. Upperparts olive-brown with greyer hindneck. Median underparts white with bold black spot on central chest, greyish sides become dusky olive-brown on flanks, undertail coverts pale cinnamon. **Juv** head and upperparts rich dusky brown with darker edgings; paler lores, ochre-flecked supercilium and sides of neck, and blackish face and throat with

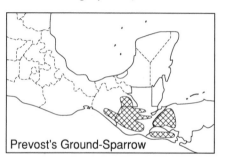

Prevost's Ground-Sparrow

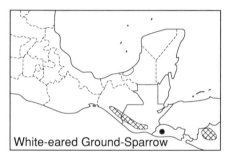

White-eared Ground-Sparrow

pale lemon auricular spot hint at adult pattern. Underparts dusky brown becoming cinnamon-brown on flanks; dirty pale lemon belly streaked dusky, undertail coverts cinnamon. *Nigrior*. Adult: *head pattern reduced to whitish lores and broken eyering, yellow sides of neck, and white auricular spot; rest of head including median crown blackish*, black extends farther down onto chest. Juv: resembles *occipitalis* but with hint of adult *nigrior* face pattern.

Voice. *Occipitalis*: a dueting, high, thin, but slightly gruff *ssyi si ssyi ssyi ssyi-ssyi ssyi-ssyi . . .*, also a single *ssiu*. Song undescribed (?). *Nigrior* group: song of explosive short phrases of staccato notes and loud, often rather hoarse or penetrating whistles: *spit-CHUR see-see-see, PSEET-seecha seecha seecha*, etc. (Stiles and Skutch 1989).

Habs. Humid to semihumid, semideciduous forest, adjacent second growth woodland and plantations. Habits similar to Rusty-crowned Ground-Sparrow. Eggs: 2, white, with cinnamon speckles (Murray 1985).

SD: F resident (600–1800 m) on Pacific Slope from Chis to El Salvador, and (*nigrior*) in N cen Nicaragua. A report from NW Guatemala (Griscom 1932) requires verification.

RA: S Mexico to El Salvador, locally in Nicaragua and cen Costa Rica.

GENUS PIPILO: Towhees

Towhees are large, primarily terrestrial sparrows with sturdy legs and feet. Their long claws are used for scratching in leaf litter. Ages differ, sexes similar or slightly different, juvs quickly attain adult-like plumage. The four 'brown' towhees all have amber eyes, dusky flesh to grey bills, and dusky flesh legs. Nest cups of grasses, twigs, rootlets, etc., low in bush or small tree, rarely on ground (Rufous-sided at least). Eggs: 2–4.

GREEN-TAILED TOWHEE
Pipilo chlorurus   Not illustrated
Rascador Coliverde

ID: 6.8–7.3 in (17–18.5 cm). Winter visitor to Mexico. Descr. Adult: bill black above, blue-grey below, legs dark flesh. *Rufous crown contrasts with grey face*, white lore

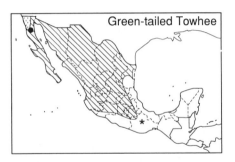

Green-tailed Towhee

spot; *throat white with dark malar stripe*. Upperparts greyish olive, *wings and tail edged bright yellow-olive. Underparts grey*, becoming whitish on belly, flanks and undertail coverts washed buff. Juv: head and upperparts sandy olive-brown, streaked dark brown, with dark malar stripe and pale moustache and lore spot; throat and underparts whitish, streaked dark brown.

Voice. A nasal mewing *neyuh* or *meew*, hollower and less complaining than Rufous-sided Towhee; also a thin *ssi* or *ssir*. Song a varied chipping series typically ending with a trill, *chu chri chi-chi-chi iirrrr*, etc.

Habs. Arid to semiarid brush, semiopen areas with scattered bushes and scrub, chaparral. Nests in low scrub in open conifer woods. Singly or, in migration, in loose flocks, on or near ground, fairly secretive but often sings from prominent perches. Eggs: 2–4, white, heavily speckled with reddish brown and greys (WFVZ).

SD: U and local breeder (May–Aug?, 2750 m) in Sierra San Pedro Martír (Howell and Webb 1992f). F to C transient and winter visitor (Sep–May; SL–2500 m) in Baja, on Pacific Slope and in interior S to cen Mexico, and on Atlantic Slope in Tamps; U to R in interior S to cen Oax (SNGH).

RA: Breeds W N.A.; winters SW USA to cen Mexico.

COLLARED TOWHEE
*Pipilo ocai**   Plate 61
Rascador Collarejo

ID: 7.8–9 in (19.5–23 cm). Endemic to cen Mexico, local. Descr. Adult: eyes amber to reddish, bill black, legs flesh. *Black face with whitish supercilium contrasts with rufous crown and white throat* (supercilium often indistinct or lacking in *alticola* of Jal and Col). Upperparts olive, *wings and tail edged bright yellow-olive. Broad black chest band below white throat.* Greyish sides become dusky cinnamon on flanks, median underparts whitish, undertail coverts pale cinnamon. **Juv:** eyes brownish. Crown, nape, and upperparts brown, streaked darker, face blackish with dull lemon supercilium. Throat and underparts pale lemon with broad necklace of dark brown streaks on upper chest, indistinct dusky streaks on belly; flanks and undertail coverts dusky cinnamon. Hybrids with Rufous-sided Towhee show intermediate characters.

Voice. Distinctive call a high, clear, usually ascending whistle, *pseeeeeu* or *teeeeeu.* Song a varied arrangement of chips and trills suggesting Rufous-sided Towhee: *chri chri ss chi-chi-chi-i,* or *ssr ssr chi-i-i-i-ir rreeu,* or *whi chu ssi-i-i-i-i-i chwee chwee,* or *chwer chee chee tsiiirrr,* etc.

Habs. Humid to semihumid pine–oak and pine–evergreen forest and edge, typically in higher, wetter, and cooler areas than Rufous-sided Towhee. Singly or in pairs, on or near ground, secretive; sings from low to mid-levels but often well hidden. Eggs: 2, pale bluish, speckled with reddish brown and greys (WFVZ).

SS: See Chestnut-capped Brushfinch; note possibility of hybrids with Rufous-sided Towhee.

SD/RA: C to F but local resident (1500–3500 m) in interior and on adjacent slopes from Jal to Mich, S Gro to S Oax, and from Pue to N Oax. Hybridizes with Rufous-sided Towhee in E Pue and adjacent Ver, and populations in E Mich appear variably intermediate between Collared and Rufous-sided.

NB: Sometimes considered conspecific with Rufous-sided Towhee.

RUFOUS-SIDED TOWHEE
Pipilo erythrophthalmus *   Plate 61
Rascador Ojirrojo

ID: 7.5–9 in (19–23 cm). Widespread. Three groups: *maculatus* group (widespread), *macronyx* group (cen volcanic belt), and *socorroensis* (Isla Socorro). **Descr. Maculatus.** ♂: eyes red, bill blackish, legs flesh. *Head and chest blackish, upperparts greyish olive, streaked blackish, to blackish, spotted and streaked white on scapulars and tertial edges; 2 white wingbars and bold white tips to outer rectrices. Sides, flanks, and undertail coverts cinnamon to cinnamon-rufous, median underparts white.* ♀: *head, chest, and upperparts blackish grey to slaty grey-brown, back with darker streaks.* **Macronyx:** sexes similar. *Upperparts olive, back mottled darker, with scapular spots, wingbars, and tailspots washed lemon; wingbars and scapular spots may be reduced or lacking, especially in ♀♀.* **Socorroensis:** *small (6–7 in, 15–17.5 cm).* Upperparts greyish olive-brown with reduced whitish markings. Head and chest of ♂ brownish black; ♀ has grey-brown head and chest, amber eyes. **Juv** (all ssp): eyes brownish. Head and upperparts dark brown with richer brown edgings. Dirty whitish throat and underparts streaked dark brown, heaviest on throat and chest; sides, flanks, and undertail coverts washed cinnamon.

Voice. Common call a slurred, nasal mewing *rreeah* or *jeree'r.* Song a varied arrangement of chips and trills, typically 2 chips and a trill, *chi chi chi-i-ir* or *weet weet diiiiir,* but other variations common. Songs can be confused with Bewick's Wren and Yellow-eyed Junco.

Habs. Brushy woodland and scrub, understory of pine and pine–oak woodland, chaparral, semiopen areas with

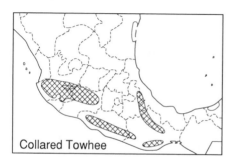

Collared Towhee

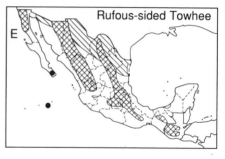

Rufous-sided Towhee

scattered bushes and brush. Habits much as Collared Towhee but often more conspicuous when singing. Eggs: 2–3, white to bluish white, densely speckled with reddish brown and grey (WFVZ).

SS: See Collared Towhee.

SD: C to F resident (1000–3300 m, to near SL in BCN) in Baja, on Isla Socorro, and in interior and on adjacent slopes from Son and Coah to Guatemala, but absent from much of SW Mexico; extirpated from Isla Guadalupe, BCN. Irregular R to U winter visitor (Nov–Mar) from N Son to N Tamps.

RA: NA to Guatemala; N populations move S in winter.

NB1: *Socorroensis* (Socorro Towhee) and (rarely) *macronyx* group (Olive-backed Towhee) have been considered specifically distinct. **NB2:** See Collard Towhee.

CALIFORNIA [BROWN] TOWHEE
Pipilo crissalis *   Plate 61
Rascador Californiano

ID: 7.5–9 in (19–23 cm). Baja. **Descr. Adult:** *head and upperparts dark grey-brown* in N to grey-brown in S, with pale broken eyering and streaking in face. *Throat buffy cinnamon, bordered and often flecked with dark spots*; underparts dusky grey becoming *cinnamon on vent and undertail coverts*. P. c. *albigula* of S BCS has rufous crown, whitish throat and belly. **Juv:** upperparts with faint dark streaking and 2 cinnamon wingbars, underparts paler with dusky streaking.

Voice. A sharp, metallic *chink* or *chik*, and a high, thin *ssi* which may run into an accelerating chatter, *ssi chuh-chuh-chuh-chuh-chuhr*, etc. Song an accelerating series of metallic notes, typically run into a slurred ending, *chink chink chink-chink-chinkchikchih-churuhchuhchuh*, etc.

Habs. Arid to semiarid brushy scrub,

semiopen areas with scattered bushes and trees. Singly or in pairs, fairly conspicuous but usually close to cover. Eggs: 2–4, pale bluish, marked with browns, greys, and black (WFVZ).

SS: Canyon and Abert's towhees allopatric.

SD: C to F resident (SL–2700 m) in Baja, except NE corner.

RA: W USA to Baja.

NB: Traditionally lumped with Canyon Towhee and known as Brown Towhee *P. fuscus*.

CANYON [BROWN] TOWHEE
Pipilo fuscus *   Plate 61
Rascador Arroyero

ID: 7.5–8.8 in (19–22.5 cm). N and cen Mexico. **Descr. Adult:** *head and upperparts grey-brown* with pale broken eyering and streaking in face, *crown rufous in N* (S to NE Jal and SLP). Wings and especially *tail darker*. *Throat buff to pale buff, bordered and often flecked with dark spots*. Chest and flanks greyish with *dark central chest spot*, belly whitish becoming cinnamon on vent and undertail coverts. **Juv:** upperparts streaked darker, throat and underparts paler, heavily streaked dusky, 2 pale cinnamon wingbars.

Voice. A slurred, nasal *cheh* and *nyeeah*, a high, thin *ssi*, and excited nasal bickering. Song a varied series of sometimes accelerating chips which may have one or more introductory notes, *chwee chwee chi-chi-chi-chi-chi-chit*, or *chihk chihk-chih-chih-chih-chihchih...*, or simply a steady *ching ching ching ching*, etc.

Habs. Much as California Towhee but more social, often in small groups. Eggs: 2–4, bluish white, marked with browns, greys, and black (WFVZ).

SS: White-throated Towhee has clean white throat and upper chest crossed by cinnamon band, 2 whitish wingbars, lacks dark

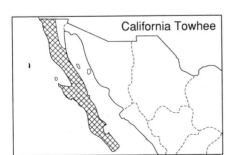
California Towhee

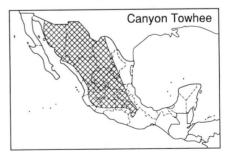
Canyon Towhee

chest spot. California and Abert's towhees allopatric.

SD: C to F resident (near SL–1500 m) on Pacific Slope from Son to N Sin, and (800–3000 m) in interior and on adjacent slopes from Chih and Coah to NW Oax.

RA: SW USA to cen Mexico.

NB: See NB under California Towhee.

ABERT'S TOWHEE
Pipilo aberti   Not illustrated
Rascador de Abert

ID: 8.3–9.8 in (21–25 cm). *Rio Colorado drainage.* **Descr. Adult:** head and upperparts grey-brown with *blackish lores and chin*; tail contrastingly darker. *Throat and underparts brownish cinnamon,* streaked dark on throat, becoming cinnamon on vent and undertail coverts. **Juv:** underparts duller and paler, streaked dusky.

Voice. A tinny, slightly nasal *beehk* or *chiink*, and excited churring and bickering. Song a series of 4–6 call notes often run into a lower series, *chiink chiink chiink chiink . . . chih-chih-chih-chih-chih,* etc.

Habs. Riparian thickets and brush in arid to semiarid country. Habits much as Canyon Towhee. Eggs: 2–4, bluish white, marked with purplish brown, greys, and black (WFVZ).

SS: California and Canyon towhees allopatric.

SD: C to F resident (around SL) in Rio Colorado drainage of NE BCN and NW Son.

RA: SW USA and extreme NW Mexico.

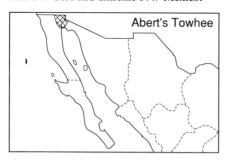

WHITE-THROATED TOWHEE
Pipilo albicollis*   Plate 61
Rascador Oaxaqueño

ID: 7.3–8.8 in (18.5–22.5 cm). *Endemic to interior SW Mexico.* **Descr. Adult:** *throat and upper chest white with dark malar*

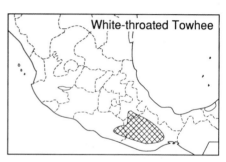

stripe, *cinnamon wash across central throat,* and dark border along lower edge of white. Chest and belly whitish, mottled dusky on lower chest, sides greyish, flanks washed cinnamon; vent and undertail coverts cinnamon. Head and upperparts grey-brown with whitish lores, indistinct eyering, and flecking in face. Wings and tail darker with *2 narrow whitish wingbars.* **Juv:** chest has heavier brownish mottling, wingbars pale cinnamon.

Voice. A slurred, slightly metallic *churenk* or *chehrrk,* a high, thin *ssi* and *siirr,* and excited, nasal, squeaky, bickering chatters. Song a series of chips, usually with one or more introductory notes, *chik tchu-chu-chu-chu chi-i-i-ir,* etc.

Habs. Arid to semiarid, semiopen country with scattered trees and bushes, brushy scrub. Habits much as Canyon Towhee but more social and often in noisy groups up to 10–15 birds. Eggs: 2–3, pale bluish to bluish white, marked with reddish brown and greys (WFVZ).

SS: See Canyon Towhee.

SD/RA: C to F resident (1000–2500 m) in interior from E Gro and S Pue to cen Oax.

BLUE-BLACK GRASSQUIT
Volatinia jacarina spendens   Plate 62
Semillero Brincador

ID: 4–4.3 in (10–11 cm). Widespread. **Descr.** Ages/sexes differ, molts need study. **Basic ♂** (Oct–May): bill black above, grey below, legs greyish. *Head and upperparts brown, slightly mottled blue-black,* wings and tail blue-black, edged brown. *Throat and underparts scalloped pale buff and blue-black,* white tufts at sides of chest usually concealed. Feather tips wear off through winter to reveal glossy blue-black bases, supplemented by a partial molt into alt. **Alt ♂** (May–Aug): *glossy blue-black*

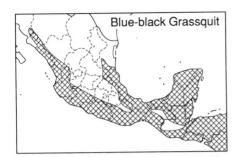

Blue-black Grassquit

overall, *white tufts at sides of chest visible in display.* ♀: bill and legs grey to greyish flesh. *Head and upperparts brown*, wings and tail darker, edged brown. *Throat and underparts pale buff, streaked dusky on throat and chest*, flanks and undertail coverts buffy brown. **Juv:** resembles ♀ but wing coverts tipped buff, forming 2 narrow wingbars, molts quickly into imm plumage. **Imm** ♂: resembles ♀ but wings and tail blue-black, head and body variably mottled blue-black in 1st summer. Attains adult plumage by 2nd prebasic molt when about 1 year old.

Voice. A high, sharp to slightly liquid *tsik* or *sip*. Song a slurred, buzzy, slightly metallic or lisping *tzzzzzu* or *tzssiiu*, often repeated over and over as ♂♂ jump up and down with each song.

Habs. Weedy fields and second growth. ♂♂ sing conspicuously from fences, grass stalks, etc. In non-breeding season forms flocks, at times to a few hundred birds, which join readily with other seed-eating birds. Nest cup of rootlets, fine grasses, etc., low in weed stalks or bush. Eggs: 2–3, pale blue to bluish white, speckled rusty (WFVZ).

SS: See ♀ Indigo Bunting.

SD: C to F resident (SL–1800 m) on both slopes from S Son and S Tamps, and in interior from Balsas drainage, to Honduras and W Nicaragua. Possibly summer resident only (Apr–Sep) at higher elevations locally in W Mexico.

RA: Mexico to N Chile and N Argentina.

SLATE-COLORED SEEDEATER

Sporophila schistacea ssp.   Plate 62
Semillero Apizarrado

ID: 4.2–4.5 in (10.5–11.5 cm). *C.A.* Bill relatively stout. **Descr.** Ages/sexes differ. ♂: *bill bright orange-yellow*, legs grey.

Head, chest, and upperparts slaty blue-grey, with white subocular crescent and white patch at sides of neck; wings and tail darker with *white flash at base of primaries* (at times indistinct) and short white upper wingbar. Belly and undertail coverts white, flanks grey. ♀: bill greyish. *Head and upperparts olive-brown*, wings and tail slightly darker, edged olive-brown. *Throat and underparts dirty buff, paler on throat, becoming pale lemon on belly and undertail coverts.* **Juv/imm** ♂: resemble ♀, ♂ may have a few grey feathers in 1st summer.

Voice. A buzzy, nasal *shih* and *shih-shih*, and a high, slightly lisping *ssik* or *siik*. Song varied, a high, slightly buzzy twittering and trilling, at times sweeter with more trilling.

Habs. Humid evergreen and semideciduous forest edge, second growth, plantations, particularly with bamboo. Notably nomadic and irregular in occurrence, movements tied to seeding bamboo. Singly, in pairs, or loose flocks, often at mid- to upper levels in open forest edge and bamboo thickets, also on or near ground in seeding fields (Honduras). Nest cup of fine plant materials at mid-levels in tree or vine. Eggs undescribed (Kiff and Hough 1985).

SS: ♂ unmistakable, ♀ told from Variable Seedeater by more brownish upperparts, dirty buff underparts, more like ♀ White-collared Seedeater (*morelleti* group) in color but lacks whitish wingbars.

SD: R and local (increasing?) presumed resident in Belize where first recorded in 1989 (Howell *et al.* 1992*b*), first recorded Honduras in 1979 (Marcus 1983), and has invaded Caribbean Slope of Costa Rica since about 1975 (Stiles and Skutch 1989). In Mexico, known from 2 ♂♂ (one labeled ♀) collected at Palomares, Oax, by Avilés pur-

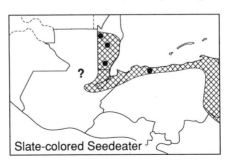

Slate-colored Seedeater

portedly in Sep 1957 and described as a new ssp, *subconcolor* (Berlioz 1959); Avilés data are notoriously suspect however, and we consider the record hypothetical.

RA: Belize (SE Mexico?) and Honduras locally to Ecuador and Brazil.

VARIABLE SEEDEATER

Sporophila aurita corvina   Plate 62
Semillero Variable

ID: 4.2–4.5 in (11–11.5 cm). Humid SE lowlands. **Descr.** Ages/sexes differ. ♂: bill blackish, legs dark grey. *Black overall with white spot at base of primaries*, white underwing coverts flash in flight. ♀/juv/imm ♂: bill dark grey. Head and upperparts *brownish olive*, underwing coverts white. Throat and underparts paler brownish olive, palest on throat, washed ochre on belly and undertail coverts. Imm ♂ may show black patches by spring, main molt into adult plumage May–Oct.
Voice. A nasal *chiyh* or *nyih*, and a fuller nasal *chieh* or *chiih*, at times run into short chatters, *cheh-cheh-cheh chi-chi*, etc. Song a complex, typically prolonged warble, often including buzzy trills suggesting a siskin.
Habs. Overgrown weedy fields, second growth and forest edge, in humid situations. In pairs or small groups, associating rarely with flocks of White-collared Seedeaters. Nest cup of rootlets, fibers, etc., at low to mid-levels in bush or tree. Eggs: 2–3, pale blue-grey, marked with browns, greys, and black (WFVZ).
SS: ♂ often mistaken for much rarer (in Mexico at least) ♂ Thick-billed Seedfinch which is slightly larger, with massive deep-based bill.
SD: F to C resident (SL–1000 m) on Atlantic Slope from N Oax to Honduras.
RA: SE Mexico to NW Peru.

NB: Sometimes considered conspecific with *Sporophila americana* of S.A.

WHITE-COLLARED SEEDEATER

*Sporophila torqueola**   Plate 62
Semillero Collarejo

ID: 4–4.5 in (10–11.5 cm). Common in lowlands. Marked geographic and individual variation in ♂ plumages. Three groups: *sharpei* (NE Mexico) and *morelleti* (S Mexico and C.A.), intergrading from S Ver through Guatemala; and *torqueola* group (W Mexico). **Descr.** Ages/sexes differ. *Sharpei.* Alt ♂ (Mar–Aug, rarely Sep–Feb): bill black, legs grey. *Head and upperparts greyish olive, mottled black,* often buff on lower rump; whitish subocular crescent. *Throat and underparts pale buff to pale cinnamon, sides of neck usually whitish, chest often mottled black.* Basic ♂ (Aug–Apr, possibly only an 'eclipse' plumage worn for short period): bill dull flesh with dusky culmen, to all dusky. *Head and upperparts olive-brown,* rump often paler, *wings and tail black with 2 white wing-bars (tinged buff when fresh) and white flash at base of primaries.* Throat and underparts rich buff (pale buff when worn). ♀: *resembles basic ♂ but wings and tail brown, 2 whitish wingbars,* flight feathers edged pale olive-brown. **Juv:** resembles ♀ but wingbars cinnamon, ♂ has blackish-brown wings, white spot at base of primaries. **Imm** ♂: resembles ♂ basic, attains variable degree of alt in 1st summer. *Morelleti.* Alt ♂: *head and upperparts more solidly black with buff to whitish rump. Throat and sides of neck* (collar) *whitish,* separated from *whitish to pale buff underparts by black chest band.* ♀, imm, and basic ♂ plumages similar to *sharpei* but throat and underparts generally paler. *Torqueola.* Alt ♂: *black head contrasts with white sides of neck,* upper-

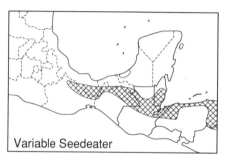

Variable Seedeater

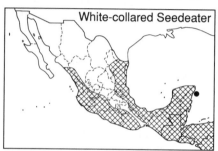

White-collared Seedeater

parts greyish olive (rarely black or mottled black) with *cinnamon rump*; wings and tail darker to blackish with white spot at base of primaries. *Throat and underparts buffy cinnamon with black chest band.* ♀: head and upperparts greyish olive, *wings and tail* slightly darker, *edged greyish olive.* *Throat and underparts dusky pale buff* to buffy lemon, palest on throat and belly. **Juv/imm/basic** ♂ (basic ♂ Oct– Apr): plumages correspond to *sharpei* group, basic ♂ has pale buffy-cinnamon rump; *note lack of white wingbars in all ages/sexes.* **Imm** ♂ (1st alt?): duller than ♂, head dark slaty olive, collar tinged buffy cinnamon, upperparts olive with a few dark spots, rump dull cinnamon; paler underparts lack black chest band, white wing spot smaller.

Voice. A nasal *cheh* or *nyeh*, and a clear, slightly piercing, often quite loud *seeu* suggesting a euphonia. Song a character-istic sound of open country on both slopes, especially in summer: a varied sweet warble, often speeding into a buzzy trill, *s'wee s'wee s'wee s'wee s'wee's-wee-il-idl-idl*, or *swee twee-wee-wee wit wit-wit-wit-wit wil-idl-idl...*, or simply *swee swee swee-swee-swee*, or buzzy trills, etc.
Habs. Weedy fields and second growth. ♂♂ sing from fences, trees, roadside wires, etc. In non-breeding season, in flocks up to a few hundred birds, often joining other seed-eating birds. Nest cup of root-lets, fibers, etc., at low to mid-levels in bush, tree, coarse weeds. Eggs: 2–3, pale blue-grey, marked with browns, greys, and black (WFVZ).
SS: Ruddy-breasted Seedeater (Pacific Slope) from *torqueola* group by smaller size (often hard to judge), bill color (flesh-orange versus dusky in White-collared); note more contrasting wing edgings and overall buffier, less olive tones of Ruddy-breasted; basic ♂ Ruddy-breasted has dull cinnamon undertail coverts and dull rufous-chestnut rump. White-collareds S of Isthmus have bold pale wingbars.
SD: C to F resident (SL–2400 m, mainly be-low 1500 m) on both slopes from Sin and Tamps, and in interior from Jal, to Hon-duras and W Nicaragua.
RA: Mexico to cen Panama.
NB: *Torqueola* group sometimes considered specifically distinct, Cinnamon-rumped Seedeater.

RUDDY-BREASTED SEEDEATER
Sporophila minuta parva   Plate 62
Semillero Pechicanelo

ID: 3.7–4 in (9.5–10 cm). *Pacific Slope.* ♀/ *imm plumages much like White-collared Seedeater.* **Descr.** Ages/sexes similar. **Alt** ♂ (Apr–Sep): bill black, legs grey. *Head and upperparts slaty blue-grey with rufous-chestnut rump.* Wings and tail blackish with *white patch on base of primaries*, underwings flash white in flight. *Throat and underparts rufous-chestnut.* **Basic** ♂ (Oct–Apr): *bill flesh-orange with dusky culmen.* Head and upperparts olive-brown, often washed greyish, with *dull rufous-chestnut rump*; wings and tail blackish with *white patch at base of primaries, tertials and wing coverts edged cinnamon-buff. Throat and underparts cinnamon-buff, becoming dull cinnamon on undertail coverts.* ♀: *bill flesh-orange with dusky culmen.* Head and upperparts olive-brown; *wings and tail dark brown, tertials and wing coverts edged tawny-brown to buff.* Throat and underparts rich buff to pale greyish buff, palest on throat and belly. **Juv:** resembles ♀ but wingbars pale buffy cinnamon, ♂ may have blackish-brown wings with white check at base of primaries. **Imm** ♂: resembles ♀ but wings and tail blackish brown with white check at base of primaries. In 1st alt, head and upperparts mottled blue-grey, rump and underparts mottled rufous-chestnut. Molts and plumages need study.
Voice. A quiet, nasal *chih* and *nyeh*, much like White-collared Seedeater. Song a sweet warble, slower and without the ten-dency to run into a trill as in White-collared Seedeater, *ssi ss-ee' si-wee si-wee s-wee-chee-chee-chee-it* or *ssi ss-see s-wee' s-wee' s-wee' s-wee' chi chi-chu'*, etc.
Habs. Overgrown weedy fields and

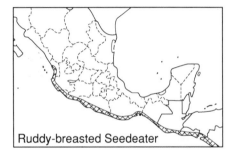

Ruddy-breasted Seedeater

second growth, especially near water, marshes. Habits much as White-collared Seedeater with which it associates readily. Nest cup of rootlets, fibers, etc., low in bush or tree. Eggs: 2–3, white to pale buff, spotted with browns.

SS: See White-collared Seedeater.

SD: F to C resident (SL–1000 m, mainly below 500 m) on Pacific Slope from Col to El Salvador, U N to Nay.

RA: W Mexico to NW Ecuador and N Argentina.

THICK-BILLED [LESSER] SEEDFINCH
Oryzoborus funereus   Plate 62
Semillero Piquigrueso

ID: 4.5–5 in (11.5–12.5 cm). Humid SE lowlands. *Bill massive and deep based with straightish culmen.* Descr. Ages/sexes differ. ♂: bill blackish, legs grey. *Black overall with white spot on base of primaries*, white underwing coverts flash in flight. ♀/juv: rich dark brown overall, paler and more rufous-brown below; underwing coverts white. Imm ♂: resembles ♀ but may show black mottling on face and throat in 1st summer.

Voice. An often quiet, slightly nasal *chihk* or *jiit*, and a full nasal *beehn*. Song a prolonged rich warble, occasionally with introductory short buzzy twittering; typically richer and less buzzy than Variable Seedeater.

Habs. Humid evergreen forest edge and second growth. Singly, in pairs, or small groups, not in flocks, often low and retiring in dense growth. ♂♂ sing from mid- to upper levels of small trees. Nest cup of rootlets, fibers, etc., at low to mid-levels in bush, tree, vine tangle. Eggs: 2, bluish white, flecked with browns and black (WFVZ).

SS: See ♂ Variable Seedeater, ♀ Blue Bunting.

SD: U to F resident (SL–1000 m, mainly below 500 m) on Atlantic Slope from S Ver to Honduras; apparently U to R in Mexico.

RA: SE Mexico to NW S.A.

BLUE SEEDEATER
Amaurospiza c. concolor   Plate 60
Semillero Azul

ID: 4.8–5.3 in (12–13.5 cm). *Local S of Isthmus. Bill stubby with straightish culmen.* Descr. Ages/sexes differ. ♂: bill and legs dark grey, bill paler grey below at base. *Dark dull bluish overall* (may look blackish in shade) *with darker lores*; lesser wing coverts edged slightly brighter bluish but usually concealed. ♀/imm: *rich cinnamon-brown overall*, throat and underparts paler cinnamon-brown, may show slightly paler lores and eyering. Juv: brighter than ♀, with dull cinnamon wingbars.

Voice. A high, sharp, slightly metallic *tswik* or *sik*, and thin twittering. Song variable, a high, slightly tinny to sweet warble suggesting *Passerina* buntings, see *see-wee-see si-si-wee* or *seet see-wee-see si-si-wee-su*, etc.

Habs. Humid to semihumid brushy woodland, thickets, forest edge, in association with bamboo. Singly or in pairs, at low to mid-levels in dense shrubby growth, especially bamboo. Sings from cover or while foraging, as well as from prominent perches. Nest and eggs undescribed (?).

SS: See Blue Bunting. Slate-blue Seedeater allopatric.

SD: Locally/sporadically U to F resident (600–2100 m) in interior and on adjacent slopes from Chis to N cen Nicaragua. Recently discovered (around SL) in Belize (Howell *et al.* 1992*b*). A report from Guatemala (Vannini 1989*b*) requires

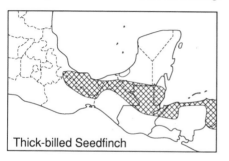
Thick-billed Seedfinch

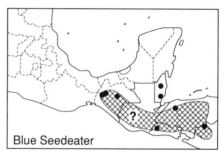
Blue Seedeater

verification. Abundance and distribution closely tied to seeding bamboo.
RA: S Mexico to NW Ecuador.

SLATE-BLUE [BLUE] SEEDEATER
Amaurospiza relicta or *A. concolor* (in part)
Semillero Azuligris   Plate 60

ID: 5–5.5 in (12.5–14 cm). *Endemic to W Mexico.* Slightly larger and longer-tailed than Blue Seedeater, with slightly stubbier bill. **Descr.** Ages/sexes differ. ♂: bill and legs dark grey, bill paler grey below at base. *Dull slaty blue overall with darker lores*; lesser wing coverts slightly brighter bluish but usually concealed. ♀: *cinnamon-brown overall*; head and upperparts more olive-brown, may show slightly paler lores and eyering, throat and underparts paler cinnamon-brown. Duller and greyer in worn plumage. **Imm ♂:** resembles ♀ but in 1st alt has slaty blue mottling on head and body, often solidly slaty on face and throat. **Juv:** undescribed (?).
Voice. Much like Blue Seedeater but song slightly higher and faster: a variable warble, ranging from high, quiet, and slightly tinny to fairly strong, bright, and sweet, *sii sii-wii-sii si-si-wi*, or *tsee tsee tsee-si-wee*, or *swiit swiit swiitsi wiit-i-wiit*, etc., suggesting *Passerina* buntings. Call a slightly metallic to slightly wet *tslik* or *tsiik*.
Habs. Brushy woodland to evergreen forest with thickets of bamboo. Habits much as Blue Seedeater. Nest cup of fine grasses, fibers, etc., at mid-levels in bush or tree. Eggs: 2, pale blue, unmarked (WFVZ).
SS: See ♀/imm Varied Bunting and Blue Bunting. Blue Seedeater allopatric.
SD/RA: Locally/sporadically U to F resident (1200–2500 m) on Pacific Slope and in interior of SW Mexico, recorded from Jal, Col (SNGH), Mor, and Sierra Madre del

Sur of Gro and Oax. A report from 'Moctum', Oax, is questionable (Binford 1989).

YELLOW-FACED GRASSQUIT
*Tiaris olivacea**   Plate 62
Semillero Oliváceo

ID: 4–4.3 in (10–11 cm). E and S. **Descr.** Ages/sexes differ. ♂: bill and legs grey. *Face and chest black with rich yellow supercilium*, subocular crescent, *and throat patch* (black restricted to upper chest in *intermedia* of Isla Cozumel); belly and undertail coverts dusky olive, often mottled black on upper belly. Crown and upperparts greenish olive, crown may be mottled black. ♀: bill paler below. *Olive overall with trace of ♂ face pattern*, typically narrow lemon supercilium and lemon chin, dusky mottling on throat sides and chest. **Juv:** resembles ♀ but duller, with dull whitish supraloral stripe and chin, quickly molts into **1st basic:** similar to juv but face pattern brighter, ♂ has slight black mottling on face. **1st alt:** supercilium and throat yellowish, ♂ with more black mottling on face and chest, much like adult ♀.
Voice. A high, slightly sharp *sik* or *tsi*. Song a rapid, high, often soft, insect-like trill, *siiiiiiiiir* and variations, at times a slower *siiiriririr*, etc.
Habs. Humid to semiarid weedy fields, second growth, and forest edge. Singly or in flocks up to 100+ birds, associates with seedeaters. Nest is a globular structure of grasses, fibers, etc., with side entrance, low in tangle, on bank, etc. Eggs: 2–4, bluish white, speckled with browns and greys (WFVZ).
SS: ♂ unmistakable; dull ♀ fairly plain, note face pattern.
SD: F to C resident (SL–2000 m, mainly above 500 m in C.A.) on Atlantic Slope

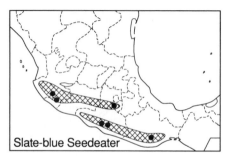

Slate-blue Seedeater

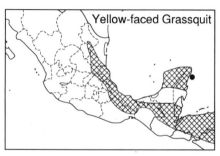

Yellow-faced Grassquit

from cen NL (SNGH) and Tamps to Honduras (but U to R in Belize), on Pacific Slope of El Salvador and Honduras.

RA: E Mexico and Greater Antilles to NW Venezuela.

SLATY FINCH
Haplospiza rustica*   Plate 60
Fringilo Plomizo

ID: *5–5.5 in (12.5–14 cm). Rare and local. Bill pointed with straight culmen.* **Descr.** Ages/sexes differ. ♂: bill blackish above, horn to greyish below, legs brownish flesh. *Slaty blue-grey overall, slightly paler below.* ♀/juv: *head and upperparts dark olive-brown, streaked darker;* wings and tail edged olive-brown to rufous-brown. *Throat and underparts dusky pale tawny to buffy brown, palest on throat, with dusky streaking on throat and chest, flanks washed olive, vent washed lemon.* **Imm** ♂: resembles ♂ but overall mottled with olive-brown, often some rufous-brown on back, retains juv flight feathers and sometimes greater wing coverts, may attain adult tail.
Voice. A high, thin, slightly shrill to buzzy, metallic *zchiiiir-i-iiir* or *zzziiiri-chi* (GHR tape, Costa Rica). Song a very high and thin medley of staccato notes and longer whistles and trills (Stiles and Skutch 1989).
Habs. Humid evergreen forest and edge, adjacent overgrown clearings and second growth. Singly or in small groups, in weedy fields and overgrown pastures in humid forest, bamboo thickets in forested ravines. Nest and eggs undescribed (?).
SS: ♀ suggests ♀ Blue-black Grassquit but more thickset with longer heavier bill.
SD: Poorly known. Locally/sporadically R to U resident (1200–3000 m): known from cen Ver (but not since the 1860s), and Pacific Slope and adjacent interior from Chis to N cen Nicaragua (Martinez-S.

1989); extralimital (?) record from cen Mich (Aug 1982 spec. per ARP).
RA: S Mexico to Peru and NW Bolivia.

CINNAMON-BELLIED FLOWERPIERCER
Diglossa baritula*   Plate 55
Picaflor Vientre-canelo

ID: *4.2–4.7 in (10.5–12 cm). Highlands from cen Mexico S. Bill slender and slightly recurved with hooked tip to upper mandible.* **Descr.** Ages/sexes differ. ♂: bill black, pale flesh below at base, legs dusky flesh. *Head and upperparts slaty grey-blue, darker on head. Chin and upper throat slaty grey-blue* (extending slightly onto upper chest in *montana,* S of Isthmus), *underparts cinnamon.* ♀: *head and upperparts greyish olive,* wings and tail darker with paler wingbars and tertial edges. *Throat, chest, and flanks dusky pale cinnamon with blurry dusky streaking on chest, belly cinnamon.* **Juv:** *resembles* ♀ *but head and upperparts brownish olive to olive,* wings and tail darker with pale cinnamon lower wingbar and tertial edgings, olive flight feather edgings. *Throat and chest dirty pale lemon, blurrily streaked dusky, belly and undertail coverts pale buffy cinnamon, washed dusky on flanks.* **Imm** ♂: resembles ♂ but upperparts may be mixed with olive, chin and median throat mixed with pale cinnamon, retains juv flight feathers.
Voice. A high, thin, sharp *tsi* or *tsik,* and a quiet mellow trill *triiiir.* Song a high, thin, slightly squeaky twittering warble, increasing in intensity then fading, *ss-ssi-ssi ssiiu-i-siin-i-siin-i,* and variations.
Habs. Semiopen weedy areas in humid to semiarid pine–oak and evergreen forest, in banks of flowers, gardens, etc. Singly or often several birds at flowering shrubs, highly active and almost constantly flitting about. ♂♂ sing from low to high

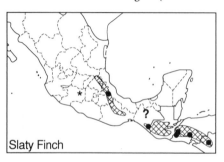

Slaty Finch

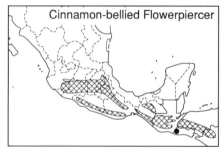

Cinnamon-bellied Flowerpiercer

perches. Nest is a bulky cup of mosses, pine needles, rootlets, etc., low in bushes. Eggs: 2–3, pale blue, speckled and flecked with browns and greys (WFVZ). Nesting: Aug–Feb (at least).

SS: ♀/juv might suggest dull *Vermivora* warblers but note distinctive bill, cinnamon wash to underparts.

SD/RA: F to C resident (1200–3500 m) in interior and on adjacent slopes from Jal and Hgo to Honduras, R (?) to N cen Nicaragua (Martinez-S. 1989).

NB1: Formerly known as Cinnamon Flower-Piercer and sometimes as Highland Honey-creeper. **NB2:** Sometimes considered conspecific with *Diglossa plumbea* of S C.A.

GRASSLAND YELLOW-FINCH
*Sicalis luteola**   Plate 59
Zacatero Amarillo

ID: 4–4.5 in (10–11.5 cm). Local in grasslands. **Descr.** Ages/sexes differ. ♂: bill black above, grey below, legs flesh. *Face and underparts bright yellow*, duskier on auriculars. *Crown and upperparts yellow-olive, streaked blackish*, with *bright yellow rump*; wings and tail darker, edged olive to pale grey. ♀: bill greyish. *Duller, face and underparts pale lemon.* Crown and upperparts brownish olive, streaked blackish, with *pale olive rump*; wings and tail darker, edged pale brown. **Imm** ♂: resembles ♀ but face and underparts brighter, rump yellow-olive. **Juv:** resembles ♀ but paler, throat whitish, chest streaked dark brown, rump streaked blackish.

Voice. A sharp *siik* or *syiik*, and a longer *sii-sii chi* and *ss-siit*, etc. Song varied, high, insect-like, buzzy trills and chipping, including a liquid *tsip tsip...* run into a trill, and a buzzy *dzzi dzzi...* or *dzzirr dzzirr...* given in display flight.

Habs. Arid to semihumid grassland, ranging from short grass savanna to rice and cane fields. Often in flocks, up to several hundred birds; perches on fences and low bushes, gathers at water holes to drink. In display, ♂♂ fly with exaggerated, deep, slow 'butterfly' wingbeats, then glide down with wings raised and tail spread. Nest (S.A.) is a cup of dry grasses on or near ground. Eggs (S.A.): 3–5, pale bluish, speckled with reddish brown and greys (WFVZ).

SD: Locally C to F resident (around SL) on Atlantic Slope from S Ver to W Camp and in the Mosquitia; in interior (1200–2200 m) in Mor and W Pue, irregularly (Mar–Jul) to DF (RGW). Also, formerly (?) in cen Guatemala, and recently in Belize (Wood and Leberman 1987). Somewhat nomadic but this and similar species are popular cage birds so extralimital occurrences should be checked carefully.

RA: Mexico locally to cen Chile and cen Argentina.

GENUS AIMOPHILA

These medium-sized sparrows are usually found in arid scrub and adjacent grassland. Bills relatively long, tails long and graduated, wings short and rounded. Ages differ, sexes similar. Juvs show faint trace of distinctive adult face patterns. Most species' juv (fledgling) plumage is lost quickly after leaving the nest, some also have a distinct imm plumage. Molts often complex and variable, in need of study. Songs only given in breeding season (mainly Apr–Sep) but several species are social and noisy much of year. Nest cups of grasses, rootlets, etc., usually placed low in bushes. Eggs: 2–5, bluish white to whitish, unmarked (WFVZ).

BRIDLED SPARROW
Aimophila mystacalis   Plate 63
Zacatonero Bigote-blanco

ID: 6–6.5 in (15–16.5 cm). *Endemic to interior SW Mexico.* **Descr. Adult:** bill black above, blue-grey below, legs dull flesh. *White loral spot and white moustache contrast with dark face and black throat*; rest of head grey, streaked blackish. Chest grey becoming whitish on belly, flanks and undertail coverts pale cinnamon. *Brown upperparts streaked black, with contrasting rufous outer scapulars and rump* often hidden at rest. Wings and tail blackish brown with brown edgings, 2 white wing-

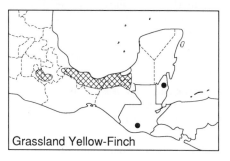

Grassland Yellow-Finch

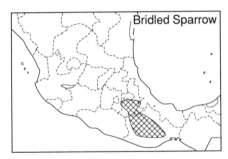

bars, and whitish outer webs to outermost rectrices. **Juv** (fledgling): head and upperparts brownish, face pattern indistinct, moustache washed lemon. Underparts pale lemon, streaked dusky on chest.

Voice. Thin, high, wheezy to slightly liquid twittering, at times becoming a rhythmic series with birds dueting; also a sharp, liquid *suik*. Song a series of hesitant, slightly sneezy, sharp, sweet chips, in full song breaking into an excited liquid chatter: *w-sik' w-sik' w-sik', w-sik' w-sik'* ... or *w-syu', w-syu'* ... run into an accelerating, rapid *seeu seeuseeuseeuseeuseeu*; at times with a shorter introduction, *s-ik' w-chi' seeuseeuseeuseeu*, etc.

Habs. Semiopen arid scrub, thickets, and adjacent overgrown grassy clearings. In pairs or small groups, usually on or near ground but often sings from fairly conspicuous perches; at times takes nectar in flowering trees. Eggs undescribed (?).

SD/RA: C to F resident (900–1800 m) in interior from S Pue to Oax; also locally in extreme E Mex, E Mor (SNGH, JKe), and W cen Ver.

FIVE-STRIPED SPARROW
Aimophila quinquestriata *   Plate 63
Zacatonero de Cinco-rayas

ID: 5.8–6.3 in (14.5–16 cm). W Mexico.
Descr. Adult: bill black above, blue-grey

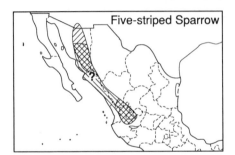

below, legs dull flesh. *Face grey with white supercilium,* subocular crescent, *and moustache; broad black malar stripe borders white throat.* Crown, nape, and upperparts brown, becoming chestnut-brown on back; tail blackish brown. *Chest and flanks grey with black median chest spot,* belly white, undertail coverts grey, tipped whitish. **Imm:** head and upperparts brownish with trace of paler supercilium and moustache. Throat and underparts dusky olive-greyish with pale lemon median throat and belly. Sequence to adult plumage unknown (?), perhaps similar to Black-chested Sparrow. **Juv** (fledgling): head and upperparts brownish, streaked darker, with narrow pale wingbars. Throat and underparts whitish, with dark streaking on chest and flanks. Quickly attains imm plumage.

Voice. A high, thin *sik* or *tsit,* a dry, soft, slightly nasal clucking chatter, *ch-chch-ch* ... which, when excited, may run into a duet of rollicking liquid twitters suggesting Black-chested Sparrow. Song varied, short chipping and trilled phrases, as often as not each differs from the preceding phrase, *sirr-it chee-chee' it, ts chi-chi-chit, seerr seerr,* ..., or *ssi ssi chi-i-chirr, ssi-i-i chir-i, chi ss-i-i,* ..., etc.

Habs. Arid to semiarid brushy scrub and thorn forest, typically on rocky slopes; also more open scrub and grassy areas in migration and winter. Singly or in pairs, on or near ground, overlooked easily; sings conspicuously from tops of bushes, but may also sing from ground while foraging. Nest may be on ground in grass clump.

SD/RA: F to C resident (Apr–Oct; 50–1700 m) on Pacific Slope in Son (also extreme SE Arizona) and SW Chih (probably also N Sin); in winter, U to F (Nov–mid-Apr; to SL) from cen Son to S Sin (and N Nay?), R to U and irregular N to SE Arizona (Phillips and Phillips 1993). Little-known resident population (F but local?) in SW Zac, W Ags, and N Jal.

NB: Sometimes placed in the genus *Amphispiza.*

BLACK-CHESTED SPARROW
Aimophila humeralis   Plate 63
Zacatonero Pechinegro

ID: 6–6.5 in (15–16.5 cm). *Endemic to SW Mexico.* **Desc. Adult:** bill black above,

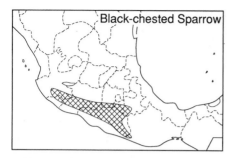

Black-chested Sparrow

blue-grey below, legs flesh. *Head blackish brown with white loral spot and moustache; white throat bordered by black malar stripe and black chest band.* Rest of underparts white, washed cinnamon on flanks and undertail coverts. *Back rufous, mottled darker,* becoming olive-brown on rump and uppertail coverts. Wings and tail blackish brown, edged brown, palest on outer webs of outermost rectrices, 2 whitish wingbars. **Imm:** duller, head and chest band greyish, crown streaked rufous-brown, attains adult plumage Mar–May. **Juv** (fledgling): head and upperparts tawny-brown, back streaked darker, with whitish moustache and dark malar stripe. Throat and underparts whitish, streaked dusky on chest and flanks. Quickly attains imm plumage.

Voice. Commonly heard call an excited, high, liquid twittering and chattering at times somewhat rhythmic, perhaps birds dueting: *swi-chi-ti swi-chi-ti swi-chi-ti . . .,* or *s'wee-chu s'wee-chu . . .,* etc., also a high, thin, sharp, slightly metallic *siik* and *si-ik.* Song typically a single note followed by a rapid, liquid series of 8–16 notes sometimes accelerating into a trill: *swiet, swieswieswieswie . . .* or *psu, susususu . . .,* etc.

Habs. Arid thorn forest and adjacent scrubby edge. Habits much as Bridled Sparrow. Nest near ground in thick weed growth.

SD/RA: C to F resident (300–1500 m, locally to near SL) in interior and on adjacent Pacific Slope from S Jal to S Pue and extreme W Oax.

STRIPE-HEADED SPARROW

Aimophila ruficauda*   Plate 63
Zacatonero Cabecirrayada

ID: 6–7.7 in (15–19.5 cm) S > N. *Pacific Slope and S interior.* Two groups: *acumi-*

nata (W Mexico) and *ruficauda* group (S Mexico and C.A.). **Descr. Adult.** *Acuminata:* bill blackish above, pale greyish to flesh below, legs flesh. *Broad white supercilium contrasts with black sides of crown and black auricular mask, median crown stripe white.* Throat and underparts whitish becoming pale grey on sides, washed pale cinnamon on flanks and undertail coverts. Greyish nape streaked dark brown; *back rufous, streaked darker* and edged sandy brown; rump and uppertail coverts sandy brown. Wings blackish brown, edged tawny-rufous; *brownish tail edged cinnamon.* **Ruficauda:** larger, *head stripes blackish brown with pale spot in auriculars, chest mottled pale grey, tail overall cinnamon* to rufous-brown. **Juv** (both groups): head stripes dark brown and creamy to pale cinnamon, chest flecked dark brown, upperparts sandy brown overall, without rufous.

Voice. Sharp, slightly nasal clucks, *chuh, chuh chuh . . .* or *kyuh . . .,* may be repeated persistently and often break into an excited, rollicking, and bickering duet, also higher and more liquid, rollicking duets, and a high *ssi.* Song a bright, sharp, fairly rapid *beeh-cha' beeh-cha' beeh-cha' . . .* or *ta'bee ta'bee ta'bee . . .,* etc., usually 5–6×, may suggest a squeezy toy.

Habs. Arid to semiarid brushy scrub and semiopen areas. In pairs or small groups, generally conspicuous and noisy.

SD: C to F resident (SL–1400 m) on Pacific Slope and locally in interior from S Dgo and Nay to W Oax and (*ruficauda* group) in Isthmus and from E Guatemala to Honduras.

RA: W Mexico to NW Costa Rica.

NB: Formerly known as Russet-tailed Sparrow.

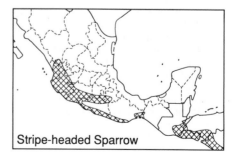

Stripe-headed Sparrow

SUMICHRAST'S (CINNAMON-TAILED) SPARROW
Aimophila sumichrasti   Plate 64
Zacatonero de Sumichrast

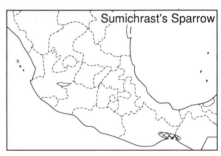

Sumichrast's Sparrow

ID: 5.8–6.5 in (14.5–16.5 cm). Endemic to Pacific Slope of Isthmus. **Descr. Adult:** bill dark grey above, horn-flesh below, legs flesh. *Face grey with* paler supercilium and dark brown eyestripe, *short dark moustache and malar stripe;* throat whitish. *Crown streaked rufous and dark brown with grey median stripe.* Chest pale grey becoming whitish on belly, flanks and undertail coverts washed pale cinnamon. Brown upperparts streaked blackish on back, anterior scapulars rufous. Wings blackish brown with *rufous lesser coverts (usually concealed)* and rufous-brown edgings; *tail cinnamon.* **Juv** (fledgling): head less distinctly marked, chest streaked dusky, upperparts lack rufous.

Voice. High, slightly sharp, clear notes, *tsit* and *tsi-tsit,* a buzzier *sssir* or *zzri,* and fairly loud, excited, rapid, liquid tinkling, at times rhythmic (dueting pairs?). Song a short, chipping series, often with 1–3 introductory notes, *bzz, bzz, bzz t-chip-ichip tip-i-see,* or *tik ssiu chi-chi-chit,* or *spi spi speen chi-chi-chi-chi,* etc., the 2nd part louder and fuller.

Habs. Arid brushy thorn forest and scrub. In pairs or small groups, often fairly conspicuous and noisy at dawn but quickly becomes retiring and elusive except when singing. Nest and eggs undescribed (?).

SS: None in range.

SD/RA: C to F resident (SL–900 m) on Pacific Slope in Isthmus (from Rio Tehuantepec basin to W Chis).

BOTTERI'S SPARROW
Aimophila botterii*   Plate 64
Zacatonero de Botteri

ID: 5.3–6 in (13.5–15 cm). Widespread. *Fairly plain overall; culmen slightly decurved.* Two groups: *botterii* group and *petenica* group (see SD below). **Descr. Botterii. Adult:** bill greyish with dark culmen, legs flesh. Face greyish with paler eyering and rufous postocular stripe; crown slightly darker, streaked dark brown to blackish. *Upperparts grey, streaked rufous and blackish, uppertail coverts with blackish shaft streaks.* Wings and tail dark brown, edged sandy grey-

brown to cinnamon-brown. *Throat and underparts pale grey, washed buff on chest,* brownish buff on flanks and undertail coverts; often shows a short dusky malar stripe. **Juv:** face washed buffy lemon, often with brighter lemon supraloral; throat and underparts washed buffy lemon, streaked dark brown on chest and sides. Crown, nape, and upperparts streaked dark brown and buffy brown. *Petenica.* Darker and greyer overall. **Adult:** face grey with dark rufous postocular stripe. *Crown, nape, and upperparts dusky grey, heavily streaked black and rufous-brown,* uppertail coverts blackish, edged pale grey-brown. Wings and tail blackish brown edged dull sandy grey-brown to dusky cinnamon-brown. *Throat pale grey becoming dusky greyish buff on chest, flanks,* and undertail coverts (flanks and undertail coverts often slightly warmer greyish buff), belly whitish. Birds in interior Chis and Guatemala are intermediate in appearance between the two groups.

Voice. Song a varied series of hesitant chips, in full song breaking into a sweet, accelerating or bouncing-ball trill: *ssi, si si-pit sirr si, si-pit see seeu weet weet-weewiwiwiwiwiwiwiwiwi,* etc. Calls include a high, thin, slightly metallic *sik* or *siik.*

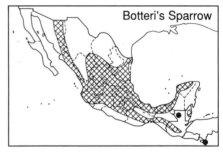

Botteri's Sparrow

Habs. Arid to semihumid brushy scrub, thorn forest with grassy clearings, savannas and fields with scattered bushes and trees. Retiring and elusive except when singing. Drops back to cover quickly when flushed; sings from bush or fence, at times in low gliding flight between perches. Nest may be on ground in grass clump.

SS: Cassin's Sparrow has straightish culmen, plainer face with less distinct cap and supercilium, dark flank streaks, upperparts with subterminal dark bars and spots appear scalloped (versus streaked in Botteri's), outer rectrices tipped whitish, central rectrices often have dark cross-barring; note song. Grasshopper Sparrow shorter-tailed; note yellow supraloral area. Juv Botteri's may recall Savannah Sparrow but note longer, graduated tail.

SD: *Botteri:* F to C breeder (Apr–Sep; SL–2500 m) on Pacific Slope from Son to Nay, in interior from Dgo and S Coah (SNGH, SW) to Guatemala, and on Atlantic Slope S to cen Ver. *Petenica:* F to C but local resident (SL–500 m) on Atlantic Slope from S Ver to the Mosquitia, and on Pacific Slope in W Nicaragua. Breeding season in some areas triggered by start of the rainy season. Resident in much of range but migratory in N Mexico (withdrawing S Sep/Oct–Apr/May) and probably locally elsewhere.

RA: SW USA to cen Mexico, locally to NW Costa Rica.

NB: *Petenica* group has been considered specifically distinct, Peten or Yellow-carpalled Sparrow.

CASSIN'S SPARROW
Aimophila cassinii   Plate 64
Zacatonero de Cassin

ID: 5.3–6 in (13.5–15 cm). N Mexico. *Plain overall, much like Botteri's Sparrow but*

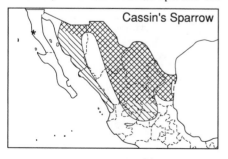

Cassin's Sparrow

culmen straightish. **Descr. Adult.** **Grey morph:** bill greyish with dark culmen, legs flesh. Face plain greyish with paler eyering and rufous-brown to rufous postocular stripe. *Crown and upperparts grey, streaked brown to rufous with darker subterminal spots and bars (including uppertail coverts).* Wings and tail dark brown, edged pale grey and pale brown, pale lemon carpal bend often concealed; *tips to outer webs of outermost rectrices whitish, central rectrices often with dark cross-barring.* Throat and underparts pale grey, usually with dusky malar stripe, chest and flanks washed pale buff to greyish buff, dark streaks on flanks often covered by wings; often also shows dark flecks on sides of chest (imms only?). **Rufous morph:** appears brighter overall, more rufous, due to broader rufous upperpart markings, rufous wing and tail edgings, and rufous uppertail coverts. **Juv:** buffier upperparts have little or no rufous, chest streaked and spotted dark brown.

Voice. Song typically a sweet, slightly quavering trill with 1–2 introductory and ending notes, *sur eeeeeeeeeeeur whee-wheet'*, or *si see seeeeeeur swee-swee*, etc.; also a varied chipping warble interspersed with sweet trills. Calls include a high, thin *sik* or *seek* and a quiet trill.

Habs. Arid to semiarid, open and semi-open brush, grassland with scattered bushes and brush, in winter also weedy field edges, etc. Habits much as Botteri's Sparrow but song flight common, climbing up to 5 m or so before a fluttering glide down. Nest may be on ground in grass clump.

SS: See Botteri's Sparrow. Brewer's Sparrow slimmer with cleft tail, smaller pinkish bill. Grasshopper Sparrow shorter-tailed; note yellow supraloral.

SD: F to C breeder (Apr–Sep; near SL–2500 m) in N Son and in interior S an undetermined distance over the Plateau, at least to Zac (Webster 1968*b*) and SLP (Webster and Orr 1954*b*), and on Atlantic Slope in Tamps. Resident in much of this range, but probably some withdraw S from NW; U to F in winter (Oct–Feb at least) on Pacific Slope from Son to Nay (SNGH, DEW), and S in interior to Gto. Vagrant to BCN (Jun 1984, REW).

RA: SW USA to cen Mexico; N USA birds withdraw S in winter.

RUFOUS-WINGED SPARROW
Aimophila carpalis*   Plate 64
Zacatonero Alirrufo

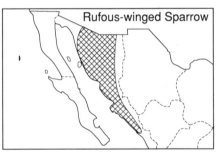

Rufous-winged Sparrow

ID: 5.3–5.8 in (13.5–14.5 cm). NW Mexico.
Descr. Adult: bill blackish above, yellow-orange below, legs flesh. *Face grey with dark lores,* narrow whitish eyering, rufous postocular stripe; *throat white with short dark moustache and malar stripe. Rufous crown streaked grey,* often with grey median stripe. Underparts pale grey, washed buffy brown on flanks. Upperparts grey-brown, streaked dark brown on back, *rufous lesser wing coverts often concealed.* Wings dark brown, edged buffy brown, palest on wingbars and becoming rufous on tertial edges. Tail dark brown, tips of outer webs of outermost rectrices whitish to pale buff. **Juv:** head and upperparts streaked dark brown and buffy brown with trace of adult face pattern. Throat and underparts whitish, streaked dark brown on chest.
Voice. A high, thin, slightly tinny *tsit* or *sik,* and a sharp, sweet *chiip* or *chip.* Song a varied series of high, thin, sharp, slightly sweet chips typically accelerating into a trill: *tiik tik tik tik ti-i-i-i-i-i,* or *tit tit ti-tit-tiiii* ..., also simpler series without a trill, *chit chit chit chi-chi-chi,* etc.
Habs. Arid to semiarid thorn forest and brushy scrub. Sings conspicuously from bushes; at other times can be elusive but in winter may form small flocks. Nest may be at mid-levels in bush or tree.
SS: Suggests a *Spizella* more than other *Aimophila;* note face pattern, bicolored bill, and *Aimophila* structure.
SD/RA: C to F resident (SL–1200 m) on Pacific Slope from Son (also SE Arizona) to N Sin.

RUFOUS-CROWNED SPARROW
Aimophila ruficeps*   Plate 64
Zacatonero Coronirrufo

ID: 5.3–6 in (13.5–15 cm). N of Isthmus.
Descr. Adult: *bill flesh-grey with dark culmen,* legs dusky flesh. *Face grey with* dusky lores, *narrow white eyering* (often broken), and indistinct rufous postocular stripe; supercilium often paler grey. *Crown rufous to chestnut, streaked grey along median line. Throat whitish with dark malar stripe,* chest and flanks grey to buffy grey, becoming whitish on belly. *Upperparts streaked grey and rufous to dark*

grey-brown. Wings and *tail dark brown, edged rufous-brown to grey-brown,* central rectrices sometimes with hint of darker cross-barring. **Juv:** head and upperparts grey-brown with trace of adult face pattern. Throat and underparts pale buff, streaked dusky brown on chest and flanks.
Voice. A nasal *tchew* or *dew,* often run into an excited twangy scolding, *tchew-tchew* ..., which may run into a drier chatter, *chi-chi* ...; also a dry churring rattle *jerrerrrr* ... or *churr-rrr* ..., and a thin *tssi.* Song varies from a rapid, jumbled, bubbling series of chips to a more structured chipping series, such as *chri-chri-chri chri-chri-chri chi-chi-chi chi-chi-chih,* at times accelerating toward the end.
Habs. Arid to semiarid, open and semi-open areas, often rocky, with scrubby bushes, scrub, chaparral. In pairs or small groups, on or near ground, fairly conspicuous and easily seen, even when not singing. Nest may be on banks, in grass clumps, etc.
SS: Rusty Sparrow (*mcleodii* group) larger and bulkier with proportionately larger bill, eyering often more distinctly broken in front of eye, postocular stripe brown; other ssp of Rusty Sparrow brighter and more boldly marked. Oaxaca Sparrow has black bill, bolder face pattern with blackish lores and surround to eye setting off

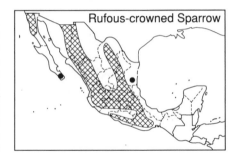

Rufous-crowned Sparrow

white eyering, darker crown typically has grey median stripe above bill, back has distinct black streaks.

SD: F to C resident (900–3000 m, to SL in BCN) in Baja and in interior and on adjacent slopes from E Son and Coah to Oax.
RA: SW USA to Mexico.

OAXACA SPARROW

Aimophila notosticta   Plate 64
Zacatonero Oaxaqueño

ID: 6–6.5 in (15–16.5 cm). *Endemic to interior Oaxaca.* **Descr. Adult:** *bill black* (in winter can be pale grey below, often with dark tip, imm?), legs dull flesh. *Face grey with black lores and surround to eye setting off white eyering* (often broken); supraloral stripe whitish, postocular stripe brown. *Crown chestnut with grey median stripe above bill* and narrow black edging above supercilium. Throat whitish with black malar stripe and often a buff wash to moustachial region. Chest greyish, flanks grey-brown to dusky buffy brown, belly whitish, undertail coverts pale cinnamon. Upperparts brown, *back streaked black*; wings dark brown with rufous secondary and tertial edgings, coverts and primaries edged rufous-brown to grey-brown (worn?). Uppertail coverts and tail dark grey-brown, edged paler, coverts with blackish shaft streaks. **Juv:** head and upperparts olive-brown, streaked darker, with trace of adult face pattern, supercilium and moustache washed buff. Throat and underparts pale buff with dark malar stripe, chest and sides streaked dark brown, flanks washed cinnamon.
Voice. A slightly nasal, dry, scolding chatter *shasha* ... or *chehcheh* ... suggesting a wren; also a harsh, more excited, often prolonged chattering *chii-i-i-i-i-ir*, accelerating and slowing, and a high, thin *tik*. Song typically a single note followed by a

varied chipping series at times run into a short warble: *swi chi-chi-chi-chi-chi chu-chu-chut*, or *ssiu chi-chi-chi chi-di-rit*, etc.
Habs. Arid to semiarid oak–thorn scrub and adjacent overgrown grassy areas. Usually on or near ground in cover and often hard to see, but in winter may form flocks up to 10+ birds which associate loosely with other sparrows. Typically sings from fairly conspicuous perches. Nest and eggs undescribed (?).
SS: See Rufous-crowned Sparrow. Rusty Sparrow (apparently *allopatric*) is larger and bulkier, proportionately larger bill grey below, face pattern less bold, primaries edged rufous, tail rufous-brown, underparts often have warmer buffy tone.
SD/RA: F to C resident (1600–1900 m) in interior of Oaxaca, possibly also in adjacent SE Pue.

RUSTY SPARROW

*Aimophila rufescens**   Plate 64
Zacatonero Rojizo

ID: 6.5–8 in (16.5–20 cm). N > S. Two groups: *rufescens* (widespread) and *mcleodii* (NW Mexico). **Descr. Rufescens. Adult:** bill black above, grey below, legs dull flesh. *Face greyish with blackish lores, dark eyestripe, and white eyering* (usually broken). *Crown chestnut with greyish median stripe, often obscure when worn,* and narrow black edging above supercilium. *Throat whitish with black malar stripe and buff wash to moustachial region.* Chest greyish to greyish buff, *flanks washed brownish to rich ochre-brown,* belly whitish, undertail coverts pale buff to cinnamon. *Upperparts rufous-brown,* back with blackish streaks and greyish edgings. Wings dark brown, coverts edged rufous-brown to grey-brown (worn?), tertials and remiges edged rufous to chestnut. Uppertail coverts and *tail rufous-brown to chestnut-brown,* coverts with dark shaft streaks. **Juv:** head and upperparts rufous-brown, streaked blackish, with trace of adult face pattern, supercilium and moustache washed lemon. Throat and underparts dirty pale lemon with black malar stripe, chest streaked dark brown. *Mcleodii: duller and browner overall.* Rufous crown often lacks blackish edging above supercilium, flanks washed dusky grey-brown, *dark back streaks indistinct, tail brown.* Juv has grey-brown head and

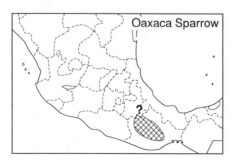
Oaxaca Sparrow

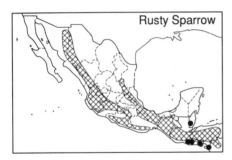

Rusty Sparrow

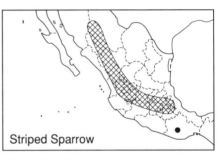

Striped Sparrow

upperparts, dirty pale buff underparts streaked dark brown.

Voice. A gruff, dry scolding *chrrr-rrr-rrr* or *grrr-grr-grrr*, at times prolonged, and other chattering calls. Song notably varied, usually short, a bright series of chips, *cheeu chik-chik* or *chree ch-cheeo* or *seeyr seeyr sit-sit-sit-churr*, etc.

Habs. Scrubby second growth, brushy woodland, forest edge, overgrown clearings, particularly in oak and pine–oak zones but also in areas of humid evergreen forest. Singly or in pairs, often at edges or roadsides; sings from fairly conspicuous perches. Nest may be on ground, in grass clump, bank, etc.

SS: See Oaxaca (allopatric) and Rufous-crowned sparrows.

SD: *Rufescens:* F to C resident (600–2700 m, locally to near SL) on both slopes from cen Sin and S Tamps, and in interior from cen Mexico, to W Nicaragua, locally (to near SL) in pine and oak savannas from N Chis to the Mosquitia. *Mcleodii:* F to C resident on Pacific Slope from Son to SW Chih and N Sin.

RA: Mexico to NW Costa Rica.

STRIPED SPARROW

Oriturus superciliosus*   Plate 63
Zacatonero Rayado

ID: 6.5–7 in (16.5–18 cm). A large, fat, flat-headed sparrow endemic to highlands of NW and cen Mexico. **Descr. Adult:** bill blackish, legs flesh. *Creamy white supercilium contrasts with blackish auricular mask and black-streaked chestnut crown sides; median crown stripe grey, streaked black.* Underparts pale grey, becoming whitish on throat, washed buff on dark-streaked flanks. *Upperparts streaked tawny-brown and black with pale grey-brown edgings.* Wings and tail blackish

brown, edged tawny-brown, becoming rufous on tertials and secondaries; central rectrices have wavy cross-barring; note contrasting buff outer webs and tips to outer rectrices. **Juv:** head pattern duller, upperparts less neatly marked, chest and flanks flecked and spotted dusky.

Voice. A sharp, metallic *tik* or *chik*, often run into excited hard chattering, *chik chik chik-chik sshi sshisshisshi* . . ., etc. Song one to several metallic chips and often 1–3 nasal beeps preceding a rattling trill that may suggest a ♀ cowbird, *tiuk, tiuk, tiuk, tiuk, beeh beeh drrrrrrrrrrrr*, etc.

Habs. Open pine woodland, fields, and meadows, especially with bunch-grass. Usually in loose groups; a noisy, social, and conspicuous sparrow often seen on roadside fence posts, rocks, etc. Nest cup of grasses, pine needles, etc., placed in bunch grass. Eggs: 3–4, whitish, marked with reddish browns and purples (MLZ).

SD/RA: C to F resident (1500–3500 m) in interior and locally on adjacent Pacific Slope from E Son to cen volcanic belt; disjunctly in cen Oax.

BLACK-THROATED SPARROW

Amphispiza bilineata*   Plate 63
Gorrión Gorjinegro

ID: 5.2–5.7 in (13–14.5 cm). N Mexican deserts. **Descr.** Ages differ, sexes similar.

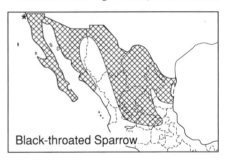

Black-throated Sparrow

Adult: bill black above, grey below, legs grey. *White supercilium contrasts with grey crown, black lores, and grey auriculars. Black throat and upper chest bordered by white moustache,* underparts otherwise whitish, washed grey to greyish buff on sides and flanks. Upperparts grey-brown, *tail blackish with white outer webs and tips to outermost rectrices.* **Juv:** grey-brown and whitish head pattern less distinct; throat and underparts whitish, streaked dusky on chest; brownish upperparts streaked darker on back.

Voice. High, sweet, tinkling twitters and warbles, also a single high *ti* or *sii*, etc. Song varied, an overall high and slightly tinkling warble, often with buzzy notes and short trills thrown in.

Habs. Arid open and semiopen areas with scattered bushes, cacti, etc. In pairs or small groups, on or near ground where runs well, often with tail cocked; sings from low bushes. Nest cup of fine twigs, grasses, etc., low in cactus or bush. Eggs: 2–4, bluish white, unmarked (WFVZ).

SS: Juv Sage and Black-throated sparrows sometimes confused with one another but note that they show a trace of the distinctive adult face patterns.

SD: C to F resident (SL–2500 m) in Baja, on Pacific Slope S to N Sin, on Plateau S to Gto, and on Atlantic Slope in Tamps.

RA: SW USA to cen Mexico.

SAGE SPARROW

Amphispiza belli*   Not illustrated
Gorrión de Artemesia

ID: 5.5–6.3 in (14–16 cm). NW Mexico. Two ssp breed in region: *belli* (BCN N of 29° N) and *cinerea* (cen Baja). **Descr.** Ages differ, sexes similar. *Belli.* **Adult:** bill dark grey above, paler below, legs blackish to dark greyish flesh. *Head greyish with white supraloral spot, dark lores, and white eye-*

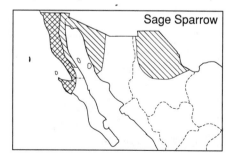

Sage Sparrow

crescents, upperparts grey-brown with indistinct dark streaks. Wings dark brown, edged paler, with 2 pale buff wingbars; *tail blackish brown,* outer webs of outermost rectrices may be pale buff, rarely noticeable in field. *Throat and underparts whitish with blackish malar stripe and central chest spot, sides and belly washed vinaceous cinnamon,* streaked dark brown on sides. **Juv:** head browner, face pattern less distinct, head and upperparts streaked dark brown, chest finely streaked dusky, wing edgings broader and brighter, underparts dirty pale buff to buffy white. *Cinerea.* **Adult:** *head and upperparts pallid sandy brown* to grey-brown with *whitish to buffy-white supraloral spot and short postocular stripe, broken whitish eyering.* Wings browner with paler edgings, 2 pale buff wingbars often wear off. *Tail contrastingly dark brown.* Throat and underparts buffy white to pale buff with dark brown malar stripe and central chest spot, sides and flanks with blurry dusky streaks. **Juv:** plainer overall, face pattern less distinct, broader sandy-buff tertial edgings and wingbars, hint of fine dusky streaks on chest; note dark tail. Wintering ssp resemble *belli* but paler, dark back streaks more distinct.

Voice. A high, thin *tik* or *ti* and *tik-tik,* and high, thin, tinkling twitters. Song varied, a pleasant to scratchy, jangling and tinkling warble.

Habs. Arid to semiarid, open to semiopen areas with scattered bushes and trees, also chaparral in NW Baja. Habits much as Black-throated Sparrow but in flocks up to 30+ birds in winter. Nest cup of fine twigs, grasses, etc., low in bushes. Eggs: 2–4, pale blue to bluish white, speckled with reddish browns and greys (WFVZ).

SS: See juv Black-throated Sparrow.

SD: C to F resident (SL–2100 m) in W Baja S to 26° N; F to U winter visitor (Oct–Mar; SL–1800 m) to NE BCN, N Son, and N Chih.

RA: W USA to N Baja; winters to NW Mexico.

GENUS SPIZELLA

These are small, relatively slender sparrows of open country and open woodland. Bills small, often pinkish, tails relatively long, usually slightly cleft. Ages differ, sexes similar. Flocks of one species do not usually mingle with flocks of another, although a few individuals of one species are often in flocks of another. Nest cups of grasses, rootlets, etc., usually low in bushes. Eggs: 2–5, pale bluish, with dark flecks and scrawls (WFVZ).

CHIPPING SPARROW

Spizella passerina*   Not illustrated
Gorrión Cejiblanco

ID: 5–5.5 in (12.5–14 cm). Widespread. **Descr.** Seasonal variation. **Alt:** bill black, legs flesh. *Whitish supercilium contrasts with rufous crown and black eyestripe,* forehead black with white median streak. Face and sides of neck grey, throat and underparts paler grey, fading to whitish on throat, belly, and undertail coverts. Upperparts brown to rufous-brown, streaked black, with *unstreaked grey rump.* Wings and tail dark brown, edged paler, with 2 whitish to pale buff wingbars and rich brown tertial edgings. *S. p. pinetorum* of C.A. darker overall, crown chestnut, underparts duskier grey. **Basic/imm:** bill pinkish. *Crown streaked brown and blackish, typically with paler median stripe and a few rufous flecks at sides,* supercilium and wingbars may be washed buff. Imm often has chest washed buff, throat with indistinct dusky malar stripe. **Juv:** face less distinctly marked than imm; supercilium streaked dusky; pale grey underparts streaked dark brown, brown upperparts (including rump) streaked dark brown.

Voice. A high, thin *tssip* or *ssip* and *tik*. Song a dry, rapid trill on one pitch, typically 2–3 s in duration, less often a slower chipping trill, *chichichi...* lasting up to 7–9 s.

Habs. Breeds in open pine and pinyon–juniper woodland, typically with grassy clearings; in winter also in open and semi-open areas with brushy woodland, grasslands with scattered bushes, etc. Sings from mid- to upper levels. In winter often in flocks to 100+ birds. Nest may be at mid- to upper levels in trees, rarely on ground.

SS: Clay-colored Sparrow usually brighter and buffier overall, more strongly patterned face lacks singularly contrasting supercilium; it shows instead buff supercilium and moustache outlining brown auricular mask; note dark malar stripe, more contrasting grey sides of neck, and sandy grey-brown rump. Brewer's Sparrow (deserts) longer-tailed, face plain without strong supercilium; note whitish eyering, dusky malar stripe, sandy grey-brown rump. Field Sparrow (winter) has rufous-brown crown with grey median stripe, plain greyish face with white eyering, rufous postocular stripe, bright pink bill; upperparts brighter rufous, rump sandy grey-brown, chest often buffier. Worthen's Sparrow has unstreaked rufous crown and grey forehead, plain grey face with white eyering, bright pink bill.

SD: F to C resident (Apr–Aug; 1000–3500 m, locally to 500 m) in BCN and in interior and on adjacent slopes from Son and S NL to N cen Nicaragua; also (near SL–1000 m) in pine savannas from Petén to the Mosquitia. C to F transient and winter visitor (mid-Sep–early May; near SL–3500 m) in Baja, and on Pacific Slope S to Oax, in interior S to cen Mexico; irregularly U on Atlantic Slope to Tamps.

RA: Breeds N.A. to Nicaragua, migratory N populations winter to N C.A.

CLAY-COLORED SPARROW

Spizella pallida   Not illustrated
Gorrión Pálido

ID: 4.8–5.3 in (12–13.5 cm). Winters N of Isthmus. **Descr.** Bill and legs flesh. *Pale*

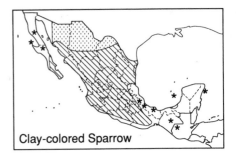

Chipping Sparrow

Clay-colored Sparrow

buff supercilium and moustache border dark-sided sandy brown auricular mask. *Crown streaked brown and blackish with pale median stripe, sides of neck grey.* Throat and underparts whitish with dark malar stripe, *chest and flanks washed warm buff* (often whitish by late spring). Upperparts sandy brown, streaked black, rump duller grey-brown. Wings and tail dark brown, edged paler, with 2 buff wingbars and buffy-rufous tertial and secondary edgings.

Voice. A high, thin *tsi* or *tsip*, and a more lisping *tssi*.

Habs. Arid to semihumid grassland and arable fields with scattered bushes and hedges, often in same areas as Lark Sparrow. Usually in flocks, feeding on ground or perched in bushes.

SS: See Chipping Sparrow. Brewer's Sparrow longer-tailed, overall sandier and greyer with less contrasting grey sides of neck, face relatively plain with whitish eyering, indistinct median crown stripe.

SD: C to F transient and winter visitor (Sep–May; near SL–2700 m) in BCS (U to R transient in rest of Baja), on Pacific Slope from cen Son to Nay, and in interior from Chih and NL to Oax; R and irregular to cen Guatemala; U to R on Atlantic Slope S to Ver. Vagrant (mid-Sep–Nov) to N Yuc Pen (KK, AMS).

RA: Breeds W N.A.; winters N and cen Mexico.

BREWER'S SPARROW

Spizella b. breweri   Not illustrated
Gorrión de Brewer

ID: 5–5.5 in (12.5–14 cm). Winters N and cen Mexico. A relatively long-tailed *Spizella*. **Descr.** Bill and legs flesh. *Face relatively plain brownish with paler greyish-buff supercilium and whitish eyering, crown streaked brown and blackish*

with hint of paler median stripe; sides of neck greyish. Throat and underparts whitish with dark malar stripe, chest and flanks washed buff in autumn and winter. *Upperparts sandy brown, streaked blackish.* Wings and tail dark brown, edged paler, with 2 pale buff wingbars, richer buffy-rufous tertial and secondary edgings.

Voice. A high, thin *tsi* or *tsip*, much like Clay-colored Sparrow.

Habs. Arid to semiarid grassland and semiopen areas with scattered trees and bushes, more often in deserts than other *Spizella*. Habits much as Clay-colored Sparrow.

SS: See Clay-colored, Chipping, and Cassin's sparrows.

SD: C to F transient and winter visitor (Sep–early May; near SL–2500 m) in Baja, on Pacific Slope in Son (probably also N Sin), and in interior from Chih and NL S over Plateau to Gto.

RA: Breeds W N.A.; winters SW USA to cen Mexico.

NB: Some authors (Sibley and Monroe 1990) consider birds breeding in NW N.A. as specifically distinct, Timberline Sparrow (*S. taverneri*), and suggest that this form may winter in NW Mexico. It tends to be slightly larger and darker than typical Brewer's Sparrows but field separation criteria are essentially unknown.

FIELD SPARROW

Spizella pusilla arenacea   Not illustrated
Gorrión Llanero

ID: 5.3–5.8 in (13.5–14.5 cm). *Winter visitor to NE Mexico.* **Descr.** Bill bright pink, legs flesh-pink. *Face grey with distinct whitish eyering, rufous postocular stripe, and often a brownish wash on auriculars. Crown greyish rufous-brown to rufous-brown with grey forehead and median stripe,* dark streaking in crown may be

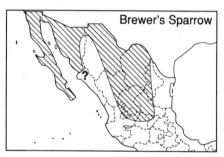

Brewer's Sparrow

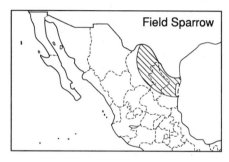

Field Sparrow

visible at close range. Sides of neck greyish, throat whitish, underparts pale grey, washed greyish buff on chest and flanks, often with *rufous patch on sides of chest.* Back sandy brownish grey, streaked blackish, and usually with a few rufous edgings, rump unstreaked grey to grey-brown. Wings and tail dark brown, wings edged paler, with broader white to whitish upper wingbar, narrower whitish to pale buff lower wingbar, buffy-rufous tertial and secondary edgings, greyish lesser coverts; tail narrowly edged pale grey to whitish.

Voice. A high, fairly sharp *tik* or *tsip.*

Habs. Arid to semihumid brush and overgrown fields, grasslands with scattered trees and bushes. Likely to be found with flocks of other sparrows, especially *Spizella.*

SS: Worthen's Sparrow relatively shorter-tailed; note solidly rufous to rufous-brown crown, lack of rufous chest side patches, less distinct wingbars, and dark legs. See Chipping Sparrow.

SD: U to R winter visitor (late Oct–Mar; near SL–500 m) on Atlantic Slope from E Coah and NL to cen Tamps. Possible vagrant to Son (Nov 1979, Terrill 1981).

RA: Breeds E N.A.; winters E USA to NE Mexico.

NB: Worthen's Sparrow has been considered conspecific with Field Sparrow.

WORTHEN'S SPARROW

Spizella wortheni*   Plate 64
Gorrión de Worthen

ID: 5–5.5 in (12.5–14 cm). *Endemic to Mexican Plateau.* **Descr.** *Bill bright pink, legs dark dusky flesh to blackish brown.* **Adult:** *face grey with distinct whitish eye-ring* and often a brownish postocular stripe and brownish wash on auriculars. *Crown rufous to rufous-brown with greyish forehead,* dark streaking in crown

may be visible at close range. Sides of neck greyish, throat whitish, underparts pale grey, washed greyish buff to buff on chest and flanks. Upperparts sandy grey-brown, streaked dark brown, with unstreaked grey rump. Wings and tail dark brown, wings edged paler, with broader whitish to pale buff upper wingbar, narrower pale buff lower wingbar, buffy-rufous tertial and secondary edgings, greyish lesser coverts; tail edged whitish. **Juv:** face and chest washed brownish buff, face streaked dusky, chest and flanks streaked dark brown, wingbars buff.

Voice. A high, thin, fairly dry *tssip* or *tsip,* at times repeated rapidly. Song a dry chipping trill of 2–3 s duration (AMS tape), suggests Chipping Sparrow or Dark-eyed Junco.

Habs. Arid to semiarid rolling grassland with trees and bushes, especially junipers. Sings from low bushes and fences. In winter locally in flocks up to 50+ birds.

SS: See Field and Chipping sparrows.

SD/RA: Poorly known. U to F but local breeder (Apr–Aug; 1200–2400 m) on Plateau from NW Zac and SE Coah to SW Tamps and S SLP, probably also cen Chih (Jun spec., MVZ). Some local movement occurs: at least in Coah and NL forms winter flocks in areas where U to R in summer (SNGH, SW). Specimens from N Pue and W cen Ver (and New Mexico?) may represent isolated (extirpated?) populations; there is no evidence that they are migrants (*contra* AOU 1983).

NB: See Field Sparrow.

BLACK-CHINNED SPARROW

Spizella atrogularis*   Plate 63
Gorrión Barbinegro

ID: 5.5–5.8 in (14–15 cm). Widespread N of Isthmus. **Descr.** Age/sex/seasonal variation. **Alt:** bill bright pink, legs flesh.

Worthen's Sparrow

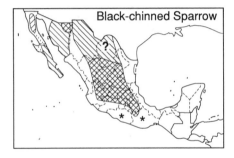
Black-chinned Sparrow

Head and underparts grey with black lores and upper throat (less extensive, greyish black in ♀). *Back brown, streaked blackish, with rufous outer scapulars, rump* and uppertail coverts *unstreaked greyish.* Wings and tail dark brown, wings edged paler, tail edged whitish. **Basic/imm:** *lores and throat grey,* crown washed brown in fresh plumage, also head and chest washed brown in fresh imm. **Juv:** head and underparts washed brownish, chest and flanks with indistinct dusky streaks, outer scapulars tawny-brown.

Voice. A high, fairly sharp *tsik* or *sik.* Song 2 or more chips run into a pleasant, sweet, bouncing-ball trill, *teu swi-swi siiiiiiiiir,* or *swee twee wee-i-iiiiiiir,* etc.

Habs. Arid to semiarid brushy scrub, chaparral, semiopen areas with bushes and scrub. Singly or in pairs, rarely in small flocks. Usually retiring and inconspicuous but often sings from prominent perches.

SD: F to C resident (1200–2500 m, lower locally in winter) in N Son and in interior and on adjacent slopes from Dgo and S Coah to cen volcanic belt, U to F and local to NW Oax. F to C breeder (Apr–Aug; 300–2400 m) in NW BCN, winters (Sep–Apr) BCS. U winter visitor (Sep–Apr) S to S Son (SNGH) and on N Plateau, irregular R to U winter visitor to Gro and cen Oax (SNGH).

RA: Breeds SW USA to Mexico; N populations withdraw S in winter.

VESPER SPARROW
Poocetes gramineus *   Not illustrated
Gorrión Coliblanco

ID: 5.5–6 in (14–15 cm). Widespread in winter N of Isthmus. **Descr.** Bill flesh with dark culmen, legs flesh. Head streaked pale

grey-brown and dark brown with short whitish supercilium and *white eyering,* dark-bordered auriculars, and *broad whitish moustache.* Throat and underparts whitish with dark-streaked malar stripe, *fine dark streaks on chest and flanks;* chest and especially flanks often washed buff to pale pinkish cinnamon. Upperparts greybrown, streaked dark brown. Wings dark brown, edged paler, with pale buff wingbars; *rufous lesser wing coverts* often concealed by scapulars. Tail dark brown with mostly *white outermost rectrices, obvious when tail spread.*

Voice. A high, thin *tssit* or *ssit.*

Habs. Wide variety of grasslands, fields, brushy second growth. Singly or in loose flocks, perches on bushes and fences although at times fairly skulking.

SS: Savannah Sparrow shorter-tailed, face has yellowish supraloral stripe and lacks contrasting eyering, also lacks rufous lesser wing coverts and white tail sides.

SD: C to F transient and winter visitor (Sep–Apr; SL–2500 m) on Pacific Slope from Son to Nay, in interior and on adjacent slopes from Son and NL to Oax, and on Atlantic Slope in Tamps; U to F in Baja; R and irregular to Yuc Pen and Guatemala.

RA: Breeds N.A.; winters S USA to Mexico.

LARK SPARROW
Chondestes grammacus *   Not illustrated
Gorrión Arlequín

ID: 6–6.5 in (15–16.5 cm). *A strikingly patterned sparrow, common in winter N of Isthmus.* **Descr.** Ages differ, sexes similar. **Adult:** bill flesh-grey with dark culmen, legs flesh. Broad whitish supercilium becomes buff behind eye and contrasts with chestnut sides of crown, black sides of forehead, and black-bordered chestnut auriculars; white subocular crescent, broad white moustache, and whitish

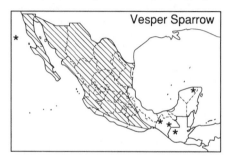

Vesper Sparrow

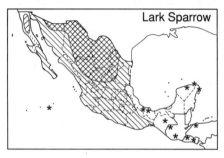

Lark Sparrow

median crown stripe complete *harlequin head pattern. Throat whitish with black malar stripe*; underparts pale grey with black central chest spot, flanks washed buff. Upperparts grey-brown, streaked black on back. Wings and tail dark brown, edged paler, with 2 pale buff wingbars and whitish flash at base of primaries; *tail boldly edged white.* **Juv:** head pattern less distinct with brown replacing chestnut, dark streaks on crown; throat, chest, and flanks spotted and streaked dark brown.

Voice. Distinctive call often given in flight, a sharp, slightly metallic *tik* or *sik.* Song a fairly full, loud, and varied series of rich chips and buzzes, *chru-chru choo-choo-choo-choo chi-chi-chi-chi,* or *chru-chru chri-chri-chri-chri tsi-i-i chu-chu',* etc.; also prolonged series of sweet chips, trills, short buzzes, and longer trills, at times given in winter.

Habs. Open grassland and stony plains with scattered bushes and trees. Singly, or more usually in flocks, at times to a few hundred birds, which often associate loosely with Clay-colored Sparrows. Flies higher than most sparrows and often flies overhead, calling. Nest cup of grasses low in bush or depression in ground. Eggs: 3–5, whitish, with black and grey scrawls and spots (WFVZ).

SD: F to U breeder (Apr–Aug; 1000–2000 m) in N BCN, and in interior from NE Son and Coah to Zac. C to F transient and winter visitor (mid-Aug–early May; SL–2500 m) in Baja (winters in S), on Pacific Slope and in interior S to Oax, and on Atlantic Slope to N Ver; irregularly U to R (Sep–Mar) in coastal Yuc Pen and S to Honduras and El Salvador.

RA: Breeds W and cen N.A. to N Mexico; winters S USA to Mexico.

LARK BUNTING

Calamospiza melanocorys   Not illustrated
Gorrión Alipálido

ID: 6.2–6.7 in (15.5–17 cm). A stocky stout-billed sparrow wintering in N and cen Mexico. **Descr.** Age/sex/seasonal variation. **Basic** ♂: bill blackish above, grey below, legs brownish flesh. Head and upperparts brown, streaked darker, with *broad, slightly teardrop-shaped whitish eyering, black mottling on lores and throat. Underparts whitish, streaked blackish.* Wings blackish with broad

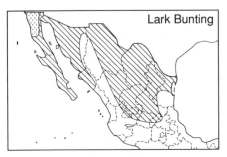

creamy-buff to whitish panel on greater and median coverts, tertials edged tawny-brown. *Tail blackish with white tips* to all but central rectrices. Plumage overall mottled black by late winter/spring in molt to alt. **Alt** ♂: black overall with bold white wing panel, white tertial edgings. ♀/imm: resemble basic ♂ but lack black mottling on face and throat; throat whitish with black malar stripe, chin mottled black, belly unstreaked. Wings and tail dark brown, wing panel reduced to broad creamy buff edgings; imm ♂ usually has outer 4 primaries contrastingly blackish.

Voice. A nasal *whiu* or *whew,* often given in flight.

Habs. Open and semiopen areas with bushes and brush, often along roadsides. Flocks up to a few hundred birds fly in tight, undulating formation.

SD: C to F transient and winter visitor (Aug–Apr, R from mid-Jul, U into May) in Baja, on Pacific Slope S to N Sin, on Plateau S to Gto, and on Atlantic Slope in Tamps.

RA: Breeds cen N.A., winters SW USA to cen Mexico.

GENUS AMMODRAMUS

These small, streaky brown sparrows inhabit grasslands and marshes. Relatively short-tailed and often flat-headed, tails notably spiky in several species. Ages differ, sexes similar, juv plumage of most species lost before arrival in the region.

BAIRD'S SPARROW

Ammodramus bairdii   Not illustrated
Gorrión de Baird

ID: 4.8–5.2 in (12–13 cm). *Winters in grasslands of N Mexico.* Tail fairly short and spiky, crown flattish. **Descr. Adult:** bill flesh with dusky culmen, legs flesh. *Face*

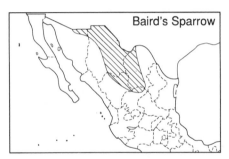

Baird's Sparrow

ochraceous buff to buff with black moustache and two dark corner marks on hind auriculars. *Crown streaked blackish and ochre with ochre median stripe*; ochre to ochraceous-buff nape streaked blackish. Upperparts cryptically patterned with blackish, chestnut-brown, and buff, typically with 2 buff braces. Brown wings edged buff to rufous-buff. Tail dark brown with paler edgings, outer rectrices contrastingly paler. *Throat and underparts whitish with blackish-brown malar stripe and fairly coarse but often sparse streaks on chest and flanks*; chest often washed buff. **Juv** (Sep–Oct): face, crown, and nape buffier, less ochraceous, median crown stripe less distinct.

Habs. Native grasslands, grassy fields. Rarely seen unless flushed when flies some distance before dropping back down; silent when flushed.

SS: Savannah Sparrow has whitish to buff supercilium washed yellow in lores, lacks bright ochre median crown stripe and nape, upperparts more distinctly streaked, less patterned. Grasshopper Sparrow has plain buff face and chest.

SD: U to R and local transient and winter visitor (Sep–Apr; 1200–2000 m) on Plateau S to N Dgo.

RA: Breeds N interior N.A.; winters SW USA and N Mexico.

GRASSHOPPER SPARROW

Ammodramus savannarum* Not illustrated
Gorrión Chapulín

ID: 4.7–5 in (12–12.5 cm). Widespread, but often local, in grasslands. Tail fairly short and spiky, crown flattish. **Descr.** Ages differ, sexes similar. **Adult:** bill greyish flesh with dusky culmen, legs flesh. *Face plain buff with paler eyering, ochre wash in lores.* Crown blackish brown with buff median stripe; *nape streaked blue-grey and*

rufous. Upperparts cryptically patterned with blackish, chestnut, blue-grey, and buff. Wings and tail brown, edged buff and pale brown, often with 2 buff wingbars, outer rectrices contrastingly paler. *Throat, chest, and flanks buff*, palest on throat, becoming whitish on belly. **Juv:** face brownish with paler eyering and supercilium, whitish underparts streaked dusky on chest, upperparts less neatly patterned.

Voice. Song a high, thin, wiry, insect-like buzz of 1–2 s duration; usually one or more introductory notes audible at close range, *ts, zzzziiir*, etc. Calls include a soft insect-like *tk* or *tik*.

Habs. Grasslands, arable fields, etc., often with scattered bushes. Habits similar to Baird's Sparrow but more often seen in open, usually silent when flushed; sings from low bushes and fences. Nest cup of grasses in depression on ground, under grass tussocks, etc. Eggs: 3–5, white, speckled reddish brown and grey (WFVZ).

SS: Savannah Sparrow often calls when flushed and perches in view before dropping back to ground. See Cassin's and Botteri's sparrows.

SD: Breeding range poorly known. F to C but local breeder (Apr–Aug; near SL–2500 m) in NW Baja and N Son, on Plateau in Zac (and almost certainly elsewhere), and probably (resident?) from Oax to Honduras (but no confirmed breeding); F to C local resident (SL–1000 m) on Atlantic Slope from S Ver to the Mosquitia. F to C transient and winter visitor (Sep–May; SL–3000 m) virtually throughout N of Isthmus; F to U in interior and on Pacific Slope to cen Honduras and El Salvador; U to R (Oct–Apr) in Yuc Pen.

RA: Breeds N.A. and locally from Mexico to Panama; N populations migratory, wintering S to C.A.

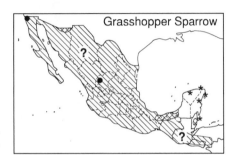

Grasshopper Sparrow

LE CONTE'S SPARROW

Ammodramus leconteii   Not illustrated
Gorrión de Le Conte

ID: 4.7–5 in (12–12.5 cm). *Vagrant to NE Mexico.* **Descr. Adult:** bill greyish to flesh with dark culmen, legs flesh. *Ochre supercilium, sides of neck, and moustache border grey auriculars* edged above with black postocular stripe. *Crown blackish brown with whitish median stripe*, nape streaked blue-grey and rufous. Upperparts cryptically patterned blackish, brown, blue-grey, and buff, with *bold striping on back*. Wings and tail dark brown, edged rufous and buff. Throat and chest ochraceous buff becoming white on belly, *sides and flanks streaked blackish.* **Juv** (Oct): lacks blue-grey on nape and upperparts. *Face buff with duskier cheeks, dark triangular postocular stripe*, buff nape streaked dusky, dark streaks on chest and flanks, upperparts striped buff and blackish.
  Habs. Grasslands, fields. Habits similar to Baird's Sparrow; usually silent when flushed.
SS: Sharp-tailed Sparrow (*nelsoni*) has broad blue-grey median crown stripe, unstreaked blue-grey nape, less contrastingly patterned upperparts overall olive with whitish streaks; usually in marshes.
SD: One record: Sabinas, Coah (23 Feb 1910, UCLA spec.). Probably R to U winter visitor (Oct–Apr?) to grasslands from N Coah to N Tamps. A Christmas Count record from Tamps in 1988 (*Am. Birds*, 42, 1147), however, is not credible (description examined).
RA: Breeds cen N.A.; winters SE USA.

Le Conte's Sparrow

SHARP-TAILED SPARROW

Ammodramus caudacutus nelsoni
Gorrión Coliagudo   Not illustrated

ID: 4.8–5.2 in (12–13 cm). *Rare winter visitor to N Mexico.* **Descr.** Bill grey, legs

Sharp-tailed Sparrow

flesh. *Ochre supercilium, sides of neck, and moustache border grey auriculars* edged above with black postocular stripe. *Blackish-brown sides of crown border broad blue-grey median stripe, nape blue-grey.* Upperparts olive-brown with blackish centers and *whitish streaks on back*. Wings and tail brown, edged buff and pale brown, becoming rufous on greater coverts, tertials, and secondaries. *Throat and chest ochre*, becoming white on belly, *sides and flanks streaked dusky*, flanks and undertail coverts buff.
  Habs. Salt-marshes with dense rushes and reeds, could also occur in freshwater marshes. Skulking and overlooked easily; when flushed, flies a short distance, silently, and often hard to relocate.
SS: See Le Conte's Sparrow. Juv Seaside Sparrow (Rio Grande) unlikely to overlap seasonally, paler buff on face and chest, upperparts streaked blackish, legs dusky.
SD: U to R winter visitor (late Oct–Apr?) to N Tamps (JCA); probably R to U vagrant (Nov–Mar?) to salt-marshes of NW Baja, but only record to date is from Feb 1961 (Northern 1962). A report from Yuc (Edwards 1978, 1989) is in error (photo examined).
RA: Breeds cen and E N.A.; winters coastal E and S USA.

SEASIDE SPARROW

Ammodramus maritimus (sennetti?)
Gorrión Marino   Not illustrated

ID: 5.2–5.7 in (13–14 cm). *Mouth of Rio Grande, NE Mexico.* **Descr. Adult:** bill greyish, legs dusky. *Head and upperparts dark olive-grey*, streaked darker, *with yellow supraloral spot; pale buff moustache, bordered below by dark malar stripe, curves up around lower auriculars.* Wings

Seaside Sparrow

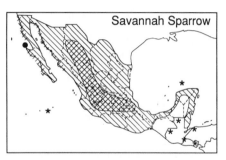

Savannah Sparrow

and tail dark brown, edged paler, becoming rufous on greater coverts and tertials, olive-rufous on secondaries. Whitish throat contrasts with dusky underparts, chest streaked buff, belly whitish. **Juv:** *pale buff supercilium, sides of neck, and moustache border dark auriculars; crown, nape, and upperparts greyish, streaked blackish,* some buff streaks on back. Throat and underparts whitish, streaked dusky on flanks, chest and flanks washed buff.

Voice. A low *chrrt* or *chrit*. Song 1–2 squeaky gurgles run into a rough rasping buzz, *chi eegle eedle zhrrrrr* or *ji cheedle uhrrr*, etc. (GWL tape).

Habs. Brackish and saltwater marshes with reeds and rushes. Often skulking and easily overlooked when not singing. Nest cup of grasses, low in marsh grasses. Eggs: 2–4, whitish, heavily marked with reddish brown and greys (WFVZ).

SS: See Sharp-tailed Sparrow.

SD: U and local resident in marshes on N (USA) side of mouth of Rio Grande, wanders to adjacent bank in N Tamps (JCA, AMS) but there is apparently insufficient habitat to support a breeding population on the Mexican side of the river.

RA: Coasts of E and SE N.A.; N birds withdraw S in winter.

SAVANNAH SPARROW
*Ammodramus sandwichensis**   Figure 41
Gorrión Sabanero

ID: 5–5.8 in (12.5–14.5 cm). *Overall brown and streaky, numerous ssp:* bill varies from small and pointed (most ssp) to large and deep (Islas San Benito, Rio Colorado delta to N Sin) or long and slender (W Baja marshes). **Descr.** Ages differ, sexes similar. **Adult:** bill flesh with dark culmen, legs flesh. *Face brown with paler supercilium*

washed yellow in supraloral region, dark eyestripe and moustache border auriculars. *Crown, nape, and upperparts streaked dark brown and pale sandy brown to grey-brown,* with darker sides of crown. Wings and tail dark brown, edged pale brown to rufous, brightest on greater coverts and tertials, often 2 paler wingbars; outer rectrices often contrastingly paler, especially in *brunnescens* of cen Mexico. Throat and underparts whitish with dark malar stripe, *dark brown streaking on chest and flanks sometimes merges to form dark central chest spot;* chest and flanks may be washed buff. Resident sspp of marshes in W Baja darker overall with olive-grey upperparts. ***Rostratus*** **group** ('Large-billed Sparrow', breeds coastal Son and Sin): overall paler and sandier greybrown with indistinct dark streaking on upperparts, yellow on lores often absent, underparts streaked sandy brown to brown. Legs orange-flesh. **Juv:** less distinctly

Fig. 41 Large-billed [Savannah] Sparrow

marked than adult, upperparts more mottled and spotted, less streaked, streaks on underparts dusky and diffuse.

Voice. A high, thin *tsi* or *tsit*, and a fairly hard, smacking *tsk* or *tsik!*; *rostratus* also gives a high *tsiip* or *tsiih*. Song (*rostratus* group similar) a few high, thin or tinkling chips run into a trilled buzz, often with 1–2 lower notes at end, *si sik-sik-sik sirrrr ji-ir*, or *si sik-sik-sik zzzzzzr zzr-i*, etc..

Habs. Grasslands in general, marshes, adjacent scrub. Somewhat secretive, at times very much so, recalling Grasshopper Sparrow. Often calls when flushed and perches up on grass stalk or fence before dropping back down; sings from fence or other conspicuous perch. May form loose flocks to 30+ birds in migration and winter. Nest cup of grasses in depression in ground, in or under grass tussock, etc. Eggs: 2–5, whitish to pale greenish blue, marked with browns (WFVZ).

SS: See Vesper and Baird's sparrows. Song Sparrow larger and longer-tailed, darker overall with grey supercilium.

SD: C to F resident (around SL) on Pacific coast of Baja, and (*rostratus* group) from Rio Colorado delta to N Sin; in interior (1200–2500 m), often locally, from Chih to cen volcanic belt; reports of breeding in Gro and SW Guatemala (AOU 1983) are unconfirmed. C to F transient and winter visitor (Sep–mid-May; SL–3000 m) virtually throughout N of Isthmus; U to F but local (Oct–Apr) in Yuc Pen and Belize, R to Honduras and El Salvador. In winter, *rostratus* group disperses to both coasts of Baja.

RA: Breeds N.A. to cen Mexico; N populations migratory, wintering to Caribbean and N C.A.

NB1: Traditionally placed in the monotypic genus *Passerculus*. **NB2:** *Rostratus* group has been considered specifically distinct, Large-billed Sparrow.

SIERRA MADRE SPARROW
Ammodramus baileyi*   Plate 64
Gorrión Serrano

ID: 4.7–5 in (12–12.5 cm). *Very local in cen Mexico*, resembles a small Song Sparrow. **Descr. Adult:** bill grey, legs flesh. *Grey supercilium contrasts with blackish-brown sides of crown, median crown streaked grey* (obscure when worn). *Greyish auricu-*

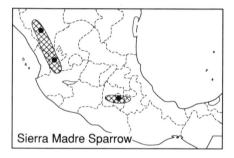

Sierra Madre Sparrow

lars bordered by blackish postocular stripe and moustache. Throat and underparts whitish with black malar stripe, black streaks on chest and flanks often merge to form black central chest spot; chest, flanks, and undertail coverts washed buff in fresh plumage. Upperparts rufous, streaked blackish, greyish to buff edgings form neat scaling in fresh plumage; outer scapulars rufous. Wings and tail dark brown, edged rufous on tertials, secondaries, and wing coverts, tail and primaries edged sandy grey-brown. **Juv:** duller overall, supercilium dirty greyish buff, washed lemon in lores, throat and underparts washed dirty buff, dark brown chest spots diffuse. Bill dull yellow-orange with dark culmen.

Voice. A nasal *nyew* or *cheh*, suggesting Song Sparrow, and a high, thin *ssi*. Song a varied arrangement of nasal or plaintive to slightly burry chips, often including a distinctive burry buzz, at times with a rapid series of 4–5 chips at or near start. Typically 7–12 chips overall, less often a simpler series of 4–6 chips: *chie tsi si-sii-r zzzzzr sir chi-i'*, or *tchiu-sir si si si-chu si-chu sier*, or *chi-chi-chi-chi-chi zzzzzzr si-ch' hu*, or *whie zzzzzzr whie-tu*, etc.

Habs. Bunch grass in open pine woodland. In Dgo reported from dried-up marshes in open pine woods (Bailey and Conover 1935), possibly because marsh grasses were taller than adjacent habitat. Nest cup of grass, rootlets, etc., low in clump of bunch grass. Eggs: 3, pale greenish blue, finely flecked blackish and dark brown (Dickerman *et al.* 1967).

SS: Song Sparrow (mainly marshes in cen Mexico) larger and longer-tailed upperparts less rufous, underparts duller with darker flanks.

SD/RA: F to U but local resident (around 3000 m) in S DF and N Mor; also known from Jal (Sierra de Bolaños) and Dgo

(2200–2700 m), but no records from these areas since 1951.

NB: Traditionally placed in the monotypic genus *Xenospiza*, we consider this species better placed in *Ammodramus* although in some ways, such as voice, it appears more like *Melospiza*.

GENUS MELOSPIZA

These are medium-sized sparrows of marshes and brush. Tails relatively long and squared to slightly rounded. When flushed, typically fly low with distinctive 'pumping' tail action. Bills greyish to flesh-grey, darker above, legs flesh to dark flesh.

SONG SPARROW

*Melospiza melodia**   Not illustrated
Gorrión Cantor

ID: 5.7–6.3 in (14.5–16 cm). *Widespread in marshes S to cen Mexico, numerous ssp.* **Descr. Adult:** *grey supercilium contrasts with rufous-brown and black sides of crown and dark postocular stripe;* median crown stripe (indistinct or lacking in cen volcanic belt) and auriculars grey to grey-brown, auriculars often bordered below by dark moustache. *Throat and underparts whitish with blackish malar stripe, coarse blackish-brown streaks (heaviest in cen volcanic belt) on chest and flanks often merge to form central chest spot;* flanks and undertail coverts buff to buffy brown. Upperparts grey-brown to rufous-brown, streaked blackish to dark brown. *Wings and tail dark brown, edged rufous to pale brown, brightest on greater coverts,* tertials, and secondaries. **Juv:** less contrastingly marked, upperparts brighter and paler, chest washed buff. *Rivularis* **group** (deserts of BCS and Rio Colorado): *much paler overall,* head with rufous sides of crown, postocular stripe, and moustache; sandy upperparts streaked rufous. Throat and underparts whitish, with rufous malar stripe and relatively sparse rufous streaks and triangular spots on chest and flanks.

Voice. A nasal *nyik* or *nyieh* a high *ssi* or *ssir*, a sibilant *ssip*, and a high , lisping, accelerating chatter, *ssi chi-chi-chichichi*, etc. Song a varied arrangement of chips and trills, usually begins with 1–3 separate notes and runs into a buzzy trill, at times ending more slowly: *bree bree si-chi chi-chi-chiiii*, or *reeh, reeh reeh chriiiir chi chi chi-chirr*, etc.

Habs. Breeds in freshwater marshes, riparian washes, less often in drier, semi-open brushy areas. In winter also weedy field borders, brushy second growth, usually near water. Singly or in pairs, often fairly skulking but sings from tops of bushes, reeds, etc. Nest cup of grasses, leaves, rootlets, etc., low in or under bush or grass tussock. Eggs: 2–5, pale blue to greenish white, heavily marked with reddish browns and greys (WFVZ).

SS: Lincoln's Sparrow smaller with neater, finer markings, buff wash across chest, smacking call. See Savannah and Sierra Madre sparrows.

SD: C to F local resident (SL–2500 m) in Baja, from Rio Colorado delta to cen Son, and in interior from Chih to Dgo and in cen volcanic belt. U to R winter visitor (Oct–Apr; near SL–2000 m) to S BCS, and in N Mexico from Son to NL (AMS); reports S to Nay and SLP (Am. Birds Christmas Counts) are not credible.

RA: Breeds N.A. to cen Mexico; N populations withdraw S in winter.

LINCOLN'S SPARROW

*Melospiza lincolnii**   Not illustrated
Gorrión de Lincoln

ID: 5.3–5.8 in (13.5–14.5 cm). *Widespread in winter.* **Descr.** Grey supercilium contrasts with rufous-and-black sides of crown and dark eyestripe; median crown stripe and auriculars grey, auriculars often bordered below by dark moustache. *Throat whitish with dark malar stripe and buff-washed submoustachial stripe. Chest, flanks, and undertail coverts finely streaked blackish on buff to ochraceous-buff wash,* belly whitish. Upperparts grey-brown, streaked blackish. Wings and tail

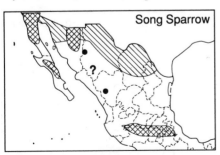
Song Sparrow

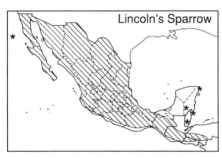
Lincoln's Sparrow

dark brown, edged rufous-brown to pale brown, brightest on greater coverts, tertials, and secondaries.

Voice. A smacking *tchik* or *tsk*, and a high, lisping *tssi*.

Habs. Weedy fields, brushy second growth and forest edge, marshes. Singly or, in migration, in loose flocks up to 20+ birds, often skulking.

SS: See Song Sparrow.

SD: C to F transient and winter visitor (Oct–early May; SL–3000 m) virtually throughout N of Isthmus, F to U in interior (above 600 m) to Honduras; R transient (Oct, Mar) in Yuc Pen and Belize.

RA: Breeds N and W N.A.; winters S USA to Panama.

SWAMP SPARROW
Melospiza georgiana   Not illustrated
Gorrión Pantanero

ID: 5.3–5.8 in (13.5–14.5 cm). Winters in marshes to cen Mexico. **Descr.** Seasonal variation. **Basic:** *grey to buffy-grey supercilium contrasts with rufous-and-black sides of crown* and dark eyestripe; median crown stripe and auriculars greyish to grey-brown, auriculars often bordered below with indistinct dark moustache. *Whitish throat with dark malar stripe contrasts with dusky-grey to greyish-buff chest*, chest often indistinctly streaked

dusky; flanks and undertail coverts buff to buffy brown, belly whitish. Sides of neck grey, upperparts rufous-brown to olive-brown, streaked blackish. *Wings and tail dark brown, edged rufous, brightest on greater coverts and secondaries*, tertials edged whitish. **Alt** (Mar–Apr): crown rufous with blackish forehead and grey median forehead streak, face cleaner grey.

Voice. A sharp *peek*, suggests Eastern Phoebe.

Habs. Freshwater marshes, especially with reeds (*Phragmites*) and adjacent bushes and scrub; usually skulking.

SS: White-throated Sparrow larger and bulkier with yellow supraloral spot, whitish or buff supercilium, whitish wingbars.

SD: F to U winter visitor (Oct–Apr; SL–1800 m), mostly in interior and on Atlantic Slope, S to Nay (PP, DEW), N Gro (AMNH spec.), and cen Ver (SNGH). R and irregular W to Rio Colorado delta (SNGH, PP).

RA: Breeds E N.A.; winters E and S USA to cen Mexico.

FOX SPARROW
Passerella iliaca*   Not illustrated
Gorrión Rascador

ID: 6.7–7.3 in (17–18.5 cm). *BCN in winter.* A large sturdy sparrow with a fairly stout bill, stout legs, and long claws. Two groups: *unalaschcensis* and *schistacea*. **Descr. *Unalaschcensis*:** bill dark grey, orangish below at base. *Head and upperparts dark brownish* with indistinct whitish eyering and flecking in face; wing edgings, uppertail coverts, and *tail brighter, rufous* to rufous-brown. *Throat and underparts whitish, marked with coarse triangular dark brown spots*, heaviest on chest and flanks. **Schistacea:** bill grey, paler below. Head and back grey to brownish-grey.

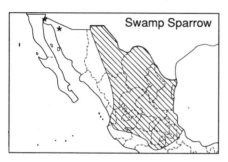
Swamp Sparrow

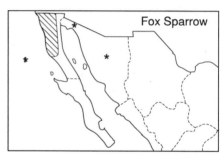
Fox Sparrow

Voice. *Unalaschcensis*: a hard, smacking *tchk!* or *tssk*, deeper than Lincoln's Sparrow, at times run into a rapid chatter. *Schistacea*: a metallic *chik*, suggesting California Towhee.

Habs. Brushy understory of woodland, chaparral, thickets. Fairly skulking, scratches in leaf litter with long claws. Usually solitary but may associate loosely with other sparrows.

SD: Both groups: U to F winter visitor (late Oct–early Apr; near SL–2700 m) in BCN. Vagrant (Dec–Jan) to N Son (ssp. unknown?).

RA: Breeds N and W N.A.; winters USA to N Baja.

GENUS ZONOTRICHIA

These are fairly large, bulky sparrows with longish squared tails and sturdy legs. Head patterns important for species recognition. Bills greyish to orange, legs flesh. Ages differ, sexes similar; migrants lose juv plumage before arriving in region.

RUFOUS-COLLARED SPARROW

Zonotrichia capensis septentrionalis   Plate 63
Gorrión Chingolo

ID: 5.5–6 in (14–15 cm). *Highlands S of Isthmus.* **Descr.** Ages differ, sexes similar. **Adult:** bill dark above, horn to grey below. *Face grey with black sides of crown, eyestripe, and moustache,* median crown stripe grey. *Rufous collar extends from sides of neck around nape. Throat white with blackish patch on sides of chest,* underparts pale grey becoming whitish on belly, washed buffy brown on flanks. Upperparts brown, streaked blackish on back. Wings and tail dark brown, edged pale brown to rufous, with 2 white wingbars. **Juv:** head pattern subdued, brown and greyish, rufous collar duller and streaked blackish. Throat and underparts whitish, streaked and spotted dark brown on chest and flanks.

Voice. A metallic *chink* or *chik* suggesting White-crowned Sparrow. Song a notably varied arrangement of plaintive, often slightly slurred whistles, *swee swee sweeer'* or *tseeoo tchoo*, or *pee-pew-seeer*, etc.

Habs. Open and semiopen areas from towns and villages to pine woods and grassy clearings with nearby brush. Singly or in loose flocks, feeding on or near ground; sings conspicuously from buildings, bushes, etc. Nest cup of grasses, rootlets, etc., in niche in wall, bank, on ground, low in bush, etc. Eggs: 2–3, pale greenish blue, heavily speckled with reddish browns and greys (WFVZ).

SD: C to F resident (1000–3500 m) in interior and on adjacent slopes from Chis to Honduras.

RA: S Mexico to Tierra del Fuego.

WHITE-THROATED SPARROW

Zonotrichia albicollis   Not illustrated
Gorrión Gorjiblanco

ID: 6–6.5 in (15–16.5 cm). Winter visitor to N Mexico. **Descr.** Bill greyish, often flesh below. *Whitish to pale buff supercilium with yellow supraloral spot* contrasts with rufous-and-blackish mottled sides of crown and dark eyestripe; median crown stripe greyish. Greyish auriculars often bordered below by dark moustache. *Whitish throat with dark malar stripe contrasts with dusky greyish chest,* chest often streaked brownish; belly whitish, flanks and undertail coverts washed buffy brown. Brown upperparts streaked blackish and rufous on back. *Wings dark brown, edged brown to rufous, especially bright on tertials and greater coverts,* with 2 whitish to buffy-white wingbars. Tail dark brown.

Voice. A slightly shrill, metallic *chink* or *chihk* and a high, slurred *tseep* or *tsiir*.

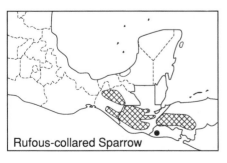

Rufous-collared Sparrow

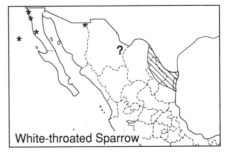

White-throated Sparrow

Habs. Brush and woodland. On or near ground close to cover, often with flocks of other sparrows but often in less open situations than White-crowned Sparrow.

SS: White-crowned Sparrow lacks contrasting whitish throat and yellow supraloral spot, tertials not bright rufous. See Swamp Sparrow.

SD: U to R winter visitor (Oct–Apr; SL–1250 m) on both slopes, recorded BCN and Son, and E NL to Tamps.

RA: Breeds N N.A.; winters S and E USA to N Mexico.

GOLDEN-CROWNED SPARROW

Zonotrichia atricapilla   Not illustrated
Gorrión Coronidorado

ID: 6.5–7.2 in (16.5–18 cm). Winters NW Mexico. **Descr. Adult:** bill dull flesh with dark culmen. *Yellow forecrown contrasts with broad black sides of crown, median hindcrown and face greyish.* Throat and underparts grey, becoming whitish on belly, washed brown on flanks and undertail coverts. Upperparts grey-brown, streaked blackish on back. Wings dark brown, edged pale brown and rufous, brightest on inner coverts and tertials, with 2 white wingbars. Tail dark brown. **Imm:** *crown brown, streaked darker, often with darker sides and yellowish patch on forecrown.*

Voice. A sharp, bright *chik* or *chiik*, flatter and less metallic than White-crowned Sparrow; also a high *tsii* or *tsi*.

Habs. Brush and woodland. Habits much as White-throated Sparrow but also forms flocks up to 20+ birds.

SS: Imm White-crowned Sparrow has dark brown sides of crown and distinct buffy-brown median crown stripe, pinkish to yellowish bill.

SD: F to U winter visitor (mid-Oct–Apr; SL–2700 m) in NW BCN, R and irregular to BCS and (Nov–Jan) to N Son. A report from Nay (AOU 1983) requires verification.

RA: Breeds NW N.A.; winters W N.A. to NW Mexico.

WHITE-CROWNED SPARROW

Zonotrichia leucophrys*   Not illustrated
Gorrión Coroniblanco

ID: 6.2–7 in (16–18 cm). Winters N and cen Mexico. **Descr. Adult:** *bill pink to orange-yellow. Broad white median crown stripe bordered by black sides of crown and black forehead, face grey with white supercilium and black postocular stripe;* black sides of crown touch front of eye in *oriantha* and *leucophrys*, supercilium unbroken in *gambelii*. Throat and chest grey, becoming whitish on belly, flanks and undertail coverts washed brown. Upperparts grey-brown, streaked dark brown on back. Wings dark brown, edged pale brown and rufous, brightest on inner coverts and tertials, with 2 white wingbars. Tail dark brown. **Imm** (Oct–Apr): *sides of crown chestnut-brown with pale buffy-brown median stripe, supercilium washed buff,* face and chest washed brownish.

Voice. A metallic *chink* or *tchik*, and a high, thin *tsir*. Winter flocks often give partial or full songs consisting of 1–2 plaintive whistles followed by a twittering warble.

Habs. Brush and woodland edge, often in more open situations than other *Zonotrichia*. Habits much as White-throated Sparrow but often in flocks, up to 100+ birds at times.

SS: See Golden-crowned and White-throated sparrows.

SD: C to F transient and winter visitor (Sep–May; SL–2500 m) in Baja, on Pacific Slope S to Nay, over Plateau S to N Mich, and on Atlantic Slope in Tamps. Irregularly R to U (Oct–Mar) in Yuc Pen (G and JC, SNGH, BM); vagrant to Belize (Oct 1988, Howell *et al.* 1992*b*.)

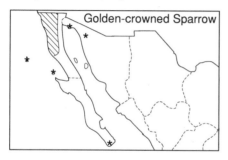
Golden-crowned Sparrow

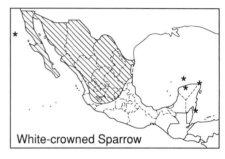
White-crowned Sparrow

RA: Breeds N and W N.A.; winters N.A. to cen Mexico.

NB: Harris' Sparrow Z. querula (6.8–7.3 in, 17–18.5 cm) may occur in N Mexico in winter. Adult basic (Nov–Apr) has pinkish bill, tawny-buff face, crown mottled black and grey, throat and underparts whitish with black mottling (may be solidly black) on throat and chest. Imm (Nov–Apr) similar but typically throat white with blackish malar stripe, inverted triangle of blackish mottling on median upper chest. Alt (Apr) has crown, lores, throat, and upper chest solidly black, contrasting with grey to greyish-tawny face.

GENUS JUNCO: Juncos

Juncos are smartly marked sparrows, typically with bright pinkish bills and striking white outer rectrices; legs flesh. From two to eight species have been recognized in Mexico, and species limits in the genus have yet to be delineated satisfactorily. Ages differ, sexes usually slightly different in dark-eyed forms, similar in yellow-eyed forms; streaked juvenal plumage lost by migrants before arriving in region.

DARK-EYED JUNCO
Junco hyemalis *   Not illustrated
Junco Ojioscuro

ID: 5.5–6 in (14–15 cm). N Mexico, mainly in winter. Four groups (see SD below): *oreganus* group (Oregon Junco), *mearnsi* group (Pink-sided Junco), *hyemalis* group (Slate-colored Junco), and *caniceps* group (Grey-headed Junco). **Descr.** Sexes similar or different. *Oreganus.* ♂: bill pinkish. *Head and chest blackish to blackish grey, sides and flanks vinaceous*, belly and undertail coverts white. *Back rufous-brown*, rump grey. Wings and tail black-

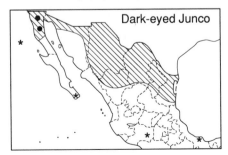

Dark-eyed Junco

ish, greater coverts edged grey-brown to rufous-brown, *mostly white outer rectrices flash in flight.* ♀: head and chest grey with blackish lores, back duller, crown often washed brown when fresh. *Mearnsi:* resembles ♀ *oreganus* but *head and chest typically a purer bluer grey.* **Hyemalis** ♂: *head, chest, flanks, and upperparts slaty grey.* ♀: duller, *crown, nape, and upperparts washed brownish, sides and flanks washed pinkish.* Intergrades with Oregon Junco show mixed characters. *Caniceps: grey overall with chestnut back*, black lores, and whitish belly and undertail coverts; *dorsalis* (may occur in N Son and N Chih) has bicolored bill, blackish above, blue-grey below. **Juv:** head and upperparts duller, streaked blackish. Throat and underparts dirty whitish, chest and brown-washed flanks streaked dusky.

Voice. Song a ringing, chipping trill *ji-ji-ji-ji-ji-ji-ji-ji-ji*, or *ji-i-i-i-i-i-i-i-i-i*, etc. Pink-sided (*townsendi*) gives a thin sharp *siik*, migrant Pink-sideds give a harder *tchehk* or *tchak*, Oregon and Slate-colored give a smacking *tsik* or *tik*, Grey-headed give a harder *tchik* or *tsik*; all make high, sharp twitters.

Habs. Breeds in arid to semiarid pine woodland and pine–oak–chaparral; in winter also semiopen brush and woodland, grassy areas. In winter often in flocks when several ssp may occur together; feeds on ground, flies into trees and bushes when flushed. Sings from prominent perches, often high in trees. Nest cup of grasses, rootlets, etc., in shallow depression with overhead protection, rarely low in bushes. Eggs: 3–5, bluish white, speckled with reddish browns and greys (WFVZ).

SS: Mexican (Yellow-eyed) Junco told from Grey-headed by yellow eyes, greater coverts and tertials broadly edged rufous, throat and underparts paler than head.

SD: *Oreganus:* C to F breeder (1800–2500 m) in Sierra Juarez, wandering to lower elevations in winter. C to F winter visitor (Oct–Apr; near SL–2500 m) in BCN, N Son and Chih; R and irregular to BCS and NL (DEW). *Mearnsi:* C to F breeder (1800–2900 m) in Sierra San Pedro Martír, BCN. F in winter (Oct–Apr) to N Son and Chih. *Hyemalis:* irregular R to U winter visitor (Oct–Mar; near SL–2000 m) in N Mexico from BCN to Tamps, vagrant to S

Ver (Nov 1973, Winker *et al.* 1992*b*). *Caniceps*: F to C winter visitor (Oct–Apr; 1000–2500 m) in interior from Son and Chih to N Sin and Dgo; vagrant S to Mich (Mar 1993, SNGH, RGW)

RA: Breeds N N.A. to NW Mexico; winters to N Mexico.

NB: The four groups have been considered separate species.

GUADALUPE [DARK-EYED] JUNCO
Junco insularis   Plate 69
Junco de Guadalupe

ID: *5.5–6 in (14–15 cm). Endemic to Isla Guadalupe, endangered. Bill relatively long and slender.* **Descr.** Sexes similar. **Adult:** *bill blue-grey. Head and chest grey* with blackish lores, *sides and flanks vinaceous*, median underparts white. *Back dull brownish* becoming greyish on rump. Wings and tail blackish, wings edged grey-brown, brighter on tertials. Mostly white outer rectrices flash in flight. **Juv:** head and upperparts duller, streaked blackish. Throat and underparts dirty whitish, chest and brown-washed flanks streaked dusky. **Voice.** A high, slightly metallic *sik*. Song a varied series of chips, often including a buzz or trill: *sui si swii si-si-si-i-i iu,* or *swi si si-i-i-i-i-iu,* or *ti-si-wee' swee-i'chu,* or an accelerating *si sit-i wi si-chi-chi-chi-chi,* etc. **Habs.** Mostly confined to remnant cypress and pine–oak groves at N end of island, also a few pairs in *Nicotiana* scrub near northeast anchorage. Gathers in flocks in winter, breeds Feb–Jun. Nest cup of grasses etc., in shallow depression in ground or low in tree. Eggs: 3–4, greenish white, speckled with reddish browns. **SS:** Dark-eyed (Oregon) Junco occasionally reaches Isla Guadalupe in winter; note pinkish bill, brighter upperparts, call. **SD/RA:** Formerly C resident on Isla Guada-

lupe, BCN. Population in Feb 1988 estimated at 50–100 birds (SNGH, SW), but rampant overgrazing by goats threatens to eradicate all trees, and hence juncos, from the island.

NB: Traditionally considered conspecific with Dark-eyed Junco.

BAIRD'S [YELLOW-EYED] JUNCO
Junco bairdi   Plate 63
Junco de Baird

ID: *5.5–6 in (14–15 cm). Endemic to Cape mountains of BCS.* **Descr. Adult:** *eyes golden yellow, bill flesh to yellowish flesh with dusky culmen. Head blue-grey with blackish lores,* throat and median underparts pale grey becoming white on undertail coverts; *sides and flanks pale cinnamon-vinaceous. Back brownish cinnamon,* rump and uppertail coverts olive-sandy. Wings and tail blackish, *inner greater coverts and tertials broadly edged brownish cinnamon,* mostly white outer rectrices flash in flight. **Juv:** eyes dusky, quickly becoming yellow. Brownish head and upperparts streaked blackish. Throat and underparts dirty whitish, chest washed buff becoming cinnamon on flanks, chest and flanks streaked dusky. **Voice.** A hard, smacking *tsk* or *tsik* and *tsk-tsk!* which, in interactions, may run into squeaky to shrill querulous chatters; also a high sibilant *tssi* or *ssip*. Song a pleasant, tinkling, and trilled warble, typically of 1.5–2 s duration, may suggest a small *Troglodytes* wren and strikingly different from mainland Yellow-eyed Juncos. **Habs.** Arid to semiarid oak and pine–oak woodland. In pairs or flocks, feeds on ground and flies up to trees when disturbed. Sings from mid- to upper levels in trees, at times also on ground. Nest cup of grasses, rootlets, etc., in shallow

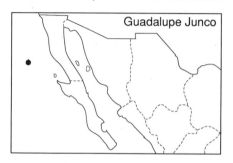
Guadalupe Junco

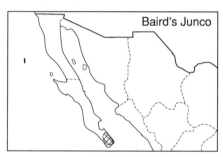
Baird's Junco

depression in ground or low in tree. Eggs: 2, whitish, speckled with reddish browns.
SS: See Dark-eyed (Oregon) Junco (R in BCS).
SD/RA: C to F resident (1200–1900 m, rarely lower in winter) in Sierra Victoria, BCS.
NB: Traditionally considered conspecific with Yellow-eyed Junco.

YELLOW-EYED JUNCO
Junco phaeonotus *   Plate 63
Junco Ojilumbre

ID: 5.7–6.7 in (14.5–17 cm). *Highlands S to Guatemala*, three groups (see SD below): *phaeonotus* group (Mexican Junco); *fulvescens* (Chiapas Junco); and *alticola* (Guatemalan Junco). **Descr. Adult. Phaeonotus:** *eyes yellow, bill blackish above, yellowish to flesh below. Head grey with black lores, throat and underparts pale grey becoming whitish on belly and undertail coverts. Rufous back* edged greyish olive when fresh, rump and upper-tail coverts greyish. Wings and tail black-ish, *inner greater coverts and tertials broadly edged rufous,* mostly white outer rectrices flash in flight. **Fulvescens:** *larger and darker* overall, rump and uppertail coverts olive-brown, underparts grey with buffy-brown flanks, often less white in outer rectrices. **Alticola:** larger (6.2–6.7 in; 15.5–17 cm) and darker. Bill yellowish be-low. *Throat and underparts dusky grey, washed buffy brown on flanks. Back dull olivaceous rufous-brown,* rump and uppertail coverts olive-brown. Wings and tail dark brown, inner greater coverts and tertials edged dark rufous, *white in outer rectrices reduced and not striking.* **Juv** (all groups): eyes dusky. Head and upperparts streaked dusky; throat and underparts paler, chest and flanks streaked dusky.
Voice. *Phaeonotus:* a high, sharp, often smacking *tsik* or *sik,* and high, sharp twit-

tering. Song a varied series of bright chips, often with trills or buzzes thrown in, typi-cally the last note rising or up-slurred: *swi swi see-i-ewi',* or *si-si-si-si-si-si-si-ssiu,* or *chiwee chiwee chiwee chiiiiiii chiwee,* or *chee-chee zzhi-zzhi zzhi-zzhi,* or *we-chu we-chu chi-irrrr we-shu,* etc.; can be confused with songs of Rufous-sided Towhee and Bewick's Wren. *Alticola* has sharp, slightly buzzy or metallic *tchik.* Song a varied series of bright chips, usually the last note rising: *ch-wee' ch-wee' ch-wee' ch-wee' ch-wee'chu,* or *ch-wee ch-wee ssi ssi si chu,* etc., much like *phaeonotus.*
Habs. Pine, pine–oak, and at times oak woodland with grassy clearings, brush. Feeds on ground and flies up into trees and bushes when flushed. Forms flocks in winter. Sings from prominent perches, often high in trees. Nest cup of grasses, moss, rootlets, etc., in shallow depression with overhead protection, rarely low in bushes. Eggs: 2–4, pale bluish, unmarked or with sparse reddish-brown and grey speckling (WFVZ).
SS: See Dark-eyed (Grey-headed) Junco.
SD/RA: *Phaeonotus:* C to F resident (1200–3500 m) in interior and on adjacent slopes from Son (also SE Arizona/SW New Mexico) and Coah to Oax. *Fulvescens:* C to F resident (2100–3500 m) in N Chis. *Alticola:* C to F resident (2000–3500 m) in Guatemala and adjacent SE Chis.
NB: See NB under Baird's Junco.

GENUS CALCARIUS: Longspurs

Longspurs are terrestrial, sparrow-like birds of open grassland. They feed low to the ground, run well, and rarely perch on bushes. Long primary projections quite different from typical sparrows, hindclaws long like pipits. Sexes differ, juvs attain adult-like plumage before migration, seasonal plumage variation marked. Bills in winter dusky flesh to greyish, legs dark flesh. Tail patterns and voice are best identification points.

McCOWN'S LONGSPUR
Calcarius mccownii   Not illustrated
Escribano de McCown

ID: 5.7–6.2 in (14.5–15.5 cm). N Mexico in winter. *Bill relatively deep and stout, tail relatively short.* **Descr. Basic ♂:** *dull pale buff supercilium and lores contrast slightly*

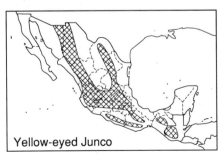

Yellow-eyed Junco

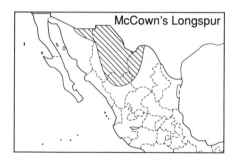

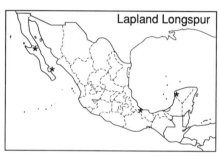

with buffy-brown auriculars and dark-streaked brown crown; may show darker moustache. *Throat and underparts dirty whitish, black chest band heavily veiled whitish to pale buff but usually noticeable,* sides and flanks washed buff. Upperparts sandy brown to grey-brown, streaked dark brown; wings dark brown, edged pale brown, with rufous-centered median coverts often concealed by scapulars. *Central rectrices blackish brown, rest of tail white, tipped dark brown,* white in spread tail striking. **Alt** ♂ (Mar–Apr): bill black. Face whiter with black crown, thick black moustache, broad black chest band. **Basic** ♀: *resembles ♂ basic but chest washed buff without veiled black band, rufous median coverts duller and paler.* **Alt** ♀: face more contrasting with hint of ♂ pattern.

Voice. A dry rattled *dri-ri-rit* and a single *chee*; flocks make a social twittering.

Habs. Open short-grass fields, dried lake beds, often more barren areas than other longspurs. Singly or in small flocks, ones and twos may join flocks of Horned Larks or Chestnut-collared Longspurs but usually keep apart.

SS: Chestnut-collared Longspur often in large flocks; note nasal chattering calls, white lesser coverts and tail pattern distinctive if visible, ♂ has buff face with dark post-ocular stripe, ♀ very plain, underparts dirty buff with indistinct dusky streaking.

SD: U to F winter visitor (Nov–Mar; 1000–2500 m), mostly on Plateau, from N Son and Chih to N Dgo.

RA: Breeds W interior N.A.; winters SW USA to N Mexico.

LAPLAND LONGSPUR
Calcarius lapponicus   Not illustrated
Escribano Ártico

ID: 6.2–6.7 in (15.5–17 cm). *Vagrant.*

Descr. Basic ♂: *buff supercilium and lores contrast with brown-streaked blackish sides of crown; buff auriculars bordered black, thickest at rear. Chestnut nape veiled with greyish buff. Throat and underparts whitish* to pale buff, *mottled black on throat and upper chest,* buff flanks streaked blackish. Upperparts sandy brown to grey-brown, streaked blackish brown. Wings dark brown, edged pale brown, with *rufous edgings to greater coverts and tertials.* Tail blackish brown, *outermost rectrices mostly white but not striking in spread tail.* **Alt** ♂ (Mar–Apr): bill horn-yellow. Face, throat, and upper chest black, bordered white, crown black, nape chestnut. **Basic** ♀: *resembles ♂ basic but rufous nape often completely veiled with greyish when fresh, black mottling mainly in malar area and on central chest.* Greater covert and tertial edgings paler and duller, cinnamon. **Alt** ♀: nape rufous, throat and chest more extensively mottled black.

Voice. A dry, rattled *dri-ri-it* and a slightly plaintive *teu*, given singly or often following the rattle call.

Habs. Grasslands, open areas. Singly or in small groups, may associate with other longspurs.

SS: Dark-bordered auriculars, dark mottling on malar area and chest, rufous edged greater coverts, and white restricted to outer rectrices distinguish Lapland from other longspurs; note call.

SD: Vagrant (Nov–Feb) to Baja (Banks 1962; Wilbur 1987), NW Yuc (Lee 1978), and cen Ver (Feb 1985, SNGH, PP, SW).

RA: Holarctic breeder; winters S in New World to cen, rarely S, USA.

CHESTNUT-COLLARED LONGSPUR
Calcarius ornatus   Not illustrated
Escribano Cuellicastaño

ID: 5.3–5.8 in (13.5–15 cm). Winters N Mexico. **Descr. Basic ♂:** *face and throat pale buff with dark postocular stripe curving down around auriculars*; crown streaked blackish and brown; *chestnut nape veiled with greyish buff. Underparts blackish, heavily veiled buff*, undertail coverts whitish. Upperparts sandy brown to grey-brown, streaked dark brown. Wings dark brown, edged pale brown, *broad white edgings to lesser coverts* often concealed by scapulars. *Central rectrices dark brown, rest of tail mostly white with dark tips broadest on inner rectrices.* **Alt ♂** (Mar–Apr): bill blue-grey with dark culmen. Face creamy white with black postocular stripe, black crown, and chestnut nape; underparts black with white undertail

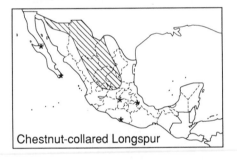

Chestnut-collared Longspur

coverts. **Basic ♀:** *face plain buff with paler supercilium, crown streaked dark brown and buff; throat paler buff, often with dusky malar stripe. Underparts dirty pale buff, indistinctly streaked dusky*, becoming paler on undertail coverts. Upperparts similar to ♂ basic but white lesser-covert edgings narrower, often hard to see. **Alt ♀:** much as basic but underparts often with dark patches, inconspicuous chestnut patch at base of nape.

Voice. A nasal *cheedl* or *chittl*, and *cheedl-ee-l* or *chittl-it*, distinct from other longspurs; flocks make social, buzzy, twittering calls.

Habs. Open grassland, lake edges, often in areas with longer grass than McCown's Longspur. Usually in flocks, locally to 100+ birds, may associate loosely with Horned Larks and American Pipits.

SS: See McCown's and Lapland longspurs. Sparrows have short primary projections, more upright and usually hopping gait, all except Vesper lack white in tail.

SD: C to F winter visitor (Oct–early Apr; 1000–2500 m) in N Son, and on Plateau from Chih and Coah (SNGH, SW) to Zac and Ags (SNGH, SW), R and irregular to cen volcanic belt, and (Oct–Jan; around SL) to Baja (Wilbur 1987). Vagrant to Gro (Mar 1990, Howell and Wilson 1990).

RA: Breeds W interior N.A.; winters SW USA to cen Mexico.

734

AMERICAN BLACKBIRDS AND ORIOLES: ICTERIDAE (32*)

Icterids are here treated as a separate family although some authors (such as AOU 1983) merge this group into the nine-primaried emberizid assemblage. Blackbirds and orioles comprise a diverse group characterized by stout, fairly long, and sharply pointed bills, and strong legs and feet. Wings often fairly long and pointed, tails generally long, squared to graduated. Ages differ; young altricial; 1st prebasic molt partial to complete. Sexes similar in most tropical species, different in temperate species; ♂♂ larger than ♀♀, often strikingly so. Plumage varies from black with various sheens in most blackbirds, to brightly and strikingly colored in orioles. Most species are conspicuous and easily seen. In winter, temperate-breeding, migratory blackbirds often form large flocks in which sexes typically segregate almost completely. Blackbirds, grackles, and cowbirds are mostly terrestrial and walk with their tails slightly cocked, whereas orioles, caciques, and oropendolas are mostly arboreal. Songs and calls usually loud and distinctive.

Blackbirds eat mostly grains and insects, orioles eat mostly fruit and take nectar. Some species nest colonially. Nests vary from bulky cups to neatly woven, often conspicuous pensile structures (Fig. 43). Eggs: 2–6, whitish to pale blue, typically marked with browns and black.

BOBOLINK

Dolichonyx oryzivorus   Not illustrated
Tordo Arrocero

ID: 6–7 in (15–17.5 cm). Transient on Caribbean coasts. *Tail spiky-tipped.* **Descr.** Age/sex/seasonal variation. **Alt ♂** (Apr–Jul): bill black, legs dark flesh. *Head and underparts black with golden-buff hindneck.* Upperparts black with *whitish scapulars*, rump, and uppertail coverts, buff striping on back, and buff tertial edgings. In spring, black often veiled with buff. ♀/imm/basic ♂ (Aug–Dec): bill flesh

Bobolink

with dark culmen, legs flesh. *Head and underparts buff with dark brown lateral crown stripe and postocular stripe,* and blackish streaking on flanks. *Upperparts buffy brown, streaked blackish;* wings and tail blackish brown, edged buff and brown.
  Voice. Flight call a nasal *inhk* or *eenk;* also a dry chatter.
  Habs. Open and semiopen grassy areas. Singly or in flocks, feeding in weedy areas; flight strong and slightly undulating.
SD: Irregularly U to F spring transient (mid-Apr–early Jun) and R to U autumn transient (Oct–Nov) from N Yuc to Honduras Bay Islands; R to Ver (May 1986, PW). Vagrant (Jun, Oct–Dec) to Baja (CAS spec., GMe photo) and Clipperton.
RA: Breeds N.A.; winters S S.A., migrates mainly across Caribbean.

RED-WINGED BLACKBIRD

*Agelaius phoeniceus**   Plate 65
Tordo Sargento

ID: ♂ 8.5–9.5 in (21.5–24 cm) ♀ 7–7.5 in (18–19 cm). Widespread in marshes and open country. Two groups: *phoeniceus* group (widespread) and *gubernator* ('Bicolored Blackbird', cen Mexico). **Descr.** Ages/sexes differ. **Phoeniceus.** ♂: bill and legs blackish. *Black overall with orange-red to red lesser wing coverts and creamy to pale ochre-yellow median wing coverts* (red often covered by scapulars). Head and body, especially upperparts, narrowly edged cinnamon-brown in fresh plumage. ♀: *head and upperparts dark brown with greyish-buff to buff supercilium and often a median crown stripe, upperparts with chestnut and buff streaking.* Wings and tail darker, wings edged rufous-brown to buff, often with 2 paler buff wingbars. *Throat and underparts whitish to pale buff, heavily streaked dark brown* on underparts, often becoming solidly dark on vent and under-tail coverts; throat and supraloral stripe may be washed pinkish. **Imm** ♂: black overall with extensive rufous-brown to buff edgings on head, body, and wings, red-and-yellow shoulder mixed with black. **Juv:** paler and buffier than ♀, without chestnut edgings on upperparts, under-parts with finer dark streaking. Intergrades with *gubernator* show intermediate characters. *Gubernator.* ♂: orange-red to *red lesser wing coverts lack pale border*, thus often looks all black at rest. ♀: *dark sooty brown overall*, often with trace of paler supercilium and median crown stripe. Most birds show whitish streaking on throat, and chest and chest may be washed pink. **Imm** ♂: resembles ♂ but with some fine buff edgings, especially on wing coverts and tertials, and some dark flecks in red shoulder. **Juv:** resembles ♀ but throat and underparts have more extensive buff streaking, upperparts with narrow buff fringes.

Voice. A low *chek* or *chwek* and a slightly metallic *tink* or *diink* in flight, also a clipped, nasal, crake-like *deek* or *biehk*. *Phoeniceus* song a slightly metallic, warbled, strangled gurgle. *Gubernator* song shorter and simpler, a strangled *chuh-krrrir* or *kyu-rreihr*, lacking warbled quality. Intergrades sing intermediate songs.

Habs. Freshwater marshes, especially with reeds, and adjacent open land, fields, and other agricultural areas. Nests and roosts colonially, usually in marshes; in Lerma marshes most *gubernator* nest in low sedge marshes versus reedbeds at other sites (Hardy and Dickerman 1965). In flocks, at times to thousands of birds. Nest cup of grasses, fibers, etc., suspended amid marsh grasses or in low bush or tree. Eggs: 2–5, pale blue to pale grey, marked with black dots and scrawls (WFVZ).

SS: *Phoeniceus:* Tricolored Blackbird (NW Baja) slightly longer and more slender-looking, ♂ has red lesser wing coverts and whitish median coverts: in flight at a distance looks black with whitish wingbars (versus black with orange-red shoulders), but some imm ♂♂ perhaps not safely identified. ♀ Tricolored darker than Red-wing, without strong chestnut and buff edgings above, underparts typically more heavily streaked, solidly dark on belly, throat lacks pink wash. *Gubernator:* cowbirds have less sharply pointed bills, lack pale streaking on face and throat. Brewer's Blackbird slimmer and longer-tailed, lacks pale streaking on face and throat, ♂ has yellow eyes.

SD: *Phoeniceus:* C to F resident (SL–2500 m) in BCN and N Son, on Pacific Slope locally from Nay to El Salvador, in interior from E cen volcanic belt to Guatemala, and on Atlantic Slope from Coah and Tamps to Honduras. More widespread in winter (Oct–Mar), especially NW Mexico where N migrants occur. *Gubernator:* C to F resident (1000–2500 m) in interior from Dgo (and S Chih?) to cen volcanic belt (Jal to Pue). The two forms breed sympatrically in several marshes, intergrading readily in Tlax and Pue but rarely in Mex (Hardy and Dickerman 1965).

RA: Breeds N.A. to NW Costa Rica; N N.A. populations withdraw S in winter.

TRICOLORED BLACKBIRD
Agelaius tricolor Not illustrated
Tordo Tricolor

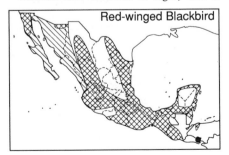

Red-winged Blackbird

ID: ♂ 8.5–9 in (21.5–23 cm), ♀ 7–7.5 in (18–19 cm). NW BCN. Overall more slender than Red-winged Blackbird. **Descr.** Ages/sexes differ. ♂: bill and legs blackish. *Black overall with blue sheen, lesser wing coverts red to dark red, median wing coverts white to creamy.* Head and body have fine buff fringes in fresh plumage. ♀: *head and upperparts dark brown, greyish-buff supercilium usually indistinct, upperparts with grey-brown to greyish-buff edgings and streaking,* lesser wing coverts often edged dull red. Wings and tail darker, wings edged buffy brown, often with 2 narrow paler buff wingbars. *Throat and chest whitish with heavy dark brown streaking becoming solidly dark on belly* and undertail coverts. **Imm** ♂: resembles ♂ but with more extensive buff edgings to head and body, lesser wing coverts orange-red, median coverts cream to creamy buff, flecked blackish. **Juv:** resembles ♀ but brighter buff to whitish streaking on underparts extends onto belly.

Voice. Varied squawking to rough, strangled, and gurgled calls; flight call *chek* or *chwek*, not hard. Song a short, low, gruff, strangled gurgle, rougher than Red-winged Blackbird.

Habs. Freshwater marshes, open agricultural areas. Usually in flocks which may include other blackbirds. Nests in dense colonies in marshes. Nest cup of grasses low in reeds, weeds, bushes, etc. Eggs: 2–5, pale blue to pale grey, marked with black dots and scrawls (WFVZ).

SS: See Red-winged Blackbird (*phoeniceus* group).

SD: Formerly described as a C breeder in marshes of NW BCN (Bryant 1889) but now apparently local (Howell and Webb 1992*f*).

RA: W USA to NW BCN.

YELLOW-HEADED BLACKBIRD
Xanthocephalus xanthocephalus
Tordo Cabeciamarillo   Not illustrated

ID: ♂ 9–10 in (23–25.5 cm), ♀ 7.5–8.5 in (19–21.5 cm). Mainly in winter, N of Isthmus. **Descr.** Ages/sexes differ. ♂: bill and legs dark greyish. *Black overall with golden-yellow head and chest* and black loral mask. *White patch on outer greater wing coverts and primary coverts striking in flight.* ♀: *dark brown overall with yellow supercilium, throat, and chest;* throat often paler with dark malar stripe, belly mottled whitish. **Imm** ♂: resembles ♀ but brighter yellow on throat and face, with darker loral mask, lacks white mottling on belly. White tips to outer greater wing coverts and primary coverts hint at ♂ wing pattern. **Juv:** head and chest cinnamon to ochraceous buff with dusky auriculars and paler throat, rest of underparts pale buff, washed cinnamon on flanks. Upperparts dark brown with dull cinnamon edgings to body feathers.

Voice. A gruff *krrek* and *chrruk* or *ch-ruk,* the overall effect of a flock being a gruff chattering; also a slightly whining *cheh-eh eh-heh-heh.* Song a rough, strangled, honking gurgle.

Habs. Nests and roosts in reed-fringed irrigation ditches and freshwater marshes with extensive reedbeds; feeds in open agricultural areas. Flocks can number millions in winter, when sexes often segregate like other blackbirds. Nest cup of grasses in reeds low over water. Eggs: 3–5, greyish to greenish white, finely and densely marked with browns (WFVZ).

SS: Flocks of ♀♀ appear much like other blackbirds at a distance but note gruffer calls, face pattern distinctive when seen.

SD: F to U but local breeder (May–Aug; SL–1000 m) in NE BCN (Howell and Webb 1992*f*) and, at least formerly, in N Tamps

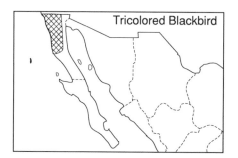

Tricolored Blackbird

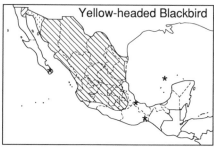

Yellow-headed Blackbird

(RR). C to F transient and winter visitor (late Jul–mid-May; SL–2500 m) on Pacific Slope S to Nay, and in interior and on adjacent slopes S to cen volcanic belt, U to F to N Gro, irregularly (?) to S Oax; U to R in Baja and on Atlantic Slope S to Ver. Vagrant to Yuc (Oct 1984, Howell 1989*a*).

RA: Breeds W N.A. to N Mexico; winters SW USA to Mexico, rarely to NW Costa Rica.

Fig. 42 Eastern Meadowlark (*Sturnella magna lilianae*)

GENUS STURNELLA:
Meadowlarks

Meadowlarks are sturdily built, stout-legged inhabitants of grasslands. Bills moderately long and pointed, blue-grey with blackish culmens, legs flesh. Ages/sexes differ, ♂♂ markedly larger and brighter than ♀♀; juvs soon undergo a complete 1st prebasic molt.

Field separation of the two species is confounded by sexual dimorphism, seasonal variation, and by geographic variation in Eastern Meadowlark. There are several tendencies in appearance but few solid field marks, and voice is often the best clue. In fresh plumage, broad buff tips obscure much of the face and underpart patterning, and feather wear may affect greatly the appearance of the upperparts.

EASTERN MEADOWLARK
*Sturnella magna**   Figure 42
Pradero Común

ID: ♂ 8.5–9.5 in (21.5–24 cm), ♀ 7.5–8.5 in (19–21.5 cm). Widespread. See genus note. **Descr.** ♂: yellow supraloral stripe merges with buff to pale greyish-buff supercilium and contrasts with blackish to brownish sides of crown and postocular stripe; median crown stripe buff. Face buffy grey, *throat and median underparts bright yellow* to orange-yellow (*pectoralis* of W-cen Mexico) *with U-shaped black chest band*, flanks and undertail coverts pale buff, streaked blackish. *Nape and upperparts cryptically streaked and barred with sandy brown to chestnut-brown, blackish, and buff.* Central rectrices barred blackish and brown, *outer 3 pairs of rectrices mostly white*, and often some white on 4th from outermost rectrix. In fresh plumage, buff to whitish edgings veil much of throat and underparts. ♀: generally duller, black chest band narrower. **Juv:** more washed-out

overall, underparts paler and duller yellow with necklace of dark streaks replacing black chest band. *S. m. **lilianae*** (NW Mexico): paler and browner overall with little or no obvious black in upperparts (similar in this respect to *S. m. hoopesi* of arid NE Mexico), and *mostly white outer 4 pairs of rectrices*.

Voice. A distinctive, full, slightly raspy *zzzrt* or *zzzrip*, also a spluttering chatter which sometimes follows on from song; in flight a nasal *sweenk* or *sweink*. Song (including *lilianae*) a plaintive, often slurred, rich, whistled *see seeur seee-u* or *see seeu see-u-seu'* and variations, at times run into a quiet warble, but typically more plaintive and less varied than Western Meadowlark song, without a strong, warbled ending.

Habs. Grasslands and open agricultural

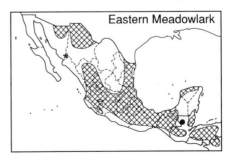

Eastern Meadowlark

areas, often near marshes. Singly or, in winter, often in flocks up to 50+ birds. Walks with high-stepping gait, flight interspersed with stiff-winged glides recalls Starling, tail usually spread on landing. May associate with Western Meadowlark in winter flocks. Nest cup of grasses usually roofed over, on or near ground, in grassy area or crops. Eggs: 3–5, whitish, speckled and spotted with reddish browns and greys (WFVZ).

SS: ♂ Western Meadowlark has yellow of throat extending up into submoustachial region (may be hard to see in fresh plumage), ♀ shows this to lesser extent; sides of crown and postocular stripe usually brownish (versus blackish in most Easterns). Most Westerns have less white in outer rectrices than Easterns (obviously less than *lilianae*); identification safe by voice and extent of yellow in submoustachial region.

SD: C to F but often local resident (SL–2500 m) in N Son and Chih and disjunctly, mostly in interior and on Atlantic Slope, from Dgo and Tamps to Honduras and the Mosquitia. More widespread in winter when N migrants may occur on N Plateau. With extensive clearing of forests, will undoubtedly continue to spread in S Mexico and probably elsewhere.

RA: Breeds E N.A. to Mexico, locally to S.A.; partially migratory in N N.A.

NB: Some authors (Sibley and Monroe 1990) suggest that populations in SW USA and NW Mexico may be specifically distinct, Lilian's Meadowlark (*S. lilianae*).

WESTERN MEADOWLARK
Sturnella neglecta   Not illustrated
Pradero Occidental

ID: ♂ 8.5–9.5 in (21.5–24 cm), ♀ 7.5–8.5 in (19–21.5 cm). NW and cen Mexico. See genus note. **Descr.** ♂: yellow supraloral

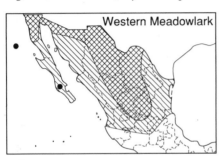

stripe merges with buff to pale greyish-buff supercilium and contrasts with brown sides of crown and postocular stripe; median crown stripe buff. Face buffy grey, *throat (including submoustachial) and median underparts bright yellow with U-shaped black chest band*, flanks and undertail coverts pale buff, streaked blackish. *Nape and upperparts cryptically streaked and barred with sandy brown to rufous-brown, blackish, and buff.* Central rectrices barred blackish and grey-brown, *outer 2–3 pairs of rectrices mostly white*, 4th from outermost rectrix usually has little or no white. In fresh plumage, buff to whitish edgings veil much of throat and underparts. ♀: generally duller, black chest band narrower. **Juv:** more washed-out overall, underparts paler and duller yellow with necklace of dark streaks replacing black chest band.

Voice. A distinctive low *chuk* or *chook*; also a spluttering rattle, a mellow *whiu* or *whew* suggesting Phainopepla and, in flight, a nasal *zweenk* or *chwink*, much like Eastern Meadowlark. Song more complex and warbled than Eastern Meadowlark, quality clear and fluty: *whee hir weedle-e whi-chee* or *chee-oo-e-lee chee-ee le-ee*, etc.

Habs. Much as Eastern Meadowlark, including nest and eggs, but rarely around marshes in interior.

SS: See Eastern Meadowlark.

SD: C to F resident (SL–2500 m) in Baja (irregularly to Isla Guadalupe, Howell and Webb 1992e), N Son, and on Plateau S to N Gto; U to R in winter (Oct–Mar) on Pacific Slope to Nay, in interior to Mich and Mex, and on Atlantic Slope from Tamps to N Ver.

RA: Breeds W and cen N.A. to cen Mexico; N populations partially migratory.

MELODIOUS BLACKBIRD
Dives dives   Plate 65
Tordo Cantor

ID: 9–11.5 in (23–29 cm). E and S tropical lowlands. *Bill fairly stout and pointed.* **Descr.** Ages differ slightly, sexes similar. **Adult:** bill and legs blackish. *Glossy black overall*, a blue sheen visible in good light. **Juv:** dull blackish overall, quickly attains adult-like plumage.

Voice. Varied, rich, sharp whistles, a sharp *weet!* or *piik!* sometimes repeated

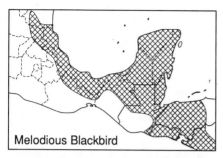

Melodious Blackbird

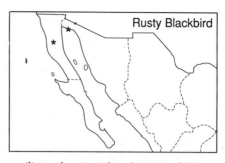

Rusty Blackbird

persistently, or preceding a characteristic piercing *whee' choo* or *chee' tweu*; also a drawn-out *tseeeuu*, and a loud, rich, whistled *cheeu*, etc.

Habs. Open and semiopen areas with hedges, scattered trees, forest edge, plantations. Feeds mostly on ground, usually in pairs, at times in flocks up to 100+ birds. Characteristically flicks its tail up sharply, often accompanied by a whistled call. Flies in jerky, hesitant manner. Nest is a deep woven cup of plant fibers at mid-levels in dense foliage of tree or bush. Eggs: 3, pale blue, spotted and scrawled with black (WFVZ).

SS: Bronzed Cowbird has red eyes, thick neck, and stouter bill, glossy sheen purplish, often in flocks. ♂ Brewer's Blackbird more slender, with purple and green sheens, pale yellow eyes. Yellow-billed Cacique skulking, with pale greenish-yellow bill, pale yellow eyes. Giant Cowbird larger and longer-tailed, appears small-headed with neck ruff, heavier bill, and red eyes visible at close range.

SD/RA: C to F resident (SL–2000 m) on Atlantic Slope from S Tamps to Honduras, on Pacific Slope and in interior from E Guatemala (Thurber *et al.* 1987) to W Nicaragua. Apparently spread into El Salvador in 1960s, and can be expected to spread with increased clearing of forest; first recorded in Costa Rica, Mar 1987 (SNGH).

RUSTY BLACKBIRD
Euphagus carolinus   Not illustrated
Tordo Canadiense

ID: 8–9 in (20.5–23 cm). *Vagrant. Descr.* Ages/sex/seasonal variation. ♂ (autumn): *eyes pale yellow*, bill and legs blackish. *Overall black plumage obscured by broad rufous edgings on crown, nape, and upperparts and by cinnamon edgings of super-*

cilium, throat, and underparts, which set off *blackish loral mask*. Wings and tail black with blue-green sheen. Edgings wear off through winter to reveal black 'breeding' plumage. ♀ (autumn): similar but underlying color dark grey-brown; note blackish loral mask in fresh plumage. By spring, appears overall dark slaty grey, wings and tail darker with blue-green sheen.

Voice. A gruff *chehk* or *chak*, similar to Brewer's Blackbird.

Habs. Agricultural land, open and semiopen areas. Associates with other blackbirds.

SS: Some autumn ♂ Brewer's Blackbirds have contrasting pale edgings to head and body (but not to tertials and wing coverts) which are, however, duller buffy brown, not rufous or cinnamon; ♀ Brewer's Blackbird has dark eyes, ♂ Brewer's glossy overall with purple and green sheens.

SD: Vagrant (can be expected Nov–Mar), 2 records: BCN (Dec 1888, Grinnell 1928) and NW Son (Dec 1989/Jan 90, DS and others).

RA: Breeds N N.A.; winter SE USA, vagrant to W USA and NW Mexico.

BREWER'S BLACKBIRD
Euphagus cyanocephalus   Not illustrated
Tordo de Brewer

ID: 8.5–9.5 in (21.5–24 cm). *Breeds BCN,*

Brewer's Blackbird

winters N and cen Mexico. **Descr.** Ages/sexes differ. ♂: *eyes yellow*, bill and legs blackish. *Black overall with purple sheen on head, blue-green sheen on body, wings, and tail.* Some autumn birds have head, back, and chest edged buffy brown. ♀: eyes dark brown. *Dark grey-brown overall, wings and tail darker with slight green sheen.* **Juv:** resembles ♀ but lacks green sheen on flight feathers.

Voice. Varied whistles and clucks, including a high, slightly piercing, whistled *tsiiw*, and a dry, slightly gruff to hard *chehk* or *chwehk*. Song a short, soft, creaky gurgle.

Habs. Open and semiopen areas, especially agricultural areas, often in towns and villages. In pairs or loose colonies when breeding, in flocks up to a few hundred birds in winter; associates with other blackbirds. Nest cup of twigs, grasses, etc., from on or near ground to mid- to upper levels in tree or bush. Eggs: 3–6, pale grey to greenish white, spotted and mottled with browns and greys (WFVZ).

SS: ♀ Brown-headed Cowbird has shorter, deeper-based bill, shorter tail, usually shows pale face pattern. See Rusty Blackbird.

SD: C to F resident (SL–1800 m) in NW BCN. C to F transient and winter visitor (mid-Sep–early May; SL–3000 m) in Baja, on Pacific Slope from Son to Nay, in interior S to cen volcanic belt, and on Atlantic Slope S to cen Ver; U and irregular to Oax, R to W Guatemala.

RA: Breeds W N.A. to BCN; winters USA to cen Mexico.

SLENDER-BILLED GRACKLE *Quiscalus palustris* Extinct. See Appendix A.

GREAT-TAILED GRACKLE
*Quiscalus mexicanus**   Plate 65
Zanate Mayor

ID: ♂ 13.5–18.5 in (34.5–47 cm), ♀ 10.5–12.5 in (26.5–31.5 cm), E > W. *Widespread and common. Tail very long, graduated, and keel-shaped.* **Descr.** Ages/sexes differ. ♂: *eyes yellow*, bill and legs blackish. *Glossy blue-black overall with purple sheen on head, chest, and back*, more greenish blue on wings and tail. ♀: *head and upperparts dark brown to blackish brown, head browner with paler super-*

cilium; scapulars, wings, and tail glossed blue-green. *Throat and underparts dusky cinnamon to tawny-brown*, palest on throat and becoming sooty brown on flanks and undertail coverts. *Q. m. nelsoni* (Son and NE BCN): *smaller and paler overall, with supercilium, throat, and underparts pale greyish buff.* **Juv:** resembles ♀ but eyes brown, underparts often brighter, ochre-brown, streaked dusky.

Voice. Varied loud shrieks, clacks, whistles, and chatters, including a bright, piercing, ascending whistle, *wheeeeu'* or *s-weeeeerk!*, bright piping to shrieking series in various combinations, *wee kee-kee-kee-keek* or *shreeih dee-dee-dee-dee-dee-dee-dee-dee*, etc., a burry note followed by a muted shriek, *shrr iik!* or *ehrr eik!*, a fuller shrieking *shree kee-keer* or *shrr kee-keek!*; flight call a gruff, dry *chek*.

Habs. Open and semiopen areas, especially agricultural and marshy, towns and villages, etc.; only absent from, or R in, deserts, heavy forest, and high mountains. Noisy, conspicuous, and gregarious. Breeds in colonies, rarely singly. Large flocks most evident near noisy roosts (which are often in town parks) and during spring migration in E Mexico. Nest is a bulky cup of grasses etc., low to high in tree or bush. Eggs: 2–4, pale grey to bluish white, with vinaceous staining and black scrawls and spots (WFVZ).

SS: Giant Cowbird in flight has thicker neck, broader-based and more pointed wings, square-tipped tail not keel-shaped, flies with strong flap-flap-flap-pause progression versus steadier, less powerful flight of grackles; at rest note stout bill, red eyes.

SD: C to F resident virtually throughout the region, but absent from Baja except NE where colonized since 1960s; less common

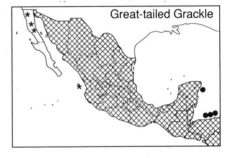

on much of Plateau and in cen volcanic belt. Range expanding locally, spread into NW Son in 1940s and 1950s; apparently colonized Valley of Mexico since 1950s (Dickerman 1965).

RA: S USA to NW Venezuela and NW Peru; N populations withdraw S in winter.

NB1: Common Grackle *Q. quiscula* (♂ 10.5–11 in, 26.5–28 cm; ♀ 10–10.5 in, 25.5–26.5 cm) has been reported without documentation from NE Mexico in winter. Markedly smaller than Great-tailed Grackle. May be confused with Brewer's Blackbird but bill markedly heavier than that species, tail fairly long and graduated. ♂ glossy black overall, head and chest have strong purple to bluish sheen, rest of body with green sheen, wings and tail glossed bronzy purple. ♀ duller than ♂, head and chest with greenish-blue sheen, rest of underparts dull dark brownish, upperparts with slight greenish gloss; wings and tail dark brown with purple sheen. **NB2: Boat-tailed Grackle** *Q. major* (♂ 15.5 in, 39.5 cm; ♀ 13 in 33 cm) of USA Gulf coast could reach NE Mexico. Much like Great-tailed Grackle but eyes dull yellowish (♂) to brownish (♀), slightly smaller but looks less small-headed (forehead rounder, less flat-headed), rarely far from salt water.

BRONZED (RED-EYED) COWBIRD
Molothrus aeneus* Plate 65
Vaquero Ojirrojo

ID: 7.5–8.5 in (19–21.5 cm). Widespread. Note neck ruff, especially in ♂, bill *relatively stout and pointed.* Two groups, readily told in ♀ plumage: *assimilis* group (W Mexico) and *aeneus* (E Mexico and C.A.), intergrading in interior S Mexico. **Descr.** Ages/sexes differ. ♂: *eyes red*, bill and legs blackish. *Glossy black overall with bronzy green sheen on head and body*

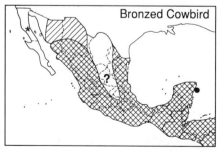

Bronzed Cowbird

becoming purplish on scapulars and rump; *wings and tail with more bluish sheen.* **Assimilis** ♀: *head and upperparts grey-brown, throat and underparts paler grey-brown, palest on throat* and often with indistinct dusky streaking on underparts. **Juv:** eyes dull; paler and browner than ♀, underparts streaked dusky and buff. **Aeneus** ♀: *brownish black overall with slight bluish sheen on upperparts*; notably duller and browner when worn. **Juv:** eyes dull, dark brown overall without sheen.

Voice. High, thin, tinny, whining trills and wheezy squeaks, thinner than calls of Brown-headed Cowbird, at times suggesting European Starling; also a spluttering chatter or rattle.

Habs. Open and semiopen areas in winter, dispersing to open woodland, forest patches, scrub, etc., in breeding season. Large flocks form in winter, associating loosely with other blackbirds and readily with Brown-headed Cowbirds. Singly or in small groups in summer. In display, ♂♂ 'inflate' themselves and, with rapid audible wing-quivering, parade around and hover over ♀♀. Brood parasite, wide range of hosts known. Eggs: pale bluish to bluish white, unmarked (WFVZ).

SS: ♀ Brown-headed Cowbird smaller, has dark eyes, smaller and less stout bill and distinct forehead (Bronzed Cowbird's mean-looking bill often appears as a continuation of the forecrown), usually sandier brown overall with trace of pale face pattern.

SD: C to F resident (SL–3000 m) on both slopes from Son and Tamps, and in interior from cen volcanic belt, to El Salvador and Honduras; F to U in interior N to W Chih. Many withdraw S and to lowlands in winter (Oct–Mar?) from NW areas. Vagrant to BCN (Jun 1980, Wilbur 1987); should be sought in Río Colorado delta.

RA: SW USA and Mexico to cen Panama.

NB: Shiny Cowbird *M. bonariensis* (♂ 7–7.5 in, 18–19 cm; ♀ 6.5–7 in, 16.5–17.5 cm) (Plate 65) is spreading rapidly through much of Caribbean and SE USA and should be sought in places such as Isla Cozumel or Cancún, QR. Bill more slender and pointed than Bronzed or Brown-headed cowbirds, lacks neck ruff, eyes dark brown. ♂ glossy purplish black overall, wings and tail glossy blue-black. ♀ dark grey-brown overall, paler below, palest on throat; underparts often tinged ochre-brown and faintly streaked dusky. Eggs: whitish to pale bluish to pale pinkish, speckled and spotted with browns and greys (WFVZ).

BROWN-HEADED COWBIRD

Molothrus ater *   Plate 65
Vaquero Cabecicafé

ID: 6.3–7.3 in (16–18.5 cm). Widespread N of Isthmus. *Bill relatively short and stubby.* **Descr.** Ages/sexes differ. ♂: bill and legs blackish. *Overall glossy blue-black* (duller and browner when worn) *with brown head.* ♀: *head and upperparts grey-brown, often with slightly paler eyering and supercilium. Throat and underparts paler grey-brown to dusky buffy brown, palest on throat, and often with faint dark malar stripe and dusky streaking on underparts.* **Juv:** overall paler than ♀, upperparts with scaly buff edgings, underparts more distinctly streaked, buff and dusky. Quickly attains adult plumage.
Voice. A high, slightly tinny or whining, slurred *ts-eeu* and *tsee' t-seeu*, high, squeaky, thin whistles and warbles in flocks, and a slightly liquid to dry spluttering chatter or rattle.
Habs. Much as Bronzed Cowbird but lacks inflation and hovering display. Eggs: white to bluish white, densely speckled with browns and greys (WFVZ).
SS: See Bronzed Cowbird, Brewer's Blackbird.
SD: F to C resident (SL–2500 m) in Baja, on Pacific Slope, and in interior S to cen Mexico, U to R to Isthmus (Rowley 1984). F to C transient and winter visitor (Aug–Apr) on Atlantic Slope S to N Ver, R to S

Ver. A report from Isla Cozumel (Paynter 1955a) is based on a Gaumer spec. and is thus questionable.
RA: N.A. to Mexico; N populations withdraw S in winter.

GIANT COWBIRD

Scaphidura oryzivora impacifica   Plate 65
Vaquero Gigante

ID: ♂ 12.5–13.5 in (31.5–34 cm), ♀ 11–12 in (28–30.5 cm). *Humid SE lowlands. Bill stout and pointed, tail fairly long and graduated; ♂ has thick neck ruff accentuating small-headed look.* **Descr.** Ages/sexes differ slightly. ♂: *eyes red,* bill and legs black. *Glossy black overall with strong purple sheen,* imm duller with slight sheen. ♀: *duller,* sooty blackish, with slight metallic sheen on head, chest and upperparts. **Juv:** eyes greyish, bill pale horn-flesh (black by winter). Duller, with little or no sheen. Attains adult plumage by 2nd prebasic molt when about 1 year old.
Voice. Usually silent, at times utters sharp clucks and chatters, including a nasal *chechk' chehk'* or *chehk-chik'*, and a sharp *chrrik' rrik-rrik-rrik-rrik-rrik-rrik'*, etc. Wings in flight make a hollow whirr.
Habs. Humid evergreen forest edge and adjacent open and semiopen areas. Often seen flying over, with steady flap-flap-flap-glide flight. Singly or in small groups, may associate loosely with other blackbirds. Brood parasite of oropendolas and often seen about active colonies. Eggs: white, unmarked (WFVZ).
SS: See Great-tailed Grackle and Melodious Blackbird.
SD: U to F resident (SL–750 m) on Atlantic Slope from S Ver to Honduras.
RA: SE Mexico to Ecuador and N Argentina.

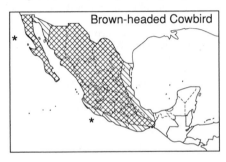

Brown-headed Cowbird

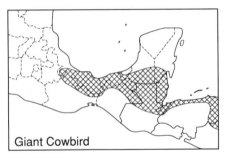

Giant Cowbird

GENUS ICTERUS: Orioles

Orioles are brightly colored 'blackbirds' that often exhibit marked geographic as well as age and sex variation; in addition, males and females in some species look different in the north of their range but become increasingly similar southward! Orioles present challenging identification problems and some birds may not be identifiable to species in the field. Typically bills are black with a variable amount of blue-grey at the base below, legs are blue-grey.

Many species have black bibs; juvs of most species lack the black lores and bib of adults. Juv plumage is lost quickly in most tropical species, usually while the young are still attended by parents, but is kept through autumn migration in most northern species. Imm plumage is retained for one year, although heavily worn feathers such as tertials and central rectrices may be replaced. Songs often rich and loud, calls harsh and grating. The nests of most species are woven cups or pouches of plant fibers slung under a branch or leaf; long and conspicuous in Altamira, Spot-breasted, and Streak-backed (Fig. 43). Eggs: 2–6 (usually 3–5), white to bluish white, marked with black and brown scrawls, flecks, and spots (WFVZ, MVZ).

BLACK-COWLED ORIOLE

Icterus dominicensis prosthemelas   Plate 66
Bolsero Capucha-negra

ID: 7.3–8.3 in (18.5–21 cm). Humid SE lowlands. *Bill relatively small and slender, culmen decurved. ♀ and imm resemble several oriole species and are a common source of confusion.* **Descr.** Ages/sexes differ. ♂: bill black with 30–60% of lower mandible blue-grey at base. *Head, chest, and back black; belly, rump, and tail coverts yellow,* usually with indistinct brownish-orange wash adjacent to black chest. Wings and tail black, yellow shoulder mostly covered by scapulars. ♀: *face, throat, and upper chest black, crown and upperparts olive-yellow.* Some ♀ (especially in S) have head and back mixed black and yellowish, *some are black headed,* and others look like ♂♂. **Imm:** *resembles ♀ but wings and tail dark brown,* shoulder and back often duller. During molt into adult plumage *can show black head contrasting with yellowish back.* **Juv:** resembles imm but duller, *black mainly on lores and throat,* wing coverts dark brown, narrowly edged olive.

Voice. A chuttering scold *chuh-chuh...* or *cheh-cheh-cheh-chek,* and varied nasal calls, *cheh* or *chek,* and a sharper *beehk* or *bihk,* etc., also a quiet *tee-u.* Song a fairly quiet, rich, scratchy warble.

Habs. Humid to semihumid forest, edges, clearings, semiopen areas with scattered trees and bushes. Singly or in pairs, at times in small groups, feeds and flocks readily with other orioles. Nests at midlevels in tree or bush. Eggs of *prosthemelas* group (see NB) undescribed (?).

SS: Yellow-backed Oriole has straighter stout bill, less black on face, typically brighter yellow on head and back. Audubon's Oriole (rarely in same range or habitat) is larger with heavier straight bill, tertials edged whitish when fresh. Imm Yellow-tailed Oriole has stouter bill, less black on face, dark olive tail with yellowish outer rectrices. Larger and longer-tailed Black-vented Oriole is a bird of dry highlands, has long straight bill, orange-yellow belly, imms have little black on face. ♀ Bar-winged Oriole (allopatric?) has paler wings and usually a pale lower wingbar.

SD: F to C resident (SL–500 m, locally to 1000 m) on Atlantic Slope from S Ver to Honduras. A report from the Pacific Slope of Guatemala (Vannini 1989*b*) is probably in error.

RA: SE Mexico to W Panama, also Caribbean.

NB: Mainland and Caribbean forms are sometimes considered specifically distinct, *I. prosthemelas* and *I. dominicensis,* respectively.

BAR-WINGED ORIOLE

Icterus maculialatus   Plate 66
Bolsero Guatemalteco

ID: 8–9 in (20.5–23 cm). *Endemic; SE Oax*

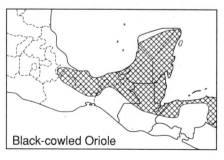

Black-cowled Oriole

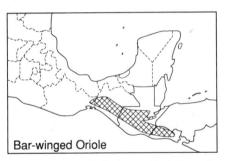

Bar-winged Oriole

to N C.A. *Bill fairly small, culmen de-curved.* **Descr.** Ages/sexes differ. ♂: bill black with 40–70% of lower mandible blue-grey at base. Head, central upper chest, and back black, rest of body *golden yellow.* Wings black with orange-yellow shoulder (mostly covered by scapulars), *broad white lower wingbar, white flash on primaries,* and narrow white tertial edgings. Tail black. ♀: *face, throat, and central chest black,* crown and upperparts yellow-olive, underparts yellow. Wings dark brown with with narrow pale edgings and indistinct whitish lower wingbar. **Imm** ♂: resembles ♀ but brighter, some-times with more extensive black on head and some black on back. **Imm** ♀: *black restricted to lores and broad black bib.* **Juv**: resembles imm ♀ but lacks black lores and bib, paler overall, wings brownish without pale wingbar.

Voice. A short, dry, gruff chatter *ahrı, ɹı grrrr,* and a longer *ah-rrrrrr.* Song a slow, rich warble, not prolonged or loud.

Habs. Arid to semiarid oak scrub and semideciduous woodland, semiopen areas with scattered trees, especially oaks. Often in pairs or small groups, associates occasionally with other species at flower-ing trees. Nest and eggs undescribed (?).

SS: Black-vented Oriole has longer straight bill, adult lacks white wingbar, has black undertail coverts, imms have less black on face and throat. Imm Yellow-backed Oriole has brighter head and back, dark brown wings and tail, and stouter straight bill. See Black-cowled Oriole (humid SE lowlands).

SD/RA: Locally/seasonally F to C resident (500–1800 m) in interior and on adjacent Pacific Slope from E Oax and Chis through Guatemala to N El Salvador (possibly SW Honduras). Engages in poorly known seasonal movements (for example, typically C to F at El Sumidero, Chis, Apr–Oct, but R to absent there Dec–Mar).

ORCHARD ORIOLE

Icterus spurius*   Plate 68
Bolsero Castaño

ID: 6–6.8 in (15–17 cm). *Small and slender; bill small, culmen decurved.* Two groups, separable in ♂ plumage: *spurius* group (breeds N and cen Mexico; including *phil-lipsi,* shown) and *fuertesi* (breeds NE Mexico). **Descr.** Ages/sexes differ. Bill black, 30–95% of lower mandible blue-grey basally. *Spurius* ♂: head, upper chest, and back black; in fresh plumage, back broadly edged buff, at times to the extend that birds appear pale-backed with black face and chest. *Underparts, rump, and uppertail coverts chestnut.* Wings black with chestnut shoulder and whitish lower wingbar and remex edgings. Tail black, outer rectrices narrowly tipped whitish. *Fuertesi* ♂: ochre replaces chestnut of *spurius.* ♀/juv: face and underparts greenish yellow to lemon-yellow, upper-parts greyish olive. Wings dark brown with 2 white wingbars and white remex edgings. Tail olive. **Imm** ♂: after autumn migration, attains black lores and bib, may also show a few patches of adult color on chest in 1st summer.

Voice. A low, slightly gruff, dry chattering *chuhchuh...* or *chuh-uh-uh,* etc., and a low gruff *chuk.* Song quite loud, a rich, scratchy warble begining with distinct rich notes and running into a jumbled, scratchy ending; at times given in flight, at least by *fuertesi.*

Habs. Riparian groves, gardens, semiopen areas with scattered trees and bushes. In flocks up to hundreds in migration and winter, often feeding in roadside hedges of flowering trees. Nests at low to mid-levels in tree or bush.

SS: ♀ and imm ♂ similar to Hooded Oriole (*californicus* group) but smaller with pro-portionately smaller bill and shorter tail,

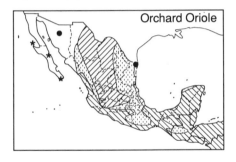

Orchard Oriole

lower wingbar usually broader, black face often less extensive; undertail coverts of Hooded usually brighter, orange-yellow.

SD: *Spurius:* F to C breeder (Apr–Jul, near SL–2000 m) in N Tamps and on Plateau from Chih and Coah to N Mich, irregularly (?) in E Son (GHR). C to F transient and winter visitor (Jul–Mar) from Balsas basin along Pacific Slope to El Salvador and Honduras, U to R on Pacific Slope N to S Sin and on Atlantic Slope from S Ver and Yuc Pen to Honduras. C transient (Feb–Apr, Jul–Sep) on Atlantic Slope N of Isthmus. Vagrant to BCS (Oct–Jan; KAR, RS). *Fuertesi:* F to C breeder (Mar–Jul, around SL) from S Tamps to Ver. Winters from Balsas Basin S on Pacific Slope at least to Chis (SNGH, SW, photo).

RA: Breeds E N.A. to cen Mexico; winters Mexico to N S.A.

NB: *Fuertesi* sometimes considered specifically distinct, Fuerte's or Ochre Oriole.

HOODED ORIOLE
*Icterus cucullatus**   Plate 67
Bolsero Cuculado

ID: 7.3–7.8 in (18.5–20 cm). A slender, medium-sized, long-tailed oriole. *Bill slender, culmen decurved. Two groups: californicus* group (breeds NW and NE Mexico; including *nelsoni* shown) and *cucullatus* group (E Mexico to Belize; including *igneus*, shown). **Descr.** Ages/sexes differ. *Californicus.* ♂: bill black with 30–40% of lower mandible blue-grey at base. Head, underparts, rump, and uppertail coverts *tawny-yellow*; lores, bib, and *back black* (in fresh plumage back broadly edged olive-buff, at times may appear pale-backed). Wings black with *2 white wingbars* (upper much broader, lower often absent in worn plumage) and white remex edgings. Tail black, outer rectrices narrowly

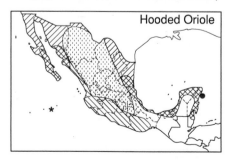

Hooded Oriole

tipped white. ♀/**juv:** 50–100% of lower mandible blue-grey. *Face and underparts greenish yellow* to lemon yellow, undertail coverts brighter orange-yellow. *Upperparts greyish olive*; wings darker with 2 white wingbars and whitish remex edgings. *Tail olive.* **Imm** ♂: resembles ♀ but lores and bib black (attained by N migrants after autumn migration), upper wingbar broader. *Cucullatus:* similar to *californicus*, but ♂: *bright flame-orange* to orange instead of tawny-yellow. ♀/**juv:** upperparts richer olive, less greyish, *face and underparts yellow to yellow-orange.*

Voice. A slightly metallic, nasal *weink* or *zwink*, a plaintive *whiet*, a quiet *cheu*, a hard dry *chek* or *tchek*, and a fairly hard, dry chatter, etc. Song a varied, fairly rapid warble, not loud or striking.

Habs. Scrubby woodland, open and semi-open areas with scattered trees and bushes; the common oriole of arid beach scrub, often in and around palms. In pairs or small groups, often with other orioles. Nests at mid- to upper levels in trees, especially palms.

SS: Streak-backed Oriole (Pacific Slope) has stouter bill with straight culmen, dark streaks on back (versus scallops of fresh ♂ Hooded); Altamira Oriole larger with stout, deep-based bill, orange shoulder; Orange Oriole (Yuc Pen) has straight culmen, less extensive black lores, ♂ has yellow-orange back, ♀ and imm have brighter yellowish head and back, ♀ has darker tail; worn plumage Orange shows more white on primaries. See ♀/imm Orchard Oriole.

SD: C to F breeder (Feb–Sep, SL–1500 m) in Baja and Son, and on Atlantic Slope from E Coah to N Ver, thence resident through Yuc Pen to Belize (including offshore islands); U and local to N Sin (SNGH, SW). NW populations winter (Sep–Apr) on Pacific Slope from BCS and S Son to Gro, NE populations winter mainly on Pacific Slope from Nay to Oax; U to R on Atlantic Slope N to S Tamps. Vagrant to Isla Socorro (Feb 1988, SNGH, SW).

RA: SW USA, Mexico, and Belize.

BLACK-VENTED ORIOLE
*Icterus wagleri**   Plate 66
Bolsero de Wagler

ID: 8–9 in (20.5–23 cm). Endemic; interior and highlands. *Bill long and narrow, cul-*

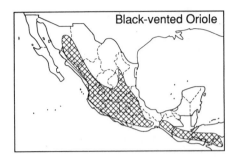

Black-vented Oriole

men straight to slightly decurved. *Tail long and strongly graduated.* **Descr.** Ages differ, sexes similar. **Adult:** bill black with 30–60% of lower mandible blue-grey at base. *Head, chest, and back black. Belly, rump, and uppertail coverts orange-yellow* with brownish-orange wash adjacent to black chest; *undertail coverts black.* Wings and tail black, orange-yellow shoulder mostly covered by scapulars. Tail black. **Imm:** crown and upperparts greyish olive, back mottled darker, face and underparts orange-yellow with *black lores and bib (bib typically restricted to chin* but sometimes extends onto upper chest). Wings dark brown with dull yellow-orange shoulder and pale greyish edgings to greater coverts and remiges. *Tail blackish brown* above, outer rectrices olive. **Juv:** resembles imm but lacks black lores and bib, crown and upperparts less greyish, underparts pale yellow, wing coverts dark brown, edged olive-yellow.

Voice. A nasal, slightly gruff *nyeh* or *yahn,* often repeated in series. Song a strong, squeaky to wheezy, gurgling warble with hard, nasal notes thrown in (JH tape).

Habs. Arid to semiarid scrub, semiopen areas with hedges and scattered trees. Usually in pairs or small groups which may be joined by other orioles. Often seen in leafless flowering trees. Nests at low to mid-levels in tree or bush.

SS: Scott's Oriole has distinct white wing-bars, yellow underparts. See Bar-winged and Black-cowled orioles.

SD/RA: F to C resident (500–2500 m; to near SL (Nov–Mar at least) in NW of range) in interior and on adjacent Pacific Slope from Son and S NL to Honduras and N cen Nicaragua. Wanders seasonally but extent of movements little known.

YELLOW-BACKED ORIOLE
Icterus chrysater *   Plate 66
Bolsero Dorsidorado

ID: 8–9.5 in (20.5–24 cm). S Mexico and C.A. *Bill fairly stout, culmen straight.* Two ssp, alike in plumage but Yuc Pen birds (*mayensis*) markedly smaller. **Descr.** Ages/sexes differ slightly. ♂: bill black with 30–50% of lower mandible blue-grey at base. *Forehead, lores, and bib black,* rest of *head and body* (except black outer scapulars) *bright golden yellow. Wings and tail black,* small yellow shoulder patch mostly covered by black scapulars, tertials and primaries edged white when fresh. ♀: head, back, and chest duller, *washed brownish orange,* especially around edge of bib. **Imm:** head and back washed olive. Wings dark brown, remiges narrowly edged pale olive when fresh. Tail dark brown with olive inner webs to outer rectrices. **Juv:** resembles imm but lacks black lores and bib, paler yellow overall.

Voice. A plaintive *choo,* a nasal *cheh* or *yehnk,* often repeated, a rich *tchew tcheeo cheoo,* and an ascending series *wheep wheep wheep wheep,* etc. Song a sad, rich, often hesitant series of whistles, rising and falling, much like Audubon's Oriole, at times with faster series of short whistles thrown in, etc.

Habs. Scrubby woodland, overgrown clearings, semiopen areas with hedges, plantations and (*chrysater* of S Ver to N C.A.) pine and pine–oak woodland. In small groups which often feed and flock with other orioles; *chrysater* often in mixed flocks with Band-backed wrens and jays. Nests at mid-levels in trees.

SS: See imm Black-cowled Oriole. Audubon's Oriole (*dickeyae* group) probably allopatric, has more black on face and head.

SD: F to C resident (500–2500 m) in S Ver (Los Tuxtlas) and in interior and on ad-

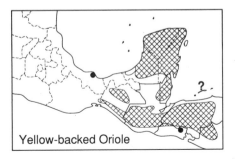

Yellow-backed Oriole

jacent slopes from E Oax (Binford 1989) and Chis to N cen Nicaragua; disjunctly in Yuc Pen (SL−900 m) S to Belize, and in the Mosquitia. Reports from Roatan Island, Honduras, may be in error (Appendix D).
RA: SE Mexico locally to N S.A.
NB: Yellow-backed and Audubon's orioles appear to be closely related, based on their strikingly similar morphology and vocalizations.

AUDUBON'S (BLACK-HEADED) ORIOLE

Icterus graduacauda *   Plate 68
Bolsero de Audubon

ID: 8.5−9.5 in (21.5−24 cm). *N of Isthmus. Bill fairly stout, culmen straight.* Two groups: *graduacauda* group (E Mexico) and *dickeyae* group (W and S Mexico). **Descr.** Ages/sexes differ slightly. *Graduacauda.* ♂: bill black with 30−50% of lower mandible blue-grey at base. *Black head and broad bib contrast with olive-yellow body,* underparts yellower. Wings black with yellow shoulder (mostly covered by scapulars), *broken white lower wingbar, and narrow white edgings to remiges.* Tail black, outer rectrices narrowly tipped white. ♀: duller, back and rump olive, white wing edgings bolder. **Imm:** resembles ♀ but sometimes (rarely) black hood less extensive and crown mostly olive. Wings grey-brown, tail dark olive. **Juv:** resembles imm but lacks black on head, paler yellow overall. *Dickeyae.* ♂: narrower black bib than *graduacauda, body bright golden yellow to yellow. Wings appear all black,* shoulder black except yellow marginal wing coverts, narrow whitish tertial edgings quickly wear off. Tail black. ♀ duller, back more olive-yellow, tertials often with narrow whitish edgings. **Imm/juv:** differ from ♀ in same respects as imm *graduacauda.*

Voice. A nasal *yehnk,* often doubled and repeated, and a clear *peu* and *hew-hoo,* the 2nd note higher. Song a sad, rich, often hesitant series of whistles, rising and falling, often with nasal calls thrown in.
Habs. Humid evergreen to semideciduous forest, clearings, also (*dickeyae*) in semiarid pine−oak woodland. In pairs or small groups, often in flocks with other species such as Green Jays, tanagers, and other orioles. Nests at mid-levels in trees.
SS: None in most of range but in Isthmus see ♀/imm Black-cowled Oriole; also see Yellow-backed Oriole (allopatric?).
SD/RA: C to F resident (500−2500 m, locally near SL in Tamps and Isthmus) on Atlantic Slope (*graduacauda*) from NL and Tamps (also S Texas) to Ver, and on Pacific Slope (*dickeyae*) of Gro and Oax; U to F on Pacific Slope from Nay to Mich and in interior Oax. May occur in W Chis; an old report from Guatemala is in error (Jenkinson and Mengel 1979).
NB: See NB under Yellow-backed Oriole.

YELLOW-TAILED ORIOLE

Icterus m. mesomelas   Plate 66
Bolsero Coliamarillo

ID: 8−9.3 in (20.5−23.5 cm). *A fairly skulking but flashy oriole of humid SE lowlands.* Culmen slightly decurved. **Descr.** Ages differ, sexes similar. **Adult:** bill black with 30−50% of lower mandible blue-grey at base. Head, underparts, rump, and uppertail coverts yellow, with black lores and broad bib. Back and wings black, yellow shoulder and inner greater coverts partially covered by black scapulars to form *broad yellow wingbar;* tertials and outer primaries narrow edged whitish. *Inner rectrices black, outer rectrices mostly yellow,* striking when tail spread. **Imm:** *back mottled olive and black* (base of feathers blackish so back becomes darker

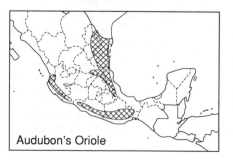

Audubon's Oriole

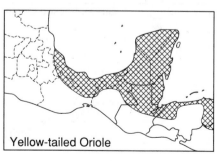

Yellow-tailed Oriole

with wear), rump and uppertail coverts olive; wings dark brown. *Tail olive, yellower inner webs of outer rectrices suggest adult pattern* (often attains much of adult tail by 1st spring). **Juv:** resembles imm but lacks black lores and bib, dark brown upperwing coverts edged yellowish.

Voice. A slightly nasal *chew* or *cheuk*, and a harder *chuk* or *chook*, both may be repeated steadily. Song often involves dueting birds, a varied, rhythmic or rollicking repetition of phrases, may suggest *Campylorhynchus* wrens: *tch wee-choo weeeep, tch wee-choo weeeep...* or *cheer ro-ror chee ro-ror...*, or *roo chee-roo roo chee-roo*, etc.

Habs. Humid to semihumid second growth, forest edge, often near water, at times in mangroves. Less conspicuous than most orioles, keeps low in dense growth and most obvious when singing from exposed perches. In pairs or small groups, may join flocks of other orioles. Nests at low to mid-levels in bush or tree.

SS: Imm told by mottled olive and black back, yellow wingbar.

SD: F to C resident (SL–300 m in Mexico, locally to 500 m in N C.A.) on Atlantic Slope from cen Ver to Honduras.

RA: SE Mexico to Peru.

STREAK-BACKED ORIOLE
Icterus pustulatus*   Plates 67, 69
Bolsero Dorsirrayado

ID: 7.5–9 in (19–23 cm). Mainly Pacific Slope. *Culmen straight.* Both sexes have black lores and bib. *Four groups: microstictus* group (breeds Son–Jal), *graysonii* (Islas Tres Marías), *pustulatus* group (Jal–Honduras; including *formosus*, shown), and *alticola* group (interior valleys of C.A.). **Descr.** Ages/sexes differ. *Microstictus.* ♂: bill black with 30–60% of lower mandible blue-grey at base. Head and chest *flame-*

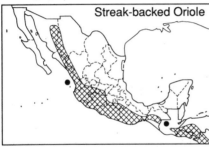

Streak-backed Oriole

orange-red, fading to bright orange on body; lores and bib black, *back narrowly streaked blackish.* Wings black, boldly edged white. Tail black, outer rectrices tipped white. ♀: duller, *upperparts washed olive, back streaks often indistinct,* face and underparts yellow-orange. Wings dark brown, tail dark brownish olive. **Imm** ♂: resembles ♀ but tail olive. **Imm** ♀: very dull, upperparts greyish olive, *dark back streaks indistinct at best, dusky lores and bib indistinct,* face and chest yellow to dull orange-yellow, underbody pale lemon, washed greyish. *Wings brown with whitish edgings,* tail olive. **Juv:** resembles imm ♀ but lacks dusky lores and bib (attained after migration, at least by some birds). *Graysonii* ♂: head, back, and underparts golden yellow overall, back with few or no black streaks. ♀: back plain greyish olive, face and underparts yellow. *Pustulatus.* ♂: slightly larger than *microstictus*, rarely as bright red on head, *back more heavily streaked black*; white wing edgings often less bold but white flash at base of primaries more contrasting. ♀: averages less bright than ♂ (bright orange to yellow-orange), tail blackish, outer rectrices olive basally and tipped whitish. **Imm:** as ♀ but tail dark olive, wings browner (especially when worn), ♂ brighter than ♀. **Juv:** resembles imm but lacks black lores and bib, paler orange-yellow overall. *Alticola:* larger than *pustulatus, back averages more heavily streaked, often solidly black when worn.* Sexes similar but ♀ averages duller. **Imm:** told from ♀ by dark olive tail, browner wings. **Juv:** resembles imm (may show dusky lores and bib) but head and body duller, more yellowish, back dull olive with darker mottled stripes. *I. p. maximus* (Rio Negro Valley, Guatemala) is bright orange-yellow, even in adult ♂.

Voice. A rough, nasal *yehr* or *yehnk*, often repeated, a nasal *cheh* or *chehk*, a sharp, slightly liquid, House Finch-like *chuwit* and *weet*, a quieter *chu* or *chew*, a hard, dry chatter, and a steady series of clear notes (typically 5–10), *sweet sweet...* or *weet weet...* Song a varied rich warble, *whee'tchi-wee-chi-wee*, or *chree chree chree chree chree-chu chee-chi'*, etc., also a jerky, discordant, jangling warble, often ending with a short dry churr, may suggest Yellow-winged Cacique.

Habs. Arid to semiarid brushy woodland and scrub, semiopen areas with scattered trees and bushes, plantations. Singly or in small groups, often associates with other orioles, nesting readily alongside Altamira and Spot-breasted. Pensile woven nest of plant fibers typically 25–40 cm (10–16 in) long (Fig. 43), hung from end of branch at mid- to upper levels in tree or bush, often thorny, also from roadside wires, etc.

SS: *Microstictus:* ♀ and imm ♂ Bullock's Oriole similar to imm ♀ but tomium of upper mandible and all lower mandible blue-grey, belly often pale greyish; see Hooded Oriole. *Pustulatus:* worn imm Spot-breasted Oriole best distinguished by slightly decurved culmen, note voice; juv Spot-breasted told by bill shape and wing pattern. *Alticola:* solidly black-backed birds and worn imms often best told from Altamira by voice; note also less heavy bill with more extensive blue-grey below, more extensive white primary edgings (useful only in fresh plumage); juv told from juv Spot-breasted and Altamira by wing pattern; worn imm from corresponding plumage of Spot-breasted by bill shape and voice.

SD: C to F resident (SL–1800 m) on Pacific Slope from Son, and in interior from Jal, to Honduras; *microstictus* partly migratory, some birds winter S to Gro.

RA: W Mexico to NW Costa Rica.

ORANGE ORIOLE
Icterus auratus   Plate 67
Bolsero Yucateco

ID: 7–7.5 in (18–19 cm). *Endemic to Yuc Pen. Culmen straight.* **Descr.** Ages/sexes differ. ♂: bill black above, blue-grey below (may have black tip). Head orange, becoming yellow-orange on body, with black

lores and bib; outer scapulars tipped black. Wings blackish (orange shoulder usually covered by black-tipped scapulars) with 2 white wingbars, upper bar fairly broad, lower bar narrow (often wears off), remiges edged white. Tail black, outer rectrices narrowly tipped white. ♀: *head and body orange-yellow overall with black lores and bib,* back duller, *outer scapulars tipped black.* Wings dark brown with 2 narrow white wingbars and edgings to remiges. *Tail dark brown.* **Imm** ♂: resembles ♀ but brighter, especially on head and back, upper wingbar wider when fresh, tail olive. **Imm** ♀: resembles ♀ but nape and back washed olive, white wing edgings absent when worn, tail olive. **Juv:** resembles imm ♀ but lacks black lores and bib, wingbars washed lemon.

Voice. A low, dry chatter, a fairly hard, nasal *nyehk,* a clear *choo,* and a slightly nasal, drawn-out *wheet.* Song a varied series of clear, slightly plaintive whistles, recalling Altamira Oriole, and a rapid, fairly mellow, whistled *chuchuchuchu . . .* or *cheechee. . . .*

Habs. Arid to semihumid brushy woodland, semiopen areas with scattered trees. Usually singly or in pairs, often associates with other orioles. Nests low to high in tree or bush (Fig. 43), at times colonially (Howell *et al.* 1992a). Eggs undescribed (Kiff and Hough 1985).

SS: See Hooded Oriole.

SD/RA: F to C resident (around SL) in Yuc Pen; also NE Belize, at least in winter (Howell *et al.* 1992b).

SPOT-BREASTED ORIOLE
Icterus pectoralis*   Plate 67
Bolsero Pechimanchado

ID: 8.5–9.5 in (21.5–24 cm). Pacific Slope. A fairly large oriole with *fairly deep-based*

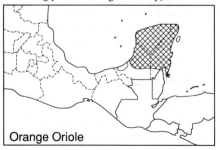

Orange Oriole

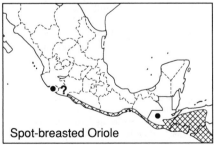

Spot-breasted Oriole

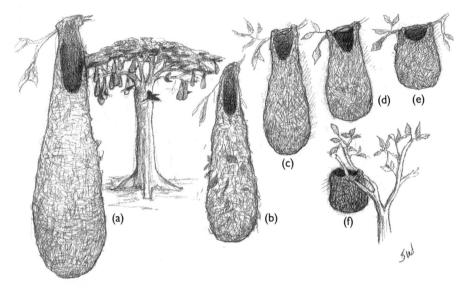

Fig. 43 Pendulous icterid nests. (a) Montezuma Oropendola and colony (Chestnut-headed Oropendola is similar); (b) Yellow-winged Cacique; (c) Altamira Oriole; (d) Spot-breasted Oriole; (e) Streak-backed Oriole; (f) Orange Oriole

bill, culmen slightly decurved. **Descr.** Ages differ, sexes differ slightly. ♂: bill black with 30–50% of lower mandible blue-grey at base. Head and underparts orange to orange-yellow with black lores and bib, *sides of chest spotted black.* Back black, rump and uppertail coverts orange to orange-yellow. Wings black (yellow shoulder often covered by scapulars) with *white triangle on tertials and inner secondaries and white flash on base of primaries.* Tail black, outer rectrices narrowly tipped white. ♀: averages duller and less heavily spotted. **Imm:** resembles adult but *back olive, mottled dark brown,* wings dark brown, tail olive. *By spring back mainly dark brown (olive edgings wear off) and chest spots may have worn off.* **Juv:** lacks black lores, bib, and chest spots, paler yellow overall than imm, back olive. *I. p. carolynae* (N of Isthmus) has head and underparts yellow with heavy black chest spots sometimes forming solid pectoral band.

Voice. A nasal *nyeh*, often repeated. Song a varied, rich, slow warble, *whi whew hi hew hew,* or *whee ch-wee'chu-u,* etc., at times more prolonged or with repetitive series, *chee-hee-oo hee-hee chee-chee-chee-chee-chee-chee-chee...*, etc.

Habs. Arid to semihumid woodland, brushy scrub and semiopen areas with scattered trees. Singly or in pairs, less often in small groups, often with other orioles. Pensile woven nest of plant fibers typically 25–45 cm (10–18 in) long (Fig. 43), hung from end of branch at mid- to upper levels in tree or bush, often thorny.

SS: Note distinctive wing pattern (may be hard to see when worn), decurved culmen. Streak-backed (*alticola*) and Altamira orioles occasionally show 1–2 small black spots at sides of chest. Worn imm orioles may not always be identified safely by plumage; note heavier bill and olive back of Altamira; see Streak-backed Oriole.

SD: F to C resident (SL–250 m in Mexico, to 1500 m in C.A.) on Pacific Slope in Col (CB, tape) and from Gro to Honduras; also in interior valleys from Guatemala S and locally on Atlantic Slope in Honduras.

RA: SW Mexico to NW Costa Rica.

ALTAMIRA ORIOLE

Icterus gularis*   Plate 67
Bolsero de Altamira

ID: 9–10 in (23–25.5 cm). *A large, heavily built oriole with very deep-based bill, culmen straight.* **Descr.** Ages differ, sexes

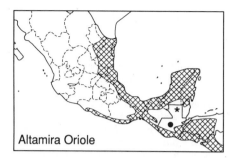

Altamira Oriole

BALTIMORE [NORTHERN] ORIOLE
Icterus galbula   Plate 68
Bolsero de Baltimore

ID: 7–8 in (18–20.5 cm). Winter visitor to E and S. *Relatively short-tailed. Culmen straight.* **Descr.** Ages/sexes differ. ♂: *bill blue-grey with blackish culmen. Head, upper chest, and back black, rest of body bright orange.* Wings black with orange shoulder mostly covered by scapulars, broad white lower wingbar, and white edgings to remiges. *Tail black, outer rectrices yellow-orange distally.* ♀: *head and back mottled olive, black, and orange.* Underparts yellow-orange (belly may be washed greyish), rump and uppertail coverts yellow-olive. Wings blackish brown with 2 *broad white wingbars* and narrow white edgings to remiges. Tail olive, outer rectrices paler and yellower distally with hint of ♂ pattern. **Imm ♂:** brighter than ♀, usually more black on head and back by spring when often shows black central rectrices. **Imm ♀:** resembles ♀ but head and back greyish olive, paler in face, wings browner, rump greyish olive, paler underparts often washed greyish on belly. Intergrades with Bullock's Oriole appear variably intermediate.

Voice. A gruff, dry chatter, *chachacha...*, often harder than Bullock's Oriole, and a soft *whuit.*

Habs. Humid to semihumid woodland, forest edge, gardens, hedges. Often in large flocks during migration (mainly Sep and Apr), in winter often in mixed-species flocks.

SS: Imm Bullock's Oriole often not safely distinguishable but tends to be greyer above and yellower below with greyer belly, paler face has hint of ♂ Bullock's pattern.

SD: C transient (Mar–May, late Aug–Oct;

similar. **Adult:** bill black with 5–25% of lower mandible blue-grey at base. Head and underparts orange with black lores and bib; color varies from intense red-orange in *yucatanensis* of Yuc Pen to orange-yellow in *flavescens* of Gro. Back black, rump and uppertail coverts orange. *Wings black with* orange to yellow shoulder (often covered by scapulars), *white lower wingbar, flash on base of primaries, and narrow edgings to remiges.* Tail black, outer rectrices tipped white. ♀ averages duller. **Imm:** orange-yellow overall rather than orange of adult, *back olive*, wings brown, tail olive. **Juv:** resembles imm but paler yellow, lacks black lores and bib.

Voice. A nasal *yehn* or *yehnk*, often doubled and repeated, a bright, slightly ringing *chiu*, a sharp *peen*, and a rich, plaintive *chu wee chu.* Song varied, a rich, often jerky series of whistles, often containing a fairly rapid series, *chee-choo' chee-choo' chree chu-chu-chu-chu-chu-chee'*, etc.; also a simple *chee'chu*, repeated and at times run into a more varied series.

Habs. Woodland, scrub, second growth and semiopen areas with scattered trees and bushes. Typically in pairs or small groups, often with other orioles. Conspicuous woven pensile nest of plant fibers typically 35–65 cm (14–26 in) long (Fig. 43), hung from end of branch at mid- to upper levels in tree or bush, often thorny, also from roadside wires, etc.

SS: See Spot-breasted, Streak-backed (*alticola*), and Hooded (*cucullatus*) orioles.

SD/RA: C to F resident (SL–1500 m) on Atlantic Slope from E NL and Tamps (also S Texas) through Yuc Pen to N Belize (where U), on Pacific Slope from cen Gro to Nicaragua; also in interior locally from Isthmus S.

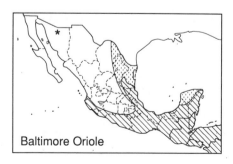

Baltimore Oriole

SL–2500 m) on Atlantic Slope of Mexico (but U in Yuc Pen) and on both slopes S of Isthmus. F to C winter visitor from S Ver and Chis S, mainly on Atlantic Slope, U to F to S Tamps, U to Nay (PA). Vagrant (Oct–Jan) to Son (Terrill 1981).

RA: Breeds E N.A.; winters Mexico to N S.A.

BULLOCK'S [NORTHERN] ORIOLE
Icterus (galbula?) bullockii   Plate 68
Bolsero de Bullock

ID: 7–8 in (18–20.5 cm). Two groups: *bullockii* group (breeds N Mexico) intergrading with *abeillei* (breeds cen Mexico). *Tail relatively short. Culmen straight.* **Descr.** Ages/sexes differ. *Bullockii.* ♂: *bill blue-grey with blackish culmen. Face and underparts orange with black eyestripe and narrow black bib.* Crown, nape, and back black, rump and uppertail coverts orange, often washed olive. Wings black with *bold white panel on median and greater coverts* and white edgings to remiges. *Tail orange with black tip and black central rectrices.* ♀: crown, nape, and upperparts olive-grey. Face, throat, and chest yellowish to orange-yellow, often with indistinct narrow black bib; belly pale grey (sometimes washed orange-yellow), undertail coverts yellowish. Wings blackish brown with *2 broad white wingbars* and narrow white edgings to remiges. Tail olive. **Imm** ♂: resembles ♀ but lores and bib black, plumage generally brighter, white wingbars often broader. **Imm ♀/ juv:** resembles ♀ but usually duller, lacks black bib. Intergrades with Baltimore Oriole show intermediate characters. *Abeillei.* ♂: resembles *bullockii* but yellow-orange overall, *face black with narrow orange spectacles; sides, rump, and uppertail coverts black.* ♀/imm: resemble *bullockii* but head and often sides dusky,

hinting at ♂ pattern; face pattern least distinct in imm ♀.

Voice. A gruff, nasal *cheh* or *shehh*, and a chattering *shehkshehk...* or *cheh-cheh...*, etc. Song a varied rich warble often introduced by, or containing, gruff scratchy notes.

Habs. Arid to semihumid wooded areas, pine–oak forest, parks, orchards, riparian groves. Singly or in small groups, larger flocks during migration; both ssp occur together in winter flocks. Nests at mid- to upper levels in tree or bush.

SS: See imm Baltimore and Streak-backed (*microstictus*) orioles.

SD: *Bullockii:* F to C breeder (Apr–Aug; near SL–2500 m) in N BCN, N Son, and on Plateau S to Dgo where intergrades with *abeillei.* C to F transient and winter visitor (Aug–Apr; near SL–3000 m) on Pacific Slope from Sin (U to S Son), and in interior from cen Mexico, to Oax; U to F on Atlantic Slope from S Tamps to Isthmus and in interior to cen Guatemala, possibly to El Salvador (Thurber *et al.* 1987). C to F transient (Jul–Sep) in Son, U in BCS. *Abeillei:* F to C breeder (Apr–Aug; 1500– 3000 m) in interior and on adjacent slopes from Dgo and S NL to Mich and cen Ver; winters (Aug–Apr; 1000–3000 m) from cen volcanic belt to Oax (RAB, SNGH).

RA: Breeds W N.A. to cen Mexico; winters SW USA to Guatemala.

NB: *Abeillei* has been considered a separate species, Abeille's or Black-backed Oriole.

SCOTT'S ORIOLE
Icterus parisorum   Plate 68
Bolsero Tunero

ID: 7.5–8.3 in (19–21 cm). *Mainly arid country N of Isthmus. Bill long and narrow, culmen straight.* **Descr.** Ages/ sexes differ. ♂: bill black with 30–65% of

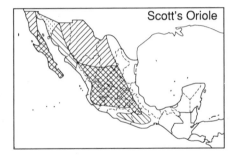

Bullock's Oriole   Scott's Oriole

lower mandible blue-grey at base. Head, chest, and back black, rest of body *bright yellow*, rump and uppertail coverts often washed olive. Wings black with yellow shoulder (mostly covered by scapulars), 2 *broad white wingbars*, and white edgings to remiges. *Tail black, outer rectrices yellow basally.* ♀: variable. *Face, throat, and chest yellowish, often mottled blackish creating a hooded effect.* Crown, nape, and back greyish olive with dusky streaking, rump and uppertail coverts yellower; underparts yellow. Wings dark brown with 2 *broad white wingbars* and white edgings to remiges. Tail dark olive, outer rectrices paler and yellower basally. **Imm ♂**: resembles heavily marked ♀ but usually more solidly black on head and chest. **Imm ♀/juv**: *resemble ♀ with no black on head and chest.*
  Voice. A harsh, nasal scolding *cheh-cheh* ..., or *chuhk*, and a quiet *huit*. Song a rich mellow warble, often with a fluty quality suggesting Western Meadowlark.
  Habs. Arid to semihumid, open and semiopen areas with scattered trees and bushes (especially Joshua trees and agaves), locally in pine and pine–oak woodland, at least in winter. In pairs or small groups up to 20+ birds, often joins mixed-species flocks in winter. Nests at mid- to upper levels in tree, bush, yucca, etc.
  SS: See Black-vented Oriole.
  SD: F to C breeder (SL–3000 m, highest in S of range) in Baja and in interior and on adjacent slopes from Son and Coah to Jal and Gto; U to F and local to NW Oax. Winters (Sep–Mar) from cen Baja, Son, and S Plateau S, and F to U to Sierra Madre del Sur of Gro and Oax.
  RA: Breeds SW USA to cen Mexico; winters to S Mexico.

YELLOW-BILLED CACIQUE
Amblycercus h. holosericeus   Plate 65
Cacique Piquiclaro

ID: 8.5–10 in (21.5–25.5 cm). E and S lowlands, *skulking*. **Descr.** Ages differ slightly, sexes similar. **Adult:** *eyes pale lemon, bill pale greenish yellow*, legs blue-grey. *Glossy black overall.* **Juv:** eyes dusky. Dull black overall, quickly attains adult-like plumage.
  Voice. A gruff *shehr shehr* ..., a gruff

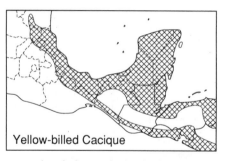
Yellow-billed Cacique

nasal *yahnk* or *ah'nk ah'nk*, a slightly gruff, crowing to laughing *yeh-yeh-yeh* ... or *ahrr ruh-ruh-ruh* ..., and descending *kyow kyow kyow* ..., and a growling *ah-rrrr*, etc. Song, often heard late in the day, a loud, rich *ch-wee* or *wheeu' hu* or *ch-wee-hu*, etc., usually 4–6×, often in duet with a rich whistle followed by a rough descending chatter, *wheeee irrrrrrrrrrrrr* or *hew chrrrrrrrrrrrr* from the other member of the pair.
  Habs. Thickets, second growth, forest edge, often near water. In pairs or singly, skulks in tangles where often hard to see, at times attends ant swarms. Nest is a thick-walled cup of vines, fibers, etc., at low to mid-levels in thickets. Eggs: 2, pale blue, with black dots.
  SS: See Melodious Blackbird.
  SD: C to F resident (SL–1500 m) on Atlantic Slope from S Tamps (Arvin 1990) to Honduras, on Pacific Slope from E Oax to Honduras.
  RA: Mexico to Peru and Bolivia.
  NB: Sometimes placed in the genus *Cacicus*.

YELLOW-WINGED CACIQUE
Cacicus melanicterus   Plate 65
Cacique Mexicano

ID: ♂ 12–13 in (30.5–33 cm), ♀ 10.5–11.5 in (26.5–29 cm). *Endemic to Pacific Slope lowlands.* Wispy crest often raised. **Descr.**

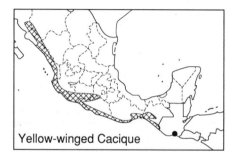
Yellow-winged Cacique

Ages/sexes differ. ♂: *bill pale greenish yellow*, legs dark grey. *Glossy black overall with broad yellow bar across upperwing coverts, yellow rump* and tail coverts, and *all except central rectrices mostly yellow*. ♀: duller, overall blackish slate above, slaty grey below, sides of forehead flecked yellow. **Juv/imm:** duller and browner than ♀, imm ♂ may be slightly mottled black in 1st summer, attains adult plumage by 2nd prebasic molt when about 1 year old.

Voice. Varied. A nasal, crowing, upslurred *rrahnk* and *raah*, a mellow whistled *tyoo* or *tiyih*, a hollow, slightly plaintive *wheeoo?*, a clipped *ch-tewk*, a ringing *chehnk*, a short, hard, churred rattle, often followed by bell-like notes, *ki-errr ink-ink-ink*. Song (?) a short rattle followed by quiet notes and ending with a discordant, slightly mechanical phrase inflected upward at the end, *rrah uh-uu uh-uu raahn'ee raahn-ee*. Wings make a hollow rushing or drumming sound in flight.

Habs. Forest and edge, plantations, semi-open areas with hedges and scattered trees. Arboreal, in pairs or small groups, roosts communally, at times in hundreds. Nests singly or in small colonies. Conspicuous pensile nest of plant fibers 61–76 cm (24–30 in) long (Fig. 43), slung at mid- to upper levels in trees, on roadside wires, etc. Eggs: 2–4, pale blue, sparsely flecked with black and grey (WFVZ).

SD/RA: C to F resident (SL–1500 m) on Pacific Slope from S Son (at least formerly) to Chis, and along Balsas drainage to SW Mex (SNGH, SW, RGW); disjunctly near La Avellena in E Guatemala (RWD spec.) where may have colonized recently.

CHESTNUT-HEADED (WAGLER'S) OROPENDOLA

Psarocolius w. wagleri　　　　Plate 65
Oropéndola Cabecicastaña

ID: ♂ 13–14 in (33–35.5 cm), ♀ 9.5–11 in (24–28 cm). *Humid SE forest.* Wispy crest often visible. **Descr.** Ages differ slightly, sexes similar. **Adult:** *eyes pale bluish, bill and shield pale greenish yellow*, legs dark grey. *Head and chest dark purplish chestnut, body and wings metallic blue-black overall* (duller in ♀ and imm ♂), becoming

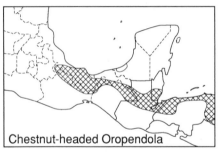

Chestnut-headed Oropendola

dark chestnut on rump and uppertail coverts; vent and undertail coverts chestnut. *Central rectrices blackish, rest of tail yellow except dark outer webs of outermost rectrices.* **Juv:** eyes dusky, bill duller. Duller, back and underparts dark sooty chestnut without metallic sheen, quickly resembles adult.

Voice. Varied throaty clucks and chatters: a low throaty *chuk-uk-luk* or *k-loh-uh*, a gurgled *g-lok*, a nasal *kyah*, and a mellower *wheuk*, etc. Wings make a hollow rushing sound in flight.

Habs. Humid evergreen forest and edge. Arboreal, in small groups or singly. Flies with fairly fast, deep wingbeats. Nests colonially, usually in a single tree, up to 60+ nests per colony. Conspicuous pensile nests of plant fibers 60–100 cm (24–40 in) long, at upper levels in trees (see Fig. 43). Eggs: 2, pale blue, marked with browns and black (WFVZ).

SS: Montezuma Oropendola larger with slower flight, clearly different if seen well.

SD: F to C resident (SL–1200 m) on Atlantic Slope from S Ver and N Oax (RAB, SNGH) to Honduras.

RA: SE Mexico to N Ecuador.

MONTEZUMA OROPENDOLA

Psarocolius montezuma　　　　Plate 65
Oropéndola de Moctezuma

ID: ♂ 18–20 in (45.5–50.5 cm), ♀ 15–16 in (38–40.5 cm). *Humid SE lowlands.* **Descr.** Ages differ slightly, sexes similar. **Adult:** *bill and shield blue-black with orange tip*, orbital skin grey, *naked patch below eye pale blue*, malar stripe pale pink, legs dark grey. *Chestnut overall, becoming blackish on head. Central rectrices blackish, rest of tail yellow.* **Juv:** duller, head and underparts mainly sooty blackish, quickly resembles adult.

Voice. A squeaky *woik*, a low *chuck* or

whek, often given in flight, a gruff, sneezy *rruh*, a cooing, slightly clucking *whi-t-wuu*, and a gruff clucking *kyuk kyuk* ..., etc. Song an unforgettable, bizarre gurgling and hollow popping series.

Habs. Humid evergreen forest and edge,

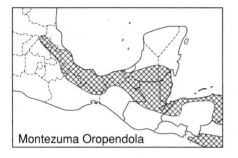

Montezuma Oropendola

adjacent open areas with hedges and scattered trees. Arboreal, in small groups or singly, at times in flocks up to 100+ birds; in full display ♂ swings upside down and flashes yellow in his raised tail. Flight slow and steady. Nests colonially, usually in a single tree, up to 140+ nests per colony. Conspicuous pensile nests of plant fibers 60–180 cm (24–72 in) long, at upper levels in trees (Fig. 43). Eggs: 2, whitish with coarse blackish spots and scrawls, to pale pinkish buff with dense dusky flecking (WFVZ).

SD: C to F resident (SL–1000 m) on Atlantic Slope from SE SLP (SNGH, PP) to Honduras.

RA: E Mexico to cen Panama.

CARDUELINE FINCHES AND ALLIES: FRINGILLIDAE (12)

Fringillids are finches of temperate origin with highest diversity in the Old World; this family formerly included the Emberizinae. Unlike emberizids and icterids, however, fringillids have ten primaries. Bills typically stout, very much so in *Coccothraustes* which can crack open hard seeds; legs fairly short. Wings usually long and pointed, tails short to medium length, often forked. Ages differ; young altricial; juvs usually resemble ♀♀, 1st prebasic molt partial to complete. Sexes differ or similar. Plumage often brightly colored, especially in ♂♂. Their preference for, or restriction to, temperate habitats is illustrated by the fact that only three species in the region range south of Guatemala, and two of these have their southernmost limit in Nicaragua.

Fringillids eat mostly seeds, but also insects and fruit. Nests typically are compact cups of fine materials placed low to high in bushes or trees. Eggs: 2–7, varied.

GENUS CARPODACUS

Three members of this genus, appropriately known in the Old World as rosefinches, occur in the New World. Ages/sexes differ. ♂♂ have rosy heads and chests, ♀♀ overall brown and streaked. Bills dull horn, darker above, legs greyish to dull flesh.

PURPLE FINCH
Carpodacus purpureus californicus
Fringílido Purpureo   Not illustrated

ID: 5.5–6 in (14–15 cm). *NW Baja in winter. Tail forked, culmen straightish.* **Descr.** ♂: *face, throat, and chest pinkish red,* with paler pinkish supercilium and dusky auriculars, crown brighter pinkish red; flanks pinkish red, becoming whitish on belly and undertail coverts, flanks indistinctly streaked dusky. Nape and *upperparts brownish red,* streaked darker, with

brighter pinkish-red rump. *Wings and tail dark brown, edged brownish red.* ♀/imm ♂: *head and upperparts brown to olive-brown,* streaked darker, with *contrasting whitish supercilium and broad moustache* flecked dark brown. Wings and tail dark brown, edged paler olive-brown. *Throat and underparts whitish, coarsely streaked dark brown.*

Voice. A dry *pit* given in flight, suggests Red Crossbill; also a slightly burry, rolled *bree-ah* and *ch-ree-ah* when perched. Song a mellow, rapid, rolled warble, suggests Warbling Vireo.

Habs. Deciduous and pine–oak woodland and edge, riparian groves. Singly or in small groups, feeds from high in trees to low in weedy fields.

SS: Cassin's Finch has longer pointed bill with straight culmen, long primary projection, nape often slightly crested, ♀ colder grey-brown overall with neater and finer dark streaking, ♂ has brighter red crown contrasting with grey-brown back, wings edged pale pink. House Finch has stubbier bill with decurved culmen, squared tail, ♂ streaked dusky on underparts, ♀ has plain face.

SD: Irregularly U to R winter visitor (Oct–Apr; SL–2500 m) to NW BCN; reports of breeding in Sierra Juarez (Huey 1926, and all subsequent authors) rest on a second-hand report of a spec. and require confirmation.

RA: N.A.; winters to NW Baja.

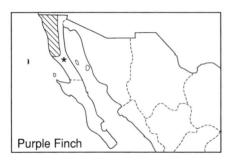

Purple Finch

CASSIN'S FINCH
Carpodacus cassinii   Not illustrated
Fringílido de Cassin

ID: 5.8–6.3 in (14.5–15.5 cm). NW Mexico. *Tail forked, culmen straight.* **Descr.** ♂: *face, throat, and chest pinkish,* with paler supercilium and dusky auriculars, *crown bright pinkish red;* flanks pinkish, becoming whitish on belly and undertail coverts, flanks and undertail coverts with indistinct dark streaks. Nape and *upperparts grey-brown* (back washed red), streaked darker, with pinkish-red rump. *Wings and tail dark brown, edged pale pink.* ♀/imm ♂: *head and upperparts grey-brown,* streaked darker, *with contrasting whitish supercilium and broad moustache* flecked dark brown. Wings and tail dark brown, edged paler grey-brown. *Throat and underparts whitish, neatly streaked dark brown.* **Juv:** resembles ♀ but less distinctly marked, head fairly plain, wingbars brighter, pale cinnamon.
Voice. A liquid *p-lip,* and a slightly plaintive *hweet.* Song a rich, fairly rapid warble, burrier than Purple Finch, not as scratchy as House Finch.
Habs. Arid to semiarid conifer and pine–oak woods. Singly or in small groups, feeds from trees to ground, sings from high in trees. Nest cup of fine twigs, grasses, rootlets, etc., at mid- to upper levels in trees. Eggs: 2–5, pale blue, speckled brown and black (WFVZ).
SS: See Purple Finch. House Finch smaller with stubby bill, squared tail, ♂ streaked on underparts, ♀ has plain face.
SD: F resident (2500–3000 m, lower in winter) in Sierra San Pedro Martír, BCN. Irregular R winter visitor (Oct–Apr; 2000–3500 m) to Sierra Juarez, BCN, and irregularly F to R in interior from NE Son (per GM) and Coah to N Mich (Nov 1984,

COIBUNAM specs) and Mex. Vagrant (May) to Islas Tres Marías (CAS spec.). A Jun 1864 spec. from Ver (Loetscher 1941) may be mislabeled.
RA: W N.A. to N Baja; winters to cen Mexico.

HOUSE FINCH
*Carpodacus mexicanus**   Not illustrated
Fringílido Mexicano

ID: 5.5–6.3 in (14–15.5 cm). Widespread N of Isthmus. *Tail squared, bill relatively stubby with decurved culmen.* **Descr.** ♂: *head and chest red* (rarely orange to yellowish) *with brownish auriculars* or (in much of S and cen Mexico) red restricted to forehead, supercilium, and throat. *Rest of underparts whitish, streaked dusky brown.* Nape and upperparts grey-brown, indistinctly streaked darker, with *red rump;* back sometimes washed reddish. Wings and tail dark brown, edged pale brown. ♀: *head and upperparts grey-brown,* indistinctly streaked darker, *with little or no contrasting face pattern,* rump often tinged red. Wings and tail dark brown, edged pale brown. *Throat and underparts whitish, blurrily streaked dusky brown.* **Juv:** much like ♀ but with paler wingbars, soon resembles adult.
Voice. A nasal *chee* or *chieh* and *chie-chiet,* all in flight; also a nasal *rrea-h* or *nyee-ah* and a short *wiet.* Song varied, a rich to scratchy warble, at times prolonged.
Habs. Arid to semiarid, open and semi-open areas, towns, villages. Singly or more often in pairs or flocks, feeds mostly on or near ground. Nest cup of fine twigs, grasses, rootlets, etc., low to high in tree, bush, cactus, building, etc. Eggs: 2–5, bluish white, sparsely scrawled and flecked with brown and blackish (WFVZ).
SS: See Purple and Cassin's finches. Pine

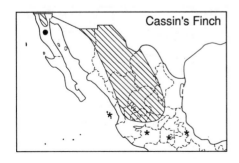

Cassin's Finch

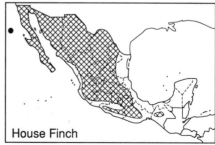

House Finch

Siskin smaller with finer pointed bill, forked tail, yellow flash in wings and tail.

SD: C to F resident in Baja (including many islands), on Pacific Slope from Son to N Sin, and in interior (and on adjacent slopes S to cen volcanic belt) to Oax. Also apparently F to U resident (at least since 1980s) in interior Chis (Tuxtla Gutierrez, GCH; San Cristobal, SNGH, SW), perhaps derived from escaped caged birds.

RA: W N.A. (introduced in E) to Mexico.

NB: The large sspp. on Islas San Benitos (*mcgregori*, extinct) and Isla Guadalupe (*amplus*, still fairly common in 1988, SNGH, SW) have been considered specifically distinct.

RED CROSSBILL

*Loxia curvirostra**   Not illustrated
Picotuerto Rojo

ID: 5.5–6.3 in (14–16 cm). Pines in highlands. *Mandibles conspicuously crossed at tips*; wings long, tail fairly short and forked. **Descr.** Ages/sexes differ. ♂: bill dark grey, legs dusky flesh. *Reddish to orange-red overall*, mottled darker on back, *with brighter red rump, blackish wings and tail.* Plumage often mottled yellowish. ♀/imm ♂: *overall greyish olive to yellow-olive, darker on back, rump brighter yellowish, wings and tail blackish.* Plumage may be flecked reddish. **Juv:** mandible tips shorter and only slightly crossed. Head and upperparts greyish olive, heavily streaked dark brown, with paler buffy-lemon rump. Throat and underparts whitish, often suffused lemon on underparts, heavily streaked dark brown. Wings and tail dark brown. Soon attains imm plumage.

Voice. Calls vary in intensity and pitch. Usually in series of 2 or 3, including a bright *kiip kiip kiip*, a lower *cheh cheh*

cheh, and a more House Finch-like *chee chee*. Song in region undescribed (?).

Habs. Pine and pine–oak woodland. Usually in pairs or small groups, often seen overhead in undulating flight. Feeds in conifers; crossed mandibles used to extract seeds from cones. Nest cup of twigs, bark strips, rootlets, etc., at mid- to upper levels in conifers. Eggs (N.A.): 3–4, bluish white to greenish white, flecked with browns and blackish (WFVZ).

SS: Crossed bill distinctive; in flight note thickset shape, shortish forked tail, long wings, and call.

SD: U to F resident (900–3500 m) in BCN and in interior, and on adjacent slopes from Chih and S Tamps to N cen Nicaragua, also (near SL–900 m) in pine savannas of Belize and the Mosquitia. Irregular U winter visitor (rarely to SL) to BCN, including Isla Guadalupe where may have bred (Howell and Cade 1954).

RA: Holarctic, S in New World to Nicaragua.

NB: There may be more than one species of Red Crossbill in the New World and, should several species be recognized, Mexican (and C.A.?) birds would be *L. stricklandi*.

GENUS CARDUELIS

Siskins and goldfinches are fairly small finches with pointed bills (notably slender in siskins) and with mostly yellow, green, and blackish plumage. Tails cleft to forked and wings fairly long. Ages/sexes usually differ. Twittering and warbling songs may include mimicry.

PINE SISKIN

*Carduelis pinus**   Plate 59
Dominico Pinero

ID: 4.5–4.8 in (11.5–12 cm). Two groups: *spinus* group (N of Isthmus) and *perplexus*

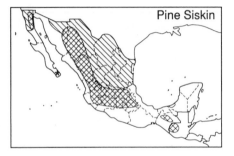
Red Crossbill   Pine Siskin

(S of Isthmus). **Descr.** Ages differ, sexes similar. *Spinus.* **Adult:** bill and legs greyish. *Head and upperparts sandy grey-brown, streaked darker.* Wings and tail dark brown, edged paler with 2 pale buff wingbars, and *yellow flash across base of remiges and outer rectrices.* Throat and underparts whitish, finely streaked dark brown. **Juv:** upperparts brighter with buff wingbars. *Perplexus.* **Adult:** face and underparts pale grey with dusky auriculars, *indistinct dusky streaking on throat, chest, and flanks,* vent whitish. Crown, nape, and upperparts dusky brownish grey with indistinct darker streaking. Wings and tail as *spinus* group but covert edgings duller. **Juv:** brighter, upperparts browner with black streaking, 2 distinct pale cinnamon wingbars, throat and underparts tinged buff with dark streaking.

Voice. *Spinus:* a nasal *zzheeu* or *sveea,* a dry twittering *ch-cht* or *chi-chit chi-chi-chit,* etc., and a drier chatter *ji-ji-ji-ji-jink,* etc.; song a varied, often prolonged, scratchy, twittering warble. *Perplexus:* probably similar but undescribed (?).

Habs. Arid to semihumid pine−oak and conifer forests, open areas with scattered trees, also deciduous woodland in winter. Often in flocks, at times to 100+ birds; feeds in trees and on ground. Nest cup of twigs, grasses, rootlets, etc., at mid- to upper levels in trees. Eggs: 2−4, greenish white or bluish white, flecked with brown and black (WFVZ).

SS: Juv Black-capped Siskin (Chis and Guatemala) has more slender bill, contrastingly darker cap. Hybrid *perplexus* × Black-capped Siskin show intermediate plumage characters, but bill usually like Pine Siskin. See House Finch.

SD: *Spinus:* F to C resident (1800−3500 m, lower in winter) in BCN and in interior and on adjacent slopes from Son and Chih to cen volcanic belt. Irregular U to C winter visitor (late Oct−Apr; SL−2500 m) to N Mexico, from BCN to Tamps, S to Dgo and Tamps. Vagrant (Jan−early May) to BCS (Unitt *et al.* 1992). *Perplexus:* U (?) resident (2000−3500 m) in highlands of N Chis and W Guatemala (Van Rossem 1938*b*).

RA: N.A. to W Guatemala.

NB1: *Perplexus* is quite distinct in appearance from *pinus* group, and studies are needed to determine its taxonomic status.

Hybrids between Black-capped and Pine (*perplexus*) siskins not uncommon (at least formerly, BM specs), and Black-capped sometimes considered conspecific with Pine Siskin.

BLACK-CAPPED SISKIN
Carduelis atriceps   Plate 59
Dominico Coroninegro

ID: 4.5−4.8 in (11.5−12 cm). *Endemic to highlands of Chis and Guatemala.* Bill sharply pointed. **Descr.** Ages/sexes differ. ♂: bill flesh, legs dusky. *Crown black,* throat mottled sooty, *rest of head and body olive-green,* brighter on underparts and rump, undertail coverts lemon, streaked dark. *Wings and tail blackish, egded yellow-olive, with 2 olive wingbars, yellow flash across base of remiges and outer rectrices.* ♀: *duller* overall, yellow flash at base of primaries reduced or lacking. **Juv:** face and underparts pale lemon, streaked dark brown. Crown, nape, and upperparts greyish olive, heavily streaked blackish (heaviest on crown); 2 buffy-lemon wingbars. Quickly attains adult-like plumage but throat and underparts paler, yellowish, into mid-winter.

Voice. Much like Pine Siskin. A nasal *zwee* or *dzee,* a dry *ch-ch-cht* or *cheh-cheh-cheht,* and a drawn-out, buzzy *djeeeerrr,* etc. Song a prolonged, scratchy, twittering warble.

Habs. Open and semiopen areas with scattered trees, streamside alders, etc., in humid to semihumid pine−oak and oak woodland. Habits much as Pine Siskin, in flocks up to 30+ birds in autumn and winter. Nest and eggs undescribed (?).

SS: Distinctive; compare juv with Pine Siskin. Hybridizes with Pine Siskin.

SD/RA: F but local resident (2000−3500 m)

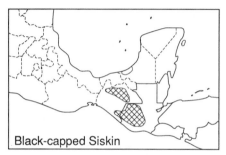

Black-capped Siskin

in interior and on adjacent Pacific Slope of Chis and Guatemala.

NB: See Pine Siskin.

BLACK-HEADED SISKIN

Carduelis notata*   Plate 59
Dominico Cabecinegro

ID: 4.2–4.5 in (11–11.5 cm). Endemic; highlands, mainly with oaks. **Descr.** Ages/sexes differ. ♂: bill blue-grey, legs grey. *Black head and chest contrast with golden-yellow underparts.* Nape and upperparts yellow-olive, back often mottled dusky, rump brighter yellowish, uppertail coverts black. *Wings black with* yellow lower wing-bar and *broad yellow patch across base of remiges.* Tail black, outer rectrices yellow at base of inner webs. ♀: duller, upperparts olive. **Imm** (Jan–Dec): *head and upperparts olive,* slightly paler on rump. Wings blackish with lemon lower wingbar, *yellow flash across base of remiges.* Tail blackish, outer rectrices edged yellow basally. *Throat and underparts dirty pale lemon.* May show black patches on head after 1st prebasic molt; most attain adult plumage Apr–Aug. **Juv:** resembles imm but underparts washed buff, lower wing-bar buff. Quickly attains imm plumage.

Voice. A nasal *teu* much like Lesser Goldfinch, a drawn-out *tseeeu* or *djeein,* a dry *jeh-jeht,* and a nasal *ti-chie,* etc. Song a varied, rapid, jangling, twittering warble, often prolonged, with repetition of phrases and odd nasal and metallic notes thrown in.

Habs. Arid to semihumid oak and pine–oak woodland, adjacent open areas. Feeds in trees and on ground, at times in flocks up to 20+ birds. Nest and eggs (2) (WFVZ) much like Pine Siskin.

SS: Lesser Goldfinch has deeper-based bill, white flash across base of primaries.

SD/RA: F to C resident (600–3000 m, lower locally) on both slopes from S Son and S Tamps (Arvin 1990), and in interior from cen Mexico, to N cen Nicaragua, and (around SL) the Mosquitia.

LESSER GOLDFINCH

Carduelis psaltria*   Plate 59
Dominico Dorsioscuro

ID: 4–4.5 in (10–11.5 cm). Widespread. Two groups: *psaltria* group (widely) and *hesperophilus* (NW Mexico). **Descr.** Ages/sexes differ. *Psaltria.* ♂: bill and legs grey, bill may be flesh in winter. *Head and upperparts black with bold white flash at base of primaries, white tertial tips* and, in N and cen Mexico, narrow white lower wingbar; white tail-spots on outer rectrices. *Throat and underparts yellow,* richer in *colombianus* (S of Isthmus). ♀: head and upperparts olive to greyish olive, wings blackish to dark brown with whitish to pale grey lower wingbar, *white flash on base of primaries* (indistinct or lacking in imm), and white tertial tips. Tail blackish, inner webs of outer rectrices edged white basally. *Throat and underparts yellowish to dirty whitish, tinged lemon.* **Imm** ♂: resembles ♀ but head and upperparts mottled black, typically all black in 1st summer. **Juv:** resembles ♀ but throat and underparts washed buff, 2 buff wingbars, and tertial edges buff to buffy white. *Hesperophilus:* much as *psaltria* but ♂ has *olive head and upperparts with black crown,* ♂ and ♀ have whitish lower wing-bar and tertial edgings.

Voice. A drawn-out, nasal *zweeir,* a plaintive *teu* or *tew,* and a dry *ch-ch-cht* or *ch-cht,* etc. Song varied and complex, a pleasant twittering warble, often prolonged and including mimicry.

Habs. Open and semiopen areas with

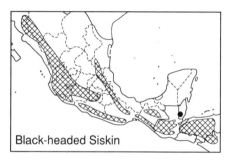
Black-headed Siskin

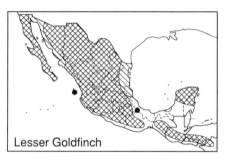
Lesser Goldfinch

scattered trees. In pairs or small groups, up to 30+ birds. Habits much as Pine Siskin. Nest cup of bark strips, rootlets, lichens, etc., at low to mid-levels in tree or bush. Eggs: 3–4, bluish white, unmarked (WFVZ).

SS: American Goldfinch has blackish wings with whitish to buff wingbars, lacks white flash across base of primaries, basic plumage brownish above, undertail coverts whitish. See imm Black-headed Siskin.

SD: C to F resident (SL–3000 m) in Baja and, mostly in interior and on Pacific Slope, from Son and NL to Oax, U to F to N cen Nicaragua; U to F in N Yuc Pen. F to C in winter (Sep–Apr) in NW lowlands and on Atlantic Slope S to N Ver.

RA: W USA to Peru and Venezuela.

LAWRENCE'S GOLDFINCH
Carduelis lawrencei   Not illustrated
Dominico de Lawrence

ID: 4.2–4.7 in (11–12 cm). NW Mexico. Bill small. **Descr.** Ages/sexes differ. ♂: bill pink to greyish flesh, legs dark flesh. *Black crown, lores, and throat contrast with grey face and yellow chest*, rest of underparts pale grey. Upperparts grey, washed olive on back, with ochre-yellow rump. *Wings and tail blackish, broadly edged yellow* except white tertial edgings, white tail-spots on outer rectrices. ♀: *head and upperparts greyish*, washed brown on back, rump dull ochre-yellow. *Wings and tail blackish brown, patterned as ♂ but duller. Throat and underparts pale grey, washed yellow on chest.* **Juv:** bill dusky. Resembles ♀ but browner overall, upperparts often with indistinct dusky streaks, wingbars pale cinnamon, yellow reduced or absent on chest.
Voice. A high *tinkl* or *dink* given in flight, also a high, thin *dee*, a drier *dri-i-iit*, a quiet, nasal *wheeh* or *whieh*, and a

rougher, soft mewing *rrehr*. Song varied, a pleasant, tinkling and twittering warble, often prolonged and including mimicry.
Habs. Arid to semiarid, open deciduous and pine–oak woodland, often near water. In pairs or small groups. Habits much as Pine Siskin. At times loosely colonial. Nest cup of fine plant material, wool, feathers, etc., at low to mid-levels in tree or bush. Eggs: 3–5, whitish, unmarked (WFVZ).

SD: U to F resident (near SL–2500 m) in NW BCN, irregularly U to R winter visitor (Oct–Apr) to cen Son and probably N Chih. Vagrant (Nov) to BCS (CAS spec.).
RA: Breeds California and NW Baja; winters to SW USA and Son.

AMERICAN GOLDFINCH
*Carduelis tristis**   Not illustrated
Dominico Americano

ID: 4.5–5 in (11.5–12.5 cm). Winters in N Mexico. **Descr.** Ages/sex/seasonal variation. **Basic** ♂: (Oct–Mar): legs dusky flesh, bill dusky flesh to grey with some orange-pink below at base. *Head and upperparts tawny-brown to grey-brown, face washed yellow, lower rump and uppertail coverts pale grey. Wings and tail black with pale lemon lesser and median wing coverts, whitish to pale buff lower wingbar and tertial and tail edgings. Throat and upper chest yellowish to dusky lemon, rest of underparts dusky buffy brown* to dusky buffy grey, becoming whitish *on undertail coverts*. ♀: wings dark brown, face, throat, and chest with less yellow. **Alt.** ♂: bill and legs pink. *Yellow overall with black crown and black wings with white wingbars*, white tertial and tail edgings. ♀: *resembles basic but throat and underparts yellowish*, becoming white on vent and undertail coverts, *head and upperparts washed olive*, wings and tail edgings narrower, whitish.

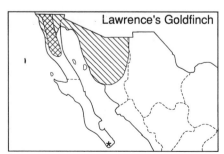

Lawrence's Goldfinch

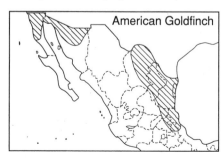

American Goldfinch

Voice. Common call a pleasant chipping *chih-tih* and *chih tih-tih-tih*, etc., often given in flight; also a slightly plaintive *chween*, and sweet to mellow warbling and twittering phrases. Song a varied, sweet, twittering warble, often prolonged.

Habs. Open and semiopen areas, especially weedy fields. Usually in small groups, feeding mostly low or on ground. Nest cup of bark strips, fibers, other fine plant material, etc., at low to mid-levels in tree or bush. Eggs: 4–5, bluish white, unmarked (WFVZ).

SS: See ♀ Lesser Goldfinch.

SD: Irregularly U to F winter visitor (Oct–Apr; near SL–2000 m) in N BCN and N Son, F to C from Coah and Tamps to N Ver, mainly on Atlantic Slope. The often-cited 'breeding' in BCN refers to a June 1925 sighting (Huey 1926); no nest has been found in Mexico and there are no subsequent breeding season reports.

RA: N.A.; winters to N Mexico.

HOODED GROSBEAK
Coccothraustes abeillei *   Plate 59
Pepitero Encapuchado

ID: 6.7–7.2 in (17–18.5 cm). *Endemic; highlands S to Guatemala.* **Descr.** Ages/sexes differ. ♂: bill pale greenish lemon, legs flesh. *Black head and throat contrast with yellow underparts. Upperparts yellow-olive* to olive, becoming pale grey on outer scapulars, brighter yellowish on rump, uppertail coverts black. *Wings black with silvery-grey tertials*; tail black, outer rectrices tipped white on inner webs in imm. ♀: *black crown contrasts with greyish-olive face, upperparts olive to greyish olive*, brighter on rump, pale grey on outer scapulars, uppertail coverts black, edged white. *Wings black with silvery-grey tertials*, white flash across base of inner

primaries. Tail black, outer rectrices tipped white on inner webs. Throat and underparts pale greyish vinaceous, paler on belly. *C. a. pallida* (NW Mexico) paler and greyer overall, undertail coverts may be whitish to pale lemon. **Juv:** undescribed (?).

Voice. A ringing note, usually doubled, *beehn beehn* or *beenk beenk*, and a nasal *whew* or *kew*. Song 1–2 clipped, slightly metallic, burry, rolled phrases often preceded by ringing calls, *beenk beenk eihrr-r*, or *wheirr whrr*, or *beehn beehn bee-beihr*, etc.

Habs. Humid to semihumid evergreen and pine–oak forest and edge. In pairs or small groups, locally (Tamps and SLP at least) in flocks to 50+ birds in winter; often seen flying high overhead, calling. Feeds quietly at mid- to upper levels and overlooked easily. Nest at mid- to upper levels in trees (RGW). Eggs undescribed (?).

SS: ♀ Evening Grosbeak lacks black crown, greyer overall.

SD/RA: U to F resident (1000–3500 m) on Pacific Slope from S Chih to Dgo, in Gro, and from Chis to El Salvador (Thurber *et al.* 1987); and in cen volcanic belt from Mich to Ver; F to C on Atlantic Slope in S Tamps and SE SLP, U to F to Chis and Guatemala.

EVENING GROSBEAK
Coccothraustes vespertinus montana
Pepitero Norteño   Not illustrated

ID: 6.5–7 in (16.5–18 cm). Highlands N of Isthmus. **Descr.** Ages/sexes differ. ♂: bill pale greenish lemon, legs flesh. *Yellow forehead and short supercilium contrast with blackish crown and brownish face.* Nape and back ochre-brown becoming yellowish on scapulars and rump, uppertail coverts black. Wings and tail black

Hooded Grosbeak

Evening Grosbeak

with *bold white wing panel on inner coverts and tertials.* Throat and upper chest ochre-brownish becoming ochre-yellow on underparts. ♀: *head and upperparts grey-brown, head slightly darker, with ochre-yellow wash on sides of neck,* rump paler; uppertail coverts black, boldly spotted white. Wings and tail black with *whitish wing panel on inner coverts and tertials,* white flash across base of inner primaries, and white tips to outer rectrices. Throat and underparts pale greyish to greyish vinaceous, with whitish undertail coverts, trace of black malar stripe. **Juv:** resembles ♀ but duller, less distinctly marked, soon attains adult-like plumage.

Voice. A whining, slightly nasal *dein* or *djein,* often doubled, a short trilled *dji-i-i-ir,* and a sharp *kyew.* Song in region undescribed (?).

Habs. Arid to semihumid pine and pine–oak woodland. Habits much as Hooded Grosbeak but not in large flocks in the region. Nest cup of twigs, rootlets, etc., at mid- to upper levels in tree or bush. Eggs (N.A.): 2–4, pale blue, spotted and scrawled blackish and brown (WFVZ).

SS: See ♀ Hooded Grosbeak.

SD: U to R resident (1500–3000 m) in interior and on adjacent Pacific Slope from Chih to Dgo, and in cen volcanic belt from Mich to Ver, R to N Oax (Binford 1989; KK).

RA: N.A. to cen Mexico.

OLD WORLD SPARROWS: PASSERIDAE (I)

One member of this Old World family (also known as Weaver Finches) occurs in the region; House Sparrows were first introduced to N.A. in New York in 1850 and have since spread quickly and successfully. Superficially resembling New World buntings or sparrows, Old World Sparrows have 10 primaries. Ages/sexes differ; young altricial; 1st prebasic molt complete.

Food mostly seeds and other vegetable matter. Nests in cavities in buildings, trees, etc., may also build ball-shaped nest of grasses, etc., placed in trees. Eggs: 4–6, whitish with dark brown spots and mottling (WFVZ).

HOUSE SPARROW

Passer d. domesticus　　　Not illustrated
Gorrión Domestico

ID: 5.5–6 in (14–15 cm). Introduced; widespread in towns. **Descr.** ♂: bill black (summer) to dusky horn with dark culmen (winter), legs dull flesh. *Crown grey with broad chestnut postocular stripe curving around paler grey auriculars; lores, throat, and central upper chest black* (bib broadly veiled with grey in fresh basic). Rest of underparts dusky grey. Upperparts chestnut-brown, streaked black, with plain olive-grey rump and uppertail coverts. Wings blackish brown with chestnut lesser coverts, white upper wingbar, and broad rufous-brown edges to greater coverts and secondaries. Tail blackish brown, edged olive-brown. ♀: bill horn to dusky flesh with dark culmen. *Head grey-brown with pale buffy-brown postocular stripe* and slightly paler and greyer lores and auriculars. Throat and underparts smoky grey, washed dirty brownish when fresh. Upperparts brown, broadly streaked blackish, with uniform olive-brown rump and uppertail coverts. Wings blackish brown with pale buff upper wingbar, greater coverts and secondaries edged dull cinnamon-brown. **Juv:** resembles adult ♀ but throat whitish in ♀, dusky in ♂.

Voice. Varied chirps and chips, singly or in series, *chip cheer chrip cheer...*, etc., and dry, churring and chattering calls.

Habs. Cities to villages, parks, farmland, human habitation in general. Singly or in flocks, feeding on ground, in trees, etc.; noisy and social.

SS: ♀ fairly nondescript but note habits, voice.

SD: F to C resident (SL–3000 m) virtually throughout region except Yuc Pen, and U to R and local in Belize (Howell *et al.* 1992*b*).

RA: Native to Eurasia; introduced to New World where has spread from N.A. to cen Panama; disjunctly in S.A.

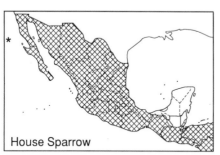

House Sparrow

APPENDIX A: EXTINCT SPECIES

Six species are considered or presumed recently extinct in the region, four of them endemic to Mexico. Two others (Trumpeter Swan, Whooping Crane) are extirpated as regular winter visitors due to declines of the N.A. breeding populations. In addition, Townsend's Shearwater, Socorro Mockingbird, and Guadalupe Junco, all island endemics, seem on the road to extinction unless effective conservation measures are taken now.

GUADALUPE STORM-PETREL

Oceanodroma macrodactyla Mexican endemic, presumed extinct; last reported 1912 (Davidson 1928). Bred commonly in burrows in pine–oak and cypress groves at N end of Isla Guadalupe, BCN. Habits at sea unknown. Demise considered due to feral cat predation. It is conceivable that some birds still survive, Guadalupe being a rugged island that has yet to be explored thoroughly.

A large (9 in; 23 cm), fork-tailed species, resembling an intermediate morph Leach's Storm-Petrel. Blackish brown overall with pale grey upperwing bar. White uppertail-covert patch tipped black, mottled dark at base and with a partial dusky median stripe; white extends onto lateral tail coverts.

CALIFORNIA CONDOR *Gymnogyps californianus* Extinct in the wild; last recorded in Mexico, where known from the mountains of BCN, in 1937 (Koford 1953). The sad recent history of this species in adjacent California is well documented: the last wild birds were taken into captivity in 1987 as part of a controversial attempt to produce a viable captive breeding population.

SOCORRO (GRAYSON'S) DOVE

Zenaida graysoni Plate 69 Mexican endemic, presumed extinct in the wild; last reported 1958 (Villa 1960). Formerly a common resident on Isla Socorro, Islas Revillagigedo, this species apparently vanished between 1958 and 1978, presumably due to cat predation in conjunction with human settlement of the island in 1957. Viable breeding populations exist in N.A. aviaries and reintroduction may be possible if introduced mammals can be eliminated from Socorro.

Larger (10.5–12 in; 26.5–30.5 cm) and more heavily built than Mourning Dove and probably mainly terrestrial. Ages/sexes differ as in closely related Mourning Dove. Orbital ring pale blue, bill dark grey with reddish-

pink base, legs pinkish. ♂: head and underparts deep cinnamon with black streak on lower auriculars, blue-grey hindcrown and nape, and iridescent pink neck patch. Upperparts dark brownish with bold black spots on scapulars, tertials, and inner wing coverts. Remiges dark grey; central rectrices dark brown, outer rectrices grey with black subterminal band, broadly tipped pale grey. ♀: duller overall, blue-grey on head reduced to small nape patch, iridescent neck patch smaller. Juv: resembles ♀ but duller, upperparts with cinnamon-buff feather tips, chest with coarse dusky streaks. Song a hoarse cooing *wah-ah ah ah ah, ahh-ah*', quite different from Mourning Dove (LFB tape).

PASSENGER PIGEON *Ectopistes migratorius* Extinct; last recorded in the wild in 1900 (AOU 1983). A classic case of a once abundant bird wiped out by habitat destruction and over-hunting. Breeding in N.A., apparently it occurred as a winter visitor to the Atlantic Slope of Mexico (Friedmann *et al.* 1950).

IMPERIAL WOODPECKER *Campephilus imperialis* Figure 44 Mexican endemic, presumed extinct. Last reported 1956 (Tanner 1964). Formerly common resident (1800–3000 m) in open pine and pine–oak woodland of Sierra Madre Occidental from Chih to N Jal, and in W mountains of Central Volcanic Belt from Jal to Mich. Demise apparently due to hunting by humans for food, in combination with habitat loss. May survive in remotest areas of Sierra Madre Occidental.

Largest woodpecker in the world (22–24 in; 56–61 cm), ♂ with long, tufted crest, ♀ crest recurved. Eyes pale yellow, bill pale horn. ♂: black overall with black-centered red crest and narrow white braces running into white tertials and secondaries. ♀ and juv have recurved black crests. Voice described

Fig. 44 Imperial Woodpecker

as 'nasal penny-trumpet-like notes' (Nelson 1898).

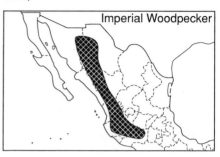

SLENDER-BILLED GRACKLE *Quiscalus palustris* Mexican endemic, extinct. Last recorded 1910 (Dickerman 1965). Formerly resident in marshes at headwaters of Río Lerma, Mex. Little known, may have nested in reedbeds (Hardy 1965), and probably colonial. Demise likely due to habitat loss, the Lerma marshes having been drained steadily and mercilessly over the years.

Resembled small Great-tailed Grackle in shape, tail strongly graduated and slightly keel-shaped. ♂ (13–14 in; 33–35.5 cm): purplish black overall; in fresh plumage, crown, nape, and back edged cinnamon-brown, less so on throat and underparts, with black loral mask reminiscent of Rusty Blackbird. ♀ (11–11.5 in; 28–29 cm): head and underparts cinnamon-brown, paler buffy cinnamon below, with darker lores and auriculars and pale buff supercilium. Upperparts dark rich brown, wings and tail glossy brownish black. **Juv** like ♀ but paler, throat and underparts buffier.

APPENDIX B: SPECIES OF HYPOTHETICAL OCCURRENCE

Several species have, at various times, been attributed to Mexico and northern Central America on the basis of what we consider to be unsatisfactory grounds. These are roughly divisible into old published (usually specimen) records with doubtful data, and recent sight reports. See Acknowledgements for discussion of acceptance criteria for records.

The following species reported from Mexico and northern Central America are considered unacceptable; although many of the recent sight records are undoubtedly valid, they lack sufficient documentation. Where hypothetical reports have been discussed elsewhere, we simply cite the relevant references. When we consider the species' occurrence probable, this is also mentioned with a short identification summary as a Note (NB) under a similar or closely related species in the main species accounts. Identification notes are given for the non-native species discussed below.

RED-NECKED GREBE *Podiceps grisegena* Documentation for a Dec 1974 report from Son (*Am. Birds*, **29**, 594) apparently has been lost (R. Ryan, personal communication) and we consider the record hypothetical. See Horned Grebe.

SOUTHERN FULMAR *Fulmarus glacialoides* While still accepted for Mexico (and thus the northern hemisphere) by AOU (1983), Banks (1988a) has correctly pointed out that this report (a specimen) is unacceptable.

CAPE PETREL *Daption capense* An old report from Guerrero is considered hypothetical (Miller *et al.* 1957).

WHITE-NECKED PETREL *Pterodroma cervicalis* Reported without documentation from waters S of Islas Revillagigedo and around Clipperton (Pitman 1986). Identification criteria separating this species from Juan Fernandez Petrel have been clarified only recently (Spear *et al.* 1992) and we consider all reports from waters in the region hypothetical. See Juan Fernandez Petrel.

BLACK-WINGED PETREL *Pterodroma nigripennis* Reported from waters off southern Mexico and Clipperton (Pitman 1986); while these reports are almost certainly correct, apparently no documentation exists (R. L. Pitman, personal communication) and we consider the records hypothetical. A re-

port from Islas Revillagigedo (Santaella and Sada 1991b) is not credible (description examined). See Cook's Petrel.

WHITE-WINGED PETREL *Pterodroma leucoptera* Reported without documentation from waters S of Clipperton (Pitman 1986) and from Mexico (Howell 1986); the latter record was due to a plotting error and should be omitted. Identification criteria for separating White-winged and Stejneger's petrels (the latter unrecorded by Pitman 1986) have only recently become clear and we consider reports of the former species in the region hypothetical. See Stejneger's Petrel.

BULWER'S PETREL *Bulweria bulwerii* Reported from waters S of Clipperton (Pitman 1986) but no documentation apparently exists for this record (R. L. Pitman, personal communication). Probably also occurs N to vicinity of Islas Revillagigedo (D. Roberson, personal communication) but there are no documented records within 200 miles of the islands. We consider the species' status in the region hypothetical. See Parkinson's Petrel.

LITTLE SHEARWATER *Puffinus assimilis* Reported from Colima as Dusky Shearwater *P. assimilis* (King and Pyle 1957). Given the confusion formerly surrounding the taxonomy and identification of small shearwaters, and present knowledge of their distributions, we agree with R. L. Pyle (personal communication) that this report should be considered hypothetical.

WILSON'S STORM-PETREL   *Oceanites oceanicus*   The occurrence of this species in Mexico has rested for a long time on an old late-Dec sight report from Veracruz (see Loetscher 1941). We consider this unacceptable as a first Mexican record. AOU (1983) cited the species as 'recorded from ... Oaxaca ... also sight records from Michoacan, Guatemala ...', based on Jehl (1974). The Oaxaca record, however, is also only a sight record (see Binford 1989) and, as no conclusive written documentation exists (J. R. Jehl, personal communication), we consider the reports hypothetical. A report from Islas Revillagigedo (Santaella and Sada 1991*b*) is not credible (description examined). See Leach's Storm-Petrel.

FORK-TAILED STORM-PETREL   *Oceanodroma furcata*   A report from BCS (Crossin, in King 1974) lacks documentation and is thus unacceptable (Wilbur 1987). See Ashy Storm-Petrel.

RUFESCENT TIGER-HERON   *Tigrisoma lineatum*   Reports of this species from Mexico (Chiapas, Campeche) and Guatemala (El Golfete) remain unconfirmed and we consider this species' occurrence in the guide region hypothetical. The closest confirmed occurrence is in extreme E Honduras (see Appendix E). See Bare-throated Tiger-Heron.

WESTERN REEF-HERON   *Egretta gularis*   Two dark morph birds were reported flying N off the coast at Tulum, QR, in Apr 1986 (DFD). Not all key identification points were noted, however, and we consider the record hypothetical. See Little Egret.

SCARLET IBIS   *Eudocimus ruber*   An old report from Tamaulipas is considered hypothetical (Miller *et al.* 1957).

BLACK DUCK   *Anas (platyrhynchos?) rubripes*   One individual was reported, with Mottled Ducks, 50 km S of Matamoros, Tamps, in Jan 1979 (JRG, PRG). Not all key identification points were noted, however, and we consider the record hypothetical. See Mottled Duck.

HAWAIIAN DUCK   *Anas (platyrhynchos?) wyvilliana*   An old record for Sinaloa (Friedmann *et al.* 1950) has been questioned although it is still included by AOU (1983).

We consider that the record is almost certainly based on a mislabeled specimen and relegate it to hypothetical status.

BARROW'S GOLDENEYE   *Bucephala islandica*   Saunders and Saunders (1981) gave two reports from Mexico: one, a band recovery, has been questioned (Wilbur 1987); the other, a sight report, lacks documentation. See Common Goldeneye.

GYR FALCON   *Falco rusticolus*   An old specimen reportedly from the Gulf of California is considered hypothetical (Grinnell 1928).

CHUKAR   *Alectoris chukar*   Native to SE Europe and Asia, introduced to N.A. Reported as 'introduced successfully in mts. [=mountains] of N Baja California' (Peterson and Chalif 1973), presumably based on Leopold (1959). We are aware of no recent reports and consider the species' occurrence in the guide region hypothetical, unless a viable population can be shown to exist. Wilbur (1987) overlooked this species in his summary of Baja California birds.

Description: 13–15 in (33–38 cm). **Adult:** bill and orbital ring bright red, legs reddish. Bold black border to creamy lower face and throat. Crown, nape, and chest blue-grey, washed vinaceous on sides of chest, rest of underparts vinaceous cinnamon with bold black bars, often edged chestnut and blue-grey, on flanks. Upperparts olive grey-brown with vinaceous wash on back and outer scapulars. Gruff, dry, chattering calls carry well, *cha-cha-cha* or *ch-ch-ka*, etc. Inhabits arid rocky slopes and open country. In pairs or small flocks.

COMMON (RING-NECKED) PHEASANT *Phasianus colchicus*   Native to Asia, introduced in many parts of the world, including N.A. While Peterson and Chalif (1973) reported this species as 'established in Mexicali valley in extreme NE Baja California', which was presumably based on Leopold (1959), Wilbur (1987) reported only an unsuccessful introduction in W Baja in the 1940s (see Hill and Wiggins 1948) and stated that pheasants had not been recorded in NE Baja. Howell and Pyle (1988), however, saw 5–6 pheasants just S of Mexicali, BCN, in 1983 and suggested that the species may be a wary resident there. We have subsequently learned of other recent reports from NE Baja but consider the status of Common Pheasant in

Mexico hypothetical until a viable population can be shown to exist.

Description: ♂ 30–36 in, ♀ 20–25 in (♂ 76–91 cm, ♀ 51–63.5 cm). Tail long and pointed, much longer in ♂. ♂: bill pale horn, large red facial wattle red, eyes pale yellow, legs greyish. Plumage variable, commonest type as follows. Head and upper neck iridescent blue-green with short ear tufts and white collar. Foreneck, chest, and underparts burnished rufous with violet highlights and fine black scaling and spots, flanks buff to burnished golden with bolder black spots. Upperparts cryptically patterned buff, chestnut, black, and grey. Attenuated rectrices greybrown to rufous-brown with black bars. ♀: buff overall with rufous to blackish mottling, spots, and scallops, mostly on upperparts; tail brown with black bars. Flushes abruptly with rattling whirr of wings. Voice a raucous crowing *ehr-ehk!* Inhabits farmland, open and semiopen country. Singly or in small groups.

PURPLE SWAMPHEN   *Porphyrio porphyrio*   An old specimen from BCN is considered hypothetical (Grinnell 1928).

BRISTLE-THIGHED CURLEW   *Numenius tahitiensis*   Old references to this species from Baja are considered hypothetical (Grinnell 1928).

ROCK SANDPIPER   *Calidris ptilocnemis*   Reports from BCN by Hubbs (1960) rightly have been questioned (Wilbur 1987) and are considered hypothetical.

CHILEAN SKUA   *Catharacta chilensis*   A sight report off the Gulf of Tehuantepec in Nov 1956 (Murphy 1958) has been questioned by Devillers (1977). We consider the record hypothetical. See South Polar Skua.

TUFTED PUFFIN   *Fratercula cirrhata*   An old report from Baja is considered hypothetical (Grinnell 1928).

HORNED PUFFIN   *Fratercula corniculata*   Sightings from BCN (J. Butler, in McCaskie 1975) lack documentation (J. Butler, personal communication) and thus we consider them hypothetical. See Rhinoceros Auklet.

SPOTTED DOVE   *Streptopelia chinensis*   Native to cen and E Asia; introduced to some Pacific islands, and to S California. Reported to occur in 'Tijuana area', BCN, by AOU (1983); not mentioned by Peterson and

Chalif (1973), and overlooked by Wilbur (1987). Spotted Doves are uncommon residents in the San Diego area (Unitt 1984) and have been noted recently in Tijuana (Howell and Webb 1992*f*), but we consider this species' place on the Mexican list hypothetical until a viable population can be shown to exist.

Description: 11–12 in (28–30.5 cm). Long, graduated tail looks squared in flight. **Adult:** eyes amber, bill blackish, legs reddish. Crown blue-grey, rest of head and underparts ruddy-vinaceous with black neck patch densely spotted white. Upperparts greybrown with narrow paler scaly edgings, bluegrey bend of wing, and darker remiges. Central rectrices dark brown, outer rectrices black with bold white tips. Voice a burry cooing *whoo h'whoo* or *wh whoo hoor*. Inhabits urban parks, gardens, etc. Singly or in pairs, strong flight suggests Mourning Dove but heavier.

GREATER ANI   *Crotophaga major*   The inclusion of this species by AOU (1983) is based on two specimens reportedly collected in 1960 in S Tamps, along 'the Rio Tamesi, about 56 km SW of Ciudad Mante' (Olsen 1978). This location, however, is far from the Rio Tamesi, and may be a lapsus for 56 km SE. Recent searches along the upper Rio Tamesi (Howell and Webb 1990*c*; A. M. Sada, personal communication) suggest that, if Greater Anis did occur there, they probably have been extirpated by habitat loss. It is also of interest that F. M. Chapman and L. A. Fuertes did not find this species in their travels along the Rio Tamesi in 1910 (see Loetscher 1941). We consider the species' occurrence in the guide region hypothetical.

DOWNY WOODPECKER   *Picoides pubescens*   One individual was seen in the foothills of the Sierra San Pedro Martír, BCN, in Feb 1984 (RS). Not all key identification points were noted, however, and we consider the record hypothetical. Wilbur (1987) listed two other hypothetical records from the same region. See Hairy Woodpecker.

PILEATED WOODPECKER   *Dryocopus pileatus*   One individual reported flying upriver along the Rio Grande, Texas, 'probably violated Mexican airspace' (Webster 1983); we consider the record hypothetical. See Lineated Woodpecker.

GIANT KINGBIRD *Tyrannus cubensis* The often-cited old specimen from Isla Mujeres was collected by Gaumer (BM spec. examined by Howell) and thus should be considered hypothetical.

CARIBBEAN MARTIN *Progne dominicensis* A BM specimen reportedly from Isla Cozumel (Phillips 1986; spec. examined by Howell) was collected by Gaumer and thus should be considered hypothetical. Reports of *P. dominicensis* from Nay (Escalante 1988) are based on a misunderstanding of Phillips (1986). See Sinaloa Martin.

BLUE JAY *Cyanocitta cristata* A report in Edwards (1978) lacks documentation and should be considered hypothetical; Edwards (personal communication) was of no help in tracing the record and omitted this species from a later work (Edwards 1989). See Steller's Jay.

BLACK-BACKED WAGTAIL *Motacilla (alba?) lugens* Wilbur (1987) reported three sightings (Nov–Apr) thought to pertain to this species, all of which he considered hypothetical; while the April record should be identifiable, no written notes were made and a potentially useful photograph cannot be found (B. Goodhart, personal communication). See White Wagtail.

KIRTLAND'S WARBLER *Dendroica kirtlandii* A report from Mexico (AOU 1983) is based on a barely documented sight report (description examined) and is unacceptable.

RUFOUS-WINGED TANAGER *Tangara lavinia* A report from Guatemala (cited in Griscom 1932; Land 1970) is based on Lantz (1899). This record is in error and refers to a mislabeled *T. larvata* (Jenkinson and Mengel 1979). Reports from Belize (Wood *et al.* 1986) are not credible. See Appendix E.

COMMON GRACKLE *Quiscalus quiscula* An old report from Tamaulipas is unsatisfactory (Miller *et al.* 1957). See Great-tailed Grackle.

APPENDIX C: BIRDS OF PACIFIC ISLANDS

The avifauna of Mexico's offshore Pacific islands is still poorly known, largely due to their relative inaccessibility. As a baseline for future work, the status of species that have occurred on Isla Guadalupe, Alijos Rocks, Las Islas Revillagigedo, and Clipperton Atoll is given below. We also include the Islas Tres Marías, closer inshore but still far from well known. For offshore islands we include surrounding waters out to 200 miles from habitable land.

In general, only the most recent summaries and subsequent observations are given here. Our status definitions may differ from those used in works below; this usually results from different interpretation of earlier data. Our sources of information are as follow. Isla Guadalupe: Jehl and Everett (1985), Oberbauer *et al.* (1989), Mellink and Palacios (1990), Howell and Webb (1992*d*). Alijos Rocks: Pitman (1985), Howell and Webb (personal observations). Islas Revillagigedo: Brattstrom and Howell (1956), Jehl and Parkes (1982), Pitman (1986, and personal communication), Everett (1988*b*), Howell and Webb (1989*b*, 1990*a*, 1992*g*), Santaella and Sada (1991*b*), specimens from UBC, and unpublished data of S. F. Bailey, L. F. Baptista, and H. Gomez de Silva. Clipperton Atoll: Stager (1964), Ehrhardt (1971), Pitman (1986), and Howell *et al.* (1993). Islas Tres Marías: Stager (1957), Grant (1964*a*), Northern (1965), Contreras B. (1987), and Konrad (unpublished observations).

Birds recorded from: Isla Guadalupe, Alijos Rocks, Islas Revillagigedo, Clipperton Atoll, Islas Tres Marías. R: confirmed or suspected resident breeder. S: breeds, present only seasonally, usually in summer. T: transient migrant. W: winter visitor, at least irregularly. V: non-breeding visitor (such as oversummering waterbirds). X: vagrant. E: extinct or extirpated. H: hypothetical occurrence (usually refers to published records which probably are in error or for which documentation is unavailable). ?: when used alone, indicates possible occurrence (mostly unpublished records); when used with another symbol, indicates probable status (suspected status may be given in parentheses). Endemic species and subspecies are indicated in bold in the relevant columns. An asterisk (*) indicates non-native or introduced stock.

	Isla Guadalupe	Alijos Rocks	Islas Revillagigedo	Clipperton Atoll	Islas Tes Marías
Pacific Loon	W				
Pied-billed Grebe	X		X		?
Eared Grebe	W?	X			W
Western/Clark's Grebe	W?				
Short-tailed Albatross	V		X		
Black-footed Albatross	V	V	V		
Laysan Albatross	S	V	S		
Northern Fulmar	V	V			

	Isla Guadalupe	Alijos Rocks	Islas Revillagigedo	Clipperton Atoll	Islas Tes Marías
Juan Fernandez Petrel			V	V	
White-necked Petrel			H	H	
Dark-rumped Petrel			V	V	
Kermadec Petrel			V	V	
Herald Petrel			X (V?)	V?	
Tahiti Petrel			V	V	
Cook's Petrel	V	V	V (T?)	V (T?)	
White-winged Petrel				H	
Black-winged Petrel			H	H	
Bulwer's Petrel			?	H	
Pink-footed Shearwater	H		V	V	
Flesh-footed Shearwater			V		
Wedge-tailed Shearwater			S	V	
Buller's Shearwater				V	
Sooty Shearwater	V	T (V?)	T	T	
Christmas Shearwater				V	
Black-vented Shearwater	S	V			
Townsend's Shearwater		V	S	?	
Newell's Shearwater				?	
Audubon's Shearwater			H	V	
Leach's Storm-Petrel	S	V (S?)	V	V	
Guadalupe Storm-Petrel	E?				
Harcourt's Storm-Petrel			V	V	
Galapagos Storm-Petrel	V	V	V	V	
Black Storm-Petrel	H	V	V		
Markham's Storm-Petrel				V	
Least Storm-Petrel			V		
White-tailed Tropicbird				X	
Red-billed Tropicbird	V	R	R	V	R (S?)
Red-tailed Tropicbird	V		R?	R?	
Masked Booby		R	R	R	
Blue-footed Booby			H		R
Brown Booby			R	R	R
Red-footed Booby			R	R	
Brown Pelican	X		X		R
Double-crested Cormorant	H		H		H
Brandt's Cormorant	R	?			
Pelagic Cormorant	H				
Magnificent Frigatebird	V	V	R		R
Great Frigatebird			R	V	
American Bittern				X	
Least Bittern				X	
Great Blue Heron	W		W	X	W
Great Egret			V	X	W
Snowy Egret			V	X	W
Little Blue Heron					V
Tricolored Heron			X		V
Cattle Egret	V	V	V	X	W

	Isla Guadalupe	Alijos Rocks	Islas Revillagigedo	Clipperton Atoll	Islas Tes Marías
Green Heron			V	X	T (W?)
Black-crowned Night-Heron				X	
Yellow-crowned Night-Heron			R	X	R
White-faced Ibis			V	X	V
White Ibis				X	
White-fronted Goose	X				
Brant	X				
Mallard	X				
Northern Pintail	X			X	
Blue-winged Teal	X		X	X	
Cinnamon Teal	X			X	
Northern Shoveler	X			X	
Eurasian Wigeon				X	
American Wigeon				X	
Redhead					H
Canvasback				X	
Ring-necked Duck				X	
Lesser Scaup				X	
Scaup sp.	X			X	
Red-breasted Merganser	X				X
Black Vulture					H
Turkey Vulture					R
Osprey	V (W?)		W	V (W?)	R
Hook-billed Kite					V (R?)
Sharp-shinned Hawk			X		
Northern Harrier					T
Common Black Hawk					V (R?)
Red-tailed Hawk	E		R		R
Crested Caracara	E				R
American Kestrel	R		R?		W
Merlin			W?		W
Bat Falcon					V
Peregrine Falcon	W?		W	W?	W
Prairie Falcon	H				
Elegant Quail*					R
Virginia Rail				X	
Sora				X	
Purple Gallinule				X	
Common Moorhen				R	
American Coot			X	R	W
Black-bellied Plover			V (W?)	W?	W
American Golden Plover					T
Pacific Golden Plover	W		W	W?	
Snowy Plover					T?
Semipalmated Plover			W	W?	W
Killdeer	V (W?)		W	W?	W
American Oystercatcher					R
Black-necked Stilt			X		T

	Isla Guadalupe	Alijos Rocks	Islas Revillagigedo	Clipperton Atoll	Islas Tes Marías
Greater Yellowlegs				X	
Lesser Yellowlegs			X	X	T
Solitary Sandpiper				X	T
Willet	V		X		W
Wandering Tattler	W		W	W?	T (W?)
Spotted Sandpiper			W	W?	W
Whimbrel			W	W?	W
Long-billed Curlew					T
Marbled Godwit					T
Ruddy Turnstone	W		W	W?	W
Black Turnstone	W				H
Surfbird					T
Sanderling	W		W	W	T
Western Sandpiper	T (W?)		W	W?	T (W?)
Least Sandpiper			W	W?	T
Baird's Sandpiper					T
Pectoral Sandpiper				X	T
Dunlin					T
Short-billed Dowitcher	X				T
Long-billed Dowitcher			X	X	H
Common Snipe	X				T
Wilson's Phalarope				X	T
Red-necked Phalarope			H		
Red Phalarope	W	W	W	W	
Pomarine Jaeger	W	W	W	W	
Parasitic Jaeger	T	W?	W	W	
Long-tailed Jaeger	T		T	T	
South Polar Skua			V	V	
Laughing Gull	X	V	W?	W?	V
Franklin's Gull			X	X	
Bonaparte's Gull			X		X
Heermann's Gull	V		X		V
Ring-billed Gull	V				W
California Gull	W	W	X		
Herring Gull	W	W	W?		W
Thayer's Gull	W				
Western Gull	R	W	H		
Glaucous-winged Gull	W		X		
Black-legged Kittiwake	W	W			
Sabine's Gull	T				
Royal Tern	H				W
Elegant Tern					T
Common Tern				X	
Arctic Tern	T	T	T	T	
Grey-backed Tern				X	
Sooty Tern		S	S	R (S?)	
Black Tern				X	
Brown Noddy			S	R (S?)	

	Isla Guadalupe	Alijos Rocks	Islas Revillagigedo	Clipperton Atoll	Islas Tes Marías
Black Noddy				R (S?)	
White Tern			X	R (S?)	
Xantus' Murrelet	S				
Cassin's Auklet	R	V			
Rhinoceros Auklet	W				
Feral Pigeon (Rock Dove)*	R		R?		R
Red-billed Pigeon					R
White-winged Dove	X				R
Mourning Dove	R		R	X	T
Socorro Dove			E		
Inca Dove					H
Common Ground-Dove			R		R
White-tipped Dove					R
Socorro Parakeet			R		
Mexican Parrotlet					R
Yellow-headed Parrot					R
Yellow-billed Cuckoo				X	
Mangrove Cuckoo					R
Dark-billed Cuckoo				X	
Groove-billed Ani					R
Barn Owl			X (R?)		V (R?)
Great Horned Owl	H				
Elf Owl			R (E?)		T
Burrowing Owl	R		R		V
Short-eared Owl			X		
Lesser Nighthawk					R? (S?)
Common Nighthawk				X	
Pauraque					R
Vaux's/Chimney Swift	X (T?)		X		
White-throated Swift	E				
Lawrence's [Broad-billed] Hummingbird					R
Cinnamon Hummingbird					R
Anna's Hummingbird	R				
Rufous Hummingbird					H
Elegant Trogon					R
Ringed Kingfisher					V
Belted Kingfisher	W		W	W	W
Green Kingfisher					V
Yellow-bellied Sapsucker			X		
Ladder-backed Woodpecker					R
Northern Flicker	E? (W?)				
Northern Beardless Tyrannulet					R
Greenish Elaenia					R
Western Pewee					T
Eastern Pewee				X	
Eastern/Western Pewee				X	
Least Flycatcher	X				
Western (Pacific Slope?) Flycatcher					W

	Isla Guadalupe	Alijos Rocks	Islas Revillagigedo	Clipperton Atoll	Islas Tes Marías
Empidonax sp.				X	
Say's Phoebe	W				
Vermilion Flycatcher					T
Dusky-capped Flycatcher					R (S?)
Brown-crested Flycatcher					S (R?)
Tropical Kingbird					S (R?)
Rose-throated Becard					R
Purple Martin				X	
Martin sp.				X	
Tree Swallow				X	
Mangrove Swallow					H
Northern Rough-winged Swallow	X (T?)		X		
Bank Swallow				X	
Cliff Swallow				X	
Barn Swallow	T?		X	X	
Clark's Nutcracker	X				
Northern Raven			R		
Red-breasted Nuthatch	R (E?)		X		
Rock Wren	R		E		
Happy Wren					R
Bewick's Wren	E				
Socorro Wren			R		
Clarion Wren			R		
Wren sp.				X	
Ruby-crowned Kinglet	E? (W?)				
Mountain Bluebird	X				
Townsend's Solitaire	X				
Brown-backed Solitaire					R
Grey-cheeked Thrush				X	
Swainson's Thrush					W
Hermit Thrush	W				
Grayson's Thrush					R
American Robin	X (W?)				
Varied Thrush	X				
Northern Mockingbird	W?		R		R
Socorro Mockingbird			R		
Sage Thrasher	X				
Blue Mockingbird					R
Americn Pipit	W?		W?	X	T
Cedar Waxwing	X				T
Phainopepla	X				
Loggerhead Shrike	X				H
European Starling*	W (R?)				
White-eyed Vireo			X		
Solitary Vireo					T
Golden Vireo					R
Warbling Vireo					W
Yellow-green Vireo					S

	Isla Guadalupe	Alijos Rocks	Islas Revillagigedo	Clipperton Atoll	Islas Tes Marías
Golden-winged Warbler				X	
Tennessee Warbler			X	X	
Orange-crowned Warbler					T (W?)
Tropical Parula			R		R
Yellow Warbler			W?	X	T (W?)
Magnolia Warbler				X	
Cape May Warbler			X		
Yellow-rumped Warbler	W		W		W
Black-throated Grey Warbler					W
Townsend's Warbler	W?		W?		T
Black-throated Green Warbler			X	X	
Yellow-throated Warbler				X	
Prairie Warbler				X	
Palm Warbler			X	X	
Bay-breasted Warbler			X	X	
Blackpoll Warbler				X	
Black-and-white Warbler	X				W
American Redstart			X	X	W
Prothonotary Warbler				X	
Ovenbird	X			X	W
Northern Waterthrush			X (W?)	X	W
Louisiana Waterthrush					T
Connecticut Warbler				X	
MacGillivray's Warbler					T (W?)
Common Yellowthroat	X (T?)		X (W?)	X	
Wilson's Warbler	X (T?)				T (W?)
Canada Warbler				X	
Red-breasted Chat					R
Summer Tanager	X			X	
Scarlet Tanager				X	
Western Tanager					T
Flame-colored Tanager					R
Northern Cardinal					R
Rose-breasted Grosbeak	X		X		
Black-headed Grosbeak	X				T
Indigo Bunting			X		X
Varied Bunting					T (W?)
Painted Bunting					W
Dickcissel				X	T
Rufous-sided Towhee	E		R		
White-collared Seedeater					V
Chipping Sparrow	X				
Brewer's Sparrow					H
Vesper Sparrow	X				T
Lark Sparrow			X		T
Fox Sparrow	X				
Lincoln's Sparrow	X (T?)				T
White-throated Sparrow	X				

	Isla Guadalupe	Alijos Rocks	Islas Revillagigedo	Clipperton Atoll	Islas Tes Marías
Golden-crowned Sparrow	X				
White-crowned Sparrow	W?				X
Guadalupe Junco	R				
Bobolink				X	
Western Meadowlark	R?				
Brewer's Blackbird	X				
Great-tailed Grackle					V
Brown-headed Cowbird	X (W?)		X		T
Hooded Oriole			X		H
Streak-backed Oriole					R
Scott's Oriole	H				
Oriole sp.				X	
Cassin's Finch					X
House Finch	R				
Red Crossbill	E				
Lesser Goldfinch					R
House Sparrow*	R?				R?

APPENDIX D: BIRDS OF GULF AND CARIBBEAN ISLANDS

As with Mexico's Pacific islands, the avifaunas of islands in the Gulf of Mexico and off the Caribbean coast of the guide region remain poorly known. This may be explained by the relative inaccessibility of some of the islands. It is strange, however, that a list of the birds of the highly visited Isla Cozumel, for example, has not been available previously. Here we give the status of species that have occurred on and in waters surrounding the Campeche Bank, Isla Cozumel, Isla Holbox, the Belize Cays, and the Honduras Bay Islands. We also include Ambergris Cay, Belize, no more than a peninsula of the Yucatan mainland cut off by a canal, since it is visited frequently by birders. (Note that Ambergris has many mainland species not found on any of the other islands.) Isla Contoy, home to breeding Brown Boobies, Brown Pelicans, and Magnificent Frigatebirds, and Isla Cancún (effectively part of mainland QR) are excluded. The relatively inaccessible Swan Islands, claimed by Honduras, are included in Bond's *Birds of the West Indies* (1985).

In general, only the most recent summaries and subsequent observations are given here. Our sources of information are as follow. Campeche Bank: Paynter (1953), Fosberg (1962), Boswall (1978*b*), Tunnell and Chapman (1988), Howell (1989*a*), Sada (1989*a*), and B. M. de Montes (personal communication). Isla Holbox: Paynter (1955*a*), Howell and Johnston (1993). Isla Cozumel: Paynter (1955*a*), Bond (1961), Klaas (1968), Parkes (1970), Paulson (1986), Howell and de Montes (1989), Howell *et al.* (1990), Howell (personal observations), and unpublished data of R. A. Behrstock, J. Coons, H. Gomez de Silva, S. Johnston, K. Kaufman, B. M. de Montes, D. R. Paulson, P. Pyle, R. G. Wilson, and D. E. Wolf. Belize Cays: Russell (1964), Pelzl (1969), Barlow *et al.* (1972), Fosberg and Pelzl (1973), Wood *et al.* (1986), Wood and Leberman (1987), Howell *et al.* (1992*b*), and Howell (personal observations). Honduras Bay Islands: Monroe (1968), Udvardy *et al.* (1973), Udvardy (1976), Marcus (1983), Howell and Webb (1992*b*), and Howell (unpublished observations). Many of the published records for Holbox, Cozumel, and the Honduras Bay Islands are based on Gaumer specimens; we consider hypothetical any species' occurrence based *solely* on a Gaumer specimen since, although some of the species in question (such as northern migrants) certainly occur on these islands, others (such as Plain Chachalaca, Lineated Woodpecker, and Green Jay, all supposedly from Cozumel) are clearly in error (see Parkes 1970). Of the island lists given below, that for Cozumel is probably the most complete. Any other presumed resident breeding species' occurrence there should be documented thoroughly.

Birds recorded from: **Campeche Bank, Isla Holbox, Isla Cozumel, Belize Cays** (* = **Ambergris only**), **Honduras Bay Islands.** R: confirmed or suspected resident breeder. S: breeds, present only seasonally, usually in summer. T: transient migrant. W: winter visitor, at least irregularly. V: non-breeding visitor (such as oversummering waterbirds). X: vagrant. E: extinct or extirpated. H: hypothetical occurrence (usually refers to published records which probably are in error or for which documentation is not available; HG = species known only by Gaumer specimens). ?: when used alone, indicates possible occurrence (mostly unpublished records); when used with another symbol, indicates probable status; suspected status may be given in parentheses. Endemic species and subspecies are indicated in bold in the relevant columns. An asterisk (*) indicates non-native or introduced stock.

We consider two species (Least Grebe and Neotropic Cormorant) reported from the Belize Cays by Pelzl (1969) to be of marginal occurrence in the Cays and thus omitted them from this table.

	Campeche Bank	Isla Holbox	Isla Cozumel	Belize Cays	Honduras Bay Islands
Least Grebe			S?		
Pied-billed Grebe			W		
Audubon's Shearwater			V		
White-tailed Tropicbird			X	H	
Masked Booby	R (S?)			V	
Brown Booby	S	T		V	V
Red-footed Booby	S			R (S?)	
American White Pelican		W			
Brown Pelican		R?	V (R?)	R	V (R?)
Double-crested Cormorant		R	W	R	
Neotropic Cormorant		R?	V (R?)		
Anhinga		W (R?)	W (R?)	V (R*?)	
Magnificent Frigatebird	R	R?	V (R?)	R	V
Least Bittern	T		W	R?	
Bare-throated Tiger-Heron			HG	R*?	
Great Blue Heron	T	R?	V (R?)	R	W (V?)
Great Egret	T	R?	V (R?)	V	W (V?)
Snowy Egret	T	R?	W (R?)	R?	W (V?)
Little Blue Heron	T	R?	W (R?)	R	W (V?)
Tricolored Heron		R?	W (R?)	R	W (V?)
Reddish Egret		R?	W (R?)	R	
Cattle Egret	T	W (V?)	W (V?)	W (V?)	W (V?)
Green Heron	T	R?	R	R	R?
Black-crowned Night-Heron	T				
Yellow-crowned Night-Heron	T	R?	W (R?)	R	R?
Boat-billed Heron			R?	R*?	
Glossy Ibis			V		
White Ibis	T	R?	W (R?)	R	R (E?)

	Campeche Bank	Isla Holbox	Isla Cozumel	Belize Cays	Honduras Bay Islands
Roseate Spoonbill		W	W (R?)	V	V
Wood Stork		W	HG	V	
American Flamingo		V	V	V	
Black-bellied Whistling-Duck			HG		
Muscovy Duck			HG		
Northern Pintail			W		
Blue-winged Teal	T	T (W?)	W		T (W?)
Northern Shoveler			W		
Ring-necked Duck			W		
Lesser Scaup			W	H	
Masked Duck			S?		
Black Vulture		R?	R		
Turkey Vulture		R?	R	V (R?)	R
Lesser Yellow-headed Vulture		W (R?)		V*(R?)	
Osprey	T	W (V?)	R?	R	V
Hook-billed Kite			V (R?)		
Swallow-tailed Kite			T	T*	
White-tailed Kite				V* (R?)	X
Snail Kite		X			
Double-toothed Kite				X* (R?)	
Plumbeous Kite					X (T?)
Black-collared Hawk		V (R?)			
Northern Harrier	T				
Sharp-shinned Hawk	T		HG		
Common Black Hawk		R?	R?	R	R
Great Black Hawk		R?			
Grey Hawk				R*	R
Roadside Hawk		HG	R	R*?	R (E?)
Broad-winged Hawk				X*	
Short-tailed Hawk		R?	R		
White-tailed Hawk		X			
Laughing Falcon				V* (R?)	
American Kestrel	T		W	T	W
Merlin	T	W	W	W	W
Bat Falcon			V		V
Peregrine Falcon	T	W	W	W	W
Plain Chachalaca		HG	HG	R*	R
Great Curassow			R (E?)		
Ruddy Crake			R	R*	
Clapper Rail		R		R	
Grey-necked Wood-Rail			HG		
Rufous-necked Wood-Rail				R	R
Sora	T	T	W	W	W
Purple Gallinule	T		T	W*	HG
Common Moorhen	T		R		
American Coot			W		

	Campeche Bank	Isla Holbox	Isla Cozumel	Belize Cays	Honduras Bay Islands
Limpkin			HG		
Black-bellied Plover	T (W?)	W	W	V	W
American Golden Plover		T			
Collared Plover				H (V?)	
Snowy Plover	T	R?	T (W?)		
Wilson's Plover		R?	W (R?)	R	
Semipalmated Plover		W	W	W	W
Piping Plover		W			
Killdeer		W	W	W	W
American Oystercatcher		R	W (R?)		
Black-necked Stilt		R	R	S (R?)	HG
Northern Jacana			R?		
Greater Yellowlegs	T	T	W	W	T
Lesser Yellowlegs	T	T	T	T	T
Solitary Sandpiper	T	T	W	T (W*)	HG
Willet	T (W?)	W	W	V	
Spotted Sandpiper	T	W	W	W	W
Upland Sandpiper	T	T		T	HG
Whimbrel		W			
Long-billed Curlew		T	HG		
Marbled Godwit		T (W?)	HG		
Ruddy Turnstone	T (W?)	W	V	V	W
Red Knot	T	W?	T		
Sanderling	T (W?)	W	W	W	
Semipalmated Sandpiper		T (W?)	HG	T	
Western Sandpiper	T	W	W	T (W?)	T (W?)
Least Sandpiper	T	W	W	T (W?)	W
White-rumped Sandpiper		T	T	T	
Pectoral Sandpiper		T		T	HG
Buff-breasted Sandpiper					HG
Short-billed Dowitcher	T	W	W		
Common Snipe	T		W	T	
Great Skua				X*	
Pomarine Jaeger	W?	W			
Parasitic Jaeger		W			
Jaeger sp.	T		T		
Laughing Gull	R	V	V	R	V
Ring-billed Gull		W			
Herring Gull	W	W	W		W
Caspian Tern		W	W	H	
Royal Tern	R (S?)	V	V	V	V
Sandwich Tern	R (S?)	V	V	R	
Roseate Tern			T (S?)	S	S
Common Tern	H		V		
Forster's Tern	T	W			X
Least Tern		T (S?)	S?	S	S
Bridled Tern	H		S	S	

	Campeche Bank	Isla Holbox	Isla Cozumel	Belize Cays	Honduras Bay Islands
Sooty Tern	S		T	S	V
Black Tern	T	T		T	
Brown Noddy	S		V	S	V
Black Noddy				E	X
Black Skimmer	T	W	W		
Feral Pigeon (Rock Dove)*	R	R	R*	R	
Pale-vented Pigeon				H	
White-crowned Pigeon			S (R?)	R	R
White-winged Dove	T	R (S?)	R (S?)	R	
Zenaida Dove	X	R	X (R?)		
Mourning Dove	T	W	W		
Common Ground-Dove		R	R	R	R
Ruddy Ground-Dove			R		
Blue Ground-Dove			V (S?)		
Caribbean Dove		H	R	R*	R
Aztec Parakeet		HG		R*	
Yucatan Parrot			R		E?
Red-lored Parrot					R
Yellow-headed Parrot					R
Black-billed Cuckoo		HG			
Yellow-billed Cuckoo	T	T	T	T	HG
Mangrove Cuckoo		R	R	R	R?
Squirrel Cuckoo		HG		R*	
Smooth-billed Ani		HG	R	R	R
Groove-billed Ani		R	R	R*	
Barn Owl			R?		R
Vermiculated Screech-Owl			R		
Ferruginous Pygmy-Owl			HG		
Stygian Owl			V (R?)		
Lesser Nighthawk		S	S	W	T
Common Nighthawk		T	T		
Chordeiles sp.	T	T	T		
Pauraque		R	R	R*?	
Chuck-will's-widow	T				
Yucatan Nightjar			R	X (T?)	
Northern Potoo					R
Chimney Swift	T		T	T*	
Vaux's Swift			R		
Green-breasted Mango		S	R (S?)	R (S?)	R
Canivet's Emerald		R (S?)		R*(S?)	R
Cozumel Emerald		HG	**R**		
Cinnamon Hummingbird		R		R	
Mexican Sheartail		X (R?)			
Ruby-throated Hummingbird	H	T	H	T (W*)	T (W?)
Black-headed Trogon				R*?	
Turquoise-browed Motmot			HG		
Belted Kingfisher	T (W?)	W	W	W	W
Pygmy Kingfisher		?	R		R

	Campeche Bank	Isla Holbox	Isla Cozumel	Belize Cays	Honduras Bay Islands
Yucatan Woodpecker			**R**	R*	**R**
Golden-fronted Woodpecker		R?	**R**	R	**R**
Yellow-bellied Sapsucker			W	W	T (W?)
Ladder-backed Woodpecker		V (R?)	HG		
Lineated Woodpecker			HG	R*	
Ruddy Woodcreeper			HG		
Barred Antshrike			HG	R*	
Northern Beardless Tyrannulet			R		
Greenish Elaenia			R		R?
Caribbean Elaenia		HG	R	R?	
Yellow-bellied Elaenia		S	?	R*	
Common Tody-Flycatcher		X (R?)		R*	
Sulphur-rumped Flycatcher			HG		
Eastern Pewee	T	T	T	T	T
Tropical Pewee			W (R?)		
Acadian Flycatcher			T	T	HG
Least Flycatcher	T	HG	W	W*	W
Empidonax sp.	T	T			
Eastern Phoebe	H				
Vermilion Flycatcher			HG		
Bright-rumped Attila		HG	R		
Yucatan Flycatcher			R	R*?	
Dusky-capped Flycatcher			S	R*?	HG
Great Crested Flycatcher	T		T	T	
Brown-crested Flycatcher		S	S	S	R (S?)
Kiskadee			HG	R*	
Social Flycatcher		R?	HG	R*	
Sulphur-bellied Flycatcher				H	
Piratic Flycatcher				X	
Tropical Kingbird		R	R	R	
Couch's Kingbird				R*	
Eastern Kingbird	T	T	T	T	T
Scissor-tailed Flycatcher	T	T	T		
Fork-tailed Flycatcher				X	
Masked Tityra				R*	
Rose-throated Becard		HG	HG		
Purple Martin	T	T	T	T	T
Grey-breasted Martin			S?		
Tree Swallow		T	W		T (W?)
Mangrove Swallow		R	R	R	
Northern Rough-winged Swallow	T	T	T	T*	
Bank Swallow	T	T	T	T	T
Cliff Swallow	T	T	T	T	T
Cave Swallow			?	H	

	Campeche Bank	Isla Holbox	Isla Cozumel	Belize Cays	Honduras Bay Islands
Barn Swallow	T	T	T	T	T
Green Jay			HG		
Cozumel Wren			R		
Southern House Wren					X
Ruby-crowned Kinglet	X				
Blue-grey Gnatcatcher			R		HG
White-lored Gnatcatcher			HG		
Eastern Bluebird			X		
Veery	T				HG
Grey-cheeked Thrush			T	T	HG
Swainson's Thrush	T	T	T		T
Wood Thrush	T		W	T	W
Clay-colored Thrush			HG		
American Robin		X	X		
Grey Catbird	T	T	W	W	W
Black Catbird		HG	R	R	
Tropical Mockingbird		R	R	R	X
Cozumel Thrasher			R		
Cedar Waxwing			W	W	
White-eyed Vireo	T	T	W	W	W
Mangrove Vireo		R		R	R
Cozumel Vireo			R		
Yellow-throated Vireo			W	W	W
Philadelphia Vireo	T		W		W
Red-eyed Vireo	T	T	T	T	T
Black-whiskered Vireo			H	T	
Yucatan Vireo		HG	R	R	R
Rufous-browed Peppershrike			R	R*	
Blue-winged Warbler			W	T*	W
Golden-winged Warbler	T		T	T	
Tennessee Warbler	T	T	W	W	W
Nashville Warbler	T				
Northern Parula	T	W	W	W	W
Yellow Warbler	T	T	R	W	W
Mangrove [Yellow] Warbler		R		R	R
Chestnut-sided Warbler	T		T	T	T
Magnolia Warbler	T		W	W	W
Cape May Warbler	T		W	W	W
Black-throated Blue Warbler	T	W	W	W	
Myrtle [Yellow-rumped] Warbler	H	T	W	W	W
Black-throated Green Warbler	T	HG	W	W	W
Blackburnian Warbler	T		T	T	T
Yellow-throated Warbler	T	W	W	W	W
Prairie Warbler		T	W	W	W
Palm Warbler	T	W	W	W	W
Bay-breasted Warbler			T	T	T

	Campeche Bank	Isla Holbox	Isla Cozumel	Belize Cays	Honduras Bay Islands
Blackpoll Warbler		T	T		
Cerulean Warbler	T			T	
Black-and-white Warbler	T	W	W	W	W
American Redstart	T	W	W	W	W
Prothonotary Warbler	T	T	T (W?)	W	HG
Worm-eating Warbler			W	W	W
Swainson's Warbler	T		W	T (W?)	
Ovenbird	T	HG	W	W	W
Northern Waterthrush	T	W	W	W	W
Louisiana Waterthrush			H	T	HG
Kentucky Warbler	T		W	T	W
Connecticut Warbler				T	
Mourning Warbler				T	T
Common Yellowthroat	T	W	W	W	W
Grey-crowned Yellowthroat			?	H*	
Hooded Warbler	T	T	W	W	W
Wilson's Warbler	T			T	
Canada Warbler	T			T	
Yellow-breasted Chat			W	W	W
Bananaquit		HG	R		
Red-legged Honeycreeper			X		
Scrub Euphonia			HG		
Stripe-headed Tanager			R		
Red-throated Ant-Tanager				R*	
Rose-throated Tanager			R		
Summer Tanager	T	T	W	W	W
Scarlet Tanager	T		T	T	HG
Black-headed Saltator				R*?	
Northern Cardinal		R	R		
Rose-breasted Grosbeak	T	HG	W	W	W
Blue Grosbeak	T	T	W	T	
Indigo Bunting	T	W?	W	W	W
Painted Bunting		HG	W	W	
Dickcissel	T	T	HG	T	T
White-collared Seedeater			R	R	
Yellow-faced Grassquit		R	R		
Chipping Sparrow				H	HG
Clay-colored Sparrow	X		X		
Lark Sparrow		T*			
Savannah Sparrow	T	W	HG	W	
Grasshopper Sparrow		W	W	T	
Lincoln's Sparrow				T*	
White-crowned Sparrow	X	X		X*	
Bobolink	T	T	HG	T	T
Red-winged Blackbird		R?	HG		
Yellow-headed Blackbird	X				

	Campeche Bank	Isla Holbox	Isla Cozumel	Belize Cays	Honduras Bay Islands
Melodious Blackbird		R		H*	
Great-tailed Grackle		R	R	R	R
Bronzed Cowbird		R?	S?	R*?	
Orchard Oriole		T	T	T	T
Hooded Oriole		R	R	R	
Orange Oriole				W* (R?)	
Yellow-backed Oriole				R*	HG
Altamira Oriole			HG		
Baltimore Oriole	T	T	T	T	T
Yellow-billed Cacique				R*	

APPENDIX E: ADDITIONAL SPECIES OF EASTERN HONDURAS

While we have chosen to delimit the area covered by this guide along bio-geographic lines, we realize that visitors to Honduras are unlikely to restrict their explorations accordingly! The following, therefore, is an annotated list of the 50 species recorded from the Atlantic Slope lowlands (SL–500 m, unless stated otherwise) and foothills of eastern Honduras (Monroe 1968; Marcus 1983; R. S. Ridgely, personal communication). The rain forests of SE Honduras are often referred to as the Olancho (including rain forests in *both* the departments of Olancho and Gracias a Dios). When species described below are similar to species in the main text, their differences are discussed under SS of the latter. Distinct ssp occurring in E Honduras (such as White Hawk, Mealy Parrot) are described in the main text. Calls and songs described below are from Stiles and Skutch (1989) or SNGH's field work in Honduras, Costa Rica, and Panama.

Of the 50 E Honduras species, 35 are known only from the Olancho, while the remaining 15 range N to the coastal strip E of the Sula Valley, between Tela and Trujillo. We consider two species (Fulvous-bellied Antpitta, accepted by Monroe (1968); and Palm Tanager, *Thraupis palmarum*, accepted by AOU (1983) as hypothetical; the source of the Palm Tanager record is unclear (B. L. Monroe, personal communication). The birds of the Olancho are still poorly known and several other S C.A. species can be expected to occur there, such as Yellow-crowned Euphonia (*Euphonia luteicapilla*) (see Monroe 1968).

RUFESCENT TIGER-HERON *Tigrisoma l. lineatum* 25–29 in (63.5–73.5 cm). A solitary, *retiring species of small forest pools and streams*, rarely if ever in open situations; for description see NB under Bare-throated Tiger-Heron and Plate 1.

U resident in the Olancho; a report from Lake Yojoa, W Honduras, appears to be in error (BLM, personal communication). E Honduras to W Ecuador and N Argentina. See Appendix B.

GREEN IBIS *Mesembrinibis cayennensis* 22–25 in, B 4–5 in (56–63.5 cm, B 10–12.5 cm). A *heavy-bodied, broad-winged, retiring ibis of forest swamps* and pools, river bars, etc.; usually in pairs or small groups. Bill and legs dull pale greenish, bare facial skin grey. **Adult:** *looks dark overall.* Head, neck, and underparts sooty blackish with metallic green sheen on sides of neck and hindneck, dull bronzy-green sheen to underparts. Upperparts metallic bronzy green. Juv duller overall, head and neck with little gloss. Deep,

rolling calls, perched or in flight may suggest Limpkin; often calls loudly when flushed. See Glossy Ibis (unlikely in same habitat).

U resident in the Olancho. E Honduras to Brazil.

SEMIPLUMBEOUS HAWK *Leucopternis semiplumbea* 13.5–15 in, W 22–25 in (34–38 cm, W 56–63 cm). A fairly chunky, relatively short-tailed and broad-winged hawk of *forest canopy and understory; does not soar.* **Adult:** eyes yellow, *cere and legs orange. Head and upperparts slaty grey*, becoming black on distal uppertail coverts; *tail black with white median band. Throat and underparts creamy white*, sides of neck with a few dark streaks. Underwings: coverts creamy white, remiges barred dark grey and pale grey. **Juv:** throat and chest have sparse, fine dark streaks which may extend to belly. Calls include a plaintive squealing *kreee-eeoo* and *ree-a*, a clear, whistled *h-h-hoo-lee*, and a thin, drawn-out, hissing squeal.

U to F resident in the Olancho. E Honduras to NW Ecuador.

GREY-HEADED CHACHALACA *Ortalis cinereiceps* 20–24 in (51–61 cm). *Replaces Plain Chachalaca in the Olancho*, and much like that species in habits. Bill and legs greyish, naked throat skin reddish. Head greyish, neck, upper chest, and upperparts olive-brown; mostly *rufous-chestnut primaries striking in flight.* Underparts pale buffy grey becoming dusky olive-grey on undertail coverts. Tail blackish with green sheen and broad, dirty buffy-white tips to all but central rectrices. Raucous calls and rollicking choruses suggest allopatric Plain Chachalaca.

F to C resident in the Olancho. E Honduras to NW Colombia.

BLACK-EARED [RUFOUS-FRONTED] WOOD-QUAIL *Odontophorus melanotis** 9–10 in (23–25.5 cm). *Replaces Spotted Wood-Quail in lowlands E of Sula Valley,* and habits, shape, etc., much like that species. **Adult:** bill black, legs grey, orbital skin dark. *Crown and erectile crest deep rufous, face and throat black.* Upperparts dark grey-brown, scapulars and wing coverts with indistinct black, grey-brown, and buff barring. Underparts rufous, thighs barred dusky, undertail coverts barred dark brown and greyish buff. **Juv:** back streaked white, throat pale brown, underparts sparsely barred and spotted black. Voice a gabbling *kooLAWik kooLAWik, kooKLAWK kooK-LAWK* (Stiles and Skutch 1989). See Spotted Wood-Quail (mostly at higher elevations in E Honduras).

U (to F?) resident (SL–500 m) E of Sula Valley. E Honduras to Panama.

TAWNY-FACED QUAIL *Rhynchortyx cinctus pudibundus* 7.5–8 in (19–20.5 cm). *A small, stub-tailed, elusive forest quail.* In pairs or small groups, rarely flies, prefers to run or freeze when alarmed. ♂: bill and legs grey. *Face bright orange-tawny, crown and eyestripe dark brown.* Upperparts greybrown, scapulars and wing coverts cryptically patterned with grey, buff, black, and rufous. Chest grey, underbody ochraceous buff, thighs and undertail coverts barred dark brown. ♀: *supercilium and throat whitish*, chest tawny-brown; rest of underparts whitish, barred black. Voice a clear, sad, ventriloquial, dove-like *cooo* or *toot* (Stiles and Skutch 1989).

U resident E of Sula Valley. E Honduras to Ecuador.

WHITE-THROATED CRAKE *Laterallus albigularis cinereiceps* 5.5–6 in (14–15.5 cm). *Counterpart of Ruddy Crake in S C.A.,* inhabiting wet grassy fields, roadside ditches, marshes, etc. **Adult:** eyes red, bill blackish with pale green base, legs olive-grey. *Resembles ♂ Ruddy Crake but crown, hindneck, and back washed brown, throat whitish; belly, flanks, and undertail coverts barred black and white.* **Juv:** resembles juv Ruddy Crake but often shows a trace of whitish flank barring; quickly attains adult-like plumage. Prolonged churring trill much like Ruddy Crake; also quiet contact-call notes, *kik* or *kek.* See Ruddy Crake; area of overlap little known, allopatric (?).

Probably F to C resident in the Olancho, but few records. E Honduras to Ecuador.

GREAT GREEN MACAW *Ara a. ambigua* 28–31 in (71–78.5 cm). In pairs or small groups, usually seen flying over humid evergreen forest or adjacent areas. *Resembles allopatric Military Macaw* (bill blackish) but slightly paler overall, averages larger, face more strongly washed reddish. Deep, raucous cries, much as Military Macaw.

U resident in the Olancho. E Honduras to Ecuador. See NB under Military Macaw.

VIOLET-HEADED HUMMINGBIRD *Abeillia guimeti merrittii* 3–3.2 in (7.5–8 cm). Perches and feeds at low to mid-levels in forest, at edges, etc.; sings from exposed twigs at mid-levels. *Black bill very short and straight,* 1× head, tail broad and squared. ♂: *crown and throat glittering violet, auriculars blue-green with bold white postocular spot.* Nape blue-green, upperparts emerald green becoming blue-green on uppertail coverts. Tail deep blue-green, to blue-black distally, with narrow whitish tips to outer rectrices. ♀: *median crown glittering blue,* crown sides, nape, and upperparts emerald green. *Face dusky with bold white postocular spot,* throat and underparts pale grey to smoky grey, mottled green on sides. White tips of outer rectrices often bolder than ♂. **Imm ♂:** resembles ♀ but crown duller, greenish blue, throat with some violet feathers. Calls suggest Emerald-chinned Hummingbird but harder; song a high, slightly buzzy *tsi-s-see...* or *tiz-i-it...,* often repeated insistently. See ♀ Emerald-chinned Hummingbird.

U to F resident (near SL–750 m) E of Sula Valley. E Honduras to Brazil. Traditionally placed in the monotypic genus *Klais*.

SNOWCAP *Microchera albocoronata* 2.6–2.8 in (6.5–7 cm). A tiny hummingbird of forest and edge; perches and feeds low to high. Flight fast, not slow and bee-like; often flashes tail open while feeding. Black bill straight, 1× head. ♂: *deep glittering purple overall with snow-white crown*, sooty blackish throat, white thighs. *Central rectrices bronzy, outer rectrices white*, narrowly edged and broadly tipped dusky bronze. ♀: crown, nape, and upperparts emerald green. Face dusky with small whitish postocular spot, *throat and underparts pale grey*. Central rectrices bronzy green, outer rectrices blackish, broadly tipped white and with whitish bases obvious in flight. Imm ♂: resembles ♀ but soon (?) shows purplish patches on underparts, whitish supraloral stripe. Older imms mostly purple below, upperparts mottled purple and bronzy green with white patches in crown. Mostly silent; high twitters when fighting. Song a soft, sputtering warble *tsitsup tsitsup tsitsup tsew tttt-tsew*, etc. (Stiles and Skutch 1989).

U to F resident in the Olancho. E Honduras to Panama.

BRONZE-TAILED PLUMELETEER *Chalybura urochrysia* cf. *melanorrhoa* 4.7–4.9 in (12–12.5 cm). Favors humid second growth, old plantations, *Heliconia* thickets, etc. *Bill stout, straightish*, 1.3× head, *tail long, broad, and cleft*. ♂: black *bill pinkish below at base, feet pinkish red*. Head and upperparts deep emerald green becoming bronzy on rump, dark purplish on uppertail coverts. *Tail dark bronzy green*. Throat and chest glittering emerald-green to blue-green, belly sooty, mottled green, undertail coverts deep purple. ♀: face dusky with slight whitish postocular spot, throat and underparts pale grey, mottled green from sides of neck to flanks. Outer rectrices tipped pale grey. Fairly loud, rippling chips when perched. Song a soft, scratchy, warbled *ter-pleeleeleelee ter-pleeleeleelee ter-pleeleeleelee ter-plee* (Stiles and Skutch 1989).

U resident in the Olancho. E Honduras to NW Ecuador. Sometimes known as Red-footed Plumeleteer.

BLACK-THROATED TROGON *Trogon rufus tenellus* 10–10.5 in (25.5–26.5 cm).

A yellow-bellied trogon of humid forests, usually at low to mid-levels in understory. ♂: *bill greenish horn, orbital ring pale blue*. Face and throat blackish, *crown, nape, and upperparts green*; green chest separated from yellow belly by white band. Wing panel vermiculated white and black (looks grey), blackish primaries white on outer webs. Uppertail blue-green, tipped black. *Undertail barred black and white, outer rectrices broadly tipped white*. ♀: bill greyish horn, orbital ring blue-grey. Head, chest, and upperparts brown, white crescent in front of and behind eye. Uppertail rufous, tipped black. Wing panel vermiculated brown and black (looks paler brown). Yellow belly below greyish chest band. Undertail as ♂. Imm ♂: resembles ♂ but wing panel browner, chest may be mottled brown, undertail has narrower black bars so appears whiter overall, uppertail rufous distally. A plaintive *kyow kyow* or *caow caow*, and *kyow kyow-kyow*, recalling Collared Trogon. See Violaceous and Black-headed trogons.

F resident E of Sula Valley. E Honduras to Ecuador and N Argentina.

RUFOUS MOTMOT *Baryphthengus ruficapillus* cf. *costaricensis* 18–20 in (45.5–50.5 cm). A large motmot of humid evergreen forest; tail racket-tipped. **Adult:** *head, chest, and upper belly cinnamon with black mask, small black chest spot*; lower belly and undertail coverts blue-green. Upperparts green, flight feathers bluer, rackets tipped black. **Juv:** duller, without black on chest, tail lacks rackets. Song a resonant, fairly rapid *whu-whu-whup* or *hoo-oo-oo*, suggesting a sheet of aluminum being flexed quickly. See Broad-billed Motmot.

F resident in the Olancho. E Honduras to Brazil and N Argentina. Some authors consider populations in S S.A. as a separate species.

BROAD-BILLED MOTMOT *Electron platyrhynchum minor* 12–13 in (30.5–33 cm). A forest species, habits much like closely related Keel-billed Motmot. Tail racket-tipped (often not strongly so), blackish bill very broad. **Adult:** *head and chest cinnamon with green chin, black mask, 2 large black chest spots*; belly and undertail coverts green. Upperparts green, flight feathers bluer, rackets tipped black. **Juv:** duller, rufous underparts replaced largely by green, crown washed olive, short blue-green streak above eye (Stiles and Skutch 1989). Voice much like Keel-billed Motmot but slightly higher, less rough. See Rufous and Keel-billed motmots.

Apparently U resident E of Sula Valley. Honduras to Ecuador and Brazil. See NB under Keel-billed Motmot.

WHITE-FRONTED NUNBIRD   *Monasa morphoeus grandior*   11–12 in (28–30.5 cm). A fairly slender, long-tailed puffbird of forest and edge, clearings, etc.; often in groups of 3–4 birds which associate with mixed-species flocks. **Adult:** *bill long, bright pinkish red,* feet dark grey. *Slaty blue-grey overall, head blackish with bushy white forehead and chin;* flight feathers blackish. **Juv:** duller, bill paler, facial tufts pale cinnamon. Calls include a loud, screaming *sheek sheek-sheek,* which may run into a chatter; other loud cries often given by groups.

U to F resident in the Olancho. E Honduras to Brazil.

YELLOW-EARED TOUCANET   *Selenidera spectabilis*   14.5–16 in (37–41.5 cm). In pairs or small groups in forest canopy, clearings with fruiting trees, etc. ♂: orbital skin blue, green, and yellow, *bill blackish with broad greenish-yellow stripe on culmen,* legs grey-blue. *Head and underparts black with yellow auricular patch,* fiery-yellow flash on flanks, chestnut thighs, and red undertail coverts. *Upperparts olive-green,* remiges slaty grey. ♀: crown and nape chestnut-brown, black auriculars lack yellow patch. A dry, rasping crackle, rather frog-like; may recall Keel-billed Toucan.

U to F resident E of Sula Valley. E Honduras to NW Colombia.

CHESTNUT-MANDIBLED (SWAIN-SON'S) TOUCAN   *Ramphastos swainsonii* 21–24 in (53.5–61 cm). Humid evergreen forest and edge. *Resembles Keel-billed Toucan but bill deep chestnut (looks blackish) with broad yellow stripe along culmen,* red border to yellow chest slightly wider. Calls include a loud, squeaky, jerky *kyew kik'* or, more fully, *k-yew ki-di' ki-dit',* at times repeated incessantly. See Keel-billed Toucan.

F resident in the Olancho. E Honduras to N Colombia. Sometimes considered conspecific with *R. ambiguus* of S.A.

RUFOUS-WINGED WOODPECKER *Piculus simplex**   7–7.5 in (18–19 cm). Like other *Piculus,* an easily overlooked forest woodpecker; mostly at mid- to upper levels, may join mixed-species flocks. ♂: *eyes pale greyish. Red crown, nape, and moustache border olive auriculars.* Upperparts golden olive, *remiges boldly barred rufous,* lateral uppertail coverts barred buffy lemon. Tail blackish. Throat and chest dusky olive, flecked buffy lemon, rest of underparts barred buffy lemon and dark olive. ♀: crown and moustache greyish olive, thus head mostly dark with red nape band. **Juv:** resembles ♀ but ♂ has forecrown and moustache mottled red. Call a sharp *keeeah,* suggesting Golden-olive Woodpecker. Also a loud, emphatic series of down-slurred, slightly nasal notes, *heew heew heew...* (Stiles and Skutch 1989).

U to F resident E of Sula Valley. Honduras to W Panama. Sometimes considered conspecific with the forms *callopterus* and *leucolaemus* of E Panama and S.A.

SLATY SPINETAIL   *Synallaxis brachyura nigrofumosa*   6–6.5 in (15–16.5 cm). Skulks in humid second growth, overgrown fields, and clearings; usually detected by voice. **Adult:** *dark brownish slate overall,* greyer below, *with bright rufous crown, nape, and panel from sides of chest across wing coverts* to base of remiges. Chin flecked whitish, lower throat blackish. **Juv:** crown, nape, and upperparts dusky brownish, face, throat, and underparts dusky grey, palest on throat, washed brown on flanks and undertail coverts; rufous wing panel duller. Common call a gruff, slightly chattering *chukr-r-r-r* or *churr.* See Rufous-breasted Spinetail (sympatric E to around Lancetilla).

F resident E of Sula Valley. Honduras to NW Peru.

PLAIN-BROWN WOODCREEPER   *Dendrocincla fuliginosa ridgwayi*   8–8.5 in (20.5–21.5 cm). A typical *Dendrocincla* (see p. 467) of humid evergreen forest. Ages/sexes similar. *Stout bill blackish. Dull tawny-brown overall,* darker above, *with slightly paler and greyer face and throat, and dark malar stripe.* Remiges rufous, tail rufous-chestnut. Calls include a sharp, buzzy, slightly nasal *cheehrr,* and a rising, metallic screech *shriek.* See Tawny-winged Woodcreeper.

U resident in the Olancho. E Honduras to Ecuador and N Argentina.

LONG-TAILED WOODCREEPER   *Deconychura longicauda typica*   7.5–8 in

(19–20 cm). A slender, fairly long-tailed woodcreeper with a *slender, medium-long, straight bill.* Feeds low to high, spiraling up trunks; retiring but often joins mixed-species flocks. Ages/sexes similar. Bill blackish above, grey below, legs greyish. Head dark brown with *inconspicuous buff streaks on crown*, paler lores, buff supercilium, and buff-streaked auriculars. Upperparts rufous-brown becoming rufous on remiges and rufous-chestnut on tertials, uppertail coverts, and tail. *Throat buff, underparts dusky brown, spotted buff on chest and fading to dull buff streaks on upper belly*; undertail coverts rufous. A long series of soft but resonant *chip* or *chrit* notes starts slowly, speeds up into a slow trill, then slows into a jerky rhythm (Stiles and Skutch 1989).

U resident in the Olancho. E Honduras to Brazil.

FASCIATED ANTSHRIKE *Cymbilaimus lineatus fasciatus* 6.8–7.3 in (17–18 cm). A fairly stocky antshrike with bushy crest, *very stout, hooked, blackish bill.* Skulks in humid second growth, forest edge, etc.; typically in pairs. ♂: *eyes red. Crown, nape, and upperparts black with narrow white barring on nape and upperparts. Face, throat, and underparts densely barred black and white.* ♀: *crown rufous, forehead, nape, and upperparts blackish, barred buffy cinnamon. Face, throat, and underparts pale buffy cinnamon, narrowly barred blackish.* **Juv**: resembles ♀ but crown barred black, underparts paler. Song a series of 6–8 hollow, laughing notes, rising and falling, *ca-ca-ca-ca ca-ca*, etc., may suggest Laughing Falcon; also a gruff, chattering scold. See Barred Antshrike.

U resident in the Olancho. E Honduras to Brazil.

STREAK-CROWNED ANTVIREO
Dysithamnus striaticeps 4.2–4.5 in (10.5–11.5 cm). Replaces Plain Antvireo at lower elevations in E Honduras, and much like it in shape, habits, etc., but bill heavier. ♂: resembles Plain Antvireo but *crown coarsely streaked black and blue-grey, throat and chest coarsely streaked blackish and white, flanks rich tawny.* ♀: crown striped black and rufous, upperparts olive-brown, underparts more extensively washed rich tawny. Calls a mellow whistled *wheu heu* and *heu*. Song an accelerating series of rich whistled notes that rises, grows louder, then falls, ending in a roll (Stiles and Skutch 1989). See Plain Antvireo.

U to F resident in the Olancho. E Honduras to Costa Rica.

CHECKER-THROATED ANTWREN
Myrmotherula fulviventris costaricensis
4–4.3 in (10–11 cm). Active at low to mid-levels of forest understory, usually in pairs or small groups with mixed-species flocks. ♂: *eyes pale*, bill black above, grey below, legs grey. Head and upperparts dusky olive-brown becoming rufous on uppertail coverts, scalloped blackish on crown and nape. Blackish-brown *wing coverts boldly tipped buffy cinnamon*, remiges edged rufous-brown, tail rufous-brown. *Throat boldly checkered black and white*, underparts dusky cinnamon-buff, becoming brighter cinnamon on flanks and undertail coverts. ♀: throat and upper chest buffy cinnamon, rest of underparts duskier, typically becoming dusky brownish on flanks. Calls include sharp hummingbird-like chips, *chip-chip...* or *siip-siip...*, often repeated insistently. ♀ White-flanked Antwren has dull eyes, duller wingbars, whitish throat, and brighter underparts.

F resident in the Olancho. E Honduras to Ecuador.

WHITE-FLANKED ANTWREN *Myrmotherula axillaris albigula* 3.7–4 in (9.5–10 cm). Active at low to mid-levels of forest understory, usually in pairs with mixed-species flocks. ♂: bill blackish, legs grey. *Velvety black overall with bold white dots on wing coverts, fluffy white flanks*; white scapular flashes usually mostly concealed. Tail tipped white. ♀: bill grey below. Crown, nape, and upperparts dusky olive-brown becoming brighter on uppertail coverts. *Wing coverts dark with dull cinnamon tips*, flight feathers edged tawny-brown. *Throat whitish becoming ochraceous buff on underparts.* **Imm** ♂: slaty grey overall, mottled black on throat and chest, wing dots often duller, tinged cinnamon, white flanks less striking. Calls include a nasal *chew tew* and *chee-u*, and a rapid, squeaky, nasal chatter, *chi-ri-ri-ri-ri*. See Dot-winged Antwren, ♀ Checker-throated Antwren.

U to F resident in the Olancho. E Honduras to Ecuador and Brazil.

CHESTNUT-BACKED ANTBIRD *Myrmeciza exsul* cf. *exsul* 5.7–6 in (14.5–15 cm). A fairly heavy-bodied antbird found singly or in pairs, near ground in shady forest understory. ♂: *orbital skin pale blue*, bill and legs blackish. *Head and chest slaty blackish* becoming dark slaty grey on belly, hind flanks and undertail coverts chestnut. *Upperparts chestnut.* ♀: nape washed brown, upperparts duller. Throat sooty brownish grey, underparts dusky chestnut. **Juv:** sooty blackish overall with some chestnut wing edgings; quickly attains adult-like plumage. Song imitated easily, a whistled *pu, pu pu* or *whu hu*, suggesting Black-faced Antthrush.

Probably U resident in the Olancho, only one record. E Honduras to W Ecuador.

SPOTTED ANTBIRD *Hylophylax naevioides capnitis* 4.3–4.7 in (11–12 cm). A plump, short-tailed antbird, usually low in shady forest understory; often attends army-ant swarms. ♂: bill blackish, legs grey. Crown and auriculars brownish grey, scalloped black, upperparts chestnut; white back patch usually concealed. *Wings blackish with bold chestnut wingbars and tertial edgings*, rufous flash across remiges. Tail dusky tawny-brown with black subterminal band, cinnamon tips. *Throat black, chest and median underparts white with necklace of bold black spots* across chest, flanks greyish. ♀: head browner, upperparts duller, rufous-chestnut, wingbars and tertial spots paler, cinnamon. Throat and median underparts whitish with necklace of bold dusky spots across chest. **Juv:** resembles ♀ but throat and underparts dirty whitish, mottled dusky brown. Calls include a sharp *schip* or *cheet*, often doubled; song a slightly eerie, accelerating, then slowing series of whistles, *wheew wheew wheew wheeu*, etc.

F to U resident E of Sula Valley. E Honduras to Ecuador.

BICOLORED ANTBIRD *Gymnopithys leucaspis olivascens* 5.5–6 in (14–15 cm). A fairly stout-bodied, shortish-tailed antbird; habits much as Spotted Antbird with which it often occurs at ant swarms. **Adult:** broad orbital ring pale blue, bill blackish, legs blue-grey. Crown, nape, and *upperparts chestnut-brown*, wing and tail edgings slightly brighter. *Black mask sets off blue orbital skin. Throat and median chest white*, sides, flanks, belly, and undertail coverts dark sooty tawny-brown with black mottling at sides of chest. **Juv:** duller, with sooty throat and chest, greyish orbital skin. Common call a whining *chirr*; song a slightly querulous series of whistles, accelerating then fading to-ward end, *whee-whee-whee-whee-whee-hee-hee-hee-hee-hee-heer*, etc.

F to U resident E of Sula Valley. E Honduras to Brazil. Birds from W of the Andes sometimes considered specifically distinct, *G. bicolor.*

OCELLATED ANTBIRD *Phaenostictus mcleannani saturatus* 8–8.5 in (20.5–21.5 cm). Fairly large and long-tailed; an unmistakable and striking antbird of humid evergreen forest interior; often with Spotted and Bicolored antbirds at ant swarms. **Adult:** *bold orbital ring bright blue*, bill blackish, legs flesh. Crown grey-brown, nape rufous, extending around below black bib to rufous chest band. *Upperparts blackish with bold, scaly pale cinnamon edgings;* tail blackish, outer rectrices narrowly tipped white. *Underparts rufous with bold black spots* mostly on chest. **Juv:** crown feathers blackish, edged pale grey, upperparts with cinnamon tips forming bars and scallops; throat and underparts sooty blackish brown, spotted rufous on chest. Calls include a reedy trill and a slightly buzzy *seeirrr*; song a series of high, thin, piping notes accelerating, at times into a trill, then ending abruptly, *pi-pi-pi-pi* ...

U to F resident in the Olancho. E Honduras to Ecuador.

BLACK-FACED ANTTHRUSH *Formicarius analis umbrosus* 7–7.5 in (18–19.5 cm). *Lowland counterpart of Mexican Antthrush in E Honduras*, which there occurs in foothills. Ages differ, sexes similar. **Adult:** bill black, legs dusky flesh. *Face and throat black, orbital ring pale blue*, lore spot whitish, sides of neck rufous-chestnut. Upperparts rich dark brown, crown blacker, uppertail coverts chestnut. Tail blackish. *Underparts sooty grey without rufous-chestnut forecollar*, flanks washed olive, undertail coverts cinnamon. Song a sharp whistle followed by 2–4 (rarely more) slow-paced whistles, *pee, piu piu*, etc., 1st note higher; suggests Chestnut-backed Antbird. See Mexican Antthrush.

F to C resident in the Olancho. E Honduras to Bolivia. Traditionally considered conspecific with Mexican Antthrush, *F. moniliger.*

SPECTACLED ANTPITTA *Hylopezus perspicillatus* cf. *intermedius* 5.3–5.7 in (13.5–14.5 cm). Plump, stub-tailed, and long-legged; a retiring inhabitant of relatively open forest floor. Moves by hopping like other antpittas. Bill blackish above, pale below, legs grey. Crown, nape, and upperparts slaty grey becoming olive-brown on tertials, back sparsely streaked cinnamon; wings browner with *bold cinnamon spots on coverts*, cinnamon outer webs to primaries. Face dusky buff with *bold pale cinnamon spectacles*, blackish moustache. Throat and median underparts white with cinnamon wash and bold black stripes across chest, flanks and thighs bright orange-cinnamon. Song a far-carrying series of 6–9 hollow, mellow whistles, easily imitated, the 1st 3 notes rising, the rest of song fading away, *whew whew whew whew   whew, whew,   whew,   whew.*

F resident in the Olancho. E Honduras to Ecuador. **Fulvous-bellied Antpitta** *H. fulviventris* (5.3–5.7 in; 13.5–14.5 cm) has been reported from the Olancho on basis of vocalizations (Monroe 1968) and probably is an U to F resident there; lacks bold spectacles and black moustache, face dusky cinnamon with brighter supraloral stripe, dusky-flecked auriculars, wings lack contrasting edgings and spots. Chest orange-cinnamon with narrow dusky streaking. Song an ascending series of mellow whistles suggesting Spectacled Antpitta but intensifying overall and ending abruptly.

SCALE-CRESTED PYGMY-TYRANT *Lophotriccus pileatus* cf. *luteiventris* 3.7–4 in (9.5–10 cm). An inconspicuous flycatcher of low to mid-levels of forest interior, best detected by voice. Blackish bill flesh below at base, legs flesh, *eyes whitish.* Face olive-brown with pale grey supraloral stripe, cinnamon eyering, crown *and shaggy crest rufous with scaly black feather centers.* Upperparts olive-green, wings and tail dark greyish, edged yellow-olive to olive. Throat and upper chest whitish, heavily streaked dusky, belly and undertail coverts pale lemon. Calls sharp and often loud, including a reedy, frog-like *prriep* or *pirrip*, often repeated in ascending series.

Probably F resident in foothills of the Olancho, but only one record. E Honduras to Peru and N Venezuela.

GOLDEN-CROWNED SPADEBILL *Platyrinchus coronatus superciliaris* 3.3–3.7 in (8.5–9.5 cm). Like many forest flycatchers, easily overlooked; singly or in pairs at low to mid-levels of shady forest understory.

Bill blackish above, pale greyish flesh below, legs flesh-grey. *Rufous-orange crown bordered by black sides,* yellow crown patch usually concealed. *Face pale buffy lemon with blackish auricular mark and* stripe below eye. Upperparts greyish olive, wings and tail darker, edged dull tawny-brown. Throat pale buffy lemon, chest and flanks dusky, belly and undertail coverts pale lemon. Call a thin, insect-like, buzzy, trilled *ssssirrrrrit* or *sirrrrrir.*

U to F resident E of Sula Valley. Honduras to Ecuador and Brazil.

LONG-TAILED TYRANT *Colonia colonus leuconota* 4.7–5.2 in (12–13.5 cm); narrow, projecting central rectrices add 3–5 in (7.5–12.5 cm) to length. A conspicuous bird of forest edge, clearings, etc., perching atop prominent twigs and snags. **Adult:** bill and legs blackish. *Blackish overall,* paler on belly, with *sooty grey-brown cap bordered by frosty whitish band;* also a *greyish-white stripe down back.* **Juv:** sooty blackish overall with broader central rectrices extending up to 1 in (2.5 cm) beyond tail. Calls include a high, rising *sweet* or *pweet.*

C to F resident in the Olancho. E Honduras to Ecuador and N Argentina.

WHITE-RINGED FLYCATCHER *Conopias albovittata* ssp 6–6.3 in (15–16 cm). In pairs or small groups in forest canopy at edges, along rivers, etc. **Adult:** bill and legs black. *Broad white supercilia surround blackish crown,* yellow crown patch usually concealed; auriculars blackish, throat white, underparts yellow. Upperparts greyish olive, wings and tail blackish with indistinct narrow paler edgings, whitish tertial edgings. Calls mostly shrill, scolding trills, *kree-ee-ee-eer* or *wheeerr*, etc. See Social Flycatcher.

F to U resident in the Olancho. E Honduras to NW Ecuador. Birds E of the Andes, *C. parva*, sometimes considered conspecific.

GREY-CAPPED FLYCATCHER *Myiozetetes g. granadensis** 6–6.5 in (15–16.5 cm). In pairs or small groups in forest edge, second growth, semiopen areas with scattered trees, etc. **Adult:** bill and legs black. *Crown and nape grey with whitish supercilium barely extending behind eye,* a flame-orange crown patch usually concealed; auriculars slaty grey, throat white, underparts yellow. Upperparts olive, wings and tail blackish, wings narrowly edged yellow-olive, tail narrowly edged rufous. **Juv:** greyish-olive crown lacks flame patch, wings and tail edged cinnamon-rufous becoming whitish on tertial edgings. Common calls a sharp *pic*, suggesting Western or Couch's kingbirds, and a sharp *kyew-kyew!*; also burry and bickering calls. See Social Flycatcher.

F resident in the Olancho. E Honduras to NW Peru and Brazil.

SNOWY COTINGA *Carpodectes nitidus* 8–8.5 in (20–21.5 cm). Singly or in small groups at fruiting trees or perched conspicuously in forest canopy, trees in clearings, etc. ♂: bill black above, grey below, legs blackish. *Gleaming snowy white overall, crown, nape, and upperbody clouded blue-grey.* ♀: *crown, nape, and upperparts grey,* washed mauve on back, *wings and tail darker with bold white edgings* to wing coverts, secondaries, and tertials. *Face, throat, chest, and flanks smoky mauve-grey with white eyering,* belly and undertail coverts white. **Imm** ♂: crown, nape, and upperparts pale blue-grey, washed mauve on back, scapulars and innerwing coverts whitish, outerwing coverts and remiges blackish brown, boldly edged white on tertials and secondaries; tail grey-brown. Throat and underparts white, clouded bluish grey on chest. Mostly silent.

U to F resident E of Sula Valley. Honduras to W Panama.

WHITE-RUFFED MANAKIN *Corapipo leucorrhoa altera* 3.7–4 in (9.5–10 cm). A typical, stocky, short-tailed manakin, found singly at low to mid-levels of shady forest understory; at times joins mixed-species flocks. ♂: bill and legs blackish. *Blue-black overall with bushy white throat,* distal undertail coverts pale lemon. ♀: *head and upperparts bright olive-green.* Throat bib pale blue-grey, chest and flanks dirty olive-green becoming whitish, washed yellow-green, on belly and undertail coverts. **Imm** ♂: resembles ♀ but at least median throat white, body may be mottled blue-black. Calls include a fairly full *ssip* or *pssih.*

U resident in the Olancho. E Honduras to N Colombia and NW Venezuela. Birds in C.A. have been considered specifically distinct, *C. altera.*

SOUTHERN ROUGH-WINGED SWALLOW *Stelgidopteryx ruficollis uropygialis* 5–5.3 in (12.5–13.5 cm). Southern counterpart of Northern Rough-winged Swallow which it resembles in shape, habits, voice, etc. **Adult:** head and upperparts dark brown with *contrasting buffy-grey rump. Throat pinkish cinnamon,* chest and flanks dusky grey-brown, belly and undertail coverts lemon-white with black tips to distal undertail coverts. **Juv:** cinnamon of throat extends as wash over chest, wingbars and tertial edgings pinkish cinnamon.

F resident (SL–1200 m) in the Olancho. E Honduras to NW Peru and cen Argentina.

SONG WREN *Cyphorhinus phaeocephalus richardsoni* 4.5–5 in (11.5–12.5 cm). An antbird-like wren; note ridged culmen base. In pairs or small groups, low in forest understory. **Adult:** *orbital skin pale blue,* bill and legs blackish. *Dark rich brown overall,* wings and tail barred black, *with rufous-chestnut face, throat, and upper chest,* belly washed greyish. **Juv:** dark scaling on upperparts, rufous face bordered by pale buff line. Calls and song are deep, gurgling churrs and chatters, often interspersed with sweet, whistled notes.

Probably F resident in the Olancho, few records. E Honduras to W Ecuador.

OLIVE-CROWNED YELLOWTHROAT *Geothlypis semiflava bairdi* 5–5.2 in (12.5–13 cm). In humid overgrown fields, brushy second growth, marshy areas, etc. Habits much as other *Geothlypis.* ♂: bill blackish, legs yellowish flesh. *Forecrown and mask black, rest of head and upperparts olive.* Throat and underparts yellow, washed olive on sides and flanks. ♀: head olive with paler, olive-yellow supraloral stripe and often slight spectacles. Call a sharp *chip*; song a rich warble ending with a slight chatter.

Probably F resident in the Olancho, few records. E Honduras to W Ecuador.

BUFF-RUMPED WARBLER *Basileuterus fulvicauda leucopygia* 4.8–5.3 in (12–13 cm). An active warbler of forested streams and rivers; fans and wags its broad tail while hopping around. **Adult:** bill blackish, legs dull flesh. Head and upperparts dark olive-brown, sootier on crown, with *dirty pale buffy-grey supercilium* fading behind eye, and *striking, bright pale cinnamon-buff uppertail coverts and tail base.* Throat and median underparts whitish, sides and flanks brownish, washed ochraceous buff, with brown mottling on chest; undertail coverts pale cinnamon. **Juv**: sooty brown overall with cinnamon-buff uppertail coverts and tail base, buffy-white belly. Calls include a fairly hard, smacking *chik* which may be run into rapid series suggesting Green Kingfisher, and a sharp, slightly metallic *ch-tiik*. Song begins quietly, then runs into a crescendo of rich notes which carries over the sound of rushing water.

F resident E of Sula Valley. Honduras to NW Peru and Brazil. Sometimes placed in the genus *Phaeothlypis*.

RUFOUS-WINGED TANAGER   *Tangara lavinia cara*   4.7–5 in (12–13 cm). A striking tanager of humid evergreen forest and edge, found in pairs or small groups in the canopy, especially at fruiting trees; at times with mixed-species flocks ♂: bill horn with dark culmen, legs greyish. *Brilliant green overall with reddish-chestnut head*, gold sheen on mantle, turquoise belly and vent, and *mostly golden-chestnut wing coverts; remiges dark, edged chestnut.* ♀: *brilliant green overall*, head with golden sheen, belly mottled turquoise, *wing coverts greenish gold; remiges dark, edged greenish gold.* **Juv**: resembles ♀ but duller. **Imm** ♂: head flecked golden chestnut, may attain full chestnut head in 1st prebasic molt (?). Calls include a sharp *tseeup* or *tseep* and a piercing *zeek* (Stiles and Skutch 1989). Compare with ♀ Blue-crowned Chlorophonia, ♂ Green Honeycreeper. See Appendix B.

U resident (near SL–750 m; mostly foothills?) E of Sula Valley. C.A. to W Ecuador.

BLUE DACNIS   *Dacnis cayana* cf. *callaina*   4.8–5 in (12–12.5 cm). Resembles a honeycreeper with a straight pointed bill. In pairs or small groups in forest canopy, often with mixed-species flocks. ♂: eyes reddish, bill blackish, flesh below at base, legs flesh. *Head and underparts turquoise-blue with black loral mask and median throat.* Upperparts black with outer scapulars, wings, and tail edged turquoise-blue, turquoise-blue rump and uppertail coverts. ♀: *head and throat turquoise-blue* with dark loral mask. *Body emerald green*, wings and tail blackish, edged emerald-green with some blue in wing edgings. Mostly silent, at times gives high *tsit* notes.

Probably F resident in the Olancho, few records. E Honduras to Ecuador and N Argentina.

WHITE-THROATED SHRIKE-TANAGER
Lanio l. leucothorax   7.8–8.3 in (20–21 cm). *Replaces Black-throated Shrike-Tanager in the Olancho; voice and habits similar to that species.* ♂: bill and legs blackish. *Head black, throat white above pale vinaceous band across upper chest, rest of underparts bright yellow.* Upperparts yellow with black outer scapulars and distal uppertail coverts; *wings and tail black*, white lesser wing coverts usually covered by scapulars. ♀: *head greyish olive-brown, throat and upper chest paler, dusky vinaceous-brown, rest of underparts ochre-yellow* to yellow, washed olive. Upperparts rich *tawny-brown becoming bright tawny on rump* and uppertail coverts, wings and tail darker, edged tawny. See Black-throated Shrike-Tanager (apparently allopatric).

U to F resident in the Olancho. E Honduras to W Panama. See Black-throated Shrike-Tanager.

WHITE-SHOULDERED TANAGER
Tachyphonus luctuosus axillaris   5.5–5.8 in (14–15 cm). Typically in pairs at mid- to upper levels of forest, often with mixed-species flocks. ♂: bill black with silvery base below, legs grey. *Glossy black overall with white lesser and median upperwing coverts*; white underwing coverts flash in flight. ♀: head and upperparts olive, throat and underparts paler, dirty olive-yellow, washed dusky on flanks. Calls include a thin *ssit*, and a nasal *cheh-cheh-chet*, often combined. Song a squeaky *tseek chur-chur-chur* (Stiles and Skutch 1989).

U to F resident in the Olancho. E Honduras to Ecuador and Brazil.

TAWNY-CRESTED TANAGER   *Tachyphonus delatrii*   5.8–6.3 in (15–16 cm). Usually in small, active and noisy flocks at low to mid-levels of forest. ♂: bill black with silvery base below, legs blackish. *Black overall with erectile tawny-orange crown feathers.* ♀: bill dark. Sooty olive-brown overall, brighter and tawnier on throat and chest. Calls include a metallic *tchit* or *zick*, a higher *tsip* or *tchewp*, and a high, thin, sibilant or slightly scratchy *zeeet* or *pseeet* (Stiles & Skutch 1989).

Probably U to F resident in the Olancho, only one record. E Honduras to W Ecuador.

SLATE-COLORED GROSBEAK   *Pitylus grossus* cf. *saturatus*   7.5–8 in (19–20.5 cm).

Singly or in pairs at mid- to upper levels of forest, at edges, etc.; may join mixed-species flocks. ♂: *stout bill bright pinkish red*, legs blackish. *Black face and chest surround white throat, rest of plumage slaty blue-grey*, darker on flight feathers. ♀: slaty grey overall with white throat, head and upperparts bluer, underparts with dusky-olive cast. Calls include a sharp *plik* and a nasal *yeahr*. Song a rich, warbled *twee tee-wee tee-weechee wer*, etc., may suggest peppershrike.

Probably U to F resident in the Olancho, only one record. E Honduras to Ecuador and Brazil.

BLACK-STRIPED SPARROW *Arremonops conirostris richmondi* 6.5–7.2 in (16.5–18 cm). *A large, bulky counterpart of Green-backed Sparrow*, inhabiting forest edge, plantations, gardens, etc.; often in more open and conspicuous situations than other *Arremonops*. **Adult:** bill black, paler below at base, legs dusky flesh. *Head grey with black lateral crown stripe and eyestripe; upperparts olive-green.* Chest and flanks greyish with whitish throat and belly, undertail coverts buff. **Juv:** unlike other *Arremonops*, juv plumage kept several weeks. Head and upperparts olive with blackish lateral crown stripe and eyestripe, back streaked dark brown. Lemon throat contrasts with dark-streaked olive underparts, belly and undertail coverts lemon. Calls include a nasal, fairly sharp *cheuk* and a softer, slightly gruff *cheuh, cheuh ...*, etc. Song begins with a hesitant series of rich and burry chips which, sooner or later, typically break into an accelerating and slowing series of nasal *chuh* or *choo* notes: *chreeu, cheu wh-eep, cheuh, wh-eep, tuu, ... chuh chuh chuh-chuh-chuh ...*, often ending with an abrupt, nasal note, *chieh*. Green-backed Sparrow allopatric.

C to F resident E of Sula Valley. E Honduras to W Ecuador and N Brazil. See NB2 under Green-backed Sparrow.

SCARLET-RUMPED CACIQUE *Cacicus uropygialis microrhynchus* 8.5–9.5 in (21.5–24 cm). Often noisy and conspicuous in groups up to 10 birds in forest canopy, may join flocks of oropendolas, etc. **Adult:** eyes pale blue, bill pale horn, legs black. *Glossy black overall with flaming-scarlet rump* (often covered by wings). ♀ duller. **Juv:** duller, rump cinnamon-orange. Loud, ringing whistles attract attention, *clee-oo-woo*, and *pleeoo*, etc.

Probably F resident in the Olancho, few records. E Honduras to Ecuador and Peru. Taxonomic status of S.A. forms unclear, perhaps more than one species involved.

BIBLIOGRAPHY

Abbreviations

AMNH	American Museum of Natural History	PRBO	Point Reyes Bird Observatory
ANSP	Academy of Natural Sciences, Philadelphia	RSMHN	Revista de la Sociedad Mexicana de Historia Natural
AOU	American Ornithologists' Union	SDSNH	San Diego Society of Natural History
BBOC	Bulletin of the British Ornithologists' Club	SEDUE	Secretaria de Desarrollo Urbano y Ecología
CAS	California Academy of Sciences	UCPZ	University of California, Publications in Zoology
CM	Carnegie Museum		
FMNH	Field Museum of Natural History	UKPMNH	University of Kansas Publication, Museum of Natural History
ICACH	Instituto de Ciencias y Artes de Chiapas	UNAM	Universidad Autonoma Nacional de Mexico
ICBP	International Council for Bird Preservation	USFWS	U.S. Fish and Wildlife Service
INIREB	Instituto Nacional de Investigaciónes Sobre Recursos Bióticos	USNM	U.S. National Museum
		WFVZ	Western Foundation of Vertebrate Zoology
MBA	Mexican Birding Association		
MCZ	Museum of Comparative Zoology		
MZLSU	Museum of Zoology, Louisiana State University		
MZUM	Museum of Zoology, University of Michigan		Fieldiana (Chicago Natural History Museum); Nemouria (Occasional Papers, Delaware Museum of Natural History); Notulae Naturae (Academy of Natural Sciences, Philadelphia); Novitates (American Museum of Natural History); Postilla (Peabody Museum of Natural History, Yale University).
NAS	National Audubon Society		
NMFS	National Marine Fisheries Service		
PBSW	Proceedings of the Biological Society of Washington		

Aldrich, J. W. and Baer, K. P. (1970). Status and speciation in the Mexican Duck (*Anas diazi*). *Wilson Bull.*, **82**, 63–73.

Aldrich, J. W. and Bole, B. P. (1937). The birds and mammals of the western slope of the Azuero Peninsula. *Cleveland Mus. Nat. Hist.*, **7**, 3–196.

Aldrich, J. W. and Duvall, A. J. (1955). *Distribution of American gallinaceous game birds*. US Dept. Int. Circular 34.

Allen, J. A. (1893). List of mammals and birds collected in northeastern Sonora and northwestern Chihuahua, Mexico, on the Lumholtz Expedition, 1890–92. *Bull. AMNH*, **5**, 27–42.

Alvarez del Toro, M. (1949). Striped Owl in southern Mexico. *Condor*, **51**, 232.

Alvarez del Toro, M. (1950). The English Sparrow in Chiapas. *Condor*, **52**, 166.

Alvarez del Toro, M. (1952*a*). Contribución al conocimiento de la oología y nidología de las aves Chiapanecas. *Ateneo*, **4**, 11–21.

Alvarez del Toro, M. (1952*b*). New records of birds from Chiapas, Mexico. *Condor*, **54**, 112–14.

Alvarez del Toro, M. (1954). Notes on the occurrence of birds in Chiapas, Mexico. *Condor*, **56**, 365.

Alvarez del Toro, M. (1955). The Rufescent Mourner in Chiapas, Mexico. *Condor*, **57**, 370–1.

Alvarez del Toro, M. (1958). Lista de las especies de aves que habitan en Chiapas, endemicas, emigrantes y de paso. *RSMHN*, **19**, 73–113.

Alvarez del Toro, M. (1963). *Miscelanea ornitologica*. ICACH. Tuxtla Gutierrez, Chis.

Alvarez del Toro, M. (1964). *Lista de las aves de Chiapas, endemicas, emigrantes y de paso*. ICACH. Tuxtla Gutierrez, Chis.

Alvarez del Toro, M. (1965). The nesting of the Belted Flycatcher. *Condor*, **67**, 339–43.

Alvarez del Toro, M. (1970). Notas para la biología del Pájaro Cantil (*Heliornis fulica*). *ICACH*, **1**, 7–13.

Alvarez del Toro, M. (1971*a*). On the biology of the American Finfoot in southern Mexico. *Living Bird*, **10**, 79–88.

Alvarez del Toro, M. (1971*b*). El Bienparado o Pájaro Estaca (*Nytibius griseus mexicanus* Nelson). *ICACH*, **2**, 7–13.

Alvarez del Toro, M. (1971*c*). *Las aves de*

Chiapas. Gobierno del Estado de Chiapas. Tuxtla Gutierrez.

Alvarez del Toro, M. (1976). Datos biologicas del Pavón (*Oreophasis derbianus* G. R. Gray). *Rev. Univ. Chis.*, 1, 43–54.

Amadon, D. (1961). Remarks on the genus *Buteogallus*. *Nov. Colombianas*, 1, 358–60.

Amadon, D. (1970). Taxonomic categories below the level of genus: theoretical and practical aspects. *J. Bombay Nat. Hist. Soc.*, 67, 1–13.

Amadon, D. and Eckelbery, D. R. (1955). Observations on Mexican birds. *Condor*, 57, 65–80.

Amadon, D. and Phillips, A. R. (1947). Notes on Mexican birds. *Auk*, 64, 576–81.

Amadon, D. and Short, L. L. (1976). Treatment of subspecies approaching species status. *Syst. Zool.*, 25, 161–7.

Amadon, D., Bull, J., Marshall, J. T., and King, B. F. (1988). Hawks and owls of the world: a distributional and taxonomic list. *Proc. WFVZ*, 3, 295–357.

AOU (1957). *Checklist of North American birds*, 5th edition. AOU.

AOU (1973). Bird conservation in Middle America. Report of the AOU Conservation Committee, 1972–73. *Auk*, 90, 877–87.

AOU (1983). *Checklist of North American birds*, 6th edition. AOU.

AOU (1985). Thirty-fifth supplement to the AOU Checklist of North American Birds. *Auk*, 102, 680–6.

AOU (1987). Thirty-sixth supplement to the AOU Checklist of North American Birds. *Auk*, 104, 591–6.

AOU (1989). Thirty-seventh supplement to the AOU Checklist of North American Birds. *Auk*, 106, 532–8.

AOU (1991). Thirty-eighth supplement to the AOU Checklist of North American Birds. *Auk*, 108, 750–4.

Anderson, D. W. and Gress, F. (1983). Status of northern populations of California Brown Pelicans. *Condor*, 85, 79–88.

Anderson, D. W., Deweese, L. R., and Tiller, D. V. (1977). Passive dispersal of California Brown Pelicans. *Bird-Banding*, 48, 228–38.

Andrle, R. F. (1961). Cattle Egrets in Mexico. *Wilson Bull.*, 73, 280.

Andrle, R. F. (1966). North American migrants in the Sierra de Tuxtla (sic) of southern Veracruz, Mexico. *Condor*, 68, 177–84.

Andrle, R. F. (1967a). The Horned Guan in Mexico and Guatemala. *Condor*, 69, 93–109.

Andrle, R. F. (1967b). Birds of the Sierra de Tuxtla (sic) in Veracruz, Mexico. *Wilson Bull.*, 79, 163–87.

Andrle, R. F. (1967c). Notes on the Black

Chachalaca (*Penelopina nigra*). *Auk*, 84, 169–72.

Anthony, A. W. (1925). Expedition to Guadalupe Island, Mexico, in 1922. The birds and mammals. *Proc. CAS*, 14, 277–320.

Arnold, K. A. (1971). Three additional specimens of the Eared Poor-will from the state of Guerrero, Mexico. *Condor*, 73, 475.

Arnold, K. A. and Maxwell, T. C. (1970). The Great Swallow-tailed Swift (*Panyptila sanctihieronymi*) from the state of Guerrero, Mexico. *Condor*, 72, 108.

Arvin, J. C. (1990). *A checklist of the birds of the Gomez Farias region, southwestern Tamaulipas, Mexico*. Publ. by author. Austin, TX.

Ash, J. S. and Watson, G. E. (1980). Great Shearwater *Puffinus gravis* new to Mexico. *BBOC*, 100, 194–5.

Atkinson, P. W., Whittingham, M. J., Gomez de Silva G., H., Kent, A. M., and Maier, R. T. (1993). The taxonomic status of the genus *Hylorchilus* with notes on its ecology and conservation. *Bird Conserv. Int.*, 3, 75–85.

Atwood, J. L. (1988). *Speciation and geographic variation in Black-tailed Gnatcatchers*. Ornithological Monographs No. 42. AOU.

Baepler, D. H. (1962). The avifauna of the Soloma region in Huehuetenango, Guatemala. *Condor*, 64, 140–53.

Bailey, A. M. and Conover, H. B. (1935). Notes from the state of Durango, Mexico. *Auk*, 52, 421–4.

Baker, R. H. (1958). Nest of the Military Macaw in Durango. *Auk*, 75, 98.

Baker, R. H. and Fleming, R. L. (1962). Birds near La Pesca, Tamaulipas, Mexico. *SW Nat.*, 7, 253–61.

Baltosser, W. H. (1989). Costa's Hummingbird: its distribution and status. *W. Birds*, 20, 41–62.

Bancroft, G. (1927). Notes on breeding coastal and insular birds of central Lower California. *Condor*, 29, 188–95.

Bangs, O. and Peters, J. L. (1927). Birds from the rain forest region of Vera Cruz (sic). *Bull. MCZ*, 67, 471–87.

Bangs, O. and Peters, J. L. (1928). A collection of birds from Oaxaca. *Bull. MCZ*, 68, 385–404.

Banks, R. C. (1962). Lapland Longspur in Mexico. *Condor*, 64, 329.

Banks, R. C. (1963a). Birds of Cerralvo Island, Baja California. *Condor*, 65, 300–12.

Banks, R. C. (1963b). New birds from Cerralvo Island, Baja California, Mexico. *Occas. Papers CAS*, 37.

Banks, R. C. (1965). An unusual habitat for Purple Martins. *Auk*, 82, 271–3.

Banks, R. C. (1967a). Birds and mammals of La

Laguna, Baja California. *Trans. SDSNH*, **14** 205–32.

Banks, R. C. (1967*b*). Some supplementary records of birds in Baja California, Mexico. *Condor*, **69**, 318.

Banks, R. C. (1974). Clapper Rail in Tamaulipas, Mexico. *Wilson Bull.*, **86**, 76–7.

Banks, R. C. (1975). Plumage variation in the Masked Bobwhite. *Condor*, **77**, 486–7.

Banks, R. C. (1988*a*). Supposed northern records of the Southern Fulmar. *W. Birds*, **19**, 121–4.

Banks, R. C. (1988*b*). Geographic variation in the Yellow-billed Cuckoo. *Condor*, **90**, 473–7.

Banks, R. C. (1990*a*). Taxonomic status of the Coquette Hummingbird of Guerrero, Mexico. *Auk*, **107**, 191–2.

Banks, R. C. (1990*b*). Taxonomic status of the Rufous-bellied Chachalaca (*Ortalis wagleri*). *Condor*, **92**, 749–53.

Banks, R. C. and Dickerman, R. W. (1978). Mexican nesting records of the American Bittern. *W. Birds*, **9**, 130.

Banks, R. C. and Tomlinson, R. E. (1974). Taxonomic status of certain Clapper Rails of south-western United States and north-western Mexico. *Wilson Bull.*, **86**, 325–35.

Baptista, L. F. (1978). A revision of the Mexican *Piculus* (Picidae) complex. *Wilson Bull.*, **90**, 159–81.

Baptista, L. F., Boarman, W. I., and Kandianidis, P. (1983). Behaviour and taxonomic status of Grayson's Dove. *Auk*, **100**, 907–19.

Barlow, J. C. and James, R. D. (1975). Aspects of the biology of the Chestnut-sided Shrike-Vireo. *Wilson Bull.*, **87**, 320–34.

Barlow, J. C. and Johnson, R. R. (1969). The Grey Vireo *Vireo vicinior* Coues (Passeriformes: Vireonidae), in the Sierra del Carmen, Coahuila, Mexico. *Can. J. Zool.*, **47**, 151–2.

Barlow, J. C., Dick, J. A., Baldwin D. H., and Davis, R. A. (1969). New records of birds from British Honduras. *Ibis*, **111**, 399–402.

Barlow, J. C., Dick, J. A., and Pendergast, E. (1970). Additional records of birds from British Honduras (Belize). *Condor*, **72**, 371–2.

Barlow, J. C., Dick, J. A., Weyer, D., and Young, W. F. (1972). New records of birds from British Honduras (Belize), including a skua. *Condor*, **74**, 486–7.

Beavers, R. A. (1992). *The Birds of Tikal. An annotated checklist for Tikal National Park and Peten, Guatemala*. Texas A & M University Press, College Station.

Beavers, R. A., Delaney, D. J., Leahy, C. W., and Oatman, G. F. (1991). New and noteworthy bird records from Peten, Guatemala, including Tikal National Park. *BBOC*, **111**, 77–90.

Beck, R. H. (1907). Notes from Clipperton and Cocos islands. *Condor*, **9**, 109–10.

Beehle, W. H. (1950). Clines in the Yellow-throats of western North America. *Condor*, **52**, 193–219.

Behrstock, R. A. (1977). Notes on some northern birds wintering on the Pacific coast of Guatemala. *Am. Birds*, **31**, 382.

Bent, A. C. (1919–1968). *Life histories of North American birds* (26 vols). USNM. Washington, DC.

Berlioz, J. (1959). Un oiseau nouveau du Mexique. *L'Oiseau*, **29**, 40–2.

Berrett, D. G. (1962). The birds of the Mexican state of Tabasco; a dissertation. Unpubl. PhD thesis. Louisiana State Univ., Baton Rouge.

Berrett, D. G. (1963). First records for the Ruddy-tailed Flycatcher and Purple-crowned Fairy Hummingbird from Mexico. *Condor*, **65**, 163.

Bierregaard, R. O., Jr. (1984). Observations of the nesting biology of the Guiana Crested Eagle (*Morphnus guianensis*). *Wilson Bull.*, **96**, 1–5.

Binford, L. C. (1965). Two new subspecies of birds from Oaxaca, Mexico. *Occas. Papers MZLSU*, **30**.

Binford, L. C. (1970). Audubon's Shearwater, Hudsonian Godwit and Long-tailed Jaeger in Oaxaca, Mexico. *Condor*, **72**, 366.

Binford, L. C. (1973). Virginia Rail and Cape May Warbler in Chiapas, Mexico. *Condor*, **75**, 350–1.

Binford, L. C. (1985). Re-evaluation of the 'hybrid' hummingbird *Cynanthus sordidus* × *C. latirostris* from Mexico. *Condor*, **87**, 148–50.

Binford, L. C. (1989). *A distributional survey of the birds of the Mexican state of Oaxaca*. Ornithological Monographs No. 43. AOU.

Birkenstein, L. R. and Tomlinson, R. E. (1981). Native names of Mexican birds. *USFWS Resource Publ.*, **139**, 1–156.

Bjelland, A. D. and Ray, J. C. (1977). Birds collected in the state of Hidalgo, Mexico. *Occas. Papers Museum Texas Tech. University.*

Blake, E. R. (1949*a*). First Mexican records of tropical American birds. *Nat. Hist. Misc., Chicago Acad. Sci.*, **42**.

Blake, E. R. (1949*b*). Distribution and variation of *Caprimulgus maculicaudus*. *Fieldiana, Zool.*, **31**, 207–13.

Blake, E. R. (1950*a*). Report on a collection of birds from Guerrero, Mexico. *Fieldiana, Zool.*, **31**, 375–93.

Blake, E. R. (1950*b*). Report on a collection of birds from Oaxaca, Mexico. *Fieldiana, Zool.*, **31**, 395–419.

Blake, E. R. (1953). *Birds of Mexico*. Univ. Chicago Press, Chicago, IL.

Blake, E. R. (1977). *Manual of neotropical birds*. Vol. 1. Univ. Chicago Press. Chicago.

Blake, E. R. and Hanson, H. C. (1942). Notes on a collection of birds from Michoacan, Mexico. *Zool. Series, FMNH*, 22, 513–51.

Bond, J. (1936). Resident birds of the Bay Islands of Spanish Honduras. *Proc. ANSP*, 88, 353–64.

Bond, J. (1954). Birds of Turneffe and Northern Two Cays, British Honduras. *Notulae Naturae*, 260.

Bond, J. (1961). Notes on birds of Cozumel Island, Quintana Roo, Mexico. *Carib. J. Sci.*, 1, 41–7.

Bond, J. (1985). *Birds of the West Indies.* 5th edition. Houghton Mifflin, Boston.

Boswall, J. (1978a). Notes on the breeding birds of Isla Raza, Baja California. *W. Birds*, 9, 93–108.

Boswall, J. (1978b). The birds of Alacran Reef, Gulf of Mexico. *BBOC*, 98, 99–109.

Boswall, J. (1978c). The birds of the San Benito Islands, Lower California, Mexico. *Bristol Ornithol.* 11, 23–30.

Brand, D. D. *et al.* (1960). *Coalcoman and Motines del Oro, An 'ex-distrito' of Michoacan, Mexico.* Inst. Latin Amer. Studies. Austin, TX.

Brattstrom, B. H. and Howell, T. R. (1956). The birds of the Revilla Gigedo (sic) Islands, Mexico. *Condor*, 58, 107–20.

Braun, M. J., Braun, D., and Terrill, S. B. (1986). Winter records of the Golden-cheeked Warbler (*Dendroica chrysoparia*) from Mexico. *Am. Birds*, 41, 564–6.

Briggs, M. A. (1954). Apparent neotcony in the Saw-Whet Owls of Mexico and Central America. *PBSW*, 67, 179–82.

Brodkin, H. and Brodkin, P. (1981). Summer birds at Polol, Guatemala. *Cont. Birdlife*, 2, 111–17.

Brodkorb, P. (1938). New birds from the district of Soconusco, Chiapas. *Occas. Papers MZUM*, 369.

Brodkorb, P. (1939). Rediscovery of *Heleodytes chiapensis* and *Tangara cabanisi*. *Auk*, 56, 447–50.

Brodkorb, P. (1940). New birds from southern Mexico. *Auk*, 57, 542–9.

Brodkorb, P. (1943a). Birds from the gulf lowlands of southern Mexico. *Misc. Publ. MZUM*, 55.

Brodkorb, P. (1943b). Notes on two rare birds in Chiapas, Mexico. *Auk*, 60, 280–3.

Brodkorb, P. (1943c). The Rufous-browed Wrens of Chiapas, Mexico. *Occas. Papers MZUM*, 480, 1–3.

Brodkorb, P. (1948). Some birds from the lowlands of central Veracruz, Mexico. *J. Florida Acad. Sci.*, 10, 31–8.

Brown, B. T. and Warren, P. L. (1985). Winter-

ing Bald Eagles along the Rio Yaqui, Sonora, Mexico. *Wilson Bull.*, 97, 224–6.

Brown, B. T., Warren, P. L., and Anderson L. S. (1987). First Bald Eagle nesting record from Sonora, Mexico. *Wilson Bull.*, 99, 279–80.

Brown, H. C. and Monroe, B. L., Jr. (1974). Bird records from Honduras. *Condor*, 76, 348–9.

Brown, J. L. (1958). A nesting record of the Scissor-tailed Flycatcher in Nuevo Leon, Mexico. *Condor*, 60, 193–4.

Brown, L. and Amadon, D. (1968). *Eagles, hawks and falcons of the world.* McGraw-Hill, New York.

Bryant, W. E. (1889). A catalogue of the birds of Lower California, Mexico. *Proc. CAS, Ser. 2*, 2, 237–320.

Bubb, P. (1991). *The current status of the cloud forest in northern Chiapas, Mexico.* Publ. by author. Amersham, Bucks, UK.

Buchanan, O. M. (1964a). The Mexican races of the Least Pygmy Owl. *Condor*, 66, 103–12.

Buchanan, O. M. (1964b). Another Pacific record of the Black Swift off Mexico. *Condor*, 66, 161–2.

Burleigh, T. D. and Lowery G. H., Jr. (1942). Notes on the birds of south-eastern Coahuila. *Occas. Papers MZLSU*, 12, 185–212.

Cade, T. J., Enderson, J. H., Thelander, C. G., and White, C. M. (ed.) (1988). *Peregrine Falcon populations, their management and recovery.* Peregrine Fund Inc., Boise, ID.

Carriker, M. A., Jr. and Meyer de Schauensee, R. (1935). An annotated list of two collections of Guatemalan birds in the Academy of Natural Sciences of Philadelphia. *Proc. ANSP*, 87, 411–55.

Castellanos, A. and Rodriguez-E., R. (1993). Current status of the Socorro Mockingbird. *Wilson Bull.*, 105, 167–71.

CENREN (1987). Elaboracion del plan y estrategia del sistema nacional de areas silvestres protegidas de El Salvador. Soyapango, El Salvador.

Chapman, F. M. (1898). Notes on birds observed at Jalapa and Las Vigas, Vera Cruz (sic), Mexico. *Bull. AMNH*, 10, 15–43.

Clapp, R. B. *et al.* (1982). *Marine birds in the southern United States and Gulf of Mexico. Part I: Gaviiformes through Pelecaniformes.* USFWS. FWS/OBS-82/01.

Clapp, R. B. *et al.* (1983). *Marine birds in the southern United States and Gulf of Mexico. Part III: Charadriiformes.* USFWS. FWS/OBS-83/30.

Clark, T. O. (1984). Notable records of birds from eastern Sonora, Mexico. *W. Birds*, 15, 134–6.

Clark, W. H. and Kendall, S. S. (1986). First record of the Blackburnian Warbler, *Den-*

droica fusca, from the central desert of Baja California. *SW Naturalist*, **31**, 544.

Clark, W. S. and Wheeler, B. K. (1987). *A field guide to hawks, North America*. Houghton Mifflin Co., Boston, MA.

Clow, B. (1976). A new look at the San Blas area. *Mexican Birds Newsletter*, **1**(2), 12–15.

Cobb, J. (1990). A nest of the Collared Forest-Falcon (*Micrastur semitorquatus*). *Aves Mexicanas*, **2**(90–1), 8.

Coffey, B. B. (1948). Southward migration of herons. *Bird Banding*, **19**, 1–5.

Coffey, B. B. (1959). The Starling in eastern Mexico. *Condor*, **61**, 299.

Coffey, B. B. (1960). Late North American spring migrants in Mexico. *Auk*, **77**, 288–97.

Coffey, B. B. (1961). Some shorebird records from Mexico. *Wilson Bull.*, **73**, 207–8.

Coffey, B. B., Jr., and Coffey, L. C. (1989). *Songs of Mexican birds*. ARA 13–1. ARA Records. Gainesville, FL.

Cole, L. J. (1906). Vertebrata from Yucatan: Introduction. *Aves. Bull. MCZ*, **50**, 101–47.

Collar, N. J., Gonzaga, L. P., Krabbe, N., Madroño N., A., Naranjo, L. G., Parker, T. A., III, and Wege, D. C. (1992). *Threatened birds of the Americas: the ICBP/IUCN Red Data Book*, 3rd edition, Part 2. ICBP, Cambridge, UK.

Collins, C. T., Delaney, D., and Atwood, J. L. (1990). First record of the Great Kiskadee in Baja California, Mexico. *W. Birds*, **21**, 73–4.

Connors, P. G. (1983). Taxonomy, distribution, and evolution of Golden Plovers (*Pluvialis dominica* and *Pluvialis fulva*). *Auk*, **100**, 607–20.

Contreras B., A. J. (1977). *Ornitofauna comparativa de tres areas fisiográficas del sur de Nuevo León, México*. Mem. 1ª Congreso Nac. de Zoología. Chapingo, Mexico.

Contreras B., A. J. (1984). Birds from Cuatro Cienegas, Coahuila, Mexico. *J. Arizona–Nevada Acad. Sci.*, **19**, 77–9.

Contreras B., A. J. (1987). Nuevos registros de aves para la Isla María Madre, Nayarit, México. *Rev. Biol. Trop.*, **35**, 353–4.

Contreras B., A. J. (1988). New records from Nuevo León, Mexico. *SW Nat.*, **33**, 251–2.

Contreras B., A. J. (1992). Status of Clark's Nutcrackers on Cerro Potosí, Nuevo León, Mexico. *W. Birds*, **23**, 181.

Cotera-C., M. and Contreras B., A. J. (1985). Ornitofauna de un transecto ecologico de Cañon de la Boca, Santiago, Nuevo León. Publ. Biol. Inst. Cient. *UANL*, **2**(1), 31–49.

Cracraft, J. (1981). Toward a phylogenetic classification of the recent birds of the world (Class Aves). *Auk*, **98**, 681–714.

Cramp, S. and Simmons, K. E. L. (1977–1990). *Handbook of the birds of Europe, the Middle East and North Africa*. Vols 1–5. Oxford University Press.

Crosby, G. T. (1972). Spread of the Cattle Egret in the Western Hemisphere. *Bird-Banding*, **43**, 205–12.

Crossin, R. S. (1967). The breeding biology of the Tufted Jay. *Proc. WFVZ*, **1**, 265–99.

Crossin, R. S. and Ely, C. A. (1973). A new race of Sumichrast's Wren from Chiapas, Mexico. *Condor*, **75**, 137–9.

Danemann, G. D. and Guzman P., J. R. (1992). Notes on the birds of San Ignacio Lagoon, Baja California Sur, Mexico. *W. Birds*, **23**, 11–19.

Davidson, M. E. McL. (1928). On the present status of the Guadalupe Petrel. *Condor*, **30**, 355–6.

Davis, D. E. (1940). Social nesting habits of the Smooth-billed Ani. *Auk*, **57**, 179–218.

Davis, J. (1951). Distribution and variation of the Brown Towhees. *UCPZ*, **52**, 1–120.

Davis, J. (1953). Birds of the Tzitzio region, Michoacan, Mexico. *Condor*, **55**, 90–8.

Davis, J. (1959a). The Sierra Madrean element of the avifauna of the Cape District, Baja California. *Condor*, **61**, 75–84.

Davis, J. (1959b). A new race of the Mexican Potoo from western Mexico. *Condor*, **61**, 300–1.

Davis, J. (1960). Notes on the birds of Colima, Mexico. *Condor*, **62**, 215–19.

Davis, J. (1962). Notes on some birds of the state of Michoacan, Mexico. *Condor*, **64**, 325.

Davis, J. (1965). Natural history, variation and distribution of the Strickland's Woodpecker. *Auk*, **82**, 537–90.

Davis, J. and Miller, A. H. (1962). Further information on the Caribbean Martin in Mexico. *Condor*, **64**, 237–9.

Davis, L. I. (1952). Winter bird census at Xilitla, San Luis Potosi, Mexico. *Condor*, **54**, 345–55.

Davis, L. I. (1957). Observations on Mexican birds. *Wilson Bull.*, **69**, 364–5.

Davis, L. I. (1958). Acoustic evidence for relationship in North American Crows. *Wilson Bull.*, **70**, 151–67.

Davis, L. I. (1962). Acoustic evidence for relationship in *Caprimulgus*. *Tex. J. Sci.*, **14**, 72–106.

Davis, L. I. (1965). Acoustic evidence for relationship in *Ortalis* (Cracidae). *SW Nat.*, **10**, 288–301.

Davis, L. I. (1972). *A field guide to the birds of Mexico and Central America*. Univ. Texas Press. Austin.

Davis, L. I. (1978). Acoustic evidence for relationship in Potoos. *Pan Amer. Studies*, **1**, 4–21.

Davis, W. B. (1944). Notes on summer birds of Guerrero, Mexico. *Condor*, **46**, 9–14.

Davis, W. B. (1945). Notes on Veracruzean birds. *Auk*, **62**, 272–86.

Davis, W. B. (1950). Summer range of the Scissor-tailed Flycatcher. *Condor*, **52**, 138.

Davis, W. B. and Russell, R. J. (1953). Aves y mamíferos del estado de Morelos. *RSMHN*, **14**, 77–121.

Dawn, W. (1964). Nest and eggs of the Cabanis Tiger Heron in Chiapas, Mexico. *Auk*, **81**, 230–1.

DeSante, D. F. and Pyle, P. (1986). *Distributional checklist of North American birds*. Artemesia Press. Lee Vining, CA.

Devillers, P. (1977). The skuas of the North American Pacific coast. *Auk*, **94**, 417–29.

Devillers, P., McKaskie, G., and Jehl, J. R., Jr. (1971). The distribution of certain large gulls (*Larus*) in southern California and Baja California. *Calif. Birds*, **2**, 11–26.

DeWeese, L. R. and Anderson, D. W. (1976). Distribution and breeding biology of Craveri's Murrelet. *Trans. SDSNH*, **18**, 155–68.

Dickerman, R. W. (1958). The nest and eggs of the White-throated Flycatcher. *Condor*, **60**, 259–60.

Dickerman, R. W. (1963a). The Song Sparrows of the Mexican Plateau. *Occas. Papers Minn. Mus. Nat. Hist.*, **9**.

Dickerman, R. W. (1963b). The grebe *Aechmophorus occidentalis* as a nesting bird of the Mexican plateau. *Condor*, **65**, 66–7.

Dickerman, R. W. (1963c). A critique of 'Birds from Coahuila, Mexico'. *Condor*, **65**, 330–2.

Dickerman, R. W. (1965). The juvenal plumage and distribution of *Cassidix palustris* (Swainson). *Auk*, **82**, 268–70.

Dickerman, R. W. (1966a). A new subspecies of the Boat-tailed Grackle from Mexico. *Wilson Bull.*, **78**, 129–31.

Dickerman, R. W. (1966b). A new subspecies of Virginia Rail from Mexico. *Condor*, **68**, 215–16.

Dickerman, R. W. (1968). Notes on the Red Rail (*Laterallus ruber*). *Wilson Bull.*, **80**, 94–9.

Dickerman, R. W. (1969). Nesting records of the Eared Grebe in Mexico. *Auk*, **86**, 144.

Dickerman, R. W. (1970). A systematic revision of *Geothlypis speciosa*, the Black-polled Yellowthroat. *Condor*, **72**, 95–8.

Dickerman, R. W. (1971). Notes on various rails in Mexico. *Wilson Bull.*, **83**, 49–56.

Dickerman, R. W. (1972). Further notes on the Pinnated Bittern in Mexico and Central America. *Wilson Bull.*, **84**, 90.

Dickerman, R. W. (1973a). Further notes on the Western Grebe in Mexico. *Condor*, **75**, 131–2.

Dickerman, R. W. (1973b). A review of the White-breasted Wood-Wrens of Mexico and Central America. *Condor*, **75**, 361–3.

Dickerman, R. W. (1973c). The Least Bittern in Mexico and Central America. *Auk*, **90**, 689–91.

Dickerman, R. W. (1973d). A review of the Boat-billed Heron *Cochlearius cochlearius*. *BBOC*, **93**, 111–14.

Dickerman, R. W. (1974). Review of the Red-winged Blackbirds (*Agelaius phoeniceus*) of eastern, west-central and southern Mexico and Central America. *Novitates*, **2538**.

Dickerman, R. W. (1975a). Revision of the Short-billed Marsh-Wrens (*Cistothorus platensis*) of Central America. *Novitates*, **2569**.

Dickerman, R. W. (1975b). Nine new specimen records for Guatemala. *Wilson Bull.*, **87**, 412–13.

Dickerman, R. W. (1977). Three more new specimen records for Guatemala. *Wilson Bull.*, **89**, 612–13.

Dickerman, R. W. (1980). Preliminary review of the Clay-colored Robin *Turdus grayi* with redesignation of the type locality of the nominate form and description of a new subspecies. *BBOC*, **101**, 285–9.

Dickerman, R. W. (1981a). A taxonomic review of the Spotted-breasted Oriole. *Nemouria*, **26**.

Dickerman, R. W. (1981b). Geographic variation in the juvenal plumage of the Lesser Nighthawk (*Chordeiles acutipennis*). *Auk*, **98**, 619–21.

Dickerman, R. W. (1985). Taxonomy of the Lesser Nighthawks (*Chordeiles acutipennis*) of North and Central America. In *Neotropical ornithology* (ed. P. A. Buckley *et al.*). Ornithological Monographs No. 36. AOU.

Dickerman, R. W. (1987). Two new subspecies of birds from Guatemala. *Occas. Papers WFVZ*, **3**.

Dickerman, R. W. (1993). Birds of the southern Pacific lowlands of Guatemala. *Trans. Linn. Soc. N.Y.* (in press).

Dickerman, R. W. and Haverschmidt F. (1971). Further notes on the juvenal plumage of the Spotted Rail (*Rallus maculatus*). *Wilson Bull.*, **83**, 444–6.

Dickerman, R. W. and Juarez, L. C. (1969). Juvenal plumage of the Spotted Rail (*Rallus maculatus*). *Wilson Bull.*, **81**, 207–9.

Dickerman, R. W. and Juarez, L. C. (1971). Nesting studies of the Boat-billed Heron *Cochlearius cochlearius* at San Blas, Nayarit, Mexico. *Ardea*, **59**, 1–16.

Dickerman, R. W. and Phillips, A. R. (1966). A new subspecies of the Boat-tailed Grackle from Mexico. *Wilson Bull.*, **78**, 129–31.

Dickerman, R. W. and Phillips, A. R. (1967). Botteri's Sparrows of the Atlantic coastal lowlands of Mexico. *Condor*, 69, 596–600.

Dickerman, R. W. and Phillips, A. R. (1970). Taxonomy of the Common Meadowlark (*Sturnella magna*) in central and southern Mexico and Caribbean Central America. *Condor*, 72, 305–9.

Dickerman, R. W. and Warner, D. W. (1961). Distribution records from Tecolutla, Veracruz, with the first record of *Porzana flaviventer* for Mexico. *Wilson Bull.*, 73, 336–40.

Dickerman, R. W. and Warner, D. W. (1962). A new Orchard Oriole from Mexico. *Condor*, 64, 311–14.

Dickerman, R. W., Phillips, A. R., and Warner, D. W. (1967). On the Sierra Madre Sparrow *Xenospiza baileyi* of Mexico. *Auk*, 84, 49–60.

Dickerman, R. W., Zink, R. M., and Frye, S. L. (1980). Migration of the Purple Martin in southern Mexico. *W. Birds*, 11, 203–4.

Dickerman, R. W., Parkes K. C., and Bell, J. (1981). Notes on the plumages of the Boat-billed Heron. *Living Bird*, 19, 115–20.

Dickey, D. R. and Van Rossem, A. J. (1927). The spotted Rock Wrens of Central America. *PBSW*, 40, 25–8.

Dickey, D. R. and Van Rossem, A. J. (1929). The races of *Lampornis viridipallens* (Bourcier and Mulsant). *PBSW*, 42, 209–12.

Dickey, D. R. and Van Rossem, A. J. (1938). The Birds of El Salvador. *Zool. Series FMNH*, 23.

Dixon, K. L. and Davis, W. B. (1958). Some additions to the avifauna of Guerrero, Mexico. *Condor*, 60, 407.

Donagho, W. R. (1965). The Starling in Guanajuato, Mexico. *Condor*, 67, 447.

Dunlap, E. (1988). Laysan Albatross nesting on Guadalupe Island, Mexico. *Am. Birds*, 42, 180–1.

Easterla, D. A. (1964). Bird records from San Luis Potosi, Mexico. *Condor*, 66, 514.

Easterla, D. A. (1975). The Chestnut-collared Swift in the lowlands of Sinaloa, Mexico. *Condor*, 77, 352.

Eaton, S. W. and Edwards, E. P. (1947). The Mangrove Cuckoo in interior Tamaulipas, Mexico. *Wilson Bull.*, 59, 110–11.

Eaton, S. W. and Edwards, E. P. (1948). Notes on the birds of the Gomez Farias region of Tamaulipas. *Wilson Bull.*, 60, 109–14.

Edwards, E. P. (1957). Yellow-billed Cuckoo nesting in Yucatan. *Condor*, 59, 69–70.

Edwards, E. P. (1959). Nesting of the Lesser Swallow-tailed Swift *Panyptila cayennensis* in Guatemala. *Auk*, 76, 358.

Edwards, E. P. (1968). *Finding birds in Mexico*, 2nd edition, E. P. Edwards, Sweet Briar, VA.

Edwards, E. P. (1972). *A field guide to the birds of Mexico*. E. P. Edwards, Sweet Briar, VA.

Edwards, E. P. (1978). *Appendix for 'A field guide to the birds of Mexico'*. E. P. Edwards, Sweet Briar, VA.

Edwards, E. P. (1985). *Supplement to 'Finding birds in Mexico'*. E. P. Edwards, Sweet Briar, VA.

Edwards, E. P. (1989). *A field guide to the birds of Mexico*, 2nd edition. E. P. Edwards, Sweet Briar, VA.

Edwards, E. P. and Hilton, F. K. (1956). *Streptoprocne semicollaris* in the lowlands of Sinaloa and Nayarit. *Auk*, 73, 138.

Edwards, E. P. and Lea, R. B. (1955). Birds of the Monserrate area, Chiapas, Mexico. *Condor*, 57, 31–54.

Edwards, E. P. and Martin, P. S. (1955). Further notes on birds of the Lake Patzcuaro Region, Mexico. *Auk*, 72, 174–8.

Edwards, E. P. and Morton, E. S. (1963). Occurrence of the Starling in Baja California. *Condor*, 65, 530.

Edwards, E. P. and Tashian, R. E. (1956). The Prothonotary and Kentucky warblers on Cozumel Island, Quintana Roo, Mexico. *Wilson Bull.*, 68, 73.

Edwards, E. P. and Tashian, R. E. (1959). Avifauna of the Catemaco Basin of Southern Veracruz, Mexico. *Condor*, 61, 325–37.

Edwards, E. P. and Tashian, R. E. (1963). Occurrence of the Starling in Baja California, Mexico. *Condor*, 65, 530.

Ehrhardt, J. P. (1971). Census of the birds of Clipperton Island, 1968. *Condor*, 73, 476–80.

Eisenmann, E. (1955a). The species of Middle American birds. *Trans. Linn. Soc. N.Y.*, 7.

Eisenmann, E. (1955b). Status of the Black-polled, Bay-breasted and Connecticut warblers in Middle America. *Auk*, 72, 206–7.

Eisenmann, E. (1962). Notes on nighthawks of the genus *Chordeiles* in southern Middle America, with a description of a new race of *Chordeiles minor* breeding in Panama. *Novitates*, 2094.

Eisenmann, E. (1963). Breeding nighthawks in Central America. *Condor*, 65, 165–6.

Eisenmann, E. (1971). Range expansion and population increase in North and Middle America of the White-tailed Kite (*Elanus leucurus*). *Am. Birds*, 25, 529–36.

Eitniear, J. C. (1986). Status of the large forest eagles of Belize. *Birds of Prey Bull.*, 3, 107–10.

Eitniear, J. C. (1991). The Solitary Eagle *Harpyhaliaetus solitarius*: a new threatened species. *Birds of Prey Bull.*, 4, 81–5.

Elliot, B. G. (1965). The nest of the Red Warbler. *Condor*, 67, 540.

Elliot, B. G. (1969). Life history of the Red Warbler. *Wilson Bull.*, 81, 184–95.

Ellis, D. H. and Whaley, W. H. (1981). Three Crested Eagle records for Guatemala. *Wilson Bull.*, 93, 284–5.

Ely, C. A. (1962). The birds of southeastern Coahuila, Mexico. *Condor*, 64, 34–9.

Ely, C. A. (1973). Returns of North American birds to their wintering grounds in southern Mexico. *Bird-Banding*, 44, 228–9.

Ely, C. A. and Crossin, R. S. (1972). A northern wintering record of the Elf Owl (*Micrathene whitneyi*). *Condor*, 74, 215.

Ely, C. A., Latas, P. J., and Lohoefener, R. R. (1977). Additional returns and recoveries of North American birds banded in southern Mexico. *Bird-Banding*, 48, 275–7.

Erickson, R. (1977). First record of a Knot *Calidris canutus*, and other records, from Belize (British Honduras). *BBOC*, 97, 78–81.

Erickson, R. A., Barron, A. D., and Hamilton, R. A. (1992). A recent Black Rail record for Baja California. *Euphonia*, 1, 19–21.

Ericsson, S. (1981). Loggerhead Shrike in Guatemala in December 1979. *Dutch Birding*, 3, 27–8.

Escalante P., B. P. (1988). Aves de Nayarit. Universidad Autonoma de Nayarit.

Escalante P., P. and Peterson, A. T. (1992). Geographic variation and species limits in Middle American Woodnymphs (*Thalurania*). *Wilson Bull.*, 104, 205–19.

Espino-B., R. and Baldassare, G. A. (1989). Numbers, migration chronology, and activity patterns of nonbreeding Caribbean Flamingos in Yucatan, Mexico. *Condor*, 91, 592–7.

Estrada, R. C. and Estrada, A. (1985). *Lista de las aves de la Estación de Biología Los Tuxtlas*. Inst. de Biol. UNAM.

Evenden, F. G. (1952). Notes on Mexican bird distribution. *Wilson Bull.*, 64, 112–13.

Evenden, F. G., Argante, A. J., and Argante, L. B. (1965). Southerly occurrence of Clark's Nutcracker. *Wilson Bull.*, 77, 86.

Everett, W. T. (1988*a*). Biology of the Black-vented Shearwater. *W. Birds*, 19, 89–104.

Everett, W. T. (1988*b*). Notes from Clarion Island. *Condor*, 90, 512–13.

Everett, W. T. and Teresa, S. (1988). A Masked Booby at Islas Los Coronados, Baja California, Mexico. *W. Birds*, 19, 173–4.

Farrand, J. (ed.) (1985). *The Audubon Society master guide to birding* (3 vols). Alfred A. Knopf, New York.

Feduccia, A. (1976). New bird records for El Salvador. *Wilson Bull.*, 88, 150–1.

Feduccia, J. A. (1970). Natural history of the avian families Dendrocolaptidae (Woodhewers) and Furnariidae (Ovenbirds). *J. Grad. Res. Center*, 38, 1–26.

Feerer, J. L. (1977). Niche partitioning by Western Grebe polymorphs. Unpubl. M.S. thesis. Humboldt State Univ., Arcata, CA.

Feltner, T. B. (1976). A Scaly-breasted Hummingbird in the Republic of Mexico. *Mexican Birds Newsletter*, 1(2), 11.

ffrench, R. (1973). *A guide to the birds of Trinidad and Tobago*. Livingston, Wynnewood, PA.

Fjeldsa, J., Krabbe, N., and Ridgely, R. S. (1987). Great Green Macaw *Ara ambigua* collected in northwest Ecuador, with taxonomic notes on *Ara militaris*. *BBOC*, 107, 28–31.

Fleming, R. L. and Baker, R. H. (1963). Notes on the birds of Durango, Mexico. *Publ. Mus. Mich. State Univ. Biol.*, Series 2, 273–304.

Forshaw, J. M. (1981). *Parrots of the world*, 2nd rev. edn, reprinted with corrections. Lansdowne Press, Melbourne, Australia.

Fosberg, F. R. (1962). *A brief survey of the cays of Arrecife Alacrán, a Mexican atoll*. Atoll Research Bull. No. 93. Pacific Science Board, Washington, D.C.

Fosberg, F. R. and Pelzl, H. W. (1973). Ornithological observations. In *Investigations of marine shallow water ecosystems, progress report, terrestrial investigations* (ed. F. R. Fosberg), pp. 16–21. Environmental Sci. Program, Smithsonian Inst., Washington D.C.

Foster, M. S. (1976). Nesting biology of the Long-tailed Manakin. *Wilson Bull.*, 88, 400–20.

Fowler, J. M. and Cope, J. B. (1964). Notes on the Harpy Eagle in British Guiana. *Auk*, 81, 257–73.

Freese, C. H. (1975). Notes on the nesting of the Double-striped Thick-knee (*Burhinus bistriatus*) in Costa Rica. *Condor*, 77, 353–4.

Friedmann, H. (1934). The hawks of the genus *Chondrohierax*. *J. Wash. Acad. Sci.*, 24, 310–18.

Friedmann, H. (1947). The Spotted Rail, *Pardirallus maculatus*, in southern Mexico. *Auk*, 64, 460.

Friedmann, H., Griscom, L., and Moore, R. T. (1950). *Distributional checklist of the birds of Mexico. Part 1*. Pacific Coast Avifauna No. 29. Cooper Ornithol. Soc.

Galluci, T. (1981) Summer bird records from Sonora, Mexico. *Am. Birds*, 35, 243–7.

Gardner, A. L. (1972). The occurrence of *Streptoprocne zonaris albicincta* and *Ara militaris* in Chiapas, Mexico. *Condor*, 74, 480–1.

Garrett, K. and Dunn, J. (1981). *Birds of southern California*. Los Angeles Audubon Soc., Los Angeles, CA.

Garza de León, A. (1987). Unusual records from Coahuila, Mexico. *Condor*, 89, 672–3.

Gatz, T., Gent, P., Jakle, M., Otto, R., Otto, W.,

and Ellis, B. (1985). Spotted Rail, Brant and Yellow-breasted Crake—records from the Yucatan. *Am. Birds*, **39**, 871–2.

Gaviño de La Torre, G., Monroy M., R., Soria, G., and Arellano A., N. (1974). Notas sobre la biología de *Sula leucogaster nesiotes* (Aves: Sulidae) en la Bahia de Chamela, Jalisco, México. *Analecta*, 1974–**12**, 1–15.

Gaviño de La Torre, G., Martinez, A., Uribe P., Z., and Santillan A., S. (1979). Vertebrados terrestres y vegetación dominante de la Isla Ixtapa, Guerrero. UNAM. *Anales Inst. Biol.*, 50, Ser. Zool. (1), 701–19.

Gehlbach, F. R., Dillon, D. O., Harrell, H. L., Kennedy, S. E., and Wilson, K. R. (1976). Avifauna of the Rio Corona, Tamaulipas, Mexico: northeastern limit of the Tropics. *Auk*, **93**, 53–65.

Gisiner, B., Riedman, M., Keith, E., and LeBoeuf, B. J. (1979). *Report of a Scripps Institute of Oceanography expedition to the Gulf of California and the Pacific Ocean west of Baja California, 21 June to 21 July 1979.* Univ. Calif., Santa Cruz.

Goldman, E. A. (1951). Biological investigations in Mexico. *Smithson. Misc. Coll.*, **115**.

Goldman, E. A. and Moore, R. T. (1945). The biotic provinces of Mexico. *J. Mam.*, **26**, 347–60.

Graber, R. R. and Graber, J. W. (1954a). Yellow-headed Vulture in Tamaulipas, Mexico. *Condor*, **56**, 165–6.

Graber, R. R. and Graber, J. W. (1954b). Comparative notes on Fuertes and Orchard orioles. *Condor*, **56**, 274–82.

Grabowski, G. L. (1979). Vocalizations of the Rufous-backed Thrush (*Turdus rufopalliatus*) in Guerrero, Mexico. *Condor*, **81**, 409–16.

Grant, P. J. (1986). *Gulls: a guide to identification*, 2nd edition. Buteo Books. SD.

Grant, P. R. (1964a). Nuevos datos sobre las aves de Jalisco y Nayarit, México. *Anales Inst. Biol.*, **35**, 123–6.

Grant, P. R. (1964b). The birds of the Tres Marietas Islands, Nayarit, Mexico. *Auk*, **81**, 514–19.

Grant, P. R. (1965). A systematic study of the birds of the Tres Marias Islands, Mexico. *Postilla*, **90**.

Grant, P. R. (1966). The coexistence of two wren species of the genus *Thryothorus*. *Wilson Bull.*, **78**, 266–78.

Grant, P. R. and Cowan, I. McT. (1964). A review of the avifauna of the Tres Marias Islands, Nayarit, Mexico. *Condor*, **66**, 221–8.

Grinnell, J. (1928). A distributional summation of the ornithology of Lower California. *UCPZ*, **32**, 1–300.

Grinnell, J. and Lamb, C. C. (1927). New bird records from Lower California. *Condor*, **29**, 124–6.

Griscom, L. (1926a). The ornithological results of the Mason–Spinden expedition to Yucatan, Part I: Introduction; birds of the mainland of eastern Yucatan. *Novitates*, **235**.

Griscom, L. (1926b). The ornithological results of the Mason–Spinden expedition to Yucatan, Part II: Chinchorro and Cozumel Islands. *Novitates*, **236**.

Griscom, L. (1929). Studies from the Dwight collection of Guatemala birds. I. *Novitates*, **379**.

Griscom, L. (1930a). Critical notes on Central American Birds. *Proc. New England Zool. Club*, **12**, 1–8.

Griscom, L. (1930b). Studies from the Dwight collection of Guatemala birds. II. *Novitates*, **414**.

Griscom, L. (1931). Notes on rare and little known Neotropical pygmy-owls. *Proc. New England Zool. Club*, **12**, 37–43.

Griscom, L. (1932). The distribution of bird-life in Guatemala. *Bull. AMNH*, **64**.

Griscom, L. (1934). The ornithology of Guerrero, Mexico. *Bull. MCZ*, **75**, 367–422.

Griscom, L. (1935). The birds of the Sierra de Las Minas, eastern Guatemala. *Ibis*, **77**, 801–37.

Griscom, L. (1937). A collection of birds from Omilteme, Guerrero. *Auk*, **54**, 192–9.

Gutierrez, C. (ed.) (1990). *Propuesta de restructuración del Instituto Nicaraguense de Recursos Naturales y del Ambiente.* IRENA. Managua, Nicaragua.

Hainebach, K. (1992). First records of Xantus' Hummingbird in California. *W. Birds*, **23**, 133–6.

Haney, J. C. (1983). First sight record of Orange-breasted Falcon for Belize. *Wilson Bull.*, **95**, 314–15.

Hanson, D. A. (1982). Distribution of the Quetzal in Honduras. *Auk*, **99**, 385.

Hardy, J. W. (1964). Behaviour, habitat and relationships of jays of the Genus *Cyanolyca*. *Occas. Papers C. C. Adams Cen. Ecol. Studies*, **11**.

Hardy, J. W. (1965). Evolutionary and ecological relationships between three species of blackbirds (Icteridae) in central Mexico. *Evolution*, **21**, 196–7.

Hardy, J. W. (1969). A taxonomic revision of the New World jays. *Condor*, **71**, 360–75.

Hardy, J. W. (1971). Habitat and habits of the Dwarf Jay *Aphelocoma nana*. *Wilson Bull.*, **83**, 5–30.

Hardy, J. W. (1973). Age and sex differences in the black-and-blue jays of Middle America. *Bird-Banding*, **44**, 81–90.

Hardy, J. W. (1976). Comparative breeding behaviour and ecology of the Bushy-crested and Nelson San Blas jays. *Wilson Bull.*, 88, 96–120.

Hardy, J. W. (1979). Social behaviour, habitat, and food of the Beechey Jay. *Wilson Bull.*, 91, 1–15.

Hardy, J. W. and Delaney, D. J. (1987). The vocalizations of the Slender-billed Wren (*Hylorchilus sumichrasti*): who are its closest relatives? *Auk*, 104, 528–30.

Hardy, J. W. and Dickerman, R. W. (1965). Relationships between two forms of the Red-winged Blackbird in Mexico. *Living Bird*, 4, 107–29.

Hardy, J. W., Raitt, R. J., Orjuela, J., Webber, T., and Edinger, B. (1975). First observation of the Orange-breasted Falcon in the Yucatan Peninsula of Mexico. *Condor*, 77, 512–13.

Hardy, J. W., Parker, T. A., III, and Coffey, B. B., Jr. (1991). *Voices of the woodcreepers*. Ara Records, Gainesville, FL.

Harris, A., Tucker, L., and Vinicombe, K. (1989). *The Macmillan field guide to bird identification*. Macmillan, London.

Harrison, E. N. and Kiff, L. F. (1977). The nest and egg of the Black Solitary Eagle. *Condor*, 79, 132–3.

Hartman, F. A. (1956). A nest of the Striped Horned Owl. *Condor*, 58, 73.

Heath, M. and Long, A. (1991). Habitat, distribution and status of the Azure-rumped Tanager *Tangara cabanisi* in Mexico. *Bird Conserv. International*, 1, 223–54.

Helbig, A. (1983). Notes on the distribution of seabirds in western Mexico. *Gerfaut*, 73, 147–60.

Hellebuyck, V. (1983). Three new specimen records of birds for El Salvador. *Wilson Bull.*, 95, 662–4.

Henny, C. J., Anderson, D. W., and Knoder, C. E. (1978). Bald Eagles nesting in Mexico. *Auk*, 95, 424.

Hill, H. M. and Wiggins, I. L. (1948). Ornithological notes from Lower California. *Condor*, 50, 155–61.

Hilty, S. L. and Brown, W. L. (1986). *A guide to the birds of Colombia*. Princeton Univ. Press, Princeton, NJ.

Holt, E. G. (1926). On a Guatemalan specimen of *Progne sinaloae* Nelson. *Auk*, 43, 550–1.

Howell, S. N. G. (1986). *A field checklist to the birds of Mexico*. Golden Gate Audubon Soc., Berkeley, CA.

Howell, S. N. G. (1987). Little Gull (*Larus minutus*) in Mexico. *MBA Bulletin Board*, 1(87–1), 2–3.

Howell, S. N. G. (1989a). Additional information on the birds of the Campeche Bank, Mexico. *J. Field Ornithol.*, 60, 504–9.

Howell, S. N. G. (1989b). Scaly-breasted Hummingbird (*Phaechroa cuvierii*) nesting in Mexico. *Aves Mexicanas*, 2(89–2), 8–9.

Howell, S. N. G. (1989c). Short-tailed Nighthawk (*Lurocalis 'semitorquatus'*) in Mexico. *Aves Mexicanas*, 2(89–2), 9–10.

Howell, S. N. G. (1989d). Hummingbird discoveries. *PRBO Newsletter*, 85, 8–9.

Howell, S. N. G. (1990a). Songs of Mexican birds by B. B. and L. C. Coffey. Review. *Wilson Bull.*, 102, 184–5.

Howell, S. N. G. (1990b). A distributional survey of the birds of the Mexican state of Oaxaca by L. C. Binford. Review. *Wilson Bull.*, 102, 185–7.

Howell, S. N. G. (1990c). Identification of White and Black-backed wagtails in alternate plumage. *W. Birds*, 21, 41–9.

Howell, S. N. G. (1992a). The Short-crested Coquette: Mexico's least-known endemic. *Birding*, 24, 86–91.

Howell, S. N. G. (1992b). Recent records of Maroon-chested Ground-Dove (*Claravis mondetoura*) in Mexico. *Euphonia*, 1, 39–41.

Howell, S. N. G. (1993a). Taxonomy and distribution of the hummingbird genus *Chlorostilbon* in Mexico and northern Central America. *Euphonia*, 2, 25–37.

Howell, S. N. G. (1993b). A taxonomic review of the Green-fronted Hummingbird. *BBOC*, 113, 179–87.

Howell, S. N. G. (1994). The specific status of Black-faced Antthrushes in Middle America. *Cotinga*, 1, 21–5.

Howell, S. N. G. and Cannings, R. J. (1992). Songs of two Mexican populations of the Western Flycatcher *Empidonax difficilis* complex. *Condor*, 94, 785–7.

Howell, S. N. G. and de Montes, B. M. (1989). Status of the Glossy Ibis in Mexico. *Am. Birds*, 43, 43–5.

Howell, S. N. G. and de Montes, B. M. (1990). Additional information on the Wood Duck (*Aix sponsa*) in Mexico: first records for the Yucatan Peninsula. *Aves Mexicanas*, 2(90–1), 2–3.

Howell, S. N. G. and Engel, S. J. (1993). Seabird observations off western Mexico. *W. Birds*, 24, 167–81.

Howell, S. N. G. and Johnston, S. (1993). The birds of Isla Hobox, Mexico. *Euphonia*, 2, 1–18.

Howell, S. N. G. and Prairie, L. J. (1989). Notable gull records from the Yucatan Peninsula. *Aves Mexicanas*, 2(89–2), 3–4.

Howell, S. N. G. and Pyle, P. (1987). Oldsquaw (*Clangula hyemalis*) in Jalisco. *MBA Bulletin Board*, 1(87–3), 2–3.

Howell, S. N. G. and Pyle, P. (1988). Additional

notes on birds in Baja California. *Aves Mexicanas*, 1(88–1), 2–3.

Howell, S. N. G. and Pyle, P. (1989). Another vagrant location in Baja California. *Aves Mexicanas*, 2(89–1), 3–4.

Howell, S. N. G. and Pyle, P. (1990). Additional notes on birds in Baja California, May 1989. *Aves Mexicanas*, 2(90–1); 6–7.

Howell, S. N. G. and Pyle, P. (1993). New and noteworthy bird records from Baja California, October 1991. *W. Birds*, 24, 57–62.

Howell, S. N. G. and Webb, S. (1987). Birding at Tecolutla, Veracruz. *MBA Bulletin Board*, 1(87–3), 4.

Howell, S. N. G. and Webb, S. (1989a). Red-throated Pipit (*Anthus cervinus*) in Michoacan. *Aves Mexicanas*, 2(89–1), 2.

Howell, S. N. G. and Webb, S. (1989b). Additional notes from Isla Clarión, Mexico. *Condor*, 91, 1007–8.

Howell, S. N. G. and Webb, S. (1989c). Notes on the Honduran Emerald. *Wilson Bull.*, 101, 642–3.

Howell, S. N. G. and Webb, S. (1990a). The seabirds of Las Islas Revillagigedo, Mexico. *Wilson Bull.*, 102, 140–6.

Howell, S. N. G. and Webb, S. (1990b). A site for Buff-collared Nightjar (*Caprimulgus ridgwayi*) and Mexican Sheartail (*Calothorax* [= *Doricha*] *eliza*) in Veracruz. *Aves Mexicanas*, 2(90–1), 1–2.

Howell, S. N. G. and Webb, S. (1990c). Comments on trying to find the Greater Ani (*Crotophaga major*) in Tamaulipas. *Aves Mexicanas*, 2(90–2), 1–2.

Howell, S. N. G. and Webb, S. (1990d). Notes on migrants in eastern Mexico, spring 1990. *Aves Mexicanas*, 2(90–2), 5–7.

Howell, S. N. G. and Webb, S. (1992a). Southernmost records of Western and Yellow-footed gulls. *W. Birds*, 23, 31–2.

Howell, S. N. G. and Webb, S. (1992b). New and noteworthy bird records from Guatemala and Honduras. *BBOC*, 112, 42–9.

Howell, S. N. G. and Webb, S. (1992c). Changing status of the Laysan Albatross in Mexico. *Am. Birds*, 46, 220–3.

Howell, S. N. G. and Webb, S. (1992d). Observations of birds from Isla Guadalupe, Mexico. *Euphonia*, 1, 1–6.

Howell, S. N. G. and Webb, S. (1992e). A little-known cloud forest in Hidalgo, Mexico. *Euphonia*, 1, 7–11.

Howell, S. N. G. and Webb, S. (1992f). Noteworthy bird observations from Baja California, Mexico. *W. Birds*, 23, 153–63.

Howell, S. N. G. and Webb, S. (1992g). Observations of northern migrant birds on the Revillagigedo Islands. *Euphonia*, 1, 27–33.

Howell, S. N. G. and Wilson, R. G. (1990).

Chestnut-collared Longspur (*Calcarius ornatus*) and other migrants of note in Guerrero, Mexico. *Aves Mexicanas*, 2(90–1), 7–8.

Howell, S. N. G., Webb, S., and de Montes, B. M. 1990. Notes on tropical terns in Mexico. *Am. Birds*, 44: 381–3.

Howell, S. N. G., Webb, S. and de Montes, B. M. (1992a). Colonial nesting of the Orange Oriole. *Wilson Bull.*, 104, 189–90.

Howell, S. N. G., Dowell, B. A., James, D. A., Behrstock, R. A., and Robbins, C. S. (1992b). New and noteworthy bird records from Belize. *BBOC*, 112, 235–44.

Howell, S. N. G., Pyle, P., Spear, L. B., and Pitman, R. L. (1993). North American migrant birds on Clipperton Atoll. *W. Birds*, 24, 73–80.

Howell, T. R. (1952). Natural history and differentiation in the Yellow-bellied Sapsucker. *Condor* 54, 237–82.

Howell, T. R. (1955). A southern hemisphere migrant in Nicaragua. *Condor*, 57, 188–9.

Howell, T. R. (1957). Birds of a second growth rain forest in Nicaragua. *Condor*, 59, 73–111.

Howell, T. R. (1959). Land birds from Clipperton Island. *Condor*, 61, 155–6.

Howell, T. R. (1969). Avian distribution in Central America. *Auk*, 86, 293–326.

Howell, T. R. (1972). Birds of the lowland pine savanna of north-eastern Nicaragua. *Condor*, 74, 316–40.

Howell, T. R. and Cade, T. J. (1954). The birds of Guadalupe Island in 1954. *Condor*, 56, 283–94.

Hubbard, J. P. (1967). Notes on some Chiapas birds. *Wilson Bull.*, 79, 236.

Hubbard, J. P. (1970). Geographic variation in the *Dendroica coronata* complex. *Wilson Bull.*, 82, 355–69.

Hubbard, J. P. (1972). Palm Warbler in Guerrero and comments on Audubon's Warbler in Costa Rica. *Auk*, 89, 885–6.

Hubbard, J. P. (1974). Avian evolution in the aridlands of North America. *Living Bird*, 12, 155–96.

Hubbard, J. P. (1977). The status of Cassin's Sparrow in New Mexico and adjacent states. *Am. Birds*, 31, 933–41.

Hubbard, J. P. (1987). Regional reports: New Mexico, northern Chihuahua. *Am. Birds*, 41, 314–16.

Hubbard, J. P. and Crossin, R. S. (1974). Notes on northern Mexican birds. *Nemouria*, 14, 1–41.

Hubbs, C. L. (1955). Black Scoters reported from Baja California. *Condor*, 57, 121–2.

Hubbs, C. L. (1956). Off-season, southern occurrence of the Black Scoter on the Pacific coast. *Condor*, 58, 448–9.

Hubbs, C. L. (1960). The Rock Sandpiper,

another northern bird recorded from the cool coast of northwestern Baja California. *Condor*, **62**, 68–9.

Huey, L. M. (1926). Notes from northwestern Lower California, with the description of an apparently new race of the Screech Owl. *Auk*, **43**, 347–62.

Huey, L. M. (1927). Birds recorded in spring at San Felipe, northeastern Lower California, Mexico, with a description of a new woodpecker from that locality. *Trans. SDSNH*, **5**, 13–40.

Huey, L. M. (1928). Some bird records from northern Lower California. *Condor*, **30**, 158–9.

Huey, L. M. (1954). Notes from southern California and Baja California, Mexico. *Condor*, **56**, 51–2.

Humphrey, P. S. and Parkes, K. C. (1959). An approach to the study of molts and plumages. *Auk*, **76**, 1–31.

Humphrey, P. S. and Parkes, K. C. (1963). Comments on the study of plumage succession. *Auk*, **80**, 496–503.

Hunn, E. (1971). Noteworthy bird observations from Chiapas, Mexico. *Condor*, **73**, 483.

Hunt, W. G., Enderson, J. H., Lanning, D., Hitchcock, M. A., and Johnson, B. S. (1988). Nesting Peregrines in Texas and northern Mexico. In *Peregrine falcon populations* (ed. T. J. Cade *et al.*). Peregrine Fund Inc., Boise, ID.

Hunter, L. A. (1988). Status of the endemic Atitlan Grebe of Guatemala: is it extinct? *Condor*, **90**, 906–12.

Hussel, D. J. T. (1980). The timing of fall migration and molt in Least Flycatchers. *J. Field Ornithol.*, **51**, 65–71.

Hussel, D. J. T. (1982). The timing of fall migration in Yellow-bellied Flycatchers. *J. Field Ornithol.*, **53**, 1–6.

Iñigo-E., E., Ramos, M., and Gonzalez, F. (1987). Two recent records of Neotropical eagles in southern Veracruz, Mexico. *Condor*, **89**, 671–2.

ICBP. (1992). *Putting biodiversity on the map: priority areas for global conservation*. ICBP, Cambridge, UK.

Jehl, J. R., Jr. (1970). A Mexican specimen of the Yellow-billed Loon. *Condor*, **72**, 376.

Jehl, J. R. Jr. (1971). The status of *Carpodacus mcgregori*. *Condor*, **73**, 375–6.

Jehl, J. R. Jr. (1974). The near-shore avifauna of the Middle American west coast. *Auk*, **91**, 681–99.

Jehl, J. R. Jr. (1976). The northernmost colony of Heermann's Gull. *W. Birds*, **7**, 25–6.

Jehl, J. R. Jr. (1977). An annotated list of birds of Islas Los Coronados, Baja California, and adjacent waters. *W. Birds*, **8**, 91–101.

Jehl, J. R. Jr. (1982). The biology and taxonomy of the Townsend's Shearwater. *Gerfaut*, **72**, 121–35.

Jehl, J. R., Jr. and Bond, S. I. (1975). Morphological variation and species limits in murrelets of the genus *Endomychura*. *Trans. SDSNH*, **18**, 9–23.·

Jehl, J. R., Jr. and Everett, W. T. (1985). History and status of the avifauna of Isla Guadalupe, Mexico. *Trans. SDSNH*, **20**, 313–36.

Jehl, J. R., Jr. and Parkes, K. C. (1982). The status of the avifauna of the Revillagigedo Islands, Mexico. *Wilson Bull.*, **94**, 1–19.

Jenkinson, M. A. and Mengel, R. M. (1979). Notes on an important nineteenth century collection of Central and North American birds made by N. S. Goss. *Occas. Papers Mus. Nat. Hist. Univ. Kansas*, **81**, 1–10.

Jenny, J. P. and Cade, T. J. (1986). Observations on the biology of the Orange-breasted Falcon *Falco deiroleucus*. *Proc. ICBP W. Hemisphere World Working Group for Birds of Prey. Birds of Prey Bull.* **3**, 119–24.

Johnsgard, P. A. (1973). *Grouse and quails of North America*. Univ. Nebraska Press, Lincoln, NE.

Johnson, K. W., Johnson, J. E., Albert, R. O., and Albert, T. R. (1988). Sightings of Golden-cheeked Warblers (*Dendroica chrysoparia*) in northeastern Mexico. *Wilson Bull.*, **100**, 130–1.

Johnson, N. K. (1970). Fall migration and winter distribution of the Hammond Flycatcher. *Bird-Banding*, **41**, 169–90.

Johnson, N. K. (1980). Character variation and evolution of sibling species flycatchers of the *Empidonax difficilis-flavescens* complex (Aves: Tyrannidae). *UCPZ*, **112**, 1–151.

Kantak, G. E. (1979). Observations on some fruit-eating birds in Mexico. *Auk*, **96**, 183–6.

Kaufman, K. and Witzeman, J. (1979). A Harlequin Duck reaches Sonora, Mexico. *Cont. Birdlife*, **1**, 16–17.

Keith, A. R. and Stejskal, D. (1987). Two outstanding sites for vagrants in Baja California. *MBA Bulletin Board*, 1(87–3), 3.

Kiff, L. F. and Hough, D. J. (1985). *Inventory of bird egg collections in North America, 1985*. AOU and Oklahoma Biol. Survey, Norman, OK.

King, J. E. and Pyle, R. L. (1957). Observations of sea birds in the tropical Pacific. *Condor*, **59**, 27–39.

King, W. B. (ed.) (1974). *Pelagic studies of seabirds in the central and eastern Pacific*. Smithson. Contr. Zool. No. 158. Smithsonian, Washington.

Klass, E. E. (1968). Summer birds from the Yucatan Peninsula, Mexico. *UKPMNH*, **17**, 579–611.

Knoder, C. E., Plaza, P. D., and Sprunt, A., IV (1980). Status and distribution of the Jabiru Stork and other water birds in western Mexico. In *The birds of Mexico, their ecology and conservation* (ed. P. P. Schaeffer and S. M. Ehlers). Proc. NAS symposium, Tiburon, CA.

Koford, C. B. (1953). The California Condor. *Nat. Aud. Soc. Res. Report*, 4.

Kramer, G. W. (1982). Oldsquaw record from Sinaloa, Mexico. *Condor*, 84, 243.

Kramer, G. W. (1983). Some winter birds of Bahia de San Quintin, Baja California. *Am. Birds*, 37, 270–2.

Kratter, A. W. (1991). First nesting record for Williamson's Sapsucker (*Sphyrapicus thyroides*) in Baja California, Mexico, and comments on the biogeography of the fauna of the Sierra San Pedro Martir. *SW Nat.*, 36, 247–50.

Kroodsma, D. E. (1988). Two species of Marsh Wren in Nebraska? *Birding*, 20, 371–4.

LaBastille, A. (1974). *Ecology and management of the Atitlan Grebe, Lake Atitlan, Guatemala*. Wildlife Monographs No. 37. Wildlife Society.

Lamb, C. C. (1925a). Observations on the Xantus' Hummingbird. *Condor* 27, 89–92.

Lamb, C. C. (1925b). Some birds new to the Cape San Lucas region. *Condor*, 27, 117–18.

Lancaster, D. A. (1964). Life history of the Boucard Tinamou in British Honduras. Part II: breeding biology. *Condor*, 66, 253–76.

Land, H. C. (1961). Birding on eastern Guatemala's Highest Mountain. *Fl. Field Nat.*, Apr. 1961.

Land, H. C. (1962a). A collection of birds from the arid interior of eastern Guatemala. *Auk*, 79, 1–11.

Land, H. C. (1962b). A collection of birds from the Sierra de Las Minas, Guatemala. *Wilson Bull.*, 74, 267–83.

Land, H. C. (1963). A collection of birds from the Caribbean lowlands of Guatemala. *Condor*, 65, 49–65.

Land, H. C. (1970). *Birds of Guatemala*. Livingstone, Wynnewood, PA.

Land, H. C. and Kiff, L. F. (1965). The Band-tailed Barbthroat, *Threnetes ruckeri* (Trochilidae) in Guatemala. *Auk*, 82, 286.

Land, H. C. and Schultz, W. L. (1963). A proposed subspecies of the Great Potoo, *Nyctibius grandis* (Gmelin). *Auk*, 80, 195–6.

Land, H. C. and Wolf, L. L. (1961). Additions to the Guatemala bird list. *Auk*, 78, 94–5.

Lanning, D. V. and Hitchcock, M. A. (1991). Breeding distribution and habitat of Prairie Falcons in northern Mexico. *Condor*, 93, 762–5.

Lanning, D. V. and Shiflett, J. T. (1983). Nesting ecology of Thick-billed Parrots. *Condor*, 85, 66–73.

Lanning, D. V., Marshall, J. T., and Shiflett, J. T. (1990). Range and habitat of the Colima Warbler. *Wilson Bull.*, 103, 1–13.

Lantz, D. E. (1899). List of birds collected by Col. N. S. Goss in Mexico and Central America. *Kansas Acad. Sci. Trans.*, 16, 218–24.

Lanyon, W. E. (1960a). Relationship of the House Wren (*Troglodytes aedon*) of North America and the Brown-throated Wren (*T. brunneicollis*) of Mexico. *Proc. 12th Int. Ornith. Congress 1958*, 450–8.

Lanyon, W. E. (1960b). The Middle American populations of the Crested Flycatcher, *Myiarchus tyrannulus*. *Condor*, 62, 341–50.

Lanyon, W. E. (1961). Specific limits and distribution of Ash-throated and Nutting's flycatchers. *Condor*, 63, 421–49.

Lanyon, W. E. (1962). Specific limits and distribution of meadowlarks of the desert grassland. *Auk*, 79, 183–207.

Lanyon, W. E. (1965a). Specific limits in the Yucatan Flycatcher *Myiarchus yucatanensis*. *Novitates*, 2229.

Lanyon, W. E. (1965b). Correction of erroneous records of the Ash-throated Flycatcher for northern Guatemala and Yucatan, Mexico. *Condor*, 67, 354.

Lanyon, W. E. (1969). Vocal characters and avian systematics. In *Bird vocalizations* (ed. R. A. Hinde). Cambridge Univ. Press, Cambridge.

Lanyon, W. E. (1982a). Behaviour, morphology and systematics of the Flammulated Flycatcher of Mexico. *Auk*, 99, 414–23.

Lanyon, W. E. (1982b). The subspecies concept: then, now, and always. *Auk*, 99, 603–4.

Lanyon, W. E. (1984). A phylogeny of the kingbirds and their allies. *Novitates*, 2797.

Lanyon, W. E. (1985). A phylogeny of the Myiarchine flycatchers. In *Neotropical ornithology* (ed. P. A. Buckley *et al.*), pp. 361–80. Ornithological Monographs No. 36. AOU.

Lanyon, W. E. (1986). A phylogeny of the thirty-three genera in the *Empidonax* assemblage of Tyrant-Flycatchers. *Novitates*, 2846.

Lanyon, W. E. (1988a). A phylogeny of the thirty-two genera in the *Elaenia* assemblage of Tyrant-Flycatchers. *Novitates*, 2914.

Lanyon, W. E. (1988b). The phylogenetic affinities of the flycatcher genera *Myiobius* Darwin and *Terrenotriccus* Ridgway. *Novitates*, 2915.

Lanyon, W. E. (1988c). A phylogeny of the Flatbill and Tody-tyrant assemblage of Tyrant-Flycatchers. *Novitates*, 2923.

Lasley, G. W. (1987). The first records of the Lesser Black-backed Gull (*Larus fuscus*) in Mexico. *MBA Bulletin Board*, 1(87–1), 2.

Lasley, G. W. and DeSante, D. F. (1988).

Thayer's Gull in Tamaulipas. *Aves Mexicanas*, **1**(88–1), 1.

Lawson, P. W. and Lanning, D. V. (1980). Nesting and status of the Maroon-fronted Parrot (*Rhynchopsitta terrisi*). In *Conservation of New World parrots*, (ed. R. F. Pasquier) ICBP Tech. Pub. No. 1. Smithson. Inst. Press.

Lea, R. B. and Edwards, E. P. (1950). Notes on birds of the Lake Patzcuaro region, Michoacan, Mexico. *Condor*, **52**, 260–71.

Lea, R. B. and Edwards, E. P. (1951). A nest of the Rufous-breasted Spinetail in Mexico. *Wilson Bull.*, **63**, 337–8.

Leber, K. (1975). Notes on the life history of the Spot-bellied Bobwhite, *Colinus leucopogon dickeyi*, Conover. *Brenesia*, **5**, 7–21.

Lee, J. C. (1978). Lapland Longspur in southeastern Mexico. *Condor*, **80**, 452–3.

Le Febvre, E. A. and Le Febvre, J. H. (1958). Notes on the ecology of *Dactylortyx thoracicus*. *Wilson Bull.*, **70**, 372–7.

Le Febvre, E. A. and Warner, D. W. (1959). Molts, plumages and age groups in *Piranga bidentata* in Mexico. *Auk*, **76**, 208–17.

Leopold, A. S. (1946). Clark Nutcracker in Nuevo Leon, Mexico. *Condor*, **48**, 278.

Leopold, A. S. (1948). The wild turkeys of Mexico. *Trans. 13th North Amer. Wildlife Conference*.

Leopold, A. S. (1950). Vegetation zones of Mexico. *Ecology*, **31**, 507–18.

Leopold, A. S. (1959). *Wildlife of Mexico*. Univ. Cal. Press, Berkeley.

Leopold, A. S. and McCabe, R. A. (1957). Natural history of the Montezuma Quail in Mexico. *Condor*, **59**, 3–26.

Leopold, A. S., Gutierrez, R. G., and Bronson, M. T. (1984). *North American game birds and mammals*. Charles Scribner's Sons. New York.

Lewis, J. C. (ed.) (1982). *Proceedings of the 1981 crane workshop*. NAS.

Ligon, J. D. (1968a). Observations on Strickland's Woodpecker (*Dendrocopus stricklandi*). *Condor*, **70**, 83–4.

Ligon, J. D. (1968b). The biology of the Elf Owl *Micrathene whitneyi*. *Misc. Publ. MZUM*, **136**.

Loetscher, F. W., Jr. (1941). Ornithology of the Mexican state of Veracruz, with an annotated list of the birds. Unpubl. PhD Thesis. Cornell Univ.

Loetscher, F. W., Jr. (1955). North American migrants in the state of Veracruz, Mexico: a summary. *Auk*, **72**, 14–54.

Loomis, L. M. (1918). Expedition of the California Academy of Sciences to the Galapagos Islands, 1905–1906. XII. A review of the albatrosses, petrels, and diving petrels. *Proc. CAS* (4th series), **2** (2), 1–187.

Lopez, O., A., Lynch, J. F., and de Montes, B. M. (1989). New and noteworthy records of birds from the eastern Yucatan Peninsula. *Wilson Bull.*, **101**, 390–409.

Lopez, O., A. and Ramo, C. (1992). Colonial waterbird populations in the Sian Ka'an Biosphere Reserve (Quintana Roo, Mexico). *Wilson Bull.* **104**, 501–15.

Lousada, S. (1989). *Amazona auropalliata caribaea*: a new subspecies of parrot from the Bay Islands, northern Honduras. *BBOC*, **109**, 232–5.

Lowery, G. H., Jr. and Berrett, D. G. (1963). A new Carolina Wren (Aves: Troglodytidae) from southern Mexico. *Occas. Papers MZLSU*, **24**.

Lowery, G. H., Jr. and Dalquest, W. W. (1951). Birds from the state of Veracruz, Mexico. *UKPMNH*, **3**, 531–649.

Lowery, G. H., Jr. and Newman, R. J. (1949). New birds from the state of San Luis Potosi and the Tuxtla Mountains of Veracruz, Mexico. *Occas. Papers MZLSU*, **22**.

Lowery, G. H. Jr. and Newman, R. J. (1951). Notes on the ornithology of south-eastern San Luis Potosi. *Wilson Bull.*, **63**, 315–22.

Lynch, J. F. (1989). Distribution of overwintering Nearctic migrants in the Yucatan Peninsula, I: general patterns of occurrence. *Condor*, **91**, 515–44.

Lyon, B. and Kuhnigk, A. (1985). Notes on nesting Ornate Hawk-Eagles in Guatemala. *Wilson Bull.*, **97**, 141–7.

McCaskie, G. (1975). The spring migration, Southern Pacific Coast Region. *Am. Birds*, **29**, 907–12.

McCaskie, G. (1983). Another look at the Western and Yellow-footed gulls. *W. Birds*, **14**, 85–107.

McCaskie, G. (1988). Southern Pacific Coast Region. *Am. Birds*, **42**, 480–3.

MacKinnon Vda. de Montes, B. (1990). *100 common birds of the Yucatan Peninsula*. Amigos de Sian Ka'an, A. C. Cancún, QR, Mexico.

MacKinnon H., B. (1992). *Check-list of the birds of the Yucatan Peninsula*. Amigos de Sian Ka'an, Cancun, Mexico.

McKitrick, M. and Zink, R. M. (1988). Species concepts in ornithology. *Condor*, **90**, 1–14.

McLellan, M. E. (1926). Expedition to the Revillagigedo Islands, Mexico, in 1925. The birds and mammals. *Proc. CAS*, **15**, 279–322.

McLellan, M. E. (1927). Notes on birds of Sinaloa and Nayarit, Mexico, in the fall of 1925. *Proc. CAS*, **16**, 1–51.

Madge, S. C. (1987). Field identification of Radde's and Dusky warblers. *Brit. Birds*, **80**, 595–603.

Maillard, J. (1923). Expedition of the California Academy of Sciences to the Gulf of California in 1921. *Proc. CAS*, **12**, 443–56.

Marcus, M. J. (1983). Additions to the avifauna of Honduras. *Auk*, **100**, 621–9.

Marin A., M. (1989). Notes on the breeding of Chestnut-bellied Herons (*Agamia agami*) in Costa Rica. *Condor*, **91**, 215–17.

Marin A., M and Schmitt, N. J. (1991). Nests and eggs of some Costa Rican birds. *Wilson Bull.*, **103**, 506–9.

Marshall, J. T., Jr. (1943). Additionl information concerning the birds of El Salvador. *Condor*, **45**, 21–33.

Marshall, J. T., Jr. (1956). Summer birds of the Rincon Mountains, Saguaro National Monument, Arizona. *Condor*, **58**, 81–97.

Marshall, J. T., Jr. (1957). *Birds of pine–oak woodland in southern Arizona and adjacent Mexico.* Pacific Coast Avifauna No. 32. Cooper Ornithol. Soc.

Marshall, J. T., Jr. (1967). *Parallel variation in North and Middle American Screech Owls.* WFVZ Monographs No. 1.

Marshall, J. T., Behrstock, R. A., and Konig, C. (1991). Voices of the New World Nightjars and their allies, Voices of the New World Owls, by J. W. Hardy, B. B. Coffey, and G. B. Reynard. Review. *Wilson Bull.*, **103**, 311–15.

Martin, P. S. (1951a). Bicolored Hawk in Tamaulipas, Mexico. *Wilson Bull.*, **63**, 339.

Martin, P. S. (1951b). Black Robin in Tamaulipas, Mexico. *Wilson Bull.*, **63**, 340.

Martin, P. S., Robins, C. R., and Heed, W. B. (1954). Birds and biogeography of the Sierra de Tamaulipas, an isolated pine-oak habitat. *Wilson Bull.*, **66**, 38–57.

Martinez-S., J. C. (1988). Deforestación y conservación de cracidos en Nicaragua; un informe preliminar. II International Symposium on the biology and conservation of the family Cracidae.

Martinez-S., J. C. (1989). Records of new or little known birds for Nicaragua. *Condor*, **91**, 468–9.

Mason, C. R. (1976). Cape May Warblers in Middle America. *Auk*, **93**, 167–9.

Massey, B. W. (1977). Occurrence and nesting of the Least Tern and other endangered species in Baja California, Mexico. *W. Birds*, **8**, 67–70.

Mayfield, H. (1948). Boat-billed Heron in east-central Tamaulipas. *Condor*, **50**, 228.

Mayfield, H. (1968). Nests of the Red Warbler and Crescent-chested Warbler in Oaxaca, Mexico. *Condor*, **70**, 271–2.

Mays, N. M. (1985). Ants and foraging behaviour of the Collared Forest-Falcon. *Wilson Bull.*, **97**, 231–2.

Mellink, E. and Palacios, E. (1990). Observations on Isla Guadalupe in November 1989. *W. Birds*, **21**, 177–80.

Mengel, R. M. and Warner, D. W. (1948). Golden Eagles in Hidalgo, Mexico. *Wilson Bull.*, **60**, 122.

Meyer de Schauensee, R. and Phelps, W. H. (1978). *A guide to the birds of Venezuela.* Princeton Univ. Press, Princeton, NJ.

Michener, M. C. (1964). A breeding colony of Agami Herons in Veracruz. *Condor*, **66**, 77.

Miller, A. H. (1948). A new subspecies of Eared Poorwill from Guerrero, Mexico. *Condor*, **50**, 224–5.

Miller, A. H. (1955). The avifauna of the Sierra del Carmen of Coahuila, Mexico. *Condor*, **57**, 154–78.

Miller, A. H. and Moore, R. T. (1954). A further record of the Slaty Finch in Mexico. *Condor*, **56**, 310–11.

Miller, A. H. and Ray, M. S. (1944). Discovery of a new vireo of the genus *Neochloe* in south-western Mexico. *Condor*, **46**, 41–5.

Miller, A. H., Friedmann, H., Griscom, L., and Moore, R. T. (1957). *Distributional checklist of the birds of Mexico. Part 2.* Pacific Coast Avifauna No. 33. Cooper Ornithol. Soc.

Miller, B. W. and Miller, C. M. (1992a). Checklist—Birds of Belize. Publ. by authors. Gallon Jug, Belize.

Miller, B. W. and Miller, C. M. (1992b). Distributional notes and new species records for birds in Belize. *Occas. Papers Belize Nat. Hist. Soc.*, **1**, 6–25.

Miller, W. DeW. (1906). List of birds collected in Durango, Mexico, by J. H. Batty, during 1913. *Bull. AMNH*, **22**, 161–83.

Mills, E. D. and Rogers, D. T., Jr. (1988). First record of the Blue-throated Goldentail (*Hylocharis eliciae*) in Belize. *Wilson Bull.*, **100**, 510.

Mills, G. S., Silliman, J. R., Groschupf, K. D., and Speich, S. M. (1979). Life history of the Five-striped Sparrow. *Living Bird*, **18**, 95–110.

Mirsky, E. N. (1976). Song divergence in hummingbird and junco populations on Guadalupe Island. *Condor*, **78**, 230–5.

Monroe, B. L., Jr. (1963a). Three new subspecies of birds from Honduras. *Occas. Papers MZLSU*, **26**.

Monroe, B. L., Jr. (1963b). A revision of the *Lampornis viridipallens* complex (Aves: Trochilidae). *Occas. Papers MZLSU*, **27**.

Monroe, B. L., Jr. (1963c). Notes on the avian genus *Arremonops* with description of a new subspecies from Honduras. *Occas. Papers MZLSU*, **28**.

Monroe, B. L., Jr. (1968). *A distributional survey of the birds of Honduras.* Ornithological Monographs No. 7. AOU.

Monroe, B. L., Jr., and Howell, T. R. (1966). Geographic variation in Middle American parrots of the *Amazona ochrocephala* complex. *Occas. Papers MZLSU*, **34**.

Monson, G. (1986). Gray-collared Becard in Sonora. *Am. Birds*, **40**, 562.

Moore, J. V. (1992). *A bird walk at Chan Chich*. Publ. by author. San Jose, CA.

Moore, R. T. (1938). Unusual birds and extensions of ranges in Sonora, Sinaloa and Chihuahua, Mexico. *Condor*, **40**, 23–8.

Moore, R. T. (1939). A new race of *Cynanthus latirostris* from Guanajuato. *PBSW*, **52**, 57–60.

Moore, R. T. (1940). Notes on Middle American *Empidonaces*. *Auk*, **57**, 349–89.

Moore, R. T. (1944). Nesting of the Brown-capped Leptopogon in Mexico. *Condor*, **46**, 6–8.

Moore, R. T. (1945a). Sinaloa Martin nesting in western Mexico. *Auk*, **62**, 308–9.

Moore, R. T. (1945b). The transverse volcanic biotic province of central Mexico and its relationship to adjacent provinces. *Trans. SDSNH*, 217–36.

Moore, R. T. (1949). A new hummingbird of the genus *Lophornis* form Guerrero, Mexico. *PBSW* 62: 103–4.

Moore, R. T. (1951). Records of two North American corvids in Mexico. *Condor*, **53**, 101.

Moore, R. T. (1953). Notes on two rare Tyrannids in Mexico. *Auk*, **70**, 210–11.

Moore, R. T. and Medina, D. R. (1957a). A record of the Slaty Finch for Honduras. *Condor*, **59**, 67.

Moore, R. T. and Medina, D. R. (1957b). The status of the chachalacas of western Mexico. *Condor*, **59**, 230–4.

Mora, M. A. (1989). Predation by a Brown Pelican at a mixed-species heronry. *Condor*, **91**, 742–3.

Morlan, J. (1981). Status and identification of forms of White Wagtail in western North America. *Cont. Birdlife*, **2**, 37–50.

Murphy, R. C. (1958). The vertebrates of SCOPE. In *Physical, chemical, and biological oceanographic observations obtained on expedition SCOPE in the eastern tropical Pacific, November–December 1956* (ed. R. W. Holmes). USFWS Special Scientific Report—Fisheries No. 279. Washington, DC.

Murphy, R. C. and Pennoyer, J. M. (1952). Large petrels of the genus *pterodroma*. *Novitates*, **1580**.

Murray, K. W. (1985). First reported nest of the White-eared Ground-Sparrow (*Melozone leucotis*). *Condor*, **87**, 554.

Murray, K. W., Murray, K. G., and Busby,

W. H. (1988). Two nests of the Azure-hooded Jay with notes on nest attendance. *Wilson Bull.*, **100**, 134–5.

Navarro S., A. G. (1986). Distribución altitudinal de las aves en La Sierra de Atoyac, Guerrero. Tesis Profesional, UNAM. Mexico D.F.

Navarro S., A. G. (1992). Altitudinal distribution of birds in the Sierra Madre del Sur, Guerrero, Mexico. *Condor*, **94**, 29–39.

Navarro S., A. G., Torres C., M. G., and Escalante P., B. P. (1991). Catálogo de Aves. Museo de Zoolgía, Facultad de Ciencias, UNAM.

Navarro S., A. G., Peterson, A. T., and Escalante P., P. (1992a). New distributional information on Mexican birds. I. The Sierra de Atoyac, Guerrero. *BBOC*, **112**, 6–11.

Navarro S., A. G., Peterson, A. T., Escalante P., B. P., and Benitez D., H. (1992b). *Cypseloides storeri*, a new species of swift from Mexico. *Wilson Bull.*, **104**, 55–64.

Nelson, E. W. (1898). The Imperial Ivory-billed Woodpecker, *Campephilus imperialis* (Gould). *Auk*, **15**, 217–23.

Nelson, E. W. (1922). Lower California and its natural resources. *Memoirs Nat. Acad. Sci.*, **16**.

Nelson, J. B. (1978). *The Sulidae*. Oxford University Press.

Newcomer, M. W. and Silber, G. K. (1989). Sightings of the Laysan Albatross in the northern Gulf of California, Mexico. *W. Birds*, **20**, 134–5.

Newman, R. J. (1950). A nest of the Mexican Ptilogonys. *Condor*, **52**, 157–8.

Nordhoff, C. B. (1922). Notes on some waterfowl. *Condor*, **24**, 64–5.

Northern, J. R. (1962). Noteworthy bird records from Baja California. *Condor*, **64**, 240.

Northern, J. R. (1965). Notes on the owls of the Tres Marias Islands, Mexico. *Condor*, **67**, 358.

Oates, E. W. (1901). *Catalogue of the collection of bird's eggs in the British Museum*. Vol. 1. Brit. Mus. Nat. Hist., London.

Oberbauer, T. A., Cibit, C., and Lichtwardt, E. (1989). Notes from Isla Guadalupe. *W. Birds*, **20**, 89–90.

Oberholser, H. C. (1923). Bird banding as an aid to the study of migration. *Auk*, **40**, 436–41.

Oberholser, H. C. (1974). *The bird life of Texas* (2 vols). Univ. Texas Press, Austin.

Olsen, K. M. (1989). Field identification of the smaller skuas. *Brit. Birds*, **82**, 143–76.

Olson, S. L. (1978). Greater Ani (*Crotophaga major*) in Mexico. *Auk*, **95**, 766–7.

Olson, S. L. (1982). The distribution of fused phalanges of the inner toe in Accipitridae. *BBOC*, **102**, 8–12.

Ornelas, J. F. (1987). Rediscovery of the Rufous-crested Coquette (*Lophornis delattrei brachylopha*) in Guerrero, Mexico. *Wilson Bull.*, 99, 719–21.

Ornelas, F., Navarijo, L., and Chavez, N. (1988). Análisis avifaunistico de la localidad de Temascaltepec, estado de México, México. *Anales. Inst. Biol. UNAM, Ser. Zool.* 58 (1), 373–88.

Orr, R. T. and Ray, M. S. (1945). Critical comments on seed-eaters of the genus *Amaurospiza*. *Condor*, 47, 225–8.

Orr, R. T. and Webster, J. D. (1968). New subspecies of birds from Oaxaca (Aves: Phasianidae, Turdidae and Parulidae). *PBSW*, 81, 37–40.

Palacios, E. and Alfaro, L. (1991). Breeding birds of Laguna Figueroa and La Pinta Pond, Baja California, Mexico. *W. Birds*, 22, 27–32.

Palacios, E. and Alfaro, L. (1992a). First breeding records of the Caspian Tern in Baja California, (Norte), Mexico. *W. Birds*, 23, 143–4.

Palacios, E. and Alfaro, L. (1992b). Records of Clark's Nutcracker and Grey Catbird in the Vizcaino Desert, Baja California. *Euphonia*, 1, 14–16.

Palacios, E. and Alfaro L. (1992c). Occurrence of Black Skimmers in Baja California. *W. Birds*, 23, 173–6.

Palacios, E. and Mellink, E. (1992). Breeding bird records from Montague Island, northern Gulf of California. *W. Birds*, 23, 41–4.

Palmer, R. S. (ed.) (1962–1988). *Handbook of North American Birds*. Vols 1–4. Yale Univ. Press, New Haven, CT.

Panza, R. K. and Parkes, K. C. (1992). Baja California specimens of *Parus gambeli baileyae*. *W. Birds*, 23, 87–9.

Parfitt, B. D. (1976). Cattle Egrets in central Coahuila. *Condor*, 78, 273.

Parker, J. W. (1977). Second record of the Mississippi Kite from Guatemala. *Auk*, 94, 168–9.

Parker, T. A., Hilty, S., and Robbins, M. (1976). Birds of El Triunfo cloud forest, Mexico. *Am. Birds*, 30, 779–82.

Parkes, K. C. (1954). A revision of the Neo- .tropical finch *Atlapetes brunnei-nucha* (sic). *Condor*, 56, 129–38.

Parkes, K. C. (1957). The juvenal plumages of the finch genera *Atlapetes* and *Pipilo*. *Auk*, 74, 499–502.

Parkes, K. C. (1970). On the validity of some supposed 'first state records' from Yucatan. *Wilson Bull.*, 82, 92–5.

Parkes, K. C. (1974a). Variation in the Olive Sparrow in the Yucatan Peninsula. *Wilson Bull.*, 86, 293–5.

Parkes, K. C. (1974b). Systematics of the White-

throated Towhee (*Pipilo albicollis*). *Condor*, 76, 457–9.

Parkes, K. C. (1976). Geographic variation in the Common Tody-Flycatcher (*Todirostrum cinereum*) with special reference to Middle America. *Nemouria*, 18.

Parkes, K. C. (1982). Parallel geographic variation in three *Myiarchus* flycatchers in the Yucatan Peninsula and adjacent areas (Aves: Tyrannidae). *An. CM*, 51, 1–16.

Parkes, K. C. (1990a). Additional records of birds from Oaxaca, Mexico. *Occas. Papers WFVZ*, 5.

Parkes, K. C. (1990b). A critique of the description of *Amazona auropalliata caribaea* Lousada, 1989. *BBOC*, 110, 175–9.

Parkes, K. C. (1991). A revision of the Mangrove Vireo (*Vireo pallens*) (Aves: Vireonidae). *An. CM*, 59, 49–60.

Parkes, K. C. and Dickerman, R. W. (1967). A new subspecies of Mangrove Warbler (*Dendroica petechia*) from Mexico: *An. CM*, 39, 85–9.

Parkes, K. C., Kibbe, D. P., and Roth, E. L. (1978). First records of the Spotted Rail (*Pardirallus maculatus*) for the United States, Chile, Bolivia and western Mexico. *Am. Birds*, 32, 295–9.

Pashley, D. (1987). New duck records for southern Veracruz. *MBA Bulletin Board*, 1(87–2), 2–3.

Patten, M. A., Radamaker, K., and Wurster, T. E. (1993). Noteworthy observations from northeastern Baja California. *W. Birds*, 24, 89–93.

Paulson, D. R. (1986). *Bird records from the Yucatan Peninsula, Tabasco, and Chiapas, Mexico*. Burke Mus. Contr. in Anthropology and Nat. Hist. No. 3.

Paynter, R. A., Jr. (1951). Autumnal trans-Gulf migrants and a new record for the Yucatan Peninsula. *Auk*, 68, 113–14.

Paynter, R. A., Jr. (1952). Birds from Popocatepetl and Ixtaccihuatl, Mexico. *Auk*, 69, 293–230.

Paynter, R. A., Jr. (1953). Autumnal migrants on the Campeche Bank. *Auk*, 70, 338–49.

Paynter, R. A., Jr. (1954). Two species new to the Mexican avifauna. *Auk*, 71, 204.

Paynter, R. A., Jr. (1955a). *The ornithogeography of the Yucatan Peninsula*. Peabody Mus. Nat. Hist. Bull. No. 9. New Haven, CT.

Paynter, R. A., Jr. (1955b). Additions to the ornithogeography of the Yucatan Peninsula. *Postilla*, 22.

Paynter, R. A., Jr. (1956). Avifauna of the Jorullo region, Michoacan, Mexico. *Postilla*, 25.

Paynter, R. A., Jr. (ed.) (1957a). Biological investigations in the Selva Lacandona, Chiapas, Mexico. *Bull. MCZ*, **116**, 193–298.

Paynter, R. A., Jr. (1957*b*). Rough-winged Swallows of the race *stuarti* in Chiapas and British Honduras. *Condor*, **59**, 212–13.

Paynter, R. A., Jr. (1978). Biology and evolution of the avian genus *Atlapetes* (Emberizinae). *Bull. MCZ*, **148**, 323–69.

Paynter, R. A., Jr. and Alvarez del Toro, M. (1957). Blue-and-White Swallow in Mexico. *Condor*, **59**, 268.

Pelzl, H. W. (1969). *Birds of the British Honduras Keys.* Publ. by author, St. Louis, MO.

Perales, F., L. E. and Contreras B., A. J. (1986). Aves acuaticas y semiacuaticas de La Laguna Madre, Tamaulipas, México. *Universidad y Ciencia*, **3**, 39–46.

Peters, J. L. (1913). List of birds collected in the territory of Quintana Roo, Mexico, in the winter and spring of 1912. *Auk*, **30**, 367–80.

Peterson, A. T. (1991). New distributional informational on the *Aphelocoma* jays. *BBOC*, **111**, 28–33.

Peterson, R. T. and Chalif, E. L. (1973). *A field guide to Mexican birds.* Houghton Mifflin Co., Boston, MA.

Peterson, R. T. and Chalif, E. L. (1989). *Aves de México.* Editorial Diana, Mexico, D.F.

Phillips, A. R. (1942). The races of *Empidonax affinis. Auk*, **59**, 424–8.

Phillips, A. R. (1949). Further notes on *Empidonax affinis. Auk*, **66**, 92–3.

Phillips, A. R. (1950). The Great-tailed Grackles of the South-West. *Condor*, **52**, 78–81.

Phillips, A. R. (1959). The nature of avian species. *J. Ariz. Acad. Sci.*, **1**, 22–30.

Phillips, A. R. (1962*a*). Notas sistemáticas sobre aves Mexicanas I. *Anales Inst. Biol.*, **32**, 333–81.

Phillips, A. R. (1962*b*). Notas sobre la chuparosa *Thalurania* y ciertos plumajes de otras aves Mexicanas. *Anales Inst. Biol.*, **32**, 383–90.

Phillips, A. R. (1962*c*). Notas sistemáticas sobre aves Mexicanas II. *Anales Inst. Biol.*, **33**, 331–72.

Phillips, A. R. (1964). Notas sistemáticas sobre aves Mexicanas III. *RSMHN*, **25**, 217–42.

Phillips, A. R. (1966*a*). Further systematic notes on Mexican birds. *BBOC*, **86**, 86–94.

Phillips, A. R. (1966*b*). Further systematic notes on Mexican birds. *BBOC*, **86**, 103–12.

Phillips, A. R. (1966*c*). Further systematic notes on Mexican birds. *BBOC*, **86**, 125–31.

Phillips, A. R. (1966*d*). Further systematic notes on Mexican birds. *BBOC*, **86**, 148–59.

Phillips, A. R. (1968). A notable specimen of *Vireo nelsoni. Condor*, **70**, 90.

Phillips, A. R. (1969). An ornithological comedy of errors: *Catharus occidentalis* and *C. frantzii. Auk*, **86**, 605–23.

Phillips, A. R. (1971*a*). Avian breeding cycles: are they related to photoperiods? *An. Inst. Biol. UNAM*, **42**, 87–98.

Phillips, A. R. (1971*b*). What is *Amazilia microrhyncha? Auk*, **88**, 679.

Phillips, A. R. (1975). The migrations of Allen's and other hummingbirds. *Condor*, **77**, 196–205.

Phillips, A. R. (1981). Subspecies vs. forgotten species: the case of Grayson's Robin (*Turdus graysoni*). *Wilson Bull.*, **93**, 301–9.

Phillips, A. R. (1986). *The known birds of North and Middle America. Part 1.* A. R. Phillips. Denver, CO.

Phillips, A. R. (1991). *The known birds of North and Middle America. Part 2.* A. R. Phillips. Denver, CO.

Phillips, A. R. and Amadon, D. (1952). Some birds of northwestern Sonora, Mexico. *Condor*, **54**, 163–8.

Phillips, A. R. and Dickerman, R. W. (1957). Notes on the Song Sparrows of the Mexican plateau. *Auk*, **74**, 376–82.

Phillips, A. R. and Hardy, J. W. (1965). *Tanagra minuta*, an addition to the Mexican List. *Wilson Bull.*, **77**, 89.

Phillips, A. R. and Phillips F., R. (1993). Distribution, migration, ecology, and relationships of the Five-striped Sparrow, *Aimophila quinquestriata. W. Birds*, **24**, 65–72.

Phillips, A. R. and Rook, W. (1965). A new race of the Spotted Nightingale-Thrush from Oaxaca, Mexico. *Condor*, **67**, 3–5.

Phillips, A. R. and Schaldach, W. J., Jr. (1960). New records of raptors from Jalisco, Mexico. *Condor*, **62**, 295.

Phillips, A. R. and Webster, J. D. (1957). The Vaux Swift in western Mexico. *Condor*, **59**, 140–1.

Phillips, A., Marshall, J., and Monson, G. (1964). *The birds of Arizona.* Univ. Ariz. Press, Tucson.

Pierson, J. E. (1986). Notes on the vocalizations of the Yucatan Poorwill (*Nyctiphrynus yucatanicus*) and Tawny-collared Nightjar (*Caprimulgus salvini*). *MBA Bulletin Board*, **1**(86–1), 3–4.

Pitelka, F. A. (1947). Taxonomy and distribution of the Mexican sparrow, *Xenospiza baileyi. Condor*, **49**, 199–203.

Pitelka, F. A. (1948). Notes on the distribution and taxonomy of Mexican game birds. *Condor*, **50**, 113–23.

Pitelka, F. A. (1951*a*). Speciation and ecological distribution in American jays of the genus *Aphelocoma. UCPZ*, **50**, 195–464.

Pitelka, F. A. (1951*b*). The Tyrannid *Aechmolophus mexicanus* in Guerrero. *Condor*, **53**, 300.

Pitelka, F. A., Selander, R. K., and Alvarez del Toro, M. (1956). A hybrid jay from Chiapas, Mexico. *Condor*, 58, 98–106.

Pitman, R. L. (1985). The marine birds of Alijos Rocks, Mexico. *W. Birds*, 16, 81–92.

Pitman, R. L. (1986). *Atlas of seabird distribution and relative abundance in the eastern tropical Pacific*. NMFS Admin. Report LJ-86-02C.

Pitman, R. L., Newcomer, M., Butler, J., Cotton, J., and Friedrichsen, G. (1983). A Crested Auklet from Baja California. *W. Birds*, 14, 47–8.

Porter, R. D., Jenkins, M. A., Kirven, M. N., Anderson, D. W., and Keith, J. O. (1988). Status and reproductive performance of marine Peregrines in Baja California and the Gulf of California, Mexico. In *Peregrine falcon populations* (ed. T. J. Cade *et al.*). Peregrine Fund Inc., Boise, ID.

Prum, R. O. and Lanyon, W. E. (1989). Monophyly and phylogeny of the *Schiffornis* group (Tyrannoidea). *Condor*, 91, 444–61.

Pyle, P. and Howell, S. N. G. (1993). An Arctic Warbler in Baja California, Mexico. *W. Birds*, 24, 53–6.

Pyle, P., Howell, S. N. G., Yunick, R. P., and DeSante, D. F. (1987). *Identification guide to North American passerines*. Slate Creek Press, Bolinas, CA.

Raffaele, H. A. (1989). *A guide to the birds of Puerto Rico and the Virgin Islands*. Princeton Univ. Press, Princeton, NJ.

Raitt, R. J. (1967). Relationships between black-eared and plain-eared forms of Bushtits (*Psaltriparus*). *Auk*, 84, 503–28.

Raitt, R. J. and Hardy, J. W. (1970). Relationships between two partly sympatric species of thrushes (*Catharus*) in Mexico. *Auk*, 87, 20–57.

Raitt, R. J. and Hardy, J. W. (1976). Behavioural ecology of the Yucatan Jay. *Wilson Bull.*, 88, 529–54.

Ramo, C. and Busto, B. (1988). Observations at a King Vulture (*Sarcoramphus papa*) nest in Venezuela. *Auk*, 105, 195–6.

Ramos, M. A. (1985). Problems hindering the conservation of tropical forest birds in Mexico and Central America, and steps towards a conservation strategy. *ICBP tech. publ.*, 4, 67–76.

Rangel-S., J. L. and Vega-R., J. H. (1989). Two new records of birds for southern Mexico. *Condor*, 91, 214–15.

Rangel-S., J. L. and Vega-R., J. H. (1991). The Great Potoo (*Nyctibius grandis*) as a probable resident in southern Mexico. *Ornitología Neotropical*, 2, 38–9.

Ratti, J. T. (1979). Reproductive separation and isolating mechanisms between sympatric dark- and light-phased Western Grebes. *Auk*, 96, 573–86.

Rea, A. M. and Weaver, K. L. (1990). The taxonomy, distribution, and status of coastal California Cactus Wrens. *W. Birds*, 21, 81–126.

Rettig, N. L. (1978). Breeding behaviour of the Harpy Eagle (*Harpia harpyja*). *Auk*, 95, 629–43.

Reynard, G. B. (1981). *Bird songs of the Dominican Republic*. Cornell Lab. of Ornithol., Ithaca, NY.

Richmond, C. W. (1895). Partial list of birds collected at Alta Mira (sic), Mexico, by Mr. Frank B. Armstrong. *Proc. USNM*, 18, 627–32.

Ridgely, R. S. and Gwynne, J. (1989). *A guide to the birds of Panama*, 2nd edition. Princeton Univ. Press, Princeton, NJ.

Ridgely, R. S. and Tudor, G. (1989). *The birds of South America*. Vol. 1. Univ. Texas Press, Austin.

Ridgway, R. (1883a). On the probable identity of *Motacilla ocularis* Swinhoe and *M. amurensis* Seebohm, with remarks on an allied supposed species, *M. blakistoni* Seebohm. Proc. USNM, 6 144–7.

Ridgway, R. (1883b). *Anthus cervinus* (Pallas) in Lower California. *Proc. USNM*, 6, 156–7.

Ridgway, R. (and Friedmann, H., Parts 9–11) (1901–1950). *The birds of North and Middle America*. Bull. USNM. Parts 1–11.

Ripley, C. D. (1977). *Rails of the world*. Godine. Boston, MA.

Rising, J. D. (1988). Phenetic relationships among the warblers in the *Dendroica virens* complex and a record of *D. virens* from Sonora, Mexico. *Wilson Bull.*, 100, 312–16.

Robins, C. R. and Heed, W. B. (1951). Bird notes from La Joya de Salas, Tamaulipas. *Wilson Bull.*, 63, 263–70.

Robles Gil, P., Eccardi, F., and Robles Gil., J. (1989). *El libro de las aves de México*. Centro de Arte Vitro, Mexico, D.F.

Rodewald, P. (1989). Scissor-tailed Flycatcher (*Tyrannus forficatus*) in Baja California Sur. *Aves Mexicanas*, 2(89–2), 4–5.

Rodriguez-E., R., Llinas-G., J., and Cancino, J. (1991). New Golden Eagle records from Baja California. *J. Raptor Res.*, 25, 68–71.

Rodriguez-E., R., Mata, E., and Rivera, L. (1992). Ecological notes on the Green Parakeet of Isla Socorro, Mexico. *Condor*, 94, 523–5.

Rogers, D. T., Garcia B., J., and Rogel B., A. (1986). Additions to records of North American avifauna in Yucatan, Mexico. *Wilson Bull.*, 98, 163–7.

Rogers, D. T., Jr., Hicks, D. L., Wischusen, E.

W., and Parrish, J. R. (1982). Repeats, returns, and estimated flight ranges of some North American migrants in Guatemala. *J. Field Ornithol.*, **53**, 133–8.

Rosenberg, K. V. and Rosenberg, G. H. (1979). First documented record of Sparkling-tailed Hummingbird from Sinaloa, Mexico. *Cont. Birdlife*, **1**, 57–61.

Rowley, J. S. (1962). Nesting of the birds of Morelos, Mexico. *Condor*, **64**, 253–72.

Rowley, J. S. (1963). Notes on the life history of the Pileated Flycatcher. *Condor*, **65**, 318–23.

Rowley, J. S. (1966). Breeding records of birds of the Sierra Madre del Sur, Oaxaca, Mexico. *Proc. WFVZ*, **1**, 107–204.

Rowley, J. S. (1968). Geographic variation in four species of birds in Oaxaca, Mexico. *Occas. Papers WFVZ*, **1**.

Rowley, J. S. (1984). Breeding records of land birds in Oaxaca, Mexico. *Proc. WFVZ*, **2**, 74–224.

Rowley, J. S. and Orr, R. T. (1962a). The nest and eggs of the Slaty Vireo. *Condor*, **64**, 88–90.

Rowley, J. S. and Orr, R. T. (1962b). The nesting of the White-naped Swift. *Condor*, **64**, 361–7.

Rowley, J. S. and Orr, R. T. (1964a). A new hummingbird from southern Mexico. *Condor*, **66**, 81–4.

Rowley, J. S. and Orr, R. T. (1964b). The status of Frantzius' Nightingale Thrush. *Auk*, **81**, 308–14.

Rowley, J. S. and Orr, R. T. (1965). Nesting and feeding habits of the White-collared Swift. *Condor*, **67**, 449–56.

Ruiz-C., G. and Quintana-B., L. (1991). First mainland record of Red-breasted Nuthatch from Baja California, Mexico. *W. Birds*, **22**, 189–90.

Russell, S. M. (1964). *A distributional survey of the birds of British Honduras*. Ornithological Monographs No. 1. AOU.

Russell, S. M. (1966). Status of the Black Rail and Grey-breasted Crake in British Honduras. *Condor*, **68**, 105–7.

Russell, S. M. and Lamm, D. W. (1978). Notes on the distribution of birds in Sonora, Mexico. *Wilson Bull.*, **90**, 123–31.

Sada, A. M. (1989a). Birds from Cayo Arcas, Campeche, 14–16 September 1986. *Aves Mexicanas*, **89–1**, 4.

Sada, A. M. (1989b). *Mexican birds, codes for ease of observation*. Publ. by author. Monterrey, NL.

Sada, A. M., Phillips A. R., and Ramos, M. A. (1987). *Nombres en Castellano para las aves Mexicanas*. INIREB. Xalapa, Ver.

Salvin, O. (1861). A list of species to be added to the ornithology of Central America. *Ibis*, **3**, 351–7.

Salvin, O. (1888). A list of the birds of the islands of the coast of Yucatan and of the Bay of Honduras. *Ibis*, **1888**, 241–65.

Salvin, O. (1889). A list of the birds of the islands of the coast of Yucatan and of the Bay of Honduras (contd.). *Ibis*, **1889**, 359–79.

Salvin, O. (1890). A list of the birds of the islands of the coast of Yucatan and of the Bay of Honduras (contd.). *Ibis*, **1890**, 84–95.

Salvin, O. and Godman, F. D. (1879–1904). *Biologia Centrali-Americana. Aves*. Vols. 1–3. Taylor & Francis, London.

Santaella, L. and Sada, A. M. (1991a). A Short-tailed Albatross near San Benedicto Island, Revillagigedo Islands, Mexico. *W. Birds*, **22**, 33–4.

Santaella, L. and Sada, A. M. (1991b). The avifauna of the Revillagigedo Islands, Mexico: additional data and observations. *Wilson Bull.*, **103**, 668–75.

Santaella, L. and Sada, A. M. (1992). A Cory's Shearwater off Isla Cozumel, Mexico. *Euphonia*, **1**, 17–18.

Saunders, G. B. (1953). The Tule Goose (*Anser albifrons gambelli*), Blue Goose (*Chen caerulescens*), and Mottled Duck (*Anas fulvigula mexicana*) added to the list of the birds of Mexico. *Auk*, **70**, 84–5.

Saunders, G. B. and Saunders, D. C. (1981). *Waterfowl and their wintering grounds in Mexico, 1937–64*. USFWS Res. Publ. No. 138. Washington D.C.

Schaeffer, P. P. and Ehlers, S. M. (ed.) (1980). *The birds of Mexico, their ecology and conservation*. Proc. NAS symposium, Tiburon, CA.

Schaldach, W. J., Jr. (1960). Occurrence of the Slaty and Dwarf vireos in Jalisco, Mexico. *Condor*, **62**, 139.

Schaldach, W. J., Jr. (1963). The avifauna of Colima and adjacent Jalisco, Mexico. *Proc. WFVZ*, **1**, 1–100.

Schaldach, W. J., Jr. (1969). Further notes on the avifauna of Colima and adjacent Jalisco, Mexico. *Anales Inst. Biol. UNAM*, **40**, 299–316.

Schaldach, W. J., Jr. and Phillips, A. R. (1961). The Eared Poor-Will. *Auk*, **78**, 567–72.

Schonwetter, M. (1960). *Handbuch der Oologie*. Vol. 1. Zool. Staatsinstitut und. Zool. Mus. Hamburg, Akademie-Verlag, Berlin.

Scott, P. E., Andrews D. D., and de Montes, B. M. (1985). Spotted Rail: first record from the Yucatan Peninsula, Mexico. *Am. Birds*, **39**, 852–3.

SEDUE. (1989). *Información básica sobre las areas naturales protegidas de México*. Directorio de areas naturales protegidas administradas por la Secretaria de Desarrollo Urbano y Ecología. SEDUE.

Selander, R. K. (1955). Great Swallow-tailed Swift in Michoacan, Mexico. *Condor*, 57, 123–5.

Selander, R. K. (1956). Additional record of the Pinyon Jay in Mexico. *Condor*, 58, 239.

Selander, R. K. (1959). Polymorphism in the Mexican Brown Jay. *Auk*, 76, 385–417.

Selander, R. K. (1964). Speciation in wrens of the genus *Campylorhynchus*. *UCPZ*, 74, 1–305.

Selander, R. K. (1965). Hybridization of the Rufous-naped Wrens in Chiapas, Mexico. *Auk*, 82, 206–14.

Selander, R. K. and Alvarez del Toro, M. (1953). The breeding distribution of *Chordeiles minor* in Mexico. *Condor*, 55, 160–1.

Selander, R. K. and Alvarez del Toro, M. (1955). A new race of Booming Nighthawk from southern Mexico. *Condor*, 57, 144–7.

Selander, R. K. and Giller, D. R. (1959). The avifauna of the Barranca de Oblatos, Jalisco, Mexico. *Condor*, 61, 210–22.

Selander, R. K. and Giller, D. R. (1963). Species limits in the woodpeckers of the genus *Centurus* (Aves). *Bull. AMNH*, 124, 217–73.

Selander, R. K., Johnston, R. F., Wilks, B. J., and Raun, G. G. (1962). Vertebrates from the barrier islands of Tamaulipas, Mexico. *UKPMNH*, 12, 309–45.

Short, L. L., Jr. (1967a). Taxonomic aspects of avian hybridization. *Auk*, 86, 84–105.

Short, L. L., Jr. (1967b). Variation in Central American flickers. *Wilson Bull.*, 79, 5–21.

Short, L. L., Jr. (1971). Systematics and behaviour in some North American woodpeckers of the genus *Picoides* (Aves). *Bull. AMNH*, 145, 1–118.

Short, L. L., Jr. (1972). Hybridization, taxonomy and avian evolution. *An. Missouri Bot. Gar.*, 59, 447–53.

Short, L. L., Jr. (1974a). Nesting of southern Sonoran birds during the summer rainy season. *Condor*, 76, 21–32.

Short, L. L., Jr. (1974b). The contribution of external morphology to avian classification. *Proc. 16th Int. Ornith. Congress*, 185–95.

Short, L. L., Jr. (1982). *Woodpeckers of the world*. Delaware Mus. Nat. Hist, Greenville, DE.

Short, L. L., Jr. and Banks, R. C. (1965). Notes on the birds of north-western Baja California. *Trans. SDSNH*, 14, 41–52.

Short, L. L., Jr. and Crossin, R. S. (1967). Notes on the avifauna of north-western Baja California. *Trans. SDSNH*, 14, 281–300.

Sibley, C. G. (1950). Species formation in the Red-eyed Towhees of Mexico. *UCPZ*, 50, 109–94.

Sibley, C. G. (1961). Hybridization and isolating mechanisms. In *Vertebrate speciation* (ed. W. F. Blair). Univ. Texas Press, Austin.

Sibley, C. G. and Monroe, B. L., Jr. (1990). *Distribution and taxonomy of birds of the world*. Yale Univ. Press, New Haven, CT.

Sibley, C. G. and Sibley F. (1964). Hybridization in the Red-eyed Towhees of Mexico: populations of the south-eastern plateau region. *Auk*, 81, 479–504.

Sibley, C. G. and West, D. A. (1958). Hybridization in the Red-eyed Towhees of Mexico: the eastern plateau populations. *Condor*, 60, 85–104.

Sibley, F. C., Barrowclough G. F., and Sibley, C. G. (1980). Notes on the birds of Honduras. *Wilson Bull.*, 92, 125–6.

Simon, D. and Simon, W. F. (1974). A Yellow-billed Loon in Baja California, Mexico. *W. Birds*, 5, 23.

Skutch, A. F. (1940). Social and sleeping habits of Central American wrens. *Auk*, 57, 293–312.

Skutch, A. F. (1942). Life history of the Mexican Trogon. *Auk*, 59, 341–63.

Skutch, A. F. (1944). Life history of the Quetzal. *Condor*, 46, 213–35.

Skutch, A. F. (1945a). Incubation and nestling periods of Central American birds. *Auk*, 62, 8–37.

Skutch, A. F. (1945b). Life history of the Allied Woodcreeper. *Condor*, 47, 85–94.

Skutch, A. F. (1945c). Life history of the Blue-throated Green Motmot. *Auk*, 62, 489–517.

Skutch, A. F. (1948). Life history of the Citreoline Trogon. *Condor*, 50, 137–47.

Skutch, A. F. (1949). Life history of the Ruddy Quail-Dove. *Condor*, 51, 3–19.

Skutch, A. F. (1950). Life history of the White-breasted Blue Mockingbird. *Condor*, 52, 220–7.

Skutch, A. F. (1952). Life history of the Chestnut-tailed Automolus. *Condor*, 54, 93–100.

Skutch, A. F. (1953). Life history of the Southern House Wren. *Condor*, 55, 121–49.

Skutch, A. F. (1954a). Life history of the White-winged Becard. *Auk*, 71, 113–29.

Skutch, A. F. (1954b). *Life histories of the Central American birds*. Pacific Coast Avifauna No. 31. Cooper Ornithol. Soc.

Skutch, A. F. (1957). Life history of the Amazon Kingfisher. *Condor*, 59, 217–29.

Skutch, A. F. (1959a). Life history of the Black-headed Trogon. *Wilson Bull.*, 71, 5–18.

Skutch, A. F. (1959b). Life history of the Blue Ground Dove. *Condor*, 61, 65–74.

Skutch, A. F. (1959c). Life history of the Groove-billed Ani. *Auk*, 76, 281–317.

Skutch, A. F. (1960). *Life histories of Central American birds II*. Pacific Coast Avifauna No. 34. Cooper Ornithol. Soc.

Skutch, A. F. (1962a). Life histories of honey-creepers. *Condor*, 64, 92–116.

Skutch, A. F. (1962b). On the habits of the Queo, *Rhodinocichla rosea*. *Auk*, 79, 633–9.

Skutch, A. F. (1963). Life history of the Little Tinamou. *Condor*, 65, 224–31.

Skutch, A. F. (1964a). Life histories of Hermit hummingbirds. *Auk*, 81, 5–25.

Skutch, A. F. (1964b). Life history of the Scaly-breasted Hummingbird. *Condor*, 66, 186–98.

Skutch, A. F. (1964c). Life histories of Central American pigeons. *Wilson Bull.*, 76, 211–47.

Skutch, A. F. (1965). Life history notes on two tropical American kites. *Condor*, 67, 235–46.

Skutch, A. F. (1967). *Life histories of Central American highland birds*. Publ. Nuttall Ornithol. Club No. 7, Cambridge, MA.

Skutch, A. F. (1969). *Life histories of Central American birds III*. Pacific Coast Avifauna No. 35. Cooper Ornithol. Soc.

Skutch, A. F. (1971). Life history of the Keel-billed Toucan. *Auk*, 88, 381–424.

Skutch, A. F. (1972). *Studies of tropical American birds*. Publ. Nuttall Ornithol. Club No. 10, Cambridge, MA.

Skutch, A. F. (1989). Courtship of the Rufous Piha *Lipaugus unirufus*. *Ibis*, 131, 303–4.

Smith, J. I. (1987). Evidence of hybridization between Red-bellied and Golden-fronted woodpeckers. *Condor*, 89, 377–86.

Smith, N. G. (1970). Nesting of King Vulture and Black Hawk-Eagle in Panama. *Condor*, 72, 247–8.

Smith, T. B. (1982). Nests and young of two rare raptors from Mexico. *Biotropica*, 14, 79–80.

Smithe, F. B. (1960). First records of Cattle Egrets (*Bubulcus ibis*) in Guatemala. *Auk*, 77, 218.

Smithe, F. B. (1966). *The birds of Tikal*. Natural History Press, New York.

Smithe, F. B. and Paynter, R. A., Jr. (1963). Birds of Tikal, Guatemala. *Bull. MCZ*, 128, 245–324.

Snell, R. R. (1989). Status of *Larus* gulls at Home Bay, Baffin Island. *Col. Waterbirds*, 12, 12–23.

Snyder, D. E. (1957). A recent Colima Warbler's nest. *Auk*, 74, 97–8.

Spear, L. B., Howell, S. N. G., and Ainley, D. G. (1992). Notes on the at-sea identification of some gadfly petrels (genus: *Pterodroma*). *Col. Waterbirds*, 15, 202–18.

Sprunt, A., IV and Knoder, C. E. (1980). Populations of wading birds and other colonial nesting species on the Gulf and Caribbean coasts of Mexico. In *The birds of Mexico, their ecology and conservation* (ed. P. P. Schaeffer and S. M. Ehlers). Proc. NAS symposium, Tiburon, CA.

Stager, K. E. (1954). Birds of the Barranca de Cobre region of southwestern Chihuahua, Mexico. *Condor*, 56, 21–32.

Stager, K. E. (1957). The avifauna of the Tres Marias Islands, Mexico. *Auk*, 74, 413–32.

Stager, K. E. (1964). The birds of Clipperton Island, eastern Pacific. *Condor*, 66, 357–71.

Stiles, F. G. (1981a). Notes on the Uniform Crake in Costa Rica. *Wilson Bull.*, 93, 107–8.

Stiles, F. G. (1981b). The taxonomy of Rough-winged Swallows (*Stelgidopteryx*; Hirundinidae) in southern Central America. *Auk*, 98, 282–93.

Stiles, F. G. and Hespenheide, H. A. (1972). Observations on two rare Costa Rican finches. *Condor*, 74, 99–101.

Stiles, F. G. and Skutch, A. F. (1989). *A guide to the birds of Costa Rica*. Cornell Univ. Press, Ithaca, NY.

Storer, R. W. (1951). A preliminary report on the summer bird life of south-western Michoacan. In *Coalcoman and Motines del Oro, An 'ex-Distrito' of Michoacan, Mexico* (ed. D. D. Brand *et al.*). Inst. Latin Amer. Studies, Austin, TX.

Storer, R. W. (1952). Variation in the resident Sharp-shinned Hawks of Mexico. *Condor*, 54, 283–9.

Storer, R. W. (1955). A preliminary survey of the sparrows of the genus *Aimophila*. *Condor*, 57, 193–201.

Storer, R. W. (1961). Two collections of birds from Campeche, Mexico. *Occas. Papers MZUM*, 621.

Storer, R. W. (1962). Variation in the Red-tailed Hawks of southern Mexico and Central America. *Condor*, 64, 77–8.

Storer, R. W. (1972). The juvenal plumage and relationships of *Lophostrix cristatus*. *Auk*, 89, 452–5.

Storer, R. W. (1978). Systematic notes on the loons (Gaviidae: Aves). *Brevoria*, 448.

Storer, R. W. and Zimmerman, D. A. (1959). Variation in the Blue Grosbeak (*Guiraca caerulea*) with special reference to the Mexican populations. *Occas. Papers MZUM*, 609

Straneck, R., Ridgely, R., Rumboll, M., and Herrera, J. (1987). El nido del Atajacaminos Castaño *Lurocalis nattereri* (Temminck) (Aves, Caprimulgidae). *Com. del Museo Argentino de Ciencias Naturales "Bernadino Rivadavia", Zool.*, 4(17), 133–6.

Strauch, J. G., Jr. (1975). Observations at a nest of the Black-and-white Hawk-Eagle. *Condor*, 77, 512.

Stresemann, E. (1954). Ferdinand Deppe's

travels in Mexico, 1824–1829. *Condor*, 56, 86–92.

Sumichrast, F. (1869). The geographical distribution of the native birds of the department of Vera Cruz (sic), with a list of the migratory species. *Mem. Boston. Soc. Nat. Hist.*, 1, 542–63.

Sutton, G. M. (1941). A new race of *Chaetura vauxi* from Tamaulipas. *Wilson Bull.*, 53, 231–3.

Sutton, G. M. (1948a). The nest and eggs of the White-bellied Wren. *Condor*, 50, 101–12.

Sutton, G. M. (1948b). White-throated or Bat Falcon in Nuevo Leon, Mexico. *Auk*, 65, 603.

Sutton, G. M. (1951a). *Mexican birds, first impressions*. Univ. Oklahoma Press, Norman.

Sutton, G. M. (1951b). *Empidonax albigularis* in southwestern Tamaulipas. *Wilson Bull.*, 63, 339.

Sutton, G. M. (1952). New birds for the state of Michoacan, Mexico. *Wilson Bull.*, 64, 221–3.

Sutton, G. M. (1954). Blackish Crane-Hawk. *Wilson Bull.*, 66, 237–42.

Sutton, G. M. (1955). Great Curassow. *Wilson Bull.*, 67, 75–7.

Sutton, G. M. and Burleigh, T. D. (1939a). A list of birds observed on the 1938 Semple Expedition to north-eastern Mexico. *Occas. Papers MZLSU*, 3, 15–46.

Sutton, G. M. and Burleigh, T. D. (1939b). A new Abeille's Grosbeak from Tamaulipas. *PBSW*, 52; 145–6.

Sutton, G. M. and Burleigh, T. D. (1940a). Birds of Tamazunchale, San Luis Potosi. *Wilson Bull.*, 52, 221–33.

Sutton, G. M. and Burleigh, T. D. (1940b). Birds recorded in the state of Hidalgo, Mexico, by the Semple Expedition of 1939. *An. CM*, 28, 169–86.

Sutton, G. M. and Burleigh, T. D. (1940c). Birds of Valles, San Luis Potosi, Mexico. *Condor*, 42, 259–62.

Sutton, G. M. and Burleigh, T. D. (1941). Some birds recorded in Nuevo Leon, Mexico. *Condor*, 43, 158–60.

Sutton, G. M. and Burleigh, T. D. (1942). Birds recorded in the Federal District and states of Puebla and Mexico by the 1939 Semple Expedition. *Auk*, 59, 418–23.

Sutton, G. M. and Pettingill, O. S. Jr. (1942). Birds of the Gomez Farias Region, South-Western Tamaulipas. *Auk*, 59, 1–34.

Sutton, G. M. and Pettingill, O. S. Jr. (1943). Birds of Linares and Galeana, Nuevo Leon, Mexico. *Occas. Papers MZLSU*, 16, 273–91.

Sutton, G. M., Pettingill, O. S., and Lea, R. B. (1942). Notes on the birds of the Monterrey district of Nuevo Leon, Mexico. *Wilson Bull.*, 54, 199–203.

Sutton, G. M., Lea, R. B., and Edwards, E. P. (1950). Notes on the ranges and breeding habits of certain Mexican birds. *Bird-Banding*, 21, 45–59.

Swarth, H. S. (1933). Off-shore migrants over the Pacific. *Condor*, 35, 39–41.

Taibel, A. M. (1955). Uccelli del Guatemala con speciale riguardo alla region del Peten raccolti dal Maggio al Setembre 1932. *Atti Societa Italiana di Scienze Naturali*, 94, 15–84.

Tanner, J. T. (1964). The decline and present status of the Imperial Woodpecker in Mexico. *Auk*, 81, 74–81.

Tashian, R. E. (1952). Some birds from the Palenque region of north-eastern Chiapas, Mexico. *Auk*, 69, 60–6.

Tashian, R. E. (1953). The Birds of southeastern Guatemala. *Condor*, 55, 198–210.

Taylor, R. C. (1986). *Checklist to the birds of Sonora and the Sea of Cortez*. Borderland Productions, Portal, AZ.

Terrill, S. B. (1981). Notes on the winter avifauna of two riparian sites in northern Sonora, Mexico. *Cont. Birdlife*, 2, 11–18.

Tershy, B. R., Van Gelder, E., and Breese, D. (1993). Relative abundance and seasonal distribution of seabirds in the Canal de Ballenas, Gulf of California. *Condor*, 95 458–64.

Thayer, J. E. (1925). The nesting of the Worthen Sparrow in Tamaulipas, Mexico. *Condor*, 27, 34.

Thompson, M. C. (1962). Noteworthy records of birds from the Republic of Mexico. *Wilson Bull.*, 74, 173–6.

Thompson, W. L. (1968). The songs of five species of *Passerina* buntings. *Behaviour*, 31, 261–87.

Thorstrom, R. K., Turley, C. W., Gutierrez R., F. and Gilroy, B. A. (1990). Descriptions of nests, eggs, and young of the Barred Forest-Falcon (*Micrastur ruficollis*) and of the Collared Forest-Falcon (*M. semitorquatus*). *Condor*, 92, 237–9.

Thurber, W. A. (1972). House Sparrows in Guatemala. *Auk*, 89, 200.

Thurber, W. A. (1980). Hurricane Fifi and the 1974 autumn migration in El Salvador. *Condor*, 82, 212–18.

Thurber, W. A. and Serrano, J. F. (1972). Status of the White-tailed Kite in El Salvador. *Condor*, 74, 489–91.

Thurber, W. A. and Villeda, A. (1980). Notes on parasitism by Bronzed Cowbird in El Salvador. *Wilson Bull.*, 92, 112–13.

Thurber, W. A., Serrano, J. F., Sermeño, A. and Benitez, M. (1987). Status of uncommon or previously unreported birds of El Salvador. *Proc. WFVZ*, 3, 109–293.

Tilley, F. C., Hoffman, S. W., and Tilly, C. R. (1990). Spring hawk migration in southern Mexico, 1989. *Hawk Migration Studies*, **15**, 21–9.

Todd, W. E. C. (1929). Revision of the genus *Basileuterus. Proc. USNM*, **74**, 1–95.

Townsend, C. H. (1923). Birds collected in Lower California. *Bull. AMNH*, **48**, 1–26.

Traylor, M. A. (1941). Birds from the Yucatan Peninsula. *Zool. Series, FMNH*, **24**, 195–225.

Traylor, M. A. (1949). Notes on some Veracruz Birds. *Fieldiana*, **31**, 269–75.

Traylor, M. A. (1979). Two sibling species of *Tyrannus* (Tyrannidae). *Auk*, **96**, 221–33.

Tunnell, J. W. and Chapman, B. R. (1988). First record of Red-footed Boobies nesting in the Gulf of Mexico. *Am. Birds*, **42**, 380–1.

Tyler, J. D. (1978). Elegant Quail in the Barranca del Cobre, Chihuahua. *W. Birds*, **9**, 134.

Udvardy, M. D. F. (1976). Contributions to the avifauna of the Bay Islands of Honduras, Central America. *Ceiba*, **20**, 80–5.

Udvardy, M. D. F., de Beausset, C. S., and Ruby, M. (1973). New tern records from Caribbean Honduras. *Auk*, **90**, 440–2.

Unitt, P. (1984). *The birds of San Diego County*. SDSNH Memoir No. 13.

Unitt, P. (1987). *Empidonax traillii extimus*: an endangered subspecies. *W. Birds*, **18**, 137–62.

Unitt, P., Rodriguez E., R., and Castallanos V., A. (1992). Ferruginous Hawk and Pine Siskin in the Sierra de la Laguna, Baja California Sur; subspecies of the Pine Siskin in Baja California. *W. Birds*, **23**, 171–2.

Urban, E. K. (1959). Birds from Coahuila, Mexico. *UKPMNH*, **11**, 443–516.

Vanderwerf, E. A. (1988). Observations on the nesting of the Great Potoo (*Nyctibius grandis*) in central Venezuela. *Condor*, **90**, 948–50.

Vannini, J. P. (1989a). Neotropical raptors and deforestation: notes on diurnal raptors at Finca el Faro, Quetzaltenango (sic), Guatemala. *J. Raptor Res.*, **23**, 27–38.

Vannini, J. P. (1989b). *Preliminary checklist to the birds of Finca El Faro, Quetzaltenango (sic), Guatemala*. Fundación Interamericana de Investigación Tropical. Publ. Ocas. No. 2.

Van Hoose, S. G. (1955). Distributional and breeding records of some birds from Coahuila. *Wilson Bull.*, **67**, 302–3.

Van Rossem, A. J. (1929). Nesting of the American Merganser in Chihuahua. *Auk*, **46**, 380.

Van Rossem, A. J. (1931). Report on a collection of land birds from Sonora, Mexico. *Trans. SDSNH*, **6**, 237–304.

Van Rossem, A. J. (1934). Critical notes on Middle American birds. *Bull. MCZ*, **77**, 387–490.

Van Rossem, A. J. (1938a). A Mexican race of the Goshawk (*Accipiter gentilus* [Linnaeus]). *PBSW*, **51**, 99–100.

Van Rossem, A. J. (1938b). Descriptions of twenty-one new races of Fringillidae and Icteridae from Mexico and Guatemala. *BBOC*, **58**, 124–38.

Van Rossem, A. J. (1945). A distributional survey of the birds of Sonora, Mexico. *Occas. Papers MZLSU*, **21**, 1–379.

Van Rossem, A. J. (1947). A synopsis of the Savannah Sparrows of northwestern Mexico. *Condor*, **49**, 97–107.

Van Tyne, J. (1935). The birds of northern Petén, Guatemala. *MZUM Misc. Publ.* No. 27.

Van Tyne, J. and Trautman, M. B. (1941). New birds from Yucatan. *Occas. Papers MZUM*, **439**.

Vaurie, C. (1965). Systematic notes on the bird family Cracidae, No. 2. Relationships and geographical variation of *Ortalis vetula*, *O. poliocephala* and *O. leucogastra. Novitates*, **2222**.

Villa, R. B. (1960). Vertebrados terrestres. La Isla Socorro. *Monogr. Inst. Geofísica, UNAM*, **2**, 203–16.

Villada, D. M. M. (1875). Troquilideos del Valle de Mexico. *La Naturaleza*, **2**, 339–69.

Voous, K. H. (1964). Wood owls of the genera *Strix* and *Ciccaba. Zool. Meded. Rijksmus Nat. Hist. Leiden*, **39**, 471–8.

Vuillemier, F. and Williams, J. E. (1964). Notes on some birds of the Rio Conchos, Chihuahua, Mexico. *Condor*, **66**, 515–16.

Wagner, H. O. (1945a). Observaciones sobre el comportamiento de *Chiroxiphia linearis* durante su propogación. *Anales Inst. Biol.*, **16**, 539–46.

Wagner, H. O. (1945b). Notes on the life history of the Mexican Violet-ear. *Wilson Bull.*, **57**, 165–87.

Wagner, H. O. (1946a). Food and feeding habits of Mexican hummingbirds. *Wilson Bull.*, **58**, 69–93.

Wagner, H. O. (1946b). Observaciones sobre la vida de *Calothorax lucifer. Anales Inst. Biol.*, **17**, 283–99.

Wagner, H. O. (1953). Beitrag zur Biologie und Domestizierungsmöglichkeit des Pfauentruthuhues (*Agriornis ocellata* Cuvier). *Veröffentlichung Mus. Bremen A*, **2**, 2.

Wagner, H. O. (1955). The molt of hummingbirds. *Auk*, **72**, 286–91.

Wagner, H. O. (1957). The molting periods of Mexican hummingbirds. *Auk*, **74**, 251–7.

Walsh, J. M. (1990). Estuarine habitat use and age-specific foraging behavior of Wood Storks (*Mycteria americana*). Unpubl. M.S. thesis, University of Georgia, Athens.

Warner, D. W. (1959). The song, nest, eggs and young of the Long-tailed Wood-Partridge. *Wilson Bull.*, 71, 307–12.

Warner, D. W. and Beer, J. R. (1957). Birds and mammals of the Mesa de San Diego, Puebla, Mexico. *Act. Zool. Mex.*, 2(4–5), 1–17.

Warner, D. W. and Dickerman, R. W. (1959). The status of *Rallus elegans tenuirostris* in Mexico. *Condor*, 61, 49–51.

Warner, D. W. and Harrell, B. E. (1957). The systematics and biology of the Singing Quail, *Dactylortyx thoracicus*. *Wilson Bull.*, 69, 123–48.

Warner, D. W. and Mengel, R. M. (1951). Notes on the birds of the Veracruz Coastal Plain. *Wilson Bull.*, 63, 288–95.

Warnock, N., Griffin, S., and Stenzel, L. E. (1989). Results of the 22–23 Apr 1989 shorebird census in coastal wetlands of San Diego Country and northern Baja California. PRBO Report.

Webb, S. and Howell, S. N. G. (1993). New records of Hawk-Eagles from Guerrero, Mexico. *Euphonia*, 2, 19–21.

Webster, F. S. (1974). Resident birds of the Gomez Farias⁻ region, Tamaulipas, *Mexico Am. Birds*, 28, 3–10.

Webster, F. S. (1983). The winter season, South Texas region. *Am. Birds*, 37, 317–18.

Webster, J. D. (1958a). Further ornithological notes from Zacatecas, Mexico. *Wilson Bull.*, 70, 243–56.

Webster, J. D. (1958b). Systematic notes on the Olive Warbler. *Auk*, 75, 469–73.

Webster, J. D. (1959). A Revision of Botteri's Sparrow. *Condor*, 61, 136–46.

Webster, J. D. (1959b). Another collection from Zacatecas, Mexico. *Auk*, 76, 365–7.

Webster, J. D. (1961). A revision of Grace's Warbler. *Auk*, 78, 554–66.

Webster, J. D. (1962). Systematic and ecologic notes on the Olive Warbler. *Wilson Bull.*, 74, 417–25.

Webster, J. D. (1963). A revision of the Rose-throated Becard. *Condor*, 65, 383–99.

Webster, J. D. (1968a). A revision of the Tufted Flycatchers of the genus *Mitrephanes*. *Auk*, 85, 287–303.

Webster, J. D. (1968b). Ornithological notes from Zacatecas, Mexico. *Condor*, 70, 395–7.

Webster, J. D. (1973). Richardson's Zacatecas collection, I. *Condor*, 75, 239–41.

Webster, J. D. (1984). Richardson's Mexican collection: birds from Zacatecas and adjoining states. *Condor*, 86, 204–7.

Webster, J. D. and Orr, R. T. (1952). Notes on Mexican birds from the states of Durango and Zacatecas. *Condor*, 54, 309–13.

Webster, J. D. and Orr, R. T. (1954a). Summering birds of Zacatecas, Mexico, with a description of a new race of Worthen's Sparrow. *Condor*, 56, 155–60.

Webster, J. D. and Orr, R. T. (1954b). Miscellaneous notes on Mexican birds. *Wilson Bull.*, 66, 267–9.

Wendelken, P. W. and Martin, R. F. (1986). Recent data on the distribution of birds in Guatemala. *BBOC*, 106, 16–21.

Wendelken, P. W. and Martin, R. F. (1989). Recent data on the distribution of birds in Guatemala, 2. *BBOC*, 109, 31–6.

Wetmore, A. (1939). Birds from Clipperton Island collected on the presidential cruise of 1938. *Smithson. Misc. Coll.*, 98, 1–6.

Wetmore, A. (1941a). New forms of birds from Mexico and Colombia. *PBSW*, 54, 203–10.

Wetmore, A. (1941b). Notes on the birds of the Guatemala highlands. *Proc. USNM*, 89, 523–81.

Wetmore, A. (1943). The birds of southern Veracruz, Mexico. *Proc. USNM*, 93, 215–340.

Wetmore, A. (1965–1984). The birds of the Republic of Panama. Parts 1–4. Smithson. Misc. Coll. Vol. 150 (Part 4 completed by R. F. Pasquier and S. L. Olson).

Wetmore, A. (1974). The egg of a Collared Forest-Falcon. *Condor*, 76, 103.

Weyer, D. (1984). Diurnal birds of prey of Belize. *The Hawk Trust Annual Report*, 14, 22–39.

Whitacre, D. F. (1989). Conditional use of nest structure by White-naped and White-collared swifts. *Condor*, 91, 813–15.

Whitaker, L. M. (1960). Nests of the Lesser Swallow-tailed Swift in Mexico. *Wilson Bull.*, 72, 288.

Whitney, B. and Kaufman, K. (1986). The *Empidonax* challenge. Part IV. *Birding*, 18, 315–27.

Widrig, R. S. (1983). December nesting of the Collared Plover in west Mexico. *Am. Birds*, 37, 273–4.

Wilbur, S. R. (1987). *Birds of Baja California*. Univ. California Press, Berkeley.

Wilbur, S. R. and Kiff L. F. (1980). The California Condor in Baja California, Mexico. *Am. Birds*, 34, 856–9.

Williams, S. O. III (1975). Redhead breeding in the state of Jalisco, Mexico. *Auk*, 92, 152–3.

Williams, S. O. III (1977). Notes on the waterbirds of the Laguna de Santiaguillo region, Durango, Mexico. Rept. to USA-Mexico Joint Committee on Wildlife Conservation, August 1977.

Williams, S. O. III (1978). Colonial waterbirds on the Mexican Plateau. *Proc. 1977 Conf. Colonial Waterbird Group*, 1, 44–7.

Williams, S. O. III (1982a). Black Skimmers on the Mexican Plateau. *Am. Birds*, 36, 255–7.

Williams, S. O. III (1982*b*). Notes on the breeding and occurrence of Western Grebes on the Mexican plateau. *Condor*, **84**, 127–30.

Williams, S. O. III (1983). Distribution and migration of the Black Tern in Mexico. *Condor*, **85**, 376–8.

Williams, S. O. III (1987). The changing status of the Wood Duck (*Aix sponsa*) in Mexico. *Am. Birds*, **41**, 372–5.

Williams, S. O. III (1989). Notes on the rail *Rallus longirostris tenuirostris* in the highlands of central Mexico. *Wilson Bull.*, **101**, 117–20.

Williams, S. O. III and Skaggs, R. W. (1993). Distribution of the Mexican Spotted Owl (*Strix occidentalis lucida*) in Mexico. Report to Endangered Species Program, NM Dept. of Game and Fish. Santa Fe, NM.

Willis, E. O. (1961*a*). Prairie Warbler off the Pacific coast of Guatemala. *Condor*, **63**, 419.

Willis, E. O. (1961*b*). A study of nesting anttanagers in British Honduras. *Condor*, **63**, 479–503.

Willis, E. O. (1982). The behaviour of Blackbanded Woodcreepers (*Dendrocolaptes picumnus*). *Condor*, **84**, 272–85.

Wilson, R. G. (1992). Parasitism of Yellow-olive Flycatcher by the Pheasant Cuckoo. *Euphonia*, **1**, 34–6.

Wilson, R. G. and Ceballos-L., H. (1986). *The birds of Mexico City*. BBC Printing & Graphics, Burlington, Ontario, Canada.

Wilson, R. G., Hernandez, C., and Melendez, A. (1988). Eared Grebes nesting in the Valley of Mexico. *Am. Birds*, **42**, 29.

Winker, K., Warner, D. W., and Dickerman, R. W. (1992*a*). Additional bird records from Oaxaca, Mexico. *Ornitologia Neotropical*, **3**, 69–70.

Winker, K., Ramos, M. A., Rappole, J. H., and Warner, D. W. (1992*b*). A note on *Campylopterus excellens* in southern Veracruz, with a guide to sexing captured individuals. *J. Field Ornithol.*, **63**, 339–43.

Winker, K., Oehlenschlager, R. J., Ramos, M. A., Zink, R. M., Rappole, J. H., and Warner, D. W. (1992*c*). Avian distribution and abundance records for the Sierra de Los Tuxtlas, Veracruz, Mexico. *Wilson Bull.*, **104**, 699–718.

Wolf, D. E. (1978). First sighting of a Woodcock in Mexico. *Mexican Birds Newsletter*, **3**, 10–11.

Wolf, L. L. (1964). Nesting of the Fork-tailed Emerald in Oaxaca, Mexico. *Condor*, **66**, 51–5.

Wolf, L. L. (1967). Notes on the taxonomy and plumages of the Slaty Vireo. *Condor*, **69**, 82–4.

Wolf, L. R. (1954). Nesting of the Laughing Falcon. *Condor*, **56**, 161–2.

Wood, D. S. and Leberman, R. C. (1987). Results of the Carnegie Museum of Natural History expedition to Belize. III. Distributional notes on the birds of Belize. *An. CM*, **56**, 137–60.

Wood, D. S., Leberman, R. C., and Weyer, D. (1986). *Checklist of the birds of Belize*. CM Special Publ. No. 12.

Wurster, T. E. and Radamaker, K. (1992). Semipalmated Sandpiper records for Baja California. *Euphonia*, **1**, 37–8.

Zimmerman, D. A. (1957*a*). Some remarks on the behavior of the Yucatan Cactus Wren. *Condor*, **59**, 53–8.

Zimmerman, D. A. (1957*b*). Spotted-tailed Nightjar nesting in Veracruz, Mexico. *Condor*, **59**, 124–7.

Zimmerman, D. A. (1957*c*). Notes on Tamaulipan birds. *Wilson Bull.*, **69**, 273–7.

Zimmerman, D. A. (1969). New records of wood warblers from New Mexico. *Auk*, **86**, 346–7.

Zimmerman, D. A. (1973*a*). Range expansion of Anna's Hummingbird. *Am. Birds*, **27**, 827–35.

Zimmerman, D. A. (1973*b*). Cattle Egrets in northern Mexico. *Condor*, **75**, 480–1.

Zimmerman, D. A. (1978). Eared Trogon — immigrant or visitor? *Am. Birds*, **32**, 135–9.

Zimmerman, D. A. and Harry, G. B. (1951). Summer Birds of Autlan, Jalisco. *Wilson Bull.*, **63**, 302–14.

INDEX OF ENGLISH NAMES

Pages with figures on them are indicated by (Fig.). (NB) indicates that the species is mentioned only in a note. Numbers following Pl. are plate numbers.

INDEX OF SCIENTIFIC NAMES

Pages with figures on them are indicated by (Fig.). (NB) indicates that the species is mentioned only in a note. Numbers following Pl. are plate numbers.

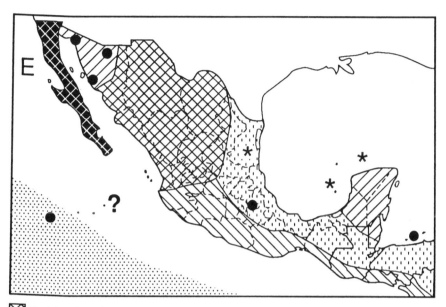

⊠ Resident breeder

⊘ Summer resident (breeder)

◩ Winter (non breeding) visitor

⬖ Former (resident breeder) range

⬚ Transient migrant

⬚ Non-breeding visitor

● Breeding colony (mainly used for water birds), or a presumed resident breeding record in an extensively interpolated range (mainly land birds), or a disjunct, presumed or proven breeding population.

★ Migrant (non-breeding) occurrence, includes transient, vagrant, and winter records (see text). If superimposed on transient range indicates rare winter occurrence.

? Status uncertain (see text)

E Extirpated population (islands)